Handbook of Geomathematics

Willi Freeden, M. Zuhair Nashed, Thomas Sonar (Eds.)

Handbook of Geomathematics

Volume 1

With 360 Figures and 62 Tables

Springer

Willi Freeden
Geomathematics Group
Technische Universität Kaiserslautern
P. O. Box 3049
67653 Kaiserslautern
Germany

M. Zuhair Nashed
Department of Mathematics
University of Central Florida
Orlando, FL 32816
USA

Thomas Sonar
Computational Mathematics
Technische Universität Braunschweig
Pockelsstraße 14
38106 Braunschweig
Germany

Library of Congress Control Number: 2010928832

ISBN: 978–3–642–01545–8
This publication is available also as:
Electronic publication under ISBN 978–3–642–01546–5 and
Print and electronic bundle under ISBN 978–3–642–01547–2

Springer is part of Springer Science+Business Media

springer.com

Editor: Clemens Heine, Heidelberg, Germany
Development Editor: Sylvia Blago, Heidelberg, Germany
Production: SPI-Publishing, Pondicherry, India
Cover Design: Frido Steinen-Broo, Girona, Spain

Printed on acid-free paper

Preface

Mathematics concerned with problems of geoscientifical relevance, i.e., *geomathematics*, is becoming increasingly important. Surprisingly, there is no authoritative mathematical forum offering appropriate means of assimilating, assessing, and reducing to comprehensible form the readily increasing flow of data from geochemical, geodetic, geological, geophysical, and satellite sources and providing an objective basis for scientific interpretation, classification, testing of concepts, modeling, simulation, and solution of problems. Therefore, it seems that the stage is set for a "Handbook of Geomathematics" as a central reference work for academics, policymakers, and practitioners internationally.

The handbook will fill the gap of a basic reference work on geomathematics; it consolidates the current knowledge by providing succinct summaries of concepts and theories, definitions of terms, biographical entries, organizational profiles, a guide to sources of information, and an overview of the landscapes and contours of today's geomathematics. Contributions are written in an easy-to-understand and informative style for a general readership, typically from areas outside the particular research field.

The "Handbook of Geomathematics" will comprise the following scientific fields:

1. Key issues, historical background and future perspectives
2. Observational and measurement key technologies
3. Modeling of the system Earth (geosphere, cryosphere, hydrosphere, atmosphere, biosphere, anthroposphere)
4. Analytic, algebraic, and operator-theoretical methods
5. Statistical and stochastic methods
6. Computational and numerical methods

We as editors wish to express our particular gratitude to the people who not only made this work possible, but also made it an extremely satisfying project:

- The *contributors to the Handbook*, who dedicated much time, effort, and creative energy to the project. The Handbook evolved continuously throughout the recruitment period, as more and more new facets became apparent, many of them were entirely new at the time of recruitment. Eventually, in order to move the Handbook into production, we reluctantly had to stop adding new topics. Moreover, during the process of reviewing and editing, authors as well as editors became exposed to exciting new information that, in some cases, strongly increased activity and creativity. At the end, as far as we are aware, this Handbook breaks a new geomathematical ground in dealing generically with fundamental problems and "key technologies" as well as exploring the wide range of consequences and interactions for (wo)mankind. The Handbook opens new opportunities to contribute significantly to the understanding of our planet, its climate, its environment, and about an expected shortage of natural resources. Even more, the editors are convinced that it offers the key parameters for the study of the Earth's dynamics such that interactions of its solid part with ice, oceans, atmosphere, etc. become more and more accessible.

- The *folks at Springer*, particularly Clemens Heine, who initiated the whole work and gave a lot of advice. Sylvia Blago and Simone Giesler, who were the production interface with authors and editors. They did all the manuscript processing. The editors have had a number of editorial experiences during a long time, but this was certainly the most enjoyable with these professionals.

Thank you very much for all your exceptional efforts and support in creating a work summarizing exciting discoveries and impressive research achievements. We hope that the "Handbook of Geomathematics" will stimulate and inspire new research efforts and the intensive exploration of very promising directions.

Willi Freeden, Kaiserslautern
M. Zuhair Nashed, Orlando
Thomas Sonar, Braunschweig March 2010

About the Editors

Willi Freeden
Geomathematics Group
Technische Universität Kaiserslautern
P.O. Box 3049
67653 Kaiserslautern
Germany

Studies in Mathematics, Geography, and Philosophy at the RWTH Aachen, 1971 'Diplom' in Mathematics, 1972 'Staatsexamen' in Mathematics and Geography, 1975 PhD in Mathematics with distinction, 1979 'Habilitation' in Mathematics at the RWTH Aachen, 1981/1982 Visiting Research Professor at the Ohio State University, Columbus (Department of Geodetic Science and Surveying), 1984 Professor of Mathematics at the RWTH Aachen (Institute of Pure and Applied Mathematics), 1989 Professor of Technomathematics (Industrial Mathematics), 1994 Head of the Geomathematics Group, 2002-2006 Vice-President for Research and Technology at the University of Kaiserslautern, Published over 130 papers, several book chapters, and 4 books. Editor of the Springer "International Journal on Geomathematics (GEM)" and member of editorial board of 4 journals. Organized several international conferences.

M. Zuhair Nashed
Department of Mathematics
University of Central Florida
Orlando, FL 32816
USA

S.B. and S.M. degrees in electrical engineering from MIT and Ph.D. in Mathematics from the University of Michigan. Served for many years as a Professor at Georgia Tech and the University of Delaware. Held visiting professor position at the University of Michigan, University of

Wisconsin, AUB and KFUPM, and distinguished visiting scholar positions at various universities worldwide. Recipient of the Lester Ford Award of the Mathematical Association of America, the Sigma Xi Faculty Research Award and sustained Research Award in Science, Dr. Zakir Husain Award of the Indian Society of Industrial and Applied Mathematics, and several international awards. Published over 130 papers and several expository papers and book chapters. Editor of two journals and member of editorial board of 23 journals, including 4 Springer journals. Organized several international conferences. Editor or co-editor of 10 books. Executive editor of a book series, CRC Press.

Thomas Sonar
Computational Mathematics
Technische Universität Braunschweig
Pockelsstraße 14
38106 Braunschweig
Germany

Study of Mechanical Engineering at the Fachhochschule Hannover, 'Diplom' in 1980. Work as a Laboratory Engineer at the Fachhochschule Hannover from 1980 to 1981. Founding of an engineering company and consulting work from 1981 to 1984. Studies in Mathematics and Computer Science at the University of Hannover, 'Diplom' with distinction in 1987. Research Scientist at the DLR in Braunschweig from 1987 to 1989, then PhD studies in Stuttgart and Oxford, 1991 PhD in Mathematics with distinction, from 1991 to 1996 'Hausmathematiker' at the Institute for Theoretical Fluid Dynamics of DLR at Göttingen. 1995 'Habilitation' in Mathematics at the TH Darmstadt, 1996 Professor of Applied Mathematics at the University of Hamburg. Since 1999 Professor of Technomathematics at the TU Braunschweig. Head of the Group 'Partial Differential Equations'. Member of the 'Braunschweigische Wissenschaftliche Gesellschaft' and corresponding member of the Hamburg Academy of Sciences. Published more than 100 papers, several book chapters, and 10 books. Member of editorial board of 4 journals and responsible book review editor in 2 journals. Organized several international conferences.

Table of Contents

Volume 2 Modeling the System Earth, Analytical and Statistical Methods, Computational and Numerical Algorithms

Contributors

Amir Z. Averbuch
School of Computer Science
Tel Aviv University
Tel Aviv 69978
Israel
amir@math.tau.ac.il

Jörn Behrens
KlimaCampus
Numerical Methods in Geosciences
University of Hamburg
Grindelberg 5
20144 Hamburg
Germany
Joern.Behrens@zmaw.de

Joachim Benz
Faculty of Organic Agricultural Sciences
Work-Group Data-Processing and Computer Facilities
University of Kassel
Witzenhausen
Germany

Jef Caers
Department of Energy Resources Engineering
Stanford University
Stanford
CA 94305-2220
USA
jcaers@stanford.edu

Charles K. Chui
Department of Mathematics
University of Missouri
St. Louis
MO
USA
and
Department of Statistics
Stanford University
Stanford
CA 94305-2220
USA
charleskchui@yahoo.com

Stephan Dahlke
FB 12 Mathematics and Computer Sciences
Philipps–University of Marburg
Hans–Meerwein–Str.
Lahnberge
35032 Marburg
Germany
dahlke@mathematik.uni-marburg.de

Sergey Danilov
Alfred Wegener Institute for Polar and Marine Research
P.O. 1201
27515 Bremerhaven
Germany

Nils Dorband
Meteomedia, AG
Schwäbrig 833
9056 Gais
Schweiz

Paul N. Eggermont
Food and Resource Economics
University of Delaware
224 Townsend Hall
Newark
DE
USA
eggermon@udel.edu

Martin Fengler
Meteomedia, AG
Schwäbrig 833
9056 Gais
Schweiz
mfengler@meteomedia.ch
martin.fengler@gmail.com

Luciana Fenoglio-Marc
Institute of Physical Geodesy
Technical University Darmstadt
Petersenstr. 13
64287 Darmstadt
Germany
fenoglio@jpg.tu-darmstadt.de

Bernd Flemisch
Department of Hydromechanics
and Modeling of Hydrosystems
Institute of Hydraulic Engineering
Universität Stuttgart
Pfaffenwaldring 61
70569 Stuttgart
Germany

Willi Freeden
Geomathematics Group
Technische Universität Kaiserslautern
P.O. Box 3049
67653 Kaiserslautern
Germany
freeden@mathematik.uni-kl.de

Jochen Fritz
Department of Hydromechanics
and Modeling of Hydrosystems
Institute of Hydraulic Engineering
Universität Stuttgart
Pfaffenwaldring 61
70569 Stuttgart
Germany

Christian Gerhards
Geomathematics Group
Department of Mathematics
University of Kaiserslautern
P.O. Box 3049
67653 Kaiserslautern
Germany
gerhards.christian@googlemail.com

Erik W. Grafarend
Department of Geodesy and
Geoinformatics
Stuttgart University
Geschwister Scholl-Straße 24 D
70174 Stuttgart
Germany
Grafarend@gis.uni-stuttgart.de

Markus Grasmair
Computational Science Center
University of Vienna
Nordbergstr. 15
1090 Vienna
Austria

Erwin Groten
Institute of Physical Geodesy
Technical University Darmstadt
Petersenstr. 13
64287 Darmstadt
Germany
groten@ipgs.ipg.verm.tu-darmstadt.de

Martin Grothaus
Functional Analysis Group
University of Kaiserslautern
Gottlieb-Daimler-Str.
67663 Kaiserlautern
Germany
grothaus@mathematik.uni-kl.de

Andreas Gumann
Meteomedia, AG
Schwäbrig 833
9056 Gais
Schweiz

Markus Haltmeier
Computational Science Center
University of Vienna
Nordbergstr. 15
1090 Vienna
Austria

Helmut Harder
Institut für Geophysik
Westfälische Wilhelms-Universität
Schlossplatz 2
48149 Münster
Germany
harder@earth.uni-muenster.de

Rainer Helmig
Department of Hydromechanics
and Modeling of Hydrosystems
Institute of Hydraulic Engineering

Universität Stuttgart
Pfaffenwaldring 61
70550 Stuttgart
Germany
Rainer.Helmig@iws.uni-stuttgart.de

Kerstin Hesse
Department of Mathematics
University of Sussex
Falmer
Brighton
UK
k.hesse@sussex.ac.uk

Ralf Hielscher
Applied Functional Analysis
Technical University Chemnitz
Reichenhainerstr
09126 Chemnitz
Germany
ralf.hielscher@helmholtz-muenchen.de

Gauthier Hulot
Equipe de Géomagnétisme
Institut de Physique du Globe de Paris
Institut de recherche associé au
CNRS et à
l'Université Paris 7
4, Place Jussieu
75252, Paris, Cedex 05
France
gh@ipgp.jussieu.fr

Maxim Ilyasov
Fraunhofer ITWM
Fraunhofer-Platz 1
67663 Kaiserslautern
Germany
ilyasov@itwm.fhg.de

Christopher Jekeli
Division of Geodetic Science
School of Earth Sciences
275 Mendenhall Laboratory
125 South Oval Mall
Ohio State University
Columbus, 43210-1398
OH
USA
jekeli.1@osu.edu

Matthias Klapp
Wunnensteinstr. 9
71679 Asperg
Germany
matthias_klapp@gmx.de

Rupert Klein
FB Mathematik & Informatik
Freie Universität Berlin
Institut für Mathematik
Arnimallee 6
14195 Berlin
Germany
rupert.klein@math.fu-berlin.de

Dan D. Kosloff
Department of Earth and Planetary Sciences
Tel Aviv University
Tel Aviv 69978
Israel

Inna Kozlov
Department of Computer Science
Holon Institute of Technology
Holon
Israel

Jürgen Kusche
Astronomical, Physical and
Mathematical Geodesy Group
Bonn University
Nussallee 17
53115 Bonn
Germany
jkusche@geod.uni-bonn.de

Hansjörg Kutterer
Geodätisches Institut
Leibniz Universität Hannover
Schneiderberg 50
30167 Hannover
Germany
kutterer@gih.uni-hannover.de

Stefan Laps
Meteomedia, AG
Schwäbrig 833
9056 Gais
Schweiz

Vincent LaRiccia
Food and Resource Economics
University of Delaware
224 Townsend Hall
Newark
DE
USA

Henri Laur
European Space Agency (ESA)
Head of Earth Observation Mission
Management Office
ESRIN
Via Galileo Galilei-Casella Postale 64
00044 Frascati
Italy
henri.laur@esa.int

Volker Liebig
European Space Agency (ESA)
Director of Earth Observation Programmes
ESRIN
Via Galileo Galilei-Casella Postale 64
00044 Frascati
Italy
Volker.Liebig@esa.int

Shuai Lu
Johann Radon Institute for Computational
and Applied Mathematics
Austrian Academy of Sciences
Altenbergstrasse 69
4040 Linz
Austria

David Mainprice
Géosciences UMR CNRS 5243
Université Montpellier 2
Montpellier
France
David.Mainprice@gm.univ-montp2.fr

Zdeněk Martinec
Dublin Institute for Advanced Studies
5 Merrion Square
Dublin 2
Ireland
zdenek@cp.dias.ie
and
Department of Geophysics
Faculty of Mathematics and Physics
Charles University
V Holešovičkách 2
180 00 Prague 8
Czech Republic
zdenek@hervam.troja.mff.cuni.cz

Andreas Meister
Work-Group of Analysis and
Applied Mathematics
Department of Mathematics
University of Kassel
Heinrich Platt Str. 40
34132 Kassel
Germany
meister@mathematik.uni-kassel.de

Liqiu Meng
Institute for Photogrammetry and
Cartography
Technische Universität München
Arcisstr. 21
8333 Munich
Germany
meng@bv.tum.de

Volker Michel
Geomathematics Group
University of Siegen
Walter-Flex Str. 3
57068 Siegen
Germany
michel@mathematik.uni-siegen.de

Helmut Moritz
Institut für Navigation und
Satellitengeodäsie
Graz University of Technology
Steyrergasse 30/II
8010 Graz
Austria
helmut.moritz@tugraz.at

M. Zuhair Nashed
Department of Mathematics
University of Central Florida
Orlando
FL 32816
USA
znashed@mail.ucf.edu

Jennifer Niessner
Department of Hydromechanics and Modeling of Hydrosystems
Institute of Hydraulic Engineering
Universität Stuttgart
Pfaffenwaldring 61
70569 Stuttgart
Germany

Helga Nutz
Center for Mathematical and Computational Modelling ((CM)^2)
University of Kaiserslautern
P.O. Box 3045
67653 Kaiserslautern
Germany
nutz@mathematik.uni-kl.de

Nils Olsen
DTU Space
Juliane Maries Vej 30
2100 Copenhagen
Denmark
nio@space.dtu.dk

Isabel Ostermann
Fraunhofer ITWM
Fraunhofer-Platz 1
67653 Kaiserslautern
Germany
osterma@mathematik.uni-kl.de

Kwangwon Park
Department of Energy Resources Engineering
Stanford University
Stanford
CA 94305-2220
USA

Sergei V. Pereverzev
Johann Radon Institute for Computational
and Applied Mathematics
Austrian Academy of Sciences
Altenbergstrasse 69
4040 Linz
Austria
sergei.pereverzyev@oeaw.ac.at

Alexander Petukhov
Department of Mathematics
University of Georgia
Athens 30602
GA
USA
petukhov@hotmail.com

Helmut Pruscha
Mathematical Institute
University of Munich
Theresienstr. 39
80333 Munich
Germany
pruscha@math.lmu.de

Alessandro Punzi
Fraunhofer ITWM
Fraunhofer-Platz 1
67653 Kaiserslautern
Germany
alessandropunzi@gmail.com

Thomas Raskop
Functional Analysis Group
University of Kaiserslautern
Gottlieb-Daimler-Str.
67663 Kaiserlautern
Germany

Reiner Rummel
Institut für Astronomische und Physikalische Geodäsie
Technische Universität München
80290 München
Germany
rummel@bv.tum.de

Terence J. Sabaka
Planetary Geodynamics Laboratory
Code 698
NASA Goddard Space Flight Center
Greenbelt
MD 20771
USA
Terence.J.Sabaka@nasa.gov

Helmut Schaeben
Geoscience Mathematics and Informatics
Technical University Bergakademie

Gustav-Zeiner-Str. 12
09596 Freiberg
Germany
schaeben@geo.tu-freiberg.de

Céline Scheidt
Department of Energy Resources Engineering
Stanford University
Stanford
CA 94305-2220
USA

Otmar Scherzer
Computational Science Center
University of Vienna
Nordbergstr. 15
1090 Vienna
Austria
otmar.scherzer@univie.ac.at

Jens Schröter
Alfred Wegener Institute for Polar and Marine Research
P.O. 1201
27515 Bremerhaven
Germany
jschroeter@AWI-Bremerhaven.de

Michael Schreiner
Institute for Computational Engineering
University of Buchs
Werdenbergstr. 4
9471 Buchs
Switzerland
schreiner@ntb.ch

Monika Sester
Institute of Cartography and Geoinformatics
Leibniz Universität Hannover
Appelstr. 9a
30167 Hannover
Germany
Monika.Sester@ikg.uni-hannover.de

Frederik J. Simons
Department of Geosciences
Princeton University
Guyot Hall 321B
Princeton 08544-1003
NJ
USA
fjsimons@alum.mit.edu

Ian H. Sloan
School of Mathematics and Statistics
University of New South Wales
2052 Sydney
Australia
I.Sloan@unsw.edu.au

Thomas Sonar
Computational Mathematics
Technische Universität Braunschweig
Pockelstr. 14
38106 Braunschweig
Germany
t.sonar@tu-bs.de

Stephan Stellmach
Institut für Geophysik
Westfälische Wilhelms-Universität
Schlossplatz 2
48149 Münster
Germany
stellma@earth.uni-muenster.de

Peter J.G. Teunissen
Department of Spatial Sciences
Curtin University of Technology
Perth
Australia
and
Department of Earth Observation and Space Systems
Delft University of Technology
Delft
The Netherlands
P.J.G.Teunissen@tudelft.nl

Jianzhong Wang
Department of Mathematics
Sam Houston State University
Huntsville 77341
TX
USA
mth_jxw@shsu.edu
jzwang2004@yahoo.com

Yanfei Wang
Key Laboratory of Petroleum Geophysics
Institute of Geology and Geophysics
Chinese Academy of Sciences
P.O. Box 9825
Beijing, 100029
People's Republic of China
yfwang@mail.iggcas.ac.cn
yfwang_ucf@yahoo.com

Johannes Wicht
Max-Planck Intitut für
Sonnensystemforschung
Max-Planck-Str. 2
37191 Kaltenburg-Lindau
Germany
wicht@mps.mpg.de

Kerstin Wolf
Center for Mathematical and
Computational Modelling ($(CM)^2$)
University of Kaiserslautern
Kaiserslautern
Germany
kerstin-wlf@gmx.de

Detlef Wolf
Department of Geodesy and
Remote Sensing
German Research Center for
Geosciences GFZ
Telegrafenberg
14473 Potsdam
Germany
dasca@gfz-potsdam.de

Markus Wolff
Department of Hydromechanics
and Modeling of Hydrosystems
Institute of Hydraulic Engineering
Universität Stuttgart
Stuttgart
Germany

Robert S. Womersley
School of Mathematics and Statistics
University of New South Wales
2052 Sydney
Australia

Peiliang Xu
Disaster Prevention Research Institute
Kyoto University
Uji
Kyoto 611-0011
Japan
pxu@rcep.dpri.kyoto-u.ac.jp

Valery A. Zheludev
School of Computer Science
Tel Aviv University
Tel Aviv 69978
Israel

Part 1

General Issues, Historical Background, and Future Perspectives

1 Geomathematics: Its Role, Its Aim, and Its Potential

Willi Freeden
Geomathematics Group, Technische Universität Kaiserslautern, Kaiserslautern, Germany

W. Freeden, M.Z. Nashed, T. Sonar (Eds.), *Handbook of Geomathematics*, DOI 10.1007/978-3-642-01546-5_1,

Abstract During the last decades geosciences and geoengineering were influenced by two essential scenarios: First, the technological progress has completely changed the observational and measurement techniques. Modern high-speed computers and satellite-based techniques more and more enter all geodisciplines. Second, there is a growing public concern about the future of our planet, its climate, its environment, and about an expected shortage of natural resources. Obviously, both aspects, namely efficient strategies of protection against threats of a changing Earth and the exceptional situation of getting terrestrial, airborne as well as spaceborne data of better and better quality explain the strong need of new mathematical structures, tools, and methods, i.e., *geomathematics.*

This chapter deals with geomathematics, its role, its aim, and its potential. Moreover, the "circuit" geomathematics is exemplified by two classical problems involving the Earth's gravity field, namely gravity field determination from terrestrial deflections of the vertical and ocean flow modeling from satellite (altimeter measured) ocean topography.

1 Introduction

Geophysics is an important branch of physics; it differs from the other physical disciplines due to its restriction to objects of geophysical character. Why shouldn't the same hold for mathematics what physics regards as its canonical right since the times of Emil Wiechert in the late nineteenth century? More than ever before, there are significant reasons for a well-defined position of geomathematics as a branch of mathematics and simultaneously of geosciences. These reasons are based on the one hand intrinsically in the self-conception of mathematics; on the other hand, they are explainable from the current situation of geosciences.

In the following, I would like to explain my thoughts on geomathematics in more detail. My objective is to convince not only the geoscientists, but also a broader audience: "Geomathematics is essential and important. As geophysics is positioned within physics, it is an adequate forum within mathematics, and it should have a fully acknowledged position within geosciences!"

2 Geomathematics as Cultural Asset

In ➲ Chap. 2 of this handbook, T. Sonar begins his contribution with the sentence: "Geomathematics in our times is thought of being a very young science and a modern area in the realms of mathematics. Nothing is farer from the truth. Geomathematics began as man realized that he walked across a sphere-like Earth and that this observation has to be taken into account in measurements and calculations." In consequence, we can only do justice to Geomathematics if we look at its historic importance, at least shortly.

According to the oldest evidence which has survived in written form, geomathematics was developed in Sumerian Babylon on the basis of practical tasks concerning measuring, counting, and calculation for reasons of agriculture and stock keeping.

In the ancient world, mathematics dealing with problems of geoscientific relevance flourished for the first time, e.g., when Eratosthenes (276–195 BC) of Alexandria calculated the radius of the Earth. We also have evidence that the Arabs carried out an arc measurement northwest of Bagdad in the year 827 AD. Further key results of geomathematical research lead us from the Orient across the occidental Middle Ages to modern times. N. Copernicus (1473–1543) successfully made the transition from the Ptolemaic geocentric system to the heliocentric system. J. Kepler (1571–1630) determined the laws of planetary motion. Further milestones from a historical point of view are, e.g., the theory of geomagnetism developed by W. Gilbert (1544–1608), the development of triangulation methods for the determination of meridians by T. Brahe (1547–1601) and W. Snellius (1580–1626), the laws of falling bodies by G. Galilei (1564–1642), and the basic theory on the propagation of seismic waves by C. Huygens (1629–1695). The laws of gravitation formulated by the English scientist I. Newton (1643–1727) have taught us that gravitation decreases with an increasing distance from the Earth. In the seventeenth and eighteenth century, France took over an essential role through the foundation of the Academy in Paris (1666). Successful discoveries were the theory of the isostatic balance of mass distribution in the Earth's crust by P. Bouguer (1698–1758), the calculation of the Earth's shape and especially of the pole flattening by P. L. Maupertuis (1698–1759) and A. C. Clairaut (1713–1765), and the development of the calculus of spherical harmonics by A. M. Legendre (1752–1833) and P. S. Laplace (1749–1829). The nineteenth century was essentially characterized by C. F. Gauß (1777–1855). Especially important were the calculation of the lower Fourier coefficients of the Earth's magnetic field, the hypothesis of electric currents in the ionosphere, as well as the definition of the level set of the geoid (however, the term "geoid" was defined by J. B. Listing (1808–1882), a disciple of C.F. Gauß). At the end of the nineteenth century, the basic idea of the dynamo theory in geomagnetics was developed by B. Stewart (1851–1935), etc. This very incomplete list (which does not even include the last century) already shows that Geomathematics is one of the large achievements of mankind from a historic point of view.

3 Geomathematics as Task and Objective

Modern geomathematics deals with the qualitative and quantitative properties of the current or possible structures of the Earth system. It guarantees appropriate concepts of scientific research concerning the Earth system, and it is simultaneously the force behind it.

The system Earth consists of a number of elements which represent individual systems themselves. The complexity of the entire Earth system is determined by interacting physical, biological, and chemical processes transforming and transporting energy, material, and information (cf. Emmermann and Raiser 1997). It is characterized by natural, social, and economic processes influencing one another. In most instances, a simple theory of cause and effect is therefore completely inappropriate if we want to understand the system. We have to think in terms of dynamical structures. We have to account for multiple, unforeseen, and of course sometimes even undesired effects in the case of interventions. Inherent networks must be recognized and made use of, and self-regulation must be accounted for. All these aspects require a type of mathematics which must be more than a mere collection of theories and numerical methods. Mathematics dedicated to geosciences, i.e., geomathematics deals with nothing more than the organization of the complexity of the Earth system. Descriptive thinking is required to clarify abstract complex situations. We also need a correct simplification of complicated interactions, an appropriate system of mathematical concepts for their description, and exact thinking and formulations. Geomathematics has thus become the key science of the complex Earth system. Wherever there are data and observations to be processed, e.g., the diverse scalar, vectorial, and tensorial clusters of satellite data, we need mathematics. For example, statistics serves for noise reduction, constructive approximation for compression and evaluation, the theory of special function systems yield georelevant graphical and numerical representations – there are mathematical algorithms everywhere. The specific task of geomathematics is to build a bridge between mathematical theory and geophyiscal as well as geotechnical applications.

The special attraction of this branch of mathematics is therefore based on the vivid communication between applied mathematicians more interested in model development, theoretical foundation, and the approximate as well as computational solution of problems, and geoengineers and physicists more familiar with measuring technology, methods of data analysis, implementation of routines, and software applications.

There is a very wide range of modern geosciences on which geomathematics is focused (see ➲ *Fig. 1*), not least because of the continuously increasing observation diversity. Simultaneously, the mathematical "toolbox" is becoming larger. A special feature is that geomathematics primarily deals with those regions of the Earth which are only insufficiently or not at all accessible for direct measurements (even by remote sensing methods (as discussed, e.g., in ➲ Chap. 26 of this work)). Inverse methods (see, for instance, ➲ Chaps. 5, ➲ 8, ➲ 17, ➲ 24–26, and ➲ 32 of this handbook) are absolutely essential for mathematical evaluation in these cases. Mostly, a physical field is measured in the vicinity of the Earth's surface, and is then continued downward or upward by mathematical methods until one reaches the interesting depths or heights (see, e.g., ➲ Chaps. 7, ➲ 9, ➲ 19, ➲ 27, and ➲ 32 for more details on downward and upward continuation).

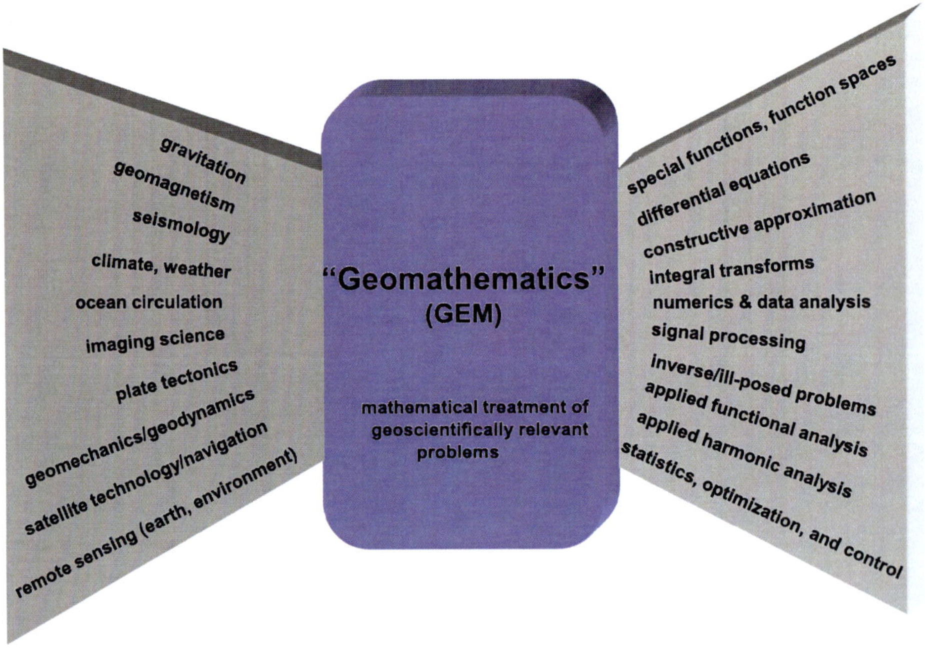

Fig. 1
Geomathematics, its range of fields, and its tools

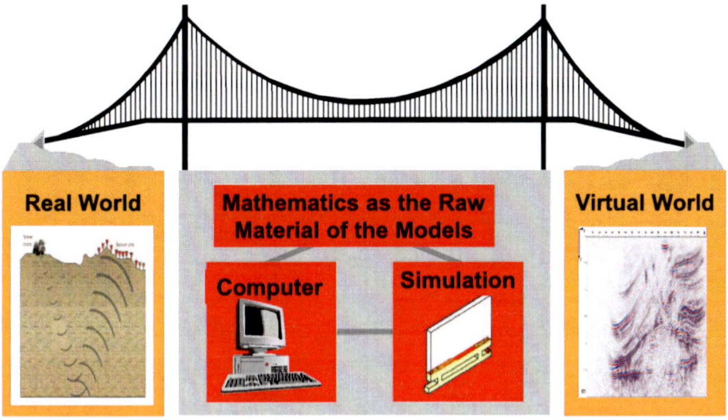

Fig. 2
Geomathematics as key technology bridging the real and virtual world (see Chaps. 22, 25, and 43)

4 Geomathematics as Interdisciplinary Science

It should be mentioned once more that, today, computers and measurement technology have resulted in an explosive propagation of mathematics in almost every area of society. Mathematics as an interdisciplinary science can be found in almost every area of our lives. Consequently, mathematics is closely interacting with almost every other science, even medicine and parts of the arts (*mathematization of sciences*). The use of computers allows for the handling of complicated models for real data sets. Modeling, computation, and visualization yield reliable simulations of processes and products.

Mathematics is the "raw material" for the models and the essence of each computer simulation. As the key technology, it translates the images of the real world to models of the virtual world, and vice versa (cf. *Fig. 2*).

The special importance of mathematics as an interdisciplinary science (see also Pesch 2002) has been acknowledged increasingly within the last few years in technology, economy, and commerce. However, this process does not remain without effects on mathematics itself. New mathematical disciplines, such as scientific computing, financial and business mathematics, industrial mathematics, biomathematics, and also geomathematics have complemented the traditional disciplines.

Interdisciplinarity also implies the interdisciplinary character of mathematics at school (cf. Sonar 2001). Relations and references to other disciplines (especially informatics, physics, chemistry, biology, and also economy and geography) become more important, more interesting, and more expandable. Problem areas of mathematics become explicit and observable, and they can be visualized. Of course, this undoubtedly also holds for the Earth system.

5 Geomathematics as Challenge

From a scientific and technological point of view, the past twentieth century was a period with two entirely different faces concerning research and its consequences. The first two thirds of the century were characterized by a movement toward a seemingly inexhaustible future of science and technology; they were marked by the absolute belief in technical progress which would make everything achievable in the end. Up to the 1960s, mankind believed to have become the master of the Earth (note: in geosciences as well as other sciences, to master is also a synonym for to understand). Geoscience was able to understand plate tectonics on the basis of Wegener's theory of continental drift, geoscientific research began to deal with the Arctic and Antarctic, and man started to conquer the universe by satellites, so that for the first time in mankind's history the Earth became measurable on a global scale, etc. Then, during the last third of the past century, there was a growing skepticism as to the question whether scientific and technical progress had really brought us forth and whether the effects of our achievements were responsible. As a consequence of the specter of a shortage in raw materials (mineral oil and natural gas reserves), predicted by the Club of Rome, geological/geophysical research with the objective of exploring new reservoirs was stimulated during the 1970s (cf. Jakobs and Meyer 1992). Moreover, the last two decades of the century have sensitized us for the global problems resulting from our behavior with respect to climate and environment. Our senses have been sharpened as to the dangers caused by the forces of nature, from earthquakes and volcanic eruptions, to the temperature development and the hole in the ozone layer, etc. Man has become aware of his environment.

The image of the Earth as a potato drenched by rainfall (which is sometimes drawn by oceanographers) is not a false one (see ➋ *Fig. 3*). The humid layer on this potato, maybe only a fraction of a millimeter thick, is the ocean. The entire atmosphere hosting the weather and climate events is only a little bit thicker. Flat bumps protruding from the humid layer represent the continents. The entire human life takes place in a very narrow region of the outer peel (only a few kilometers in vertical extension). However, the basically excellent comparison of the Earth with a huge potato does not give explicit information about essential ingredients of the Earth system, e.g., gravitation, magnetic field, deformation, wind and heat distribution, ocean currents, internal structures, etc.

In our twenty-first century, geoproblems currently seem to overwhelm the scientific programs and solution proposals. *How much more will our planet Earth be able to take?* has become an appropriate and very urgent question.

Indeed, there has been a large number of far reaching changes during the last few decades, e.g., species extinction, climate change, formation of deserts, ocean currents, structure of the atmosphere, transition of the dipole structure of the magnetic field to a quadrupole structure, etc. These changes have been accelerated dramatically. The main reasons for most of these phenomena are the unrestricted growth in the industrial societies (population and

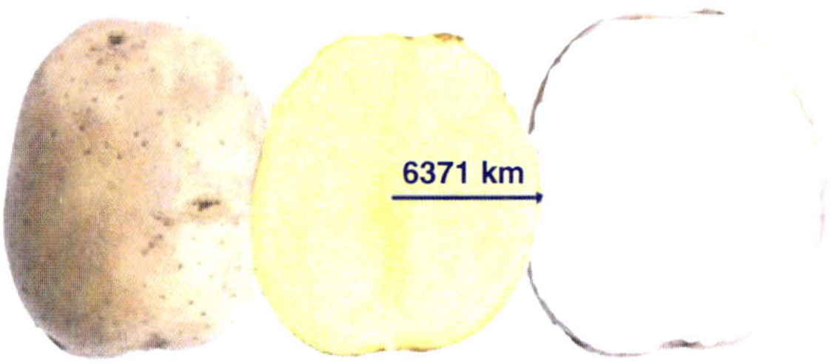

Fig. 3
Potato "Earth"

consumption, especially of resources, territory, and energy) and severe poverty in the developing and newly industrialized countries. The dangerous aspect is that extreme changes have taken place within a very short time; there has been no comparable development in the dynamics of the Earth system in the past. Changes brought about by man are much faster than changes due to natural fluctuations. Besides, the current financial crisis shows that our western model of affluence (which holds for approximately 1 billion people) cannot be transferred globally to 5–8 billion people. Massive effects on mankind are inevitable. The appalling résumé is that the geoscientific problems collected over the decades must now all be solved simultaneously. Interdisciplinary solutions including intensive mathematical research are urgently required as answers to an increasingly complex world. Geomathematics is absolutely essential for a sustainable development in the future.

However, the scientific challenge does not only consist of increasing the leading role of mathematics within the current "scientific consortium Earth." The significance of the subject "Earth" must also be acknowledged (again) within mathematics itself, so that mathematicians will become more enthusiastic about it. Up to now, it has become usual and good practice in application-oriented mathematical departments and research institutions to present applications in technology, economy, finances, and even medicine as being very career-enhancing for young scientists. Geomathematics can be integrated smoothly into this phalanx with current subjects like exploration, geothermal research, navigation, and so on. Mathematics should be the leading science for the solution of these complex and economically very interesting problems, instead of fulfilling mere service functions. Of course, basic research is indispensable. We should not hide behind the other geosciences! Neither should we wait for the next horrible natural disaster! Now is the time to turn expressly toward georelevant applications. The Earth as a complex, however limited system (with its global problems concerning climate, environment, resources, and population) needs new political strategies. Consequently, these will step by step also demand changes in research and teaching at the universities due to our modified concept of "well being" (e.g., concerning milieu, health, work, independence, financial situation, security, etc). This will be a very difficult process. We dare to make the prognosis that, finally, it will also result in a modified appointment practice at the universities. Chairs in the field of geomathematics will increase in number and importance, in order to promote attractiveness, but also to accept a general responsibility for society. Additionally, the curricular standards and models for school lessons in mathematics (see Bach et al. 2004; Sonar 2001) will also change. We will not be able to afford any jealousies or objections on our way in that direction.

6 Geomathematics as Solution Potential

Current methods of applied measurement and evaluation processes vary strongly, depending on the examined measurement parameters (gravity, electric or magnetic field force, temperature and heat flow, etc), the observed frequency domain, and the occurring basic "field characteristic" (potential field, diffusion field, or wave field, depending on the basic differential equations). In particular, the differential equation strongly influences the evaluation processes. The typical mathematical methods are therefore listed here according to the respective "field characteristic" – as it is usually done in geomathematics.

- *Potential methods* (potential fields, elliptic differential equations) in geomagnetics, geoelectrics, gravimetry, geothermal research, etc

- *Diffusion methods* (diffusion fields, parabolic differential equations) in magnetotellurics, geoelectromagnetics, etc
- *Wave methods* (wave fields, hyperbolic differential equations) in seismics, georadar, etc.

The diversity of mathematical methods will increase in the future due to new technological developments in computer and measurement technology. More intensively than before, we must aim for the creation of models and simulations for combinations and networks of data and observable structures. The process (i.e., the "circuit") for the solution of practical problems usually has the following components:

- **Mathematical modeling**: the practical problem is translated into the language of mathematics, requiring the cooperation between application-oriented scientists and mathematicians.
- **Mathematical analysis**: the resulting mathematical problem is examined as to its "well-posedness" (i.e., existence, uniqueness, dependence on the input data).
- **Development of a mathematical solution method**: appropriate analytical, algebraic, and/or numerical methods and processes for a specific solution must be adapted to the problem; if necessary, new methods must be developed. The solution process is carried out efficiently and economically by the decomposition into individual operations, usually on computers.
- **"Back-transfer" from the language of Mathematics to applications**: the results are illustrated adequately to ensure their evaluation. The mathematical model is validated on the basis of real data and modified, if necessary. We aim for good accordance of model and reality.

Usually, the circuit must be applied several times in an iterative way in order to get a sufficient insight into the Earth system. Nonetheless, the advantage and benefit of the mathematical processes are a better, faster, cheaper, and more secure problem solution on the basis of the mentioned means of simulation, visualization, and reduction of large amounts of data. So, what is it exactly that enables mathematicians to build a bridge between the different disciplines? The mathematics' world of numbers and shapes contains very efficient tokens by which we can describe the rule-like aspect of real problems. This description includes a simplification by abstraction: essential properties of a problem are separated from unimportant ones and included into a solution scheme. Their "eye for similarities" often enables mathematicians to recognize a posteriori that an adequately reduced problem may also arise from very different situations, so that the resulting solutions may be applicable to multiple cases after an appropriate adaptation or concretization. Without this second step, abstraction remains essentially useless (cf. Bach et al. 2004).

The interaction between abstraction and concretization characterizes the history of mathematics and its current rapid development as a common language and independent science. A problem reduced by abstraction is considered as a new "concrete" problem to be solved within a general framework, which determines the validity of a possible solution. The more examples one knows, the more one recognizes the causality between the abstractness of mathematical concepts and their impact and cross-sectional importance.

Of course, geomathematics is closely interconnected with geoinformatics, geoengineering, and geophysics. However, geomathematics basically differs from these disciplines (cf. Kümmerer 2002). Engineers and physicists need the mathematical language as a tool. In contrast to this, geomathematics also deals with the further development of the language itself. Geoinformatics concentrates on the design and architecture of processors and computers, data bases and programming languages, etc, in a georeflecting environment. In geomathematics, computers do not represent the objects to be studied, but instead represent technical auxiliaries for the solution of mathematical problems of georeality.

7 Geomathematics as Solution Method

Up to now, ansatz functions for the description of geoscientifically relevant parameters have been frequently based on the almost spherical shape of the Earth. By modern satellite positioning methods, the maximum deviation of the actual Earth's surface from the average Earth's radius (6371 km) can be proved as being less than 0.4%. Although a mathematical formulation in a spherical context may be a restricted simplification, it is at least acceptable for a large number of problems (see ➋ Chap. 31 of this handbook for more details).
Usually, in geosciences, when we are interested in modeling on the spherical Earth's surface, we consider a separable Hilbert space with a (known) polynomial basis as reference space for *ansatz functions*. A standard approximation method for discrete data since the times of C. F. Gauß (1835) has been the Fourier series in an orthogonal basis (i.e., spherical basis functions). It is characteristic for such an approach that the polynomial ansatz functions (usually called spherical harmonics) do not show any localization in space (cf. ➋ *Fig. 4*). In the momentum domain (here called frequency domain), each spherical function corresponds to exactly one single Fourier coefficient. We call this ideal frequency localization. Due to the ideal frequency localization and the simultaneous dispensation with localization in space, local data modifications influence all the Fourier coefficients (that have to be determined by global integration). Consequently, this also leads to global modifications of the data representations in local space.

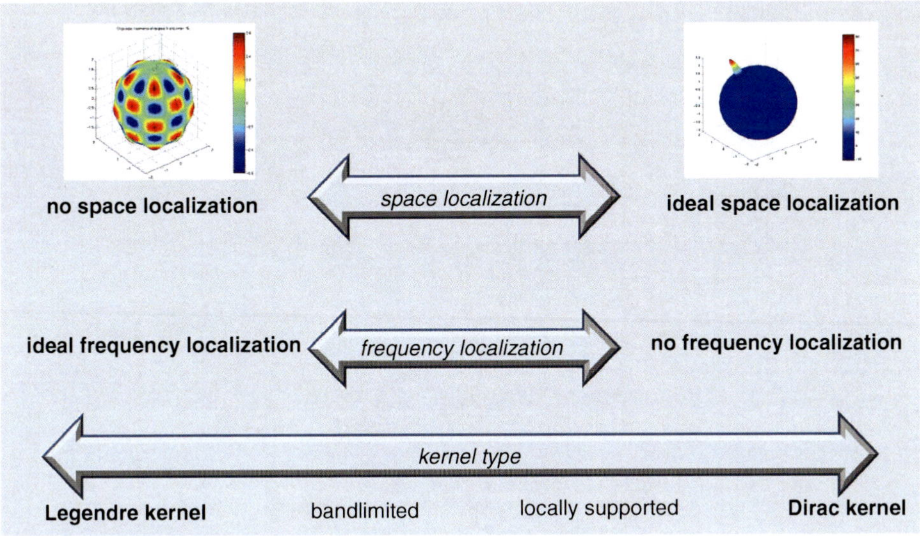

Fig. 4
Uncertainty principle and its consequences for space-frequency localization (see also ➋ Chap. 31)

Nevertheless, we may state that ideal frequency localization has proved to be extraordinarily advantageous due to the important physical interpretability (as multipole moments) of the model and due to the simple comparability of the Fourier coefficients for observables in geophysical interrelations. From a mathematical and physical point of view, however, certain kinds of ansatz functions would be desirable which show ideal frequency localization and localization in space. Such an ideal system of ansatz functions would allow for models of highest resolution in space; simultaneously, individual frequencies would remain interpretable. However, the principle of uncertainty, which connects frequency localization and localization in space quantitatively and qualitatively, teaches us that both properties are mutually exclusive (except of the trivial case).

Extreme ansatz functions (cf. ➋ *Fig. 4*) in the sense of such an *uncertainty principle* are, on the one hand, spherical polynomials, i.e., spherical harmonics (no space localization, ideal frequency localization) and, on the other hand, the Dirac (kernel) function(als) (ideal space localization, no frequency localization).

In consequence (see also ➋ Chap. 31 of this handbook, and for more details Freeden et al. 1998, Freeden and Schreiner 2009), (spherical harmonic) Fourier methods are surely well-suited to resolve low and medium frequency phenomena, while their application is critical to obtain high resolution models. This difficulty is also well known to theoretical physics, e.g., when describing monochromatic electromagnetic waves or considering the quantum-mechanical treatment of free particles. In this case, plane waves with fixed frequencies (ideal frequency localization, no space localization) are the solutions of the corresponding differential equations, but do certainly not reflect the physical reality. As a remedy, plane waves of different frequencies are superposed to so-called wave-packages, which gain a certain amount of space localization while losing their ideal frequency (spectral) localization.

A suitable *superposition of polynomial ansatz functions* (see Freeden and Schreiner 2009) leads to so-called kernel functions/kernels with a reduced frequency – but increased space localization (cf. ➋ *Fig. 5*).

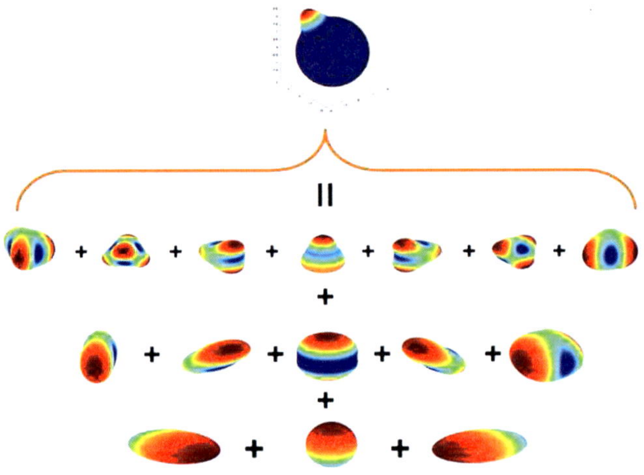

Fig. 5
Summation of spherical harmonics for the generation of a band-limited zonal kernel

These kernels can be constructed as to cover various spectral bands and, hence, can show all intermediate stages of frequency – and space localization. The width of the corresponding frequency – and space localization is usually controlled using a so-called *scale-parameter*. If the kernel is given by a finite superposition of polynomial ansatz-functions, it is said to be bandlimited, while in the case of infinitely many ansatz functions the kernel is called non-bandlimited. It turns out that, due to their higher frequency localization (fine frequency band), the bandlimited kernels show less space localization than their non-bandlimited counterparts (infinite frequency band). This leads to the following characterization of ansatz functions: Fourier methods with polynomial trial-functions are the canonical point of departure for approximations of low-frequent phenomena (global to regional modeling). Because of their excellent localization properties in the space domain, bandlimited and non-bandlimited kernels with increasing space localization properties can be used for stronger and stronger modeling of short-wavelength phenomena (local modeling). Using kernels of different scales the modeling approach can be adapted to the localization properties of the physical process under consideration. By use of sequences of scale-dependent kernels tending to the Dirac-kernel, i.e., so-called Dirac-sequences, a multiscale approximation (i.e., "zooming in" process) can be established appropriately. Later on in this treatise, we deal with simple (scalar and/or vectorial) wavelet-techniques, i.e., with multiscale techniques based on special kernel functions: (spherical) scaling functions and wavelets. Typically, the generating-functions of scaling functions have the characteristics of low-pass filters, i.e., the polynomial basis-functions of higher frequencies are attenuated or even completely left out. The generating-functions of wavelets, however, have the typical properties of band-pass filters, i.e., polynomial basis functions of low and high frequency are attenuated or even completely left out when constructing the wavelet. Thus, wavelet-techniques usually lead to a multiresolution of the Hilbert space under consideration, i.e., a certain two-parameter splitting with respect to scale and space. To be more concrete, the Hilbert space under consideration can be decomposed into a nested sequence of approximating subspaces – the scale spaces – corresponding to the scale parameter. In each scale space a model of the data–function can be calculated usually using the respective scaling functions and, thus, leading to an approximation of the data at a certain resolution. For increasing scales, the approximation improves and the information obtained on coarse levels is contained in all levels of approximation above. The difference between two successive approximations is called the detail information, and it is contained in the so-called detail spaces. The wavelets constitute the basis-functions of the detail spaces and, summarizing the subject, every element of the Hilbert space (i.e., the "continuous data") can be represented as a structured linear combination of scaling functions and wavelets of different scales and at different positions ("multiscale approximation") (cf. Freeden et al. 1998; Freeden and Michel 2004 and the references therein). Hence we have found a possibility to break up complicated functions like the geomagnetic field, electric currents, gravitational field, deformation field, oceanic currents, etc, into single pieces of different resolutions and to analyze these pieces separately. This helps to find adaptive methods that take into account the structure of the data, i.e., in areas where the data show only a few coarse spatial structures, the resolution of the model can be chosen to be rather low; in areas of complicated data structures the resolution can be increased accordingly. In areas, where the accuracy inherent in the measurements is reached, the solution process can be stopped by some kind of threshholding, i.e., using scaling functions and wavelets at different scales, the corresponding approximation techniques can be constructed as to be suitable for the particular data situation. Consequently, although most data show correlation in space as well as in frequency, the kernel functions with their simultaneous space and frequency localization allow for the

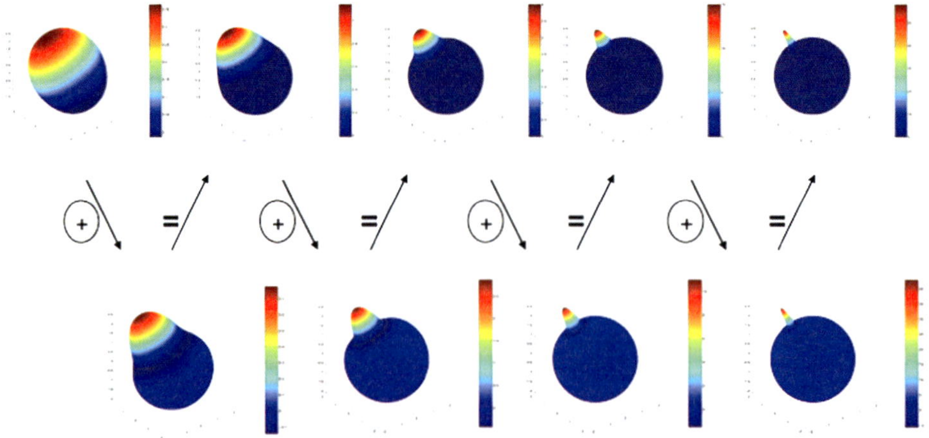

◘ Fig. 6
Scaling functions (upper row) and wavelet functions (lower row) in mutual relation ("tree structure") within a multiscale approximation

efficient detection and approximation of essential features in the data structure by only using fractions of the original information (decorrelation) (➋ *Fig. 6*).

Finally it is worth mentioning that future spaceborne observation combined with terrestrial and airborne activities will provide huge datasets of the order of millions of data (see ➋ Chaps. 3–5, ➋ 11, ➋ 13, and ➋ 18–22 concerning different fields of observation). Standard mathematical theory and numerical methods are not at all adequate for the solution of data systems with a structure such as this, because these methods are simply not adapted to the specific character of the spaceborne problems. They quickly reach their capacity limit even on very powerful computers. In our opinion, a reconstruction of significant geophysical quantities from future data material requires much more: e.g., it requires a careful analysis, dimensionality reduction (see ➋ Chap. 34), fast solution techniques, and a proper stabilization of the solution, usually including procedures of regularization (see Freeden 1999 and the references therein), sparsity in various aspects (see, e.g., ➋ Chaps. 25, ➋ 29, and ➋ 42), new models for ill posed problems (see ➋ Chap. 24), etc. In order to achieve these objectives various strategies and structures must be introduced reflecting different aspects. As already pointed out, while global long-wavelength modeling can be adequately done by the use of polynomial expansions, it becomes more and more obvious that splines and/or wavelets are most likely the candidates for medium- and short-wavelength approximation. But the multiscale concept of wavelets demands its own nature which – in most cases – cannot be developed from the well-known theory in Euclidean spaces. In consequence, the stage is also set to present the essential ideas and results involving a multiscale framework to the geoscientific community (see ➋ Chaps. 11, ➋ 18, ➋ 30, ➋ 32, ➋ 41, and ➋ 43 for different realizations).

8 Geomathematics: Two Exemplary "Circuits"

In the sequel, the "circuit" geomathematics as solution potential will be demonstrated with respect to contents, origin, and intention on two completely different geoscientifically relevant examples, namely the determination of the gravity field from terrestrial deflections of the vertical and the computation of (geostrophic) oceanic circulation from altimeter satellite data (i.e., RADAR data). We shall restrict the calculation of these geophysical quantities to an Earth's model that is for simplicity supposed to be a spherical one. Our mathematical analysis will demonstrate that both illustrations are closely related via the Helmholtz Theorem for decomposing a spherical vector field into a radial and a tangential part, where the tangential part is split into a divergence and a curl free vector field. The mathematical background coincides for both geoscientifically distinct problems with respect to notation and terminology. In fact, their solution can be performed within the same mathematical context; it can be described actually within the same apparatus of formulas. However, the numerical methods must be specifically adapted to the concrete solution of the problems, they are basically dissimilar. In the first example (i.e., the gravity field determination) we are confronted with a process of integration, in the second example (i.e., modeling of the oceanic circulation) we have to execute a process of differentiation – in both cases to discretely given data. The first example uses vectorial data on the Earth's surface, the second example uses scalar satellite altimetry data.

8.1 Circuit: Gravity Field from Deflections of the Vertical

The modeling of the gravity field and its equipotential surfaces, especially the geoid of the Earth, is essential for many applications (see ➋ *Fig. 7* for a graphical illustration), from which we only mention some significant examples (cf. Rummel 2002):

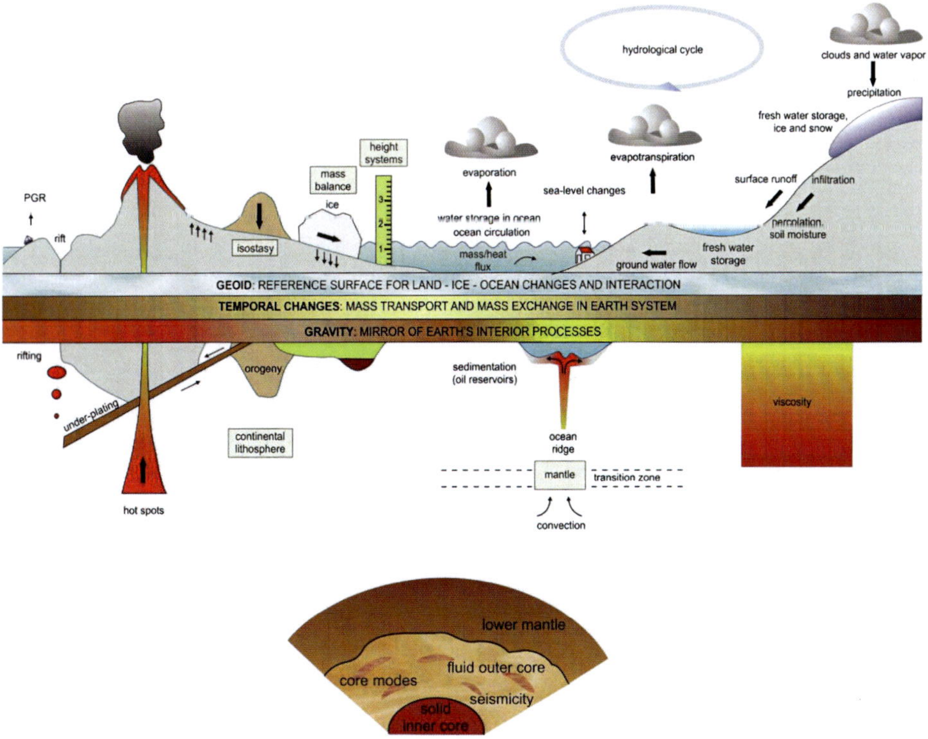

◘ Fig. 7
Gravity involved processes (from "German Priority Research Program: Mass transport and mass distribution in the Earth system, DFG–SPP 1257")

Earth System: there is a growing awareness of global environmental problems (e.g., the CO_2-question, the rapid decrease of rain forests, global sea-level changes, etc). What is the role of the future terrestrial, airborne methods, and satellite missions in this context? They do not tell us the reasons for physical processes, but it is essential to bring the phenomena into one system (e.g., to make sea-level records comparable in different parts of the world). In other words, the geoid, i.e., the equipotential surface at sea level is viewed as an almost static reference for many rapidly changing processes and at the same time as a "frozen picture" of tectonic processes that evolved over geological time spans.

Solid Earth Physics: the gravity anomaly field has its origin mainly in mass inhomogenities of the continental and oceanic lithosphere. Together with height information and regional tomography, a much deeper understanding of tectonic processes should be obtainable in the future.

Physical Oceanography: the altimeter satellites in combination with a precise geoid will deliver global dynamic ocean topography. Global surface circulation can be computed resulting in a completely new dimension of ocean modeling. Circulation allows the determination of transport processes of, e.g., polluted material.

Satellite Orbits: for any positioning from space the uncertainty in the orbit of the spacecraft is the limiting factor. The future spaceborne techniques will basically eliminate all gravitational uncertainties in satellite orbits.

Geodesy and Civil Engineering: accurate heights are needed for civil constructions, mapping, etc. They are obtained by leveling, a very time consuming and expensive procedure. Nowadays geometric heights can be obtained fast and efficiently from space positioning (e.g., GPS, GLONASS, and (future) GALILEO). The geometric heights are convertable to leveled heights by subtracting the precise geoid, which is implied by a high-resolution gravitational potential. To be more specific, in those areas where good gravity information is already available, the future data information will eliminate all medium- and long-wavelength distortions in unsurveyed areas. GPS, GLONASS, and GALILEO together with the planned explorer satellite missions for the past 2010 time frame will provide extremely high quality height information at the global scale.

Exploration Geophysics and Prospecting: airborne gravity measurements have usually been used together with aeromagnetic surveys, but the poor precision of air-borne gravity measurements has hindered a wider use of this type of measurements. Strong improvements can be expected by combination with terrestrial and spaceborne observations in the future scenario. The basic interest in gravitational methods in exploration is based on the small variations in the gravitational field anomalies in relation to an ellipsoidal reference model, i.e., the so-called "normal" gravitational field.

8.1.1 Mathematical Modeling of the Gravity Field

Gravity as observed on the Earth's surface is the combined effect of the gravitational mass attraction and the centrifugal force due to the Earth's rotation. The force of gravity provides a directional structure to the space above the Earth's surface. It is tangential to the vertical plumb lines and perpendicular to all level surfaces. Any water surface at rest is part of a level surface. As if the Earth were a homogeneous, spherical body, gravity turns out to be constant all over the Earth's surface, the well-known quantity 9.8 ms^{-2}. The plumb lines are directed toward the

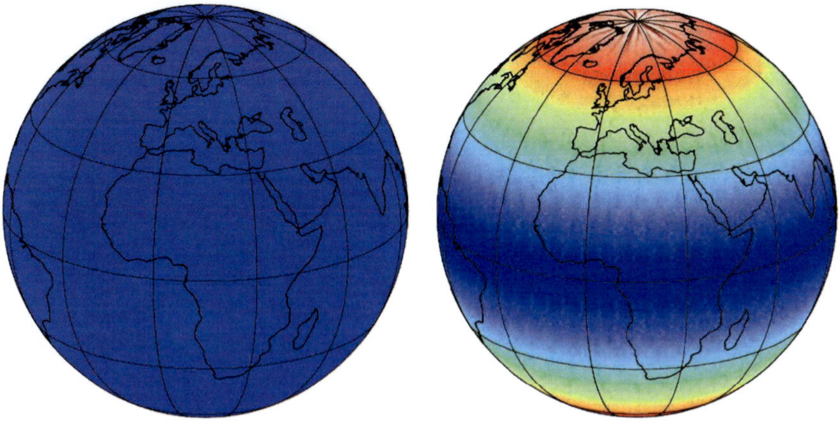

Fig. 8
Illustration of the gravity intensity: constant, i.e., 9.8 ms^{-2} (left), decreasing to the poles by about 0.05 ms^{-2} (right)

Earth's center-of-mass, and this implies that all level surfaces are nearly spherical too. The gravity decreases from the poles to the equator by about 0.05 ms^{-2} (see *Fig. 8*). This is caused by the flattening of the Earth's figure and the negative effect of the centrifugal force, which is maximal at the equator.

High mountains and deep ocean trenches (cf. *Fig. 9*) cause the gravity to vary. Materials within the Earth's interior are not uniformly distributed. The irregular gravity field shapes the geoid as virtual surface. The level surfaces are ideal reference surfaces, e.g., for heights (cf. Helmert 1881).

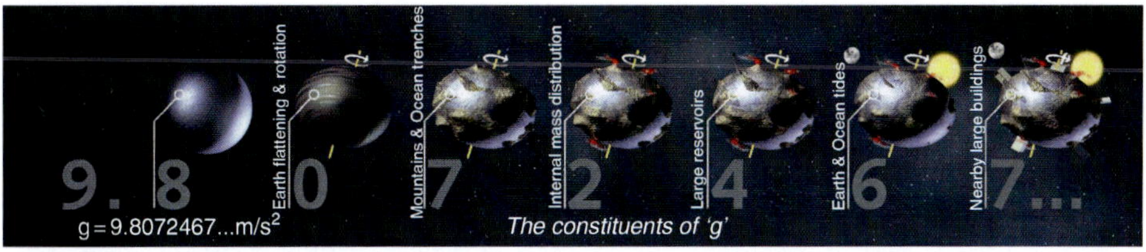

Fig. 9
Illustration of the constituents of the gravity intensity (ESA medialab, ESA communication production SP-1314)

The *gravity acceleration (gravity)* w is the resultant of gravitation v and centrifugal acceleration c (see Chaps. 6–8)

$$w = v + c. \tag{1}$$

The centrifugal force c arises as a result of the rotation of the Earth about its axis. Here, we assume a rotation of constant angular velocity Ω around the rotational axis x_3, which is further assumed to be fixed with respect to the Earth. The centrifugal acceleration acting on a unit mass is directed outward, perpendicularly to the spin axis (see *Fig. 10*). If the ϵ^3-axis of an Earth-fixed coordinate system coincides with the axis of rotation, then we have $c = \nabla C$, where

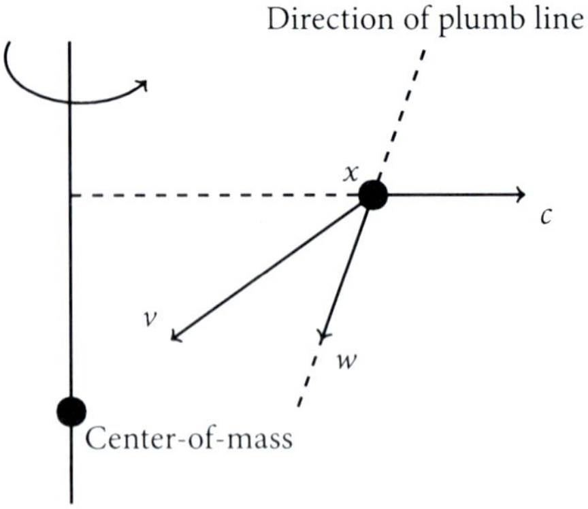

Fig. 10
Gravitation v, centrifugal acceleration c, gravity acceleration w

C is the so-called *centrifugal potential* (with $\{\epsilon^1, \epsilon^2, \epsilon^3\}$ the canonical orthonormal system in Euclidean space $\mathbb{R}^3$). The direction of the gravity w is known as the direction of the *plumb line*, the quantity $g = |w|$ is called the *gravity intensity* (often just *gravity*). The *gravity potential of the Earth* can be expressed in the form:

$$W = V + C. \tag{2}$$

The (vectorial) gravity acceleration w (see Chap. 6 for more details) is given by

$$w = \nabla W = \nabla V + \nabla C. \tag{3}$$

The surfaces of constant gravity potential $W(x) = \text{const}, x \in \mathbb{R}^3$, are designated as *equipotential (level, or geopotential) surfaces of gravity*. The *gravity potential* W of the Earth is the sum of the *gravitational potential* V and the *centrifugal potential* C, i.e., $W = V + C$. In an Earth-fixed coordinate system the centrifugal potential C is explicitly known. Hence, the determination of equipotential surfaces of the potential W is strongly related to the knowledge of the gravitational potential V. The gravity vector w given by $w = \nabla W$ is normal to the equipotential surface passing through the same point. Thus, equipotential surfaces intuitively express the notion of tangential surfaces, as they are normal to the plumb lines given by the direction of the gravity vector (cf. *Fig. 11*).

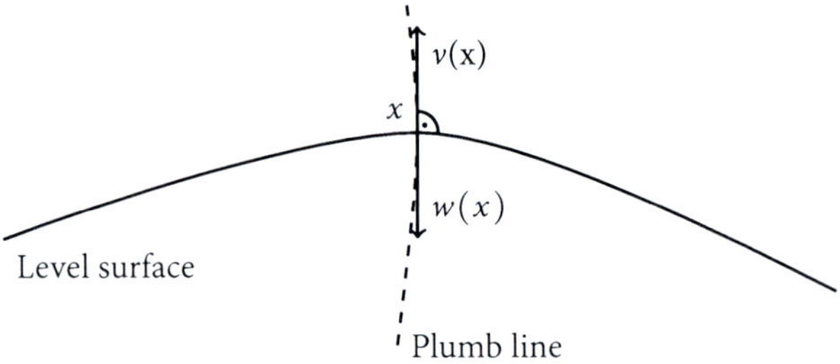

Fig. 11
Level surface and plumb line

The traditional concept in gravitational field modeling is based on the assumption that all over the Earth the position (i.e., latitude and longitude) and the scalar gravity g are available. Moreover, it is common practice that the gravitational effects of the sun, moon, Earth's atmosphere, etc, are accounted for by means of corrections. The gravitational part of the gravity potential can then be regarded as a harmonic function in the exterior of the Earth. A classical approach to gravity field modeling was conceived by Stokes 1849. He proposed to reduce the given gravity accelerations from the Earth's surface to the geoid. As the geoid is a level surface, its potential value is constant. The difference between the reduced gravity on the geoid and the reference gravity on the so-called normal ellipsoid is called the gravity anomaly. The disturbing potential, i.e., the difference between the actual and the reference potential, can be obtained from a (third) boundary value problem of potential theory. Its solution is representable in integral form, i.e., by the Stokes integral. The disadvantage of the Stokes approach is that the reduction to the geoid requires the introduction of assumptions concerning the unknown mass distribution between the Earth's surface and the geoid (for more details concerning the classical theory the reader is referred, e.g., to ➋ Chap. 6 and the references therein).

In this chapter, we briefly recapitulate the classical approach to global gravity field determination by formulating the differential/integral relations between gravity disturbance, gravity anomaly, and deflections of the vertical on the one hand and the disturbing potential and the geoidal undulations on the other. The representation of the disturbing potential in terms of gravity disturbances, gravity anomalies, and deflections of the vertical are written in terms of well-known integral representations over the geoid. For practical purposes the integrals are replaced by approximate formulas using certain integration weights and knots within a spherical framework (see ➋ Chap. 40 for a survey paper on numerical integration on the sphere).

Equipotential surfaces of the gravity potential W allow in general no simple representation. This is the reason why a reference surface – in physical geodesy usually an ellipsoid of revolution – is chosen for the (approximate) construction of the geoid. As a matter of fact, the deviations of the gravity field of the Earth from the normal field of such an ellipsoid are small. The remaining parts of the gravity field are gathered in a so-called *disturbing gravity field* ∇T corresponding to the *disturbing potential* T. Knowing the gravity potential, all equipotential surfaces – including the geoid – are given by an equation of the form $W(x) = \text{const}$. By introducing U as the normal gravity potential corresponding to the ellipsoidal field and T as the disturbing potential we are led to a decomposition of the gravity potential in the form

$$W = U + T \tag{4}$$

such that

(C1) the center of the ellipsoid coincides with the center of gravity of the Earth.
(C2) the difference of the mass of the Earth and the mass of the ellipsoid is zero.

According to the classical Newton Law of Gravitation (1687), knowing the density distribution of a body, the gravitational potential can be computed everywhere in $\mathbb{R}^3$ (see ➋ Chap. 8 for more information in time-space dependent relation). More explicitly, the gravitational potential V of the Earth's exterior is given by (cf. ➋ Chaps. 8 and ➋ 32)

$$V(x) = G \int_{\text{Earth}} \frac{\rho(y)}{|x-y|}\, dV(y), \quad x \in \mathbb{R}^3 \backslash \text{Earth}, \tag{5}$$

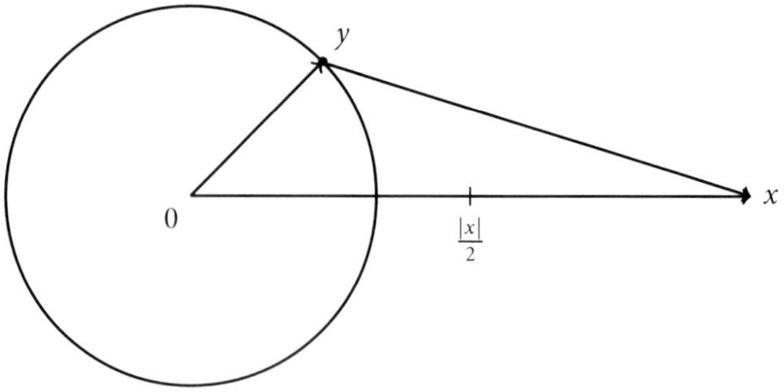

Fig. 12
Regularity at infinity

where G is the gravitational constant ($G = 6.6742 \cdot 10^{-11} \mathrm{m}^3\,\mathrm{kg}^{-1}\,\mathrm{s}^{-2}$) and ρ is the density function. The properties of the gravitational potential (5) in the Earth's exterior are easily described as follows: V is harmonic in $x \in \mathbb{R}^3 \backslash \mathrm{Earth}$, i.e., $\Delta V(x) = 0,\ x \in \mathbb{R}^3 \backslash \mathrm{Earth}$.
Moreover, the gravitational potential V is *regular at infinity*, i.e.,

$$|V(x)| = O\left(\frac{1}{|x|}\right), \quad |x| \to \infty, \tag{6}$$

$$|\nabla V(x)| = O\left(\frac{1}{|x|^2}\right), \quad |x| \to \infty. \tag{7}$$

Note that, for suitably large values $|x|$ (see *Fig. 12*), we have $|y| \leq \frac{1}{2}|x|$, hence, $|x - y| \geq ||x| - |y|| \geq \frac{1}{2}|x|$.

However, the actual problem (cf. *Fig. 2*) is that in reality the density distribution ρ is very irregular and known only for parts of the upper crust of the Earth. Actually, geoscientists would like to know it from measuring the gravitational field. Even if the Earth is supposed to be spherical, the determination of the gravitational potential by integrating Newton's potential is not achievable (see Chap. 32 for inversion methods of Newton's law).

Remark. As already mentioned, the classical remedy avoiding any knowledge of the density inside the Earth are boundary-value problems of potential theory to determine the external gravitational potential from terrestrial data. For reasons of demonstration, however, here we do not follow the standard (Vening-Meinesz) approach in physical geodesy of obtaining the disturbing potential from deflections of the vertical as terrestrial data (as explained in Chap. 6). Our approach is based on the new context developed in Freeden and Schreiner 2009 (see also the references therein).

Furthermore, it should be noted that the determination of the Earth's gravitational potential under the assumptions of nonspherical geometry and terrestrial oblique derivatives in form of a (stochastic) boundary-value problem of potential theory is discussed in Chap. 35. Further basic details concerning oblique derivative problems can be found in Freeden and Michel 2004.

A point x of the geoid is projected onto the point y of the ellipsoid by means of the ellipsoidal normal (see *Fig. 13*). The distance between x and y is called the *geoidal height*, or *geoidal undulation*. The *gravity anomaly vector* is defined as the difference between the gravity vector $w(x)$ and the normal gravity vector $u(y)$, $u = \nabla U$, i.e., $\alpha(x) = w(x) - u(y)$ (see *Fig. 13*). It is also possible to distinguish the vectors w and u at the same point x to get the *gravity disturbance vector* $\delta(x) = w(x) - u(x)$.

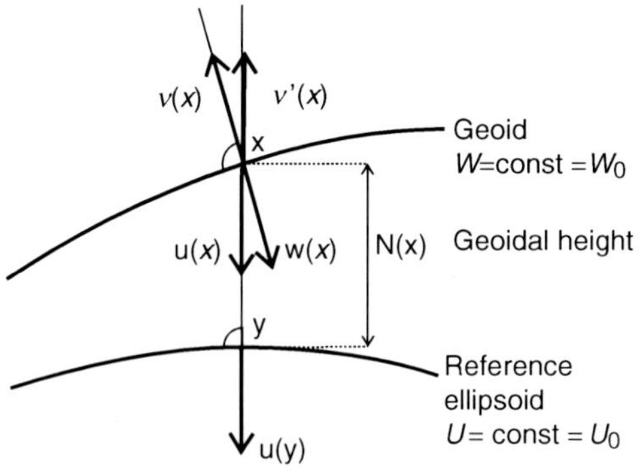

Fig. 13
Illustration of the definition of the gravity anomaly vector $\alpha(x) = w(x) - u(y)$ and the gravity disturbance vector $\delta(x) = w(x) - u(x)$

Of course, several basic mathematical relations between the quantities just mentioned are known. In what follows, we only heuristically describe the fundamental relations. We start by observing that the gravity disturbance vector at the point x can be written as

$$\delta(x) = w(x) - u(x) = \nabla(W(x) - U(x)) = \nabla T(x). \tag{8}$$

Expanding the potential U at x according to Taylor's theorem and truncating the series at the linear term we get

$$U(x) \doteq U(y) + \frac{\partial U}{\partial v'}(y)N(x) \tag{9}$$

($\doteq$ means approximation in linearized sense). Here, $v'(y)$ is the ellipsoidal normal at y, i.e., $v'(y) = -u(y)/\gamma(y)$, $\gamma(y) = |u(y)|$, and the geoid undulation $N(x)$, as indicated in ➋ *Fig. 14*, is the aforementioned distance between x and y, i.e., between the geoid and the reference ellipsoid. Using

$$\gamma(y) = |u(y)| = -v'(y) \cdot u(y) = -v'(y) \cdot \nabla U(y) = -\frac{\partial U}{\partial v'}(y) \tag{10}$$

we arrive at

$$N(x) = \frac{T(x) - (W(x) - U(y))}{|u(y)|} = \frac{T(x) - (W(x) - U(y))}{\gamma(y)}. \tag{11}$$

Letting $U(y) = W(x) = \text{const} = W_0$ we obtain the so-called *Bruns' formula* (cf. Bruns 1878)

$$N(x) = \frac{T(x)}{\gamma(y)}. \tag{12}$$

It should be noted that Bruns' formula (12) relates the physical quantity T to the geometric quantity N.

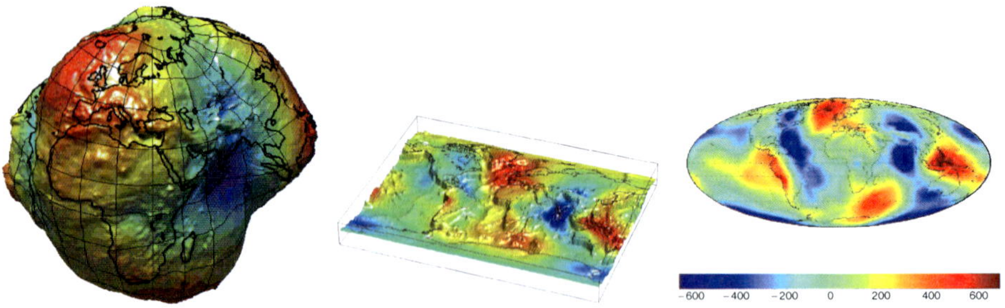

Fig. 14
Three graphical illustrations of the geoidal surface (from EGM96)

In what follows we are interested in introducing the deflections of the vertical of the gravity disturbing potential T. For this purpose, let us consider the vector field $\nu(x) = -w(x)/|w(x)|$. This gives us the identity (with $g(x) = |w(x)|$ and $\gamma(x) = |u(x)|$)

$$w(x) = \nabla W(x) = -|w(x)|\, \nu(x) = -g(x)\nu(x). \tag{13}$$

Furthermore, we have

$$u(x) = \nabla U(x) = -|u(x)|\, \nu'(x) = -\gamma(x)\nu'(x). \tag{14}$$

The *deflection of the vertical* $\Theta(x)$ at the point x on the geoid is defined to be the angular (i.e., tangential) difference between the directions $\nu(x)$ and $\nu'(x)$, i.e., the plumb line and the ellipsoidal normal through the same point:

$$\Theta(x) = \nu(x) - \nu'(x) - \big((\nu(x) - \nu'(x)) \cdot \nu(x)\big)\, \nu(x). \tag{15}$$

Clearly, because of (15), $\Theta(x)$ is orthogonal to $\nu(x)$, i.e., $\Theta(x) \cdot \nu(x) = 0$. Since the plumb lines are orthogonal to the level surfaces of the geoid and the ellipsoid, respectively, the deflections of the vertical give briefly spoken a measure of the gradient of the level surfaces. This aspect will be described in more detail below: From (13) we obtain, in connection with (15),

$$w(x) = \nabla W(x) = -|w(x)| \big(\Theta(x) + \nu'(x) + ((\nu(x) - \nu'(x)) \cdot \nu(x))\nu(x)\big). \tag{16}$$

Altogether we get for the gravity disturbance vector

$$\begin{aligned} w(x) - u(x) = \nabla T(x) = &-|w(x)| \big(\Theta(x) + \big((\nu(x) - \nu'(x)) \cdot \nu(x)\big)\, \nu(x)\big) \\ &- (|w(x)| - |u(x)|)\, \nu'(x). \end{aligned} \tag{17}$$

The magnitude

$$D(x) = |w(x)| - |u(x)| = g(x) - \gamma(x) \tag{18}$$

is called the *gravity disturbance*, while

$$A(x) = |w(x)| - |u(y)| = g(x) - \gamma(y) \tag{19}$$

is called the *gravity anomaly*.

Since the vector $\nu(x) - \nu'(x)$ is (almost) orthogonal to $\nu'(x)$, it can be neglected in (17). Hence, it follows that

$$w(x) - u(x) = \nabla T(x) \doteq -|w(x)|\Theta(x) - (|w(x)| - |u(x)|)\, \nu'(x). \tag{20}$$

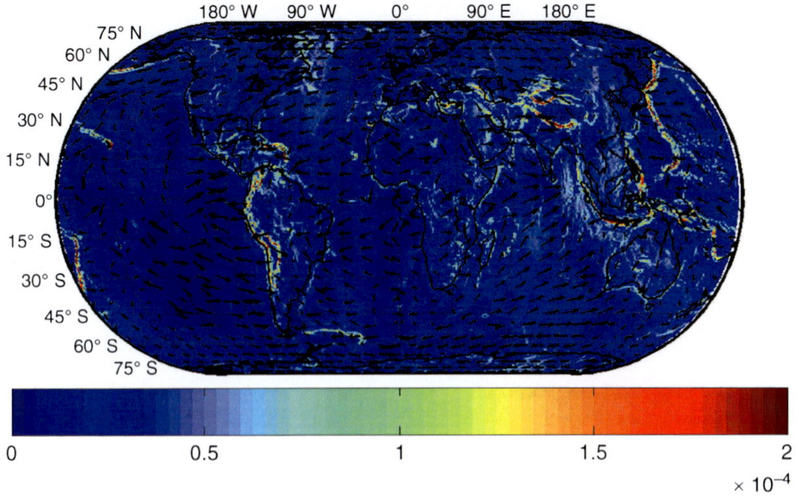

Fig. 15
Absolute values of the deflections of the vertical and their directions computed from EGM96 (cf. Lemoine et al. 1996) from degree 2 up to degree 360 (reconstructed by use of space-limited (locally supported) scaling functions, Geomathematics Group, TU Kaiserslautern, Wolf 2009) (min = 0, max = $3.0 \cdot 10^{-4}$)

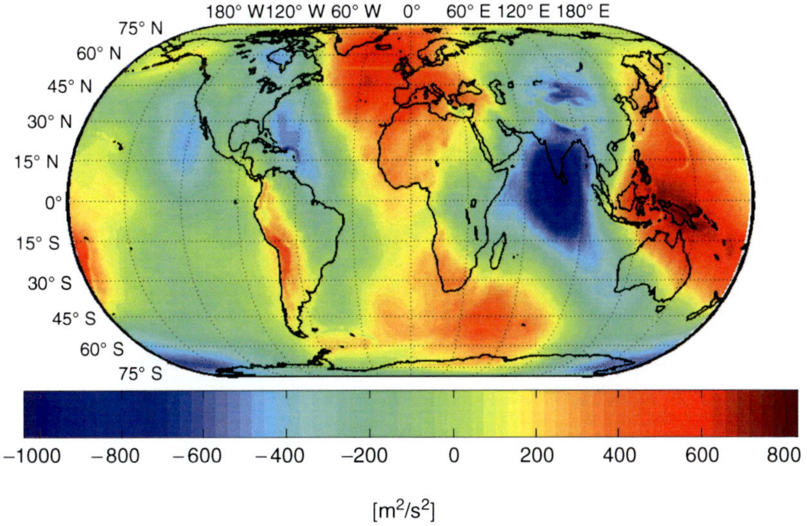

Fig. 16
Disturbing potential computed from EGM96 from degree 2 up to degree 360 (reconstructed by use of space-limited (locally supported) scaling functions, Geomathematics Group, TU Kaiserslautern, Wolf 2009) (min ≈ $-1038\,m^2/s^2$, max ≈ $833\,m^2/s^2$)

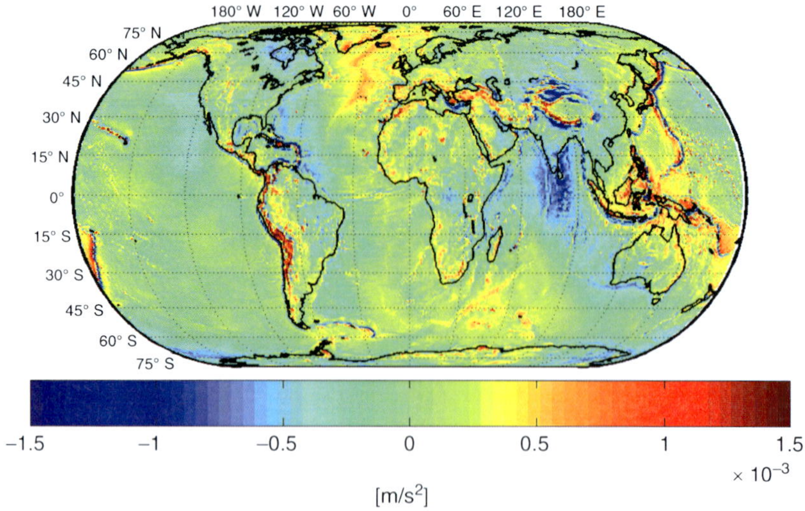

Fig. 17
Gravity disturbances computed from EGM96 from degree 2 up to degree 360 (reconstructed by use of space-limited (locally supported) scaling functions, Geomathematics Group, TU Kaiserslautern, Wolf 2009) (min = $-3.6 \cdot 10^{-3}$ m/s^2, max = $4.4 \cdot 10^{-3}$ m/s^2)

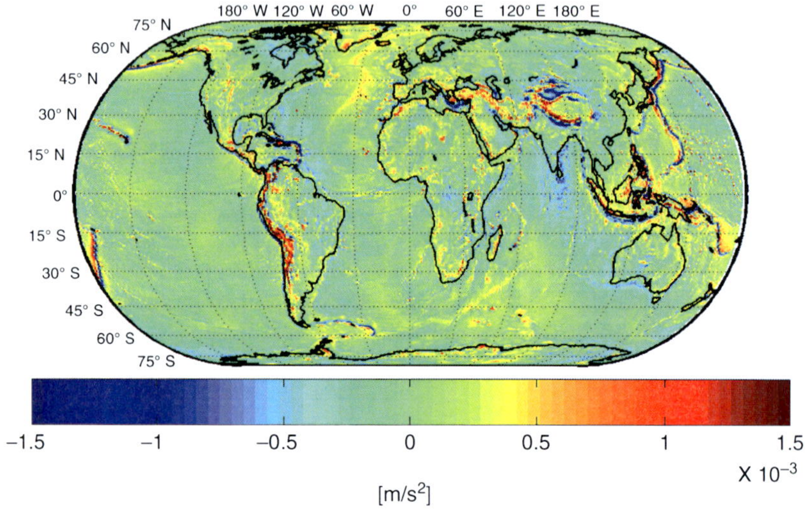

Fig. 18
Gravity anomalies computed from EGM96 from degree 2 up to degree 360 (reconstructed by use of space-limited (i.e., locally supported) scaling functions, Geomathematics Group, TU Kaiserslautern, Wolf 2009) (min = $-3.4 \cdot 10^{-3}$ m/s^2, max = $4.4 \cdot 10^{-3}$ m/s^2)

The gradient $\nabla T(x)$ can be split into a normal part (pointing in the direction of $\nu(x)$) and an angular (tangential) part (characterized by the surface gradient ∇^*). It follows that

$$\nabla T(x) = \frac{\partial T}{\partial \nu}(x)\nu(x) + \frac{1}{|x|}\nabla^* T(x). \tag{21}$$

By comparison of (20) and (21), we therefore obtain

$$D(x) = g(x) - \gamma(x) = |w(x)| - |u(x)| = -\frac{\partial T}{\partial \nu'}(x), \tag{22}$$

i.e., the gravity disturbance, beside being the difference in magnitude of the actual and the normal gravity vector, is also the normal component of the gravity disturbance vector. In addition, we are led to the angular, i.e., (tangential) differential equation

$$\frac{1}{|x|}\nabla^* T(x) = -|w(x)|\,\Theta(x). \tag{23}$$

Remark. Since $|\Theta(x)|$ is a small quantity, it may be (without loss of precision) multiplied either by $-|w(x)|$ or by $-|u(x)|$, i.e., $-g(x)$ or by $-\gamma(x)$.

Remark. The reference ellipsoid deviates from a sphere only by quantities of the order of the flattening. Therefore, in numerical calculations, we treat the reference ellipsoid as a sphere around the origin with (certain) mean radius R. This may cause a relative error of the same order (for more details the reader is referred to standard textbooks of physical geodesy). In this way, together with suitable pre-reduction processes of gravity, formulas are obtained that are rigorously valid for the sphere.

Remark. An ellipsoidal approach to the gravity field determination including the theory of ellipsoidal functions (ellipsoidal harmonics) is given in ➋ Chap. 7 of this handbook. This approach certainly is more realistic from geodetic view, but it amounts to much more numerical efforts.

If now – as explained above – the relative error between normal ellipsoid and mean sphere of radius R is permissible, we are allowed to go over to the spherical nomenclature $x = R\xi$, $R = |x|$, $\xi \in \mathbb{S}^2$ with $\mathbb{S}^2$ being the unit sphere in $\mathbb{R}^3$. Replacing $|u(R\xi)|$ by its spherical approximation GM/R^2 we find

$$\nabla_\xi^* T(R\xi) = -\frac{GM}{R}\Theta(R\xi), \quad \xi \in \mathbb{S}^2, \tag{24}$$

where G is the gravitational constant and M is the mass.

By virtue of Bruns' formula we finally find the relation between geoidal undulations and deflections of the vertical

$$\frac{GM}{R^2}\nabla_\xi^* N(R\xi) = -\frac{GM}{R}\Theta(R\xi), \quad \xi \in \mathbb{S}^2, \tag{25}$$

i.e.,

$$\nabla_\xi^* N(R\xi) = -R\,\Theta(R\xi), \quad \xi \in \mathbb{S}^2. \tag{26}$$

In other words, the knowledge of the geoidal undulations allows the determination of the deflections of the vertical by taking the surface gradient on the unit sphere.

From the identity (22) it follows that

$$-\frac{\partial T}{\partial \nu'}(x) = D(x) = |w(x)| - |y(x)| \doteq |w(x)| - |\gamma(y)| - \frac{\partial \gamma}{\partial \nu'}(y)\, N(x)$$
$$= A(x) - \frac{\partial \gamma}{\partial \nu'}(y)\, N(x), \tag{27}$$

where A represents the scalar gravity anomaly as defined by (19). Observing Bruns' formula, we get

$$A(x) = -\frac{\partial T}{\partial \nu'}(x) + \frac{1}{\gamma(y)}\frac{\partial \gamma}{\partial \nu'}(y)\, T(x). \tag{28}$$

In well-known *spherical approximation* we have

$$\gamma(y) = |u(y)| = \frac{GM}{|y|^2}, \tag{29}$$

$$\frac{\partial \gamma}{\partial \nu'}(y) = \frac{y}{|y|} \cdot \nabla_y \gamma(y) = -2\frac{GM}{|y|^3} \tag{30}$$

and

$$\frac{1}{\gamma(y)}\frac{\partial \gamma}{\partial \nu'}(y) = -\frac{2}{|y|}. \tag{31}$$

This leads us to the basic relations

$$-D(x) = \frac{x}{|x|} \cdot \nabla T(x), \quad |x| = R, \tag{32}$$

and

$$-A(x) = \frac{x}{|x|} \cdot \nabla T(x) + \frac{2}{|x|} T(x), \quad |x| = R, \tag{33}$$

as so-called *fundamental equations of physical geodesy*. In the sense of physical geodesy, the meaning of the spherical approximation should be carefully kept in mind. It is used only for expressions relating to small quantities of the disturbing potential, the geoidal undulations, the gravity disturbances, the gravity anomalies, etc. Clearly, as already noted, together with the Laplace equation in the exterior of the sphere around the origin with radius R and the regularity at infinity, the (boundary) relations (32) and (33) form boundary-value problems of potential theory (cf. Neumann 1887, Stokes 1849), which are known as Neumann and Stokes problems, respectively (see ➋ Chap. 6 for more details).

➋ *Figures 15–18* give rise to the following remarks: the gravity disturbances, which enable a physically oriented comparison of the real Earth with the ellipsoidal Earth model, are consequences of the imbalance of forces inside the Earth according to Newton's law of gravitation. It leads to the suggestion of density anomalies. In analogy the difference between the actual level surfaces of the gravity potential and the level surfaces of the model body form a measure for the deviation of the Earth from a hydrostatic status of balance. In particular, the geoidal undulations (geoidal heights) represent the deflections from the equipotential surface on the mean level of the ellipsoid. The geoidal anomalies generally show no essential correlation to the distribution of the continents (see ➋ *Fig. 19*).

It is conjecturable that the geoidal undulations mainly depend on the reciprocal distance of the density anomaly. They are influenced by lateral density variations of large vertical extension, from the core-mantle-layer to the crustal layers. In fact, the direct geoidal signal, which

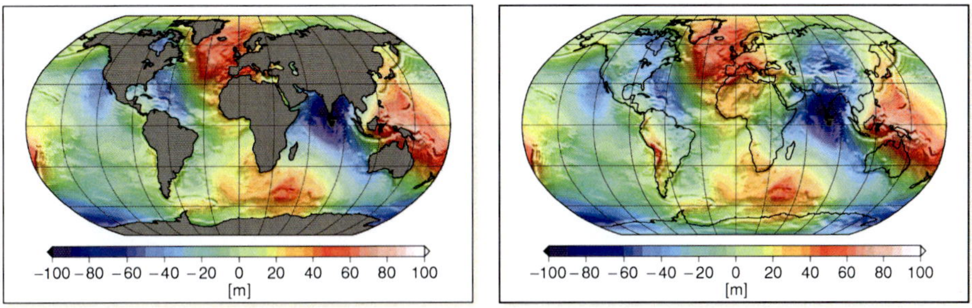

Fig. 19
Geoidal surface over oceans (left) and over the whole Earth (right), modeled by smoothed Haar wavelets (with $k = 5$)

would result from the attraction of the continental and the oceanic bottom, would be several hundreds of meters (for more details see Rummel 2002 and the references therein). In consequence, the weight of the continental and oceanic masses in the Earth's interior are almost perfectly balanced. This is the phenomenon of isostatic compensation that already was observed by P. Bouguer in the year 1750.

8.1.2 Mathematical Analysis

Equation (26) under consideration is of vectorial tangential type with the unit sphere $\mathbb{S}^2$ in Euclidian space $\mathbb{R}^3$ as the domain of definition. More precisely, we are confronted with a surface gradient equation $\nabla^* P = p$ with p as given continuous vector field (i.e., $p(\xi) = -R\Theta(R\xi)$, $\xi \in \mathbb{S}^2$) and P as continuously differentiable scalar field that is desired to be reconstructed. In the so far formulated abstraction, the determination of the *surface potential function* P via the equation $\nabla^* P = p$ is certainly not uniquely solvable. We are able to add an arbitrary constant to P without changing the equation. For our problem, however, this argument is not valid, since we have to observe additional integration relations resulting from the conditions (C1) and (C2), namely

$$\int_{\mathbb{S}^2} P(\xi)(\varepsilon^i \cdot \xi)^k dS(\xi) = 0; \quad k = 0, 1; \; i = 1, 2, 3. \tag{34}$$

Hereby, we are able to guarantee uniqueness. The solution theory is based on the Green theorem (cf. Freeden et al. 1998; Freeden and Schreiner 2009; see also ➔ Chap. 18 for its role in geomagnetism)

$$P(\xi) = \frac{1}{4\pi} \int_{\mathbb{S}^2} P(\eta)\, dS(\eta) - \int_{\mathbb{S}^2} \nabla^*_\eta G(\xi \cdot \eta) \cdot \nabla^*_\eta P(\eta)\, dS(\eta), \tag{35}$$

where

$$t \mapsto G(t) = \frac{1}{4\pi} \ln(1 - t) + \frac{1}{4\pi}(1 - \ln 2), \quad t \in [-1, 1) \tag{36}$$

is the *Green function (i.e., the fundamental solution) with respect to the (Laplace)–Beltrami operator on the unit sphere* $\mathbb{S}^2$. This leads us to the following result: suppose that p is a given continuous tangential, curl free vector field on the unit sphere $\mathbb{S}^2$. Then

$$P(\xi) = \frac{1}{4\pi} \int_{\mathbb{S}^2} \frac{1}{1 - \xi \cdot \eta} (\xi - (\xi \cdot \eta)\eta) \cdot p(\eta)\, dS(\eta), \quad \xi \in \mathbb{S}^2 \tag{37}$$

is the uniquely determined solution of the *surface gradient equation* $\nabla^* P = p$ with

$$\int_{\mathbb{S}^2} P(\eta)\, dS(\eta) = 0. \tag{38}$$

Consequently, the existence and uniqueness of the equation is assured. Even more, the solution admits a representation in form of the singular integral (37).

8.1.3 Development of a Mathematical Solution Method

Precise terrestrial data of the deflections of the vertical are not available in dense distribution. They exist, e.g., on continental areas in much larger density than on oceanic ones. In order to exhaust the existent scattered data reservoir we are not allowed to apply a Fourier technique in terms of spherical harmonics (note that, according to Weyl 1916, the integrability is equivalently interrelated to the equidistribution of the datapoints). Instead we have to use an appropriate zooming in procedure which starts globally from (initial) rough data width (low scale) and proceeds to more and more finer local data width (higher scales). A simple solution for a multiscale approximation consists of appropriate "regularization" of the singular kernel $\Phi : t \mapsto \Phi(t) = (1-t)^{-1}$, $-1 \le t < 1$, in (35) by the continuous kernel $\Phi_\rho : t \mapsto \Phi_\rho(t)$, $t \in [-1, 1]$, given by

$$\Phi_\rho(t) = \begin{cases} \dfrac{1}{1-t}, & 1-t > \rho, \\ \dfrac{1}{\rho}, & 1-t \le \rho. \end{cases} \tag{39}$$

In fact, it is not difficult to verify (see Freeden and Schreiner 2009) that

$$P_\rho(\xi) = \frac{1}{4\pi} \int_{\mathbb{S}^2} \Phi_\rho(\xi\cdot\eta)(\xi - (\xi\cdot\eta)\eta)\cdot p(\eta)\, dS(\eta), \quad \xi \in \mathbb{S}^2, \tag{40}$$

where the *surface curl-free space-regularized Green vector scaling kernel* (see ➋ *Fig. 20*) is given by

$$(\xi,\eta) \mapsto \Phi_\rho(\xi\cdot\eta)(\xi - (\xi\cdot\eta)\eta), \quad \xi,\eta \in \mathbb{S}^2, \tag{41}$$

satisfies the following limit relation

$$\lim_{\substack{\rho\to 0\\ \rho>0}} \sup_{\xi\in\mathbb{S}^2} |P(\xi) - P_\rho(\xi)| = 0. \tag{42}$$

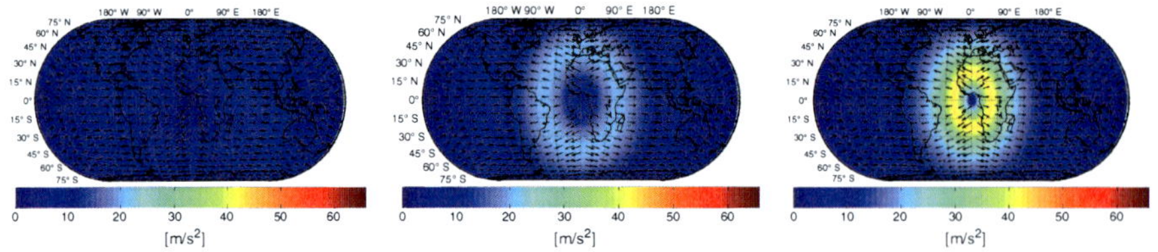

◘ Fig. 20

Absolute values and the directions of the surface curl-free space-regularized Green vector scaling kernel $\phi_j(\xi,\eta)$, $\xi,\eta \in \mathbb{S}^2$, as defined by (43) with ξ fixed and located at 0° N 0° W, for scales $j = 1, 2, 3$, respectively

Furthermore, in the scale discrete formulation using a strictly monotonically decreasing sequence $(\rho_j)_{j\in\mathbb{N}_0}$ converging to zero with $\rho_j \in (0,2)$ (for instance, $\rho_j = 2^{1-j}$ or $\rho_j = 1 - \cos(2^{-j}\pi)$, $j \in \mathbb{N}_0$), we are allowed to write the *surface curl-free space-regularized Green vector scaling kernel* as follows

$$\phi_j(\xi,\eta) = \Phi_{\rho_j}(\xi\cdot\eta)(\xi - (\xi\cdot\eta)\eta), \quad \xi,\eta \in \mathbb{S}^2. \tag{43}$$

The *surface curl-free space-regularized Green vector wavelet kernel* then reads as

$$\psi_j(\xi,\eta) = \Psi_{\rho_j}(\xi\cdot\eta)(\xi - (\xi\cdot\eta)\eta), \quad \xi,\eta \in \mathbb{S}^2, \tag{44}$$

where

$$\Psi_{\rho_j} = \Phi_{\rho_{j+1}} - \Phi_{\rho_j} \tag{45}$$

is explicitly given by

$$\Psi_{\rho_j}(t) = \begin{cases} 0, & 1-t > \rho_j, \\ \dfrac{1}{1-t} - \dfrac{1}{\rho_j}, & \rho_j \geq 1-t > \rho_{j+1}, \\ \dfrac{1}{\rho_{j+1}} - \dfrac{1}{\rho_j}, & \rho_{j+1} > 1-t. \end{cases} \tag{46}$$

With W_{ρ_j} defined by

$$W_{\rho_j}(\xi) = \frac{1}{4\pi}\int_{\mathbb{S}^2} \psi_j(\xi,\eta)\cdot p(\eta)\, dS(\eta), \quad \xi \in \mathbb{S}^2, \tag{47}$$

we are led to the recursion

$$P_{\rho_{j+1}} = P_{\rho_j} + W_{\rho_j}. \tag{48}$$

Hence, we immediately get by elementary manipulations for all $m \in \mathbb{N}$

$$P_{\rho_{j+m}} = P_{\rho_j} + \sum_{k=0}^{m-1} W_{\rho_{j+k}}. \tag{49}$$

The functions P_{ρ_j} represent low-pass filters of P, P_{ρ_j} is improved by the band-pass filter W_{ρ_j} in order to obtain $P_{\rho_{j+1}}$, $P_{\rho_{j+1}}$ is improved by the band-pass filter $W_{\rho_{j+1}}$ in order to obtain $P_{\rho_{j+2}}$, etc.

Summarizing our results we are allowed to formulate the following conclusion: three features are incorporated in our way of thinking about multiscale approximation by the use of locally supported wavelets, namely basis property, decorrelation, and fast computation. More concretely, our vector wavelets are "building blocks" for huge discrete data sets. By virtue of the *basis property* the function P can be better and better approximated from p with increasing scale j. Our wavelets have the power to decorrelate. In other words, the representation of data in terms of wavelets is somehow "more compact" than the original representation. We search for an accurate approximation by only using a small fraction of the original information of a potential. Typically, our *decorrelation* is achieved by vector wavelets which have a compact support (localization in space) and which show a decay toward high frequencies. The main calamity in multiscale approximation is how to *decompose* the function under consideration into wavelet coefficients, and how to efficiently reconstruct the function from the coefficients. There is a "tree algorithm" (cf. also ➋ *Fig.* 6) that makes these steps simple and fast:

$$\begin{array}{ccccccccc} W^{\rho_j} & & & & W^{\rho_{j+1}} & & & & \\ & \searrow & & & & \searrow & & & \\ P^{\rho_j} & \longrightarrow & \oplus & \longrightarrow & P^{\rho_{j+1}} & \longrightarrow & \oplus & \longrightarrow & P^{\rho_{j+2}} \quad \ldots. \end{array}$$

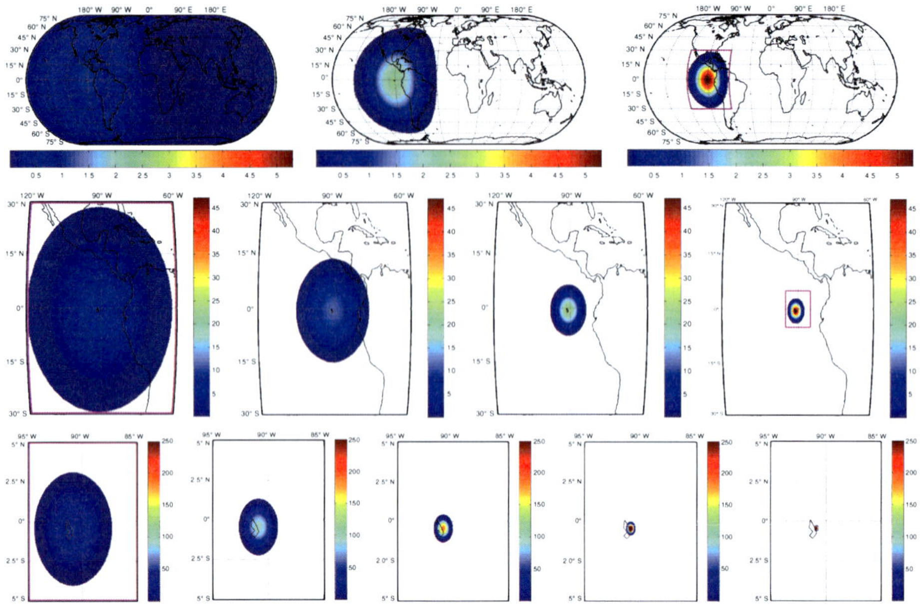

Fig. 21
"Zooming-in"-strategy with a hotspot (here, Galapagos (0° N ,91°W)) as target area (the colored areas illustrate the local support of the wavelets for increasing scale levels, Geomathematics Group, Kaiserslautern, Wolf 2009)

The fast decorrelation power of wavelets is the key to applications such as data compression, fast data transmission, noise cancellation, signal recovering, etc. With increasing scales $j \to \infty$ the supports become smaller and smaller. This is the reason why the calculation of the integrals has to be extended over smaller and smaller caps ("zooming-in"). Of course, downsizing spherical caps and increasing data widths are in strong correlation (➲ *Fig. 21*). Thus, the variable width of the caps with increasing scale parameter j enables the integration of data sets of heterogenous data width for local areas without violating Weyl's law of equidistribution.

8.1.4 "Back-transfer" to Application

The multiscale techniques as presented here will be used to investigate the anomalous gravity field particularly for areas, in which mantle plumes and hotspots occur. In this respect it should be noted that *mantle plume* is a geoscientifical term which denotes an upwelling of abnormally hot rocks within the Earth's mantle (cf. ➲ *Fig. 7*). Plumes are envisioned to be vertical conduits in which the hot mantle material rises buoyantly from the lower mantle to the lithosphere at velocities as large as $1\,\text{m}\,\text{yr}^{-1}$ and these quasicylindrical regions have a typical diameter of about 100–200 km. In mantle convection theory, mantle plumes are related to *hotspots* which describe centers of surface volcanism that are not directly caused by plate tectonic processes. A hotspot is a long-term source of volcanism which is fixed relative to the plate overriding it. A classical example is Hawaii (see Ritter and Christensen 2007 for more details). Due to the local nature of plumes and hotspots such as Hawaii we have to use high-resolution

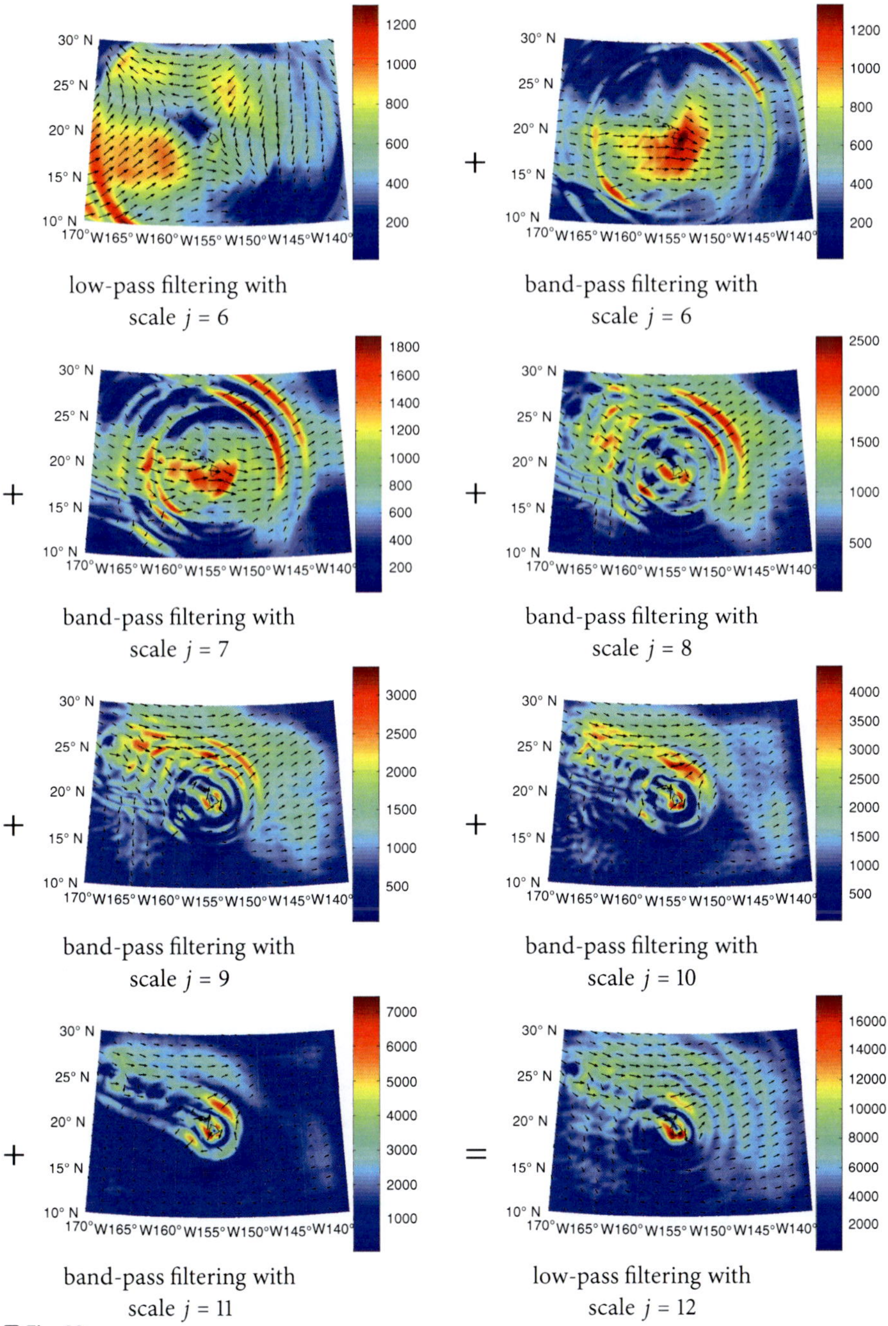

Fig. 22

Approximation of the vector valued vertical deflections Θ in $[ms^{-2}]$ of the Hawaiian region with smoothed Haar wavelets (a rough low-pass filtering at scale 6 is improved with several band-pass filters of scale $j = 6, \ldots, 11$, the last picture shows the multiscale approximation at scale $j = 12$, numerical and graphical illustration with Fehlinger 2008

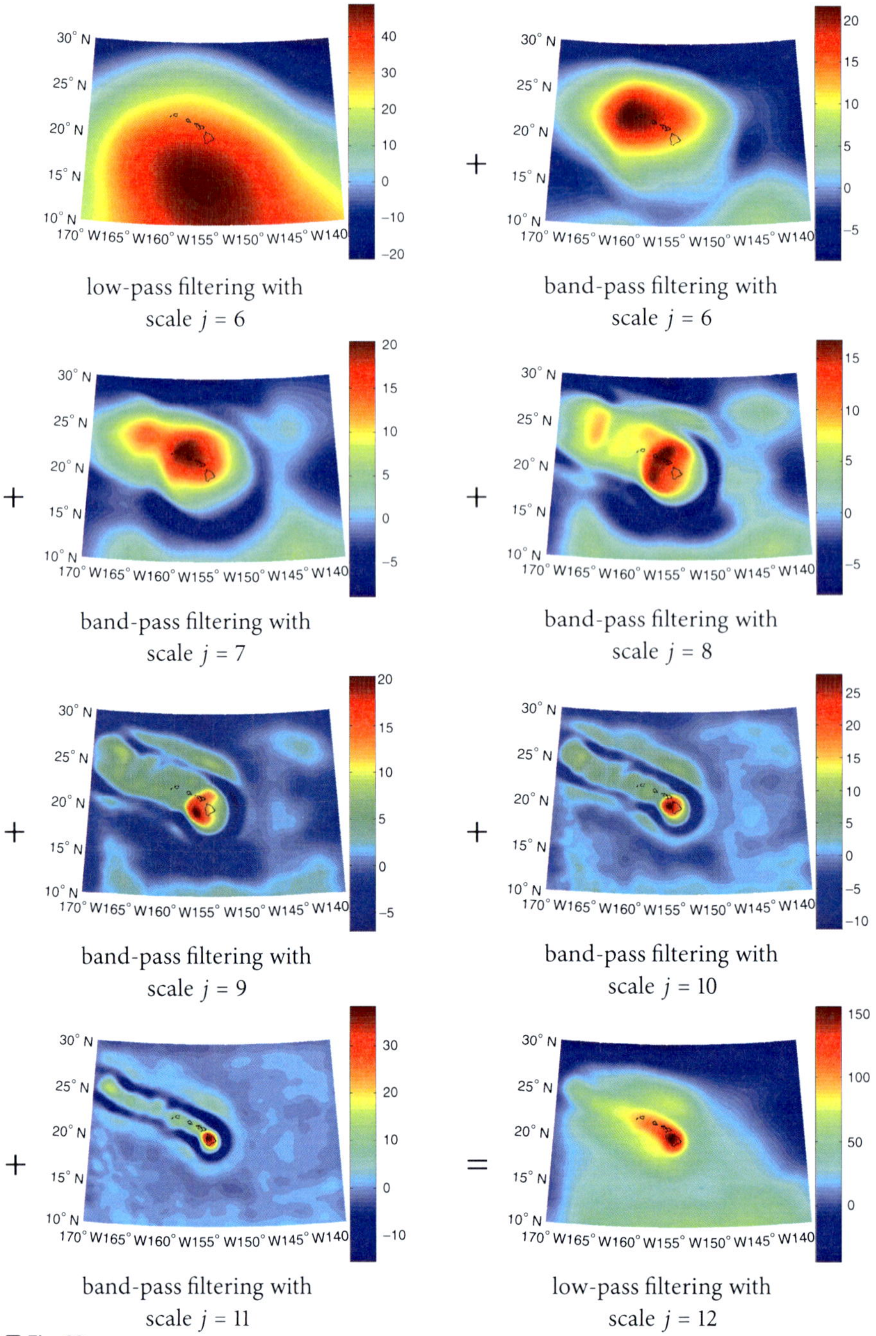

Fig. 23

Multiscale reconstruction of the disturbing potential T in $[m^2 s^{-2}]$ from vertical deflections for the Hawaiian (plume) area using regularized vector Green functions (a rough low-pass filtering at scale $j = 6$ is improved with several band-pass filters of scale $j = 6, \ldots, 11$, the last illustration shows the approximation of the disturbing potential T at scale $j = 12$, numerical and graphical illustration with Fehlinger 2008

gravity models. Because of the lack of terrestrial – only data the GFZ-combined gravity model EIGEN-GLO4C is used, that consists of satellite data, gravimetry, and altimetry surface data. The "zooming-in" property of our analysis is of great advantage. Especially the locally compact wavelet turns out to be an essential tool of the (vectorial) multiscale decomposition of the deflections of the vertical and correspondingly the (scalar) multiscale approximation of the disturbing potential.

In the multiscale analysis (cf. ➋ *Figs. 22* and ➋ *23*) several interesting observations can be detected. By comparing the different positions and different scales, we can see that the maximum of the "energy" contained in the signal of the disturbing potential – measured in so-called wavelet variances (see Freeden and Michel 2004 for more details) – starts in the North West of the Hawaiian islands for scale $j = 2$ and travels in eastsouthern direction with increasing scale. It ends up, for scale $j = 12$, in a position at the geologically youngest island under which the mantle plume is assumed to exist. Moreover, in the multiscale resolution with increasing scale, more and more local details of the disturbing potential appear. In particular, the structure of the Hawaiian island chain is clearly reflected in the scale and space decomposition. Obviously, the "energy-peak" observed at the youngest island of Hawaii is highly above the "energy-intensity-level" of the rest of the island chain. This seems to strongly corroborate the belief of a stationary mantle plume, which is located beneath the Hawaiian islands and that is responsible for the creation of the Hawaii-Emperor seamount chain, while the oceanic lithosphere is continuously passing over it. An interesting area is the southeastern part of the chain, situated on the Hawaiian swell, a 1,200 km broad anomalously shallow region of the ocean floor, extending from the island of Hawaii to the Midway atoll. Here a distinct geoidal anomaly occurs that has its maximum around the youngest island that coincides with the maximum topography, and both decrease in northwestern direction.

8.2 Circuit: Oceanic Circulation from Ocean Topography

Ocean flow (see also ➋ Chaps. 12 and ➋ 13) has a great influence on mass transport and heat exchange. By modeling oceanic currents we therefore gain, for instance, a better understanding of weather and climate. In what follows we devote our attention to the geostrophic oceanic circulation on bounded regions on the sphere (and in a first approximation the oceanic surfaces under consideration may be assumed to be parts of the boundary of a spherical Earth model), i.e., to oceanic flow under the simplifying assumptions of stationarity, spherically reflected horizontal velocity, and strict neglect of inner frictions. This leads us to inner-oceanic long-scale currents, which still give meaningful results – as, e.g., for the phenomenon of El Niño.

8.2.1 Mathematical Modeling of Ocean Flow

The numerical simulation of ocean currents is based on the Navier-Stokes equation. Its formulation (see, e.g., Ansorge, Sonar 2009) is well-known: Let us consider a fluid occupying an arbitrary (open and bounded) subdomain $G_0 \subset \mathbb{R}^3$ at time $t = 0$. The vector function $v : [0, t_{\text{end}}] \times G_0 \to G_t \subset \mathbb{R}^3$ describes the motion of the particle positions $\pi \in G_0$ with time, so that, at times $t \geq 0$ the fluid occupies the domain $G_t = \{v(t;\pi) | \pi \in G_0\}$, respectively. Hence, G_t is a closed system in the sense that no fluid particle flows across its boundaries. The path of a particle $\pi \in G_0$ is given by the graph of the function $t \mapsto v(t;\pi)$, and the velocity of the fluid at a fixed location $x = v(t;\pi) \in G_t$ by the derivative $u(t;x) = \frac{\partial}{\partial t} v(t;\pi)$. The derivation of the governing equations relies on the conservation of mass and momentum. The essential tool is the transport theorem, which shows how the time derivative of an integral over a domain changing with the time may be computed. The mass of a fluid occupying a domain is determined by the integral over the density of the fluid ρ. Since the same amount of fluid occupying the domain at time 0 later occupies the domain at time $t > 0$, we must have that $\int_{G_0} \rho(0;x)\, dV(x)$ coincides with $\int_{G_t} \rho(t;x)\, dV(x)$ for all $t \in (0, t_{\text{end}}]$. Therefore, the derivative of mass with respect to time must vanish, upon which the transport theorem yields for all t and G_t

$$\int_{G_t} \left(\frac{\partial}{\partial t} \rho(t;x) + \text{div}\,(\rho u)(t;x) \right) dV(x) = 0. \tag{50}$$

Since this is valid for arbitrary regions (in particular, for arbitrarily small ones), this implies that the integrand itself vanishes, which yields the *continuity equation for compressible fluids*

$$\frac{\partial}{\partial t} \rho + \text{div}\,(\rho u) = 0. \tag{51}$$

The momentum of a solid body is the product of its mass with its velocity

$$\int_{G_t} \rho(t;x) u(t;x)\, dV(x). \tag{52}$$

According to Newton's second law, the same rate of change of (linear) momentum is equal to the sum of the forces acting on the fluid. We distinguish two types of forces, viz. body forces k (e.g., gravity, Coriolis force), which can be expressed as $\int_{G_t} \rho(t;x) k(t;x)\, dV(x)$ with a given force density k per unit volume and surface forces (e.g., pressure, internal friction) representable as $\int_{\partial G_t} \boldsymbol{\sigma}(t;x) \nu(x)\, dS(x)$, which includes the stress tensor $\boldsymbol{\sigma}(t;x)$. Thus Newton's law reads

$$\frac{\mathrm{d}}{\mathrm{dt}} \int_{G_t} \rho(t;x) u(t;x)\, dV(x) = \int_{G_t} \rho(t;x) k(t;x)\, dV(x) + \int_{\partial G_t} \boldsymbol{\sigma}(t;x) \nu(x)\, dS(x). \tag{53}$$

If we now apply the product rule and the transport theorem componentwise to the term on the left and apply the divergence theorem to the second term on the right, we obtain the momentum equation

$$\frac{\partial}{\partial t} (\rho u) + (u \cdot \nabla)(\rho u) + (\rho u) \nabla \cdot u = \rho k + \nabla \cdot \boldsymbol{\sigma}. \tag{54}$$

The nature of the oceanic flow equation depends heavily on the model used for the stress tensor. In the special case of incompressible fluids (here, ocean water) that is characterized by a density $\rho(t;x) = \rho_0 =$ const dependent neither on space nor on time, we find $\nabla \cdot u = 0$, i.e., u is divergence-free. When modeling an inviscid fluid, internal friction is neglected and the stress tensor is determined solely by the pressure $\boldsymbol{\sigma}(t;x) = -P(t;x)\mathbf{i}$ ($\mathbf{i}$ is the unit matrix). In the absence of inner friction (in consequence of, e.g., effects of wind and surface influences) we are able to ignore the derivative $\frac{\partial u}{\partial t}$ and, hence, the dependence of time. As relevant volume forces k, the gravity field w and the coriolis force $c = 2u \wedge \omega$ remain valid; they have to be observed. Finally, for large-scale currents of the ocean, the nonlinear part does not play any role, i.e., the term $(u \cdot \nabla) u$ is negligible. Under all these very restrictive assumptions the equation of motion (54) reduces to the following identity

$$2\omega \wedge u = -\frac{\nabla P}{\rho_0} - w. \tag{55}$$

Even more, we suppose a velocity field of a spherical layer model. For each layer, i.e., for each sphere around the origin 0 with radius $r(\leq R)$, the velocity field u can be decomposed into a normal field u_{nor} and a tangential field u_{tan}. The normal part is negligibly small in comparison with the tangential part (see the considerations in Pedlovsky 1979). Therefore, we obtain with $\omega = \Omega(\xi \cdot \epsilon^3)\xi$ (the expression $C(\xi) = 2\Omega(\epsilon^3 \cdot \xi)$ is called *Coriolis parameter*) the following separations of Eq. (55) (observe that $\xi \cdot u_{\text{tan}}(r\xi) = 0$, $\xi \in \mathbb{S}^2$)

$$C(\xi)\xi \wedge u_{\text{tan}}(r\xi) = -\frac{1}{\rho_0 r}\nabla^*_\xi P(r\xi) \tag{56}$$

and

$$(C(\xi)\xi \wedge u_{\text{tan}}(r\xi)) \cdot \xi = -\frac{1}{\rho_0}\frac{\partial}{\partial r}P(r\xi) + g_r. \tag{57}$$

Equation (56) essentially tells us that the tangential surface gradient is balanced by the Coriolis force. For simplicity, in our approach, we regard the gravity acceleration as a normal field: $w(r\xi) = -g_r\xi$, $\xi \in \mathbb{S}^2$ (with g_r as mean gravity intensity). Moreover, the vertical Coriolis acceleration in comparison with the tangential motion is very small, i.e., we are allowed to assume that $(2\Omega(\xi \cdot \epsilon^3)\xi \wedge u_{\text{tan}}(R\xi)) \cdot \xi = 0$. On the surface of the Earth (here, $r = R$), we then obtain from (57) a direct relation of the product of the mean density and the mean gravity acceleration to the normal pressure gradient (hydrostatic approximation): $\frac{\partial}{\partial r}P(R\xi) = \rho_0 g_r$. This is the reason why we obtain the pressure by integration (see also ➋ Chaps. 13 and ➋ 20 of this handbook) as follows $P(R\xi) = \rho_0 g_R\, \Xi(R\xi) + P_{\text{Atm}}$, where P_{Atm} denotes the mean atmospheric pressure. The quantity $\Xi(R\xi)$ (cf. ➋ *Fig. 24*) is the difference between the heights of the ocean surface and the geoid at the point $\xi \in \mathbb{S}^2$. The scalar function $\xi \mapsto \Xi(R\xi)$, $\xi \in \mathbb{S}^2$, is called *ocean topography*. By use of altimeter satellites, we are able to measure the difference H between the known satellite height H_{Sat}, and the (unknown) height of the ocean surface H_{Ocean}: $H = H_{\text{Sat}} - H_{\text{Ocean}}$. After calculation of H_{Ocean} we then get the ocean topography $\Xi = H_{\text{Ocean}} - H_{\text{Geoid}}$ from the

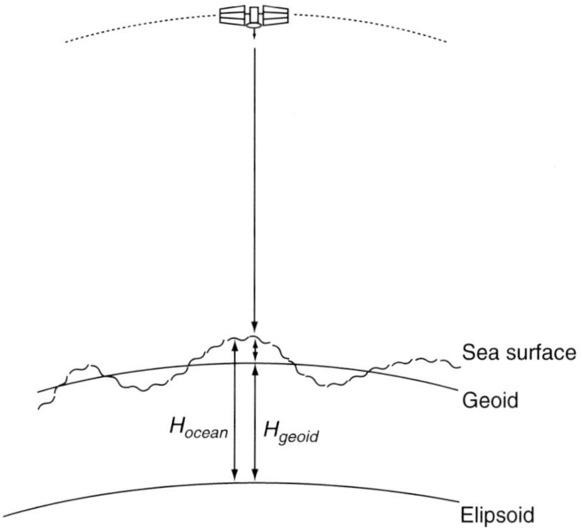

Fig. 24
Ocean topography

known geoidal height H_{Geoid}. In connection with (56) this finally leads us to the equation

$$2|\Omega|(\epsilon^3 \cdot \xi)\xi \wedge u_{\tan}(R\xi) = \frac{g_R \rho_0}{R} \nabla^*_\xi \Xi(R\xi). \tag{58}$$

Remembering the surface curl gradient $\mathrm{L}^*_\xi = \xi \wedge \nabla^*_\xi$ we are able to conclude

$$\xi \wedge (\omega \wedge u_{\tan}(R\xi)) = \frac{\rho_0 g_R}{2R} \mathrm{L}^*_\xi \Xi(R\xi), \tag{59}$$

i.e.,

$$C(\xi) u_{\tan}(R\xi) = \frac{g_R}{R} \mathrm{L}^*_\xi \Xi(R\xi). \tag{60}$$

This is the *equation of the geostrophic oceanic flow.*

8.2.2 Mathematical Analysis

Again we have an equation of vectorial tangential type given on the unit sphere $\mathbb{S}^2$. This time, however, we have to deal with an *equation of the surface curl gradient* $\mathrm{L}^* S = s$ (with $s(\xi) = \frac{2R}{g_R}\Omega(\epsilon^3 \cdot \xi) u_{\tan}(R\xi)$ and $S = \Xi$). The solution theory providing the *surface stream function* S from the surface divergence free vector field s would be in accordance with our considerations above (with ∇^* replaced by L^*). The computation of the geostrophic oceanic flow simply is a problem of differentiation, namely the computation of the derivative $\mathrm{L}^*_\xi S(\xi) = \xi \wedge \nabla^*_\xi S(\xi), \ \xi \in \mathbb{S}^2$.

8.2.3 Development of a Mathematical Solution Method

The point of departure for our intention to determine the geostrophic oceanic flow (as derived above from the basic hydrodynamic equation) is the ocean topography which is obtainable via satellite altimetry (see ➋ Chap. 12 for observational details). As a scalar field on the spherical Earth, the ocean topography consists of two ingredients. First, on an Earth at rest, the water masses would align along the geoid related to a (standard) reference ellipsoid. Second, satellite measurements provide altimetric data of the actual ocean surface height which is also used in relation to the (standard) reference ellipsoid. The difference between these quantities is understood to be the actual ocean topography. In other words, the ocean topography is defined as the deviation of the ocean surface from the geoidal surface, which is here assumed to be due to the geostrophic component of the ocean currents. The data used for our demonstration are extracted from the French CLS01 model (in combination with the EGM96 model).

The calculation of the derivative $\mathrm{L}^* S$ is not realizable, at least not directly. Also in this case we are confronted with discrete data material that, in addition, is only available for oceanic areas. In the geodetic literature a spherical harmonic approach is usually used in form of a Fourier series. The vectorial isotropic operator L^* (cf. Freeden and Schreiner 2009 for further details) is then applied – unfortunately under the leakage of its vectorial isotropy when decomposed in terms of scalar components – to the resulting Fourier series. The results are scalar components of the geostrophic flow. The serious difficulty with global polynomial structures such as spherical harmonics (with $\Xi = 0$ on continents!) is the occurrence of the Gibbs phenomenon close to the coast lines (see, e.g., Nerem and Koblinski 1994; Albertella et al. 2008). In this respect it should be mentioned additionally that the equation of the geostrophic oceanic flow cannot be regarded as adequate for coastal areas, hence, our modeling fails for coastal areas.

An alternative approach avoiding numerically generated oscillations in coastal areas is the application of kernels with local support, as, e.g., smoothed Haar kernels (see Freeden et al. 1998 and the references therein)

$$\Phi_\rho^{(k)}(t) = \begin{cases} 0, & \rho < 1-t \le 2, \\ \dfrac{k+1}{2\pi}\dfrac{(t-(1-\rho))^k}{\rho^{k+1}}, & 0 \le 1-t \le \rho. \end{cases} \tag{61}$$

For $\rho > 0$, $k \in \mathbb{N}$, the function $\Phi_\rho^{(k)}$ as introduced by (61) is $(k-1)$-times continuously differentiable on the interval $[0,2]$. $\left(\Phi_{\rho_j}^{(k)}\right)_{j\in\mathbb{N}_0}$ is a sequence tending to the Dirac function(al), i.e., a Dirac sequence (cf. *Figs. 25* and *26*).

For a strictly monotonically decreasing sequence $(\rho_j)_{j\in\mathbb{N}_0}$ satisfying $\rho_j \to 0$, $j \to \infty$, we obtain for the convolution integrals (low-pass filters)

$$H_{\rho_j}^{(k)}(\xi) = \int_{\mathbb{S}^2} \Phi_{\rho_j}^{(k)}(\xi\cdot\eta)H(\eta)\,dS(\eta), \quad \xi \in \mathbb{S}^2, \tag{62}$$

the limit relation

$$\lim_{j\to\infty}\sup_{\xi\in\mathbb{S}^2}\left|H(\xi) - H_{\rho_j}^{(k)}(\xi)\right| = 0. \tag{63}$$

An easy calculation yields

$$L_\xi^*\Phi_{\rho_j}^{(k)}(\xi\cdot\eta) = \begin{cases} 0, & \rho_j < 1-\xi\cdot\eta \le 2, \\ \dfrac{k(k+1)}{2\pi}\dfrac{((\xi\cdot\eta)-(1-\rho_j))^{k-1}}{\rho_j^{k+1}}\,\xi\wedge\eta, & 0 \le 1-\xi\cdot\eta \le \rho_j. \end{cases} \tag{64}$$

It is not hard to show (cf. Freeden and Schreiner 2009) that

$$\lim_{j\to\infty}\sup_{\xi\in\mathbb{S}^2}\left|L_\xi^* H(\xi) - L_\xi^* H_{\rho_j}^{(k)}(\xi)\right| = 0. \tag{65}$$

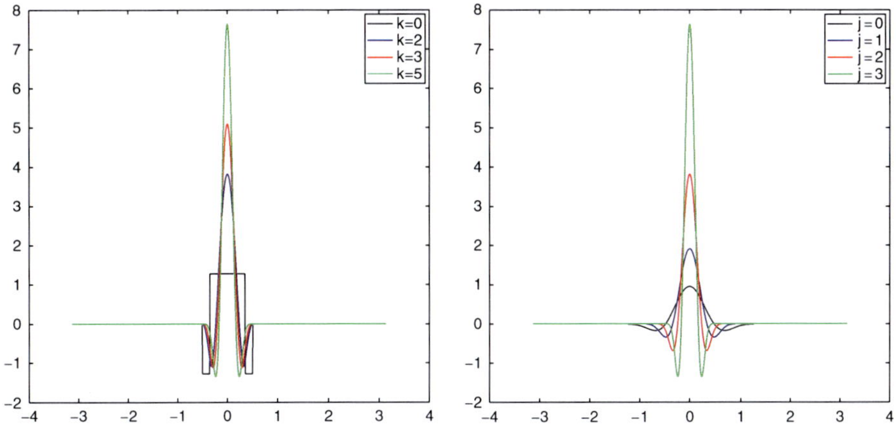

Scale $j = 3$, $k = 0, 2, 3, 5$ Scale $j = 0, 1, 2, 3$, $k = 5$

Fig. 25

Sectional illustration of the smoothed Haar wavelets $\Phi_{\rho_j}^{(k)}(\cos(\theta))$, $(\rho_j = 2^{-j})$

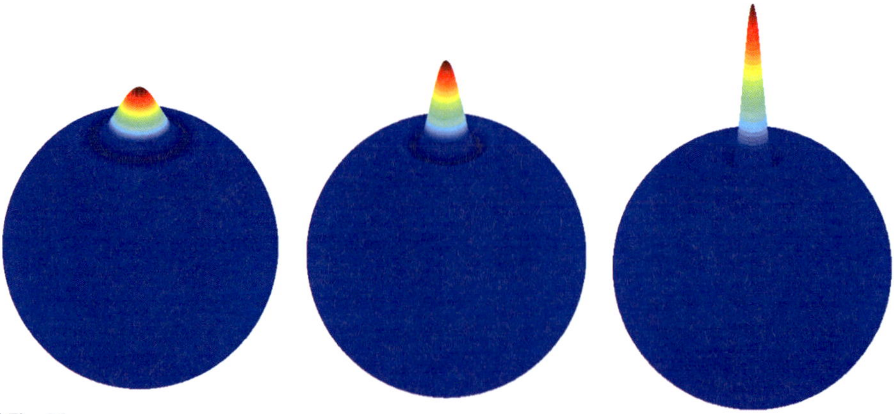

Fig. 26
Illustration of the first members of the wavelet sequence for the smoothed Haar scaling function on the sphere ($\rho_j = 2^{-j}, j = 2, 3, 4, k = 5$)

The multiscale approach by smoothed Haar wavelets can be formulated in a standard way (➋ *Fig. 27*). For example, the Haar wavelets can be understood as differences of two successive scaling functions. In doing so, an economical and efficient algorithm in a tree structure (Fast Wavelet Transform (FWT)) can be implemented appropriately.

8.2.4 "Back-transfer" to Application

Ocean currents are subject to different influence factors, such as wind field, warming of the atmosphere, salinity of the water, etc, which are not accounted for in our modeling. Our approximation must be understood in the sense of a geostrophic balance. An analysis shows that its validity may be considered as given on spatial scales of an approximate expansion of a little more than 30 km, and on time scale longer than approximately 1 week. Indeed, the geostrophic velocity field is perpendicular to the tangential gradient of the ocean topography (i.e., perpendicular to the tangential pressure gradient). This is a remarkable property. The water flows along curves of constant ocean topography (i.e., along isobars). Despite the essentially restricting assumptions necessary for the modeling, we obtain instructive circulation models for the internal ocean surface current for the northern or southern hemisphere, respectively (however, difficulties for the flow computation arise from the fact that the Coriolis parameter vanishes on the equator) (➋ *Fig. 28*).

An especially positive result is that the modeling of the ocean topography has made an essential contribution to the research in exceptional phenomena of internal ocean currents, such as *El Niño*. El Niño is an anomaly of the ocean-atmosphere system. It causes the occurrence of modified currents in the equatorial Pacific, i.e., the surface water usually flowing in western direction suddenly flows to the east. Geographically speaking, the cold Humboldt Current is weakened and finally ceases. Within only a few months, the water layer moves from South East Asia to South America. Water circulation has reversed. As a consequence, the Eastern Pacific is warmed up, whereas the water temperature decreases off the shores of Australia and Indonesia. This phenomenon has worldwide consequences on the weather, in the form of extreme droughts

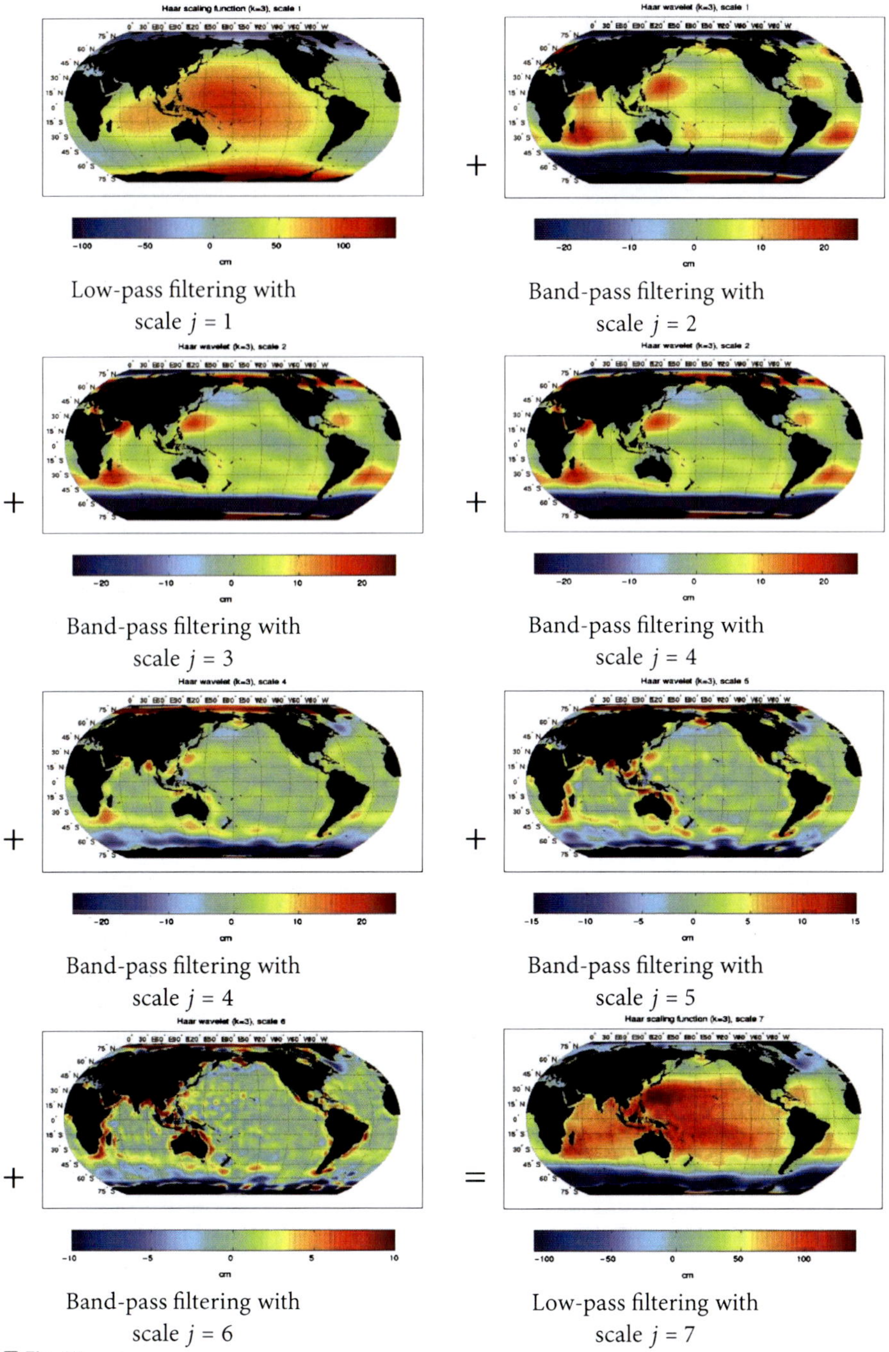

Fig. 27

Multiscale approximation of the ocean topography [cm] (a rough low-pass filtering at scale $j = 1$ is improved by several band-pass filters of scale $j = 1, \ldots, 6$, where the last picture shows the multiscale approximation at scale $j = 7$, numerical and graphical illustration in cooperation with D. Michel and V. Michel 2006

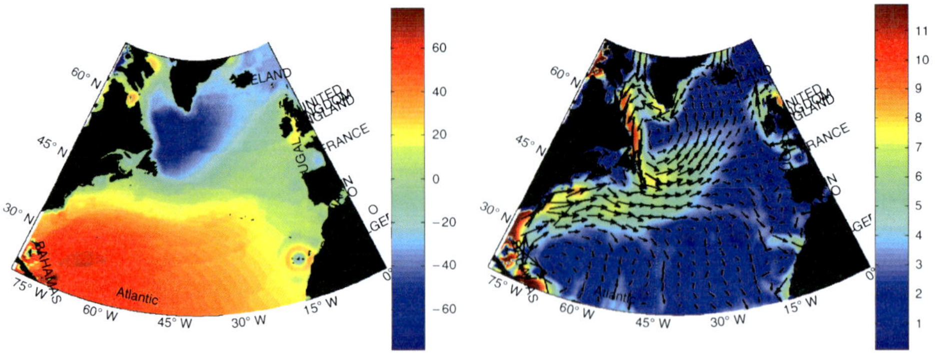

■ **Fig. 28**
Ocean topography [cm] (left) and geostrophic oceanic flow [cm/s] (right) of the gulf stream computed by the use of smoothed Haar wavelets ($j = 8, k = 5$), numerical and graphical illustration in cooperation with D. Michel and V. Michel 2006

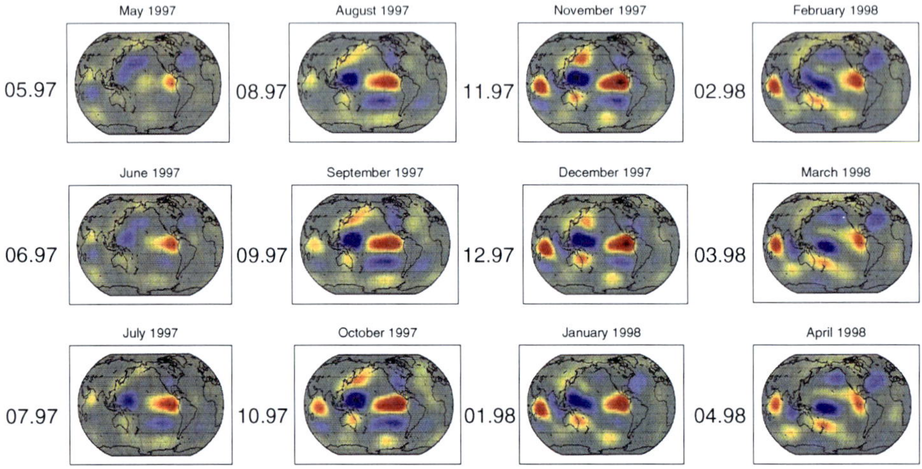

■ **Fig. 29**
Ocean topography during the El Niño period (May 1997–April 1998) from data of the satellite CLS01 model, numerical realization and graphical illustration in cooperation with V. Michel and Maßmann 2006

and thunderstorms. Our computations do not only help to visualize these modifications graphically, they also offer the basis for future predictions of El Niño characteristics and effects (➋ *Fig. 29*).

9 Final Remarks

The Earth is a dynamic planet in permanent change, due to large-scale internal convective material and energy rearrangement processes, as well as manifold external effects. We can, therefore, only understand the Earth as our living environment if we consider it as a complex system of all its interacting components. The processes running on the Earth are coupled with one another, forming ramified chains of cause and effect which are additionally influenced by man who intervenes into the natural balances and circuits. However, knowledge of these chains of cause and effect has currently still remained incomplete to a large extent. In adequate time, substantial improvements can only be reached by the exploitation of new measurement and observation methods, e.g., by satellite missions and by innovative mathematical concepts of modeling and simulation, all in all by *geomathematics*. As far as data evaluation is concerned in the future, traditional mathematical methods will not be able to master the new amounts of data neither theoretically nor numerically – especially considering the important aspect of a more intensively localized treatment with respect to space and time, embedded into a global concept. Instead, geoscientifically relevant parameters must be integrated into constituting modules; the integration must be characterized by three essential characteristics: good approximation property, appropriate decorrelation ability, and fast algorithms. These characteristics are the key for a variety of abilities and new research directions.

10 Acknowledgments

This introductory chapter is based on the German note "Freeden 2009: Geomathematik, was ist das überhaupt?, Jahresbericht der Deutschen Mathematiker Vereinigung (DMV), JB.111, Heft 3, 125–152". I am obliged to the publisher Vieweg+Teubner for giving the permission for an English translation of essential parts of the original version.

Furthermore, I would like to thank my Geomathematics Group, Kaiserslautern, for the assistance in numerical implementation and calculation as well as graphical illustration.

References

Albertella A, Savcenko R, Bosch W, Rummel R (2008) Dynamic ocean topography – the geodetic approach. IAPG/FESG Mittelungen, vol. 27, TU München

Ansorge R, Sonar T (2009) Mathematical models of fluid dynamics, 2nd edn. Wiley-VCH, Weinheim

Bach V, Fraunholz W, Freeden W, Hein F, Müller J, Müller V, Stoll H, von Weizsächer H, Fischer H (2004) Curriculare Standards des Fachs Mathematik in Rheinland-Pfalz (Vorsitz: W. Freeden). Studie: Reform der Lehrerinnen- und Lehrerausbildung, MWWFK Rheinland-Pfalz

Bruns EH (1878) Die Figur der Erde. Publikation Königl. Preussisch. Geodätisches Institut. P. Stankiewicz, Berlin

Emmermann R, Raiser B (1997) Das System Erde-Forschungsgegenstand des GFZ. Vorwort des GFZ-Jahresberichts 1996/1997, V-X, GeoForschungsZentrum, Potsdam

Freeden W (1999) Multiscale modelling of spaceborne geodata. B.G. Teubner, Stuttgart, Leipzig

Freeden W, Michel V (2004) Multiscale potential theory (with applications to geoscience). Birkhäuser Verlag, Boston, Basel, Berlin

Freeden W, Schreiner M (2009) Spherical functions of mathematical geosciences – a scalar, vectorial, and tensorial setup. Springer, Berlin

Freeden W, Gervens T, Schreiner M (1998) Constructive approximation on the sphere (with applications to geomathematics). Oxford Science Publications, Clarendon, Oxford

Helmert FR (1881) Die mathematischen und physikalischen Theorien der Höheren Geodäsie 1+2, B.G. Teubner, Verlagsgesellschaft Leipzig

Jakobs F, Meyer H (1992) Geophysik – Signale aus der Erde. Teubner, Leipzig

Kümmerer B (2002) Mathematik. Campus, Spektrum der Wissenschaftsverlagsgesellschaft, 1–15

Lemoine FG, Kenyon SC, Factor JK, Trimmer RG, Pavlis NK, Shinn DS, Cox CM, Klosko SM, Luthcke SB, Torrence MH, Wang YM, Williamson RG, Pavlis EC, Rapp RH, Olson

TR (1998) The development of the joint NASA GSFC and NIMA geopotential model EGM96. NASA/TP-1998-206861, NASA Goddard Space Flight Center, Greenbelt, MD

Listing JB (1873) Über unsere jetzige Kenntnis der Gestalt und Größe der Erde. Dietrichsche Verlagsbuchhandlung, Göttingen

Nerem RS, Koblinski CJ (1994) The geoid and ocean circulation (geoid and its geophysical interpretations). In: Vanicek P, Christon NT (eds.). CRC Press, Boca Ruton, FL, pp 321–338

Neumann F (1887) Vorlesungen über die Theorie des Potentials und der Kugelfunktionen. Teubner, Leipzig, pp 135–154

Pedlovsky J (1979) Geophysical fluid dynamics. Springer, New York

Pesch HJ (2002) Schlüsseltechnologie Mathematik. Teubner, Stuttgart, Leipzig, Wiesbaden

Ritter JRR, Christensen UR (eds) (2007) Mantle plumes – a multidisciplinary approach. Springer, Heidelberg

Rummel R (2002) Dynamik aus der Schwere – Globales Gravitationsfeld. An den Fronten der Forschung (Kosmos, Erde, Leben), Hrsg. R. Emmermann u.a., Verhandlungen der Gesellschaft Deutscher Naturforscher und Ärzte, 122. Versammlung, Halle

Sonar T (2001) Angewandte Mathematik, Modellbildung und Informatik: Eine Einführung für Lehramtsstudenten, Lehrer und Schüler. Vieweg, Braunschweig, Wiesbaden

Stokes GG (1849) On the variation of gravity at the surface of the Earth. Trans. Cambr Phil Soc 8:672–712. In: Mathematical and physical papers by George Gabriel Stokes, Vol. II. Johanson Reprint Corporation, New York, pp 131–171

Weyl H (1916) Über die Gleichverteilung von Zahlen mod. Eins Math Ann 77:313–352

2 Navigation on Sea: Topics in the History of Geomathematics

Thomas Sonar
Computational Mathematics, Technische Universität Braunschweig, Braunschweig, Germany

W. Freeden, M.Z. Nashed, T. Sonar (Eds.), *Handbook of Geomathematics*, DOI 10.1007/978-3-642-01546-5_2,

Abstract In this essay we review the development of the magnet as a means for navigational purposes. Around 1600, knowledge of the properties and behavior of magnetic needles began to grow in England mainly through the publication of William Gilbert's influential book *De Magnete*. Inspired by the rapid advancement of knowledge on one side and of the English fleet on the other, scientists associated with Gresham College began thinking of using magnetic instruments to measure the degree of latitude without being dependent on a clear sky, a quiet sea, or complicated navigational tables. The construction and actual use of these magnetic instruments, called dip rings, is a tragic episode in the history of seafaring since the latitude does not depend on the magnetic field of the Earth but the construction of a table enabling seafarers to take the degree of latitude from is certainly a highlight in the history of geomathematics.

1 General Remarks on the History of Geomathematics

Geomathematics in our times is thought of being a very young science and a modern area in the realms of mathematics. Nothing is farer from the truth. Geomathematics began as man realized that he walked across a sphere-like Earth and that this had to be taken into account in measurements and computations. Hence, Eratosthenes can be seen as an early geomathematician when he tried to determine the circumference of the Earth by measurements of the sun's position from the ground of a well and the length of shadows farther away at midday. Other important topics in the history of geomathematics are the struggles for an understanding of the true shape of the Earth which led to the development of potential theory and much of multidimensional calculus, see Greenberg (1995), the mathematical developments around the research of the Earth's magnetic field, and the history of navigation.

2 Introduction

The history of navigation is one of the most exciting stories in the history of mankind and one of the most important topics in the history of Geomathematics. The notion of navigation thereby spans the whole range from the ethnomathematics of polynesian stick charts via the compass to modern mathematical developments in understanding the earth's magnetic field and satellite navigation via GPS. We shall concentrate here on the use of the magnetic needle for navigational purposes and in particular on developments having taken place in early modern England. However, we begin our investigations with a short overview on the history of magnetism following Balmer (1956).

3 The History of the Magnet

The earliest sources on the use of magnets for the purpose of navigation stem from China. During the *Han* epoche 202–220 we find the description of carriages equipped with compass-like devices so that the early Chinese imperators were able to navigate on their journeys through their enormous empire. These carriages were called *tschinan-tsche*, meaning "carriages that show noon." The compass-like devices consisted of little human-like figures which swam on water in a bowl, the finger of the stretched arm pointing always straight to the south. We do not know nowadays why the ancient Chinese preferred the southward direction instead of a northbound one. In a book on historical memoirs written by *Sse-ma-tsien* (or *Schumatsian*) we find a report dating back to the first half of the second century on a present that imperator *Tsching-wang* gave 1100 before Christ to the ambassadors of the cities of Tonking and Cochinchina. The ambassadors received five "magnetic carriages" in order to guide them safely back to their cities even through sand storms in the desert. Since the ancient Chinese knew about the attracting forces of a magnet they called them 'loving stones.' In a work on natural sciences written by Tschin-tsang-ki from the year 727 we read:

- The loving stone attracts the iron like an affectionate mother attracts their children around her; this is the reason where the name comes from.

It was also very early known that a magnet could transfer its attracting properties to iron when it was swept over the piece of iron. In a dictionary of the year 121 the magnet is called a stone "with which a needle can be given a direction" and hence it is not surprising that a magnetic needle mounted on a piece of cork and swimming in a bowl of water belonged to the standard equipment of larger chinese ship as early as the fourth century. Such simple devices were called "bussola" by the Italians and are still known under that name.

A magnetic needle does not point precisely to the geographic poles but to the magnetic ones. The locus of the magnetic poles moves in time so that a deviation has always to be taken into account. Around the year 1115 the problem of deviation was known in China.

The word "magnet" comes from the greek word magnes describing a sebacious rock which, according to the greek philosopher and natural scientist Theophrastos (ca. 371–287 BC), was a forgeable and silver white rock. The philosopher Plato (428/27–348/47 BC) called magnetic rock the "stone of Heracles" and the poet Lucretius (ca. 99–55 BC) used the word "magnes" in the sense of attracting stone. He attributed the name to a place named "Magnesia" where this rock could be found. Other classical greek anecdotes call a shepherd named "Magnes" to account for the name. It is said that he wore shoes with iron nails and while accompanying his sheep suddenly could no longer move because he stood on magnetic rock.

Homer wrote about the force of the magnet as early as 800 BC. It seems typical for the ancient greek culture that one sought for an explanation of this force fairly early on. Plato thought of this force as being simply "divine." Philosopher Epicurus (341–270) had the hypothesis that magnets radiate tiny particles – atoms. Eventually Lucretius exploited this hypothesis and explained the attracting force of a magnet by the property of the radiated atoms to clear the space between the magnet and the iron. Into the free space then iron atoms could penetrate and since iron atoms try hard to stay together (says Lucretius) the iron piece would follow them. The pressure of the air also played some minor role in this theory. Lucretius knew that he had to answer the question why iron would follow but other materials would not. He simply declared that gold would be too heavy and timber would show too large porosities so that the atoms of the magnet would simply go through.

The first news on the magnet in Western Europe came from Paris around the year 1200. A magnetic needle was used to determine the orientation. We do not know how the magnet came to Western Europe and how it was received but it is almost certainly true that the crusades and the associated contact with the peoples in the mediterranean played a crucial role. Before William Gilbert around 1600 came up with a "magnetick philosophy" it was the crusader, astronomer, chemist, and physician Peter Peregrinus De Maricourt who developed a theory of the magnet in a famous "letter on the magnet" dating back to 1269. He describes experiments

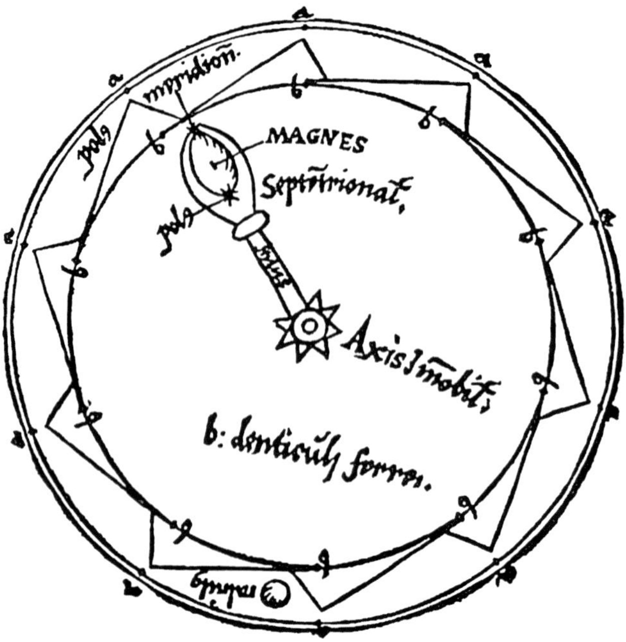

Fig. 1
The magnetic perpetuum mobile of Peregrinus

with magnetic stones which are valid even nowadays. Peregrinus grinds a magnetic stone in the form of a sphere, places it in a wooden plate and puts this plate in a bowl with water. Then he observes that the sphere moves according to the poles. He develops ideas of magnetic clocks and describes the meaning of the magnet with respect to the compass. He also develops a magnetic *perpetuum mobile* according to ➋ *Fig. 1.* A magnet is mounted at the tip of a hand which is periodically moving (says Peregrinus) beacuse of iron nails on the circumference.

Peregrinus' work was so influential in Western Europe that even 300 years after his death he is still accepted as the authority on the magnet.

4 Early Modern England

In the sixteenth century Spain and Portugal developed into the leading sea powers. Currents of gold, spices, and gemstones regorge from the South Americans into the home countries. England had missed connection. When Henry VIII died in 1547, only a handful of decaying ships were lying in the English sea harbors. His successor, his son Edward VI, could only rule for 6 years before he died young. Henry's daughter Mary, a devout catholic, tried hard to re-catholize the country her father had steered into protestantism and married Philipp II, King of Spain. Mary was fairly brutal in the means of the re-catholization and many of the protestant intelligentsia left the country in fear of their lives. "Bloody Mary" died in 1558 at the age of 47 and the way opened to her stepsister Elizabeth. Within one generation itself Elizabeth I transformed rural England to the leading sea power on Earth. She was advised very well by Sir Walter Raleigh who clearly saw the future of England on the seas. New ships were built for the navy and in 1588

Fig. 2
Elizabeth I (Armada portrait)

the small English fleet was able to drown the famous Spanish armada – by chance and with good luck; but this incident served to boost not only the feeling of self-worth of a whole nation but also the realization of the need of a navy and the need of efficient navigational tools.

English mariners realized on longer voyages that the magnetic needle inside a compass lost its magnetic power. If that was detected the needle had to be magnetized afresh – it had to be "loaded" afresh. However, this is not the reason why the magnet is called *loadstone* in the English language but, only a mistranslation. The correct word should be *lodstone* – "leading stone" – but that word was actually never used (Pumfrey, 2002).

5 The Gresham Circle

In 1592, Henry Briggs (1561–1639), chief mathematician in his country, was elected Examiner and Lecturer in Mathematics at St John's College, Cambridge, which nowadays corresponds to a professorship. In the same year he was elected *Reader of the Physics Lecture founded by Dr Linacre* in London. One hundred years before the birth of Briggs, Thomas Linacre was horrified by the pseudomedical treatment of sick people by hair dressers and vicars who did not shrink back from chirurgical operations without a trace of medical instruction. He founded the *Royal College of Physicians of London* and Briggs was now asked to deliver lectures with medical contents. The Royal College of Physicians was the first important domain for Briggs to make contact with men outside the spheres of the two great universities and, indeed most important, he met William Gilbert (1544–1603) who was working on the wonders of the magnetical forces and who revolutionized modern science only a few years later.

While England was on its way to become the worlds leading sea force the two old English universities Oxford and Cambridge were in an alarming state of sleepyness (Hill 1997, p. 16ff). Instead of working and teaching on the forefront of modern research in important topics like navigation, geometry, astronomy, the curricula were directly rooted in the ancient greek tradition. Mathematics included reading of the first four or five books of Euclid, medicine was read after Galen and Ptolemy ruled in astronomy. When the founder of the English stock exchange (Royal Exchange) in London, Thomas Gresham, died, he left in his last will money and buildings in order to found a new form of university, the Gresham College, which is still in function. He ordered the employment of seven lecturers to give public lectures in theology, astronomy, geometry, music, law, medicine, and rhetorics mostly in English language. The salary of the Gresham professors was determined to be £50 a year which was an enormous sum as compared to the salary of the Regius professors in Oxford and Cambridge (Hill 1997, p. 34). The only conditions on the candidates for the Gresham professorships were brilliance in their field and an unmarried style of life.

Briggs must have been already well known as a mathematician of the first rank since he was chosen to be the first Gresham professor of Geometry in 1596. Modern mathematics was needed badly in the art of navigation and public lectures on mathematics were in fact already given in 1588 on behalf of the East India Company, the Muscovy Company, and the Virginia Company. Even before 1588 there were attempts by Richard Hakluyt to establish public lectures and none less than Francis Drake had promised £20 (Hill 1997, p. 34), but it needed the national shock of the attack of the Armada in 1588 to make such lectures come true.

During his time in Gresham College Briggs became the center of what we can doubtlessly call the Briggsian circle. Hill writes, (Hill 1997, p. 37):

> ▸ He [i.e. BRIGGS] was a man of the first importance in the intellectual history of his age, ... Under him Gresham at once became a centre of scientific studies. He introduced there the modern method of teaching long division, and popularized the use of decimals.

The Briggsian circle consisted of true copernicans; men like William Gilbert who wrote *De Magnete*, the able applied mathematician Edward Wright who is famous for his book on the errors in navigation, William Barlow, a fine instrument maker and men of experiments, and the great popularizer of scientific knowledge, Thomas Blundeville. Gilbert and Blundeville were protégés of the Earl of Leicester and we know about connections with the circle of Ralegh in which the brilliant mathematician Thomas Harriot worked. Blundeville held contacts with John Dee who introduced modern continental mathematics and the Mercator maps in England, (Hill 1997, p. 42). Hence, we can think of a scientific sub-net in England in which important work could be done which was impossible to do in the great universities. It was this time in Gresham College in which Briggs and his circle were most productive in the calculation of tables of astronomical and navigational importance. In the center of their activities was Gilbert's "Magnetick Philosophy."

6 William Gilberts Dip Theory

The role of William Gilbert in shaping modern natural sciences cannot be overestimated and a recent biography of Gilbert (Pumfrey 2002) emphasizes his importance in England and abroad. Gilbert, a physician and member of the Royal College of Physisicans in London, became interested in navigational matters and the properties of the magnetic needle in particular, by his contacts to seamen and famous navigators of his time alike. As a result of years of experiments,

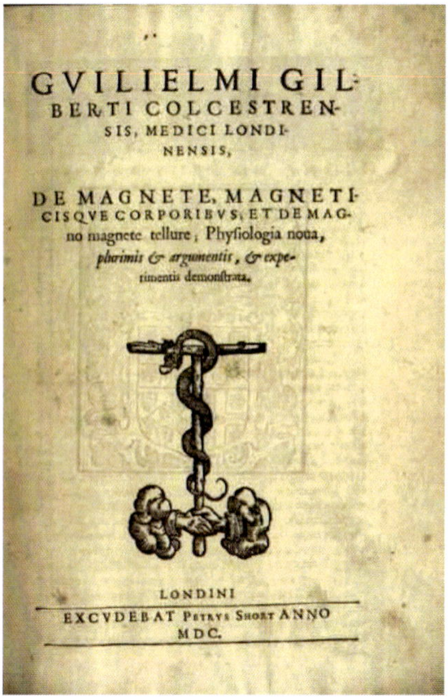

Fig. 3
De Magnete

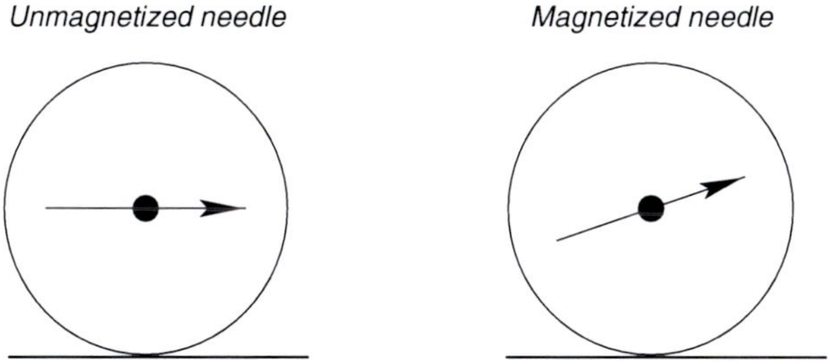

Fig. 4
Norman's discovery of the magnetic dip

thought, and discussions with his Gresham friends, the book *De Magnete, magneticisque corporibus, et de magno magnete tellure; Physiologia nova, plurimis & argumentis, & experimentis demonstrata* was published in 1600. (I refer to the English translation Gilbert (1958) by P. Fleury Mottelay which is a reprint of the original of 1893. There is a better translation by Sylvanus P. Thompson from 1900 but while the latter is rare the former is still in print.) It contained many magnetic experiments with what Gilbert called his *terrella* – the little Earth – which was a magnetical sphere. In the spirit of the true copernican Gilbert deduced the rotation of the Earth from the assumption of it being a magnetic sphere.

Concerning navigation, Briggs, and the Gresham circle the most interesting chapter in *De Magnete* is Book V: *On the dip of the magnetic needle*. Already in 1581 the instrument maker

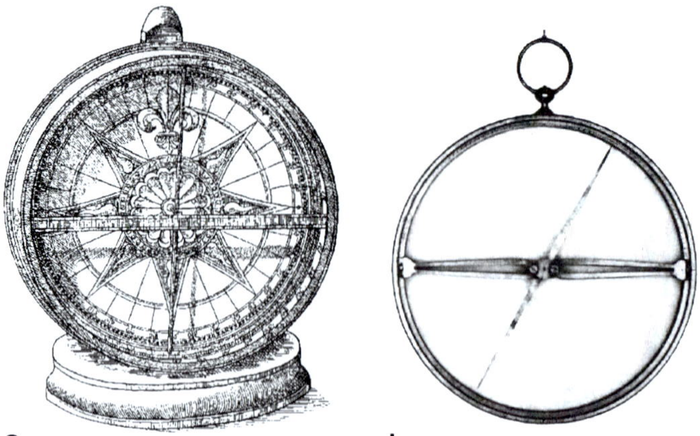

a b

Fig. 5
Dip rings: (a)The dip ring after Gilbert in *De Magnete* (b) A dip ring used in the 17th century

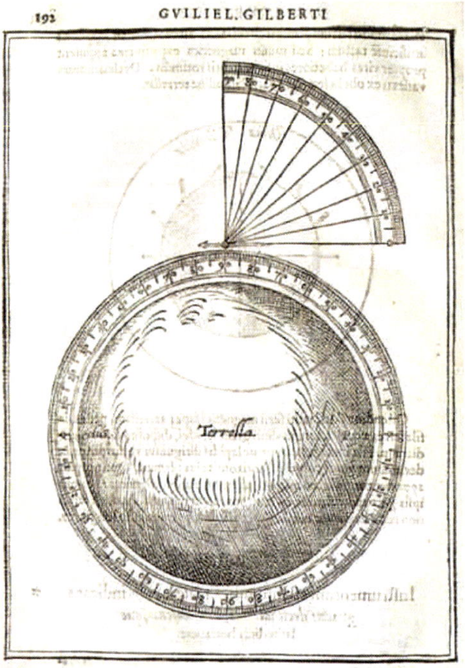

Fig. 6
Measuring the dip on the terrella in *De Magnete*

Robert Norman had discovered the magnetic dip in his attempts to straighten magnetic needles in a fitting on a table. He had observed that an unmagnetized needle could be fitted in a parallel position with regard to the surface of a table but when the same needle was magnetized and fitted again it made an angle with the table. Norman published his results already in 1581 (R. Norman — The New Attractive *London, 1581.* Even before Norman the dip was reported by

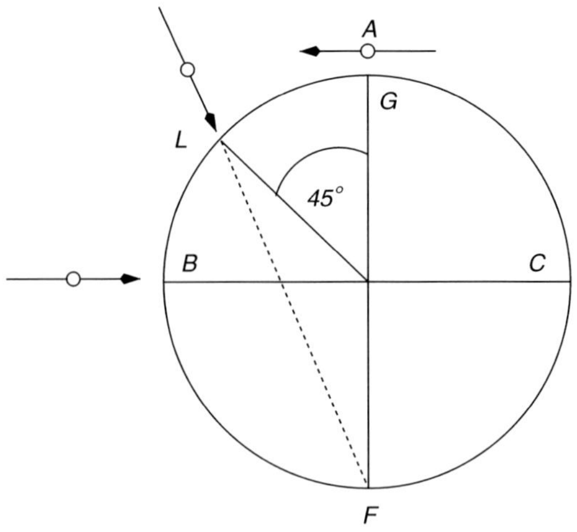

Fig. 7
Third hypothesis

the german astronomer and instrument maker Georg Hartmann from Nuremberg in a letter to the Duke Albrecht of Prussia from 4th of March, 1544, see Balmer 1995, p. 290–292.) but he was not read. In modern notion the phenomenon of the dip is called *inclination* in contrast to the *declination* or *variation* of the needle. A word of warning is appropriate here: in Gilbert's time many authors used the word *declination* for the *inclicnation*. Anyway, Norman was the first to build a *dip ring* in order to measure the inclination. This ring is nothing else but a vertical compass. Already Norman had discovered that the dip varied with time!

However, Gilbert believed that he had found the secrets of magnetic navigation. He explained the variation of the needle by land masses acting on the compass which fitted nicely with the measurements of seamen but is wrong, as we now know.

Concerning the dip let me give a summary of Gilbert's work in modern terms. Gilbert must have measured the dip on his terrella many, many times before he was led to his

First hypothesis: *There is an invertible mapping between the lines of latitude and the lines of constant dip.*
Hence, Gilbert believed to have found a possibility of determining the latitude on Earth from the degree of the dip. Let β be the latitude and α the dip. He then formulated his

Second Hypothesis: *At the equator the needle is parallel to the horizon, i.e.* $\alpha = 0^\circ$. *At the north pole the needle is perpendicular to the surface of the Earth, i.e.* $\alpha = 90^\circ$.
He then draws a conlcusion but in our modern eyes this is nothing but another

Third hypothesis: *If* $\beta = 45^\circ$ *then the needle points exactly to the second equatorial point.*

What he meant by this is best described in *Fig. 7*. Gilbert himself writes

- … *points to the equator F as the mean of the two poles.* (Gilbert 1958, p. 293)

Note that in *Fig. 7* the equator is given by the line A–F and the poles are B (north) and C so that our implicit (modern) assumption that the north pole is always shown on top of a figure is not satisfied.

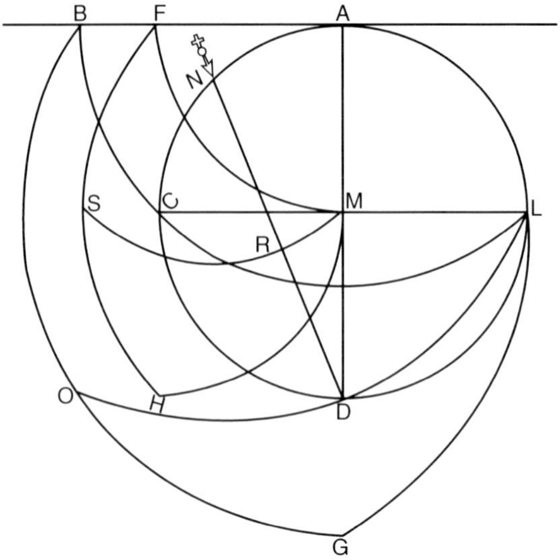

Fig. 8
Gilbert's geometrical construction of the mapping in *De Magnete*

From his three hypothesises Gilbert concludes correctly:

First Conclusion: *The* rotation *of the needle has to be faster on its way from A to L than from L to B*. Or, in Gilbert's words,

▸ … the movement of this rotation is quick in the first degrees from the equator, from *A* to *L*, but slower in the subsequent degrees, from *L* to *B*, that is, with reference to the equatorial point *F*, toward *C*. (Gilbert 1958, p. 293)

And on the same page we read:

▸ … it dips; yet, not in ratio to the number of degrees or the arc of the latitude does the magnetic needle dip so many degrees or over a like arc; but over a very different one, for this movement is in truth not a dipping movement, but really a revolution movement, and it describes an arc of revolution proportioned to the arc of latitude.

This is simply the lengthy description of the following

Second Conclusion: *The mapping between latitude and dip cannot be linear.*

Now that Gilbert had made up his mind concerning the behavior of the mapping at $\beta = 0^\circ, 45^\circ, 90^\circ$ a construction of the general mapping was sought. It is exactly here where *De Magnete* shows strange weaknesses, in fact, we witness a qualitative jump from a geometric construction to a *dip instrument*. Gilbert's geometrical description can be seen in ❯ *Fig. 8*. I do not intend to comment on this construction because this was done in detail elsewhere, (Sonar 2001), but it is *not* possible to understand the construction from Gilbert's writings in *De Magnete*. Even more surprising, while all figures in *De Magnete* are raw wood cuts in the quality shown before, suddenly there is a fine technical drawing of the resulting construction

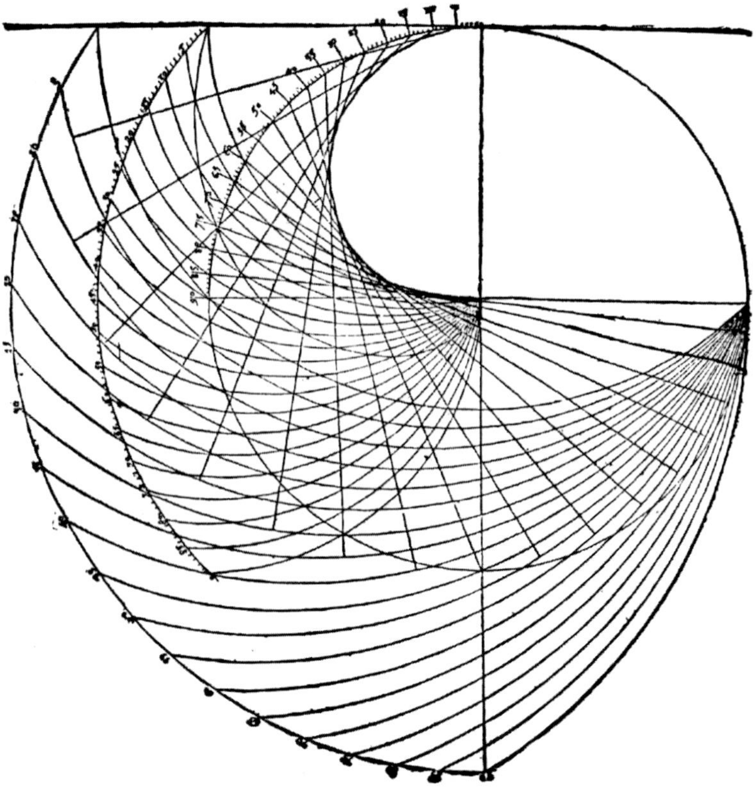

Fig. 9
The fine drawing in *De Magnete*

as shown in Fig. 9. The difference between this drawing and all other figures in *De Magnete* and the weakness in the description of the construction of the mapping between latitudes and dip angles suggest that at least this part was not written by Gilbert alone but by some of his friends in the Gresham circle. Pumfrey speaks of *the dark secret of De Magnete* (Pumfrey 2002, p. 173ff), and gives evidence that Edward Wright, whose *On Certain Errors in Navigation* had appeared a year before *De Magnete*, had his hands in some parts of Gilberts book. In Parsons (1939, pp. 61–67), we find the following remarks:

> Wright, and his circle of friends, which included Dr. W. Gilbert, Thomas Blundeville, William Barlow, Henry Briggs, as well as Hakluyt and Davis, formed the centre of scientific thought at the turn of the century. Between these men there existed an excellent spirit of co-operation, each sharing his own discoveries with the others. In 1600 Wright assisted Gilbert in the compilation of *De Magnete*. He wrote a long preface to the work, in which he proclaimed his belief in the rotation of the earth, a theory which Gilbert was explaining, and also contributed chapter 12 of Book IV, which dealt with the method of finding the amount of the variation of the compass. Gilbert devoted his final chapters to practical problems of navigation, in which he knew many of his friends were interested.

There is no written evidence that Briggs was involved too but it seems very unlikely that the chief mathematician of the Gresham circle should not have been in charge in so important a

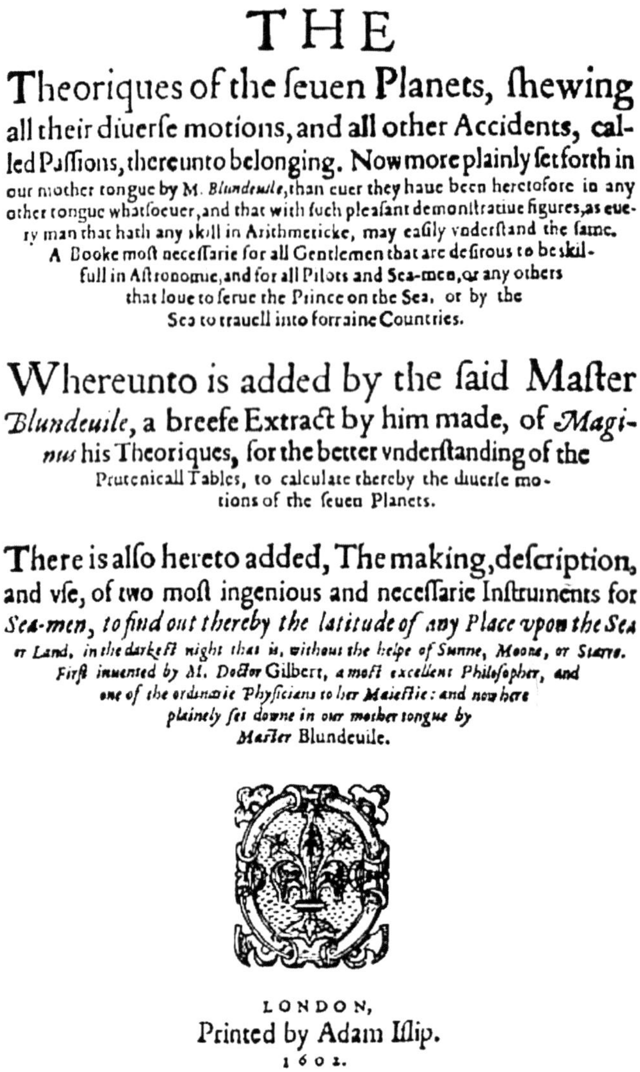

THE
Theoriques of the ſeuen Planets, ſhewing all their diuerſe motions, and all other Accidents, called Paſſions, thereunto belonging. Now more plainly ſet forth in our mother tongue by M. *Blundeuile*, than euer they haue been heretofore in any other tongue whatſoeuer, and that with ſuch pleaſant demonſtratiue figures, as euery man that hath any skill in Arithmeticke, may eaſily vnderſtand the ſame.
A Booke moſt neceſſarie for all Gentlemen that are deſirous to be skilfull in Aſtronomie, and for all Pilots and Sea-men, or any others that loue to ſerue the Prince on the Sea, or by the Sea to trauell into forraine Countries.

Whereunto is added by the ſaid Maſter *Blundeuile*, a breefe Extract by him made, of *Maginus* his Theoriques, for the better vnderſtanding of the Prutenicall Tables, to calculate thereby the diuerſe motions of the ſeuen Planets.

There is alſo hereto added, The making, deſcription, and vſe, of two moſt ingenious and neceſſarie Inſtruments for *Sea-men, to find out thereby the latitude of any Place vpon the Sea or Land, in the darkeſt night that is, without the helpe of Sunne, Moone, or Starre. Firſt inuented by M. Doctor* Gilbert, *a moſt excellent Philoſopher, and one of the ordinarie Phyſicians to her Maieſtie: and now here plainely ſet downe in our mother tongue by Maſter* Blundeuile.

LONDON,
Printed by Adam Iſlip.
1602.

Fig. 10
Title page

development as the dip theory. We shall see later on that the involvement of Briggs is highly likely when we study his contributions to dip theory in books of other authors.

7 The Briggsian Tables

If we trust (Ward, (1740), pp. 120–129), the first published table of Henry Briggs is the table which represents Gilbert's mapping between latitude and dip angles in Thomas Blundevilles book

- The Theoriques of the seuen Planets, shewing all their diuerse motions, and all other Accidents, called Passions, thereunto belonging.

Whereunto is added by the said Master Blundeuile, a breefe Extract by him made, of Magnus his Theoriques, for the better vnderstanding of the Prutenicall Tables, to calculate thereby the diuerse motions of the seuen Planets.

There is also hereto added, The making, description, and vse, of the two most ingenious and necessarie Instruments for Sea-men, to find out therebye the latitude of any place vpon the Sea or Land, in the darkest night that is, without the helpe of Sunne, Moone, or Starre. First inuented by M. Doctor Gilbert, a most excellent Philosopher, and one of the ordinarie Physicians to her Maiestie: and now here plainely set down in our mother tongue by Master Blundeuile.

London

Printed by Adam Islip.

1602.

Blundeville is an important figure in his own right, see Waters (1958, pp. 212–214), and Taylor (1954, p. 173). He was one of the first and most influencial *popularizers* of scientific knowledge. He did not write for the expert, but for the layman, i.e., the *young gentlemen* interested in so diverse questions of science, writing of history, map making, logic, seamenship, or horse riding. We do not know much about his life (Campling 1921–1922), but his role in the Gresham circle is apparent through his writings. In *The Theoriques* Gilbert's dip theory is explained in detail and a step-by-step description of the construction of the dip instrument is given. I have followed Blundeville's instructions and constructed the dip instrument again elsewhere, see Sonar (2002). See also Sonar (2001). The final result is shown in Blundeville's book as in ➲ *Fig. 11*. In order to understand the geometrical details it is necessary to give a condensed description of the actual construction in ➲ *Fig. 8* which is given in detail in *The Theoriques*. We start with a circle $ACDL$ representing planet Earth as in ➲ *Fig. 12*. Note that A is an equatorial point while C is a pole. The navigator (and hence the dip instrument) is assumed to be in point N which corresponds to the latitude $\beta - 45^\circ$. In the first step of the construction a horizon line is sought, i.e., the line from the navigator in N to the horizon. Now a circle is drawn around A with radius AM (the Earth's radius). This marks the point F on a line through A parallel to CL. A circle around M through F now gives the arc $\overset{\frown}{FH}$. The point H is constructed by drawing a circle around C with radius AM. The point of intersection of this circle with the outer circle through F is H. *If the dip instrument is at* A*, the navigator's horizon point is* F*. If it is in* C*, the navigator's horizon will be in* H. Correspondingly, drawing a circle through N with radius AM gives the point S, hence, S *is the point at the horizon seen from* N. Hence to every position N of the needle there is a *quadrant of dip* which is the arc from M to a corresponding point on the outer crircle through F. If N is at $\beta = 45^\circ$ latitude as in our example we know from Gilbert's third hypotheses that the needle points to D. The angle between S and the intersection point of the quadrant of dip with the direction of the needle is the dip angle. The remaining missing information is the point to which the needle points for a general latitude β. This is accomplised by *Quadrants of rotation* which implement Gilbert's idea of the needle *rotating* on its way from A to C. The construction of these quadrants is shown in ➲ *Fig. 13*. We need a second outer circle which is constructed by drawing a circle around A through L. The intersection point of this circle with the line AF is B and the second outer circle is then the circle through B around M. Drawing a circle around C through L defines the point G on the second outer circle. These arcs, $\overset{\frown}{GL}$ and $\overset{\frown}{BL}$, are the *quadrants of*

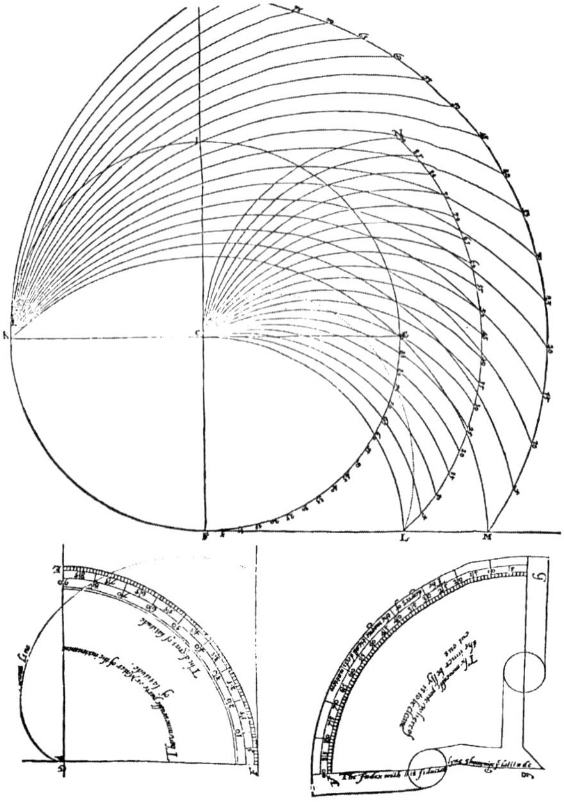

Fig. 11
The dip instrument in *The Theoriques*

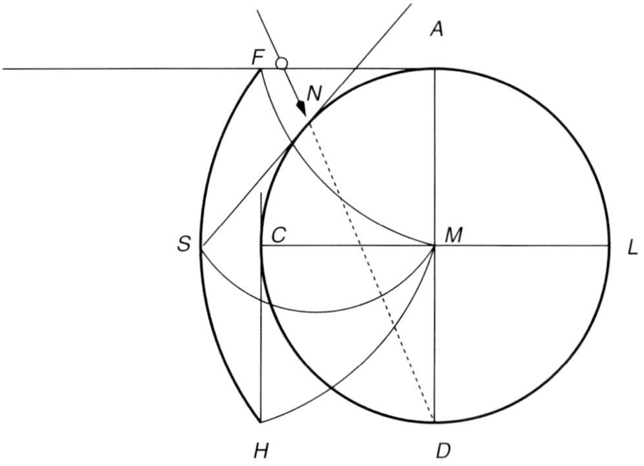

Fig. 12
The *Quadrant of dip*

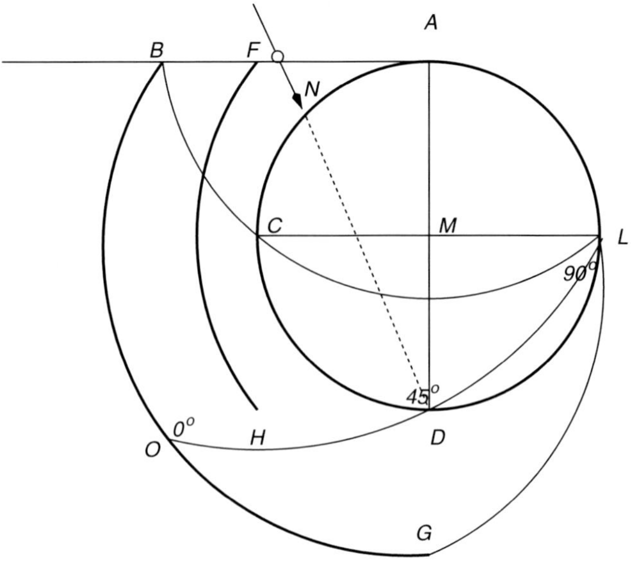

Fig. 13
The *Quadrant of rotation*

rotation corresponding to the positions C and A, of the needle respectively. Assuming again the dip instrument in N at $\beta = 45^\circ$. Then the corresponding quadrant of rotation is constructed by a circle around N through L and is the arc $\overset{\frown}{OL}$. This arc is now divided in 90 parts, starting from the second outer circle (0°) and ending at L (90°). Obviously, in our example, according to Gilbert, the 45° mark is exactly at D.

Now we are ready for the final step. Putting together all our quadrants and lines, we arrive at ➲ *Fig. 14*. The needle at N points to the mark 45° on the arc of rotation $\overset{\frown}{OL}$ and hence intersects the quadrant of dip (arc $\overset{\frown}{SM}$) in the point S. *The angle of the arc* $\overset{\frown}{SR}$ *is the dip angle* δ.

We can now proceed in this manner for all latitudes from $\beta = 0^\circ$ to $\beta = 90^\circ$ in steps of 5°. Each latitude gives a new quadrant of dip, a new quadrant of rotation, and a new intersection point R. The final construction is shown in ➲ *Fig. 15*. However, in ➲ *Fig. 15* the construction is shown in the lower right quadrant instead of in the upper left and uses already the notation of Blundeville instead of those of William Gilbert.

The main goal of the construction, however, is a spiral line which appears, after the removal of all the construction lines, as in ➲ *Fig. 16* and can already be seen in the upper left picture in Blundeville's drawing in ➲ *Fig. 11*. The spiral line consists of all intersection points R.

Together with a quadrant which can rotate around the point C of the *mater* the instrument is ready to use. In order to illustrate its use we give an example. Consider a seamen who has used a dip ring and measured a dip angle of 60°. Then he would rotate the quadrant until the spiral line intersects the quadrant at the point 60° on the inner side of the quadrant. Then the line A–B on the quadrant intersects the scale on the mater at the degree of latitude; in our case 36°, see ➲ *Fig. 18*. However, accurate reading of the scales becomes nearly impossible for angles of dip larger than 60° and the reading depends heavily on the accuracy of the construction of the spiral line. Therefore, Henry Briggs was asked to compute a table in order to replace the dip

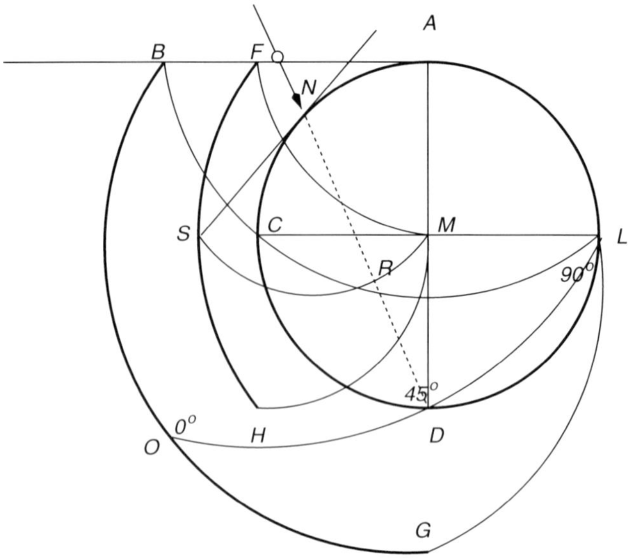

Fig. 14
The final steps

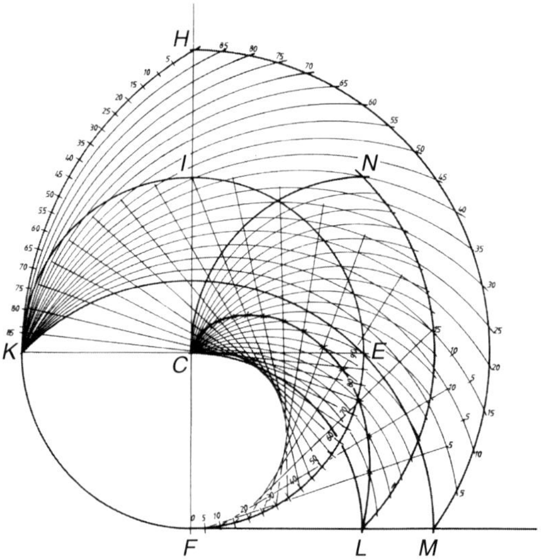

Fig. 15
The construction of Gilbert's mapping

instrument by a simple table look-up. At the very end of Blundeville's *The Theoriques* we find the following appendix, see *Fig. 19*:

- A short Appendix annexed to the former Treatise by Edward Wright, at the motion of the right Worshipfull M. Doctor Gilbert

  Because of the making and using of the foresaid Instrument, for finding the latitude by the declination of the Magneticall Needle, will bee too troublesome for the most part of Sea-men, being

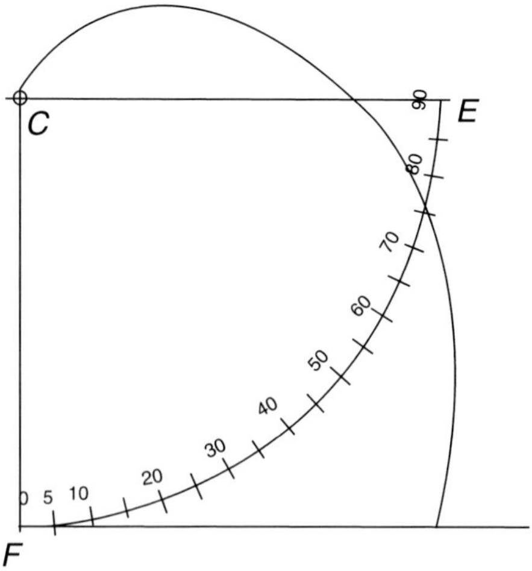

Fig. 16
The *mater* of the dip instrument in *The Theoriques*

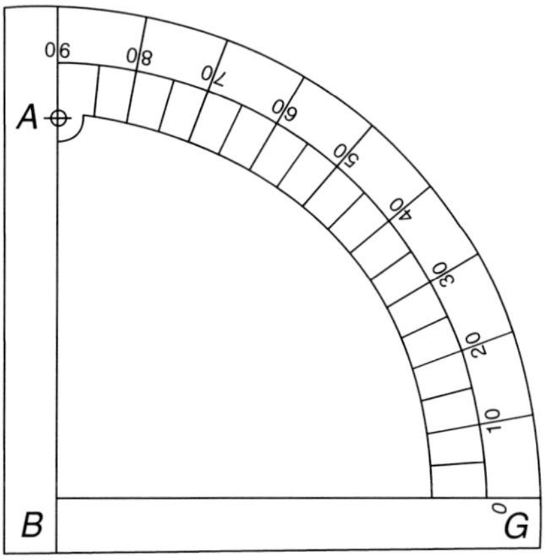

Fig. 17
The quadrant

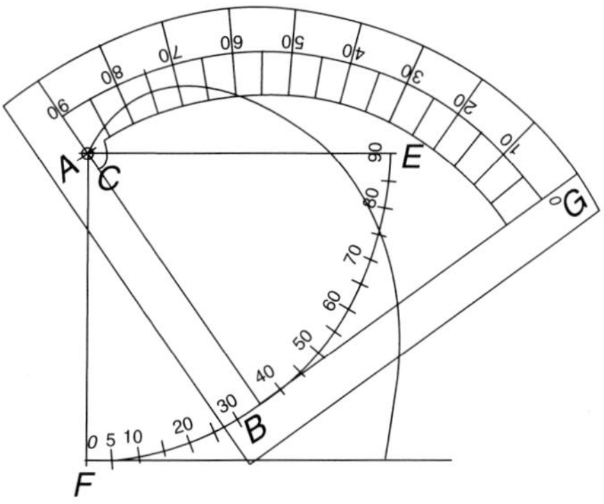

Fig. 18
Determining the latitude for 60° dip

A short Appendix annexed to the former Treatise by *Edward Wright*, at the motion of the right Worshipfull M. Doctor *Gilbert*.

BEcause the making and vsing of the foresaid Instrument, for finding the latitude by the declination of the Magneticall Needle, will bee too troublesome for the most part of Sea-men, being notwithstanding a thing most worthie to be put in daily practise, especially by such as vndertake long voyages: it was thought meet by my worshipfull friend M. Doctor Gilbert, *that (according to M.* Blundeuiles *earnest request) this Table following should be hereunto adioined; which M.* Henry Brigs *(professor of Geometrie in* Gresham *Colledge at London) calculated and made out of the doctrine and tables of Triangles, according to the Geometricall grounds and reason of this Instrument, appearing in the 7 and 8 Chapter of M. Doctor* Gilberts *fift booke of the Loadstone. By helpe of which Table, the Magneticall declination being giuen, the height of the Pole may most easily be found, after this manner.*

With the Instrument of Declination before described, find out what the Magneticall declination is at the place where you are: Then looke that Magneticall declination in the second Collumne of this Table, and in the same line immediatly towards the left hand, you shall find the height of the Pole at the same place, vnlesse there be some variation of the declination, which must be found out by particular obseruation in euery place.

The Table followes on the next Page.

First Collumn.	Second Collumn.		First Collumn.	Second Collumn.		First Collumn.	Second Collumn.	
Height of the Pole.	Magneticall declination.		Height of the Pole.	Magneticall declination.		Height of the Pole.	Magneticall declination.	
Degrees.	Deg.	Min.	Degrees.	Deg	Min	Degrees.	Deg.	Min.
1	2	11	31	52	27	61	79	29
2	4	20	32	53	41	62	10	4
3	6	27	33	54	53	63	80	38
4	8	31	34	56	4	64	81	31
5	10	34	35	57	13	65	81	43
6	12	34	36	58	11	66	82	13
7	14	32	37	59	28	67	82	43
8	16	28	38	60	33	68	83	12
9	18	22	39	61	37	69	83	40
10	20	14	40	62	39	70	84	7
11	22	4	41	63	40	71	84	32
12	23	52	42	64	39	72	84	57
13	25	38	43	65	38	73	85	21
14	27	22	44	66	35	74	85	41
15	29	4	45	67	30	75	86	7
16	30	45	46	68	24	76	86	28
17	32	24	47	69	17	77	86	48
18	34	0	48	70	9	78	87	8
19	35	36	49	70	59	79	87	26
20	37	9	50	71	48	80	87	44
21	38	41	51	72	36	81	88	1
22	40	11	52	73	23	82	88	17
23	41	39	53	74	8	83	88	33
24	43	6	54	74	52	84	88	47
25	44	30	55	75	35	85	89	1
26	45	54	56	76	17	86	89	14
27	47	15	57	76	57	87	89	27
28	48	36	58	77	37	88	89	39
29	49	54	59	78	15	89	89	50
30	51	11	60	78	13	90	90	0

A 318.

Fig. 19
The Appendix

notwithstanding a thing most worthie to be put in daily practise, especially by such as undertake long voyages: it was thought meet by my worshipfull friend M. Doctor Gilbert, that (according to M. Blundeuiles earnest request) this Table following should be hereunto adioned; which M. Henry Briggs (professor of Geometrie in Gresham Colledge at London) calculated and made out of the doctrine and tables of Triangles, according to the Geometricall grounds and reason of this Instrument, appearing in the 7 and 8 Chapter of M. Doctor Gilberts fift[h] booke of the Loadstone. By helpe of which Table, the Magneticall declination being giuen, the height of the Pole may most easily be found, after this manner.

With the Instrument of Declination before described, find out what the Magneticall declination is at the place where you are: Then look that Magneticall declination in the second Collum[n]e of this Table, and in the same line immediatly towards the left hand, you shall find the height of the Pole at the same place, unleße there be some variation of the declination, which must be found out by particular obseruation in euery place.

The next page (which is the final page of *The Theoriques*) indeed shows the Table. *Fig. 19* shows the Appendix. In order to make the numbers in the table more visible I have retyped the table.

First Column. Heighs of the Pole.	Second Column. Magnetical declination.		First Column. Heighs of the Pole.	Second Column. Magnetical declination.		First Column. Heighs of the Pole.	Second Column. Magnetical declination.	
Degrees.	Deg.	Min.	Degrees.	Deg.	Min.	Degrees.	Deg.	Min.
1	2	11	31	52	27	61	79	29
2	4	20	32	53	41	62	80	4
3	6	27	33	54	53	63	80	38
4	8	31	34	56	4	64	81	11
5	10	34	35	57	13	65	81	43
6	12	34	36	58	21	66	82	13
7	14	32	37	59	28	67	82	43
8	16	28	38	60	33	68	83	12
9	18	22	39	61	37	69	83	40
10	20	14	40	62	39	70	84	7
11	22	4	41	63	40	71	84	32
12	23	52	42	64	39	72	84	57
13	25	38	43	65	38	73	85	21
14	27	22	44	66	35	74	85	44
15	29	4	45	67	30	75	86	7
16	30	45	46	68	24	76	86	28
17	32	24	47	69	17	77	86	48
18	34	0	48	70	9	78	87	8
19	35	36	49	70	59	79	87	26
20	37	9	50	71	48	80	87	44
21	38	41	51	72	36	81	88	1
22	40	11	52	73	23	82	88	17
23	41	39	53	74	8	83	88	33
24	43	6	54	74	52	84	88	47
25	44	30	55	75	35	85	89	1
26	45	54	56	76	17	86	89	14
27	47	15	57	76	57	87	89	27
28	48	36	58	77	37	88	89	39
29	49	54	59	78	15	89	89	50
30	51	11	60	78	53	90	90	0

We shall not discuss this table in detail but it is again worthwhile to review the relations between Gilbert, Briggs, Blundeville, and Wright (Hill 1997), p. 36.:

- Briggs was at the center of Gilbert's group. At Gilbert's request he calculated a table of magnetic dip and variation. Their mutual friend Edward Wright recorded and tabulated much of the information which Gilbert used, and helped in the production of *De Magnete*. Thomas Blundeville, another member of Brigg's group, and - like Gilbert - a former protégé of the Earl of Leicester, popularized Gilbert's discoveries in *The Theoriques of the Seven Planets* (1602), a book in which Briggs and Wright again collaborated.

It took Blundeville's *The Theoriques* to describe the construction of the dip instrument accurately which nebulously appeared in Gilbert's *De Magnete*. However, even Blundeville does not say a word concerning the computation of the table. Another friend in the Gresham circle, famous

Edward Wright, included all of the necessary details in the second edition of his *On Errors in Navigation* (Wright 1610), the first edition of which appeared 1599. Much has been said about the importance of Edward Wright, see for instance (Parsons 1939), and he was certainly one of the first – if not the first – who was fully aware of the mathematical background of Mercator's mapping, see Sonar 2001, p. 131ff).

It is in ➋ Chap. 14 where Wright and Briggs explain the details of the computation of the dip table which was actually computed by Briggs showing superb mastership of trigonometry. We shall now turn to this computation.

8 The Computation of the Dip Table

In the second edition of Wright's book *On Errors in Navigation* we find in chapter XIIII:

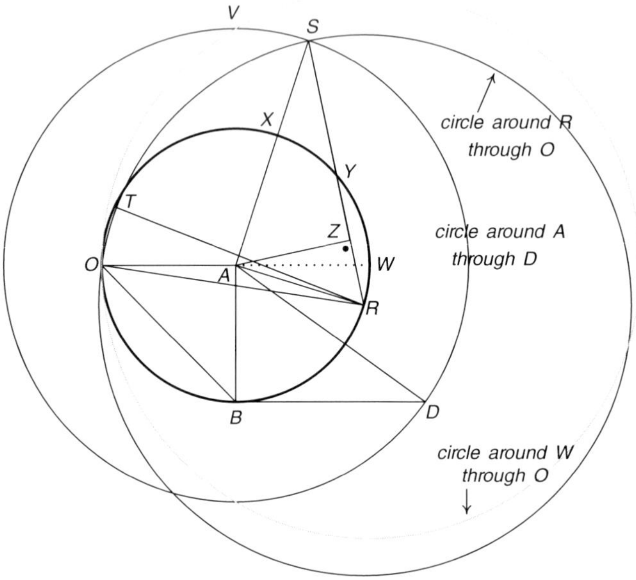

Figure 20

CHAP. XIIII To finde the inclination or dipping of the magneticall needle under the Horizon

- Let *OBR* be a meridian of the earth, wherein let *O* be the pole, *B* the æquinoctal, *R* the latitude (suppose 60 degrees) let *BD* be perpendicular to *AB* in *B*, and equall to the subtense *OB*; and drawing the line *AD*, describe therwith the arch *DSV*. Then draw the subtense *OR*, wherewith (taking *R* for the center) draw the lines *RS* equall to *RO*, and *AS* equall to *AD*. Also because *BR* is assumed to be 60 deg. therefore let *ST* be $\frac{60}{90}$ parts of the arch *STO*, and draw the line *RT*, for the angle *ART* shall be the cõplement of the magnetical needles inclinatiõ vnder the Horizon. Which may be found by the solution of the two triangles *OAR* and *RAS*, after this manner:

Although here again other notation is used as in Blundeville's book as well as in *De Magnete* we can easily see the situation as described by Gilbert. Now the actual computation starts:

- First the triangle *OAR* is given because of the arch *OBR*, measuring the same 150 degr. and consequently the angle at *R* 15 degr. being equall to the equall legged angle at *O*; both which together are 30 degr. because they are the complement of the angle *OAR* (150 degr.) to a semicircle of 180 degr.

The first step in the computation hence concerns the triangle OAR in *Figure 21*. Since point R lies at 60° (measured from B) the arc OBR corresponds to an angle of $90^\circ + 60^\circ = 180^\circ - 30^\circ = 150^\circ$. Hence the angle at A in the triangle OAR is just $90^\circ + (90^\circ - 30^\circ) = 150^\circ$. Since OAR is isosceles the angles at O and R are identical and each is 15°.

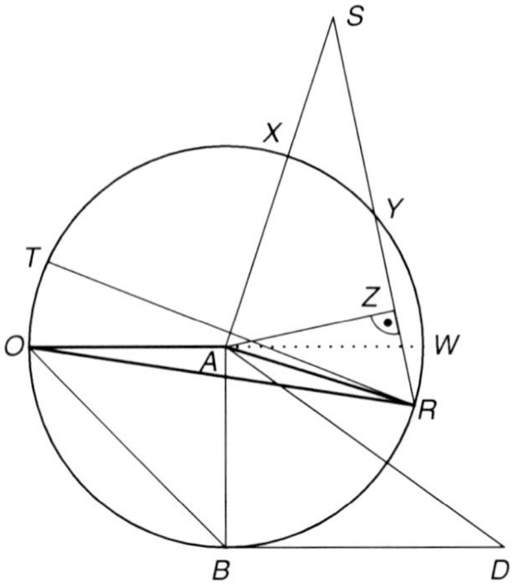

Figure 21
The first step

Let us go on with Wright:

- Secondly, in the triangle *ARS* all the sides are given *AR* the Radius or semidiameter 10,000,000: *RS* equal to *RO* the subtense of 150 deg. 19,318,516: and *AS* equall to *AD* triple in power to *AB*, because it is equal in power to *AB* and *BD*, that is *BO*, which is double in power to *AB*.

The triangle ARS in *Figure 21* is looked at where S lies on the circle around A with radius AD *and* on the circle around R with radius OR. The segment AR is the radius of the Earth or the "whole sine." Wright takes this value to be 10^7. We have to clarify what is meant by *subtense* and where the number 19318516 comes from. Employing the law of sines in triangle OAR we get

$$\frac{OR}{AR} = \frac{\sin 150^\circ}{\sin 15^\circ}$$

and therefore it follows that

$$OR = RO = AR \cdot \frac{OR}{AR} = AR \cdot \frac{\sin 150^\circ}{\sin 15^\circ} = 19318516(.5257...).$$

Since O lies on the circle around R with radius OR as S does we also have $RO = RS$. Furthermore $AS = AD$ since D as S lies on the circle around A with radius AD. *Per constructionem* we have $BD = OB$ and using the Theorem of Pythagoras we conclude

$$OB^2 = BD^2 = 2AB^2$$

as well as

$$AD^2 = AB^2 + BD^2 = AB^2 + 2AB^2 = 3AB^2.$$

This reveals the meaning of the phrase *triple in power to AB*: "the square is three times as big as *AB*." Hence it follows for *AS*:

$$AS = AD0\sqrt{3} \cdot AB = 17320508(.0757...).$$

It is somewhat interesting that Wright does not compute the square root but gives an alternative mode of computation as follows:

- Or else thus: The arch *OB* being 90 degrees, the subtense therof *OB*, that is, the tangent *BD* is 14,142,136, which sought in the table of Tangents, shall giue you the angle *BAD* 54 degr. 44 min. 8 sec. the secant whereof is the line *AD* that is *AS* 17,320,508.

In the triangle *ABD* we know the lenghts of the segments *AB* and $BD = OB = \sqrt{2} \cdot AB = 14142135(.6237...)$. Hence for the angle at *A* we get

$$\tan \angle A = \frac{BD}{AB} = \frac{\sqrt{2} \cdot AB}{AB} = \sqrt{2},$$

which results in $\angle A = 54.7356...^\circ = 54^\circ 44' 8''$. Using this value it follows from

$$\sin \angle A = \frac{BD}{AD} = \frac{OB}{AD}$$

that

$$AD = AS = \frac{OB}{\sin \angle A} = \sqrt{2} \cdot \frac{AB}{\sin 54^\circ 44' 8''} = 17320508(.0757...).$$

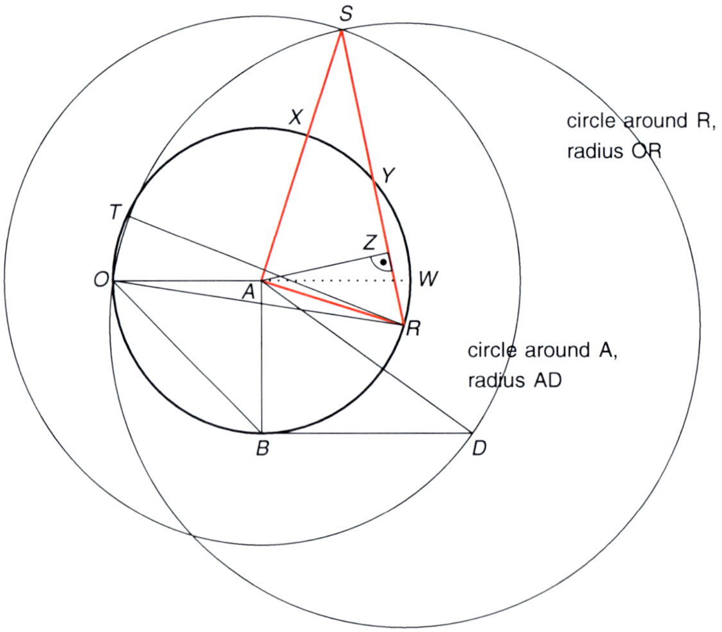

Figure 22
Second step

Wright goes on:

- Now then by 4 Axiom of the 2 booke of Pitisc.[1] as the base or greatest side *SR* 19,318,516 is to y[e] summe of the two other sides *SA* and *AR* 27,320,508; so is the difference of them *SX* 7,320,508 to the segment of the greatest side *SY* 10,352,762; which being taken out of *SR* 19,318,516, there remaineth *YR* 8,965,754, the halfe whereof *RZ* 4,482,877, is the Sine of the angle *RAZ* 26 degr. 38 min. 2 sec. the complement whereof 63 degr. 21 min. 58 sec. is the angle *ARZ*, which added to the angle *ARO* 15 degr. maketh the whole angle *ORS*, 78 degr. 21 min. 58 sec. wherof $\frac{60}{90}$ make 52 degr. 14 min. 38 sec. which taken out of *ARZ* 63 degr. 21 min. 58 sec. there remaineth the angle *TRA* 11 deg. 7 min. 20 sec. the cõplement whereof is the inclination sought for 78 degrees, 52 minutes, 40 seconds.

The "Axiom 4" mentiod is nothing but the Theorem of chords:

- If two chords in a circle intersect then the product of the segments of the first chord equals the product of the segments of the other.

Looking at ➋ *Figure 23* the Theorem of chords is

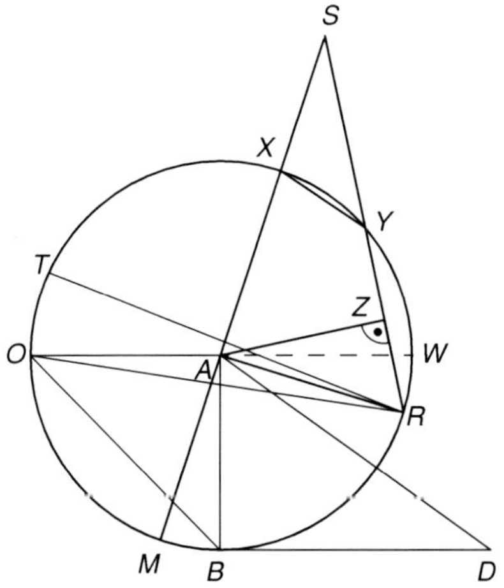

Figure 23

$$MS \cdot SX = SR \cdot SY$$

and since $MS = AS + AB$ it follows

$$(AS + AB) \cdot SX = SR \cdot SY,$$

resulting in

$$\frac{SR}{AS + AB} = \frac{SX}{SY}.$$

[1] The Silesian Bartholomäus Pitiscus (1561-1613) authored the first useful text book on trigonometry: *Trigonometriae sive dimensione triangulorum libre quinque*, Frankfurt 1595, which was published as an appendix to a book on astronomy by Abraham Scultetus. First independent editions were published in Frankfurt 1599, 1608, 1612 and in Augsburg 1600. The first English translation appeared in 1630.

Now the computations should be fully intelligible. Given are $AS = 17320508$, $AB = 10^7$, $SR = OR = 19318516$, and $SX = AS - AX = 7320508$. Hence,

$$SY = \frac{SX \cdot (AS + AB)}{SR} = 10352762.$$

The segment YR has length $YR = SR - SY = 8965754$. *Per constructionem* the point Z is the midpoint of YR. Half of YR is $RZ = 4482877$. From $\sin \angle RAZ = RZ/AR = 4482877/10^7 = 0.4482877$ we get $\angle RAZ = 26.6339^\circ = 26^\circ 38' 2''$. In the right-angled triangle ARZ we see from ➋ *Figure 24* that $\angle ARZ = 90^\circ - 26^\circ 38' 2'' = 63^\circ 21' 58''$.

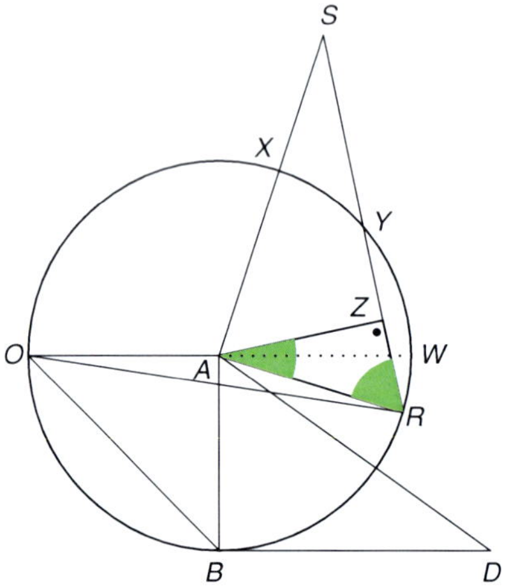

Figure 24

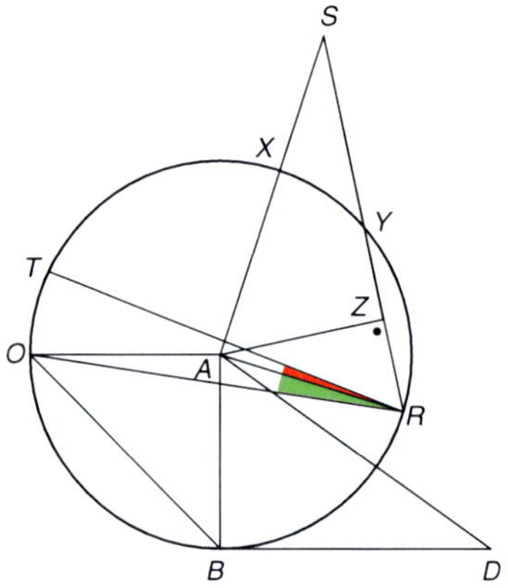

Figure 25

At a degree of latitude of 60° the angle ARO at R is 15° since the obtuse angle in the isosceles triangle ORA is $90^\circ + 60^\circ = 150^\circ$. Therefore, $\angle ORS = \angle ARO + \angle ARZ = 78^\circ 21' 58''$. The part TRS of this angle is 60/90 of it, hence $\angle TRS = 52^\circ 14' 38''$. We arrive at

$$\angle TRA = \angle ARZ - \angle TRA = 63^\circ 21' 58'' - 52^\circ 14' 38'' = 11^\circ 7' 20''.$$

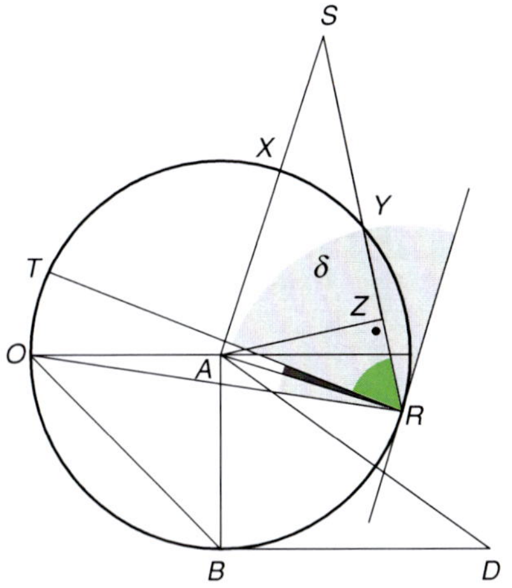

Figure 26

The dip angle δ is the complement of the angle TRA,

$$\delta = 90^\circ - \angle TRA = 78^\circ 52' 40''.$$

Although the task of computing the dip if the degree of latitude is given is now accomplished we find a final remark on saving of labor:

- The Summe and difference of the sides *SA* and *AR* being alwaies the same, viz. 27,320,508 and 7,320,508, the product of them shall likewise be alwaies the same, viz. 199,999,997,378,064 to be diuided by ye side *SR*, that is *RO* the subtense of *RBO*. Therefore there may be some labour saued in making the table of magneticall inclination, if in stead of the said product you take continually but ye halfe thereof, that is 99,999,998,689,032, and so diuide it by halfe the subtense *RO*, that is, by the sine of halfe the arch *OBR*. Or rather thus: As halfe the base *RS* (that is, as the sine of halfe the arch *OBR*) is to halfe the summe of the other two sides *SA* & *AR* 13,660,254, so is half the difference of theẽ 3,660,254 to halfe of the segment *SY*, which taken out of half the base, there remaineth *RZ* ye sine of *RAZ*, whose cõplement to a quadrãt is ye angle sought for *ARZ*.

  According to this Diagramme and demonstration was calulated the table here following; the first columne whereof conteineth the height of the pole for euery whole degree; the second columne sheweth the inclination or dipping of the magnetical needle answerable thereto in degr. and minutes.

Although we have taken these computations from Edward Wright's book there is no doubt that the author was Henry Briggs as is also clear from the foreword of Wright.

9 Conclusion

The story of the use of magnetic needles for the purposes of navigation is fascinating and gives deep insight into the nature of scientific inventions. Gilbert's dip theory and the unhappy idea to link latitude to dip is a paradigm of what can go wrong in mathematical modeling. The computation of the dip table is, however, a brillant piece of mathematics and shows clearly the mastery of Henry Briggs.

References

Balmer H (1956) Beiträge zur Geschichte der Erkenntnis des Erdmagnetismus. Verlag H.R. Sauerländer & Co. Aarau

Campling A (1922) Thomas Blundeville of Newton Flotman, co. Norfolk (1522-1606). Norfolk Archaeol 21: 336–360

Gilbert W (1958) De Magnete. Dover, New York

Greenberg JL (1995) The problem of the Earth's shape from Newton to Clairault. Cambridge University Press, Cambridge, UK

Hill Ch (1997) Intellectual origins of the English revolution revisited. Clarendon Press, Oxford

Parsons EJS, Morris WF (1939) Edward Wright and his work. Imago Mundi 3: 61–71

Pumfrey S (2002) Latitude and the Magnetic Earth. Icon Books, UK

Sonar Th (2001) Der fromme Tafelmacher. Logos Verlag, Berlin

Sonar Th (2002) William Gilberts Neigungsinstrument I: Geschichte und Theorie der magnetischen Neigung. Mitteilungen der Math. Gesellschaft in Hamburg, Band XXI/2, 45-68

Taylor EGR (1954) The mathematical practioneers of Tudor & Stuart England. Cambridge University Press, Cambridge, UK

Ward J (1740) The lives of the professors of Gresham College. Johnson Reprint Corporation London

Waters DJ (1958) The art of navigation. Yale University Press, New Haven

Wright E (1610) Certaine errors in navigation detected and corrected with many additions that were not in the former edition as appeareth in the next pages. London

Part 2

Observational and Measurement Key Technologies

3 Earth Observation Satellite Missions and Data Access

Henri Laur[1] · *Volker Liebig*[2]
[1]European Space Agency (ESA), Head of Earth Observation Mission Management Office, ESRIN, Frascati, Italy
[2]European Space Agency (ESA), Director of Earth Observation Programmes, ESRIN, Frascati, Italy

W. Freeden, M.Z. Nashed, T. Sonar (Eds.), *Handbook of Geomathematics*, DOI 10.1007/978-3-642-01546-5_3,

Abstract This chapter provides an overview on Earth Observation (EO) satellites, describing the end-to-end elements of an EO mission, then focusing on the European EO programmes. Some significant results obtained using data from European missions (ERS, Envisat) are provided. Finally the access to EO data through the European Space Agency (ESA), mostly free of charge, is described.

1 Introduction

Early pictures of the Earth seen from space became icons of the Space Age and encouraged an increased awareness of the precious nature of our common home. Today, images of our planet from orbit are acquired continuously and have become powerful scientific tools to enable better understanding and improved management of the Earth. Satellites provide clear and global views of the various components of the Earth system—its land, ice, oceans and atmosphere—and how these processes interact and influence each other.

Space-derived information about the Earth provides a whole new dimension of knowledge and services which can benefit our lives on a day-to-day basis. Earth Observation (EO) satellites supply a consistent set of continuously updated global data which can offer support to policies related to environmental security by providing accurate information on various environmental issues, including global change. Meteorological satellites have radically improved the accuracy of weather forecasts and have become a crucial part of our daily life. EO data are gradually integrated within many economic activities, including exploitation of natural resources, land use efficiency, or transport routing.

The European Space Agency (ESA) has been dedicated to observing the Earth from space ever since the launch of its first meteorological mission Meteosat in 1977. Following the success of this first mission, the subsequent series of Meteosat satellites, the ERS and Envisat missions have been providing a growing number of users with a wealth of valuable data about the Earth, its climate and changing environment.

It is critical, however, to continue learning about our planet. As our quest for scientific knowledge continues to grow, so does our demand for accurate satellite data to be used for numerous practical applications related to protecting and securing the environment. Responding to these needs, ESA's EO programmes comprise a science and research element, which includes the Earth Explorer missions, and an applications element, which is designed to facilitate the delivery of EO data for use in operational services, including the well-established meteorological missions with the European Organisation for the Exploitation of Meteorological Satellites (EUMETSAT). In addition, the Global Monitoring for Environment and Security (GMES) Sentinel Missions, which form part of the GMES Space Component, will collect robust, long-term relevant datasets. Together with other satellites, their combined data archives are used to produce Essential Climate Variables for climate monitoring, modeling, and prediction.

Mathematics is of crucial importance in EO: on-board data compression and signal processing are common features of many EO satellites, inverse problems for data applications (e.g., the determination of the gravitational field from ESA Gravity field and steady-state Ocean Circulation Explorer [GOCE] satellite measurements), data assimilation to combine EO data with numerical models, etc.

For this chapter many texts and information have been used which are found on ESA Web sites.

2 End-to-End Earth Observation Satellite Systems

In this chapter the end-to end structure of a typical satellite system shall be explained. This will enable the user of EO data to better understand the process of gathering the information he or she is using.

EO satellites usually fly on a so-called *Low Earth Orbit* (LEO) which means an orbit altitude of some 250–1,000 km above the Earth surface and an orbit inclination close to 90°, i.e., a polar orbit. Another orbit used mainly for weather satellites like Meteosat is the *Geostationary Orbit* (GEO). This is the orbit in which the angular velocity of the satellite is exactly the same as the one of Earth (360° in 24 h). If the satellite is positioned exactly over the Equator (inclination 0°) then the satellite has always the same position relative to the Earth surface. This is why the orbit is called geostationary. Using this position the satellite's instruments can always observe the same area on Earth. The price to pay for this position is that the orbit altitude is approx. 36000 km which is very far compared to a LEO.

A special LEO is the *sun-synchronous orbit* which is used for all observation systems which need the same surface illumination angle of the sun. This is important for imaging systems like optical cameras. For this special orbit, typically 600–800 km in altitude, the angle between the satellite orbit and the line between sun and Earth is kept equal. As the Earth rotates once per year around the sun the satellite orbit has also to rotate by 360° in 365 days. This can be reached by letting the rotational axis of the satellite orbit precess, i.e., rotate, by 360°/365° per day to keep exactly pace with the Earth rotation around the sun. As the Earth is not a perfect sphere the excess mass at the Equator creates an angular momentum which lets a rotating system precess like a gyroscope. The satellite has to fly with an inclination of approx. 98° which is close to an orbit flying over the poles. As the Earth is rotating under this orbit the satellite almost "sees" the Earth's entire surface during several orbits. Typical LEO satellites are ERS, Envisat, or SPOT (see ❯ Sect. 3).

All EO Satellite Systems consist of a Space Segment (S/S) and a Ground Segment (G/S). If we regard the whole chain of technical infrastructures to acquire, downlink, process and distribute the EO data we speak about an end-to-end system.

2.1 Space Segment

The S/S mainly consists of the spacecraft, i.e., the EO satellite which has classically a modular design and is subdivided into satellite *platform* (or *bus*) and satellite *payload*. ❯ *Figure 1* shows as example the Envisat satellite. The satellite bus offers usually all interfaces (mechanical, thermal, energy, data handling) which are necessary to run the payload instruments mounted on it. In addition it contains usually all housekeeping, positioning, and attitude control functions as well as a propulsion system. It should be mentioned that small satellites often have a more integrated approach.

In some cases the S/S can consist of more than one satellite, e.g., if two or more satellites fly in tandem, such as Swarm (see ❯ Sect. 3). The S/S may also include a data relay satellite (DRS), positioned in a geostationary orbit and used to relay data between the EO satellite and the Earth when the satellite is out of visibility of the ground stations, such as ESA Artemis satellite data relay for the Envisat satellite.

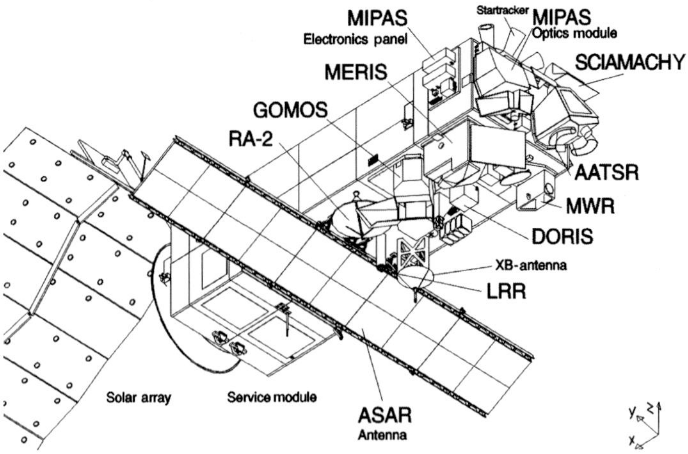

Fig. 1
Envisat satellite with its platform (service module and solar array) and the payload composed of 10 instruments

2.2 Ground Segment

The G/S provides the means and resources to manage and control the EO satellite, to receive and process the data produced by the payload instruments, and to disseminate and archive the generated products. In general the G/S can be split into two major elements (see example in *Fig. 2* for CryoSat mission):

- The Flight Operations Segment (FOS), which is responsible for the command and control of the satellite
- The Payload Data Segment (PDS), which is responsible for the exploitation of the instruments data

Both G/S elements have different communication paths: satellite control and telecommand uplink use normally S-band whereas instrument data is downlinked to ground stations via X-band either directly or after on-board recording. In addition, Ka-band or laser links with very high bandwidth are used for inter-satellite links needed by data relay satellites like Artemis or the future European Data Relay Satellite (EDRS).

2.2.1 Flight Operations Segment

The FOS at ESA is under the responsibility of the European Space Operation Centre (ESOC) located in Darmstadt, Germany. The mandate of ESOC is to conduct mission operations for ESA satellites and to establish, operate and maintain the related G/S infrastructure. Most of ESA EO satellites are controlled and commanded from ESOC control rooms.

ESOC's involvement in a new mission normally begins with the analysis of possible operational orbits or trajectories and the calculation of the corresponding launch windows—selected to make sure that the conditions encountered in this early phase remain within the spacecraft

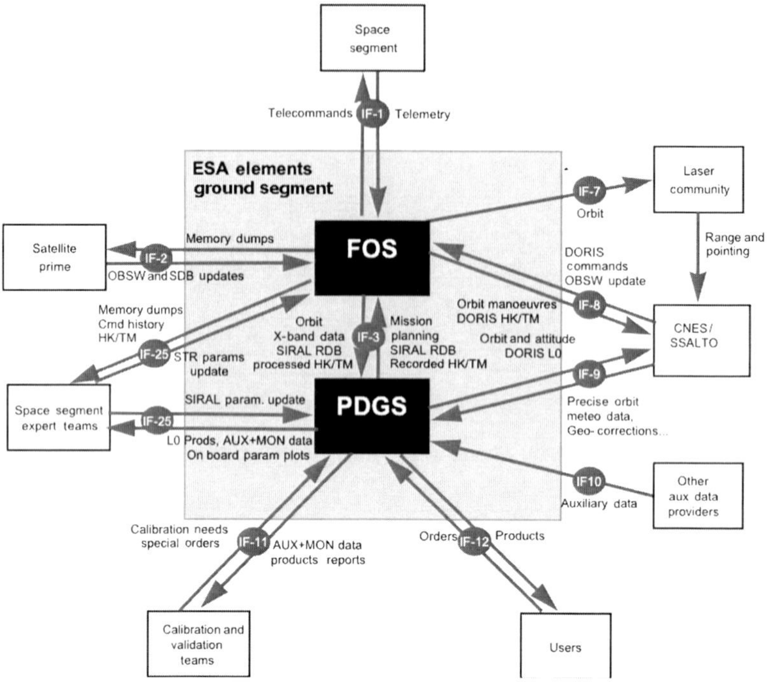

Fig. 2

Example of an EO G/S for ESA CryoSat mission including the FOS and the Payload Data Ground Segment (PDGS) as well as the various interfaces (IF) around the G/S

capabilities. The selection of the operational orbit is a complex task with many trade-offs involving the scientific objectives of the mission, the launch vehicle, the spacecraft, and the ground stations. ESOC's activity culminates during the Launch and Early Orbit Phase (LEOP), with the first steps after the satellite separates from the launcher's upper-most stage, including the deployment of antennas and solar arrays as well as critical orbit and attitude control manoevres.

After the LEOP, ESOC starts the operations of the FOS, including generally the command and control of the satellite, the satellite operations uploading—based on the observation plans prepared by the PDS, the satellite configuration and performance monitoring, the orbit prediction, restitution and maintenance, the contingency and recovery operations following satellite anomalies. ESOC does also provide a valuable service for avoidance of collision with space debris, a risk particularly high for LEO satellites orbiting around 800 km altitude.

More information about ESOC activities can be found at: http://www.esa.int/esoc.

2.2.2 Payload Data Segment

The EO PDS at ESA is under the responsibility of ESRIN, located in Frascati, Italy, and known as the ESA Centre for Earth Observation, because it also includes the ESA activities for developing the EO data exploitation.

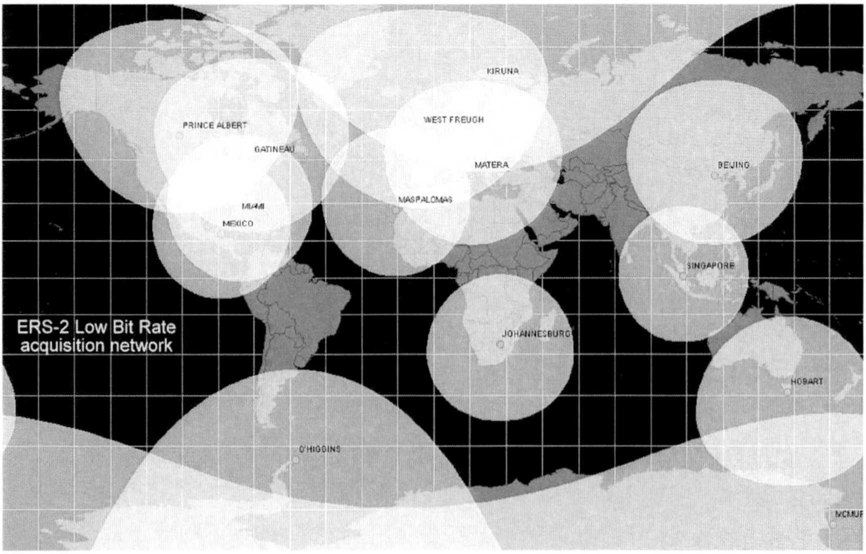

Fig. 3

Example of a network of acquisition stations for the low bit rate instruments on-board the ERS-2 satellite (status in 2009)

The PDS provides all services related to the exploitation of the data produced by the instruments embarked on-board the ESA EO satellites and is therefore the gateway for EO data users. The activities performed within a PDS are:

- The payload data acquisition using a network of worldwide acquisition stations as well as data relay satellites such as Artemis
- The processing of products either in near real time from the above data acquisitions, or on demand from the archives
- The monitoring of products quality and instrument performance, as well as the regular upgrade of data processing algorithms
- The archiving and long-term data preservation, and the data re-processing following the upgrade of processing algorithms
- The interfaces to the user communities from user order handling to product delivery, including planning of instrument observations requested by the users
- The availability of data products on Internet servers, or through dedicated satellite communication (see Sect. 5 for data access)
- The development of new data products and new services in response to evolving user demand

The ESA PDS is based on a decentralized network of acquisition stations (*Fig. 3*) and archiving centers. The data acquisition stations located close to the poles can "see" most of the LEO satellite passes and are therefore used for acquiring the data recorded on-board during the previous orbit. Acquisition stations located away from polar areas are used essentially to acquire data collected over their station mask (usually 4000 km diameter). The archiving centers are generally duplicated to avoid data loss in case of fire or accidents.

In carrying out the PDS activities, ESA works closely with other space agencies as well as with coordination and standardization bodies. This includes a strong inter-agency collaboration to acquire relevant EO data following a natural disaster.

More information about ESRIN activities can be found at: http://www.esa.int/esrin .

3 Overview on ESA Earth Observation Programmes

3.1 Background

The ESA is Europe's gateway to space. Its mission is to shape the development of Europe's space capability and ensure that investment in space continues to deliver benefits to the citizens of Europe and the world. ESA is an international organization with 18 Member States. By coordinating the financial and intellectual resources of its members, it can undertake programmes and activities far beyond the scope of any single European country. ESA's programmes are designed to find out more about Earth, its immediate space environment, our Solar System and the Universe, as well as to develop satellite-based technologies and services, and to promote European industries. ESA works closely with space organizations inside and outside Europe.

ESA's EO programme, known as the ESA Living Planet programme, embodies the fundamental goals of EO in developing our knowledge of Earth, preserving our planet and its environment, and managing life on Earth in a more efficient way. It aims to satisfy the needs and constraints of two groups, namely the *scientific community* and also the *providers of operational satellite-based services*. The ESA Living Planet Programme is therefore composed of two main components: a science and research element in the form of Earth Explorer missions, and an operational element known as Earth Watch designed to facilitate the delivery of EO data for use in operational services. The Earth Watch element includes the future development of meteorological missions in partnership with EUMETSAT and also new missions under the European Union's GMES initiative, where ESA is the partner responsible for developing the space component. The past and current ERS and Envisat missions contribute both to the science and to the applications elements of the ESA Living Planet programme.

General information on ESA EO programmes can be found at: http://www.esa.int/esaEO.

3.2 ERS-1 and ERS-2 Missions

The ERS-1 (European Remote Sensing) satellite, launched in 1991, was ESA's first EO satellite on LEO; it carried a comprehensive payload including an imaging C-band Synthetic Aperture Radar (SAR), a radar altimeter and other instruments to measure ocean surface temperature and winds at sea. ERS-2, which overlapped with ERS-1, was launched in 1995 with an additional sensor for atmospheric ozone research.

At their time of launch the two ERS satellites were the most sophisticated EO spacecraft ever developed and launched in Europe. These highly successful ESA satellites have collected a wealth of valuable data on the Earth's land surfaces, oceans, and polar caps and have been called upon to monitor natural disasters such as severe flooding or earthquakes in remote parts of the world.

ERS-1 was unique in its systematic and repetitive global coverage of the Earth's oceans, coastal zones and polar ice caps, monitoring wave heights and wavelengths, wind speeds and directions, sea levels and currents, sea surface temperatures and sea ice parameters. Until ERS-1 appeared, such information was sparse over the polar regions and the southern oceans.

The ERS missions were both an experimental and a preoperational system, since it has had to demonstrate that the concept and the technology had matured sufficiently for successors such as Envisat, and that the system could routinely deliver to end users some data products such as reliable sea-ice distribution charts and wind maps within a few hours of the satellite observations.

The experimental nature of the ERS missions was outlined shortly after the launch of ERS-2 in 1995 when ESA decided to link the two spacecrafts in the first ever "tandem" mission which lasted for nine months. During this time the increased frequency and level of data available to scientists offered a unique opportunity to observe changes over a very short space of time, as both satellites orbited Earth only 24 h apart and to experiment innovative SAR measurement techniques.

The ERS-1 satellite ended its operations in 2000, far exceeding its 3 years planned lifetime. The ERS-2 satellite, now a veteran in space, is expected to continue operating until 2011, i.e., 16 years after its launch, maximizing the benefit of the past investment.

More information on ERS missions can be found at: http://earth.esa.int/ers/.

3.3 Envisat Mission

Envisat, ESA's second generation remote sensing satellite, launched in 2002, not only provides continuity of many ERS observations—notably the ice and ocean elements—but adds important new capabilities for understanding and monitoring our environment, particularly in the areas of atmospheric chemistry and ocean biological processes.

Envisat is the largest and most complex satellite ever built in Europe. Its package of 10 instruments is making major contributions to the global study and monitoring of the Earth and its environment, including global warming, climate change, ozone depletion, and ocean and ice monitoring. Secondary objectives are more effective monitoring and management of the Earth's resources, and a better understanding of the solid Earth.

As a total package, Envisat capabilities exceed those of any previous or planned Earth observation satellite. The payload includes three new atmospheric sounding instruments designed primarily for atmospheric chemistry, including measurement of ozone in the stratosphere. The advanced C-band SAR collects high-resolution images with a variable viewing geometry, with wide swath and selectable dual polarization capabilities. A new imaging spectrometer is included for ocean color and vegetation monitoring, and there are improved versions of the ERS radar altimeter, microwave radiometer, and visible/near-infrared radiometers, together with a new very precise orbit measurement system.

Combined with ERS-1 and ERS-2 missions, the Envisat mission is an essential element in providing long-term continuous datasets that are crucial for addressing environmental and climatological issues. In addition, the Envisat mission further promotes the gradual transfer of applications of EO data from experimental to preoperational and operational exploitation.

Thanks to its current good technical status, ESA envisages to operate Envisat satellite until 2013, i.e., up to 11 years after its launch.

More information on Envisat mission can be found at: http://envisat.esa.int.

3.4 Proba-1

Launched in 2001, the small Project for On-Board Autonomy (Proba) satellite was intended as a 1 year ESA technology demonstrator. Once in orbit, however, its unique capabilities and performance made it evident that it could make big contributions to science so its operational lifetime is currently extended until 2012 to serve as an EO mission. Its main payload is a spectrometer (CHRIS), designed to acquire hyperspectral images with up to 63 spectral bands. Also aboard is the high resolution camera (HRC), which acquires 5 m black and white images.

3.5 Earth Explorers

The Earth Explorer missions form the science and research element of ESA's Living Planet Programme and focus on the atmosphere, biosphere, hydrosphere, cryosphere and the Earth's interior with the overall emphasis on learning more about the interactions between these components and the impact that human activity is having on natural Earth processes.

The Earth Explorer missions are designed to address key scientific challenges identified by the science community whilst demonstrating breakthrough technology in observing techniques. By involving the science community right from the beginning in the definition of new missions and introducing a peer-reviewed selection process, it is ensured that a resulting mission is developed efficiently and provides the exact data required by the user. The process of mission selection has given the Earth science community an efficient tool for advancing the understanding of the Earth system. The science questions addressed also form the basis for development of new applications of EO. This approach also gives Europe an excellent opportunity for international cooperation, both within the wide scientific domain and also in the technological development of new missions (ESA, 2006).

The family of Earth Explorer missions is a result of this strategy. Currently there are six Earth Explorer missions and a further three undergoing feasibility study:

3.5.1 GOCE (Gravity Field and Steady-State Ocean Circulation Explorer)

Launched in March 2009, GOCE is dedicated to measuring the Earth's gravity field and modeling the geoid with unprecedented accuracy and spatial resolution to advance our knowledge of ocean circulation, which plays a crucial role in energy exchanges around the globe, sea-level change, and Earth interior processes. GOCE will also make significant advances in the field of geodesy and surveying. (This handbook also includes ➋ Chaps. 4 and ➋ 9 dedicated to gradiometry).

3.5.2 SMOS (Soil Moisture and Ocean Salinity)

Launched in November 2009, SMOS will observe soil moisture over the Earth's landmasses and salinity over the oceans. Soil moisture data are urgently required for hydrological studies and data on ocean salinity are vital for improving our understanding of ocean circulation patterns (for more details on hydrology and ocean circulation see ➋ Chaps. 11–13, respectively).

3.5.3 CryoSat

Due for launch towards early 2010, CryoSat-2 will acquire accurate measurements of the thickness of floating sea-ice so that seasonal to inter-annual variations can be detected, and will also survey the surface of continental ice sheets to detect small elevation changes. It will determine regional trends in Arctic perennial sea-ice thickness and mass, and determine the contribution that the Antarctic and Greenland ice sheets are making to mean global rise in sea level. CryoSat-2 replaces CryoSat which was lost in 2005 after a launch failure.

3.5.4 Swarm

Due for launch in 2011, Swarm is a constellation of three satellites that will provide high-precision and high-resolution measurements of the strength and direction of the Earth's magnetic field. The geomagnetic field models resulting from the Swarm mission will provide new insights into the Earth's interior, further our understanding of atmospheric processes related to climate and weather, and will also have practical applications in many different areas such as space weather and radiation hazards (for more aspects on geomagnetism the reader is referred to the ❯ Chaps. 5 and ❯ 17–19).

3.5.5 ADM-Aeolus (Atmospheric Dynamics Mission)

Due for launch in 2012, ADM-Aeolus will be the first space mission to measure wind profiles on a global scale. It will improve the accuracy of numerical weather forecasting and advance our understanding of atmospheric dynamics and processes relevant to climate variability and climate modeling. ADM-Aeolus is seen as a mission that will pave the way for future operational meteorological satellites dedicated to measuring the Earth's wind fields.

3.5.6 EarthCARE (Earth Clouds Aerosols and Radiation Explorer)

Due for launch in 2013, EarthCARE is being implemented in cooperation with the Japanese Aerospace Exploration Agency (JAXA). The mission addresses the need for a better understanding of the interactions between cloud, radiative, and aerosol processes that play a role in climate regulation.

3.5.7 Future Earth Explorers

In 2005, ESA released the latest opportunity for scientists from ESA Member States to submit proposals for ideas to be assessed for the next in the series of Earth Explorer missions. As a result, 24 proposals were evaluated and a shortlist of six missions underwent assessment study. Subsequently, three missions—BIOMASS, CoReH2O, and PREMIER have been selected in 2009 for the next stage of development (feasibility study). This process will lead to the launch of ESA's seventh Earth Explorer mission in the 2016 timeframe. In parallel, ESA is starting the process for selecting the eighth Earth Explorer mission.

More information on Earth Explorer missions can be found at: http://www.esa.int/esaLP/LPearthexp.html.

3.6 GMES and Sentinels

The GMES programme has been established to fulfill the growing need amongst European policy-makers to access accurate and timely information services to better manage the environment, understand and mitigate the effects of climate change and ensure civil security.

Under the leadership of the European Commission, GMES relies largely on data from satellites observing the Earth. Hence, ESA is developing and managing the *GMES Space Component*. The European Commission, acting on behalf of the European Union (EU), is responsible for the overall initiative, setting requirements, and managing the *GMES services*.

To ensure the operational provision of EO data, the Space Component includes a series of five space missions called "Sentinels," which are being developed by ESA specifically for GMES. In addition, data from satellites that are already in orbit, or are planned will also be used for the initiative. These "Contributing Missions" include both existing and new satellites, whether owned and operated at European level by the EU, ESA, EUMETSAT, or on a national basis. They also include data acquired from non-European partners. The GMES Space Component forms part of the European contribution to the worldwide Global Earth Observation System of Systems (GEOSS).

The acquisition of reliable information and the provision of services form the backbone of Europe's GMES initiative. Services are based on data from a host of existing and planned EO satellites from European and national missions, as well as a wealth of measurements taken *in situ* from instruments carried on aircraft, floating in the oceans or positioned on the ground. Services provided through GMES fall so far into five main categories: services for land management, services for the marine environment, services relating to the atmosphere, services to aid emergency response, and services associated with security. The services component of GMES is under the responsibility of the European Commission.

More information on GMES can be found at: http://www.esa.int/esaLP/LPgmes.html

3.6.1 Sentinel Missions

The success of GMES will be achieved largely through a well-engineered Space Component for the provision of EO data to turn them into services for continuous monitoring the environment.

The GMES Space Component comprises five types of new satellites called *Sentinels* that are being developed by ESA specifically to meet the needs of GMES. The Sentinel Missions include radar and super-spectral imaging for land, ocean, and atmospheric monitoring. The so-called a and b models of the first three Sentinels (Sentinel-1a/1b, Sentinel-2a/2b, and Sentinel-3a/3b) are currently under industrial development, with the first satellite planned to launch in 2012.

Sentinel-1 will provide all-weather, day and night C-band radar imaging for land and ocean services and will allow to continue SAR measurements initiated with ERS and Envisat missions, Sentinel-2 will provide high-resolution optical imaging for land services, Sentinel-3 will carry an altimeter, optical and infrared radiometers for ocean and global land monitoring as a continuation of Envisat measurements, Sentinel-4 and Sentinel-5 will provide data for atmospheric composition monitoring (Sentinel-4 from geostationary orbit and Sentinel-5 from low-Earth polar orbit). It is envisaged that Sentinel-4 and Sentinel-5 will be instruments carried on the next generation of EUMETSAT meteorological satellites—Meteosat Third Generation (MTG) and post- EUMETSAT Polar System (EPS), respectively. However a dedicated Sentinel-5 precursor mission is planned to be launched in 2014 to avoid any data gap between Envisat and Sentinel-5.

3.6.2 GMES Contributing Missions

Before data from the Sentinel satellites is available, missions contributing to GMES play a crucial role ensuring that an adequate dataset is provided for the GMES Services. The role of the Contributing Missions will, however, continue to be essential once the Sentinels are operational by complementing Sentinel data and ensuring that the whole range of observational requirements is satisfied.

Contributing Missions are operated by national agencies or commercial entities within ESA's Member States, EUMETSAT or other third parties. GMES Contributing Missions data initially address services for land and ice and also focus on ocean and atmosphere. Current services mainly concentrate on the following observation techniques:

- SAR sensors, for all weather day/night observations of land, ocean and ice surfaces,
- Medium-low resolution optical sensors for information on land cover, for example, agriculture indicators, ocean monitoring, coastal dynamics, and ecosystems
- High-resolution and medium-resolution optical sensors—panchromatic and multispectral—for regional and national land monitoring activities
- Very high resolution optical sensors for targeting specific sites, especially in urban areas as for security applications
- High accuracy radar altimeter systems for sea-level measurements and climate applications
- Radiometers to monitor land and ocean temperature
- Spectrometer measurements for air quality and atmospheric composition monitoring

A G/S, facilitating access to both Sentinel and Contributing Missions data, complements the GMES S/S.

More information on GMES data is available can be found at: http://gmesdata.esa.int .

3.7 Meteorological Programmes

3.7.1 Meteosat

With the launch of the first Meteosat satellite into a geostationary orbit in November 1977, Europe gained the ability to gather weather data over its own territory with its own satellite. Meteosat began as a research programme for a single satellite by the European Space Research Organisation, a predecessor of ESA. Once the satellite was in orbit, the immense value of the data it provided led to the move from a research to an operational mission. ESA launched three more Meteosat satellites before the founding of EUMETSAT, organization partner of ESA for the meteorological programmes. Launched in 1997, Meteosat-7 was the last of the first generation of Meteosat satellites.

The first generation of seven Meteosat satellites brought major improvements to weather forecasting. But technological advances and increasingly sophisticated weather forecasting requirements created demand for more frequent, more accurate and higher resolution space observation. To meet this demand, EUMETSAT started the Meteosat Second Generation

(MSG) programme, in coordination with ESA. In 2002, EUMETSAT launched the first MSG satellite, renamed Meteosat-8 when it began routine operations to clearly maintain the link to earlier European weather satellites. It was the first of four MSG satellites, which are gradually replacing the original Meteosat series (aspects of weather forecasting are contained in ➋ Chap. 21).

The MTG will take the relay in 2015 from Meteosat-11, the last of a series of four MSG satellites. MTG will enhance the accuracy of forecasts by providing additional measurement capability, higher resolution, and more timely provision of data. Like its predecessors, MTG is a joint project between EUMETSAT and ESA that followed the success of the first generation Meteosat satellites.

3.7.2 MetOp

Launched in 2006, in partnership between ESA and EUMETSAT, MetOp is Europe's first polar-orbiting satellite dedicated to operational meteorology. It represents the European contribution to a new cooperative venture with the United States providing data to monitor climate and improve weather forecasting. MetOp is a series of three satellites to be launched sequentially over 14 years, forming the S/S of EPS.

MetOp carries a set of "heritage" instruments provided by the United States and a new generation of European instruments that offer improved remote sensing capabilities to both meteorologists and climatologists. The new instruments augment the accuracy of temperature humidity measurements, readings of wind speed and direction, and atmospheric ozone profiles.

Preparations have started for the next generation of this EPS, the so-called Post-EPS.

More information can be found at: http://www.eumetsat.int.

3.8 ESA Third Party Missions

ESA uses its multi-mission ground systems to acquire, process, distribute, and archive data from other satellites—known as Third Party Missions (TPM). The data from these missions are distributed under specific agreements with the owners or operators of those missions, which can be either public or private entities outside or within Europe.

ESA TPM are addressing most of the existing observation techniques:

- SAR sensors, e.g., L-band instrument (*PALSAR*) on-board the Japanese Space Agency *ALOS* satellite
- Low resolution optical sensors, e.g., *MODIS* sensors on-board US Terra and Aqua satellites, *SeaWIFS* sensor on-board US OrbView-2 satellite, etc.
- High resolution optical sensors, e.g., *Landsat* imagery (15–30 m), *SPOT-4* data (10 m), a 10 m radiometer instrument (AVNIR-2) on-board ALOS satellite, medium resolution (20–40 m) sensors on-board the Disaster Monitoring Constellation (*DMC*), etc.
- Very high resolution optical sensors, e.g., US *Ikonos-2* (1–4 m), Korean *Kompsat-2* (1–4 m), Taiwan *Formosat-2* (2–8 m), Japanese *PRISM* (2.5 m) on-board ALOS, etc.
- Atmospheric chemistry sensors, e.g., Swedish *Odin*, Canadian *SciSat-1*, and Japanese *GOSAT* satellites

The complete list of TPM currently supported by ESA is available at: http://earth.esa.int/thirdpartymissions/.

4 Some Major Results of ESA Earth Observation Missions

Since 1991, the flow of data provided by the consecutive ERS-1, ERS-2, and Envisat missions has been converted in extensive results, giving new insights into our planet. These results encompass many fields of Earth science, including land, ocean, ice, and atmosphere studies and have shown the importance of long-term data collection to identify trends such as those associated to climate change.

The ERS and Envisat missions are valuable tools not only for Earth scientists, but gradually also for public institutions providing operational services such as sea ice monitoring for ship routing or UV index forecast. The ERS and Envisat data have also stimulated the emergence of new analysis techniques such as SAR Interferometry.

4.1 Land

One of the most striking results of the ERS and Envisat missions for land studies is the development of the *interferometry* technique using SAR instruments. SAR instruments are microwave imaging systems. Besides their cloud penetrating capabilities and day and night operational capabilities, they have also 'interferometric' capabilities, i.e., capabilities to measure very accurately the travel path of the transmitted and received radiation. SAR Interferometry (InSAR) exploits the variations of the radiation travel path over the same area observed at two or more acquisition times with slightly different viewing angles. Using InSAR, scientists are able to measure millimetric surface deformations of the terrain, like the ones associated with earthquakes movements, volcanoes deformation, landslides or terrain subsidence (ESA, 2007).

Since the launch of ERS-1 in 1991, InSAR has advanced the fields of tectonics and volcanology by allowing scientists to monitor the terrain deformation anywhere in the world at any time. Some major *earthquakes* such as Landers, California in 1992, Bam, Iran in 2003, or the 2009 L' Aquila earthquake in central Italy, were "imaged" by the ERS and Envisat missions, allowing geologists to better understand the fault rupture mechanisms. Using ERS and Envisat data, scientists have been able to monitor the long-term behavior of some *volcanoes* such as Mt. Etna, providing crucial information for understanding how the volcano's surface deformed during the rise, storage, and eruption of magma. Changes in surface deformation, such as sinking, bulging, and rising, are indicators of different stages of volcanic activity, which may result in eruptions. Thus, precise monitoring of a volcano's surface deformation could lead to predictions of eruptions.

The InSAR technique was also exploited by merging SAR data acquired by different satellites. For 9 months in 1995/1996, the two ERS-1 and ERS-2 satellites undertook a "tandem" mission, in which they orbited Earth only 24 h apart. The acquired image pairs provide much greater interferometric coherence than is normally possible, allowing scientists to generate detailed *digital elevation maps* and observe changes over a very short space of time. The same "tandem" approach is currently followed by the ERS-2 and Envisat satellites, adding to the ever growing set of SAR interferometric data.

Because SAR instruments are able "to see" through clouds or at night time, their ability to map *river flooding* was quickly recognized and has been gradually exploited by civil protection authorities. Similarly other ERS and Envisat instruments such as the ATSR infrared radiometers have been able to provide relevant *forest fires* statistics.

Using data of the MERIS spectrometer instrument on-board Envisat, the most detailed Earth *global land cover map* was created. This global map, which is ten times sharper than any previous global satellite map, was derived by an automatic and regionally adjusted classification of the MERIS data global composites using land cover classes defined according to the UN Land Cover Classification System (LCCS). The map and its various composites support the international community in modeling climate change impacts, studying ecosystems, and plotting worldwide land-use trends.

4.2 Ocean

Through the availability of satellite data from ERS-1 up until now, scientists have gained an understanding of the ocean and its interaction with the entire Earth system that they would not have otherwise been able to do. Thanks to their almost 20 year's extent, the time series of the ERS and Envisat missions' data allow scientists to investigate the effects of climate change, in particular on the oceans. Those long-term data measurements allow removing the yearly variability of most of the geophysical parameters, providing results of fundamental significance in the context of climate change.

Radar altimeters on-board satellites play an important role in those long-term measurements. They work by sending thousands of separate radar pulses down to Earth per second then recording how long their echoes take to bounce back to the satellite platform. The sensor times its pulses' journey down to under a nanosecond to calculate the distance to the planet below to a maximum accuracy of 2 cm. The consecutive availability of altimeters on-board ERS-1, ERS-2 then Envisat allowed establishing that the global *mean sea level* raised by about 2–3 mm per year since the early 1990s, with important regional variations (cf. ➲ Chaps. 12 and ➲ 13).

Sea surface temperature (SST) is one of the most stable of several geographical variables which, when determined globally, helps diagnose the state of the Earth's climate system. Tracking SST over a long period is a reliable way researchers know of measuring the precise rate at which global temperatures are increasing and improves the accuracy of our climate change models and weather forecasts. There is evidence from measurements made from ERS and Envisat ATSR radiometer instruments that there is a distinctive upward trend in global sea surface temperatures. The ATSR instruments produced data of unrivalled accuracy on account of its unique dual view of the Earth's surface, whereby each part of the surface is viewed twice, through two different atmospheric paths. This not only enables scientists to correct for the effects of dust and haze, which degrade measurements of surface temperature from space, but also enables scientists to derive new measurements of the actual dust and haze, which are needed by climate scientists.

One of the main assets of the Envisat mission is its multi sensor capability, which allows observing geophysical phenomena from various "viewpoints." A good example is the observation of *cyclones*: the data returned by Envisat includes cloud structure and height at the top of the cyclone, wind and wave fields at the bottom, sea surface temperature and even sea height anomalies, indicative of upper ocean thermal conditions that influence its intensity.

The ERS and Envisat SAR data did also stimulate the development of maritime applications. They include monitoring of illegal fisheries or monitoring of oil slicks (➲ *Fig. 4*) which can be natural or the results of human activities.

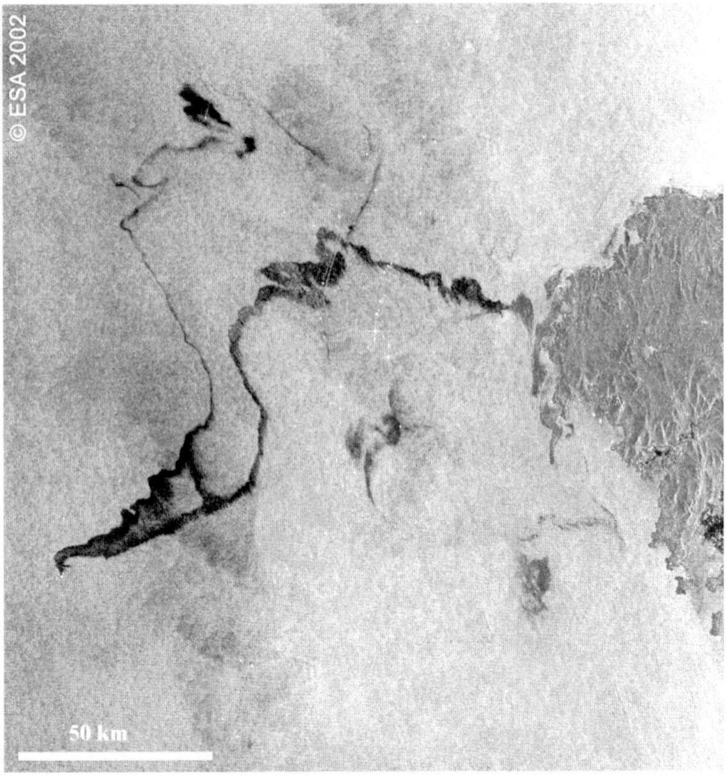

Fig. 4
This Envisat's ASAR image acquired on November 17, 2002 shows a double-headed oil spill originating from the stricken Prestige tanker, lying 100 km off the Spanish coast

The level of shipping and offshore activities occurring in and around icebound regions is growing steadily, and with it the demand for reliable sea ice information. Traditionally, ice-monitoring services were based on data from aircraft, ships, and land stations. But the area coverage available from such sources is always limited, and often impeded further by bad weather. Satellite data has begun to fill this performance gap, enabling continuous wide-area ice surveillance. SAR instruments of the type flown on ERS and Envisat are able to pierce through clouds and darkness and therefore are particularly adapted for generating high-quality images of sea ice. Ice classification maps generated from radar imagery are now being supplied to users at sea.

4.3 Cryosphere

The cryosphere is both influenced by and has a major influence on climate. Because any increase in the melt rate of ice sheets and glaciers has the potential to greatly increase sea level, researchers are looking to the cryosphere to get a better idea of the likely scale of the impact of climate change. In addition, the melting of sea-ice will increase the amount of solar radiation that will be absorbed by ice-free polar oceans rather than reflected by ice-covered oceans, increasing the ocean temperature.

The remoteness, darkness, and cloudiness of Earth's polar regions make them difficult to study. Being microwave active instruments, the radar on-board ERS and Envisat missions allow seeing through clouds and darkness.

Since about 30 years, satellites have been observing the Arctic and have witnessed reductions in the *minimum Arctic sea ice extent*—the lowest amount of ice recorded in the area annually—at the end of summer from around 8 million km^2 in the early 1980s to the historic minimum of 4 million km^2 in 2007, changes widely viewed as a consequence of greenhouse warming. ERS and Envisat radar instruments, i.e., the imaging radar (SAR) and the radar altimeter instruments witnessed this sharp decline, providing detailed measurements respectively on sea ice areas and sea ice thickness. The CryoSat mission shall add accurate sea ice thickness measurements to complement the sea ice extent measurements.

In addition to mapping sea ice, scientists have used repeat-pass SAR image data to map the *flow velocities of glaciers*. Using SAR data collected by ERS-1 and ERS-2 during their tandem mission in 1995 and Envisat and Canada's Radarsat-1 in 2005, scientists discovered that the Greenland glaciers are melting at a pace twice as fast as previously thought. Such a rapid pace of melting was not considered in previous simulations of climate change, therefore showing the important role of EO in advancing our knowledge of climate change and improving climate models.

Similar phenomena also take place in Antarctica, with some large glaciers such as the Pine Island glacier, thinning at a constantly accelerating rate as suggested by ERS and Envisat altimetry data. In Antarctica, the stability of the glaciers is also related to their floating terminal platform, the ice shelves. The break-up of large *ice shelves* (e.g., Larsen ice shelf, Wilkins ice shelf) has been observed by Envisat SAR, and is likely a consequence of both sea and air temperature increase around the Antarctica peninsula and West Antarctica.

4.4 Atmosphere

ERS-2 and Envisat are equipped with several atmospheric chemistry instruments, which can look vertically or sideways to map the atmosphere in three dimensions, producing high-resolution horizontal and vertical cross sections of trace chemicals stretching from ground level to a 100 km in the air, all across a variety of scales. Those instruments can detect holes in the thinning ozone layer, plumes of aerosols and pollutants hanging over major cities or burning forests, and exhaust trails left in the atmosphere by commercial airliners.

ERS-2 and Envisat satellites have been maintaining a regular census of global stratospheric *ozone levels* since 1995, mapping yearly Antarctica ozone holes as they appear. The size and precise time of the ozone hole is dependent on the year-to-year variability in temperature and atmospheric dynamics, as established by satellite measurements.

Envisat results benefit from improved sensor capabilities. As an example, the high spatial resolution delivered by the Envisat atmospheric instruments means precise maps of atmospheric trace gasses, even resolving individual city sources such than the high-resolution global atmospheric map of *nitrogen dioxide* (NO_2), an indicator of air pollution (➲ *Fig. 5*). By making a link with the measurements started with ERS-2, the scientists could note a significant increase

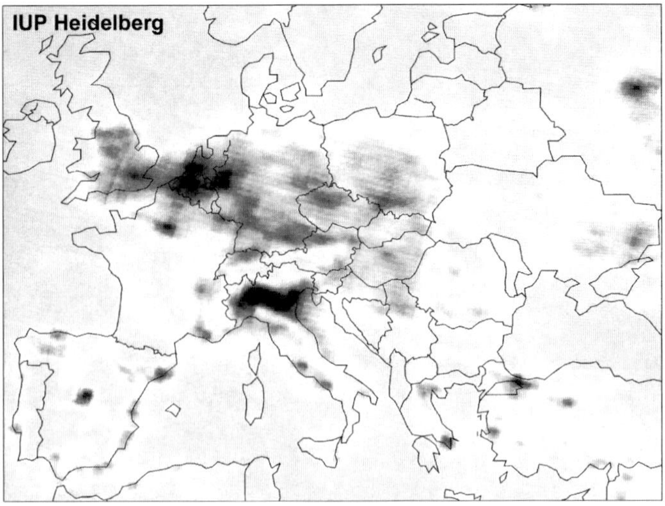

Fig. 5
Nitrogen dioxide (NO_2) concentration map over Europe derived from several years of SCIAMACHY instrument data on-board the Envisat satellite (Courtesy Institute of Environmental Physics, University of Heidelberg, Heidelberg)

in the NO_2 emissions above Eastern China, a tangible sign of the fast economical growth of China.

Based on several years of Envisat observations, scientists have produced global distribution maps of the most important *greenhouse gases*—carbon dioxide (CO_2) and methane (CH_4)—that contribute to global warming. The importance of cutting emissions from these "anthropogenic," or manmade, gases has been highlighted with EU leaders endorsing binding targets to cut greenhouse gases in the mid-term future. Careful monitoring is essential to ensuring these targets are met, and space-based instruments are new means contributing to this. The SCIAMACHY instrument on-board Envisat satellite was the first space sensor capable of measuring the most important greenhouse gases with high sensitivity down to the Earth's surface because it observes the spectrum of sunlight shining through the atmosphere in "nadir" looking operations.

Envisat atmospheric chemistry data are useful for helping build scenarios of greenhouse gas emissions, such as methane—the second most important greenhouse gas after carbon dioxide. Increased methane concentrations induced mainly by human activities were observed. By comparing model results with satellite observations, the model is continually adjusted until it is able to reproduce the satellite observations as closely as possible. Based on this, scientists continually improve models and their knowledge of nature.

4.5 ESA Programmes for Data Exploitation

EO is an inherently multipurpose tool. This means there is no typical EO user: it might be anyone who requires detailed characterization of any given segment of our planet, across a wide variety of scales from a single city block to a region, or continent, right up to coverage of the entire globe. EO is already employed by many thousands of users worldwide. However ESA works to further increase EO take-up by encouraging development of new science, new applications and services centered on user needs.

New applications usually emerge from scientific research. ESA supports scientific research either by providing *easy access to high quality data* (see ➋ Sect. 5), by organizing dedicated workshops and symposia, by training users or by taking a proactive role in the formulation of new mission concepts and by providing support to science.

Converting basic research and development into an operational service requires the fostering of partnership between research institutions, service companies, and user organizations. ESA's *Data User Element* (DUE) programme addresses institutional users tasked with collecting specific geographic or environmental data. The DUE aims to raise such institutions' awareness of the applicability of EO to their day-to-day operations, and develop demonstration products tailored to increase their effectiveness. The intention is then to turn these products into sustainable services provided by public or private entities. More information on ESA's DUE programme can be found at: http://dup.esrin.esa.it.

Complementing DUE objectives is ESA's EO *Market Development* programme. This provides a supportive framework within which to organise end-to-end service chains capable of leveraging scientific EO data into commercial tools supplied by self-supporting businesses. More information on ESA EO Market Development programme can be found at: http://www.eomd.esa.int.

5 User Access to ESA Data

ESA endeavors to maximize the beneficial use of EO data. It does this by fostering the use of this valuable information by as many people as possible, in as many ways as possible. For users, and therefore for ESA, *easy access to EO data* is of paramount importance. However the challenges for easy EO data access are many:

1. The *ESA EO data policy* shall be beneficial to various categories of users, ranging from global change scientists to operational services, and shall have the objective to stimulate a balanced development of Earth science, public services and value-adding companies. ESA has always pursued an approach of low cost fees for its satellite data, trying to provide *free of charge* the maximum amount of data. This approach will continue and even be reinforced in the future, by further increasing the amount of data available on Internet and by reducing the complexity of EO missions.
2. The *volume of data* transmitted to the ground by ESA EO satellites is particularly high: the Envisat satellite transmits about 270 GB of data every day; the future Sentinel-1 satellite will transmit about 900 GB of data every day. Once acquired, the data shall be transformed (i.e., processed) into products in which the information is related either to an engineering calibrated parameter (so-called Level 1 products, e.g., a SAR image) or to a geophysical parameter (so-called Level 2 products, e.g., sea surface temperature). Despite the high volume of data, the processing into Level 1 and Level 2 products shall be as fast as possible to serve increasingly demanding operational services. Finally, the data products shall be almost immediately available with users either through a broad use of Internet or through dedicated communication links.

3. The *quality* of EO data products shall be high so that user can effectively rely on the delivered information content. This means that the processing algorithms (i.e., the transformation of raw data into products) as well as the products calibration and validation shall be given strong attention. ESA has constantly given such attention for their EO missions, investing large efforts in algorithms development, particularly for innovative instruments such as the ones flying on board the Earth Explorer missions. Of equal importance for the credibility of EO data are the validation activities aiming at comparing the geophysical information content of EO data products with similar measurements collected through, e.g., airborne campaigns or ocean buoys set-up.
4. Finally, *EO data handling tools* shall be offered to users. Besides the general assistance given through a centralized ESA EO user service (eohelp@esa.int), the ESA user tools include
   - *On-line data information* (http://earth.esa.int) providing EO missions news, data products description, processing algorithms documentation, workshops proceedings, etc.
   - Data collections visualization through *on-line catalogs*, including request for product generation when the product does not yet exist (e.g., for future data acquisition) or direct download when the product is already available in online archives
   - *EO data software tools*, aiming to facilitate the utilization of data products by provision of, e.g., viewing capabilities, innovative processing algorithms, format conversion, etc.

5.1 How to Access the EO Data at ESA

Internet is the main way to access to the EO satellite services and products at ESA. Besides the general description on data access described below, further assistance can be provided by the ESA EO help desk (eohelp@esa.int).

1. For most of the ESA EO data, *open and free of charge* access is granted after a simple *user registration*. The data products are those that are systematically acquired, generated, and available online. This includes all altimetry, sea surface temperature, atmospheric chemistry, and future Earth Explorers data, but also large collections of optical (MERIS, Landsat) and SAR datasets.

   User registration is done at: **http://eopi.esa.int** .

   The detailed list of open and free of charge data products is available on this Web site.

   For users who are only interested in basic imagery (i.e., false color jpg images), ESA provides access to large galleries of free of charge Earth images (http://earth.esa.int/satelliteimages), including near real-time Earth images (http://miravi.eo.esa.int).
2. Some ESA EO data and services cannot be provided free of charge or openly either because of restrictions in the distribution rights granted to ESA (e.g., some ESA TPM), either because the data/service is restrained by technical capacities and therefore is on-demand (i.e., not systematically provided). The *restrained dataset* essentially includes on-demand SAR data acquisition and production.

   In this case, users shall describe the intended use of the data within a *project proposal*. The project proposals are collected at: http://eopi.esa.int. ESA will analyze the project proposal to review its scientific objectives, to assess its feasibility and to establish project quotas for the requested products and services (e.g., instrument tasking). The products and services are provided free of charge or with a small fee to cover production costs or instrument tasking. Products can be provided either on Internet or on media (DVD-Rom).

3. Finally some few ESA EO data can also be obtained from *commercial distributors*. This essentially includes SAR data needed for operational purposes and therefore requiring higher priority for acquisition and delivery. The commercial distributors may use a specific network of acquisition stations essentially dedicated to operational services.

6 Conclusion

During the last three decades, EO satellites have gradually taken on a fundamental role with respect to understanding and managing our planet. Contributing to this trend, the EO programme of the ESA addresses a growing number of scientific issues and operational services, thanks to the continuous development of new satellites and sensors and to a considerable effort in stimulating the use of EO data. Partnerships between satellite operators and strong relations with user organizations are essential at ESA for further improving information retrieval from EO satellite data.

The coming decade will see an increasing number of orbiting EO satellites, not only in Europe. This is a natural consequence of the growing user demands and expectations, but also of the gradual decrease of the costs for satellite manufacturing. The main challenge in EO will therefore be to maximize the synergies between existing satellites, both with respect to combining their respective observations as well as optimizing their operation concepts.

References

ESA (2006) The changing Earth—new scientific challenges for ESA's Living Planet Programme, ESA SP-1304

ESA (2007) InSAR principles—guidelines for SAR interferometry processing and interpretation, ESA TM-19

4 GOCE: Gravitational Gradiometry in a Satellite

Reiner Rummel
Institut für Astronomische und Physikalische Geodäsie,
Technische Universität München, München, Germany

W. Freeden, M.Z. Nashed, T. Sonar (Eds.), *Handbook of Geomathematics*, DOI 10.1007/978-3-642-01546-5_4,

Abstract Spring 2009 the satellite Gravity and steady-state Ocean Circulation Explorer (GOCE), equipped with a gravitational gradiometer, has been launched by European Space Agency (ESA). Its purpose is the detailed determination of the spatial variations of the Earth's gravitational field, with applications in oceanography, geophysics, geodesy, glaciology, and climatology. Gravitational gradients are derived from the differences between the measurements of an ensemble of three orthogonal pairs of accelerometers, located around the center of mass of the spacecraft. Gravitational gradiometry is complemented by gravity analysis from orbit perturbations. The orbits are thereby derived from uninterrupted and three-dimensional GPS tracking of GOCE.

The gravitational tensor consists of the nine second-derivatives of the Earth's gravitational potential. These nine components can also be interpreted in terms of the local curvature of the field or in terms of components of the tidal field generated by the Earth inside the spacecraft. Four of the nine components are measured with high precision ($10^{-11}s^{-2}$ per square-root of Hz), the others are less precise. Several strategies exist for the determination of the gravity field at the Earth's surface from the measured tensor components at altitude. The analysis can be based on one, several or all components; the measurements may be regarded as time series or as spatial data set; recovery may take place in terms of spherical harmonics or other types of global or local base functions. After calibration GOCE entered into its first measurement phase in fall 2009. First results are expected to become available in summer 2010.

1 Introduction: GOCE and Earth Sciences

On March 17, 2009 the European Space Agency (ESA) launched the satellite Gravity and steady-state Ocean Circulation Explorer (GOCE). It is the first satellite of ESA's Living Planet Programme (see ESA 1999a, 2006). It is also the first one that is equipped with a gravitational gradiometer. The purpose of the mission is to measure the spatial variations of the Earth's gravitational field globally with maximum resolution and accuracy. Its scientific purpose is essentially twofold. First, the gravitational field reflects the density distribution of the Earth's interior. There are no direct ways to probe the deep Earth interior, only indirect ones in particular seismic tomography, gravimetry, and magnetometry. The gravimetry part is now been taken care of by GOCE. Also in the field of space magnetometry an ESA mission is in preparation; it is denoted Swarm and consists of three satellites. Seismic tomography is based on a worldwide integrated network of seismic stations. From a joint analysis of all seismic data, a tomographic image of the spatial variations in the Earth's interior of the propagation velocity of seismic waves is derived. The three methods together establish the experimental basis for the study of solid Earth physics, or more specifically, of phenomena such as core–mantle topography, mantle convection, mantle plumes, ocean ridges, subduction zones, mountain building, and mass compensation. Inversion of gravity alone is nonunique but joint inversion together with seismic tomography, magnetic field measurements and in addition with surface data of plate velocities and topography and models from mineralogy will lead to a more and more comprehensive picture of the dynamics and structure of the Earth's interior (see, e.g., Hager and Richards 1989; Lithgow-Bertelloni and Richards 1998; Bunge et al. 1998; Kaban et al. 2004). Second, the gravitational field and therefore the mass distribution of the Earth determines the geometry of level surfaces, plumb lines, and lines of force. This geometry constitutes the natural reference in our physical and technical world. In particular in cases where small potential differences matter such as in ocean dynamics and large civil constructions, precise knowledge of this reference is an important source of information. The most prominent example is ocean circulation. Dynamic ocean topography, the small one up to 2 m deviation of the actual ocean surface from an equipotential surface can be directly translated into ocean surface circulation. The equipotential surface at mean ocean level is referred to as geoid and it represents the hypothetical surface of the world oceans at complete rest. GOCE, in conjunction with satellite altimetry missions, like Jason will allow for the first time direct and detailed measurement of ocean circulation (see discussions in Wunsch and Gaposchkin 1980; Ganachaud et al. 1997; LeGrand and Minster 1999; Losch et al. 2002; Albertella and Rummel 2009; Maximenko et al. 2009). GOCE is an important satellite mission for oceanography, solid Earth physics, geodesy, and climate research (compare ESA 1999b; Rummel et al. 2002; Johannessen et al. 2003).

2 GOCE Gravitational Sensor System

In the following, the main characteristics of the GOCE mission which is unique in several ways will be summarized (see also ➋ Chap. 3). The mission consists of two complementary gravity sensing systems. The large-scale spatial variations of the Earth's gravitational field will be derived from its orbit, while the medium to short scales are measured by a so-called gravitational gradiometer. Even though satellite gravitational gradiometry has been proposed already in the late 1950s in Carroll and Savet (1959) (see also Wells 1984), the GOCE gradiometer is the first instrument of its kind to be put into orbit. The principles of satellite gradiometry will be described in ➋ Chap. 3. The purpose of gravitational gradiometry is the measurement of the second derivatives of the gravitational potential. In total there exist nine second-derivatives in the orthogonal coordinate system of the instrument. The GOCE gradiometer is a three-axis instrument and its measurements are based on the principle of differential acceleration measurement. It consists of three pairs of orthogonally mounted accelerometers, each accelerometer with three axes (see ➋ *Fig. 1*). The gradiometer baseline of each one-axis gradiometer is 50 cm. The precision of each accelerometer is about $2 \cdot 10^{-12}$ m/s^2 per square-root of Hz along two sensitive axes; the third axis has much lower sensitivity. This results in a precision of the gravitational gradients of 10^{-11} s^{-2} or 10 mE per square-root of Hz (1 E = 10^{-9} s^{-2} = 1 Eötvös Unit). From the measured gravitational acceleration differences the three main diagonal terms and one off-diagonal term of the gravitational tensor can be determined with high precision. These are the three diagonal components $\Gamma_{xx}, \Gamma_{yy}, \Gamma_{zz}$ as well as the off-diagonal component Γ_{xz}, while the components Γ_{xy} and Γ_{yz} are less accurate. Thereby the coordinate axes of the instrument are pointing in flight direction (x), cross direction (y), and radially toward the Earth (z). The extremely high gradiometric performance of the instrument is confined to the so-called measurement bandwidth (MBW), while outside the measurement bandwidth noise is increasing.

Strictly speaking, the derivation of the gradients from accelerometer differences does only hold if all six accelerometers (three pairs) are perfect twins and if all accelerometer test masses are perfectly aligned. In real world small deviations from such an idealization exist. Thus, the calibration of the gradiometer is of high importance. Calibration is essentially the process of determination of a set of scale, misalignment, and angular corrections. They are the parameters

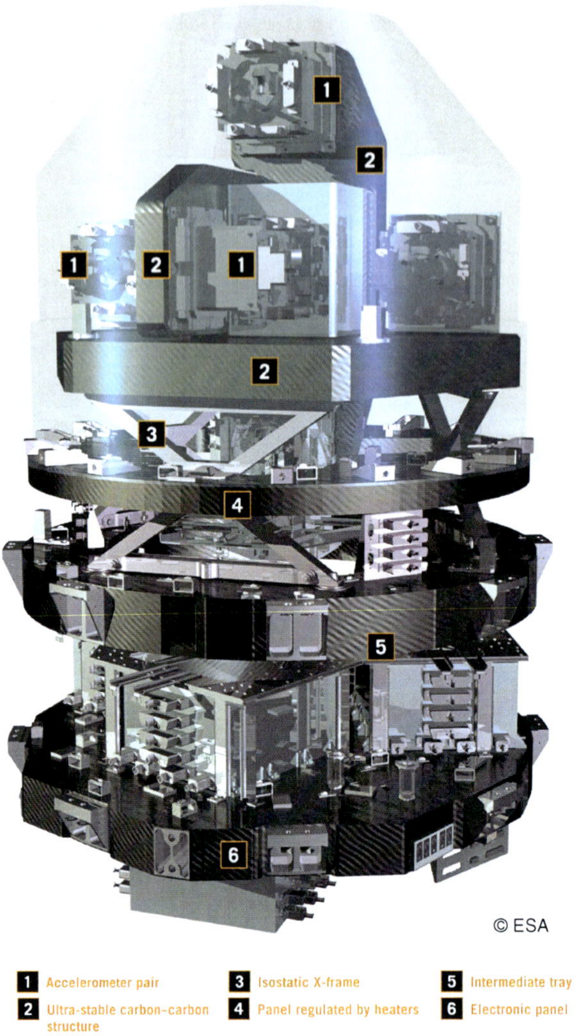

Fig. 1
GOCE gravitational gradiometer (courtesy ESA)

of an affine transformation between an ideal and the actual set of six accelerometers. Calibration in orbit requires random shaking of the satellite by means of a set of cold gas thrusters and comparison of the actual output with the theoretically correct one. Before calibration the nonlinearities of each accelerometer are removed electronically; in other words, the proof mass of each accelerometer inside the electrodes of the capacitive electronic feedback system is brought into its linear range.

The gravitational signal is superimposed by the effects of angular velocity and angular acceleration of the satellite in space. Knowledge of the latter is required for the removal of the angular effects from the gradiometer data and for angular control. The separation of angular acceleration from the gravitational signal is possible from a particular combination of the measured nine acceleration differences. The angular rates (in MBW) as derived from the gradiometer data in combination with those deduced from the star sensor readings are used for attitude control of the spacecraft. The satellite has to be well controlled and guided smoothly around the Earth. It is Earth pointing, which implies that it performs one full revolution in inertial space per full orbit cycle. Angular control is attained via magnetic torquers, i.e., using the Earth's magnetic field lines for orientation. This approach leaves uncontrolled one-directional degree of freedom at any moment. In order to prevent nongravitational forces, in particular atmospheric drag, to "sneak" into the measured differential accelerations as secondary effect, the satellite is kept "drag-free" in along-track direction by means of a pair of ion thrusters. The necessary control signal is derived from the available "common-mode" accelerations (sum instead of differences of the measured accelerations) along the three orthogonal axes of the accelerometer pairs of the gradiometer. Some residual angular contribution may also add to the common-mode acceleration, due to the imperfect symmetry of the gradiometer relative to the spacecraft's center of mass. This effect has to be modeled.

The second gravity sensor device is a newly developed European GPS receiver. From its measurements, the orbit trajectory is computed to within a few centimeters, either purely geometrically, the so-called kinematic orbit, or by the method of reduced dynamic orbit determination (compare Švehla and Rothacher 2004; Jäggi 2007). As the spacecraft is kept in an almost drag-free mode (at least in along-track direction), the orbit motion can be regarded as purely gravitational. It complements the gradiometric gravity field determination and covers the long wavelength part of the gravity signal.

The orbit altitude is extremely low, only about 255 km. This is essential for a high gravitational sensitivity. No scientific satellite has been flown at such low altitude so far. Its altitude is maintained through the drag-free control and additional orbit maneuvers, which are carried out at regular intervals. As said above, this very low altitude results in high demands on drag-free and attitude control. Finally, any time-varying gravity signal of the spacecraft itself, the so-called self-gravitation, must be excluded. This results in extremely tight requirements on metrical stiffness and thermal control.

In summary, GOCE is a technologically very complex and innovative mission. The gravitational field sensor system consists of a gravitational gradiometer and GPS receiver as core instruments. Orientation in inertial space is derived from star sensors. Common-mode and differential-mode accelerations from the gradiometer and orbit positions from GPS are used together with ion thrusters for drag-free control and together with magneto-torquers for angular control. The satellite and its instruments are shown in ➲ *Fig. 2*. The system elements are summarized in ➲ *Table 1*.

3 Gravitational Gradiometry

Gravitational gradiometry is the measurement of the second derivates of the gravitational potential V. Its principles are described in textbooks such as Misner et al. (1970), Falk and Ruppel (1974), and Ohanian and Ruffini (1994) or in articles like Rummel (1986), compare also Colombo (1989) and Rummel (1997). Despite the high precision of the GOCE gradiometer instrument the theory can still be formulated by classical Newton mechanics. Let us denote the

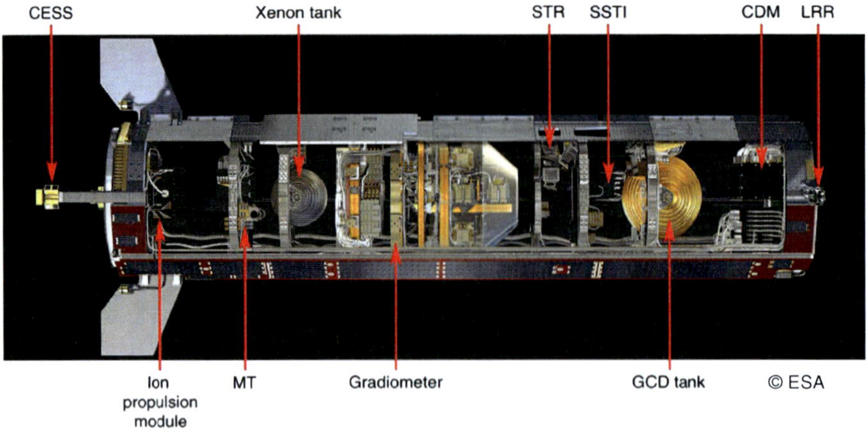

Fig. 2
GOCE satellite and main instruments (courtesy ESA) (CESS Coarse earth and Sun Sensor, MT Magneto torquer, STR star tracker, SSTI satellite to satellite tracking instrument, CDM Command and data management unit, LRR laser retro reflector)

Table 1
Sensor elements and type of measurement delivered by them (approximate orientation of the instrument triad: x = along track, y = out-of-orbit-plane, z = radially downward)

Sensor	Measurements
Three-axis gravity gradiometer	Gravity gradients Γ_{xx}, Γ_{yy}, Γ_{zz}, Γ_{xz} in instrument system and in MBW (measurement bandwidth)
	Angular accelerations (highly accurate around y-axis, less accurate around x, z axes)
	Common-mode accelerations
Star sensors	High-rate and high-precision inertial orientation
GPS receiver	Orbit trajectory with centimeter precision
Drag control with two ion thrusters	Based on common-mode accelerations from gradiometer and GPS orbit
Angular control with magnetic torquers	Based on angular rates from star sensors and gradiometer
Orbit altitude maintenance	Based on GPS orbit
Internal calibration (and quadratic factors removal) of gradiometer	Calibration signal from random shaking by cold gas thrusters (and electronic proof mass shaking)

gravitational tensor, expressed in the instrument frame as

$$\Upsilon = V_{ij} = \begin{pmatrix} V_{xx} & V_{xy} & V_{xz} \\ V_{yx} & V_{yy} & V_{yz} \\ V_{zx} & V_{zy} & V_{zz} \end{pmatrix} = \begin{pmatrix} \frac{\partial^2 V}{\partial x^2} & \frac{\partial^2 V}{\partial x \partial y} & \frac{\partial^2 V}{\partial x \partial z} \\ \frac{\partial^2 V}{\partial y \partial x} & \frac{\partial^2 V}{\partial y^2} & \frac{\partial^2 V}{\partial y \partial z} \\ \frac{\partial^2 V}{\partial z \partial x} & \frac{\partial^2 V}{\partial z \partial y} & \frac{\partial^2 V}{\partial z^2} \end{pmatrix}, \tag{1}$$

where the gravitational potential represents the integration over all Earth masses (cf. Chaps. 8 and 32)

$$V_P = G \iiint \frac{\rho_Q}{\ell_{PQ}} d\Sigma_Q, \tag{2}$$

where G is the gravitational constant, ρ_Q the density, ℓ_{PQ} the distance between the mass element in Q and the computation point P, and $d\Sigma$ is the infinitesimal volume. If we assume the atmosphere to be, taken away, then the space outside of the Earth is empty and it holds $\nabla \cdot \nabla V = 0$ (source free) apart from $\nabla \times \nabla V = 0$ (vorticity free). This corresponds to saying in Eq. 1 $V_{ij} = V_{ji}$ and $\sum_i V_{ii} = 0$. It leaves only five independent components in each point and offers important cross checks between the measured components. If the Earth were a homogenous sphere, the off-diagonal terms would be zero and in a local triad {north, east, radial} one would find

$$\Upsilon = V_{ij} = \begin{pmatrix} V_{xx} & V_{xy} & V_{xz} \\ V_{yx} & V_{yy} & V_{yz} \\ V_{zx} & V_{zy} & V_{zz} \end{pmatrix} = \frac{GM}{r^3} \begin{pmatrix} -1 & 0 & 0 \\ 0 & -1 & 0 \\ 0 & 0 & 2 \end{pmatrix}, \tag{3}$$

where M is the mass of the spherical Earth. This simplification gives an idea about the involved orders of magnitude. At satellite altitude, it is $V_{zz} = 2740\,\text{E}$. This also implies that at a distance of 0.5 m from the spacecraft's center of mass, the maximum gravitational acceleration is about $1.5 \cdot 10^{-6} \text{m/s}^2$.

In an alternative interpretation one can show that the V_{ij} express the local geometric curvature structure of the gravitational field, i.e.,

$$\Upsilon = V_{ij} = -g \begin{pmatrix} k_1 & t_1 & f_1 \\ t_2 & k_2 & f_2 \\ f_1 & f_2 & -H \end{pmatrix}, \tag{4}$$

where g is the gravity, k_1 and k_2 express the local curvature of the level surfaces in north and east directions, t_1 and t_2 are the torsion, f_1 and f_2 the north and east components of the curvature of the plumb line, and H the mean curvature. For a derivation refer to Marussi (1985). This interpretation of gravitational gradients in terms of gravitational geometry provides a natural bridge to Einstein's general relativity, where gravitation is interpreted in terms of space–time curvature for it holds

$$R^i_{0j0} = \frac{1}{c^2} V_{ij} \tag{5}$$

for the nine components of the tidal force tensor, which are components of $R^{\mu}_{\ \nu\alpha\beta}$ the Riemann curvature tensor with its indices running from 0, 1, 2, 3 (Ohanian and Ruffini 1994, p. 41; Moritz and Hofmann-Wellenhof 1993, Chap. 5).

A third interpretation of gravitational gradiometry is in term of tides. Sun and moon produce a tidal field on Earth. It is zero at the Earth's center of mass and maximum at its surface. Analogously, the Earth is producing a tidal field in every Earth-orbiting satellite. At the center of mass of a satellite, the tidal acceleration is zero; i.e., the acceleration relative to the center of mass is zero; this leads to the terminology "zero-g." The tidal acceleration increases with distance from the satellite's center of mass like

$$a_i = V_{ij} dx^j. \tag{6}$$

with the measurable components of tidal acceleration a_i and of relative position dx^j, both taken in the instrument reference frame. Unlike sun and moon relative to the Earth, GOCE is always

Earth pointing with its z-axis. This implies that the gradiometer measures permanently "high-tide" in z-direction and "low-tide" in x- and y-directions. The gradient components are deduced from taking the difference between the acceleration at two points along one gradiometer axis and symmetrically with respect to the satellite's center of mass

$$V_{ij} = \frac{a_i(1) - a_i(2)}{2dx^j}. \tag{7}$$

Remark: In addition to the tidal acceleration of the Earth, GOCE is measuring the direct and indirect tidal signal of sun and moon. This signal is much smaller, well known and taken into account.

If the gravitational attraction of the atmosphere is taken care of by an atmospheric model, the Earth's outer field can be regarded source free and Laplace equation holds. It is common practice to solve Laplace equation in terms of spherical harmonic functions. The use of alternative base functions is discussed, for example, in Schreiner (1994) or Freeden et al. (1998). For a spherical surface, the solution of a Dirichlet boundary value problem yields the gravitational potential of the Earth in terms of normalized spherical harmonic functions $Y_{nm}(\Omega_P)$ of degree n and order m as

$$V(P) = \sum_n \left(\frac{R}{r_P}\right)^{n+1} \sum_m t_{nm} Y_{nm}(\Omega_P) = \underline{Y}\,\underline{t} \tag{8}$$

with $\{\Omega_P, r_P\} = \{\theta_P, \lambda_P, r_P\}$ the spherical coordinates of P and t_{nm} the spherical harmonic coefficients. The gravitational tensor at P is then

$$V_{ij}(P) = \sum_n \sum_m t_{nm} \partial_{ij} \left\{ \left(\frac{R}{r_P}\right)^{n+1} Y_{nm}(\Omega_P) \right\} = \underline{Y}\{ij\}\,\underline{t}. \tag{9}$$

The spherical harmonic coefficients t_{nm} are derived from the measured gradiometric components V_{ij} by least-squares adjustment.
Remarks: In the case of GOCE gradiometry the situation is as follows.

Above degree and order $n = m = 200$, noise starts to dominate signal. Thus, the series has to be truncated in some intelligent manner, minimizing aliasing and leakage effects.

GOCE will cover our globe in 60 days with a dense pattern of ground tracks. Altogether, current planning of its mission life time assumes three times such 60-day cycles.

The orbit inclination is 96°. This leaves the two polar areas (opening angle 6°) free of observations, the so-called polar gaps. Various strategies have been suggested for minimization of the effect of the polar gaps on the determination of the global field (compare, e.g., Baur et al. 2009).

Instead of dealing in an least-squares adjustment with the analysis of the individual tensor components one could consider the study of particular combinations. A very elegant approach is the use of the invariants of the gravitational tensor. They are independent of the orientation of the gradiometer triad. It is referred to Rummel (1986) and in particular to Baur and Grafarend (2006) and Baur (2007). While the first invariant cannot be used for gravity field analysis, it is the Laplace trace condition, the two others can be used. They are nonlinear and lead to an iterative adjustment.

Let us assume for a moment the gradiometer components V_{ij} would be given in an Earth-fixed spherical {north, east, radial}-triad. In that case the tensor can be expanded in tensor spherical harmonics and decomposed into the irreducible radial, mixed normal–tangential, and pure tangential parts with the corresponding eigenvalues (Rummel 1997; Rummel and van Gelderen 1992; Schreiner 1994; Nutz 2002; Martinec 2003),

$$(n+1)(n+2) \quad \text{for} \quad \Gamma_{zz} \quad \text{and} \quad \Gamma_{xx}+\Gamma_{yy},$$
$$-(n+2)\sqrt{n(n+1)} \quad \text{for} \quad \{\Gamma_{xz},\, \Gamma_{yz}\}, \frac{(n+2)!}{(n-2)!} \quad \text{for} \quad \{\Gamma_{xx}-\Gamma_{yy},\, -2\Gamma_{xy}\}.$$

All eigenvalues are of the order of n^2. In Schreiner (1994), Freeden et al. (1998) as well as ➋ Chap. 9 of this handbook, it is also shown that $\{\Gamma_{xz}, \Gamma_{yz}\}$ and $\{\Gamma_{xx}-\Gamma_{yy}, -2\Gamma_{xy}\}$ are insensitive to degree zero, the latter combination also to degree one. In the case of GOCE, the above properties cannot be employed in a straight forward manner, because (1) not all components are of comparable precision and, more importantly, (2) the gradiometric components are measured in the instrument frame, which is following in its orientation the orbit and the attitude control commands.

There exist various competing strategies for the actual determination of the field coefficients t_{nm}, depending on whether the gradients are regarded in situ measurements on a geographical grid, along the orbit tracks, as a time series along the orbit or as Fourier-coefficients derived from the latter (compare, e.g., Migliaccio et al. 2004; Brockmann et al. 2009; Pail and Plank 2004; Stubenvoll et al. 2009). These methods take into account the noise characteristics of the components and their orientation in space. Different "philosophies" will also be followed concerning the choice of the system of base functions, the boundary surface, or the use of prior information (see, e.g., ➋ Chap. 9 for a tensorial approach). Ultimately, any gravity model from GOCE has to be based on a combination with gravity deduced from precise orbits, as derived from GPS tracking. Whether a stable solution can be computed without regularization needs to be seen (Kusche and Klees 2002). It will mainly depend on how one is dealing with the polar gaps. In a second step, local refinements will be attempted and combinations with terrestrial data sets (e.g., Eicker et al. 2009; Stubenvoll et al. 2009).

4 Conclusions: GOCE Status

Currently—fall 2009—GOCE is in its calibration phase. Calibration consists of two steps. First, the six accelerometers are "linearized." This is done by electronically shaking them, deriving from this shaking the off-sets of each accelerometer from linearity and correcting the test masses for these deviations. In the second step, the gradiometer instrument is calibrated as a whole. By applying random pulses to the gradiometer by cold gas thrusters, all departures of the actual gradiometer axes from an ideal set are estimated. It is a set of scale, misalignment, and orientation corrections. After the completion of calibration, GOCE will enter the first mission operational phase, most likely at an altitude of below 255 km. The expectation is, that 6 months later the first data can be delivered to a wide user community in geodesy, oceanography, solid Earth physics, and glaciology.

References

Albertella A, Rummel R (2009) On the spectral consistency of the altimetric ocean and geoid surface: a one-dimensional example. J Geodesy 83(9):805–815

Baur O (2007) Die Invariantendarstellung in der Satellitengradiometrie, DGK, Reihe C, Beck, München

Baur O, Grafarend EW (2006) High performance GOCE gravity field recovery from gravity gradient tensor invariants and kinematic orbit information. In: Flury J, Rummel R, Reigber Ch, Rothacher M, Boedecker G, Schreiber U (eds) Observation of the earth system from space. Springer, Berlin, pp 239–254

Baur O, Cai J, Sneeuw N (2009) Spectral approaches to solving the polar gap problem. In: Flechtner F, Mandea M, Gruber Th, Rothacher M, Wickert J, Güntner A, Schöne T (eds) System earth via geodetic-geophysical space techniques. Springer, Berlin

Brockmann JM, Kargoll B, Krasbutter I, Schuh WD, Wermuth M (2009) GOCE data analysis: from calibrated measurements to the global earth gravity field. In: Flechtner F, Mandea M, Gruber Th, Rothacher M, Wickert J, Güntner A, Schöne T (eds) System earth via geodetic-geophysical space techniques. Springer, Berlin

Bunge H-P, Richards MA, Lithgow-Bertelloni C, Baumgardner JR, Grand SP, Romanowiez BA (1998) Time scales and heterogeneous structure in geodynamic earth models. Science 280:91–95

Carroll JJ, Savet PH (1959) Gravity difference detection. Aerospace Eng 18:44–47

Colombo O (1989) Advanced techniques for high-resolution mapping of the gravitational field. In: Sansò F, Rummel R (eds) Theory of satellite geodesy and gravity field determination. Lecture notes in earth sciences, vol 25. Springer, Heidelberg, pp 335–369

Eicker A, Mayer-Gürr T, Ilk KH, Kurtenbach E (2009) Regionally refined gravity field models from in-situ satellite data. In: Flechtner F, Mandea M, Gruber Th, Rothacher M, Wickert J, Güntner A, Schöne T (eds) System earth via geodetic-geophysical space techniques. Springer, Berlin

ESA (1999a) Introducing the "Living Planet" Programme—The ESA strategy for earth observation. ESA SP-1234. ESA Publication Division, ESTEC, Noordwijk, the Netherlands

ESA (1999b): Gravity field and steady-state ocean circulation mission. Reports for mission selection, SP-1233 (1). ESA Publication Division, ESTEC, Noordwijk, the Netherlands. http://www.esa.int./livingplanet/goce

ESA (2006) The changing earth—New scientific challenges for ESA's Living Planet Programme. ESA SP-1304. ESA Publication Division, ESTEC, Noordwijk, the Netherlands

Falk G, Ruppel W (1974) Mechanik, Relativität, Gravitation. Springer, Berlin

Freeden W, Gervens T, Schreiner M (1998) Constructive approximation on the sphere. Oxford Science Publications, Oxford

Ganachaud A, Wunsch C, Kim M-Ch, Tapley B (1997) Combination of TOPEX/POSEIDON data with a hydrographic inversion for determination of the oceanic general circulation and its relation to geoid accuracy. Geophys J Int 128: 708–722

Hager BH, Richards MA (1989) Long-wavelength variations in Earth's geoid: physical models and dynamical implications, Phil Trans R Soc Lond A 328:309–327

Jäggi A (2007) Pseudo-stochastic orbit modelling of low earth satellites using the global positioning system. Geodätisch – geophysikalische Arbeiten in der Schweiz, 73

Johannessen JA, Balmino G, LeProvost C, Rummel R, Sabadini R, Sünkel H, Tscherning CC, Visser P, Woodworth P, Hughes CH, LeGrand P, Sneeuw N, Perosanz F, Aguirre-Martinez M, Rebhan H, Drinkwater MR (2003) The European gravity field and steady-state ocean circulation explorer satellite mission: its impact on geophysics. Surv Geophys 24:339–386

Kaban MK, Schwintzer P, Reigber Ch (2004) A new isostatic model of the lithosphere and gravity field. J Geodesy 78:368–385

Kusche J, Klees R (2002) Regularization of gravity field estimation from satellite gravity gradients. J Geodesy 76:359–368

LeGrand P, Minster J-F (1999) Impact of the GOCE gravity mission on ocean circulation estimates. Geophys Res Lett 26(13):1881–1884

Lithgow-Bertelloni C, Richards MA (1998) The dynamics of cenozoic and mesozoic plate motions. Rev. Geophys. 36(1):27–78

Losch M, Sloyan B, Schröter J, Sneeuw N (2002) Box inverse models, altimetry and the geoid; problems with the omission error. J Geophys Res 107(C7):10.1029

Martinec Z (2003) Green's function solution to spherical gradiometric boundary-value problems. J Geodesy 77:41–49

Marussi A (1985) Intrinsic geodesy. Springer, Berlin

Maximenko N, Niiler P, Rio M-H, Melnichenko O, Centurioni L, Chambers D, Zlotnicki V, Galperin B (2009) Mean dynamic topography of the ocean derived from satellite and drifting buoy data using three different techniques. J Atmos Ocean Technol 26:1910–1919

Migliaccio F, Reguzzoni M, Sansò F (2004) Space-wise approach to satellite gravity field determination in the presence of coloured noise. J Geodesy 78:304–313

Misner CW, Thorne KS, Wheeler JA (1970) Gravitation. Freeman, San Francisco

Moritz H, Hofmann-Wellenhof B (1993) Geometry, relativity, geodesy. Wichmann, Karlsruhe

Nutz H (2002) A unified setup of gravitational observables. Dissertation, Shaker Verlag, Aachen

Ohanian HC, Ruffini R (1994) Gravitation and spacetime. Norton & Comp., New York

Pail R, Plank R (2004) GOCE gravity field processing strategy. Stud Geophys Geod 48:289–309

Rummel R (1986) Satellite gradiometry. In: Sünkel H (ed) Mathematical and numerical techniques in physical geodesy. Lecture notes in earth sciences. vol 7. Springer, Berlin, pp 317–363, ISBN (Print) 978-3-540-16809-6, doi:10.1007/BFb0010135

Rummel R, van Gelderen M (1992) Spectral analysis of the full gravity tensor. Geophys J, Int 111:159–169

Rummel R (1997) Spherical spectral properties of the earth's gravitational potential and its first and second derivatives. In: Sansò F, Rummel R (eds) Geodetic boundary value problems in view of the one centimeter geoid. Lecture notes in earth sciences, vol 65. Springer, Berlin, pp 359–404, ISBN 3-540-62636-0

Rummel R, Balmino G, Johannessen J, Visser P, Woodworth P (2002) Dedicated gravity field missions—principles and aims. J Geodyn 33/1–2: 3–20

Schreiner M (1994) Tensor spherical harmonics and their application in satellite gradiometry. Dissertation, Universität Kaiserslautern

Stubenvoll R, Förste Ch, Abrikosov O, Kusche J (2009) GOCE and its use for a high-resolution global gravity combination model. In: Flechtner F, Mandea M, Gruber Th, Rothacher M, Wickert J, Güntner A, Schöne T (eds) System earth via geodetic-geophysical space techniques. Springer, Berlin

Švehla D, Rothacher M (2004) Kinematic precise orbit determination for gravity field determination. In: Sansò F (ed) The proceedings of the international association of geodesy: a window on the future of geodesy. Springer, Berlin, 181–188

Wells W C (ed) (1984) Spaceborne gravity gradiometers. NASA conference publication 2305, Greenbelt, MD

Wunsch C, Gaposchkin EM (1980) On using satellite altimetry to determine the general circulation of the oceans with application to geoid improvement. Rev Geophys 18:725–745

5 Sources of the Geomagnetic Field and the Modern Data That Enable Their Investigation

Nils Olsen[1] · *Gauthier Hulot*[2] · *Terence J. Sabaka*[3]
[1]Copenhagen, Denmark
[2]Equipe de Géomagnétisme, Institut de Physique du Globe de Paris, Institut de recherche associé au CNRS et *à* l'Université Paris 7, Paris, France
[3]Planetary Geodynamics Laboratory, Code 698, Greenbelt, MD, USA

W. Freeden, M.Z. Nashed, T. Sonar (Eds.), *Handbook of Geomathematics*, DOI 10.1007/978-3-642-01546-5_5,

Abstract The geomagnetic field one can measure at the Earth's surface or on board satellites is the sum of contributions from many different sources. These sources have different physical origins and can be found both below (in the form of electrical currents and magnetized material) and above (only in the form of electrical currents) the Earth's surface. Each source happens to produce a contribution with rather specific spatiotemporal properties. This fortunate situation is what makes the identification and investigation of the contribution of each source possible, provided appropriate observational data sets are available and analyzed in an adequate way, to produce the so-called geomagnetic field models. Here a general overview of the various sources that contribute to the observed geomagnetic field, and of the modern data that enable their investigation via such procedures is provided.

1 Introduction

The Earth has a large and complicated magnetic field, the major part of which is produced by a self-sustaining dynamo operating in the fluid outer core. What is measured at or near the surface of the Earth, however, is the superposition of the core field and of additional fields caused by magnetized rocks in the Earth's crust, by electric currents flowing in the ionosphere, magnetosphere, and oceans, and by currents induced in the Earth by the time-varying external fields. The sophisticated separation of these various fields and the accurate determination of their spatial and temporal structure based on magnetic field observations is a significant challenge, which requires advanced modeling techniques (see, e.g., Hulot et al. 2007). These techniques rely on a number of mathematical properties which are reviewed in the chapter entitled "Mathematical properties relevant to geomagnetic field modeling" by Sabaka et al. in this handbook. But as many of those properties have been derived by relying an assumptions motivated by the nature of the various sources of the Earth's magnetic field and of the available observations, it is important that a general overview of those sources and observations be given. This is precisely the purpose of this chapter. First, the various sources that contribute to the Earth's magnetic field are described (➋ Sect. 2) and next the observations currently available to investigate them (➋ Sect. 3) are discussed. Special emphasis is given on data collected by satellites, since these are extensively used for modeling the present magnetic field. The chapter concludes with a few words with respect to the way the fields these sources produce can be identified and investigated thanks to geomagnetic field modeling.

2 Sources of the Earth's Magnetic Field

Several sources contribute to the magnetic field that is measured at or above the surface; the most important ones are sketched in ➋ *Fig. 1*. The main part of the field is due to electrical currents in the Earth's fluid outer core at depths larger than 2,900 km; this is the so-called core field. Its strength at the Earth's surface varies from less than 30.000 nT near the equator to about 60.000 nT near the poles, which makes the core field responsible for more than 95% of the observed field at ground. Magnetized material in the crust (the uppermost few km of Earth) causes the crustal field; it is relatively weak and accounts on average only for a few percent of the observed field at ground. Core and crustal fields together make the internal field (since their sources are internal to the Earth's surface). External magnetic field contributions are caused by

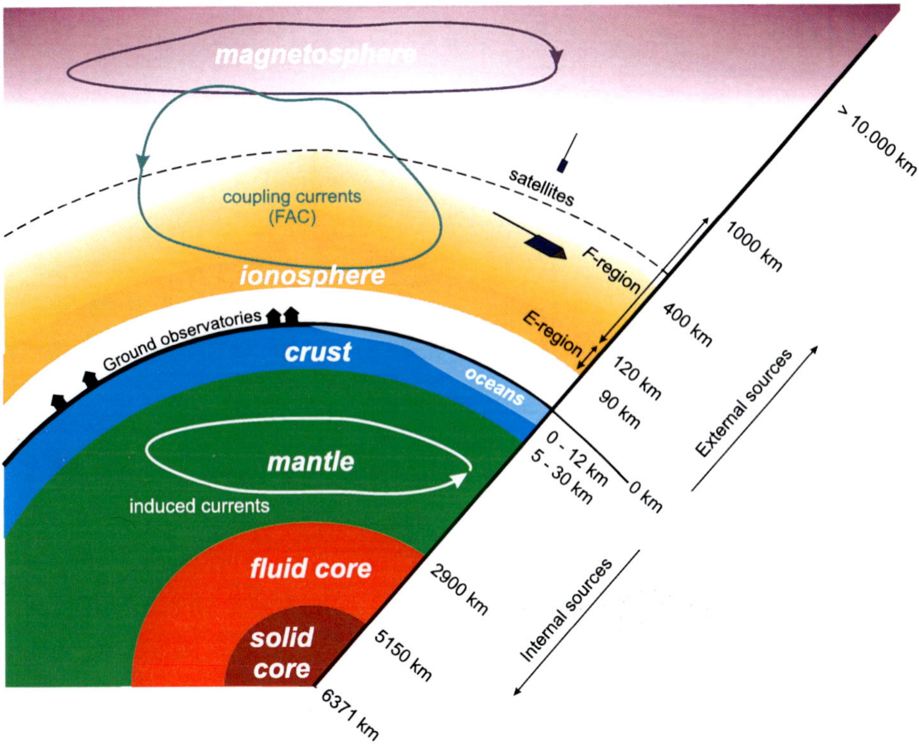

Fig. 1
Sketch of the various sources contributing to the near-Earth magnetic field

electric currents in the ionosphere (at altitudes 90–1000 km) and magnetosphere (at altitudes of several Earth radii). On average their contribution is also relatively weak—a few percent of the total field at ground during geomagnetic quiet conditions. However, if not properly considered, they disturb the precise determination of the internal field. It is therefore of crucial importance to account for external field (by data selection, data correction, and/or field coestimation) in order to obtain reliable models of the internal fields. Finally, electric currents induced in the Earth's crust and mantle by the time-varying fields of external origin, and the movement of electrically conducting seawater, cause magnetic field contributions that are of internal origin like the core and crustal field; however, typically only core and crustal field is meant when speaking about "internal sources."

A useful way of characterizing the spatial behavior of the geomagnetic field is to make use of the concept of spatial power spectra (e.g., Lowes 1966 and Sect. 4 of Sabaka et al. in this handbook). *Figure 2* shows the spectrum of the field of internal origin, often referred to as the *Lowes–Mauersberger* spectrum, which gives the mean square magnetic field at the Earth's surface due to contributions with horizontal wavelength λ_n corresponding to spherical harmonic degree n. The spectrum of the observed magnetic field (based on a combination of the recent field models derived by Olsen et al. (2009) and Maus et al. (2008)) is shown by black dots, while theoretical spectra describing core, resp. crustal, field spectra (Voorhies et al. 2002) are shown as blue, resp. magenta, curves. Each of these two theoretical spectra has two free parameters

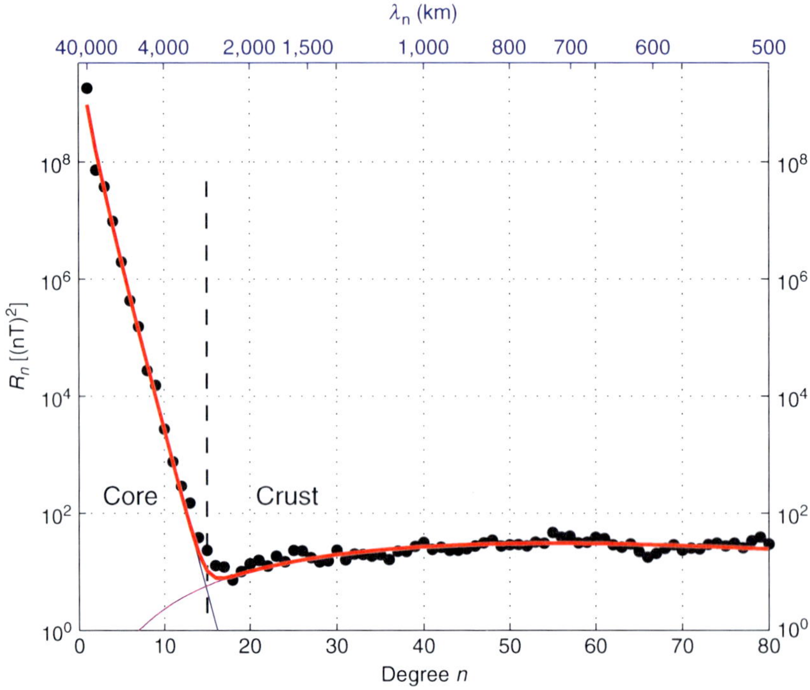

Fig. 2
Spatial power spectrum of the geomagnetic field at the Earth's surface. Black dots represent the spectrum of a recent field model (Olsen et al. 2009; Maus et al. 2008). Also shown are theoretical spectra (Voorhies et al., 2002) for the core (*blue*) and crustal (*magenta*) part of the field, as well as their superposition (*red curve*)

which have been fitted to the observed spectra; their sum (red curve) provides a remarkable good fit to the observed spectrum. There is a sharp "knee" at about degree $n = 14$ which indicates that contributions from the core field are dominant at large scales ($n < 14$) while those of the crustal field dominate for the smaller scales ($n > 14$).

A more detailed overview of the various field sources and their characteristics is now presented.

2.1 Internal Field Sources: Core and Crust

2.1.1 Core Field

Although the Earth's magnetic field has been known for at least several thousands of years (see, e.g., Merrill et al. 1998), the nature of its sources has eluded scientific understanding for a very long time. It was not until the nineteenth century that its main source was finally proven to be internal to the Earth. It is now known that this main source is most likely a self-sustaining dynamo within the Earth's core (see, e.g., Roberts 2007; the ➲ Chap. 16 "Theory and modeling of planetary dynamos" by Wicht et al. in this handbook)

This dynamo is the result of the fact that the liquid electrically conducting outer core (consisting of a Fe–Ni alloy) is cooling down and convecting vigorously enough to maintain electrical currents and a magnetic field. The basic process is one whereby the convective motion of the conducting fluid within the magnetic field induces electromotive forces which maintain electrical currents, and therefore also the magnetic field, against Ohmic dissipation.

The Earth's core dynamo has several specific features. First, the core contains a solid, also conducting, inner core. This inner core is thought to be the solidified part of the core (Jacobs 1953), the growth of which is the result of the cooling of the core. Because the Fe–Ni alloy that makes the core must in fact contain additional light elements, the solidification of the inner core releases the so-called compositional buoyancy at the inner-core boundary. This buoyancy will add up to the thermal buoyancy and is thought to be a major source of energy for the convection (see, e.g., Nimmo 2007). Second, the core, together with the whole Earth, is rotating fast, at a rate of one rotation per day. This leads Coriolis forces to play a major role in organizing the convection, and the way the dynamo works. In particular, spherical symmetry is dynamically broken, and a preferential axial (north–south) dipole field can be produced (see, e.g., Gubbins and Zhang 1993). At any given instant however, the field produced cannot be too *simple* (a requirement that has been formalized in terms of antidynamo theorems, starting with the best-known Cowling theorem (Cowling 1957)). In particular, no fully axisymmetric field can be produced by a dynamo. In effect, all dynamo numerical simulations run so far with conditions approaching that of the core dynamo produce quite complex fields in addition to a dominant axial dipole. This complexity not only affects the so-called *toroidal* component of the magnetic field which remains for the most part within the fluid core (such toroidal components are nonzero only where their sources lie, cf. ➋ Sect. 3 of Sabaka et al. in this handbook, and the poorly conducting mantle forces those to essentially remain within the core). It also affects the *poloidal* component of the field which can escape the core by taking the form of a *potential* field (such poloidal components can indeed escape their source region in the form of a Laplacian potential field, see again Sabaka et al. in this handbook), reach the Earth's surface and make the core field one can observe. The core field thus has a rich spatial spectrum beyond a dominant axial dipole component. It also has a rich temporal spectrum (with typical time scales from decades to centuries, see e.g., Hulot and Le Mouël 1994) directly testifying for the turmoil of the poloidal field produced by the dynamo at the core surface.

However, what is observable at and above the Earth's surface is only part of the core field that reaches it. Spatially, its small-scale contributions are masked by the crustal field, as shown in ➋ *Fig. 2*, and therefore only its largest scales (corresponding to spherical harmonic degrees smaller than 14) can be recovered. And temporally, the high-frequency part of the core field (corresponding to periods shorter than a few months) is screened by the finite conductivity of the mantle (see, e.g., Alexandrescu et al. 1999). This puts severe limitations on the possibility to recover the spatiotemporal structure of the core field, regardless of the quality of the magnetic field observations.

More information on the present knowledge of the core field based on recent data can be found in, e.g., Hulot et al. (2007) and Jackson and Finlay (2007).

2.1.2 Crustal Field

The material that makes the mantle and crust contains substantial amounts of magnetic minerals. Those minerals can become magnetized in the presence of an applied magnetic field. To produce any significant magnetic signals, this magnetism must however be of ferromagnetic type, which also requires the material to be at a low enough temperature (below the so-called Curie temperature of the minerals, see, e.g., Dunlop and Özdemir 2007). Those conditions are only met within the Earth's upper layers, above the so-called Curie-isotherm. Its depth can vary between zero (such as at mid-oceanic ridges) and several tens of kilometers, with a typical value on the order of 20 km in continental regions. Magnetized rocks can thus only be found in those layers.

Magnetized rocks essentially carry two types of magnetization, *induced* magnetization and *remanent* magnetization. Induced magnetization is one that is proportional, both in strength and direction, to the ambient field within which the rock is embedded. The ability of such a rock to acquire this magnetization is a function of the nature and proportion of the magnetic minerals it contains. It is measured in terms of a proportionality factor known as the *magnetic susceptibility*. Were the core field (or more correctly the local field experienced by the rock) to disappear, this induced magnetization would also disappear. Then, only the second type of magnetization, remanent magnetization, would remain. This remanent magnetization may have been acquired by the rock in many different ways (see, e.g., Dunlop and Özdemir 2007). For instance at times of deposition for a sedimentary rock, or via chemical transformation, if the rock has been chemically altered. The most ubiquitous process however, which also usually leads to the strongest remanent magnetization, is thermal. It is the way igneous and metamorphic rocks acquire their remanent magnetization when they cool down below their Curie temperature. The rock becomes magnetized in proportion, both in strength and direction, to the ambient magnetic field that the rocks experiences at the time it cools down (the proportionality factor being again a function of the magnetic minerals contained in the rock). Remanent magnetization from a properly sampled rock thus can provide information about the ancient core field (see, e.g., Hulot et al. 2010).

There is no way one can identify the signature of the present core field without taking the crustal field into account (in fact modern ways of modeling the core field from satellite data often also involves modeling the crustal field). It is therefore important to also mention some of the most important spatiotemporal characteristics of the field produced by magnetized sources.

It should first be recalled that not all magnetized sources will produce observable fields at the Earth's surface. In particular, if the upper layers of the Earth consisted in a spherical shell of uniform magnetic properties magnetized within the core field at a given instant, they would produce no observable field at the Earth's surface. This is known as Runcorn's theorem (Runcorn 1975), an important implication of which is that the magnetic field observed at the Earth's surface is not sensitive to the induced magnetization due to the average susceptibility of a spherical shell best describing the upper magnetic layers of the Earth. It will only sense the departure of those layers from sphericity (see, e.g., Lesur and Jackson 2000), either because of the Earth's flattening, because of the variable depth of the Curie isotherm, or because of the contrasts in magnetization due to the variable nature and susceptibility of rocks within those layers (although even such contrasts can sometimes also fail to produce observable fields (Maus and Haak 2003)).

Another issue of importance is that of the relative contributions of induced and remanent magnetization. Induced magnetization is most likely the main source of large-scale magnetization, while remanent magnetization plays a significant role only at regional (especially in oceans) and local scales (e.g., Purucker et al. 2002; Hemant and Maus 2005).

Finally, it is important to briefly mention the poorly known issue of possible temporal changes in crustal magnetization on a human time scale. At a local scale, any dynamic process that can alter the magnetic properties of the rocks, or change the geological setting (such as an active volcano) would produce such changes. On a planetary scale, by contrast, significant changes can only occur in the induced magnetization because of the slowly time-varying core field, as has been recently demonstrated by Hulot et al. (2009) and Thebault et al. (2009).

A recent review of the crustal field is given by Purucker and Whaler (2007).

2.2 Ionospheric, Magnetospheric, and Earth-Induced Field Contributions

Ionospheric and magnetospheric currents (which produce the field of external origin), as well as Earth-induced currents (which produce externally induced internal fields) contribute in a nonnegligible way to the observed magnetic field, both at ground and at satellite altitude. It is therefore important to consider them in order to properly identify and separate their signal from that of the field of internal origin.

2.2.1 Ionospheric Contributions

Geomagnetic daily variations at nonpolar latitudes (known as *Sq* variations) are caused by diurnal wind systems in the upper atmosphere: Heating at the dayside and cooling at the nightside generates tidal winds which drive ionospheric plasma against the core field, inducing electric fields and currents in the ionospheric *E*-region dynamo region between 90 and 150 km altitude (Richmond 1989; Campbell 1989; Olsen 1997b). The currents are concentrated at an altitude of about 110–115 km and hence can be represented by a sheet current at that altitude (cf. ➋ Sects. 3.4 and ➋ 3.5 of Sabaka et al. in this handbook). They remain relatively fixed with respect to the Earth–Sun line and produce regular daily variations which are directly seen in the magnetograms of *magnetically quiet* days. On *magnetically disturbed* days there is an additional variation which includes superimposed magnetic storm signatures of magnetospheric and high-latitude ionospheric origin. Typical peak-to-peak *Sq* amplitudes at middle latitudes are 20–50 nT; amplitudes during solar maximum are about twice as large as those during solar minimum. *Sq* variations are restricted to the dayside (i.e., sunlit) hemisphere, and thus depend mainly on local time. Selecting data from the nightside when deriving models of the internal field is therefore useful to minimize field contributions from the nonpolar ionospheric *E*-region.

Because the geomagnetic field is strictly horizontal at the dip equator, there is about a fivefold enhancement of the effective (Hall) conductivity in the ionospheric dynamo region, which results in about a fivefold enhanced eastward current, called the *Equatorial Electrojet (EEJ)*, flowing along the dayside dip equator (Rastogi 1989). Its latitudinal width is about 6°–8°.

In addition, *auroral electrojets (AEJ)* flow in the auroral belts (near $\pm(65^\circ$–$70^\circ)$ magnetic latitude) and vary widely in amplitude with different levels of magnetic activity from a few 100s nT during quiet periods to several thousand nT during major *magnetic storms*. As a general rule, ionospheric fields at polar latitudes are present even at magnetically quiet times and on the nightside (i.e., dark) hemisphere, which makes it difficult to avoid their contribution by data selection.

Electric currents at altitude above 120 km, i.e., in the so-called ionospheric *F*-region (up to 1000 km altitude), cause magnetic fields that are detectable at satellite altitude as nonpotential (e.g., *toroidal*) magnetic fields (Olsen 1997a; Richmond 2002; Maus and Lühr 2006). Their contributions in nonpolar regions are also important during local nighttime, when the *E*-region conductivity vanishes and therefore contributions from *Sq* and the Equatorial Electrojet are absent.

2.2.2 Magnetospheric Contributions

The field originating in Earth's magnetosphere is due primarily to the *ring-current* and to currents on the magnetopause and in the magnetotail (Kivelson and Russell 1995). Currents flowing on the outer boundary of the magnetospheric cavity, the *magnetopause currents*, cancel the Earth's field outside and distend the field within the cavity. This produces an elongate tail in the antisolar direction within which the so-called neutral-sheet currents are established in the equatorial plane. Interaction of these currents with the *radiation belts* near the Earth produces a ring-current in the dipole equatorial plane which partially encircles the Earth, but achieves closure via *field-aligned currents (FAC)* (currents which flow along core field lines) into and out of the ionosphere. These resulting fields have magnitudes on the order of 20–30 nT near the Earth during magnetically quiet periods, but can increase to several hundreds of nT during disturbed times.

In polar regions (poleward of, say, $\pm 65^\circ$ dipole latitude), the auroral ionosphere and magnetosphere are coupled by field-aligned currents. The fields from these FAC have magnitudes that vary with the magnetic disturbance level. However, they are always present, on the order of 30–100 nT during quiet periods and up to several thousand nT during substorms. There are also currents which couple the ionospheric *Sq* currents systems in the two hemispheres that flow, at least in part, along magnetic field lines. The associated magnetic fields are generally 10 nT or less. Finally, there exists a meridional current system which is connected to the EEJ with upward directed currents at the dip equator and field-aligned downward directed currents at low latitudes. These currents result in magnetic fields of about a few tens of nT at 400 km altitude.

2.2.3 Induction in the Solid Earth and the Oceans

Time-varying external fields produce secondary, induced, currents in the oceans and the Earth's interior; this contribution is what we refer to as externally induced fields, which are the topic of *electromagnetic induction* studies (see Parkinson and Hutton 1989; Constable 2007). In addition, the motion of electrically conducting seawater through the core field, via a process referred to as *motional induced induction*, also produces secondary currents (e.g., Tyler et al. 2003; Kuvshinov and Olsen 2005; Maus 2007b). The oceans thus contribute twofold to the observed magnetic field: by secondary currents induced by primary current systems in the ionosphere and magnetosphere; and by motion-induced currents due to the movement of seawater, for instance by tides.

The amplitude of induced contributions generally decreases with the period. As an example, about one third of the observed daily *Sq* variation in the horizontal components is of induced origin (Schmucker 1985). But induction effects also depend on the scale of the source (i.e., the ionospheric and magnetospheric current systems); as a result, the induced contribution due to the daily variation of the Equatorial Electrojet is, for instance, much smaller than the above-mentioned one third, typical of the large-scale *Sq* currents (Olsen 2007b).

3 Modern Geomagnetic Field Data

3.1 Definition of Magnetic Elements and Coordinates

Measurements of the geomagnetic field taken at ground or in space form the basis for modeling the Earth's magnetic field.

Observations taken at the Earth's surface are typically given in a local *topocentric (geodetic)* coordinate system (i.e., relative to a reference ellipsoid as approximation for the geoid). The magnetic elements X, Y, Z are the components of the field vector $\mathbf{B}$ in an orthogonal right-handed coordinate system, the axis of which are pointing toward geographic North, geographic East, and vertically down, as shown in ➋ *Fig. 3*. Derived magnetic elements are the angle between geographic North and the (horizontal) direction in which a compass needle is pointing, denoted as *declination* $D = \arctan Y/X$; the angle between the local horizontal plane and the field vector, denoted as *inclination* $I = \arctan Z/H$; *horizontal intensity* $H = \sqrt{X^2 + Y^2}$; and *total intensity* $F = \sqrt{X^2 + Y^2 + Z^2}$.

In contrast to magnetic observations taken at or near ground, satellite data are typically provided in the *geocentric coordinate* system as spherical components B_r, B_θ, B_ϕ where r, θ, ϕ are radius, colatitude and longitude, respectively. Equations for transforming between geodetic components X, Y, Z and geocentric components B_r, B_θ, B_ϕ can be found in, e.g., Sect. 5.02.2.1.1 of Hulot et al. (2007).

The distribution in space of the observations at a given time determines the spatial resolution to which the field can be determined for that time. Internal sources are often fixed with respect to the Earth (magnetic fields due to induced currents in the Earth's interior are an exception) and thus follow its rotation. Internal sources are therefore best described in an Earth-centered-Earth-fixed (ECEF) coordinate frame like that given by the geocentric

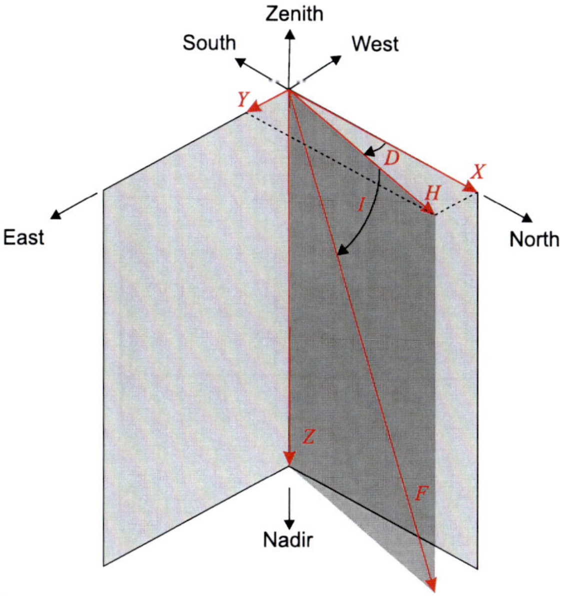

Fig. 3
The magnetic elements in the local topocentric coordinate system, seen from North-East

coordinates r, θ, ϕ. In contrast, many external fields are fixed with respect to the position of the Sun, and therefore the use of a coordinate frame that follows the (apparent) movement of the Sun is advantageous. Solar magnetospheric (SM) coordinates for describing near magnetospheric currents like the ring-current, and geocentric solar magnetospheric (GSM) coordinates for describing far magnetospheric current systems like the tail currents have turned out to be useful when determining models of Earth's magnetic field (Maus and Lühr 2005; Olsen et al. 2006, 2009).

3.2 Ground Data

Presently about 150 geomagnetic observatories monitor the time changes of the Earth's magnetic field. Their global distribution, shown in the right part of ➋ *Fig. 4*, is very uneven, with large uncovered areas especially over the oceans. Yellow symbols indicate sites that provide data (regardless of the observation time instant and the duration of the time series), while the red dots show observatories that have provided hourly mean values for the recent years. These observatories provide data of different temporal resolution which are distributed through the World-Date-Center system (e.g., http://www.ngdc.noaa.gov/wdc, http://wdc.kugi.kyoto-u.ac.jp, http://www.wdc.bgs.ac.uk). Traditionally, annual mean values have been used for deriving field models, but the availability of hourly mean values (or even one minute values) in digital form for the recent years allow for a better characterization of external field variations. The left part of ➋ *Fig. 4* shows the distribution in time of observatory data of various sampling rate. International campaigns, like the *Göttingen Magnetic Union*, the first and second *International Polar Year (IPY)*, the *International Geophysical Year (IGY/C)*, the *International Quiet Solar Year (IQSY)*, the *International Magnetospheric Study (IMS)*, and the preparation of the Ørsted satellite mission have stimulated observatory data processing and the establishment of new observatories.

Geomagnetic observatories aim at measuring the magnetic field in the geodetic reference frame with an absolute accuracy of 1 nT (Jankowski and Sucksdorff 1996). However, it is presently not possible to take advantage of that measurement accuracy due to the (unknown) contribution from near-by crustal sources. When using observatory data for field modeling it is therefore common practice to either use first time differences of the observations (thereby

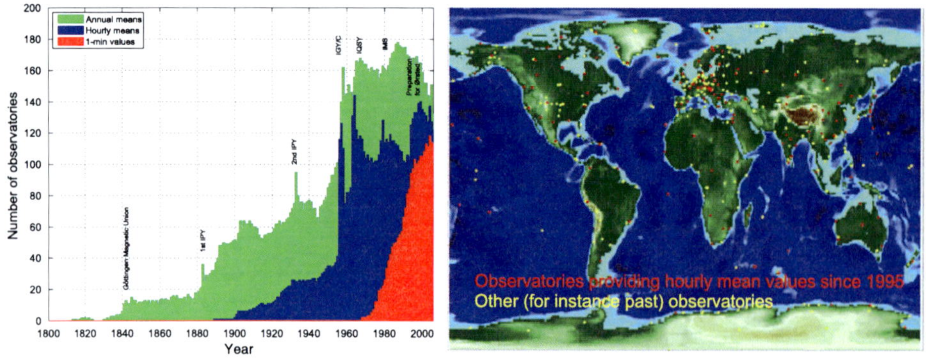

◘ Fig. 4
Distribution of ground observatory magnetic field data in time (*left*) and space (*right*)

eliminating the static crustal field contribution) or to coestimate together with the field model an "observatory bias" for each site and element, following a procedure introduced by Langel et al. (1982). A joint analysis of observatory and satellite data allows one to determine these observatory biases. The need for knowledge of the absolute baseline is therefore less important during periods for which satellite data are available. Recognizing this will simplify the observation practice, especially for ocean-bottom magnetometers for which the exact determination of true north is very difficult and expensive.

In addition to geomagnetic observatories (which monitor the time changes of the geomagnetic field at a given location), magnetic *repeat stations* are sites where high-quality magnetic measurements are taken every few years for a couple hours or even days (Newitt et al. 1996; Turner et al. 2007). The main purpose of repeat stations is to measure the time changes of the core field (secular variation); they offer better spatial resolution than observatory data but do not provide continuous time series.

3.3 Satellite Data

The possibility to measure the Earth's magnetic field from space has revolutionized geomagnetic field modeling. Magnetic observations taken by low-Earth-orbiting (LEO) satellites at altitudes below 1000 km form the basis of recent models of the geomagnetic field.

There are several advantages of using satellite data for field modeling:

1. Satellites sample the magnetic field over the entire Earth (apart from the polar gap, a region around the geographic poles that is left unsampled if the satellite orbit is not perfectly polar).
2. Measuring the magnetic field from an altitude of 400 km or so corresponds roughly to averaging over an area of this dimension. Thus the effect of local heterogeneities, for instance caused by local crustal magnetization, is reduced.
3. The data are obtained over different regions with the same instrumentation, which helps to reduce spurious effects.

There are, however, some points to consider when using satellite data instead of ground based data:

1. Since the satellite moves (with a velocity of about 8 km/s at 400 km altitude) it is not possible to decide whether an observed magnetic field variation is due to a temporal or spatial change of the field. Thus there is risk for time-space aliasing.
2. It is necessary to measure the magnetic field with high accuracy—not only regarding resolution, but also regarding orientation and absolute values.
3. Due to the Earth's rotation, the satellite revisits a specific region after about 1 day.[1] Hence the magnetic field in a selected region is modeled from time series with a sampling rate of 1 day. However, since the measurements were not low-pass filtered before "resampling," aliasing may occur.
4. Satellites usually acquire data not at one fixed altitude, but over a range of altitudes. The decay of altitude through mission lifetime often leads to time series that are unevenly distributed in altitude.

[1] Actually the satellite revisits that region already after about 12 h, but this will be for a different local time. Because of external field contributions—which heavily depend on local time—it is safer to rely on data taken at similar local time conditions, which results in the above-stated sampling recurrence of 24 h.

Table 1
Satellite missions of relevance for geomagnetic field modeling

Satellite	Years	Inclination i	Altitude (km)	Accuracy (nT)	Remarks
Cosmos 49	1964	50°	261–488	22	Scalar only
POGO					
OGO-2	1965–1967	87°	413–1510	6	Scalar only
OGO-4	1967–1969	86°	412–908	6	Scalar only
OGO-6	1969–1971	82°	397–1098	6	Scalar only
Magsat	1979–1980	97°	350–550	6	Vector and scalar
DE-1	1981–1991	90°	568–23	290 ?	Vector (spinning)
DE-2	1981–1983	90°	309–1012	$\approx 30(F)/100$	Low accuracy vector
POGS	1990–1993	90°	639–769	?	Low accuracy vector, timing problems
UARS	1991–1994	57°	560	?	Vector (spinning)
Ørsted	1999–	97°	650–850	4	Scalar and vector
CHAMP	2000–	87°	310–450	3	Scalar and vector
SAC-C	2001–2004	97°	698–705	4	Scalar only
Swarm	2012–2016	88°/87°	530/ < 450	2	Scalar and vector

5. Finally, the satellite move through an electric plasma, and the existence of electric currents at satellite altitude does, in principle, not allow to describe the observed field as the gradient of a Laplacian potential.

An overview of previous and present satellites that have been used for geomagnetic field modeling is given in ➲ *Table 1* (see also Table 3.3 of Langel and Hinze 1998).

The quality of the magnetic field measurements is rather different for the listed satellites, and before the launch of the Ørsted satellite in 1999, the POGO satellite series (Cain 2007) that flew in the second half of the 1960s, and Magsat (Purucker 2007), which flew for 6 months around 1980, were the only high-precision magnetic satellites. A timeline of high-precision missions is shown in ➲ *Fig. 5*. After a gap of almost 20 years with no high-precision satellites in orbit, the launch of the Danish Ørsted satellite (Olsen 2007a) in February 1999 marked the beginning of a new epoch for exploring the Earth's magnetic field from space. Ørsted was followed by the German CHAMP satellite (Maus 2007a) and the US/Argentinian/Danish SAC-C satellite, launched in July and November 2000, respectively. All three satellites carry essentially the same instrumentation and provide high-quality and high-resolution magnetic field observations from space. They sense the various internal and external field contributions differently, due to their different altitudes and drift rates through local time.

A closer look at the characteristics of satellite data sampling is helpful. A satellite moves around Earth in elliptical orbits. However, ellipticity of the orbit is small for many of the satellites used for field modeling, and for illustration purposes circular orbits are considered. As sketched on the left side of ➲ *Fig. 6*, orbit inclination i is the angle between the orbit plane and the equatorial plane. A perfectly polar orbit implies $i = 90°$, but for practical reasons most satellite orbits have inclinations that are different from 90°. This results in *polar gaps*, which are regions around the geographic poles that are left unsampled. The right part of ➲ *Fig. 6* shows the ground track of 1 day (January 2, 2001) of Ørsted satellite data. It is obvious that the coverage in latitude

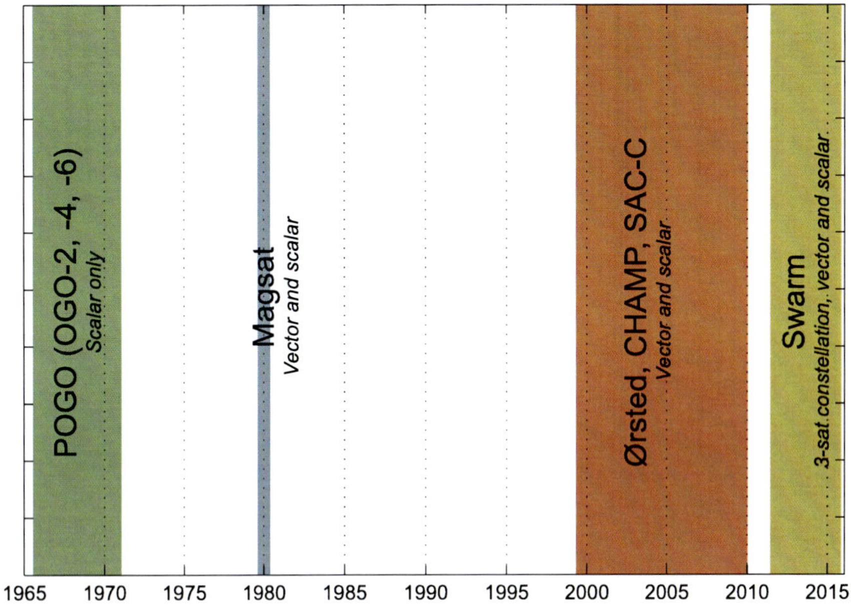

Fig. 5
Distribution of high-precision satellite missions in time

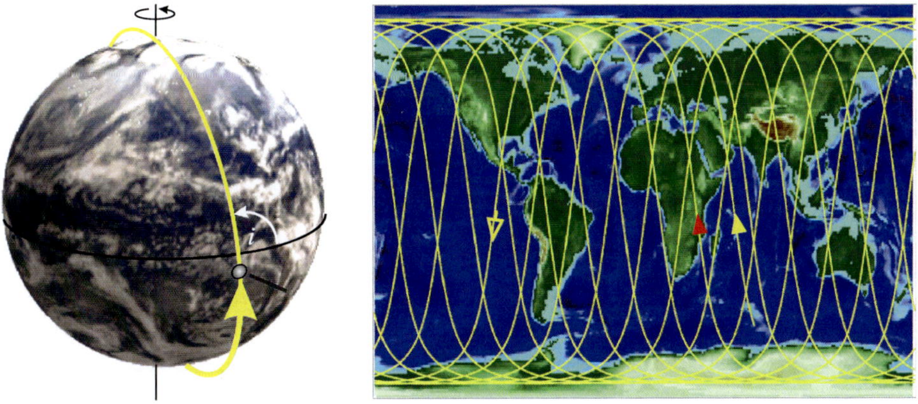

Fig. 6
Left: The path of a satellite at inclination *i* in orbit around the Earth. *Right*: Ground track of 24 h of the Ørsted satellite on January 2, 2001 (*yellow curve*). The satellite starts at −57°N, 72°E at 00 UT, moves northward on the morning side of the Earth, and crosses the Equator at 58°E (*yellow arrow*). After crossing the polar cap it moves southward on the evening side and crosses the equator at 226°E (*yellow open arrow*) 50 min after the first equator crossing. The next Equator crossing (after additional 50 minutes) is at 33°E (*red arrow*), 24° westward of the first crossing 100 minutes earlier, while moving again northward

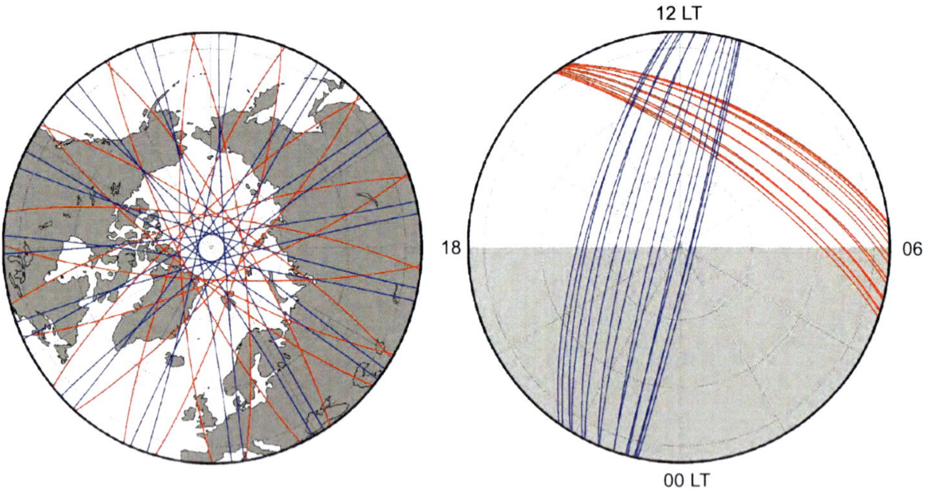

Fig. 7
Left: **Ground track of one day (January 2, 2001) of the Ørsted (*red*) and CHAMP (*blue*) satellites in dependence on geographic coordinates. *Right*: Orbit in the solar-magnetospheric (SM) reference frame**

and longitude provided even by only a few days of satellites data is much better than that of the present ground based observatory network (cf. *Fig. 4*).

The polar gaps, the regions of half-angle $|90^\circ - i|$ around the geographic poles, are obvious when looking at the orbits in a polar view in an Earth-fixed coordinate system, as done in the left part of *Fig. 7* for the Ørsted (red), resp. CHAMP (blue), satellite tracks of January 2, 2001. The polar gaps are larger for Ørsted (inclination $i = 97^\circ$) compared to CHAMP ($i = 87^\circ$), as confirmed by the figure.

As mentioned before, internal magnetic field sources are often fixed (or slowly changing, in the case of the core field) with respect to the Earth while most external fields have relatively fixed geometries with respect to the Sun. A good description of the various field contributions requires a good sampling of the data in the respective coordinate systems.

Good coverage in latitude and longitude is essential for modeling the internal field. There are, however, pitfalls due to peculiarities of the satellite orbits, which may result in less optimal sampling. The top panel of *Fig. 8* shows the longitude of the ascending node (the equator crossing of the satellite going from south to north) of the Ørsted (left) and CHAMP (right) satellite orbits. Depending on orbit altitude (shown in the middle part of the figure) there are periods with pronounced "revisiting patterns": in June and July 2003 the CHAMP satellite samples for instance only the field near the equator at longitudes $\phi = 7.6^\circ, 19.2^\circ, 30.8^\circ \ldots 356.0^\circ$. This longitudinal sampling of $\Delta\phi = 11.6^\circ$ hardly allows to resolve features of the field of spatial scale corresponding to spherical harmonic degree $n > 15$.

Another issue that has to be considered when deriving field models from satellite data concerns satellite altitude. The middle panel of *Fig. 8* shows the altitude evolvement for the Ørsted and CHAMP satellites. Various altitude maneuvers are the reason for the sudden increase of altitude of CHAMP. At lower altitudes the magnetic signal of small-scale features of the internal magnetic field (corresponding to higher spherical harmonic degrees) is relatively more amplified compared to large-scale features (represented by low-degree spherical

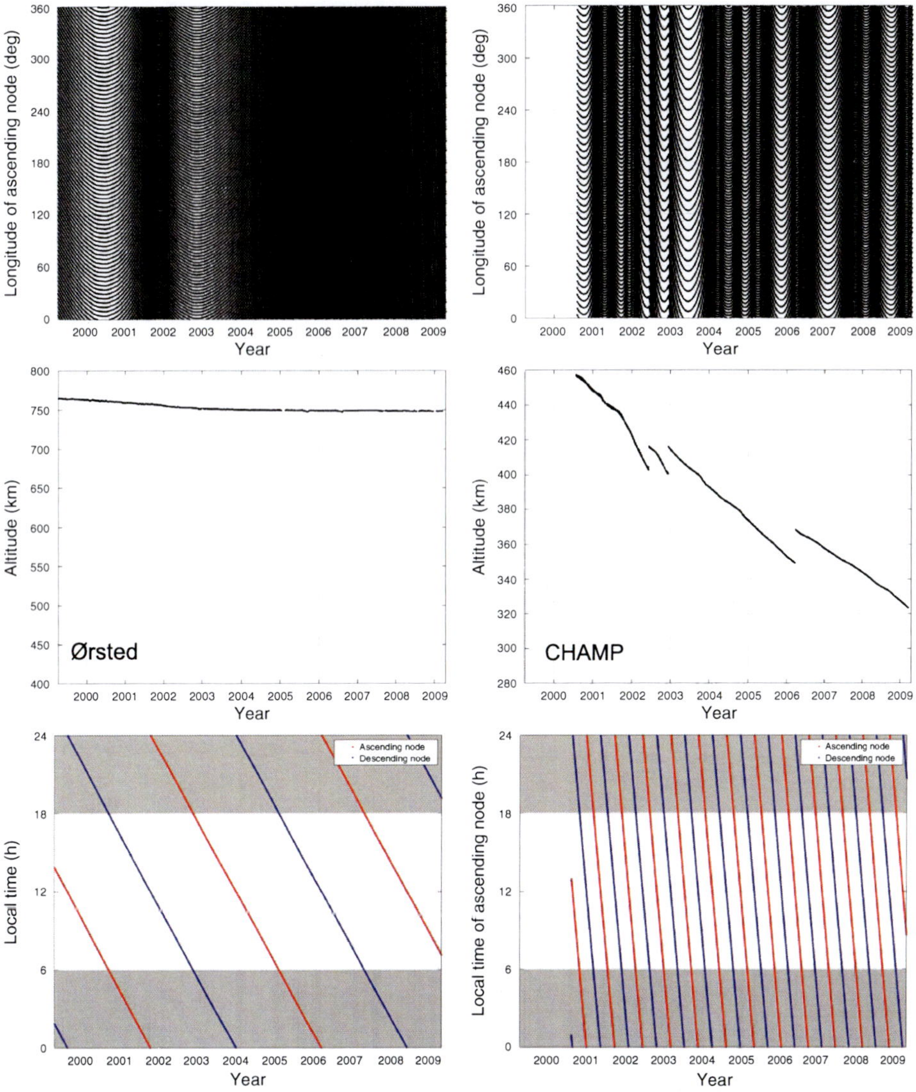

Fig. 8

Some orbit characteristics for the Ørsted (*left*), resp. CHAMP (*right*) satellite in dependence on time. *Top*: Longitude of the ascending node, illustrating longitudinal "revisiting patterns." *Middle*: Mean altitude. *Bottom*: Local time of ascending (*red*), resp. descending (*blue*) node

harmonics). The crustal field signal measured by a satellite is thus normally stronger toward the end of the mission lifetime due to the lower altitude. However, if the crustal field is not properly accounted for, the decreasing altitude may hamper the determination of the core field time changes.

Good sampling in the Earth-fixed coordinate system, which is essential for determining the internal magnetic field, can, at least in principle, be obtained from a few days of satellite data. However, good sampling in sun-fixed coordinates is required for a reliable determination of the external field contributions. The right part of ➋ *Fig. 7* shows the distribution of the Ørsted, resp. CHAMP, satellite data of January 2, 2001 in the SM coordinate system, i.e., in dependence on the distance from the geomagnetic North pole (which is in the center of the plot) and magnetic local time LT. Despite the rather good sampling in the geocentric frame (left panel), the distribution in the sun-fixed system (right panel) is rather coarse, especially when only data from one satellite are considered. Data obtained at different local times are essential for a proper description of external fields. The bottom panel of ➋ *Fig. 8* shows how local time of the satellite orbits change through mission lifetime. The Ørsted satellite scans all local times within 790 days (2.2 years), while the local time drift rate of CHAMP is much higher: CHAMP covers all local times within 130 days.

Combining observations from different spacecraft flying at different local times helps to improve data coverage in the various coordinate systems. Especially the upcoming *Swarm* satellite constellation mission (Friis-Christensen et al. 2006, 2009) consisting of three satellites has specifically been designed to reduce the time-space ambiguity that is typical of single-satellite missions.

4 Making the Best of the Data to Investigate the Various Field Contributions: Geomagnetic Field Modeling

Using the observations of the magnetic field described in the previous section to identify the various magnetic fields described in ➋ Sect. 2 is the main purpose of geomagnetic field modeling. This requires the use of mathematical representations of such fields in both space and time. The mathematical tools that make such a representation in space possible are described in the accompanying chapter by Sabaka et al. in this handbook. But because the fields vary in time, some temporal representations are also needed. Using such spatiotemporal representations formally makes it possible to represent all the fields that contribute to the observed data in the form of a linear superposition of elementary functions. The set of numerical coefficients that define this linear combination is then what one refers to as a geomagnetic field model. It can be recovered from the data via inverse theory, which next makes it possible to identify the various field contributions. In practice however, one has to face many pitfalls, not the least because the data are limited in number and not ideally distributed. In particular, although numerous, the usefulness of satellite data is limited by the time needed for satellites to complete an orbit, during which some of the fields can change significantly. This can then translate into some ambiguity in terms of the spatial/temporal representations. But advantage can be taken of the known spatio temporal properties of the various fields described in ➋ Sect. 2 and of the combined use of ground and satellite data.

Fast changing fields, with periods up to typically a month, are for instance known to mainly be of external origin (both ionospheric and magnetospheric), but with some electrically induced internal fields. Those are best identified with the help of observatory data, which can be temporally band-filtered, and next spatially analyzed with the help of the tools described in Sabaka et al. in this handbook. This then makes it possible to identify the contribution from sources above and below the Earth's surface. The relative magnitude and temporal phase shifts between the (induced) internal and (inducing) external fields can then be computed, which provides very useful information with respect to the distribution of the electrical conductivity within the solid Earth (see, e.g., Constable 2007; Kuvshinov 2008). Satellite data can also be used for similar purposes, but this is a much more difficult endeavor since, as we already noted, one then has to deal with additional space/time separation issues related to the fact that satellites sense both changes due to their motion over stationary sources (such as the crustal field) and true temporal field changes. Much efforts are currently devoted to deal with those issues and make the best of such data for recovering the solid Earth electrical conductivity distribution, with encouraging preliminary successes (see, e.g., Kuvshinov and Olsen 2006), especially in view of the upcoming *Swarm* satellite constellation mission (Kuvshinov et al. 2006).

As a matter of fact, and for the time being, satellite data turn out to be most useful for the investigation of the field of internal origin (the core field and the crustal field). But even recovering those fields requires considerable care and advanced modeling strategies. In principle, and as explained in Sabaka et al. in this handbook, full vector measurements can be combined with ground-based vector data to infer both the field of internal origin, the *E*-region ionospheric field, the local *F*-region ionospheric field, and the magnetospheric field. But there are many practical limits to this possibility, again because satellites do not provide instantaneous sets of measurements on a sphere at all times, and also because the data distribution at the Earth's surface is quite sparse. This sets a limit on the quality of the *E*-region ionospheric field one can possibly hope to recover by a joint use of ground-based and satellite data. But appropriate knowledge of the spatiotemporal behavior of each type of sources can again be used. This basically leads to the two following possible strategies to infer the contribution of each field from satellite data.

A first strategy consists in acknowledging that the field due to nonpolar ionospheric *E*-region is weak at night, especially at the so-called magnetic quiet time (as may be inferred from ground-based magnetic data), and selecting satellite data in this way, so as to minimize contributions from the ionosphere. Those satellite data can then be used alone to infer both the field of internal origin and the field of external (then only magnetospheric) origin (though this usually still requires some care when dealing with polar latitude data, because those are always, also at night and during quiet conditions, affected by some ionospheric and local field-aligned currents). This is a strategy that can be used to focus on the field of internal origin, and in particular the crustal field (see, e.g., Maus et al. 2008).

A second strategy consists in making use of both observatory data and satellite data, and to simultaneously parameterizing the spatial and temporal behavior of as many sources as possible. This strategy has been used in particular to improve the recovery of the core field and its slow secular changes (see, e.g., Thomson and Lesur 2007; Lesur et al. 2008; Olsen et al. 2009), but can more generally be used to try and recover all field sources simultaneously, using the so-called comprehensive modelling approach (Sabaka et al. 2002, 2004), to investigate the temporal evolution of all fields over long periods of times when satellite data are available. This strategy is one that looks particularly promising in view of the upcoming *Swarm* satellite constellation mission (Sabaka and Olsen 2006). Finally, it is worth pointing out that whatever strategy is being used, residuals from the modeled fields may then also always be used to investigate additional non-modeled sources such as local ionospheric *F*-region sources (Lühr et al. 2002; Lühr and Maus 2006) to which satellite data are very sensitive.

Considerable more details about all those strategies are given in, e.g., Hulot et al. (2007), to which the reader is referred, and where many more references can be found.

Acknowledgment

This is IPGP contribution 2595.

References

Alexandrescu MM, Gibert D, Le Mouël JL, Hulot G, Saracco G (1999) An estimate of average lower mantle conductivity by wavelet analysis of geomagnetic jerks. J Geophys Res 104: 17735–17745

Cain JC (2007) POGO (OGO-2, -4 and -6 spacecraft). In: Gubbins D, Herrero-Bervera E (eds) Encyclopedia of geomagnetism and paleomagnetism. Springer, Heidelberg

Campbell WH (1989) The regular geomagnetic field variations during quiet solar conditions. In: Jacobs JA (ed) Geomagnetism, vol 3. Academic, London, pp 385–460

Constable S (2007) Geomagnetic induction studies. In: Kono M (ed) Treatise on geophysics, vol 5. Elsevier, Amsterdam, pp 237–276

Cowling TG (1957) Magnetohydrodynamics. Wiley Interscience, New York

Dunlop D, Özdemir Ö (2007) Magnetizations in rocks and minerals. In: Kono M (ed) Treatise on geophysics, vol 5. Elsevier, Amsterdam, pp 277–336

Friis-Christensen E, Lühr H, Hulot G (2006) *Swarm*: a constellation to study the Earth's magnetic field. Earth Planets Space 58:351–358

Friis-Christensen E, Lühr H, Hulot G, Haagmans R, Purucker M (2009) Geomagnetic research from space. Eos 90:25

Gubbins D, Zhang K (1993) Symmetry properties of the dynamo equations for paleomagnetism and geomagnetism. Physics of the Earth and Planetary Interior 75:225–241

Hemant K, Maus S (2005) Geological modeling of the new CHAMP magnetic anomaly maps using a geographical information system technique. J Geophys Res 110:B12103. doi:10.1029/2005JB003837

Hulot G, Le Mouël JL (1994) A statistical approach to the Earth's main magnetic field. Physics of the Earth and Planetary Interior 82:167–183. doi: 10.1016/0031-9201(94)90070-1

Hulot G, Sabaka TJ, Olsen N (2007) The present field. In: Kono M (ed) Treatise on geophysics, vol 5. Elsevier, Amsterdam

Hulot G, Finlay C, Constable C, Olsen N, Mandea M (2010) The magnetic field of planet Earth. Space Sci Rev (in press), doi:10.1007/s11214-010-9644-0

Hulot G, Olsen N, Thebault E, Hemant K (2009) Crustal concealing of small-scale core-field secular variation. Geophys J Int 177(2): 361–366

Jackson A, Finlay CC (2007) Geomagnetic secular variation and its application to the core. In: Kono M (ed) Treatise on geophysics, vol 5. Elsevier, Amsterdam

Jacobs JA (1953) The earth's inner core. Nature 172: 297–300

Jankowski J, Sucksdorff C (1996) IAGA guide for magnetic measurements and observatory practice. IAGA, Warszawa

Kivelson MG, Russell CT (1995) Introduction to space physics. Cambridge University Press, Cambridge

Kuvshinov A (2008) 3-D global induction in the oceans and solid earth: recent progress in modeling magnetic and electric fields from sources of magnetospheric, ionospheric and oceanic origin. Surv Geophys 29(2):139–186

Kuvshinov AV, Olsen N (2005) 3D modelling of the magnetic field due to ocean flow. In: Reigber C, Lühr H, Schwintzer P, Wickert J (eds) Earth observation with CHAMP, results from three years in orbit. Springer, Berlin

Kuvshinov AV, Olsen N (2006) A global model of mantle conductivity derived from 5 years of CHAMP, Ørsted, and SAC-C magnetic data. Geophys Res Lett 33:L18301. doi: 10.1029/2006GL027083

Kuvshinov AV, Sabaka TJ, Olsen N (2006) 3-D electromagnetic induction at *Swarm* constellation. Mapping the conductivity anomalies in the mantle. Earth Planets Space 58(4): 417–427

Langel RA, Hinze WJ (1998) The magnetic field of the Earth's lithosphere: the satellite perspective. Cambridge University Press, Cambridge

Langel RA, Estes RH, Mead GD (1982) Some new methods in geomagnetic field modelling applied to the 1960–1980 epoch. J Geomagn Geoelectr 34:327–349

Lesur V, Jackson A (2000) Exact solution for internally induced magnetization in a shell. Geophys J Int 140:453–459

Lesur V, Wardinski I, Rother M, Mandea M (2008) GRIMM: the GFZ reference internal magnetic model based on vector satellite and observatory data. Geophys J Int 173:382–294

Lowes FJ (1966) Mean-square values on sphere of spherical harmonic vector fields. J Geophys Res 71:2179

Lühr H, Maus S (2006) Direct observation of the F region dynamo currents and the spatial structure of the EEJ by CHAMP. Geophys Res Lett 33:L24102. doi: 10.1029/2006GL028374

Lühr H, Maus S, Rother M (2002) First in-situ observation of night-time *F* region currents with the

CHAMP satellite. Geophys Res Lett 29(10):1489. doi:10.1029/2001 GL 013845

Maus S (2007a) CHAMP magnetic mission. In: Gubbins D, Herrero-Bervera E (eds) Encyclopedia of geomagnetism and paleomagnetism. Springer, Heidelberg

Maus S (2007b) Electromagnetic ocean effects. In: Gubbins D, Herrero-Bervera E (eds) Encyclopedia of geomagnetism and paleomagnetism. Springer, Heidelberg

Maus S, Haak V (2003) Magnetic field annihilators: invisible magnetization at the magnetic equator. Geophys J Int 155:509–513

Maus S, Lühr H (2005) Signature of the quiet-time magnetospheric magnetic field and its electromagnetic induction in the rotating Earth. Geophys J Int 162:755–763

Maus S, Lühr H (2006) A gravity-driven electric current in the earth's ionosphere identified in CHAMP satellite magnetic measurements. Geophys Res Lett 33:L02812. doi:10.1029/2005GL024436

Maus S, Yin F, Lühr H, Manoj C, Rother M, Rauberg J, Michaelis I, Stolle C, Müller R (2008) Resolution of direction of oceanic magnetic lineations by the sixth-generation lithospheric magnetic field model from CHAMP satellite magnetic measurements. Geochem Geophys Geosyst 9(7):Q07021

Merrill R, McFadden P, McElhinny M (1998) The magnetic field of the Earth: paleomagnetism, the core, and the deep mantle. Academic, San Diego

Newitt LR, Barton CE, Bitterly J (1996) Guide for magnetic repeat station surveys. International Association of Geomagnetism and Aeronomy, Boulder

Nimmo F (2007) Energetics of the core. In Treatise on geophysics, vol 8. Elsevier, Amsterdam, pp 31–65

Olsen N (1997a) Ionospheric *F* region currents at middle and low latitudes estimated from Magsat data. J Geophys Res 102(A3):4563–4576

Olsen N (1997b) Geomagnetic tides and related phenomena. In: Wilhelm H, Zürn W, Wenzel HG (eds) Tidal phenomena. Lecture notes in Earth sciences, vol 66. Springer, Berlin

Olsen N (2007a) Ørsted. In: Gubbins D, Herrero-Bervera E (eds) Encyclopedia of geomagnetism and paleomagnetism. Springer, Heidelberg

Olsen N (2007a) Natural sources for electromagnetic induction studies. In: Gubbins D, Herrero-Bervera E (eds) Encyclopedia of geomagnetism and paleomagnetism. Springer, Heidelberg

Olsen N, Lühr H, Sabaka TJ, Mandea M, Rother M, Tøffner-Clausen L, Choi S (2006) CHAOS—a model of Earth's magnetic field derived from CHAMP, Ørsted, and SAC-C magnetic satellite data. Geophys J Int 166:67–75. doi: 10.1111/j.1365-246X.2006.02959.x

Olsen N, Mandea M, Sabaka TJ, Tøffner-Clausen L (2009) CHAOS-2—A geomagnetic field model derived from one decade of continuous satellite data. Geophys J Int 179:1477–1487. doi: doi: 10.1111/j.1365-246X.2009.04386.x

Parkinson WD, Hutton VRS (1989) The electrical conductivity of the earth. In: Jacobs JA (ed) Geomagnetism, vol 3. Academic, London, pp 261–321

Purucker M, Whaler K (2007) Crustal magnetism. In: Kono M (ed) Treatise on geophysics, vol 5. Elsevier, Amsterdam, pp 195–235

Purucker M, Langlais B, Olsen N, Hulot G, Mandea M (2002) The southern edge of cratonic north america: evidence from new stallite magnetometer observations. Geophys Res Lett 29(15):8000. doi: 10.1029/2001GL013645

Purucker ME (2007) Magsat. In: Gubbins D, Herrero-Bervera E (eds) Encyclopedia of geomagnetism and paleomagnetism. Springer, Heidelberg

Rastogi RG (1989) The equatorial electrojet: magnetic and ionospheric effects. In: Jacobs JA (ed) Geomagnetism, vol 3. Academic, London, pp 461–525

Richmond AD (2002) Modeling the geomagnetic perturbations produced by ionospheric currents, above and below the ionosphere. J Geodynamics 33:143–156

Richmond AD (1989) Modeling the ionospheric wind dynamo: a review. In: Campbell WH (ed) Quiet daily geomagnetic fields. Birkhäuser, Basel

Roberts PH (2007) Theory of the geodynamo. In: Treatise on geophysics, vol 8. Elsevier, Amsterdam, pp 67–106

Runcorn SK (1975) On the interpretation of lunar magnetism. Physics of the Earth and Planetary Interior 10:327–335

Sabaka TJ, Olsen N (2006) Enhancing comprehensive inversions using the *Swarm* constellation. Earth Planets Space 58:371–395

Sabaka TJ, Olsen N, Langel RA (2002) A comprehensive model of the quiet-time near-Earth magnetic field: phase 3. Geophys J Int 151:32–68

Sabaka TJ, Olsen N, Purucker ME (2004) Extending comprehensive models of the Earth's magnetic field with Ørsted and CHAMP data. Geophys J Int 159:521–547. doi:10.1111/j.1365-246X.2004.02421.x

Schmucker U (1985) Magnetic and electric fields due to electromagnetic induction by external sources. In Landolt–Börnstein, new-series, 5/2b. Springer, Berlin–Heidelberg, pp 100–125

Thebault E, Hemant K, Hulot G, Olsen N (2009) On the geographical distribution of induced time-varying crustal magnetic fields. Geophys Res Lett 36:L01307. doi: 10.1029/2008GL036416

Thomson AWP, Lesur V (2007) An improved geomagnetic data selection algorithm for global geomagnetic field modelling. Geophys J Int 169 (3):951–963

Turner GM, Rasson JL, Reeves CV (2007) Observation and measurement techniques. In: Kono M (ed) Treatise on geophysics, vol 5. Elsevier, Amsterdam

Tyler RH, Maus S, Lühr H (2003) Satellite observations of magnetic fields due to ocean tidal flow. Science 299:239–241

Voorhies CV, Sabaka TJ, Purucker M (2002) On magnetic spectra of Earth and Mars. J Geophys Res 107(E6):5034

Part 3

Modeling of the System Earth (Geosphere, Cryosphere, Hydrosphere, Atmosphere, Biosphere, Anthroposphere)

6 Classical Physical Geodesy

Helmut Moritz
Institut für Navigation und Satellitengeodäsie, Graz University of Technology, Graz, Austria

W. Freeden, M.Z. Nashed, T. Sonar (Eds.), *Handbook of Geomathematics*, DOI 10.1007/978-3-642-01546-5_6,

Abstract Geodesy can be defined as the science of the figure of the Earth and its gravitational field, as well as their determination. Even though today the figure of the Earth, understood as the visible Earth's surface, can be determined purely geometrically by satellites, using global positioning system (GPS) for the continents and satellite altimetry for the oceans, it would be pretty useless without gravity. One could not even stand upright or walk without being "told" by gravity where the upright direction is. So as soon as one likes to work with the Earth's surface, one does need the gravitational field. (Not to speak of the fact that, without this gravitational field, no satellites could orbit around the Earth.)

To be different from the existing textbooks, a working knowledge of professional mathematics can be taken for granted. In some areas where professors of geodesy are hesitant to enter too deeply, afraid of losing their students, some fundamental problems can be studied.

Of course, there is a brief introduction to terrestrial gravitation as treated in the first few chapters of every textbook of geodesy, such as gravitation and gravity (gravitation plus the centrifugal force of the Earth's rotation), the geoid, and heights above the ellipsoid (now determined directly by GPS) and above the sea level (a surprisingly difficult problem!).

But then, as accuracies rise from 10^{-6} in 1960 (about <10 m globally) to 10^{-8} to 10^{-9} (a few centimeters globally) one has to rethink the fundamentals, and make use of the new powerful measuring devices, not to forget the computers that are able to handle all this stuff.

At the new accuracies, Newtonian mechanics is no longer sufficient. Einstein's General Relativity is needed. Fortunately these "relativistic corrections" are small, and Newtonian mechanics and Euclidean geometry still provide a classical basis to which these corrections can be applied.

Einstein's relativity has put into focus an old ingenious technique of measuring the gravity field, gradiometry, which was invented around 1890 by Roland Eötvös. His torsion balance measured the second-order gradients of the gravitational potential, rather than the three first-order gradients, which form the gravity vector (with the centrifugal forces included). If one wants to measure gravity in a satellite, one would get zero, because the centrifugal force exactly balances gravitation (this is the essence of weightlessness already recognized by Jules Verne). So one has to one step further to measure the second-order gradients, which leads to satellite gradiometry. The newest dedicated satellite mission, launched in 2009, is GOCE, and there is a long way of more than some 120 years to go from Eötvös to GOCE. On the way one has Einstein and then, in 1960, Synge who showed that Eötvös' gradients are nothing else but components of the mysterious Riemann curvature tensor so prominent in General Relativity. Since 1957, of course, this was done in artificial satellites, after not so spectacular results in terrestrial and aerial gradiometry.

Since the Earth's rotation was, and still is, a fundamental measure of time, and the Earth is not rotating uniformly due to tidal effects, it is not surprising that geodesists became involved in precise time measurements. Time, however, is also affected by gravity according to Einstein.

Immediately after the Sputnik of 1957, satellites were used to measure the global features of the external geopotential and to bring it down to the earth by "downward continuation," analytical continuation of the harmonic potential, at least to the Earth's surface, but still better, down to the geoid, to sea level.

The old problem of the geodesist, to "reduce" their data to sea level, is not solvable exactly because the density of the masses above the geoid is not known to sufficient accuracy. If it were, then one could apply the classical boundary-value problems formulas by Stokes in 1849 and Neumann in 1887 (the latter is particularly appropriate in the GPS era).

In 1945, the Russian geodesist and geophysicist M.S. Molodensky devised a highly ingenious and absolutely novel approach to overcome this problem. His idea was to forget about the geoid and to directly determine the Earth's surface. Only, the boundary-value problem becomes much more difficult! Using the language of modern mathematics it is a "hard" problem of nonlinear functional analysis. Its existence and uniqueness was first shown on the basis of Krarup's exact linearization by the well-known mathematician Lars Hörmander in 1976, but with presupposing a considerable amount of smoothing the topography.

However, Molodensky and several others found approximate solutions, which seem to be practically sufficient and did not require the rock density. One of the best solutions, found and rejected by Molodensky and rediscovered by several others, uses again analytical continuation!

Still, one cannot get rid of the rock density altogether in a very practical engineering problem: tunnel surveying. Here one is inside the rock masses and GPS cannot be used. If these masses are disregarded, GPS and ISS (inertial survey systems) may have an unpleasant encounter at the ends of the tunnel.

A well-known practical and theoretical tool is the use of series of spherical harmonics, both for satellite determination of the gravitational field and for the study of analytical continuation. Harmonic functions are a three-dimensional analogue to complex functions in the plane, for which a well-known approximation theorem by Runge guarantees, loosely speaking, analytical continuability to any desired accuracy, as pointed out by Krarup. This chapter contains a comprehensive review of this problem.

Since relativistic effects and analytical continuation are not easily found in books on geodesy, they are relatively broadly treated here.

A method of data combination for arbitrary data to determine the geopotential in three-dimensional space is the least-squares collocation developed as an extension of least-squares gravity interpolation together with least-squares adjustment by Krarup and others. As is extensively used and well documented, a brief account will be given here.

Open current problems such as an adequate treatment of ellipticity of the reference ellipsoid (already studied by Molodensky!), nonrigidity of the Earth, and relevant inverse problems are pointed out finally.

1 Introduction

1.1 Preliminary Remarks

This chapter presupposes a normal knowledge of modern technical mathematics, including some elementary nonlinear function analysis for ➋ Sect. 4.1.1, as well as some notions of General Relativity. As basic references one can use Hofmann-Wellenhof and Moritz (2005) and Moritz (1980) (reprint available from Ohio State University Press, c/o Christopher Jekeli, jekeli.1@osu.edu). Articles or other references are mainly quoted only if they are not contained in one of these books. However, this chapter should be self-contained.

The emphasis is on contemporary problems and techniques. New research not adequately contained in the two references above is emphasized, such as application of the *General Theory of Relativity* to gradiometry, and the treatment of geodetic boundary problems as *"hard" inverse problems* of modern nonlinear functional analysis (Hörmander 1976), as well as Krarup's density theorem, which should show that in practical computation, convergence and divergence of harmonic functions are as meaningless as the question whether a given measured number is rational or irrational.

1.2 What Is Geodesy?

Geodesy is considered to be the science of the figure of the Earth and its gravitational field, according to Bruns (1878) and Helmert (1884). This holds for the Earth's figure, both in its basic general form, which is a flattened rotational ellipsoid, and in its local features caused by gravitational attraction of irregular local masses. Thus geometric and physical aspects are inextricably interrelated. If the gravitational field elements are measured on the Earth's surface, then one can speak of (classical) physical geodesy (in German: Erdmessung); equally important is the external gravitational field at satellite altitudes, which defines and is determined by the orbits of artificial satellites. Also, here one has geometric aspects, expressed par excellence by GPS, and physical aspects, expressed by various gravimetric satellites such as GRACE or GOCE. It is therefore appropriate to speak of satellite geodesy.

To repeat: no satellite data, also with GPS, is purely geometrical, and all satellite orbits are affected by gravitational irregularities. In fact, geodetic satellites are the most powerful tools for determining the Earth's surface and its external gravity field, such as mapping the Earth and oceans.

Physical geodesy is mainly concerned with the theory of gravitational field, and this is the way in which one understands geodesy in this chapter.

So far, the Earth has been considered as a rigid body. This is permissible to a relative accuracy of 10^{-6}. At present one may obtain relative accuracies of 10^{-8} to 10^{-9}. This requires taking into account the effects of Special and General Theory of Relativity, as well as the effects of a nonrigid Earth, such as tides and irregularities in the Earth's rotation.

1.3 Reference Systems

First one must define the coordinate or reference system one will be using. This fundamental Earth-fixed rectangular coordinate system xyz is usually defined in the following way. The origin is the Earth's center of mass (the *geocenter*); the z-axis coincides with the (*mean*) axis of rotation; the x- and y-axes lie in a plane normal to the z-axis and pass through the geocenter (the *equatorial plane*); the x-axis passes through the (*mean*) meridian of Greenwich, and the y-axis is normal to the xz-plane and directed such that the xyz-system is right-handed. It is thus an "Earth-fixed coordiate system."

Since the Earth is not rigid, the expression "mean" is conventional, and thus the rigid system xyz (Moritz and Mueller 1987) is also conventional but well defined. The elastic effects must be disregarded here and this book must be referred for details.

2 Basic Principles

2.1 Gravitational Potential and Gravity Field

The gravitational potential V may be expressed by the formula

$$V(P) = V(x, y, z) = G \iiint \frac{1}{\ell} \rho(Q) \mathrm{d}v(Q). \tag{1}$$

Here P is a given point having coordinates (x, y, z) and usually lying outside the volume V of the Earth, over which the volume integral is extended. The point Q *is* the center of any element dv of volume v, and ℓ denotes the distance between P and Q. G is the Newtonian gravitational constant, whose numerical value is irrelevant in the context of this chapter and will usually be set equal to 1. The integral is to be extended over the whole Earth.

The gravitational potential is theoretically of fundamental importance, but it cannot be evaluated directly because the distribution of density ρ inside the Earth is unknown (see also Chaps. 8 and 32 of this work for more information).

If one assigns the point Q the coordinates (ξ, η, ζ), then the distance ℓ may be expressed by

$$\ell^2 = (x-\xi)^2 + (y-\eta)^2 + (z-\zeta)^2.$$

For large distances ℓ, Eq. 1 may be written as

$$V = \frac{GM}{r} + O\left(\frac{1}{r^2}\right) \quad \text{as} \quad r \to \infty, \tag{2}$$

where $r^2 = x^2 + y^2 + z^2$. Here, r denotes the distance of P from the origin and is not to be confused with ℓ, M is the total mass of the Earth, and $O(\,)$ denotes terms of order r^{-2} or smaller. The physical sense of this equation is that, at large distances and approximately, any body acts gravitationally as a point mass.

The *gravity potential* W is the sum of V and the *potential of centrifugal force*,

$$V_c = \frac{1}{2}\omega^2(x^2 + y^2), \tag{3}$$

which is a formula well known from elementary mechanics, ω being the angular velocity of the Earth's rotation, regarded as a constant. (Check: centrifugal force = grad V_c). Thus,

$$W(x,y,z) = V(x,y,z) + V_c(x,y,z) \tag{4}$$

is the gravity potential and its *gradient*

$$\boldsymbol{g} = \text{grad } W = (W_x, W_y, W_z) = \left(\frac{\partial W}{\partial x}, \frac{\partial W}{\partial y}, \frac{\partial W}{\partial z}\right) \tag{5}$$

denotes the *gravity vector*. Its norm $|\boldsymbol{g}| = g$ is *gravity*. The field with potential V is the *gravitational field*, and the field with potential W is the *gravity field*.

The surfaces W = const are *equipotential surfaces* or *level surfaces*. Their orthogonal trajectories are the (slightly curved) *plumb lines*, and the gravity vector $\boldsymbol{g}$ is tangent to a plumb line since it is normal to the level surfaces.

The unit vector of $\boldsymbol{g}$, $\boldsymbol{n} = -\boldsymbol{g}/g$, may be expressed by *geographical coordinates* Φ and Λ (*astronomical latitude and longitude*) defining the direction of the unit vector of the plumb line $\boldsymbol{n}$

$$\boldsymbol{n} = \frac{-\boldsymbol{g}}{g} = (\cos\Phi\cos\Lambda, \cos\Phi\sin\Lambda, \sin\Phi). \tag{6}$$

The minus sign indicates that the normal $\boldsymbol{n}$ is directed upward and the gravity vector $\boldsymbol{g}$ points downward.

2.2 The Normal Field

The *geoid* is a level surface $W = W_0$ defined in such a way that the sea surface (idealized in some way) is part of it. It can be approximated by a *reference ellipsoid* $U = U_0$. The real gravity field

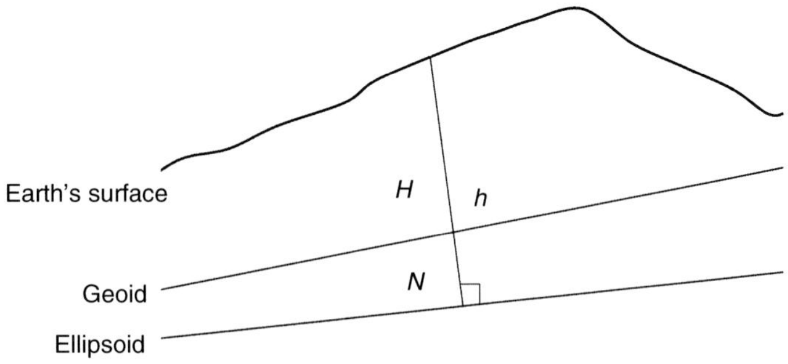

Fig. 1
Various surfaces and heights

is thus referred to as an ellipsoidal field, with the anomalous gravity field being the difference between these two fields, in the sense of *linearization*

	Real Field	Normal Field	Anomalous Field
Basic surface	Geoid $W = W_o$	Ellipsoid $U = U_o$	Deviation of the two surfaces
Potential	Gravity potential W	Normal potential U	Perturbing potential $T = W - U$
Gravity	Gravity $\mathbf{g}$, g	Normal gravity $\mathbf{Y}$, γ,	Gravity anomaly $\Delta g = g - Y$
Plumb line	Φ, Λ	φ, λ	Deflections of the vertical: $\xi = \Phi - \varphi, \eta = (\Lambda - \lambda) \cos\varphi$
Heights	Orthometric height H	Ellipsoidal height h	Geoidal height $N = h - H$

These definitions and symbols are largely self-explanatory; see also Eq. 1 and *Fig. 1.*

The reference ellipsoid is an ellipsoid of revolution of semi-axes a and b. All quantities of the normal field can be rigorously computed once the four constants $a, b, U_o = W_o, \omega$ have been adopted (cf. Hofmann-Wellenhof and Moritz 2005, sect. 2.11).

The quantities of the reference field thus are given. The quantities of the real field are interrelated by difficult *nonlinear* equations, but *linearization* gives the quantities of the anomalous field, which are related by linear equations that are manageable. Linearization will be essential throughout this chapter, particularly in Sect. 4.

2.3 The Geoid and Height Systems

The geoid has just been defined as an equipotential surface $W(x, y, z) = W_o$ = const. at sea level. The classical definition of height above sea level gives the *orthometric height H* measured along the plumb lines defined in Sect. 2.1. (The fact that the plumb lines are slightly curved is irrelevant in this context.) By GPS one gets the *ellipsoidal height h*, measured along the normal to the reference ellipsoid.

If the geoidal heights are denoted by N, then

$$h = H + N. \tag{7}$$

(This relation is not rigorous but very accurate because the plumb line and the ellipsoidal normal almost coincide.)

Thus ➋ *Fig. 1* and Eq. 7 give a simple but fundamental relation between the two kinds of height systems: the orthometric height H above sea level and the ellipsoidal height h above the ellipsoid. The classic way of determining sea-level heights H is by spirit leveling, which is very time consuming. The ellipsoidal height h is determined rapidly by GPS, so the orthometric height H is found by

$$H = h - N \tag{8}$$

provided the geoidal height N is available, a goal, which is being rapidly attained in many countries.

2.4 Gravity Gradients and General Relativity

2.4.1 Separability of Gravitation and Inertia

The centrifugal force, due to the Earth's rotation, is inseparably superimposed on gravitational attraction to form the force of gravity. This is expressed by Eq. 4, $W(x, y, z) = V(x, y, z) + V_c(x, y, z)$, where W is the gravity potential, V is the gravitational potential, and V_c is the potential of centrifugal force.

If one looks at the reason for this inseparability, one finds it in the equivalence between gravitational mass, entering in Newton's law of universal attraction, and inertial mass, entering in Newton's law of motion: mass times acceleration equals force. In fact, classical mechanics disdains centrifugal force as a pseudoforce due to the reference system not being inertial, and refuses to consider it as a real force. It is a so-called inertial force, and thus is related to inertial mass. Within the frame of classical mechanics, the equivalence between gravitational and inertial mass is a fact but remains a mystery. There is no stronger reason for the equivalence of the two than, say, for an equivalence of electric charge and inertial mass, which clearly does not hold. Still, the mass equivalence was confirmed by all mechanical experiments. Already in 1889, Roland Eötvös in Hungary had shown experimental equivalence to a few parts in 10^9, and in 1964, R.H. Dicke and associates at Princeton reduced this number to less than one part in 10^{11}. The explanation for the equivalence of gravitation and inertia had to wait until Einstein. In his General Theory of Relativity he showed that gravitational and inertial forces are intrinsically identical, both being due to the geometrical properties of space-time *and* of the reference system used. All reference systems are equally admissible, and as a result of this democracy, there is no theoretical basis for distinguishing gravitation and inertia: both are equally real (or equally apparent) forces.

Geodesists have had no motive to worry about this situation as long as they used only measuring instruments at rest with respect to the Earth. What counted for them was gravity, the combined effect of gravitation and centrifugal force. They had no reason to be excessively interested in the separation of the two effects; for them gravitation and inertia were inseparable friends. (Actually, in geodetic slang, speaking of "geopotential" one frequently does not distinguish between "gravity" and "gravitation"; the exact meaning is clear from the context, since the difference is the obvious and user-friendly centrifugal force.)

2.4.2 Separability in First-Order Gradients

A different situation arose first in aerial gravimetry (gravity measurement) in the early 1960s (It is true that marine gravimetry had already presented essentially the same theoretical problems before, but it did so in a much more harmless way since ship motion is less irregular.) There occurred inertial forces of a much different nature than the friendly predictable centrifugal force of Earth's rotation. They are due to the aircraft motion, which exerts a great and unpredictable effect on aerial gravity measurements. It was tried to separate gravitation and inertia by statistical filtering, but the results were not satisfactory, and aerial gravimetry was put on the shelf at that time. Gravitation and inertia had become inseparable enemies.

Now, one also has a somewhat similar situation: in inertial surveying one should like to separate the inertial "signal" from the gravitational "noise," just as in aerial gravimetry one should have liked to separate the gravitational "signal" from the inertial "noise." One person's signal is the other person's noise!

2.4.3 Separability in Second-Order Gradients

Around 1965 the author tried to do some theoretical work on aerial gravimetry. When he talked to a physicist about this, the physicist declared solemnly that a separation of gravitation and inertia was impossible since "Einstein had proved this." In fact, in the beginning, relativists were so enchanted by the recently discovered geometric nature of gravitation and the equivalence of all systems, so well corresponding to an ideal democracy, to a classless society, that similar statements can still be found in most texts on general relativity.

But no democracy is perfect, not even in physics. The Soviet physicist V. Fock (1959) tried to show, with convincing arguments, that some reference systems are more equal than others. Even in general relativity, one could introduce a privileged class of coordinate systems, something like inertial systems of classical physics and special relativity, which would provide a theoretical basis for a meaningful, though somewhat conventional, separation between gravitation and inertia.

But the decisive moment for the author came when he hit on the book on general relativity by J. L. Synge (1960). The author tries to explain the basic argument, in a simple way.

Consider the line element ds on a curved surface. It was first discovered by C. F. Gauss. In the modern tensor notation, it is

$$ds^2 = g_{ij}dx_i dx_j. \tag{9}$$

In Eq. 9, the summation over i and j from 1 to n is implied as usual, where n denotes the dimension of the manifold under consideration; in the case of the surface, of course, one has $n = 2$.

The same form (Eq. 9) holds for the line element in the plane referred to curvilinear coordinates, e.g., polar coordinates. How is it possible to tell whether a given line element (Eq. 9) refers to a curved surface or to a plane? The answer was also given already by Gauss:

Form a certain combination of the g_{ij} and of its first and second derivatives, called the *Gaussian curvature K*.

If $K \neq 0$, then one has a genuine curved surface; but if $K = 0$, then the surface is a plane (or developable on a plane), and by a simple coordinate transformation one may reduce Eq. 9 to the Cartesian form

$$ds^2 = dx^2 + dy^2. \tag{10}$$

Now, according to the general theory of relativity, space-time represents a four-dimensional manifold, whose line element is again given by Eq. 9, but now with $n = 4$, of course. Gravitation is nothing else but a manifestation of space-time curvature.

For a flat space-time without curvature, the special theory of relativity holds and inertial systems, corresponding to rectangular or Cartesian coordinates in the plane, exist. Then, one has

$$ds^2 = dx_1^{\ 2} + dx_2^{\ 2} + dx_3^{\ 2} + dx_4^{\ 2}, \tag{11}$$

where x_1, x_2, x_3 are rectangular spatial coordinates and $x_4 = ct\sqrt{-1}$ is related to time t via light velocity c. Clearly, (11) is a four-dimensional analogue of (10). (This hides a minus sign, which makes the metric nonpositive definite, a kind of "*diabolus in physica*.")

Also in flat space-time, in the absence of gravitational field, noninertial reference systems exist, for which the general form (Eq. 9), with $n = 4$, holds rather than (11). They correspond to rotating and/or accelerated reference systems. Any system of three axes rigidly attached to a moving automobile or an airplane, but also any terrestrial reference frame, is such a noninertial system, in which "inertial forces" such as centrifugal, Coriolis or Euler forces occur.

But for any system in *flat* space-time, whether it is inertial or not, the four-dimensional equivalent of the Gaussian curvature K, the *Riemann curvature tensor* R_{ijkl} must vanish. Do not be scared by this quantity: like the Gaussian curvature, it consists of g_{ik} together with their first and second derivatives. Thus,

$$R_{ijkl} = 0 \tag{12}$$

in the absence of a true gravitational field, no matter how strong inertial forces occur. What happens if $R_{ijkl} \neq 0$? This means a nonvanishing curvature of space-time, or in Einstein's terms, a true gravitational field. Thus, J. L. Synge (1960) was justified to write

$$R_{ijkl} = \text{Gravitational field.} \tag{13}$$

But this means that even within the frame of general relativity it is possible to distinguish inertial and gravitational forces, namely, through the Riemann tensor!

As it has been said above, even in general relativity there are privileged coordinate systems, equivalent to inertial systems. In such a system one may write to a good approximation (Moritz and Hofmann-Wellenhof 1993, p. 226):

$$R_{i4j4} \sim \frac{\partial^2 V}{\partial x_i \partial x_j}, \tag{14}$$

restricting i and j to spatial indices 1, 2, 3; as usual, 4 denotes the time-like index. Thus, certain important components of the Riemann tensor are simply proportional to second derivatives, or *(second) gradients*, of the gravitational potential V!

This furnishes a fundamental theoretical reason for the importance of gradiometry, which measures just these second derivatives. And one should not forget that it was Roland Eötvös, who in 1887 has constructed the first gradiometer, which he called *torsion balance*.

Synge (1960) was well aware of the geodetic importance of these facts: in passing, he mentioned a "*relativistically valid geodetic survey*" without, however, going into details.

Soon the author took up the subject under the heading "*kinematical geodesy*" (Moritz 1967). Now he was not too happy about this name: as there are already too many "geodesies" around: three-dimensional geodesy, physical geodesy, satellite geodesy, theoretical geodesy, and the like. But the damage has been done, and the term is being used, fortunately not too often. Taken as such, the term "kinematical geodesy" seems to be rather appropriate, though. It denotes the geodetic use of data obtained by instruments in motion; furthermore, general relativity reduces the dynamics of gravitation to kinematics as the *geometry* of the space-time continuum.

To continue with the story, the well-known relativity book *Gravitation* by C.W. Misner, K.S. Thorne and J.A. Wheeler (Misner et al. 1973) even describes with considerable respect the gradiometer designed by R.L. Forward. So the relation between the Riemann curvature tensor, the second-order gradients, and the separability of gravitation and inertia is now a much less exotic subject than it was some 50 years ago.

The situation may be summarized in a quite simple way. Separation is indeed impossible if the force vector is considered *at one point* only. Instead, this vector must be given in an (arbitrarily small) *three-dimensional region*. Then a spatial differentiation can be performed, and the derivatives are just the second-order gradients. A purely inertial field, being due to the translation and/or rotation of a rigid frame (whether it is Einstein's famous elevator or an airplane) *has a relatively simple structure*, and this very structural simplicity of the field is expressed by the vanishing Riemann tensor (12). Truly gravitational fields have an intrinsically more complicated structure, and the Riemann curvature tensor no longer vanishes.

So a *separation between gravitation and inertia is indeed possible*, but only with *second* derivatives of the potential, and not with the force vector consisting of *first* derivatives. This is one of the reasons for the importance of *gradiometry*, the measurement of second derivatives (Eq. 14) of the potential V.

2.4.4 Satellite Orbits

Since the results of General Relativity slightly differ from the results of classical mechanics, also the orbits of artificial satellites in Einstein's mechanics will differ from those of Newtonian theory.

Since these differences are very small, they are best be taken into account by small "relativistic corrections" to the values computed by classical mechanics. As these relativistic corrections are very small, one may approximate the Earth by a sphere and further by a mass point.

So one needs a particular metric for the space-time of General Relativity *having spherical symmetry*. Such a solution was found by Karl Schwarzschild already in 1916, in the same year in which Einstein published his general relativity.

It has a rather simple form

$$\mathrm{d}s^2 = -\left(1-\frac{2m}{r}\right)c^2\mathrm{d}t^2 + \left(1-\frac{2m}{r}\right)^{-1}\mathrm{d}r^2 + r^2(\mathrm{d}\vartheta^2 + \sin^2\vartheta\,\mathrm{d}\lambda^2). \tag{15}$$

As a check, m representing the central point mass, for $m \to 0$, the Schwarzschild solution becomes

$$\begin{aligned}\mathrm{d}s^2 &= -c^2\mathrm{d}t^2 + \mathrm{d}r^2 + r^2(\mathrm{d}^2\vartheta + \sin^2\vartheta\,\mathrm{d}\lambda^2)\\ &= -c^2\mathrm{d}t^2 + (\mathrm{d}x^2 + \mathrm{d}y^2 + \mathrm{d}z^2),\end{aligned} \tag{16}$$

and one is back at the flat Minkowski space-time of special relativity, as it should be. One also recognizes r, ϑ, λ as spatial spherical coordinates.

(Many people nowadays are interested in what happens if $r \to 0$, at the so-called Schwarzschild singularity, more popularly known as a *black hole*. The interested reader can learn more about these mysterious black holes by surfing the Internet. So far it seems that black holes are not needed in physical geodesy.)

One is thus interested in the space outside the Earth, hence in the *outside* of any possible black hole. In this external gravitational field of a point mass, the equations of motion of a satellite in this exterior Schwarzschild field can be found as described in Moritz and Hofmann-Wellenhof (1993, p. 234–241).

The result is a Kepler ellipse with a very small perihelion precession

$$r = \frac{p}{1 + e\cos\left[(1 - 3m/p)\lambda\right]}, \tag{17}$$

where the relativistic correction $3m/p$ depends on the Schwarzschild mass m in Eq. 15. *For practical formulas, useful for satellite geodesy cf. Moritz and Hofmann-Wellenhof (1993).*

2.4.5 Applications

The second-order gradients are important to geodesy for several reasons:

- *Terrestrial gradiometry.* The *Eötvös torsion balance* was of great theoretical importance but, because the second-order gradients are very irregular on the Earth's surface and it is slow in application, the torsion balance is hardly used any more.
- *Aerial gradiometry.* As seen in ❯ Sects. 2.4.2 and ❯ 2.4.3, the inertial noise is very inconvenient in the first-order gradients $\partial V/\partial x_i$ but that it vanishes automatically in the second-order gradients by Eqs. 13 and 14, which provide gravitational second-order gradients $\partial^2 V/\partial x_i \partial x_j$, unaffected by inertial effects. (The reader will have noticed that in geodesy, "gradients" are usually used for the "second-order gradients"). They form a square matrix, which is frequently called "Marussi tensor" (Marussi 1985).
- *Satellite gradiometry.* This is by far the most important application. *The first-order gradient vector $\partial W/\partial x_i$ is zero* because, for a satellite, the gravitational force exactly balances the inertial force. (This causes the well-known weightlessness inside a satellite!) There remains the Marussi tensor of second-order gradients, which is purely gravitational by Eqs. 13 and 14. The second-order gradients are more irregular gravitational signals than the first gradients are; therefore, they are better for describing the local gravitational irregularities. They are not affected by irregular inertial noise, so that they are ideal data for gravity satellite missions, of which the most recent is *GOCE* (Rummel et al. 2002). See also the contribution of Roland Pail in Hofmann-Wellenhof and Moritz (2005).
- *Other applications.* In ❯ Sect. 2.4.4, relativistic corrections of satellite orbits are mentioned, and in the following section relativistic effects in exact time definition will be discussed.

2.4.6 Gravitation and Time

The relation between Newton's gravitation and Einstein's gravitation is, in a first but for these purposes completely sufficient approximation, given by

$$ds^2 = dx^2 + dy^2 + dz^2 - \left(\frac{1 - 2V}{c^2}\right) c^2 dt^2. \tag{18}$$

This is nothing else but Minkowski's formula (16), with a factor $(1 - 2V/c^2)$, which is small (divided by the square of the light velocity c, which is fantastically large!) but of basic importance. Comparison with Eq. 9 shows that

$$g_{44} = c^2\left(\frac{1-2V}{c^2}\right), \quad \text{with} \quad x_4 = t \tag{19}$$

is the 44-component of the metric tensor g_{ij}, the good friend from Eq. 9, which, of course, now is a 4×4 matrix. So, the (Newtonian!) gravitational potential V appears again in Einstein's theory. Thus, the gravitational potential V enters General Relativity directly into the metric tensor, primarily into the component g_{44}, but it influences also the other g_{ij} although in a way far below the present measuring accuracies. (Only accuracies of about 1 in 100 million are considered here.)

If a *clock is at rest with respect to its environment*, then the velocity components $dx/dt = dy/dt = dz/dt = 0$, and the clock shows "proper time" τ. Hence the line element ds (Eq. 18) reduces to

$$ds_0 = i\sqrt{g_{44}}\,dt = i\sqrt{g_{44}}\,d\tau \quad \text{with "proper time" } \tau = t. \tag{20}$$

In fact, the line element can be used with "signature" $+++-$, as in Eq. 18 and generally so far, if emphasis was on space. There is no reason why one should not use the alternative signature $---+$ if emphasis is on time t. Thus one can apply the "proper time"

$$d\tau = i\,c^{-1} ds_0\,, \tag{21}$$

using the factor c to get the dimension right (time vs. length). Then one gets from (19)

$$d\tau^2 = -c^{-2} ds_0{}^2 = -c^{-2} g_{44}\,dt^2\,. \tag{22}$$

For a stationary object (no motion with respect to the environment) one can thus have

$$d\tau = c^{-1}\sqrt{g_{44}}dt = \left(\frac{1-V}{c^2}\right) dt. \tag{23}$$

Einstein has called τ "Eigenzeit" or "*proper time*" because it is the *local* time measured by a very precise clock at rest in its *proper* environment. The elementary fact that $\sqrt{1-2\varepsilon} \doteq 1-\varepsilon$ has been also used.

The formula can be generalized in a very simple and evident way. First, a clock at rest on the rotating Earth's surface is subject to the centrifugal force, which has a potential, in the usual geocentric coordinate system

$$V_c = \frac{1}{2}\omega^2(x^2 + y^2), \tag{24}$$

where ω is the angular velocity of the Earth's rotation. The total *gravity potential* is then $W = V + V_c$, familiar from ❯ Sect. 2.1. Thus, the generalization of (23) to the case of a rotating Earth is straightforward, replacing V by W (the clock is affected by total W rather than by V only)

$$d\tau = \left(\frac{1-W}{c^2}\right) dt. \tag{25}$$

This formula, however, permits another generalization, which is perfectly natural, although surprising for a moment. The *total energy* E in mechanics is E = potential energy + kinetic energy = $U + T$. The potential energy is simply the potential W acting on a unit mass $m = 1$. The kinetic energy T (also for $m = 1$)

$$T = \frac{mv^2}{2} \tag{26}$$

of a measuring system moving with velocity v can be "smuggled" additively into (17) to give

$$\frac{d\tau}{dt} = 1 - \frac{W}{c^2} - \frac{v^2}{2c^2}, \tag{27}$$

which beautifully shows the effects of the general theory of relativity (W) and the special theory of relativity (v/c).

Coordinate time. So far, the emphasis has been mainly on the *proper time.* It is measurable by an atomic clock, which depends on the local frequency and hence on the location of the observation station. The *coordinate time t* has no physical significance but it is a uniform time unaffected by gravitation and motion. Therefore, it is useful as a time standard.

Using the well-known approximation $(1-\varepsilon)^{-1} = 1+\varepsilon$ one gets

$$\frac{dt}{d\tau} = 1 + \frac{W}{c^2} + \frac{v^2}{2c^2} = 1 + \frac{V}{c^2} + \frac{V_c}{c^2} + \frac{v^2}{2c^2}. \tag{28}$$

This formula is fundamental for time synchronization by transportable clocks and for the comparison of satellite and station clocks. It very clearly shows the symmetric dependence on gravitational potential V, on the potential of the centrifugal force V_c, and on the potential corresponding on the local movement with velocity v.

It is instructive to interpret the elements on the second part of Eq. 28. The "1" is classical, W/c^2 is general-relativistic (gravity!), and $v^2/2c^2$ is special-relativistic (velocity!).

The standard precise time systems take relativistic corrections into account (Moritz and Mueller 1987, p. 392; Moritz and Hofmann-Wellenhof 1993, Chap. 5).

3 Key Issues of Theory: Harmonicity, Analytical Continuation, and Convergence Problems

3.1 Harmonic Functions and Spherical Harmonics

The gravitational potential V outside the Earth, essential for the computation of satellite orbits is now dealt with. There V is well known to be a *harmonic function* satisfying *Laplace's equation*

$$\Delta V = 0, \tag{29}$$

where Δ is the Laplace operator

$$\Delta = \frac{\partial^2}{\partial x^2} + \frac{\partial^2}{\partial y^2} + \frac{\partial^2}{\partial z^2}. \tag{30}$$

Transforming (29) to spherical coordinates r, ϑ, λ one obtain a linear partial differential equation $\Delta F = 0$, which one tries to solve by the trial substitution

$$F(r,\vartheta,\lambda) = f(r)g(\vartheta)h(\lambda). \tag{31}$$

The result is

$$\begin{aligned} f(r) &= r^n \quad \text{or} \quad 1/r^{n+1}, \\ g(\vartheta) &= P_{nm}(\cos\vartheta), \\ h(\lambda) &= \cos\lambda \quad \text{or} \quad \sin\lambda, \end{aligned} \tag{32}$$

where

$$n = 0, 1, 2, 3, 4, \ldots, \infty$$
$$m = 0, 1, \ldots, n \tag{33}$$

n is called the *degree* and m is the *order* of the function under consideration.

The functions $f(r)$ and $h(\lambda)$ are simple. The functions P_{nm} are less elementary. They are called *Legendre functions* and defined by

$$P_{nm}(t) = \frac{1}{2^n\, n!}(1-t^2)^{m/2}\frac{\mathrm{d}^{n+m}}{\mathrm{d}t^{n+m}}(t^2-1)^m \quad \text{with } t = \cos\vartheta. \tag{34}$$

The infinite series may be formed with coefficients A_{nm} and B_{nm}

$$V(r,\vartheta,\lambda) = \sum_{n=0}^{\infty}\sum_{m=0}^{n}\frac{P_{nm}(\cos\vartheta)}{r^{n+1}}(A_{nm}\cos m\lambda + B_{nm}\sin m\lambda), \tag{35}$$

which converges *outside a certain sphere* σ and satisfies $\Delta V = 0$ there. This is why the elements of series (35) are called spherical harmonics. They play a basic role in physical geodesy, both theoretically and for numerical computation.

The "external convergence sphere" σ_E is the smallest sphere (with origin at the geocenter O), which contains the whole Earth in its interior. The series (35) definitely cannot be used *inside* the Earth because there it satisfies, rather than Laplace's equation, *Poisson's equation*

$$\Delta V = -4\pi G\rho, \tag{36}$$

which is a simple consequence of the definition Eq. 1 of V and reduces to the Laplace equation (29) only for $\rho = 0$. The symbol ρ denotes the mass density at the "computation point" $P(x, y, z)$ at which V, and therefore ΔV, refers. *The potential V is not harmonic inside the Earth !*

3.2 Convergence and Analytical Continuation

Since spherical-harmonic series are the standard mathematical tool for representing the Earth's gravitational field as determined from satellite methods, their convergence has been thoroughly investigated.

A first fact is that the spherical harmonic series (35) converges *outside* a sphere σ_E, the "external convergence sphere," defined in ➋ Sect. 3.1. This is a nice result, easy to prove (Kellogg 1929, p. 142), but it is not sufficient for geodetic purposes because the Earth's surface lies completely *inside* this sphere, only touching it at one point. So one does not know its convergence behavior on the Earth's surface. Anyway it cannot apparently be used on the Earth's surface S because the geopotential V inside the Earth is not harmonic.

Here one may remember from complex function theory that an analytic function f in two dimensions also satisfies Laplace's equation $\Delta f = 0$ and is therefore harmonic (in two dimensions). There the "real convergence circle" may be smaller than the "external convergence circle." This is the case if the *analytical continuation* of f exists and is regular. The circle of convergence is the largest circle outside of which there are no singularities; it may be smaller than what is called the "external convergence circle," which is defined in ➋ Sect. 3.1.

It is tempting to extend this well-known property of analytic harmonic functions from two to three dimensions. This property is being tried to illustrate by ➋ *Fig. 2*. The

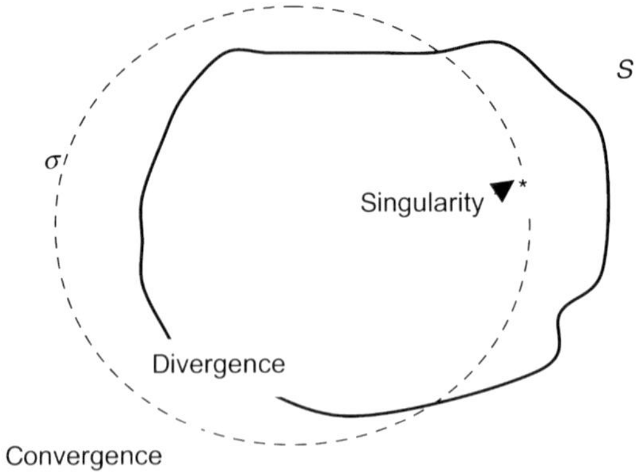

Fig. 2
A singularity separates regions of convergence and divergence

spherical-harmonic series converges outside the sphere of convergence σ, concentric to σ_E, but in general *smaller*, which is important. It diverges inside σ and converges for any point of the Earth's surface S, which lies *above* σ (right side of the figure) and represents the geopotential V there. Unfortunately it diverges on the remaining part of S, which is below σ (left, largely lower and upper part of the figure). (In exceptional cases, however, there may be other surfaces of convergence, see Sect. 3.3). The reader is requested not to confuse the external convergence sphere σ_E with the convergence sphere σ!

Wherever the spherical-harmonic converges, it is harmonic. Inside the mountain on the right-hand side of the picture it represents the "analytical continuation" V_{cont} of the potential into the masses down to the convergence sphere σ. The external potential V together with its analytic continuation V_{cont} forms a *single harmonic function*, everywhere satisfying Laplace equation $\Delta f = 0$. It is clear that, inside the masses, V_{cont} is in general not equal to V because $\Delta V_{\text{cont}} = 0$, whereas $\Delta V = -4\pi G\rho$ there.

It is very important to understand this. It may happen that the only singularity of a harmonic function is the geocenter O, for which $r = 0$. This is the case if the series (35) contains only a finite number of terms. (This one knows to happen, for instance, after the originally infinite series (35) has been truncated after a certain order n). Each spherical-harmonic term has the geocenter $r = 0$ as its only singularity, where this term $\to \infty$. (This is the reason why functions $f(r)$ of type $1/r^{n+1}$ have been used rather than r^n from (32): the functions r^n would have a bad singularity at infinity!) The finite spherical-harmonic sum then also has the origin O as its only singularity; it is regular and harmonic down to the geocenter !

This simple case is by no means far-fetched: all of the spherical-harmonic series determined by satellite methods are truncated for practical reasons. This would close the discussion: forget about convergence and use such truncated series down to the Earth surface (and even deeper if necessary). For these applications, the geopotential is harmonic and regular right down to (but exempting) the geocenter.

It is worthwhile to look more closely into a problem that has been considered one of the apparently greatest mysteries of theoretical geodesy, about which so much nonsense has been written (also by the author; for an amusing history see Moritz (1978)). The reader is requested to write about analytical continuation only if he or she understands the situation thoroughly!

Imagine such a well-behaved potential has been determined. Its convergence sphere σ coincides with the geocenter O. Now put a grain of sand, or more mathematically, a point of mass m somewhere on the Earth's surface or hide it in its interior. It has a singularity of type $1/r_{grain}$, where r_{grain} denotes the distance from the sand grain to the origin. The singularity persists however small the mass of the sand grain is: its mass m may be 1 g, 1 mg or 10^{-33} g. It should only be only >0. However, much the modest sand grain tries to be nonobtrusive, it will cause a singularity into the originally regular geopotential model and thus introduce a convergence sphere (➋ *Fig. 2*)!

This shows that the possibility of regular analytic continuation of the geopotential and convergence is a highly complicated and unstable problem. A single sand grain may change convergence into divergence, and think of how many mass points, rocks, molecules, and electrons make up the Earth's body and cause the most fanciful singularities.

A counterexample. This would indicate that for all solid bodies, consisting of many mass points, the analytical continuation of the outer potential into its interior is automatically singular. A counterexample is the homogeneous sphere of radius R. Its external potential is well known to be that of a point mass M situated at its center:

$$V_{\text{ext}} = \frac{GM}{r}, \qquad r > R, \tag{37}$$

whereas the potential inside the Earth is (int = interior)

$$V_{\text{int}} = 2\pi G\rho\left(R^2 - \frac{1}{3}r^2\right), \tag{38}$$

which satisfies Poisson's equation $\Delta V_{\text{int}} = -4\pi G\rho$.

The analytical continuation of V_{ext} is evidently the harmonic function

$$V_{\text{cont}} = \frac{GM}{r}, \qquad 0 < r < R \tag{39}$$

with $V_{\text{cont}} = V_{\text{ext}} = V_{\text{int}}$ at the spherical surface $r = R$.

This counterexample is refuted by the simple remark that solid bodies do not consist of mass points but have continuous density. So the analytical continuation may well exist inside solid bodies.

3.3 More About Convergence

Nobody has understood convergence and analytical continuation better than Krarup did.

Torben Krarup (1919–2005) was a Danish mathematician and geodesist whose impact on geodesy was tremendous. He did not like publishing; however, his booklet (Krarup 1969) became a classic. It is the great merit of Kai Borre to have made is easily accessible by including it, together with other pioneering papers of Krarup, in the book Borre (2006, pp. 29–90).

Is the convergence sphere in three dimensions universal? The circle of convergence for complex analytical functions in two dimensions *is* universal. In three dimensions, the sphere of convergence, *seemed* also universal: convergence outside, divergence inside this sphere (➋ *Fig. 2*). Krarup has given an elegant counterexample where the surface of convergence is a *torus* rather than a sphere (Borre 2006, p. 63).

This seems to be an exception. In most cases, a convergence sphere does seem to apply, so that the use of a convergence sphere in ➋ *Fig. 2* continues to be appropriate. Of eminently

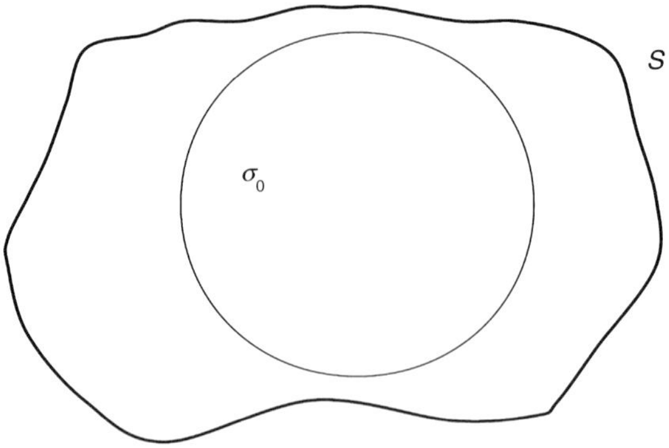

Fig. 3
Earth's surface and Krarup's sphere σ_0

practical value, however, is Krarup's application of a theorem by Runge (see also Chap. 9 and the references therein). One can quote Krarup (Borre 2006, p.67)

- *Runge's theorem*: Given any potential regular outside the surface of the Earth and any sphere in the interior of the Earth. For every closed surface surrounding the Earth (which surface can be arbitrarily near the surface of the Earth) there exists a sequence of potentials regular in the whole space outside the given sphere and uniformly converging to the given potential on and outside the given surface.

Krarup (1969 p. 83) also gave a rigorous proof of this theorem for three dimensions, because Runge had given it only for the two-dimensional case of analytical complex functions.

So to speak, it is the converse of the sand grain case of Sect. 3.2, which indicated that an arbitrarily small grain of sand can convert a harmonic function (potential), regular (for instance) right down to the geocenter (of course, excluding this point), to a potential that converges only to a certain sphere σ_0 (*Fig. 3*). Consider a potential (here used synonymously to the concept of a regular harmonic function) *on the Earth surface's S* that *cannot be analytically continued* into the Earth's body. This is bad, but one can approximate by a potential that *can be so continued*. This can be done to an arbitrary accuracy. In terms of gravity the accuracy may be 1 mgal (milligal), or 0.001 mgal, or less (1 gal = 1 cm s^{-2}).

Casually speaking, a sand grain can change convergence into divergence, and Krarup's theorem can change divergence into convergence.

The reader is requested not to confuse the *convergence sphere* σ (convergence outside, divergence inside; see *Fig. 2*), the "*external convergence sphere*" σ_E, defined in Sect. 3.1, and "*Runge's sphere*" σ_0, which is arbitrary but must lie completely inside the Earth (*Fig. 3*). All three spheres are concentric.

3.4 Krarup's Density Theorem

One may say (mathematical rigor should be added by the reader):

- *Krarup's density theorem*: the set of downward-continuable functions is *dense* in the set of not-continuable harmonic functions, in rather the same way as the set of rational functions is dense in the set of real functions.

For numerical computations one can use only rational numbers because the number of digits used in these computations is always finite. Also, numbers obtained by measurement are also always *rational* because of small measuring errors. Thus, in most of experimental physics, the distinction between rational and irrational numbers is meaningless. Practically all numbers are rational. The author likes to formulate this fact as the

▸ *Pleasant theorem.* The question of convergence or divergence of spherical harmonics in physical geodesy is practically meaningless. The series may always be considered to be convergent.

Thus one can practically always assume that one can continue the external gravitational potential to the inside of the Earth, at least *cum grano salis*, with a grain of salt (or of sand. . .). As remarked before, however, the downward continuation is an improperly posed problem: it may be instable! (In practical cases, the instability might be negligible.)

After Krarup's insightful work, it has turned out that Runge-type theorems, or density theorems, have been found before, for instance Frank and von Mises (1930, pp. 760–762), with references to G. Szegö and to J.L. Walsh.

More details can be found in Krarup (1969) and Moritz (1980); for improperly posed problems one of the earliest sources is Courant and Hilbert (1962, pp. 130–231); see also Anger et al. (1993) or, easiest to find, the paper by Anger and Moritz in http://www.helmut-moritz.at.

4 Key Issues of Applications

4.1 Boundary-Value Problems of Physical Geodesy

4.1.1 Nonlinear Inverse Problems of Functional Analysis

The details of these nonlinear equations, their linearization, and their solutions are rather complex. It is, however, quite easy to understand the general principles, following Hofmann-Wellenhof and Moritz (2005, sect. 8.3). Here the language of modern nonlinear functional analysis tremendously improves the understanding of the structure of the problem.

Recall that the gravity vector $\boldsymbol{g}$ and the gravity potential W are related by Eq. 5. Let S be the Earth's topographic surface (the "physical Earth surface" on which one can walk, drive, and occasionally swim), let now W and $\boldsymbol{g}$ be the gravity potential and the gravity vector *on this surface* S. Then, the gravity vector $\boldsymbol{g}$ on S is a function of the surface S and the geopotential W on S:

$$\boldsymbol{g} = f(S, W). \tag{40}$$

This can be seen in the following way. Let the surface S and the geopotential W on S be given. The gravitational potential V is obtained by subtracting the potential of the centrifugal force, which is simple and well-known (Eq. 4):

$$V = W - V_c. \tag{41}$$

The potential V outside the Earth is a solution of Laplace's equation $\Delta V = 0$ and is consequently *harmonic*. Thus, knowing V *on* S, one can obtain V outside S by solving Dirichlet's well-known boundary-value problem for the exterior of S, the first boundary-value problem of potential theory, which is practically always uniquely solvable at least if V is sufficiently smooth on S. After having found V as a function in space outside S, one can get the gravitational force grad

V on S. Adding the well-known and simple vector of the centrifugal force, one obtains the gravity vector $\boldsymbol{g}$ outside and, by continuity, on S.

This is precisely what Eq. 40 means. The modern general concept of a function can be explained as a rule of computation, indicating that given S and W on S, one can uniquely calculate $\boldsymbol{g}$ on S. Note that f is not a function in the elementary sense but rather a "nonlinear operator," but one can disregard this. (In computer slang, everything the computer does, is an operator! It you add $3+4=7$, the sign "+" is an operator. This is at least psychologically helpful.) Therefore one may formulate the following:

- *Molodensky's boundary-value problem* is the task to determine S, the Earth's surface, if $\boldsymbol{g}$ and W on it are given. Formally, one has to solve for S:
$$S = F_1(\boldsymbol{g}, W), \tag{42}$$
that is, one gets geometry from gravity.
- *GPS boundary-value problem.* Since one has GPS at one's disposal, one can consider S as known, or at least determinable by GPS. In this case, the geometry S is known, and one can solve for W:
$$W = F_2(S, \boldsymbol{g}), \tag{43}$$
that is, one gets potential from gravity. As shall be seen, this is far from being trivial: one has now a method *to replace leveling*, a tedious and time-consuming old-fashioned method, *by GPS leveling*, a fast and modern technique. (An approximate formula for this purpose is (8).)

In spite of all similarities, one should bear in mind a fundamental difference: (43) solves a *fixed-boundary problem* (boundary S given), whereas (42) solves a *free-boundary problem*: the boundary S is a priori unknown ("free"). Fixed-boundary problems are usually simpler than free ones.

This is only the principle of both solutions. The formulation is quite easy to understand intuitively. The direct implementation of these formulas is extremely difficult, however, because that would imply the solution of "hard inverse function theorems" of nonlinear functional analysis. So one can use the usual iterative method, staring from a *linearization* and get successive approximations (Newton iteration) applied for nonlinear inverse or implicit functions (Sternberg 1969, sect. 3.11). This idea was used by the well-known mathematician Hörmander (1976) (using a Nash-type iteration) to provide a rigorous proof of existence and uniqueness of the nonlinear Molodensky problem, to be sure, for unrealistically smooth conditions. A simplified account is found in Moritz (1980, sects. 50 and 51).

Another approach, the gravity space by Fernando Sansò, is also described in this reference (Moritz 1980, sect. 52). This approach turns the free-boundary problem of Molodensky into an easier fixed boundary-value problem in gravity space.

In practice, the linearized problem is sufficient. It may be solved by other series solutions proposed by Molodensky, Brovar, Marych, and others (see Moritz 1980, sects. 43–47). See also Sects. 4.1.2–4.1.5.

The Simplest Possible Example of a Molodensky-Type Problem

Let the boundary surface S be a sphere of radius R. The Earth is represented by this sphere, which is considered homogeneous and nonrotating. The potential W is then identical to the gravitational potential V, so that on the surface S one can have constant values:

$$W = \frac{GM}{R}, \quad g = \frac{GM}{R^2}. \tag{44}$$

Knowing W and g, one can have

$$R = \frac{W}{g} \tag{45}$$

the radius of the sphere S. Thus, Molodensky's problem has been solved in this trivial but instructive example. One has indeed got geometry (i.e., R) from physics (i.e., $\mathbf{g}$ and W)!

4.1.2 The Standard Classical Model

The solution of the boundary-value problems shall be now illustrated by simplifying the situation. The model may first appear unrealistic and oversimplified but it underlies all practical solutions: most present practical solutions use it as their basic conceptual structure.

This *standard classical model* (SCM) has the following properties:

- It replaces the reference ellipsoid by a sphere whose radius R is considered known. Its radius has the value $R = 6371$ km, which is conventionally fixed and may be interpreted as a mean radius of the Earth, for instance $R = (a + a + b)/3$ (arithmetic mean) or $R^3 = a^2 b$ (geometric mean) of the semimajor axis a and the semiminor axis b of the reference ellipsoid. It is called the *geosphere.*
- The Earth's surface S is replaced by an equipotential surface $W = W_0 =$ const called *geoid.* It represents the *sea level.* The orthometric heights, heights above sea level, are identically zero.

This model, unrealistic and unfriendly to tourism as it is, is what lies at the base of almost every book and every course on physical geodesy. The real world is taken into account later by ellipsoidal corrections and corrections for height above sea level. This has been working since Stokes in 1849 and will probably work still for some time.

4.1.3 Linearization

It should be first emphasized that a linearization in the sense of ➲ Sect. 2.2 and ➲ 4.1.3 is being dealt with. The following notations are used (see ➲ *Fig. 4*).

P is any point of the geoid, and Q is the corresponding point of the geosphere, separated by the vertical distance N, the *geoidal height.* W is the *gravity potential* or *geopotential* $W = W_0 =$ const in the case of our model *SCM.* $W(Q)$ means W at Q, etc.

Thus one has by the definitions of ➲ Sect. 2.2

$$W_0 = W(P) = U(Q) = U_0 = W_0. \tag{46}$$

This means that the geopotential W is constant by definition of the *geopotential*, whereas the spheropotential U is constant by the definition of the *geosphere*; both constants in (46) are defined to be equal. Gravity is denoted by $\mathbf{g}$ (vector) or g (norm), and normal gravity is denoted by gamma (vector $\boldsymbol{\gamma}$, norm γ).

There is

$$\mathbf{g} = \text{grad}\, W \quad \text{and} \quad \boldsymbol{\gamma} = \text{grad}\, U, \tag{47}$$

where

$$\gamma = -\frac{\partial U}{\partial r} \tag{48}$$

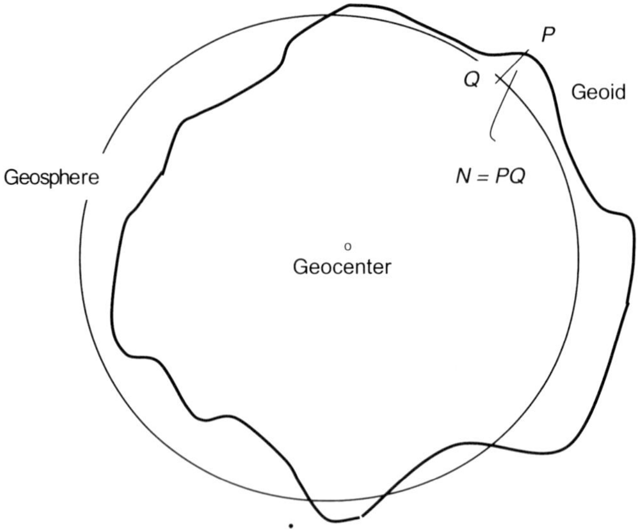

Fig. 4
Geoid and geosphere

is the (negative) radial derivative of the spheropotential U

$$U = \frac{GM}{R} \tag{49}$$

the external potential of a sphere (44), R being the radius of the geosphere.

By Sect. 2.2, the *perturbing potential*

$$T(P) = W(P) - U(P) \tag{50}$$

and the *gravity disturbance* is its (negative) radial derivative (in the spherical approximation)

$$\delta g = -\frac{\partial T}{\partial r} = g(P) - \gamma(P). \tag{51}$$

The classical *gravity anomaly*, however, is defined slightly differently:

$$\Delta g = g(P) - \gamma(Q) \tag{52}$$

for historical reasons. Before the satellite era, the point Q in (52) was basic since the reference ellipsoid (here approximated by the geosphere) was known. Now the geometry of S and hence the point P can be considered known by GPS, so that the input datum is δg rather than Δg.

(*Note*: This spherical approximation does not mean that the reference ellipsoid is replaced, in any geometrical sense, by a sphere. It only means that Taylor series with respect to a small elliptic parameter are used, and only the first term is retained, which formally is spherical. For present accuracies, higher ellipsoidal terms must, of course, be taken into account. This is done for instance in the computation of normal gravity by closed ellipsoidal formulas, and also known ellipsoidal corrections should be added to Stokes formula if a "cm-geoid" is to be computed; cf. Moritz (1980, sect. 39).)

One can further have the following as the first term of Taylor expansions:

$$\gamma(P) = \gamma(Q) + \frac{\partial\gamma}{\partial r}N, \tag{53}$$

$$U(P) = U(Q) + \frac{\partial U}{\partial r}N \tag{54}$$

with

$$\frac{\partial U}{\partial r} = -\gamma \tag{55}$$

Combining all these equations one gets (see also the approach presented in Chap. 1)

$$T(P) = W(P) - U(P) = W(P) - U(Q) - \frac{\partial U}{\partial r}N = W_0 - W_0 - \frac{\partial U}{\partial r}N = -\frac{\partial U}{\partial r}N,$$

that is, the famous Bruns equation

$$N = \frac{T}{\gamma}. \tag{56}$$

Proceeding similarly with (51) and (52) one can find

$$\frac{\partial T}{\partial r} - \frac{1}{\gamma}\frac{\partial\gamma}{\partial r}T + \Delta g = 0. \tag{57}$$

With

$$\gamma = \frac{GM}{r^2}, \quad \frac{1}{\gamma}\frac{\partial\gamma}{\partial r} = -\frac{2}{r} \tag{58}$$

this reduces for the geosphere $r = R$ to

$$\frac{\partial T}{\partial r} + \frac{2}{R}T = \Delta g \tag{59}$$

for the Molodensky boundary-value problem, and (51) becomes simply

$$\frac{\partial T}{\partial r} = \delta g \tag{60}$$

for the GPS boundary-value problem.

4.1.4 Solution by Spherical Harmonics: A Useful Formula for Spherical Harmonics

From Sect. 3.1 the spherical-harmonic expansion (35) can be taken for any function V, which is harmonic outside a sphere:

$$V(r, \vartheta, \lambda) = \sum_{n=0}^{\infty}\sum_{m=0}^{n}\frac{P_{nm}(\cos\vartheta)}{r^{n+1}}(A_{nm}\cos m\lambda + B_{nm}\sin m\lambda). \tag{61}$$

One can put

$$V_n(\vartheta, \lambda) = \sum_{m=0}^{n} R^{-(n+1)} P_{nm}(\vartheta, \lambda)(A_{nm}\cos m\lambda + B_{nm}\sin m\lambda) \tag{62}$$

to get

$$V(r, \vartheta, \lambda) = \sum_{n=0}^{\infty}\left(\frac{R}{r}\right)^{n+1} V_n(\vartheta, \lambda). \tag{63}$$

For the geosphere $r = R$ this becomes

$$V(\vartheta, \lambda) = \sum_{n=0}^{\infty} V_n(\vartheta, \lambda) \equiv \sum V_n. \tag{64}$$

Vice versa, given the function $V(\vartheta, \lambda)$ on the geosphere in form (64), Eq. 63 expresses it in space $r > R$, thus solving the first boundary-value problem (*Dirichlet problem*, mentioned in ➋ Sect. 4.1.1) for an easy, but nontrivial case. The function $V_n(\vartheta, \lambda)$ is called the nth Laplace (surface) spherical harmonic of V.

4.1.5 Solution by Stokes-Type Integral Formulas

All the functions T, Δg, and δg can be developed into Laplace spherical harmonics:

$$T = \sum T_n, \quad \Delta g = \sum \Delta g_n, \quad \delta g = \sum \delta g_n. \tag{65}$$

From (59) and (60) one gets

$$T_n = R \frac{\Delta g_n}{n-1}, \tag{66}$$

$$T_n = R \frac{\delta g_n}{n+1}, \tag{67}$$

originating from radial differentiation in (59) and (60).

The summation of (66) for n from 2 (why?) to infinity, and of (67) from 0 to infinity gives

$$\begin{aligned} T &= R \sum \frac{\Delta g_n}{n-1}, \\ T &= R \sum \frac{\delta g_n}{n+1}, \end{aligned} \tag{68}$$

and further

$$T = \frac{R}{4\pi} \iint_{\sigma} S(\psi)\, \Delta g \, d\sigma, \tag{69}$$

the classical formula of Stokes from 1849 for *gravity anomalies*, and

$$T = \frac{R}{4\pi} \iint_{\sigma} K(\psi)\, \delta g \, d\sigma, \tag{70}$$

the recent solution (Hotine 1969; Koch 1971) of the classical second boundary value problem or the Neumann problem for *gravity disturbances* (radial derivatives of T !).

The Stokes function is

$$S(\psi) = \frac{1}{\sin(\psi/2)} - 6 \sin \frac{\psi}{2} + 1 - 5 \cos \psi - 3 \cos \psi \ln \left(\sin \frac{\psi}{2} + \sin^2 \frac{\psi}{2} \right) \tag{71}$$

and the Neumann–Koch function is

$$K(\psi) = \frac{1}{\sin(\psi/2)} - \ln \left(1 + \frac{1}{\sin(\psi/2)} \right). \tag{72}$$

From T, one can find the geoidal heights N from Bruns' formula (56). For more details see Hofmann and Moritz (2005, pp. 102–108, 302–303).

4.1.6 Remarks

The standard classical model can be made more realistic, by considering the topography and heights above sea level, in two ways:

Classical: Gravity reduction to sea level by computationally removing the topography (remove–restore process)
Molodensky: Working directly on the Earth's surface rather than on the geoid

These two approaches are not so different as they appear; for instance, a kind of free-air reduction holds practically in both approaches.

Ellipticity: The effects of ellipticity are taken into account by ellipsoidal corrections. Of course, this is not very elegant. Ideally, *ellipsoidal harmonics* should be used consistently instead of spherical harmonics, preferably by closed expressions, following work by Grafarend (Martinec and Grafarend 1997); see also ❯ Chap. 7 in this handbook). This may be a task for the future.

Gravity disturbances vs. gravity anomalies. The difference between the "classical" Δg and the "modern" and more convenient δg is that GPS is now available to determine P rather than Q as reference point for use in future gravity data archives, so probably the future is in favor of gravity disturbances. This leads to the question:

Is the boundary-problem solution necessary at all to determine the Earth's surface if one now has GPS? The answer is *The gravity field* is also needed, so that the *geopotential* must be determined by formulas such as by (70) or by dedicated satellite missions such as GOCE (contribution by R. Rummel in this handbook; see also Rummel et al. (2002)), which, however, cannot be expected to give exact point values.

The complete answer is *both approaches* are needed for a good gravity field and are complementary to each other: satellites give global and regional aspects, whereas terrestrial gravimetry supplies the local details. Thus one needs the combination of terrestrial and satellite data by a method such as the least-squares collocation, which will be outlined in ❯ Sect. 4.2.

4.2 Collocation

4.2.1 Principles

One works with quantities of the linear perturbing field only, as defined in ❯ Sect. 4.1.3. The anomalous potential T outside the Earth is harmonic, that is, it satisfies Laplace's differential equation

$$\Delta T = \frac{\partial^2 T}{\partial x^2} + \frac{\partial^2 T}{\partial y^2} + \frac{\partial^2 T}{\partial z^2} = 0. \tag{73}$$

An approximate analytical representation of the external potential T is obtained by

$$T(P) = f(P) = \sum_{k=1}^{q} b_k \varphi_k(P), \tag{74}$$

as a linear combination f of suitable base functions $\varphi_1, \varphi_2, \varphi_3, \ldots, \varphi_q$ with appropriate coefficients b_k. All these as functions of the space point P are under consideration.

As T is harmonic outside the Earth's surface, it is natural to choose *base functions* φ_k, which are likewise harmonic (but this time down to a Krarup sphere wholly inside the Earth, see ➋ *Fig. 3*), so that

$$\Delta\varphi_k = 0, \tag{75}$$

in correspondence to Eq. 73

There are many simple systems of functions satisfying the harmonicity condition (75), and thus there are many possibilities for a suitable choice of base functions φ_k. One might, for instance, choose *spherical harmonics* but also local functions; see the review by Willi Freeden (➋ Chap. 31) (*Special functions in mathematical geosciences—an attempt at a categorization*) in this handbook.

The coefficients b_k may be chosen such that *the given observational values are reproduced exactly*. This means that the assumed approximating function f in (74) gives the same values at the observation stations as the actual potential T and hence may well be a suitable approximation for T. These ideas can be now put into a mathematical form, following Hofmann-Wellenhof and Moritz (2005, Chap. 10).

Interpolation

Let errorless values of T be given at q spatial points $P_1, P_2, P_3, \ldots..P_q$. One can put

$$T(P_i) = f_i, \quad i = 1, 2, \ldots, q, \tag{76}$$

and postulate that, in approximating $T(P)$ by $f(P)$, the observations (76) will be exactly reproduced. The condition for this is

$$\sum_{k=1}^{q} A_{ik}\, b_k = f_i \quad \text{with } A_{ik} = \varphi_k(P_i) \tag{77}$$

or in matrix notation

$$\mathbf{A}\,\mathbf{b} = \mathbf{f}. \tag{78}$$

If the square matrix $\mathbf{A}$ is regular, then the coefficients b_k are uniquely determined by

$$\mathbf{b} = \mathbf{A}^{-1}\mathbf{f}. \tag{79}$$

Least-Squares Interpolation

Consider a function

$$K = K(P, Q) \tag{80}$$

in which two points P and Q are the independent variables. Let this function K be

- Symmetric with respect to P and Q
- Harmonic with respect to both points, everywhere outside a certain sphere
- *Positive definite* (the positive-definiteness of a function is defined similarly as in the case of a matrix)

Then the function $K(P, Q)$ is called a (harmonic) *kernel function* (Moritz 1980, p. 205). A kernel function $K(P, Q)$ may serve as "building material" from which one can construct base functions. Taking for the base functions the form

$$\varphi_k(P) = K(P, P_k) \tag{81}$$

where P denotes the variable point and P_k is a fixed point in space, the *least-squares interpolation* can be obtained.

This well-known term originates from the statistical interpretation of the kernel function as a *covariance function* (cf. Hofmann-Wellenhof and Moritz 2005 sect. 9.2). Then, the least-squares interpolation has some minimum properties (least-error variance, similarly as in least-squares adjustment). This interpretation is not essential; however, one may also work with arbitrary analytical kernel functions, regarding the procedure as a purely analytical mathematical approximation technique. Normally one tries to combine both aspects in a reasonable way.

Substituting (81) into (77), one gets

$$A_{ik} = K(P_i, P_k) \equiv C_{ik} \tag{82}$$

this matrix is symmetric and positive-definite because of the corresponding properties of the function $K(P, Q)$. Then the coefficients b_k follow from (82) and may be substituted into (79). With the notation

$$\varphi_k(P) = K(P, P_k) = C_{Pk}, \tag{83}$$

the result may be written in the form

$$f(P) = \begin{bmatrix} C_{P1} C_{P2} \dots C_{Pq} \end{bmatrix} \begin{bmatrix} C_{11} C_{12} \dots C_{1q} \\ C_{21} C_{22} \dots C_{2q} \\ \dots\dots\dots\dots \\ C_{q1} C_{q2} \dots C_{qq} \end{bmatrix}^{-1} \begin{bmatrix} f_1 \\ f_2 \\ \dots \\ f_q \end{bmatrix} \tag{84}$$

which is a linear combination of harmonic base functions and thus itself a harmonic function.

4.2.2 Least-Squares Collocation

Formulas like (84) are used for interpolation, under the condition that measured values f_k of a function $f(P)$ are reproduced. In *collocation*, the measured data are assumed to be *linear functionals* $L_1 T$, $L_2 T$,..., $L_q T$, of the anomalous potential $T(P)$. A linear functional is usually a derivative $\partial T / \partial x_i$ multiplied by a constant factor. For instance, in a local coordinate system, let x, y, z denote north, east, and vertical. Then,

$$\xi = -\frac{1}{\gamma}\frac{\partial T}{\partial x}, \quad \eta = -\frac{1}{\gamma}\frac{\partial T}{\partial y} \tag{85}$$

are the north–south and east–west components of the deflection of the vertical (Hofmann-Wellenhof and Moritz 2005, p.91)

$$\delta g = -\frac{\partial T}{\partial z}, \quad \Delta g = -\frac{\partial T}{\partial z} - \frac{2T}{R}. \tag{86}$$

They are gravity disturbance and gravity anomaly, as it is seen above; the vertical derivatives are with respect to z locally.

All these are linear functionals of T and can be used to compute T. Putting

$$L_i f = L_i T = \ell_i, \tag{87}$$

and inserting this into (77) one gets

$$\sum_{k=1}^{q} C_{ik}\, b_k = \ell_i \quad \text{with} \quad C_{ik} = \varphi_k(P_i). \tag{88}$$

Compared to (77) the matrix **A** is replaced by **B**. In agreement with (87), the base functions $\varphi_k(P)$ must be derived from the basic kernel function $K(P, Q)$ by

$$\varphi_k(P) = L_k^Q K(P, Q) = C_{Pk}, \tag{89}$$

where L_k^Q means that the functional L_k is applied to the variable Q, so that the result no longer depends on Q (since the application of a functional results in a definite number).

Thus, in (88) one must put

$$C_{ik} = L_i^P L_k^Q K(P, Q), \tag{90}$$

which gives a matrix that again is symmetric. Solving (88) for b_k and inserting into (74) gives with

$$\varphi_k(P) = L_k^Q K(P, Q) = C_{Pk}, \tag{91}$$

the formula

$$f(P) = \left[C_{P1} C_{P2} \ldots C_{Pq}\right] \begin{bmatrix} C_{11} C_{12} \ldots C_{1q} \\ C_{21} C_{22} \ldots C_{2q} \\ \ldots\ldots\ldots\ldots\ldots \\ C_{q1} C_{q2} \ldots C_{qq} \end{bmatrix}^{-1} \begin{bmatrix} \ell_1 \\ \ell_2 \\ \ldots \\ \ell_q \end{bmatrix}, \tag{92}$$

which is a linear combination of harmonic base functions and thus *itself a harmonic function.* This statement is repeated because it is such an important fact.

4.2.3 Concluding Remarks

Least-squares collocation gives a method to combine *all* geodetic measures, terrestrial and satellite data. Several or even all of the ℓ's might be satellite data (Lerch et al. 1979; Moritz 1980, pp. 138–139).

The simple formula (92) can be extended in several ways:

- *Errors.* The data ℓ may contain observational errors.
- *Systematic parameters* may be introduced by combining collocation with *least-squares adjustment.*
- Collocation gives not only adjusted values but also their *error covariance matrix* expressing the *accuracy* of the results.
- Collocation is not restricted to the standard classical model (SCM) of ➋ Sect. 4.1.2 but can also take height into account.
- The topography is considered by a "remove–restore" process, not unlike the boundary-value problem (➋ Sect. 4.1.4).

Practical results can be found in the proceedings of geoid-relevant symposia of the International Association of Geodesy published by Springer; Kühtreiber (2002), Erker et al. (2003), and Pail et al. (2008) deserves a mention here.

5 Future Directions

5.1 The Earth as a Nonrigid Body

The Earth is not a rigid body but undergoes elastic deformations on the order of several decimeters. These are tides of the solid earth, as well as oscillations of the Earth's axis on the same order of magnitude. These variations are routinely measured and should be used to reduce the measurements to a solid standard (Moritz and Mueller 1987) if the well-advertised "cm-geoid" is to be reached.

5.2 The Smoothness of the Earth's Surface

In practicality all methods of geodesy, the Earth's surface is considered to be smooth: continuous and differentiable as often as it is needed. A glance through the window shows that this is not realistic: there are trees, houses, and similar objects in high mountains such as rugged cliffs, vertical slopes, etc. They are usually neglected in the formulas of geodesy, but cartographers think differently. Highly irregular curves and surfaces are treated in *chaos theory*, but it is not obvious how this is included in geodetically useful formulas (Moritz 2009) although the remote-sensing people are interested in fractal terrain models (see the paper "Doppelsummation von Zufallsgrössen: von Aerotriangulation und Inertialnavigation zu digitalen Geländemodellen" on the author's science page http://www.helmut-moritz.at).

5.3 Inverse Problems

The geodetic boundary-value problems basically are nonlinear "hard" inverse problems (➋ Sect. 4.1.1). Work on these problems is an exciting challenge to mathematicians following Hörmander (1976).

The principal direct linear problem of physical geodesy is the computation of the gravitational potential from the mass density (see Eq. 1). The *inverse problem*, to compute the mass from the density, is much more difficult (Anger et al. 1993; Anger and Moritz in http://www.helmut-moritz.at).

6 Conclusion

In Moritz (2009) mathematics applied to geodesy within the frame of geosciences have been divided into three parts.

A. Classical physics
B. Complexity
C. Inverse problems

Physical geodesy is indeed mainly based on *classical potential theory* sufficient to a relative accuracy of 10^{-6}, up to 1970 (typically 10 m with respect to the geocenter and 1 mgal in absolute gravimetry). With highly precise absolute gravimetry and satellite laser ranging, accuracy requirements quickly jumped to 10^{-8} to 10^{-9}. This is the order of smallness where *relativistic effects* must be considered, also in the precise measurement of time, for which geodesy is frequently responsible. Main relativistic measurements are the second-order gravitational gradients; see Sect. 2.4.5; cf. Synge's equation (Eq. 13).

For geomathematics, *complexity theory* may offer a great challenge, but this theory is still in its beginning stage (Turcotte 1997). (The use of chaos theory has already been fully established in meteorology (Lorenz 1993)). An interesting problem would be the consideration of the Earth's surface as a random boundary, but it might be of theoretical rather than practical importance; cf. Sect. 5.2.

Concerning *improperly posed inverse problems*, the geodetic boundary value is pretty well solved computationally (see Sect. 4.1), but apart from investigations by Hörmander and Sansò, the area is still wide open with respect to mathematical treatment of existence, uniqueness, and stability.

Already Gauss has invented the least-squares adjustment for the solution of improperly posed overdetermined problems. The least-squares collocation has been developed to solve underdetermined problems and combined with adjustment into a unified method. Theoretical research seems to be more or less finished, and the method is being used in all relevant problems. New mathematical research would be desirable to treat nonlinear adjustment and collocation.

Considerable research on the treatment of inverse gravimetric problems has been done by Anger and others, but it requires methods of modern potential theory that are extremely difficult. They are not only challenging mathematically but also of great practical importance for geophysical prospecting.

6.1 The Geoid

The *geoid* is the most typical feature of the Earth's gravitational field: it is an equipotential surface. However, difficulties begin with this very definition. It can be called the "geoid dilemma."

The definition $W = W_0 =$ constant is natural and well intended but very difficult to implement. When the external gravity field is determined by a spherical harmonic series from satellite measurements, and this series is used at sea level, one does not really get "the geoid," because under the continents, the "real" geopotential is not harmonic but satisfies Poisson's equation (36), which contains the density ρ inside the masses. What is really got in this way is *the analytical downward continuation* through the mass down to sea level, which is a nontrivial inverse problem. Therefore, this problem is treated in detail. It *may* not be of great practical importance, but it is vital to know the reason. Any *truncated* spherical harmonic expression has the origin O as its only singularity (if every term is harmonic down to O, also any *finite* sum of them has the same property because of the linearity of Laplace's equation).

Also any field obtained by the least-squares collocation is harmonic inside the masses because it is a *finite* linear combination of harmonic base function; cf. Eq. 92. It cannot therefore be used to obtain the inner geopotential W, which is not harmonic. So why not use a geoid of downward continuation, hoping that the error can be managed, rather than the real geoid? Since downward continuation is not unique, this approach does not seem quite natural. Although it could work practically, heights above that "geoid" are not what the author understand by "heights above sea level." (Other people might think differently.)

Molodensky's solution of the geoid dilemma was to work only with the *external* potential, disregarding the geoid at all, without bothering about internal masses and the internal potential. This "Molodensky postulate," *to disregard the internal masses*, has had a tremendous, profound, and irreversible influence on geodetic thinking after 1945, but at present again it is not completely sufficient. Furthermore, his "quasigeoid" (not to be confused with the above-mentioned "geoid of downward continuation") and "normal heights" seem artificial and unsatisfactory.

In fact, curiously enough, a new problem, typically non-Molodenskian, is posed by engineering geodesy: precise *tunnel surveying*. The gravity field *in the tunnel* is needed for correct positioning and orientation. This is implicit with the classical method of direction measurements by gyrotheodolites and laser distances. It gets new significance with inertial surveying (Hofmann-Wellenhof et al. 2003). Inertial surveying (ISS) depends on gravity. Basically, the problem is the *separation of gravity and inertia*. Theoretically this is possible even in Einstein's relativity; see sec. 2.4.3, especially Synge' theorem, Eq. 13.

This does occur with ISS. It can be used for positioning. Then acceleration is the signal and gravitation is the noise, or for gravimetry. Then gravity is the signal and inertial position is the noise. In any case, the *internal* gravitational field is needed, and again the "geoid dilemma" comes up. It can be solved if one knows the density ρ, but how? A partial answer is gravity reduction, discussed at various places in Hofmann-Wellenhof and Moritz (2005).

Unsolved problems abound. When in tunnel surveying, topographic reduction is disregarded, GPS and ISS may have an unpleasant encounter at the ends of the tunnel.

It is hoped that this chapter provides suggestions for future research. Most urgently needed and most difficult may be a detailed investigation of the geoid dilemma.

Acknowledgment

The author thanks his colleagues in the Institute of Navigation and Satellite Geodesy at TU Graz for constant support and help in the wonderful atmosphere in the institute, especially to B. Hofmann-Wellenhof, S. Berghold, F. Heuberger, N. Kühtreiber, R. Mayrhofer, and R. Pail, who has carefully read the manuscript.

References

Monographs

Bruns H (1878) Die Figur der Erde. Publikation des Preussischen Geodätischen Instituts, Berlin

Courant R, Hilbert D (1962) Methods of mathematical physics, vol 2. Wiley-Interscience, New York

Fock V (1959) The theory of space-time and gravitation. Pergamon, London

Frank P, Mises R von (eds) (1930) Die Differential- und Integralgleichungen der Mechanik und Physik, 2nd edn, Part 1: Mathematischer Teil. Vieweg, Braunschweig (reprint 1961 by Dover, New York and Vieweg, Braunschweig)

Helmert FR (1884) Die mathematischen und physikalischen Theorien der Höheren Geodäsie, Part 2. Teubner, Leipzig (reprint 1962)

Hofmann-Wellenhof B, Legat K, Wieser M (2003) Navigation—principles of positioning and guidance. Springer, Wien

Hofmann-Wellenhof B, Moritz H (2005) Physical geodesy. Springer, Wien

Hotine M (1969) Mathematical geodesy. ESSA Monograph 2, U.S. Department of Commerce, Washington (reprint 1992 by Springer)

Kellogg OD (1929) Foundations of potential theory. Springer, Berlin (reprint 1954 by Dover, New York, and 1967 by Springer)

Krarup T (1969) A contribution to the mathematical foundation of physical geodesy, vol 44. Danish Geodetic Institute, Copenhagen (reprinted in (Borre 2006))

Lorenz E (1993) The essence of chaos. University of Washington, Seattle

Marussi A (1985) Intrinsic geodesy. Springer, Berlin

Misner CW, Thorne KS, Wheeler JA (1973) Gravitation. Freemann, San Francisco

Moritz H (1967) Kinematical geodesy. Report 92. Department of Geodetic Science, Ohio State University, Columbus

Moritz H (1980) Advanced physical geodesy, Wichmann, Karlsruhe

Moritz H, Hofmann-Wellenhof B (1993) Geometry, relativity, geodesy. Wichmann, Karlsruhe

Moritz H, Mueller II (1987) Earth rotation—theory and observation. Ungar, New York

Neumann F (1887) Vorlesungen über die Theorie des Potentials und der Kugelfunktionen, Neumann C (ed). Teubner, Leipzig

Sternberg S (1969) Celestial mechanics, vol 2. Benjamin, New York

Synge JL (1960) Relativity: the general theory. North-Holland, Amsterdam

Turcotte DL (1997) Fractals and chaos in geology and geophysics, 2nd edn. Cambridge University, Cambridge

Collections

Anger G, Gorenflo R, Jochmann H, Moritz H, Webers W (1993) Inverse problems: principles and applications in geophysics, technology, and medicine. Mathematical Research 74. Akademie Verlag, Berlin

Borre K (ed) (2006) Mathematical foundation of geodesy (Selected papers by Torben Krarup). Springer, Berlin

Journal Articles

Erker E, Höggerl N, Imrek E, Hofmann-Wellenhof B, Kühtreiber N (2003) The Austrian geoid—recent steps to a new solution. Österreichische Zeitschrift für Vermessung und Geoinformation 91(1):4–13

Hörmander L (1976) The boundary problems of physical geodesy. Arch Rational Mech Anal 62:1–52

Koch KR (1971) Die geodätische Randwertaufgabe bei bekannter Erdoberfläche. Zeitschrift für Vermessungswesen 96:218–224

Kühtreiber N (2002) High precision geoid determination of Austria using heterogeneous data. In: Tziavos IN (ed) Gravity and geoid 2002. Proceedings of the third meeting of the international gravity and geoid commission, Thessaloniki, Greece, 26–30 August 2002. http://olimpia.topo.auth.gr/GG2002/SESSION2/kuehtreiber.pdf or http://olimpia.topo.auth.gr/GG2002/SESSION2/session2.html

Lerch FJ, Klosko SM, Laubscher RE, Wagner CA (1979) Gravity model improvement using Geos 3 (GEM 9 and 10). J Geophys Res 84(B8):3897–3916

Martinec Z, Grafarend EW (1997) Solution to the Stokes boundary-value problem on an ellipsoid of revolution. Studia geoph et geod 41:103–129

Moritz H (1978) On the convergence of the spherical-harmonic expansion for the geopotential at the Earth's surface. Bollettino de geodesia e scienze affini 37:363–381

Moritz H (2009) Grosse Mathematiker und die Geowissenschaften: Von Leibniz und Newton bis Einstein und Hibert, Sitzungsberichte Leibniz-Sozietät der Wissenschaften 104: 115–130

Pail R, Kühtreiber N, Wiesenhofer B, Hofmann-Wellenhof B, Of G, Steinbach O, Höggerl N, Imrek E, Ruess D, Ullrich C (2008) Ö.Z. Vermessung und Geoinformation 96 (1):3–14

Rummel R, Balmino G, Johannessen J, Visser P, Woodworth P (2002) Dedicated gravity field missions – principles and aims. J Geodynamics 33:3–20

7 Spacetime Modeling of the Earth's Gravity Field by Ellipsoidal Harmonics

Erik W. Grafarend[1] · *Matthias Klapp*[2] · *Zdeněk Martinec*[3,4]
[1]Department of Geodesy and Geoinformatics, Stuttgart University, Stuttgart, Germany
[2]Asperg, Germany
[3]Dublin Institute for Advanced Studies, Dublin, Ireland
[4]Department of Geophysics, Faculty of Mathematics and Physics, Charles University, Prague, Czech Republic

W. Freeden, M.Z. Nashed, T. Sonar (Eds.), *Handbook of Geomathematics*, DOI 10.1007/978-3-642-01546-5_7,

Abstract All planetary bodies like the Earth rotate causing centrifugal effect! The result is an equilibrium figure of ellipsoidal type. A natural representation of the planetary bodies and their gravity fields has therefore to be in terms of ellipsoidal harmonics and ellipsoidal wavelets, an approximation of its gravity field which is three times faster convergent when compared to the "ruling the world" spherical harmonics and spherical wavelets. Freeden et al. (1998, 2004). Here, various effects are treated when considering the Earth to be "ellipsoidal": ➋ Sections 2 and ➋ 3 start the chapter with the celebrated ellipsoidal Dirichlet and ellipsoidal Stokes (to first order) boundary-value problems. ➋ Section 4 is devoted to the definition and representation of the ellipsoidal vertical deflections in gravity space, extended in ➋ Sect. 5 to the representation in geometry space. The potential theory of horizontal and vertical components of the gravity field, namely, in terms of ellipsoidal vector fields, is the target of ➋ Sect. 6. ➋ Section 7 is concentrated on the reference potential of type Somigliana–Pizzetti field and its tensor-valued derivatives. ➋ Section 8 illustrates an ellipsoidal harmonic gravity field for the Earth called SEGEN (Gravity Earth Model), a set-up in ellipsoidal harmonics up to degree/order 360/360. Five plates are shown for the West–East/North–South components of type vertical deflections as well as gravity disturbances refering to the International Reference Ellipsoid 2000. The final topic starts with a review of the curvilinear datum problem refering to ellipsoidal harmonics. Such a datum transformation from one ellipsoidal representation to another one in ➋ Sect. 9 is a seven-parameter transformation of type (i) translation (three parameters), (ii) rotation (three parameters) by Cardan angles, and (iii) dilatation (one parameter) as an action of the conformal group in a three-dimensional Weitzenbäck space W(3) with seven parameters. Here, the chapter is begun with an example, namely, with a datum transformation in terms of spherical harmonics in ➋ Sect. 10. The hard work begins with ➋ Sect. 11 to formulate the datum transformation in ellipsoidal coordinates/ellipsoidal harmonics! The highlight is ➋ Sect. 12 with the characteristic example in terms of ellipsoidal harmonics for an ellipsoid of revolution transformed to another one, for instance, polar motion or gravitation from one ellipsoid to another ellipsoid of reference. ➋ Section 13 reviews various approximations given in the previous three sections.

1 Introduction

It is a common tradition in *Geosciences* to represent its various effects in terms of spherical functions, namely, spherical harmonics and spherical wavelets. Freeden et al. (1998). Here, the intention is to generalize to a less-symmetric representation which is more close to *reality*, namely, to represent the *spacetime gravity field and its tensor fields* in terms of *ellipsoidal functions*, namely, *ellipsoidal harmonics* and *ellipsoidal wavelets*. This alternative generalization has been developed as early as the nineteenth century, for instance, by Jacobi (1834), and generalized, for instance, by Thong and Grafarend (1989) by *four alternatives*. Due to the *rotation* of the Earth or any other celestial body, the ellipsoidal representation of its gravity field converges *three times faster* than in terms of a spherical representation. The introduction to *zero-order derivatives*, namely, gravity potential—the sum of the *gravitational potential* and the *centrifugal potential* due to the rotation of celestial bodies—and to *first-order derivatives*, namely, *gravity disturbance* and *vertical deflections* is specialized here. For *second-order derivatives* and its ellipsoidal spectral representation in terms of *tensor-valued ellipsoidal harmonics* refer to Bölling and Grafarend (2005). A lot of work has been done already to present numerical computations in terms of *ellipsoidal harmonics*, for instance, the spectral representation to degree/order 360/360 in Grafarend et al. (2006).

Why should the spacetime gravity field in terms of *ellipsoidal harmonics* be considered? To this end, refer to the great works being done in the context of *ellipsoidal figures of equilibrium*, the origin of Jacobi's contribution of 1834, generalized by Neumann (1848), Cayley (1875a ,b), Niven (1891), Hobson (1896, 1931), Louis (1957), Dyson (1991), Sona (1995).

An excellent introduction into *ellipsoidal harmonics* representing *ellipsoidal figures of equilibrium* is the book from Chandrasekhar (1969), namely, pages 38–63. The *Nobel Laureat in Physics*, together with Lebovitz (1961), introduced the method of *virial equations* (up to order four) *in a rotating frame of reference*: *Biaxial* ellipsoidal figures of equilibrium were already known to Maclaurin (1742). Dependent on the rotational speed, Jacobi (1834) found that *triaxial ellipsoids* are sometimes dynamically (*angular momentum*) preferable. Mayer (1942) was the first to study the *Jacobian sequence "bifurcating"* (in the terminology of *H. Poincare*) from the *Maclaurin sequence* at a point where the eccentricity reaches 0.81267. The fact that there is no figure of equilibrium for *uniformly rotating bodies*, when the angular velocity *exceeds a certain limit*, raises the question: What happens when the angular speed exceeds this limit? *L. Dirichlet* adressed this question during the years 1856–1857. He included his results in his lectures, but did not publish any detailed account. *Dirichlet's results* were collected in some papers left by him and were published by Dedekind (1860). These results opened a new way to study in more detail the motion of a *self-gravitating, homogeneous, and inhomogeneous ellipsoid*. The complete solution of the problem of stationary figures admissible under *Dirichlet's general assumption* was given by Riemann (1892). He has proven that under the restriction of motions *which are linear in the coordinates* the most general type of motion is compatible with an ellipsoidal figure of equilibrium consisting of a superposition of a *uniform rotation* and the *internal motion of a uniform vorticity* in the rotating frame of reference. Three conditions have to be fulfilled:

1. The case of uniform rotation with *no internal motions*
2. The case when the direction of the uniform rotation vector and the uniform vorticity vector coincide with a principal axis of the ellipsoid
3. The case when the directions of the uniform rotation vector and the uniform vorticity vector lie in a principal plane of the ellipsoidal

B. Riemann also studied the stability; his results were more precisely written by Lebovitz and Lifschitz (1998) and Fasso and Louis (2001). The whole research on ellipsoidal figures of equilibrium got a strong push when Poincare (1885) discussed that the *Jacobian sequence* has also points of *bifurcation* similar to the *Maclaurin sequence* due to linear harmonics. He has also argued, that a *planet with two satellites on each side is unstable due to the four zonal harmonics*. Finally, Cartan (1922, 1928) established that the *Jacobi ellipsoid* becomes *unstable* at its first point of bifurcation and behaves differently from the *Maclaurin spheroid*. As the recent number of papers show, there are still many open questions, namely, how to deal with *inhomogeneous bodies* and their series of internal stability. For instance, *M.E. Roche* considered the equilibrium of an *infinitesimal satellite* rotating about a spherical planet in a *circular Keplerian orbit* and showed, in general, that no equilibrium figures are possible if the orbital rotation exceeds a certain limit. The lower limit is called the *Roche limit*. In a particular paper, Darwin (1906) studied the *figure and stability of a liquid satellite*. An interesting case was finally analyzed by Jeans (1917, 1919) when in *Roche's problem tidal forces* act. At this end, refer to the review paper of Lebovitz (1996) on the *virial method* and the *classical ellipsoids*.

Since the concern is also with the *time-dependence of the planetary gravity fields*, refer to the *review of* Grafarend et al. (1997). In addition, the *Habilitation Thesis of* Engels (2006) on *ellipsoidal models for Earth rotation studies*, namely, polar motion, polar wandering, and length-of-day variations should also the mentioned.

Here, the chapter proceeds as follows: The *first two Sections* (➋ Sects. 2 and ➋ 3) are devoted to the *Dirichlet* as well as the *Stokes boundary-value problem*, both on the *ellipsoid of revolution*, which is important in all branches of *Geosciences*. In all detail, the *ellipsoidal Poisson kernel* as well as series expansion in terms of ellipsoidal harmonics and closed formulae is constructed. For the *Stokes boundary-value problem* on the *ellipsoid of revolution* the highlight is on the importance of the *zero-degree harmonics*. ➋ Section 4 is concerned with the definition and representation of *ellipsoidal vertical deflection in gravity space* extended in ➋ Sect. 5 by an analysis of deflections of the vertical and *gravity disturbance in geometry space*, both in ellipsoidal coordinates. The *potential theory of horizontal and vertical components* of the gravity field is presented, namely, based on the ellipsoidal *reference potential of type Somigliana–Pizzetti and its first derivative*. At the end, the study is on the *time changes* of the *Somigliana–Pizzetti* reference field according to *best estimates* for the Earth gravity field. ➋ Section 7 reviews the *ellipsoidal reference potential of type Somigliana–Pizzetti* in terms of first derivatives, namely, vector-valued ellipsoidal components $\{S_{kl}, R_{kl}\}$ characterizing ellipsoidal vertical deflections and ellipsoidal gravity disturbances. ➋ Section 8 illustrates an ellipsoidal harmonic *Gravity Earth Model* (SEGEN) up to *degree/order 360/360*, namely, by *global maps* of ellipsoidal vertical deflections and ellipsoidal gravity disturbances. *Plates I to V* show West–East/South–North components of type $\{\eta, \xi\}$ as well as *gravity disturbances* refering to the *International Reference Ellipsoid* both in *ellipsoidal Mollweide projection* and *ellipsoidal equal-area azimutal projection* for polar regions.

The final topic in ➋ Sect. 9 is curvilinear datum transformation by changing seven parameters: (i) *translation* (three parameters), (ii) *rotation* (three parameters) by *Cardan angles*, and (iii) *dilatation* (one parameter). Under the action of the *conformal group* (similarity transformation) angles and distance ratios are left equivariants (invariant). They locate a point in the *three-dimensional Weitzenbäck space* W_7^3. An example is the datum transformation in terms of *spherical harmonic coefficients* in ➋ Sect. 10. The *highlight* of ➋ Section 11 is the *datum transformation of ellipsoidal harmonic coefficients*. ➋ Section 12 gives characteristic examples in eight tables. The conclusion of the final 3 sections is given in ➋ Sect. 13.

2 Dirichlet Boundary-Value Problem on the Ellipsoid of Revolution

Green's function to the external Dirichlet boundary-value problem for the Laplace equation with data distributed on an ellipsoid of revolution has been constructed in a closed form. The ellipsoidal Poisson kernel describing the effect of the ellipticity of the boundary on the solution of the investigated boundary-value problem has been expressed as a finite sum of elementary functions which describe analytically the behavior of the ellipsoidal Poisson kernel at the singular point $\psi = 0$. It has shown that the degree of singularity of the ellipsoidal Poisson kernel in the vicinity of its singular point is of the same degree as that of the original spherical Poisson kernel.

To be able to work with a harmonic gravitational potential outside the *Geoid*, it is necessary to replace the gravitational effect of the topographical masses (extending between the

Geoid and the surface) by the gravitational effect of an auxiliary body situated below or on the *Geoid*. To carry out this "mass redistribution" mathematically, a number of more or less idealized so-called compensation models have been proposed. Although the way of compensation might originally be motivated by geophysical ideas, for the purpose of *Geoid* computation, one may, in principle, employ any compensation model generating a harmonic gravitational field outside the *Geoid*. For instance, the *topographic-isostatic compensation* models (for instance, Rummel et al. 1988; Moritz 1990) are based on the anomalies of density distribution in a layer between the *Geoid* and the compensation level. In the limiting case, the topographical masses may be compensated by a mass surface located on the *Geoid*, i.e., by a layer whose thickness is infinitely small. This kind of compensation is called Helmert's second compensation (Helmert 1884).

Redistributing the topographical masses and subtracting the centrifugal potential from the *"observed" surface geopotential*, one gets the gravitational potential induced by the masses below the *Geoid*, which is harmonic in the space above the *Geoid*. To find it on the *Geoid*, a fixed boundary-value problem governed by the *Laplace equation* with the *Dirichlet boundary condition* on the Earth's surface (determined by GPS observations) can be formulated.

The problem can be solved by employing the *Poisson integral* (Kellogg 1929, Sect. IX.4) provided that the *Geoid* in the boundary condition for a potential is approximated by a mean sphere. Such an approximation may produce significant errors in geoidal heights. The only possibility to reduce them is to approximate the *Geoid* by a surface fitting it in a closer way. An ellipsoid of revolution, best fitting the *Geoid*, represents an alternative to approximate the *Geoid* closer to the reality. This motivated the question: Can *Green's function* to the external *Dirichlet boundary-value problem* for the *Laplace equation* with data distributed on an ellipsoid of revolution be constructed? *Green's function* to the *Stokes problem* for gravity data distributed on an ellipsoid of revolution (Martinec and Grafarend 1997) has already been constructed. Here, the intention is to apply the same analytical approach to solve the raised question.

The construction of this *Green's function* (later on, called the ellipsoidal *Poisson kernel*) enables one to avoid making the assumption on a spherical approximation of the *Geoid* in the problem of the upward and downward continuation of gravitation (for instance, P. Vaniček et al. 1996). Even pessimistic estimates show that ellipsoidal approximation of the *Geoid* in this type of boundary-value problem is sufficient to determinate a very accurate *Geoid* with an error not exceeding 1 cm. Such an accuracy is the ultimate goal of scientists dealing with the determination of a very precise *Geoid*.

2.1 Formulation of the Dirichlet Boundary-Value Problem on an Ellipsoid of Revolution

To begin with, the ellipsoidal coordinates $\{u,\vartheta,\lambda\}$ through the transformation relations into cartesian coordinates $\{x,y,z\}$ (for details refer to P. Moon and Spencer 1961; Heiskanen and Moritz 1967; N.C Thong and E.W Grafarend 1989) are introduced:

$$\begin{aligned} x &= \sqrt{u^2 + E^2}\sin\vartheta\cos\lambda \\ y &= \sqrt{u^2 + E^2}\sin\vartheta\,\sin\lambda \\ z &= u\cos\vartheta \end{aligned} \tag{1}$$

where ϑ is the reduced co-latitude, λ is geometric longitude, and $E = \sqrt{a^2 - b^2}$ (= constant) is the linear eccentricity, also called the radius of the focal circle, of the set of confocal ellipsoidal coordinate surfaces u = constant. The problem to be dealt with is to determine a function $T(u, \Omega)$, $\Omega = \{\vartheta, \lambda\}$, outside the reference ellipsoid $u = b_0$ such that

$$\nabla^2 T = 0 \text{ for } u > b_0 \tag{2}$$

$$T = f \text{ for } u = b_0 \tag{3}$$

$$T \sim O\left(\frac{1}{u}\right) \text{ for } u \to \infty \tag{4}$$

where $f(\Omega)$ is assumed to be a known square-integrable function, $f(\Omega) \in L_2(\Omega)$. The asymptotic condition in Eq. 4 means that the harmonic function T approaches zero at infinity.

The solution of the *Laplace equation* 2 can be written in terms of ellipsoidal harmonics as follows (Moon and Spencer 1961; Heiskanen and Moritz 1967; Thong and Grafarend 1989):

$$T(u, \Omega) = \sum_{j=0}^{\infty} \sum_{m=-j}^{j} T_{jm} \frac{Q_{jm}\left(i\frac{u}{E}\right)}{Q_{jm}\left(i\frac{b_0}{E}\right)} Y_{jm}(\Omega) \tag{5}$$

where $Q_{jm}\left(i\frac{u}{E}\right)$ are *Legendre's function of the second kind*, $Y_{jm}(\Omega)$ are complex spherical harmonics of degree j and order m, and T_{jm} are coefficients to be determined from the boundary condition of Eq. 3. Substituting Eq. 5 into Eq. 3, and expanding the function $f(\Omega)$ in a series of spherical harmonics,

$$f(\Omega) = \sum_{j=0}^{\infty} \sum_{m=-j}^{j} \int_{\Omega_0} f(\Omega') Y_{jm}^{*}(\Omega') d\Omega' Y_{jm}(\Omega) \tag{6}$$

where Ω_0 is the full solid angle and $d\Omega = \sin\vartheta d\vartheta d\lambda$, and comparing the coefficients at spherical harmonics $Y_{jm}(\Omega)$ in the result, one gets

$$T_{jm} = \int_{\Omega_0} f(\Omega') Y_{jm}^{*}(\Omega') d\Omega' \tag{7}$$

for $j = 0, 1, \ldots,$ and $m = -j, \ldots, j$. Furthermore, substituting coefficients T_{jm} into Eq. 5, interchanging the order of summation over j and m and integration over Ω' due to the uniform convergence of the series expansion given by Eq. 5, the solution to the *ellipsoidal Dirichlet boundary-value problem* Eqs. 2–4, finally reads

$$T(u, \Omega) = \int_{\Omega_0} f(\Omega') \sum_{j=0}^{\infty} \sum_{m=-j}^{j} T_{jm} \frac{Q_{jm}\left(i\frac{u}{E}\right)}{Q_{jm}\left(i\frac{b_0}{E}\right)} Y_{jm}^{*}(\Omega') Y_{jm}(\Omega) d\Omega' \tag{8}$$

From a practical point of view, the spectral form of Eq. 8 of the solution to the *Dirichlet problem* given by Eqs. 2–4 may often become inconvenient, since the construction of $Q_{jm}(z)$ functions and their summation up to high degrees and orders ($j \sim 10^4$–10^5) is time consuming and numerically unstable (Sona 1995). Moreover, in the case that the level ellipsoid $u = b_0$ deviates from a sphere by only a tiny amount, which is the case for the Earth, the solution of the problem should be close to the solution to the same problem but formulated on a sphere. One should thus attempt to rewrite $T(u, \Omega)$ as a sum of the well-known *Poisson integral*

(Kellogg 1929, Sect. IX.4), which solves the *Dirichlet problem* on a sphere, plus the corrections due to the ellipticity of the boundary. An evident advantage of such a decomposition is that existing theories as well as numerical codes for solving the *Dirichlet problem* on a sphere can simply be corrected for the ellipticity of the boundary.

2.2 Power-Series Representation of the Integral Kernel

Thong 1993 and Martinec and Grafarend 1997 showed that the *Legendre function of the second kind*, $Q_{jm}\left(i\frac{u}{E}\right)$, can be developed in an infinite power series of the first eccentricity e.

$$Q_{jm}\left(i\frac{u}{E}\right) = (-1)^{m-(j+1)/2}\frac{(j+m)!}{(2j+1)!!}e^{j+1}\sum_{k=0}^{\infty}a_{jmk}e^{2k} \tag{9}$$

where coefficients a_{jmk} can, for instance, be defined by the recurence relation as

$$a_{jm0} = 1 \tag{10}$$

$$a_{jmk} = \frac{(j+2k-1)^2 - m^2}{2k(2j+2k+1)}a_{jmk-1} \text{ for } k \geq 1 \tag{11}$$

Throughout this chapter it is assumed that the eccentricity e_0 of the reference ellipsoid $u = b_0$

$$e_0 = \frac{E}{\sqrt{b_0^2 + E^2}} \tag{12}$$

is less than 1. Then, for points (u,Ω) being outside or on the reference ellipsoid, i.e., when $u = b_0$, the series in Eq. 9 is convergent.

By Eq. 9, the ratio of the *Legendre functions of the second kind* in Eq. 8 reads

$$\frac{Q_{jm}\left(i\frac{u}{E}\right)}{Q_{jm}\left(i\frac{b_0}{E}\right)} = \left(\frac{e}{e_0}\right)^{j+1}\frac{\sum_{k=0}^{\infty}a_{jmk}e^{2k}}{\sum_{k=0}^{\infty}a_{jmk}e_0^{2k}} \tag{13}$$

Dividing the polynomials in Eq. 13 term by term, one can write

$$\frac{\sum_{k=0}^{\infty}a_{jmk}e^{2k}}{\sum_{k=0}^{\infty}a_{jmk}e_0^{2k}} = 1 + \sum_{k=1}^{\infty}b_{jmk} \tag{14}$$

where the explicit forms of the first few constituents read

$$b_{jm1} = a_{jm1}\left(e^2 - e_0^2\right) \tag{15}$$

$$b_{jm2} = a_{jm2}\left(e^4 - e_0^4\right) - a_{jm1}^2 e_0^2\left(e^2 - e_0^2\right) \tag{16}$$

$$b_{jm3} = a_{jm3}\left(e^6 - e_0^6\right) - a_{jm2}a_{jm1}e_0^2\left(e^4 + e^2e_0^2 - 2e_0^4\right) + a_{jm1}^3 e_0^4\left(e^2 - e_0^2\right) \tag{17}$$

Generally,

$$b_{jmk} \sim O\left(e^{2r}e_0^{2s}\right),\ r + s = k \tag{18}$$

To get an analytical expression for the kth term of the series in Eq. 14 some cumbersome algebraic manipulations have to be performed. To avoid them, confine to the case where the computation point ranges in a limited layer above the reference ellipsoid (for instance, *topographical layer*), namely, $b_0 < u < b_0 + 9{,}000$ m, which includes all the actual topographical masses of the Earth. For this restricted case, which is, however, often considered when geodetic boundary-value problems are solved, express the first eccentricity e of the computation point by means of the first eccentricity e_0 of the reference ellipsoid and a quantity $\varepsilon, \varepsilon > 0$, as

$$e = e_0(1-\varepsilon) \tag{19}$$

Assuming $b_0 < u < b_0 + 9{,}000$ m means, that $\varepsilon < 1.4 \times 10^{-3}$, and one can put approximately

$$e^{2k} = e_0^{2k}(1-2k\varepsilon) \tag{20}$$

$$b_{jm1} = -2\varepsilon e_0^2 a_{jm1} \tag{21}$$

This allows one to write

$$\frac{\sum\limits_{k=0}^{\infty} a_{jmk} e^{2k}}{\sum\limits_{k=0}^{\infty} a_{jmk} e_0^{2k}} = 1 - 2\varepsilon \frac{\sum\limits_{k=1}^{\infty} k a_{jmk} e_0^{2k}}{\sum\limits_{k=0}^{\infty} a_{jmk} e_0^{2k}} \tag{22}$$

Expand the fraction on the right-hand side of the last equation (divided by a_{jm1}) into an infinite power series e_0^2

$$\frac{1}{a_{jm1}} \frac{\sum\limits_{k=1}^{\infty} k a_{jmk} e_0^{2k}}{\sum\limits_{k=0}^{\infty} a_{jmk} e_0^{2k}} = \sum_{k=1}^{\infty} \beta_{jmk} e_0^{2k} \tag{23}$$

To find the coefficients β_{jmk}, rewrite the last equation in the form

$$\frac{1}{a_{jm1}} \sum_{k=1}^{\infty} k a_{jmk} e_0^{2k} = \sum_{k=0}^{\infty} a_{jmk} e_0^{2k} \sum_{l=1}^{\infty} \beta_{jml} e_0^{2l} \tag{24}$$

Since both the series on the right-hand side are absolutely convergent, their product may be rearranged as

$$\sum_{k=0}^{\infty} a_{jmk} e_0^{2k} \sum_{l=1}^{\infty} \beta_{jml} e_0^{2l} = \sum_{k=1}^{\infty} \sum_{l=1}^{k} \beta_{jml} a_{jm,k-l} e_0^{2k} \tag{25}$$

Substituting Eq. 25 into Eq. 24 and equating coefficients at e_0^{2k} on both sides of Eq. 24, one obtains

$$k \frac{a_{jmk}}{a_{jm1}} = \sum_{l=1}^{k} \beta_{jml} a_{jmk-l}, k = 1, 2 \ldots \tag{26}$$

which yields the recurrence relation for β_{jmk}:

$$\beta_{jmk} = k \frac{a_{jmk}}{a_{jm1}} - \sum_{l=1}^{k-1} \beta_{jml} a_{jm,k-l}, k = 2, 3 \ldots \tag{27}$$

with the starting value

$$\beta_{jm1} = 1 \tag{28}$$

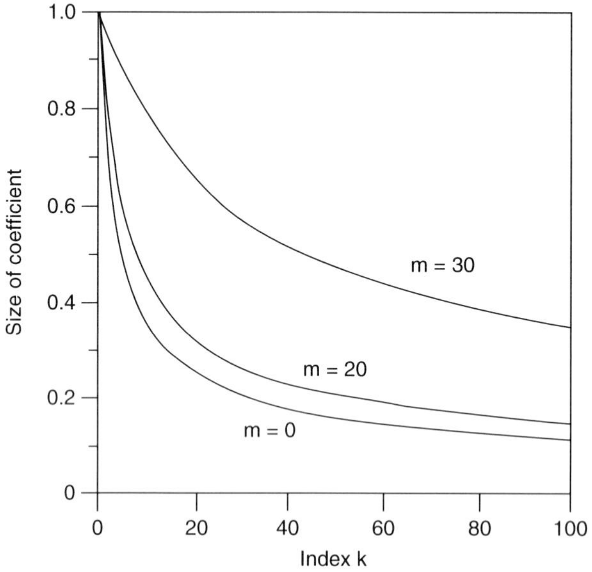

Fig. 1
Coefficients β_{jmk} for $j = 30$, $m = 0,20,30$ and $k = 1,\ldots,100$

With the recurrence relation in Eq. 27 one may easily construct the higher coefficients β_{jmk}:

$$\beta_{jm2} = 2\frac{a_{jm2}}{a_{jm1}} - a_{jm1} \tag{29}$$

$$\beta_{jm3} = 3\frac{a_{jm3}}{a_{jm1}} - 3a_{jm2} + a_{jm1}^2 \tag{30}$$

This process may be continued indefinitely. Important properties of the coefficients β_{jmk} are:

$$1 = \beta_{jm1} > \beta_{jm2} > \beta_{jm3} > \ldots \geq 0 \tag{31}$$

$$\beta_{jjk} > \beta_{jj-1k} > \ldots > \beta_{j0k} \tag{32}$$

Figure 1 demonstrates these properties for $j = 30$.

Finally, substituting the series Eq. 23 into Eq. 22 and using Eqs. 21 and 28 for b_{jm1} and β_{jm1}, respectively, yields

$$\frac{\sum\limits_{k=0}^{\infty} a_{jmk} e^{2k}}{\sum\limits_{k=0}^{\infty} a_{jmk} e_0^{2k}} = 1 + b_{jm1}\left(1 + \sum_{k=2}^{\infty} \beta_{jmk} e_0^{2k-2}\right) \tag{33}$$

2.3 The Approximation of $O\left(e_0^2\right)$

The harmonic upward or downward continuation of the potential or the gravitation between the *Geoid* to the Earth's surface is an example of the practical application of the boundary-value problem Eqs. 2–4 (Engels et al. 1993; Martinec 1996; P. Vaniček et al. 1996). To make the theory as simple as possible but still matching the requirements on *Geoid* height accuracy, one should keep throughout the following derivations the terms of magnitudes of the order of the first eccentricity e_0^2 of the Earth's level ellipsoid and neglect the term of higher powers of e_0^2. This approximation is justifiable because the error introduced by this approximation is at most 1.5×10^{-5}, which then causes an error of at most 2 mm in the geoidal heights.

Keeping in mind the inequalities in Eq. 31, and assuming that $e_0^2 \ll 1$, the magnitude of the last term in Eq. (33) is of the order of e_0^2,

$$\sum_{k=2}^{\infty} \beta_{jmk} e_0^{2k-2} \sim \sigma\left(e_0^2\right) \tag{34}$$

Hence, Equation 13 becomes

$$\frac{Q_{jm}\left(i\frac{u}{E}\right)}{Q_{jm}\left(i\frac{b_0}{E}\right)} = \left(\frac{e}{e_0}\right)^{j+1} \left(1 + a_{jm1}\left(e^2 - e_0^2\right)\left[1 + O\left(e_0^2\right)\right]\right) \tag{35}$$

Substituting Eq. 35 into Eq. 8, evaluating a_{jm1} according to Eq. 11 for $k = 1$, bearing in mind the *Laplace addition theorem* for spherical harmonics (for instance, Varshalovich et al. 1989, p. 164),

$$P_j(\cos\psi) = \frac{4\pi}{2j+1} \sum_{m=-j}^{j} Y_{jm}(\Omega) Y_{jm}^*(\Omega') \tag{36}$$

where $P_i(\cos\psi)$ is the *Legendre polynomial* of degree j and ψ is the angular distance between directions Ω and Ω', one gets

$$T(u,\Omega) = \frac{1}{4\pi} \int_{\Omega_0} f(\Omega') \left[K^{\text{sph}}(t,\cos\psi) - 2e_0^2 k^{\text{ell}}(t,\Omega,\Omega')\left(1 + O\left(e_0^2\right)\right)\right] d\Omega' \tag{37}$$

where

$$t := \frac{e}{e_0} \tag{38}$$

$K^{sph}(t, \cos\psi)$ is the *spherical Poisson kernel* (Kellogg 1929; Heiskanen and Moritz 1967; Pick et al. 1973),

$$K^{\text{sph}}(t,\cos\psi) = \sum_{j=0}^{\infty} (2j+1) t^{j+1} P_j(\cos\psi) = \frac{t(1-t^2)}{g^3} \tag{39}$$

with

$$g \equiv g(t,\cos\psi) := \sqrt{1 + t^2 - 2t\cos\psi} \tag{40}$$

and $k^{\text{ell}}(t,\Omega,\Omega')$ stands for

$$k^{\text{ell}}(t,\Omega,\Omega') = 4\pi(1-t^2) \sum_{j=0}^{\infty} \sum_{m=-j}^{j} t^{j+1} \frac{(j+1)^2 - m^2}{2(2j+3)} Y_{jm}(\Omega) Y_{jm}^*(\Omega') \tag{41}$$

Moreover, it holds that ($t < 1$):

$$\begin{aligned} |k^{\text{ell}}(t,\Omega,\Omega')| &\leq k^{\text{ell}}(t,\Omega,\Omega) \leq k^{\text{ell}}(t,\Omega,\Omega)|_{\vartheta=0} \\ &= \frac{1-t^2}{2}\sum_{j=0}^{\infty} t^{j+1}(j+1)^2\frac{2j+1}{2j+3} \\ &< \frac{1-t^2}{2}\sum_{j=0}^{\infty} t^{j+1}(j+1)^2 \\ &= \frac{t(1+t)(1-t^2)}{2(1-t)^3} < \frac{t(1-t^2)}{(1-t)^3} \\ &= K^{\text{sph}}(t,1) \end{aligned} \tag{42}$$

where the following relations have been used:

$$Y_{jm}(0,\lambda) = \delta_{m0}\sqrt{\frac{2j+1}{4\pi}} \tag{43}$$

and

$$\sum_{j=0}^{\infty} t^j (j+1)^2 = \frac{1+t}{(1-t)^3} \tag{44}$$

borrowed from Varshalovich et al. (1989, Chap. 5) and Mangulis (1965, p. 72), respectively.

The estimate in Eq. 42 helps us to restrict to terms of magnitude of $O\left(e_0^2\right)$ at most. Within this accuracy, Eq. 37 reads

$$T(u,\Omega) = \frac{1}{4\pi}\int_{\Omega_0} f(\Omega')K^{\text{ell}}(t,\Omega,\Omega')d\Omega' \tag{45}$$

where

$$K^{\text{ell}}(t,\Omega,\Omega') := K^{\text{sph}}(t,\cos\psi) - e_0^2 k^{\text{ell}}(t,\Omega,\Omega') \tag{46}$$

2.4 The Ellipsoidal Poisson Kernel

The function $K^{\text{ell}}(t,\Omega,\Omega')$ is called the *ellipsoidal Poisson kernel* since it solves *Dirichlet's problem* Eqs. 2–4 on an ellipsoid of revolution. The next effort will be devoted to convert the spectral form Eq. 41 of $k^{\text{ell}}(t,\Omega,\Omega')$ to a spatial representation.

One of the trickiest steps in such a conversion is to sum up the series

$$\sum_{m=-j}^{j} m^2 Y_{jm}(\Omega)Y_{jm}^*(\Omega') \tag{47}$$

Martinec and Grafarend (1997) presented a method showing how to sum up this series. They obtained

$$\begin{aligned}
\frac{4\pi}{2j+1}\sum_{m=-j}^{j}\left[(j+1)^2-m^2\right]Y_{jm}(\Omega)Y_{jm}^*(\Omega') \\
\frac{\sin\vartheta\cos\alpha}{\sin\psi}(\cos\vartheta-\cos\vartheta'\cos\psi) \\
\times\left[2\cos\psi\frac{dP_{j+1}(\cos\psi)}{d\cos\psi}-(j+1)(j+2)P_{j+1}(\cos\psi)\right] \\
+\sin\vartheta(\sin\vartheta-\cos\vartheta'\sin\psi\cos\alpha)\frac{dP_{j+1}(\cos\psi)}{d\cos\psi} \\
+(j+1)(1-\sin^2\vartheta\sin^2\alpha)\sin^2\psi\frac{dP_{j+1}(\cos\psi)}{d\cos\psi} \\
+(j+1)^2\cos\vartheta\cos\vartheta' P_{j+1}(\cos\psi)
\end{aligned} \tag{48}$$

where α is the azimuth between directions Ω and Ω' (Heiskanen and Moritz 1967).

Later on, one will see that Eq. 48 ensures that the *ellipsoidal Poisson kernel* $K^{\text{ell}}(t,\Omega,\Omega')$ has the same degree of singularity at the point $t = 1$ and $\Omega = \Omega'$ as the original *spherical Poisson kernel* $K^{\text{sph}}(t,\cos\psi)$. Such a result will be acceptable, since it can now never happen that the ellipsoidal correction to the *Dirichlet problem on an ellipsoid* with a small flattening becomes larger than the contribution due to the *spherical Poisson kernel*. Substituting Eq. 48 into Eq. 41, the integral kernel $k^{\text{ell}}(t,\Omega,\Omega')$ can be composed from the different terms

$$k^{\text{ell}}(t,\Omega,\Omega') = \frac{1-t^2}{2}\sum_{i=1}^{4}k_i^{\text{ell}}(t,\Omega,\Omega') \tag{49}$$

where

$$k_1^{\text{ell}}(t,\Omega,\Omega') = \frac{\sin\vartheta\cos\alpha}{\sin\psi}(\cos\vartheta-\cos\vartheta'\cos\psi)\left[\cos\psi L_1(t,\cos\psi)-L_2(t,\cos\psi)\right] \tag{50}$$

$$k_2^{\text{ell}}(t,\Omega,\Omega') = \sin\vartheta(\sin\vartheta-\cos\vartheta'\sin\psi\cos\alpha)L_1(t,\cos\psi) \tag{51}$$

$$k_3^{\text{ell}}(t,\Omega,\Omega') = (1-\sin^2\vartheta\sin^2\alpha)L_3(t,\cos\psi) \tag{52}$$

$$k_4^{\text{ell}}(t,\Omega,\Omega') = \cos\vartheta\cos\vartheta' L_4(t,\cos\psi) \tag{53}$$

The *isotropic parts* $L_i(t, \cos\psi)$ of functions $k_i^{\text{ell}}(t,\Omega,\Omega')$ are given by infinite series of the *Legendre polynomials*,

$$L_1(t,x) = \sum_{j=1}^{\infty}\frac{2j-1}{2j+1}t^j\frac{dP_j(x)}{dx} \tag{54}$$

$$L_2(t,x) = \sum_{j=1}^{\infty}\frac{j(j+1)(2j-1)}{2j+1}t^jP_j(x) \tag{55}$$

$$L_3(t,x) = (1-x^2)\sum_{j=1}^{\infty}\frac{j(2j-1)}{2j+1}t^j\frac{dP_j(x)}{dx} \tag{56}$$

$$L_4(t,x) = \sum_{j=1}^{\infty}\frac{j^2(2j-1)}{2j+1}t^jP_j(x) \tag{57}$$

where x abbreviates the expression:

$$x := \cos\psi = \cos\vartheta\cos\vartheta' + \sin\vartheta\sin\vartheta'\cos(\lambda - \lambda') \tag{58}$$

2.5 Spatial Forms of Kernels $L_i(t,x)$

It is now attempted to express infinite sums for $L_i(t,x)$ as finite combinations of elementary functions depending on t and x. Simple manipulations with Eqs. 54–57 result in fact that $L_i(t,x)$ can be, besides others, expressed in terms of sums

$$\sum_{j=1}^{\infty} \frac{t^j}{2j+1} P_j(x) \text{ and/or } \sum_{j=1}^{\infty} \frac{t^j}{2j+1} \frac{dP_j(x)}{dx} \tag{59}$$

Both the sums are expressible in terms of the incomplete elliptic integrals (Abramowitz and Stegun 1970, ➋ Chap. 32) which cannot be evaluated analytically but only numerically by numerical quadrature (Press et al. 1989, Sect. 6.11). Hence, the method of summation will follow another idea. Since the kernels $L_i(t,x)$ are singular at the point $(t = 1, x = 1)$, take out those terms from the sums in Eqs. 54–57, which are responsible for this singular behavior. This step will be performed analytically. After subtracting the singular contributions, the rest will be represented by quickly convergent infinite series, which are bounded over the whole interval $-1 \le x \le 1$ and $t = 1$. Prescribing an error of computation, this can simply be summed up numerically.

To sum up the infinite series of Eqs. 54–57, use the summation formulae for infinite series of *Legendre polynomials* and their derivatives listed in Appendix A (see Thong and Grafarend 1989). In order to use them, the fractions occurring in Eqs. 54–57 need to be arranged as follows:

$$\begin{aligned}
\frac{2j-1}{2j+1} &= 1 - \frac{15}{8j} + \frac{5}{4(j-1)} - \frac{3}{8(j-2)} + \frac{15}{4(j-2)(j-1)j(2j+1)} \qquad (j \ge 3) \\
\frac{j(j+1)(2j-1)}{2j+1} &= j^2 - \frac{1}{2} + \frac{1}{4j} - \frac{1}{4j(2j+1)} \\
\frac{j(2j-1)}{2j+1} &= j - \frac{1}{2} - \frac{2j-1}{2(2j+1)} \\
\frac{j^2(2j-1)}{2j+1} &= 2j^2 - j - \frac{j(j+1)(2j-1)}{2j+1}
\end{aligned} \tag{60}$$

These expressions allows one to write Eqs. 54–57 as

$$L_1(t,x) = \frac{\partial S_3(t,x)}{\partial x} - \frac{15}{8}\frac{\partial S_4(t,x)}{\partial x} + \frac{5}{4}\frac{\partial S_5(t,x)}{\partial x} - \frac{3}{8}\frac{\partial S_6(t,x)}{\partial x} + \frac{29}{24}t - \frac{171}{80}t^2 x + R_1(t,x) \tag{61}$$

$$L_2(t,x) = S_1(t,x) - \frac{1}{2}S_3(t,x) + \frac{1}{4}S_4(t,x) + R_2(t,x) \tag{62}$$

$$L_3(t,x) = (1-x^2)\left[\frac{\partial S_2(t,x)}{\partial x} - \frac{1}{2}\frac{\partial S_3(t,x)}{\partial x} - \frac{1}{2}L_1(t,x)\right] \tag{63}$$

$$L_4(t,x) = 2S_1(t,x) - S_2(t,x) - L_2(t,x) \tag{64}$$

where residuals $R_i(t,x)$, $i = 1,2$, are discussed in the next section and the sums $S_i(t,x)$, $i = 1,\ldots,6$, and their derivatives with respect to a variable x are listed in *Appendix A* (see Thong and Grafarend 1989). Substituting them into Eqs. 61–64, the final formulae for the integral kernels $L_i(t,x)$, $i = 1,\ldots,4$ can be obtained. Instead of writing down these final formulae explicitly, the

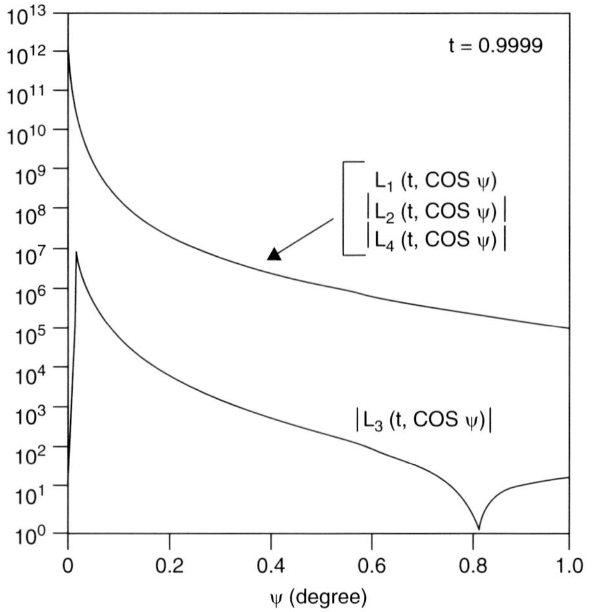

Fig. 2
The kernels $L_i(t, \cos\psi)$ and $|L_i(t, \cos\psi)|$, $i = 2,3,4$, for $t = 0.9999$ and $0 \le \psi \le 1$

subroutine KERL (see *Appendix B*, see Grafarend et al. 2006), which computes the kernels $L_i(t, x)$, $i = 1,\ldots,4$, and demonstrates the simplicity of the computation is presented. Note that KERL does not consider the residuals $R_i(t,x)$, $i = 1,2$, because of their negligible size; they can usually be neglected in practical implementations of the presented theory. However, a simple way of computing them is shown in the next section.

Figure 2 plots the integral kernel $L_1(t,x)$ and the absolute values of kernels $L_i(t,x)$, $i = 2,3,4$ (in order to use the logarithmic scale on the vertical axis), for $t = 0.9999$ and $x = \cos\psi$, $0 \le \psi \le 1°$. A huge increase can be observed in the amplitude of $L_i(t, \cos\psi)$, $i = 1,2,4$ when $\psi \to 0$ (but $L_i(t,1)$ are still bounded when $t < 1$), while $L_3(t, \cos\psi) = 0$ when $\psi \to 0$.

2.6 Residuals $R_i(t, x)$

The residuals $R_i(t,x)$, $i = 1,2$, appearing in Eqs. 61–64, are of the form

$$R_1(t,x) = \sum_{j=3}^{\infty} \frac{15}{4(j-2)(j-1)(2j+1)} t^j \frac{dP_j(x)}{dx} \tag{65}$$

$$R_2(t,x) = -\sum_{j=1}^{\infty} \frac{15}{4j(2j+1)} t^j P_j(x) \tag{66}$$

It has to be shown that the $R_i(t,x)$ are bounded for $0 < t \le 1$ and $-1 \le x \le 1$. Since

$$\left| \frac{dP_j(x)}{dx} \right| \le j^2 \tag{67}$$

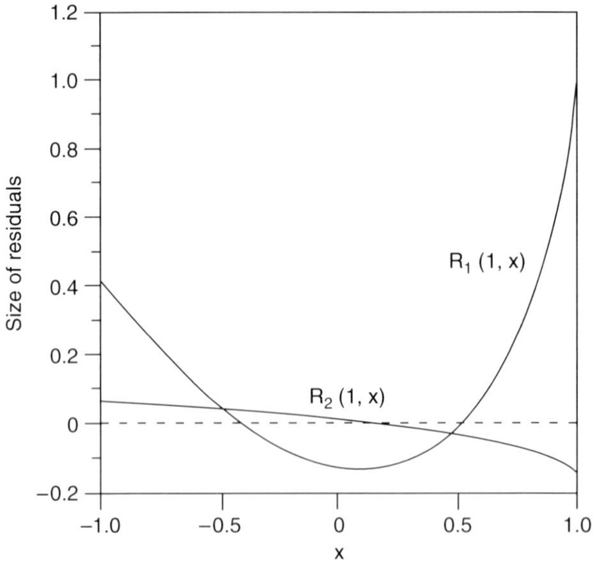

Fig. 3
The residuals $R_i(1,x), i = 1,2$ for $x \in (-1, 1)$

for $-1 \le x \le 1$ one can write

$$|R_1(t,x)| \le |R_1(1,x)| \le \frac{15}{4}\sum_{j=3}^{\infty}\frac{j}{(j-2)(j-1)(2j+1)} < \frac{15}{8}\sum_{j=3}^{\infty}\frac{1}{(j-2)(j-1)} = \frac{15}{8} \tag{68}$$

where the summation formula has been borrowed from Mangulis (1965, p. 53):

$$\sum_{j=1}^{\infty}\frac{1}{j(j+1)} = 1 \tag{69}$$

Similarly, the size of residual $R_2(t,x)$ can be estimated by realizing that $|P_j(x)| \le 1$

$$|R_2(t,x)| \le |R_2(1,x)| \le \frac{1}{4}\sum_{j=1}^{\infty}\frac{1}{j(2j+1)} < \frac{1}{8}\sum_{j=1}^{\infty}\frac{1}{j^2} = \frac{\pi^2}{48} \approx 0.206 \tag{70}$$

where the summation formula has been used

$$\sum_{j=1}^{\infty}\frac{1}{j^2} = \frac{\pi^2}{6} \tag{71}$$

taken from Mangulis (1965, p. 50). *Figure 3* plots the residuals $R_1(1,x)$ and $R_2(1,x)$ for $-1 \le x \le 1$. It can be seen that $|R_1(1,x)| < 1.04$ and $|R_2(1,x)| < 0.16$, which are more precise estimates than those given by Eqs. 68 and 70, respectively.

It is important to realize that the residuals $R_i(t,x)$, i = 1,2, are bounded at any point of the domain $(0,1)\otimes(-1, 1)$, so that the singularities of integral kernels $L_i(t,x)$, i = 1,…,4, at the point (1,1) are expressed analytically by the first terms on the right-hand sides of Eqs. 61–64. In addition, the function $R_i(t,x)$ can simply be evaluated numerically, since it is represented by a quickly convergent series. *Figure 4* demonstrates this fact in a transparent way; it plots the decay of

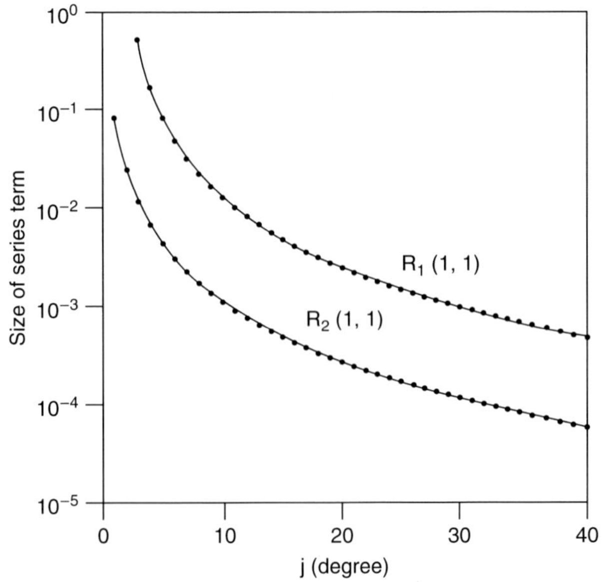

Fig. 4
The sizes of particular terms creating the infinite series for R_i (t,x),i = 1,2 at the point (1, 1)

magnitudes of series terms Eqs. 65 and 66 at the point (1, 1) with increasing degree j. Inspecting *Fig. 4* one can estimate, that it is sufficient to sum up the infinite series in Eq. 65 for $R_1(t,x)$ up to $j \approx 11$ and that in Eq. 66 for $R_2(t,x)$ up to $j \approx 3$ in order to achieve an absolute accuracy of order 0.01. This accuracy is sufficient for the evaluation of the ellipsoidal Poisson kernel $K^{\mathrm{ell}}(t, \Omega, \Omega')$ in the frame of the $O\left(e_0^2\right)$-approximation.

2.7 The Behavior at the Singularity

Though the *spherical Poisson integral kernel* (Kellogg 1929, Sect. IX.4) as well as the *ellipsoidal Poisson integral* in Eq. 45 are regular, which means that there is no singularity of the integral kernel inside the integration domain, it may simply happen that the distance of the computation point from the reference sphere or ellipsoid becomes small. Then the singularities of the integral kernels lie in a close neighborhood of the integration domain and the integrals turn out to be nearly singular. As a matter of fact, the integration error of a quadrature method for computing nearly singular integral depends, besides other factors, on the degree of singularity of an integral kernel (Hackbusch 1995, p. 330). Hence, one may raise the question, whether the degree of singularity of the original *spherical Poisson kernel* is the same as the degree of singularity of the *ellipsoidal Poisson kernel*. The answer to this question gives one a hint as to whether one can apply a numerical quadrature method of the same precision for computing both the spherical and the *ellipsoidal Poisson integral*.

The singularity of the *Poisson kernels* $K^{\mathrm{sph}}(t, \cos\psi)$ and $K^{\mathrm{ell}}(t,\Omega,\Omega')$ lies at the point ($t = 1$, $\psi = 0$), or, equivalently,($t = 1,\Omega'=\Omega$). To determine the degree of singularity of $K^{\mathrm{ell}}(t,\Omega,\Omega')$, start to study the behavior of the kernels $L_i(t, \cos\psi)$ at the singular point. Using Eqs. 61–64, one can, after some algebra, derive that

$$L_i(t, \cos\psi)\Big|_{\substack{t\to 1\\ \psi\to 0}} \sim \frac{1}{d^3} \text{ for } i = 1, 2, 4 \tag{72}$$

$$L_3(t, \cos\psi)\Big|_{\substack{t\to 1\\ \psi\to 0}} \sim \frac{1}{d^2} \tag{73}$$

where

$$d = \sqrt{(1-t)^2 + \psi^2} \tag{74}$$

Consequently, the particular constituents $k_i^{\text{ell}}(t, \Omega, \Omega')$, see Eqs. (50–53), of the kernel $k^{\text{ell}}(t, \Omega, \Omega')$, behave as follows:

$$k_i^{\text{ell}}(t, \Omega, \Omega')\Big|_{\substack{t\to 1\\ \Omega'\to\Omega}} \sim \frac{1}{d^2} \text{ for } i = 1, 3 \tag{75}$$

$$k_i^{\text{ell}}(t, \Omega, \Omega')\Big|_{\substack{t\to 1\\ \Omega'\to\Omega}} \sim \frac{1}{d^3} \text{ for } i = 2, 4 \tag{76}$$

Substituting the last estimates into Eq. 49, one has

$$k^{\text{ell}}(t, \Omega, \Omega')\Big|_{\substack{t\to 1\\ \Omega'\to\Omega}} \sim \frac{1-t^2}{d^3} \tag{77}$$

Realizing that

$$k^{\text{sph}}(t, \cos\psi)\Big|_{\substack{t\to 1\\ \psi\to 0}} \sim \frac{1-t^2}{d^3} \tag{78}$$

one can conclude that the *ellipsoidal Poisson kernel* $k^{\text{ell}}(t, \Omega, \Omega')$ has the same degree of singularity at the point ($t = 1, \Omega' = \Omega$) as the original *spherical Poisson kernel* at the same point; the estimate in Eq. 42 confirms this conclusion.

2.8 Conclusion

The chapter was motivated by the question as to whether *Green's function* to the *external Dirichlet boundary-value problem* for the *Laplace equation* with data distributed on an ellipsoid of revolution can be constructed. To answer this question, the solution of this boundary-value problem in terms of ellipsoidal harmonics was first constructed. The fact that the solution is represented by slowly convergent series of ellipsoidal harmonics prevents its use for regional *Geoid* computation. That is why it is confined in the next step to the $O\left(e_0^2\right)$-approximation, meaning that the terms of the magnitudes of the order of e_0^2 are kept; the terms of order $O\left(e_0^4\right)$ and of higher powers have not been considered. Nevertheless, the accuracy of the order of $O\left(e_0^2\right)$ is fairly good for today's requirement on *Geoid* height computation. Within this accuracy, it has been shown that the solution can be written as an integral taken over the full solid angle from data distributed on an ellipsoid of revolution which are multiplied by a kernel consisting of the traditional *spherical Poisson kernel* and the correction due to the elliptical shape of the boundary; the whole resolving kernel is called the *ellipsoidal Poisson kernel*. It has been managed to express the *ellipsoidal Poisson kernel*, originally represented in the form of an infinite sum of ellipsoidal harmonics, as a finite combination of elementary functions analytically describing the behavior of the *ellipsoidal Poisson kernel* at the singular point. Such an expression is suitable for a numerical solution to the harmonic upward and downward continuation of gravitation. Moreover, it has been shown, that the degree of singularity of the *ellipsoidal Poisson kernel* in the vicinity of its singular point is of the same degree as that of the original *spherical Poisson kernel*.

3 Stokes Boundary-Value Problem on the Ellipsoid of Revolution

Green's function to Stokes's boundary-value problem with the gravity data distributed over an ellipsoid of revolution has been constructed. It has been shown that the problem has an unique solution provided that the first eccentricity e_0 *of the ellipsoid of revolution is less than 0.65041. The ellipsoidal Stokes function describing the effect of ellipticity of the boundary is expressed in the* $O\left(e_0^2\right)$*-approximation as a finite sum of elementary functions which describe analytically the behavior of the ellipsoidal Stokes function at the singular point* $\psi = 0$*. It is proved that the degree of singularity of the ellipsoidal Stokes function in the vicinity of its singular point is the same as that of the spherical Stokes function.*

Stokes's integral for the gravimetric determination of the *Geoid* requires, besides other assumptions, that the gravity anomalies are referred to a sphere in geometry and gravity space. The relative error introduced by this spherical approximation is, according to Heiskanen and Moritz (1967, Sect. 2.14), of the order of 3×10^{-3} which causes an absolute error of 0·5 m in terms of geoidal heights. A simple error analysis reveals that this error can be even larger and that it may reach 2 m. Such an error is unacceptable at a time when "the 1 cm geoid" is the target, i.e., observations of the coordinates of topographic points by means of the Global Positioning System can provide ellipsoidal heights with a relative accuracy of a few centimetres. The question arises whether the geoid can be determined with a comparable relative accuracy in order, for instance, to be able to define a system of precise orthometric heights.

To develop the theory of geoidal height determination as precisely as possible is one step toward obtaining the geoid with a relative accuracy of a few centimetres. In this chapter, the aim is to solve the *Stokes boundary-value problem* for gravity anomalies distributed on an ellipsoid of revolution in geometry and gravity space. The ellipsoidal approximation of the geoid reflects reality much better than a spherical approximation since the actual shape of the geoid deviates from an ellipsoid of revolution by 100 m at the most. Treating the geoid with respect to an ellipsoid of revolution in an appropriate normal gravity field, in particular as a boundary condition prescribed on the approximate geoid surface—the reference ellipsoid of revolution—may cause relative errors in the order of 1.5×10^{-5}. The absolute error in geoidal heights computed from the *ellipsoidal Stokes integral* does not exceed 2 mm.

3.1 Formulation of the Stokes Problem on an Ellipsoid of Revolution

To begin with, ellipsoidal coordinates u, ϑ, λ are introduced through the transformation relations into Cartesian coordinates $\{x,y,z\}$ (for details see, e.g., Moon and Spencer 1961, Table 1.07, p. 31;

Heiskanen and Moritz, 1967, Sect. 1.19; Thong and Grafarend 1989, pp. 290–295):

$$x = \sqrt{u^2 + E^2} \sin \vartheta \cos \lambda$$
$$y = \sqrt{u^2 + E^2} \sin \vartheta \sin \lambda \tag{79}$$
$$z = u \cos \vartheta$$

where $E = \sqrt{a^2 - b^2}$(= constant) is the linear eccentricity, also called the radius of the focal circle, of the set of confocal ellipsoidal coordinate surfaces u = constant. The problem to be dealt is to determine the potential $T(u,\Omega)$, $\Omega = (\vartheta,\lambda)$, on and outside the reference ellipsoid of revolution $u = b_0$ such that

$$\nabla^2 T = 0 \quad \text{for } u>b_0, \tag{80}$$

$$\frac{\partial T}{\partial u} + \frac{2}{u} = -f \quad \text{for } u = b_0 \tag{81}$$

$$T = \frac{c}{u} + O\left(\frac{1}{u^3}\right) \quad \text{for } u \to \infty, \tag{82}$$

where $f(\Omega)$ is assumed to be a known square-integrable function, i.e., $f(\Omega) \in L_2(\Omega)$, and c is a constant. A limiting case $E = 0$ when the reference ellipsoid of revolution, $u = b_0$, reduces to a sphere $x^2 + y^2 + z^2 = b_0$, and the boundary-value problem Eqs. (80–82) reduces to the *spherical Stokes problem* (Heiskanen and Moritz 1967, Sect. 2.16) will also be considered. To guarantee the existence of a solution in this particular case, the first-degree harmonics of $f(\Omega)$ have to be removed by means of the postulate (Holota 1995)

$$\int f(\Omega) Y_{1m}(\Omega) d\Omega \text{ for } m = -1, 0, 1 \tag{83}$$

Here, $Y_{1m}(\Omega)$ are complex spherical harmonics of the first degree and order m, the asterisk denotes a complex conjugate, Ω_0 is the full solid angle, and $d\,\Omega = sin\vartheta\; d\vartheta d\lambda$. Throughout the chapter it is assumed that conditions (5) are satisfied. Moreover, to guarantee the uniqueness of the solution, the first-degree harmonics have to be removed from the potential as the asymptotic condition (82) indicates. The question of the uniqueness of the problem (80–82) for a general case $E \neq 0$ will e examined later in ➲ Sect. 3.5.

The problem (80–82) will be called the *ellipsoidal Stokes boundary-value problem* since it generalizes the traditional *Stokes boundary-value problem* formulated on a sphere. It approximates the boundary-value problem for geoid determination (Martinec and Vaniček 1996) or the *Molodensky scalar boundary-value problem* (Heck 1991, Sect. 6.3) by maintaining the two largest terms in boundary condition (81) and omitting two small ellipsoidal correction terms. (The effect of the ellipsoidal correction terms on the solution of the *ellipsoidal Stokes boundary-value problem* will be treated in a separate chapter.) Alternatively, the boundary-value problem for geoid determination may be formulated in geodetic coordinates (Otero 1995) which results in a boundary condition of a form slightly different from Eq. (81). However, ellipsoidal coordinates and formulation (80–82) will be used since the *Laplace operator* is separable in ellipsoidal coordinates (Moon and Spencer 1961) which will substantially help in finding a solution to the problem.

The solution of the *Laplace equation* (80) can be written in terms of ellipsoidal harmonics as follows (Moon and Spencer 1961, p. 32; Heiskanen and Moritz 1967, Sect. 1.20; Thong and Grafarend 1989, p. 302):

$$T(u,\Omega) = \sum_{j=0}^{\infty} \sum_{m=-j}^{j} T_{jm} \frac{Q_{jm}\left(i\frac{u}{E}\right)}{Q_{jm}\left(i\frac{b_0}{E}\right)} Y_{jm}(\Omega) \tag{84}$$

where $Q_{jm}\left(i\frac{u}{E}\right)$ are *Legendre's function of the second kind*, and T_{jm} are coefficients to be determined from boundary condition (81). Equation A.11, derived in *Appendix A* (see Thong and Grafarend 1989), shows that

$$Q_{jm}\left(i\frac{u}{E}\right) \approx O\left(\frac{1}{u^{j+1}}\right) \text{ for } u \to \infty \tag{85}$$

To satisfy asymptotic condition (82), the term with $j = 1$ must be eliminated from the sum over j in Eq. 84, i.e.,

$$T(u,\Omega) = \sum_{\substack{j=0\\ j\neq 1}}^{\infty}\sum_{m=-j}^{j} T_{jm}\frac{Q_{jm}\left(i\frac{u}{E}\right)}{Q_{jm}\left(i\frac{b_0}{E}\right)} Y_{jm}(\Omega) \tag{86}$$

Substituting expansion (86) into boundary conditions (81) yields

$$\sum_{\substack{j=0\\ j\neq 1}}^{\infty}\sum_{m=-j}^{j}\frac{1}{Q_{jm}\left(i\frac{b_0}{E}\right)}\left[\frac{dQ_{jm}\left(i\frac{u}{E}\right)}{du}\Big|_{u=b_0} + \frac{2}{b_0}Q_{jm}\left(i\frac{b_0}{E}\right)\right] T_{jm}Y_{jm}(\Omega) = -f(\Omega). \tag{87}$$

Moreover, expanding function $f(\Omega)$ in a series of spherical harmonics

$$f(\Omega) = \sum_{j=0}^{\infty}\sum_{m=-j}^{j}\int_{\Omega_0} f(\Omega')Y^*_{jm}(\Omega')d\Omega' Y_{jm}(\Omega) \tag{88}$$

where the first-degree spherical harmonics of $f(\Omega)$ are equal to zero by assumption (83), substituting expansion (88) into Eq. 87, and comparing the coefficients at spherical harmonics $Y_{jm}(\Omega)$ in the result yields

$$T_{jm} = -\frac{\int_{\Omega_0} f(\Omega')Y_{jm}(\Omega')d\Omega'}{\frac{1}{Q_{jm}\left(i\frac{b_0}{E}\right)}\left[\frac{dQ_{jm}\left(i\frac{u}{E}\right)}{du}\Big|_{u=b_0} + \frac{2}{b_0}Q_{jm}\left(i\frac{b_0}{E}\right)\right]} \tag{89}$$

for $j = 0, 2, \ldots,$ and $m = -j, \ldots, j$. Substituting coefficients T_{jm} into Eq. 86 and changing the order of summation over j and m and of integration over Ω', due to the uniform convergence of the series, one obtains

$$T(u,\Omega) = \int_{\Omega_0} f(\Omega')\sum_{j=0}^{\infty}\sum_{m=-j}^{j}\alpha_{jm}(u)Y_{jm}(\Omega')Y_{jm}(\Omega)d\Omega' \tag{90}$$

where functions $\alpha_{jm}(u)$ have been introduced

$$\alpha_{jm}(u) = -\frac{Q_{jm}\left(i\frac{u}{E}\right)}{\frac{dQ_{jm}\left(i\frac{u}{E}\right)}{du}\Big|_{u=b_0} + \frac{2}{b_0}Q_{jm}\left(i\frac{b_0}{E}\right)} \tag{91}$$

to abbreviate notations.

3.2 The Zero-Degree Harmonic of T

The zero-degree harmonic of potential $T(u,\Omega)$ is to be calculated. One first has:

$$\alpha_{00}(u) = \frac{Q_{00}\left(i\frac{u}{E}\right)}{\frac{dQ_{00}\left(i\frac{u}{E}\right)}{du}\Big|_{u=b_0} + \frac{2}{b_0}Q_{00}\left(i\frac{u}{b_0}\right)} \tag{92}$$

where $Q_{00}\left(i\frac{u}{E}\right)$ can be expressed as (Arfken 1968, Eq. 12.222)

$$Q_{00}\left(i\frac{u}{E}\right) = -i\arctan\left(\frac{E}{u}\right) \tag{93}$$

with $i = \sqrt{-1}$. Taking the derivatives of the last equation with respect to u, one gets

$$\frac{dQ_{00}\left(i\frac{u}{E}\right)}{du} = \frac{iE}{u^2 + E^2} \tag{94}$$

By substituting Eqs. 93 and 94 into 92,

$$\alpha_{00}(u) = \frac{\arctan\left(\frac{E}{u}\right)}{\frac{E}{b_0^2+E^2} - \frac{2}{b_0}\arctan\left(\frac{E}{b_0}\right)} \tag{95}$$

Since $\arctan x \doteq x$ for $x \ll 1$, one can see that

$$\alpha_{00}(u) = \frac{c}{u} \text{ for } u \to \infty \tag{96}$$

with

$$c = \frac{\arctan\left(\frac{E}{u}\right)}{\frac{E}{b_0^2+E^2} - \frac{2}{b_0}\arctan\left(\frac{E}{b_0}\right)} \tag{97}$$

3.3 Solution on the Reference Ellipsoid of Revolution

Solution (90) can now be expressed as

$$T(u,\Omega) = \frac{\alpha_{00}(u)}{4\pi}\int_{\Omega_0} f(\Omega')d\Omega' + \int_{\Omega_0} f(\Omega')\sum_{j=2}^{\infty}\sum_{m=-j}^{j}\alpha_{jm}(u)Y_{jm}^*(\Omega')Y_{jm}(\Omega)d\Omega' \tag{98}$$

In particular, the interest is in finding potential $T(u,\Omega)$ on the reference ellipsoid of revolution $u = b_0$, i.e., function $T(b_0,\Omega)$. In this case, the general formulae (91) and (98) reduce to

$$T(b_0,\Omega) = \frac{\alpha_{00}(u)}{4\pi}\int_{\Omega_0} f(\Omega')d\Omega' + \int_{\Omega_0} f(\Omega')\sum_{j=2}^{\infty}\sum_{m=-j}^{j}\alpha_{jm}(b_0)Y_{jm}^*(\Omega')Y_{jm}(\Omega)d\Omega' \tag{99}$$

where $\alpha_{00}(b_0)$ are given by Eq. 95 for $u = b_0$, and the other $\alpha_{jm}(b_0)$ read

$$\alpha_{jm}(b_0) = -\frac{Q_{jm}\left(i\frac{b_0}{E}\right)}{\frac{dQ_{jm}\left(i\frac{u}{E}\right)}{du}\Big|_{u=b_0} + \frac{2}{b_0}Q_{jm}\left(i\frac{b_0}{E}\right)} \tag{100}$$

From the practical point of view, the spectral form Eq. 99 of the solution of the *Stokes boundary-value problem* Eqs. (80–82) is often inconvenient, since constructing the spectral components of $f(\Omega)$ and summing them up to high degrees and orders may become time consuming and numerically unstable. Moreover, for the case in which the reference ellipsoid of revolution $u = b_0$ deviates only slightly from a sphere, which is the case of the Earth, the solution to the problem should be close to the solution for the same problem but formulated for a sphere. It is thus attempted to rewrite $T(b_0,\Omega)$ as a sum of the well-known *Stokes integral* plus the corrections due to the ellipticity of the boundary. An evident advantage of such a decomposition is that existing theories as well as numerical codes for geoidal height computations can simply be corrected for the ellipticity of the geoid.

To make the theory as simple as possible, but still maintain the requirements for geoidal height accuracy throughout the following derivations, one shall retain the terms of magnitudes of the order of the first eccentricity e_0^2 of an ellipsoid of revolution, and neglect the term of higher power of e_0^2. This approximation is justifiable, because the error introduced by this approximation is $1 \cdot 5 \times 10^{-5}$ at the most which then causes an error of no more than 2 mm in the geoidal heights.

3.4 The Derivative of the Legendre Function of the Second Kind

Now, look for the derivative of the *Legendre function* $Q_{jm}\left(i\frac{u}{E}\right)$ with respect to variable u. Equation A.11, derived in Appendix A (see Thong and Grafarend 1989), shows that $Q_{jm}\left(i\frac{u}{E}\right)$ can be expressed as an infinite power series of first eccentricity e,

$$e = \frac{E}{\sqrt{u^2 + E^2}} \tag{101}$$

in the form

$$Q_{jm}\left(i\frac{u}{E}\right) = (-1)^{m-(j+1)/2} \frac{(j+m)!}{(2j+1)!!} e^{j+1} \sum_{k=0}^{\infty} \alpha_{jmk} e^{2k} \tag{102}$$

where coefficient α_{jmk} can, for instance, be defined by recurrence relations (A 14). In particular, one shall need

$$\alpha_{jm0} = 1, \tag{103}$$

and

$$\alpha_{jm1} = \frac{(j+1)^2 - m^2}{2(2j+3)}, \quad \alpha_{jm2} = \frac{(j+3)^2 - m^2}{4(2j+5)} \alpha_{jm1} \tag{104}$$

Throughout this chapter it is assumed that the eccentricity e_0 of the reference ellipsoid of revolution $u = b_0$,

$$e_0 = \frac{E}{\sqrt{b_0^2 + E^2}} \tag{105}$$

is less than 1. For points (u,Ω) outside, or on the reference ellipsoid of revolution, i.e., when $u \geq b_0$, series Eq. 102 then converges. One can take the derivative of this series with respect to u and change the order of integration and summation since the resulting series Eq. 106 is uniformly convergent. Consequently,

$$\frac{dQ_{jm}\left(i\frac{u}{E}\right)}{du} = (-1)^{m-(j+1)/2} \frac{(j+m)!}{(2j+1)!!} e^{j+1} \sum_{k=0}^{\infty} (2k+j+1) \alpha_{jmk} e^{2k} \frac{de}{du} \tag{106}$$

The derivative of eccentricity e with respect to u can be easily obtained from Eq. 101:

$$\frac{de}{du} = -(1 - e^2)\frac{e}{u} \tag{107}$$

Substituting Eq. 107 into Eq. 106 yields

$$\frac{dQ_{jm}\left(i\frac{u}{E}\right)}{du} = (-1)^{m-(j+1)/2}\frac{(j+m)!}{(2j+1)!!}(1-e^2)\frac{e^{j+1}}{u}\sum_{k=0}^{\infty}(-2k-j-1)\alpha_{jmk}e^{2k} \tag{108}$$

3.5 The Uniqueness of the Solution

Now, one is ready to express the sum in the denominator of Eq. 100,

$$\frac{dQ_{jm}\left(i\frac{u}{E}\right)}{du}\Big|_{u=b_0} + \frac{2}{b_0}Q_{jm}\left(i\frac{b_0}{E}\right)$$
$$= (-1)^{m-(j+1)/2}\frac{(j+m)!}{(2j+1)!!}\frac{e_0^{j+1}}{b_0}\sum_{k=0}^{\infty}\alpha_{jmk}\left[\left(1-e_0^2\right)(-2k-j-1)+2\right]e_0^{2k} \tag{109}$$

Substituting Eqs. 102 and 109 into Eq. 100 yields

$$\alpha_{jm}(b_0) = b_0\frac{\sum_{k=0}^{\infty}\alpha_{jmk}e_0^{2k}}{\sum_{k=0}^{\infty}\alpha_{jmk}\left[\left(1-e_0^2\right)(2k+j+1)-2\right]e_0^{2k}} \tag{110}$$

The last formula enables one to examine the uniqueness of the solution of boundary-value problem Eqs. (80–82). Let the denominator of expression Eq. 110 for α_{jm} be denoted by $d_{jm}\left(e_0^2\right)$,

$$d_{jm}\left(e_0^2\right) = \sum_{k=0}^{\infty}\alpha_{jmk}\left[\left(1-e_0^2\right)(2k+j+1)-2\right]e_0^{2k} \tag{111}$$

$j = 2,\ldots;\ m = 0,1,\ldots,j$, and investigate the dependence of $d_{jm}\left(e_0^2\right)$ on e_0^2. Numerical examinations have resulted in ➋ *Fig. 1* showing the behavior of $d_{jm}\left(e_0^2\right)$ for three values of e_0^2 as functions of angular degree j and order m, $j = 2,\ldots,8$; $m = 0,2,\ldots,j$; the combined index $jm{:}= j(j+1)/2 + m + 1$. If $e_0^2 < 0.42303$, then all coefficients $d_{jm}\left(e_0^2\right)$ are positive which means that the solution of boundary-value problem Eqs. (80–82) is unique.

Once, $e_0^2 \cong 0.42303$, $d_{22}\left(e_0^2\right) = 0$, and the solution of the problem is not unique. If the size of e_0^2 is increased further, see, for instance, the curve denoted by triangles in ➋ *Fig. 5* for $e_0^2 = 0.7$, other $d_{jm}\left(e_0^2\right)$ may then vanish and problem Eqs. 80–82 have a nonunique solution. One can conclude that the uniqueness of the solution can only be guaranteed if the square of the first eccentricity is less than 0.42303. Fortunately, this condition is satisfied for the Earth since $e_0^2 = 0.006\,694\,380$ (Moritz 1980) for the ellipsoid best fitting the Earth's figure (the corresponding $d_{jm}\left(e_0^2\right)$ are plotted in ➋ *Fig. 5* as black dots).

3.6 The Approximation up to $O\left(e_0^2\right)$

Arrange formula Eq. 110 into a form that is more suitable for highlightning the Stokes contribution. After some cumbersome but straightforward algebra, one arrives at

$$\alpha_{jm}(b_0) = \frac{b_0}{j-1}\left[1-\frac{\sum_{k=1}^{\infty}\left[\frac{2k}{j-1}\alpha_{jmk}-\left(1+\frac{2k}{j-1}\right)\alpha_{jm,k-1}\right]e_0^{2k}}{1+\sum_{k=1}^{\infty}\left(1+\frac{2k}{j-1}\right)\left(\alpha_{jmk}-\alpha_{jm,k-1}\right)e_0^{2k}}\right] \tag{112}$$

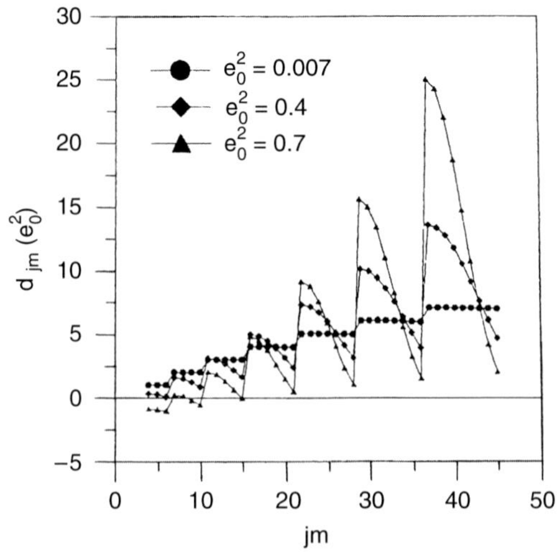

Fig. 5
Functions $d_{jm}\left(e_0^2\right)$ for various e_0^2 as functions of degree j and order m; the combined index $jm := j(j+1)/2 + m + 1$

where α_{jm0} has been substituted from Eq. 103. The ratio of the two power series in Eq. 112 can further be expanded as a power series of eccentricity e_0^2. The explicit forms of the first two terms of such a series are as follows:

$$\frac{\sum\limits_{k=1}^{\infty}\left[\frac{2k}{j-1}\alpha_{jmk} - \left(1+\frac{2k}{j-1}\right)\alpha_{jm,k-1}\right]e_0^{2k}}{1+\sum\limits_{k=1}^{\infty}\left(1+\frac{2k}{j-1}\right)\left(\alpha_{jmk}-\alpha_{jm,k-1}\right)e_0^{2k}} = d_{jm1}e_0^2 + d_{jm2}e_0^4 + R \tag{113}$$

where

$$d_{jm1} = \frac{2}{j-1}\alpha_{jm1} - \frac{j+1}{j-1} \tag{114}$$

$$d_{jm2} = \frac{2}{(j-1)^2}\left[2(j-1)\alpha_{jm2} - (j+1)\alpha_{jm1}^2 + (j+3)\alpha_{jm1}\right] - \left(\frac{j+1}{j-1}\right)^2 \tag{115}$$

and R is the residual of the series expansion.

It shall now be attempted to estimate the sizes of particular terms on the right-hand side of Eq. 113. Substituting for α_{jm1} and α_{jm2} from Eq. 104 into Eqs. 114 and 115 and after some more algebra, one gets

$$d_{jm1} = -\frac{j^2+3j+2+m^2}{(j-1)(2j+3)} \tag{116}$$

and

$$d_{jm2} = \frac{10j^3+50j^2+100j+58+2m^2(3j+4)}{(j-1)^2(2j+3)(2j+5)}\alpha_{jm1} - \left(\frac{j+1}{j-1}\right)^2 \tag{117}$$

Hence,

$$|d_{jm1}| \le \frac{2j^2+3j+2}{(j-1)(2j+3)} = 1 + \frac{1}{j-1} + \frac{2}{(j-1)(2j+3)} \le \frac{16}{7} \tag{118}$$

for $j \ge 2$ and $|m| \le j$.

It shall continue estimating the maximum size of the first constituent creating coefficients $d_j m2$ for $j \ge 2$ and $|m| \le j$

$$\begin{aligned}
0 &< \frac{10j^3+50j^2+100j+58+2m^2(3j+4)}{(j-1)^2(2j+3)(2j+5)}\alpha_{jm1} \le \frac{16j^3+58j^2+100j+58}{(j-1)^2(2j+3)(2j+5)}\alpha_{jm1} \\
&\le \frac{16j^3+58j^2+100j+58}{(j-1)^2(2j+3)(2j+5)}\alpha_{j01} = \frac{(j+1)^2(16j^3+58j^2+100j+58)}{2(j-1)^2(2j+3)^2(2j+5)} \\
&< \frac{(j+1)^2(16j^3+58j^2+101j+75)}{2(j-1)^2(2j+3)^2(2j+5)} = \frac{(j+1)^2(8j^2+17j+25)}{2(j-1)^2(2j+3)(2j+5)} \\
&< \frac{(j+1)^2(8j^2+30j+27)}{2(j-1)^2(2j+3)(2j+5)} = \frac{(j+1)^2(4j+9)}{2(j-1)^2(2j+5)} < \left(\frac{j+1}{j-1}\right)^2
\end{aligned} \tag{119}$$

Therefore, coefficients d_{jm2}, Eq. 117, can be estimated as

$$|d_{jm2}| < \left(\frac{j+1}{j-1}\right)^2 = \left(1+\frac{2}{j-1}\right)^2 \le 9 \tag{120}$$

which is again valid for $j \ge 2$ and $|m| \le j$.

It remains to estimate the size of the residual R of series (113). Since it is fairly problematical to find an analytical expression for R, it will be estimated numerically. For $e_0^2 < 1$, the power series over k in the ratio on the left-hand side of Eq. 113 are convergent and may be summed up. It has been done numerically for the Earth's eccentricity, $e_0^2 = 0.006\,694\,380$ (Moritz 1980), and for $j = 2, \ldots, 10^4$, and $m = 0, 1, \ldots, j$. Note that the larger the j, the slower the decrease in the magnitude of series terms. For instance, to achieve the relative accuracy of 10^{-8} in evaluating the fraction in Eq. 113 for $j = 10^4$, the power series must be summed up to $k = 60$. The numerical investigations have shown that the magnitude of the residual R in Eq. 113 is of the order e_0^6 at the most

$$R \sim O\left(e_0^6\right) \tag{121}$$

Finally, retaining only the term of order $O\left(e_0^2\right)$ and dropping out higher-order terms, Eq. 113 becomes

$$\frac{\sum_{k=1}^{\infty}\left[\frac{2k}{j-1}\alpha_{jmk} - (1+\frac{2k}{j-1})\alpha_{jm,k-1}\right]e_0^{2k}}{1+\sum_{k=1}^{\infty}(1+\frac{2k}{j-1})(\alpha_{jmk}-\alpha_{jm,k-1})e_0^{2k}} = \left(\frac{2}{j-1}\alpha_{jm1} - \frac{j+1}{j-1}\right)e_0^2 + O\left(9\times e_0^4\right) \tag{122}$$

Therefore, if it is restricted only to terms of the magnitude of $O\left(e_0^2\right)$ at the most, coefficients $\alpha_{jm}(b_0)$, Eq. 112, can then be expressed as

$$\alpha_{jm}(b_0) = \frac{b_0}{j-1}\left[1 - \left(\frac{2}{j-1}\alpha_{jm1} - \frac{j+1}{j-1}\right)e_0^2\right] \tag{123}$$

Substituting Eq. 123 into Eq. 99, and bearing in mind the Laplace addition theorem for spherical harmonics (for instance, Varshalovich et al. 1988, p. 164, Eq. 9),

$$P_j(\cos\psi) = \frac{4\pi}{2j+1}\sum_{m=-j}^{j} Y_{jm}(\Omega)Y^*_{jm}(\Omega') \tag{124}$$

where $P_j(\cos\psi)$ is the *Legendre polynomial* of degree j, and ψ is the angular distance between directions Ω and Ω', one gets

$$T_{jm}(b_0,\Omega) = \frac{\alpha_{00}(b_0)}{4\pi}\int_{\Omega_0} f(\Omega')d\Omega' + \frac{b_0}{4\pi}\int_{\Omega_0} f(\Omega')\left[S(\psi) - e_0^2 S^{ell}(\Omega,\Omega')\right]d\Omega' \tag{125}$$

where $S(\psi)$ is the *spherical Stokes function* (Stokes 1849; Heiskanen and Moritz 1967, Eq. 2-169), and

$$S^{ell}(\Omega,\Omega') = 4\pi\sum_{j=2}^{\infty}\sum_{m=-j}^{j}\frac{1}{(j-1)^2}\left[\frac{(j+1)^2-m^2}{2j+3} - (j+1)\right]Y_{jm}(\Omega)Y^*_{jm}(\Omega') \tag{126}$$

In carrying out the last operation, α_{jml} has been substituted from Eq. 104.

3.7 The Ellipsoidal Stokes Function

The function $S^{ell}(\Omega,\Omega')$ shall be called the ellipsoidal Stokes function because it describes the effect of the ellipticity of the boundary on the solution of Stokes's problem. Subsequent effort will be devoted to converting the spectral form Eq. 126 of $S^{ell}(\Omega,\Omega')$ to a spatial representation.

One of the trickiest steps in such a conversion is to sum up the series

$$\sum_{m=-j}^{j} m^2 Y_{jm}(\Omega)Y_{jm}(\Omega') \tag{127}$$

Thong 1993 used a simple way of finding the expression for this sum. He took the second-order derivative of the *Laplace addition theorem* Eq. 124 with respect to longitude Λ, and obtained sum Eq. 127 as a linear combination of the zero-, first- and second-order derivatives of the *Legendre polynomials* $P_j(\cos\psi)$. To obtain the spatial form of the *Hotine function*, he summed this result from $j = 2$ up to infinity. Unfortunately, this procedure leads to a strongly singular *ellipsoidal Hotine function* growing as $1/\psi^2$ as $\Omega' \to \Omega$. (Note that the original spherical Hotine function grows as $1/\psi$ as $\psi \to 0$). Evidently, this result cannot be accepted since, in particular, as $\Omega' \to \Omega$, the ellipsoidal correction to the *spherical Hotine problem* may become larger (even for a very small eccentricity of an ellipsoid of revolution) than the contribution due to the *spherical Hotine function*.

In Appendix B (see Grafarend et al. 2006), another way of summing up series (127) is shown. Later on, it can be seen that this approach ensures that the *ellipsoidal Stokes function* $S^{ell}(\Omega,\Omega')$ has the same degree of singularity at computation point $\psi = 0$ as the original *spherical Stokes function* $S(\psi)$. This result will be acceptable because, for an ellipsoid of revolution with a small flattening, the ellipsoidal correction to the *Stokes boundary-value problem* cannot become larger than the contribution due to the *spherical Stokes function*. Substituting Eq. B.21 into Eq. 126, the *ellipsoidal Stokes function* $S^{ell}(\Omega,\Omega')$ can be composed of four different terms,

$$\begin{aligned} S^{ell}(\Omega,\Omega') &= \sin\vartheta(\cos\vartheta\sin\psi\cos\psi\cos\alpha - \sin\vartheta\cos^2\psi\cos^2\alpha \\ &\quad + \sin\vartheta\sin^2\alpha)K_1(\cos\psi) + (1-\sin^2\vartheta\sin^2\alpha)K_2(\cos\psi) \\ &\quad - \sin\vartheta\cos\alpha(\cos\vartheta\sin\psi - \sin\vartheta\cos\psi\cos\alpha)K_3(\cos\psi) - K_4(\cos\psi) \end{aligned} \tag{128}$$

where the isotropic parts $K_i \cos\psi$, $i = 1,\ldots,4$ of function $S^{ell}(\Omega,\Omega')$ are given by infinite series of *Legendre polynomials*

$$K_1(\cos\psi) = \sum_{j=3}^{\infty} \frac{2j-1}{(j-2)^2(2j+1)} \frac{dP_j(\cos\psi)}{d\cos\psi} \tag{129}$$

$$K_2(\cos\psi) = \sum_{j=2}^{\infty} \frac{(j+1)^2(2j+1)}{(j-1)^2(2j+3)} P_j(\cos\psi) \tag{130}$$

$$K_3(\cos\psi) = \sum_{j=3}^{\infty} \frac{j(2j-1)}{(j-2)^2(2j+1)} P_j(\cos\psi) \tag{131}$$

$$K_4(\cos\psi) = \sum_{j=2}^{\infty} \frac{(j+1)(2j+1)}{(j-1)^2} P_j(\cos\psi) \tag{132}$$

3.8 Spatial Forms of Functions $K_i(\cos\psi)$

It shall now be attempted to express the infinite sums for $K_i(\cos\psi)$ as finite combinations of elementary functions depending on $\cos\psi$. Following simple algebra applied to expressions Eqs. (129–132), $K_i(\cos\psi)$ can be expressed in terms of sums

$$\sum_{j=3}^{\infty} \frac{P_j(\cos\psi)}{2j+1} \tag{133}$$

and

$$\sum_{j=3}^{\infty} \frac{P_j(\cos\psi)}{(j-1)^2} \tag{134}$$

The first sum can be expressed in full elliptic integrals (Pick et al. 1973, Appendix 18) which can be calculated only approximately by a method of numerical quadrature (Press et al. 1989, Sect. 6.11). The second sum is equal to a definite integral, the primitive function of which cannot be expressed in closed analytical form (Pick et al. 1973, Appendix 18) but again only numerically. It can thus be seen that sums Eqs. (129–132) cannot be expressed in closed analytical forms. Therefore, this method of summation will be based on the following idea. Since kernels $K_i(\cos\psi)$ are singular at point $\psi = 0$, the contributions that are responsible for the singular behavior at point $\psi = 0$ shall be removed from sums Eqs. 129–132. These contributions will be expressed in closed analytical forms. Having removed the singular contributions, the rest will be represented by quickly convergent infinite series, which are bounded on the whole interval $0 \le \psi \le \pi$. Prescribing an error of computation, they can be simply summed up numerically.

In a preparatory step, the series of derivatives of the *Legendre polynomials* will be summed up. Making use of the recurrence relation (for instance, Arfken 1968, Eq. 12.23)

$$\frac{dP_{j+1}(\cos\psi)}{d\cos\psi} - \frac{dP_{j-1}(\cos\psi)}{d\cos\psi} = (2j+1)P_{j+1}(\cos\psi) \tag{135}$$

one can readily derive that

$$\sum_{j=0}^{\infty} \frac{4}{(2j-1)(2j+3)} \frac{dP_j(\cos\psi)}{d\cos\psi} = \sum_{j=0}^{\infty} P_j(\cos\psi) \tag{136}$$

and

$$\sum_{j=0}^{\infty} \frac{2(4j+1)}{(j^2-1)(2j-1)(2j+3)} \frac{dP_j(\cos\psi)}{d\cos\psi} = \sum_{j=1}^{\infty} \frac{P_j(\cos\psi)}{j} + \frac{1}{10} \tag{137}$$

Furthermore, borrowing two formulae from Pick et al. 1973, Eqs. D.18.1 and D.18.1,

$$\sum_{j=0}^{\infty} P_j(\cos\psi) = \frac{1}{2\sin\frac{\psi}{2}} \tag{138}$$

and

$$\sum_{j=1}^{\infty} \frac{P_j(\cos\psi)}{j} = -\ln\left(\sin\frac{\psi}{2} + \sin^2\frac{\psi}{2}\right) \tag{139}$$

one gets two useful formulae

$$\sum_{j=1}^{\infty} \frac{4}{(2j-1)(2j+3)} \frac{dP_j(\cos\psi)}{d\cos\psi} = \frac{1}{2\sin\frac{\psi}{2}} \tag{140}$$

$$\sum_{j=2}^{\infty} \frac{2(4j+1)}{(j^2-1)(2j-1)(2j+3)} \frac{dP_j(\cos\psi)}{d\cos\psi} = -\ln\left(\sin\frac{\psi}{2} + \sin^2\frac{\psi}{2}\right) + \frac{1}{10} \tag{141}$$

Start summing up infinite series Eq. 129. The fraction occuring in this series can be decomposed as

$$\frac{2j-1}{(j-2)^2(2j+1)} = \frac{4}{(2j-1)(2j+3)} + \frac{4(4j+1)}{(j^2-1)(2j-1)(2j+3)} + \frac{3(26j^3-27j^2-18j-1)}{(j-2)^2(j^2-1)(2j-1)(2j+1)(2j+3)} \tag{142}$$

In view of the last equation and Eqs. 140 and 141, function $K_1(\cos\psi)$ becomes

$$K_1(\cos\psi) = \frac{1}{2\sin\frac{\psi}{2}} - 2\ln\left(\sin\frac{\psi}{2} + \sin^2\frac{\psi}{2}\right) - \frac{3}{5} - \frac{16}{7}\cos\psi + R_1(\cos\psi) \tag{143}$$

where

$$R_1(\cos\psi) = \sum_{j=3}^{\infty} \frac{3(26j^3-27j^2-18j-1)}{(j-2)^2(j^2-1)(2j-1)(2j+1)(2j+3)} \frac{dP_j(\cos\psi)}{d\cos\psi} \tag{144}$$

It has to be proved that function $R_1(\cos\psi)$ is bounded for $\psi \in \langle 0, \pi \rangle$. For these ψ's, the derivatives of the *Legendre polynomials* can be estimated as

$$\left| \frac{dP_j(\cos\psi)}{d\cos\psi} \right| \leq j^2 \tag{145}$$

which yields

$$|R_1(\cos\psi)| \leq 3\sum_{j=3}^{\infty}\frac{j^2(26j^3-27j^2-18j-1)}{(j-2)^2(j^2-1)(2j-1)(2j+1)(2j+3)}$$
$$< 3\sum_{j=3}^{\infty}\frac{j(26j^3-27j^2-18j-1)}{(j-2)^2(j-1)(2j-1)(2j+1)(2j+3)} \tag{146}$$

Using the identity

$$j(26j^3-27j^2-18j-1) = 7j^2(2j-1)(2j+1) - 2j^4 - 27j^3 - 11j^2 - j \tag{147}$$

one has

$$|R_1(\cos\psi)| < 21\sum_{j=3}^{\infty}\frac{j^2}{(j-2)^2(j-1)(2j+3)} \tag{148}$$

since $2j^4+27j^3+11j^2+j>0$ for any $j>0$. Finally, the sum in inequality Eq. 148 may be estimated as

$$\sum_{j=3}^{\infty}\frac{j^2}{(j-2)^2(j-1)(2j+3)} < \frac{1}{2}\sum_{j=1}^{\infty}\frac{j+2}{j^2(j+1)} < \frac{1}{2}\sum_{j=1}^{\infty}\left(\frac{1}{j^2}+\frac{2}{j^3}\right) = \frac{\pi^2}{12}+\zeta(3) \tag{149}$$

where $\zeta(.)$ is the Riemann zeta function (for instance, Arfken 1968, Sect. 5.8). In addition, we have used (mangulis, 1965, p. 50)

$$\sum_{j=1}^{\infty}\frac{1}{j^2} = \frac{\pi^2}{6} \tag{150}$$

Substituting $\zeta(3) \cong 1.20205$ (for instance, Arfken 1968, p. 359) yields

$$R_1(\cos\psi) < 21.4 \tag{151}$$

It should be pointed out that the last estimate is rather weak; ➲ *Fig. 7* shows that $R_1(\cos\psi) < 5.5$.

Nevertheless, it is important that function $R_1(\cos\psi)$ is bounded at any point in the interval $0 \leq \psi \leq \pi$, so that the singularity of function $K_1(\cos\psi)$ at point $\psi = 0$ is expressed analytically by the first two terms on the right-hand side of Eq. 143. In addition, function $R_1(\cos\psi)$ can simply be calculated numerically since it is represented by a quickly convergent series. ➲ *Figure 6* demonstrates this fact in a transparent way; for $\psi = 0$ it plots the decay of magnitudes of series terms Eq. 143 with increasing degree j. Inspecting ➲ *Fig. 6* one can estimate that it is sufficient to sum up infinite series Eq. 144 for $R_1(\cos\psi)$ up to $j = 25$ in order to achieve an absolute accuracy of the order of 0.01. This accuracy is sufficient for calculating the *ellipsoidal Stokes function* $S^{ell}(\Omega,\Omega')$ in the frame of the $O\left(e_0^2\right)$-approximation.

The spatial forms of the other integral kernel $K_1(\cos\psi)$, $i = 2,\ldots,4$, can be expressed in a similar fashion as the kernel $K_1(\cos\psi)$; after some cumbersome but straightforward algebra one can arrive at

$$K_2(\cos\psi) = \frac{1}{2\sin\frac{\psi}{2}} - 3\cos\psi\ln\left(\sin\frac{\psi}{2}+\sin^2\frac{\psi}{2}\right) + 2 - 4\cos\psi - 6\sin\frac{\psi}{2} + R_2(\cos\psi)$$
$$K_3(\cos\psi) = -\ln\left(\sin\frac{\psi}{2}+\sin^2\frac{\psi}{2}\right) - \cos\psi - \frac{1}{2}P_2(\cos\psi) + R_3(\cos\psi)$$
$$K_4(\cos\psi) = \frac{1}{\sin\frac{\psi}{2}} - 7\cos\psi\ln\left(\sin\frac{\psi}{2}+\sin^2\frac{\psi}{2}\right) + 5 - 9\cos\psi - 14\sin\frac{\psi}{2} + R_4(\cos\psi) \tag{152}$$

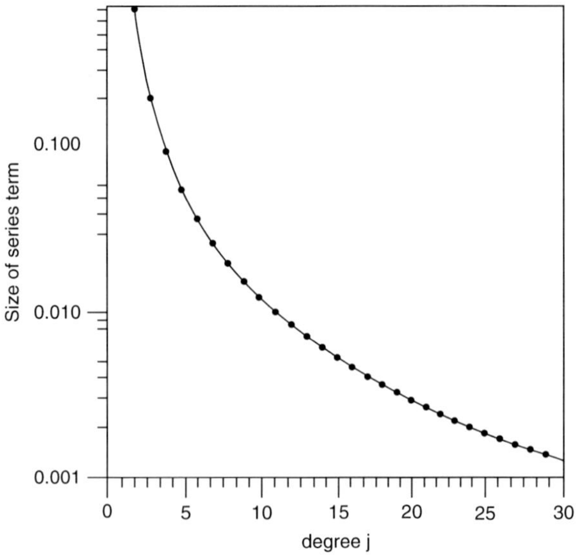

Fig. 6
The sizes of particular terms creating the infinite series for $R_1(1)$

where the residuals $R_i(\cos\psi)$ are of the forms

$$R_2(\cos\psi) = \sum_{j=2}^{\infty} \frac{5j+7}{(j-1)^2(2j+3)} P_j(\cos\psi)$$
$$R_3(\cos\psi) = \sum_{j=3}^{\infty} \frac{6j^2-4j-4}{j(j-2)^2(2j+1)} P_j(\cos\psi)$$
$$R_4(\cos\psi) = \sum_{j=2}^{\infty} \frac{6}{(j-1)^2} P_j(\cos\psi) \tag{153}$$

Figure 7 plots the residuals $R_i(\cos\psi)$, $i = 1,\ldots,4$ within the interval $0 \le \psi \le \pi$. One can see that $R_i(\cos)$ are "reasonably" smooth functions bounded for all angles ψ. This is the consequence of the fact that the magnitudes of the series terms in infinite sums (153) decrease quickly with increasing summation index j. In order to achieve an absolute accuracy of the order of 0.01, which is sufficient in the framework of the $O\left(e_0^2\right)$-approximation, series Eq. 153 may be truncated at degree $j \approx 25$.

Moreover, the above formulae make it possible to study the behavior of functions $K_i(\cos\psi)$ in the vicinity of point $\psi = 0$. It can readily be seen that

$$\lim_{\psi\to 0} K_i(\cos\psi) \approx \frac{1}{\psi},\ i = 1,2$$
$$\lim_{\psi\to 0} K_3(\cos\psi) \approx -\ln\frac{\psi}{2} \tag{154}$$
$$\lim_{\psi\to 0} K_4(\cos\psi) \approx \frac{2}{\psi}$$

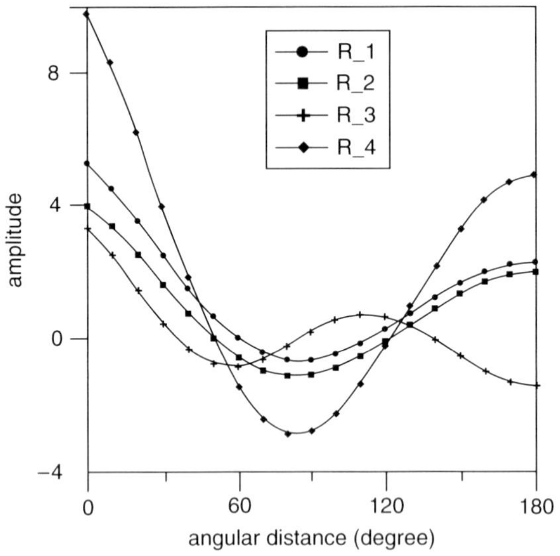

Fig. 7
Function $R_i(\cos\psi), i = 1, \ldots, 4$ for $\psi \in \psi_t < 0, \pi >$

Consequently, when integration point Ω' lies in the vicinity of computation point Ω, the *ellipsoidal Stokes function* $S^{ell}(\Omega,\Omega')$, see Eq. 128, may be approximated as

$$S^{ell}(\Omega,\Omega') = -\frac{1}{\psi}(1+\sin^2\vartheta\cos^2\alpha) \text{ for } \psi \ll 1 \tag{155}$$

This also means, that the *ellipsoidal Stokes function* $S^{ell}(\Omega,\Omega')$ has the same degree of singularity at point $\psi = 0$, namely, $1/\psi$, as the *original Stokes function.*

3.9 Conclusion

This work was motivated by the question whether the solution of the *Stokes boundary-value problem* with the boundary condition prescribed on an ellipsoid of revolution can be expressed in a closed spatial form, suitable for numerical computations. To answer this question, the solution of the *Stokes boundary-value problem* in terms of ellipsoidal harmonics was first found. The fact that this solution is represented by slowly convergent series of ellipsoidal harmonics prevents its use for regional geoid computations. That is why, in the next step, it was confined to the $O\left(e_0^2\right)$-approximation, meaning that the terms of magnitudes of the order of e_0^2 were retained; the terms of the order of $O\left(e_0^4\right)$ and of higher powers were not considered. Nevertheless, the accuracy of the order of $O\left(e_0^2\right)$ is fairly good for today's requirement of geoidal height computations. Within this accuracy, it has been shown that the solution of the *Stokes boundary-value problem* can be expressed as an integral taken over the full solid angle, and applied to gravity anomalies multiplied by a kernel consisting of the traditional *spherical Stokes function* and a correction due to the elliptical shape of the boundary; this additional term is called the *ellipsoidal Stokes function.* It has been managed to express the *ellipsoidal Stokes function,* originally represented in the form of an infinite sum of ellipsoidal harmonics, as a finite combination of elementary functions analytically describing the singular behavior of the *ellipsoidal Stokes function* at the point $\psi = 0$. This expression is suitable for the numerical solution of the *Stokes boundary-value problem* on an ellipsoid of revolution. The most important result is that the *ellipsoidal Stokes function* can be approximated by function $1/\psi$ in the vicinity of its singular point $\psi = 0$. Thus, the degree of singularity of the *ellipsoidal Stokes function* in the vicinity of point $\psi = 0$ is the same as that of the original *spherical Stokes function.*

4 Vertical Deflections in Gravity Space

In this chapter a definition of the vertical which relates to astronomical longitude and astronomical latitude as spherical coordinates in gravity space is presented. Vertical deflections and gravity disturbances relate to a reference gravity potential. In order to refer the horizontal and vertical components of the disturbing gravity field to a reference gravity field, which is physically meaningful, the Somigliana–Pizzetti gravity potential as well as its gradient has been chosen.

The local gravity vector $\mathbf{\Gamma}$ as well as a reference gravity vector $\boldsymbol{\gamma}$, both in a global frame of reference, are introduced in order to be able to construct the vertical deflection vector $\{\mathbf{\Gamma} \div \|\mathbf{\Gamma}\|^2 - \boldsymbol{\gamma} \div \|\boldsymbol{\gamma}\|^2\}$ in gravity space. In order to relate the vertical deflection vector to observables as elements of an observation space one has to transform the incremental gravity vector $\{\mathbf{\Gamma} \div \|\mathbf{\Gamma}\|^2 - \boldsymbol{\gamma} \div \|\boldsymbol{\gamma}\|^2\}$ from a global frame of reference to a local frame of reference, also called "natural", "horizontal," or "local level."

4.1 Representation of the Actual Gravity Vector as Well as the Reference Gravity Vector Both in a Global and a Local Frame of Reference

Conventionally, the local gravity vector Γ is represented in a global orthonormal frame of reference $\{\boldsymbol{E}_1, \boldsymbol{E}_2, \boldsymbol{E}_3|O\}$ either in Cartesian coordinates ($\Gamma_x = \Gamma_1, \Gamma_y = \Gamma_2, \Gamma_z = \Gamma_3$), or in spherical coordinates (Λ_Γ, Φ_Γ,Γ) subject to $\Gamma = \|\mathbf{\Gamma}\|$, its Euclidean length ($\ell^2$-norm). Λ_Γ is called astronomical longitude, Φ_Γ denotes astronomical latitude and Γ the "modulus of the actual gravity vector." Consult Eqs. 156 and 157 for such a representation. The global orthonormal frame of reference $\{\boldsymbol{E}_1, \boldsymbol{E}_2, \boldsymbol{E}_3|O\}$ is attached to the origin O, the mass center of the Earth. Since the actual gravity vector Γ is attached to the position P (placement vector from O to P) one may think of a parallel transport of $\{\boldsymbol{E}_1, \boldsymbol{E}_2, \boldsymbol{E}_3\}$ from O to P, namely, parallel transport in the Euclidean sense. The base vectors $\{\boldsymbol{E}_1, \boldsymbol{E}_2, \boldsymbol{E}_3|O\}$ may be materialized by the definitions of the Equatorial Frame of Reference or quasi-Earth fixed reference frame (International Terrestrial Reference Frame [ITRF]) distributed by the International Earth Rotation Service (IERS).

Alternatively, by means of the derivatives $\{D_{\Lambda_\Gamma}\mathbf{\Gamma}, D_{\Phi_\Gamma}\mathbf{\Gamma}, D_\Gamma\mathbf{\Gamma}\}$ one is able to construct a local orthonormal frame of reference, namely, $\boldsymbol{E}_{\Lambda_\Gamma} := D_{\Lambda_\Gamma}\mathbf{\Gamma} \div \|D_{\Lambda_\Gamma}\mathbf{\Gamma}\|$, $\boldsymbol{E}_{\Phi_\Gamma} := D_{\Phi_\Gamma}\mathbf{\Gamma} \div \|D_{\Phi_\Gamma}\mathbf{\Gamma}\|$, $\boldsymbol{E}_\Gamma := D_\Gamma\mathbf{\Gamma} \div \|D_\Gamma\mathbf{\Gamma}\|$, in short $\{\boldsymbol{E}_{\Lambda_\Gamma}, \boldsymbol{E}_{\Phi_\Gamma}, \boldsymbol{E}_\Gamma|P\}$, also called {astronomical East, astronomical North, astronomical Vertical} at P. Equations (160–162) outline the various operations in detail.

A similar procedure applies to the reference gravity vector γ. It is represented in a global orthonormal frame of reference $\{\boldsymbol{e}_1, \boldsymbol{e}_2, \boldsymbol{e}_3|O\}$, either in cartesian coordinates ($\gamma_x = \gamma_1, \gamma_y = \gamma_2, \gamma_z = \gamma_3$) or in spherical coordinates ($\lambda_\gamma, \varphi_\gamma, \gamma$) subject to $\gamma = \gamma^2$, its Euclidean length (ℓ^2-norm). λ_γ is called reference longitude, φ_γ reference latitude, and γ "modulus of the reference gravity vector" in gravity space. Consult Eqs. 158 and 159 for such a representation. The global orthonormal reference frame $\{\boldsymbol{e}_1, \boldsymbol{e}_2, \boldsymbol{e}_3|O\}$ is attached to the origin O, the mass center of the Earth.

Again by Euclidean parallelism $\{e_1, e_2, e_3\}$ is transported from O to P such that $\{e_1, e_2, e_3|P\} = \{e_1, e_2, e_3|O\}$. The reference frame of type $\{e_1, e_2, e_3|O\}$ is chosen by means of the definition of the reference gravity field, here gauged to ITRF, namely, $\{E_1, E_2, E_3|O\} = \{e_1, e_2, e_3|O\}$.

Local and global representation of the actual gravity vector and the reference gravity vector.
Actual gravity vector

$$E_\Gamma := -\boldsymbol{\Gamma}/\|\boldsymbol{\Gamma}\| = -\Gamma/\Gamma \tag{156}$$

$$-\Gamma = E_\Gamma \boldsymbol{\Gamma} = \boldsymbol{E}_1 \cos\Phi_\Gamma \cos\Lambda_\Gamma + \boldsymbol{E}_2 \cos\Phi_\Gamma \sin\Lambda_\Gamma + \boldsymbol{E}_3 \sin\Phi_\Gamma \tag{157}$$

Λ_Γ—astronomical longitude, Φ_Γ—astronomical latitude, $\boldsymbol{\Gamma}$—modulus of the actual gravity vector

Reference gravity vector

$$\boldsymbol{e}_\gamma := -\boldsymbol{\gamma}/\|\boldsymbol{\gamma}\| = -\boldsymbol{\gamma}/\boldsymbol{\gamma} \tag{158}$$

$$-\boldsymbol{\gamma} = \boldsymbol{e}_\gamma \boldsymbol{\gamma} = \boldsymbol{e}_1 \cos\varphi_\gamma \cos\lambda_\gamma + \boldsymbol{e}_2 \cos\varphi_\gamma \sin\lambda_\gamma + \boldsymbol{e}_3 \sin\varphi_\gamma \tag{159}$$

Λ_γ—astronomical longitude, φ_γ—astronomical latitude, $\boldsymbol{\gamma}$—modulus of the actual gravity vector

In contrast, by means of the derivatives $\{D_{\lambda_\gamma}\boldsymbol{\gamma}, D_{\varphi_\gamma}\boldsymbol{\gamma}, D_\gamma\boldsymbol{\gamma}\}$ one can compute a local orthonormal frame of reference, namely, $\boldsymbol{e}_{\lambda_\gamma} := D_{\lambda_\gamma}\boldsymbol{\gamma} \div \|D_{\lambda_\gamma}\boldsymbol{\gamma}\|$, $\boldsymbol{e}_{\varphi_\gamma} := D_{\varphi_\gamma}\boldsymbol{\gamma} \div \|D_{\varphi_\gamma}\boldsymbol{\gamma}\|$, $\boldsymbol{e}_\gamma := D_\gamma\boldsymbol{\gamma} \div \|D_\gamma\boldsymbol{\gamma}\|$, also called {reference East, reference North, reference Vertical} at P. Equations 163–165 outline the various relations in detail.

Both local orthonormal frames of reference $\{E_{\Lambda_\Gamma}, E_{\Phi_\Gamma}, E_\Gamma|P\}$ and $\{e_{\Lambda_\Gamma}, e_{\Phi_\Gamma}, e_\Gamma|P\}$ are related by means of an orthonormal matrix $R_E\left(\Lambda_\Gamma + \frac{\pi}{2}, \frac{\pi}{2} - \Phi_\Gamma, 0\right) := R_3(0)R_2\left(\frac{\pi}{2} - \Phi_\Gamma\right) R_3\left(\Lambda_\Gamma + \frac{\pi}{2}\right)$, Eq. 162 to the global orthonormal frame of reference $\{E_1, E_2, E_3|O\}$ as well as, Eq. 164, to the global orthonormal frame of reference $\{e_1, e_2, e_3|O\}$. R_E contains as an index the symbol of a rotation matrix which is parameterized by Euler angles. R_3 denotes a rotation around the three-axis (or in the (1,2)-plane), R_2 around the two-axis (or in the (3,1)-plane).

Construction of the local reference frame $\{e_{\Lambda_\Gamma}, e_{\Phi_\Gamma}, e_\Gamma|P\}$ from the spherical coordinates of the actual gravity vector

$$\begin{aligned}
\boldsymbol{E}_{\Lambda_\Gamma} &= \frac{D_{\Lambda_\Gamma}\Gamma}{\|D_{\Lambda_\Gamma}\Gamma\|} = \left(\boldsymbol{e}_1\frac{\partial\Gamma_1}{\partial\Lambda_\Gamma} + \boldsymbol{e}_2\frac{\partial\Gamma_2}{\partial\Lambda_\Gamma} + \boldsymbol{e}_3\frac{\partial\Gamma_3}{\partial\Lambda_\Gamma}\right) \div \left[\left(\frac{\partial\Gamma_1}{\partial\Lambda_\Gamma}\right)^2 + \left(\frac{\partial\Gamma_2}{\partial\Lambda_\Gamma}\right)^2 + \left(\frac{\partial\Gamma_3}{\partial\Lambda_\Gamma}\right)^2\right]^{1/2} \\
\boldsymbol{E}_{\Phi_\Gamma} &= \frac{D_{\Phi_\Gamma}\Gamma}{\|D_{\Phi_\Gamma}\Gamma\|} = \left(\boldsymbol{e}_1\frac{\partial\Gamma_1}{\partial\Phi_\Gamma} + \boldsymbol{e}_2\frac{\partial\Gamma_2}{\partial\Phi_\Gamma} + \boldsymbol{e}_3\frac{\partial\Gamma_3}{\partial\Phi_\Gamma}\right) \div \left[\left(\frac{\partial\Gamma_1}{\partial\Phi_\Gamma}\right)^2 + \left(\frac{\partial\Gamma_2}{\partial\Phi_\Gamma}\right)^2 + \left(\frac{\partial\Gamma_3}{\partial\Phi_\Gamma}\right)^2\right]^{1/2} \\
\boldsymbol{E}_\Gamma &= \frac{D_\Gamma\Gamma}{\|D_\Gamma\Gamma\|} = \left(\boldsymbol{e}_1\frac{\partial\Gamma_1}{\partial\Gamma} + \boldsymbol{e}_2\frac{\partial\Gamma_2}{\partial\Gamma} + \boldsymbol{e}_3\frac{\partial\Gamma_3}{\partial\Gamma}\right) \div \left[\left(\frac{\partial\Gamma_1}{\partial\Gamma}\right)^2 + \left(\frac{\partial\Gamma_2}{\partial\Gamma}\right)^2 + \left(\frac{\partial\Gamma_3}{\partial\Gamma}\right)^2\right]^{1/2}
\end{aligned} \tag{160}$$

$$\begin{bmatrix} \boldsymbol{E}_{\Lambda_\Gamma} \\ \boldsymbol{E}_{\Phi_\Gamma} \\ \boldsymbol{E}_\Gamma \end{bmatrix} = R_E\left(\lambda_\gamma + \frac{\pi}{2}, \frac{\pi}{2} - \phi_\gamma, 0\right) \begin{bmatrix} \boldsymbol{E}_1 \\ \boldsymbol{E}_2 \\ \boldsymbol{E}_3 \end{bmatrix} \tag{161}$$

$$\begin{bmatrix} \boldsymbol{E}_{\Lambda_\Gamma} \\ \boldsymbol{E}_{\Phi_\Gamma} \\ \boldsymbol{E}_\Gamma \end{bmatrix} = \begin{bmatrix} -\sin_{\Lambda_\Gamma} & \cos_{\Lambda_\Gamma} & 0 \\ -\cos_{\Lambda_\Gamma}\sin_{\Phi_\Gamma} & \sin_{\Lambda_\Gamma}\sin_{\Phi_\Gamma} & \cos_{\Phi_\Gamma} \\ \cos_{\Lambda_\Gamma}\cos_{\Phi_\Gamma} & \sin_{\Lambda_\Gamma}\cos_{\Phi_\Gamma} & \sin_{\Phi_\Gamma} \end{bmatrix} \begin{bmatrix} \boldsymbol{E}_1 \\ \boldsymbol{E}_2 \\ \boldsymbol{E}_3 \end{bmatrix} \tag{162}$$

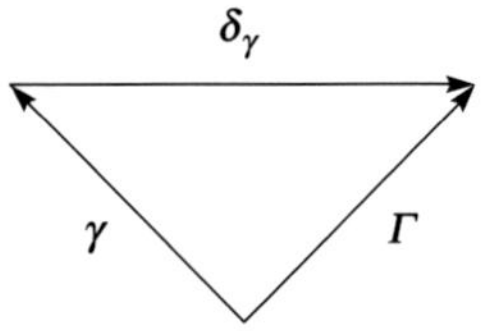

Fig. 8
Decomposition of the actual gravity vector $\Gamma = \gamma + \delta\gamma$

Construction of the local reference frame $\{e_{\Lambda_\Gamma}, e_{\Phi_\Gamma}, e_\Gamma | P\}$ from the spherical coordinates of the reference gravity vector.

$$e_{\lambda_\gamma} = \frac{D_{\lambda_\gamma}\boldsymbol{\gamma}}{\|D_{\lambda_\gamma}\boldsymbol{\gamma}\|} = \left(e_1\frac{\partial\gamma_1}{\partial\lambda_\gamma} + e_2\frac{\partial\gamma_2}{\partial\lambda_\gamma} + e_3\frac{\partial\gamma_3}{\partial\lambda_\gamma}\right) \div \left[\left(\frac{\partial\gamma_1}{\partial\lambda_\gamma}\right)^2 + \left(\frac{\partial\gamma_2}{\partial\lambda_\gamma}\right)^2 + \left(\frac{\partial\gamma_3}{\partial\lambda_\gamma}\right)^2\right]^{1/2}$$

$$e_{\varphi_\gamma} = \frac{D_{\varphi_\gamma}\boldsymbol{\gamma}}{\|D_{\varphi_\gamma}\boldsymbol{\gamma}\|} = \left(e_1\frac{\partial\gamma_1}{\partial\varphi_\gamma} + e_2\frac{\partial\gamma_2}{\partial\varphi_\gamma} + e_3\frac{\partial\gamma_3}{\partial\varphi_\gamma}\right) \div \left[\left(\frac{\partial\gamma_1}{\partial\varphi_\gamma}\right)^2 + \left(\frac{\partial\gamma_2}{\partial\varphi_\gamma}\right)^2 + \left(\frac{\partial\gamma_3}{\partial\varphi_\gamma}\right)^2\right]^{1/2}$$

$$e_{\gamma} = \frac{D_{\gamma}\boldsymbol{\gamma}}{\|D_{\gamma}\boldsymbol{\gamma}\|} = \left(e_1\frac{\partial\gamma_1}{\partial\gamma} + e_2\frac{\partial\gamma_2}{\partial\gamma} + e_3\frac{\partial\gamma_3}{\partial\gamma}\right) \div \left[\left(\frac{\partial\gamma_1}{\partial\gamma}\right)^2 + \left(\frac{\partial\gamma_2}{\partial\gamma}\right)^2 + \left(\frac{\partial\gamma_3}{\partial\gamma}\right)^2\right]^{1/2} \tag{163}$$

$$\begin{bmatrix} e_{\lambda_\gamma} \\ e_{\phi_\gamma} \\ e_\gamma \end{bmatrix} = R_E\left(\lambda_\gamma + \frac{\pi}{2}, \frac{\pi}{2} - \phi_\gamma, 0\right)\begin{bmatrix} e_1 \\ e_2 \\ e_3 \end{bmatrix} \tag{164}$$

$$\begin{bmatrix} e_{\lambda_\gamma} \\ e_{\phi_\gamma} \\ e_\gamma \end{bmatrix} = \begin{bmatrix} -\sin_{\lambda_\gamma} & \cos_{\lambda_\gamma} & 0 \\ -\cos_{\lambda_\gamma}\sin_{\phi_\gamma} & \sin_{\lambda_\gamma}\sin_{\phi_\gamma} & \cos_{\phi_\gamma} \\ \cos_{\lambda_\gamma}\cos_{\phi_\gamma} & \sin_{\lambda_\gamma}\cos_{\phi_\gamma} & \sin_{\phi_\gamma} \end{bmatrix}\begin{bmatrix} e_1 \\ e_2 \\ e_3 \end{bmatrix} \tag{165}$$

4.2 The Incremental Gravity Vector

The incremental gravity vector $\delta\gamma$ is defined as the vector-valued difference of the actual gravity vector Γ and the reference gravity vector γ, computed at the same point in space: $\delta\gamma := \Gamma - \gamma$.

Figure 8 is an illustration of the local incremental gravity vector, also called "disturbing gravity vector" $\delta\boldsymbol{\gamma}$. Indeed, the choice of the reference gravity vector which leads to the unit reference gravity vector $-\boldsymbol{\gamma}/\parallel\boldsymbol{\gamma}\parallel$ has to be specified.

Next, computing the rotation matrix R for the transformation between the local orthonormal frames of reference $\{\boldsymbol{E}_{\Lambda_\Gamma}, \boldsymbol{E}_{\Phi_\Gamma}, \boldsymbol{E}_\Gamma\} \mapsto \{e_{\lambda_\gamma}, e_{\varphi_\gamma}, e_\gamma\}$, according to *Fig. 9* is aimed at. At first the moving frames $\{e_{\lambda_\gamma}, e_{\varphi_\gamma}, e_\gamma | P\}$ and $\{\boldsymbol{E}_{\Lambda_\Gamma}, \boldsymbol{E}_{\Phi_\Gamma}, \boldsymbol{E}_\Gamma | P\}$ are rotated by means of Eulerian rotation matrices $\boldsymbol{R}_E$ of type Eqs. 162 and 165 to the fixed frame $\{e_1, e_2, e_3 | O\} = \{\boldsymbol{E}_1, \boldsymbol{E}_2, \boldsymbol{E}_3 | O\}$. Second, in closing the commutative diagram of *Fig. 9* one is led to the unknown matrix R of Eq. 167. All rotation matrices belong to the special orthogonal group $SO(3) := \{\boldsymbol{R} \in \boldsymbol{R}^{3\times3} | \boldsymbol{R}^T\boldsymbol{R} = I_3 \text{ and } |\boldsymbol{R}| = 1\}$. Its inverse is just the transpose.

The compound rotation matrix R depends on the two sets of parameters Λ_γ, φ_γ, and Λ_γ, Φ_γ, respectively. Since the differences $\delta\lambda_\gamma := \Lambda_\gamma - \lambda_\gamma$, $\delta\varphi_\gamma := \Phi_\gamma - \varphi_\gamma$ between astronomical longitude/latitude and reference longitude/latitude in gravity space are small to first order, $\delta\lambda_\gamma$,

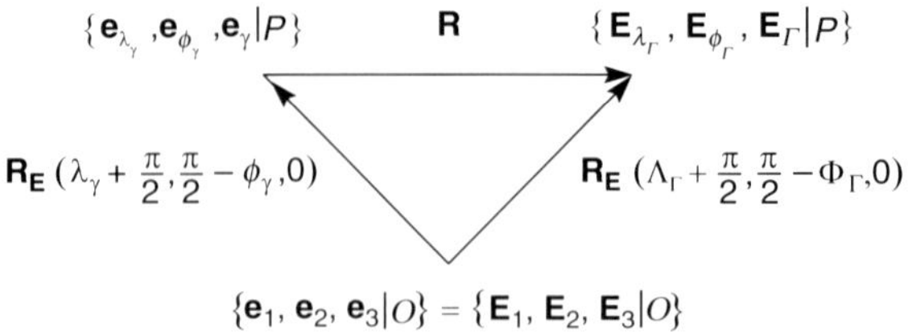

Basis transformation in gravity space.

Fig. 9
Basis transformation in gravity space

$\delta\varphi_\gamma$ are called Euler increments. As soon as one implements $\Lambda_\gamma := \lambda_\gamma + \delta\lambda_\gamma$, $\Phi_\gamma := \varphi_\gamma + \delta\varphi_\gamma$ into the *Euler rotation* matrix $\boldsymbol{R}_E\left(\Lambda_\Gamma + \frac{\pi}{2}, \frac{\pi}{2} - \Phi_\Gamma, 0\right)$ and linearizes it to the first order, for instance, $\cos(\lambda_\gamma + \delta\lambda_\gamma) \doteq \cos\lambda_\gamma - \sin\Lambda_\gamma\delta\lambda_\gamma$, $\sin(\varphi_\gamma + \delta\varphi_\gamma) \doteq \sin\varphi_\gamma + \cos\varphi_\gamma\delta\varphi_\gamma$, one is able to compute the compound rotation matrix $\boldsymbol{R} = \boldsymbol{I} + \boldsymbol{\delta A}$ of type Eqs. 169 and 170 decomposed into the unit matrix I_3 and the incremental antisymmetric matrix δA if one neglects terms of second order. *Basic transformation between both local bases*

$$\begin{bmatrix} \boldsymbol{E}_{\Lambda_\Gamma} \\ \boldsymbol{E}_{\Phi_\Gamma} \\ \boldsymbol{E}_\Gamma \end{bmatrix} = \boldsymbol{R} \begin{bmatrix} \boldsymbol{e}_{\lambda_\gamma} \\ \boldsymbol{e}_{\phi_\gamma} \\ \boldsymbol{e}_\gamma \end{bmatrix} \tag{166}$$

$$\boldsymbol{R} = \boldsymbol{R}_E\left(\Lambda_\Gamma + \frac{\pi}{2}, \frac{\pi}{2} - \Phi_\Gamma, 0\right)\boldsymbol{R}_E^T\left(\lambda_\gamma + \frac{\pi}{2}, \frac{\pi}{2} - \varphi_\gamma, 0\right) \tag{167}$$

$$\boldsymbol{R} = \begin{bmatrix} 1 & \sin\varphi_\gamma\delta\lambda_\gamma & -\cos\varphi_\gamma\delta\lambda_\gamma \\ -\sin\varphi_\gamma\delta\lambda_\gamma + O(\delta\lambda_\gamma\delta\varphi_\gamma) & 1 & -\delta\varphi_\gamma \\ \cos\varphi_\gamma\delta\lambda_\gamma + O(\delta\lambda_\gamma\delta\varphi_\gamma) & \delta\varphi_\gamma & 1 \end{bmatrix} \tag{168}$$

$$\boldsymbol{R} = \boldsymbol{I}_3 + \boldsymbol{\delta A} \tag{169}$$

$$g\boldsymbol{\delta A} = \begin{bmatrix} 0 & \sin\varphi_\gamma\delta\lambda_\gamma & -cos\varphi_\gamma\delta\lambda_\gamma \\ -\sin\varphi_\gamma\delta\lambda_\gamma + O(\delta\lambda_\gamma\delta\varphi_\gamma) & 0 & -\delta\varphi_\gamma \\ \cos\varphi_\gamma\delta\lambda_\gamma + O(\delta\lambda_\gamma\delta\varphi_\gamma) & \delta\varphi_\gamma & 0 \end{bmatrix} \tag{170}$$

Finally by means of Eqs. (171–178) the incremental gravity vector $-\boldsymbol{\delta\gamma} = -(\Gamma - \gamma)$ in the local reference frame $\left\{\boldsymbol{e}_{\lambda_\gamma}, \boldsymbol{e}_{\varphi_\gamma}, \boldsymbol{e}_\gamma | P\right\}$ at P is represented. While the actual gravity vector Γ is originally given in the local base E_γ, by means of Eq. 172 one succeeds to represent Γ in the local reference base $\left\{\boldsymbol{e}_{\lambda_\gamma}, \boldsymbol{e}_{\varphi_\gamma}, \boldsymbol{e}_\gamma | P\right\}$ by Eq. 171. As soon as one combines Eq. 171 and Eq. 172 one finally represents the negative incremental gravity vector in the local reference frame $(-\delta\gamma)$ of type Eq. 173. The modulus of gravity $\Gamma := \|\Gamma\|^2 \doteq \|\gamma\|^2 + \|\delta\gamma\|^2 = \gamma + \delta\gamma$ is finally approximated by the modulus of reference gravity $\gamma := \|\gamma\|^2$ and the modulus of incremental gravity $\delta\gamma := \|\delta\gamma\|^2$. In this way, from Eq. 174 one is led to Eq. 175 and Eq. 176, just defining the horizontal components of the incremental gravity vector as vertical deflections "East" $\eta := \cos\varphi_\gamma\delta\lambda_\gamma$ and "North" $\xi := \delta\varphi_\gamma$ up to order $O(2)$. *Figure 10* illustrates those vertical deflections in the tangent space $T_P S_\gamma^2$ of the sphere S_γ^2 of radius $\gamma := \|\boldsymbol{\gamma}\|^2$ spanned by the unit vectors East, North$P := \left\{e_{\lambda_\gamma}, e_{\varphi_\gamma} | P\right\}$ at P.

Incremental gravity vector within the local reference frame

$$\left\{\boldsymbol{e}_{\lambda_\gamma}, \boldsymbol{e}_{\varphi_\gamma}, \boldsymbol{e}_\gamma | P\right\}.$$

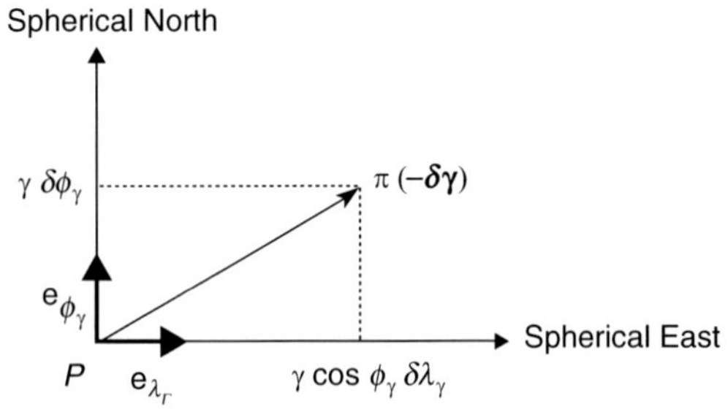

Fig. 10
Negative incremental gravity vector $-\delta\gamma$ in the tangent space at point P, projection $\pi(-\delta\gamma)$ of the negative incremental gravity vector onto the tangent plane at point P

actual gravity vector

$$-\boldsymbol{\Gamma} = E_\Gamma \Gamma = \left[\boldsymbol{e}_{\lambda_\gamma}, \boldsymbol{e}_{\varphi_\gamma}, \boldsymbol{e}_\gamma\right] \Gamma \begin{bmatrix} \cos\varphi_\gamma \delta\lambda_\gamma \\ \delta\varphi_\gamma \\ 1 \end{bmatrix} \tag{171}$$

reference gravity vector

$$-\boldsymbol{\gamma} = e_\gamma \gamma = \left[\boldsymbol{e}_{\lambda_\gamma}, \boldsymbol{e}_{\varphi_\gamma}, \boldsymbol{e}_\gamma\right] \gamma \begin{bmatrix} 0 \\ 0 \\ 1 \end{bmatrix} \tag{172}$$

incremental gravity vector

$$-\boldsymbol{\delta\gamma} = -(\boldsymbol{\Gamma} - \boldsymbol{\gamma}) = \left[\boldsymbol{e}_{\lambda_\gamma}, \boldsymbol{e}_{\varphi_\gamma}, \boldsymbol{e}_\gamma\right] \begin{bmatrix} \Gamma\cos\varphi_\gamma \delta\lambda_\gamma \\ \Gamma\delta\varphi_\gamma \\ \Gamma - \gamma \end{bmatrix} \tag{173}$$

$$-\boldsymbol{\delta\gamma} = \left[\boldsymbol{e}_{\lambda_\gamma}, \boldsymbol{e}_{\varphi_\gamma}, \boldsymbol{e}_\gamma\right] \begin{bmatrix} (\gamma + \delta\gamma)\cos\varphi_\gamma \delta\lambda_\gamma \\ (\gamma + \delta\gamma)\delta\varphi_\gamma \\ \delta\gamma \end{bmatrix} \tag{174}$$

$$-\boldsymbol{\delta\gamma} = \left[\boldsymbol{e}_{\lambda_\gamma}, \boldsymbol{e}_{\varphi_\gamma}, \boldsymbol{e}_\gamma\right] \begin{bmatrix} \gamma\cos\varphi_\gamma \delta\lambda_\gamma + O(\delta\gamma\delta\lambda_\gamma) \\ \gamma\delta\varphi_\gamma + O(\delta\gamma\delta\lambda_\gamma) \\ \delta\gamma \end{bmatrix} \tag{175}$$

$$-\boldsymbol{\delta\gamma} = \left[\boldsymbol{e}_{\lambda_\gamma}, \boldsymbol{e}_{\varphi_\gamma}, \boldsymbol{e}_\gamma\right] \begin{bmatrix} \gamma\eta \\ \gamma\xi \\ \delta\gamma \end{bmatrix} \tag{176}$$

vertical deflections

$$\eta := \cos\varphi_\gamma \delta\lambda_\gamma + O(\delta\gamma\delta\lambda_\gamma) \tag{177}$$
$$\xi := \delta\varphi_\gamma + O(\delta\gamma\delta\lambda_\gamma) \tag{178}$$

Gauss surface normal coordinates (L,B,H):

(i) Forward transformation

$$\mathbf{X} = \mathbf{E}_1 X + \mathbf{E}_2 Y + \mathbf{E}_3 Z \tag{179}$$

$$X = \left[A_1/\sqrt{1-E^2\sin^2 B} + H(L,B)\right]\cos B\cos L$$

$$Y = \left[A_1/\sqrt{1-E^2\sin^2 B} + H(L,B)\right]\cos B\sin L \tag{180}$$

$$Z = \left[A_1(1-E^2)/\sqrt{1-E^2\sin^2 B} + H(L,B)\right]\sin B$$

relative eccentricity of the ellipsoid of revolution $E^2_{A_1,A_1,A_2}$

$$E^2 := \left(A_1^2 - A_2^2\right)/A_1^2, E \in R^+ \tag{181}$$

(ii) Backward transformation – see review paper by *Grafarend (2001)*

(iii) Jacobi matrix

$$\boldsymbol{J} = \begin{bmatrix} D_L X & D_B X & D_H X \\ D_L Y & D_B Y & D_H Y \\ D_L Z & D_B Z & D_H Z \end{bmatrix} \tag{182}$$

$$\boldsymbol{J} = \begin{bmatrix} \cos B\left[H_L\cos L - (N+H)\sin L\right] & \cos L\left[H_B\cos B - (M+H)\sin B\right] & \cos B\cos L \\ \cos B\left[H_L\sin L - (N+H)\cos L\right] & \sin L\left[H_B\cos B - (M+H)\sin B\right] & \cos B\sin L \\ \sin B H_L & H_B\sin B + (M+H)\cos B & \sin B \end{bmatrix} \tag{183}$$

(iv) Metric

$$dS^2 = \left[dL\, dB\, dH\right]\boldsymbol{J}^T\boldsymbol{J}\begin{bmatrix} dL \\ dB \\ dH \end{bmatrix} \tag{184}$$

$$\mathbf{G} = \boldsymbol{J}^T\boldsymbol{J} \tag{185}$$

$$\mathbf{G} = \begin{bmatrix} (N+H)^2\cos^2 B + H_L^2 & H_L H_B & H_L \\ H_L H_B & (M+H)^2 + H_B^2 & H_B \\ H_L & H_B & 1 \end{bmatrix} \tag{186}$$

$$M(B) = A_1(1-E^2)/\sqrt[3]{1-E^2\sin^2 B} \tag{187}$$

$$N(B) = A_1/\sqrt{1-E^2\sin^2 B} \tag{188}$$

$$H_L = 0, H_B = 0 \Rightarrow dS^2 = (N+H)^2\cos^2 B dL^2 + (M+H)^2 dB^2 + dH^2 \tag{189}$$

5 Vertical Deflections and Gravity Disturbance in Geometry Space

Next, two different definitions of the vertical which relate to (i) *Gauss surface normal coordinates* (also called geodetic coordinates) of type ellipsoidal longitude and ellipsoidal latitude and (ii) *Jacobi ellipsoidal coordinates* of type spheroidal longitude and spheroidal latitude in geometry space are presented. Up to terms of second order those vertical deflections agree to each other.

5.1 Ellipsoidal Coordinates of Type Gauss Surface Normal

With reference to the extensive review on ellipsoidal coordinates by Thong and Grafarend (1989) Gauss surface normal coordinates (LBH) called ellipsoidal longitude L, ellipsoidal latitude B, and ellipsoidal height H which relate to Cartesian coordinates $\{X, Y, Z\}$ in the global orthonormal frame of reference $\{E_1, E_2, E_3|O\}$ by Eqs. (179–181) are introduced.

The derivatives $\{D_L x, D_B x, D_H x\}$ will enable one to derive the local orthonormal frame of reference $E_L := D_L x \div \|D_L x\|$, $E_B := D_B x \div \|D_B x\|$, $E_H := D_H x \div \|D_H x\|$, in short also called {ellipsoidal East, ellipsoidal North, ellipsoidal Vertical} at *P*. The local orthonormal frame of reference $\{E_L, E_B, E_H|P\}$ at *P* is related to the global orthonormal frame of reference $\{E_1, E_2, E_3|O\}$ at *O* by the orthonormal matrix T_1 as outlined in Eqs. 190 and 191.

Construction of the local reference frame $\{e_L, e_B, e_H|Q\}$ from the geodetic coordinates of point Q on the ellipsoid of revolution.

$$
\begin{aligned}
\mathbf{E}_L &:= \frac{D_L x}{\|D_L x\|} = \left(\mathbf{E}_1\frac{\partial x_1}{\partial L} + \mathbf{E}_2\frac{\partial x_2}{\partial L} + \mathbf{E}_3\frac{\partial x_3}{\partial L}\right) \div \left[\left(\frac{\partial x_1}{\partial L}\right)^2 + \left(\frac{\partial x_2}{\partial L}\right)^2 + \left(\frac{\partial x_3}{\partial L}\right)^2\right]^{1/2}\\
\mathbf{E}_B &:= \frac{D_B x}{\|D_B x\|} = \left(\mathbf{E}_1\frac{\partial x_1}{\partial B} + \mathbf{E}_2\frac{\partial x_2}{\partial B} + \mathbf{E}_3\frac{\partial x_3}{\partial B}\right) \div \left[\left(\frac{\partial x_1}{\partial B}\right)^2 + \left(\frac{\partial x_2}{\partial B}\right)^2 + \left(\frac{\partial x_3}{\partial B}\right)^2\right]^{1/2}\\
\mathbf{E}_H &:= \frac{D_H x}{\|D_H x\|} = \left(\mathbf{E}_1\frac{\partial x_1}{\partial H} + \mathbf{E}_2\frac{\partial x_2}{\partial H} + \mathbf{E}_3\frac{\partial x_3}{\partial H}\right) \div \left[\left(\frac{\partial x_1}{\partial H}\right)^2 + \left(\frac{\partial x_2}{\partial H}\right)^2 + \left(\frac{\partial x_3}{\partial H}\right)^2\right]^{1/2}\\
&= \mathbf{E}_L \times \mathbf{E}_B = *(\mathbf{E}_L \wedge \mathbf{E}_B)
\end{aligned}
\tag{190}
$$

$$
\begin{bmatrix}\mathbf{E}_L\\ \mathbf{E}_B\\ \mathbf{E}_H\end{bmatrix} = T_1\begin{bmatrix}\mathbf{E}_1\\ \mathbf{E}_2\\ \mathbf{E}_3\end{bmatrix} = \begin{bmatrix}-\sin L & \cos L & 0\\ -\cos L\sin B & \sin L\sin B & \cos B\\ \cos L\cos B & \sin L\cos B & \sin B\end{bmatrix}\begin{bmatrix}\mathbf{E}_1\\ \mathbf{E}_2\\ \mathbf{E}_3\end{bmatrix}
\tag{191}
$$

Basis transformation from geometry space to gravity space.

$$
\begin{bmatrix}\mathbf{e}_{\lambda_\gamma}\\ \mathbf{e}_{\phi_\gamma}\\ \mathbf{e}_\gamma\end{bmatrix} = \mathbf{T}_2\mathbf{T}_1\begin{bmatrix}\mathbf{E}_L\\ \mathbf{E}_B\\ \mathbf{E}_H\end{bmatrix} = \mathbf{T}\begin{bmatrix}\mathbf{E}_L\\ \mathbf{E}_B\\ \mathbf{E}_H\end{bmatrix}
\tag{192}
$$

subject to

$$
\mathbf{T}_1 = \mathbf{T}_1(L, B) \tag{193}
$$

$$
\mathbf{T}_2 = \mathbf{T}_2(\lambda_\gamma, \varphi_\gamma) = R_E\left(\lambda_\gamma + \frac{\pi}{2}, \frac{\pi}{2} - \varphi_\gamma, 0\right) \tag{194}
$$

$$
\mathbf{T} := \mathbf{T}_2\mathbf{T}_1^T = \mathbf{T}(L, B; \lambda_\gamma, \varphi_\gamma), \mathbf{T}_1, \mathbf{T}_2 \in SO(3) \tag{195}
$$

$$
\mathbf{T} = \begin{bmatrix}
-\cos(L-\lambda_\gamma) & -\sin(L-\lambda_\gamma)\sin B & \sin(L-\lambda_\gamma)\cos B\\
\sin(L-\lambda_\gamma)\sin\varphi_\gamma & \cos(L-\lambda_\gamma)\sin B\sin\varphi_\gamma + \cos B\cos\varphi_\gamma & \cos(L-\lambda_\gamma)\cos B\sin\varphi_\gamma + \sin B\cos\varphi_\gamma\\
-\sin(L-\lambda_\gamma)\cos\varphi_\gamma & \cos(L-\lambda_\gamma)\sin B\cos\varphi_\gamma + \cos B\sin\varphi_\gamma & \cos(L-\lambda_\gamma)\cos\varphi_\gamma + \sin B\sin\varphi_\gamma
\end{bmatrix}
\tag{196}
$$

additive decomposition

$$L = \Lambda_\gamma + \delta L \Leftrightarrow \Lambda_\gamma = L - \delta L \tag{197}$$

$$B = \Lambda_\gamma + \delta B \Leftrightarrow \Lambda_\gamma = B - \delta B \tag{198}$$

$$\sin(L - \delta L) \doteq \sin L - \cos L\delta L \tag{199}$$

$$\sin(B - \delta B) \doteq \cos B + \sin B\delta B \tag{200}$$

linearized basis transformation from geometry space to gravity space

$$\begin{bmatrix} \boldsymbol{e}_{\lambda_\gamma} \\ \boldsymbol{e}_{\phi_\gamma} \\ \boldsymbol{e}_\gamma \end{bmatrix} = \begin{bmatrix} 1 & -\sin B\delta L & \cos B\delta L \\ -\sin B\delta L & 1 & \delta B \\ -\cos B\delta L & -\delta B & 1 \end{bmatrix} \begin{bmatrix} \boldsymbol{E}_L \\ \boldsymbol{E}_B \\ \boldsymbol{E}_H \end{bmatrix} \tag{201}$$

subject to

$$\boldsymbol{T} = \boldsymbol{I}_3 + \boldsymbol{\delta A} \tag{202}$$

$$\boldsymbol{\delta A} = \begin{bmatrix} 0 & -\sin \mathrm{B}\delta L & \cos \mathrm{B}\delta L \\ \sin \mathrm{B}\delta L & 0 & \delta B \\ -\cos \mathrm{B}\delta L & -\delta B & 0 \end{bmatrix} \tag{203}$$

By means of the transformation has succeeded already succeeded in establishing the orthonormal basis $T_2\ (\Lambda_\gamma,\varphi_\gamma)$, Eq. 194. Such a relation will be used to transform the ellipsoidal orthonormal frame of reference $\{E_L, E_B, E_H|Q\}$, $Q \in E^2_{A_1,A_1,A_2}$ to the reference "East, North, Vertical" frame $\left\{\mathbf{e}_{\lambda_\gamma}, \mathbf{e}_{\varphi_\gamma}, \mathbf{e}_\gamma|P\right\}$. Indeed, $[\mathbf{E_L}, \mathbf{E_B}, \mathbf{E_H}]^T = \mathbf{T}_1 [\mathbf{e}_1, \mathbf{e}_2, \mathbf{e}_3]^T = \mathbf{T}_1\mathbf{T}_2^T \left[\mathbf{e}_{\lambda_\gamma}, \mathbf{e}_{\varphi_\gamma}, \mathbf{e}_\gamma\right]^T$ or $[\boldsymbol{e}_1, \boldsymbol{e}_2, \boldsymbol{e}_3]^T = \boldsymbol{T}_2\boldsymbol{T}_1^T [\boldsymbol{E}_L, \boldsymbol{E}_B, \boldsymbol{E}_H]^T$ is the compound transformation of T of Eqs. (192–204), linearized with respect to the antisymmetric matrix δA, Eq. 203.

Based upon the linearized version of the transformation $[E_L, E_B, E_H] \rightarrow \left[e_{\lambda_\gamma}, e_{\varphi_\gamma}, e_\gamma\right]$, Eqs. (202–204), one can take advantage of the representation of the incremental gravity vector $-\boldsymbol{\delta\gamma}$, Eq. 176, in terms of $\left[\boldsymbol{e}_{\lambda_\gamma}, \boldsymbol{e}_{\varphi_\gamma}, \boldsymbol{e}_\gamma\right]$: Eqs. 204–207 illustrate the representation of the incremental gravity vector $-\boldsymbol{\delta\gamma}$ in the basis $\{\boldsymbol{E_L}, \boldsymbol{E_B},\boldsymbol{E_H}\}$. The first basic result has to be interpreted as following. The second representation of the incremental gravity vector $-\delta\gamma$ contains horizontal components as well as a vertical component which are all functions of $\{\eta,\xi,\delta\gamma\}$. For instance, the East component $\boldsymbol{E_L}$ is a function of $\gamma\eta$ (first order) and of $(\gamma\xi,\delta\gamma)$ (second order). Or the vertical component $\boldsymbol{E_H}$ is a function of $\delta\gamma$ (first order) and of $(\gamma\eta,\gamma\xi)$ (second order). If one concentrates on first-order terms only, Eqs. 206 and 207 prove the identity of the first and second definition of vertical deflections.

Vertical deflections with respect to a basis in geometry space: Gauss surface normal coordinates {L,B,H}

$$-\boldsymbol{\delta\gamma} = \left[\boldsymbol{e}_{\lambda_\gamma}\ \boldsymbol{e}_{\varphi_\gamma}\ \boldsymbol{e}_\gamma\right] = \begin{bmatrix} \gamma\eta \\ \gamma\xi \\ \delta\gamma \end{bmatrix} = \left[\boldsymbol{E_L}\ \boldsymbol{E_B}\ \boldsymbol{E_H}\right] T \begin{bmatrix} \gamma\eta \\ \gamma\xi \\ \delta\gamma \end{bmatrix} \tag{204}$$

$$-\boldsymbol{\delta\gamma} \doteq \left[\boldsymbol{E_L}\ \boldsymbol{E_B}\ \boldsymbol{E_H}\right] \begin{bmatrix} \gamma\eta + \sin B\delta L\gamma\xi - -\cos B\delta L\delta\gamma \\ - -\sin B\delta L\gamma\eta + \gamma\xi - -\delta B\delta\gamma \\ \cos B\delta L\gamma\eta + \delta B\gamma\xi + \delta\gamma \end{bmatrix} \tag{205}$$

vertical deflections

$$\eta := \cos\varphi_\gamma\delta\Lambda_\gamma \doteq \cos B\delta L \tag{206}$$

$$\xi := \delta\varphi_\gamma \doteq \delta B \tag{207}$$

The potential theory of the horizontal and vertical components of the gravity field is reviewed in Eqs. (208–219). First, the reference gravity vector $\boldsymbol{\gamma}$ as the gradient of the gravity potential, the incremental gravity vector $\boldsymbol{\delta\gamma}$ as the gradient of the incremental gravity potential, also called disturbing potential, both represented in the ellipsoidal frame of reference $\{\boldsymbol{e}_L,\boldsymbol{e}_B,\boldsymbol{e}_H\}$, neglecting $H_L = 0, H_B = 0$ are presented. The elements $\{g_{LL}, g_{BB}, g_{HH}\}$ of the matrix of metric G have to be implemented. Second, the highlight is the first-order potential representation of $\{\eta,\xi,\delta\gamma\}$, Eqs. (214–216), as functionals of type $\{D_L\delta w, D_B\delta w, D_H\delta w\}$.

Potential theory of horizontal and vertical components of the gravity field.

$$\boldsymbol{\gamma} = \text{grad}w = e_L\frac{1}{\sqrt{g_{LL}}}D_Lw + e_B\frac{1}{\sqrt{g_{BB}}}D_Bw + e_H\frac{1}{\sqrt{g_{HH}}}D_Hw \tag{208}$$

$$\boldsymbol{\delta\gamma} = \text{grad}\delta w = \boldsymbol{e}_L\frac{1}{\sqrt{g_{LL}}}D_L\delta w + \boldsymbol{e}_B\frac{1}{\sqrt{g_{BB}}}D_B\delta w + \boldsymbol{e}_H\frac{1}{\sqrt{g_{HH}}}D_H\delta w(H_L = 0, H_B = 0) \tag{209}$$

functionals of the disturbing potential δw

$$\begin{bmatrix} (1/\sqrt{g_{LL}})D_L\delta w \\ (1/\sqrt{g_{BB}})D_B\delta w \\ (1/\sqrt{g_{HH}})D_H\delta w \end{bmatrix} = -T\begin{bmatrix} \gamma\eta \\ \gamma\xi \\ \delta\gamma \end{bmatrix} \tag{210}$$

$$\eta = -\frac{1}{\gamma}\left(\frac{1}{\sqrt{g_{LL}}}D_L\delta w + \sin B\delta L\frac{1}{\sqrt{g_{BB}}}D_B\delta w - \cos B\delta L\frac{1}{\sqrt{g_{HH}}}D_H\delta w\right) \tag{211}$$

$$\xi = -\frac{1}{\gamma}\left(-\sin B\delta L\frac{1}{\sqrt{g_{LL}}}D_L\delta w + \frac{1}{\sqrt{g_{BB}}}D_B\delta w - \delta B\frac{1}{\sqrt{g_{HH}}}D_H\delta w\right) \tag{212}$$

$$\delta\gamma = -\cos B\delta L\frac{1}{\sqrt{g_{LL}}}D_L\delta w + \delta B\frac{1}{\sqrt{g_{BB}}}D_B\delta w - \frac{1}{\sqrt{g_{HH}}}D_H\delta w \tag{213}$$

$$\frac{\sin B\delta L}{\sqrt{g_{BB}}}D_B\delta w \ll 1, \frac{\cos B\delta L}{\sqrt{g_{HH}}}D_H\delta w \ll 1, \frac{\delta B}{\sqrt{g_{BB}}}D_B\delta w \ll 1,$$
$$\frac{\sin B\delta L}{\sqrt{g_{LL}}}D_L\delta w \ll 1, \frac{\cos B\delta L}{\sqrt{g_{LL}}}D_L\delta w \ll 1, \frac{\delta B}{\sqrt{g_{HH}}}D_H\delta w \ll 1$$

then

$$\eta \doteq -\frac{1}{\gamma}\frac{1}{(N+H)\cos B}D_L\delta w \tag{214}$$

$$\xi \doteq -\frac{1}{\gamma}\frac{1}{M+H}D_B\delta w \tag{215}$$

$$\delta\gamma \doteq -D_H\delta w \tag{216}$$

5.2 Jacobi Ellipsoidal Coordinates

With reference to the extensive review on ellipsoidal coordinates by Thong and Grafarend 1989, Jacobi ellipsoidal coordinates (λ,φ,u), also called "mixed elliptic trigonometric-elliptic coordinates," namely, spheroidal longitude λ, spheroidal latitude φ, and semi-minor axis u which relate to Cartesian coordinates (X,Y,Z) in the global orthonormal frame of reference $\{\boldsymbol{E}_1, \boldsymbol{E}_2, \boldsymbol{E}_3|O\}$ by Eqs. 217 and 219 are introduced. Those elliptic coordinates have been introduced to physical geodesy since on contrast to Gauss surface normal ellipsoidal coordinates the *Laplace differential equation*, which governs the *Newtonian gravitational field*, separates in these coordinates (see, e.g., Grafarend 1988). The *Jacobi ellipsoidal coordinates* are generated by the intersection of

(i) A confocal, oblate ellipsoid
(ii) A confocal half hyperboloid
(iii) A half plane

in a unique way.

Jacobi ellipsoidal coordinates $\{\Lambda, \varphi, u\}$.
(i) Forward transformation

$$\mathbf{X} = \mathbf{E}_1 X + \mathbf{E}_2 Y + \mathbf{E}_3 Z \tag{217}$$

$$\begin{aligned} X &= \sqrt{u^2+\varepsilon^2}\cos\varphi\cos\lambda \\ Y &= \sqrt{u^2+\varepsilon^2}\cos\varphi\cos\lambda \\ Z &= u\sin\varphi \end{aligned} \tag{218}$$

absolute eccentricity of the ellipsoid of revolution $E^2_{A_1,A_1,A_2}$

$$\varepsilon^2 := \left(A_1^2 - A_2^2\right)/A_1^2, \varepsilon \in R^+ \tag{219}$$

(ii) Backward transformation

$$\lambda = \begin{cases} \arctan(Y/X) & \text{for } (X>0) \text{ and } (Y \geq) \\ \arctan((Y/X)+\pi)l & \text{for } (X<0) \text{ and } (Y \neq 0) \\ \arctan((Y/X)+2\pi) & \text{for } (X>0) \text{ and } (Y<0) \\ \pi/2 & \text{for } (X=0) \text{ and } (Y>0) \\ 3\pi/2 & \text{for } (X=0) \text{ and } (Y<0) \\ \text{undefined} & \text{for } (X=0) \text{ and } (Y=0) \end{cases} \tag{220}$$

$$\varphi = (\operatorname{sgn} Z)\arcsin\left[\frac{1}{2\varepsilon^2}\left(\varepsilon^2 - (X^2+Y^2+Z^2) + \sqrt{(X^2+Y^2+Z^2-\varepsilon^2)^2 + 4\varepsilon^2 Z^2}\right)\right]^{1/2} \tag{221}$$

$$u = \left[\frac{1}{2}\left(X^2+Y^2+Z^2-\varepsilon^2 + \sqrt{(X^2+Y^2+Z^2-\varepsilon^2)^2 + 4\varepsilon^2 Z^2}\right)\right]^{1/2} \tag{222}$$

(iii) Jacobi-matrix

$$\boldsymbol{J} := \begin{bmatrix} D_\lambda X \; D_\varphi X \; D_u X \\ D_\lambda Y \; D_\varphi Y \; D_u Y \\ D_\lambda Z \; D_\varphi Z \; D_u Z \end{bmatrix} \tag{223}$$

$$\boldsymbol{J} := \begin{bmatrix} -\sqrt{u^2+\varepsilon^2}\cos\varphi\sin\lambda & -\sqrt{u^2+\varepsilon^2}\sin\varphi\cos\lambda & \frac{u}{\sqrt{u^2+\varepsilon^2}}\cos\varphi\cos\lambda \\ \sqrt{u^2+\varepsilon^2}\cos\varphi\cos\lambda & -\sqrt{u^2+\varepsilon^2}\sin\varphi\sin\lambda & \frac{u}{\sqrt{u^2+\varepsilon^2}}\cos\varphi\sin\lambda \\ 0 & u\cos\varphi & \sin\varphi \end{bmatrix} \tag{224}$$

(iv) Metric

$$dS^2 = \begin{bmatrix} d\lambda \, d\varphi \, du \end{bmatrix} \boldsymbol{J}^T \boldsymbol{J} \begin{bmatrix} d\lambda \\ d\varphi \\ du \end{bmatrix} \tag{225}$$

$$\boldsymbol{G} := \boldsymbol{J}^T \boldsymbol{J} \tag{226}$$

$$\boldsymbol{G} := \begin{bmatrix} (u^2+\varepsilon^2)\cos^2\varphi & 0 & 0 \\ 0 & u^2+\varepsilon^2\sin^2\varphi & 0 \\ 0 & 0 & (u^2+\varepsilon^2\sin^2\varphi)/(u^2+\varepsilon^2) \end{bmatrix} \tag{227}$$

$$dS^2 = (u^2+\varepsilon^2)\cos^2\varphi d\lambda^2 + (u^2+\varepsilon^2\sin^2\varphi)d\varphi^2 + \frac{u^2+\varepsilon^2\sin^2\varphi}{u^2+\varepsilon^2}du^2 \tag{228}$$

The derivatives $\{D_\Lambda x, D_\varphi x, D_u x\}$ are presented in order to construct in Eqs. 229 and 230, a local orthonormal frame of reference, namely, $\boldsymbol{E}_\lambda := D_\lambda x \div \|D_\lambda x\|$, $\boldsymbol{E}_\varphi := D_\varphi x \div \|D_\varphi x\|$, $\boldsymbol{E}_u := D_u x \div \|D_u x\|$, in short $\{\boldsymbol{E}_\lambda, \boldsymbol{E}_\varphi, \boldsymbol{E}_u | P\}$, also called {ellipsoidal East, ellipsoidal North, ellipsoidal Vertical} at P. The local orthonormal frame of reference $\{\boldsymbol{E}_\lambda, E_\varphi, E_u | P\}$ at P is related to the global orthonormal frame of reference $\{\boldsymbol{E}_1, \boldsymbol{E}_2, \boldsymbol{E}_3 | O\}$ at O by the orthonormal matrix T_1.

Thanks to the transformation Eq. 165 $\{\boldsymbol{e}_1, \boldsymbol{e}_2, \boldsymbol{e}_3\} \rightarrow \{e_\Lambda, e_\varphi, e_u\}$; one has already succeeded to establish the orthonormal matrix $T_2(\lambda_\gamma, \varphi_\gamma)$, Eq. 233. Such a relation will again be used to transform the ellipsoidal orthonormal frame of reference $\{\boldsymbol{E}_\lambda, E_\varphi, E_u | P\}$ to the reference "East, North, Vertical" frame $\{\boldsymbol{e}_\lambda, e_\varphi, e_u | P\}$. Indeed $[\boldsymbol{E}_\lambda, E_\varphi, E_u]^T = T_1[e_1, e_2, e_3]^T = T_1 T_2^T [\boldsymbol{e}_\lambda, \boldsymbol{e}_\varphi, \boldsymbol{e}_u]^T$ or $[\boldsymbol{e}_\lambda, \boldsymbol{e}_\varphi, \boldsymbol{e}_u]^T = T_2 T_1^T [\boldsymbol{E}_\lambda, E_\varphi, E_u]^T$ is the compound transformation T of Eqs. 231–235. As soon as one develops T close to the identity, namely, by means of the decomposition Eqs. 236–239 as well as of a special case Eq. 240 of the binomial series, $(\varepsilon u) < 1$, one gains the elements of the matrix $\boldsymbol{T}$ (case study t_{23}). The matrix $\boldsymbol{T}$ is finally decomposed into the unit matrix $\boldsymbol{I}_3$ and the incremental antisymmetric matrix $\boldsymbol{A}$, Eqs. 248 and 249.

Construction of the local reference frame $\{\boldsymbol{E}_\lambda, \boldsymbol{E}_\varphi, \boldsymbol{E}_u | P\}$ from the Jacobi ellipsoidal coordinates of point P on the topography.

$$\begin{aligned}
\boldsymbol{E}_\lambda &:= \frac{\boldsymbol{D}_\lambda \boldsymbol{x}}{\|\boldsymbol{D}_\lambda \boldsymbol{x}\|} = \left(\boldsymbol{E}_1 \frac{\partial x_1}{\partial \lambda} + \boldsymbol{E}_2 \frac{\partial x_2}{\partial \lambda} + \boldsymbol{E}_3 \frac{\partial x_3}{\partial \lambda}\right) \div \left[\left(\frac{\partial x_1}{\partial \lambda}\right)^2 + \left(\frac{\partial x_2}{\partial \lambda}\right)^2 + \left(\frac{\partial x_3}{\partial \lambda}\right)^2\right]^{1/2} \\
\boldsymbol{E}_\varphi &= \frac{\boldsymbol{D}_\varphi \boldsymbol{x}}{\|\boldsymbol{D}_\varphi \boldsymbol{x}\|} = \left(\boldsymbol{E}_1 \frac{\partial x_1}{\partial \varphi} + \boldsymbol{E}_2 \frac{\partial x_2}{\partial \varphi} + \boldsymbol{E}_3 \frac{\partial x_3}{\partial \varphi}\right) \div \left[\left(\frac{\partial x_1}{\partial \varphi}\right)^2 + \left(\frac{\partial x_2}{\partial \varphi}\right)^2 + \left(\frac{\partial x_3}{\partial \varphi}\right)^2\right]^{1/2} \\
\boldsymbol{E}_u &= \frac{\boldsymbol{D}_u \boldsymbol{x}}{\|\boldsymbol{D}_u \boldsymbol{x}\|} = \left(\boldsymbol{E}_1 \frac{\partial x_1}{\partial u} + \boldsymbol{E}_2 \frac{\partial x_2}{\partial u} + \boldsymbol{E}_3 \frac{\partial x_3}{\partial u}\right) \div \left[\left(\frac{\partial x_1}{\partial u}\right)^2 + \left(\frac{\partial x_2}{\partial u}\right)^2 + \left(\frac{\partial x_3}{\partial u}\right)^2\right]^{1/2}
\end{aligned} \tag{229}$$

$$\begin{bmatrix} \boldsymbol{E}_\lambda \\ \boldsymbol{E}_\varphi \\ \boldsymbol{E}_u \end{bmatrix} = \boldsymbol{T}_1 \begin{bmatrix} \boldsymbol{E}_1 \\ \boldsymbol{E}_2 \\ \boldsymbol{E}_3 \end{bmatrix} \tag{230}$$

$$\boldsymbol{T}_1 := \begin{bmatrix} -\sin\lambda & \cos\lambda & 0 \\ \frac{-\sqrt{u^2+\varepsilon^2}}{\sqrt{u^2+\varepsilon^2\sin^2\varphi}}\cos\lambda\sin\varphi & \frac{-\sqrt{u^2+\varepsilon^2}}{\sqrt{u^2+\varepsilon^2\sin^2\varphi}}\sin\lambda\sin\varphi & \frac{u}{\sqrt{u^2+\varepsilon^2\sin^2\varphi}}\cos\varphi \\ \frac{u}{\sqrt{u^2+\varepsilon^2\sin^2\varphi}}\cos\lambda\cos\varphi & \frac{u}{\sqrt{u^2+\varepsilon^2\sin^2\varphi}}\sin\lambda\cos\varphi & \frac{\sqrt{u^2+\varepsilon^2}}{\sqrt{u^2+\varepsilon^2\sin^2\varphi}}\sin\varphi \end{bmatrix}$$

Basis transformation from geometry space to gravity space

$$\begin{bmatrix} \boldsymbol{e}_{\lambda_\gamma} \\ \boldsymbol{e}_{\varphi_\gamma} \\ \boldsymbol{e}_\gamma \end{bmatrix} = \boldsymbol{T}_2 \boldsymbol{T}_1^T \begin{bmatrix} \boldsymbol{E}_\lambda \\ \boldsymbol{E}_\varphi \\ \boldsymbol{E}_u \end{bmatrix} = \boldsymbol{T} \begin{bmatrix} \boldsymbol{E}_\lambda \\ \boldsymbol{E}_\varphi \\ \boldsymbol{E}_u \end{bmatrix} \tag{231}$$

subject to

$$\boldsymbol{T}_1 \doteq \boldsymbol{T}_1(\Lambda, \varphi, u) \tag{232}$$

$$\boldsymbol{T}_2 \doteq \boldsymbol{T}_2(\lambda_\gamma, \varphi_\gamma) = \boldsymbol{R}_E\left(\lambda_\gamma + \frac{\pi}{2}, \frac{\pi}{2} - \varphi_\gamma, 0\right) \tag{233}$$

$$\boldsymbol{T} := \boldsymbol{T}_2 \boldsymbol{T}_1^T = \boldsymbol{T}(\Lambda, \varphi, u; \lambda_\gamma, \varphi_\gamma), \boldsymbol{T}_1, \boldsymbol{T}_2 \in SO(3) \tag{234}$$

$$\boldsymbol{T} = \begin{bmatrix} \cos(\lambda - \lambda_\gamma) & \dfrac{-\sqrt{u^2+\varepsilon^2}\sin(\lambda-\lambda_\gamma)\sin\varphi}{\sqrt{u^2+\varepsilon^2\sin^2\varphi}} & \dfrac{u\sin(\lambda-\lambda_\gamma)\cos\varphi}{\sqrt{u^2+\varepsilon^2\sin^2\varphi}} \\ \sin(\lambda-\lambda_\gamma)\sin\varphi_\gamma & \begin{array}{c}\dfrac{\sqrt{u^2+\varepsilon^2}\cos(\lambda-\lambda_\gamma)\sin\varphi_\gamma}{\sqrt{u^2+\varepsilon^2\sin^2\varphi}} + \\ + \dfrac{u\cos\varphi\cos\varphi_\gamma}{\sqrt{u^2+\varepsilon^2\sin^2\varphi}}\end{array} & \begin{array}{c}\dfrac{-u\cos(\lambda-\lambda_\gamma)\cos\varphi\sin\varphi_\gamma}{\sqrt{u^2+\varepsilon^2\sin^2\varphi}} + \\ + \dfrac{\sqrt{u^2+\varepsilon^2}\sin\varphi\cos\varphi_\gamma}{\sqrt{u^2+\varepsilon^2\sin^2\varphi}}\end{array} \\ -\sin(\lambda-\lambda_\gamma)\cos\varphi_\gamma & \begin{array}{c}\dfrac{\sqrt{u^2+\varepsilon^2}\cos(\lambda-\lambda_\gamma)\sin\varphi\cos\varphi_\gamma}{\sqrt{u^2+\varepsilon^2\sin^2\varphi}} + \\ + \dfrac{u\cos\varphi\sin\varphi_\gamma}{\sqrt{u^2+\varepsilon^2\sin^2\varphi}}\end{array} & \begin{array}{c}\dfrac{u\cos(\lambda-\lambda_\gamma)\cos\varphi\cos\varphi_\gamma}{\sqrt{u^2+\varepsilon^2\sin^2\varphi}} + \\ + \dfrac{\sqrt{u^2+\varepsilon^2}\sin\varphi\sin\varphi_\gamma}{\sqrt{u^2+\varepsilon^2\sin^2\varphi}}\end{array} \end{bmatrix} \tag{235}$$

additive decomposition

$$\lambda = \lambda_\gamma + \delta\lambda \Leftrightarrow \lambda_\gamma = \varphi - -\delta\lambda \tag{236}$$

$$\varphi = \varphi_\gamma + \delta\varphi \Leftrightarrow \varphi_\gamma = \varphi - -\delta\varphi \tag{237}$$

$$\sin(\Lambda - -\delta\lambda) \doteq \sin\Lambda - -\cos\Lambda\delta\lambda \tag{238}$$

$$\sin(\varphi - -\delta\varphi) \doteq \cos\varphi + \sin\varphi\delta\varphi \tag{239}$$

special case of binomial series

$$(1+x)^{-1/2} = 1 - \frac{1}{2}x + \frac{3}{8}x^2 - \frac{15}{48}x^3 + O(4) \tag{240}$$

condition: $u > \varepsilon = \sqrt{a^2 - b^2}$

$$\begin{aligned} \frac{u}{\sqrt{u^2+\varepsilon^2\sin^2\varphi}} &= \left(1 + \frac{\varepsilon^2\sin^2\varphi}{u^2}\right)^{-1/2} \\ &= 1 - \frac{1}{2}\left(\frac{\varepsilon\sin\varphi}{u}\right)^2 + \frac{3}{8}\left(\frac{\varepsilon\sin\varphi}{u}\right)^4 + O_6\left(\frac{\varepsilon\sin\varphi}{u}\right) \end{aligned} \tag{241}$$

$$\begin{aligned} \sqrt{\frac{u^2+\varepsilon^2}{u^2+\varepsilon^2\sin^2\varphi}} &= \left(1 + \frac{\varepsilon^2\cos^2\varphi}{u^2+\varepsilon^2}\right)^{-1/2} \\ &= 1 - \frac{1}{2}\frac{\varepsilon^2\cos^2\varphi}{u^2+\varepsilon^2} + \frac{3}{8}\left(\frac{\varepsilon^2\cos^2\varphi}{u^2+\varepsilon^2}\right)^2 + O_6\left(\frac{\varepsilon\cos\varphi}{\sqrt{u^2+\varepsilon^2}}\right) \end{aligned} \tag{242}$$

case study

$$t_{23} = \left[1 - \frac{\varepsilon^2 \sin^2\varphi}{u^2} + O(4)\right](-\sin\varphi\cos\varphi + \cos^2\varphi\,\delta\varphi)$$
$$+\left[1 - \frac{\varepsilon^2\cos^2\varphi}{u^2+\varepsilon^2} + O(4)\right](\sin\varphi\cos\varphi + \sin^2\varphi\,\delta\varphi) \tag{243}$$

$$t_{23} \doteq \delta\varphi + \frac{\varepsilon^2 \sin\varphi\cos\varphi}{2u^2(u^2+\varepsilon^2)}$$
$$\times\left[(u^2+\varepsilon^2)\sin^2\varphi + u^2\cos^2\varphi - (u^2+\varepsilon^2)\sin\varphi\cos\varphi\,\delta\varphi + u^2\sin\varphi\cos\varphi\,\delta\varphi\right] \tag{244}$$

$$t_{23} \doteq \delta\varphi + \frac{\varepsilon^2\sin\varphi\cos\varphi}{2u^2}$$
$$\times\left[\sin^2\varphi + \frac{u^2}{u^2+\varepsilon^2}\cos^2\varphi - \sin\varphi\cos\varphi\,\delta\varphi + \frac{u^2}{u^2+\varepsilon^2}\sin\varphi\cos\varphi\,\delta\varphi\right] \tag{245}$$

$$t_{23} \doteq \delta\varphi + (\varepsilon^2/4u^2)\sin 2\varphi \tag{246}$$

linearized basis transformation from geometry space to gravity space

$$\begin{bmatrix} \boldsymbol{e}_{\lambda_y} \\ \boldsymbol{e}_{\varphi_y} \\ \boldsymbol{e}_{y} \end{bmatrix} = \begin{bmatrix} 1 & -\sin\varphi\,\delta L & \cos\varphi\,\delta L \\ \sin\varphi\,\delta L & 1 & \delta\varphi + (\varepsilon/4u^2)\sin 2\varphi \\ -\cos\varphi\,\delta L & -\delta\varphi - (\varepsilon/4u^2)\sin 2\varphi & 1 \end{bmatrix} \begin{bmatrix} \boldsymbol{E}_\lambda \\ \boldsymbol{E}_\varphi \\ \boldsymbol{E}_u \end{bmatrix} \tag{247}$$

subject to

$$\boldsymbol{T} = \boldsymbol{I}_3 + \boldsymbol{\delta A} \tag{248}$$

$$\delta\boldsymbol{A} = \begin{bmatrix} 0 & -\sin\varphi\,\delta L & \cos\varphi\,\delta L \\ \sin\varphi\,\delta L & 0 & \delta\varphi + (\varepsilon/4u^2)\sin 2\varphi \\ -\cos\varphi\,\delta L & -\delta\varphi - (\varepsilon/4u^2)\sin 2\varphi & 0 \end{bmatrix} \tag{249}$$

Based upon the linearized version of the transformation $\{\boldsymbol{E}_\Lambda,\boldsymbol{E}_\varphi,\boldsymbol{E}_u\} \to \{\boldsymbol{e}_\Lambda,\boldsymbol{e}_\varphi,\boldsymbol{e}_u\}$, Eqs. (231–249), one can build-up the representation of the incremental gravity vector $-\boldsymbol{\delta y}$, Eq. 176, in terms of $[\boldsymbol{e}_\Lambda,\boldsymbol{e}_\varphi,\boldsymbol{e}_u]$: Eqs. 250–253 illustrate the third representation of the incremental gravity vector $-\boldsymbol{\delta y}$, now in the basis $[\boldsymbol{E}_\Lambda,\boldsymbol{E}_\varphi,\boldsymbol{E}_u]$. The second basic result has to be interpreted as following. The third representation of the incremental gravity vector $-\boldsymbol{\delta y}$ contains horizontal components as well as a vertical component which are all functions of $(\eta,\xi,\delta y)$. For instance, the East component E_Λ is a function of $y\eta$ (first order) and of $(y\xi,\delta y)$ (second order). Or the vertical component E_u is a function of $\delta\gamma$ (first order) and of $(\gamma\eta, \gamma\xi)$ (second order). If one concentrates on first-order terms only, Eqs. 252 and 253 prove the identity of the first and third definition of vertical deflections.

Vertical deflections with respect to a basis in geometry space: Jacobi ellipsoidal coordinates (Λ, φ, u).

$$-\boldsymbol{\delta\gamma} = \left[\boldsymbol{e}_{\lambda_\gamma}, \boldsymbol{e}_{\varphi_\gamma}, \boldsymbol{e}_\gamma\right] \tag{250}$$

$$-\boldsymbol{\delta\gamma} \doteq \left[\boldsymbol{E}_\lambda \quad \boldsymbol{E}_\varphi \quad \boldsymbol{E}_u\right] \begin{bmatrix} \gamma\eta + \sin\varphi\delta\lambda\gamma\xi - \cos\varphi\delta\lambda\delta\gamma \\ -\sin\varphi\delta\lambda\gamma\eta + \gamma\xi - \left(\delta\varphi + \frac{\varepsilon^2}{4u^2}\sin 2\varphi\right)\delta\lambda \\ \cos\varphi\delta\lambda\gamma\eta + \left(\delta\varphi + \frac{\varepsilon^2}{4u^2}\sin 2\varphi\right)\gamma\xi + \delta\lambda \end{bmatrix} \tag{251}$$

vertical deflections

$$\eta := \cos\varphi_\gamma \delta\lambda_\gamma \doteq \cos\varphi\delta\lambda \tag{252}$$

$$\xi := \delta\varphi_\gamma \doteq \delta\varphi + \frac{\varepsilon^2}{4u^2}\sin 2\varphi \tag{253}$$

The potential theory of the horizontal and vertical components of the gravity field is reviewed in Eqs. (254–262). First, the reference gravity vector γ as the gradient of the gravity potential, the incremental gravity vector $\delta\gamma$ as the gradient of the incremental gravity potential, also called disturbing potential, newly represented in the Jacobi ellipsoidal frame of reference $\{\boldsymbol{E}_\lambda, \boldsymbol{E}_\varphi, \boldsymbol{E}_u | P\}$ are presented. The elements $(g_{\lambda\lambda}, g_{\varphi\varphi}, g_{uu})$ of the matrix of the metric G have to be implemented. Second, the updated highlight is the first-order potential representation $(\eta, \xi, \delta\gamma)$, Eqs. (257–259), as functionals of type $(D_\Lambda\delta w, D_\varphi\delta w, D_u\delta w)$.

Potential theory of horizontal and vertical components of the gravity field.

$$\boldsymbol{\gamma} = \text{grad}w = \boldsymbol{E}_\lambda \frac{1}{\sqrt{g_{\lambda\lambda}}} D_\lambda w + \boldsymbol{E}_\varphi \frac{1}{\sqrt{g_{\varphi\varphi}}} D_\varphi w + \boldsymbol{E}_u \frac{1}{\sqrt{g_{uu}}} D_u w \tag{254}$$

$$\boldsymbol{\delta\gamma} = \text{grad}\delta w = \boldsymbol{E}_\lambda \frac{1}{\sqrt{g_{\lambda\lambda}}} D_\lambda \delta w + \boldsymbol{E}_\varphi \frac{1}{\sqrt{g_{\varphi\varphi}}} D_\varphi \delta w + \boldsymbol{E}_u \frac{1}{\sqrt{g_{uu}}} D_u \delta w \tag{255}$$

functionals of the disturbing potential δw

$$\begin{bmatrix} (1/\sqrt{g_{\lambda\lambda}})D_\lambda\delta w \\ (1/\sqrt{g_{\varphi\varphi}})D_\varphi\delta w \\ (1/\sqrt{g_{uu}})D_u\delta w \end{bmatrix} = -\boldsymbol{T} \begin{bmatrix} \gamma\eta \\ \gamma\xi \\ \delta\gamma \end{bmatrix} \tag{256}$$

$$\eta = -\frac{1}{\gamma}\left(\frac{1}{\sqrt{g_{\lambda\lambda}}} D_\lambda\delta w + \sin\varphi\delta\lambda \frac{1}{\sqrt{g_{\varphi\varphi}}} D_\varphi\delta w - \cos\varphi\delta\lambda \frac{1}{\sqrt{g_{uu}}} D_u\delta w\right) \tag{257}$$

$$\xi = -\frac{1}{\gamma}\left(-\sin\varphi\delta\lambda \frac{1}{\sqrt{g_{\lambda\lambda}}} D_\lambda\delta w + \frac{1}{\sqrt{g_{\varphi\varphi}}} D_\varphi\delta w - \left(\delta\varphi + \frac{\varepsilon^2}{4u^2}\sin 2\varphi\right)\frac{1}{\sqrt{g_{uu}}} D_u\delta w\right) \tag{258}$$

$$\delta\gamma = -\left(-\cos\varphi\delta\lambda \frac{1}{\sqrt{g_{\lambda\lambda}}} D_\lambda\delta w + \left(\delta\varphi + \frac{\varepsilon^2}{4u^2}\sin 2\varphi\right)\frac{1}{\sqrt{g_{\varphi\varphi}}} D_\varphi\delta w + \frac{1}{\sqrt{g_u}} D_u\delta w\right) \tag{259}$$

if

$$\frac{\sin\varphi\delta\lambda}{\sqrt{g_{\lambda\lambda}}} D_\lambda\delta w \ll 1, \frac{\cos\varphi\delta\lambda}{\sqrt{g_{\lambda\lambda}}} D_\lambda\delta w \ll 1, \frac{\delta\varphi + (\varepsilon^2/4u^2)\sin 2\varphi}{\sqrt{g_{\varphi\varphi}}} D_\varphi\delta w \ll 1,$$

$$\frac{\sin\varphi\delta\lambda}{\sqrt{g_{\varphi\varphi}}} D_\varphi\delta w \ll 1, \frac{\cos\varphi\delta\lambda}{\sqrt{g_{uu}}} D_u\delta w \ll 1, \frac{\delta\varphi + (\varepsilon^2/4u^2)\sin 2\varphi}{\sqrt{g_{uu}}} D_u\delta w \ll 1,$$

then

$$\eta \doteq -\frac{1}{\gamma}\frac{1}{\sqrt{u^2+\varepsilon^2}\cos\varphi}D_\lambda \delta w \tag{260}$$

$$\xi \doteq -\frac{1}{\gamma}\frac{1}{\sqrt{u^2+\varepsilon^2\sin^2\varphi}}D_\varphi \delta w \tag{261}$$

$$\delta\gamma \doteq -\sqrt{\frac{u^2+\varepsilon^2\sin^2\varphi}{u^2+\varepsilon^2}}D_u \delta w \tag{262}$$

6 Potential Theory of Horizontal and Vertical Components of the Gravity Field: Gravity Disturbance and Vertical Deflections

Now the gravitational disturbing potential in terms of *Jacobi ellipsoidal harmonics* is represented. As soon as one takes reference to a normal potential of *Somigliana–Pizzetti* type, the ellipsoidal harmonics of degree/order (0,0), (1,0), (1,–1), (1,1) and (2,0) are eliminated from the gravitational disturbing potential.

In order to present the potential theory of the horizontal and vertical components of the gravity field, namely, ellipsoidal vertical deflections and ellipsoidal gravity disturbance, one has to make a decision about:

- What is the proper choice of the ellipsoidal potential field of reference $w(\Lambda,\varphi,u)$,
- And about the related ellipsoidal incremental potential field $\delta w(\Lambda,\varphi,u)$, also called "disturbing potential."

6.1 Ellipsoidal Reference Potential of Type Somigliana–Pizzetti

There has been made three proposals for an ellipsoidal potential field of reference. The first choice is the zero-degree term $[\text{arccot}(u/\varepsilon)](\text{GM}/\varepsilon)$ of the external ellipsoidal harmonic expansion. Indeed, it would correspond to the zero-order term GM/r of the external spherical harmonic expansion. As proven in Grafarend et al. (1999), the equipotential surface $w = [\text{arccot}(u/\varepsilon)](\text{GM}/\varepsilon) = \text{constant}$ where GM is the geocentric gravitational constant. Unfortunately, such an equipotential reference surface does not include the rotation of the Earth, namely, its centrifugal potential $\Omega^2\varepsilon^2[1+(u/\varepsilon)^2]/3+\Omega^2\varepsilon^2[1+(u/\varepsilon)^2]P_2(\sin\varphi)/3$. Accordingly, there has been made the proposal for a second choice, namely, to choose $[\text{arccot}(u/\varepsilon)]\,\text{GM}/\varepsilon + \Omega^2\varepsilon^2(u^2+\varepsilon^2)\cos^2\varphi/2$. In this approach the zero-degree term of the gravitational potential and the centrifugal potential is superimposed. Unfortunately, the level surface $[\text{arccot}(u/\varepsilon)]\,\text{GM}/\varepsilon + (1+P_2(\sin\varphi))\Omega^2\varepsilon^2[1+(u/\varepsilon)^2]/3 = \text{constant}$ is not an ellipsoid of revolution. It is for this reason that the proposal for the third choice has been chosen. Superimpose the gravitational potential, which is externally expanded in ellipsoidal harmonics, and the centrifugal potential represented also in ellipsoidal base functions (the centrifugal potential is not a harmonic function) and postulate an equipotential reference surface to be a level ellipsoid. Such a level ellipsoid should be an ellipsoid of revolution. Such an ellipsoidal reference field has been developed by *Somigliana* (1929) and *Pizzetti* (1894) and is properly called Somigliana–Pizzetti reference potential. The Euclidean length of its gradient is referred to as the

International Gravity Formula, which recently has been developed to the sub-nano Gal level by Ardalan and Grafarend (2001). Here, the recommendations followed are that of the International Association of Geodesy, namely, Moritz 1984, to use the Somigliana–Pizzetti potential of a level ellipsoid as the reference potential summarized in Eqs. (263–267).

Reference gravity potential field of type Somigliana–Pizzetti.
Reference Level Ellipsoid
semi-major axis a, semi-minor axis b
absolute eccentricity $\varepsilon = \sqrt{a^2 - b^2}$

$$\mathbf{E}^2_{a,b} := \mathbf{X} \in R^3 | (X^2 + Y^2)/a^2 + Z^2/b^2 = 1 \tag{263}$$

the first version of the reference potential field

$$w(\varphi, u) = \frac{GM}{\varepsilon}\operatorname{arccot}\left(\frac{u}{\varepsilon}\right) + \frac{1}{6}\Omega^2 a^2 \frac{\left(3\left(\frac{u}{\varepsilon}\right)^2 + 1\right)\operatorname{arccot}\left(\frac{u}{\varepsilon}\right) - 3\frac{u}{\varepsilon}}{\left(3\left(\frac{b}{\varepsilon}\right)^2 + 1\right)\operatorname{arccot}\left(\frac{b}{\varepsilon}\right) - 3\frac{b}{\varepsilon}}(3\sin^2\varphi - 1)$$
$$+ \frac{1}{2}\Omega^2(u^2 + \varepsilon^2)\cos^2\varphi \tag{264}$$

input
4 parameters: $GM, \Omega, a, \varepsilon = \sqrt{a^2 - b^2}$ or b
Legendre polynomials of the first and second kind

$$\operatorname{arccot}(u/\varepsilon) = Q^*_{00}(u/\varepsilon)$$
$$\frac{\left(3\left(\frac{u}{\varepsilon}\right)^2 + 1\right)\operatorname{arccot}\left(\frac{u}{\varepsilon}\right) - 3\frac{u}{\varepsilon}}{\left(3\left(\frac{b}{\varepsilon}\right)^2 + 1\right)\operatorname{arccot}\left(\frac{b}{\varepsilon}\right) - 3\frac{b}{\varepsilon}} = \frac{Q^*_{20}(u/\varepsilon)}{Q^*_{20}(b/\varepsilon)} \tag{265}$$

$$3\sin^2\varphi - 1 = \frac{2}{\sqrt{5}}P^*_{20}(\sin\varphi) \tag{266}$$

the second version of the reference potential field

$$w(\varphi, u) = \frac{GM}{\varepsilon}Q^*_{00}(u/\varepsilon) + \frac{\sqrt{5}}{15}\Omega^2 a^2 \frac{Q^*_{20}(u/\varepsilon)}{Q^*_{20}(b/\varepsilon)}P^*_{20}(\sin\varphi) + \frac{1}{2}\omega^2(u^2 + \varepsilon^2)\cos^2\varphi \tag{267}$$

Constraints: reference gravity potential field of type Somigliana–Pizzetti.
Conditions for the ellipsoidal terms of degree/order (0,0) and (2,0).
the first version

$$(0,0) : u_{00} + \frac{1}{3}\Omega^2 a^2 = W_0 \tag{268}$$

$$(2,0) : u_{20} - \frac{\sqrt{5}}{15}\Omega^2 a^2 = 0 \tag{269}$$

the first condition in multipole expansion

$$u_{00} = \frac{GM}{\varepsilon}\operatorname{arccot}\left(\frac{b}{\varepsilon}\right) = \frac{GM}{\sqrt{a^2 - b^2}}\operatorname{arccot}\left(\frac{b}{\sqrt{a^2 - b^2}}\right) \tag{270}$$

$$u_{00} = \frac{GM}{\varepsilon}Q^*_{00}\left(\frac{b}{\varepsilon}\right) \tag{271}$$

Corollary:

$$\frac{GM}{\sqrt{a^2-b^2}}\operatorname{arccot}\left(\frac{b}{\sqrt{a^2-b^2}}\right)+\frac{1}{3}\Omega^2a^2-W_0=0 \tag{272}$$

the second condition in multipole expansion

$$u_{20}=\frac{G}{\varepsilon}\frac{1}{2}\left[\left(3\left(\frac{b}{\varepsilon}\right)^2+1\right)\operatorname{arccot}\left(\frac{b}{\varepsilon}\right)-3\frac{b}{\varepsilon}\right]J_{20} \tag{273}$$

$$u_{20}=\frac{G}{\varepsilon}Q^*_{20}\left(\frac{b}{\varepsilon}\right)J_{20} \tag{274}$$

Ellipsoidal multipole of degree/order (2,0)

$$J_{20}:=\int_0^{2\pi}d\lambda'\int_{-\pi/2}^{\pi/2}d\varphi'\cos\varphi'\int_0^{u'(\lambda',\varphi')}du'(u'^2+\varepsilon^2\sin^2\varphi')\frac{1}{2}\left(3\left(\frac{u'}{\varepsilon}\right)^2+1\right)$$
$$\times\frac{\sqrt{5}}{2}(3\sin^2\varphi'-1)\rho(\lambda',\varphi',u') \tag{275}$$

functional of the mass density field $\rho(\Lambda',\varphi', u')$

$$J_{20}:=\int_0^{2\pi}d\lambda'\int_{-\pi/2}^{\pi/2}d\varphi'\cos\varphi'\int_0^{u'(\lambda,\varphi')}du'(u'^2+\varepsilon^2\sin^2\varphi')P^*_{20}\left(\frac{u'}{\varepsilon}\right)P^*_{20}(\sin\varphi')\rho(\lambda,\varphi,u') \tag{276}$$

Cartesian multipole of degree two versus ellipsoidal multipole of degree/order (2,0)

$$J_{11}=:A,\ J_{22}=:B,\ J_{33}=:C \tag{277}$$

$$J_{pq}:=\int dw_3(\|X\|^2\delta_{pq}-X_pX_q)\rho(X,Y,Z)\ \forall\ p,q\in\{1,2,3\} \tag{278}$$

$$\begin{aligned}J_{11}&:=\int dw_3(Y^2+Z^2)\rho(X,Y,Z)\\ J_{22}&:=\int dw_3(X^2+Z^2)\rho(X,Y,Z)\\ J_{33}&:=\int dw_3(Y^2+Y^2)\rho(X,Y,Z)\end{aligned} \tag{279}$$

$$J_{20}=\frac{1}{2}\sqrt{5}\frac{1}{\varepsilon^2}\left[3\left(\frac{A+B}{2}-C\right)+M\varepsilon^2\right] \tag{280}$$

the second condition in Cartesian multipole expansion

$$u_{20}=\frac{3\sqrt{5}}{8}\frac{1}{\varepsilon^3}\left[\left(3\left(\frac{b}{\varepsilon}\right)^2+1\right)\operatorname{arccot}\left(\frac{b}{\varepsilon}\right)-3\frac{b}{\varepsilon}\right]\left[G(A+B-2C)+\frac{2}{3}GM\varepsilon^2\right] \tag{281}$$

$$u_{20}=\frac{3\sqrt{5}}{4}\frac{1}{\varepsilon^3}Q^*_{20}\left(\frac{b}{\varepsilon}\right)\left[G\left(\frac{A+B}{2}-C\right)+\frac{1}{3}GM\varepsilon^2\right] \tag{282}$$

Corollary:

$$\frac{1}{\varepsilon^3}\left[\frac{1}{4}\frac{GM}{\sqrt{a^2-b^2}}+\frac{3}{8}\frac{G}{\sqrt[3]{a^2-b^2}}(A+B-2C)\right]$$
$$\times\left[\left(3\frac{b^2}{a^2-b^2}+1\right)\operatorname{arccot}\left(\frac{b}{\sqrt{a^2-b^2}}\right)-3\frac{b}{\sqrt{a^2-b^2}}\right]-\frac{1}{15}\Omega^2a^2=0 \tag{283}$$

Corollary (second ellipsoidal condition):

$$u_{20} = \frac{\mathrm{GM}}{\varepsilon}\left(\operatorname{arccot}\frac{b}{\varepsilon}\right) C_{20} \tag{284}$$

is a transformation of the dimensionless ellipsoidal coefficient of degree (2,0) C_{20} to the nondimensionless ellipsoidal coefficient of degree/order (2,0), then the second condition in the ellipsoidal harmonic coefficient C_{20} is

$$\frac{\mathrm{GM}}{\sqrt{a^2-b^2}}\operatorname{arccot}\left(\frac{b}{\sqrt{a^2-b^2}}\right) C_{20} - \frac{1}{3\sqrt{5}}\Omega^2 a^2 = 0 \tag{285}$$

Lemma (World Geodetic Datum, Grafarend and Ardalan 1999)

If the parameters $\{W_0, GM, C_{20}, \Omega\}$ are given, the Newton iteration of the nonlinear two condition equations is contractive and leads to $a = a\,(W_0, GM, C_{20}, \Omega)$, $b = b\,(W_0, GM, C_{20}, \Omega)$, and $\varepsilon = \varepsilon\,(W_0, GM, C_{20}, \Omega)$.

6.2 Ellipsoidal Reference Gravity Intensity of Type Somigliana–Pizzetti

Vertical deflections $\{\eta(\Lambda,\varphi,u),\xi(\Lambda,\varphi,u)\}$ as defined as the longitudinal and lateral derivatives of the disturbing potential are normalized by means of reference gravity intensity $\gamma(\lambda,\varphi,u) =: \|\mathrm{grad}(\lambda,\varphi,u)\|$ (Eqs. 254–262, 285, and 286). Here the aim is at computing the modulus of reference gravity with respect to the ellipsoidal reference potential of type Somigliana–Pizzetti.

The detailed computation of $\|\mathrm{grad}(\lambda,\varphi,u)\|$ is presented by means of Eqs. (285–291), two lemmas, and two corollaries.

Since the reference potential of type Somigliana–Pizzetti $w(\varphi, u)$ depends only on spheroidal latitude φ and spheroidal height u, the modulus of the reference gravity vector Eq. 286 is a nonlinear operator based upon the lateral derivative $D_\varphi w$ and the vertical derivative $D_u w$. As soon as one departs from the standard representation of the gradient operator in orthogonal coordinates, namely, $\mathrm{grad}w = e_\lambda(g_{\lambda\lambda})^{-1/2}D_\lambda w + e_\varphi(g_{\varphi\varphi})^{-1/2}D_\varphi w + e_u(g_{uu})^{-1/2}D_u w$, one arrives at the standard form of $\|\mathrm{grad}w\|$ of type Eq. 286. Here, for the near field computation, one shall assume $x := (u^2+\varepsilon^2)^{-1}(D_\varphi w/D_u w)$. Accordingly by means of Eq. 287, $\sqrt{1+x}$ is expanded in binomial series and is led to the first-order approximation of $\gamma =: \|\mathrm{grad}w\|$ by Eq. 288. Obviously up to $O(2)$, it is sufficient to compute the vertical derivative $|D_u w|$. An explicit version of $D_u w$ is given by Eqs. 291 and 292, in the form of ellipsoidal base functions by Eq. 293.

Reference gravity intensity of type Somigliana–Pizzetti, International Gravity Formula.

$$\begin{aligned}\gamma &= \|\mathrm{grad}w(\varphi,u)\| = \sqrt{\langle \mathrm{grad}w(\varphi,u)|\mathrm{grad}w(\varphi,u)\rangle}\\ &= \sqrt{(u^2+\varepsilon^2\sin^2\varphi)^{-1}(D_\varphi w)^2 + (u^2+\varepsilon^2)(u^2+\varepsilon^2\sin^2\varphi)^{-1}(D_u w)^2}\\ &= \sqrt{(u^2+\varepsilon^2)/(u^2+\varepsilon^2\sin^2\varphi)}\sqrt{(D_u w)^2+(u^2+\varepsilon^2)^{-1}(D_\varphi w)^2}\end{aligned} \tag{286}$$

$$\gamma = \sqrt{(u^2+\varepsilon^2)/(u^2+\varepsilon^2\sin^2\varphi)}\,|D_u w|\sqrt{1+(u^2+\varepsilon^2)^{-1}(D_\varphi w/D_u w)^2} \tag{287}$$

$$x := (u^2+\varepsilon^2)^{-1}(D_\varphi w/D_u w)^2, \sqrt{1+x} = 1 + \frac{1}{2}x - \frac{1}{8}x^2 + O(3)\forall |x| < 1$$

if $x = (u^2 + \varepsilon^2)^{-1}(D_\varphi w / D_u w)^2 < 1$ then

$$\gamma = \sqrt{(u^2 + \varepsilon^2)/(u^2 + \varepsilon^2 \sin^2 \varphi)} |D_u w| + O(2) \tag{288}$$

the first version of $D_u w$

$$D_u w = -\frac{\text{GM}}{u^2 + \varepsilon^2} + \frac{1}{6}\Omega^2 a^2 (3\sin^2\varphi - 1) \frac{\frac{6u}{\varepsilon^2}\text{arccot}\left(\frac{u}{\varepsilon}\right) - \left(3\left(\frac{u}{\varepsilon}\right)^2 + 1\right)\frac{\varepsilon}{u^2+\varepsilon^2} - \frac{3}{\varepsilon}}{\left(3\left(\frac{b}{\varepsilon}\right)^2 + 1\right)\text{arccot}\left(\frac{b}{\varepsilon}\right) - 3\frac{b}{\varepsilon}} + \Omega^2 u \cos^2\varphi \tag{289}$$

$$D_u w = -\frac{\text{GM}}{u^2 + \varepsilon^2} - \frac{1}{6}\Omega^2 a^2 \frac{1}{u^2 + \varepsilon^2} \frac{6\left(\left(\frac{u}{\varepsilon}\right)^2 + 1\right) u\text{arccot}\left(\frac{u}{\varepsilon}\right) - 2\varepsilon\left(3\left(\frac{u}{\varepsilon}\right)^2 + 2\right)}{\left(3\left(\frac{b}{\varepsilon}\right)^2 + 1\right)\text{arccot}\left(\frac{b}{\varepsilon}\right) - 3\frac{b}{\varepsilon}}$$
$$+ \frac{1}{2}\Omega^2 a^2 \frac{\sin^2\varphi}{u^2 + \varepsilon^2} \frac{6\left(\left(\frac{u}{\varepsilon}\right)^2 + 1\right) u\text{arccot}\left(\frac{u}{\varepsilon}\right) - 2\varepsilon\left(3\left(\frac{u}{\varepsilon}\right)^2 + 2\right)}{\left(3\left(\frac{b}{\varepsilon}\right)^2 + 1\right)\text{arccot}\left(\frac{b}{\varepsilon}\right) - 3\frac{b}{\varepsilon}} + \Omega^2 u \cos^2\varphi \tag{290}$$

"the second version of $D_u w$*"*

$$D_u w = \frac{\text{GM}}{\varepsilon}\left[Q^*_{00}(u/\varepsilon)\right]' + \frac{\sqrt{5}}{15}\Omega^2 a^2 P^*_{20}(\sin\varphi)\frac{\left[Q^*_{20}(u/\varepsilon)\right]'}{Q^*_{20}(b/\varepsilon)} + \Omega^2 \text{u}\cos^2\varphi \tag{291}$$

A more useful closed-form representation of the reference gravity intensity of type Somigliana–Pizzetti will be given by two lemmas and two corollaries. First, if one collects the coefficients of $\{1, \cos^2\varphi, \sin^2\varphi\}$ by $\{\gamma_0, \gamma_c, \gamma_s\}$, one is led to the representation of γ by Eqs. 292–295 expressed in the Lemma 1. Second, if one takes advantage of the (a,b) representation of $\sqrt{u^2 + \varepsilon^2}/\sqrt{u^2 + \varepsilon^2 \sin^2\varphi}$ and decompose γ_0 according to $\gamma_0(\sin^2\varphi + \cos^2\varphi)$, one is led to the alternative elegant representation of γ by Eqs. (296–303), presented in the Lemma 2.

Lemma 1

Reference gravity intensity of type Somigliana–Pizzetti, International Gravity Formula:

If $x := (u^2 + \varepsilon^2)^{-1}(D_\varphi w / D_u w)^2 \ll 1$ holds, where $w = w(\varphi,u)$ is the reference potential field of type Somigliana–Pizzetti, then its gravity field intensity $\gamma(\varphi,u)$ can be represented up to the order $O(2)$ by

$$\gamma = \sqrt{(u^2 + \varepsilon^2)/(u^2 + \varepsilon^2 \sin^2\varphi)} |\gamma_0 + \gamma_c \cos^2\varphi + \gamma_s \sin^2\varphi| \tag{292}$$

subject to

$$\gamma_0 := -\frac{\text{GM}}{a^2 + (u+b)(u-b)} - \frac{1}{3}\Omega^2 a^2 \frac{1}{a^2 + (u+b)(u-b)}$$
$$\times \frac{3((u/\varepsilon)^2 + 1)\text{uarccot}(u/\varepsilon) - \varepsilon(3(u/\varepsilon)^2 + 2)}{(3(b/\varepsilon)^2 + 1)\text{arccot}(b/\varepsilon) - 3(b/\varepsilon)} \tag{293}$$

$$\gamma_c := \Omega^2 u \tag{294}$$

$$\gamma_s := \Omega^2 a^2 \frac{1}{a^2 + (u+b)(u-b)} \frac{3((u/\varepsilon)^2 + 1)\text{uarccot}(u/\varepsilon) - \varepsilon(3(u/\varepsilon)^2 + 2)}{(3(b/\varepsilon)^2 + 1)\text{arccot}(b/\varepsilon) - 3(b/\varepsilon)} \tag{295}$$

Lemma 2

Reference gravity intensity of type Somigliana–Pizzetti, International Gravity Formula:

If $x := (u^2+\varepsilon^2)^{-1}(D_\varphi w/D_u w)^2 << 1$ holds, where $w = w(\varphi,u)$ is the reference potential field of type *Somigliana–Pizzetti*, then its gravity field intensity $\gamma(\varphi,u)$ can be represented up to the order $O(2)$ by

$$\sqrt{\frac{u^2+\varepsilon^2}{u^2+\varepsilon^2\sin^2\varphi}} = \frac{\sqrt{a^2+(u+b)(u-b)}}{\sqrt{a^2\sin^2\varphi+b^2\cos^2\varphi+(u+b)(u-b)}} \tag{296}$$

$$\gamma_0+\gamma_C\cos^2\varphi+\gamma_S\sin^2\varphi = \gamma_a\cos^2\varphi+\gamma_b\sin^2\varphi \tag{297}$$

$$\gamma_a := \gamma_0+\gamma_C \tag{298}$$

$$\gamma_b := \gamma_0+\gamma_S \tag{299}$$

$$\Gamma_a := (\gamma_0+\gamma_c)\sqrt{1+(\varepsilon/u)^2} \tag{300}$$

$$\Gamma_b := \gamma_b = \gamma_0+\gamma_S \tag{301}$$

$$\gamma(u,\varphi) = \sqrt{a^2+(u+b)(u-b)}\frac{|\gamma_b\sin^2\varphi+\gamma_a\cos^2\varphi|}{\sqrt{a^2\sin^2\varphi+b^2\cos^2\varphi+(u+b)(u-b)}} \tag{302}$$

$$\gamma(u,\varphi) = \frac{|\sqrt{a^2+(u+b)(u-b)}\Gamma_b\sin^2\varphi+u\Gamma_a\cos^2\varphi|}{\sqrt{a^2\sin^2\varphi+b^2\cos^2\varphi+(u+b)(u-b)}} \tag{303}$$

Two special cases of the modulus of the reference gravity vector of type Somigliana–Pizzetti will be finally presented. By means of Corollary 1, it is specialized to $u = b$, namely, assuming a location of points on the level ellipsoid $E^2_{a,b}$, the International Reference Ellipsoid. In contrast, in Corollary 2 results of a computation of the International Gravity Formula at the equator $\varphi = 0$ and the poles $\varphi = \pm\frac{\pi}{2}$ are presented. On the one side the characteristic formulae Eqs. (304–306) for $u = b$, on the other side Eqs. (307–310) for $\varphi = \pm\frac{\pi}{2}, u \neq b$ and $u = b$ are found. ➋ *Figure 11* outlines the gravity's lateral dependency at $u = b$, while ➋ *Fig. 12* demonstrates its radial behavior at $\varphi = 0$ in the exterior domain.

Corollary 1

Reference gravity intensity of type Somigliana–Pizzetti, special case: level ellipsoid $u = b$.

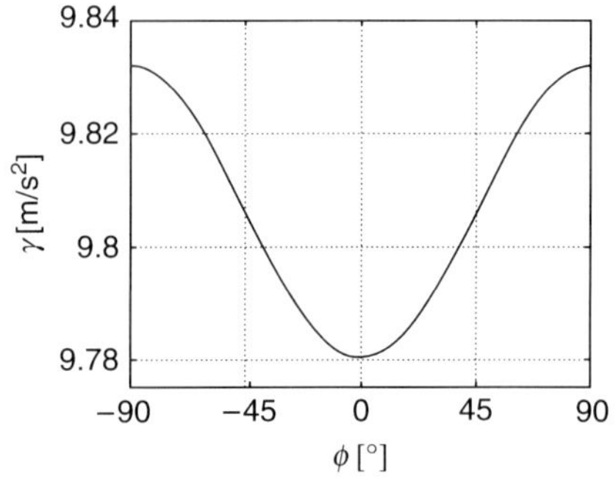

Fig. 11

Reference gravity intensity of type Somigliana–Pizzetti $\gamma(u = b, \varphi)$

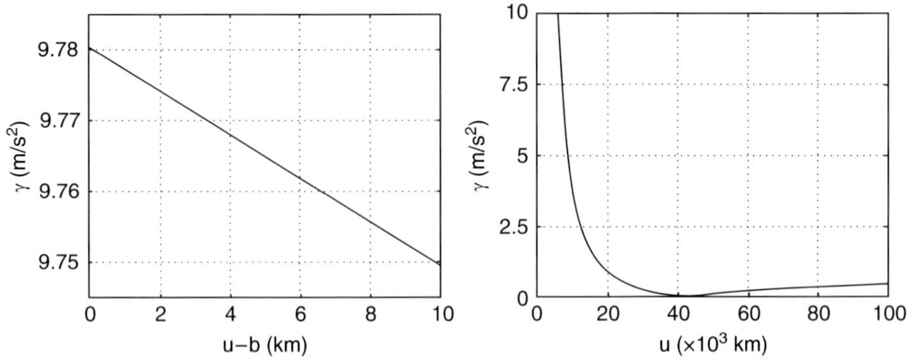

Fig. 12
Reference gravity intensity of type Somigliana–Pizzetti $\gamma(u, \varphi = 0)$ above the ellipsoid, near zone (horizontal axis $u - b$), far zone (horizontal axis u)

If $(D_\varphi w/D_u w)^2/a^2 \ll 1$ holds, where $w = w(\varphi,u)$ is the reference potential field of type Somigliana–Pizzetti on the surface of the level ellipsoid $E^2_{a,b}$, then its gravity field intensity $\gamma(\varphi,b)$ can be represented up to the order $O(2)$ by

$$\gamma(b,\varphi) = |a\Gamma_b \sin^2\varphi + b\Gamma_a \cos^2\varphi| / \sqrt{a^2\sin^2\varphi + b^2\cos^2\varphi} \tag{304}$$

subject to

$$\Gamma_a := (\gamma_0 + \gamma_c)\sqrt{1 + (\varepsilon/b)^2} = -\frac{GM}{ab} - \frac{1}{3}\Omega^2 a^2 \times \frac{3((b/\varepsilon)^2 + 1)b\operatorname{arccot}(b/\varepsilon) - \varepsilon(3(b/\varepsilon)^2 + 2)}{(3(b/\varepsilon)^2 + 1)\operatorname{arccot}(b/\varepsilon) - 3(b/\varepsilon)} + \Omega^2 a\sqrt{1 + (\varepsilon/b)^2} \tag{305}$$

$$\Gamma_b := \gamma_0 + \gamma_s = -\frac{GM}{a^2} + \frac{2}{3}\Omega^2 \frac{3((b/\varepsilon)^2 + 1)b\operatorname{arccot}(b/\varepsilon) - \varepsilon(3(b/\varepsilon)^2 + 2)}{(3(b/\varepsilon)^2 + 1)\operatorname{arccot}(b/\varepsilon) - 3(b/\varepsilon)} \tag{306}$$

Corollary 2

Reference gravity intensity of type Somigliana–Pizzetti, special cases: (i) $\varphi = 0$ and (ii) $\varphi = \pm\frac{\pi}{2}$.

If $(D_\varphi w/D_u w)^2/(u^2 + \varepsilon^2) \ll 1$ holds, where $w = w(\varphi,u)$ is the reference potential field of type Somigliana–Pizzetti, then its gravity field intensity $\gamma(\varphi,b)$ can be represented up to the order $O(2)$ by

(i) At the equator $\varphi = 0$:

$$\gamma(u, \varphi = 0) = |\Gamma_a| \tag{307}$$

(ii) At the poles $\varphi = \pm\frac{\pi}{2}$:

$$\gamma\left(u, \varphi = \pm\frac{\pi}{2}\right) = |\Gamma_b| \tag{308}$$

(iii) At the equator/level ellipsoid:

$$\gamma(b, \varphi = 0) = |\Gamma_a(u = b)| \tag{309}$$

(iv) At the poles/level ellipsoid:

$$\gamma\left(b, \varphi = \pm\frac{\pi}{2}\right) = |\Gamma_b(u = b)| \tag{310}$$

6.3 Expansion of the Gravitational Potential in Ellipsoidal Harmonics

So far the representation of the external gravity field has been found namely, potential and gravity field intensity, in the space external to a level ellipsoid, in particular the International Reference Ellipsoid. On the way to find the proper representation of the ellipsoidal disturbing potential, the expansion of the potential field in ellipsoidal harmonics (van Asche 1991; Cruz 1986; Gleason 1988, 1989; Skrzipek 1998; Sona 1996) has to be presented.

Equations 311 and 312 review two versions of the external representation of the gravitational field in terms of *associated Legendre polynomials of the first and second kind* of orthonormal type called $P^*_{kl}(\sin\varphi)$ and $Q^*_{kl}(u/\varepsilon)$. The star at $P^*_{kl}(\sin\varphi)$ highlights weighted orthonormality defined in Eq. 313. In contrast, the star at $Q^*_{kl}(u/\varepsilon)$ emphasizes the real-valued *associated Legendre polynomials of the second kind*, namely, $Q^*_{kl}(u/\varepsilon) = i^{k+1}Q_{kl}(iu/\varepsilon)$. The advantage of the first version is in terms of functional analysis, namely, illustrating the corresponding span of a *Hilbert space*. Its disadvantage is the units of the coefficients u_{kl}, for instance, cm^2s^{-2}. In addition, the gravitational constant G as well as gravitational mass GM are not represented causing great variations of the coefficients. In contrast, by means of the normalization factor $(GM/\varepsilon)\text{arccot}(b/\varepsilon)$ the coefficients $\{C_{kl}, S_{kl}\}$ of the second version are dimensionless and normalized by $C_{00} = 1$.

External representation of the gravitational potential field in terms of ellipsoidal harmonics (ellipsoid of revolution).

first version

$$U(\lambda, \varphi, u) = \sum_{k=0}^{\infty}\sum_{l=-k}^{k}\frac{Q^*_{kl}(u/\varepsilon)}{Q^*_{kl}(b/\varepsilon)}e_{kl}(\lambda, \varphi)u_{kl} \tag{311}$$

The coefficients u_{kl} are given in CGS units (cm s^{-2}).

second version

$$\begin{aligned} U(\lambda, \varphi, u) &= \tfrac{GM}{\varepsilon}\text{arccot}\left(\tfrac{b}{\varepsilon}\right) \\ &\sum_{k=0}^{\infty}\sum_{l=-k}^{k}\tfrac{Q^*_{kl}(u/\varepsilon)}{Q^*_{kl}(b/\varepsilon)}P^*_{kl}(\sin\varphi)(C_{kl}\cos l\lambda + S_{kl}\sin l\lambda) \end{aligned} \tag{312}$$

C_{kl}, S_{kl} are dimensionless coefficients: $C_{00} = 1$.

Surface ellipsoidal harmonics

$k \in \{0, 1, \ldots, \infty\}$, $l \in \{-k, -k+1, \ldots, -1, 0, 1, \ldots k-1, k\}$

$$e_{kl}(\lambda, \varphi) := P^*_{kl}(\sin\varphi)\begin{Bmatrix} \cos l\lambda \ \forall\ l \geq 0 \\ \sin|l|\lambda \ \forall\ l < 0 \end{Bmatrix} \tag{313}$$

orthonormality of the base functions $e_{kl}(\lambda, \varphi)$ with respect to the weighted scalar product

$$\langle e_{k_1l_1}|e_{k_2l_2}\rangle := \frac{1}{S}\int_{E^2_{a,b}} dS\kappa(\varphi)e_{k_1l_1}(\lambda, \varphi)e(\lambda, \varphi)_{k_2l_2} = \delta_{k_1k_2}\delta_{l_1l_2} \tag{314}$$

local area element

$$dS = d\left(\text{area } E^2_{a,b}\right) = \sqrt{\det G}\,d\lambda d\varphi = a\sqrt{b^2 + \varepsilon^2\sin^2\varphi}\cos\varphi\, d\lambda d\varphi \tag{315}$$

where G in Eq. 315 is the matrix of metric tensors.

global area element

$$S = \text{area } \mathbb{E}^2_{a,b} = 4\pi a \left\{ \frac{1}{2} + \frac{b^2}{4a\varepsilon} \ln \frac{a+\varepsilon}{a-\varepsilon} \right\} \tag{316}$$

weight function

$$\kappa(\varphi) := 4\pi a \frac{a}{\sqrt{b^2 + \varepsilon^2 \sin^2 \varphi}} \left(\frac{1}{2} + \frac{b^2}{4a\varepsilon} \ln \frac{a+\varepsilon}{a-\varepsilon} \right) \tag{317}$$

Those ellipsoidal coefficients $\{u_{kl}\}$ and $\{C_{kl}, S_{kl}\}$ for degree/order 360/360 (130 321 coefficients) for the terrestrial gravitational potential can be taken from the address (http://www.uni-stuttgart.de/gi/research/paper/coefficients/ coefficients.zip). The corresponding synthesis program for Eqs. 311 and 312, in particular the regularized computation of the associated Legendre functions of the second kind up to degree/order 1,000/1,000 has been described by *Thong and Grafarend* (1989) and has been positively evaluated *by Balmino et al.* (1991). For a more recent study of the ellipsoidal harmonic modulator, refer to ELGRAM (*Sansò and Sona* 2001).

By means of Eqs. 318 and 319, a short review of normalized associated Legendre functions of the first kind is given, while Eq. 320 does the same to normalized associated Legendre functions of the second kind.

Normalized associated Legendre functions of the first kind $P^*_{kl}(\sin\varphi)$ and $P^*_{kl}(u/\varepsilon)$.

k	l	$P^*_{kl}(\sin\varphi)$	$P^*_{kl}(u/\varepsilon)$
0	0	1	1
1	0	$\sqrt{3}\sin\varphi$	u/ε
1	1	$\sqrt{3}\cos\varphi$	$\sqrt{\left(\frac{u}{\varepsilon}\right)^2 + 1}$
2	0	$\frac{\sqrt{5}}{2}(3\sin^2\varphi - 1)$	$\frac{1}{2}\left(3\left(\frac{u}{\varepsilon}\right)^2 - 1\right)$
2	1	$\sqrt{15}\sin\varphi\cos\varphi$	$3\frac{u}{\varepsilon}\sqrt{\left(\frac{u}{\varepsilon}\right)^2 + 1}$
2	2	$\frac{\sqrt{5}}{2}\cos^2\varphi$	$\frac{1}{3}\sqrt{\left(\frac{u}{\varepsilon}\right)^2 + 1}$

recursion formula

$$P^*_{kl}(\sin\varphi) = \frac{\sqrt{4k^2 - 1}}{\sqrt{k^2 - l^2}} \sin\varphi P^*_{k-1,l}(\sin\varphi) - \frac{\sqrt{(2k+1)(k+l-1)(k-l-1)}}{\sqrt{(k^2-l^2)(2k-3)}} \times P^*_{k-2,l}(\sin\varphi) \forall k \in [3, \infty],\ 1 \in [0, k-2] \tag{318}$$

integral representation

$$P^*_{kl}(u/\varepsilon) = \frac{(k+l)!}{\pi k!} \int_0^{\pi} \left(\frac{u}{\varepsilon} + \sqrt{\left(\frac{u}{\varepsilon}\right)^2 + 1} \cos\tau \right)^k \cos(l\tau)\, d\tau \tag{319}$$

Normalized associated Legendre functions of the second kind $Q^*_{kl}(u/\varepsilon)$.

k	l	$Q^*_{kl}(u/\varepsilon)$
0	0	$\operatorname{arccot}(u/\varepsilon)$
1	0	$1 - (u/\varepsilon)\operatorname{arccot}(u/\varepsilon)$
2	2	$\frac{1}{2}\left[\left(3\left(\frac{u}{\varepsilon}\right)^2 + 1\right)\operatorname{arccot}\left(\frac{u}{\varepsilon}\right) - 3\frac{u}{\varepsilon}\right]$

integral representation

$$Q_{kl}^{*}\left(\frac{u}{\varepsilon}\right)=(-1)^{l}\frac{2^{k}(k+l)!l!}{(k-l)!(2l)!}\int_{0}^{\infty}\frac{\sin^{2l}\tau d\tau}{\left(\frac{u}{\varepsilon}+\sqrt{\left(\frac{u}{\varepsilon}\right)^{2}+1}\cosh\tau\right)^{k+l+1}} \tag{320}$$

6.4 External Representation of the Incremental Potential Relative to the Somigliana–Pizzetti Potential Field of Reference

Finally, the incremental potential $\delta w(\Lambda,\varphi,u)$ can be defined, also called *Somigliana–Pizzetti reference disturbing potential field.* Its first version, reviewed in Eqs. 321–326, uses the implicit representation of the *Somigliana–Pizzetti potential* $w\ (\varphi,u)$, while its second version is based upon the notion of the function space. Comments on the result: The ellipsoidal incremental potential δw, defined with respect to the gravity field of a level ellipsoid, does not contain any term of degree/order (0,0), (1,0), (1,–1), (1,1), (2,0). The term of degree/order (2,0), namely, C_{20}, is "*Somigliana–Pizzetti reduced*" by $\frac{\sqrt{15}}{5}\Omega^2 a^2 = 0$, a relict of the level ellipsoid. The centrifugal term $\frac{1}{2}\Omega^2(u^2+\varepsilon^2)\cos^2\varphi$ cancels. Accordingly, $\delta w(\Lambda,\varphi,u)$ is a harmonic function. For a series expansion of Eq. 326, refer to Appendix B (see Grafarend et al. 2006).

External representation of the incremental potential field $\delta\ w$ *relative to the Somigliana–Pizzetti potential of reference* w.

first version

$$\begin{aligned}\delta w(\lambda,\varphi,u) &= W(\lambda,\varphi,u)-w(\lambda,\varphi,u)\\ &= \frac{GM}{\varepsilon}\operatorname{arccot}\left(\frac{b}{\varepsilon}\right)\lim_{K\to\infty}\sum_{k=0}^{K}\sum_{l=0}^{k}\frac{Q_{kl}^{*}(u/\varepsilon)}{Q_{kl}^{*}(b/\varepsilon)}P_{kl}^{*}(\sin\varphi)(C_{kl}\cos l\lambda+S_{kl}\sin l\lambda)\\ &\quad -\frac{GM}{\varepsilon}\operatorname{arccot}\left(\frac{b}{\varepsilon}\right)-\frac{1}{6}\Omega^{2}a^{2}\frac{(3(u/\varepsilon)^{2}+1)\operatorname{arccot}(u/\varepsilon)-3u/\varepsilon}{(3(b/\varepsilon)^{2}+1)\operatorname{arccot}(b/\varepsilon)-3b/\varepsilon}(3\sin^{2}\varphi-1)\end{aligned} \tag{321}$$

subject to

$$\operatorname{arccot}(u/\varepsilon)=Q_{00}^{*}(u/\varepsilon)=\operatorname{arccot}(b/\varepsilon)\frac{Q_{kl}^{*}(u/\varepsilon)}{Q_{kl}^{*}(b/\varepsilon)} \tag{322}$$

$$\frac{(3(u/\varepsilon)^{2}+1)\operatorname{arccot}(u/\varepsilon)-3u/\varepsilon}{(3(b/\varepsilon)^{2}+1)\operatorname{arccot}(b/\varepsilon)-3b/\varepsilon}=\frac{Q_{20}^{*}(u/\varepsilon)}{Q_{20}^{*}(b/\varepsilon)} \tag{323}$$

$$3\sin^{2}\varphi-1=\frac{2}{\sqrt{5}}P_{20}^{*}(\sin\varphi) \tag{324}$$

second version (note $k\in\{2,K\}$*)*

$$\begin{aligned}\delta w(\lambda,\varphi,u) &= W(\lambda,\varphi,u)-w(\lambda,\varphi,u)=\frac{GM}{\varepsilon}\operatorname{arccot}\left(\frac{b}{\varepsilon}\right)\\ &\lim_{K\to\infty}\sum_{k=2}^{K}\sum_{l=0}^{k}\frac{Q_{kl}^{*}(u/\varepsilon)}{Q_{kl}^{*}(b/\varepsilon)}P_{kl}^{*}(\sin\varphi)(C_{kl}\cos l\lambda+S_{kl}\sin l\lambda)-\frac{\sqrt{5}}{15}\Omega^{2}a^{2}\frac{Q_{20}^{*}(u/\varepsilon)}{Q_{20}^{*}(b/\varepsilon)}P_{20}^{*}(\sin\varphi)\end{aligned} \tag{325}$$

Table 1

Time-evolution of the *World Geodetic Datum 2000* based on the current best estimates of the time variations of the Geoid gravity potential W_0,W_0 and the zonal spherical harmonic coefficients J_2,J_2 according to Ardalan and Grafarend (2001)

Best estimate of time evolution of fundamental parameters		Form and shape parameters		
J_2	$W_0(m^2/s^2)$	*amm/year*	*bmm/year*	*εmm/year*
$(-2.7 \pm 0.4) \times 10^{-11}$	-0.002 ± 0.0001	0.3 ± 0.02	-0.0002 ± 0.00008	3.7 ± 0.3

degree (2,0)-term: $\frac{\mathrm{GM}}{\varepsilon} \operatorname{arccot}\left(\frac{b}{\varepsilon}\right) C_{20} - \frac{\sqrt{5}}{15}\Omega^2 a^2 = 0$

$$\delta w(\lambda, \varphi, u) = \frac{\mathrm{GM}}{\varepsilon} \operatorname{arccot}\left(\frac{b}{\varepsilon}\right) \tag{326}$$

6.5 Time-Evolution

For the time-evolution of the *World Geodetic Datum 2000* of the Earth, Ardalan and Grafarend (2001) gave based on current best estimates time-evolutions of the Geoid gravity potential $\dot{W}_0$ and the zonal spherical harmonic coefficient, namely, $\dot{J}_2$. Such an analysis is based on data of type

$$\left\{GM, (\dot{GM}), J_2, \dot{J}_2, \Omega, \dot{\Omega}_0, W_0, \dot{W}_0\right\} \tag{327}$$

and

$$\{a, \dot{a},\ b, \dot{b},\ \varepsilon, \dot{\varepsilon}_0\} \tag{328}$$

being presented in ➲ *Table 1*.

7 Ellipsoidal Reference Potential of Type Somigliana–Pizzetti

In this chapter, the gradient of the gravitational disturbing potential, in particular in orthonormal ellipsoidal vector harmonics, is computed in all detail. Proper weighting functions for orthonormality on the *International Reference Ellipsoid* are constructed and tabulated. In this way, one finally arrives at an ellipsoidal harmonic representation of vertical deflections and gravity disturbances.

First, the intention is to represent vertical deflections and gravity disturbance in terms of vector-valued ellipsoidal surface harmonics. In particular, the weighting functions which make the spheroidal ellipsoidal component S_{kl} as well as the radial ellipsoidal component R_{kl} weighted orthonormal are constructed. Indeed, the spheroidal ellipsoidal component S_{kl} matches the ellipsoidal vertical deflections while the radial ellipsoidal component R_{kl} matches the gravity disturbance. Second, the aim is to provide an explicit ellipsoidal harmonic representation of vertical deflections and gravity disturbance with respect to the *Somigliana–Pizzetti reference potential* and reference gravity. It is verified that (i) ellipsoidal harmonic terms of degree/order (0,0), (1, −1), (1,0), 1,1), (2,0) and (ii) the centrifugal potential are removed from a *Somigliana–Pizzetti disturbing potential.*

7.1 Vertical Deflections and Gravity Disturbance in Vector-Valued Ellipsoidal Surface Harmonics

The gradient of the incremental potential $\delta w\,(\Lambda,\varphi,u)$, which is given in scalar-valued ellipsoidal harmonics (first version) has to be analyzed in terms of vector-valued ellipsoidal harmonics. There are two fundamental vector-valued base functions, namely, $S_{kl}(\Lambda,\varphi)$ and $R_{kl}(\Lambda,\varphi)$ respectively, called ellipsoidal surface as well as radial base functions. They are generated as follows: first, the gradient of the incremental potential $\delta w(\Lambda,\varphi,u)$ is decomposed into the tangential component (surface gradient) and into the normal component. Such a decomposition is outlined in Eqs. (333–336). Second, one has to orthonormalize those tangential and normal components with respect to the ellipsoid of revolution $E^2_{a,b}$, in particular the International Reference Ellipsoid. Such a procedure is not an easy one since it will turn out that weighted orthonormality has to be implemented.

For gaining a first insight, in Eqs. (329–332) the basic orthonormality conditions for vector-valued spherical surface harmonics here been collected Equation 329 defines the range of the indices k_1, l_1,k_2,l_2. In contrast, Eqs. (330–332) define spherical orthonormality of the components (i) tangential–tangential, (ii) normal–normal and (iii) tangential–normal. Note the normalization factor $k(k+1) = \sqrt{k(k+1)}\sqrt{k(k+1)}$.

Orthonormality of vector-valued spherical surface harmonics: tangential versus normal components.

$$k_1, k_2 \in \{0,\ldots,K\}, l_1, l_2 \in \{-k, -k+1,\ldots,-1,0,1,\ldots,k-1,k\} \tag{329}$$

(tangential–tangential)

$$\begin{aligned}\delta_{k1l1}\delta_{k2l2} &= \frac{1}{4\pi}\int_0^{2\pi} d\lambda \int_{-\pi/2}^{\pi/2} d\varphi_s \cos\varphi_s \\ &\left\langle \left(e_\lambda \frac{1}{\cos\varphi_s} D_\lambda + e_{\varphi_s} D_{\varphi_s}\right) \frac{e_{k1l1}(\lambda,\varphi_s)}{\sqrt{k(k+1)}} \;\middle|\; \left(e_\lambda \frac{1}{\cos\varphi_s} D_\lambda + e_{\varphi_s} D_{\varphi_s}\right) \frac{e_{k2l2}(\lambda,\varphi_s)}{\sqrt{k(k+1)}} \right\rangle\end{aligned} \tag{330}$$

(normal–normal)

$$\delta_{k1l1}\delta_{k2l2} = \frac{1}{4\pi}\int_0^{2\pi} d\lambda \int_{-\pi/2}^{\pi/2} d\varphi_s \cos\varphi_s \langle e_r e_{k1l1}(\lambda,\varphi_s) | e_r e_{k2l2}(\lambda,\varphi_s)\rangle \tag{331}$$

(tangential–normal)

$$0 = \frac{1}{4\pi}\int_0^{2\pi} d\lambda \int_{-\pi/2}^{\pi/2} d\varphi_s \cos\varphi_s \left\langle \left(e_\lambda \frac{1}{\cos\varphi_s} D_\lambda + e_{\varphi_s} D_{\varphi_s}\right) \frac{e_{k1l1}(\lambda,\varphi_s)}{\sqrt{k(k+1)}} \middle| e_r e_{k2l2}(\lambda,\varphi_s)\right\rangle \tag{332}$$

In order to continue ellipsoidal orthonormality one has to depart from the metric coefficients Eqs. 337 and 338 which are specialized for $u = b$. The longitudinal metric coefficient amounts to $a^2\cos^2\varphi$, the lateral metric coefficient to $b^2 + \varepsilon^2\sin^2\varphi = a^2 - (a^2 - b^2)\cos^2\varphi$, and the "radial coefficient" to $(b^2 + \varepsilon^2\sin^2\varphi)/a^2 = 1 - (a^2 - b^2)\cos^2\varphi/a^2$. Those metric coefficients define the areal element dS and the integral area S of the ellipsoid of revolution by means of Eqs. (339–341).

Orthonormality of vector-valued ellipsoidal surface harmonics: tangential versus normal components.

scalar-valued ellipsoidal harmonics

$$\delta w(\lambda,\varphi,u) = \sum_{\substack{k=2\\(k,l)\neq(2,0)}}^{\infty}\sum_{l=-k}^{+k} \frac{Q^*_{kl}(u/\varepsilon)}{Q^*_{kl}(b/\varepsilon)} e_{kl}(\lambda,\varphi)\delta w^{kl} \tag{333}$$

vector-valued ellipsoidal harmonics

$$\operatorname{grad}\delta w = \boldsymbol{e}_\lambda \frac{1}{\sqrt{g_{\lambda\lambda}(b)}} D_\lambda \delta w + \boldsymbol{e}_\varphi \frac{1}{\sqrt{g_{\varphi\varphi}(b)}} D_\varphi \delta w + \boldsymbol{e}_u \frac{1}{\sqrt{g_{uu}(b)}} D_u \delta w \tag{334}$$

tangential component

$$\left(\boldsymbol{e}_\lambda \frac{1}{\sqrt{g_{\lambda\lambda}(b)}} \frac{Q^*_{kl}(u/\varepsilon)}{Q^*_{kl}(b/\varepsilon)} D_\lambda + \boldsymbol{e}_\varphi \frac{1}{\sqrt{g_{\varphi\varphi}(b)}} \frac{Q^*_{kl}(u/\varepsilon)}{Q^*_{kl}(b/\varepsilon)} D_\varphi\right) e_{kl}(\lambda,\varphi) \tag{335}$$

normal component

$$\boldsymbol{e}_u \frac{1}{\sqrt{g_{uu}} Q^*_{kl}(b/\varepsilon)} e_{kl}(\lambda,\varphi) D_u Q^*_{kl}(u/\varepsilon) \tag{336}$$

$$\begin{aligned} g_{\lambda\lambda} &= (u^2+\varepsilon^2)\cos^2\varphi \\ g_{\varphi\varphi} &= u^2+\varepsilon^2\sin^2\varphi \\ g_{uu} &= (u^2+\varepsilon^2\sin^2\varphi)/(u^2+\varepsilon^2) \end{aligned} \tag{337}$$

$$\begin{aligned} g_{\lambda\lambda}(u=b) &= a^2\cos^2\varphi \\ g_{\varphi\varphi}(u=b) &= b^2+\varepsilon^2\sin^2\varphi \\ g_{uu}(u=b) &= (b^2+\varepsilon^2\sin^2\varphi)/a^2 \end{aligned} \tag{338}$$

local and global area element of the ellipsoid of revolution $E^2_{a,b}$"

$$dS = d\left\{\text{area}\left(E^2_{a,b}\right)\right\} = \sqrt{g_{\lambda\lambda}g_{\varphi\varphi}}\,d\lambda d\varphi \tag{339}$$

$$dS = d\left\{\text{area}\left(E^2_{a,b}\right)\right\} = a\cos\varphi\sqrt{b^2+\varepsilon^2\sin^2\varphi}\,d\lambda d\varphi \tag{340}$$

$$dS = d\left\{area\left(E^2_{a,b}\right)\right\} = 4\pi a^2\left\{\frac{1}{2}+\frac{b^2}{4}a\varepsilon\ln\frac{a+\varepsilon}{a}-\varepsilon\right\} \tag{341}$$

Third, orthonormality with respect to the weighting function $\kappa_0(\varphi)$ applied to the scalar-valued ellipsoidal base functions $e_{kl}(\Lambda,\varphi)$ is defined. Indeed, the weighting function $\kappa_0(\varphi)$ within Eq. 343 reflects the integral area S as well as the areal element dS, namely, the function $a\cos^2\varphi\sqrt{b^2+\varepsilon^2\sin^2\varphi} = a\cos^2\varphi\left[a^2-(a^2-b^2)\cos^2\varphi\right]^{1/2}$ within Eq. 342.

orthonormality of vector-valued ellipsoidal surface harmonics: tangential vs. normal components

$$\langle e_{k1l1}|e_{k2l2}\rangle := \frac{1}{S}\int dS\kappa_0(\varphi)e_{k1l1}(\lambda,\varphi)e_{k2l2}(\lambda,\varphi) = \delta_{k1l1}\delta_{k2l2} \tag{342}$$

ansatz

$$\kappa_0(\varphi) := \frac{a}{\sqrt{b^2+\varepsilon^2\sin^2\varphi}}\left(\frac{1}{2}+\frac{b^2}{4a\varepsilon}\ln\frac{a+\varepsilon}{a-\varepsilon}\right) \tag{343}$$

$\Rightarrow$

$$\langle e_{k1l1}|e_{k2l2}\rangle := \frac{1}{4\pi}\int_0^{2\pi} d\lambda \int_{-\pi/2}^{\pi/2} d\varphi \cos\varphi\, e_{k1l1}(\lambda,\varphi)e_{k2l2}(\lambda,\varphi) = \delta_{k1l1}\delta_{k2l2}$$

Fourth, the weighting functions $\kappa_1(\varphi)$ and $\kappa_2(\varphi)$ for the orthonormality postulate of the tangential vector-valued component $S_{kl}(\Lambda,\varphi)$ of type Eqs. 344 and 345 have to be determined. The "ansatz," Eqs. 346 and 347 of the weighting functions $\kappa_1(\varphi)$ and $\kappa_2(\varphi)$, namely, $\kappa_1 = a^2\kappa_0$, $\kappa_2 = (b^2 + \varepsilon^2 \sin^2\varphi)\kappa_0$, indeed fulfil the weighted orthonormality Eq. 348, $\langle S_{k1l1}|S_{k2l2}\rangle$, with respect to the ellipsoid of revolution. The proof runs through Eqs. (348–350).

Orthonormality of vector-valued ellipsoidal surface harmonics: weighting functions. orthonormality of the tangential vector-valued ellipsoidal harmonics

$$S_{kl}(\lambda,\varphi) := \frac{1}{\sqrt{k(k+1)}}\left\{e_\lambda \frac{1}{\sqrt{g_{\lambda\lambda}(u=b)}}D_\lambda + e_\varphi \frac{1}{\sqrt{g_{\varphi\varphi}(u=b)}}D_\varphi\right\} e_{kl}(\lambda,\varphi) \tag{344}$$

$$\begin{aligned}\langle S_{k1l1}|S_{k2l2}\rangle := \frac{1}{S}\int dS\kappa_3(\varphi)\Bigg\langle &\left(e_\lambda\sqrt{\kappa_1}\frac{1}{\sqrt{g_{\lambda\lambda}(b)}}D_\lambda + e_\varphi\sqrt{\kappa_2}\frac{1}{\sqrt{g_{\varphi\varphi}(b)}}D_\varphi\right)\frac{e_{k1l1}(\lambda,\varphi)}{\sqrt{k(k+1)}}\Bigg| \\ &\Bigg|\left(e_\lambda\sqrt{\kappa_1}\frac{1}{\sqrt{g_{\lambda\lambda}(b)}}D_\lambda + e_\varphi\sqrt{\kappa_2}\frac{1}{\sqrt{g_{\varphi\varphi}(b)}}D_\varphi\right)\frac{e_{k2l2}(\lambda,\varphi)}{\sqrt{k(k+1)}}\Bigg\rangle\end{aligned} \tag{345}$$

ansatz

$$\kappa_1 = a^2\kappa_0 = a^2\left(\frac{a}{\sqrt{b^2+\varepsilon^2\sin^2\varphi}}\left(\frac{1}{2}+\frac{b^2}{4a\varepsilon}\ln\frac{a+\varepsilon}{a-\varepsilon}\right)\right) \tag{346}$$

$$\kappa_2 = (b^2+\varepsilon^2\sin^2\varphi)\kappa_0 = a\sqrt{b^2+\varepsilon^2\sin^2\varphi}\left(\frac{1}{2}+\frac{b^2}{4a\varepsilon}\ln\frac{a+\varepsilon}{a-\varepsilon}\right) \tag{347}$$

$\Rightarrow$

$$\begin{aligned}\langle S_{k1l1}|S_{k2l2}\rangle := \frac{1}{S}\int dS\kappa_3(\varphi)\Big\{&\kappa_1\frac{1}{g_{\lambda\lambda}(b)}D_\lambda e_{k1l1}(\lambda,\varphi)D_\lambda e_{k2l2}(\lambda,\varphi)\frac{1}{k(k+1)} \\ &+ \kappa_2\frac{1}{g_{\varphi\varphi}(b)}D_\varphi e_{k1l1}(\lambda,\varphi)D_\varphi e_{k2l2}(\lambda,\varphi)\frac{1}{k(k+1)}\Big\}\end{aligned}$$

$$\begin{aligned}\langle S_{k1l1}|S_{k2l2}\rangle = &\frac{1}{4\pi}\frac{1}{a^2\left(\frac{1}{2}+\frac{b^2}{4a\varepsilon}\ln\frac{a+\varepsilon}{a-\varepsilon}\right)\int_0^{2\pi}d\lambda\int_{-\pi/2}^{\pi/2}d\varphi\cos\varphi\, a\sqrt{b^2+\varepsilon^2\sin^2\varphi}} \\ &\times\Bigg\{\frac{a}{\sqrt{b^2+\varepsilon^2\sin^2\varphi}}\left(\frac{1}{2}+\frac{b^2}{4a\varepsilon}\ln\frac{a+\varepsilon}{a-\varepsilon}\right)\frac{D_\lambda e_{k1l1}}{\cos\varphi}\frac{D_\lambda e_{k2l2}}{\cos\varphi}\frac{1}{k(k+1)} \\ &+ a\sqrt{b^2+\varepsilon^2\sin^2\varphi}\left(\frac{1}{2}+\frac{b^2}{4a\varepsilon}\ln\frac{a+\varepsilon}{a-\varepsilon}\right)\frac{D_\varphi e_{k1l1}}{\sqrt{b^2+\varepsilon^2\sin^2\varphi}} \\ &\times\frac{D_\varphi e_{k2l2}}{\sqrt{b^2+\varepsilon^2\sin^2\varphi}}\frac{1}{k(k+1)}\Bigg\}\end{aligned} \tag{348}$$

$$\langle S_{k1l1}|S_{k2l2}\rangle = \frac{1}{4\pi}\int_0^{2\pi} d\lambda \int_{-\pi/2}^{\pi/2} d\varphi \cos\varphi \frac{1}{\cos\varphi}\frac{D_\lambda e_{k1l1}}{\sqrt{k(k+1)}}\frac{1}{\cos\varphi}\frac{D_\lambda e_{k2l2}}{\sqrt{k(k+1)}}$$
$$+ \frac{1}{4\pi}\int_0^{2\pi} d\lambda \int_{-\pi/2}^{\pi/2} d\varphi \cos\varphi \frac{D_\varphi e_{k1l1}}{\sqrt{k(k+1)}}\frac{D_\varphi e_{k2l2}}{\sqrt{k(k+1)}} = \delta_{k1l1}\delta_{k2l2} \tag{349}$$

$$\langle S_{k1l1}|S_{k2l2}\rangle = \delta_{k1l1}\delta_{k2l1} \tag{350}$$

Fifth, the weighting function $\kappa_3(\varphi)$ for the orthonormality postulate of the normal vector-valued component $R_{kl}(\Lambda,\varphi)$ of type Eqs. 351 and 352 is determined. The "ansatz," Eq. 353 of the weighting functions $\kappa_3(\varphi)$, namely, $\sqrt{b^2+\varepsilon^2\sin^2\varphi}\kappa_0/a$, correctly fulfils the weighted orthonormality Eq. 354, $\langle R_{k1l1}|R_{k2l2}\rangle$, with respect to the ellipsoid of revolution. Now, the proof runs through Eqs. (354–356).

Orthonormality of vector-valued ellipsoidal surface harmonics: weighting functions. orthonormality of the normal vector-valued ellipsoidal harmonics

$$R_{kl}(\lambda,\varphi) := e_u \frac{1}{\sqrt{g_{uu}(b)}} e_{kl}(\lambda,\varphi) \tag{351}$$

$$\langle R_{k1l1}|R_{k2l2}\rangle := \frac{1}{S}\int dS\kappa_3(\varphi) \times \left\langle e_u \frac{1}{\sqrt{g_{uu}(b)}} \middle| e_u \frac{1}{\sqrt{g_{uu}(b)}} \right\rangle e_{k1l1}(\lambda,\varphi) e_{k2l2}(\lambda,\varphi) \tag{352}$$

ansatz

$$\kappa_3(\varphi) = \frac{b^2+\varepsilon^2\sin^2\varphi}{a^2}\kappa_0(\varphi) = \frac{1}{a}\sqrt{b^2+\varepsilon^2\sin^2\varphi}\left(\frac{1}{2}+\frac{b^2}{4a\varepsilon}\ln\frac{a+\varepsilon}{a-\varepsilon}\right) \tag{353}$$

$\Rightarrow$

$$\langle R_{k1l1}|R_{k2l2}\rangle = \frac{1}{4\pi}\frac{1}{a^2\left(\frac{1}{2}+\frac{b^2}{4a\varepsilon}\ln\frac{a+\varepsilon}{a-\varepsilon}\right)}\int_0^{2\pi} d\lambda \int_{-\pi/2}^{\pi/2} d\varphi \cos\varphi a\sqrt{b^2+\varepsilon^2\sin^2\varphi}$$
$$\times \frac{\sqrt{b^2+\varepsilon^2\sin^2\varphi}}{a}\left(\frac{1}{2}+\frac{b^2}{4a\varepsilon}\ln\frac{a+\varepsilon}{a-\varepsilon}\right)\frac{a}{\sqrt{b^2+\varepsilon^2\sin^2\varphi}} e_{k1l1}(\lambda,\varphi)$$
$$\times \frac{a}{\sqrt{b^2+\varepsilon^2\sin^2\varphi}} e_{k2l2}(\lambda,\varphi) \tag{354}$$

$$\langle R_{k1l1}|R_{k2l2}\rangle = \frac{1}{4\pi}\int_0^{2\pi} d\lambda \int_{-\pi/2}^{\pi/2} d\varphi \cos\varphi e_{k1l1}(\lambda,\varphi) e_{k2l2}(\lambda,\varphi) \tag{355}$$

$$\langle R_{k1l1}|R_{k2l2}\rangle = \delta_{k1l1}\delta_{k2l2} \tag{356}$$

Sixth, by means of a detailed computation in Eqs. (357–360), the orthogonality $\langle R_{k1l1}|S_{k2l2}\rangle = 0$ of the tangential and normal components of grad$\delta w(\Lambda,\varphi,u)$ is tested. Equations 357 and 358 finally lead to the zero identity Eq. 360 which is based upon the orthogonality condition Eq. 359 of the base vectors $\{\boldsymbol{e}_\lambda, \boldsymbol{e}_\varphi, \boldsymbol{e}_u|\lambda,\varphi,u\}$, namely, $\langle \boldsymbol{e}_u|\boldsymbol{e}_\lambda\rangle = 0$, $\langle \boldsymbol{e}_u|\boldsymbol{e}_\varphi\rangle = 0$.

The various weighting functions $\kappa_0,\kappa_1,\kappa_2,\kappa_3$ needed for orthonormality constraints for scalar- and vector-valued ellipsoidal harmonics are finally collected in Eqs. (361–363).

Orthonormality of vector-valued ellipsoidal surface harmonics between $R_{kl}(\Lambda,\varphi)$ and $S_{kl}(\Lambda,\varphi)$

$$\langle R_{k1l1}|S_{k2l2}\rangle = \frac{1}{S}\int dS\left\langle e_u \frac{1}{\sqrt{g_{uu}(b)}} D_u e_{k1l1} \middle| \left(e_\lambda \frac{1}{\sqrt{g_{\lambda\lambda}(b)}} D_\lambda + e_\varphi \frac{1}{\sqrt{g_{\varphi\varphi}(b)}} D_\varphi \right) e_{k2l2} \right\rangle \tag{357}$$

$$\begin{aligned}\langle R_{k1l1}|S_{k2l2}\rangle &= \frac{1}{S}\int dS\left\langle e_u \frac{1}{\sqrt{g_{uu}(b)}} D_u e_{k1l1} \middle| e_\lambda \frac{1}{\sqrt{g_{\lambda\lambda}(b)}} D_\lambda e_{k2l2} \right\rangle \\ &+ \frac{1}{S}\int dS\left\langle \boldsymbol{e_u} \frac{1}{\sqrt{g_{uu}(b)}} D_u e_{k1l1} \middle| \boldsymbol{e_\varphi} \frac{1}{\sqrt{g_{\varphi\varphi}(b)}} D_\varphi e_{k2l2} \right\rangle = 0\end{aligned} \tag{358}$$

$$\langle \boldsymbol{e_u}|\boldsymbol{e_\lambda}\rangle = 0, \langle e_u|e_\varphi\rangle = 0 \tag{359}$$

$$\langle R_{k1l1}|S_{k2l2}\rangle = 0 \tag{360}$$

A summary of weighting functions for scalar- as well as vector-valued ellipsoidal surface harmonics.

$$\kappa_0(\varphi) := \frac{a}{\sqrt{b^2+\varepsilon^2\sin^2\varphi}}\left(\frac{1}{2} + \frac{b^2}{4a\varepsilon}\ln\frac{a+\varepsilon}{a-\varepsilon}\right) \tag{361}$$

$$\kappa_1(\varphi) = a^2\kappa_0(\varphi), \kappa_2(\varphi) = (b^2+\varepsilon^2\sin^2\varphi)\kappa_0(\varphi) \tag{362}$$

$$\kappa_3(\varphi) = \frac{b^2+\varepsilon^2\sin^2\varphi}{a^2}\kappa_0(\varphi) = \frac{\kappa_2(\varphi)}{\kappa_1(\varphi)}\kappa_0(\varphi) \tag{363}$$

Seventh, as a result of all the efforts, one succeeds by means of Eqs. (364–367) to represent grad δw in terms of vector-valued base functions $\{S_{kl}(\lambda,\varphi)|R_{kl}(\lambda,\varphi)\}$. The coordinates of grad δw in the corresponding Sobolev space W_1^2 are given by Eq. 364. The "radial component" is represented by the D_u-differentiation which is outlined by the nonrecursive Eqs. 365–367 which are numerically stable (Thong and Grafarend 1989).

Synthesis of grad δw in terms of vector-valued ellipsoidal surface harmonics.

$$\begin{aligned}\mathrm{grad}\delta w &= \sum_{\substack{k=2}}^{\infty}\sum_{\substack{l=-k \\ (k,l)\neq(2,0)}}^{+k} \boldsymbol{S}_{kl}(\lambda,\varphi)\sqrt{\boldsymbol{k}(\boldsymbol{k}+1)}\frac{Q^*_{kl}(u/\varepsilon)}{Q^*_{kl}(b/\varepsilon)}\delta w^{kl} \\ &+ \sum_{k=2}^{\infty}\sum_{l=-k}^{+k} R_{kl}(\lambda,\varphi)\frac{1}{Q^*_{kl}(b/\varepsilon)}D_u Q^*_{kl}(u/\varepsilon)\delta w^{kl}\end{aligned} \tag{364}$$

ellipsoidal components $S_{kl}(\Lambda,\varphi)$ and radial components $R_{kl}(\Lambda,\varphi)$ subject to $\forall l \in [1,k]$

$$D_u Q^*_{kl}(u/\varepsilon) = -\frac{k-|l|+1}{\sqrt{u^2+\varepsilon^2}}Q^*_{k,l-1}(u/\varepsilon) - l\frac{u}{u^2+\varepsilon^2}Q^*_{kl}(u/\varepsilon) \tag{365}$$

$$D_u Q^*_{00}(u/\varepsilon) = -\frac{a}{u^2+\varepsilon^2} \tag{366}$$

$$D_u Q^*_{k0}(u/\varepsilon) = -k\frac{u}{u^2+\varepsilon^2}Q^*_{k0}(u/\varepsilon) - (2k+1)\frac{a}{u^2+\varepsilon^2}Q^*_{k-1,0}(u/\varepsilon) \tag{367}$$

Eigth, by Eqs. (368–374) one is able to represent vertical deflections and gravity disturbance in terms of vector-valued ellipsoidal harmonics. Equation 368 is a representation of grad δw in terms of $\{\eta,\xi,\delta\gamma\}$ up to order $O(2)$. Vertical deflections depend only on "ellipsoidal" or "tangential components" of vector-valued ellipsoidal harmonics as computed by Eq. 369. In contrast, gravity disturbance, also called incremental gravity, is dependent on "radial" or "normal components" of vector-valued ellipsoidal harmonics as computed by Eq. 370. Note the "windowing" $k \in [2, K]$ which already reflects the structure of Somigliana–Pizzetti reference gravity. As soon as the elliptic functions $s_{kl}(u)$ is introduced by Eq. 373 and $r_{kl}(u)$ by Eq. 374, one is led to the final representation of the vertical deflections vector and the gravity disturbance component of type Eqs. 373 and 374 in terms of vector-valued ellipsoidal harmonics.

Synthesis of vertical deflections and gravity disturbance in terms of vector-valued ellipsoidal surface harmonics.

$$\boldsymbol{\delta\gamma} = \text{grad}\,\delta w \doteq -(e_\lambda \gamma\eta + e_\varphi \gamma\xi + e_u \delta\gamma) + O(2) \tag{368}$$

$$-(\boldsymbol{e}_\lambda \gamma\eta + \boldsymbol{e}_\varphi \gamma\xi) = \sum_{\substack{k=2\\(k,l)\neq(2,0)}}^{\infty} \sum_{l=-k}^{+k} \mathbf{S}_{kl}(\lambda,\varphi)\sqrt{k(k+1)}\frac{Q^*_{kl}(u/\varepsilon)}{Q^*_{kl}(b/\varepsilon)}\delta w^{kl} \tag{369}$$

$$-\boldsymbol{e}_u \delta\gamma = \sum_{\substack{k=2\\(k,l)\neq(2,0)}}^{\infty} \sum_{l=-k}^{+k} R_{kl}(\lambda,\varphi)\frac{1}{Q^*_{kl}(b/\varepsilon)} D_u Q^*_{kl}(u/\varepsilon)\delta w^{kl} \tag{370}$$

$$-(\boldsymbol{e}_\lambda \gamma\eta + \boldsymbol{e}_\varphi \gamma\xi) = \sum_{\substack{k=2\\(k,l)\neq(2,0)}}^{\infty} \sum_{l=-k}^{+k} \mathbf{S}_{kl}(\lambda,\varphi) s_{kl}\delta w^{kl} \tag{371}$$

$$-\boldsymbol{e}_u \delta\gamma = \sum_{\substack{k=2\\(k,l)\neq(2,0)}}^{\infty} \sum_{l=-k}^{+k} \mathbf{R}_{kl}(\lambda,\varphi) r_{kl}\delta w^{kl} \tag{372}$$

subject to

$$\mathbf{S}_{kl} := \sqrt{k(k+1)}\frac{Q^*_{kl}(u/\varepsilon)}{Q^*_{kl}(b/\varepsilon)} \tag{373}$$

$$\mathbf{R}_{kl} := \frac{1}{Q^*_{kl}(b/\varepsilon)} D_u Q^*_{kl}(u/\varepsilon) \tag{374}$$

7.2 Vertical Deflections and Gravity Disturbance Explicitly in Terms of Ellipsoidal Surface Harmonics

The gradient of the incremental potential $\delta w(\Lambda,\varphi,u)$ which is given in scalar-valued ellipsoidal harmonics (first version) has been analyzed previously in terms of vector-valued ellipsoidal harmonics. For the analysis of the underlying *Sobolev* space W^2 such a treatment has been necessary. In case one wants to prepare for a detailed computation of vertical deflections and gravity disturbance either on Earth's surface or on the International Reference Ellipsoid, one needs an explicit formulation of the disturbing potential $\delta w(\Lambda,\varphi,u)$ with respect to the Somigliana–Pizzetti reference potential (second version) and the gravitational potential(second version).

Equations (375–382) review such a representation. Equation 375 summarizes the incremental gravity potential $\delta w(\Lambda,\varphi,u)$, which is given with respect to the base functions

$\{P^*_{kl}(\sin\varphi)\cos l\lambda, P^*_{kl}(\sin\varphi)\sin l\lambda\}$. Here, the window of type degree $2 \le k < K$ is introduced. Accordingly, the ellipsoidal harmonic coefficients have been written by small letters. The quotient of the associated Legendre polynomials of the second kind $Q^*_{kl}(u/\varepsilon)/Q^*_{kl}(b/\varepsilon)$ has been denoted $q_{kl}(u)$. Again, the degree/order 2/0-term which is generated by a reduction proportional to $\Omega^2 a^2$ is noticed. Equations 376 and 377 are properly chosen definitions. The gradient of the disturbing potential Eq. 378 is computed with respect to the explicit partial derivatives Eq. 379 D_Λ, Eqs. 380 and 381 D_φ and finally Eq. 382 D_u subject to Eqs. (365–367).

Explicit representation of the disturbing potential and its gradients
scalar-valued ellipsoidal harmonics (window $2 \le k < K$)

$$\delta w(\lambda,\varphi,u) = \frac{GM}{\varepsilon}\operatorname{arccot}\left(\frac{b}{\varepsilon}\right)\sum_{\substack{k=2\\(k,l)\neq(2,0)}}^{K}\sum_{l=0}^{k} q_{kl}(u)P^*_{kl}(\sin\varphi)(c_{kl}\sin l\lambda + s_{kl}\cos l\lambda) \tag{375}$$

subject to

$$q_{kl} = \frac{Q^*_{kl}(u/\varepsilon)}{Q^*_{kl}(b/\varepsilon)} \tag{376}$$

$$c_{20} := -\frac{\varepsilon}{GM}\frac{1}{\operatorname{arccot}(b/\varepsilon)}\frac{\sqrt{15}}{5}\Omega^2 a^2 = 0 \tag{377}$$

vector-valued ellipsoidal harmonics (window $2 \le k < K$)

$$\operatorname{grad}\delta w = \boldsymbol{e}_\lambda \frac{1}{\sqrt{u^2+\varepsilon^2}\cos\varphi}D_\lambda\delta w + \boldsymbol{e}_\varphi\frac{1}{\sqrt{u^2+\varepsilon^2\sin^2\varphi}}D_\varphi\delta w + \boldsymbol{e}_u\frac{\sqrt{u^2+\varepsilon^2}}{\sqrt{u^2+\varepsilon^2\sin^2\varphi}}D_u\delta w \tag{378}$$

derivatives

$$D_\lambda(c_{kl}\cos l\lambda + s_{kl}\sin l\lambda) = -lc_{kl}\sin l\lambda + ls_{kl}\cos l\lambda \tag{379}$$

$$D_\varphi P^*_{kl}(\sin\varphi) = \begin{Bmatrix} \sqrt{2(2k+1)}\sqrt{\frac{(k-l)!}{(k+l)!}}D_\varphi P_{kl}(\sin\varphi) & \text{for } l>0 \\ \sqrt{2k+1}D_\varphi P_{kl}(\sin\varphi) & \text{for } l=0 \end{Bmatrix} \tag{380}$$

$$D_\varphi P_{kl}(\sin\varphi) = P_{k,l+1}(\sin\varphi) - l\tan\varphi P_{kl}(\sin\varphi) \tag{381}$$

$$D_u q_{kl} = \frac{1}{Q^*_{kl}(b/\varepsilon)}D_u Q^*_{kl}(u/\varepsilon) \tag{382}$$

Finally by means of Eqs. (383–386), one can derive an explicit representation of vertical deflections, in particular Eq. 384 for East $\eta(\Lambda,\varphi,u)$, Eq. 385 for North $\xi(\Lambda,\varphi,u)$ and Eq. 386 for the gravity disturbance $\delta\gamma(\Lambda,\varphi,u)$.

Synthesis of windowed vertical deflections and windowed gravity disturbance explicitely in terms of ellipsoidal harmonics.

$$\boldsymbol{\delta\gamma} = \operatorname{grad}\delta w \doteq -(\boldsymbol{e}_\lambda \gamma\eta + \boldsymbol{e}_\varphi \gamma\xi + \boldsymbol{e}_u \delta\gamma) + O(2) \tag{383}$$

$$\eta(\lambda,\varphi,u) = -\frac{1}{\gamma(\varphi,u)} \frac{\mathrm{GM}_{\mathrm{arccot}}(b/\varepsilon)}{\varepsilon\sqrt{u^2+\varepsilon^2}\cos\varphi} \times \sum_{\substack{k=2\,l=0\\(k,l)\neq(2,0)}}^{K}\sum^{k} q_{kl}(u) P^*_{kl}(\sin\varphi)(-lc_{kl}\sin l\lambda + ls_{kl}\cos l\lambda) \tag{384}$$

$$\xi(\lambda,\varphi,u) = -\frac{1}{\gamma(\varphi,u)} \frac{\mathrm{GM}\ \mathrm{arccot}(b/\varepsilon)}{\varepsilon\sqrt{u^2+\varepsilon^2\sin^2\varphi}} \times \sum_{\substack{k=2\,l=0\\(k,l)\neq(2,0)}}^{K}\sum^{k} q_{kl}(u)\left[D_\varphi P^*_{kl}(\sin\varphi)\right](c_{kl}\cos l\lambda + s_{kl}\sin l\lambda) \tag{385}$$

$$\delta\gamma(\lambda,\varphi,u) = -\frac{\mathrm{GM}\ \mathrm{arccot}(b/\varepsilon)\sqrt{u^2+\varepsilon^2}}{\varepsilon\sqrt{u^2+\varepsilon^2\sin^2\varphi}} \times \sum_{\substack{k=2\,l=0\\(k,l)\neq(2,0)}}^{K}\sum^{k} [D_u q_{kl}(u)]\, P^*_{kl}(\sin\varphi)(c_{kl}\cos l\lambda + s_{kl}\sin l\lambda) \tag{386}$$

8 Case Studies

For an ellipsoidal harmonic Gravity Earth Model (SEGEN: http://www.uni-stuttgart.de/gi/research/paper/coefficients/coefficients.zip) up to degree/order 360/360 the global maps of ellipsoidal vertical deflections and ellipsoidal gravity disturbances which transfer a great amount of geophysical information in a properly chosen equiareal ellipsoidal map projection are computed.

the potential theory of the incremental gravity vector $\delta\gamma$ in the operational Jacobi ellipsoidal frame of reference $\{\boldsymbol{E}_\Lambda, \boldsymbol{E}_\varphi, \boldsymbol{E}_u|P\}$ at a point P has been outlined. The incremental potential, also called disturbing potential δw, had been defined with respect to the Somigliana–Pizzetti reference potential $w(\varphi, u)$, which includes the ellipsoidal harmonic coefficients up to degree/order (2,0) and the centrifugal potential. Note that the Somigliana–Pizzetti reference potential $w(\varphi, u)$ is not a harmonic function, a result caused by centrifugal potential. Accordingly the incremental potential δw is a harmonic function apart from the topographic masses on top of the Somigliana–Pizzetti level ellipsoid, also called International Reference Ellipsoid E_{A_1,A_1,A_2^2}. The ellipsoidal topographic potential, namely, the ellipsoidal terrain effect, as well as topographic gravity will be introduced in a subsequent contribution. Here, the global and regional computations of vertical deflections with respect to Somigliana–Pizzetti modulus of gravity γ (closed form) and gravity disturbance are presented. It begins with a review of the input data: ➋ *Table 2* summarizes the data of the International Reference Ellipsoid E_{A_1,A_1,A_2^2}, also called World Geodetic Datum 2000 (Groten 2000; Grafarend and

Table 2

Form parameters of the ellipsoid E_{A_1,A_1,A_2^2} (tide-free system)

$A_1 = a = 6,378,136.59\,\text{m}$
$A_2 = b = 6,356,751\,\text{m}$
$\varepsilon = \sqrt{a^2 - b^2} = 521,853.56\,\text{m}$

Table 3

Ellipsoidal harmonic coefficients of SEGEN (Special Ellipsoid Earth Gravity Normal) tide-free system (Ardalan and Grafarend 2000; Ekman 1996)

k	l	C_{kl}	Skl
0	0	$+1.00000000E+00$	$+0.00000000E+00$
1	0	$+0.00000000E+00$	$+0.00000000E+00$
1	1	$+0.00000000E+00$	$+0.00000000E+00$
2	0	$+5.13772373137E-04$	$+0.00000000E+00$
2	1	$-2.02347635955E-10$	$+1.27304012031E-09$
2	2	$+2.43914352398E-06$	$-1.40016683654E-06$
3	0	$+9.53080597633E-07$	$+0.00000000E+00$
3	1	$-2.02347635955E-10$	$+1.27304012031E-09$
3	2	$+9.04627768605E-07$	$-6.19025944205E-07$
3	3	$+7.21072657057E-07$	$+1.41435626958E-06$
4	0	$-2.50440406743E-07$	$+0.00000000E+00$
4	1	$-5.36322406987E-07$	$-4.73435295583E-07$
4	2	$+3.57427905814E-07$	$+6.58806099520E-07$
4	3	$+9.90771803829E-07$	$-2.00928369177E-07$
4	4	$-1.88560802735E-07$	$+3.08853169333E-07$
5	0	$+7.41875911881E-08$	$+0.00000000E+00$
5	1	$-6.21023518983E-08$	$-9.44155466167E-08$
5	2	$+6.56698810466E-07$	$-3.26265030571E-07$
5	3	$-4.49691388810E-07$	$-2.10406407888E-07$
5	4	$-2.95301647654E-07$	$+4.96658876769E-08$
5	5	$+1.74971983203E-07$	$-6.69384278219E-07$
10	10	$+1.00538634409E-07$	$-2.40148449520E-08$
20	20	$+4.01448327968E-09$	$-1.20450644785E-08$
36	36	$+4.60146465720E-09$	$-5.94245336314E-09$
60	60	$+4.23068069789E-09$	$+3.92983780545E-10$
120	120	$-4.56798788660E-10$	$-1.59135018852E-09$
180	180	$-4.06572704272E-10$	$-5.87726119822E-10$
240	240	$-2.30780589856E-10$	$-4.60857985599E-11$
300	300	$-5.02336888312E-11$	$-1.01275530680E-10$
360	360	$-4.47516389678E-25$	$-8.30224945525E-11$

Ardalan 1999). The ellipsoidal harmonic coefficients $\{C_{kl}, S_{kl}\}$ are given at http://www.uni-stuttgart.de/gi/research/paper/coefficients/coefficients.zip (for the derivation of those ellipsoidal harmonic coefficients see *Ardalan and Grafarend* 2000). A sample set of degree/order terms up to (360/360) is given in ➋ *Table 3.*

Based on those input data of Eqs. 383–386 and ➋ *Table 2* the following have been computed:

- Plate I: West–East component η of the deflection of the vertical on the international reference ellipsoid, mollweide Projection
- Plate II: West–East component η of the deflection of the vertical on the international reference ellipsoid, polar regions, equal-area azimutal projection of E_{A_1,A_1,A_2^2}
- Plate III: South–North component ξ of the deflection of the vertical on the international reference ellipsoid, mollweide projection
- Plate IV: South–North component ξ of the deflection of the vertical on the international reference ellipsoid, polar regions, equal-area azimutal projection of E_{A_1,A_1,A_2^2}
- Plate V: Gravity disturbance $\delta\gamma$ on the international reference ellipsoid, mollweide projection

The seismotectonic structure of the Earth is strongly visible in all plates, but a detailed analysis is left to a subsequent contribution. For a study of the Mollweide projection of the biaxial ellipsoid refer to *Grafarend and Heidenreich* (1995).

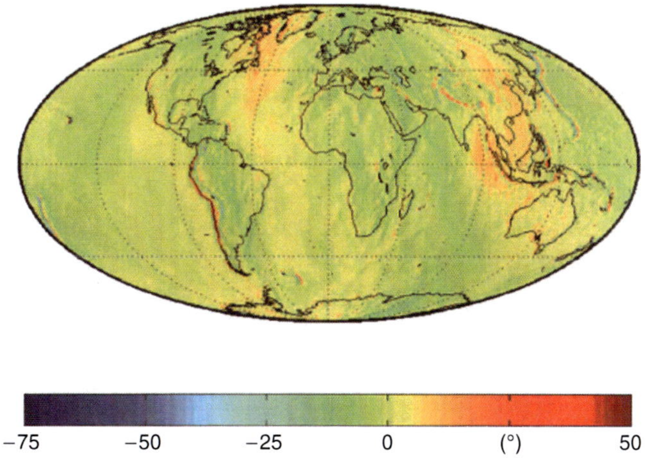

Fig. 13
Plate I: West-East component η of the deflection of the vertical on the international reference ellipsoid, mollweide projection

9 Curvilinear Datum Transformations

The *conformal group* $C_7(3)$, also called *similarity transformation*, over R^3 is introduced here. The *seven parameters* which characterize $C_7(3)$ are (i) *translation* (three parameters: t_x, t_y, t_z), (ii) *rotation* (three parameters: $SO(3) := \{\boldsymbol{R} \in \boldsymbol{R}^{3\times3} | \boldsymbol{R}^T\boldsymbol{R} =, I_3 \det \boldsymbol{R} = +1\}$, for instance, Cardan angles (α,β,γ), $\mathbf{R}(\alpha,\beta,\gamma) := \mathbf{R}_1(\alpha)\mathbf{R}_2(\beta)\mathbf{R}_3(\gamma)$), and (iii) *dilatation* (one parameter, also called *scale*: $1 + s$). Under the action of the *conformal group* (similarity transformation T) *angles and*

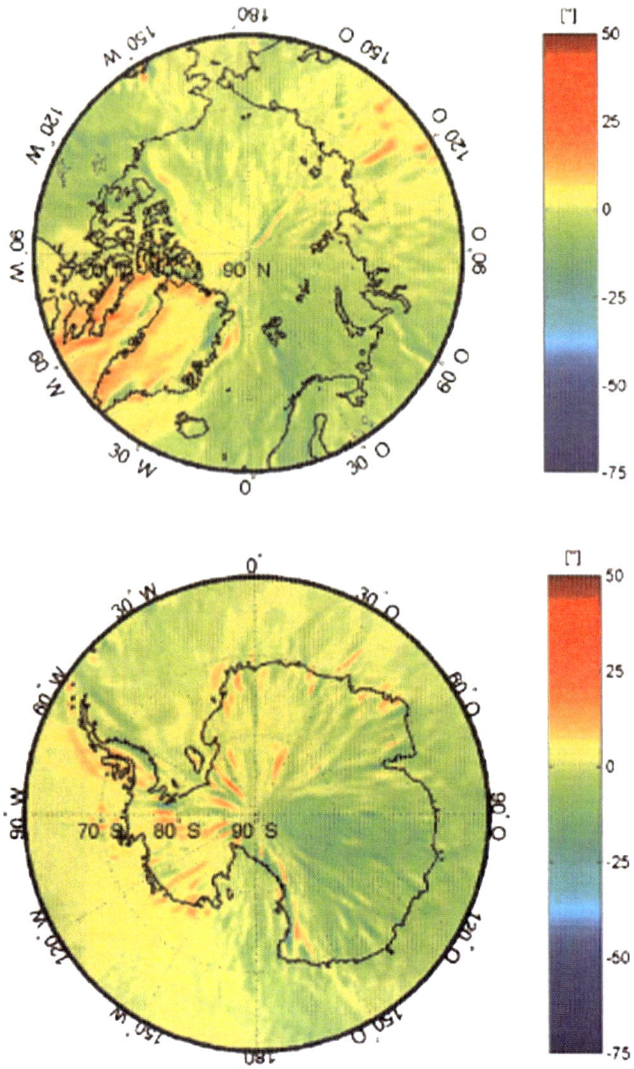

Fig. 14

Plate II: West-East component η of the deflection of the vertical on the international reference ellipsoid, polar regions, equal-area azimutal projection of E_{A_1,A_1,A_2^2}. *Top*: North pole, *Bottom*: South pole

distance ratios in an *Euclidean* R^3 are left *equivariant* (invariant). The standard representation of $T \in C_7(3)$ is given by Eqs. (387–389). {x,y,z} is a set of coordinates which locate a point in a three-dimensional *Weitzenböck space* W_l^3 called *left*. In abstract, {X,Y,Z} is a set of coordinates locating a homologous point in a three-dimensional *Weitzenböck space* W_r^3 called *right*. A *Weitzenböck space* W^3 is a three-dimensional space in which *angles* or *Euclidean distance ratios* are the structure elements.

For instance, if the *scalar product* of two relative placement vectors $\mathbf{X}_2 - \mathbf{X}_1$ and $\mathbf{X}_3 - \mathbf{X}_1$ is denoted by the bracket $\mathbf{X}_2 - \mathbf{X}_1 | \mathbf{X}_3 - \mathbf{X}_1$, the cosine of the angle $\Psi_X = <(\boldsymbol{X}_2 - \boldsymbol{X}_1), (X_3 - \boldsymbol{X}_1)>$ can

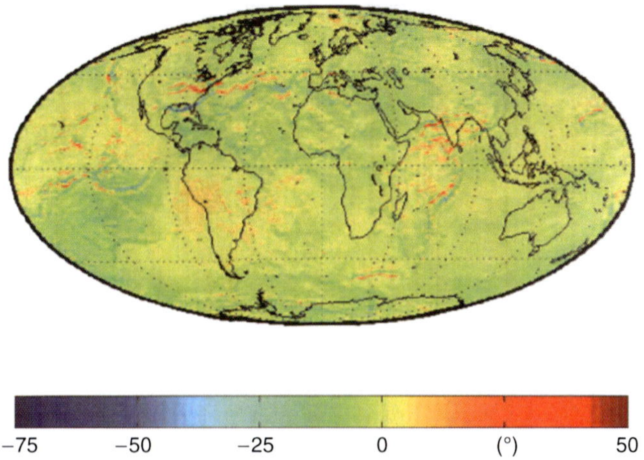

Fig. 15
Plate III: South-North component ξ of the deflection of the vertical on the international reference ellipsoid, mollweide projection

be represented by $\cos\Psi_X =< \mathbf{X}_2 - \mathbf{X}_1 \mid X_3 - \mathbf{X}_1 > \div(\|\mathbf{X}_2 - \mathbf{X}_1\|^2 \|X_3 - \mathbf{X}_1\|^2)$. $\|\mathbf{X}_2 - \mathbf{X}_1\|^2$ and $\|X_3 - \mathbf{X}_1\|^2$ denote the Euclidean length of the relative placement vector $\mathbf{X}_2 - \mathbf{X}_1$ and $\mathbf{X}_3 - \mathbf{X}_1$, respectively. The transformation T: $W_l^3 \leftrightarrow W_r^3$ leave *angles* ("space angles") and *distance ratios equivariant*, namely, $\cos\Psi_X = \cos\psi_x$ and $| |\mathbf{X}_2 - \mathbf{X}_1\|^2 / | |X_3 - \mathbf{X}_1\|^2 = | |x_2 - x_1\|^2 / | |x_3 - x_1\|^2$ a property also called *invariance under the similarity transformation.*

$$\boldsymbol{T} \in \boldsymbol{C}_7(3) := \left\{ \boldsymbol{T} \in \boldsymbol{R}_7(3) \middle| \begin{bmatrix} X \\ Y \\ Z \end{bmatrix} = T \begin{bmatrix} x \\ y \\ z \end{bmatrix}, \cos\Psi_X = \cos\psi_x, \frac{\|\mathrm{X}_3 - \mathrm{X}_1\|^2}{\|\mathrm{X}_2 - \mathrm{X}_1\|^2} = \frac{\|\mathrm{x}_3 - \mathrm{x}_1\|^2}{\|\mathrm{x}_2 - \mathrm{x}_1\|^2} \right\} \tag{387}$$

$$\boldsymbol{T} : \begin{bmatrix} x \\ y \\ z \end{bmatrix} \mapsto \begin{bmatrix} X \\ Y \\ Z \end{bmatrix} = (1+s)\mathbf{R}(\alpha,\beta,\gamma) \begin{bmatrix} x \\ y \\ z \end{bmatrix} + \begin{bmatrix} t_x \\ t_y \\ t_z \end{bmatrix} \tag{388}$$

$$\boldsymbol{R}(\alpha,\beta,\gamma) := \mathbf{R}_1(\alpha)\mathbf{R}_2(\beta)\mathbf{R}_3(\gamma) = \begin{bmatrix} \cos\beta\cos\gamma & \cos\beta\sin\gamma & -\sin\beta \\ \sin\alpha\sin\beta\cos\gamma - \cos\alpha\sin\gamma & \sin\alpha\sin\beta\sin\gamma + \cos\alpha\cos\gamma & \sin\alpha\cos\beta \\ \cos\alpha\sin\beta\cos\gamma + \sin\alpha\sin\gamma & \cos\alpha\sin\beta\sin\gamma - \sin\alpha\cos\gamma & \cos\alpha\cos\beta \end{bmatrix} \tag{389}$$

For geodetic applications, it is sufficient to consider the conformal group $C_7(3)$ *close to the identity*. *First*, the rotation matrices are expanded around $\mathbf{R}_1(0), \mathbf{R}_2(0), \mathbf{R}_3(0)$, an operation which produces the abstracted *Pauli matrices* $\mathbf{R}'_1(0), \mathbf{R}'_2(0), \mathbf{R}'_3(0)$. *Second*, advantage is taken of the scale expansion 1 + s around 1. *Third*, one realizes that the *translation parameters* t_x, t_y, t_z appear in a linear form. In this way, the linearization of the nonlinear similarity transformation "*close to the identity*" leads one via Eqs. (390–395) to the forward transformation Eqs. (396–398), to the backward transformation Eqs. (399–401) as well as to Cartesian coordinate increments $X - x =: \delta x$, $Y - y =: \delta y$, $Z - z =: \delta z$ Eq. 402, namely, functions of the transformation parameter

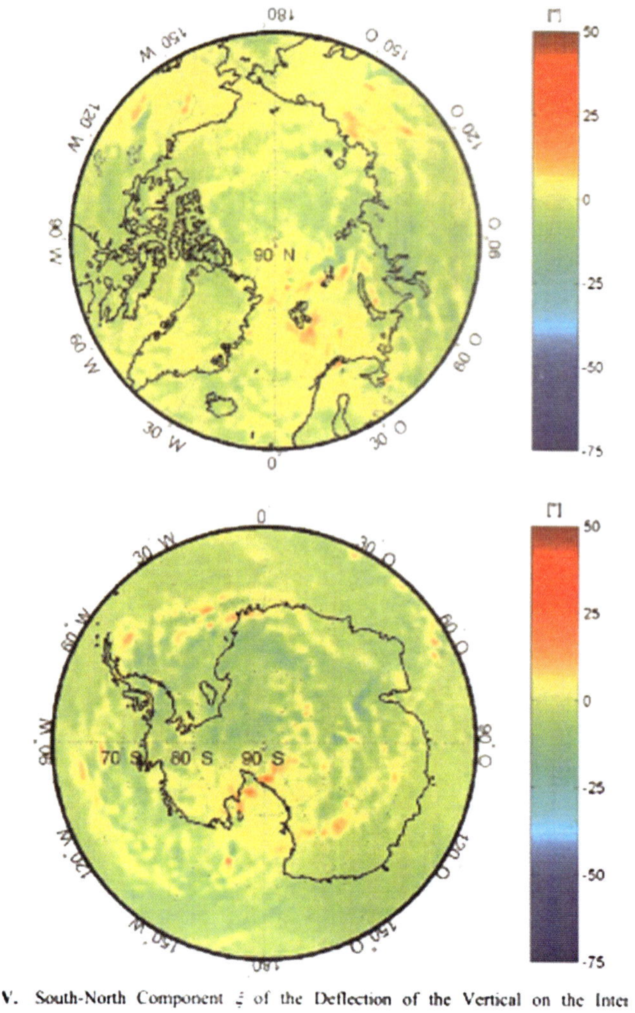

Fig. 16

Plate IV: South-North component ξ of the deflection of the vertical on the international reference ellipsoid, polar regions, equal-area azimutal projection of E_{A_1,A_1,A_2^2}. *Top*: North pole, *Bottom*: South pole

column $[t_x, t_y, t_z, \alpha, \beta, \gamma, s]'$.

$$\boldsymbol{R}_1(\alpha) = \boldsymbol{R}_1(0) + \boldsymbol{R}_1'(0)\alpha + O(\alpha^2) \tag{390}$$

$$\boldsymbol{R}_1'(0) = \begin{bmatrix} 0 & 0 & 0 \\ 0 & 0 & 1 \\ 0 & -1 & 0 \end{bmatrix} \tag{391}$$

$$\boldsymbol{R}_2(\beta) = \boldsymbol{R}_2(0) + \boldsymbol{R}_2'(0)\beta + O(\beta^2) \tag{392}$$

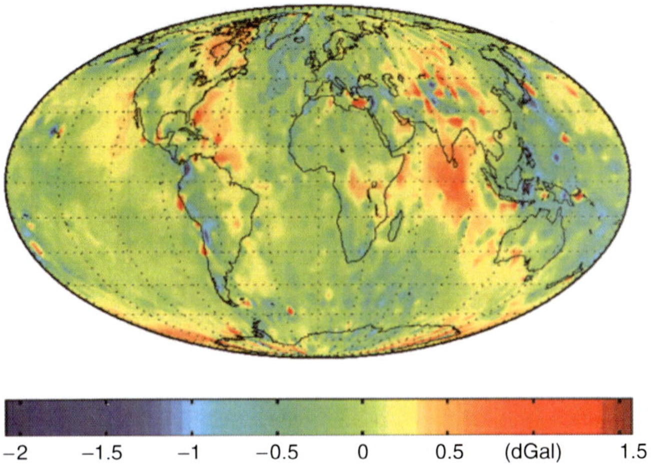

Fig. 17
Plate V: Gravity disturbance $\delta\gamma$ on the international reference ellipsoid, mollweide projection

$$\boldsymbol{R}_2'(0) = \begin{bmatrix} 0 & 0 & -1 \\ 0 & 0 & 0 \\ 1 & 0 & 0 \end{bmatrix} \tag{393}$$

$$\boldsymbol{R}_3(\gamma) = \boldsymbol{R}_3(0) + \boldsymbol{R}_3'(0)\gamma + O(\gamma^2) \tag{394}$$

$$\boldsymbol{R}_3'(0) = \begin{bmatrix} 0 & 1 & 0 \\ -1 & 0 & 0 \\ 0 & 0 & 0 \end{bmatrix} \tag{395}$$

The forward transformation close to the identity

$$X = x + t_x - -z\beta + y\gamma + xs + O_1(\alpha^2) \tag{396}$$

$$Y = y + t_y + z\alpha - -x\gamma + ys + O_2(\beta^{22}) \tag{397}$$

$$Z = z + t_z - -y\alpha + x\beta + zs + O_3(\gamma 12) \tag{398}$$

The backward transformation close to the identity

$$x = X - -t_x - -Y\gamma + Z\beta - -Xs + O_x(\alpha^2) \tag{399}$$

$$y = Y - -t_y - -Z\alpha + X\gamma - -Ys + O_y(\beta^2) \tag{400}$$

$$z = Z - -t_z - -X\beta + Y\alpha - -Zs + O_z(\gamma^2) \tag{401}$$

backward – forward

$$\begin{bmatrix} X - x \\ Y - y \\ Z - z \end{bmatrix} = \begin{bmatrix} 1 & 0 & 0 & | & 0 & -z & y & | & x \\ 0 & 1 & 0 & | & z & 0 & -x & | & y \\ 0 & 0 & 1 & | & -y & x & 0 & | & z \end{bmatrix} \begin{bmatrix} t_x \\ t_y \\ \underline{t_z} \\ \alpha \\ \beta \\ \underline{\gamma} \\ s \end{bmatrix} \tag{402}$$

As soon as is established, the Cartesian datum transformation "*close to the identity*" one departs for deriving the curvilinear datum transformation, namely, for spherical and spheroidal coordinates.

First, Eqs. 403 and 404 generate the transformation of spherical longitude, latitude, radius $(\Lambda,\Phi,R),(\Lambda,\varphi,r)$ to Cartesian coordinates (X,Y,Z), (x,y,z), respectively. Subject to Eq. 405 and Eq. 406 by means of a *Taylor expansion* up to first-order terms, one succeeds to compute the transformation Eq. 407 of incremental coordinates $(\delta\lambda,\delta\varphi\ \delta r)$ to $(\delta x,\delta y,\delta z)$ as well as the inverse transformation Eq. 408 of incremental coordinates $(\delta x,\delta y,\delta z)$ to $(\delta\lambda,\delta\varphi\ \delta r)$. Such an inverse transformation is generated by inverting the *Jacobi* matrix $\boldsymbol{J}(\boldsymbol{\Lambda},\boldsymbol{\varphi},\mathbf{r})$ "at the point (Λ,φ,r)." As soon as one transplants Eq. 402 to Eq. 408, one gains the final spherical datum transformation Eqs. (409–411). The variation of spherical coordinates $(\delta\lambda,\delta\varphi,\delta r)$ is affected by translational parameters (t_x,t_y,t_z), by rotational parameters (α,β,γ) and by incremental scale s.

$$\begin{aligned} X &= R\cos\Phi\cos\Lambda \\ Y &= R\cos\Phi\sin\Lambda \\ Z &= R\sin\Phi \end{aligned} \tag{403}$$

versus

$$\begin{aligned} x &= r\cos\varphi\cos\lambda \\ y &= r\cos\varphi\sin\lambda \\ z &= r\sin\varphi \end{aligned} \tag{404}$$

$$X = x + \delta x,\ Y = y + \delta y,\ Z = z + \delta z \tag{405}$$

$$\Lambda = \Lambda + \delta\lambda,\ \Phi = \varphi + \delta\varphi,\ R = r + \delta r \tag{406}$$

$$\begin{bmatrix} \delta x \\ \delta y \\ \delta z \end{bmatrix} = \begin{bmatrix} D_\Lambda X & D_\Phi X & D_R X \\ D_\Lambda Y & D_\Phi Y & D_R Y \\ D_\Lambda Z & D_\Phi Z & D_R Z \end{bmatrix} (\lambda,\varphi,r) \begin{bmatrix} \delta\lambda \\ \delta\varphi \\ \delta r \end{bmatrix} \tag{407}$$

$$\begin{bmatrix} \delta\lambda \\ \delta\varphi \\ \delta r \end{bmatrix} = \begin{bmatrix} D_\Lambda X & D_\Phi X & D_R X \\ D_\Lambda Y & D_\Phi Y & D_R Y \\ D_\Lambda Z & D_\Phi Z & D_R Z \end{bmatrix}^{-1} (\lambda,\varphi,r) \begin{bmatrix} \delta x \\ \delta y \\ \delta z \end{bmatrix} \tag{408}$$

(9-16) → (9-22)

$$\delta\lambda = \frac{-t_x\sin\lambda + t_y\cos\lambda + r\sin\varphi(\alpha\cos\lambda + \beta\sin\lambda)}{r\cos\varphi} - \gamma \tag{409}$$

$$\delta\varphi = \frac{\sin\varphi(t_x\cos\lambda + t_y\sin\lambda) + t_z\cos\varphi - r(\alpha\sin\lambda - \beta\cos\lambda) - rs\cos\varphi\sin\varphi}{r} \tag{410}$$

$$\delta r = \cos\varphi(t_x\cos\lambda + t_y\sin\lambda) + t_z\sin\varphi + rs \tag{411}$$

Second, Eqs. 412 and 413 generate the transformation of spheroidal longitude, latitude, semi-minor axis $(\Lambda,\Phi,U),(\Lambda,\varphi,u)$ to Cartesian coordinates (X,Y,Z), (x,y,z), respectively. Equation 414 ε^2 indicates the *absolute excentricity*, the difference of semi-major axis squared, A^2,a^2, and semi-minor axis B^2,b^2, of the *International Reference Ellipsoid*. Indeed, such an ellipsoid of reference is fixed by $A = a$, $B = b$. For a detailed introduction into *spheroidal coordinates*, namely, special ellipsoidal coordinates, refer to *Thong and Grafarend* (1989). Subject to Eqs. 415, 416 by means of a *Taylor expansion* up to first-order terms, one succeeds to compute the transformation Eq. 417 of incremental coordinates $(\delta\lambda,\delta\varphi,\delta u)$ to $(\delta x,\delta y,\delta z)$ as well as the inverse transformation Eq. 418 of incremental coordinates $(\delta x,\delta y,\delta z)$ to $(\delta\lambda,\delta\varphi,\delta u)$. Such an inverse transformation is generated by inverting the *Jacobi matrix* $\mathbf{J}(\delta\lambda,\delta\varphi,\delta u)$ "at the point $(\delta\lambda,\delta\varphi,\delta u)$." As soon as one transfers Eq. 402 to Eq. 418, one arrives at the final spherical datum transformation Eqs. (419–421). The variation of spheroidal coordinates $(\delta\lambda,\delta\varphi,\delta u)$ is caused by parameters of type translation (t_x,t_y,t_z), rotation (α,β,γ), and incremental scale s.

Spheroidal datum transformation close to the identity

$$\begin{aligned} X &= \sqrt{\varepsilon^2+U^2}\cos\Phi\cos\Lambda \\ Y &= \sqrt{\varepsilon^2+U^2}\cos\Phi\sin\Lambda \\ Z &= U\sin\Phi \end{aligned} \tag{412}$$

versus

$$\begin{aligned} x &= \sqrt{\varepsilon^2+u^2}\cos\varphi\cos\lambda \\ y &= \sqrt{\varepsilon^2+u^2}\cos\varphi\sin\lambda \\ z &= u\sin\varphi \end{aligned} \tag{413}$$

subject to

$$\varepsilon^2 = A^2 - -B^2 = a^2 - -b^2 \tag{414}$$

$$X = x+\delta x, Y = y+\delta y, Z = z+\delta z \tag{415}$$

$$\Lambda = \lambda+\delta\lambda, \Phi = \varphi+\delta\varphi, U = u+\delta u \tag{416}$$

$$\begin{bmatrix}\delta x\\ \delta y\\ \delta z\end{bmatrix} = \begin{bmatrix} D_\Lambda X & D_\Phi X & D_U X \\ D_\Lambda Y & D_\Phi Y & D_U Y \\ D_\Lambda Z & D_\Phi Z & D_U Z \end{bmatrix}(\lambda,\varphi,u)\begin{bmatrix}\delta\lambda\\ \delta\varphi\\ \delta u\end{bmatrix} \tag{417}$$

$$\begin{bmatrix}\delta\lambda\\ \delta\varphi\\ \delta u\end{bmatrix} = \begin{bmatrix} D_\Lambda X & D_\Phi X & D_U X \\ D_\Lambda Y & D_\Phi Y & D_U Y \\ D_\Lambda Z & D_\Phi Z & D_U Z \end{bmatrix}^{-1}(\lambda,\varphi,u)\begin{bmatrix}\delta x\\ \delta y\\ \delta z\end{bmatrix} \tag{418}$$

(402) → (418)

$$\delta\lambda = \frac{-t_x\sin\lambda + t_y\cos\lambda + u\sin\varphi(\alpha\cos\lambda+\beta\sin\lambda)}{\varepsilon\sqrt{1+\left(\frac{u}{\varepsilon}\right)^2}\cos\varphi} - \gamma \tag{419}$$

$$\begin{aligned}\delta\varphi = \frac{1}{\varepsilon}\frac{1}{\left(\frac{u}{\varepsilon}\right)^2+\sin^2\varphi}\Bigg(&-\sqrt{1+\left(\frac{u}{\varepsilon}\right)^2}\sin\varphi(t_x\cos\lambda+t_y\sin\lambda) + t_z\frac{u}{\varepsilon}\cos\varphi \\ &- u\sqrt{1+\left(\frac{u}{\varepsilon}\right)^2}(\alpha\sin\lambda-\beta\cos\lambda) - s\cos\varphi\sin\varphi\Bigg)\end{aligned} \tag{420}$$

$$\begin{aligned}\delta u = \varepsilon\frac{1}{\left(\frac{u}{\varepsilon}\right)^2+\sin^2\varphi}\Bigg(&\frac{u}{\varepsilon}\sqrt{1+\left(\frac{u}{\varepsilon}\right)^2}\cos\varphi(t_x\cos\lambda+t_y\sin\lambda) + t_z\left(1+\left(\frac{u}{\varepsilon}\right)^2\right)\sin\varphi \\ &- \varepsilon\sqrt{1+\left(\frac{u}{\varepsilon}\right)^2}\cos\varphi\sin\varphi(\alpha\sin\lambda-\beta\cos\lambda) - s\varepsilon\frac{u}{\varepsilon}\left(1+\left(\frac{u}{\varepsilon}\right)^2\right)\Bigg)\end{aligned} \tag{421}$$

10 Datum Transformations in Terms of Spherical Harmonic Coefficients

For the datum transformation of spherical harmonic coefficients the potential Eq. 422 in a three-dimensional *Weizenböck space* Ω_l^3 is given by known spherical harmonic coefficients $\bar{c}_{nm}$ and $\bar{s}_{nm}$. There is also given another potential Eq. 423 in a three-dimensional *Weizenböck space* Ω_r^3 with unknown spherical harmonic coefficients $\overline{C}_{nm}$ and $\overline{S}_{nm}$. With Taylor expansion Eq. 424 and a comparison of coefficients the unknown coefficients in Ω_r^3 can be determined.

Potentials in three-dimensional Weizenböck spaces Ω_l^3 and Ω_r^3 and Taylor expansion:

$$\mathrm{U}(\lambda,\varphi,r)=\frac{\mathrm{GM}}{r_0}\sum_{n=0}^{\infty}\sum_{m=0}^{n}\left(\frac{r_0}{r}\right)^{n+1}\times\overline{P}_{nm}(\sin\varphi)(\bar{c}_{nm}\cos m\lambda+\bar{s}_{nm}\sin m\lambda)\text{ in }\Omega_l^3 \tag{422}$$

$$\mathrm{U}(\Lambda,\Phi,R)=\frac{\mathrm{GM}}{R_0}\sum_{n=0}^{\infty}\sum_{m=0}^{n}\left(\frac{R_0}{R}\right)^{n+1}\times\overline{P}_{nm}(\sin\Phi)(\overline{C}_{nm}\cos m\Lambda+\overline{S}_{nm}\sin m\Lambda)\text{ in }\Omega_r^3 \tag{423}$$

$$\mathrm{U}(\Lambda,\Phi,R)=U(\lambda,\varphi,r)+D_\lambda\mathrm{U}(\lambda,\varphi,r)\delta\lambda+D_\phi\mathrm{U}(\lambda,\varphi,r)\delta\varphi+D_r\mathrm{U}(\lambda,\varphi,r)\delta r+\mathrm{O}^2 \tag{424}$$

The Taylor expansion Eq. 424 of the known potential in Ω_l^3 leads to Eq. 425, where the partial derivation to φ will be replaced by the recursive formula Eq. 426. The result is given in Eq. (427).

Equations after Taylor expansion and inserting recursive formula for partial derivation to φ:

Equation after Taylor expansion:

$$\begin{aligned}\mathrm{U}(\Lambda,\Phi,R)&=\frac{\mathrm{GM}}{r_0}\sum_{n=0}^{\infty}\sum_{m=0}^{n}\left(\frac{r_0}{r}\right)^{n+1}\overline{P}_{nm}(\sin\varphi)(\bar{c}_{nm}\cos m\lambda+\bar{s}_{nm}\sin m\lambda)\\&+\frac{\mathrm{GM}}{r_0}\sum_{n=0}^{\infty}\sum_{m=0}^{n}\left(\frac{r_0}{r}\right)^{n+1}m\overline{P}_{nm}(\sin\varphi)(-\bar{c}_{nm}\sin m\lambda+\bar{s}_{nm}\cos m\lambda)\delta\lambda\\&+\frac{\mathrm{GM}}{r_0}\sum_{n=0}^{\infty}\sum_{m=0}^{n}\left(\frac{r_0}{r}\right)^{n+1}\frac{d\overline{P}_{nm}(\sin\varphi)}{d\varphi}(\bar{c}_{nm}\cos m\lambda+\bar{s}_{nm}\sin m\lambda)\delta\varphi\\&+\frac{GM}{r_0}\sum_{n=0}^{\infty}\sum_{m=0}^{n}-\frac{n+1}{r}\left(\frac{r_0}{r}\right)^{n+1}\overline{P}_{nm}(\sin\varphi)(\bar{c}_{nm}\cos m\lambda+\bar{s}_{nm}\sin m\lambda)\delta r\end{aligned} \tag{425}$$

recursive formula for partial derivation to φ

$$\frac{d\overline{P}_{nm}(\sin\varphi)}{d\varphi}=-m\tan\varphi\overline{P}_{nm}(\sin\varphi)+\sqrt{\frac{(2-\delta_{0m})(n-m)(n+m+1)}{2}}\overline{P}_{n,m+1}(\sin\varphi) \tag{426}$$

Equation with partial derivation to φ:

$$\begin{aligned}\mathrm{U}(\Lambda,\Phi,R)&=\frac{\mathrm{GM}}{r_0}\sum_{n=0}^{\infty}\sum_{m=0}^{n}\left(\frac{r_0}{r}\right)^{n+1}\overline{P}_{nm}(\sin\varphi)(\bar{c}_{nm}\cos m\lambda+\bar{s}_{nm}\sin m\lambda)\\&+\frac{\mathrm{GM}}{r_0}\sum_{n=0}^{\infty}\sum_{m=0}^{n}\left(\frac{r_0}{r}\right)^{n+1}m\overline{P}_{nm}(\sin\varphi)(-\bar{c}_{nm}\sin m\lambda+\bar{s}_{nm}\cos m\lambda)\delta\lambda\end{aligned}$$

$$
\begin{aligned}
&-\frac{\mathrm{GM}}{r_0}\sum_{n=0}^{\infty}\sum_{m=0}^{n}\left(\frac{r_0}{r}\right)^{n+1} m\tan\varphi\overline{P}_{nm}(\sin\varphi)(\bar{c}_{nm}\cos m\lambda+\bar{s}_{nm}\sin m\lambda)\delta\varphi\\
&+\frac{\mathrm{GM}}{r_0}\sum_{n=0}^{\infty}\sum_{m=0}^{n}\left(\frac{r_0}{r}\right)^{n+1}\sqrt{\frac{(2-\delta_{0m})(n-m)(n+m+1)}{2}}\\
&\times\overline{P}_{n,m+1}(\sin\varphi)(\bar{c}_{nm}\cos m\lambda+\bar{s}_{nm}\sin m\lambda)\delta\varphi\\
&+\frac{\mathrm{GM}}{r_0}\sum_{n=0}^{\infty}\sum_{m=0}^{n}\frac{r_{n+1}}{r}\left(\frac{r_0}{r}\right)^{n+1}\overline{P}_{nm}(\sin\varphi)(\bar{c}_{nm}\cos m\lambda+\bar{s}_{nm}\sin m\lambda)\delta r
\end{aligned}
\tag{427}
$$

In Eq. 427 there are products of trigonometric functions with the arguments λ and $m\lambda$ which can be replaced by the arguments $(m+1)\lambda$ and $(m-1)\lambda$ by the application of addition theorems given in Klapp (2002). For better illustration the result in Eq. 428 is summarized by transformation parameters.

Equations summarized by transformation parameters after application of addition theorems:

$$
\mathrm{U}(\Lambda,\Phi,R)=\frac{\mathrm{GM}}{r_0}\sum_{n=0}^{\infty}\sum_{m=0}^{n}\left(\frac{r_0}{r}\right)^{n+1}\times\left(\mathrm{U}_{nm}^{\mathrm{P}}+\mathrm{U}_{nm}^{t_x,t_y}+\mathrm{U}_{nm}^{t_z}+\mathrm{U}_{nm}^{\alpha,\beta}+\mathrm{U}_{nm}^{\gamma}+\mathrm{U}_{nm}^{s}\right)
\tag{428}
$$

where

$$
\mathrm{U}_{nm}^{\mathrm{P}}=\overline{P}_{nm}(\sin\varphi)(\bar{c}_{nm}\cos m\lambda+\bar{s}_{nm}\sin m\lambda)
$$

$$
\begin{aligned}
\mathrm{U}_{nm}^{t_x,t_y}&=\frac{\left(\frac{2m}{\cos\varphi}-(n+m+1)\cos\varphi\right)\overline{P}_{nm}(\sin\varphi)-\sqrt{\frac{(2-\delta_{0m})(n-m)(n+m+1)}{2}}\sin\varphi\overline{P}_{n,m+1}(\sin\varphi)}{2r}\\
&\quad\times[(t_x\bar{c}_{nm}+t_y\bar{s}_{nm})\cos(m-1)\lambda+(t_x\bar{s}_{nm}-t_y\bar{c}_{nm})\sin(m-1)\lambda]\\
&\quad-\frac{(n+m+1)\cos\varphi\overline{P}_{nm}(\sin\varphi)+\sqrt{\frac{(2-\delta_{0m})(n-m)(n+m+1)}{2}}\sin\varphi\overline{P}_{n,m+1}(\sin\varphi)}{2r}\\
&\quad\times[(t_x\bar{c}_{nm}-t_y\bar{s}_{nm})\cos(m+1)\lambda+(t_x\bar{s}_{nm}+t_y\bar{c}_{nm})\sin(m+1)\lambda]\\
U_{nm}^{t_z}&=\frac{-(n+m+1)\sin\varphi\overline{P}_{nm}(\sin\varphi)+\sqrt{\frac{(2-\delta_{0m})(n-m)(n+m+1)}{2}}\cos\varphi\overline{P}_{n,m+1}(\sin\varphi)}{r}
\end{aligned}
$$

$$
\begin{aligned}
\mathrm{U}_{n,m}^{\alpha,\beta}&=\left(m\tan\varphi\overline{P}_{nm}(\sin\varphi)-\frac{1}{2}\sqrt{\frac{(2-\delta_{0m})(n-m)(n+m+1)}{2}}\overline{P}_{n,m+1}(\sin\varphi)\right)\\
&\quad\times[(-\beta\bar{c}_{nm}+\alpha\bar{s}_{nm})\cos(m-1)\lambda+(-\beta\bar{s}_{nm}-\alpha\bar{c}_{nm})\sin(m-1)\lambda]+\frac{1}{2}\overline{P}_{n,m+1}(\sin\varphi)\\
&\quad\times[(\beta\bar{c}_{nm}+\alpha\bar{s}_{nm})\cos(m+1)\lambda+(\beta\bar{s}_{nm}-\alpha\bar{c}_{nm})\sin(m+1)\lambda]\\
U_{nm}^{\gamma}&=m\overline{P}_{nm}(\sin\varphi)(\bar{c}_{nm}\sin m\lambda-\bar{s}_{nm}\cos m\lambda)\gamma\\
U_{nm}^{s}&=-(n+1)\overline{P}_{nm}(\sin\varphi)(\bar{c}_{nm}\cos m\lambda+\bar{s}_{nm}\sin m\lambda)s
\end{aligned}
$$

In the next step one has to replace the products of trigonometric functions with argument φ and Legendre functions of the first kind by recursive formulas given in appendix (observe all the remarks at the end of the chapter). With the result in Eq. 429 it is possible to shift the indices n and m from n to $n-1$ for $n+1$, from m to $m-1$ for $m+1$, and from m to $m+1$ for $m-1$. It is important to notice that in the case of m = 0 the index shift m to $m+1$ leads to the factor $1+\delta_{1m}$ in the terms of the index shift m to $m-1$. In Eq. 430 one has the equation to compare the coefficients, which leads to the equations for the datum transformation of spherical harmonic coefficients Eq. 431.

Equations after application of recursive formula for Legendre functions of the first kind, indices shift, and the equations for the datum transformation of spherical harmonics are as follows:

$$U(\Lambda,\Phi,R)=\frac{\mathrm{GM}}{r_0}\sum_{n=0}^{\infty}\sum_{m=0}^{n}\left(\frac{r_0}{r}\right)^{n+1}\left(\mathrm{U}^{\mathrm{P}}_{nm}+\mathrm{U}^{t_x,t_y}_{nm}+\mathrm{U}^{t_z}_{nm}+\mathrm{U}^{\alpha,\beta}_{nm}+\mathrm{U}^{\gamma}_{nm}+\mathrm{U}^{s}_{nm}\right) \quad (429)$$

where

$$U^{\mathrm{P}}_{nm}=\overline{P}_{nm}(\sin\varphi)(\bar{c}_{nm}\cos m\lambda+\bar{s}_{nm}\sin m\lambda)$$

$$U^{t_x,t_y}_{nm}=\frac{2n+1}{2r}\sqrt{\frac{(2-\delta_{0m})(n-m+1)(n-m+2)}{(2-\delta_{1m})(2n+1)(2n+3)}}$$
$$\times\begin{pmatrix}\overline{P}_{n+1,m-1}(\sin\varphi)\left[(t_x\bar{c}_{nm}+t_y\bar{s}_{nm})\cos(m-1)\lambda+(t_x\bar{s}_{nm}-t_y\bar{c}_{nm})\sin(m-1)\lambda\right]\\+\overline{P}_{n+1,m+1}(\sin\varphi)\left[(t_x\bar{c}_{nm}-t_y\bar{s}_{nm})\cos(m+1)\lambda+(t_x\bar{s}_{nm}+t_y\bar{c}_{nm})\sin(m+1)\lambda\right]\end{pmatrix}$$

$$U^{t_z}_{nm}=-\frac{2n+1}{r}\sqrt{\frac{(n-m+1)(n+m+1)}{(2n+1)(2n+3)}}\overline{P}_{n+1,m}(\sin\varphi)(\bar{c}_{nm}\cos m\lambda+\bar{s}_{nm}\sin m\lambda)t_z$$

$$U^{\alpha,\beta}_{nm}=-\frac{1}{2}\sqrt{\frac{(2-\delta_{0m})(n-m+1)(n+m)}{2-\delta_{1m}}}$$
$$\times\begin{pmatrix}\overline{P}_{n,m-1}(\sin\varphi)\left[(\beta\bar{c}_{nm}-\alpha\bar{s}_{nm})\cos(m-1)\lambda+(\beta\bar{s}_{nm}+\alpha\bar{c}_{nm})\sin(m-1)\lambda\right]\\\overline{P}_{n,m+1}(\sin\varphi)\left[(\beta\bar{c}_{nm}+\alpha\bar{s}_{nm})\cos(m+1)\lambda+(\beta\bar{s}_{nm}-\alpha\bar{c}_{nm})\sin(m+1)\lambda\right]\end{pmatrix}$$

$$U^{\gamma}_{nm}=m\overline{P}_{nm}(\sin\varphi)(\bar{c}_{nm}\sin m\lambda-\bar{s}_{nm}\cos m\lambda)\gamma$$
$$U^{s}_{nm}=-(n+1)\overline{P}_{nm}(\sin\varphi)(\bar{c}_{nm}\cos m\lambda+\bar{s}_{nm}\sin m\lambda)s$$

Equation after indices shifts $n\rightarrow n-1$, $m\rightarrow m-1$, $m\rightarrow m+1$ with $\bar{c}_{-1,m}=\bar{s}_{-1,m}=0$ and before comparison of coefficients $\overline{C}_{nm}$, $\overline{S}_{nm}$ and $\bar{c}_{nm}$, $\bar{s}_{nm}$ with $R_0=r_0$

$$\sum_{n=0}^{\infty}\sum_{m=0}^{n}\left(\frac{R_0}{R}\right)^{n+1}\overline{P}_{nm}(\sin\Phi)(\overline{C}_{nm}\cos m\Lambda+\overline{S}_{nm}\sin m\Lambda)$$
$$=\sum_{n=0}^{\infty}\sum_{m=0}^{n}\frac{R_0}{r_0}\left(\frac{r_0}{r}\right)^{n+1}\overline{P}_{nm}(\sin\varphi)\left(U^{\mathrm{P}}_{nm}+\mathrm{U}^{t_x,t_y}_{nm}+U^{t_z}_{nm}+U^{\alpha,\beta}_{nm}+\mathrm{U}^{\gamma}_{nm}+\mathrm{U}^{s}_{nm}\right) \quad (430)$$

where

$$\mathrm{U}^{\mathrm{P}}_{nm}=\bar{c}_{nm}\cos m\lambda+\bar{s}_{nm}\sin m\lambda$$

$$\mathrm{U}^{t_x,t_y}_{nm}=\frac{2n-1}{2r_0}\sqrt{\frac{2(n-m-1)(n-m)}{(2-\delta_{0m})(2n-1)(2n+1)}}$$
$$\times\left[(t_x\bar{c}_{n-1,m+1}+t_y\bar{s}_{n-1,m+1})\cos m\lambda+(t_x\bar{s}_{n-1,m+1}-t_y\bar{c}_{n-1,m+1})\sin m\lambda\right]$$
$$-\frac{(1+\delta_{1m})(2n-1)}{2r_0}\sqrt{\frac{(2-\delta_{1m})(n+m-1)(n+m)}{2(2n-1)(2n+1)}}$$
$$\times\left[(t_x\bar{c}_{n-1,m-1}-t_y\bar{s}_{n-1,m-1})\cos m\lambda+(t_x\bar{s}_{n-1,m-1}+t_y\bar{c}_{n-1,m-1})\sin m\lambda\right]$$

$$
\begin{aligned}
\mathrm{U}_{nm}^{t_z} &= -\frac{(2n-1)}{r_0}\sqrt{\frac{(n-m)(n+m)}{(2n-1)(2n+1)}}(\bar{c}_{n-1,m}\cos m\lambda + \bar{s}_{n-1,m}\sin m\lambda)t_z \\
\mathrm{U}_{nm}^{\alpha,\beta} &= -\frac{1}{2}\sqrt{\frac{2(n-m)(n+m+1)}{2-\delta_{0m}}} \\
&\quad \times [(\beta\bar{c}_{n,m+1} - \alpha\bar{s}_{n,m+1})\cos m\lambda + (\beta\bar{s}_{n,m+1} + \alpha\bar{c}_{n,m+1})\sin m\lambda] \\
&\quad + \frac{1+\delta_{1m}}{2}\sqrt{\frac{(2-\delta_{1m})(n-m+1)(n+m)}{2}} \\
&\quad \times [(\beta\bar{c}_{n,m-1} + \alpha\bar{s}_{n,m-1})\cos m\lambda + (\beta\bar{s}_{n,m-1} - \alpha\bar{c}_{n,m-1})\sin m\lambda] \\
\mathrm{U}_{nm}^{\gamma} &= m(\bar{c}_{nm}\sin m\lambda - \bar{s}_{nm}\cos m\lambda)\gamma \\
\mathrm{U}_{nm}^{s} &= -(n+1)(\bar{c}_{nm}\cos m\lambda + \bar{s}_{nm}\sin m\lambda)s
\end{aligned}
$$

Equations for the datum transformation of spherical harmonics

$$
\begin{aligned}
\begin{Bmatrix}\overline{C}_{nm}\\ \overline{S}_{nm}\end{Bmatrix} &= \frac{1}{2}(1+\delta_{1m})\sqrt{\frac{(2-\delta_{1m})(n-m+1)(n+m)}{2}} \\
&\quad \times \left(\beta\begin{Bmatrix}\bar{c}_{n,m-1}\\ \bar{s}_{n,m-1}\end{Bmatrix} \pm \alpha\begin{Bmatrix}\bar{s}_{n,m-1}\\ \bar{c}_{n,m-1}\end{Bmatrix}\right) + (1-(n+1)s)\begin{Bmatrix}\bar{c}_{nm}\\ \bar{s}_{nm}\end{Bmatrix} \mp m\gamma\begin{Bmatrix}\bar{s}_{nm}\\ \bar{c}_{nm}\end{Bmatrix} \\
&\quad - \frac{1}{2}\sqrt{\frac{2(n-m)(n+m+1)}{2-\delta_{0m}}}\left(\beta\begin{Bmatrix}\bar{c}_{n,m+1}\\ \bar{s}_{n,m+1}\end{Bmatrix} \mp \alpha\begin{Bmatrix}\bar{s}_{n,m+1}\\ \bar{c}_{n,m+1}\end{Bmatrix}\right) \\
&\quad - \frac{(2n-1)(1+\delta_{1m})}{2r_0}\sqrt{\frac{(2-\delta_{1m})(n+m-1)(n+m)}{2(2n-1)(2n+1)}} \\
&\quad \times \left(t_x\begin{Bmatrix}\bar{c}_{n-1,m-1}\\ \bar{s}_{n-1,m-1}\end{Bmatrix} \mp t_y\begin{Bmatrix}\bar{s}_{n-1,m-1}\\ \bar{c}_{n-1,m-1}\end{Bmatrix}\right) - \frac{2n-1}{2r_0}\sqrt{\frac{(n-m)(n+m)}{(2n-1)(2n+1)}}t_z\begin{Bmatrix}\bar{c}_{n-1,m}\\ \bar{s}_{n-1,m}\end{Bmatrix} \\
&\quad - \frac{2n-1}{2r_0}\sqrt{\frac{2(n-m-1)(n-m)}{(2-\delta_{0m})(2n-1)(2n+1)}} \times \left(t_x\begin{Bmatrix}\bar{c}_{n-1,m+1}\\ \bar{s}_{n-1,m+1}\end{Bmatrix} \pm t_y\begin{Bmatrix}\bar{s}_{n-1,m+1}\\ \bar{c}_{n-1,m+1}\end{Bmatrix}\right)
\end{aligned} \tag{431}
$$

11 Datum Transformations of Ellipsoidal Harmonic Coefficients

The datum transformation of spheroidal harmonic coefficients is quite similar to the datum transformation of spherical harmonic coefficients. Here, it is started with the potential Eq. 432 with known spheroidal harmonic coefficients $\bar{c}_{nm}$ and $\bar{s}_{nm}$ in a three-dimensional *Weizenböck space* Ω_{l}^3. Another potential Eq. 433 is given by unknown spheroidal harmonic coefficients $\overline{C}_{nm}$ and $\overline{S}_{nm}$ in a three-dimensional *Weizenböck space* Ω_{r}^3. With the Taylor expansion Eq. 434 and a comparison of coefficients the unknown coefficients in Ω_{l}^3 can be determined.

Potentials in three-dimensional Weizenböck spaces Ω_l^3 and Ω_r^3 and Taylor expansion

$$U(\lambda,\varphi,u)=\frac{GM}{\varepsilon}\operatorname{arccot}\left(\frac{b}{\varepsilon}\right)\sum_{n=0}^{\infty}\sum_{m=0}^{n}\frac{Q^*_{nm}\left(\frac{u}{\varepsilon}\right)}{Q^*_{nm}\left(\frac{b}{\varepsilon}\right)}\times\overline{P}_{nm}(\sin\varphi)(\bar{c}_{nm}\cos m\lambda+\bar{s}_{nm}\sin m\lambda)\text{ in }\Omega_l^3 \tag{432}$$

$$U(\Lambda,\Phi,U)=\frac{GM}{E}\operatorname{arccot}\left(\frac{B}{E}\right)\sum_{n=0}^{\infty}\sum_{m=0}^{n}\frac{Q^*_{nm}\left(\frac{U}{E}\right)}{Q^*_{nm}\left(\frac{B}{E}\right)}\times\overline{P}_{nm}(\sin\Phi)(\overline{C}_{nm}\cos m\Lambda+\overline{S}_{nm}\sin m\Lambda)\text{ in }\Omega_r^3 \tag{433}$$

$$U(\Lambda,\Phi,U)=U(\lambda,\varphi,u)+D_\lambda U(\lambda,\varphi,u)\delta\lambda+D_\phi U(\lambda,\varphi,u)\delta\varphi+D_u U(\lambda,\varphi,u)\frac{\delta u}{\varepsilon}+O^2 \tag{434}$$

The Taylor expansion of the known potential in Ω_l^3 leads to Eq. 435. After Taylor expansion the partial derivations will be replaced by the recursive formulas Eqs. 436 and 437 and the variation of spheroidal coordinates ($\delta\lambda$,$\delta\varphi$,δu) will be replaced by Eqs. (419–421). The result is given in Eq. 438. It has also to be mentioned that in all formulas in this chapter the constant factor $\frac{GM}{\varepsilon}\operatorname{arccot}\left(\frac{b}{\varepsilon}\right)$ is replaced by k.

Equations after Taylor expansion and inserting recursive formulas for partial derivations to $\frac{u}{\varepsilon}$ and φ.

$$\begin{aligned}\mathrm{U}(\Lambda,\Phi,U)&=k\sum_{n=0}^{\infty}\sum_{m=0}^{n}\frac{Q^*_{nm}\left(\frac{u}{\varepsilon}\right)}{Q^*_{nm}\left(\frac{b}{\varepsilon}\right)}\overline{P}_{nm}(\sin\varphi)(\bar{c}_{nm}\cos m\lambda+\bar{s}_{nm}\sin m\lambda)\\&+k\sum_{n=0}^{\infty}\sum_{m=0}^{n}m\frac{Q^*_{nm}\left(\frac{u}{\varepsilon}\right)}{Q^*_{nm}\left(\frac{b}{\varepsilon}\right)}\overline{P}_{nm}(\sin\varphi)(-\bar{c}_{nm}\sin m\lambda+\bar{s}_{nm}\cos m\lambda)\delta\lambda\\&+k\sum_{n=0}^{\infty}\sum_{m=0}^{n}\frac{Q^*_{nm}\left(\frac{u}{\varepsilon}\right)}{Q^*_{nm}\left(\frac{b}{\varepsilon}\right)}\frac{d\overline{P}_{nm}(\sin\varphi)}{d\varphi}(\bar{c}_{nm}\cos m\lambda+\bar{s}_{nm}\sin m\lambda)\delta\varphi\\&+k\sum_{n=0}^{\infty}\sum_{m=0}^{n}\frac{1}{Q^*_{nm}\left(\frac{b}{\varepsilon}\right)}\frac{dQ^*_{nm}\left(\frac{u}{\varepsilon}\right)}{\frac{du}{\varepsilon}}\overline{P}_{nm}(\sin\varphi)\\&\times(\bar{c}_{nm}\cos m\lambda+\bar{s}_{nm}\sin m\lambda)\frac{\delta u}{\varepsilon}\end{aligned} \tag{435}$$

recursive formula for partial derivation to φ:

$$\frac{d\overline{P}_{nm}(\sin\varphi)}{d\varphi}=-m\tan\varphi\overline{P}_{nm}(\sin\varphi)+\sqrt{\frac{(2-\delta_{0m})(n-m)(n+m+1)}{2}}\overline{P}_{nm+1}(\sin\varphi) \tag{436}$$

recursive formula for partial derivation to $\frac{u}{\varepsilon}$:

$$\frac{dQ^*_{nm}\left(\frac{u}{\varepsilon}\right)}{\frac{du}{\varepsilon}}=\frac{1}{\varepsilon}\left[\frac{m}{1+\left(\frac{u}{\varepsilon}\right)^2}\frac{u}{\varepsilon}Q^*_{nm}\left(\frac{u}{\varepsilon}\right)-\frac{n+m+1}{\sqrt{1+\left(\frac{u}{\varepsilon}\right)^2}}Q^*_{n,m+1}\left(\frac{u}{\varepsilon}\right)\right] \tag{437}$$

Equation with partial derivations to φ and $\frac{u}{\varepsilon}$:

$$
\begin{aligned}
U(\Lambda,\Phi,U) = & k\sum_{n=0}^{\infty}\sum_{m=0}^{n}\frac{Q_{nm}^{*}\left(\frac{u}{\varepsilon}\right)}{Q_{nm}^{*}\left(\frac{b}{\varepsilon}\right)}\overline{P}_{nm}(\sin\varphi)(\bar{c}_{nm}\cos m\lambda+\bar{s}_{nm}\sin m\lambda)\\
& + k\sum_{n=0}^{\infty}\sum_{m=0}^{n}\frac{Q_{nm}^{*}\left(\frac{u}{\varepsilon}\right)}{Q_{nm}^{*}\left(\frac{b}{\varepsilon}\right)} m\overline{P}_{nm}(\sin\varphi)(-\bar{c}_{nm}\sin m\lambda+\bar{s}_{nm}\cos m\lambda)\delta\lambda\\
& - k\sum_{n=0}^{\infty}\sum_{m=0}^{n}\frac{Q_{nm}^{*}\left(\frac{u}{\varepsilon}\right)}{Q_{nm}^{*}\left(\frac{b}{\varepsilon}\right)} m\tan\varphi\overline{P}_{nm}(\sin\varphi)(\bar{c}_{nm}\cos m\lambda+\bar{s}_{nm}\sin m\lambda)\delta\varphi\\
& + k\sum_{n=0}^{\infty}\sum_{m=0}^{n}\frac{Q_{nm}^{*}\left(\frac{u}{\varepsilon}\right)}{Q_{nm}^{*}\left(\frac{b}{\varepsilon}\right)}\sqrt{\frac{(2-\delta_{0m})(n-m)(n+m+1)}{2}}\overline{P}_{n,m+1}(\sin\varphi)\\
& \times(\bar{c}_{nm}\cos m\lambda+\bar{s}_{nm}\sin m\lambda)\delta\varphi\\
& + k\sum_{n=0}^{\infty}\sum_{m=0}^{n}\frac{1}{Q_{nm}^{*}\left(\frac{b}{\varepsilon}\right)}\frac{1}{\varepsilon}\frac{m}{1+\left(\frac{u}{\varepsilon}\right)^2}\frac{u}{\varepsilon}Q_{nm}^{*}\left(\frac{u}{\varepsilon}\right)\overline{P}_{nm}(\sin\varphi)\\
& \times(\bar{c}_{nm}\cos m\lambda+\bar{s}_{nm}\sin m\lambda)\delta\frac{u}{\varepsilon}\\
& - k\sum_{n=0}^{\infty}\sum_{m=0}^{n}\frac{1}{Q_{nm}^{*}\left(\frac{b}{\varepsilon}\right)}\frac{n+m+1}{\varepsilon\sqrt{1+\left(\frac{u}{\varepsilon}\right)^2}}Q_{n,m+1}^{*}\left(\frac{u}{\varepsilon}\right)\overline{P}_{nm}(\sin\varphi)\\
& \times(\bar{c}_{nm}\cos m\lambda+\bar{s}_{nm}\sin m\lambda)\delta\frac{u}{\varepsilon}
\end{aligned}
\tag{438}
$$

In Eq. 438 products of trigonometric functions with the arguments φ and $m\lambda$, which can be replaced by the arguments $(m+1)\lambda$ and $(m-1)\lambda$ by the application of addition theorems given in Klapp (2002) are obtained. In this step Eq. 439 is inserted to separate $\frac{u}{\varepsilon}$ and φ. For better illustration the result in Eq. 440 is summarized by transformation parameters.

Approximation to separate $\frac{u}{\varepsilon}$ and φ and equation after application of addition theorems summarized by transformation parameters

Approximation for $\frac{1}{\left(\frac{u}{\varepsilon}\right)^2+\sin^2\varphi}$

$$
\frac{1}{\left(\frac{u}{\varepsilon}\right)^2+\sin^2\varphi}\approx\frac{1}{1+\left(\frac{u}{\varepsilon}\right)^2}
\tag{439}
$$

$$
U(\Lambda,\Phi,U) = k\sum_{n=0}^{\infty}\sum_{m=0}^{n}\frac{1}{Q_{nm}^{*}\left(\frac{b}{\varepsilon}\right)}\begin{pmatrix} U_{nm}^{P(App)}+U_{nm}^{t_x,t_y(App)}+U_{nm}^{t_z(App)}\\ +U_{nm}^{\alpha,\beta(App)}+U_{nm}^{\gamma(App)}+U_{nm}^{s(App)}\end{pmatrix}
\tag{440}
$$

where

$$
\begin{aligned}
U_{nm}^{P(App)} &= Q_{nm}^{*}\left(\frac{u}{\varepsilon}\right)\overline{P}_{nm}(\sin\varphi)(\bar{c}_{nm}\cos m\lambda+\bar{s}_{nm}\sin m\lambda)\\
U_{nm}^{t_x,t_y(App)} &= \frac{1}{\varepsilon}\frac{1}{\sqrt{1+\left(\frac{u}{\varepsilon}\right)^2}}Q_{nm}^{*}\left(\frac{u}{\varepsilon}\right)\frac{m}{\cos\varphi}\overline{P}_{nm}(\sin\varphi)\\
&\times\left[(t_x\bar{c}_{nm}+t_y\bar{s}_{nm})\cos(m-1)\lambda+(t_x\bar{s}_{nm}-t_y\bar{c}_{nm})\sin(m-1)\lambda\right]
\end{aligned}
$$

$$-\begin{bmatrix} \frac{\sqrt{\frac{(2-\delta_{0m})(n-m)(n+m+1)}{2}}}{2\varepsilon} \frac{1}{\sqrt{1+\left(\frac{u}{\varepsilon}\right)^2}} Q^*_{nm}\left(\frac{u}{\varepsilon}\right) \sin\varphi \overline{P}_{n,m+1}(\sin\varphi) \\ +\frac{n+m+1}{2\varepsilon} \frac{\frac{u}{\varepsilon}}{1+\left(\frac{u}{\varepsilon}\right)^2} Q^*_{n,m+1}\left(\frac{u}{\varepsilon}\right) \cos\varphi \overline{P}_{nm}(\sin\varphi) \end{bmatrix}$$

$$\times \begin{bmatrix} (t_x\bar{c}_{nm} + t_y\bar{s}_{nm})\cos(m-1)\lambda + (t_x\bar{s}_{nm} - t_y\bar{c}_{nm})\sin(m-1)\lambda \\ +(t_x\bar{c}_{nm} - t_y\bar{s}_{nm})\cos(m+1)\lambda + (t_x\bar{s}_{nm} + t_y\bar{c}_{nm})\sin(m+1)\lambda \end{bmatrix}$$

$$U^{t_z(\mathrm{App})}_{nm} = \begin{bmatrix} \frac{\sqrt{\frac{(2-\delta_{0m})(n-m)(n+m+1)}{2}}}{\varepsilon} \frac{\frac{u}{\varepsilon}}{1+\left(\frac{u}{\varepsilon}\right)^2} Q^*_{nm}\left(\frac{u}{\varepsilon}\right) \cos\varphi \overline{P}_{n,m+1}(\sin\varphi) \\ -\frac{n+m+1}{\varepsilon} \frac{1}{\sqrt{1+\left(\frac{u}{\varepsilon}\right)^2}} Q^*_{n,m+1}\left(\frac{u}{\varepsilon}\right) \sin\varphi \overline{P}_{nm}(\sin\varphi) \end{bmatrix}$$
$$\times (\bar{c}_{nm}\cos m\lambda + \bar{s}_{nm}\sin m\lambda) t_z$$

$$U^{\alpha,\beta(\mathrm{App})}_{nm} = -\frac{1}{\varepsilon} \frac{\frac{u}{\varepsilon}}{\sqrt{1+\left(\frac{u}{\varepsilon}\right)^2}} Q^*_{nm}\left(\frac{u}{\varepsilon}\right) m\tan\varphi \overline{P}_{nm}(\sin\varphi)$$
$$\times [(-\alpha\bar{s}_{nm} + \beta\bar{c}_{nm})\cos(m-1)\lambda + (\alpha\bar{c}_{nm} + \beta\bar{s}_{nm})\sin(m-1)\lambda]$$
$$+ \begin{bmatrix} \frac{\sqrt{\frac{(2-\delta_{0m})(n-m)(n+m+1)}{2}}}{2} \frac{\frac{u}{\varepsilon}}{\sqrt{1+\left(\frac{u}{\varepsilon}\right)^2}} Q^*_{nm}\left(\frac{u}{\varepsilon}\right) \overline{P}_{n,m+1}(\sin\varphi) \\ -\frac{n+m+1}{2} \frac{1}{1+\left(\frac{u}{\varepsilon}\right)^2} Q^*_{n,m+1}\left(\frac{u}{\varepsilon}\right) \cos\varphi \sin\varphi \overline{P}_{nm}(\sin\varphi) \end{bmatrix}$$
$$\times \begin{bmatrix} (-\alpha\bar{s}_{nm} + \beta\bar{c}_{nm})\cos(m-1)\lambda + (\alpha\bar{c}_{nm} + \beta\bar{s}_{nm})\sin(m-1)\lambda \\ +(\alpha\bar{s}_{nm} + \beta\bar{c}_{nm})\cos(m+1)\lambda + (-\alpha\bar{c}_{nm} + \beta\bar{s}_{nm})\sin(m+1)\lambda \end{bmatrix}$$

$$U^{\gamma(\mathrm{App})}_{nm} = mQ^*_{nm}\left(\frac{u}{\varepsilon}\right)\overline{P}_{nm}(\sin\varphi)(\bar{c}_{nm}\sin m\lambda - \bar{s}_{nm}\cos m\lambda)\gamma$$

$$\mathrm{U}^{s(\mathrm{App})}_{nm} = \begin{bmatrix} mQ^*_{nm}\left(\frac{u}{\varepsilon}\right)\overline{P}_{nm}(\sin\varphi) - \frac{n+m+1}{\varepsilon} \frac{\frac{u}{\varepsilon}}{\sqrt{1+\left(\frac{u}{\varepsilon}\right)^2}} Q^*_{n,m+1}\left(\frac{u}{\varepsilon}\right) \overline{P}_{nm}(\sin\varphi) \\ -\frac{\sqrt{\frac{(2-\delta_{0m})(n-m)(n+m+1)}{2}}}{\varepsilon} \frac{1}{1+\left(\frac{u}{\varepsilon}\right)^2} Q^*_{nm}\left(\frac{u}{\varepsilon}\right) \cos\varphi \sin\varphi \overline{P}_{n,m+1}(\sin\varphi) \end{bmatrix}$$
$$\times (\bar{c}_{nm}\cos m\lambda + \bar{s}_{nm}\sin m\lambda) s$$

In the next step one has to replace the products of Legendre functions of the first kind and trigonometric functions with argument φ and the products of Legendre functions of the second kind with terms containing $\frac{u}{\varepsilon}$ by recursive formulas given in Klapp (2002). Before can be performed indices shifts one has to verify and, if necessary, adapt the degrees and orders of Legendre functions, because there should be no different degrees in a product of Legendre functions before the coefficients can be compared. The orders in a product of Legendre functions have also to be equal but also have to correspond to the factor of the argument φ in trigonometric functions. In the case of nonequal or corresponding degrees and orders one has to change them following the rules given in Eqs. 441 and 442. It is important to notice, that the factors of normalization and $Q^*_{nm}\left(\frac{b}{\varepsilon}\right)$ have also to be adapted. With Eq. 443 the indices shifts from n to $n-2$ for $n+2$, from n to $n-1$ for $n+1$, from m to $m-1$ for $m+1$, and from m to $m+1$ for $m-1$ can be done. It is important to notice that $m=0$ causes the factor $1+\delta_{1m}$ in the terms of index shift from m to $m-1$. In Eq. 444 one has the equation after indices shifts to compare the coefficients which leads to the equations for the datum transformation of spherical harmonic coefficients Eq. 445.

Equations after application of recursive formula for Legendre functions of the first and second kind and adaption of degree and order and the equations for the datum transformation of sectorial coefficients are as follows.

Rules for adaption of degree n and order m by extending the term $Q^*_{n+i,m-l}\left(\frac{u}{\varepsilon}\right)\overline{P}_{n-k,m+j}(\sin\varphi)$ with $i,j,k,l \in \{0,\ldots,\infty\}$:

Adaption of degree $n(n\text{-}k \to n+i)$: Legendre functions have the degree $n+i$:

$$\begin{matrix} Q^*_{n+i,m-l}\left(\frac{u}{\varepsilon}\right) \\ \times\overline{P}_{n-k,m+j}(\sin\varphi) \end{matrix} \quad \underset{\times\begin{pmatrix}\overline{P}_{n-k+1,m+j}(\sin\varphi)+\ldots \\ \ldots+\overline{P}_{n+i,m+j}(\sin\varphi)\end{pmatrix}}{\overset{\to}{+Q^*_{n+i,m-l}\left(\frac{u}{\varepsilon}\right)}} \quad \begin{matrix} Q^*_{n+i,m-l}\left(\frac{u}{\varepsilon}\right) \\ \times\overline{P}_{n+i,m+j}(\sin\varphi) \end{matrix} \tag{441}$$

Adaption of order $m(m-l \to m+j)$: Legendre functions have the order $m+k$:

$$\begin{matrix} Q^*_{n+i,m-l}\left(\frac{u}{\varepsilon}\right) \\ \times\overline{P}_{n+i,m+j}(\sin\varphi) \end{matrix} \quad \underset{\times\overline{P}_{n+i,m+j}(\sin\varphi)}{\overset{\to}{+\begin{pmatrix}Q^*_{n+i,m-l+1}\left(\frac{u}{\varepsilon}\right)+\ldots \\ \ldots+Q^*_{n+i,m+j}\left(\frac{u}{\varepsilon}\right)\end{pmatrix}}} \quad \begin{matrix} Q^*_{n+i,m+j}\left(\frac{u}{\varepsilon}\right) \\ \times\overline{P}_{n+i,m+j}(\sin\varphi) \end{matrix} \tag{442}$$

Equation after adaption of degree n and order m of $Q^*_{nm}\left(\frac{b}{\varepsilon}\right)$:
$n \to n+i+k,\ m \to m+j+l$

$$U(\Lambda,\Phi,U) = k\sum_{n=0}^{\infty}\sum_{m=0}^{n}\frac{1}{Q^*_{nm}\left(\frac{b}{\varepsilon}\right)}\begin{pmatrix} U_{nm}^{P(\text{Adap})} + U_{nm}^{t_x,t_y(\text{Adap})} + U_{nm}^{t_z(\text{Adap})} \\ +U_{nm}^{\alpha,\beta(\text{Adap})} + U_{nm}^{\gamma(\text{Adap})} + U_{nm}^{s(\text{Adap})} \end{pmatrix} \tag{443}$$

where

$$U_{nm}^{P(\text{Adap})} = Q^*_{nm}\left(\frac{u}{\varepsilon}\right)\overline{P}_{nm}(\sin\varphi)(\bar{c}_{nm}\cos m\lambda + \bar{s}_{nm}\sin m\lambda)$$

$$U_{nm}^{t_x,t_y(\text{Adap})} = \frac{\sqrt{2-\delta_{0m}}}{8a} \times \begin{bmatrix} -(2m+1)\sqrt{\frac{(n-m-1)(n-m)}{2(2n-1)(2n+1)}} \\ +(2m+1)\sqrt{\frac{(n+m+1)(n+m+2)}{2(2n+1)(2n+3)}} \\ +(2m-1)\sqrt{\frac{(n+m-1)(n+m)}{(2-\delta_{1m})(2n-1)(2n+1)}} \\ +(4n+2m+3)\sqrt{\frac{(n-m+1)(n-m+2)}{(2-\delta_{1m})(2n+1)(2n+3)}} \end{bmatrix} Q^*_{n+1,m-1}\left(\frac{u}{\varepsilon}\right)\overline{P}_{n+1,m-1}(\sin\varphi)$$

$$\times\left[(t_x\bar{c}_{nm}+t_y\bar{s}_{nm})\cos(m-1)\lambda+(t_x\bar{s}_{nm}-t_y\bar{c}_{nm})\sin(m-1)\lambda\right]-\frac{\sqrt{2-\delta_{0m}}}{4a}$$

$$\times\begin{bmatrix}(n+m+1)\sqrt{\frac{(n-m-1)(n-m)}{2(2n-1)(2n+1)}}\\+(n+m+1)\sqrt{\frac{(n+m-1)(n+m)}{(2-\delta_{1m})(2n-1)(2n+1)}}\\-(n+m+1)\sqrt{\frac{(n-m+1)(n-m+2)}{(2-\delta_{1m})(2n+1)(2n+3)}}\\+(n-m)\sqrt{\frac{(n+m+1)(n+m+2)}{2(2n+1)(2n+3)}}\end{bmatrix}Q^*_{n+1,m+1}\left(\frac{u}{\varepsilon}\right)\overline{P}_{n+1,m+1}(\sin\varphi)$$

$$\times\left[(t_x\bar{c}_{nm}-t_y\bar{s}_{nm})\cos(m+1)\lambda+(t_x\bar{s}_{nm}+t_y\bar{c}_{nm})\sin(m+1)\lambda\right]$$

$$U^{t_z(\mathrm{Adap})}_{nm}=-\frac{2n+1}{2a}\sqrt{\frac{(n-m+1)(n+m+1)}{(2n+1)(2n+3)}}Q^*_{n+1,m}\left(\frac{u}{\varepsilon}\right)\overline{P}_{n+1,m}(\sin\varphi)$$
$$\times(\bar{c}_{nm}\cos m\lambda+\bar{s}_{nm}\sin m\lambda)t_z$$

$$U^{\alpha,\beta(\mathrm{Adap})}_{nm}=-\frac{1}{2}\sqrt{\frac{(2-\delta_{0m})(n-m+1)(n+m)}{2-\delta_{1m}}}Q^*_{n,m-1}\left(\frac{u}{\varepsilon}\right)\overline{P}_{n,m-1}(\sin\varphi)$$
$$\times\left[(-\alpha\bar{s}_{nm}+\beta\bar{c}_{nm})\cos(m-1)\lambda+(\alpha\bar{c}_{nm}+\beta\bar{s}_{nm})\sin(m-1)\lambda\right]$$
$$+\frac{1}{2}\sqrt{\frac{(2-\delta_{0m})(n-m+1)(n+m)}{2}}Q^*_{n,m+1}\left(\frac{u}{\varepsilon}\right)\overline{P}_{n,m+1}(\sin\varphi)$$
$$\times\left[(\alpha\bar{s}_{nm}+\beta\bar{c}_{nm})\cos(m+1)\lambda+(-\alpha\bar{c}_{nm}+\beta\bar{s}_{nm})\sin(m+1)\lambda\right]$$
$$-\frac{n+m+1}{8(2n-1)(2n+3)\left(\frac{a}{\varepsilon}\right)^2}\sqrt{\frac{2-\delta_{0m}}{2-\delta_{1m}}}$$
$$\times\begin{bmatrix}(2n+3)\sqrt{\frac{(n-m)(n+m-2)(n+m-1)(n+m)}{(2n-3)(2n+1)}}\\+(2m-1)\sqrt{(n-m+1)(n+m)}\\-(2n-1)\sqrt{\frac{(n-m+1)(n-m+2)(n-m+3)(n+m+1)}{(2n+1)(2n+5)}}\end{bmatrix}$$
$$\times\begin{bmatrix}Q^*_{n+2,m-1}\left(\frac{u}{\varepsilon}\right)\overline{P}_{n+2,m-1}(\sin\varphi)\\\times\left[(-\alpha\bar{s}_{nm}+\beta\bar{c}_{nm})\cos(m-1)\lambda+(\alpha\bar{c}_{nm}+\beta\bar{s}_{nm})\sin(m-1)\lambda\right]\\+Q^*_{n+2,m+1}\left(\frac{u}{\varepsilon}\right)\overline{P}_{n+2,m+1}(\sin\varphi)\\\times\left[(\alpha\bar{s}_{nm}+\beta\bar{c}_{nm})\cos(m+1)\lambda+(-\alpha\bar{c}_{nm}+\beta\bar{s}_{nm})\sin(m+1)\lambda\right]\end{bmatrix}$$

$$\mathrm{U}^{\gamma(\mathrm{Adap})}_{nm}=mQ^*_{nm}\left(\frac{u}{\varepsilon}\right)\overline{P}_{nm}(\sin\varphi)(\bar{c}_{nm}\sin m\lambda-\bar{s}_{nm}\cos m\lambda)\gamma$$

$$\mathrm{U}^{s(\mathrm{Adap})}_{nm}=-(n+1)Q^*_{nm}\left(\frac{u}{\varepsilon}\right)\overline{P}_{nm}(\sin\varphi)(\bar{c}_{nm}\cos m\lambda+\bar{s}_{nm}\sin m\lambda)s$$
$$+\frac{1}{4(2n-1)(2n+3)\left(\frac{a}{\varepsilon}\right)^2}\begin{bmatrix}-(2n+3)(n+m+1)\\\times\sqrt{\frac{(n-m-1)(n-m)(n+m-1)(n+m)}{(2n-3)(2n+1)}}\\-(2m+1)(n-m)(n+m+1)\\+(2n-1)(n-m)\\\times\sqrt{\frac{(n-m+1)(n-m+2)(n+m+1)(n+m+2)}{(2n+1)(2n+5)}}\end{bmatrix}$$
$$\times Q^*_{n+2,m}\left(\frac{u}{\varepsilon}\right)\overline{P}_{n+2,m}(\sin\varphi)(\bar{c}_{nm}\cos m\lambda+\bar{s}_{nm}\sin m\lambda)s$$

after indices shift $n \to n-2, n \to n-1, m \to m-1, m \to m+1$ where $\bar{c}_{-n,m} = \bar{s}_{-n,m} = 0$ and before comparison of coefficients $\overline{C}_{nm}, \overline{S}_{nm}$ and $\bar{c}_{nm}, \bar{s}_{nm}$ with $\frac{B}{E} = \frac{b}{\varepsilon}$

$$\frac{E\operatorname{arccot}\left(\frac{b}{\varepsilon}\right)}{\varepsilon\operatorname{arccot}\left(\frac{B}{E}\right)\sum\limits_{n=0}^{\infty}\sum\limits_{m=0}^{n}\frac{Q^*_{nm}\left(\frac{U}{E}\right)}{Q^*_{nm}\left(\frac{B}{E}\right)}\overline{P}_{nm}(\sin\Phi)(\overline{C}_{nm}\cos m\Lambda + \overline{S}_{nm}\sin m\Lambda)}$$
$$= \sum_{n=0}^{\infty}\sum_{m=0}^{n}\frac{1}{Q^*_{nm}\left(\frac{b}{\varepsilon}\right)}\begin{pmatrix} \mathrm{U}^{\mathrm{P(Index)}}_{nm} + \mathrm{U}^{t_x,t_y(\mathrm{Index})}_{nm} + \mathrm{U}^{t_z(\mathrm{Index})}_{nm} \\ +\mathrm{U}^{\alpha,\beta(\mathrm{Index})}_{nm} + \mathrm{U}^{\gamma(\mathrm{Index})}_{nm} + \mathrm{U}^{s(\mathrm{Index})}_{nm} \end{pmatrix} \tag{444}$$

where

$$\mathrm{U}^{\mathrm{P(Index)}}_{nm} = \frac{1}{Q^*_{nm}\left(\frac{b}{\varepsilon}\right)}Q^*_{nm}\left(\frac{u}{\varepsilon}\right)\overline{P}_{nm}(\sin\varphi)(\bar{c}_{nm}\cos m\lambda + \bar{s}_{nm}\sin m\lambda)$$

$$\begin{aligned}
\mathrm{U}^{t_x,t_y(\mathrm{Index})}_{nm} &= \frac{1}{Q^*_{n-1,m+1}\left(\frac{b}{\varepsilon}\right)}\frac{1}{8a}\begin{bmatrix} -(2m+3)\sqrt{\frac{(n-m-3)(n-m-2)}{(2n-3)(2n-1)}} \\ -(2m+3)\sqrt{\frac{2(n+m-1)(n+m)}{(2-\delta_{0m})(2n-3)(2n-1)}} \\ +(2m+3)\sqrt{\frac{(n+m+1)(n+m+2)}{(2n-1)(2n+1)}} \\ +(4n+2m+1)\sqrt{\frac{2(n-m-1)(n-m)}{(2-\delta_{0m})(2n-1)(2n+1)}} \end{bmatrix} \\
&\times\left[(t_x\bar{c}_{n-1,m+1} + t_y\bar{s}_{n-1,m+1})\cos m\lambda + (t_x\bar{s}_{n-1,m+1} - t_y\bar{c}_{n-1,m+1})\sin m\lambda\right] \\
&- \frac{1}{Q^*_{n-1,m-1}\left(\frac{b}{\varepsilon}\right)}\frac{\sqrt{2-\delta_{1m}}}{4a} \\
&\times\begin{bmatrix} \left(n+m-1-\frac{\delta_{1m}}{2}(2m-3)\right)\sqrt{\frac{(n-m-1)(n-m)}{2(2n-3)(2n-1)}} \\ -\left(n+m-1-\frac{\delta_{1m}}{2}(2m-3)\right)\sqrt{\frac{(n-m+1)(n-m+2)}{(2-\delta_{2m})(2n-1)(2n+1)}} \\ +\left(n+m-1-\frac{\delta_{1m}}{2}(2m-3)\right)\sqrt{\frac{(n+m-3)(n+m-2)}{(2-\delta_{2m})(2n-3)(2n-1)}} \\ +\left(n-m-\frac{\delta_{1m}}{2}(4n-2m+1)\right)\sqrt{\frac{(n+m-1)(n+m)}{2(2n-1)(2n+1)}} \end{bmatrix} \\
&\times\left[(t_x\bar{c}_{n-1,m-1} - t_y\bar{s}_{n-1,m-1})\cos m\lambda + (t_x\bar{s}_{n-1,m-1} + t_y\bar{c}_{n-1,m-1})\sin m\lambda\right]
\end{aligned}$$

$$\begin{aligned}
U^{t_z(\mathrm{Index})}_{nm} &= -\frac{1}{Q^*_{n-1,m}\left(\frac{b}{\varepsilon}\right)}\frac{2n-1}{2a}\sqrt{\frac{(n-m)(n+m)}{(2n-1)(2n+1)}}Q^*_{nm}\left(\frac{u}{\varepsilon}\right)\overline{P}_{nm}(\sin\varphi) \\
&\times(\bar{c}_{n-1,m}\cos m\lambda + \bar{s}_{n-1,m}\sin m\lambda)t_z
\end{aligned}$$

$$\begin{aligned}
U^{\alpha,\beta(\mathrm{Index})}_{nm} &= \frac{1}{Q^*_{n,m+1}\left(\frac{b}{\varepsilon}\right)}\frac{1}{2}\sqrt{\frac{2(n-m)(n+m+1)}{2-\delta_{0m}}}Q^*_{nm}\left(\frac{u}{\varepsilon}\right)\overline{P}_{nm}(\sin\varphi) \\
&\times\left[(-\alpha\bar{s}_{n,m+1} + \beta\bar{c}_{n,m+1})\cos m\lambda + (\alpha\bar{c}_{n,m+1} + \beta\bar{s}_{n,m+1})\sin m\lambda\right]
\end{aligned}$$

$$
\begin{aligned}
&+\frac{1}{Q^*_{n,m-1}\left(\frac{b}{\varepsilon}\right)}\frac{1+\delta_{1m}}{2}\sqrt{\frac{(2-\delta_{1m})(n-m+1)(n+m)}{2}}Q^*_{nm}\left(\frac{u}{\varepsilon}\right)\overline{P}_{nm}(\sin\varphi)\\
&\times\left[(\alpha\bar{s}_{n,m-1}+\beta\bar{c}_{n,m-1})\cos m\lambda+(-\alpha\bar{c}_{n,m-1}+\beta\bar{s}_{n,m-1})\sin m\lambda\right]\\
&+\frac{1}{Q^*_{n-2,m+1}\left(\frac{b}{\varepsilon}\right)}\frac{n+m}{8(2n-5)(2n-1)\left(\frac{a}{\varepsilon}\right)^2}\\
&\times\begin{bmatrix}-(2n-1)\sqrt{\frac{2(n-m-3)(n+m-3)(n+m-2)(n+m-1)}{(2-\delta_{0m})(2n-7)(2n-3)}}\\-(2m+1)\sqrt{\frac{2(n-m-2)(n+m-1)}{2-\delta_{0m}}}\\+(2n-5)\sqrt{\frac{2(n-m-2)(n-m-1)(n-m)(n+m)}{(2-\delta_{0m})(2n-3)(2n+1)}}\end{bmatrix}Q^*_{nm}\left(\frac{u}{\varepsilon}\right)\overline{P}_{nm}(\sin\varphi)\\
&\times\left[(-\alpha\bar{s}_{n-2,m+1}+\beta\bar{c}_{n-2,m+1})\cos m\lambda+(\alpha\bar{c}_{n-2,m+1}+\beta\bar{s}_{n-2,m+1})\sin m\lambda\right]\\
&+\frac{1}{Q^*_{n-2,m-1}\left(\frac{b}{\varepsilon}\right)}\frac{\sqrt{2-\delta_{1m}}}{8(2n-5)(2n-1)\left(\frac{a}{\varepsilon}\right)^2}\\
&\times\begin{bmatrix}-(2n-1)(n+m-2)\sqrt{\frac{(n-m-1)(n+m-5)(n+m-4)(n+m-3)}{(2-\delta_{2m})(2n-7)(2n-3)}}\\+\delta_{1m}(2n-1)(n-m)\sqrt{\frac{(n-m-3)(n-m-2)(n-m-1)(n+m-3)}{(2-\delta_{0m})(2n-7)(2n-3)}}\\-(2m-3)(n+m-2)\sqrt{\frac{(n-m)(n+m-3)}{2-\delta_{2m}}}\\+\delta_{1m}(2m-1)(n-m)\sqrt{\frac{(n-m-1)(n+m-2)}{2-\delta_{0m}}}\\+(2n-5)(n+m-2)\sqrt{\frac{(n-m)(n-m+1)(n-m+2)(n+m-2)}{(2-\delta_{2m})(2n-3)(2n+1)}}\\+\delta_{1m}(2n-5)(n-m)\sqrt{\frac{(n-m)(n+m-2)(n+m-1)(n+m)}{(2-\delta_{0m})(2n-3)(2n+1)}}\end{bmatrix}\\
&\times Q^*_{nm}\left(\frac{u}{\varepsilon}\right)\overline{P}_{nm}(\sin\varphi)\\
&\times\left[(\alpha\bar{s}_{n-2,m-1}+\beta\bar{c}_{n-2,m-1})\cos m\lambda+(-\alpha\bar{c}_{n-2,m-1}+\beta\bar{s}_{n-2,m-1})\sin m\lambda\right]
\end{aligned}
$$

$$
\mathrm{U}^{\gamma(\mathrm{Index})}_{nm}=\frac{m}{Q^*_{nm}\left(\frac{b}{\varepsilon}\right)}Q^*_{nm}\left(\frac{u}{\varepsilon}\right)\overline{P}_{nm}(\sin\varphi)(\bar{c}_{nm}\sin m\lambda-\bar{s}_{nm}\cos m\lambda)\gamma
$$

$$
\begin{aligned}
\mathrm{U}^{s(\mathrm{Index})}_{nm}&=-\frac{n+1}{Q^*_{nm}\left(\frac{b}{\varepsilon}\right)}Q^*_{nm}\left(\frac{u}{\varepsilon}\right)\overline{P}_{nm}(\sin\varphi)(\bar{c}_{nm}\cos m\lambda+\bar{s}_{nm}\sin m\lambda)s\\
&+\frac{1}{Q^*_{n-2,m}\left(\frac{b}{\varepsilon}\right)}\frac{1}{4(2n-5)(2n-1)\left(\frac{a}{\varepsilon}\right)^2}\\
&\times\begin{bmatrix}-(2n-1)(n+m-1)\sqrt{\frac{(n-m-3)(n-m-2)(n+m-3)(n+m-2)}{(2n-7)(2n-3)}}\\-(2m+1)(n-m-2)(n+m-1)\\+(2n-5)(n-m-2)\sqrt{\frac{(n-m-1)(n-m)(n+m-1)(n+m)}{(2n-3)(2n+1)}}\end{bmatrix}Q^*_{nm}\left(\frac{u}{\varepsilon}\right)\overline{P}_{nm}(\sin\varphi)\\
&\times(\bar{c}_{n-2,m}\cos m\lambda+\bar{s}_{n-2,m}\sin m\lambda)s
\end{aligned}
$$

Equations for the datum transformation of spheroidal harmonics

$$
\begin{Bmatrix} \overline{C}_{nm} \\ \overline{S}_{nm} \end{Bmatrix} = \frac{E\operatorname{arccot}\left(\frac{b}{\varepsilon}\right)}{\varepsilon\operatorname{arccot}\left(\frac{B}{E}\right)} Q^*_{nm}\left(\frac{B}{E}\right)
\begin{bmatrix}
u^{\alpha,\beta}_{n,m-1}\left(\beta\begin{Bmatrix}\bar{c}_{n,m-1}\\ \bar{s}_{n,m-1}\end{Bmatrix} \pm \alpha\begin{Bmatrix}\bar{s}_{n,m-1}\\ \bar{c}_{n,m-1}\end{Bmatrix}\right) \\
+\frac{1-(n+1)s}{Q^*_{nm}\left(\frac{b}{\varepsilon}\right)}\begin{Bmatrix}\bar{c}_{nm}\\ \bar{s}_{nm}\end{Bmatrix} \mp \frac{m\gamma}{Q^*_{nm}\left(\frac{b}{\varepsilon}\right)}\begin{Bmatrix}\bar{s}_{nm}\\ \bar{c}_{nm}\end{Bmatrix} \\
-u^{\alpha,\beta}_{n,m+1}\left(\beta\begin{Bmatrix}\bar{c}_{n,m+1}\\ \bar{s}_{n,m+1}\end{Bmatrix} \mp \alpha\begin{Bmatrix}\bar{s}_{n,m+1}\\ \bar{c}_{n,m+1}\end{Bmatrix}\right) \\
-u^{t_x,t_y}_{n-1,m-1}\left(t_x\begin{Bmatrix}\bar{c}_{n-1,m-1}\\ \bar{s}_{n-1,m-1}\end{Bmatrix} \mp t_y\begin{Bmatrix}\bar{s}_{n-1,m-1}\\ \bar{c}_{n-1,m-1}\end{Bmatrix}\right) \\
-u^{t_z}_{n-1,m}t_z\begin{Bmatrix}\bar{c}_{n-1,m}\\ \bar{s}_{n-1,m}\end{Bmatrix} \\
+u^{t_x,t_y}_{n-1,m+1}\left(t_x\begin{Bmatrix}\bar{c}_{n-1,m+1}\\ \bar{s}_{n-1,m+1}\end{Bmatrix} \pm t_y\begin{Bmatrix}\bar{s}_{n-1,m+1}\\ \bar{c}_{n-1,m+1}\end{Bmatrix}\right) \\
+u^{\alpha,\beta}_{n-2,m-1}\left(\beta\begin{Bmatrix}\bar{c}_{n-2,m-1}\\ \bar{s}_{n-2,m-1}\end{Bmatrix} \pm \alpha\begin{Bmatrix}\bar{s}_{n-2,m-1}\\ \bar{c}_{n-2,m-1}\end{Bmatrix}\right) \\
-u^{s}_{n-2,m}s\begin{Bmatrix}\bar{c}_{n-2,m}\\ \bar{s}_{n-2,m}\end{Bmatrix} \\
-u^{\alpha,\beta}_{n-2,m+1}\left(\beta\begin{Bmatrix}\bar{c}_{n-2,m+1}\\ \bar{s}_{n-2,m+1}\end{Bmatrix} \mp \alpha\begin{Bmatrix}\bar{s}_{n-2,m+1}\\ \bar{c}_{n-2,m+1}\end{Bmatrix}\right)
\end{bmatrix}
\tag{445}
$$

where

$$
u^{\alpha,\beta}_{n,m-1} = \frac{1}{Q^*_{n,m-1}\left(\frac{b}{\varepsilon}\right)}\frac{1+\delta_{1m}}{2}\sqrt{\frac{(2-\delta_{1m})(n-m+1)(n+m)}{2}},
$$

$$
u^{\alpha,\beta}_{n,m+1} = \frac{1}{Q^*_{n,m+1}\left(\frac{b}{\varepsilon}\right)}\frac{1}{2}\sqrt{\frac{2(n-m)(n+m+1)}{2-\delta_{0m}}},
$$

$$
\begin{aligned}
u^{t_x,t_y}_{n-1,m-1} = {} & \frac{1}{Q^*_{n-1,m-1}\left(\frac{b}{\varepsilon}\right)}\frac{\sqrt{2-\delta_{1m}}}{4a} \\
& \times\begin{bmatrix}
\left(n+m-1-\frac{\delta_{1m}}{2}(2m-3)\right) \\
\times\left(\sqrt{\frac{(n-m-1)(n-m)}{2(2n-3)(2n-1)}} - \sqrt{\frac{(n-m+1)(n-m+2)}{(2-\delta_{2m})(2n-1)(2n+1)}} + \sqrt{\frac{(n+m-3)(n+m-2)}{(2-\delta_{2m})(2n-3)(2n-1)}}\right) \\
+n-m-\frac{\delta_{1m}}{2}(4n-2m+1)\sqrt{\frac{(n+m-1)(n+m)}{2(2n-1)(2n+1)}}
\end{bmatrix},
\end{aligned}
$$

$$
u^{t_z}_{n-1,m} = \frac{1}{Q^*_{n-1,m}\left(\frac{b}{\varepsilon}\right)}\frac{2n-1}{2a}\sqrt{\frac{(n-m)(n+m)}{(2n-1)(2n+1)}}
$$

$$
\begin{aligned}
u^{t_x,t_y}_{n-1,m+1} = {} & \frac{1}{Q^*_{n-1,m+1}\left(\frac{b}{\varepsilon}\right)}\frac{1}{4a} \\
& \times\begin{bmatrix}
-\frac{1}{2}(2m+3) \\
\times\left(\sqrt{\frac{(n-m-3)(n-m-2)}{(2n-3)(2n-1)}} + \sqrt{\frac{2(n+m-1)(n+m)}{(2-\delta_{0m})(2n-3)(2n-1)}} - \sqrt{\frac{(n+m+1)(n+m+2)}{(2n-1)(2n+1)}}\right) \\
+\frac{1}{2}(4n+2m+1)\sqrt{\frac{2(n-m-1)(n-m)}{(2-\delta_{0m})(2n-1)(2n+1)}}
\end{bmatrix},
\end{aligned}
$$

$$u^{\alpha,\beta}_{n-2,m-1} = \frac{1}{Q^*_{n-2,m-1}\left(\frac{b}{\varepsilon}\right)} \frac{\sqrt{2-\delta_{1m}}}{8(2n-5)(2n-1)\left(\frac{a}{\varepsilon}\right)^2}$$

$$\times \begin{bmatrix} -(2n-1)(n+m-2)\sqrt{\frac{(n-m-1)(n+m-5)(n+m-4)(n+m-3)}{(2-\delta_{2m})(2n-7)(2n-3)}} \\ +\delta_{1m}(2n-1)(n-m)\sqrt{\frac{(n-m-3)(n-m-2)(n-m-1)(n+m-3)}{(2-\delta_{0m})(2n-7)(2n-3)}} \\ -(2m-3)(n+m-2)\sqrt{\frac{(n-m)(n+m-3)}{2-\delta_{2m}}} \\ +\delta_{1m}(2m-1)(n-m)\sqrt{\frac{(n-m-1)(n+m-2)}{2-\delta_{0m}}} \\ +(2n-5)(n+m-2)\sqrt{\frac{(n-m)(n-m+1)(n-m+2)(n+m-2)}{(2-\delta_{2m})(2n-3)(2n+1)}} \\ +\delta_{1m}(2n-5)(n-m)\sqrt{\frac{(n-m)(n+m-2)(n+m-1)(n+m)}{(2-\delta_{0m})(2n-3)(2n+1)}} \end{bmatrix},$$

$$u^{s}_{n-2,m} = \frac{1}{Q^*_{n-2,m}\left(\frac{b}{\varepsilon}\right)} \frac{1}{4(2n-5)(2n-1)\left(\frac{a}{\varepsilon}\right)^2}$$

$$\begin{bmatrix} (2n-1)(n+m-1)\sqrt{\frac{(n-m-3)(n-m-2)(n+m-3)(n+m-2)}{(2n-7)(2n-3)}} \\ +(2m+1)(n-m-2)(n+m-1) \\ -(2n-5)(n-m-2)\sqrt{\frac{(n-m-1)(n-m)(n+m-1)(n+m)}{(2n-3)(2n+1)}} \end{bmatrix},$$

$$u^{\alpha,\beta}_{n-2,m+1} = \frac{1}{Q^*_{n-2,m+1}\left(\frac{b}{\varepsilon}\right)} \frac{n+m}{8(2n-5)(2n-1)\left(\frac{a}{\varepsilon}\right)^2}\sqrt{\frac{2}{2-\delta_{0m}}}$$

$$\times \begin{bmatrix} (2n-1)\sqrt{\frac{(n-m-3)(n+m-3)(n+m-2)(n+m-1)}{(2n-7)(2n-3)}} \\ +(2m+1)\sqrt{(n-m-2)(n+m-1)} \\ -(2n-5)\sqrt{\frac{(n-m-2)(n-m-1)(n-m)(n+m)}{(2n-3)(2n+1)}} \end{bmatrix}.$$

12 Examples

In this chapter, the examples for spherical and ellipsoidal datum transformation are not given to show a practical use but to verify the formula for ellipsoidal datum transformation, because the results of both transformations, spherical and ellipsoidal, should be similar. For this purpose one does not only compare the equations but also the transformed coefficients of spherical and ellipsoidal datum transformation and the difference between spherical and ellipsoidal transformed coefficients should decrease with increasing degree and order. At first, it is required to have parameters of two reference systems given in ➲ *Table 4*, a set of transformation parameters given in ➲ *Table 5* which are only examples for a datum transformation, and a set of coefficients $\bar{c}_{nm}$ and $\bar{s}_{nm}$ given in ➲ *Tables 6* and ➲ *7*.

With transformation formulas for spherical and ellipsoidal datum transformation given in ➲ Sects. 10 and ➲ 11, one has equations for computing transformed coefficients. For the zonal, sectorial, and tesseral coefficients the equations are given in Eqs. 446–448. The transformed

Table 4
Parameters of reference systems

Radius		
Of the reference sphere	$R = 6\,371\,000.790$ m	
Of the underlying reference sphere	$r = 6\,370\,283.158$ m	
semi-axis and linear eccentricity		
Of the reference ellipsoid	$A = 6\,378\,137.000$ m	$E = 521\,854.011$
Of the underlying reference ellipsoid	$a = 6\,377\,397.155$ m	$\varepsilon = 521\,013.137$

Table 5
Transformation parameters

Translation	t_x	100 m
	t_y	100 m
	t_z	100 m
Rotation	α	50″
	β	50″
	γ	50″
Scale variation	s	10^{-6}

Table 6
Coefficients $\bar{c}_{nm}$

	$m = 0$	1	2	3	4
$n = 0$	+1.000000+00				
1	+0.000000+00	+0.000000+00			
2	+5.137725 − 04	−1.869876−10	+2.439144−06		
3	+9.572542 − 07	+2.029989−06	+9.046278−07	+7.210727 − 07	
4	−2.504930−07	−5.363223 − 07	+3.574280 − 07	+9.907718 − 07	−1.885608−07

Table 7
Coefficients $\bar{s}_{nm}$

	$m = 0$	1	2	3	4
$n = 0$	+0.000000+00				
1	+0.000000+00	+0.000000+00			
2	+0.000000+00	−1.869876−10	+2.439144-06		
3	+0.000000+00	+2.029989−06	+9.046278-07	+7.210727−07	
4	+0.000000+00	−5.363223−07	+3.574280-07	+9.907718−07	-1.885608-07

coefficients until degree and order $n = m = 4$ are given in *Tables 8* and *10* for spherical datum transformation and in *Tables 9* and *11* for ellipsoidal datum transformation.

Zonal equations

$$\begin{aligned} \overline{C}_{00}^{\text{sph}} &= (1-s)\bar{c}_{00} \\ \overline{C}_{00}^{\text{ell}} &= k_{00}(1-s)\bar{c}_{00} \end{aligned} \tag{446}$$

$$\begin{aligned}
\begin{Bmatrix} \overline{C}_{10}^{\text{sph}} \\ \overline{S}_{10}^{\text{sph}} \end{Bmatrix} &= \begin{matrix} -\frac{\sqrt{3}}{3r} t_z \begin{Bmatrix} \bar{c}_{00} \\ \bar{s}_{00} \end{Bmatrix} + (1-2s) \begin{Bmatrix} \bar{c}_{10} \\ \bar{s}_{10} \end{Bmatrix} \\ -\beta \begin{Bmatrix} \bar{c}_{11} \\ \bar{s}_{11} \end{Bmatrix} \mp \alpha \begin{Bmatrix} \bar{s}_{11} \\ \bar{c}_{11} \end{Bmatrix} \end{matrix} \\
\begin{Bmatrix} \overline{C}_{10}^{\text{ell}} \\ \overline{S}_{10}^{\text{ell}} \end{Bmatrix} &= k_{10} \begin{bmatrix} -q_{00}^{*} \frac{\sqrt{3}}{6a} t_z \begin{Bmatrix} \bar{c}_{00} \\ \bar{s}_{00} \end{Bmatrix} + q_{10}^{*}(1-2s) \begin{Bmatrix} \bar{c}_{10} \\ \bar{s}_{10} \end{Bmatrix} \\ -q_{11}^{*} \left(\beta \begin{Bmatrix} \bar{c}_{11} \\ \bar{s}_{11} \end{Bmatrix} \mp \alpha \begin{Bmatrix} \bar{s}_{11} \\ \bar{c}_{11} \end{Bmatrix} \right) \end{bmatrix}
\end{aligned} \tag{447}$$

Tesseral equations:

$$\begin{aligned}
\begin{Bmatrix} \overline{C}_{21}^{\text{sph}} \\ \overline{S}_{21}^{\text{sph}} \end{Bmatrix} &= \begin{matrix} -\frac{3\sqrt{5}}{5r} \left(t_x \begin{Bmatrix} \bar{c}_{10} \\ \bar{s}_{10} \end{Bmatrix} \mp t_y \begin{Bmatrix} \bar{c}_{10} \\ \bar{s}_{10} \end{Bmatrix} - t_z \begin{Bmatrix} \bar{c}_{11} \\ \bar{s}_{11} \end{Bmatrix} \right) \\ +\sqrt{3} \left(\beta \begin{Bmatrix} \bar{c}_{20} \\ \bar{s}_{20} \end{Bmatrix} \pm \alpha \begin{Bmatrix} \bar{s}_{20} \\ \bar{c}_{20} \end{Bmatrix} \right) \\ +(1-3s) \begin{Bmatrix} \bar{c}_{21} \\ \bar{s}_{21} \end{Bmatrix} \mp \gamma \begin{Bmatrix} \bar{s}_{21} \\ \bar{c}_{21} \end{Bmatrix} \\ - \left(\beta \begin{Bmatrix} \bar{c}_{22} \\ \bar{s}_{22} \end{Bmatrix} \mp \alpha \begin{Bmatrix} \bar{s}_{22} \\ \bar{c}_{22} \end{Bmatrix} \right) \end{matrix} \\
\begin{Bmatrix} \overline{C}_{21}^{\text{ell}} \\ \overline{S}_{21}^{\text{ell}} \end{Bmatrix} &= k_{21} \begin{bmatrix} q_{00}^{*} \frac{\sqrt{15}}{30\left(\frac{a}{\varepsilon}\right)^2} \left(\beta \begin{Bmatrix} \bar{c}_{00} \\ \bar{s}_{00} \end{Bmatrix} \pm \alpha \begin{Bmatrix} \bar{s}_{00} \\ \bar{c}_{00} \end{Bmatrix} \right) \\ -q_{10}^{*} \frac{3\sqrt{5}}{16a} \left(t_x \begin{Bmatrix} \bar{c}_{10} \\ \bar{s}_{10} \end{Bmatrix} \mp t_y \begin{Bmatrix} \bar{c}_{10} \\ \bar{s}_{10} \end{Bmatrix} \right) \\ -q_{11}^{*} \frac{3\sqrt{5}}{10a} t_z \begin{Bmatrix} \bar{c}_{11} \\ \bar{s}_{11} \end{Bmatrix} \\ +q_{20}^{*} \sqrt{3} \left(\beta \begin{Bmatrix} \bar{c}_{20} \\ \bar{s}_{20} \end{Bmatrix} \pm \alpha \begin{Bmatrix} \bar{s}_{20} \\ \bar{c}_{20} \end{Bmatrix} \right) \\ +q_{21}^{*}(1-3s) \begin{Bmatrix} \bar{c}_{21} \\ \bar{s}_{21} \end{Bmatrix} \mp \gamma \begin{Bmatrix} \bar{s}_{21} \\ \bar{c}_{21} \end{Bmatrix} \\ -q_{22}^{*} \left(\beta \begin{Bmatrix} \bar{c}_{22} \\ \bar{s}_{22} \end{Bmatrix} \mp \alpha \begin{Bmatrix} \bar{s}_{22} \\ \bar{c}_{22} \end{Bmatrix} \right) \end{bmatrix}
\end{aligned} \tag{448}$$

with

$$k_{nm} = \frac{E \operatorname{arccot}\left(\frac{b}{\varepsilon}\right)}{\varepsilon \operatorname{arccot}\left(\frac{B}{E}\right) Q_{nm}^{*}\left(\frac{B}{E}\right)} \quad \text{and} \quad q_{nm}^{*} = \frac{1}{Q_{nm}^{*}\left(\frac{b}{\varepsilon}\right)} \tag{449}$$

For the transformed coefficients $\overline{C}_{nm}^{\text{sph}}$ and $\overline{C}_{nm}^{\text{ell}}$ the most significant differences appear for the zonal coefficient c_{00}, the sectorial coefficient c_{10}, and the tesseral coefficients c_{11} and c_{21}. The reason for these differences are the factors with the linear excentricity and the *Legendre function of the second kind* in the equation of ellipsoidal datum transformation. But for the tesseral coefficient c_{21} another reason is given by the additional terms for the ellipsoidal equation which are the result of the different approximations.

For the transformed coefficients $\overline{S}_{nm}^{\text{sph}}$ and $\overline{S}_{nm}^{\text{ell}}$ the most significant differences appear only for the tesseral coefficients c_{11} and c_{21}. The reason for these differences are the same as for the

Table 8
Transformed spherical coefficients $\overline{C}_{nm}^{sph}$

	m = 0	1	2	3	4
n = 0	+9.999990−01				
1	−9.063181−06	−9.063181−06			
2	+5.137709−04	+2.145951−07	+2.439815−06		
3	+9.357437−07	+2.013218−06	+9.059311−07	+7.199865−07	
4	−2.504193−07	−5.363879−07	+3.559571−07	+9.914496−07	−1.885526−07

Table 9
Transformed ellipsoidal coefficients $\overline{C}_{nm}^{ell}$

	m = 0	1	2	3	4
n = 0	+1.000115+00				
1	−4.534155-06	−6.796671−06			
2	+5.138355-04	+1.590650−07	+2.440108−06		
3	+9.4608580-07	+2.027601−06	+9.060808−07	+7.201633−07	
4	−2.504627-07	−5.361879−07	+3.560602−07	+9.916399−07	−1.885977−07

Table 10
Transformed spherical coefficients $\overline{S}_{nm}^{sph}$

	m = 0	1	2	3	4
n = 0	+0.000000+00				
1	+0.000000+00	−9.063181−06			
2	−4.233427−13	−2.147696−07	−1.398980−06		
3	−1.352858−09	+2.483001−07	−6.198601−07	+1.414423−06	
4	+7.908926−10	−4.739281−07	+6.587076−07	−2.001632−07	+3.082603−07

Table 11
Transformed ellipsoidal coefficients $\overline{S}_{nm}^{ell}$

	m = 0	1	2	3	4
n = 0	+0.000000+00				
1	+0.000000+00	−6.796671−06			
2	−4.235981-13	−1.592413−07	−1.399147−06		
3	−1.353573-09	+2.463250−07	−6.199595−07	+1.414581−06	
4	+7.416188-10	−4.741757 − 07	+6.587241−07	−2.001664−07	+3.082546−07

transformed coefficients $\overline{C}_{nm}^{sph}$ and $\overline{C}_{nm}^{ell}$. The zonal coefficients $\overline{S}_{00}^{sph}$ and $\overline{S}_{00}^{ell}$ are both equal to zero for spherical and ellipsoidal datum transformation because $\bar{s}_{00}$ is also equal to zero. This is the reason why in Eq. 446 there is no equation given explicitly for the transformation for the zonal coefficients $\overline{S}_{00}^{sph}$ and $\overline{S}_{00}^{ell}$.

13 Conclusions

As it is shown by the examples at last the datum transformation of spheroidal harmonics is quite similar to the spherical case. The differences which are mainly caused by the major and minor semi-axis in the spheroidal case leads to another approximation as in the spherical case. But this approximation should not be the only way to get a transformation formula because in some cases the differences of transformed spheroidal and spheroidal harmonics are quite big and for these cases it could be necessary to modify the equations of spheroidal datum transformation. It is once again important to say that the approximated solution for the datum transformation of spheroidal harmonics is only an approximation which could not be as exact as the datum transformation of spherical harmonics and that there should be searched a way to get a better result.

Due to length restrictions three appendices have been transfered to the related publications which provide equivalent compensation: Equivalent material to Appendix A relating to four alternative representations of ellipsoidal harmonics can be found in *N.C. Thong and E.W. Grafarend* (manuscipta geodaetica 14 [1989] 285–304). The power-series expansion of the associated Legendre polynomials of the second kind was the content of Appendix B. For this material the reader is referred to *Grafarend et al.* (2006, pp. 1–57). Finally, addition theorems, Jacobi matrices, normalized Legendre functions, and recursive formulae of Legendre functions, of the first and second kind were the subject of Appendix C. These topics are combined in the unpublished M.S. Thesis of the second author (see Klapp (2002a, b)). They take care of detailed representations of ellipsoidal datum transformations.

References

Abramowitz M, Stegun IA (1970) Handbook of mathematical functions. Dover, New York

Akhtar N (2009) A multiscale harmonic spline interpolation method for the inverse spheroidal gravimetric problem. Universität Siegen, Siegen

Ardalan AA (1996) Spheroidal coordinates and spheroidal eigenspace of the earth gravity field. Universität Stuttgart, Stuttgart

Ardalan AA (1999) High resolution regional geoid computation in the World Geodetic Datum 2000. Universität Stuttgart, Stuttgart

Ardalan AA, Grafarend EW (2000) Reference ellipsoidal gravity potential field and gravity intensity field of degree/order 360/360 (manual of using ellipsoidal harmonic coefficients "Ellipfree.dat" and "Ellipmean.dat"). http://www.uni-stuttgart.de/gi/research/paper/coefficients/coefficients.zip

Ardalan AA, Grafarend EW (2001) Ellipsoidal geoidal undulations (ellipsoidal Bruns formula): case studies. J Geodesy 75:544–552

Arfken G (1968) Mathematical methods for physicists, 2nd ed. Academic, New York/London

Balmino G et al (1991) Simulation of gravity gradients: a comparison study. Bull Géod 65:218–229

Bassett A (1888) A treatise on hydrodynamics. Deighton, Bell and Company, Cambridge; reprint edition in 1961 (Dover, New York)

Bölling K, Grafarend EW (2005) Ellipsoidal spectral properties of the Earth's gravitational potential and its first and second derivatives. J Geodesy 79:300–330

Cajori F (1946) Newton's principia. University of California Press, Berkeley, CA

Cartan EH (1922) Sur les petites oscillations d'une masse fluide. Bull Sci Math 46(317–352):356–369

Cartan EH (1928) Sur la stabilité ordinaire des ellipsoides de Jacobi. Proceedings of the International Mathematical Congress, Toronto 1924, 2, Toronto, University of Toronto Press, pp 2–17

Cayley A (1875a) A memoir on prepotentials. Phil Trans R Soc Lond 165:675–774

Cayley A (1875b) On the potential of the ellipse and the circle. Proc Lond Math Soc 6:38–55

Chandrasekhar S (1969) Ellipsoidal figures of equilibrium. Yale University Press, New Haven, CT

Chandrasekhar S, Roberts PH (1963) The ellipticity of a slowly rotating configuration. J Astrophys 138:801–808

Cruzy JY (1986) Ellipsoidal corrections to potential coefficients obtained from gravity anomaly data

on the ellipsoid. Rep. 371, Dept. of Geodetic Science and Surveying, The Ohio State University, Columbus
Darboux G (1910) Lecons sur les systemes orthogonaux et les cordonées curvilignes. Gauthier-Villars, Paris
Darwin GH (1906) On the figure and stability of a liquid satellite. Phil Trans R Soc Lond 206:161–248; Scietific Papers 3, Cambridge University Press, Cambridge, 1910, 436
Dedekind R (1860) Zusatz zu der vorstehenden Abhandlung. J Reine Angew Math 58:217–228
Doob JL (1984) Classical Potential theory and its probabalistic counterpart. Springer, New York
Dyson FD (1991) The potentials of ellipsoids of variable densities. Q J Pure Appl Math XXV:259–288
Eisenhart LP (1934) Separable systems of Stäckel. Ann Math 35:284–305
Ekman M (1996) The permanent problem of the permanent tide; what to do with it in geodetic reference systems. Mar Terres 125:9508–9513
Engels J (2006) Zur Modellierung von Auflastdeformationen und induzierter Polwanderung. Technical Reports Department of Geodesy and Geoinformatics University Stuttgart, Report 2006.1, Stuttgart
Engels J et al (1993) The geoid as an inverse problem to be regularized. In: Anger G et al (eds) Inverse problems: principles and applications in geophysics, technology and medicine. Akademie-Verlag, Berlin, pp 122–167
Ferrers NM (1877) On the potentials of ellipsoids, ellipsoidal shells, elliptic harmonic and elliptic rings of variable densities. Q J Pure Appl Math 14:1–22
Finn G (2001) Globale und regionale Darstellung von Lotabweichungen bezüglich des Internationalen Referenzellipsoids. Universität Stuttgart, Stuttgart
Flügge S (1979) Mathematische Methoden der Physik. Springer, Berlin
Freeden W et al (1998) Constructive approximation of the sphere. Clarendon, Oxford
Freeden W, Michel V (2004) Multiscale potential theory. Birkhäuser, Boston–Basel
Friedrich D (1998) Krummlinige Datumstransformation. Universität Stuttgart, Stuttgart
Gauss CF (1867) Werke 5, Theoria attractionis corporum sphraedicorum ellipticorum homogeneorum. Königliche Gesellschaft der Wissenschaften, Göttingen
Gleason DM (1988) Comparing corrections to the transformation between the geopotential's spherical and ellipsoidal spectrum. Manuscr Geod 13:114–129
Gleason DM (1989) Some notes on the evaluation of ellipsoidal and spheroidal harmonic expansions. Manuscr Geod 14:114–116
Gradshteyn IS, Ryzhik IM (1980) Tables of integrals, series and products. Corrected and enlarged edition, transl. by A. Jeffrey, Academic Press, New York
Grafarend EW (1988) The geometry of the earth's surface and the corresponding function space of the terrestrial gravitational field. Festschrift R. Sigl, Deutsche Geodätische Kommission, Bayerische Akademie der Wissenschaften, Report B 287, pp 76–94, München
Grafarend EW, Ardalan AA (1999) World Geodetic Datum 2000. J Geodesy 73:611–623
Grafarend EW, Awange JL (2000) Determination of vertical deflections by GPS/LPS measurements. Zeitschrift für Vermessungswesen, 125:279–288
Grafarend EW, Engels J (1998) Erdmessung und physikalische Geodäsie, Ergänzungen zum Thema Legendrefunktionen. Skript zur Vorlesung WS 1998/99. Universität Stuttgart, Stuttgart
Grafarend EW, Heidenreich A (1995) The generalized Mollweide projection of the biaxial ellipsoid. Bull Géod 69:164–172
Grafarend E, Thong NC (1989) A spheroidal harmonic model of the terrestrial gravitational field. Manuscr Geod 14:285–304
Grafarend EW, Krumm F, Okeke F (1995) Curvilinear geodetic datum transformations. Zeitschrift für Vermessungswesen 7:334–350
Grafarend EW, Engels J, Varga P (1997) The spacetime gravitational field of a deformable body. J Geodesy 72:11–30
Grafarend EW, Finn G, Ardalan AA (2006) Ellipsoidal vertical deflections and ellipsoidal gravity disturbance: case studies. Studia Geophysica et Geodaetica 50:1–57
Green G (1828) An essay on the determination of the exterior and interior attractions of ellipsoids of variable densities. In: Ferrers NM (ed) Mathematical papers of George Green. Chelsea, New York
Groten E (1979) Geodesy and the Earth's gravity field. Vol. I: Principles and Conventional Methods. Vol. II: Geodynamics and advanced methods. Dümmler Verlag, Bonn
Groten E (2000) Parameters of common relevance of astronomy, geodesy and geodynamics. The geodesist's handbook. J Geodesy 74:134–140
Hackbusch W (1995) Integral equations. Theory and numerical treatment. Birkhäuser Verlag, Basel
Hake G, Grünreich D (1994) Kartographie. Walter de Gruyter, Berlin

Heck B (1991) On the linearized boundary value problem of physical geodesy. Report 407, Department of Geodetic Science and Surveying, The Ohio State University, Columbus

Heiskanen WH, Moritz H (1967) Physical geodesy. W.H. Freeman, San Francisco, CA

Heiskanen WA, Moritz H (1981) Physical geodesy. Corrected reprint of original edition from W.H. Freeman, San Francisco, CA, 1967), order from: Institute of Physical Geodesy, TU Graz, Austria

Helmert FR (1884) Die mathematischen und physikalischen Theorien der höheren Geodäsie, Vol. 2. Leipzig, B.G. Teubner (reprinted in 1962 by Minerva GmbH, Frankfurt (Main))

Hicks WM (1882) Recent progress in hydrodynamics. Reports to the British Association, pp 57–61

Hobson EW (1896) On some general formulae for the potentials of ellipsoids, shells and discs. Proc Lond Math Soc 27:519–416

Hobson EW (1965) The theory of spherical and ellipsoidal harmonics. Second Reprint of the edition 1931 (Cambridge University Press), Chelsea, New York

Holota P (1995) Classical methods for non-spherical boundary problems in physical geodesy. Symposium 114: Geodetic Theory today. The 3rd Hotine-Massuri Symposium on Mathematical Geodesy. F. Sansó (ed) Springer, Berlin/Heidelberg

Honerkamp J, Römer H (1986) Grundlagen der klassischen theoretischen Physik. Springer, Berlin/Heidelberg/New York

Hotine M (1967) Downward continuation of the gravitational potential. General Assembly of the International Assembly of Geodesy, Luceone

Jacobi CGJ (1834) Über die Figur des Gleichgewichts. Poggendorf Annalen der Physik und Chemie 33:229–238; reprinted in Gesammelte Werke 2 (Berlin, G. Reimer, 1882), 17–72

Jeans JH (1917) The motion of tidally-distorted masses, with special reference to theories of cosmogony. Mem Roy Astron Soc Lond 62:1–48

Jeans JH (1919) Problems of cosmogony and stellar dynamics. Cambridge University Press, Cambridge, chaps. 7 and 8

Jekeli C (1988) The exact transformation between ellipsoidal and spherical harmonic expansions. Manuscr Geod 13:106–113

Jekeli C (1999) An analysis of vertical deflections derived from high-degree spherical harmonic models. J Geodesy 73:10–22

Kahle AB (1967) Harmonic analysis for a spheroidal earth. RAND Corporation Document P-3684, presented at IUGG Assembly, St. Gallen

Kahle AB, Kern JW, Vestine EH (1964) Spherical harmonic analyses for the spheroidal earth. J Geomagn Geoelectr 16:229–237

Kassir MK, Sih GC (1966) Three-dimensional stress distribution around elliptical crack under arbitrary loadings. ASME J Appl Mech 33:601–611

Kassir MK, Sih GC (1975) Three-dimensional crack problems. Mechanics of fracture, vol 2. Noordhoff International Publishing, Leyden

Kellogg OD (1929) Foundations of potential theory. Springer, Berlin/Heidelberg/New York

Klapp M (2002a) Synthese der Datumtransformation von Kugel- und Sphäroidalfunktionen zur Darstellung des terrestrischen Schwerefeldes – Beispielrechnungen zu den Transformationsgleichungen. Universität Stuttgart, Stuttgart

Klapp M (2002b) Analyse der Datumtransformation von Kugel- und Sphäroidalfunktionen zur Darstellung des terrestrischen Schwerefeldes – Herleitung der Transformationsgleichungen. Universität Stuttgart, Stuttgart

Kleusberg A (1980) The similarity transformation of the gravitational potential close to the identity. Manuscr Geod 5:241–256

Kneschke A (1965) Differentialgleichungen und Randwertprobleme. Teubner Verlag, Leipzig

Lamé G (1859) Lecons sur les cordonnées curvilignes et leurs diverses applications. Mallet-Bachelier, Paris

Lamp SH (1932) Hydrodynamics. Cambridge University Press, Cambridge, pp 722–723

Lebovitz NR (1998) The mathematical developments of the classical ellipsoids. Int J Eng Sci 36:1407–1420

Lejeune Dirichlet G (1860) Untersuchungen über ein Problem der Hydrodynamik. J Reine Angew Math 58:181–216

Lejeune Dirichlet G (1897) Gedächtnisrede auf Carl Gustav Jacob Jacobi gehalten in der Akademie der Wissenschaften am 1. Juli 1852. Gesammelte Werke, 2 (Berlin, G. Reimer), 243

Lense J (1950) Kugelfunktionen. Akademische Verlagsgesellschaft Geest-Portig, Leipzig

Lyttleton RA (1953) The stability of rotating liquid masses. Cambridge University Press, Cambridge, chap. 9

Maclaurin C (1742) A treatise on fluxions. Edinburgh

MacMillan WD (1958) The theory of the potential. Dover, New York

Mangulis V (1965) Handbook of series for scientists and engineers. Academic, New York

Martinec Z (1996) Stability investigations of a discrete downward continuation problem for geoid determination in the Canadian Rocky Mountains. J Geodesy 70:805–828

Martinec Z, Grafarend EW (1997) Solution to the Stokes boundary value problem on an ellipsoid of revolution. Studia Geophysica et Geodaetica 41:103–129

Martinec Z, Vaniček P (1996) Formulation of the boundary-value problem for geoid determination with a higher-order reference field. Geophys J Int 126:219–228

Mikolaiski HW (1989) Polbewegung eines deformierbaren Erdkörpers. PhD thesis, Deutsche Geodätische Kommission, Bayerische Akademie der Wissenschaften, Reihe C, Heft 354, München

Milne EA (1952) Sir James Jeans, a Biography. Cambridge University Press, Cambridge, chap. 9

Molodenski MS (1958) Grundbegriffe der geodÃtischen Gravimetrie. VEB Verlag Technik, Berlin

Moon P, Spencer DE (1953) Recent investigations of the Separation of Laplace's equation. Ann Math Soc Proc 4:302–307

Moon P, Spencer DE (1961) Field theory for engineers. D. van Nostrand, Princeton, NJ

Moritz H (1968a) Density distributions for the equipotential ellipsoid. Dept. of Geodetic Science and Surveying, The Ohio State University, Columbus

Moritz H (1968b) Mass distributions for the equipotential ellipsoid. Bolletino di Geofisica Teorica ed Applicata 10:59–65

Moritz H (1973) Computation of ellipsoidal mass distributions. Dept. of Geodetic Science and Surveying, The Ohio State University, Columbus

Moritz H (1980) Geodetic Reference System 1980. Bull Géod 54:395–405

Moritz H (1984) Geodetic Reference System 1980. The geodesist's handbook. Bull Géod 58:388–398

Moritz H (1990) The figure of the earth. Wichmann Verlag, Karlsruhe

Moritz H, Mueller I (1987) Earth rotation. Theory and Observation. Ungar, New York

Morse PM, Feshbach H (1953) Methods of theoretical physics, part II. McGraw Hill, New York

Natanson JP (1967) Theory of functions of a real variable. Frederick, New York

Niven WD (1891) On ellipsoidal harmonics. Phil Trans R Soc Lond A 182:231–278

Otero J (1995) A uniqueness theorem for a Robin boundary value problem of physical geodesy. Q J Appl Math

Panasyuk VV (1971) Limiting equilibrium of brittle solids with fractures. Management Information Sevices, Detroit, MI

Pflaumann E, Unger H (1974) Funktionalanalysis I. Zürich

Pick M, Picha J, Vyskočil V (1973) Theory of the Earth's gravity field. Elsevier, Amsterdam

Pizzetti P (1894) Geodesia – Sulla espressione della gravita alla superficie del geoide, supposto ellissoidico. Atti Reale Accademia dei Lincei 3: 166–172

Poincaré H (1885) Sur l'équilibre d'une masse fluide animée d'un mouvement de rotation. Acta Math 7:259–380

Polya G (1965) Mathematical discovery. On understanding, learning and teaching problem solving. Wiley, New York

Press WH et al (1989) Numerical recipes. The art of scientific computing. Cambridge University Press, Cambridge

Rapp RH, Cruzy JY (1986) Spherical harmonic expansions of the Earth's Gravitational Potential to Degree 360 Using 30′ Mean Anomalies. Report No. 376, Department of Geodetic Science and Surveying, The Ohio State University, Columbus

Riemann B (1860) Untersuchungen über die Bewegung eines flüssigen gleichartigen Ellipsoides. Abhandlung der Königlichen Gesellschaft der Wissenschaften, 9, 3–36; Gesammelte Mathematische Werke (Leipzig, B.G. Teubner, 1892), 182

Roche MEd (1847) Mémoire sur la figure d'une masse fluide (soumise á l'attraction d'un point éloigné. Acad des Sci de Montpellier 1(243–263): 333–348

Routh EJ (1902) A treatise on analytical statics, vol 2. Cambridge University Press, Cambridge

Rummel R et al (1988) Comparisons of global topographic isostatic models to the Earth's observed gravity field. Report 388, Department of Geodetic Science and Surveying, The Ohio State University, Columbus

SansÃ F, Sona G (2001) ELGRAM, an ellipsoidal gravity model manipulator. Bolletino di Geodesia e Scienze Affini 60:215–226

Sauer R, Szabo J (1967) Mathematische Hilfsmittel des Ingenieurs. Springer, Berlin

Schäfke FW (1967) Spezielle Funktionen. In: Sauer R, Szabó I (eds) Mathematische Hilfsmittel des Ingenieurs, Teil 1. Springer, Heidelberg/Berlin/New York, pp 85–232

Shah RC, Kobayashi AS (1971) Stress-intensity factor for an elliptical crack under arbitrary normal loading. Eng Fract Mech 3:71–96

Shahgholian H (1991) On the Newtonian potential of a heterogeneous ellipsoid. SIAM J Math Anal 22:1246–1255

Skrzipek MR (1998) Polynomial evaluation and associated polynomials. Numer Math 79:601–613

Smith JR (1986) From Plane to Spheroid. Landmark Enterprises. Pancho Cordova, California

Sneddon IN (1966) Mixed boundary value problems in potential theory. Wiley, New York

Somigliana C (1929) Teoria generale del campo gravitazionale dell' ellisoide di rotazione. Mem Soc Astron Ital IV

Sona G (1996) Numerical problems in the computation of ellipsoidal harmonics. J Geodesy 70:117–126

Stäckel P (1897) Über die Integration der Hamiltonschen Differentialgleichung mittels Separation der Variablen. Math Ann 49:145–147

Stokes GG (1849) On the variation of gravity on the surface of the Earth. Trans Cambridge Phil Soc 8:672–695

Thomson W, Tait PG (1883) Treatise on natural philosophie. Cambridge University Press, Cambridge, pt. 2, pp 324–335

Thong NC (1993) Untersuchungen zur Lösung der fixen gravimetrischen Randwertprobleme mittels sphäroidaler und Greenscher Funktionen. Deutsche Geodätische Kommission, Bayerische Akademie der Wissenschaften, Reihe C 399, München 1993 (in German)

Thong NC, Grafarend EW (1989) A spheroidal harmonic model of the terrestrial gravitational field. Manuscr Geod 14:285–304

Todhunter I (1966) History of the mathematical theories of attraction and the figure of the earth from the time of Newton to that of Laplace. Dover, New York

van Asche W (1991) Orthogonal polynomials, associated polynomials and functions of the second kind. J Comput Appl Math 37:237–249

Vaniček P et al (1996) Downward continuation of Helmert's gravity. J Geodesy 71:21–34

Varshalovich DA, Moskalev AN, Khersonskii VK (1989) Quantum theory of angular momentum. World Scientific, Singapore

Vijaykumar K, Atluri SN (1981) An embedded elliptical crack, in an infinite solid, subject to arbitrary crack face tractions. ASME J Appl Mech 48:88–96

Webster AG (1925) The dynamics of particles and of rigid, elastic and fluid bodies. Teubner, Leipzig

Whittaker ET, Watson GN (1935) A course of modern analysis, vol 2. Cambridge University Press, Cambridge

Yeremeev VF, Yurkina MI (1974) Fundamental equations of Molodenskii's theory for the gravitational references field. Studia Geophysica et Geodaetica 18:8–18

Yu JH, Cao HS (1996) Elliptical harmonic series and the original Stokes problem with the boundary of the reference ellipsoid. J Geodesy 70: 431–439

8 Time-Variable Gravity Field and Global Deformation of the Earth

Jürgen Kusche
Astronomical, Physical and Mathematical Geodesy Group, Bonn University, Bonn, Germany

W. Freeden, M.Z. Nashed, T. Sonar (Eds.), *Handbook of Geomathematics*, DOI 10.1007/978-3-642-01546-5_8,

The analysis of the Earth's time-variable gravity field, and its changing patterns of deformation plays an important role in Earth system research. These two observables provide, for the first time, a direct measurement of the amount of mass that is redistributed at or near the surface of the Earth by oceanic and atmospheric circulation and through the hydrological cycle. In this chapter, we will first reconsider the relations between gravity and mass change. We will in particular discuss the role of the hypothetical surface mass change that is commonly used to facilitate the inversion of gravity change to density. Then, after a brief discussion of the elastic properties of the Earth, the relation between surface mass change and the three-dimensional deformation field is considered. Both types of observables are then discussed in the framework of inversion. None of our findings is entirely new, we merely aim at a systematic compilation and discuss some frequently made assumptions. Finally, some directions for future research are pointed out.

1 Introduction

Mass transports inside the Earth, at or above its surface generate changes in the external gravity field and in the geometrical shape of the Earth. Depending on their magnitude, spatial and temporal scale, they become visible in the observations of modern space-geodetic and terrestrial techniques.

Examples for mass transport processes that are sufficiently large to become observed include changes in continental water storage in greater river basins and catchment areas, large-scale snow coverage, present-day ice mass changes occuring at Greenland and Antarctica and at the continental glacier systems, atmospheric pressure changes, sea level change and the redistribution of ocean circulation systems, land–ocean exchange of water, and glacial-isostatic adjustment (cf. ➋ *Fig. 1*). As many of these processes are directly linked to climate, an improved quantification and understanding of their present-day trends and interannual variations from geodetic data will help to separate them from the long-term evolution, typically inferred from proxy data. The order of magnitude for atmospheric, hydrological, and oceanic loads, in terms of the associated change of the geoid, an equipotential surface of the Earth's gravity field, is about 1 cm or $1 \cdot 10^{-9}$ in relative terms. For example, in the Amazon region of South America, this geoid change is largely caused by an annual basin-wide oscillation in surface and ground water that amounts to several dm.

The variety of geodetic techniques that are sufficiently sensible include intersatellite tracking as currently conducted with the Gravity Recovery and Climate Experiment (GRACE) mission, satellite laser ranging, superconducting relative and free-fall absolute gravimetry, and the monitoring of deformations with the Global Positioning System (GPS), and with the Very-Long Baseline Interferometry (VLBI) network of radiotelescopes, Interferometric Synthetic Aperture Radar (InSAR), and of the ocean surface with satellite altimetry and tide gauges.

Here, we will limit ourselves to a discussion of the fundamental relations that concern *present-day* mass redistributions and their observability in time-variable gravity (TVG) and time-variable deformation of the Earth. This is not intended to form the basis for real-data inversion schemes. Rather, we would like to point out essential assumptions and fundamental limitations in some commonly applied algorithms. To this end, it will be sufficient to consider the response of the solid Earth to the mass loading as purely elastic. In fact, this is a first assumption that will not be allowed anymore in the discussion of sea level, when geological time-scales

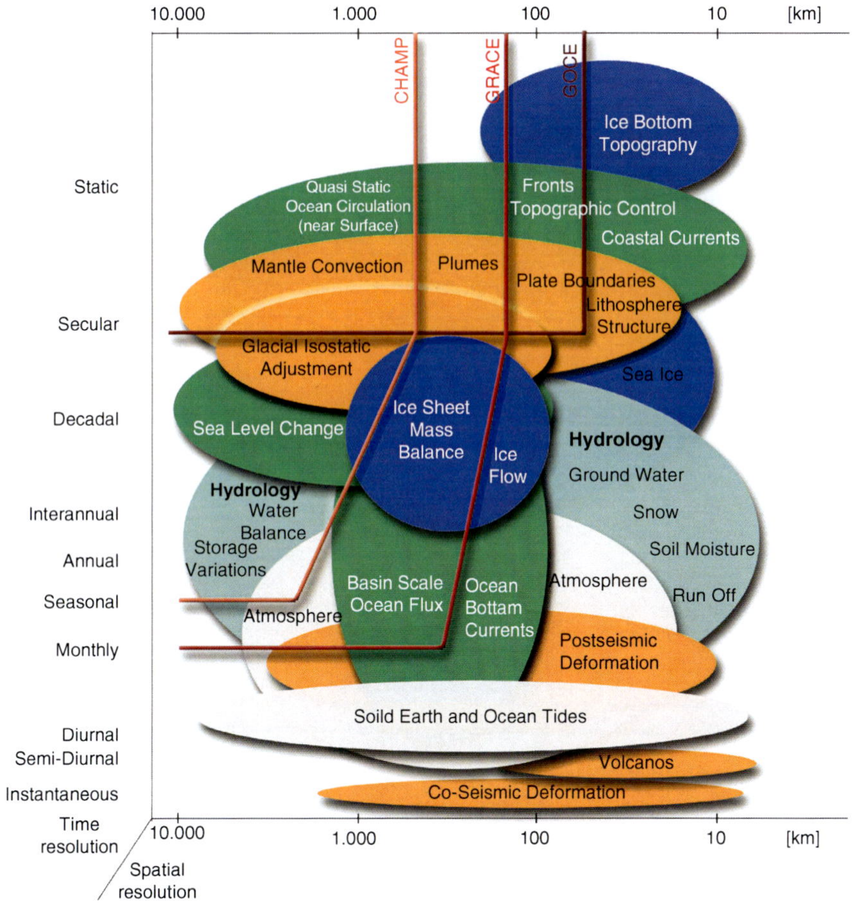

Fig. 1

Spatial and temporal scales of mass transport processes, and the resolution limits of satellite gravity missions CHAMP, GRACE, and GOCE (from Ilk et al., 2005)

are involved at which the Earth's response is driven by viscous or visco-elastic behavior. Furthermore, we will implicitly consider only that part of mass redistribution or mass transport that is actually associated with local change of density. Mass transport that does not change the local density (stationary currents in the hydrological cycle; i.e., the "motion term" in Earth's rotation analysis) is in the Nullspace of the Newton and deformation operators, it cannot be inferred from observations of time-variable gravity or from displacement data.

2 Mass and Mass Redistribution

In what follows, we will adopt an Eulerian representation of the redistribution of mass, where one considers mass density as a 4D field rather than following the path of individual particles

(Lagrangian representation). The mass density inside the Earth, including the oceans and the atmosphere, is then described by

$$\rho = \rho(\mathbf{x}', t) \,. \tag{1}$$

Mass density is the source of the external gravitational potential, described by the Newton integral

$$V(\mathbf{x}, t) = G \int_v \frac{\rho(\mathbf{x}', t)}{|\mathbf{x} - \mathbf{x}'|} \mathrm{d}v, \tag{2}$$

where v is the volume of the Earth including its fluid and gaseous envelope. The inverse distance (Poisson) kernel can be developed into

$$\frac{1}{|\mathbf{x} - \mathbf{x}'|} = \frac{1}{r'} \sum_{n=0}^{\infty} \sum_{m=-n}^{n} \left(\frac{r'}{r}\right)^{n+1} \frac{1}{(2n+1)} Y_{nm}(\mathbf{e}) Y_{nm}(\mathbf{e}'), \tag{3}$$

where $r = |\mathbf{x}|$, $r' = |\mathbf{x}'|$, and

$$\mathbf{e} = \frac{\mathbf{x}}{r} \quad \mathbf{e}' = \frac{\mathbf{x}'}{r'} \,.$$

Y_{nm} are the 4π-normalized surface spherical harmonics. Using spherical coordinates λ (spherical longitude) and θ (spherical co-latitude),

$$Y_{nm} = P_{n|m|}(\cos\theta) \begin{cases} \cos m\lambda, & m \geq 0, \\ \sin |m|\lambda, & m < 0. \end{cases}$$

Here, the $P_{nm} = \Pi_{nm} P_m^n$ are the 4π-normalized associated Legendre functions. $\Pi_{nm} = \left((2 - \delta_{0m})(2n+1)\frac{(n-m)!}{(n+m)!}\right)^{1/2}$ denotes a normalization factor that depends only on harmonic degree n and order m. The associated Legendre functions relate to the Legendre polynomials by $P_m^n(u) = (1-u^2)^{m/2} \frac{d^m P_n(u)}{du^m}$, and the Legendre polynomials may be expressed by the Rodrigues formula, $P_n(u) = \frac{1}{2^n n!} \frac{d^n (u^2-1)^n}{du^n}$. On plugging Eq. (3) into Eq. (2), the exterior gravitational potential of the Earth takes on the representation

$$V(\mathbf{x}, t) = G \sum_{n=0}^{\infty} \sum_{m=-n}^{n} \frac{1}{(2n+1) r^{n+1}} \left(\int_v \rho(\mathbf{x}', t) r'^n Y_{nm}(\mathbf{e}') \mathrm{d}v \right) Y_{nm}(\mathbf{e}), \tag{4}$$

which converges outside of a sphere that tightly encloses the Earth. On the other hand, by introducing a reference scale a, the exterior potential V can be written as a general solution of the Laplace equation,

$$V(\mathbf{x}, t) = \frac{GM}{a} \sum_{n=0}^{\infty} \sum_{m=-n}^{n} \left(\frac{a}{r}\right)^{n+1} v_{nm}(t) Y_{nm}(\mathbf{e}) \tag{5}$$

with 4π-normalized spherical harmonic coefficients

$$v_{nm} = \begin{cases} c_{nm}, & m \geq 0, \\ s_{n|m|}, & m < 0. \end{cases}$$

A direct comparison of Eq. (4) and (5) provides the source representation of the spherical harmonic coefficients (see also ➲ Chap. 32)

$$v_{nm}(t) = \frac{1}{M} \frac{1}{(2n+1)} \frac{1}{a^n} \int_v r'^n Y_{nm}(\mathbf{e}') \rho(\mathbf{x}', t) \, \mathrm{d}v. \tag{6}$$

There are $2n + 1$ coefficients for each degree n, and each coefficient follows by projecting the density on a single 3D harmonic function (solid spherical harmonic) $r^n Y_{nm}$.

An equivalent approach is to expand the inverse distance into a Taylor series,

$$\frac{1}{|\mathbf{x}-\mathbf{x}'|} = \sum_{n=0}^{\infty}\left(\sum_{i_1=1}^{3}\sum_{i_2=1}^{3}\cdots\sum_{i_n=1}^{3}\xi^{(n)}_{i_1 i_2 \dots i_n} x'_{i_1}x'_{i_2}\dots x'_{i_n}\right) \tag{7}$$

with components x'_1, x'_2, x'_3 of $\mathbf{x}'$, and

$$\xi^{(n)}_{i_1 i_2 \dots i_n} = \left(\frac{1}{n!}\right)\left[\frac{\partial^n \frac{1}{|\mathbf{x}-\mathbf{x}'|}}{\partial x'_{i_1}\partial x'_{i_2}\dots\partial x'_{i_n}}\right]_{\mathbf{x}=\mathbf{x}'} . \tag{8}$$

This leads to a representation of the exterior potential through mass moments $M^{(n)}_{i_1 i_2 \dots i_n}$ of degree n,

$$V(\mathbf{x},t) = G\sum_{n=0}^{\infty}\left(\sum_{i_1=1}^{3}\sum_{i_2=1}^{3}\cdots\sum_{i_n=1}^{3}\xi^{(n)}_{i_1 i_2 \dots i_n} M^{(n)}_{i_1 i_2 \dots i_n}\right) \tag{9}$$

with

$$M^{(n)}_{i_1 i_2 \dots i_n} = \int_v x'_{i_1}x'_{i_2}\dots x'_{i_n}\rho(\mathbf{x}',t)\mathrm{d}v' . \tag{10}$$

Note that the integrals in Eqs. (6) and (10) can be interpreted as inner products $(\cdot,\rho)$ in a Hilbert space with integrable density defined on v. However, whereas the homogeneous cartesian polynomials in Eq. (10) do provide a complete basis for the approximation of the density, the solid spherical harmonics in Eq. (6) fail to do so (in fact, they allow to represent the "harmonic density" part ρ^+ for which $\Delta\rho^+ = 0$). Due to symmetry of Eq. (10), there are $\frac{(n+1)(n+2)}{2}$ independent mass moments of degree n. Of those only $2n+1$ become visible in the gravitational potential, which means that the remaining $\frac{(n+1)(n+2)}{2} - (2n+1) = \frac{(n-1)n}{2}$ span the nullspace of the Newton integral (Chao, 2005).

Some of the low-degree coefficients and mass moments deserve a special discussion, e.g., obviously

$$v_{00} = \frac{1}{M}\int_v \rho(\mathbf{x}',t)\,\mathrm{d}v = \frac{M^{(0)}}{M}$$

corresponds to the total mass of the Earth, scaled by a reference value M. Since this is largely constant in time, inversion of geodetic data usually assume $\frac{\mathrm{d}}{\mathrm{d}t}v_{00} = 0$ (or $\delta v_{00} = 0$, see below). Moreover, the degree-1 coefficients directly correspond to the coordinates of the center of mass $\mathbf{x}^0(t)$,

$$v_{10}(t) = \frac{\sqrt{3}}{3M}\int_v \rho\, x'_3 \mathrm{d}v = \frac{M^{(1)}_3}{M} = \frac{1}{\sqrt{3}}x^0_3(t)$$

and

$$v_{11}(t) = \frac{1}{\sqrt{3}}x^0_1(t) \quad v_{1-1}(t) = \frac{1}{\sqrt{3}}x^0_2(t)$$

(in the reference system where the spherical coordinates λ, θ are referring to). Space-geodetic evidence suggests that, if the considered reference frame is fixed to the crust of the Earth, the temporal variation of the center of mass is no more than a few mm, with the x^0_3-component being largest. This is the case for realizations of the International Terrestrial Reference System (ITRS). Yet, rather large mass redistributions are required to shift the geocenter for a few mm, and the study of these effects is subject to current research.

Similarly, the degree-2 spherical harmonic coefficients v_{2m} can be related to the tensor of inertia of the Earth. However, of the $\frac{(2+1)(2+2)}{2} = 6$ mass moments $M^{(2)}_{i_1 i_2}$ only $2\cdot 2+1 = 5$

become visible in the five spherical harmonic coefficients of the gravity field, with the nullspace of the Newton integral being of dimension $6 - 5 = 1$ for degree 2. The annual variation of the "flattening" coefficient v_{20} is of the order $1 \cdot 10^{-10}$, the linear rate $\frac{d}{dt} v_{20}$ (predominantly due to the continuing rebound of the Earth in response to de-glaciation after the last ice age) at the $1 \cdot 10^{-11} \frac{1}{y}$ level. A special case is the variation of degree 2, order 1 coefficients (v_{21}, v_{2-1}), as those correspond to the position of the figure axis of the Earth (in the reference system where the λ, θ are referring to). Since the mean figure axis is close to the mean rotation axis, and since the latter can be determined with high precision from the measurement of Earth's rotation, additional observational constraints for their time-variation (of the order of $1 \cdot 10^{-11} \frac{1}{y}$) exist.

Nowadays, the v_{nm} can be observed from precise satellite tracking with global navigation systems, intersatellite ranging, and satellite accelerometry with temporal resolution of one month or below and spatial resolution of up to $n_{\max} = 120$. However, as mentioned earlier, one cannot uniquely invert gravity change into density change. The question had been raised which physically plausible assumptions nevertheless allow to somehow locate the sources of these gravity changes. A common way to regularize this "gravitational-change inverse problem" (GCIP) is to restrict the solution space to density changes within an infinitely thin spherical shell of radius a (Wahr et al., 1998). This corresponds to the determination of surface mass from an external potential. The spherical harmonic coefficients v_{nm} of the potential caused by a surface density μ on a sphere are

$$v_{nm}(t) = \frac{a^2}{M} \frac{1}{(2n+1)} \int_s \mu(\mathbf{e}', t) Y_{nm}(\mathbf{e}') \, ds . \tag{11}$$

Since only changes with respect to a *reference status*, which can be realized through the measurements (e.g., a long-term average), are observable, one defines

$$\delta\mu(\mathbf{e}', t) = \mu(\mathbf{e}', t) - \bar{\mu}(\mathbf{e}')$$

and the coefficients of potential change follow from

$$\delta v_{nm}(t) = v_{nm}(t) - \bar{v}_{nm} = \frac{a^2}{M} \frac{1}{(2n+1)} \int_s \delta\mu(\mathbf{e}', t) Y_{nm}(\mathbf{e}') \, ds, \tag{12}$$

where $\bar{v}_{nm} = \int_{t_1}^{t_2} v_{nm}(t') dt'$. Or, involving the spherical harmonic expansion of $\delta\mu$,

$$\delta v_{nm} = \frac{4\pi a^2}{M} \frac{1}{(2n+1)} \delta\mu_{nm} = \frac{3}{\rho_e} \frac{1}{(2n+1)} \delta\mu_{nm}, \tag{13}$$

where we made use of the average density of the Earth, $\rho_e = \frac{M}{v_a}$, $v_a = \frac{3}{4}\pi a^2$ being the volume of a sphere of radius a. Obviously, this relation is coefficient-by-coefficient invertible.

Corresponding to the $2n + 1$ (unnormalized) coefficients δv_{nm} at degree n, there are just $2n + 1$ surface mass moments, with components e_1', e_2', e_3' of $\mathbf{e}'$,

$$\delta A^{(n)}_{i_1 i_1 \ldots i_n} = a^n \int_s e'_{i_1} e'_{i_2} \ldots e'_{i_n} \delta\mu(\mathbf{e}', t) ds . \tag{14}$$

The integrals in Eqs. (12) and (14) can be viewed as inner products $(\cdot, \delta\mu)$ in a Hilbert space of integrable (density) functions defined on the sphere.

We will again look at low-degree terms. Obviously,

$$\delta v_{00} = \frac{4\pi a^2}{M} \delta\mu_{00} = \frac{3}{\rho_e} \delta\mu_{00} = \frac{a^2}{4\pi} \delta A^{(0)}$$

corresponds to the change in total surface mass. This should be zero as long as $\delta\mu$ considers all subsystems ($\delta\mu = \sum_s \delta\mu^{(s)}$), whereas a shift of total mass from, e.g., the ocean to the atmosphere may happen very well. For the center-of-mass shift referred to the Earth (strictly speaking, to the center of the sphere where μ is located on) with mass M,

$$\delta v_{10}(t) = \frac{4\pi a^2}{3M}\delta\mu_{10} = \frac{1}{3\rho_e}\delta\mu_{10} = \frac{1}{\sqrt{3}}x_3^0 = \frac{1}{\sqrt{3}\rho_e}\delta A_3^{(1)}$$

and similar for δv_{11} and δv_{1-1}.

Restriction of the density to a spherical shell serves for eliminating the null-space of the problem, thus allowing a unique solution for density determination from gravity. In fact, the surface layer could be located on the surface of an ellipsoid of revolution or any other sufficiently smooth surface as well. This could be implemented in Eqs. (11) and (12) by including an upward continuation integral term, however, it would not allow a simple scaling of the coefficients as with Eq. (13) anymore. On the other hand, in the light of comparing inferred surface densities to modeled densities (e.g., from ocean or hydrology models), one could determine point-wise densities on a nonspherical surface from those on a spherical shell by applying an $\left(\frac{a}{r}\right)^n$ term (Chao 2005).

The 3D density ρ and the 2D density μ are of course related to each other. This can be best seen by writing the source representation of the spherical harmonic coefficients, Eq. (6), in the following way

$$v_{nm}(t) = \frac{a^2}{M}\frac{1}{(2n+1)}\int_s \left\{\int_0^{r_{\max}} \left(\frac{r'}{a}\right)^{n+2} \rho(\mathbf{x}', t)\mathrm{d}r\right\} Y_{nm}(\mathbf{e}')\,\mathrm{d}s \tag{15}$$

and comparing the term in $\{\ldots\}$ brackets to Eq. (11). Surface mass change $\delta\mu$ can thus be considered as a column-integrating "mapping" of $\delta\rho$, say $L_2(s) \times L_2([0, r_{\max}] \to L_2(s)$ if we assume square-integrability.

A coarse approximation of the integral in Eq. (15) is used if surface density change $\delta\mu$ is transformed to "equivalent water height" change $\delta h_w = \frac{\delta\mu}{\rho_w}$ with a constant reference value ρ_w for the density of water, or if (real) water height change (e.g., from ocean model output) is transformed to surface density change $\delta\mu = \rho_w \delta h_w$. In this case, the assumption is implicitly made that $\frac{r}{a} \approx 1$, introducing an error of the order of $(n+2)f\delta\mu_{nm}$, where f is the flattening of the Earth (about 1/300).

For computing the contribution of the Earth's atmosphere to observed gravity change, it is nowadays accepted that 3D integration should be preferred to the simpler approach where surface pressure anomalies are converted to surface density. For example, in the GRACE analysis (Flechtner, 2007) $\rho(\mathbf{x}', t)\mathrm{d}r$ in Eq. (15) is replaced by $-\frac{\mathrm{d}p}{g(r)}$, assuming hydrostatic equilibrium. Then with $g(r) \approx g\left(\frac{a}{r}\right)^2$ and $r = a + N + \frac{\Phi}{1-\frac{\Phi}{a}}$,

$$v_{nm}(t) = \frac{a^2}{M}\frac{1}{(2n+1)}\frac{1}{g}\int_s \left\{\int_{p_s}^{0} \left(\frac{a}{a-\Phi} + \frac{N}{a}\right)^{n+4} \mathrm{d}p\right\} Y_{nm}(\mathbf{e}')\,\mathrm{d}s\,. \tag{16}$$

In Eq. (16), N is the height of the mean geoid above the ellipsid, p_s is surface pressure, and Φ is the geopotential height that is derived from vertical level data on temperature, pressure, and humidity.

A completely different way to solve the GIP (and the GCIP) has been suggested by Michel (2005), who derives the harmonic part ρ^+ of the density from gravity observations.

3 Earth Model

A caveat must be stated at this point, since so far we have considered the Earth as rigid. In reality, any mass redistribution at the surface or even inside the Earth is accompanied by a deformation of the solid Earth in its surroundings, which may be considered as an elastic, instantaneous response for short time scales and a visco-elastic response for longer time scales. This deformation causes an additional, "indirect" change of the gravity potential, which is not negligible and generally depends on the harmonic degree n of the load $\delta\mu_{nm}$. The linear differential equations that describe the deformation and gravity change of an elastic or visco-elastic, symmetric Earth, forced by surface loading of redistributing mass, are usually derived by considering a small perturbation of a radialsymmetric, hydrostatically pre-stressed background state.

In the linearized equation of momentum,

$$\nabla \cdot \boldsymbol{\sigma} + \nabla\left(\rho_0 g_0 \mathbf{u} \cdot \mathbf{e}_r\right) - \rho_0 \nabla \delta V - \delta\rho g_0 \mathbf{e}_r - \frac{\mathrm{d}^2}{\mathrm{d}t^2}\mathbf{u} = 0 \tag{17}$$

the first term describes the contribution of the stresses, the second term the advection of the hydrostatic pre-stress (related to the Lagrangian description of a displaced particle), the third term represents self-gravitation, expressing the change in gravity due to deformation, and the fourth term describes the change in density if one accounts for compressibility. Here, $\boldsymbol{\sigma}$ is the incremental stress tensor, $\mathbf{u} = \mathbf{u}(\mathbf{x}, t)$ the displacement of a particle at position $\mathbf{x}$, δV the perturbation of the gravitational potential, and $\delta\rho$ the perturbation of density. The fifth term in Eq. (17) is necessary when one is interested in the (free or forced) eigenoscillations of the Earth or in the body tides of the Earth caused by planetary potentials; however, the dominant time scales for external loads and the "integration times" for observing systems such as GRACE are rather long, with the consequence that one is usually confident with the quasi-static solutions.

Assuming the Earth in hydrostatic equilibrium prior to the deformation is certainly a gross simplification; it is however a necessary assumption to find "simple" solutions with the perturbation methods that are usually applied (e.g., Farrell, 1972).

Generally, the density perturbation can be expressed by the continuity equation,

$$\delta\rho = \nabla \cdot (\rho \mathbf{u}) . \tag{18}$$

For the perturbation of the potential, the Poisson equation

$$\Delta \delta V = -4\pi G \delta\rho \tag{19}$$

must hold. These equations are to be completed by a rheological law (e.g., $\boldsymbol{\sigma} = f(\boldsymbol{\epsilon})$ for elastic behavior, $\boldsymbol{\sigma} = f(\boldsymbol{\epsilon}, \dot{\boldsymbol{\epsilon}})$ for visco-elastic Maxwell behavior, $\boldsymbol{\epsilon}$ being the strain tensor), and boundary conditions for internal interfaces of a stratified Earth and for the free surface. For an elastic model consisting of Z layers, the rheology is usually prescribed by polynomial functions $\rho = \rho^{(z)}(r), \lambda = \lambda^{(z)}(r), \mu = \mu^{(z)}(r)$ of the Lamé parameters λ, μ and the density within each layer, $r_{\min}^{(z)}(r) \le r \le r_{\max}^{(z)}$. Solutions for potential change and deformation at the surface, obtained for the boundary condition of surface loading, are often expressed through Green's functions

$$\delta V(\mathbf{x}, t) = \int_{\tau=0}^{\infty} \int_s K_V(\mathbf{x}, \mathbf{x}', \tau') \delta\mu(\mathbf{x}', t - \tau') \, \mathrm{d}s \, \mathrm{d}\tau$$

($\mathbf{x}' = a\mathbf{e}'$) and

$$\mathbf{u}(\mathbf{x}, t) = \int_{\tau=0}^{\infty} \int_s K_{\mathbf{u}}(\mathbf{x}, \mathbf{x}', \tau') \delta\mu(\mathbf{x}', t - \tau') \, \mathrm{d}s \, \mathrm{d}\tau \, ,$$

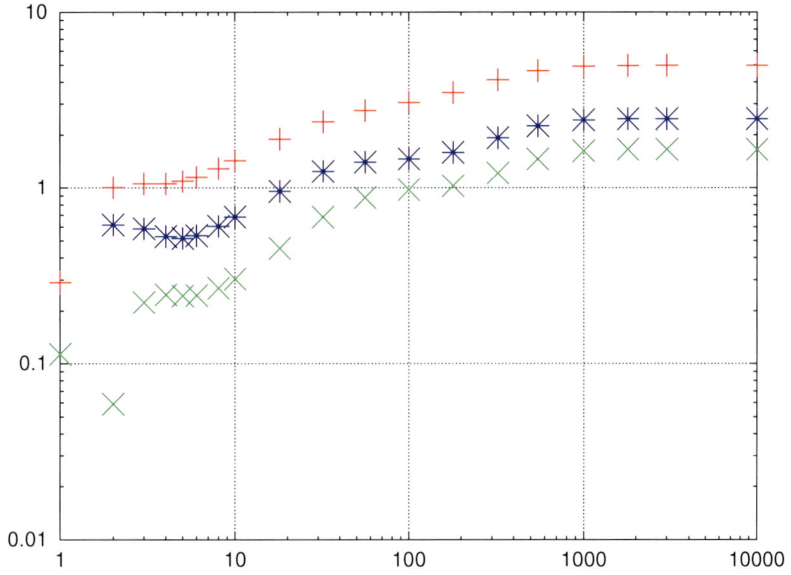

Fig. 2
Load Love numbers after Farrell (1972), $-h_n'$ (+), nl_n' (×), $-nk_n'$ (∗) versus degree *n*

where the kernels, for general (anisotropic, rotating, ellipsoidal, viscoelastic) Earth models can be represented through location- and frequency-dependent coefficients, e.g., $K(\mathbf{x}, \mathbf{x}', \tau') = \sum_{nm} k'_{nm}(\lambda, \theta, z) k'_{nm}(\lambda', \theta', z)$, where z is the Laplace transform parameter with unit $\frac{1}{s}$. Couplings between the degree-n, order-m terms of the load and potential-change and deformation harmonics of other degrees and orders have to be taken into account (e.g., Wang, 1991). It should be noted already here that the Green's functions essentially represent the impulse response of the Earth (i.e., Eqs. (17)–(19) together with a rheological law), and as such, may be determined from measurements under a known load "forcing."

For symmetric, nonrotating, elastic, and isotropic (SNREI) models of the Earth, the $k'_{nm}(\lambda, \theta, z) = k'_n$ are simply a function of the harmonic degree and the Green's function kernels depend on the spatial distance of $\mathbf{x}$ and $\mathbf{x}'$ only. Numerical solutions for the k'_n start with the observation that in Eqs. (17)–(19) the reference quantities ρ_0, g_0, and the Lame parameters that enter Hooke's law all depend on the radius r only. ➋ *Figure 2* shows the load Love numbers $-h'_n$, nl'_n, $-nk'_n$ that Farrell (1972) computed, using a Gutenberg–Bullen model for the Earth's rheology, for surface potential change and displacement (It should be noted that Love numbers are radius-dependent but usually in geodesy only the surface limits are applied).

The solution for the gravitational response to a surface mass distributed on top of an elastic Earth can be best described in spherical harmonics,

$$\delta v_{nm}(t) = \frac{a^2}{M}\left(1 + k'_n\right)\frac{1}{(2n+1)}\int_s \delta\mu(\mathbf{x}', t) Y_{nm}(\mathbf{e}')\,\mathrm{d}s \tag{20}$$

or

$$\delta v_{nm} = \frac{4\pi a^2}{M}\left(1 + k'_n\right)\frac{1}{(2n+1)}\delta\mu_{nm} = \frac{3}{\rho_e}\left(1 + k'_n\right)\frac{1}{(2n+1)}\delta\mu_{nm}, \tag{21}$$

where the "1" term is the potential change caused by $\delta\mu$, and the load Love number 'k'_n' term describes the incremental potential due to solid Earth deformation. Again, degree 0 and 1 terms deserve special attention: Due to mass conservation, k'_0 must be zero. Also, $l'_0 = 0$ (see below) is obvious since a uniform load on a spherical Earth cannot cause horizontal displacements due to symmetry reasons. In contrast, h'_0 corresponds to the average compressibility of the elastic Earth and will not vanish.

In the local spherical East–North-Up frame, the solution for **u** is

$$\mathbf{u}(\mathbf{x},t) = \begin{pmatrix} \bar{\mathbf{u}}(\mathbf{x},t) \\ h(\mathbf{x},t) \end{pmatrix} = a \sum_{n=0}^{\infty} \sum_{m=-n}^{n} \begin{pmatrix} \delta\psi_{nm}(t)\nabla^* \\ \delta h_{nm}(t) \end{pmatrix} Y_{nm}(\mathbf{e}) \tag{22}$$

with ∇^* being the tangential part of the gradient operator ∇. In spherical coordinates,

$$\nabla^* = \frac{1}{r}\frac{\partial}{\partial\theta}\mathbf{e}_\theta + \frac{1}{r\sin\theta}\frac{\partial}{\partial\lambda}\mathbf{e}_\lambda \qquad \nabla = \nabla^* + \frac{\partial}{\partial r}.$$

The radial displacement function is

$$\delta h_{nm}(t) = \frac{a^2}{gM}\frac{h'_n}{(2n+1)} \int_s \delta\mu(\mathbf{x}',t) Y_{nm}(\mathbf{e}')\,\mathrm{ds} \tag{23}$$

and at location **x** of the Earth's surface the radial displacement is

$$\delta h(\mathbf{x},t) = a \sum_{n=0}^{\infty} \sum_{m=-n}^{n} \delta h_{nm} Y_{nm}(\mathbf{e}).$$

The lateral displacement function is

$$\delta\psi_{nm}(t) = \frac{a^2}{gM}\frac{l'_n}{(2n+1)} \int_s \delta\mu(\mathbf{x}',t) Y_{nm}(\mathbf{e}')\,\mathrm{ds}. \tag{24}$$

At location **x**, the East and North displacement components are

$$\delta e(\mathbf{x}',t) = a \sum_{n=0}^{\infty} \sum_{m=-n}^{n} \delta\psi_{nm} \frac{\partial}{\partial e} Y_{nm}(\mathbf{e})$$

and

$$\delta n(\mathbf{x}',t) = a \sum_{n=0}^{\infty} \sum_{m=-n}^{n} \delta\psi_{nm} \frac{\partial}{\partial n} Y_{nm}(\mathbf{e}).$$

A reference system fixed to the center-of-mass of the solid Earth (CE system) is the natural system to compute the dynamics of solid Earth deformation and to model load Love numbers (Blewitt, 2003). This system is obviously "blind" to mass transports that shift the center-of-mass of the earth (excluding ocean, atmosphere, etc.); hence, by definition for the degree-1 potential Love number $k'_1 = 0$. Note that this is not true for the other degree-1 Love numbers; e.g., Farrell (1972) found $h'_1 = -0.290$ and $l'_1 = 0.113$. However, the CE system is difficult to realize in practice, by space-geodetic "markers." There are two principle ways to compute deformations for other reference systems: (a) first compute in CE, and subsequently apply a translation (b) transform degree-1 Love numbers, and compute in the new system. Blewitt (2003) showed that a translation of the reference system origin along the direction of the Load moment that can be absorbed in the three load Love numbers by subtracting the "isomorphic parameter" α from them (we follow the convention of Blewitt (2003)), $\tilde{k}'_1 = k'_1 - \alpha$, $\tilde{h}'_1 = h'_1 - \alpha$, and $\tilde{l}'_1 = l'_1 - \alpha$. For example, when transforming from the CE system to the CM system, which is fixed to the center of the mass of the Earth including the atmospheric and oceanographic loads, one has $\alpha = 1$. However, in reality the network shift (and rotation and scaling) might not entirely be known, as supposed for the mentioned approaches. In this situation, Kusche and Schrama (2005) and Wu et al. (2006) showed that, from 3D displacements in a realistic network and for a given set of Love numbers with $h'_1 \neq l'_1$, it is possible to separate residual unknown network translation and degree-1 deformation in an inversion approach.

For realistic loads, the theory predicts vertical deformation of up to about 1 cm, and horizontal deformation of a few mm. The solution in Eq. (22) refers to the local spherical East–North-Up frame, since it refers to an SNREI model. This must be kept in mind since geodetic displacement vectors are commonly referred to a local ellipsoidal East–North-Up frame. Otherwise a small but systematic error will be introduced from projecting height displacements erroneously onto North displacements.

In reality, the Earth is neither spherically symmetric nor purely elastic or isotropic, and it rotates in space. The magnitude of these effects are generally thought to be small. For example, Métivier et al. (2005) showed that ellipticity of the Earth, under zonal (atmospheric) loading leads to negligible amplitude and phase perturbations in low-degree geopotential harmonics. Another issue is that, even within the class of SNREI models, differences in rheology would cause differences in the response to loading. Plag et al. (1996) found for the models PREM and PEMC, differing in lithospheric properties, differences in the vertical displacement of up to 20%, in the horizontal displacement of up to 40%, and in gravity change of up to 20% in the vicinity (closer than 2° distance) of the load, but identical responses for distances greater than this value.

4 Analysis of TVG and Deformation Pattern

In many cases, the aim of the analysis of time-variable gravity and deformation patterns is, at least in an intermediate sense, the determination of mass changes within a specific volume,

$$\delta M = \int_{v^*} \left(\rho(\mathbf{x}, t) - \bar{\rho}(\mathbf{x}')\right) \mathrm{d}v \,, \tag{25}$$

e.g., an ocean basin or a hydrological catchment area. Under the assumption that the thin-layer hypothesis provides an adequate description, δM can be found by integrating $\delta\mu$ over the (spherical) surface s^* of v^* (Wahr et al., 1998). The integration is usually performed in spectral domain, with the 4π-normalized spherical harmonic coefficients s^*_{nm} of the characteristic function S^* for the area s^*

$$\delta M = \int_{s^*} \delta\mu(\mathbf{e}', t)\, \mathrm{d}s = \sum_{n=0}^{\infty} \sum_{m=-n}^{n} s^*_{nm} \delta\mu_{nm}(t) \,, \tag{26}$$

or, in brief, $\delta M = (S^*, \delta\mu)$.

Following the launch of the GRACE mission, the analysis of time-variable surface mass loads from GRACE-derived spherical harmonic coefficients via the inverse relation (Wahr et al., 1998)

$$\delta\mu_{nm} = \frac{\rho_e}{3} \frac{(2n+1)}{(1+k'_n)} \delta v_{nm} \tag{27}$$

has been applied in a fast-growing number of studies (for an overview, the publication data base of the GRACE project at http://www-app2.gfz-potsdam.de/pb1/op/grace/

might be considered). Since GRACE-derived monthly or weekly $\delta v_{nm} = v_{nm} - \bar{v}_{nm}$ are corrupted with low-frequency noise, and since they exhibit longitudinal artifacts when projected in the space domain, isotropic or anisotropic filter kernels are applied to these estimates usually. See Kusche (2007) for a review on filter methods, where in general a set of spectrally weighted coefficients $\delta\tilde{v}_{nm} = \sum_{p,q} w_{nm}^{pq} \delta v_{pq}$ is derived.

In principle, spherical harmonic models of surface load can be inferred coefficient-by-coefficient as well once a spherical harmonic model of vertical deformation has been derived,

$$\delta\mu_{nm} = \frac{\rho_e}{3} \frac{(2n+1)}{h'_n} \delta h_{nm} \tag{28}$$

or, of lateral deformation

$$\delta\mu_{nm} = \frac{\rho_e}{3} \frac{(2n+1)}{l'_n} \delta\psi_{nm} \,. \tag{29}$$

For the combination of vertical and lateral deformations in an inversion, two approaches exist: (a) either the $\delta\mu_{nm}$ are estimated directly from discrete $\mathbf{u}(\mathbf{x}_i)$, or (b) in a two-step procedure, first the h_{nm} and ψ_{nm} are estimated separatly from discrete height and lateral displacements, and the $\delta\mu_{nm}$ are subsequently inferred from those. In practice, 3D displacement vectors are available in discrete points of the Earth's surface. For the network of the International GNNS Service (IGS), a few hundred stations provide continuous data such that a maximum degree and order of well beyond degree and order 20 might be resolved. However, several studies have shown that this rather theoretical resolution cannot be reached. The reason is that spatial aliasing is present in the signal measured at a single site, which measures the sum of all harmonics up to infinity, unless the signal above the theoretical resolution can be removed from other data such as time-variable gravity.

Inversion approaches must be seen in the light of positioning accuracies of modern worldwide networks, which are currently of a similar order compared with the deformation signals (few mm horizontally and a factor of 2–3 worse in the vertical direction). Whereas random errors may significantly reduce in the spatial averaging process that is required to form estimates for low-degree harmonics, any systematic errors at the spatial scale of the signals of interest may be potentially dangerous. A particular problem is given by the presence of secular trends in the displacement data that are dominated by nonloading phenomena like GIA, plate motion, and local monument subsidence.

If the load $\delta\mu(t)$ is *known*, from independent measurements (e.g., water level measurements from gauges or surface pressure from barometric measurements and meteorological modeling) the LLN's k'_n, h'_n, l'_n could, in principle, be determined experimentally from measurements of gravity change and displacement. Plag et al. (1995) coined the term "loading inversion" (and as an alternative "loading tomography," noting that tomography is usally based on probing along rays and not by body-integrated response) and suggested a method to invert for certain spherical harmonic expansion coefficients of density $\delta\rho_{nm}$ and Lame parameter perturbations $\delta\lambda_{nm}$, $\delta\mu_{nm}$, together with polynomial parameters that describe their variation in the radial direction. This inverse approach to recover elastic properties from measurements of gravity and deformation is followed in planetary exploration, however, for the body Love numbers k_2 and h_2. There, the role of the known load is assumed by the known tidal potentials in the solar system.

On the other hand, one has the possibility to eliminate the $\delta\mu_{nm}$ from the equations and determine the ratio

$$\frac{h'_n}{1+k'_n} = \frac{\delta h_{nm}}{\delta v_{nm}} \quad \text{for each} \quad m = -n \ldots n$$

and

$$\frac{l'_n}{1+k'_n} = \frac{\delta\psi_{nm}}{\delta v_{nm}} \quad \text{for each} \quad m = -n \dots n$$

from a combination of gravity and displacement data. If all spherical harmonic coefficients up to degree $n_{\max}$ are well-determined from data, these equations provide $2n + 1$ relations per degree n. Mendes Cerveira et al. (2006) proposed that this freedom might in principle be used to uniquely solve for the ratio $\frac{h'_{nm}}{1+k'_{nm}}$ and $\frac{l'_{nm}}{1+k'_{nm}}$ of global, "azimuth-dependent" LLNs.

Other observations of gravity change and deformation of the Earth's solid and fluid surface may be considered as well. With the global network of superconducting gravimeters, it is believed that annual and short-time variations in mass loading can be observed and compared with GRACE results, after an appropriate removal of local effects (Neumeyer et al., 2006). Absolute (free-fall) gravimeters provide stable time series from which trends in gravity can be obtained, and from which time series of superconducting gravimeters can be calibrated. Gravity change $\delta g = |\nabla V| - |\nabla \overline{V}|$ as sensed by a terrestrial gravimeter, which is situated on the deforming Earth surface, reads (in spherical approximation, where the magnitude of the gradient is replaced by the radial derivative)

$$\delta g(\mathbf{x}, t) = \frac{GM}{a^2} \sum_{n=0}^{\infty} \sum_{m=-n}^{n} \left(n + 1 - 2h'_n\right) \delta v_{nm}(t) Y_{nm}(\mathbf{e}) \tag{30}$$

and when related to surface mass change,

$$\delta g(\mathbf{x}, t) = \frac{GM}{a^2} \frac{3}{\rho_e} \sum_{n=0}^{\infty} \sum_{m=-n}^{n} \frac{n + 1 - 2h'_n}{2n+1} \delta\mu_{nm}(t) Y_{nm}(\mathbf{e}) \,. \tag{31}$$

In principle, observations of the sea level might be considered in a multi-data inversion scheme as well. This requires that the steric sea-level change, which is related to changes in temperature and salinity of the ocean, can be removed from the measured total or volumetric sea level. Sea-level change can be measured using satellite altimeters, as with the current Jason-1 and Jason-2 missions, in an absolute sense, since altimetric orbits refer to an ITRS-type global reference system. Tide gauge observations, on the other hand, provide sea level in a relative sense since they refer to land benchmarks. Ocean bottom pressure recorders measure the load change directly. If one assumes that the ocean response is largely "passive" (Blewitt and Clarke, 2003), i.e., the ocean surface follows an equipotential surface, the absolute sea-level change is related to surface load as

$$\delta s(\mathbf{x}, t) = \frac{GM}{a} \frac{3}{\rho_e} \sum_{n=0}^{\infty} \sum_{m=-n}^{n} \frac{1 + k'_n}{2n+1} \delta\mu_{nm}(t) Y_{nm}(\mathbf{e}) + \delta s_0(t) \tag{32}$$

and relative sea-level as

$$\delta \check{s}(\mathbf{x}, t) = \frac{GM}{a} \frac{3}{\rho_e} \sum_{n=0}^{\infty} \sum_{m=-n}^{n} \frac{1 + k'_n - h'_n}{2n+1} \delta\mu_{nm}(t) Y_{nm}(\mathbf{e}) + \delta \check{s}_0(t) \,, \tag{33}$$

where $\delta s_0(t)$ and $\delta \check{s}_0(t)$ are spatially uniform terms that account for mass conservation (cf. Blewitt, 2003).

All spectral operators discussed in this chapter are summarized in ➲ *Fig. 3*.

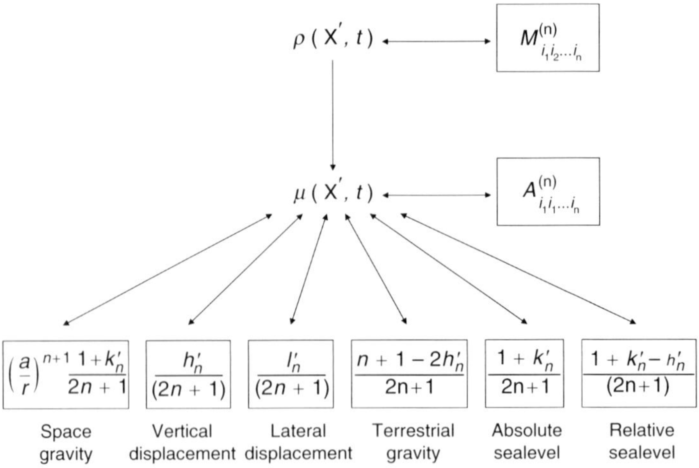

Fig. 3
Spectral operators for space and terrestrial gravity, displacement, and sea-level observables

5 Future Directions

Within the limits of accuracy and spatial resolution of current observing systems, inversions for (surface) mass appear to have almost reached their potential. Limitations are, in particular, the achievable accuracy of spherical harmonic coefficients from GRACE, systematic errors nicknamed as "striping," the presence of systematic errors in displacement vectors from space-geodetic techniques, and the spatial density and inhomogeneous distribution of global networks (space-geodetic techniques, absolute and superconducting gravimetry). However, moderate improvements in data quality and consistency can be expected from reprocessings such as the anticipated GRACE RL05 products or the IGS reprocessing project. In the long run, GRACE Follow-On missions and geometric positioning in the era of GALILEO will provide, hopefully, bright prospects.

At the time of writing, some groups focus on the combination of gravity and geometrical observations with a priori models of mass transport (so-called joint inversions, cf. Wu et al. 2006, Jansen et al. 2009, Rietbroek et al. 2009). The benefit of this strategy is that certain weaknesses of individual techniques can be covered by other techniques. For example, it has been shown that the geocenter motion or degree-1 surface load which is not observed with GRACE can be restituted to some extend by GPS and/or ocean modeling. Research is ongoing in this direction.

Another issue is that, if the space agencies fail to replace the GRACE mission in time with a follow-on satellite mission, a gap in the observation of the time-variable gravity field might occur. To some extent, the very low degrees of mass loading processes might be recovered during the gap from geometrical techniques and loading inversion, provided that their mentioned limitations can be overcome and a proper cross-calibration is facilitated with satellite gravity data in the overlapping periods (Bettadpur et al., 2008; Plag and Gross, 2008). The same situation occurs, if one tries to go back in time before the launch of GRACE, e.g., using (reprocessed) GPS solutions. Anyway, the problem remains extremely challenging.

6 Conclusions

We have reviewed concepts that are currently used in the interpretation of time-variable gravity and deformation fields in terms of mass transports and Earth system research. Potential for future research is seen in particular for combinations of different observables.

References

Bettadpur S, Ries J, Save H (2008) Time-variable gravity, low Earth orbiters, and bridging gaps. GRACE Science Team Meeting 2008, San Francisco, USA, 12–13. December 2008

Blewitt G (2003) Self-consistency in reference frames, geocenter definition, and surface loading of the solid Earth. J Geophys Res 108: B2, 2103. doi:10.1029/2002JB002082

Blewitt G, Clarke P (2003) Inversion of Earth's changing shape to weigh sea level in static equilibrium with surface mass redistribution. J Geophys Res 108: B6, 2311. doi:10.1029/2002JB002290

Chao BF (2005) On inversion for mass distribution from global (time-variable) gravity field. J Geodyn 39:223–230

Farrell W (1972) Deformation of the Earth by surface loads. Rev Geophys Space Phys 10(3): 761–797

Flechtner F (2007) AOD1B Product Description Document for Product Releases 01 to 04 (Rev. 3.1, April 13, 2007), GR-GFZ-AOD-0001, GFZ Potsdam

Ilk KH, Flury J, Rummel R, Schwintzer P, Bosch W, Haas C, Schröter J, Stammer D, Zahel W, Miller H, Dietrich R, Huybrechts P, Schmeling H, Wolf D, Götze HJ, Riegger J, Bardossy A, Güntner A, Gruber T (2005) Mass transport and mass distribution in the Earth system. Contribution of the new generation of satellite gravity and altimetry missions to geosciences, GFZ Potsdam

Jansen MWF, Gunter BC, Kusche J (2009) The impact of GRACE, GPS and OBP data on estimates of global mass redistribution. Geophys J Int doi: 10.1111/j.1365-246X.2008.04031.x

Kusche J (2007) Approximate decorrelation and non-isotropic smoothing of time-variable GRACE-type gravity fields. J Geodesy 81(11): 733– 749

Kusche J, Schrama EJO (2005). Surface mass redistribution inversion from global GPS deformation and Gravity Recovery and Climate Experiment (GRACE) gravity data. J Geophys Res 110: B09409. doi:10.1029/2004JB003556.

Métivier L, Greff-Lefftz M, Diament M (2005) A new approach to computing accurate gravity time variations for a realistic earth model with lateral heterogeneities. Geophys J Int 162:570–574

Mendes Cerveira P, Weber R, Schuh H (2006) Theoretical aspects connecting azimuth-dependent Load Love Numbers, spatiotemporal surface geometry changes, geoid height variations and Earth rotation data. In: WIGFR2006, Smolenice Castle, Slovakia, May 8-9, 2006

Michel V (2005) Regularized wavelet-based multiresolution recovery of the harmonic mass density distribution from data of the Earth's gravitational field at satellite height. Inverse Prob 21:997–1025

Neumeyer J, Barthelmes F, Dierks O, Flechtner F, Harnisch M, Harnisch G, Hinderer J, Imanishi Y, Kroner C, Meurers B, Petrovic S, Reigber C, Schmidt R, Schwintzer P, Sun H-P, Virtanen H (2006) Combination of temporal gravity variations resulting from superconducting gravimeter (SG) recordings, GRACE satellite observations and global hydrology models. J Geod 79:573–585

Plag H-P, Gross R (2008) Exploring the link between Earth's gravity field, rotation and geometry in order to extend the GRACE-determined terrestrial water storage to non-GRACE times. GRACE Science Team Meeting 2008, San Francisco, USA, 12–13. December 2008

Plag H-P, Jüttner H-U, Rautenberg V (1996) On the possibility of global and regional inversion of exogenic deformations for mechanical properties of the Earth's interior. J Geodynamics 21(3):287–308

Rietbroek R, Brunnabend S-E, Dahle C, Kusche J, Flechtner F, Schröter J, Timmermann R (2009) Changes in total ocean mass derived from GRACE, GPS, and ocean modeling with weekly resolution. J Geophys Res 114, C11004, doi:10.1029/2009JC005449

Wahr J, Molenaar M, Bryan F (1998) Time variability of the Earth's gravity field: Hydrological and oceanic effects and their possible detection using GRACE. J Geophys Res 103(B12), 30205–30229

Wang R (1991). Tidal deformations on a rotating, spherically asymmetric, visco-elastic and laterally heterogeneous Earth. Peter Lang, Frankfurt/Main

Wu X, Heflin MB, Ivins ER, Fukumori I (2006) Seasonal and interannual global surface mass variations from multisatellite geodetic data. J Geophys Res 111: B09401. doi:10.1029/2005JB004100

9 Satellite Gravity Gradiometry (SGG): From Scalar to Tensorial Solution

Willi Freeden[1] · *Michael Schreiner*[2]
[1]Geomathematics Group, Technische Universität Kaiserslautern, Kaiserslautern, Germany
[2]Institute for Computational Engineering, University of Buchs, Buchs, Switzerland

W. Freeden, M.Z. Nashed, T. Sonar (Eds.), *Handbook of Geomathematics*, DOI 10.1007/978-3-642-01546-5_9,

Abstract Satellite gravity gradiometry (SGG) is an ultra-sensitive detection technique of the space gravitational gradient (i.e., the Hesse tensor of the Earth's gravitational potential). In this note, SGG – understood as a spacewise inverse problem of satellite technology – is discussed under three mathematical aspects: First, SGG is considered from potential theoretic point of view as a continuous problem of "harmonic downward continuation." The space-borne gravity gradients are assumed to be known continuously over the "satellite (orbit) surface"; the purpose is to specify sufficient conditions under which uniqueness and existence can be guaranteed. In a spherical context, mathematical results are outlined by decomposition of the Hesse matrix in terms of tensor spherical harmonics. Second, the potential theoretic information leads us to a reformulation of the SGG-problem as an ill-posed pseudodifferential equation. Its solution is dealt within classical regularization methods, based on filtering techniques. Third, a very promising method is worked out for developing an immediate interrelation between the Earth's gravitational potential at the Earth's surface and the known gravitational tensor.

1 Introduction

Due to the nonspherical shape, the irregularities of its interior mass density, and the movement of the lithospheric plates, the external gravitational field of the Earth shows significant variations. The recognition of the structure of the Earth's gravitational potential is of tremendous importance for many questions in geosciences, for example, the analysis of present day tectonic motions, the study of the Earth's interior, models of deformation analysis, the determination of the sea surface topography, and circulations of the oceans, which, of course, have a great influence on the global climate and its change. Therefore, a detailed knowledge of the global gravitational field including the local high-resolution microstructure is essential for various scientific disciplines.

Satellite gravity gradiometry (SGG) is a modern domain of studying the characteristics, the structure, and the variation process of the Earth's graviational field. The principle of satellite gradiometry can be explained roughly by the following model (cf. ➲ *Fig. 1*): several test masses in a low orbiting satellite feel, due to their distinct positions and the local changes of the gravitational field, different forces, thus yielding different accelerations. The measurements of the relative accerlerations between two test masses provide information about the second-order partial derivatives of the gravitational potential. More concretely, measured are differences between the

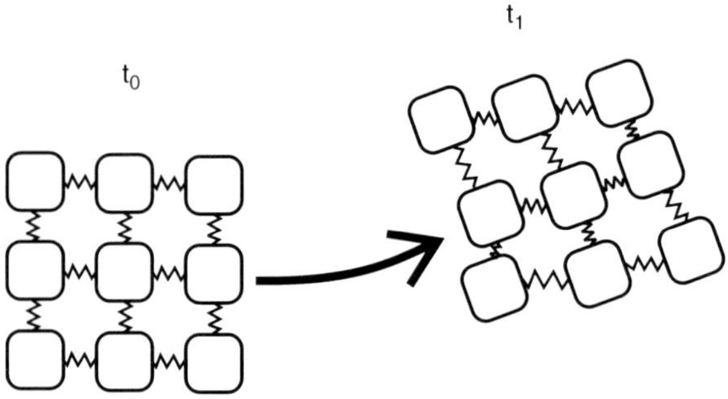

Fig. 1
The principle of a gradiometer

displacements of opposite test masses. This yields information on the differences of the forces. Since the gradiometer itself is small, these differences can be identified with differentials, so that a so-called full gradiometer gives information on the whole tensor consisting out of all second-order partial derivatives of the gravitational potential , i.e., the Hesse matrix. In an ideal case, the full Hesse matrix can be observed by an array of test masses.

On 17 March 2009, the European Space Agency (ESA) began to realize the concept of SGG with the launch of the most sophisticated mission ever to investigate the Earth's gravitational field, viz. GOCE (Gravity and Ocean Circulation Explorer). ESA's 1-ton spacecraft carries a set of six state-of-the-art high-sensitivity accelerometers to measure the components of the gravity field along all three axes (see the contribution of R. Rummel in this issue for more details on the measuring devices of this satellite). GOCE is producing a coverage of the entire Earth with measurements (apart from gaps at the polar regions). For around 20 months GOCE will be gathering gravitational data. In order to make this mission possible, ESA and its partners had to overcome an impressive technical challenge by designing a satellite that is orbiting the Earth close enough (at an altitude of only 250 km) to collect high-accuracy gravitational data while being able to filter out disturbances caused, e.g., by the remaining traces of the atmosphere.

It is not surprising that, during the last decade, the ambitious mission GOCE motivated many scientific activities such that a huge number of written material is available in different fields concerned with special user group activities, mission synergy, calibration as well as validation procedures, geoscientific progress (in fields like gravity field recovery, ocean circulation, hydrology, glacialogy, deformation, climate modeling, etc), data management, and so on. A survey about the recent status is well-demonstrated by the "ESA Living Planet Programme," which also contains a list on GOCE-publications (see also the contribution by the ESA-Frascati Group in this issue, for information from geodetic point of view the reader is referred, e.g., to the notes (Beutler et al. 2003; ESA 1999; ESA 2007; Rummel et al. 1993), too). Mathematically, the literature dealing with the solution procedures of problems related to SGG can be divided essentially into two classes: the timewise approach and the spacewise approach. The former one considers the measured data as a time series, while the second one supposes that the data are given in advance on a (closed) surface.

This chapter is part of the spacewise approach, its goal is a potential theoretically reflected approach to SGG with strong interest in the characterization of SGG-data types and tensorial oriented solution of the occuring (pseudodifferential) SGG-equations by regularization. Particular emphasis is laid on the transition from scalar data types (such as the second-order radial derivative) to full tensor data of the Hesse matrix.

2 SGG in Potential Theoretic Perspective

Gravity as observed on the Earth's surface is the combined effect of the gravitational mass attraction and the centrifugal force due to the Earth's rotation. The force of gravity provides a directional structure to the space above the Earth's surface. It is tangential to the vertical plumb lines and perpendicular to all level surfaces. Any water surface at rest is part of a level surface. As if the Earth were a homogeneous, spherical body gravity turns out to be constant all over the Earth's surface, the well-known quantity 9.8 ms^{-2}. The plumb lines are directed toward the Earth's center of mass, and this implies that all level surfaces are nearly spherical, too. However, the gravity decreases from the poles to the equator by about 0.05 ms^{-2}. This is caused by the flattening of the Earth's figure and the negative effect of the centrifugal force, which is maximal at the equator. Second, high mountains and deep ocean trenches cause the gravity to vary. Third, materials within the Earth's interior are not uniformly distributed. The irregular gravity field shapes as virtual surface the geoid. The level surfaces are ideal reference surfaces, for example, for heights. In more detail, the *gravity acceleration (gravity)* w is the resultant of gravitation v and centrifugal acceleration c, i.e., $w = v + c$. The centrifugal force c arises as a result of the rotation of the Earth about its axis. We assume here a rotation of constant angular velocity ω_0 about the rotational axis x_3, which is further assumed to be fixed with respect to the Earth. The centrifugal acceleration acting on a unit mass is directed outward perpendicularly to the spin axis. If the ε^3-axis of an Earth-fixed coordinate system coincides with the axis of rotation, then we have $c(x) = -\omega_0^2\varepsilon^3 \wedge (\varepsilon^3 \wedge x)$. Using the so-called *centrifugal potential* $C(x) = (1/2)\omega_0^2(x_1^2 + x_2^2)$ we can write $c = \nabla C$.

The direction of the gravity w is known as the direction of the *plumb line*, the quantity $|w|$ is called the *gravity intensity* (often just *gravity*). The *gravity potential of the Earth* can be expressed in the form: $W = V + C$. The gravity acceleration w is given by $w = \nabla W = \nabla V + \nabla C$. The surfaces of constant gravity potential $W(x) = \text{const}$, $x \in \mathbb{R}^3$, are designated as *equipotential (level,* or *geopotential) surfaces of gravity*. The *gravity potential* W of the Earth is the sum of the *gravitational potential* V and the *centrifugal potential* C, i.e., $W = V + C$. In an Earth's fixed coordinate system the centrifugal potential C is explicitly known. Hence, the determination of equipotential surfaces of the potential W is strongly related to the knowledge of the potential V. The gravity vector w given by $w(x) = \nabla_x W(x)$ where the point $x \in \mathbb{R}^3$ is located outside and on a sphere around the origin with Earth's radius R, is normal to the equipotential surface passing through the same point. Thus, equipotential surfaces intuitively express the notion of tangential surfaces, as they are normal to the plumb lines given by the direction of the gravity vector (for more details see, for example, Heiskanen and Moritz (1967), (Freeden and Schreiner 2009) and the contribution by H. Moritz in this issue).

According to the classical Newton's Law of Gravitation (1687), knowing the density distribution ρ of a body, the gravitational potential can be computed everywhere in $\mathbb{R}^3$. More explicitly, the gravitational potential V of the Earth's exterior is given by

$$V(x) = G\int_{\text{Earth}} \frac{\rho(y)}{|x-y|}\, dV(y), \quad x \in \mathbb{R}^3\backslash\text{Earth}, \tag{1}$$

where G is the gravitational constant ($G = 6.6742 \cdot 10^{-11}\text{m}^3\ \text{kg}^{-1}\ \text{s}^{-2}$) and dV is the (Lebesgue-) volume measure. The properties of the gravitational potential (1) in the Earth's exterior are appropriately described by the Laplace equation:

$$\Delta V(x) = 0,\ x \in \mathbb{R}^3\backslash\text{Earth}. \tag{2}$$

The gravitational potential V as defined by (1) is regular at infinity, i.e.,

$$|V(x)| = O\left(\frac{1}{|x|}\right), |x| \to \infty. \tag{3}$$

For practical purposes, the problem is that in reality the density distribution ρ is very irregular and known only for parts of the upper crust of the Earth. It is actually so that geoscientists would like to know it from measuring the gravitational field. Even if the Earth is supposed to be spherical, the determination of the gravitational potential by integrating Newton's potential is not achievable. This is the reason why, in simplifying spherical nomenclature, we first expand the so-called reciprocal distance in terms of harmonics (related to the Earth's mean radius R) as a series

$$\frac{1}{|x-y|} = \sum_{n=0}^{\infty} \sum_{j=1}^{2n+1} \frac{4\pi R}{2n+1} H^R_{-n-1,k}(x) H^R_{n,k}(y), \tag{4}$$

where $H^R_{n,k}$ is an *inner harmonic* of degree n and order k given by

$$H^R_{n,k}(x) = \frac{1}{R}\left(\frac{|x|}{R}\right)^n Y_{n,k}(\xi), \quad x = |x|\xi, \xi \in \Omega, \tag{5}$$

and $H^R_{-n-1,k}$ is an *outer harmonic* of degree n and order k given by

$$H^R_{-n-1,k}(x) = \frac{1}{R}\left(\frac{R}{|x|}\right)^{n+1} Y_{n,k}(\xi), \quad x = |x|\xi, \xi \in \Omega. \tag{6}$$

Note that the family $\{Y_{n,k}\}_{\substack{n=0,1,\ldots \\ k=1,\ldots,2n+1}}$ is an $\mathcal{L}^2(\Omega)$-orthonormal system of scalar spherical harmonics (for more details concerning spherical harmonics see, e.g., Müller (1966), Freeden et al. (1998), Freeden and Schreiner 2009). Insertion of the series expansion (4) into the Newton formula for the external gravitational potential yields

$$V(x) = G \sum_{n=0}^{\infty} \sum_{k=1}^{2n+1} \frac{4\pi R}{2n+1} \int_{\Omega_R^{\text{int}}} \rho(y)\, H^R_{n,k}(y)\, dV(y)\, H^R_{-n-1,k}(x). \tag{7}$$

The expansion coefficients of the series (7) are not computable, since their determination requires the knowledge of the density function ρ in the Earth's interior. In fact, it turns out that there are infinitely many mass distributions, which have the given gravitational potential of the Earth as exterior potential.

Nevertheless, collecting the results from potential theory on the Earth's gravitational field v for the outer space (in spherical approximation) we are confronted with the following (mathematical) characterization: v is an infinitely often differentiable vector field in the exterior of the Earth such that (v1) div $v = \nabla \cdot v = 0$, curl $v = \mathrm{L} \cdot v = 0$ in the Earth's exterior, (v2) v is regular at infinity: $|v(x)| = O\left(1/(|x|^2)\right), |x| \to \infty$. Seen from mathematical point of view, the properties (v1) and (v2) imply that the Earth's gravitational field v in the exterior of the Earth is a gradient field $v = \nabla V$, where the gravitational potential V fulfills the properties: V is an infinitely often differentiable scalar field in the exterior of the Earth such that (V1) V is harmonic in the Earth's exterior, and vice versa. Moreover, the gradient field of the Earth's gravitational field (i.e., the *Jacobi matrix field*) $\mathbf{v} = \nabla \mathbf{v}$, obeys the following properties: $\mathbf{v}$ is an infinitely often differentiable tensor field in the exterior of the Earth such that (**v1**) div $\mathbf{v} = \nabla \cdot \mathbf{v} = 0$, curl $\mathbf{v} = \mathrm{L} \cdot \mathbf{v} = 0$ in the Earth's exterior, (**v2**) $\mathbf{v}$ is regular at infinity: $|\mathbf{v}(\mathbf{x})| = O\left(1/(|x|^3)\right), |x| \to \infty$, and vice versa. Combining our identities we finally see that $\mathbf{v}$ can be represented as the *Hesse tensor of the scalar field* V, i.e., $\mathbf{v} = (\nabla \otimes \nabla)\, \mathbf{V} = \nabla^{(2)}\mathbf{V}$.

The technological SGG-principle of determining the tensor field $\mathbf{v}$ at satellite altitude is illustrated graphically in ➊ *Fig. 2*. The position of a low orbiting satellite is tracked using GPS. Inside the satellite there is a gradiometer. A simplified model of a gradiometer is sketched in ➊ *Fig. 1*. An array of test masses is connected with springs. Once more, the measured quantities are the differences between the displacements of opposite test masses. According to Hooke's law the mechanical configuration provides information on the differences of the forces. They, however, are due to local differences of ∇V. Since the gradiometer itself is small, these differences can be identified with differentials, so that a so-called full gradiometer gives information on the whole tensor consisting out of all second order partial derivatives of V, i.e., the Hesse matrix $\mathbf{v}$ of V.

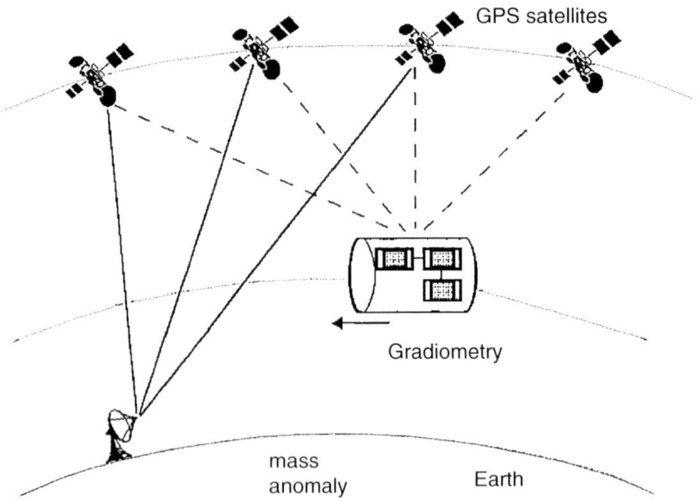

Fig. 2
The principle of satellite gravity gradiometry (from ESA (1999))

From our preparatory remarks it becomes obvious that the potential theoretic situation for the SGG problem can be formulated briefly as follows: Suppose that the satellite data $\mathbf{v} = (\nabla \otimes \nabla)\,\mathbf{V}$ are known continuously over the "orbital surface," the satellite gravity gradiometry problem amounts to the problem of determining V from $\mathbf{v} = (\nabla \otimes \nabla)\,\mathbf{V}$ at the "orbital surface."

Mathematically, SGG is a nonstandard problem of potential theory. The reasons are obvious:

- SGG is ill–posed since the data are not given on the boundary of the domain of interest, i.e., on the Earth's surface, but on a surface in the exterior domain of the Earth, i.e., at a certain height.
- Tensorial SGG-data (or scalar manifestations of them) do not form the standard equipment of potential theory (such as, e.g., Dirichlet or Neumann data). Thus, it is – at first sight – not clear whether these data ensure the uniqueness of the SGG-problem or not.
- SGG-data have its natural limit because of the strong damping of the high-frequency parts of the (spherical harmonic expansion of the) gravitational potential with increasing satellite heights. For a heuristic explanation of this calamity, let us start from the assumption that the gravitational potential outside the spherical Earth's surface Ω_R with the mean radius R is given by the ordinary expansion in terms of outer harmonics (confer the identity (7))

$$V(x) = \sum_{n=0}^{\infty} \sum_{k=1}^{2n+1} \int_{\Omega_R} V(y)\, H^R_{-n-1,k}(y)\, d\omega(y)\, H^R_{-n-1,k}(x) \tag{8}$$

($d\omega$ is the usual surface measure). Then it is not hard to see that those parts of the gravitational potential belonging to the outer harmonics $H^R_{-n-1,k}$ of order n at height H above the Earth's surface Ω_R are damped by a factor $[R/(R+H)]^{n+1}$. Just a way out of this difficulty is seen in SGG, where, e.g., second-order radial derivatives of the gravitational potential are available at a height of typically about 250 km. The second derivatives cause (roughly speaking) an amplification of the potential coefficients by a factor of order n^2. This compensates the damping effect due to the satellite's height if n is not too large. Nevertheless, in spite of the amplification, the SGG-problem still remains (exponentially) ill-posed. Altogether, the graviational potential decreases exponentially with increasing height, and therefore the process of transforming, the data down to the Earth surface (often called "downward continuation") is unstable.

The non-canonical (SGG)-situation of uniqueness within the potential theoretic framework can be demonstrated already by a simple example in spherical context: Suppose that one scalar component of the Hesse tensor is prescribed for all points x at the sphere $\Omega_{R+H} = \{x \in \mathbb{R}^3 \mid |x| = R + H\}$. Is the gravitational potential V unique on the sphere $\Omega_R = \{x \in \mathbb{R}^3 \mid |x| = R\}$? The answer is not positive, in general. To see this, we construct a counterexample: If $b \in \mathbb{R}^3$ with $|b| = 1$ is given, the second-order directional derivative of V at the point x is $b^{\mathrm{T}} \nabla \otimes \nabla V(x) b$. Given a potential V, we construct a vector field b on Ω_{R+H}, such that the second-order directional derivative $b^{\mathrm{T}} \nabla \otimes \nabla V b$. is zero: Assume that V is a solution of (2) and (3). For each $x \in \Omega_{R+H}$, we know that the Hesse tensor $\nabla \otimes \nabla V(x)$ is symmetric. Thus, there exists an orthogonal matrix $A(x)$ so that $A(x)^{\mathrm{T}} (\nabla \otimes \nabla V(x)) A(x) = \mathrm{diag}(\lambda_1(x), \lambda_2(x), \lambda_3(x))$, where $\lambda_1(x), \lambda_2(x), \lambda_3(x)$ are the eigenvalues of $\nabla \otimes \nabla V(x)$. From the harmonicity of V it is clear that $0 = \Delta V(x) = \lambda_1(x) + \lambda_2(x) + \lambda_3(x)$. Let $\mu_0 = 3^{-1/2}(1,1,1)^{\mathrm{T}}$. We define the vector field $\mu : \Omega_{R+H} \to \mathbb{R}^3$ by $\mu(x) = A(x)\mu_0$, $x \in \Omega_{R+H}$. Then we obtain

$$\begin{aligned}
\mu^{\mathrm{T}}(x)(\nabla \otimes \nabla V(x))\mu(x) &= \mu_0^{\mathrm{T}} A(x)^{\mathrm{T}} (\nabla \otimes \nabla V(x)) A(x) \mu_0 \qquad (9)\\
&= \frac{1}{3}(1\,1\,1) \begin{pmatrix} \lambda_1(x) & 0 & 0 \\ 0 & \lambda_2(x) & 0 \\ 0 & 0 & \lambda_3(x) \end{pmatrix} \begin{pmatrix} 1 \\ 1 \\ 1 \end{pmatrix} \\
&= \frac{1}{3}(\lambda_1(x) + \lambda_2(x) + \lambda_3(x)) \\
&= 0. \qquad (10)
\end{aligned}$$

Hence, we have constructed a vector field μ such that the second-order directional derivative of V in the direction of $\mu(x)$ is zero for every point $x \in \Omega_{R+H}$. It can be easily seen that, for a given V, there exist many vector fields showing the same properties for uniqueness as the vector field μ. Observing these arguments we are led to the conclusion that the function V is undetectable from the directional derivatives corresponding to μ (see also Schreiner 1994a,b).

It is, however, good news that we are not lost here: As a matter of fact, there do exist conditions under which only one quantity of the Hesse tensor yields a unique solution (at least up to low order harmonics). In order to formulate these results, a certain decomposition of the Hesse tensor is necessary, which strongly depends on the separation of the Laplace operator in terms of polar coordinates. In order to follow this path, we start to reformulate the SGG-problem more easily in spherical context. For that purpose we start with some basic facts specifically formulated on the unit sphere $\Omega = \{x \in \mathbb{R}^3 \mid |x| = 1\}$: As is well-known, any $x \in \mathbb{R}^3, x \neq 0$, can be decomposed uniquely in the form $x = r\xi$, where the directional part is an element of the unit sphere: $\xi \in \Omega$. Let $\{Y_{n,m}\} : \Omega \to \mathbb{R}^3$, $n = 0, 1, \ldots$, $m = 1, \ldots, 2n+1$, be an orthonormal set of spherical harmonics. As is well–known (see, e.g., Freeden and Schreiner 2009), the system is complete in $\mathcal{L}^2(\Omega)$, hence, each function $F \in \mathcal{L}^2(\Omega)$ can be represented by the spherical harmonic expansion

$$F(\xi) = \sum_{n=0}^{\infty} \sum_{m=1}^{2n+1} F^{\wedge}(n,m) Y_{n,m}(\xi), \quad \xi \in \Omega, \tag{11}$$

with "Fourier coefficients" given by

$$F^{\wedge}(n,m) = (F, Y_{n,m})_{\mathcal{L}^2(\Omega)} = \int_{\Omega} F(\xi) Y_{n,m}(\xi)\, d\omega(\xi). \tag{12}$$

Furthermore, the (outer) harmonics $H_{-n-1,m} : \mathbb{R}^3\backslash\{0\} \to \mathbb{R}$ related to the unit sphere Ω are denoted by $H_{-n-1,m}(x) = H^1_{-n-1,m}(x)$, where $H^1_{-n-1,m}(x) = (1/|x|^{n+1}) Y_{n,m}(x/|x|)$. Clearly, they are harmonic functions and their restrictions coincide on Ω with the corresponding spherical harmonics. Any function $F \in \mathcal{L}^2(\Omega)$ can, thus, be identified with a harmonic potential via the expansion (11), in particular, this holds true for the Earth's external gravitational potential. This motivates the following mathematical model situation of the SGG-problem to be considered next:

(i) *Isomorphism*: Consider the sphere $\Omega_R \subset \mathbb{R}^3$ around the origin with radius $R > 0$. Ω_R^{int} is the inner space of Ω_R, and Ω_R^{ext} is the outer space. By virtue of the isomorphism $\Omega \ni \xi \mapsto R\xi \in \Omega_R$ we assume functions $F : \Omega_R \to \mathbb{R}$ to be defined on Ω. It is clear that the function spaces defined on Ω admit their natural generalizations as spaces of functions defined on Ω_R. Obviously, an $\mathcal{L}^2(\Omega)$-orthonormal system of spherical harmonics forms an orthogonal system on Ω_R (with respect to $(\cdot,\cdot)_{\mathcal{L}^2(\Omega_R)}$). Moreover, with the relationship $\xi \leftrightarrow R\xi$, the differential operators on Ω_R can be related to operators on the unit sphere Ω. In more detail, the *surface gradient* $\nabla^{*;R}$, the *surface curl gradient* $\mathrm{L}^{*;\mathrm{R}}$ and the *Beltrami operator* $\Delta^{*;R}$ on Ω_R, respectively, admit the representation $\nabla^{*;R} = (1/R)\nabla^{*;1} = (1/R)\nabla^*$, $\mathrm{L}^{*;\mathrm{R}} = (1/\mathrm{R})\mathrm{L}^{*;1} = (1/\mathrm{R})\mathrm{L}^*$, $\Delta^{*;R} = (1/R^2)\Delta^{*;1} = (1/R^2)\Delta^*$, where ∇^*, L^*, Δ^* are the surface gradient, surface curl gradient, and the Beltrami operator of the unit sphere Ω, respectively. For Y_n being a spherical harmonic of degree n we have $\Delta^{*;R} Y_n = -(1/R^2)\, n(n+1) Y_n = -(1/R^2)\, \Delta^* Y_n$.

(ii) *Runge Property*: Instead of looking for a harmonic function outside and on the (real) Earth, we search for a harmonic function outside the unit sphere Ω (assuming the units are chosen in such a way that the sphere Ω with radius 1 is inside of the Earth and at the same time not too "far away" from the Earth's boundary). The justification of this simplification (see ➋ *Fig. 3*) is based on the Runge approach (see, e.g., Freeden 1980a; Freeden and Michel 2004 as well as the remarks in ➋ Chap. 6 of this handbook): To any harmonic function V

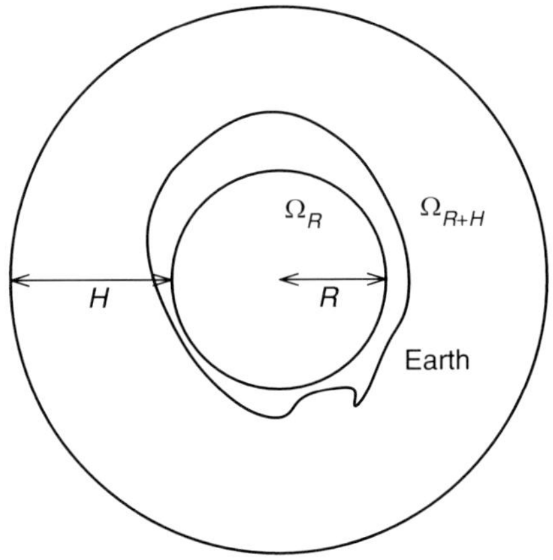

◘ Fig. 3
The role of the "Runge sphere" within the spherically reflected SGG-problem

outside of the (real) Earth and any given $\varepsilon > 0$, there exists a harmonic function U outside of the unit sphere inside the (real) Earth such that the absolute error $|V(x) - U(x)| < \varepsilon$ holds true for all points x outside and on the (real) Earth's surface.

3 Decomposition of Tensor Fields by Means of Tensor Spherical Harmonics

Let us recapitulate that any point $\xi \in \Omega$ may be represented by polar coordinates in a standard way

$$\xi = t\varepsilon^3 + \sqrt{1-t^2}(\cos\varphi\varepsilon^1 + \sin\varphi\varepsilon^2), \quad -1 \leq t \leq 1, \ 0 \leq \varphi < 2\pi, \quad t = \cos\vartheta, \tag{13}$$

($\vartheta \in [0, \pi]$: (co-)latitude, φ: longitude, t: polar distance). Consequently, any element $\xi \in \Omega$ may be represented using its coordinates (φ, t) in accordance with (13).

For the representation of vector and tensor fields on the unit sphere Ω, we are led to use a local triad of orthonormal unit vectors in the directions r, φ, and t as shown by ➋ *Fig. 4* (for more details the reader is referred to Freeden and Schreiner (2009) and the references therein).

As is well-known, the second-order tensor fields on the unit sphere, i.e., $\mathbf{f} : \Omega \to \mathbb{R}^3 \otimes \mathbb{R}^3$, can be separated into their tangential and normal parts as follows:

$$\mathbf{p}_{*,\mathrm{nor}}\mathbf{f} = (\mathbf{f}\xi) \otimes \xi, \tag{14}$$

$$\mathbf{p}_{\mathrm{nor},*}\mathbf{f} = \xi \otimes (\xi^{\mathrm{T}}\mathbf{f}), \tag{15}$$

$$\mathbf{p}_{*,\mathrm{tan}}\mathbf{f} = \mathbf{f} - \mathbf{p}_{*,\mathrm{nor}}\mathbf{f} = f - (\mathbf{f}\xi) \otimes \xi, \tag{16}$$

$$\mathbf{p}_{\mathrm{tan},*}\mathbf{f} = \mathbf{f} - \mathbf{p}_{\mathrm{nor},*}\mathbf{f} = \mathbf{f} - \xi \otimes (\xi^{\mathrm{T}}\mathbf{f}), \tag{17}$$

$$\begin{aligned} \mathbf{p}_{\mathrm{nor},\mathrm{tan}}\mathbf{f} &= \mathbf{p}_{\mathrm{nor},*}(\mathbf{p}_{*,\mathrm{tan}}\mathbf{f}) = \mathbf{p}_{*,\mathrm{tan}}(\mathbf{p}_{\mathrm{nor},*}\mathbf{f}) \\ &= \xi \otimes (\xi^{\mathrm{T}}\mathbf{f}) - (\xi^T\mathbf{f}\xi)\xi \otimes \xi. \end{aligned} \tag{18}$$

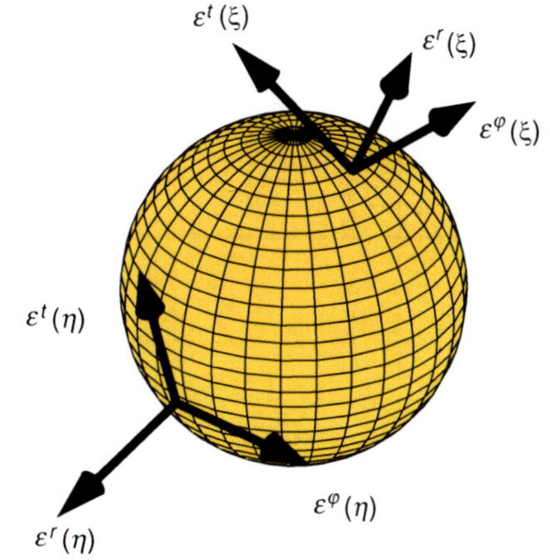

◘ Fig. 4
Local triads ε^r, ε^φ, ε^t with respect to two different points ξ and η on the unit sphere

The operators $\mathbf{p}_{\mathrm{nor,nor}}$, $\mathbf{p}_{\mathrm{tan,nor}}$, and $\mathbf{p}_{\mathrm{tan,tan}}$ are defined analogously. A vector field $\mathbf{f} : \Omega \to \mathbb{R}\otimes\mathbb{R}$ is called normal if $\mathbf{f} = \mathbf{p}_{\mathrm{nor,nor}}\mathbf{f}$ and tangential if $\mathbf{f} = \mathbf{p}_{\mathrm{tan,tan}}\mathbf{f}$. It is called left normal if $\mathbf{f} = \mathbf{p}_{\mathrm{nor},*}\mathbf{f}$, left normal/right tangential if $\mathbf{f} = \mathbf{p}_{\mathrm{nor,tan}}\mathbf{f}$, and so on.

The constant tensor fields $\mathbf{i}_{\mathrm{tan}}$ and $\mathbf{j}_{\mathrm{tan}}$ can be defined using the local triads by

$$\mathbf{i}_{\mathrm{tan}} = \varepsilon^{\varphi} \otimes \varepsilon^{\varphi} + \varepsilon^{t} \otimes \varepsilon^{t}, \quad \mathbf{j}_{\mathrm{tan}} = \xi \wedge \mathbf{i}_{\mathrm{tan}} = \varepsilon^{t} \otimes \varepsilon^{\varphi} - \varepsilon^{\varphi} \otimes \varepsilon^{t}. \tag{19}$$

Spherical tensor fields can be discussed in an elegant manner by the use of certain differential processes. Let u be a continuously differentiable vector field on Ω, i.e., $u \in c^{(1)}(\Omega)$, given in its coordinate form by

$$u(\xi) = \sum_{i=1}^{3} U_i(\xi)\varepsilon^{i}, \quad \xi \in \Omega, \quad U_i \in C^{(1)}(\Omega). \tag{20}$$

Then we define the operators $\nabla^* \otimes$ and $\mathrm{L}^* \otimes$ by

$$\nabla^*_{\xi} \otimes u(\xi) = \sum_{i=1}^{3} \left(\nabla^*_{\xi} U_i(\xi)\right) \otimes \varepsilon^{i}, \quad \xi \in \Omega, \tag{21}$$

$$\mathrm{L}^*_{\xi} \otimes u(\xi) = \sum_{i=1}^{3} \left(\mathrm{L}^*_{\xi} U_i(\xi)\right) \otimes \varepsilon^{i}, \quad \xi \in \Omega. \tag{22}$$

Clearly, $\nabla^* \otimes u$ and $\mathrm{L}^* \otimes u$ are left tangential. But it is an important fact, that even if u is tangential, the tensor fields $\nabla^* \otimes u$ and $\mathrm{L}^* \otimes u$ are generally not tangential. It is obvious, that the product rule is valid. To be specific, let $F \in \mathrm{C}^{(1)}(\Omega)$ and $u \in c^{(1)}(\Omega)$ (once more, note that the notation $u \in c^{(1)}(\Omega)$ means that the vector field u is a continuously differentiable on Ω), then

$$\nabla^*_{\xi} \otimes (F(\xi)u(\xi)) = \nabla^*_{\xi} F(\xi) \otimes u(\xi) + F(\xi)\nabla^*_{\xi} \otimes u(\xi), \quad \xi \in \Omega. \tag{23}$$

In view of the above equations and definitions we accordingly introduce operators $\mathbf{o}^{(i,k)}$: $\mathrm{C}^{(2)}(\Omega) \to \mathbf{c}^{(0)}(\Omega)$ (note that $\mathbf{c}^{(0)}(\Omega)$ is the class of continuous second-order tensor fields on the unit sphere Ω) by

$$\mathbf{o}^{(1,1)}_{\xi} F(\xi) = \xi \otimes \xi F(\xi), \tag{24}$$

$$\mathbf{o}^{(1,2)}_{\xi} F(\xi) = \xi \otimes \nabla^*_{\xi} F(\xi), \tag{25}$$

$$\mathbf{o}^{(1,3)}_{\xi} F(\xi) = \xi \otimes \mathrm{L}^*_{\xi} F(\xi), \tag{26}$$

$$\mathbf{o}^{(2,1)}_{\xi} F(\xi) = \nabla^*_{\xi} F(\xi) \otimes \xi, \tag{27}$$

$$\mathbf{o}^{(3,1)}_{\xi} F(\xi) = \mathrm{L}^*_{\xi} F(\xi) \otimes \xi, \tag{28}$$

$$\mathbf{o}^{(2,2)}_{\xi} F(\xi) = \mathbf{i}_{\mathrm{tan}}(\xi) F(\xi), \tag{29}$$

$$\mathbf{o}^{(2,3)}_{\xi} F(\xi) = \left(\nabla^*_{\xi} \otimes \nabla^*_{\xi} - \mathrm{L}^*_{\xi} \otimes L^*_{\xi}\right) F(\xi) + 2\nabla^*_{\xi} F(\xi) \otimes \xi, \tag{30}$$

$$\mathbf{o}^{(3,2)}_{\xi} F(\xi) = \left(\nabla^*_{\xi} \otimes \mathrm{L}^*_{\xi} + \mathrm{L}^*_{\xi} \otimes \nabla^*_{\xi}\right) F(\xi) + 2\mathrm{L}^*_{\xi} F(\xi) \otimes \xi, \tag{31}$$

$$\mathbf{o}^{(3,3)}_{\xi} F(\xi) = \mathbf{j}_{\mathrm{tan}}(\xi) F(\xi), \tag{32}$$

$\xi \in \Omega$.

After our preparations involving spherical second-order tensor fields it is not difficult to prove the following lemma.

Lemma 3.1
Let $F : \Omega \to \mathbb{R}$ be sufficiently smooth. Then the following statements are valid:

1. $\mathbf{o}^{(1,1)}F$ *is a normal tensor field.*
2. $\mathbf{o}^{(1,2)}F$ *and* $\mathbf{o}^{(1,3)}F$ *are left normal/right tangential.*
3. $\mathbf{o}^{(2,1)}F$ *and* $\mathbf{o}^{(3,1)}F$ *are left tangential/right normal.*
4. $\mathbf{o}^{(2,2)}F$, $\mathbf{o}^{(2,3)}F$, $\mathbf{o}^{(3,2)}F$ *and* $\mathbf{o}^{(3,3)}F$ *are tangential.*
5. $\mathbf{o}^{(1,1)}F$, $\mathbf{o}^{(2,2)}F$, $\mathbf{o}^{(2,3)}F$ *and* $\mathbf{o}^{(3,2)}F$ *are symmetric.*
6. $\mathbf{o}^{(3,3)}F$ *is skew-symmetric.*
7. $\left(\mathbf{o}^{(1,2)}F\right)^T = \mathbf{o}^{(2,1)}F$ *and* $\left(\mathbf{o}^{(1,3)}F\right)^T = \mathbf{o}^{(3,1)}F$.
8. *For* $\xi \in \Omega$

$$\operatorname{trace} \mathbf{o}_\xi^{(i,k)} F(\xi) = \begin{cases} F(\xi) & \textit{for} \quad (i,k) = (1,1) \\ 2F(\xi) & \textit{for} \quad (i,k) = (2,2) \\ 0 & \textit{for} \quad (i,k) \neq (1,1),\ (2,2) \end{cases} .$$

The tangent representation theorem (cf. Backus 1966, 1967) asserts that if $\mathbf{p}_{\tan,\tan}\mathbf{f}$ is the tangential part of a tensor field $\mathbf{f} \in \mathbf{c}^{(2)}(\Omega)$, as defined above, then there exist unique scalar fields $F_{2,2}, F_{3,3}, F_{2,3}, F_{3,2}$ such that

$$\int_\Omega F_{2,2}(\xi)\, d\omega(\xi) = \int_\Omega F_{3,3}(\xi)\, d\omega(\xi) = 0, \tag{33}$$

$$\int_\Omega F_{3,2}(\xi)(\varepsilon^i \cdot \xi)\, d\omega(\xi) = \int_\Omega F_{2,3}(\xi)(\varepsilon^i \cdot \xi)\, d\omega(\xi) = 0, \quad i = 1,2,3, \tag{34}$$

and

$$\mathbf{p}_{\tan,\tan}\mathbf{f} = \mathbf{o}^{(2,2)}F_{2,2} + \mathbf{o}^{(2,3)}F_{2,3} + \mathbf{o}^{(3,2)}F_{3,2} + \mathbf{o}^{(3,3)}F_{3,3}. \tag{35}$$

Furthermore, the following orthogonality relations may be formulated: Let $F, G : \Omega \to \mathbb{R}$ be sufficiently smooth. Then for all $\xi \in \Omega$, we have $\mathbf{o}_\xi^{(i,k)}F(\xi) \cdot \mathbf{o}_\xi^{(i',k')}F(\xi) = 0$ whenever $(i,k) \neq (i',k')$. The adjoint operators $O^{(i,k)}$ satisfying

$$\int_\Omega \mathbf{o}^{(i,k)}F(\xi) \cdot \mathbf{f}(\xi)\, d\omega(\xi) = \int_\Omega F(\xi)\, O^{(i,k)}\mathbf{f}(\xi)\, d\omega(\xi), \tag{36}$$

for all sufficiently smooth functions $F : \Omega \to \mathbb{R}$ and tensor fields $\mathbf{f} : \Omega \to \mathbb{R}^3 \otimes \mathbb{R}^3$ can be deduced by elementary calculations. In more detail, for $\mathbf{f} \in \mathbf{c}^{(2)}(\Omega)$, we find (cf. Freeden and Schreiner 2009)

$$O_\xi^{(1,1)}\mathbf{f}(\xi) = \xi^T\mathbf{f}(\xi)\xi, \tag{37}$$

$$O_\xi^{(1,2)}\mathbf{f}(\xi) = -\nabla_\xi^* \cdot p_{\tan}\left(\xi^T\mathbf{f}(\xi)\right), \tag{38}$$

$$O_\xi^{(1,3)}\mathbf{f}(\xi) = -\mathrm{L}_\xi^* \cdot p_{\tan}\left(\xi^T\mathbf{f}(\xi)\right), \tag{39}$$

$$O_\xi^{(2,1)}\mathbf{f}(\xi) = -\nabla_\xi^* \cdot p_{\tan}\left(\mathbf{f}(\xi)\xi\right), \tag{40}$$

$$O_\xi^{(3,1)}\mathbf{f}(\xi) = -L_\xi^* \cdot p_{\tan}\left(\mathbf{f}(\xi)\xi\right), \tag{41}$$

$$O_\xi^{(2,2)}\mathbf{f}(\xi) = \mathbf{i}_{\tan}(\xi)\cdot\mathbf{f}(\xi), \tag{42}$$

$$\begin{aligned} O_\xi^{(2,3)}\mathbf{f}(\xi) = {} & \nabla_\xi^* \cdot p_{\tan}\left(\nabla_\xi^* \cdot \mathbf{p}_{\tan,*}\mathbf{f}(\xi)\right) - L_\xi^* \cdot p_{\tan}\left(L_\xi^* \cdot \mathbf{p}_{\tan,*}\mathbf{f}(\xi)\right) \\ & - 2\nabla_\xi^* \cdot p_{\tan}\left(\mathbf{f}(\xi)\xi\right), \end{aligned} \tag{43}$$

$$\begin{aligned} O_\xi^{(3,2)}\mathbf{f}(\xi) = {} & L_\xi^* \cdot p_{\tan}\left(\nabla_\xi^* \cdot \mathbf{p}_{\tan,*}\mathbf{f}(\xi)\right) + \nabla_\xi^* \cdot p_{\tan}\left(L_\xi^* \cdot \mathbf{p}_{\tan,*}\mathbf{f}(\xi)\right) \\ & - 2L_\xi^* \cdot p_{\tan}\left(\mathbf{f}(\xi)\xi\right), \end{aligned} \tag{44}$$

$$O_\xi^{(3,3)}\mathbf{f}(\xi) = \mathbf{j}_{\tan}(\xi)\cdot\mathbf{f}(\xi), \tag{45}$$

$\xi \in \Omega$. Provided that $F : \Omega \to \mathbb{R}$ is sufficiently smooth we see that

$$O_\xi^{(i',k')}\mathbf{o}_\xi^{(i,k)}F(\xi) = 0 \ \text{ if } (i,k) \neq (i',k'), \tag{46}$$

whereas

$$O_\xi^{(i,k)}\mathbf{o}_\xi^{(i,k)}F(\xi) = \begin{cases} F(\xi) & \text{if } (i,k) = (1,1) \\ -\Delta^* F(\xi) & \text{if } (i,k) \in \{(1,2),(1,3) \\ & \qquad (2,1),(3,1)\} \\ 2F(\xi) & \text{if } (i,k) \in \{(2,2),(3,3)\} \\ 2\Delta^*(\Delta^*+2)F(\xi) & \text{if } (i,k) \in \{(2,3),(3,2)\}. \end{cases} \tag{47}$$

Using this set of operators we can find explicit formulas for the functions $F_{i,k}$ in the tensor decomposition theorem.

Theorem 3.2

Helmholtz decomposition theorem: Let $\mathbf{f}$ *be of class* $\mathbf{c}^{(2)}(\Omega)$*. Then there exist uniquely defined functions* $F_{i,k} \in C^{(2)}(\Omega)$, $(i,k) \in \{(1,1), (1,2), \ldots, (3,3)\}$ *with* $(F_{i,k}, Y_0)_{\mathcal{L}^2(\Omega)} = 0$ *for all spherical harmonic* Y_0 *of degree* 0*, if* $(i,k) \in \{(1,2), (1,3), (2,1), (2,3), (3,1), (3,2)\}$ *and* $(F_{i,k}, Y_1)_{\mathcal{L}^2(\Omega)} = 0$ *for all spherical harmonics* Y_1 *of degree 1 if* $(i,k) \in \{(2,3),(3,2)\}$*, in such a way that*

$$\mathbf{f} = \sum_{i,k=1}^{3} \mathbf{o}^{(i,k)} F_{i,k}, \tag{48}$$

where the functions $\xi \mapsto F_{i,k}(\xi)$, $\xi \in \Omega$*, are explicitly given by*

$$F_{1,1}(\xi) = O_\xi^{(1,1)}\mathbf{f}(\xi), \tag{49}$$

$$F_{2,2}(\xi) = \frac{1}{2}O_\xi^{(2,2)}\mathbf{f}(\xi), \tag{50}$$

$$F_{3,3}(\xi) = \frac{1}{2}O_\xi^{(3,3)}\mathbf{f}(\xi), \tag{51}$$

$$F_{1,2}(\xi) = -\int_{\Omega} G(\Delta^*; \xi, \eta) O_\eta^{(1,2)} \mathbf{f}(\eta)\, d\omega(\eta), \tag{52}$$

$$F_{1,3}(\xi) = -\int_{\Omega} G(\Delta^*; \xi, \eta) O_\eta^{(1,3)} \mathbf{f}(\eta)\, d\omega(\eta), \tag{53}$$

$$F_{2,1}(\xi) = -\int_{\Omega} G(\Delta^*; \xi, \eta) O_\eta^{(2,1)} \mathbf{f}(\eta)\, d\omega(\eta), \tag{54}$$

$$F_{3,1}(\xi) = -\int_{\Omega} G(\Delta^*; \xi, \eta) O_\eta^{(3,1)} \mathbf{f}(\eta)\, d\omega(\eta), \tag{55}$$

$$F_{2,3}(\xi) = \int_{\Omega} G(\Delta^*(\Delta^* + 2); \xi, \eta) O_\eta^{(2,3)} \mathbf{f}(\eta)\, d\omega(\eta), \tag{56}$$

$$F_{3,2}(\xi) = \int_{\Omega} G(\Delta^*(\Delta^* + 2); \xi, \eta) O_\eta^{(3,2)} \mathbf{f}(\eta)\, d\omega(\eta). \tag{57}$$

The functions $G(\Delta^*; \cdot, \cdot)$ and $G(\Delta^*(\Delta^* + 2); \cdot, \cdot)$ are the Green functions to the Beltrami operator Δ^* and its iteration $\Delta^*(\Delta^* + 2)$, respectively. For more details concerning the Green functions we refer to Freeden (1980b) and Freeden and Schreiner (2009).

The decomposition (Theorem 3.2) will be of crucial importance to verify uniqueness results for the satellite gravity gradiometry problem in spherical context.

4 Solution as Pseudodifferential Equation

Suppose that the function $H : \mathbb{R}^3 \setminus \{0\} \to \mathbb{R}$ is twice continuously differentiable. We want to show how the Hesse matrix restricted to the unit sphere Ω, i.e.,

$$\mathbf{h}(\xi) = \nabla_x \otimes \nabla_x H(x)|_{|x|=1}, \quad \xi \in \Omega, \tag{58}$$

can be decomposed according to the rules of Theorem 3.2. In order to evaluate

$$\nabla_x \otimes \nabla_x H(x) = \left(\xi \frac{\partial}{\partial r} + \frac{1}{r}\nabla_\xi^*\right) \otimes \left(\xi \frac{\partial}{\partial r} + \frac{1}{r}\nabla_\xi^*\right) H(r\xi), \tag{59}$$

we first see that

$$\xi \frac{\partial}{\partial r} \otimes \xi \frac{\partial}{\partial r} H(r\xi) = \xi \otimes \xi \left(\frac{\partial}{\partial r}\right)^2 H(r\xi), \tag{60}$$

$$\xi \frac{\partial}{\partial r} \otimes \frac{1}{r}\nabla_\xi^* H(r\xi) = -\frac{1}{r^2}\xi \otimes \nabla_\xi^* H(r\xi) + \frac{1}{r}\xi \otimes \nabla_\xi^* \frac{\partial}{\partial r} H(r\xi), \tag{61}$$

$$\frac{1}{r}\nabla_\xi^* \otimes \xi \frac{\partial}{\partial r} H(r\xi) = \frac{1}{r}\mathbf{i}_{\tan}(\xi)\frac{\partial}{\partial r} H(r\xi) + \frac{1}{r}\nabla_\xi^* \left(\frac{\partial}{\partial r} H(r\xi)\right) \otimes \xi, \tag{62}$$

$$\frac{1}{r}\nabla_\xi^* \otimes \frac{1}{r}\nabla_\xi^* H(r\xi) = \frac{1}{r^2}\nabla_\xi^* \otimes \nabla_\xi^* H(r\xi). \tag{63}$$

Summing up these terms we find (cf. Freeden and Schreiner (2009))

$$\begin{aligned}\nabla_x \otimes \nabla_x H(x)|_{|x|=1} &= \xi \otimes \xi \left(\frac{\partial}{\partial r}\right)^2 H(r\xi)|_{r=1} + \xi \otimes \nabla_\xi^* \left(\frac{\partial}{\partial r} H(r\xi)|_{r=1} - H(\xi)\right) \\ &+ \left(\nabla^* \frac{\partial}{\partial r} H(r\xi)|_{r=1}\right) \otimes \xi + \nabla_\xi^* \otimes \nabla_\xi^* H(\xi) + \mathbf{i}_{\tan}(\xi)\frac{\partial}{\partial r} H(r\xi)|_{r=1}. \end{aligned} \tag{64}$$

Using the identities (60)–(63) and the definition of the $\mathbf{o}^{(i,k)}$-operators we are able to write

$$\begin{aligned}\nabla_x \otimes \nabla_x H(x)|_{|x|=1} = {} & \mathbf{o}_\xi^{(1,1)} \left(\left(\frac{\partial}{\partial r} \right)^2 H(r\xi)|_{r=1} \right) + \mathbf{o}_\xi^{(1,2)} \left(\frac{\partial}{\partial r} H(r\xi)|_{r=1} - H(\xi) \right) \\ & + \mathbf{o}_\xi^{(2,1)} \left(\frac{\partial}{\partial r} H(r\xi)|_{r=1} - H(\xi) \right) + \mathbf{o}_\xi^{(2,2)} \left(\frac{1}{2} \Delta_\xi^* H(\xi) + \frac{\partial}{\partial r} H(r\xi)|_{r=1} \right) \\ & + \mathbf{o}_\xi^{(2,3)} \frac{1}{2} H(\xi). \end{aligned} \tag{65}$$

In particular, if we consider an outer harmonic $H_{-n-1,m} : x \mapsto H_{-n-1,m}(x)$ with $H_{-n-1,m}(r\xi) = r^{-(n+1)} Y_{n,m}(\xi)$, $r > 0$, $\xi \in \Omega$, we obtain the following decomposition of the Hesse matrix on the sphere Ω_{R+H}, i.e., for $x \in \mathbb{R}^3$ with $|x| = R + H$:

$$\begin{aligned}\nabla \otimes \nabla H_{-n-1,m}((R+H)\,\xi) = {} & (n+1)(n+2) \frac{1}{(R+H)^{n+3}} \mathbf{o}_\xi^{(1,1)} Y_{n,m}(\xi) \\ & - (n+2) \frac{1}{(R+H)^{n+3}} \left(\mathbf{o}_\xi^{(1,2)} Y_{n,m}(\xi) + \mathbf{o}_\xi^{(2,1)} Y_{n,m}(\xi) \right) \\ & - \frac{(n+1)(n+2)}{2} \frac{1}{(R+H)^{n+3}} \mathbf{o}_\xi^{(2,2)} Y_{n,m}(\xi) \\ & + \frac{1}{2} \frac{1}{(R+H)^{n+3}} \mathbf{o}_\xi^{(2,3)} Y_{n,m}(\xi). \end{aligned} \tag{66}$$

Keeping in mind, that any solution of the SGG-problem can be expressed as a series of outer harmonics and using the completeness of the spherical harmonics in the space of square-integrable functions on the unit sphere, it follows that the SGG problem is uniquely solvable (up to some low order spherical harmonics) by the $O^{(1,1)}$, $O^{(1,2)}$, $O^{(2,1)}$, $O^{(2,2)}$, and $O^{(2,3)}$ components. To be more specific, we are able to formulate the following theorem:

Theorem 4.1
Let V satisfy the following condition $V \in Pot(\mathcal{C}^{(0)}(\Omega))$, i.e.,

$$V \in \mathcal{C}^{(0)}(\overline{\Omega^{ext}}) \cap \mathcal{C}^{(2)}(\Omega^{ext}), \tag{67}$$

$$\Delta V(x) = 0, \quad x \in \Omega^{ext}, \tag{68}$$

$$|V(x)| = O\left(\frac{1}{|x|}\right), |x| \to \infty, \textit{ uniformly for all directions.} \tag{69}$$

Then the following statements are valid:

1. $O^{(i,k)} \nabla \otimes \nabla V((R+H)\xi) = 0$ *if* $(i,k) \in \{(1,3),(3,1),(3,2),(3,3)\}$.
2. $O^{(i,k)} \nabla \otimes \nabla V((R+H)\xi) = 0$ *for* $(i,k) \in \{(1,1),(2,2)\}$ *if and only if* $V = 0$.
3. $O^{(i,k)} \nabla \otimes \nabla V((R+H)\xi) = 0$ *for* $(i,k) \in \{(1,2),(2,1)\}$ *if and only if* $V|_\Omega$ *is constant.*
4. $O^{(2,3)} \nabla \otimes \nabla V((R+H)\xi) = 0$ *if and only if* $V|_\Omega$ *is linear combination of spherical harmonics of degree 0 and 1.*

This theorem gives detailed information, which tensor components of the Hesse tensor ensure the uniqueness of the SGG-problem (see also the considerations due to Schreiner (1994a), Freeden et al. (2002)). Anyway, for a potential V of class $Pot(\mathcal{C}^{(0)}(\Omega))$ with vanishing spherical harmonic moments of degree 0 and 1 such as the Earth's disturbing potential (see, e.g., Heiskanen and Moritz (1967) for its definition) uniqueness is assured in all cases (listed in Theorem 4.1).

Since we now know at least in the spherical setting, which conditions guarantee the uniqueness of an SGG-solution we can turn to the question of how to find a solution and what we mean with a solution, since we have to take into account the ill–posedness. To this end, we are interested here in analyzing the problem step by step. We start with the reformulation of the SGG-problem as pseudo differential equation on the sphere, give a short overview on regularization, and show how this ingredients can be composed together to regularize the SGG-data.

In doing so we find great help by discussing how classical boundary value problems in gravitational field of the Earth as well as modern satellite problems may be transferred into pseudodifferential equations, thereby always assuming the spherically oriented geometry. Indeed, it is helpful to treat the classical Dirichlet and Neumann boundary value problem as well as significant satellite problems such as satellite-to-satellite tracking (SST) and SGG.

4.1 SGG as Pseudodifferential Equation

Let $\Sigma \subset \mathbb{R}^3$ be a regular surface, i.e., we assume the following properties: (i) Σ divides the Euclidean space $\mathbb{R}^3$ into the bounded region Σ^{int} (inner space) and the unbounded region Σ^{ext} (outer space) so that $\Sigma^{\text{ext}} = \mathbb{R}^3 \backslash \overline{\Sigma^{\text{int}}}$, $\Sigma = \overline{\Sigma^{\text{int}}} \cap \overline{\Sigma^{\text{ext}}}$ with $\varnothing = \Sigma^{\text{int}} \cap \Sigma^{\text{ext}}$, (ii) Σ^{int} contains the origin, (iii) Σ is a closed and compact surface free of double points, (iv) Σ is locally of class $\mathcal{C}^{(2)}$ (see Freeden and Michel (2004) for more details concerning regular surfaces).

From our preparatory considerations (in particular, from the Introduction) it can be deduced that a gravitational potential of interest may be understood to be a member of the class $V \in Pot(\mathcal{C}^{(0)}(\Sigma))$, i.e.,

$$V \in \mathcal{C}^{(2)}(\overline{\Sigma^{ext}}) \cap \mathcal{C}^{(2)}(\Sigma^{ext}), \tag{70}$$

$$\Delta V(x) = 0, \quad x \in \Sigma^{ext}, \tag{71}$$

$$|V(x)| = O\left(\frac{1}{|x|}\right), \quad |x| \to \infty, \text{uniformly for all directions.} \tag{72}$$

Assume that $\Omega_R = \{x \in \mathbb{R}^3 \mid |x| = R\}$ is a (Runge) sphere with radius R around the origin, i.e., a sphere that lies entirely inside Σ, i.e. $\Omega_R \subset \Sigma^{\text{int}}$. On the class $\mathcal{L}^2(\Omega_R)$ we impose the inner product $(\cdot,\cdot)_{\mathcal{L}^2(\Omega_R)}$. Then we know that the functions $\frac{1}{R} Y_{n,m}\left(\frac{\cdot}{R}\right)$ form an orthonormal set of functions on Ω_R, i.e., given $F \in \mathcal{L}^2(\Omega_R)$, its Fourier expansion reads

$$F(x) = \sum_{n=0}^{\infty} \sum_{m=1}^{2n+1} \frac{1}{R^2} \left(F, Y_{n,m}\left(\frac{\cdot}{R}\right)\right)_{\mathcal{L}^2(\Omega_R)} Y_{n,m}\left(\frac{x}{R}\right), \quad x \in \Omega_R. \tag{73}$$

Instead of considering potentials that are harmonic outside Σ and continuous on Σ, we now consider potentials that are harmonic outside Ω_R and that are of class $\mathcal{L}^2(\Omega_R)$. In accordance with our notation we define

$$\text{Pot}(\mathcal{L}^2(\Omega_R)) = \left\{x \mapsto \sum_{n=0}^{\infty} \sum_{m=1}^{2n+1} \frac{1}{R^2} \left(F, Y_{n,m}\left(\frac{\cdot}{R}\right)\right)_{\mathcal{L}^2(\Omega_R)} \frac{R^{n+1}}{|x|^{n+1}} Y_{n,m}\left(\frac{x}{|x|}\right) \mid F \in \mathcal{L}^2(\Omega_R)\right\}. \tag{74}$$

Clearly, $\mathrm{Pot}(\mathcal{L}^2(\Omega_R))$ is a "subset" of $\mathrm{Pot}(\mathcal{C}^{(0)}(\Sigma))$ in the sense that if $V \in \mathrm{Pot}(\mathcal{L}^2(\Omega_R))$, then $V|_{\overline{\Sigma^{\mathrm{ext}}}} \in \mathrm{Pot}(\mathcal{C}^{(0)}(\Sigma))$. The "difference" of these two spaces is not "too large": Indeed, we know from the Runge approximation theorem (cf. Freeden 1980a), that for every $\varepsilon > 0$ and every $V \in \mathrm{Pot}(\mathcal{C}^{(0)}(\Sigma))$ there exists a $\hat{V} \in \mathrm{Pot}(\mathcal{L}^2(\Omega_R))$ such that $\sup_{x \in \Sigma^{\mathrm{ext}}} |V(x) - \hat{V}(x)| < \varepsilon$. Thus, in all geosciences, it is common (but not strictly consistent with the Runge argumentation) to identify Ω_R with the surface of the Earth and to assume that the restriction $V|_{\Omega_R}$ is of class $\mathcal{L}^2(\Omega_R)$. Clearly, we have a canonical isomorphism between $\mathcal{L}^2(\Omega_R)$ and $\mathrm{Pot}(\mathcal{L}^2(\Omega_R))$, which is defined via the trace operator, i.e., the restriction to Ω_R and its harmonic continuation, respectively.

4.2 Upward/Downward Continuation

Let Ω_{R+H} be the sphere with radius $R + H$. The system $\frac{1}{R+H} Y_{n,m}\left(\frac{\cdot}{R+H}\right)$ is then orthonormal in $\mathcal{L}^2(\Omega_{R+H})$. (We assume H to be the height of a satellite above the Earth's surface). Let $F \in \mathrm{Pot}(\mathcal{L}^2(\Omega_R))$ be represented in the form

$$x \mapsto \sum_{n=0}^{\infty} \sum_{m=1}^{2n+1} \frac{1}{R^2} \left(F, Y_{n,m}\left(\frac{\cdot}{R}\right)\right)_{\mathcal{L}^2(\Omega_R)} \frac{R^{n+1}}{|x|^{n+1}} Y_{n,m}\left(\frac{x}{|x|}\right). \tag{75}$$

Then the restriction of F on Ω_{R+H} reads

$$F|_{\Omega_{R+H}} : x \mapsto \sum_{n=0}^{\infty} \sum_{m=1}^{2n+1} \frac{1}{R^2} \left(F, Y_{n,m}\left(\frac{\cdot}{R}\right)\right)_{\mathcal{L}^2(\Omega_R)} \frac{R^{n+1}}{(R+H)^{n+1}} Y_{n,m}\left(\frac{x}{R+H}\right). \tag{76}$$

Hence, any element $\frac{1}{R} Y_{n,m}\left(\frac{\cdot}{R}\right)$ of the orthonormal system in $\mathcal{L}^2(\Omega_R)$ is mapped to a function $R^n/(R+H)^n\ 1/R+H\ Y_{n,m}(\cdot/R+H)$. The operation defined in such away is called *upward continuation*. It is representable by the pseudodifferential operator (for more details on pseudodifferential operators the reader should consult Svensson (1983), Schneider (1997), Freeden et al. (1998), and Freeden 1999) as well as ➋ Chap. 27 of this handbook

$$\Lambda_{\mathrm{up}}^{R,H} : \mathcal{L}^2(\Omega_R) \longrightarrow \mathcal{L}^2(\Omega_{R+H})$$

with associated symbol

$$\left(\Lambda_{\mathrm{up}}^{R,H}\right)^{\wedge}(n) = \frac{R^n}{(R+H)^n}. \tag{77}$$

In other words, we have

$$\Lambda_{\mathrm{up}}^{R,H}\left(\frac{1}{R} Y_{n,m}\left(\frac{\cdot}{R}\right)\right) = \left(\Lambda_{\mathrm{up}}^{R,H}\right)^{\wedge}(n) \frac{1}{R+H} Y_{n,m}\left(\frac{\cdot}{R+H}\right). \tag{78}$$

The image of $\Lambda_{\mathrm{up}}^{R,H}$ is given by Picard's criterion (cf. Theorem 4.4):

$$\Lambda_{\mathrm{up}}^{R,H}(\mathcal{L}^2(\Omega_R)) = \left\{ F \in \mathcal{L}^2(\Omega_{R+H}) \middle| \sum_{n=0}^{\infty} \sum_{m=1}^{2n+1} \left(\frac{(R+H)^n}{R^n}\right)^2 \right.$$
$$\left. \times \left(F, \frac{1}{R+H} Y_{n,m}\left(\frac{\cdot}{R+H}\right)\right)^2_{\mathcal{L}^2(\Omega_{R+H})} < \infty \right\}. \tag{79}$$

The inverse of $\Lambda_{\text{up}}^{R,H}$ is called the *downward continuation* operator, $\Lambda_{\text{down}}^{R,H} = \left(\Lambda_{\text{up}}^{R,H}\right)^{-1}$. It brings down the gravitational potential at height $R + H$ to the height R:

$$\Lambda_{\text{down}}^{R,H} : \Lambda_{\text{up}}^{R,H}(\mathcal{L}^2(\Omega_{R+H})) \longrightarrow \mathcal{L}^2(\Omega_R)$$

with

$$\Lambda_{\text{down}}^{R,H}\left(\frac{1}{R+H}Y_{n,m}\left(\frac{\cdot}{R+H}\right)\right) = \frac{(R+H)^n}{R^n}\frac{1}{R}Y_{n,m}\left(\frac{\cdot}{R}\right) \tag{80}$$

such that the symbol of $\Lambda_{\text{down}}^{R,H}$ is

$$\left(\Lambda_{\text{down}}^{R,H}\right)^{\wedge}(n) = \frac{(R+H)^n}{R^n}. \tag{81}$$

It is obvious that the upward continuation is well-posed, whereas the downward continuation generates an ill-posed problem.

4.3 Operator of the First-Order Radial Derivative

Let $F \in \text{Pot}(\mathcal{L}^2(\Omega_R))$ have the representation (75). If we restrict F to a sphere Ω_γ with radius γ, we have

$$\sum_{n=0}^{\infty}\sum_{m=1}^{2n+1}\frac{1}{R^2}\left(F, Y_{n,m}\left(\frac{\cdot}{R}\right)\right)_{\mathcal{L}^2(\Omega_R)}\frac{R^{n+1}}{\gamma^{n+1}}Y_{n,m}\left(\frac{x}{\gamma}\right), \quad x \in \Omega_\gamma. \tag{82}$$

Accordingly, the restriction of $\frac{\partial}{\partial r}F$ to Ω_γ amounts to

$$\sum_{n=0}^{\infty}\sum_{m=1}^{2n+1}\frac{1}{R^2}\left(F, Y_{n,m}\left(\frac{\cdot}{R}\right)\right)_{\mathcal{L}^2(\Omega_R)}\frac{-(n+1)}{\gamma}\frac{R^{n+1}}{\gamma^{n+1}}Y_{n,m}\left(\frac{x}{\gamma}\right). \tag{83}$$

Thus, the process of forming the first radial derivative at height γ constitutes the pseudodifferential operator $\Lambda_{\text{FND}}^{\gamma}$ (FND stands for first-order normal derivative) with the symbol

$$\left(\Lambda_{\text{FND}}^{\gamma}\right)^{\wedge}(n) = -\frac{n+1}{\gamma}. \tag{84}$$

4.4 Pseudodifferential Operator for SST

The principle of SST is sketched in ➋ *Fig. 5* (note that two variants of SST are discussed in satellite techniques, the so-called high–low and the low–low method. We only explain here the high–low variant, for which the GFZ–satellite CHAMP (CHAllenging Minisatellite Payload) launched in 2000 is a prototype.

The motion of a satellite in a low orbit such as CHAMP, GOCE (typical heights are in the range 200–500 km) is tracked with a GPS receiver. So the relative motion between the satellite and the GPS–satellites (the latter have a height of approximately 2,0000 km) can be measured. Assuming that the motion of the GPS–satellites is known (in fact, their orbit is very stable because of the large height), one can calculate the acceleration of the low orbiting satellite. Since the acceleration and the force acting on the satellite are proportional by Newton's law, one gets information about the gradient field $\nabla V(p)$ at the satellite's position p. Assuming that

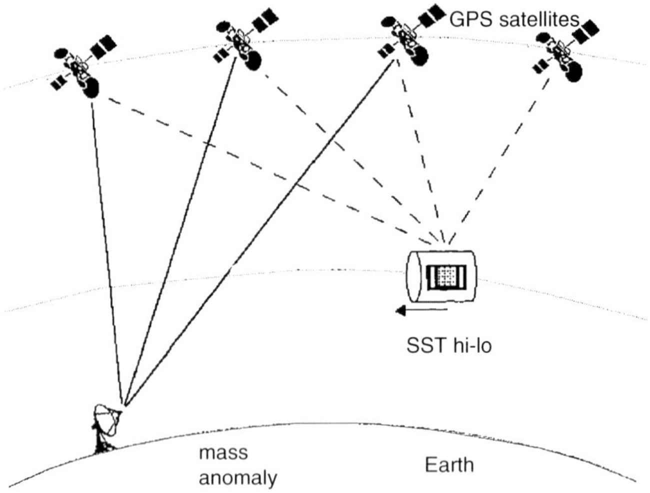

Fig. 5
The principle of satellite–to satellite tracking (from ESA (1999))

the height variations of the satellite are small, we obtain data information of ∇V at height H, that is on the sphere Ω_{R+H}. For simplicity, it is useful to consider only the radial component from these vectorial data, which is the first radial derivative.

Thus, given $F \in \mathrm{Pot}(\mathcal{L}^2(\Omega_R))$, we get the SST–data by a process of upward continuation and then taking the first radial derivative. Mathematically, SST amounts to introduce the operator

$$\Lambda_{\mathrm{SST}}^{R,H} : \mathcal{L}^2(\Omega_R) \longrightarrow \mathcal{L}^2(\Omega_{R+H})$$

via

$$\Lambda_{\mathrm{SST}}^{R,H} = -\Lambda_{\mathrm{FND}}^{R+H} \Lambda_{\mathrm{up}}^{R,H} \tag{85}$$

(we use the minus sign here, to avoid the minus in the symbol), and get

$$\left(\Lambda_{\mathrm{SST}}^{R,H}\right)^{\wedge}(n) = \frac{R^n}{(R+H)^2} \frac{n+1}{R+H}. \tag{86}$$

It is easily seen that the Picard criterion (see, e.g., Engl et al. (1997)) reads for this operator

$$\Lambda_{\mathrm{SST}}^{R,H}(\mathcal{L}^2(\Omega_R)) = \left\{ F \in \mathcal{L}^2(\Omega_{R+H}) \middle| \sum_{n=0}^{\infty} \sum_{m=1}^{2n+1} \left(\frac{(R+H)^n}{R^n} \frac{R+H}{n+1} \right)^2 \right. \\ \left. \times \left(F, \frac{1}{R+H} Y_{n,m}\left(\frac{\cdot}{R+H}\right) \right)^2_{\mathcal{L}^2(\Omega_{R+H})} < \infty \right\}. \tag{87}$$

4.5 Pseudodifferential Operator of the Second-Order Radial Derivative

Analogous considerations applied to the operator $\frac{\partial^2}{\partial r^2}$ on F in (75) at height γ yields

$$\sum_{n=0}^{\infty}\sum_{m=1}^{2n+1}\frac{1}{R^2}\left(F, Y_{n,m}\left(\frac{\cdot}{R}\right)\right)_{\mathcal{L}^2(\Omega_R)}\frac{(n+1)(n+2)}{\gamma^2}\frac{R^{n+1}}{\gamma^{n+1}}Y_{n,m}\left(\frac{x}{\gamma}\right), \quad x\in\Omega_\gamma. \tag{88}$$

Thus, the second-order radial derivative at height γ is represented by the pseudodifferential operator $\Lambda^{\gamma}_{\mathrm{SND}}$ with the symbol

$$(\Lambda^{\gamma}_{\mathrm{SND}})^{\wedge}(n) = \frac{(n+1)(n+2)}{\gamma^2}. \tag{89}$$

4.6 Pseudodifferential Operator for Satellite Gravity Gradiometry

If we restrict ourselves for the moment to the second-order radial derivative $\frac{\partial^2}{\partial r^2}V$, and assume that the height of the satellite is H, we are led to the pseudodifferential operator describing satellite gravity gradiometry by

$$\Lambda^{R,H}_{\mathrm{SGG}} = \Lambda^{R+H}_{\mathrm{SND}}\Lambda^{R,H}_{\mathrm{up}}$$

so that

$$\left(\Lambda^{R,H}_{\mathrm{SGG}}\right)^{\wedge}(n) = \frac{R^n}{(R+H)^n}\frac{(n+1)(n+2)}{(R+H)^2}. \tag{90}$$

In consequence,

$$\Lambda^{R,H}_{\mathrm{SGG}} : \mathcal{L}^2(\Omega_R) \longrightarrow \mathcal{L}^2(\Omega_{R+H})$$

with

$$\Lambda^{R,H}_{\mathrm{SGG}}(\mathcal{L}^2(\Omega_R)) = \left\{F\in\mathcal{L}^2(\Omega_{R+H})\,\middle|\, \sum_{n=0}^{\infty}\sum_{m=1}^{2n+1}\left(\frac{(R+H)^n}{R^n}\frac{(R+H)^2}{(n+1)(n+2)}\right)^2 \times\left(F,\frac{1}{R+H}Y_{n,m}\left(\frac{\cdot}{R+H}\right)\right)^2_{\mathcal{L}^2(\Omega_{R+H})}<\infty\right\}. \tag{91}$$

4.7 Survey on Pseudodifferential Operators Relevant in Satellite Technology

Until now, our purpose was to develop a class of pseudodifferential operators, which describe, in particular, important operations for actual and future satellite missions. In what follows, we are interested in a brief mathematical survey about our investigations. In order to keep the forthcoming notations as simple as possible, we use the fact that all spheres around the origin are isomorphic. Thus, we consider the resulting pseudodifferential operators on the unit sphere and ignore the different heights in the domain of definition of the functions, but not in the symbol of the operators. Hence, we can use the results of the last chapters directly for the regularization of the satellite problems. If one wants to incorporate the different heights, one has only to observe the factors R and $R+H$, respectively.

Operator	Description	Symbol	Order
$\Lambda^{R,H}_{\text{up}}$	Upward continuation operator	$\frac{R^n}{(R+H)^n}$	$-\infty$
$\Lambda^{R,H}_{\text{down}}$	Downward continuation operator	$\frac{(R+H)^n}{R^n}$	∞
Λ^{R}_{FND}	First-order radial derivative at the Earth surface	$-\frac{(n+1)}{R}$	1
Λ^{R}_{SND}	Second-order radial derivative at the Earth surface	$\frac{(n+1)(n+2)}{R^2}$	2
$\Lambda^{R,H}_{\text{SST}}$	Pseudodifferential Operator for satellite-to-satellite tracking	$\frac{R^n}{(R+H)^n}\frac{n+1}{R+H}$	$-\infty$
$\Lambda^{R,H}_{\text{SGG}}$	Pseudodifferential operator for satellite gravity gradiometry	$\frac{R^n}{(R+H)^n}\frac{(n+1)(n+2)}{(R+H)^2}$	$-\infty$

All pseudodifferential operators are then defined on $\mathcal{L}^2(\Omega)$ or on suitable Sobolev spaces (see Freeden et al. 1998; Freeden 1999). The table above gives a summary of all the aforementioned operators.

In order to show how these operators work, we give some graphical examples. We start from the disturbance potential of the NASA, GSFC, and NIMA Earth's Gravity Model EGM96 (cf. Lemoine et al. 1998). In ❯ *Figs. 6–8* we graphically show the potential at the height of the Earth surface, at the height 250 km and further more the second-order radial derivative at height 250 km.

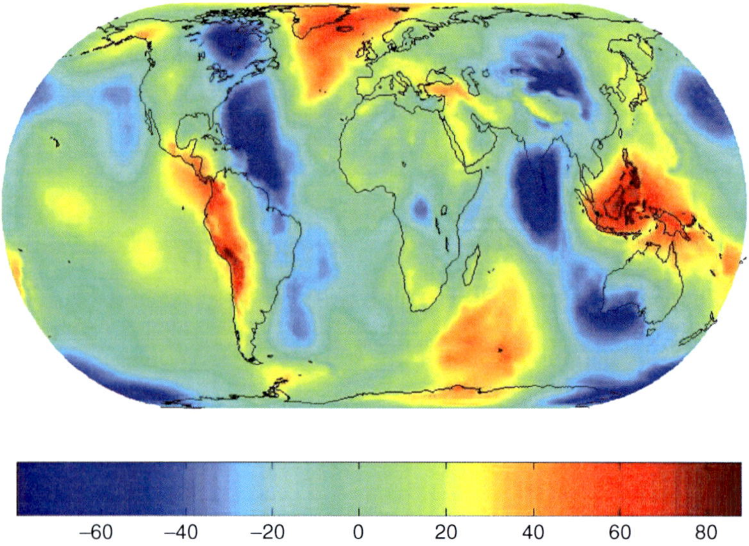

Fig. 6
The disturbance potential from EGM96 at the Earth's surface, at height 250 km, in m²/s²

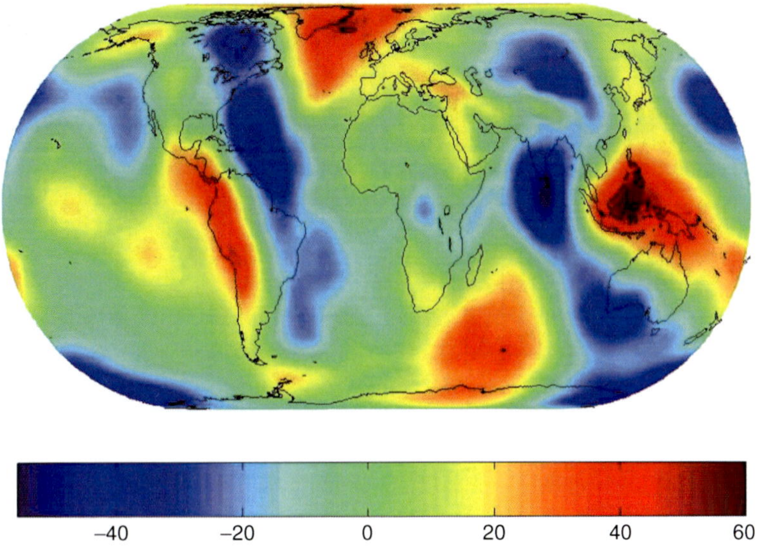

■ Fig. 7
The disturbance potential from EGM96 at height 250 km in m^2/s^2

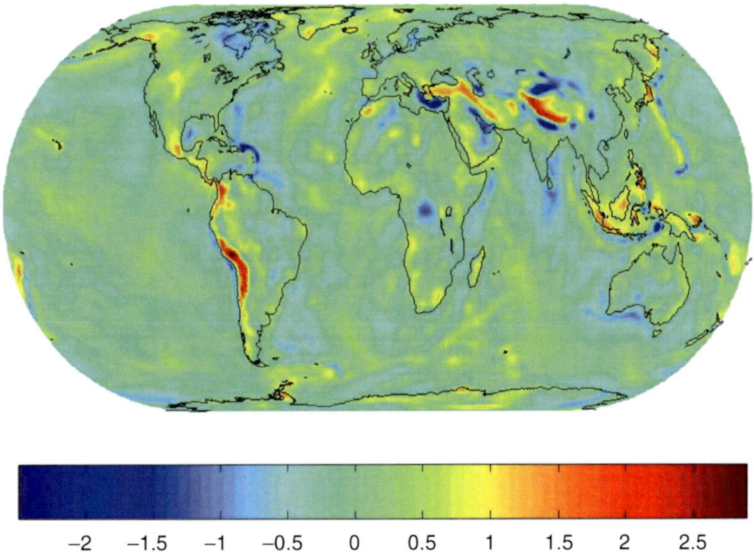

■ Fig. 8
The second-order radial derivative of the disturbance potential from EGM96 at height 250 km in $10^{-10}/s^2$

4.8 Classical Boundary Value Problems and Satellite Problems

The Neumann problem of potential theory for the outer space of the sphere Ω (based on $\mathcal{L}^2(\Omega)$ boundary data) reads as follows: Find $V \in \mathrm{Pot}(\mathcal{L}^2(\Omega))$ such that $\frac{\partial}{\partial r}V|_\Omega = G$. Since the trace of V is assumed to be a member of the class $\mathcal{L}^2(\Omega)$, the appropriate space for G is the Sobolev space $\mathcal{H}^{-1}(\Omega)$. Using pseudodifferential operators as described earlier, this problem reads in an $\mathcal{L}^2(\Omega)$-context as follows: Given $G \in \mathcal{L}^2(\Omega)$, find $F \in \mathcal{L}^2(\Omega)$ such that

$$\Lambda^R_{\mathrm{FND}} F = G \tag{92}$$

with $\left(\Lambda^R_{\mathrm{FND}}\right)^\wedge(n) = -\frac{n+1}{R}$, $n = 0, 1, \ldots$. Similar considerations show that the Dirichlet problem transfers to the trivial form $Id\, F = G$, where Id is the identity operator with $Id^\wedge(n) = 1$, $n = 0, 1, \ldots$.

Evidently, the classical problems of potential theory expressed in pseudodifferential form are well-posed in the sense that the inverse operators $\left(\Lambda^R_{\mathrm{FND}}\right)^{-1}$ and Id^{-1} are bounded in $\mathcal{L}^2(\Omega)$. In contrary, the problems coming from SST and SGG are ill-posed, as we will see in a moment. To be more concrete, SST intends to obtain information of V at the Earth's surface (radius R) from measurements of the first radial derivative at the satellite's height H. Thus, we obtain the problem: Given $G \in \mathcal{L}^2(\Omega)$, find $F \in \mathcal{L}^2(\Omega)$ so that

$$\Lambda^{R,H}_{\mathrm{SST}} F = G \tag{93}$$

with

$$\left(\Lambda^{R,H}_{\mathrm{SST}}\right)^\wedge(n) = \frac{R^n}{(R+H)^n}\,\frac{n+1}{R+H}. \tag{94}$$

Similarly, SGG is formulated as pseudodifferential equation as follows: Given $G \in \mathcal{L}^2(\Omega)$, find $F \in \mathcal{L}^2(\Omega)$ so that

$$\Lambda^{R,H}_{\mathrm{SGG}} F = G \tag{95}$$

with

$$\left(\Lambda^{R,H}_{\mathrm{SGG}}\right)^\wedge(n) = \frac{R^n}{(R+H)^n}\,\frac{(n+1)(n+2)}{(R+H)^2}. \tag{96}$$

For more detailed studies in a potential theoretic framework, the reader may wish to consult Freeden et al. (2002). The inverses of these operators possess a symbol which is exponentially increasing as $n \to \infty$. Thus, the inverse operators are unbounded, or in the jargon of regularization, these two problems are exponentially ill-posed. By a naive application of the inverse operator on the right-hand side, one cannot expect to obtain a useful solution. Thus, so-called regularization strategies have to be applied. Therefore, the basic aspects on regularization should be presented next.

4.9 A Short Introduction to the Regularization of Ill-Posed Problems

For the convenience of the reader, we present here a brief course of basic facts on regularization in a Hilbert space setting, which is useful to understand the solution strategies in the framework of pseudodifferential equations. The explanations are based on the monographs of Engl et al.

(1996) and Kirsch (1996), where much more additional material can be found even for more general reference spaces, too. .

Let $\mathcal{H}$ and $\mathcal{K}$ be two Hilbert spaces with inner products $(\cdot,\cdot)_{\mathcal{H}}$ and $(\cdot,\cdot)_{\mathcal{K}}$, respectively. Let

$$\Lambda : \mathcal{H} \longrightarrow \mathcal{K} \tag{97}$$

be a linear bounded operator. Given $y \in \mathcal{K}$, we are looking for a solution of

$$\Lambda x = y. \tag{98}$$

In accordance to Hadamard (1923), we call such a problem *well–posed*, if the following properties are valid:

- For all admissible data, a solution exists.
- For all admissible data, the solution is unique.
- The solution depends continuously on the data.

In our setting, these requirements can be translated into

- Λ is injective, i.e. $\mathcal{R}(\Lambda) = \mathcal{K}$
- Λ is surjective, i.e. $\mathcal{N}(\Lambda) = \{0\}$
- Λ^{-1} is bounded and continuous

If one of the three conditions is not fulfilled, the problem (98) is called *ill-posed*. It will turn out that the satellite problems we are concerned with are ill–posed, the largest problem being the unboundedness of the inverse operator Λ^{-1}.

Let us discuss the consequences of the violations of the above requirements for the well–posedness of (98). The lack of injectivity of Λ is perhaps the easiest problem. The space $\mathcal{H}$ can be replaced by the orthogonal complement $\mathcal{N}(\Lambda)^{\perp}$, and the restriction of the operator Λ to $\mathcal{N}(\Lambda)^{\perp}$ yields to an injective problem.

From practical point of view, one is very often confronted with the problem that $\mathcal{R}(\Lambda) \neq \mathcal{K}$, since the right-hand side is given by measurements and is, therefore, disturbed by errors. We assume now that $y \in \mathcal{R}(\Lambda)$ but only a perturbed right-hand side y^{δ} is known. We suppose

$$\|y - y^{\delta}\|_{\mathcal{K}} < \delta. \tag{99}$$

Our aim is to solve

$$\Lambda x^{\delta} = y^{\delta}. \tag{100}$$

Since y^{δ} might not be in $\mathcal{R}(\Lambda)$, the solution of this equation might not exist, and we have to generalize what is meant by a solution. x^{δ} is called *least–squares solution* of (100), if

$$\|\Lambda x^{\delta} - y^{\delta}\|_{\mathcal{K}} = \inf\{\|\Lambda z - y^{\delta}\|_{\mathcal{K}} \,|z \in \mathcal{H}\}. \tag{101}$$

The solution of (101) might not be unique, and therefore one looks for the solution of (101) with minimal norm. x^{δ} is called *best approximate solution* of $\Lambda x^{\delta} = y^{\delta}$, if x^{δ} is a least–squares solution and

$$\|x^{\delta}\|_{\mathcal{H}} = \inf\{\|z\|_{\mathcal{H}} \,|z \text{ is a least–squares solution of } \Lambda z = y^{\delta}\} \tag{102}$$

holds.

The notion of a best–approximate solution is closely related to the Moore–Penrose (generalized) inverse, of Λ (see Nashed 1976). We let

$$\tilde{\Lambda} : \mathcal{N}(\Lambda)^{\perp} \longrightarrow \mathcal{R}(\Lambda)$$

with

$$\tilde{\Lambda} = \Lambda|_{\mathcal{N}(\Lambda)^\perp} \tag{103}$$

and define the *Moore–Penrose (generalized) inverse* Λ^+ to be the unique linear extension of $\tilde{\Lambda}^{-1}$ to

$$\mathcal{D}(\Lambda^+) := \mathcal{R}(\Lambda) + \mathcal{R}(\Lambda)^\perp \tag{104}$$

with

$$\mathcal{N}(\Lambda^+) = \mathcal{R}(\Lambda)^\perp. \tag{105}$$

A standard result is provided by

Theorem 4.2
If $y \in \mathcal{D}(\Lambda^+)$, then $\Lambda x = y$ has a unique best-approximate solution which is given by

$$x^+ = \Lambda^+ y.$$

Note that the best-approximate solution is defined for all perturbed data $y^\delta \in \mathcal{K}$, whereas the last theorem requires that the right-hand side is an element of $\mathcal{D}(\Lambda^+)$.

A serious problem for ill-posed problems occurs when Λ^{-1} or Λ^+ are not continuous. This means that small errors in the data or even small numerical noise can cause large errors in the solution. In fact, in most cases the application of an unbounded Λ^{-1} or Λ^+ does not make any sense. The usual strategy to overcome this difficulty is to substitute the unbounded inverse operator

$$\Lambda^{-1} : \mathcal{R}(\Lambda) \longrightarrow \mathcal{H}$$

by a suitable bounded approximation

$$R : \mathcal{K} \longrightarrow \mathcal{H}.$$

The operator R is not chosen to be fixed, but dependent on a *regularization parameter* α. According to Kirsch (1996) we are led to introduce the following definition:

Definition 4.3
A regularization strategy is a family of linear bounded operators

$$R_\alpha : \mathcal{K} \longrightarrow \mathcal{H}, \quad \alpha > 0,$$

so that

$$\lim_{\alpha \to 0} R_\alpha \Lambda x = x \quad \textit{for all } x \in \mathcal{H},$$

i.e. the operators $\mathcal{R}_\alpha \Lambda$ converge pointwise to the identity.

From the theory of inverse problems (see, e.g., Kirsch 1996) it is also clear that if $\Lambda : \mathcal{H} \to \mathcal{K}$ is compact and $\mathcal{H}$ has infinite dimension (as it is the case for the application we have in mind), then the operators $\mathcal{R}_\alpha$ are not uniformly bounded, i.e., there exists a sequence (α_j) with $\lim_{j\to\infty} \alpha_j = 0$ and

$$\|R_{\alpha_j}\|_{\mathcal{L}(\mathcal{K},\mathcal{H})} \to \infty \quad \text{for } j \to \infty. \tag{106}$$

Note that the convergence of $\mathcal{R}_\alpha \Lambda x$ in Definition 4.3 is based on $y = \Lambda x$, i.e., on unperturbed data. In practice, the right-hand side is affected by errors and then no convergence is achieved. Instead, one is (or has to be) satisfied with an approximate solution based on a certain choice of the regularization parameter.

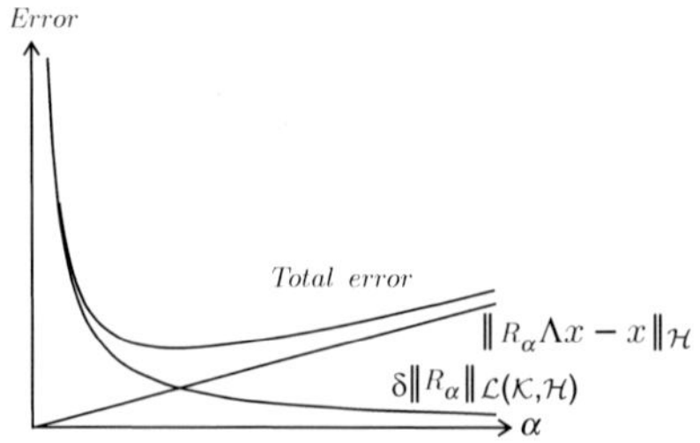

Fig. 9
Typical behavior of the total error in a regularization process

Let us discuss the error of the solution. For this purpose, we let $y \in \mathcal{R}(\Lambda)$ be the (unknown) exact right–hand side and $y^\delta \in \mathcal{K}$ be the measured data with

$$\| y - y^\delta \|_{\mathcal{K}} < \delta. \tag{107}$$

For a fixed $\alpha > 0$, we let

$$x^{\alpha,\delta} = R_\alpha y^\delta, \tag{108}$$

and look at $x^{\alpha,\delta}$ as an approximation of the solution x of $\Lambda x = y$. Then the error can be split as follows:

$$\begin{aligned} \| x^{\alpha,\delta} - x \|_{\mathcal{H}} &= \| R_\alpha y^\delta - x \|_{\mathcal{H}} \\ &\leq \| R_\alpha y^\delta - R_\alpha y \|_{\mathcal{H}} + \| R_\alpha y - x \|_{\mathcal{H}} \\ &\leq \| R_\alpha \|_{\mathcal{L}(\mathcal{K},\mathcal{H})} \, \| y^\delta - y \|_{\mathcal{K}} + \| R_\alpha y - x \|_{\mathcal{H}}, \end{aligned} \tag{109}$$

such that

$$\| x^{\alpha,\delta} - x \|_{\mathcal{H}} \leq \delta \| R_\alpha \|_{\mathcal{L}(\mathcal{K},\mathcal{H})} + \| R_\alpha \Lambda x - x \|_{\mathcal{H}}. \tag{110}$$

We see that the error between the exact and the approximate solution consists of two parts: The first term is the product of the bound for the error in the data and the norm of the regularization parameter R_α. This term will usually tend to infinity for $\alpha \to 0$ if the inverse Λ^{-1} is unbounded and Λ is compact (cf. (106)). The second term denotes the approximation error $\| (R_\alpha - \Lambda^{-1}) y \|_{\mathcal{H}}$ for the exact right–hand side $y = \Lambda x$. This error tends to zero as $\alpha \to 0$ by the definition of a regularization strategy. Thus, both parts of the error show a diametrically oriented behavior. A typical picture of the errors in dependence on the regularization parameter α is sketched in ➋ *Fig. 9*. Thus, a strategy is needed to choose α dependent on δ in order to keep the error as small as possible, i.e., we would like to minimize

$$\delta \| R_\alpha \|_{\mathcal{L}(\mathcal{K},\mathcal{H})} + \| R_\alpha \Lambda x - x \|_{\mathcal{H}}. \tag{111}$$

In principle, we distinguish two classes of parameter choice rules: If $\alpha = \alpha(\delta)$ does not depend on δ, we call $\alpha = \alpha(\delta)$ an a priori parameter choice rule. Otherwise α depends also on y^δ and we call $\alpha = \alpha(\delta, y^\delta)$ an a posteriori parameter choice rule. It is usual to say a parameter choice rule is convergent, if for $\delta \to 0$ the rule is such that

$$\lim_{\delta \to 0} \sup \{ \| R_{\alpha(\delta, y^\delta)} y^\delta - T^+ y \|_{\mathcal{H}} \, | y^\delta \in \mathcal{K}, \; \| y^\delta - y \|_{\mathcal{K}} \leq \delta \} = 0 \tag{112}$$

and

$$\lim_{\delta \to 0} \sup \{ \alpha(\delta, y^\delta) \, | y^\delta \in \mathcal{K}, \| y - y^\delta \|_{\mathcal{K}} \leq \delta \} = 0. \tag{113}$$

We stop here the discussion of parameter choice rules. For more material the interested reader is referred to, e.g., Engl et al. (1996) and Kirsch (1996).

The remaining part of this section is devoted to the case that Λ is compact, since then we gain benefits from the spectral representations of the operators. If $\Lambda : \mathcal{H} \to \mathcal{K}$ is compact, a singular system $(\sigma_n; v_n, u_n)$ is defined as follows: $\{\sigma_n^2\}_{n\in\mathbb{N}}$ are the nonzero eigenvalues of the self–adjoint operator $\Lambda^*\Lambda$ (Λ^* is the adjoint operator of Λ), written down in decreasing order with multiplicity. The family $\{v_n\}_{n\in\mathbb{N}}$ constitutes a corresponding complete orthonormal system of eigenvectors of $\Lambda^*\Lambda$. We let $\sigma_n > 0$ and define the family $\{u_n\}_{n\in\mathbb{N}}$ via $u_n = \Lambda v_n/\|\Lambda v_n\|_{\mathcal{K}}$. The sequence $\{u_n\}_{n\in\mathbb{N}}$ forms a complete orthonormal system of eigenvectors of $\Lambda\Lambda^*$, and the following formulas are valid:

$$\Lambda v_n = \sigma_n u_n, \tag{114}$$

$$\Lambda^* u_n = \sigma_n v_n, \tag{115}$$

$$\Lambda x = \sum_{n=1}^{\infty} \sigma_n (x, v_n)_{\mathcal{H}} \; u_n, \quad x \in \mathcal{H}, \tag{116}$$

$$\Lambda^* y = \sum_{n=1}^{\infty} \sigma_n (y, u_n)_{\mathcal{K}} \; v_n, \quad y \in \mathcal{K}. \tag{117}$$

The convergence of the infinite series is understood with respect to the Hilbert space norms under consideration. The identities (116) and (117) are called the *singular value expansions* of the corresponding operators. If there are infinitely many singular values, they accumulate (only) at 0, i.e., $\lim_{n\to\infty} \sigma_n = 0$.

Theorem 4.4

Let $(\sigma_n; v_n, u_n)$ *be a singular system for the compact linear operator* Λ, $y \in \mathcal{K}$. *Then we have*

$$y \in \mathcal{D}(\Lambda^+) \textit{ if and only if } \sum_{n=1}^{\infty} \frac{|(y, u_n)_{\mathcal{K}}|^2}{\sigma_n^2} < \infty, \tag{118}$$

and for $y \in \mathcal{D}(\Lambda^+)$ *it holds*

$$\Lambda^+ y = \sum_{n=1}^{\infty} \frac{(y, u_n)_{\mathcal{K}}}{\sigma_n} v_n. \tag{119}$$

The condition (118) is the *Picard criterion*. It says that a best–approximate solution of $\Lambda x = y$ exists only if the Fourier coefficients of y decay fast enough relative to the singular values.

The representation (119) of the best–approximate solution motivates a method for the construction of regularization operators, namely by damping the factors $1/\sigma_n$ in such a way that the series converges for all $y \in \mathcal{K}$. We are looking for filters

$$q : (0, \infty) \times (0, \|\Lambda\|_{\mathcal{L}(\mathcal{H},\mathcal{K})}) \longrightarrow \mathbb{R} \tag{120}$$

such that

$$R_\alpha y := \sum_{n=1}^{\infty} \frac{q(\alpha, \sigma_n)}{\sigma_n} (y, u_n)_{\mathcal{K}} \; v_n, \quad y \in \mathcal{K},$$

is a regularization strategy. The following statement is known from Kirsch (1996).

Theorem 4.5

Let $\Lambda : \mathcal{H} \to \mathcal{K}$ be compact with singular system $(\sigma_n; v_n, u_n)$. Assume that q from (120) has the following properties:

(i) $|q(\alpha, \sigma)| \leq 1$ *for all* $\alpha > 0$ *and* $0 < \sigma \leq \|\Lambda\|_{\mathcal{L}(\mathcal{H},\mathcal{K})}$.
(ii) *For every* $\alpha > 0$ *there exists a* $c(\alpha)$ *so that* $|q(\alpha, \sigma)| \leq c(\alpha)\sigma$ *for all* $0 < \sigma \leq \|\Lambda\|_{\mathcal{L}(\mathcal{H},\mathcal{K})}$.
(iii) $\lim_{\alpha \to 0} q(\alpha, \sigma) = 1$ *for every* $0 \leq \sigma \leq \|\Lambda\|_{\mathcal{L}(\mathcal{H},\mathcal{K})}$.

Then the operator $R_\alpha : \mathcal{K} \to \mathcal{H}$, $\alpha > 0$, *defined by*

$$R_\alpha y := \sum_{n=1}^{\infty} \frac{q(\alpha, \sigma_n)}{\sigma_n} (y, u_n)_{\mathcal{K}}\, v_n, \quad y \in \mathcal{K}, \tag{121}$$

is a regularization strategy with $\|R_\alpha\|_{\mathcal{L}(\mathcal{K},\mathcal{H})} \leq c(\alpha)$.

The function q is called a *regularizing filter* for Λ. Two important examples should be mentioned:

$$q(\alpha, \sigma) = \frac{\sigma^2}{\alpha + \sigma^2} \tag{122}$$

defines the *Tikhonov regularization*, whereas

$$q(\alpha, \sigma) = \begin{cases} 1, & \sigma^2 \geq \alpha, \\ 0, & \sigma^2 < \alpha, \end{cases} \tag{123}$$

leads to the regularization by *truncated singular value decomposition*.

4.10 Regularization of the Exponentially Ill-Posed SGG-Problem

We are now in position to have a closer look at the role of the regularization techniques particularly working for the SGG-problem.

In (95), the SGG-problem is formulated as pseudodifferential equation: Given $G \in \mathcal{L}^2(\Omega)$, find $F \in \mathcal{L}^2(\Omega)$ so that $\Lambda^{R,H}_{\mathrm{SGG}} F = G$ with

$$\left(\Lambda^{R,H}_{\mathrm{SGG}}\right)^{\wedge}(n) = \frac{R^n}{(R+H)^n} \frac{(n+1)(n+2)}{(R+H)^2}. \tag{124}$$

Switching now to a finite dimensional space (which is then the realization of the regularization by a singular value truncation), we are interested in a solution of the representation

$$F_N = \sum_{n=1}^{N} F^{\wedge}(n,m) Y_{n,m}. \tag{125}$$

Using a decomposition of G of the form

$$G = \sum_{n=1}^{\infty} G^{\wedge}(n,m) Y_{n,m}, \tag{126}$$

we end up with the spectral equations

$$\left(\Lambda^{R,H}_{\mathrm{SGG}}\right)^{\wedge}(n) F^{\wedge}(n,m) = G^{\wedge}(n,m), \quad n = 1, \ldots, m, \quad m = 1, \ldots, 2n+1. \tag{127}$$

In other words, in connection with (125) and (126), we find the relations

$$F^{\wedge}(n,m) = \frac{G^{\wedge}(n,m)}{\left(\Lambda_{\mathrm{SGG}}^{R,H}\right)^{\wedge}(n)}, \quad n = 1,\ldots,m, \quad m = 1,\ldots,2n+1. \tag{128}$$

For the realization of this solution we have to find the coefficients $G^{\wedge}(n,m)$. Of course, we are confronted with the usual problems of integration, aliasing, and so on.

The identity (128) also opens the perspective for SGG-applications by bandlimited regularization wavelets in Earth's gravitational field determination. For more details we refer to Schneider (1997), Freeden et al. (1997), Freeden and Schneider (1998), Glockner (2002), and Hesse (2003). The book written by Freeden (1999) contains non-bandlimited versions of (harmonic) regularization wavelets. Multiscale regularization by use of spherical up-functions is the content of the papers by Schreiner (2004) and Freeden and Schreiner (2004).

5 Future Directions

The regularization schemes described above are based on the decomposition of the Hesse tensor at satellite's height into scalar ingredients due to geometrical properties (normal, tangential, mixed) as well as to analytical properties originated by differentiation processes involving physically defined quantities (such as divergence, curl, etc). SGG-regularization, however, is more suitable and effective if it is based on algorithms involving the full Hesse tensor such as from the GOCE mission (for more insight into the tensorial decomposition of GOCE-data the reader is referred to the contribution of R. Rummel in this issue, in addition, see Rummel and van Gelderen (1992) and Rummel (1997).

Our context initiates another approach to tensor spherical harmonics. Based on cartesian operators (see Freeden and Schreiner (2009)), the construction principle starts from operators $\tilde{\mathbf{o}}_n^{(i,k)}$, $i,k \in \{1,2,3\}$ given by

$$\tilde{\mathbf{o}}_n^{(1,1)} F(x) = \left((2n+3)x - |x|^2 \nabla_x\right) \otimes \left((2n+1)x - |x|^2 \nabla_x\right) F(x), \tag{129}$$

$$\tilde{\mathbf{o}}_n^{(1,2)} F(x) = \left((2n-1)x - |x|^2 \nabla_x\right) \otimes \nabla_x F(x), \tag{130}$$

$$\tilde{\mathbf{o}}_n^{(1,3)} F(x) = \left((2n+1)x - |x|^2 \nabla_x\right) \otimes (x \wedge \nabla_x) F(x), \tag{131}$$

$$\tilde{\mathbf{o}}_n^{(2,1)} F(x) = \nabla_x \otimes \left((2n+1)x - |x|^2 \nabla_x\right) F(x), \tag{132}$$

$$\tilde{\mathbf{o}}_n^{(2,2)} F(x) = \nabla_x \otimes \nabla_x F(x), \tag{133}$$

$$\tilde{\mathbf{o}}_n^{(2,3)} F(x) = \nabla_x \otimes (x \wedge \nabla_x) F(x), \tag{134}$$

$$\tilde{\mathbf{o}}_n^{(3,1)} F(x) = (x \wedge \nabla_x) \otimes \left((2n+1)x - |x|^2 \nabla_x\right) F(x), \tag{135}$$

$$\tilde{\mathbf{o}}_n^{(3,2)} F(x) = (x \wedge \nabla_x) \otimes \nabla_x F(x), \tag{136}$$

$$\tilde{\mathbf{o}}_n^{(3,3)} F(x) = (x \wedge \nabla_x) \otimes (x \wedge \nabla_x) F(x) \tag{137}$$

for $x \in \mathbb{R}^3$ and sufficiently smooth function $F : \mathbb{R}^3 \to \mathbb{R}$.

Elementary calculations in cartesian coordinates lead us in a straightforward way to the following result.

Lemma 5.1

Let $H_n, n \in \mathbb{N}_0$, be a homogeneous harmonic polynomial of degree n. Then, $\tilde{\mathbf{o}}_n^{(i,k)} H_n$ is a homogeneous harmonic tensor polynomial of degree $\deg^{(i,k)}(n)$, where

$$\deg^{(i,k)}(n) = \begin{cases} n-2 & for \quad (i,k) = (2,2) \\ n-1 & for \quad (i,k) \in \{(2,3),(3,2)\} \\ n & for \quad (i,k) \in \{(1,2),(2,1),(3,3)\} \\ n+1 & for \quad (i,k) \in \{(1,3),(3,1)\} \\ n+2 & for \quad (i,k) = (1,1) \end{cases} . \tag{138}$$

($\deg^{(i,k)}(n) < 0$ means that $\tilde{\mathbf{o}}_n^{(i,k)} H_n = 0$).

Applying the operator $\tilde{\mathbf{o}}_n^{(1,1)}$ to the inner harmonic $x \mapsto |x|^n Y_n(x/|x|)$, we are able to deduce the following relation after some easy calculations

$$\begin{aligned} \tilde{\mathbf{o}}_n^{(1,1)} r^n Y_n(\xi)|_{r=1} = {} & (n+2)(n+1)\mathbf{o}^{(1,1)} Y_n(\xi) - (n+2)\mathbf{o}^{(1,2)} Y_n(\xi) - (n+2)\mathbf{o}^{(2,1)} Y_n(\xi) \\ & - \frac{1}{2}(n+2)(n+1)\mathbf{o}^{(2,2)} Y_n(\xi) + \frac{1}{2}\mathbf{o}^{(2,3)} Y_n(\xi). \end{aligned} \tag{139}$$

(compare with the identity (66)).

Assuming that $\{Y_{n,m}\}_{n=0,\ldots,m=1,\ldots,2n+1}$ is an orthonormal set of scalar spherical harmonics as before, we are led to introduce the following tensor spherical harmonics

$$\tilde{\mathbf{y}}_{n,m}^{(i,k)} = \left(\tilde{\mu}_n^{(i,k)}\right)^{-1/2} \tilde{\mathbf{o}}^{(i,k)} Y_{n,m}, \tag{140}$$

$n = \tilde{0}_{ik}, \ldots, m = 1, \ldots, 2n+1$, where

$$\tilde{0}_{ik} = \begin{cases} 0, & (i,k) \in \{(1,1),(2,1),(3,1)\} \\ 1, & (i,k) \in \{(1,2),(1,3),(2,3),(3,3)\} \\ 2, & (i,k) \in \{(2,2),(3,2)\} \end{cases} . \tag{141}$$

and

$$\tilde{\mu}_n^{(1,1)} = (n+2)(n+1)(2n-3)(2n-1), \tag{142}$$

$$\tilde{\mu}_n^{(1,2)} = 3n^4, \tag{143}$$

$$\tilde{\mu}_n^{(2,1)} = (n+2)(n+1)(2n-3)(2n-1), \tag{144}$$

$$\tilde{\mu}_n^{(2,2)} = n(n-1)(2n+1)(2n-1), \tag{145}$$

$$\tilde{\mu}_n^{(3,3)} = n^2(n-1)(2n+1), \tag{146}$$

$$\tilde{\mu}_n^{(1,3)} = n(n+1)^2(2n+1), \tag{147}$$

$$\tilde{\mu}_n^{(2,3)} = n^2(n+2)(n+1), \tag{148}$$

$$\tilde{\mu}_n^{(3,1)} = n^2(n+1)(2n+1), \tag{149}$$

$$\tilde{\mu}_n^{(3,2)} = n(n+1)^2(2n+1). \tag{150}$$

They are suitable for the solution of tensorial problems due to the following result involving the spaces $\mathbf{l}^2(\Omega)$ and $\mathbf{c}(\Omega)$ of square-integrable and continuous tensor fields on Ω, respectively.

Theorem 5.2

Let $\{Y_{n,m}\}_{n=0,1,\ldots,\ m=1,\ldots,2n+1}$ *be an* $\mathrm{L}^2(\Omega)$*-orthonormal set of scalar spherical harmonics. Then, the set*

$$\left\{\tilde{\mathbf{y}}_{n,m}^{(i,k)}\right\}_{i,k=1,2,3,\ n=\tilde{0}_{ik},\ldots,\ m=1,\ldots,2n+1}, \tag{151}$$

as defined by (140) *forms an* $\mathbf{l}^2(\Omega)$*-orthonormal set of tensor spherical harmonics which is closed in* $\mathbf{c}(\Omega)$ *with respect to* $\|\cdot\|_{\mathbf{c}(\Omega)}$ *and complete in* $\mathbf{l}^2(\Omega)$ *with respect to* $(\cdot,\cdot)_{\mathbf{l}^2(\Omega)}$.

Finally we introduce the tensor outer harmonics of degree n, order m, and kind (i,k) by (see Freeden and Schreiner (2009))

$$\mathbf{h}_{-n-1,m}^{(1,1);R}(x) = \frac{1}{R}\left(\frac{R}{|x|}\right)^{n+3} \tilde{\mathbf{y}}_{n,m}^{(1,1)}\left(\frac{x}{|x|}\right), \tag{152}$$

$$\mathbf{h}_{-n-1,m}^{(1,2);R}(x) = \frac{1}{R}\left(\frac{R}{|x|}\right)^{n+1} \tilde{\mathbf{y}}_{n,m}^{(1,2)}\left(\frac{x}{|x|}\right), \tag{153}$$

$$\mathbf{h}_{-n-1,m}^{(2,1);R}(x) = \frac{1}{R}\left(\frac{R}{|x|}\right)^{n+1} \tilde{\mathbf{y}}_{n,m}^{(2,1)}\left(\frac{x}{|x|}\right), \tag{154}$$

$$\mathbf{h}_{-n-1,m}^{(2,2);R}(x) = \frac{1}{R}\left(\frac{R}{|x|}\right)^{n-1} \tilde{\mathbf{y}}_{n,m}^{(2,2)}\left(\frac{x}{|x|}\right), \tag{155}$$

$$\mathbf{h}_{-n-1,m}^{(3,3);R}(x) = \frac{1}{R}\left(\frac{R}{|x|}\right)^{n+1} \tilde{\mathbf{y}}_{n,m}^{(3,3)}\left(\frac{x}{|x|}\right), \tag{156}$$

$$\mathbf{h}_{-n-1,m}^{(1,3);R}(x) = \frac{1}{R}\left(\frac{R}{|x|}\right)^{n+2} \tilde{\mathbf{y}}_{n,m}^{(1,3)}\left(\frac{x}{|x|}\right), \tag{157}$$

$$\mathbf{h}_{-n-1,m}^{(2,3);R}(x) = \frac{1}{R}\left(\frac{R}{|x|}\right)^{n} \tilde{\mathbf{y}}_{un,m}^{(2,3)}\left(\frac{x}{|x|}\right), \tag{158}$$

$$\mathbf{h}_{-n-1,m}^{(3,1);R}(x) = \frac{1}{R}\left(\frac{R}{|x|}\right)^{n+2} \tilde{\mathbf{y}}_{n,m}^{(3,1)}\left(\frac{x}{|x|}\right), \tag{159}$$

$$\mathbf{h}_{-n-1,m}^{(3,2);R}(x) = \frac{1}{R}\left(\frac{R}{|x|}\right)^{n} \tilde{\mathbf{y}}_{n,m}^{(3,2)}\left(\frac{x}{|x|}\right), \tag{160}$$

$x \in \mathbb{R}^3\backslash\{0\}$.

These definitions (in particular the one of kind $(1,1)$) offer an easy way to represent the gravitational potential V in the exterior of the sphere with radius R in terms of the gravitational tensor $\nabla \otimes \nabla V$ at the satellite's height H. We start with the observation that

$$\nabla \times \nabla \left(\frac{R}{|x|}\right)^{n+1} \frac{1}{R} Y_{n,m}\left(\frac{x}{|x|}\right) = \sqrt{(n+2)(n+1)(2n-3)(2n-1)}\, \mathbf{h}^{(1,1);R}_{-n-1,m}. \tag{161}$$

Using the orthonormal basis $\{1/R\ Y_{n,m}\}$ of the space of square-integrable functions on Ω_R and $\{1/(R+H)\ \tilde{\mathbf{y}}^{(i,k)}_{n,m}\}$ of the space of square-integrable tensor fields on Ω_{R+H}, the relation (161) can be rewritten as

$$\nabla \times \nabla \left(\frac{R}{|x|}\right)^{n+1} \frac{1}{R} Y_{n,m}\left(\frac{x}{|x|}\right) = \sqrt{(n+2)(n+1)(2n-3)(2n-1)} \frac{R+H}{R} \left(\frac{R}{|x|}\right)^{n+3} \times \frac{1}{R+H} \tilde{\mathbf{y}}^{(1,1)}_{n,m}. \tag{162}$$

In other words, the transformation of the potential at height R to the Hesse tensor at height $R+H$ can be expressed by a pseudodifferential operator $\tilde{\boldsymbol{\lambda}}^{R,H}_{\mathrm{SGG}}$ with the tensorial symbol

$$\tilde{\boldsymbol{\lambda}}^{R,H\wedge}_{\mathrm{SGG}}(n) = \begin{pmatrix} \tilde{\Lambda}^{(1,1);R,H\wedge}_{\mathrm{SGG}}(n) & \tilde{\Lambda}^{(1,2);R,H\wedge}_{\mathrm{SGG}}(n) & \tilde{\Lambda}^{(1,3);R,H\wedge}_{\mathrm{SGG}}(n) \\ \tilde{\Lambda}^{(2,1);R,H\wedge}_{\mathrm{SGG}}(n) & \tilde{\Lambda}^{(2,2);R,H\wedge}_{\mathrm{SGG}}(n) & \tilde{\Lambda}^{(2,3);R,H\wedge}_{\mathrm{SGG}}(n) \\ \tilde{\Lambda}^{(3,1);R,H\wedge}_{\mathrm{SGG}}(n) & \tilde{\Lambda}^{(3,2);R,H\wedge}_{\mathrm{SGG}}(n) & \tilde{\Lambda}^{(3,3);R,H\wedge}_{\mathrm{SGG}}(n) \end{pmatrix},$$

where

$$\tilde{\Lambda}^{(1,1);R,H\wedge}_{\mathrm{SGG}}(n) = \sqrt{(n+2)(n+1)(2n-3)(2n-1)} \left(\frac{R}{R+H}\right)^{n+2}$$

and

$$\tilde{\Lambda}^{(i,k);R,H\wedge}_{\mathrm{SGG}}(n) = 0, \quad (i,k) \neq (1,1).$$

Hence, the foreward direction of the SGG problem is described by the pseudodifferential operator $\tilde{\boldsymbol{\lambda}}^{R,H}_{\mathrm{SGG}}$, so that the SGG problem leads to the pseudodifferential equation

$$\tilde{\boldsymbol{\lambda}}^{R,H}_{\mathrm{SGG}} V = \mathbf{h}. \tag{163}$$

In order to formulate this equation more concretely, we show how the potential V is related to its Hesse tensor at height H:

$$V(x) = \sum_{n=0}^{\infty} \sum_{k=1}^{2n+1} \left(\nabla \otimes \nabla V; \mathbf{h}^{(1,1);R+H}_{-n-1,k}\right)_{\mathrm{l}^2(\Omega_{R+H})} \left(\tilde{\mu}^{(1,1)}_n\right)^{-1/2} \left(\frac{R+H}{R}\right)^{n+2} \frac{1}{|x|^{n+1}} Y_{n,m}\left(\frac{x}{|x|}\right). \tag{164}$$

Obviously, the last formula may serve as point of departure for (regularization) solution techniques to determine V at the Earth's surface from the full Hesse tensor $\mathbf{v} = \nabla \otimes \nabla V$ at the satellite altitude. Furthermore, as described in Freeden and Schreiner (2009), it is not difficult to define tensor zonal kernels in accordance with this expansion. In particular, they allow multiscale regularization (solution) schemes based on wavelet methods.

6 Conclusion

Although an impressive rate of the Earth's gravitational potential can be detected globally at the orbit of a satellite (like GOCE), the computational drawback of satellite techniques in geoscientific research is the fact that measurements must be performed at a certain altitude. Consequently, a "downward continuation" process must be applied to handle the potential at the Earth's surface, hence, a loss of information for the signal is unavoidable. Indeed, "downward continuation" causes severe problems, since the amount of amplification for the potential is not known suitably (as an a priori amount) and even small errors in the measurements may produce huge errors in the potential at the Earth's surface.

However, it is of great advantage that satellite data are globally available, at least in principle. Nevertheless, from a mathematical point of view, we are not confronted with a boundary value of potential theory. Satellite techniques such as SST and/or SGG require the solution of an inverse problem to produce gravitational information at the Earth's surface, where it is needed actually. SST/SGG can be formulated adequately as (Fredholm) pseudodifferential equation of the first kind, which is exponentially ill-posed, and this fact makes indispensable the development of suitable mathematical methods with strong relation to the nature and structure of the data.

In this respect it should be mentioned that each approximation's theoretic method has its own aim and character. Even more, it is the essence of any numerical realization that it becomes optimal only with respect to certain specified features. For example, Fourier expansion methods with polynomial trial functions (spherical harmonics) offer the canonical "trend-approximation" of low-frequency phenomena (for global modeling), they offer an excellent control and comparison of spectral properties of the signal, since any spherical harmonic relates to one frequency. This is of tremendous advantage for relating data types under spectral aspects. But it is at the price that the polynomials are globally supported such that local modeling results into serious problems of economy and efficiency. Bandlimited kernels can be used for the transition from long-wavelength to short-wavelength phenomena (global to local modeling) in the signal. Because of their excellent localization properties in the space domain, the non-bandlimited kernels can be used for the modeling of short-wavelength phenomena. Local modeling is effective and economic. But the information obtained by kernel approximations is clustered in frequency bands so that spectral investigations are laborious and time consuming. In other words, for numerical work to be done, we have to make an a priori choice. We have to reflect the different stages of space/frequency localization so that the modeling process can be adapted to the localization requirements necessary and sufficient for our geophysical or geodetic interpretation.

In conclusion, an algorithm establishing an approximate solution for the inverse SGG-problem has to reflect the intention of the applicant. Different techniques for regularization are at the disposal of the numerical analyst for global as well as local purposes. Each effort does give certain progress in the particular field of pre–defined interest. If a broad field of optimality should be covered, only a combined approach is the strategic instrument to make an essential step forward. Thus, for computational aspects of determining the Earth's gravitational potential, at least a twofold combination is demanded, viz. combining globally available satellite data (including the SGG-contribution) with local airborne and/or terrestrial data and combining tools and means of constructive approximation such as polynomials, splines, wavelets, etc. Altogether, in numerical modeling of the Earth's gravitational potential, there is no best universal method, there exist only optimized procedures with respect to certain features and the option and the feasibility for their suitable combination.

References

Backus GE (1966) Potentials for tangent tensor fields on spheroids. Arch Ration Mech Anal 22: 210–252

Backus GE (1967) Converting vector and tensor equations to scalar equations in spherical coordinates. Geophys JR Astron Soc 13: 61–101

Beutler GB, Drinkwater MR, Rummel R, von Steiger R (2003). Earth gravity field from space – from sensors to earth sciences. In the Space Sciences Series of ISSI, vol. 18. Kluwer, Dordrecht, 419–432

Engl H, Hanke M, Neubauer A (1996) Regularization of inverse problems. Kluwer, Dordrecht

Engl H, Louis AK, Rundell W (1997) Inverse problems in geophysical applications. SIAM, Philadelphia

ESA (1999) Gravity field and steady-state ocean circulation mission, ESTEC, Noordwijk, ESA SP–1233(1)

ESA (2007) Proceedings of the 3rd International GOCE User Workshop , ESTEC, Noordwijk, ESA SP-627

Freeden W (1980a) On the approximation of external gravitational potential with closed systems of (Trial) functions. Bull Géod 54: 1–20

Freeden W (1980b) On integral formulas of the (Unit) sphere and their application to numberical computation of integrals. Computing 25: 131–146

Freeden W (1999) Multiscale modelling of spaceborne geodata. B.G. Teubner, Leipzig

Freeden W, Michel V (2004) Multiscale potential theory (with applications to geoscience). Birkhäuser, Boston

Freeden W, Schneider F (1998) Regularization wavelets and multiresolution. Inverse Probl, 14: 493–515

Freeden W, Schreiner M (2004) Multiresolution analysis by spherical up functions. Constr Approx 23: 241–259

Freeden W, Schreiner M (2009) Spherical functions of mathematical geosciences. A scalar, vectorial, and tensorial setup. Springer, Berlin

Freeden W, Gervens T, Schreiner M (1998) Constructive approximation on the sphere (with applications to geomathematics). Oxford Science Publications, Clarendon, Oxford

Freeden W, Michel V, Nutz H (2002) Satellite-to-satellite tracking and satellite gravity gradiometry (advanced techniques for high-resolution geopotential field determination). J Eng Math 43: 19–56

Freeden W, Schneider F, Schreiner M (1997) Gradiometry – an inverse problem in modern satellite geodesy. In: Engl HW, Louis, A, Rundell, W (eds) Inverse Problems in Geophysical Applications, Proceedings on the GAMM-SIAM Symposium on Inverse Problems in Geophysics, Fish Lake, Yosemite, Califoria, Dec. 16–19, 1995, Society for Industrial and Applied Mathematics, 179–239

Glockner O (2002) On numerical aspects of gravitational field modelling from SST and SGG by harmonic splines and wavelets (with application to CHAMP data). Doctoral Thesis, Geomathematics Group, University of Kaiserslautern, Shaker, Aachen

Hadamard J (1923) Lectures on the Cauchy problem in linear partial differential equations. Yale University Press, New Haven

Heiskanen WA, Moritz H (1967) Physical geodesy. freeman, San Francisco, CA

Hesse K (2003) Domain decomposition methods in multiscale geopotential determination from SST and SGG. Doctoral Thesis, Geomathematics Group, University of Kaiserslautern, Shaker, Aachen

Kellogg OD (1929) Foundations of potential theory. Frederick Ungar Publishing Company, New York

Kirsch A (1996) Introduction to the mathematical theory of inverse problems. Springer, New York

Lemoine FG, Kenyon SC, Factor JK, Trimmer RG, Pavlis NK, Shinn DS, Cox CM, Klosko SM, Luthcke SB, Torrence MH, Wang YM, Williamson RG, Pavlis EC, Rapp RH, Olson TR (1998) The development of the joint NASA GSFC and NIMA geopotential model EGM96,

NASA/TP-1998-206861. NASA Goddard Space Flight Center, Greenbelt, MD, USA

Müller C (1966) Spherical harmonics. Lecture notes in mathematics, vol 17. Springer, Berlin

Nashed MZ (1976) Generalized inverses and applications. Proceedings of an Advanced Seminar Conducted by the Mathematics Research Center, University of Wisconsin, Madison

Rummel R (1997) Spherical spectral properties of the Earth's gravitational potential and its first and second derivatives. In Lecture notes in earth science, vol 65. Springer, Berlin, pp 359–404

Rummel R, van Gelderen M (1992) Spectral analysis of the full gravity tensor. Geophys J Int 111: 159–169

Rummel R, van Gelderen M, Koop R, Schrama E, Sanso F, Brovelli M, Miggliaccio F, Sacerdote F (1993) Spherical harmonic analysis of satellite gradiometry. Publications on Geodesy, Delft

Seeber G (1989) Satellitengeodäsie (Grundlagen, Methoden und Anwendungen) Walter de Gruyter, Berlin, New York

Schneider F (1996) The solution of linear inverse problems in satellite geodesy by means of spherical spline approximation. J Geodesy 71: 2–15

Schneider F (1997) Inverse problems in satellite geodesy and their approximate solution by splines and wavelets. PhD Thesis, University of Kaiserslautern, Geomathematics Group, Shaker, Aachen

Schreiner M (1994a) Tensor spherical harmonics and their application in satellite gradiometry. PhD Thesis, University of Kaiserslautern, Geomathematics Group

Schreiner M (1994b) Uniqueness problems in satellite gradiometry. In Proceedings of the 8th Conference of the European Consortium for Mathematics in Industry, Kaiserslautern

Schreiner M (2004) Wavelet approximation by spherical up functions. Habilitation Thesis, Geomathematics Group, University of Kaiserslautern, Shaker, Aachen

Svensson SL (1983) Pseudodifferential operators – a new approach to the boundary value problems of physical geodesy. Manuscr Geod 8:1–40

10 Gravitational Viscoelastodynamics

Detlef Wolf
Department of Geodesy and Remote Sensing, German Research Center for Geosciences GFZ, Telegrafenberg, Potsdam, Germany

W. Freeden, M.Z. Nashed, T. Sonar (Eds.), *Handbook of Geomathematics*, DOI 10.1007/978-3-642-01546-5_10,

Abstract We consider a compositionally and entropically stratified, compressible, rotating fluid earth and study gravitational–viscoelastic perturbations of its hydrostatic initial state. Using the Lagrangian representation and assuming infinitesimal perturbations, we deduce the incremental field equations and interface conditions of *gravitational viscoelastodynamics* (GVED) governing the perturbations. In particular, we distinguish the *material*, *material-local*, and *local* forms of the incremental equations. We also demonstrate that their short-time asymptotes correspond to generalizations of the incremental field equations and interface conditions of *gravitational elastodynamics* (GED), whereas the long-time asymptotes agree with the incremental field equations and interface conditions of *gravitational viscodynamics* (GVD). The incremental thermodynamic pressure appearing in the long-time asymptote to the incremental constitutive equation is shown to satisfy the appropriate incremental state equation. Finally, we derive approximate field theories applying to gravitational–viscoelastic perturbations of isocompositional, isentropic and compressible or incompressible fluid domains.

1 Introduction

Studies of the Earth's response to perturbing forces of short duration have conventionally been based on the assumption of elastic constitutive behavior. However, complications arise due to the presence of the gravitational initial stress in the Earth's interior. Rayleigh pointed out this problem and suggested that the *total* stress be decomposed into a hydrostatic *initial* stress and a superimposed elastic *incremental* stress (Rayleigh 1906). He also referred the displacement of a particle to its initial position and proposed relating the incremental stress to the displacement gradient using equations that formally agree with the ordinary field equations of elastodynamics valid in the absence of gravitation.

Love elaborated Rayleigh's concepts and, in particular, found it profitable to formulate the incremental field equations in terms of the *material* incremental stress experienced by a particle upon its displacement and conventionally appearing in the incremental constitutive equation of elasticity (Love 1911). However, owing to the initial stress gradient in the Earth's interior, the material incremental stress differs from the *local* incremental stress referring to a fixed location and appearing in the usual form of the incremental momentum equation. Love accounted for this difference by introducing the *advective* incremental stress into the latter equation. A consequence of Love's formulation in terms of the material incremental stress is that the initial stress associated with a particle at its current position is the hydrostatic stress at its initial position. This has sometimes been described as the particle "carrying" its hydrostatic initial stress with it while being displaced. Modern accounts of the incremental field theory of *gravitational elastodynamics* (GED) have been given by several authors (e.g., Dahlen 1974; Grafarend 1982; Vermeersen and Vlaar 1991).

The Earth's response to perturbing forces of long duration has usually been studied assuming fluid behavior. In contrast to elastic behavior, no "advective term" is required in the incremental momentum equation. This is a consequence of writing this equations in terms of the local incremental stress appearing in the conventional form of the incremental constitutive equation of fluids. The formal differences between the incremental field equations of elastic solids and fluids, respectively, in the presence of gravity have been discussed by several authors (e.g., Cathles 1975; Dahlen 1974; Grafarend 1982).

Of interest to us are viscous perturbations of an initially hydrostatic state. Such studies were initiated by Darwin and continued by others (e.g., Chandrasekhar 1961; Darwin 1879; Haskell 1935, 1936, O'Connell 1971; Parsons 1972). The investigations were usually based on the supposition of incompressibility. More recently, the modifications introduced by compressibility have been studied for internal loading (e.g., Corrieu et al. 1995; Dehant and Wahr 1991; Jarvis and McKenzie 1980; Panasyuk et al. 1996) and for surface loading (Li and Yuen 1987; Wu and Yuen 1991). We refer to the incremental field theory appropriate to viscous fluids in the presence of gravity as *gravitational viscodynamics* (GVD).

Various types of viscoelastic constitutive behavior have been suggested to explain the Earth's response to perturbing forces covering a wide period range. Comparatively recent is the formal development of the incremental field theory of viscoelastodynamics for continua in a state of initial stress. Pioneering investigations into this problem were published by Biot in a series of papers and are summarized in a monograph (Biot 1965). Relevant to us is Biot's analysis of viscoelastic perturbations, induced by surface masses, of initially hydrostatic fluids (Biot 1959). Similar to Love, Biot found it necessary to distinguish between local and material incremental stresses. However, in contrast to Love, Biot proceeded by formulating the incremental field equations in terms of the local incremental stress. On the assumption of incompressible perturbations, this allowed their formal reduction to the ordinary field equations valid in the absence of initial stress, whose solutions are well studied.

This "reduction technique" was found helpful also by Wolf in his investigations into *incompressible*, viscoelastic perturbations of initially hydrostatic planar or spherical Earth models with one or two layers (Wolf 1984, 1985a,b). Assuming *Maxwell viscoelasticity*, he could in particular show that the solution approaches that for a hydrostatic fluid at times long after the onset of forcing (Wolf 1985b). The role of the initial stress was further clarified in later studies by Wolf (1991b) and Wu (1992).

The incremental field theory of *gravitational viscoelastodynamics* (GVED) describing *compressible* gravitational–viscoelastic perturbations of initially hydrostatic fluids was largely formulated by Mitrovica et al. (1994), Peltier (1974, 1989), Wolf (1991a), and Wu and Peltier (1982). Whereas they developed the *normal-mode* form of the theory, an alternative *time-domain* approach was proposed by Hanyk et al. (1996). For the type of compressibility considered by these authors, the bulk relaxation of the material is neglected, i.e., volumetric perturbations are taken as elastic. The shear relaxation is usually assumed to correspond to Maxwell viscoelasticity. Analytical solutions for compressible Maxwell constitutive behavior have so far been obtained only for plane or spherical Earth models consisting of homogeneous layers (Hanyk et al. 1999; Vermeersen and Mitrovica 2000; Vermeersen et al. 1996; Wieczerkowski 1999; Wolf 1985c).

The objective of this review is to give a systematic and rigorous account of GVED. We begin by introducing in ➋ Sect. 2 the *Lagrangian* and *Eulerian* kinematic representations of fields and specify their *material*, *local*, and *advective* increments.

In ➋ Sect. 3, we deduce the linearized form of the incremental field equations and interface conditions of GVED for a hydrostatic initial state directly from the differential form of the fundamental principles of continuum mechanics and potential theory. The resulting equations describe infinitesimal, gravitational–viscoelastic perturbations of compositionally and entropically stratified, compressible, rotating fluids initially in hydrostatic equilibrium and apply to any type of viscoelastic constitutive behavior characterized by both bulk and shear relaxation.

In ➋ Sect. 4, we deduce two systems of equations that are asymptotically correct for short and long times after the onset of the perturbations. The short-time asymptotic equations are generalizations of the incremental field equations and interface conditions of GED. The long-time asymptotic equations agree with the incremental field equations and interface conditions of GVD.

➋ Section 5 is concerned with an example of how compressibility can be approximately accounted for in the theory. With this objective in mind, we derive a system of incremental field equations and interface conditions applying to *locally* incompressible gravitational–viscoelastic perturbations of an isocompositional, isentropic, compressible initial state. The conventional approximation of *material* incompressibility is treated as a special case.

➋ Section 6 concludes this review with a brief summary of the main results obtained.

2 Basic Concepts

In this preparatory section, we summarize the basic concepts used throughout this review. They are the *kinematic representations* underlying continuum mechanics and some essentials of perturbation theory (e.g., Eringen 1989; Krauss 1973; Malvern 1969). These concepts have also been discussed within a geophysical or geodetic context (e.g., Dahlen 1972, 1973, 1974; Dahlen and Tromp 1998; Grafarend 1982). Here, we adapt these formulations to our particular requirements. We begin by introducing the *Lagrangian* and *Eulerian* kinematic representations of fields (➋ Sect. 2.1). Following this, we specify their *material*, *local*, and *advective* increments using the Lagrangian and Eulerian representations of the fundamental perturbation equation (➋ Sect. 2.2). Finally, several notational conventions used for stating interface conditions are defined (➋ Sect. 2.3).

We restrict our study to Cartesian-tensor fields and employ for them the indicial notation and the summation convention: index subscripts i, j, ... run over $\{1,2,3\}$, respectively, and index subscripts repeated in any term imply summation over this range. The usual convention is followed, where δ_{ij} is the Kronecker symbol and ϵ_{ijk} is the Levi–Civita symbol.

2.1 Kinematic Representations

Let $\mathcal{E}$ be the unbounded 3D Euclidian space domain, $\mathcal{T}$ the time domain $[\,0,\infty)$ and consider the mapping

$$r_i = r_i(\mathbf{X}, t), \quad X_i \in \mathcal{E}, \quad t \in \mathcal{T}. \tag{1}$$

We assume that the mapping has the property $X_i = r_i(\mathbf{X}, 0)$ and is a single-valued and continuously differentiable function with respect to X_i and t, whose Jacobian determinant, $\det[\partial r_i/\partial X_j]$, does not vanish. Henceforth, t will be called current time, r_i current position, $t = 0$ initial time, and X_i initial position.

Suppose now that $\mathcal{E}$ is filled by a gravitating material continuum, whose primitive elements are called *material* points or *particles*. As a particular mapping of the type of Eq. 1 then consider

$$r_i = r_i(\mathbf{X}, t) = X_i + u_i(\mathbf{X}, t), \quad X_i \in \mathcal{E}, \quad t \in \mathcal{T}, \tag{2}$$

which refers to particles identified by their initial position, X_i. For fixed t, the mapping relates to each particle its current position, r_i, in terms of the displacement, u_i, from its initial position.

Considering the assumptions to be satisfied by the forward mapping, the inverse mapping of Eq. 2 exists and is given by

$$X_i = X_i(\mathbf{r}, t) = r_i - U_i(\mathbf{r}, t) \quad r_i \in \mathcal{E}, \ t \in \mathcal{T}. \tag{3}$$

In contrast to Eq. 2, this equation refers to *local* points or *places* identified by their position, r_i. For fixed t, the mapping relates to the place of the initial position, X_i, of the particle currently at r_i by means of the displacement, U_i, from its initial position.

Next, we seek kinematic representations equivalent to Eqs. 2 and 3 for arbitrary fields. Since such fields may have discontinuities on particular interfaces, this requires that the domains where the mappings are defined be specified more precisely. Hence, we decompose $\mathcal{E}$ into three fixed subdomains:

$$\mathcal{E} = \mathcal{X}_- \cup \mathcal{X}_+ \cup \partial\mathcal{X}, \tag{4}$$

where $\mathcal{X}_-$ denotes a simply connected *internal* 3D domain, $\mathcal{X}_+$ the complementary *external* 3D domain, and $\partial\mathcal{X}$ the 2D interface between them. The kinematic representation of fields with possible discontinuities for $X_i \in \partial\mathcal{X}$ corresponding to Eq. 2 is, therefore,

$$f_{ij\ldots} = f_{ij\ldots}(\mathbf{X}, t), \quad X_i \in \mathcal{X}_- \cup \mathcal{X}_+, \quad t \in \mathcal{T}. \tag{5}$$

For fixed t, this equation relates to each particle identified by its initial position, X_i, the current value, $f_{ij\ldots}$, of the field at this particle. Equation 5 is commonly referred to as *Lagrangian representation* of the field. To obtain the corresponding generalization of Eq. 3, we define

$$\mathcal{R}_-(t) = \{r_i(\mathbf{X}, t) \,|\, X_i \in \mathcal{X}_-, \, t \in \mathcal{T}\}, \tag{6}$$
$$\mathcal{R}_+(t) = \{r_i(\mathbf{X}, t) \,|\, X_i \in \mathcal{X}_+, \, t \in \mathcal{T}\}, \tag{7}$$
$$\partial\mathcal{R}(t) = \{r_i(\mathbf{X}, t) \,|\, X_i \in \partial\mathcal{X}, \, t \in \mathcal{T}\}, \tag{8}$$

where $\mathcal{R}_-(t)$, $\mathcal{R}_+(t)$, and $\partial\mathcal{R}(t)$ are the domains currently occupied by those particles initially in $\mathcal{X}_-$, in $\mathcal{X}_+$, and on $\partial\mathcal{X}$, respectively. With these definitions, the kinematic representation of fields with possible discontinuities for $r_i \in \partial\mathcal{R}(t)$ corresponding to Eq. 3 has the form

$$F_{ij\ldots} = F_{ij\ldots}(\mathbf{r}, t), \quad r_i \in \mathcal{R}_-(t) \cup \mathcal{R}_+(t), \quad t \in \mathcal{T}. \tag{9}$$

For fixed t, this equation relates to each place identified by its position, r_i, the current value, $F_{ij\ldots}$, of the field at this place. Equation 9 is commonly called *Eulerian representation* of the field.

The mappings defined by Eqs. 5 and 9 are assumed to be single-valued and continuously differentiable with respect to X_i, r_i, and t. Provided both kinematic representations of the field are defined, they are related by

$$f_{ij\ldots}(\mathbf{X}, t) = F_{ij\ldots}[\mathbf{r}(\mathbf{X}, t), t], \tag{10}$$
$$F_{ij\ldots}(\mathbf{r}, t) = f_{ij\ldots}[\mathbf{X}(\mathbf{r}, t), t]. \tag{11}$$

As in the preceding equations, we continue to use lower-case symbols for the Lagrangian representation and upper-case symbols for the Eulerian representation. The gradients of fields in the two kinematic representations are abbreviated according to

$$f_{ij\ldots,k}(\mathbf{X},t) = \frac{\partial}{\partial X_k} f_{ij\ldots}(\mathbf{X},t), \tag{12}$$

$$F_{ij\ldots,k}(\mathbf{r},t) = \frac{\partial}{\partial r_k} F_{ij\ldots}(\mathbf{r},t). \tag{13}$$

No ambiguity can arise from this notation, because a subscript preceded by a comma is to be understood as indicating the gradient with respect to the spatial argument of the field in the kinematic representation considered, i.e., with respect to X_i for $f_{ij\ldots}$ and with respect to r_i for $F_{ij\ldots}$. In view of Eqs. 12 and 13, differentiation of Eqs. 10 and 11 yields

$$f_{ij\ldots,k}(\mathbf{X},t) = F_{ij\ldots,\ell}[\mathbf{r}(\mathbf{X},t),t]\, r_{\ell,k}(\mathbf{X},t), \tag{14}$$

$$F_{ij\ldots,k}(\mathbf{r},t) = f_{ij\ldots,\ell}[\mathbf{X}(\mathbf{r},t),t]\, X_{\ell,k}(\mathbf{r},t). \tag{15}$$

2.2 Total, Initial, and Incremental Fields

We now assume that the current value of an arbitrary field represents a perturbation of its initial value. Allowing for discontinuities of the field values on $\partial\mathcal{R}(t)$, the Eulerian representation of the perturbation equation is then straightforward only for places that are initially in $\mathcal{R}_-(0)$ *and* currently in $\mathcal{R}_-(t)$ or that are initially in $\mathcal{R}_+(0)$ *and* currently in $\mathcal{R}_+(t)$. We call such places *strictly* internal or *strictly* external. For conciseness, we define

$$\mathcal{R}_\ominus(t) = \mathcal{R}_-(0) \cap \mathcal{R}_-(t), \tag{16}$$

$$\mathcal{R}_\oplus(t) = \mathcal{R}_+(0) \cap \mathcal{R}_+(t). \tag{17}$$

On account of these equations, the necessary and sufficient condition for strictly internal or strictly external places is therefore $r_i \in \mathcal{R}_\ominus(t) \cup \mathcal{R}_\oplus(t)$.

The Lagrangian and Eulerian representations of the fundamental perturbation equation can now be stated as follows:

$$f_{ij\ldots}(\mathbf{X},t) = f_{ij\ldots}(\mathbf{X},0) + \delta f_{ij\ldots}(\mathbf{X},t), \quad X_i \in \mathcal{X}_- \cup \mathcal{X}_+, t \in \mathcal{T}, \tag{18}$$

$$F_{ij\ldots}(\mathbf{r},t) = F_{ij\ldots}(\mathbf{r},0)\Delta F_{ij\ldots}(\mathbf{r},t), \quad r_i \in \mathcal{R}_\ominus(t) \cup \mathcal{R}_\oplus(t), t \in \mathcal{T}. \tag{19}$$

We refer to the left-hand sides of these equations as *total fields,* to the first terms on the right-hand sides as *initial fields*, and to the second terms on the right-hand sides as *incremental fields*. In particular, $\delta f_{ij\ldots}(\mathbf{X},t)$ is called *material increment* and $\Delta F_{ij\ldots}(\mathbf{r},t)$ *local increment*. Since $X_i = r_i(\mathbf{X},0)$ and $r_i = X_i(\mathbf{r},0)$, it follows by comparison of Eqs. 2 and 3 with Eqs. 18 and 19, respectively, that

$$\delta r_i(\mathbf{X},t) = u_i(\mathbf{X},t), \tag{20}$$

$$\Delta X_i(\mathbf{r},t) = -U_i(\mathbf{r},t). \tag{21}$$

Equations 18–21 are valid for *finite perturbations*. The remainder of this section, however, will be restricted to *infinitesimal perturbations*, for which the increments and their gradients are taken as "infinitely small."

In the vicinity of $\partial\mathcal{R}(t)$, some places are initially in $\mathcal{R}_-(0)$ and currently in $\mathcal{R}_+(t)$, or *vice versa*. Since the field values are not necessarily continuous on $\partial\mathcal{R}(t)$, such hybrid places require special attention. In order that this be avoided, we need the Lagrangian representation of Eq. 19. Putting $r_i = r_i(\mathbf{X}, t)$, this equation becomes

$$F_{ij\ldots}[\mathbf{r}(\mathbf{X},t),t] = F_{ij\ldots}[\mathbf{r}(\mathbf{X},t),0] + \Delta F_{ij\ldots}[\mathbf{r}(\mathbf{X},t),t], \quad X_i \in \mathcal{X}_- \cup \mathcal{X}_+, \quad t \in \mathcal{T}. \tag{22}$$

In view of Eqs. 2 and 10 and on the assumption of infinitesimal perturbations, the first term on the right-hand side of Eq. 22 can be expanded into

$$F_{ij\ldots}[\mathbf{r}(\mathbf{X},t),0] = F_{ij\ldots}[\mathbf{r}(\mathbf{X},0),0] + F_{ij\ldots,k}[\mathbf{r}(\mathbf{X},0),0]\, u_k(\mathbf{X},t). \tag{23}$$

Substiting this equation into Eq. 22 and using Eqs. 10, 14 and $r_{i,j}(\mathbf{X},0) = \delta_{ij}$ then yields

$$f_{ij\ldots}(\mathbf{X},t) = f_{ij\ldots}(\mathbf{X},0) + \Delta f_{ij\ldots}(\mathbf{X},t) + f_{ij\ldots,k}(\mathbf{X},0)\, u_k(\mathbf{X},t), \quad X_i \in \mathcal{X}_- \cup \mathcal{X}_+, \quad t \in \mathcal{T}. \tag{24}$$

We now adopt several notational simplifications: (i) the arguments X_i, r_i, and t are suppressed; (ii) the argument $t = 0$ is indicated by the label superscript 0 appended to the function symbol; (iii) the material and local increments are indicated by the label superscripts δ and Δ appended to the function symbol. With these conventions, we obtain from the Lagrangian representations of the perturbation equation, Eqs. 18 and 24, the relations

$$\left.\begin{aligned} f_{ij\ldots} &= f_{ij\ldots}^{(0)} + f_{ij\ldots}^{(\delta)} && (25)\\ f_{ij\ldots} &= f_{ij\ldots}^{(0)} + f_{ij\ldots}^{(\Delta)} + f_{ij\ldots,k}^{(0)} u_k && (26)\\ f_{ij\ldots}^{(\delta)} &= f_{ij\ldots}^{(\Delta)} + f_{ij\ldots,k}^{(0)} u_k, && (27) \end{aligned}\right\} \quad X_i \in \mathcal{X}_- \cup \mathcal{X}_+, \quad t \in \mathcal{T}.$$

We refer to Eqs. 25 and 26, respectively, as material and local forms of the Lagrangian perturbation equation. The second term on the right-hand side of Eq. 27 is called *advective increment*. It accounts for the increment resulting from the component of the displacement parallel to the gradient of the initial field. In the particular case of a spatially homogeneous initial field, we have $f_{ij\ldots,k}^{(0)} = 0$ and, therefore, $f_{ij\ldots}^{(\delta)} = f_{ij\ldots}^{(\Delta)}$. Henceforth, only the Lagrangian representation will be employed. Conventionally, the Langrangian representation is used whenever particles can be identified and the domains initially occupied by them are prescribed, but the domains currently occupied not known a priori. Elsewhere, the *Newtonian* representation and the associated *isopotential* increment of fields have been defined and employed in the formulation of the field theory of GVED (Wolf 1997).

2.3 Interface Conditions

We briefly return the behavior of field values on $\partial\mathcal{R}$. To formulate a condition expressing this behavior, we locally assign to $\partial\mathcal{R}$ the Lagrangian representation of the unit normal directed outward into $\mathcal{R}_+$. Denoting this normal by n_i and assuming $\epsilon > 0$, we define

$$\left.\begin{aligned} [f_{ij\ldots}]_\pm &= \lim_{\epsilon \to 0+} f_{ij\ldots}(\mathbf{X} \pm \epsilon \mathbf{n}^{(0)}) && (28)\\ [f_{ij\ldots}]_-^+ &= [f_{ij\ldots}]_+ - [f_{ij\ldots}]_- && (29) \end{aligned}\right\} \quad X_i \in \partial\mathcal{X}, \quad t \in \mathcal{T}.$$

The *interface condition* for $f_{ij\ldots}$ can then be written as

$$[f_{ij\ldots}]_-^+ = f_{ij\ldots}^{\pm}, \quad X_i \in \partial\mathcal{X}, \quad t \in \mathcal{T}, \tag{30}$$

where $f_{ij\ldots}^{\pm}$ is the increase of $f_{ij\ldots}$ in the direction of n_i. For convenience, we extend field values onto $\partial\mathcal{R}$ using

$$f_{ij\ldots} = \tfrac{1}{2}\{[f_{ij\ldots}]_- + [f_{ij\ldots}]_+\}, \quad X_i \in \partial\mathcal{X}, \quad t \in \mathcal{T}. \tag{31}$$

3 Field Equations and Interface Conditions

In this section, we deduce the incremental field equations and interface conditions of GVED describing infinitesimal, gravitational–viscoelastic perturbations of compositionally and entropically stratified, compressible, rotating fluids initially in hydrostatic equilibrium. In deducing the equations, we suppose that the perturbations are isocompositional and isentropic. Our assumption is justified if the characteristic times of diffusive processes are long compared with those of viscoelastic relaxation. The modifications of the theory required to include phase changes have been studied elsewhere (e.g., Johnston et al. 1997).

We use the Lagrangian representation developed in ➋ Sect. 2, i.e., the field values refer to the current position, r_i, of a particle whose initial position, X_i, is taken as the spatial argument, the temporal argument is the current time, t. The field equations and interface conditions are defined for $X_i \in \mathcal{X}_- \cup \mathcal{X}_+$ and $X_i \in \partial\mathcal{X}$, respectively, and for $t \in \mathcal{T}$. We begin by collecting the field equations and interface conditions for the total fields (➋ Sect. 3.1) and the initial fields (➋ Sect. 3.2), from which those for the incremental fields are obtained (➋ Sect. 3.3). After this, the continuity and state equations involving the density and thermodynamic pressure are given (➋ Sect. 3.4). These fields are useful when studying the large-t asymptotic behavior of the incremental field equations and interface conditions (➋ Sect. 4.3.2) and when considering locally incompressible perturbations (➋ Sect. 5.1).

3.1 Equations for the Total Fields

We follow the standard monographs on continuum mechanics (e.g., Eringen 1989; Malvern 1969). In particular, we are concerned with the relationship between the Cauchy stress, t_{ij}, and the (non-symmetric) Piola–Kirchhoff stress, τ_{ij}:

$$t_{ij}\,\mathrm{d}^2 r_j = \tau_{ij}\,\mathrm{d}^2 r_j^{(0)}, \tag{32}$$

where $\mathrm{d}^2 r_i$ is an arbitrary differential area currently at r_i and $\mathrm{d}^2 r_i^{(0)}$ the associated initial differential area at $r_i^{(0)}$. Using the transformation formula

$$r_{j,i}\,\mathrm{d}^2 r_j = j\,\mathrm{d}^2 r_i^{(0)}, \tag{33}$$

with the Jacobian determinant, j, given by

$$j = \det[r_{i,j}], \tag{34}$$

it then follows that

$$r_{i,k}\tau_{jk} = j t_{ij}. \tag{35}$$

We suppose in this review that the continuum is without couple stresses, volume couples, and spin angular momentum so that t_{ij} is symmetric.

Consider now a gravitating, rotating continuum undergoing perturbations of some initial state. Assuming that its angular velocity, Ω_i, is prescribed, the *momentum equation* relative to a co-rotating reference frame is

$$\tau_{ij,j} + \rho^{(0)}(g_i + 2\epsilon_{ijk}\Omega_k \, \mathrm{d}_t r_j) = \rho^{(0)} \, \mathrm{d}_t^2 r_i, \tag{36}$$

where g_i is the gravity force per unit mass, $2\epsilon_{ijk}\Omega_k \, \mathrm{d}_t r_j$ the Coriolis force per unit mass and $\rho^{(0)} \geq 0$ the initial volume-mass density. The symbols d_t and d_t^2 denote the first- and second-order material-derivative operators with respect to t. The field g_i, henceforth simply referred to as gravity, is given by

$$g_i = (\phi + \chi + \psi)_{,j} X_{j,i}, \tag{37}$$

with ϕ the gravitational potential, χ the centrifugal potential, and ψ the tidal potential. The *gravitational-potential equation* can be written as

$$j(\phi_{,ij} X_{i,k} X_{j,k} + \phi_{,i} X_{i,jj}) = -4\pi G \rho^{(0)}, \tag{38}$$

where G is Newton's gravitational constant. Using the relations between the Lagrangian and Eulerian gradients (➲ Sect. 2.1) and the continuity equation (➲ Sect. 3.4), Eq. 38 can be interpreted as the Lagrangian representation of the (Eulerian) *Poisson equation* (e.g., Ramsey 1981). The *rotational-potential equation* is

$$2\chi = \Omega_i \Omega_i r_j r_j - \Omega_i \Omega_j r_i r_j, \tag{39}$$

which implies that the origin of the coordinate system is on the spin axis. The *constitutive equation* is assumed to be of the form

$$t_{ij} = t_{ij}^{(0)} + \mathcal{M}_{ij}[r_{m,k}(t - t') \, r_{m,\ell}(t - t') - \delta_{k\ell}], \tag{40}$$

where $\mathcal{M}_{ij}$ is the anisotropic relaxation functional (assumed to be linear) transforming the strain *history* given by the term in brackets into the *current* incremental stress and t' is the excitation time. Clearly, $t' \in [\,0, t\,]$ must hold as a consequence of the causality principle. With $\mathcal{M}_{ij}$, $t_{ij}^{(0)}$, $\rho^{(0)}$, ψ, and Ω_i prescribed parameters and in view of $X_{(i,j)} r_{(j,i)} = 1$ (no summation), Eqs. 34–40 constitute the system of total field equations for g_i, j, r_i, t_{ij}, τ_{ij}, ϕ, and χ.

Next, the interface conditions to be satisfied by the fields will be collected. We assume here that $\partial\mathcal{R}$ coincides with a material sheet whose interface-mass density is σ. Considering Eqs. 28–30 and the direction of n_i agreed on, the following interface conditions result from Eqs. 34–40:

$$[r_i]_-^+ = 0, \tag{41}$$

$$[\phi]_-^+ = 0, \tag{42}$$

$$[n_i \phi_{,j} X_{j,i}]_-^+ = -4\pi G \sigma, \tag{43}$$

$$[n_j t_{ij}]_-^+ = -g_i \sigma. \tag{44}$$

Note that the conditions apply on the supposition that $\partial\mathcal{R}$ is a "welded" interface not admitting slip or cavitation. The value of g_i for $X_i \in \partial\mathcal{X}$ is defined according to Eq. 31 and σ is supposed to be a prescribed function of $X_i \in \partial\mathcal{X}$ and $t \in \mathcal{T}$.

3.2 Equations for the Initial Fields

We now assume that (i) the continuum is a *fluid* and (ii) the initial state applying at $t = 0$ is a *static* equilibrium state. Since a fluid at rest cannot maintain deviatoric stresses, the initial state must be a *hydrostatic equilibrium* state. With the mechanical pressure defined by $p = -t_{ii}/3$, we then have

$$t_{ij}^{(0)} = -\delta_{ij} p^{(0)} \tag{45}$$

and, with $r_{i,j}^{(0)} = \delta_{ij}$, Eqs. 34 and 35 reduce to

$$j^{(0)} = 1, \tag{46}$$

$$\tau_{ij}^{(0)} = -\delta_{ij} p^{(0)}. \tag{47}$$

Using Eqs. 45–47, $X_{i,j}^{(0)} = \delta_{ij}$ and $(d_t r_i)^{(0)} = (d_t^2 r_i)^{(0)} = 0$, Eqs. 36–40 become

$$-p_{,i}^{(0)} + \rho^{(0)} g_i^{(0)} = 0, \tag{48}$$

$$g_i^{(0)} = (\phi^{(0)} + \chi^{(0)} + \psi^{(0)})_{,i}, \tag{49}$$

$$\phi_{,ii}^{(0)} = -4\pi G \rho^{(0)}, \tag{50}$$

$$2\chi^{(0)} = \Omega_i \Omega_i r_j^{(0)} r_j^{(0)} - \Omega_i \Omega_j r_i^{(0)} r_j^{(0)}, \tag{51}$$

$$p^{(0)} = \xi(\rho^{(0)}, \lambda^{(0)}, \varphi^{(0)}). \tag{52}$$

The last expression is the form of the *state equation* assumed in this review, where $\lambda^{(0)}$ is a field representing the initial composition and $\varphi^{(0)}$ the initial entropy density. With the state function, ξ, known and $\lambda^{(0)}$, $\varphi^{(0)}$, $\psi^{(0)}$, and Ω_i prescribed, Eqs. 48–52 constitute the (nonlinear) system of initial field equations of *gravitational hydrostatics* (GHS) for $g_i^{(0)}$, $p^{(0)}$, $\rho^{(0)}$, $\phi^{(0)}$, and $\chi^{(0)}$. We point out the relationship $\epsilon_{ijk} \rho_{,j}^{(0)} g_k^{(0)} = 0$ following from Eqs. 48 and 49, whence these equations require that the level surfaces of $p^{(0)}$, $\rho^{(0)}$ and $\phi^{(0)} + \chi^{(0)} + \psi^{(0)}$ coincide. However, since Eqs. 48 and 49 represent three scalar equations, respectively, the system consisting of Eqs. 48–52 is overdetermined and solutions for the level surfaces are severely restricted.

Supposing $\sigma^{(0)} = 0$ and using Eq. 45 and $X_{i,j}^{(0)} = \delta_{ij}$, the following initial interface conditions are obtained from Eqs. 41–44:

$$\left[r_i^{(0)}\right]_-^+ = 0, \tag{53}$$

$$[\phi^{(0)}]_-^+ = 0, \tag{54}$$

$$\left[n_i^{(0)} \phi_{,i}^{(0)}\right]_-^+ = 0, \tag{55}$$

$$[p^{(0)}]_-^+ = 0. \tag{56}$$

Since solutions to Eqs. 48–52 admit a jump discontinuity of $\rho^{(0)}$ for $X_i \in \partial\mathcal{X}$, we also have

$$[\rho^{(0)}]_-^+ = \rho^{\pm}. \tag{57}$$

3.3 Equations for the Incremental Fields

3.3.1 Material Form

Using Eqs. 20, 21, and 25, we decompose the total fields in Eqs. 34–40 into initial and incremental parts. Considering also Eqs. 45–47, 52, $r_{i,j}^{(0)} = X_{i,j}^{(0)} = \delta_{ij}$ and $(\mathrm{d}_t r_i)^{(0)} = \left(\mathrm{d}_t^2 r_i\right)^{(0)} = 0$, we get

$$(1 + j^{(\delta)}) = \det[\delta_{ij} + u_{i,j}], \tag{58}$$

$$(\delta_{ik} + u_{i,k})\left(-\delta_{jk} p^{(0)} + \tau_{jk}^{(\delta)}\right) = (1 + j^{(\delta)})\left(-\delta_{ij} p^{(0)} + t_{ij}^{(\delta)}\right), \tag{59}$$

$$-p_{,i}^{(0)} + \tau_{ij,j}^{(\delta)} + \rho^{(0)}\left(g_i^{(0)} + g_i^{(\delta)} + 2\epsilon_{ijk}\Omega_k \,\mathrm{d}_t u_j\right) = \rho^{(0)}\,\mathrm{d}_t^2 u_i, \tag{60}$$

$$g_i^{(0)} + g_i^{(\delta)} = (\phi^{(0)} + \phi^{(\delta)} + \chi^{(0)} + \chi^{(\delta)} + \psi^{(0)} + \psi^{(\delta)})_{,j}(\delta_{ji} - U_{j,i}), \tag{61}$$

$$(1 + j^{(\delta)})[(\phi^{(0)} + \phi^{(\delta)})_{,ij}(\delta_{ik} - U_{i,k})(\delta_{jk} - U_{j,k}) - (\phi^{(0)} + \phi^{(\delta)})_{,i} U_{i,jj}] = -4\pi G\rho^{(0)}, \tag{62}$$

$$2(\chi^{(0)} + \chi^{(\delta)}) = \Omega_i\Omega_i\left(r_j^{(0)} + u_j\right)\left(r_j^{(0)} + u_j\right) - \Omega_i\Omega_j\left(r_i^{(0)} + u_i\right)\left(r_j^{(0)} + u_j\right), \tag{63}$$

$$\begin{aligned} -\delta_{ij} p^{(0)} + t_{ij}^{(\delta)} &= -\delta_{ij}\,\xi(\rho^{(0)}, \lambda^{(0)}, \varphi^{(0)}) \\ &+ \mathcal{M}_{ij}\{[\delta_{mk} + u_{m,k}(t - t')][\delta_{m\ell} + u_{m,\ell}(t - t')] - \delta_{k\ell}\}. \end{aligned} \tag{64}$$

We note that no restriction on the magnitude of the perturbations has been imposed so far, i.e., Eqs. 58–64 are valid for *finite perturbations*. Since we are only concerned with *infinitesimal perturbations*, this allows us to *linearize* the field equations. Accordingly, we have

$$U_{i,j} = u_{i,j} \tag{65}$$

and Eq. 58 reduces to

$$j^{(\delta)} = u_{i,i}, \tag{66}$$

by which Eq. 59 can be rewritten as

$$\tau_{ij}^{(\delta)} = t_{ij}^{(\delta)} + p^{(0)}(u_{j,i} - \delta_{ij} u_{k,k}). \tag{67}$$

Considering Eqs. 48–52 and 65–67, Eqs. 60–64 then become

$$t_{ij,j}^{(\delta)} + p_{,j}^{(0)} u_{j,i} - p_{,i}^{(0)} u_{j,j} + \rho^{(0)}\left(g_i^{(\delta)} + 2\epsilon_{ijk}\Omega_k\,\mathrm{d}_t u_j\right) = \rho^{(0)}\,\mathrm{d}_t^2 u_i, \tag{68}$$

$$g_i^{(\delta)} = (\phi^{(\delta)} + \chi^{(\delta)} + \psi^{(\delta)})_{,i} - (\phi^{(0)} + \chi^{(0)} + \psi^{(0)})_{,j} u_{j,i}, \tag{69}$$

$$\phi_{,ii}^{(\delta)} - 2\phi_{,ij}^{(0)} u_{i,j} - \phi_{,i}^{(0)} u_{i,jj} = 4\pi G\rho^{(0)} u_{i,i}, \tag{70}$$

$$\chi^{(\delta)} = \chi_{,i}^{(0)} u_i, \tag{71}$$

$$t_{ij}^{(\delta)} = \mathcal{M}_{ij}[u_{k,\ell}(t - t') + u_{\ell,k}(t - t')]. \tag{72}$$

With $\mathcal{M}_{ij}$, $\psi^{(\delta)}$ and Ω_i prescribed parameters and the initial fields given as the special solution to the initial field equations and interface conditions, Eqs. 68–72 constitute the material form of the incremental field equations of GVED for $g_i^{(\delta)}$, $t_{ij}^{(\delta)}$, u_i, $\phi^{(\delta)}$, and $\chi^{(\delta)}$.

Next, the linearized form of the associated incremental interface conditions is derived. For this purpose, we decompose the total fields in Eqs. 41–44 into initial and incremental parts. Using Eqs. 20, 21, 25, 45, and $r^{(0)}_{i,j} = X^{(0)}_{i,j} = \delta_{ij}$, we get

$$\left[r^{(0)}_i + u_i\right]^+_- = 0, \tag{73}$$

$$[\phi^{(0)} + \phi^{(\delta)}]^+_- = 0, \tag{74}$$

$$\left[\left(n^{(0)}_i + n^{(\delta)}_i\right)(\phi^{(0)} + \phi^{(\delta)})_{,j}(\delta_{ji} - U_{j,i})\right]^+_- = -4\pi G\sigma, \tag{75}$$

$$\left[\left(n^{(0)}_j + n^{(\delta)}_j\right)\left(-\delta_{ij}p^{(0)} + t^{(\delta)}_{ij}\right)\right]^+_- = -\left(g^{(0)}_i + g^{(\delta)}_i\right)\sigma. \tag{76}$$

In view of Eqs. 53–56 and 65 and on the assumption of infinitesimal perturbations, the material form of the incremental interface conditions follow as

$$[u_i]^+_- = 0, \tag{77}$$

$$[\phi^{(\delta)}]^+_- = 0, \tag{78}$$

$$\left[n^{(0)}_i\left(\phi^{(\delta)}_{,i} - \phi^{(0)}_{,j}u_{j,i}\right)\right]^+_- = -4\pi G\sigma, \tag{79}$$

$$\left[n^{(0)}_j t^{(\delta)}_{ij}\right]^+_- = -g^{(0)}_i\sigma. \tag{80}$$

Since $n^{(0)}_i$ is normal to $\partial\mathcal{R}^{(0)}$, which is a surface of constant $\phi^{(0)} + \chi^{(0)} + \psi^{(0)}$, we put on this surface

$$g^{(0)}_i = -\gamma n^{(0)}_i. \tag{81}$$

3.3.2 Material–Local Form

The material–local form of the incremental field equations and interface conditions results if we use Eq. 27 to express $g^{(\delta)}_i$, $\phi^{(\delta)}$ and $\psi^{(\delta)}$ in terms of the respective local increments:

$$g^{(\delta)}_i = g^{(\Delta)}_i + g^{(0)}_{i,j}u_j, \tag{82}$$

$$\phi^{(\delta)} = \phi^{(\Delta)} + \phi^{(0)}_{,i}u_i, \tag{83}$$

$$\psi^{(\delta)} = \psi^{(\Delta)} + \psi^{(0)}_{,i}u_i. \tag{84}$$

In view of Eqs. 48–50, 71, and 82–84, Eqs. 68–72 take the form

$$t^{(\delta)}_{ij,j} + \left(p^{(0)}_{,j}u_j\right)_{,i} - g^{(0)}_i(\rho^{(0)}u_j)_{,j} + \rho^{(0)}\left(g^{(\Delta)}_i + 2\epsilon_{ijk}\Omega_k\,\mathrm{d}_t u_j\right) = \rho^{(0)}\,\mathrm{d}^2_t u_i, \tag{85}$$

$$g^{(\Delta)}_i = (\phi^{(\Delta)} + \psi^{(\Delta)})_{,i}, \tag{86}$$

$$\phi^{(\Delta)}_{,ii} = 4\pi G(\rho^{(0)}u_i)_{,i}, \tag{87}$$

$$t^{(\delta)}_{ij} = \mathcal{M}_{ij}[u_{k,\ell}(t-t') + u_{\ell,k}(t-t')]. \tag{88}$$

With $\mathcal{M}_{ij}$, $\psi^{(\Delta)}$ and Ω_i prescribed parameters and the initial fields given as the special solution to the initial field equations and interface conditions, Eqs. 86–88 constitute the material–local form of the incremental field equations of GVED for $g^{(\Delta)}_i$, $t^{(\delta)}_{ij}$, u_i, and $\phi^{(\Delta)}$.

Equations 86–87 agree with the incremental momentum equation and the incremental gravitational-potential equation given by Love (1911) and Dahlen (1974). Love used the Eulerian representation, i.e., his incremental equations are functions of the current particle position, r_i. Since the difference between the Lagrangian and Eulerian representations is of second order in the incremental quantities, it may be ignored in linearized field theory. In contrast to Love, Dahlen used the Lagrangian representation in terms of the initial particle position, X_i, which has also been adopted here.

The associated incremental interface conditions follow upon substituting Eq. 83 into Eqs. 78 and 79, yielding

$$\left[\phi^{(\Delta)} + \phi^{(0)}_{,i} u_i\right]_-^+ = 0, \tag{89}$$

$$\left[n_i^{(0)}\left(\phi^{(\Delta)}_{,i} + \phi^{(0)}_{,ij} u_j\right)\right]_-^+ = -4\pi G\sigma. \tag{90}$$

Observing the constraints imposed by Eq. 55 on the continuity of the components of $n_j^{(0)}\phi^{(0)}_{,ij}$, Eq. 90 can be shown to be equivalent to

$$\left[n_i^{(0)}\left(\phi^{(\Delta)}_{,i} + \phi^{(0)}_{,jj} u_i\right)\right]_-^+ = -4\pi G\sigma. \tag{91}$$

Upon consideration of Eqs. 50, 77, 80, and 81, the incremental interface conditions are found to be

$$[u_i]_-^+ = 0, \tag{92}$$

$$[\phi^{(\Delta)}]_-^+ = 0, \tag{93}$$

$$\left[n_i^{(0)}\left(\phi^{(\Delta)}_{,i} - 4\pi G\rho^{(0)} u_i\right)\right]_-^+ = -4\pi G\sigma, \tag{94}$$

$$\left[n_j^{(0)} t_{ij}^{(\delta)}\right]_-^+ = \gamma n_i^{(0)} \sigma. \tag{95}$$

The material–local form of the incremental field equations and interface conditions of GVED is reconsidered below when deriving the small-t asymptotes to these equations (➋ Sect. 4.3.1).

3.3.3 Local Form

We consider Eqs. 27 and 45, giving

$$t_{ij}^{(\delta)} = t_{ij}^{(\Delta)} - \delta_{ij} p^{(0)}_{,k} u_k. \tag{96}$$

On account of Eqs. 48 and 96, the material–local form of the incremental field equations and interface conditions, Eqs. 86–88 and 92–95, transforms into

$$t^{(\Delta)}_{ij,j} - g_i^{(0)}(\rho^{(0)} u_j)_{,j} + \rho^{(0)}\left(g_i^{(\Delta)} + 2\epsilon_{ijk}\Omega_k \, \mathrm{d}_t u_j\right) = \rho^{(0)} \, \mathrm{d}_t^2 u_i, \tag{97}$$

$$g_i^{(\Delta)} = (\phi^{(\Delta)} + \psi^{(\Delta)})_{,i}, \tag{98}$$

$$\phi^{(\Delta)}_{,ii} = 4\pi G(\rho^{(0)} u_i)_{,i}, \tag{99}$$

$$t_{ij}^{(\Delta)} = \delta_{ij} p_{,k}^{(0)} u_k + \mathcal{M}_{ij}[u_{k,\ell}(t-t') + u_{\ell,k}(t-t')], \tag{100}$$

$$[u_i]_-^+ = 0, \tag{101}$$

$$[\phi^{(\Delta)}]_-^+ = 0, \tag{102}$$

$$\left[n_i^{(0)} \left(\phi_{,i}^{(\Delta)} - 4\pi G \rho^{(0)} u_i\right)\right]_-^+ = -4\pi G \sigma, \tag{103}$$

$$\left[n_j^{(0)} \left(t_{ij}^{(\Delta)} - \delta_{ij} \rho^{(0)} g_k^{(0)} u_k\right)\right]_-^+ = \gamma n_i^{(0)} \sigma. \tag{104}$$

With $\mathcal{M}_{ij}$, $\psi^{(\Delta)}$ and Ω_i prescribed parameters and the initial fields given as the special solution to the initial field equations and interface conditions, Eqs. 97–100 constitute the local form of the incremental field equations of GVED for $g_i^{(\Delta)}$, $t_{ij}^{(\Delta)}$, u_i, and $\phi^{(\Delta)}$, whose solution must satisfy the associated incremental interface conditions, Eqs. 101–104.

The term $\left[n_i^{(0)} \rho^{(0)} g_j^{(0)} u_j\right]_-^+$ in Eq. 104 is sometimes referred to as "*buoyancy term*." Note that its appearance is solely a consequence of formulating the incremental field equations and interface conditions in terms of the local incremental stress, $t_{ij}^{(\Delta)}$. In the material–local form of the equations, where the material incremental stress, $t_{ij}^{(\delta)}$, is used, no buoyancy term can therefore arise in the corresponding interface condition, Eq. 95. Conversely, the material–local momentum Eq. 86, contains the "*advective term*" $\left(p_{,j}^{(0)} u_j\right)_{,i}$, which is absent from the local momentum equation, Eq. 97. The local form of the incremental field equations and interface conditions of GVED are used when deducing the large-t asymptotes to the incremental equations (➋ Sect. 4.3.2) and when considering the approximations of local incompressibility (➋ Sect. 5.1) and material incompressibility (➋ Sect. 5.2).

3.3.4 Constitutive Equation

To obtain an expression for $\mathcal{M}_{ij}$, we use the continuous differentiability of the strain history. On this assumption, $\mathcal{M}_{ij}$ may be written as a convolution integral:

$$\mathcal{M}_{ij} = \int_0^t m_{ijk\ell}(t-t')\, \mathrm{d}_{t'}[u_{k,\ell}(t') + u_{\ell,k}(t')]\, \mathrm{d}t', \tag{105}$$

with $m_{ijk\ell}(t - t')$ the anisotropic *relaxation function* (e.g., Christensen 1982). Supposing *isotropic* viscoelasticity from now on and exploiting the usual symmetry properties of $m_{ijk\ell}$, this simplifies to

$$\begin{aligned}\mathcal{M}_{ij} = \delta_{ij} \int_0^t \left[m_1(t-t') - \tfrac{2}{3} m_2(t-t')\right] \mathrm{d}_{t'}[u_{k,k}(t')]\, \mathrm{d}t' \\ + \int_0^t m_2(t-t')\, d_{t'}[u_{i,j}(t') + u_{j,i}(t')]\, \mathrm{d}t'.\end{aligned} \tag{106}$$

We refer to this relation as incremental constitutive equation of viscoelasticity. The independent functions $m_1(t - t')$ and $m_2(t - t')$ are defined for $t - t' \in [0, \infty)$ and are called bulk- and shear-relaxation functions, respectively. For convenience, we also use $m_\nu(t - t')$ with $\nu \in \{1, 2\}$. We assume that $m_\nu(t - t')$ is continuously differentiable for $X_i \in \mathcal{X}_- \cup \mathcal{X}_+$ but may have a jump discontinuity for $X_i \in \partial\mathcal{X}$. Furthermore, we take $m_\nu(t - t')$ as continuously differentiable with respect to $t - t'$. From thermodynamic principles, it follows that

(e.g. Christensen 1982; Golden and Graham 1988)

$$m_v(t-t') \geq 0, \tag{107}$$

$$\mathrm{d}_{t-t'} m_v(t-t') \leq 0, \tag{108}$$

$$\mathrm{d}^2_{t-t'} m_v(t-t') \geq 0. \tag{109}$$

To obtain an additional constraint on $m_2(t-t')$, we recall that the assumption has been made for the continuum to be a fluid. A necessary condition of fluid constitutive behavior is that deviatoric stresses can relax completely (e.g., Christensen 1982; Golden and Graham 1988). In view of Eq. 106, this is formally expressible as the *fluidity condition*:

$$\lim_{t-t'\to\infty} m_2(t-t') = 0. \tag{110}$$

3.4 Continuity and State Equations

So far, the incremental density and incremental thermodynamic pressure have not appeared explicitly in the equations. This is in accordance with the adoption of the Lagrangian representation, where the displacement, u_i, is preferentially used. However, the incremental density and incremental thermodynamic pressure are required when interpreting the large-t asymptotes to the incremental field equations and interface conditions of GVED (➋ Sect. 4.3.2) and when studying the approximation of local incompressibility (➋ Sect. 5.1).

The current value of the density, ρ, can be related to its initial value, $\rho^{(0)}$, by means of

$$j\rho = \rho^{(0)}, \tag{111}$$

which is the *continuity equation* (e.g., Dahlen 1974; Malvern 1969). For a fluid not necessarily in hydrostatic equilibrium, the *thermodynamic* pressure, ϖ, is introduced with the aid of a *state equation* whose functional relation is identical to that governing the *mechanical* pressure, $p = -t_{ii}/3$, in the case of hydrostatic equilibrium (e.g., Dahlen 1974; Malvern 1969). In view of Eq. 52, we therefore have

$$\varpi = \xi(\rho, \lambda, \varphi), \tag{112}$$

with ϖ, in general, different from p. However, at $t = 0$, Eq. 112 reduces to

$$\varpi^{(0)} = \xi(\rho^{(0)}, \lambda^{(0)}, \varphi^{(0)}), \tag{113}$$

which, by comparison with Eq. 52, yields

$$\varpi^{(0)} = p^{(0)}. \tag{114}$$

A direct consequence of Eqs. 113 and 114 is

$$p^{(0)}_{,i} = \left(\frac{\partial \xi}{\partial \rho}\right)^{(0)} \rho^{(0)}_{,i} + \left(\frac{\partial \xi}{\partial \lambda}\right)^{(0)} \lambda^{(0)}_{,i} + \left(\frac{\partial \xi}{\partial \varphi}\right)^{(0)} \varphi^{(0)}_{,i}, \tag{115}$$

where the partial derivatives $(\partial\xi/\partial\rho)^{(0)} = [\partial\xi/\partial\rho]_{\rho=\rho^{(0)}}$ etc. are functions of $X_i \in \mathcal{X}_- \cup \mathcal{X}_+$.

Next, we use Eq. 25 to decompose the total fields in Eqs. 111 and 112 into initial and material incremental parts. Using also Eqs. 46 and 66, Eq. 111 becomes

$$(1 + u_{i,i})(\rho^{(0)} + \rho^{(\delta)}) = \rho^{(0)}. \tag{116}$$

Since, by assumption, we have *isocompositional and isentropic perturbations*, $\lambda^{(\delta)} = \varphi^{(\delta)} = 0$ applies and the decomposition of Eq. 112 takes the form

$$\varpi^{(0)} + \varpi^{(\delta)} = \xi(\rho^{(0)}, \lambda^{(0)}, \varphi^{(0)}) + \left(\frac{\partial \xi}{\partial \rho}\right)^{(0)} \rho^{(\delta)}. \tag{117}$$

Considering Eq. 113 and retaining only terms that are linear in the incremental quantities, Eqs. 116 and 117 reduce to

$$\rho^{(\delta)} = -\rho^{(0)} u_{i,i}, \tag{118}$$

$$\varpi^{(\delta)} = \left(\frac{\partial \xi}{\partial \rho}\right)^{(0)} \rho^{(\delta)}, \tag{119}$$

which constitute the material forms of the incremental continuity and state equations, respectively. Due to Eqs. 27 and 114, we, however, have

$$\rho^{(\delta)} = \rho^{(\Delta)} + \rho_{,i}^{(0)} u_i, \tag{120}$$

$$\varpi^{(\delta)} = \varpi^{(\Delta)} + p_{,i}^{(0)} u_i, \tag{121}$$

whence the material forms can be replaced by

$$\rho^{(\Delta)} = -(\rho^{(0)} u_i)_{,i}, \tag{122}$$

$$\varpi^{(\Delta)} = \left(\frac{\partial \xi}{\partial \rho}\right)^{(0)} \left(\rho^{(\Delta)} + \rho_{,i}^{(0)} u_i\right) - p_{,i}^{(0)} u_i. \tag{123}$$

These relations represent the local forms of the incremental continuity and state equations, respectively. Equation 123 takes a more familiar form upon substituting for $p_{,i}^{(0)}$ from Eq. 115, giving

$$\varpi^{(\Delta)} = \left(\frac{\partial \xi}{\partial \rho}\right)^{(0)} \rho^{(\Delta)} - \left(\frac{\partial \xi}{\partial \lambda}\right)^{(0)} \lambda_{,i}^{(0)} u_i - \left(\frac{\partial \xi}{\partial \varphi}\right)^{(0)} \varphi_{,i}^{(0)} u_i. \tag{124}$$

Explicit expressions for the partial derivatives are stated below (➋ Sects. 4.3.2 and ➋ 5.1).

4 Asymptotic Incremental Field Theories

We proceed by supposing perturbations whose limits exist for both $t \to 0$ and $t \to \infty$. Obviously, these limits correspond to the *initial* and *final* hydrostatic equilibrium states of the fluid. This means that the small- and large-t asymptotes to the incremental field equations and interface conditions of GVED also exist. We determine them by finding suitable asymptotic approximations to the Laplace transform of the incremental constitutive equation of viscoelasticity. Upon substitution of Eq. 106 into Eqs. 88 and 100, respectively, and use of Eqs. 207, 209, 211 (Appendix A) and $u_i^{(0)} = 0$, the Laplace-transformed material and local forms of the incremental constitutive equation can be written as

$$\widetilde{t}_{ij}^{(\delta)} = \delta_{ij} \left(\widetilde{m}_1 - \tfrac{2}{3}\widetilde{m}_2\right) s\widetilde{u}_{k,k} + \widetilde{m}_2 s(\widetilde{u}_{i,j} + \widetilde{u}_{j,i}), \tag{125}$$

$$\widetilde{t}_{ij}^{(\Delta)} = \delta_{ij} \left[p_{,k}^{(0)} \widetilde{u}_k + \left(\widetilde{m}_1 - \tfrac{2}{3}\widetilde{m}_2\right) s\widetilde{u}_{k,k} \right] + \widetilde{m}_2 s(\widetilde{u}_{i,j} + \widetilde{u}_{j,i}), \tag{126}$$

where $\widetilde{f}_{ij...}(\mathbf{X}, s)$ denotes the Laplace transform of $f_{ij...}(\mathbf{X}, t)$ and $s \in \mathcal{S}$ the inverse Laplace time (Appendix A). As in Eqs. 125 and 126, we continue to suppress the argument, s, of Laplace-transformed quantities. Before expanding the equations, it is necessary to specify $\widetilde{m}_\nu$. This is achieved by expressing $m_\nu(t-t')$ in terms of the associated relaxation spectrum (➋ Sect. 4.1). Laplace transformation then supplies a formula for $s\widetilde{m}_\nu$, from which asymptotic approximations for large and small s can be derived (➋ Sect. 4.2). Substituting these approximations into Eqs. 125 and 126 and applying the generalized initial- and final-value theorems finally gives the small- and large-t asymptotes to the incremental constitutive equation of viscoelasticity (➋ Sect. 4.3).

4.1 Relaxation Functions

For $\nu = 1, 2$, suppose that $m_\nu(t-t')$ can be expressed as

$$m_\nu(t-t') = m_{\nu\infty} + \int_0^\infty \overline{m}_\nu(\alpha')\, e^{-\alpha'(t-t')}\, \mathrm{d}\alpha', \tag{127}$$

where $\overline{m}_\nu(\alpha')$ is the relaxation spectrum, α' is the inverse spectral time, and $m_\nu(t-t')$ satisfies the restrictions expressed by Eqs. 107–110 (e.g., Christensen 1982; Golden and Graham 1988). Equation 127 implies

$$m_{\nu\infty} = \lim_{t-t'\to\infty} m_\nu(t-t'), \tag{128}$$

whence, by Eq. 110, it follows that

$$m_{2\infty} = 0. \tag{129}$$

Defining

$$m_{\nu 0} = m_\nu(0), \tag{130}$$

we also get

$$\int_0^\infty \overline{m}_\nu(\alpha')\, \mathrm{d}\alpha' = m_{\nu 0} - m_{\nu\infty}. \tag{131}$$

A consequence of Eqs. 108, 128, 130, and 131 is that $\int_0^\infty \overline{m}_\nu(\alpha')\, \mathrm{d}\alpha' \geq 0$. Here, we impose the more stringent condition that $\overline{m}_\nu(\alpha') \geq 0$ for $\alpha' \in [0, \infty)$. We furthermore require $\overline{m}_\nu(\alpha')$ to vanish sufficiently rapidly as $\alpha' \to 0$ and $\alpha' \to \infty$ so that, for $0 < \alpha_1 < \infty$, the integrals $\int_0^{\alpha_1} \overline{m}_\nu(\alpha')/\alpha'\, \mathrm{d}\alpha'$, and $\int_{\alpha_1}^\infty \alpha' \overline{m}_\nu(\alpha')\, \mathrm{d}\alpha'$ converge. These assumptions are of sufficient generality to include conventional mechanical and molecular models of viscoelasticity.

Next, we apply Eqs. 212 and 213 (Appendix A) to obtain the s-multiplied Laplace transform of Eq. 127 with respect to $t - t'$:

$$s\widetilde{m}_\nu = m_{\nu\infty} + \int_0^\infty \frac{s\, \overline{m}_\nu(\alpha')}{s + \alpha'}\, \mathrm{d}\alpha'. \tag{132}$$

Being interested in asymptotic approximations to $s\widetilde{m}_\nu$ for large and small s, we decompose the integral in Eq. 132 in the following way:

$$\int_0^\infty \frac{s\, \overline{m}_\nu(\alpha')}{s + \alpha'}\, \mathrm{d}\alpha' = \int_0^{s-0} \frac{\overline{m}_\nu(\alpha')}{1 + \frac{\alpha'}{s}}\, \mathrm{d}\alpha' + \int_{s+0}^\infty \frac{s}{\alpha'} \frac{\overline{m}_\nu(\alpha')}{1 + \frac{s}{\alpha'}}\, \mathrm{d}\alpha'. \tag{133}$$

Note that, on the right-hand side, $\alpha'/s < 1$ in the first integrand, whereas $s/\alpha' < 1$ in the second.

4.2 Asymptotic Relaxation Functions

4.2.1 Large-*s* Asymptotes

For sufficiently large s, the second integral on the right-hand side of Eq. 133 may be neglected. Due to the convergence of $\int_{s+0}^{\infty} \alpha' \overline{m}_\nu(\alpha')\,\mathrm{d}\alpha'$, we get the asymptotic approximation

$$\int_0^\infty \frac{s\,\overline{m}_\nu(\alpha')}{s+\alpha'}\,\mathrm{d}\alpha' \simeq \int_0^\infty \overline{m}_\nu(\alpha')\left(1-\frac{\alpha'}{s}\right)\mathrm{d}\alpha'. \tag{134}$$

Since $\overline{m}_\nu(\alpha') \geq 0$ for $\alpha' \geq 0$, we can apply the mean-value theorem of integral calculus and obtain the following estimate:

$$\int_0^\infty \alpha' \overline{m}_\nu(\alpha')\,\mathrm{d}\alpha' = \alpha_{\nu 0} \int_0^\infty \overline{m}_\nu(\alpha')\,\mathrm{d}\alpha', \tag{135}$$

where $\alpha_{\nu 0} \geq 0$. In view of Eqs. 131, 134, and 135, Eq. 132 takes the form

$$s\widetilde{m}_\nu \simeq m_{\nu 0} - (m_{\nu 0} - m_{\nu\infty})\frac{\alpha_{\nu 0}}{s}, \tag{136}$$

which is correct to the first order in $\alpha_{\nu 0}/s$. Using the abbreviations

$$\kappa_e = m_{10}, \tag{137}$$

$$\kappa_e' = (m_{10} - m_{1\infty})\alpha_{10}, \tag{138}$$

$$\mu_e = m_{20}, \tag{139}$$

$$\mu_e' = (m_{20} - m_{2\infty})\alpha_{20}, \tag{140}$$

we finally obtain

$$s\widetilde{m}_1 \simeq \kappa_e - \frac{\kappa_e'}{s}, \tag{141}$$

$$s\widetilde{m}_2 \simeq \mu_e - \frac{\mu_e'}{s}, \tag{142}$$

which are asymptotically correct for large s. From the properties of $m_\nu(t-t')$ specified above, it follows that the values of κ_e, κ_e', μ_e, and μ_e' are non-negative and continuously differentiable for $X_i \in \mathcal{X}_- \cup \mathcal{X}_+$, with jump discontinuities admitted for $X_i \in \partial\mathcal{X}$.

4.2.2 Small-*s* Asymptotes

For sufficiently small s, the first integral on the right-hand side of Eq. 133 may be neglected. As a result of the convergence of $\int_0^{s-0} \overline{m}_\nu(\alpha')/\alpha'\,\mathrm{d}\alpha'$, we arrive at the asymptotic approximation

$$\int_0^\infty \frac{s\,\overline{m}_\nu(\alpha')}{s+\alpha'}\,\mathrm{d}\alpha' \simeq \int_0^\infty \overline{m}_\nu(\alpha')\,\frac{s}{\alpha'}\,\mathrm{d}\alpha'. \tag{143}$$

Applying the mean-value theorem of integral calculus, we now have

$$\int_0^\infty \frac{\overline{m}_\nu(\alpha')}{\alpha'}\,\mathrm{d}\alpha' = \frac{1}{\alpha_{\nu\infty}}\int_0^\infty \overline{m}_\nu(\alpha')\,\mathrm{d}\alpha', \tag{144}$$

where $\alpha_{\nu\infty} \geq 0$. From Eqs. 131, 143, and 144, Eq. 132 becomes

$$s\widetilde{m}_\nu \simeq m_{\nu\infty} + (m_{\nu 0} - m_{\nu\infty}) \frac{s}{\alpha_{\nu\infty}}, \tag{145}$$

which is correct to the first order in $s/\alpha_{\nu 0}$. We simplify this by means of

$$\kappa_h = m_{1\infty}, \tag{146}$$

$$\kappa_h' = \frac{m_{10} - m_{1\infty}}{\alpha_{1\infty}}, \tag{147}$$

$$\mu_h = m_{2\infty}, \tag{148}$$

$$\mu_h' = \frac{m_{20} - m_{2\infty}}{\alpha_{2\infty}}. \tag{149}$$

Since $\mu_h = 0$ by Eqs. 129 and 148, we obtain

$$s\widetilde{m}_1 \simeq \kappa_h + \kappa_h' s, \tag{150}$$

$$s\widetilde{m}_2 \simeq \mu_h' s, \tag{151}$$

which are asymptotically correct for small s. From the properties of $m_\nu(t-t')$ described above, it follows that the values of κ_h, κ_h', and μ_h' are non-negative and continuously differentiable for $X_i \in \mathcal{X}_- \cup \mathcal{X}_+$, with jump discontinuities admitted for $X_i \in \partial\mathcal{X}$.

4.3 Asymptotic Incremental Field Equations and Interface Conditions

4.3.1 Small-*t* Asymptotes: Field Theory of GED

By the generalized initial-value theorem for Laplace transforms (Appendix A), the small-t asymptote to Eq. 88 corresponds to the large-s asymptote to Eq. 125. However, for s sufficiently large, $s\widetilde{m}_1$ and $s\widetilde{m}_2$ may be approximated by Eqs. 141 and 142, respectively, whose substitution into Eq. 125 provides

$$\widetilde{t}_{ij}^{(\delta)} = \delta_{ij}\left(\kappa_e - \tfrac{2}{3}\mu_e\right)\widetilde{u}_{k,k} + \mu_e(\widetilde{u}_{i,j} + \widetilde{u}_{j,i}) - \delta_{ij}\left(\kappa_e' - \tfrac{2}{3}\mu_e'\right)\frac{\widetilde{u}_{k,k}}{s} - \mu_e' \frac{\widetilde{u}_{i,j} + \widetilde{u}_{j,i}}{s}. \tag{152}$$

In view of Eqs. 208, 210, and 214 (Appendix A), inverse Laplace transformation from the s domain to the t domain gives

$$t_{ij}^{(\delta)} = \delta_{ij}\left(\kappa_e - \tfrac{2}{3}\mu_e\right)u_{k,k} + \mu_e(u_{i,j} + u_{j,i}) - \delta_{ij}\left(\kappa_e' - \tfrac{2}{3}\mu_e'\right)\int_0^t u_{k,k}(t')\,\mathrm{d}t'$$
$$- \mu_e' \int_0^t [u_{i,j}(t') + u_{j,i}(t')]\,\mathrm{d}t', \tag{153}$$

with κ_e called elastic bulk modulus, μ_e elastic shear modulus, κ_e' anelastic bulk modulus, and μ_e' anelastic shear modulus. Equation 153 is to be complemented by the remaining incremental field equations, Eqs. 86–87, and the associated incremental interface conditions, Eqs. 92–95. Together, they constitute the material–local form of the small-t asymptotes to the incremental field equations of GVED in terms of $g_i^{(\Delta)}$, $t_{ij}^{(\delta)}$, u_i, and $\phi^{(\Delta)}$. We refer to the equations also as *generalized incremental field equations and interface conditions* of GED.

If the integrals in Eq. 153 are neglected, it simplifies to the incremental constitutive equation of elasticity. In this case, the small-t asymptotes to the incremental field equations and interface conditions of viscoelastodynamics agree with the *ordinary incremental field equations and interface conditions* of GED (e.g., Love, 1911; Dahlen 1974; Grafarend 1982).

4.3.2 Large-*t* Asymptotes: Field Theory of GVD

By the generalized final-value theorem for Laplace transforms (Appendix A), the large-t asymptote to Eq. 100 corresponds to the small-s asymptote to Eq. 126. However, with s sufficiently small, $s\widetilde{m}_1$ and $s\widetilde{m}_2$ may be replaced by Eqs. 150 and 151, respectively, whose substitution into Eq. 126 leads to

$$\widetilde{t}_{ij}^{(\Delta)} = \delta_{ij}\left(p_{,k}^{(0)}\widetilde{u}_k + \kappa_h \widetilde{u}_{k,k}\right) + \delta_{ij}\left(\kappa_h' - \frac{2}{3}\mu_h'\right) s\widetilde{u}_{k,k} + \mu_h' s(\widetilde{u}_{i,j} + \widetilde{u}_{j,i}). \tag{154}$$

Considering Eqs. 208, 209, 214 (Appendix A), and $u_i^{(0)} = 0$, inverse Laplace transformation from the s domain to the t domain results in

$$t_{ij}^{(\Delta)} = \delta_{ij}\left(p_{,k}^{(0)}u_k + \kappa_h u_{k,k}\right) + \delta_{ij}\left(\kappa_h' - \frac{2}{3}\mu_h'\right) d_t u_{k,k} + \mu_h' \mathrm{d}_t(u_{i,j} + u_{j,i}), \tag{155}$$

with κ_h referred to as hydrostatic bulk modulus, κ_h' as viscous bulk modulus (bulk viscosity), and μ_h' as viscous shear modulus (shear viscosity). We reduce the large-t asymptote to an expression for $\varpi^{(\Delta)}$ by recalling that, for a fluid not necessarily in hydrostatic equilibrium, $\varpi^{(\Delta)}$ is related to the other state variables by the same function that relates $p^{(\Delta)} = -t_{ii}^{(\Delta)}/3$ to these variables in the case of hydrostatic equilibrium (e.g., Malvern 1969). Putting $d_t = 0$ in Eq. 155 thus yields

$$\varpi^{(\Delta)} = -p_{,i}^{(0)}u_i - \kappa_h u_{i,i}. \tag{156}$$

To replace this by a more familiar expression, we compare Eqs. 122, 123, and 156, giving

$$\left(\frac{\partial \xi}{\partial \rho}\right)^{(0)} = \frac{\kappa_h}{\rho^{(0)}}, \tag{157}$$

and put

$$\left(\frac{\partial \xi}{\partial \lambda}\right)^{(0)} = \frac{l}{\lambda^{(0)}}, \tag{158}$$

$$\left(\frac{\partial \xi}{\partial \varphi}\right)^{(0)} = \frac{\nu}{\varphi^{(0)}}, \tag{159}$$

where l and ν are the compositional and entropic moduli, respectively. Upon substitution of Eqs. 157–159, Eq. 124 takes the form

$$\varpi^{(\Delta)} = \frac{\kappa_h}{\rho^{(0)}}\rho^{(\Delta)} - \frac{l}{\lambda^{(0)}}\lambda_{,i}^{(0)}u_i - \frac{\nu}{\varphi^{(0)}}\varphi_{,i}^{(0)}u_i, \tag{160}$$

which is the incremental state equation of a fluid whose total state equation is given by Eq. 112.

Considering Eqs. 122, 155, 156, 160 and the assumption of isocompositional and isentropic perturbations, $\lambda^{(\delta)} = \varphi^{(\delta)} = 0$, the local form of the incremental field equations of GVED,

Eqs. 97–100, reduces to

$$t_{ij,j}^{(\Delta)} + g_i^{(0)}\rho^{(\Delta)} + \rho^{(0)}\left(g_i^{(\Delta)} + 2\epsilon_{ijk}\Omega_k\,\mathrm{d}_t u_j\right) = \rho^{(0)}\,\mathrm{d}_t^2 u_i, \tag{161}$$

$$g_i^{(\Delta)} = (\phi^{(\Delta)} + \psi^{(\Delta)})_{,i}, \tag{162}$$

$$\phi_{,ii}^{(\Delta)} = 4\pi G(\rho^{(0)}u_i)_{,i}, \tag{163}$$

$$t_{ij}^{(\Delta)} = -\delta_{ij}\varpi^{(\Delta)} + \delta_{ij}\left(\kappa_h' - \tfrac{2}{3}\mu_h'\right)\mathrm{d}_t u_{k,k} + \mu_h'\,\mathrm{d}_t(u_{i,j} + u_{j,i}), \tag{164}$$

$$\varpi^{(\Delta)} = \frac{\kappa_h}{\rho^{(0)}}\rho^{(\Delta)} + \frac{l}{\lambda^{(0)}}\lambda^{(\Delta)} + \frac{\nu}{\varphi^{(0)}}\varphi^{(\Delta)}, \tag{165}$$

$$\rho^{(\Delta)} = -(\rho^{(0)}u_i)_{,i}, \tag{166}$$

$$\lambda^{(\Delta)} = -\lambda_{,i}^{(0)}u_i, \tag{167}$$

$$\varphi^{(\Delta)} = -\varphi_{,i}^{(0)}u_i. \tag{168}$$

These equations are completed by the associated incremental interface conditions, Eqs. 101–104. Together, they constitute the local form of the large-t asymptotes to the incremental equations of GVED in terms of $g_i^{(\Delta)}$, $t_{ij}^{(\Delta)}$, u_i, $\lambda^{(\Delta)}$, $\varpi^{(\Delta)}$, $\rho^{(\Delta)}$, $\phi^{(\Delta)}$, and $\varphi^{(\Delta)}$; in particular, they agree with the incremental field equations and interface conditions of GVD (e.g., Backus 1967; Jarvis and McKenzie 1980).

5 Approximate Incremental Field Theories

This section is concerned with simplified field theories. We suppose that the fluid is isocompositional and isentropic in each of the domains $\mathcal{X}_-$ and $\mathcal{X}_+$:

$$\lambda_{,i}^{(0)} = \varphi_{,i}^{(0)} = 0, \tag{169}$$

that rotational and tidal effects are neglegible:

$$\Omega_i = 0, \tag{170}$$

$$\chi = \psi = 0, \tag{171}$$

that the approximation of *quasi-static perturbations* applies:

$$\mathrm{d}_t^2 u_i = 0, \tag{172}$$

and that the bulk relaxation is negligible:

$$m_1(t - t') = \kappa_h. \tag{173}$$

We note that, in the absence of tidal forces, the perturbations are now solely due to the incremental interface-mass density, σ. The field theories of GHS and GVED satisfying Eqs. 169–173 are referred to as *approximate* incremental field theories. Upon introducing further restrictions, the analysis is divided into the case of *local incompressibility* (➋ Sect. 5.1), which accounts for an initial density gradient due to self-compression, and the case of *material incompressibility* (➋ Sect. 5.2), where the initial state is also taken as incompressible.

5.1 Local Incompressibility

5.1.1 Equations for the Initial Fields

Upon consideration of Eqs. 169–171 and elimination of $g_i^{(0)}$, the initial field equations of GHS, Eqs. 48–52, simplify to

$$-p_{,i}^{(0)} + \rho^{(0)} \phi_{,i}^{(0)} = 0, \tag{174}$$

$$\phi_{,ii}^{(0)} = -4\pi G \rho^{(0)}, \tag{175}$$

$$p^{(0)} = \xi_b(\rho^{(0)}), \tag{176}$$

with ξ_b the barotropic state function. The initial interface conditions, Eqs. 53–57, continue to apply. Solutions to the approximate equations for the initial fields can be shown to exist for the level surfaces of $p^{(0)}$, $\rho^{(0)}$, and $\phi^{(0)}$ being concentric spheres, coaxial cylinders, or parallel planes (e.g., Batchelor 1967). To eliminate $p^{(0)}$, consider the gradient of Eq. 176:

$$p_{,i}^{(0)} = \left(\frac{\mathrm{d}\xi_b}{\mathrm{d}\rho}\right)^{(0)} \rho_{,i}^{(0)}, \tag{177}$$

where $(\mathrm{d}\xi_b/\mathrm{d}\rho)^{(0)}$ must be constant on the level surfaces. Comparing Eq. 174 with 177 then yields

$$\left(\frac{\mathrm{d}\xi_b}{\mathrm{d}\rho}\right)^{(0)} \rho_{,i}^{(0)} = \rho^{(0)} \phi_{,i}^{(0)}, \tag{178}$$

which is the *Williamson–Adams equation* (e.g., Bullen 1975; Williamson and Adams 1923). With $(\mathrm{d}\xi_b/\mathrm{d}\rho)^{(0)}$ prescribed, Eqs. 175 and 178 are to be solved for $\rho^{(0)}$ and $\phi^{(0)}$.

5.1.2 Equations for the Incremental Fields: Local Form

Using Eq. 173 and

$$m_2(t-t') = \mu(t-t'), \tag{179}$$

substitution of Eq. 106 into Eq. 100 leads to

$$\begin{aligned} t_{ij}^{(\Delta)} = {} & \delta_{ij} \left(p_{,k}^{(0)} u_k + \kappa_h u_{k,k} \right) - \tfrac{2}{3} \delta_{ij} \int_0^t \mu(t-t') \, \mathrm{d}_{t'} [u_{k,k}(t')] \, \mathrm{d}t' \\ & + \int_0^t \mu(t-t') \, \mathrm{d}_{t'} [u_{i,j}(t') + u_{j,i}(t')] \, \mathrm{d}t'. \end{aligned} \tag{180}$$

Since, by setting $d_t = 0$ in Eq. 180, we find

$$\varpi^{(\Delta)} = -p_{,i}^{(0)} u_i - \kappa_h u_{i,i} \tag{181}$$

and, by comparing Eqs. 122, 123, 181 and observing $\mathrm{d}\xi_b/\mathrm{d}\rho = \partial\xi/\partial\rho$, we obtain

$$\left(\frac{\mathrm{d}\xi_b}{\mathrm{d}\rho}\right)^{(0)} = \frac{\kappa_h}{\rho^{(0)}}, \tag{182}$$

it follows from Eqs. 177, 181, and 182 that

$$\varpi^{(\Delta)} = -\frac{\kappa_h}{\rho^{(0)}} (\rho^{(0)} u_i)_{,i}. \tag{183}$$

Note that, with $p^{(\Delta)} = -t_{ii}^{(\Delta)}/3$ *per definitionem*, Eqs. 180 and 181 yield

$$\varpi^{(\Delta)} = p^{(\Delta)}, \tag{184}$$

which will henceforth be implied.

Suppose now that Eq. 183 can be replaced by the simultaneous conditions

$$\kappa_h \to \infty, \tag{185}$$

$$(\rho^{(0)} u_i)_{,i} \to 0, \tag{186}$$

$$p^{(\Delta)} = \text{finite}. \tag{187}$$

The significance of Eq. 186 becomes evident, if we note that, by Eqs. 120 and 122, the condition $(\rho^{(0)} u_i)_{,i} = 0$ is equivalent to the condition $\rho^{(\delta)} = \rho_{,i}^{(0)} u_i$ or $\rho^{(\Delta)} = 0$. Equation 186 thus states that the compressibility of a displaced particle is constrained to the extent that the material incremental density "follows" the prescribed initial density gradient so that the local incremental density vanishes. For this reason, we refer to Eq. 186 as *local* incremental incompressibility condition.

Taking into account Eqs. 170–172, 178, 180, 182, and 186 and eliminating $g_i^{(\Delta)}$, the local form of the incremental field equations and interface conditions of GVED, Eqs. 97–104, reduce to

$$t_{ij,j}^{(\Delta)} + \rho^{(0)} \phi_{,i}^{(\Delta)} = 0, \tag{188}$$

$$\phi_{,ii}^{(\Delta)} = 0, \tag{189}$$

$$t_{ij}^{(\Delta)} = -\delta_{ij} p^{(\Delta)} + \delta_{ij} \frac{2\rho^{(0)} \phi_{,k}^{(0)}}{3\kappa_h} \int_0^t \mu(t-t')\, \mathrm{d}_{t'}[u_k(t')]\, \mathrm{d}t' + \int_0^t \mu(t-t')\, \mathrm{d}_{t'}[u_{i,j}(t') + u_{j,i}(t')]\, \mathrm{d}t', \tag{190}$$

$$u_{i,i} = -\frac{\rho^{(0)} \phi_{,i}^{(0)}}{\kappa_h} u_i, \tag{191}$$

$$[u_i]_-^+ = 0, \tag{192}$$

$$[\phi^{(\Delta)}]_-^+ = 0, \tag{193}$$

$$\left[n_i^{(0)} \left(\phi_{,i}^{(\Delta)} - 4\pi G \rho u_i\right)\right]_-^+ = -4\pi G \sigma, \tag{194}$$

$$\left[n_j^{(0)} \left(t_{ij}^{(\Delta)} - \delta_{ij} \rho^{(0)} \phi_{,k}^{(0)} u_k\right)\right]_-^+ = \gamma n_i^{(0)} \sigma. \tag{195}$$

The approximate incremental equations are to be solved for $p^{(\Delta)}$, $t_{ij}^{(\Delta)}$, u_i, and $\phi^{(\Delta)}$, where $\rho^{(0)}$ and $\phi^{(0)}$ must satisfy the appropriate initial field equations and interface conditions. We observe that the hydrostatic bulk modulus, κ_h, remains finite in Eqs. 190 and 191. This is because it enters into these equations as a consequence of substituting Eq. 182 into Eq. 178 for the *initial* fields, for which the approximation given by Eq. 185 does not apply. If $\phi^{(0)}$ is prescribed and $\phi^{(\Delta)}$ is neglected, the mechanical and gravitational effects decouple. In this case, solutions for the initial state are readily found. The *decoupled* incremental equations were integrated for a Newton–viscous spherical Earth model (Li and Yuen 1987; Wu and Yuen 1991) and for a

Maxwell–viscoelastic planar earth model (Wolf and Kaufmann 2000). The solution to the *coupled* incremental equations for a Maxwell–viscoelastic spherical earth model has recently been derived by Martinec et al. (2001).

5.2 Material Incompressibility

We proceed using the supposition that the *material* is incompressible. As a result, the initial state is incompressible, whence $\rho^{(0)}$ = constant replaces Eq. 176 and $\kappa_h \to \infty$ applies also in Eqs. 190 and 191, the latter reducing to the conventional *material* incremental incompressibility condition.

5.2.1 Equations for the Initial Fields

With these additional restrictions, the approximate initial field equations of GHS for local incompressibility, Eqs. 174–176, further simplify to those applying to material incompressibility:

$$-p_{,i}^{(0)} + \rho^{(0)}\phi_{,i}^{(0)} = 0, \tag{196}$$

$$\phi_{,ii}^{(0)} = -4\pi G\rho^{(0)}, \tag{197}$$

$$\rho^{(0)} = \text{constant}. \tag{198}$$

However, the initial interface conditions, Eqs. 53–57, continue to apply.

5.2.2 Equations for the Incremental Fields: Local Form

Owing to the additional assumptions, the approximate incremental field equations and interface conditions of GVED for local incompressibility, Eqs. 188–195, further reduce to the conventional form valid for material incompressibility:

$$t_{ij,j}^{(\Delta)} + \rho^{(0)}\phi_{,i}^{(\Delta)} = 0, \tag{199}$$

$$\phi_{,ii}^{(\Delta)} = 0, \tag{200}$$

$$t_{ij}^{(\Delta)} = -\delta_{ij}p^{(\Delta)} + \int_0^t \mu(t-t')\,\mathrm{d}_{t'}[u_{i,j}(t') + u_{j,i}(t')]\,\mathrm{d}t', \tag{201}$$

$$u_{i,i} = 0, \tag{202}$$

$$[u_i]_-^+ = 0, \tag{203}$$

$$[\phi^{(\Delta)}]_-^+ = 0, \tag{204}$$

$$\left[n_i^{(0)}\left(\phi_{,i}^{(\Delta)} - 4\pi G\rho^{(0)}u_i\right)\right]_-^+ = -4\pi G\sigma, \tag{205}$$

$$\left[n_j^{(0)}\left(t_{ij}^{(\Delta)} - \delta_{ij}\rho^{(0)}\phi_{,k}^{(0)}u_k\right)\right]_-^+ = \gamma n_i^{(0)}\sigma. \tag{206}$$

Deductions of *analytical solutions* to these equations for one- or two-layer Earth models are given elsewhere (e.g., Amelung and Wolf 1994; Rümpker and Wolf 1996; Wolf 1984, 1994; Wu and Ni 1996). In addition, a number of analytical or semi-analytical solutions for multi-layer Earth models have been obtained (e.g., Martinec and Wolf 1998; Sabadini et al. 1982; Spada et al.

1992; Vermeersen and Sabadini 1997; Wieczerkowski 1999; Wolf 1985d; Wu 1990; Wu and Peltier 1982). An instructive solution to a simplified form of the above equations has been derived by Wolf (1991b).

All these solutions refer to lateral homogeneity. Recently, the derivation of solutions for laterally heterogeneous Earth models has also received attention. Wheras Kaufmann and Wolf (1999) and Tromp and Mitrovica (1999a, b, 2000) limited the theory to small perturbations of the parameters in the lateral direction, D'Agostino et al. (1997), Martinec (1998, 1999), and Martinec and Wolf (1999) developed solution techniques valid for arbitrarily large perturbations.

6 Summary

The results of this review can be summarized as follows:

1. We have defined the Lagrangian and Eulerian representations of arbitrary fields and provided expressions for the relationship between the fields and their gradients in these kinematic representations. In correspondence with the Lagrangian and Eulerian representations, we have also defined the material and local increments of the fields. Using the relation between the kinematic representations, this has allowed us to establish the material and local forms of the fundamental perturbation equation.
2. Postulating only the differential form of the fundamental principles of continuum mechanics and potential theory in the Lagrangian representation, we have then presented a concise derivation of the material, material–local, and local forms of the incremental field equations and interface conditions of GVED. These equations describe infinitesimal, gravitational–viscoelastic perturbations of compositionally and entropically stratified, compressible, rotating fluids initially in hydrostatic equilibrium.
3. Following this, we have obtained, as the short-time asymptotes to the incremental field equations and interface conditions of GVED, a system of equations referred to as generalized incremental field equations and interface conditions of GED. The long-time asymptotes agree with the incremental field equations and interface conditions of GVD. In particular, we have shown that the incremental thermodynamic pressure entering into the long-time asymptote to the incremental constitutive equation of viscoelasticity satisfies the incremental state equation appropriate to viscous fluids.
4. Finally, we have adopted several simplifying assumptions and developed approximate field theories applying to gravitational–viscoelastic perturbations of isocompositional, isentropic, and compressible or incompressible fluid domains.

Appendix 1: Laplace Transform

Forward Transform

The Laplace transform, $\mathcal{L}[f(t)]$, of a function, $f(t)$, is defined by

$$\mathcal{L}[f(t)] = \int_0^\infty f(t)\, e^{-st} \mathrm{d}t, \qquad s \in \mathcal{S}, \tag{207}$$

where s is the inverse Laplace time and $\mathcal{S}$ is the complex s domain (e.g., LePage 1980). We assume here that $f(t)$ is continuous for all $t \in \mathcal{T}$ and of exponential order as $t \to \infty$, which are sufficient conditions for the convergence of the Laplace integral in Eq. 207 for Re s larger than some value, s_R. Defining $\mathcal{L}[f(t)] = \widetilde{f}(s)$ and assuming the same properties for $g(t)$, elementary consequences are then

$$\mathcal{L}[a\,f(t) + b\,g(t)] = a\,\widetilde{f}(s) + b\,\widetilde{g}(s), \qquad a, b = \text{constant}, \tag{208}$$

$$\mathcal{L}[d_t f(t)] = s\,\widetilde{f}(s) - f(0), \tag{209}$$

$$\mathcal{L}\left[\int_0^t f(t')\,\mathrm{d}t'\right] = \frac{\widetilde{f}(s)}{s}, \tag{210}$$

$$\mathcal{L}\left[\int_0^t f(t-t')\,g(t')\,\mathrm{d}t'\right] = \widetilde{f}(s)\,\widetilde{g}(s), \tag{211}$$

$$\mathcal{L}[1] = \frac{1}{s}, \tag{212}$$

$$\mathcal{L}[e^{-s_0 t}] = \frac{1}{s + s_0}, \qquad s_0 = \text{constant}. \tag{213}$$

Inverse Transform

If $\mathcal{L}[f(t)]$ is the forward Laplace transform of $f(t)$, then $f(t)$ is called inverse Laplace transform of $\mathcal{L}[f(t)]$. This is written as $\mathcal{L}^{-1}\{\mathcal{L}[f(t)]\} = f(t)$. Since $\mathcal{L}[f(t)] = \widetilde{f}(s)$, it follows that

$$\mathcal{L}^{-1}[\widetilde{f}(s)] = f(t), \qquad t \in \mathcal{T}, \tag{214}$$

which admits the immediate inversion of the forward transforms listed above.

Generalized Initial- and Final-Value Theorems

Some useful consequences of Eqs. 207 and 214 are the *generalized initial- and final-value theorems*. Assuming that the appropriate limits exist, the first theorem states that an asymptotic approximation, $p(t)$ to $f(t)$ for small t corresponds to an asymptotic approximation, $\widetilde{p}(s)$ to $\widetilde{f}(s)$ for large s. Similarly, according to the second theorem, an asymptotic approximation, $q(t)$ to $f(t)$ for large t corresponds to an asymptotic approximation, $\widetilde{q}(s)$ to $\widetilde{f}(s)$ for small s.

Appendix 2: List of Important Symbols

Latin symbols

Symbol	*Name*	*First reference*
$\mathrm{d}^2 r_i$	differential area at r_i	➲ Sect. 3.1
d_t^n	nth-order material-derivative operator with respect to t	➲ Sect. 3.1

$\widetilde{f}$	Laplace transform of f	➲ Sect. 4
$F_{ij\ldots}$	Eulerian representation of Cartesian tensor field	➲ Sect. 2.1
$f_{ij\ldots}$	Lagrangian representation of Cartesian tensor field	➲ Sect. 2.1
$F_{ij\ldots,k}$	gradient of $F_{ij\ldots}$ with respect to r_k	➲ Sect. 2.1
$f_{ij\ldots,k}$	gradient of $f_{ij\ldots}$ with respect to X_k	➲ Sect. 2.1
$f_{ij\ldots}^{(\Delta)}$	local increment of $f_{ij\ldots}$	➲ Sect. 2.2
$f_{ij\ldots}^{(\delta)}$	material increment of $f_{ij\ldots}$	➲ Sect. 2.2
$f_{ij\ldots}^{(0)}$	initial value of $f_{ij\ldots}$	➲ Sect. 2.2
$f_{ij\ldots}^{\pm}$	increase of $f_{ij\ldots}$ across $\partial\mathcal{R}$ in direction of n_i	➲ Sect. 2.3
G	Newton's gravitational constant	➲ Sect. 3.1
g_i	gravity force per unit mass	➲ Sect. 3.1
$i, j, \ldots$	index subscripts of Cartesian tensor	➲ Sect. 2
j	Jacobian determinant	➲ Sect. 3.1
l	compositional modulus	➲ Sect. 4.3.2
m_ν	relaxation function	➲ Sect. 3.3.4
$\overline{m}_\nu$	relaxation spectrum	➲ Sect. 4.1
m_1	bulk-relaxation function	➲ Sect. 3.3.4
m_2	shear-relaxation function	➲ Sect. 3.3.4
$m_{\nu 0}$	small-t limit of relaxation function	➲ Sect. 4.1
$m_{\nu\infty}$	large-t limit of relaxation function	➲ Sect. 4.1
$m_{ijk\ell}$	anisotropic relaxation function	➲ Sect. 3.3.4
n_i	outward unit normal with respect to $\partial\mathcal{R}$	➲ Sect. 2.3
p	mechanical pressure	➲ Sect. 3.2
r_i	position of place, current position of particle	➲ Sect. 2.1
s	inverse Laplace time	➲ Sect.4
t	current time	➲ Sect. 2.1
t'	excitation time	➲ Sect. 3.1
t_{ij}	Cauchy stress	➲ Sect. 3.1
u_i	displacement	➲ Sect. 2.1
v	entropic modulus	➲ Sect. 4.3.2
X_i	initial position of material point	➲ Sect. 2.1

Greek symbols

Symbol	*Name*	*First reference*
α'	inverse spectral time	➲ Sect. 4.1
γ	magnitude of $g_i^{(0)}$ on $\partial\mathcal{R}^{(0)}$	➲ Sect. 3.3.1
δ_{ij}	Kronecker symbol	➲ Sect. 2
∂	partial-derivative operator	➲ Sect. 2.1
ϵ_{ijk}	Levi–Civita symbol	➲ Sect. 2
κ_e	elastic bulk modulus	➲ Sect. 4.2.1
κ_e'	anelastic bulk modulus	➲ Sect. 4.2.1
κ_h	hydrostatic bulk modulus	➲ Sect. 4.2.2
κ_h'	viscous bulk modulus	➲ Sect. 4.2.2
λ	composition	➲ Sect. 3.2

μ	shear-relaxation function	➋ Sect. 5.1.2
μ_e	elastic shear modulus	➋ Sect. 4.2.1
μ'_e	anelastic shear modulus	➋ Sect. 4.2.1
μ'_h	viscous shear modulus	➋ Sect. 4.2.2
ξ	state function	➋ Sect. 3.2
ξ_b	barotropic state function	➋ Sect. 5.1.1
ϖ	thermodynamic pressure	➋ Sect. 3.4
ρ	volume-mass density	➋ Sect. 3.1
σ	(incremental) interface-mass density	➋ Sect. 3.1
τ_{ij}	Piola–Kirchhoff stress	➋ Sect. 3.1
ϕ	gravitational potential	➋ Sect. 3.1
φ	entropy density	➋ Sect. 3.2
χ	centrifugal potential	➋ Sect. 3.1
ψ	tidal potential	➋ Sect. 3.1
Ω_i	angular velocity	➋ Sect. 3.1

Calligraphic symbols

Symbol	*Name*	*First reference*
$\mathcal{E}$	Euclidian space domain	➋ Sect. 2.1
$\mathcal{L}$	Laplace-transformation functional	➋ Sect. A.1
$\mathcal{L}^{-1}$	inverse Laplace-transformation functional	➋ Sect. A.2
$\mathcal{M}_{ij}$	anisotropic relaxation functional	➋ Sect. 3.1
$\mathcal{R}_-$	internal r_i domain	➋ Sect. 2.1
$\mathcal{R}_+$	external r_i domain	➋ Sect. 2.1
$\mathcal{S}$	s domain	➋ Sect. 4

Symbol	*Name*	*First reference*
$\mathcal{T}$	t domain	➋ Sect. 2.1
$\mathcal{X}_-$	internal X_i domain	➋ Sect. 2.1
$\mathcal{X}_+$	external X_i domain	➋ Sect. 2.1
$\partial\mathcal{R}$	interface between $\mathcal{R}_-$ and $\mathcal{R}_+$	➋ Sect. 2.1
$\partial\mathcal{X}$	interface between $\mathcal{X}_-$ and $\mathcal{X}_+$	➋ Sect. 2.1

References

Amelung F, Wolf D (1994) Viscoelastic perturbations of the earth: significance of the incremental gravitational force in models of glacial isostasy. Geophys J Int 117: 864–879

Backus GE (1967) Converting vector and tensor equations to scalar equations in spherical coordinates. Geophys J R Astron Soc 13: 71–101

Batchelor GK (1967) An introduction to fluid dynamics. Cambridge University Press, Cambridge

Biot MA (1959) The influence of gravity on the folding of a layered viscoelastic medium under compression. J Franklin Inst 267: 211–228

Biot MA (1965) Mechanics of incremental deformations. Wiley, New York

Bullen KE (1975) The Earth's density. Chapman and Hall, London

Cathles LM (1975) The viscosity of the Earth's mantle. Princeton University Press, Princeton

Chandrasekhar S (1961) Hydrodynamic and hydromagnetic stability. Clarendon Press, Oxford

Christensen RM (1982) Theory of viscoelasticity, 2nd edn. Academic Press, New York

Corrieu V, Thoraval C, Ricard Y (1995) Mantle dynamics and geoid Green functions, Geophys J Int 120: 516–532

D'Agostino G, Spada G, Sabadini R (1997) Postglacial rebound and lateral viscosity variations: a semi-analytical approach based on a spherical model with Maxwell rheology, Geophys J Int 129: F9–F13

Dahlen FA (1972) Elastic dislocation theory for a self-gravitating elastic configuration with an initial static stress field. Geophys J R Astron Soc 28: 357–383

Dahlen FA (1973) Elastic dislocation theory for a self-gravitating elastic configuration with an initial static stress field II: energy release. Geophys J R Astron Soc 31: 469–484

Dahlen FA (1974) On the static deformation of an earth model with a fluid core. Geophys J R Astron Soc 36: 461–485

Dahlen FA, Tromp J (1998) Theoretical global seismology. Princeton University Press, Princeton

Darwin GH (1879) On the bodily tides of viscous and semi-elastic spheroids, and on the ocean tides upon a yielding nucleus. Phil Trans R Soc London Part 1 170: 1–35

Dehant V, Wahr JM (1991) The response of a compressible, non-homogeneous earth to internal loading: theory. J Geomag Geoelectr 43: 157–178

Eringen AC (1989) Mechanics of continua, 2nd edn. R. E. Krieger, Malabar

Golden JM, Graham GAC (1988) Boundary value problems in linear viscoelasticity. Springer, Berlin

Grafarend EW (1982) Six lectures on geodesy and global geodynamics. Mitt Geodät Inst Techn Univ Graz 41: 531–685

Hanyk L, Matyska C, Yuen DA (1999) Secular gravitational instability of a compressible viscoelastic sphere. Geophys Res Lett 26: 557–560

Hanyk L, Yuen DA, Matyska C (1996) Initial-value and modal approaches for transient viscoelastic responses with complex viscosity profiles. Geophys J Int 127: 348–362

Haskell NA (1935) The motion of a viscous fluid under a surface load. Physics 6: 265–269

Haskell NA (1936) The motion of a viscous fluid under a surface load, 2. Physics 7: 56–61

Jarvis GT, McKenzie DP (1980) Convection in a compressible fluid with infinite Prandtl number. J Fluid Mech 96: 515–583

Johnston P, Lambeck K, Wolf D (1997) Material versus isobaric internal boundaries in the earth and their influence on postglacial rebound. Geophys J Int 129: 252–268

Kaufmann G, Wolf D (1999) Effects of lateral viscosity variations on postglacial rebound: an analytical approach. Geophys J Int 137: 489–500

Krauss W (1973) Methods and results of theoretical oceanography, *vol. I*: Dynamics of the homogeneous and the quasihomogeneous ocean. Bornträger, Berlin

LePage WR (1980) Complex variables and the Laplace transform for engineers. Dover, New York

Li G, Yuen DA (1987) Viscous relaxation of a compressible spherical shell. Geophys Res Lett 14: 1227–1230

Love AEH (1911) Some problems of geodynamics. Cambridge University Press, Cambridge

Malvern LE (1969) Introduction to the mechanics of a continuous medium. Prentice-Hall, Englewood Cliffs

Martinec Z (1999) Spectral, initial value approach for viscoelastic relaxation of a spherical earth with three-dimensional viscosity–I. Theory. Geophys J Int 137: 469–488

Martinec Z (2000) Spectral–finite element approach to three-dimensional viscoelastic relaxation in a spherical earth. Geophys J Int 142: 117–141

Martinec Z, Wolf D (1998) Explicit form of the propagator matrix for a multi-layered, incompressible viscoelastic sphere. Sci Techn Rep GFZ Potsdam. STR98/08, 13 pp

Martinec Z, Wolf D (1999) Gravitational–viscoelastic relaxation of eccentrically nested spheres. Geophys J Int 138: 45–66

Martinec Z, Thoma M, Wolf D (2001) Material versus local incompressibility and its influence on glacial-isostatic adjustment. Geophys J Int, (144) 136–156

Mitrovica JX, Davis JL, Shapiro II (1994) A spectral formalism for computing three-dimensional deformation due to surface loads 1. Theory. J Geophys Res 99: 7057–7073

O'Connell RJ (1971) Pleistocene glaciation and the viscosity of the lower mantle. Geophys J R Astron Soc, 23: 299–327

Panasyuk SV, Hager BH, Forte AM (1996) Understanding the effects of mantle compressibility on geoid kernels. Geophys J Int 124: 121–133

Parsons BE (1972) Changes in the Earth's shape. Ph. D. thesis, Cambridge University, Cambridge

Peltier WR (1974) The impulse response of a Maxwell earth. Rev Geophys Space Phys. 12: 649–669

Peltier WR (ed) (1989) Mantle convection, plate tectonics and geodynamics. Gordon and Breach, New York

Ramsey AS (1981) Newtonian attraction. Cambridge University Press, Cambridge

Rayleigh Lord (1906) On the dilatational stability of the earth. Proc R Soc London, Ser A 77: 486–499

Rümpker G, Wolf D (1996) Viscoelastic relaxation of a Burgers half space: implications for the interpretation of the Fennoscandian uplift. Geophys J Int 124: 541–555

Sabadini R, Yuen DA, Boschi E (1982) Polar wandering and the forced responses of a rotating, multilayered, viscoelastic planet. J Geophys Res, 87: 2885–2903

Spada G, Sabadini R, Yuen DA, Ricard Y (1992) Effects on post-glacial rebound from the hard rheology in the transition zone. Geophys J Int 109: 683–700

Tromp J, Mitrovica JX (1999a) Surface loading of a viscoelastic earth–I. General theory. Geophys J Int 137: 847–855

Tromp J, Mitrovica JX (1999b) Surface loading of a viscoelastic earth–II. Spherical models. Geophys J Int 137: 856–872

Tromp J, Mitrovica JX (2000) Surface loading of a viscoelastic planet–III. Aspherical models. Geophys J Int 140: 425–441

Vermeersen LLA, Mitrovica JX (2000) Gravitational stability of spherical self-gravitating relaxation models. Geophys J Int 142: 351–360

Vermeersen LLA, Sabadini R (1997) A new class of stratified viscoelastic models by analytical techniques. Geophys J Int 129: 531–570

Vermeersen LLA, Vlaar NJ (1991) The gravito-elastodynamics of a pre-stressed elastic earth. Geophys J Int 104: 555–563

Vermeersen LLA, Sabadini R, Spada G (1996) Compressible rotational deformation. Geophys J Int 126: 735–761

Wieczerkowski K (1999) Gravito-Viskoelastodynamik für verallgemeinerte Rheologien mit Anwendungen auf den Jupitermond Io und die Erde. Publ Deut Geod Komm, Ser C 515: 130

Williamson ED, Adams LH (1923) Density distribution in the earth. J Washington Acad Sci 13: 413–428

Wolf D (1984) The relaxation of spherical and flat Maxwell earth models and effects due to the presence of the lithosphere. J Geophys 56: 24–33

Wolf D (1985a) Thick-plate flexure re-examined. Geophys J R Astron Soc 80: 265–273

Wolf D (1985b) On Boussinesq's problem for Maxwell continua subject to an external gravity field. Geophys J R Astron Soc 80: 275–279

Wolf D (1985c) The normal modes of a uniform, compressible Maxwell half-space. J Geophys 56: 100–105

Wolf D (1985d) The normal modes of a layered, incompressible Maxwell half-space. J Geophys 57: 106–117

Wolf D (1991a) Viscoelastodynamics of a stratified, compressible planet: incremental field equations and short- and long-time asymptotes. Geophys J Int 104: 401–417

Wolf D (1991b) Boussinesq's problem of viscoelasticity. Terra Nova 3: 401–407

Wolf D (1994) Lamé's problem of gravitational viscoelasticity: the isochemical, incompressible planet. Geophys J Int 116: 321–348

Wolf D (1997) Gravitational viscoelastodynamics for a hydrostatic planet. Publ Deut Geod Komm Ser C 452: 96

Wolf D, Kaufmann G (2000) Effects due to compressional and compositional density stratification on load-induced Maxwell-viscoelastic perturbations. Geophys J Int 140: 51–62

Wu J, Yuen DA (1991) Post-glacial relaxation of a viscously stratified compressible mantle. Geophys J Int 104: 331–349

Wu P (1990) Deformation of internal boundaries in a viscoelastic earth and topographic coupling between the mantle and the core. Geophys J Int 101: 213–231

Wu P (1992) Viscoelastic versus viscous deformation and the advection of prestress. Geophys J Int 108: 136–142

Wu P, Ni Z (1996) Some analytical solutions for the viscoelastic gravitational relaxation of a two-layer non-self-gravitating incompressible spherical earth. Geophys J Int 126: 413–436

Wu P, Peltier WR (1982) Viscous gravitational relaxation. Geophys J R Astron Soc 70: 435–485

11 Multiresolution Analysis of Hydrology and Satellite Gravitational Data

Helga Nutz · Kerstin Wolf
Center for Mathematical and Computational Modelling ($(CM)^2$), University of Kaiserslautern, Kaiserslautern, Germany

W. Freeden, M.Z. Nashed, T. Sonar (Eds.), *Handbook of Geomathematics*, DOI 10.1007/978-3-642-01546-5_11,

Abstract We present a multiresolution analysis of temporal and spatial variations of the Earth's gravitational potential by the use of tensor product wavelets which are built up by Legendre and spherical wavelets for the time and space domain, respectively. The multiresolution is performed for satellite and hydrological data, and based on these results we compute correlation coefficients between both data sets, which help us to develop a filter for the extraction of an improved hydrology model from the satellite data.

1 Introduction

The twin satellite gravity mission GRACE (**G**ravity **R**ecovery **A**nd **C**limate **E**xperiment) (see Tapley and Reigber 2001; Tapley et al. 2004a, b) provides a huge amount of data, which enables for the first time to quantify spatial and temporal variations of the Earth's gravity field caused by mass transport and mass distribution with sufficient accuracy (see Swenson et al. 2003; Swenson and Wahr 2006). Most of the measured gravitational variations belong to hydrological mass distribution, and the determination of the continental water changes from the GRACE data is possible with a resolution of 1 cm water column in monthly resolution. This gives us the opportunity to analyze the hydrological information at different scales in time and space with respect to topics as, e.g., global water balance and water transfer, large–scale spatial and temporal variations of terrestrial water storage, water balances in difficult accessible regions, long-term trends of continental water storage, identification of hydrological problem zones with respect to water management and the availability of water resources.

Hydrological data, as, e.g., WGHM (**W**aterGAP **G**lobal **H**ydrology **M**odel) used for our computations, are given in the form of a time series of monthly equivalent water column heights or surface density variations. These data can be directly transformed to the corresponding gravitational potential by numerical integration over the underlying grid. The classical approach for modeling the gravitational field of the Earth is to use a truncated Fourier series based on spherical harmonics where the accuracy of the approximation is given by the maximum degree. A fundamental disadvantage of the spherical harmonic expansion is the localization of the basis functions (spherical harmonics) in the frequency domain, which leads to a smearing of the spatial detail information over the whole globe. The need of a possibility to locally analyze the gravitational potential led to the development of spherical wavelets in the Geomathematics Group of the University of Kaiserslautern (Freeden et al. 1998; Freeden and Schneider 1998a, b, 2009; Freeden 1999). The spherical wavelets are kernel functions, which are constructed using clusters of a finite number of spherical harmonics, and by this means guarantee a good localization in the space domain. The uncertainty principle reveals that localization in both frequency and space domain are mutually exclusive.

Based on the one-dimensional spherical multiresolution analysis we derive a two-dimensional multiresolution analysis for both time and space domain. This is performed by transferring the spherical theory to the time domain by the use of Legendre wavelets instead of spherical wavelets and then by applying the theory of tensor product wavelets known from the classical multidimensional wavelet analysis. Note that in our case the two dimensions are the time and space domain (sphere), whereas in the classical wavelet analysis the two dimensions correspond to two directions in space. This method allows us to reveal both temporal and spatial detail information of time series of gravitational data (hydrological or satellite data). Finally, we compare the resulting temporal and spatial detail information and the scale-depending approximations of both data sets by computing local and global correlation coefficients. These comparisons reveal the temporal and spatial regions of bad correlation of the GRACE and WGHM data. With the objective of improving the existing hydrological models we finally derive a filter by weighting the detail information of different scales subject to the local correlation coefficients.

The layout of the chapter is as follows: in ➋ Sect. 2 we give a short presentation of the multiresolution for Hilbert spaces in order to explain the wavelet concept because this theory is fundamental for the further course of this chapter. The combined time–space multiresolution analysis for reconstructing a signal in the temporal and spatial domain and the theory of correlation coefficients is then introduced in ➋ Sect. 3. ➋ Section 4 is concerned with the numerical computations based on the theory which is presented in the foregoing section. All computations are performed with data from the satellite mission GRACE and with hydrological data from WGHM. A first idea for an "optimal" extraction of a hydrological model from satellite data is presented in ➋ Sect. 5 and finally some conclusions are drawn in the last section.

2 Scientific Relevance of Multiresolution

The concept of multiresolution has been developed by Mallat (1989a, b) and Meyer (1992) for fast and stable wavelet analysis and synthesis of functions in $L^2(\mathbb{R})$ and has been transferred to the spherical case by Freeden (see Freeden et al. (1998) and the references therein).

2.1 Preliminaries

We start with a short recapitulation of some notation and symbols which will be important within this chapter. Additional information can be found, e.g., in Müller (1966) and Freeden et al. (1998) and the references therein. The sets of positive integers, non–negative integers, integers, and real numbers are represented by $\mathbb{N}$, $\mathbb{N}_0$, $\mathbb{Z}$, and $\mathbb{R}$, respectively. The Hilbert space of all real, square–integrable functions F on Ω, where Ω denotes the unit sphere, is called $\mathrm{L}^2(\Omega)$ with the scalar product given by $(F,G)_{\mathrm{L}^2(\Omega)} = \int_\Omega F(\xi)G(\xi)\,d\omega(\xi)$, $F, G \in \mathrm{L}^2(\Omega)$. The space of all scalar spherical harmonics $Y_n : \Omega \to \mathbb{R}$ of degree n is of dimension $2n+1$ and the set $\{Y_{n,k} : \Omega \to \mathbb{R}, n \in \mathbb{N}_0, k = 1, \ldots, 2n+1\}$ of spherical harmonics of degree n and order k forms an orthonormal basis of $\mathrm{L}^2(\Omega)$. Thus $F \in \mathrm{L}^2(\Omega)$ can be uniquely represented by a Fourier series $F = \sum_{n=0}^{\infty}\sum_{k=1}^{2n+1} F^\wedge(n,k)Y_{n,k}$ (in $\mathrm{L}^2(\Omega)$–sense) with the Fourier coefficients $F^\wedge(n,k) = (F, Y_{n,k})_{\mathrm{L}^2(\Omega)}$. Closely related to the spherical harmonics are the Legendre polynomials $P_n : [-1,1] \to \mathbb{R}$ of degree n, $n \in \mathbb{N}_0$. Considering the space $\mathrm{L}^2([-1,1])$ with scalar product $(F,G)_{\mathrm{L}^2([-1,1])} = \int_{-1}^{1} F(t)G(t)\,dt$, $F, G \in \mathrm{L}^2([-1,1])$, the $\mathrm{L}^2([-1,1])$–orthonormal Legendre polynomials $P_n^* : [-1,1] \to \mathbb{R}$ defined by $P_n^* = \left(\frac{2n+1}{2}\right)^{\frac{1}{2}} P_n$, $n \in \mathbb{N}_0$, form an orthonormal basis in $\mathrm{L}^2([-1,1])$. Thus, every $F \in \mathrm{L}^2([-1,1])$ can be represented by a Legendre expansion $F = \sum_{n=0}^{\infty} F^\wedge(n)P_n^*$ with the Legendre coefficients $F^\wedge(n) = \left(F, P_n^*\right)_{\mathrm{L}^2([-1,1])}$. We conclude this section mentioning the addition theorem, which states the relation between the Legendre polynomial of degree n and the spherical harmonics of degree n: $\sum_{k=1}^{2n+1} Y_{n,k}(\xi)Y_{n,k}(\eta) = \frac{2n+1}{4\pi}P_n(\xi\cdot\eta)$, $\xi, \eta \in \Omega$.

2.2 Multiresolution in Hilbert Spaces

Within this subsection we briefly present the multiresolution analysis in Hilbert spaces as developed in the Geomathematics Group of the University of Kaiserslautern (see, e.g., Freeden and Schneider (1998b) and the references therein). This theory is fundamental for the understanding of the time–space multiresolution in ➋ Sect. 3.1.

With $\mathcal{H}$ we denote a real separable Hilbert space over a certain domain $\Sigma \subset \mathbb{R}^m$ with scalar product $(\cdot,\cdot)_{\mathcal{H}}$. Let $\{U_n^*\}_{n\in\mathbb{N}_0}$ be an orthonormal system which is complete in $(\mathcal{H}, (\cdot,\cdot)_{\mathcal{H}})$ and $\Gamma : \Sigma \times \Sigma \to \mathbb{R}$ an $\mathcal{H}$–product kernel given by $\Gamma(x,y) = \sum_{n=0}^{\infty} \Gamma^\wedge(n) U_n^*(x) U_n^*(y)$, $x, y \in \Sigma$, with symbol $\{\Gamma^\wedge(n)\}_{n\in\mathbb{N}_0}$. Γ is called $\mathcal{H}$–admissible if the following two conditions are satisfied:

(i) $\sum\limits_{n=0}^{\infty} \left(\Gamma^\wedge(n)\right)^2 < \infty$,
(ii) $\sum\limits_{n=0}^{\infty} \left(\Gamma^\wedge(n) U_n^*(x)\right)^2 < \infty, \quad \forall x \in \Sigma$.

These admissibility conditions ensure that the functions $\Gamma(x,\cdot) : \Sigma \to \mathbb{R}$ and $\Gamma(\cdot,x) : \Sigma \to \mathbb{R}$, $x \in \Sigma$ fixed, are elements of $\mathcal{H}$. Furthermore, they guarantee that the convolution of an admissible kernel function Γ and a function $F \in \mathcal{H}$ is again in $\mathcal{H}$, where the convolution is defined by $(\Gamma * F)(x) = \int_\Sigma F(y)\Gamma(x,y)\,dy = \sum_{n=0}^{\infty} \Gamma^\wedge(n) F^\wedge(n) U_n^*(x)$. Fundamental for the multiresolution analysis are the so-called $\mathcal{H}$-scaling functions which are defined in such a way that we can interpret them as low-pass filters for functions in $\mathcal{H}$. We start with the definition of the mother $\mathcal{H}$-scaling function. Let $\{(\Phi_0)^\wedge(n)\}_{n\in\mathbb{N}_0}$ be the symbol of an $\mathcal{H}$–admissible kernel function which additionally satisfies the following two conditions:

(i) $(\Phi_0)^\wedge(0) = 1$,
(ii) if $n > k$ then $(\Phi_0)^\wedge(n) \le (\Phi_0)^\wedge(k)$.

Then $\{(\Phi_0)^\wedge(n)\}_{n\in\mathbb{N}_0}$ is called the generating symbol of the mother $\mathcal{H}$–scaling function given by

$$\Phi_0(x,y) = \sum_{n=0}^{\infty} (\Phi_0)^\wedge(n) U_n^*(x) U_n^*(y), \quad x, y \in \Sigma.$$

For the definition of the $\mathcal{H}$–scaling function we have to extend this definition by defining the dilated versions of Φ_0 in the following way: let $\{(\Phi_J)^\wedge(n)\}_{n\in\mathbb{N}_0}$, $J \in \mathbb{Z}$, be an $\mathcal{H}$–admissible symbol satisfying in addition the following properties:

(i) $\lim\limits_{J\to\infty} (\Phi_J)^\wedge(n) = 1, \quad n \in \mathbb{N}$,
(ii) $(\Phi_J)^\wedge(n) \ge (\Phi_{J-1})^\wedge(n), \quad J \in \mathbb{Z}, n \in \mathbb{N}$,
(iii) $\lim\limits_{J\to-\infty} (\Phi_J)^\wedge(n) = 0, \quad n \in \mathbb{N}$,
(iv) $(\Phi_J)^\wedge(0) = 1, \quad J \in \mathbb{Z}$.

Then $\{(\Phi_J)^\wedge(n)\}_{n\in\mathbb{N}_0}$, $J \in \mathbb{Z}$, is called the generating symbol of an $\mathcal{H}$–scaling function and J is called the scale. The corresponding family $\{\Phi_J\}_{J\in\mathbb{Z}}$ of kernel functions given by

$$\Phi_J(x,y) = \sum_{n=0}^{\infty} (\Phi_J)^\wedge(n) U_n^*(x) U_n^*(y), \quad x, y \in \Sigma,$$

is called $\mathcal{H}$–scaling function. The symbols of the associated $\mathcal{H}$–wavelets are defined with the help of the refinement equation

$$((\Psi_J)^\wedge(n))^2 = ((\Phi_{J+1})^\wedge(n))^2 - ((\Phi_J)^\wedge(n))^2, \quad n \in \mathbb{N}_0. \tag{1}$$

Then the family $\{\Psi_J\}_{J\in\mathbb{Z}}$ of $\mathcal{H}$–product kernels defined by

$$\Psi_J(x,y) = \sum_{n=0}^{\infty} (\Psi_J)^\wedge(n) U_n^*(x) U_n^*(y), \quad x, y \in \Sigma,$$

is called $\mathcal{H}$–wavelet associated to the $\mathcal{H}$–scaling function $\{\Phi_J\}$, $J \in \mathbb{Z}$. The corresponding mother wavelet is denoted by Ψ_0.

Our numerical calculations are all performed with the so-called cubic polynomial wavelet. The corresponding cubic polynomial scaling function is composed by the symbol

$$(\Phi_J)^\wedge(n) = \begin{cases} \left(1 - 2^{-J}n\right)^2 \left(1 + 2^{1-J}n\right), & 0 \le n < 2^J, \\ 0, & n \ge 2^J. \end{cases}$$

➋ *Fig. 1* shows the scaling function and the wavelet for different scales. The corresponding symbols are shown in ➋ *Fig. 2* where the wavelet symbols are calculated with the help of the refinement Eq. (1).

With the help of the $\mathcal{H}$-scaling functions and $\mathcal{H}$-wavelets we introduce the scale spaces $\mathcal{V}_J = \{\Phi_J * \Phi_J * F \mid F \in \mathcal{H}\}$ and the corresponding detail spaces $\mathcal{W}_J = \{\Psi_J * \Psi_J * F \mid F \in \mathcal{H}\}$. The operator $T_J(F) = \Phi_J * \Phi_J * F$ can be interpreted as a low-pass filter and the corresponding scale space represents the approximation (reconstruction) of F at scale J. The operator $R_J(F) = \Psi_J * \Psi_J * F$ can be interpreted as a band-pass filter and the corresponding detail spaces $\mathcal{W}_J$ represent the wavelet approximation (detail information) of F at scale J. For these scale and detail spaces we have the decomposition $\mathcal{V}_{J+1} = \mathcal{V}_J + \mathcal{W}_J$. With increasing scale J, the scale spaces provide a better and better approximation of the function F, that is we have the limit relation (in $\mathcal{H}$-sense) $\lim_{J\to\infty} \Phi_J * \Phi_J * F = F$. Thus we end up in a *multiresolution analysis*

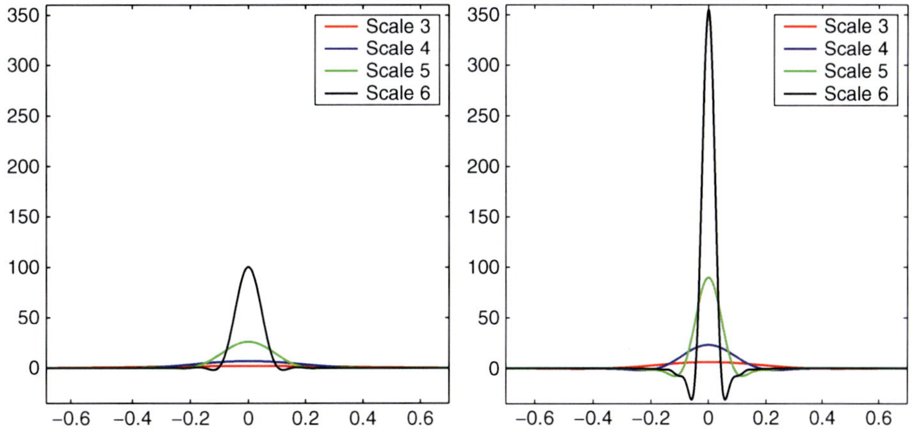

a Scaling functions: $\vartheta \mapsto \Phi_j(\vartheta)$ **b** Wavelet functions: $\vartheta \mapsto \Psi_j(\vartheta)$

Fig. 1
Cubic polynomial scaling function and wavelet for $\vartheta \in [-\pi, \pi]$; scale $j = 3, 4, 5, 6$

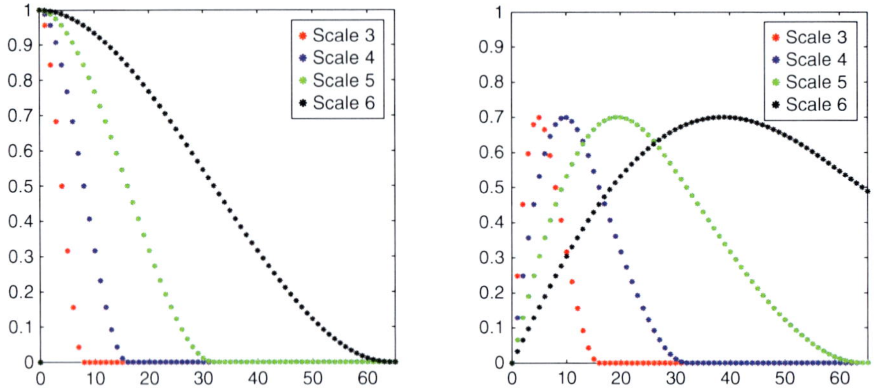

a Symbols of the scaling function: $n \mapsto (\Phi)_j^\wedge(n)$ **b** Symbols of the wavelet: $n \mapsto (\Psi)_j^\wedge(n)$

Fig. 2
Symbols of the cubic polynomial scaling function and wavelet for $n = 0, \ldots, 65$; scale $j = 3, 4, 5, 6$

given by the nested sequence of scale spaces

$$\cdots \subset \mathcal{V}_J \subset \mathcal{V}_{J+1} \subset \cdots \subset \mathcal{H},$$

and

$$\mathcal{H} = \overline{\bigcup_{J=-\infty}^{\infty} \mathcal{V}_J}^{\|\cdot\|_{\mathcal{H}}}.$$

In particular we can decompose the space $\mathcal{V}_J$ for each scale $J \in \mathbb{Z}$ in one "basic" scale space and several detail spaces: $\mathcal{V}_J = \mathcal{V}_{J_0} + \sum_{j=J_0}^{J-1} \mathcal{W}_j$.

2.3 Wavelets for the Time and Space Domain

As a matter of fact most of the functions in geophysics and geodesy are of bounded energy and thus we conclude this section with the Hilbert spaces $L^2([-1,1])$ used for the time domain (Legendre wavelets) and $L^2(\Omega)$ used for the space domain (spherical wavelets).

2.3.1 Legendre Wavelets

Let $\mathcal{H} = L^2([-1,1])$ be the space of square-integrable functions $F : [-1,1] \to \mathbb{R}$, i.e., we let $\Sigma = [-1,1]$. This choice leads to the so-called Legendre wavelets (cf. Beth and Viell (1998)). We already defined the scalar product $(F,G)_{L^2([-1,1])}$ and the orthonormal system of Legendre polynomials P_n^*. The $L^2([-1,1])$-admissible product kernels then are given by

$$\Gamma(s,t) = \sum_{n=0}^{\infty} \Gamma^\wedge(n) P_n^*(s) P_n^*(t), \quad s,t \in [-1,1],$$

and the convolution of Γ against F is given by

$$(\Gamma * F)(t) = \sum_{n=0}^{\infty} \Gamma^\wedge(n) F^\wedge(n) P_n^*(t), \quad t \in [-1,1].$$

2.3.2 Spherical Wavelets

In case of the scalar spherical wavelet theory we let $\Sigma = \Omega$ and consider the Hilbert space $\mathcal{H} = \mathrm{L}^2(\Omega)$. As an $\mathrm{L}^2(\Omega)$-orthonormal system we choose the system $\{Y_{n,k}\}_{n\in\mathbb{N}_0;\, k=1,\dots,2n+1}$ of spherical harmonics of degree n and order k. The $\mathrm{L}^2(\Omega)$-product kernels have the following representation

$$\Gamma(\xi,\eta) = \sum_{n=0}^{\infty} \sum_{k=1}^{2n+1} \Gamma^\wedge(n) Y_{n,k}(\xi) Y_{n,k}(\eta), \quad \xi,\eta \in \Omega,$$

and the convolution of Γ against F is given by

$$(\Gamma * F)(\xi) = \sum_{n=0}^{\infty} \sum_{k=1}^{2n+1} \Gamma^\wedge(n) F^\wedge(n,k) Y_{n,k}(\xi), \quad \xi \in \Omega.$$

3 Key Issues for the Comparison of GRACE and WGHM Data

In view of an improvement of existing hydrological models, as e.g., WGHM, by comparing them with measurements based on GRACE data we first perform a multiscale analysis and then compute correlation coefficients. The first part of this section (➋ Sect. 3.1) is therefore dedicated to the tensorial time–space multiresolution which is a method for the detection of temporal and spatial variations on different scales, i.e. sizes of the details. In the second part (➋ Sect. 3.2) we compute the local and global correlation coefficients between GRACE and WGHM data and thus we are able to quantify the resemblance of both data sets at different scales.

3.1 Tensorial Time–Space Multiresolution

For the combination of the temporal multiresolution based on Legendre wavelets with the spatial multiresolution based on spherical wavelets we apply the theory of tensor product wavelets (see, e.g., Louis et al. (1998)). This technique allows the transmission of the one-dimensional multiscale analysis to higher dimensions. ➋ *Fig. 3* shows the tensorial time–space multiresolution which provides a unique scale for both space and time domain and three detail parts for each scale, namely two hybrid and one pure detail part. A detailed introduction to this theory can be found in Freeden (1999), Maier (2003), Nutz and Wolf (2008), and the references therein.

Starting point of our considerations is the Hilbert space $\mathrm{L}^2([-1,1]\times\Omega)$ where without loss of generality we assume the time interval to be normalized to the interval $[-1,1]$. The scalar product of $F,G \in \mathrm{L}^2([-1,1]\times\Omega)$ is given by $(F,G)_{\mathrm{L}^2([-1,1]\times\Omega)} = \int_{-1}^{1}\int_\Omega F(t;\xi)G(t;\xi)\,d\omega(\xi)dt$. We presume that the time dependency is completely described by the spherical harmonic coefficients and we have the representation $F(t;\xi) = \sum_{n=0}^{\infty}\sum_{k=1}^{2n+1} F^\wedge(n,k)(t)Y_{n,k}(\xi)$, with $F^\wedge(n,k)(t) = \sum_{n'=0}^{\infty} F^\wedge(n';n,k)P_{n'}^*(t)$, where $F^\wedge(n';n,k) = \left(F, P_{n'}^* Y_{n,k}\right)_{\mathrm{L}^2([-1,1]\times\Omega)}$. For

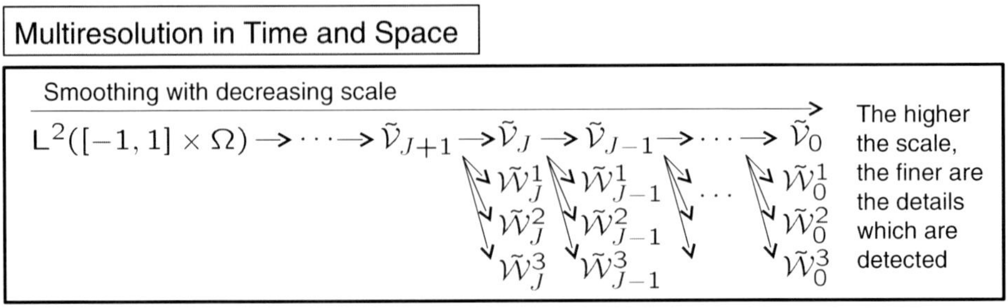

Fig. 3
Multiresolution of $L^2([-1,1]\times\Omega)$ with tensor product wavelets

notational reasons in the following text n' will always denote the summation index in the time domain (Legendre polynomials), whereas n will be used in the space domain (spherical harmonics). We finally arrive at

$$F = \sum_{n'=0}^{\infty} \sum_{n=0}^{\infty} \sum_{k=1}^{2n+1} F^{\wedge}(n'; n, k) P_{n'}^{*} Y_{n,k}$$

in $L^2([-1,1]\times\Omega)$-sense. A multiresolution of the space $L^2([-1,1]\times\Omega)$ is given by a subset of scale spaces of the form

$$\cdots \subset \tilde{\mathcal{V}}_J \subset \tilde{\mathcal{V}}_{J+1} \subset \cdots \subset L^2([-1,1]\times\Omega)$$

and

$$L^2([-1,1]\times\Omega) = \overline{\bigcup_{J=-\infty}^{\infty} \tilde{\mathcal{V}}_J}^{\|\cdot\|_{L^2([-1,1]\times\Omega)}}.$$

In the following section we follow the presentation of the tensorial time–space multiresolution analysis in Nutz and Wolf (2008) for the definition of the scaling function and the wavelets. Our starting point is the definition of the generating symbol of a time–space scaling function. Let $\{(\Phi'_J)^{\wedge}(n')\}_{n'\in\mathbb{N}_0}$ and $\{(\Phi_J)^{\wedge}(n)\}_{n\in\mathbb{N}_0}$, $J\in\mathbb{Z}$, be the generating symbols of a temporal scaling function and a spatial scaling function, respectively. Then the generating symbol of the time–space (tensor product) scaling function is given by the sequence $\{(\tilde{\Phi}_J)^{\wedge}(n';n)\}_{n',n\in\mathbb{N}_0}$, with the scaling function symbol $(\tilde{\Phi}_J)^{\wedge}(n';n) = (\Phi'_J)^{\wedge}(n')(\Phi_J)^{\wedge}(n)$. The family of kernel functions $\{\tilde{\Phi}_J\}_{J\in\mathbb{Z}}$ defined by

$$\tilde{\Phi}_J(s,t;\xi,\eta) = \sum_{n'=0}^{\infty} \sum_{n=0}^{\infty} \sum_{k=1}^{2n+1} (\tilde{\Phi}_J)^{\wedge}(n';n) P_{n'}^{*}(s) P_{n'}^{*}(t) Y_{n,k}(\xi) Y_{n,k}(\eta),$$

$s,t\in[-1,1]$, $\xi,\eta\in\Omega$, denotes the time–space (tensor product) scaling functions. Since we have two refinement equations

$$\left((\Psi'_J)^{\wedge}(n')\right)^2 = \left((\Phi'_{J+1})^{\wedge}(n')\right)^2 - \left((\Phi'_J)^{\wedge}(n')\right)^2,$$
$$\left((\Psi_J)^{\wedge}(n)\right)^2 = \left((\Phi_{J+1})^{\wedge}(n)\right)^2 - \left((\Phi_J)^{\wedge}(n)\right)^2,$$

which have to be fulfilled simultaneously we get

$$\left(\left(\Phi'_{J+1}\right)^\wedge(n')\right)^2\left(\left(\Phi_{J+1}\right)^\wedge(n)\right)^2 = \left(\left(\Phi'_J\right)^\wedge(n')\right)^2\left(\left(\Phi_J\right)^\wedge(n)\right)^2 + \left(\left(\Psi'_J\right)^\wedge(n')\right)^2\left(\left(\Phi_J\right)^\wedge(n)\right)^2 + \left(\left(\Phi'_J\right)^\wedge(n')\right)^2\left(\left(\Psi_J\right)^\wedge(n)\right)^2 + \left(\left(\Psi'_J\right)^\wedge(n')\right)^2\left(\left(\Psi_J\right)^\wedge(n)\right)^2.$$

This leads to the definition of two hybrid wavelets $\tilde{\Psi}_J^1$ and $\tilde{\Psi}_J^2$ and one pure wavelet $\tilde{\Psi}_J^3$:

$$\tilde{\Psi}_J^i(s,t;\xi,\eta) = \sum_{n'=0}^{\infty}\sum_{n=0}^{\infty}\sum_{k=1}^{2n+1}\left(\tilde{\Psi}_J^i\right)^\wedge(n';n)\,P_{n'}^*(s)P_{n'}^*(t)Y_{n,k}(\xi)Y_{n,k}(\eta),$$

$i = 1,2,3$, with the hybrid and pure wavelet symbols

$$\begin{aligned}
\left(\tilde{\Psi}_J^1\right)^\wedge(n';n) &= \left(\Phi'_J\right)^\wedge(n')(\Psi_J)^\wedge(n),\\
\left(\tilde{\Psi}_J^2\right)^\wedge(n';n) &= (\Psi'_J)^\wedge(n')(\Phi_J)^\wedge(n),\\
\left(\tilde{\Psi}_J^3\right)^\wedge(n';n) &= (\Psi'_J)^\wedge(n')(\Psi_J)^\wedge(n).
\end{aligned}$$

We now introduce the time–space convolution of a function $F \in \mathrm{L}^2([-1,1]\times\Omega)$ and a kernel function of the form

$$\Gamma(s,t;\xi,\eta) = \sum_{n'=0}^{\infty}\sum_{n=0}^{\infty}\sum_{k=1}^{2n+1}\Gamma^\wedge(n';n)\,P_{n'}^*(s)P_{n'}^*(t)Y_{n,k}(\xi)Y_{n,k}(\eta).$$

The time–space convolution of Γ against F is defined by

$$\begin{aligned}
(\Gamma \star F)(t;\eta) &= \int_{-1}^{1}\int_{\Omega}\Gamma(s,t;\xi,\eta)F(s;\xi)\,d\omega(\xi)\,ds\\
&= \sum_{n'=0}^{\infty}\sum_{n=0}^{\infty}\sum_{k=1}^{2n+1}\Gamma^\wedge(n';n)\,F^\wedge(n';n,k)\,P_{n'}^*(t)Y_{n,k}(\eta).
\end{aligned}$$

The convolution of two kernel functions is defined in analogous manner. Now let $\{\tilde{\Phi}_J\}$ be the time–space scaling functions and $\{\tilde{\Psi}_J^i\}$, $i = 1,2,3$, be the associated hybrid and pure time–space wavelets at scale J. Then the pure time–space scale spaces are defined by

$$\tilde{\mathcal{V}}_J = \left\{\tilde{\Phi}_J \star \tilde{\Phi}_J \star F \mid F \in \mathrm{L}^2([-1,1]\times\Omega)\right\},$$

and the first hybrid, the second hybrid, and the pure time–space detail spaces are given by

$$\tilde{\mathcal{W}}_J^i = \left\{\tilde{\Psi}_J^i \star \tilde{\Psi}_J^i \star F \mid F \in \mathrm{L}^2([-1,1]\times\Omega)\right\},$$

$i = 1,2,3$.

We conclude this section with the following two important properties which guarantee the time–space multiresolution based on tensor product wavelets: let $\{\tilde{\Phi}_J\}$, $J \in \mathbb{Z}$, be a time–space scaling function and $\{\tilde{\Psi}_J^i\}$, $i = 1,2,3$, $J \in \mathbb{Z}$, be the associated hybrid and pure time–space wavelets. Suppose that $F \in \mathrm{L}^2([-1,1] \times \Omega)$. Then

$$F = \lim_{J\to\infty} \left(\tilde{\Phi}_J \star \tilde{\Phi}_J \star F\right)$$

$$= \lim_{J\to\infty} \left(\tilde{\Phi}_{J_0} \star \tilde{\Phi}_{J_0} \star F + \sum_{j=J_0}^{J} \sum_{i=1}^{3} \tilde{\Psi}_j^i \star \tilde{\Psi}_j^i \star F\right),$$

$J_0 \in \mathbb{Z}$, holds true in the sense of the $\mathrm{L}^2([-1,1] \times \Omega)$-metric. Accordingly, for the time–space scale spaces and detail spaces we have

$$\tilde{\mathcal{V}}_J = \tilde{\mathcal{V}}_{J_0} + \sum_{j=J_0}^{J-1} \sum_{i=1}^{3} \tilde{\mathcal{W}}_j^i$$

with $J, J_0 \in \mathbb{Z}$, and $J_0 \leq J$. In ➋ *Fig. 4* a graphical illustration of the time–space multiscale analysis calculated with GRACE-data is shown.

3.2 Correlation Analysis Between GRACE and WGHM

By the use of the time–space multiresolution analysis we are in the position to locally measure spatial and temporal changes in the data. With respect to the application of the theory to real data sets as, e.g., hydrological or GRACE data, we need an instrument to compare these results, i.e. we must perform a correlation analysis. The correlation coefficient is a gauge for the variation of two data sets and, thus, helps us to interpret the changes of the data at different scales. Based on the different corresponding detail parts and reconstructions we compute the local correlation coefficients on the continents which reflect the good and bad accordance of the two time series. In addition we compute global correlation coefficients by averaging the local correlation coefficients over the continents.

For the definition of the local and global correlation coefficients of time series given on the sphere we use the following notation: we assume that we have $T \in \mathbb{N}$ points in time and $M \in \mathbb{N}$ grid points. The points in time are denoted by $t_i \in [-1,1]$, $i = 1,\ldots,T$, whereas all grid points are given by $\xi_m \in \Omega$, for $m = 1,\ldots,M$. Since we must take into account the latitude dependance of the grid points we let $p_m = \cos(\varphi_m)$ be the weight, where φ_m denotes the latitude of the corresponding grid point. We want to compare the values of two different time series which we denote by F and G, $F, G \in \mathrm{L}([-1,1] \times \Omega)$. Then, the local correlation coefficient c at some location $\xi \in \Omega$ is given by

$$c(\xi) = \frac{\sum_{i=1}^{T} \left(F(t_i;\xi) - \overline{F}(\xi)\right)\left(G(t_i;\xi) - \overline{G}(\xi)\right)}{\sqrt{\sum_{i=1}^{T} \left(F(t_i;\xi) - \overline{F}(\xi)\right)^2 \sum_{i=1}^{T} \left(G(t_i;\xi) - \overline{G}(\xi)\right)^2}},$$

where the weighted mean values are defined by $\overline{F}(\xi) = \frac{1}{T}\sum_{i=1}^{T} F(t_i;\xi)$ and $\overline{G}(\xi) = \frac{1}{T}\sum_{i=1}^{T} G(t_i;\xi)$. For the global correlation coefficient gc we average the local correlation

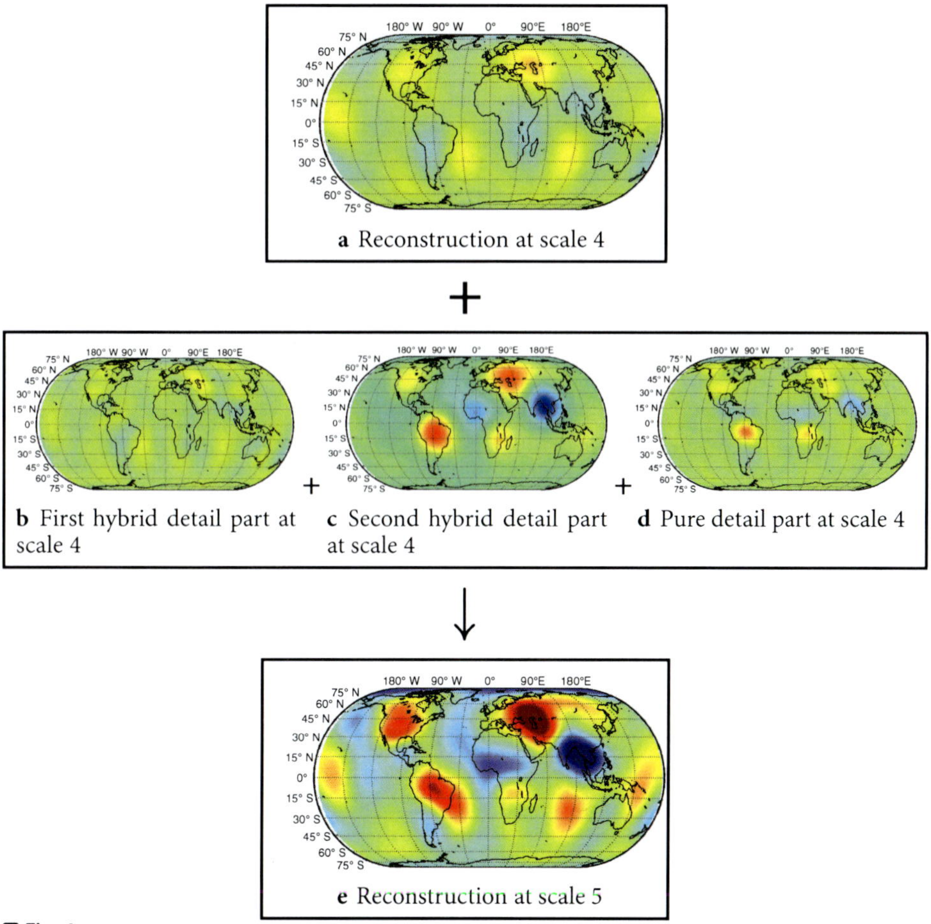
a Reconstruction at scale 4
b First hybrid detail part at scale 4
c Second hybrid detail part at scale 4
d Pure detail part at scale 4
e Reconstruction at scale 5

Fig. 4

Graphical illustration of the time–space multiresolution analysis computed from a time series of GRACE data and exemplarily shown for April 2005

coefficients over the corresponding grid points, such that we obtain

$$gc = \frac{\sum\limits_{i=1}^{T}\sum\limits_{m=1}^{M} p_m \left(F(t_i;\xi_m) - \overline{F}(\xi_m)\right)\left(G(t_i;\xi_m) - \overline{G}(\xi_m)\right)}{\sqrt{\sum\limits_{i=1}^{T}\sum\limits_{m=1}^{M} p_m \left(F(t_i;\xi_m) - \overline{F}(\xi_m)\right)^2 \sum\limits_{i=1}^{T}\sum\limits_{m=1}^{M} p_m \left(G(t_i;\xi_m) - \overline{G}(\xi_m)\right)^2}}.$$

4 Fundamental Results

This section is dedicated to numerical results for the tensorial time–space multiresolution and the correlation analysis between the GRACE and WGHM data. The computations have been carried out on the basis of 62 monthly data sets of spherical harmonic coefficients up to degree

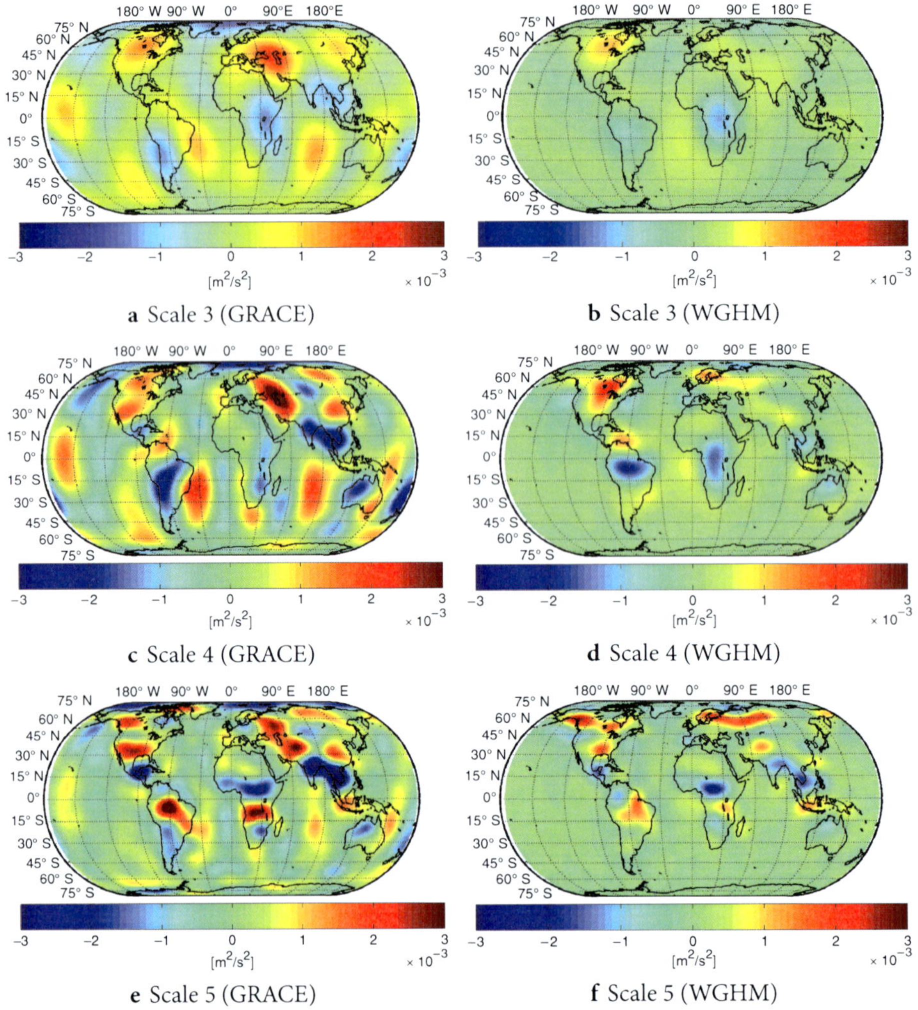

Fig. 5

First hybrid parts of the potentials of April 2005 calculated with cubic polynomial wavelet in time and space with different scales

and order 70 from GRACE and WGHM for the period of August 2002 till September 2007. These data sets have been made available to us from GeoForschungszentrum Potsdam, Department 1, Geodesy and Remote Sensing within the German Ministry of Education and Research (BMBF) project "Observation System Earth from Space."

In case of the spatial analysis we exemplarily present the results of the first hybrid parts of April 2005. The left column of ➋ *Fig. 5* shows the results based on the GRACE data and the right one shows the corresponding results in case of WGHM data. Note that in case of WGHM measurements have only been achieved on the continents whereas in case of GRACE data we also have measurements on the oceans. At scale 3 (see ➋ *Fig. 5a, b*) large-area regions are visible. With increasing scale we have better and better space localization.

In case of the temporal analysis the time dependent courses of the second hybrid parts for selected locations, more precisely for Dacca and Kaiserslautern on the Northern hemisphere and Manaus and Lilongwe on the Southern hemisphere, are shown (see ➲ *Fig. 6*). Note that Kaiserslautern has moderate seasonal variations in the water balance, whereas the other three cities are selected exemplarily for well-known regions of great changes (Ganges and Amazonas basin, region around Lake Malawi). In ➲ *Fig. 6* on the left column the time dependent courses for the GRACE data and on the right column of ➲ *Fig. 6* the time dependent courses for the corresponding results based on WGHM data are shown. The seasonal variations can be recognized best at scales 4 and 5. Even in case of Kaiserslautern located in a region with moderate variations the course of the second hybrid parts clarifies the seasonal course.

In ➲ *Fig. 7* the local correlation coefficients and, additionally, the global correlation coefficients on the continents between GRACE and WGHM are shown which are computed from the original data sets (see ➲ *Fig. 7a*) and some low-pass and band-pass filtered parts for scales 3 and 4 (see ➲ *Fig. 7b–f*). Red regions correspond to a good correlation of the two underlying time series, whereas the blue regions show the locations with more differences. In space domain, scale 3 and scale 4 correspond to a region of influence of about 8,000 km and 4,000 km, respectively, whereas in time domain we have a time of influence of about 9 months and 4 months.

We now exemplarily consider the results for North and South America in detail because these regions show very different correlation coefficients for the coarse reconstruction at scale 3. In case of North America the reconstruction at scale 4 (➲ *Fig. 7c*) shows a much better correlation than the reconstruction at scale 3 (➲ *Fig. 7b*). This is traced back to the fact that the details of the size of scale 3 (8,000 km, 9 months) are better correlated than the reconstruction at scale 3 which leads to an improvement of the correlation coefficients in case of the reconstruction at scale 4 (4,000 km, 4 months). In South America we have an excellent correlation coefficient for the coarse reconstruction at scale 3 which is slightly degraded turning to scale 4. The reason is that the detail parts at scale 3 are somewhat worse correlated than the reconstruction at scale 3.

5 Future Directions

In the previous sections we have presented some mathematical tools for the spatial and temporal analysis of hydrological and satellite data. We have also demonstrated how to compare the results of the multiresolution analysis by the use of correlation coefficients. In order to clarify the local differences between the hydrological model WGHM and the satellite measurements of GRACE, future research must now concentrate on the possibilities of how to take advantage of this knowledge for the improvement of existing hydrological models. In this section, we therefore try to give a first idea of how to interpret the results achieved from the multiscale analysis with the aid of the correlation coefficients in view of a correction of hydrology models. To this end we propose a filter based on the correlation coefficients and we assume that we have an improvement if the (global and local) correlation coefficients of the filtered GRACE and WGHM data are better than those of the original data. Furthermore we demand that a very large part of the original signal is reconstructed in the filtered data. Note that the improvement of the correlation coefficients and the increase of the percentage of the filtered signal from the original signal cannot be optimized simultaneously. To find out an optimal filter we start with computing the local correlation coefficients of GRACE and WGHM $c_J^{(i)}$, $J \in \mathbb{N}$, $i = 1, 2, 3$, on

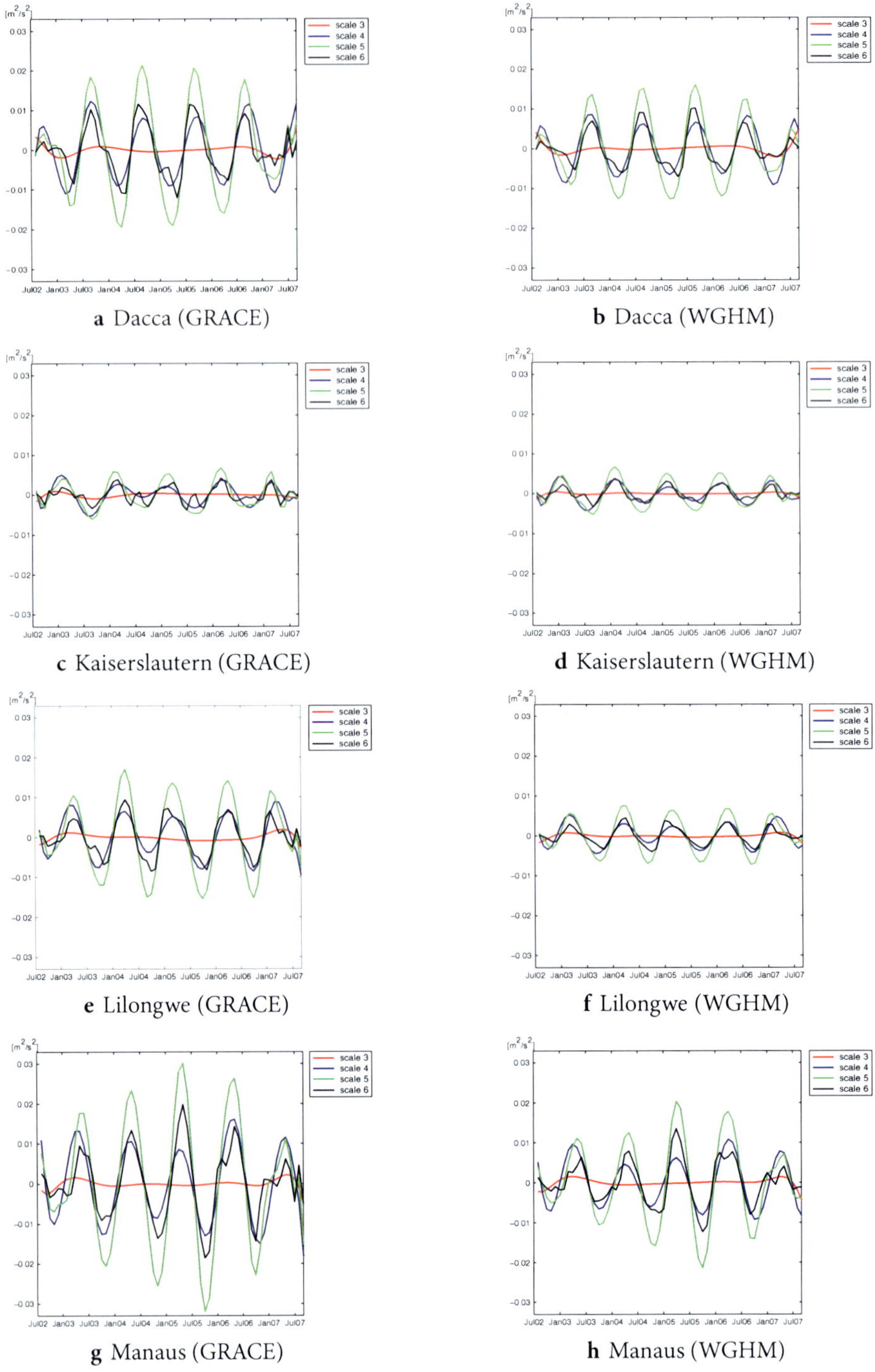

Fig. 6
Time-dependent courses of the second hybrid parts of the potentials at different locations calculated with cubic polynomial wavelet in time and space with different scales

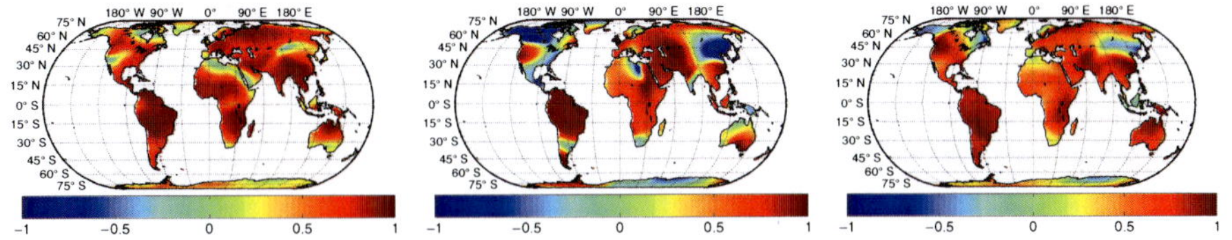

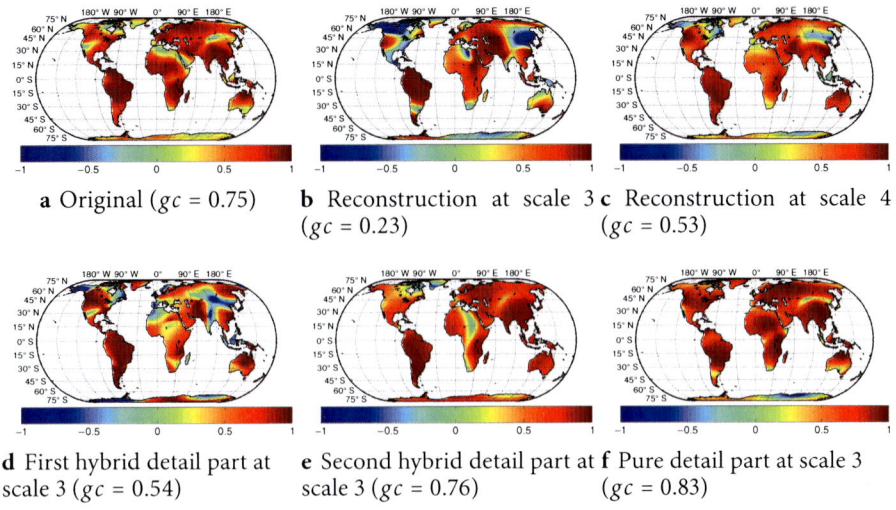

a Original ($gc = 0.75$) **b** Reconstruction at scale 3 ($gc = 0.23$) **c** Reconstruction at scale 4 ($gc = 0.53$)

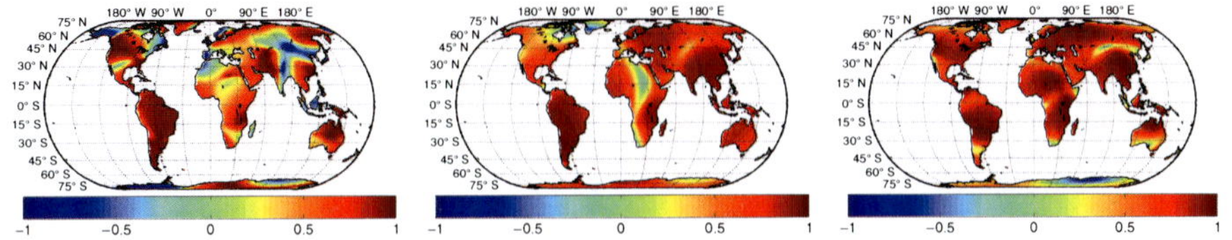

d First hybrid detail part at scale 3 ($gc = 0.54$) **e** Second hybrid detail part at scale 3 ($gc = 0.76$) **f** Pure detail part at scale 3 ($gc = 0.83$)

Fig. 7
Local correlation coefficients and in brackets the corresponding global correlation coefficients (on the continents) between GRACE and WGHM data computed from the original potential and some low-pass and band-pass parts

the continents for the corresponding detail parts $F * \Psi_J^{(i)} * \Psi_J^{(i)}$ and the local correlation coefficients of GRACE and WGHM c_J, $J \in \mathbb{N}$, of the reconstructions $F * \Phi_J * \Phi_J$. In addition, we compute the corresponding global correlation coefficients on the continents $gc_J^{(i)}$, gc_J. Using these correlation coefficients we derive a weight function $w : [-1,1] \to [0,1]$ defined by

$$w(k) = \begin{cases} 0, & k \le G_1 \\ \frac{1}{G_2 - G_1} k - \frac{G_1}{G_2 - G_1}, & G_1 < k < G_2 \\ 1, & k \ge G_2 \end{cases},$$

for controlling the influence of the corresponding parts on the resulting reconstructed signal: if the local correlation coefficient $c_J^{(i)}$ and c_J, respectively, is smaller than G_1 the corresponding part is not added in the reconstruction formula (2), whereas in case of a correlation coefficient greater than G_2 we add the entire corresponding part. In case of $G_1 < k < G_2$ we weight the corresponding part in the reconstruction formula (2) (the higher the correlation coefficient the higher the weights). We finally arrive at the following formula for a reconstruction of the signal:

$$\begin{aligned} F_{J_{\max}}^{\mathrm{rec}} &= (\Phi_{J_0} * \Phi_{J_0} * F)(\xi) w(c_{J_0}(\xi)) \\ &\quad + \sum_{j=J_0}^{J_{\max}-1} \sum_{i=1}^{3} \left(\Psi_j^{(i)} * \Psi_j^{(i)} * F \right)(\xi) w\left(c_j^{(i)}(\xi) \right). \end{aligned} \tag{2}$$

In order to get the percentage of the reconstructed signal $F^{\mathrm{rec}} = F_{J_{\max}}^{\mathrm{rec}}$ from the original signal $F^{\mathrm{orig}} = F$ we need the energy of a signal $F \in L^2([-1,+1] \times \Omega)$, which is given by $\|F\|_{L^2([-1,+1]\times\Omega)}^2 = \sum_{n'=0}^{\infty} \sum_{n=0}^{\infty} \sum_{m=1}^{2n+1} \left(F^{\wedge}(n'; n, m) \right)^2$, where $F^{\wedge}(n'; n, m)$ are the time–space

Fourier coefficients. The percentage $p(F^{\text{rec}}, F^{\text{orig}})$ is then given by

$$p(F^{\text{rec}}, F^{\text{orig}}) = \frac{\|F^{\text{rec}}\|_{L^2([-1,+1]\times\Omega)}}{\|F^{\text{orig}}\|_{L^2([-1,+1]\times\Omega)}}.$$

Table 1 shows the percentage of the reconstruction from GRACE data to the original GRACE data and the correlation coefficients for the corresponding reconstructions between GRACE and WGHM data for different values of G_1 and G_2. Note that all values of the percentage greater than 88% and all correlation coefficients greater than 0.83 are in bold numbers. As expected we realize that with decreasing percentage the correlation coefficient goes up. In dependance of the parameters G_1 and G_2 we have to optimize both percentage and correlation coefficient. In *Table 1* we have the best percentage for $G_1 = -0.3$ and $G_2 = 0.0$ if we claim a correlation coefficient greater than 0.83.

In *Fig. 8* we show the correlation coefficients of the reconstructions of GRACE and WGHM data for $G_1 = -0.3$ and $G_2 = 0.0$. Especially the regions with very bad correlation of the original data and the optimal reconstruction show differences in the local correlation

Table 1

Percentage (third column) of the reconstruction with details up to scale 9 from GRACE data to the original GRACE data and correlation coefficients (fourth column) for the corresponding reconstructions between GRACE and WGHM data for different values of G_1 and G_2

G_1	G_2	Percentage (in %)	Corr. Coeff.
-	-	100 (original)	0.75
−0.9	−0.8	**95.4**	0.79
−0.9	−0.4	**94.4**	0.80
−0.9	0.0	**92.7**	0.81
−0.9	0.4	**89.2**	0.82
−0.9	0.8	82.6	**0.83**
−0.7	−0.4	**94.0**	0.81
−0.7	0.0	**92.1**	0.82
−0.7	0.4	**88.3**	**0.83**
−0.7	0.8	81.2	**0.84**
−0.5	−0.4	**93.4**	0.81
−0.5	0.0	**91.4**	0.82
−0.5	0.4	87.3	**0.83**
−0.5	0.8	79.7	**0.84**
−0.3	0.0	**90.5**	**0.83**
−0.3	0.4	86.0	**0.84**
−0.3	0.8	77.8	**0.85**
−0.1	0.0	**89.3**	**0.83**
−0.1	0.4	84.4	**0.85**
−0.1	0.8	75.6	**0.86**
0.1	0.4	82.4	**0.85**
0.1	0.8	72.9	**0.87**
0.3	0.4	79.8	**0.86**
0.3	0.8	69.6	**0.88**

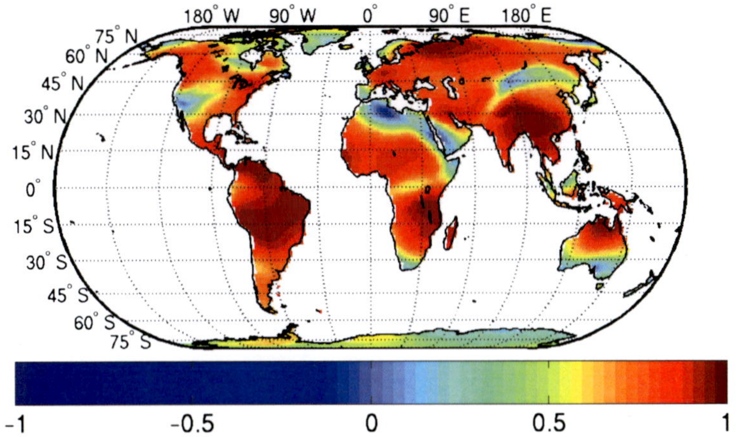

Fig. 8
Correlation coefficients of the reconstructions of GRACE and WGHM data for $G_1 = -0.3$ and $G_2 = 0.0$

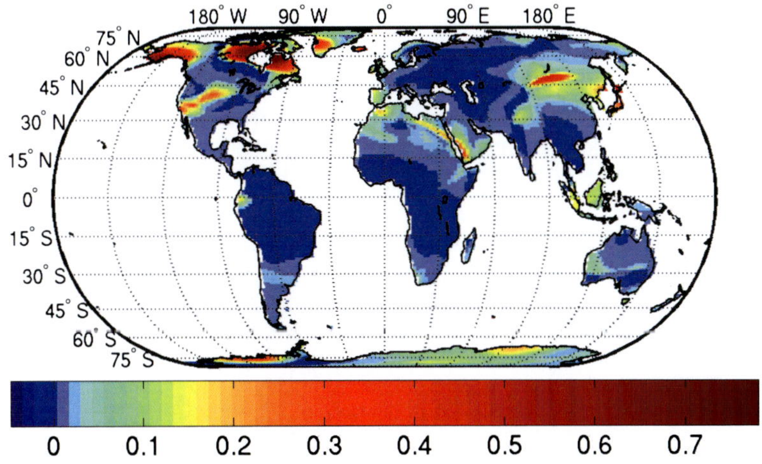

Fig. 9
Difference of the correlation coefficients of the optimal reconstruction ($G_1 = -0.3$ and $G_2 = 0.0$) and the correlation coefficients of the original GRACE and WGHM data

coefficients. To make this more evident, we additionally in ➲ *Fig. 9* show the differences of the correlation coefficients of the optimal reconstruction ($G_1 = -0.3$ and $G_2 = 0.0$) and the correlation coefficients of the original GRACE and WGHM data. Blue regions in ➲ *Fig. 9* correspond to regions of good correlation of the hydrological model with the satellite data because in our reconstruction process (see Formula (2)) we did not have to do much corrections. Red regions correspond to those regions with bad correlation coefficient between GRACE and WGHM data. In this case we had to give up much of the detail information in Formula (2) due to the bad correlation coefficient. We want to emphasize that the approach presented in this section is a first idea of how to make use of the information achieved by the multiresolution analysis in view of improving the existing hydrological models. Research in cooperation of geoscientists and mathematicians is necessary for further progress in the field of multiresolution of hydrology data.

6 Conclusion

The huge amount of data which is provided by the satellite mission GRACE allows to quantify both spatial and temporal variations of the Earth's gravity field. For this reason a time–space multiresolution analysis is presented in this chapter. The basic idea of this method is to transfer the one-dimensional multiscale analysis to higher dimensions, more precisely, a tensor product wavelet analysis using Legendre wavelets for the time domain and spherical wavelets for the space domain is realized. With the corresponding tensor product wavelets we are able to locally analyze (in one step) a time series of the gravitational potential. Particularly, the spatial detail information is not smeared over the whole Earth, which is a disadvantage of the classical approach based on spherical harmonics. Based on the results of the tensor product wavelet analysis we are interested in an extraction of a global hydrological model from the satellite data, and, thus, in an improvement of already existing hydrological models. Therefore the time series of the GRACE data and those of the existing hydrological model WGHM are compared using a correlation analysis. To this end, local and global correlation coefficients between the original data sets, the detail information, and the reconstructions of GRACE and WGHM are computed. With the aid of these correlation coefficients we are looking for an "optimal" reconstruction of the GRACE data, i.e., the aim is to find out an optimal filter. This is done by constructing a weight function which controls the influence of the corresponding detail parts on the resulting reconstructed signal. For future research it is necessary to interprete these results not only from the mathematical point of view but also with geoscientifical knowledge in order to extract a reasonable optimal global hydrology model from the GRACE satellite data.

Acknowledgments

The authors gratefully acknowledge the support by the German Ministry of Education and Research (BMBF) and German Research Foundation (DFG) within the R&D–Programme Geotechnologies Special Programme "Observation System Earth from Space", 03F0424D, (publication number GEOTECH-317). We are also much obliged to GFZ Potsdam for providing us with all GRACE and WGHM data.

References

Beth S, Viell M (1998) Uni- und multivariate Legendre-Wavelets und ihre Anwendung zur Bestimmung des Brechungsindexgradienten. In: Freeden W (ed) Progress in Geodetic Science at GW 98, Shaker, pp 25–33

Freeden W (1999) Multiscale modelling of spaceborne geodata. Teubner, Stuttgart, Leipzig

Freeden W, Schneider F (1998a) An integrated wavelet concept of physical geodesy. J Geod 72:259–281

Freeden W, Schneider F (1998b) Regularization wavelets and multiresolution. Inverse Prob 14:225–243

Freeden W, Schreiner M (2009) Spherical functions of mathematical geosciences. A scalar, vectorial, and tensorial setup. Springer, Heidelberg

Freeden W, Gervens T, Schreiner M (1998) Constructive approximation on the sphere (with applications to geomathematics). Oxford Science Publication, Clarendon Press, Oxford

Louis AK, Maaß P, Rieder A (1998) Wavelets: Theorie und Anwendungen. Teubner, Stuttgart

Maier T (2003) Multiscale geomagnetic field modelling from satellite data: theoretical aspects and numerical applications. PhD Thesis, University of Kaiserslautern, Geomathematics Group

Mallat S (1989a) Multiresolution approximations and wavelet orthonormal bases of $L^2(\mathbb{R})$. Trans Am Math Soc 315:69–87

Mallat S (1989b) A theory for multiresolution signal decompostion. IEEE Trans Pattern Anal Machine Intell 11:674–693

Meyer Y (1992) Wavelets and operators. Cambridge University Press

Müller C (1966) Spherical harmonics, vol 17. Springer, Berlin

Nutz H, Wolf K (2008) Time-space multiscale analysis by use of tensor product wavelets and its application to hydrology and GRACE data. Studia Geophysica et Geodaetica 52: 321–339

Swenson S, Wahr J (2006) Post-processing removal of correlated errors in GRACE data. Geophys Res Lett 33: L08402. doi:10.1029/2005GL025285

Swenson S, Wahr J, Milly PCD (2003) Estimated accuracies of regional water storage variations inferred from the gravity recovery and climate experiment (GRACE). Water Resour Res 39(8):1223. doi:10.1029/2002WR001808

Tapley BD, Reigber C (2001) The GRACE mission: status and future plans. EOS Trans AGU 82(47): Fall Meet Suppl G41, C-02

Tapley BD, Bettadpur S, Ries JC, Thompson PF, Watkins MM (2004a) GRACE measurements of mass variability in the Earth system. Science 305:503–505

Tapley BD, Bettadpur S, Watkins MM, Reigber C (2004b) The gravity recovery and climate experiment: mission overview and early results. Geophys Res Lett 31: L09607. doi:10.1029/2004GL019920

12 Time Varying Mean Sea Level

Luciana Fenoglio-Marc · Erwin Groten
Institute of Physical Geodesy, Technical University Darmstadt, Darmstadt, Germany

W. Freeden, M.Z. Nashed, T. Sonar (Eds.), *Handbook of Geomathematics*, DOI 10.1007/978-3-642-01546-5_12,

Abstract After a general theoretical consideration of basic mathematical aspects, numerical and physical details, which are related to a specific epoch and areas where sufficient and reliable data are available, are addressed. The concept of a "mean sea level" is in itself rather artificial, because it is not possible to determine a figure for mean sea level for the entire planet, and it varies on a much smaller scale. This is because the sea is in constant motion, affected by the high- and low-pressure zones above it, the tides, local gravitational differences, and so forth. What is possible to do is to calculate the mean sea level at that point and use it as a reference datum. Coastal and global sea level variability are analyzed between January 1993 and June 2008. The coastal variability is estimated from tide gauges, altimeter data colocated to the tide gauges and from altimeter data along the world coasts, while the global variability is estimated from altimeter data. Regionally there is a good agreement between coastal and open-ocean sea level variability from altimeter data. The sea level trends are regionally dependent, with positive average of 3.1 ± 0.4 mm/year. The variability of the trend of coastal sea level appears to be related to the interannual time scale of variability of the Northern Atlantic Oscillation and of the El Nino- Southern Atlantic Oscillation climatic indices.

Keywords: Altimetry, Global change, Sea level, Tide gauge

1 Introduction

The study of transient phenomena differs essentially from usual geodetic investigations where in the past stationary solutions prevailed. Whenever time-dependent boundary value problems (BVPs) are considered, harmonic function solutions have to be carefully selected, because nonstationary potentials are often nonharmonic and, therefore, cannot be handled by geodetic techniques deduced from classical potential theory. Consequently, the investigation of global vertical reference frames in connection with rising sea level, global warming, and coastal deformations for Earth models affected by plate tectonics does not lead to straightforward classical solutions but rather to stepwise and similar approaches where redundant data sets involving a large number of various types of observations can yield reliable results. Applications of that kind are not totally uncommon in geodesy, as geodetic systems had always been evaluated for models of the Earth perturbed by tides and other temporal variations where, however, appropriate reduction schemes had to be applied. The study of time varying mean sea level is a rather complex topic in that connection which benefits from modern satellite techniques. They make precise solutions possible. In so far geodetic approaches are typically approximation approaches.

Sea level rise is an important aspect of climate change. At least 50 year records are needed to separate secular, decadal, and interannual variations (Douglas 2001). Over this long time interval only few tide gauges along the world coastlines are available for the analysis (Church et al. 2008). The spatial distribution of tide gauges as well as the existence of interannual and low-frequency signals affects the recovery of secular trends in short records. Thus, it is important to develop techniques for the estimation of sea level trends "cleaned" from decadal variability.

The analysis based on tide gauges alone reflects however only the coastal sea level, while the global sea level change is derived from altimeter data that are available over the past 15 years. A coastal global averaged sea level rise over the past century has been observed from tide gauge data in the range of 1–2 mm/year (Cazenave and Nerem 2004; Holgate 2007; Church et al. 2004; Church and White 2006; Domingues et al. 2008).

Over the last decade, both altimetry and tide gauges have been used to estimate global coastal and open sea changes. Estimates for the rate of the global sea level from altimetry, without accounting for the Glacial Isostatic Adjustment (GIA) are around 2.8 ± 0.4 mm/year over 1993–2003 (Bindoff et al. 2007; Lombard et al. 2005), 3.1±0.4 in 1993–2006 (Beckley et al. 2007), and 3.1 ± 0.1 mm/year in 1993–2007 (Prandi et al. 2009).

Holgate and Woodworth (2004) give in 1993–2002 a rate of 4 mm/year and suggest that the coastal sea level is rising faster than the global mean. Following White et al. (2005), the difference between the rates from global altimetry and coastal tide gauges is due to the different sampling. Finally Prandi et al. (2009) show that the differences are mainly an artifact due to the interannual variability.

In addition to the uncertainty of the rate due to the fitting procedure, measurement errors and omission error are involved. The calibration of altimeter data using colocated altimetry and tide gauge stations gives an error of 0.4 mm/year for the altimetric sea level change (Mitchum 2000; Leuliette et al. 2004). Jevrejeva et al. (2006) assign an error of 1 mm/year to the global sea level change derived from tide gauges, due to the nonuniform data distribution.

The tide gauge stations used in the various studies are different, from more than 1000 stations in Jevrejeva et al. (2006) to 91 stations in Prandi et al. (2009). The filling of the incomplete time-series reduces the effective number of stations.

Fenoglio-Marc and Tel (2010) investigate if the coastal sea level observed by tide gauges conveniently represents the global sea level obtained using both altimetry and tide gauge stations. The study differs from Prandi et al. (2009) in the selected tide gauge stations and in considering the sea level averages both globally and regionally. Distinction is made between (a) global sea level change derived from altimetry (GSL), (b) costal sea level change derived from altimetry (CGSL), and (c) costal sea level change derived from tide gauges (CGTG). First, the global and regional basin averages of sea level are considered and long-term trends are identified, then the relationship between climatic indices and the main components of the interannual and interdecadal variability is investigated.

2 Theoretical Considerations

One of the key problems related to mean sea level is the global unification of vertical datums. Regionally, e.g., in Europe, the various national leveling systems have in the past been interconnected by different geodetic techniques. One of the pilot products was the UELN (Unified European Levelling Network) or REUN (Reseau Européen Unifié de Nivellement). In crossing smaller oceanic areas, hydrostatic leveling had been applied, e.g., between Denmark and Northern Germany. With the advent of GPS techniques, satellite methods played a significant role, e.g., in the construction of the railway tunnel between France and Great Britain. For precise navigation purposes the connection of vertical data between continents gained increasing interest. In this context satellite altimetry together with recent precise EGM, e.g., the Earth Gravity Model 2008, played a major role. With data from GRACE and GOCE at hand, the previous obstacles in creating a global vertical datum are substantially diminished. The traditional argument that there is no unique solution of the third "mixed" (in the sense Dirichlet (first) plus Neumannian (second) BVP) boundary value problem of potential theory in terms of altimetric data at sea and gravity data on land is long-time obsolete. With GPS available all over the Earth, the BVP is considered to be a "fixed" problem. Meanwhile, the number of altimetric satellites is so large and the quality of global gravimetric data is so high that even time-variability of global mean sea level, i.e., of a global absolute geoid, is no longer a serious problem. The transition from "observed instantaneous" to "mean sea level" data is a rather complex and numerically complicated process which depends, to some extent, on the validity of involved modeling processes and on the availability and accuracy of related data sets. The exact determination of the N(zero)-term in geoid evaluation, i.e., its scale constant, and the sufficient control of time-varying sea tide contributions have meanwhile been solved by appropriate and well-documented modeling procedures. Also temporal changes in tide gauge positions along the coasts where sea level variations are of utmost practical importance do no longer pose problems in geoid computations. Precise calibrations of altimeters, exact positioning of satellite altimeters in space via satellite-to-satellite (SST) techniques using GPS, as well as dedicated satellites such as CHAMP, GRACE, and GOCE (see ➋ Chap. 3) render traditional arguments obsolete.

In this way geodesy can substantially contribute to the long-time (controversial) discussion and associated solutions of rising mean sea level and related ecological questions. Without highly precise monitoring of global mean sea level and its combination with reliable information on time-varying global gravity, there is no solution.

Not only secular changes of mean sea surface in terms of steric and circulation variations, which vary significantly between oceanic regions, can be investigated, but also periodic and aperiodic effects. In analyses of secular effects these different types of time-varying contributions can be separated only over relatively long time spans. A typical example is the El Nino and La Nina effects in the Pacific between Peru and Australia affecting climate perturbations almost all over the globe. To what extent frequency and amplitude variations of such effects are really interdependent (global warming affects frequency and amplitudes and vice versa) has not yet been answered reliably in the past. Recent announcements by NOAA indicate the possibility that the 2007–2008 cool period of relevant La Nina activities is over since June 2009. If higher (by one degree above yearly average) temperatures in the tropical Pacific are going to prevail for longer time, say half a year or so as usually, this could indicate the return of El Nino associated with higher global average temperatures, milder winter in North America, droughts in Indonesia, and floods along the western coasts of the Americas. Moreover, California could be affected by strong winter storms, in the Atlantic a relatively mild hurricane season could be expected as well as a significant general change in wind, air pressure, and precipitation patterns. As a consequence, 2010 could become a relatively warm year all over the globe.

M. Bursa (Bursa et al. 2001) is one of the pioneers in investigating the global vertical datum. His first attempts led to accuracy estimates of about a decimeter or so, as far as the intercontinental connection of vertical datums is concerned. His various publications since 1997 reveal clearly that the determination of an absolute global geoid is basically a question of uniform global, i.e., unified scaling including time scales. This is obvious, because today the length scale is defined via the time scale given by an interval of travel time of a photon in vacuo. Moreover, atomic time is directly related to a specific value of the geopotential and thus to the dimension of the Earth. Consequently, the definition and implementation of a global vertical datum is a topic of broad practical and theoretical implications. Formally, the atomic time scale is related to "mean sea level geopotential," but this is just one of different possible definitions. At present, this quantity is identified with the (constant) geoidal potential W° (Bursa et al. 2007). It is a consequence of the classical definition of the geoid as the global approximation of mean sea level in least-squares sense. In a transient system it needs a revision. This remark gains importance with increasing precision, i.e., relative accuracy, of time standards, as, e.g., "cold fountains," which is presumably of the order of 10 (exp − 18), on the one hand and significant secular changes of global gravity and mean sea level on the other hand, where the aforementioned different types of mean sea level changes (steric, etc.) are of interest. The prime topic is the existence of a global geocentric vertical reference system.

Meanwhile the situation has been essentially improved by much better data coverage, higher accuracies and substantially improved modeling. But still the crucial areas are shallow coastal zones where altimeter data show higher noise and worse coverage than in other off-shore zones, which are basically the boundaries between land and sea in the aforementioned mixed boundary value problem. With recent data at hand, Bursa (Bursa et al. 2007), Ardalan and others improved these results numerically as well as from the very principle, where the dedicated new geodetic satellites play a dominant role, as in the aforementioned coastal zones also shipborne gravity data coverage is weak. Airborne gravity cannot completely supplement and resolve those deficits. Moreover, airborne coverage cannot fully replace spaceborne coverage, as it is less frequent and thus cannot detect time-varying gravity. Even at elevations above 350 km the extremely high resolution of space methods turns out to be, in general, superior to aircraft results. Intercontinental air and sea traffic will strongly benefit from new improved determinations of such a global vertical datum. Open questions such as the melting processes of glaciers, particularly in Arctic and Antarctic regions, will benefit as well. There is no precise solution of such questions without dependable information about the time-varying Earth's gravity field. The key to these solutions is the transition from steady-state to transitional solutions. Therefore, besides oceanography, glaciology, and precise measuring techniques, geodesy plays a dominant role for exact solutions of related problems by providing the precise global reference systems. The existing results on sea level rise (secular and periodic) as well as the investigations of underlying processes leading to reliable prediction methods are promising but still extremely controversial. This is not surprising, because GOCE was basically designed as a tool for investigation of steady-state phenomena whereas GRACE's aim is primarily the study of temporal changes within limited areas. Both differ basically in their spatial resolution. Moreover, their life times are relatively short so that secular global gravity changes are difficult to be detected by stand-alone methods. Consequently, (numerically) ill-posed and (physically) improperly posed procedures are involved in the hybrid data processes, such as analytical downward continuation of gradiometric satellite results to sea level. Whereas the conversion of accelerations, i.e., gravity, to geopotential differences is stable, in general, the opposite procedure in converting altimetric results into sea surface gravity anomalies is not well posed. Only skilful combinations of the available data sets can, e.g., lead to useful results on secular variations, where the different aforementioned types (steric, etc.) of sea level changes are of interest. The final (stationary) product of a unified global vertical datum would be a value of the geopotential $W^{\circ}(x)$ associated with a set of geocentric coordinates (x) related to the equatorial terrestrial system at epoch (t). With present data at hand, the accuracy of the numerical values is assumed to be of the order of 10^{-9}. With new results as expected from GOCE, the transition to a time-dependent transient geoid $W^{\circ}(x,t)$ including secular temporal changes would be a challenging goal. A more exact definition of the geoid would be necessary in that case, as is seen from $W^{\circ}(x,t)$ because also $W^{\circ}(x(t))$ would make sense by keeping W° constant. The decision in that case depends on the coupling between volume and potential of the Earth. It is easily seen that, for instance, steric mean sea level changes can lead to significant changes of the Earth's volume without affecting the geopotential seriously. Therefore, only modeling processes based on gravity and geometric information, for instance satellite altimetry, yield satisfying results.

The analysis and separation of the different components of sea level variation are now investigated by several authors. They were first initiated in areas where GPS-controlled tide gauge data of high data density were available, due to the particular interest in coastal areas. Also here the application of space techniques appears superior to terrestrial approaches.

3 Scientific Relevance

Sea level is a very sensitive indicator of climate change and variability. It responds to change of several components of the climate system. Sea level rises due to global warming, as sea waters expand and fresh water comes from melted mountain glaciers. Variations in the mass balance of the ice sheets in response to global warming has also direct effect on sea level. The modification of the land hydrological cycle due to climate variability and anthropogenic forcing leads to change in runoff in river basins, hence ultimately to sea level change. Even solid Earth affects sea level through small changes of the shape of ocean basins. Coupled atmosphere–ocean perturbations, like El Nino-Southern Oscillation (ENSO) and North Atlantic Oscillation (NAO) also affect sea level in a rather complex manner and contribute to sea level variability.

While sea level had remained almost stable during the last two millennia, subsequently to the last deglaciation that started ~18,000 years ago, tide gauge measurements available since the late nineteenth century have indicated significant sea level rise during the twentieth century, at a mean rate of about 1.7 mm/year (e.g., Church et al. 2004; Church and White 2006; Holgate 2007).

Since early 1993, sea level variations are accurately measured by satellite altimetry from Topex/Poseidon, Jason-1, and Jason-2 missions (Cazenave and Nerem 2004; Leuliette et al. 2004; Beckley et al. 2007). This 15 year long data set shows that, in terms of global mean, sea level is presently rising at a rate of 3.5 ± 0.4 mm/year, with the GIA correction applied (Peltier 2004; Peltier 2009). The altimetry-based rate of sea level rise is therefore significantly higher than the mean rate recorded by tide gauges over the previous decades, suggesting that sea level rise is currently accelerating in response to global warming. Owing to its global coverage, altimetry also reveals considerable regional variability in the rates of sea level change. In some regions, such as the Western Pacific or North Atlantic around Greenland, sea level rates are several times faster than the global mean, while in other regions, e.g., Eastern Pacific, sea level has been falling during the past 15 years.

There is an urgent need to understand the causes for the observed sea level rise and its possible development. Over the 1993–2003 time span, sea level rise was about equally caused by two main contributions: ~50% from warming of the oceans (through thermal expansion) and ~50% from glaciers melting and ice mass loss from Greenland and Antarctica (Bindoff et al. 2007). Since 2003 thermal expansion has negligibly contributed to sea level (e.g., Willis et al., 2008) although meanwhile satellite altimetry-based sea level has continued to rise. Accelerated ice mass loss from the Greenland and West Antarctica ice sheets (evidenced from various remote sensing techniques such as radar and laser altimetry, INSAR and GRACE space gravimetry) as well as increased in glacier melting (e.g., Meier et al., 2007, Rignot et al., 2008) appear to account alone for last 5 years sea level rise (Cazenave et al., 2009).

4 Data and Methodology of Numerical Treatment

We use altimeter data from January 1993 to December 2008 of the Topex/Poseidon, Jason-1, and Envisat altimeter missions. Standard corrections are applied to the altimetry data with exception of the inverse barometric correction, for consistency with the monthly tide gauge data that are not corrected either. This correction, given by the MOG2D model, is applied for the comparison between global and coastal sea level change derived from altimeter data only. Monthly 1° grids are constructed by using a simple Gaussian weighted average method (half-weight equal to 1 and search radius equal to 1.5°).

Tide gauge data are available from the Permanent Service for Mean Sea Level (PSMSL, http://www.pol.ac.uk/psmsl/) database. We correct the sea level at the tide gauge station for the GIA using the SELEN software (Spada and Stocchi 2007) forced with the ICE5-G glaciation history and a viscoelastic Maxwell Earth derived from VM2 (Paulson et al. 2007). Also sea level change derived from the altimeter is corrected for the GIA effect and this correction increases the satellite estimates of global sea surface rates by 0.3 mm/year (Church et al. 2004).

Selection criteria or the tide gauges are based on the time length of the time-series and on the data gaps. The stations available in each of the three intervals 1900–2001, 1950–2001, and 1993–2001 are 1158, 1103, and 738, respectively. A station is used if it is available over 90% of the interval with gaps shorter than 2 years (availability criteria), stations fulfilling these criteria are 15, 117, and 365, respectively, and mostly located in the northern hemisphere along the European and North-American coasts.

A further criterion for the selection of tide gauge is based on proximity and agreement with satellite altimetry. The chosen stations will represent the large-scale open sea level variability, and stations describing the local sea level variability are eliminated (Fenoglio-Marc et al. 2004). A station is eliminated if the minimum distance from a point of the altimeter grid is greater than 2°. For each tide gauge station we consider the nearest node of the altimeter grid within 2° radius and four parameters: (1) correlation, (2) the trend of the difference, (3) the standard error of the trend of the difference, and (4) the standard deviation of the difference are used as indicators of the agreement between altimetry and tide gauge data. The criteria are correlation >0.5, the trend and standard error of the trend <5 mm/year, and the standard deviation of the differences <80 mm. Not fulfillment of the selection criteria can arise from a jump in the record or from local variability (location in an harbor or near to an estuary). In this last case, the station, also if correctly recording, will not be used for sea level change analysis.

Only complete time-series are then used to estimate the empirical models. To increase the number of tide gauge stations, the gaps in the records are filled by linear regression with the highest correlated time-series. For each of the time-series of anomalies, the trend and seasonal signal are subtracted and the correlation matrix of the residuals computed. For each of the time-series with gaps, the gaps are filled by linear regression of the residuals with the complete time-series that realizes the highest correlation. The process is repeated until all the gaps are filled. The filling is not done for a time-serie if the correlation coefficient with any of the complete tide gauges is lower than 0.5. The time-series are then reconstructed by adding trends and seasonal cycles.

For the analysis of the sea level trend, the monthly time-series are deseasoned and the trend is evaluated by linear regression. The probable uncertainty of the linear term of the regression is estimated accounting for the temporal correlation of the residuals. The effective degree of freedom and the probably uncertainty of the linear term are derived from the lag-1 autocorrelation of the detrended time-series and the standard error obtained from the linear regression (Wilks 1995; Fenoglio-Marc et al. 2004).

For the analysis of the long-term trends, we compute monthly spatial averages over the complete globe and over selected basins. The seasonal variation is removed and a linear regression model is fitted to the residuals after application of a moving average with a window of 12 months.

For the analysis of interannual and interdecadal signals, the trend and seasonal component are removed from each time-series and the residuals time-series are low-pass filtered. For interannual analysis, the filtering consists in averaging the deseasoned monthly time-series over 1 year of data with a time spacing of 0.5. For interdecadal analysis, the filtering consists in averaging monthly data over 5 years with a time sampling of 2.5 years.

The statistical method of principal component analysis (PCA) (Preisendorfer 1988) is applied to the interannual and interdecadal anomaly fields to detect the principal modes of variability. A significance test based on a Monte Carlo technique (Overland and Preisendorfer 1982) is applied to find the number, of principal components, k, to be retained.

The statistical method of canonical correlation analysis (CCA) in the PCA basis (PCA-CCA) is used to analyze the coupled variability of altimetric and tide gauge data, the decomposition is obtained maximizing the correlation between the temporal patterns.

The interannual variability is reconstructed in the past from altimetry and tide gauge data using the climatic indices (Trenberth 1984; Trenberth and Hurrel 1994; Woolf et al. 2005). The method consists in first applying separately the PCA method to the standardized anomaly fields of altimetry and of tide gauges. The modes maximizing the temporal correlation with the same climatic index are selected to build the empirical model of the sea level variability by using the corresponding spatial patterns from altimetry and the temporal coefficients from tide gauges.

In conclusion, relatively simple mathematical procedures have been used, namely gridding, interpolation, filtering, and trend estimation.

5 Fundamental Results

The trend of sea level is regional and has time-dependent characteristics (➋ *Fig. 1*).

Tide gauge stations suitable for large-scale sea level change analysis are reduced to 365 by the "time-length" condition, to 339 by the "proximity" condition, and finally to 290 by the comparison to the nearest altimeter point (➋ *Fig. 2*).

The variability of the averaged coastal sea level measured by tide gauge and altimetry at a colocated altimetric point are in good agreement with correlation of 0.96, RMS of 3.4 mm between the smoothed time-series. The difference in the trends is within the errors.

The interannual and interdecadal variability of global sea level change derived from altimetry, costal sea level change derived from altimetry, and costal sea level change derived from colocated altimetry at tide gauge locations have slightly different characteristics and similar positive trends (➋ *Fig. 3*).

Trends are evaluated from January 1993. For intervals longer than 8 years all trends are positive and in the range 2–5 mm/year. The trend of global sea level is almost constant over the last 10 years (3.0 ± 0.4 mm/year from January 1993 to December 2007). The trend of coastal global sea level is monotonic decreasing (2.28 ± 0.59 mm/year in 1993–2007), while the trend of global sea level at tide locations is more variable (2.68 ± 0.64 mm/year in 1993–2007). The best agreement between all trends is reached in 2002–2004 (➋ *Fig. 4*).

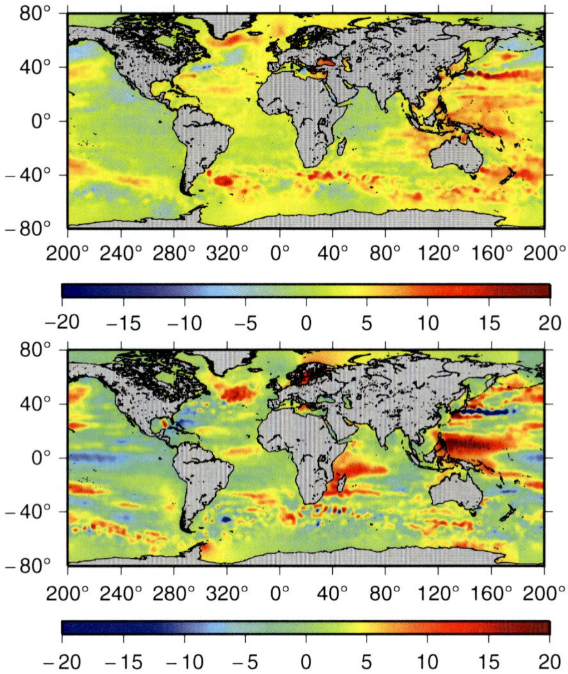

Fig. 1
Sea surface height trends from Topex-Poseidon in 1993–2005 (*above*) and from Jason-1 in 2002–2008 (*below*)

The omission error of the coastal sea level trend from tide gauges σ_{miss}, due to the incomplete coastal coverage, is evaluated to be 0.3 mm/year from the difference in trends of coastal altimetry and coastal altimetry at the tide gauges. The error σ_{miss} is smaller than the adopted measurement error σ_{meas} of 0.4 mm/year (Mitchum 2000).

The interannual variability is stronger in the coastal averaged time-series. Regionally there is a good agreement between coastal and open-ocean sea level variability, trends are positive in the main world basins. The location, number, and distribution of the tide gauge stations and the extension of the coasts explain part of the differences observed. The Tropical Pacific dominates the global variability, the highest agreement between the coastal sea level variability from altimetry and tide gauge is found there.

There is a strong correlation between the climatic indices and the patterns of interannual variability. The first four dominant modes of the PCA altimeter decomposition account for more than 60% of the total variability. The Southern Ocean Index (SOI) and the Northern Atlantic Index (NAO) are significantly correlated to the dominant modes over 1993–2001 (*Fig. 5*). The variability of the tide gauge datasets is biased to the North Hemisphere coasts. (*Fig. 6*) as most of the tide gauges are located along the European and North-American coasts. This explains the different order of the dominant modes obtained from the tide gauge and from the altimeter datasets.

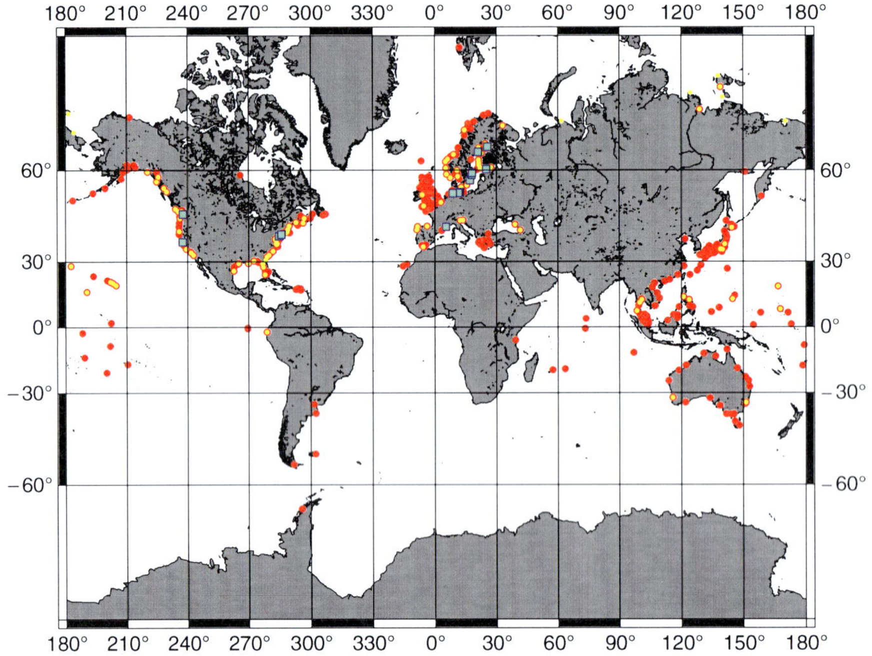

Fig. 2
Tide gauge stations of the PSMSL dataset satisfying availability criteria. Number of stations are 365 in 1993–2001 (*red*), 117 in 1950–2001 (*yellow*), and 15 in 1900–2001 (*blue*)

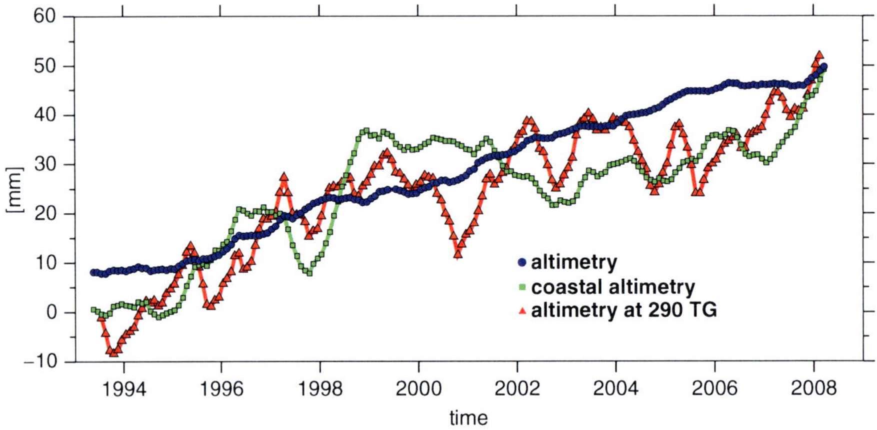

Fig. 3
Sea level average from Topex and Jason-1 altimeter data: Global sea level (*circle*), coastal global sea level at 100 km from the coast (*square*), and coastal sea level at the nearest point to each of the 290 selected tide gauges (*triangle*)

The empirical model based on NAO and SOI for the period 1993–2001 reproduces about 50% of the global averaged interannual variability observed by altimetry. The corresponding average time-series of global interannual mean sea level is estimated to have an accuracy of

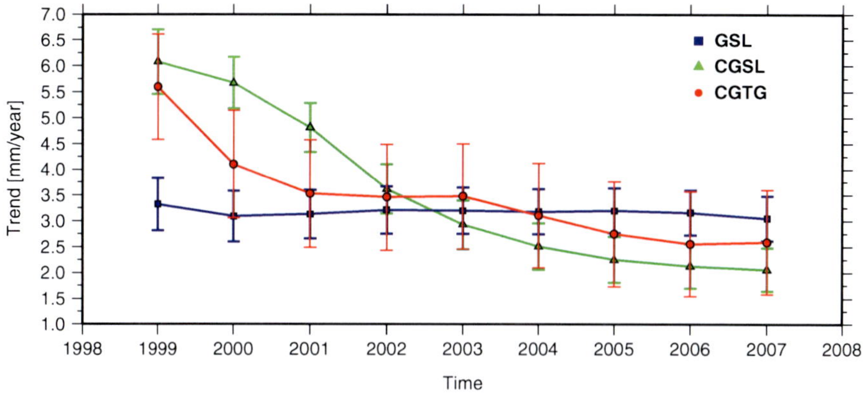

Fig. 4
Trends and error bars of global sea level (*square*), coastal sea level (*triangle*), and coastal sea level at 290 tide gauge stations (*circle*) computed over different time intervals starting from 1994

2–3 mm. Over 1950–2001 the correlation between the SOI and NAO climatic indices and the temporal patterns of the interannual and interdecadal models is conserved. We conclude that the variability of the trends is related to the time-scales of the NAO and SOI indices.

The study confirms the regional characteristics of sea level variability and the importance of the a priori selection of suitable tide gauges. Being the signal of large-scale, the results do not considerably change using suitable subsets of the original selected set.

6 Future Directions

The space technologies allow the estimation of sea level change at global scale. For the past century, estimates of sea level variations have been based on tide gauges data, that suffer from some limitations, as incomplete coverage and contamination by vertical motions of the ground from GIA, tectonics, volcanism, and local subsidence. Satellite altimetry gives new challenges.

6.1 Accuracy of Sea Level Observation

Since the early 1990s, satellite altimetry has become the major tool for precisely measuring sea level with quasi-global coverage and short revisit time. This objective has pushed the altimetry systems toward their ultimate performance limit. To measure global mean sea level rise with 5% uncertainty, a precision at the 0.1 mm/yr level or better is needed.

Errors affecting altimetry-derived sea level estimates fall into two main categories: (1) orbit errors (through force model and reference frame errors), and (2) errors in geophysical corrections applied to sea surface height measurements. The latter result either from drifts and bias of the instruments used to measure these corrections, e.g., onboard radiometers used to measure atmospheric water content, or from model errors. The precision of a single sea surface height measurement from Topex/Poseidon (1 s along track average) is estimated to be 4 cm (Chelton et al. 2001). Recent progress in data processing has decreased this error level to 2 cm for both Topex/Poseidon and Jason-1, averaging over a 10 day interval over the oceanic domain leads to

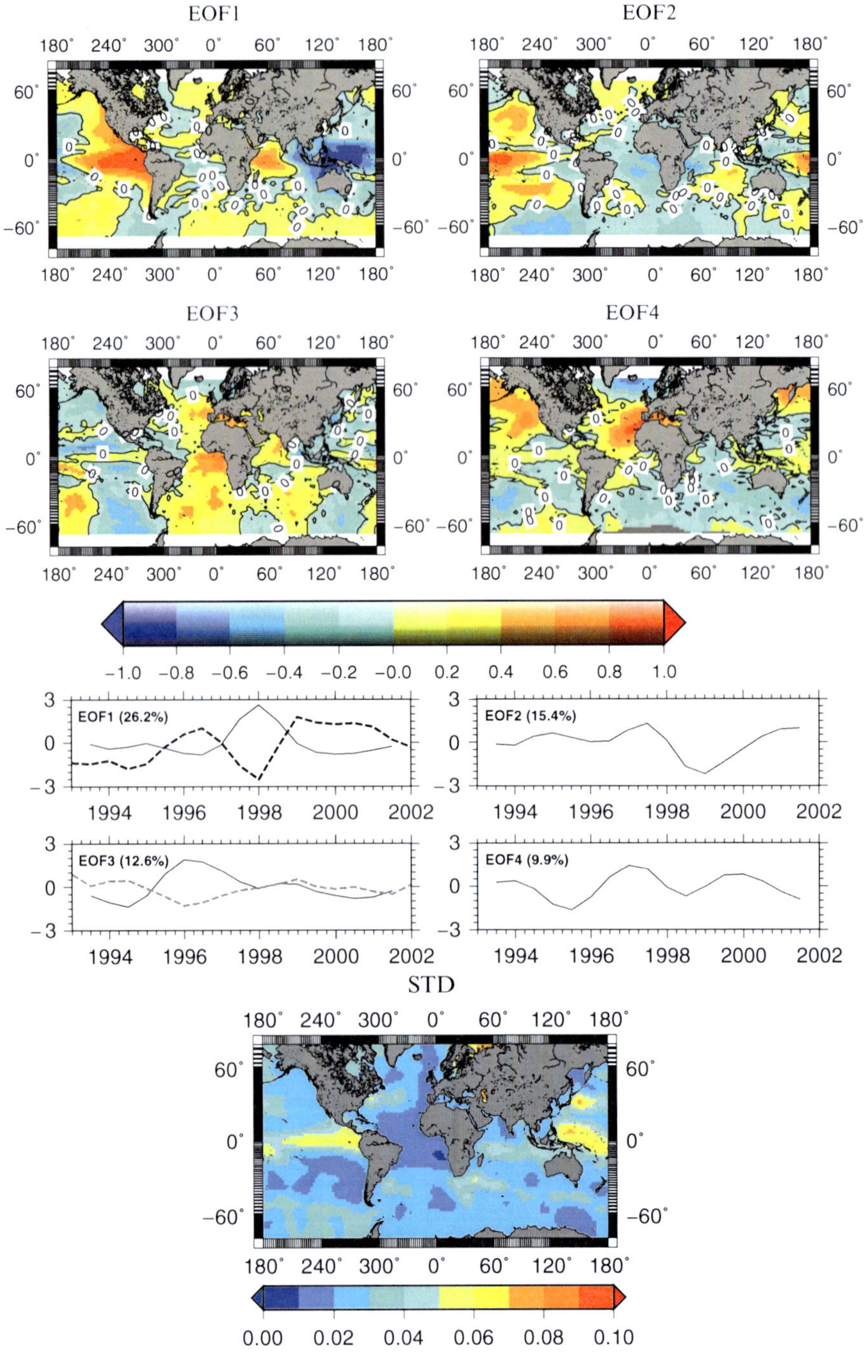

Fig. 5

Dominant spatial patterns (*top*), temporal EOF coefficients and cumulative percentage of variance (*centre*), standard deviation (*bottom*) of interannual variability from altimetry in 1993–2002. Interannual time-series of SOI (*dashed black*) and NAO (*dashed grey*) are plotted with the correlated temporal coefficients

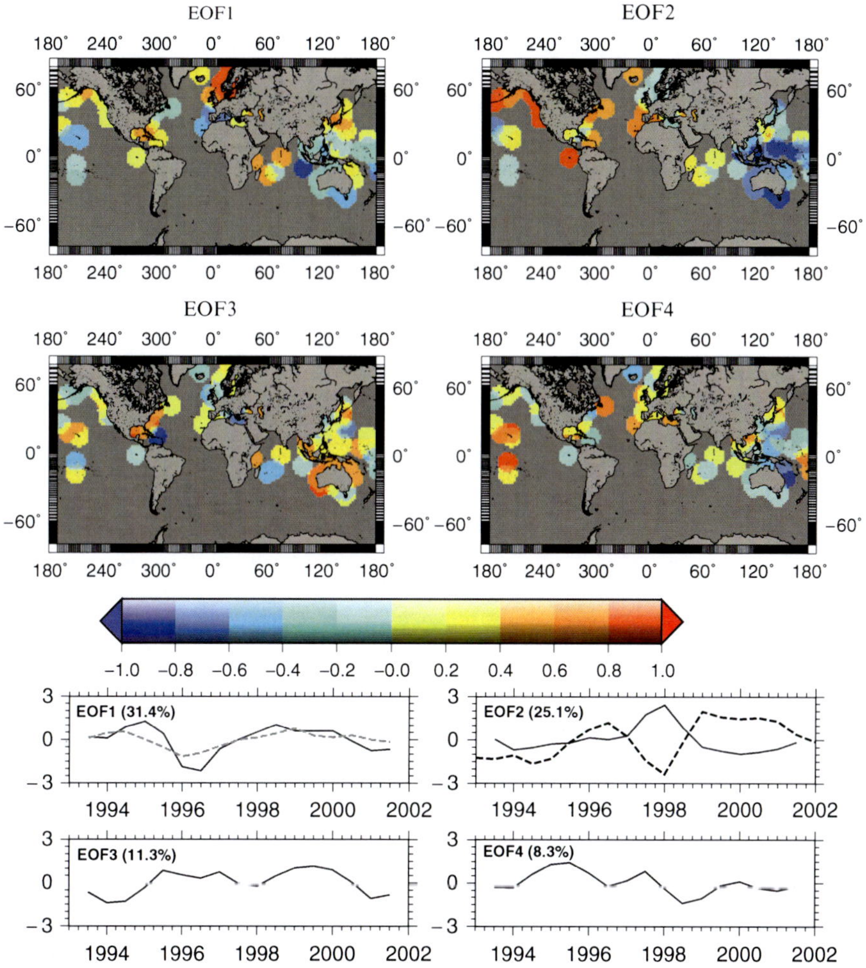

Fig. 6

Dominant spatial patterns (*top*) and temporal EOF coefficients and cumulative percentage of variance (*bottom*) of interannual variability from 197 tide gauge stations in 1993–2001. Interannual time-series of SOI (*dashed black*) and NAO (*dashed grey*) are plotted with the correlated temporal coefficients

a global mean sea level precision of about 4 mm, that translates into 0.1 mm/year uncertainty on the global mean sea level rate. Differences up to 0.5 mm/year in altimetry-based rates of sea level rise, commonly seen between investigators, result from differences in data processing and geophysical corrections used.

A robust procedure to assess the precision/accuracy of altimetry-derived sea level change consists in an external calibration using high-quality tide gauges records (Mitchum 2000). The tide gauge data need to be corrected for vertical crustal motions using GPS and so far, errors on the global mean sea level rate of rise based on tide gauge calibration are about 0.4 mm/year, because of uncertainty in crustal motions at the tide gauge sites.

6.2 Observation at the Coast and in Open Ocean

A key issue is if the coastal sea level observed by tide gauges conveniently represent the global sea level using both altimetry and tide gauge stations.

Global mean sea level rises by an estimated 3 mm/year. But how important is the global figure if we only see the impact at the coast line? It is hence prudent to ask: What is the sea level rise near the coast? Is it different from the global number, and if so, why?

Preliminary analyses show that the sea level in a 200 km radius from the coast is actually rising faster, about 4 mm/year, but this value is dominated by the large 5 mm/year rise the Indonesian Sea, which value is largely influenced by El Nino-Southern Oscillation (ENSO) variability. Some other contributors to increased coastal sea level rise would include warming of the coastal waters.

Another analysis that can shed some light on what components are currently the largest contributors to sea level change is the separation of the regional time-series into empirical orthogonal functions (EOFs). The largest EOFs show the predominant features: seasonal heating, ENSO, PDO, and North Atlantic Oscillation (NAO) and very little power that is purely coastal. That makes us conclude that eustatic and steric sea level rise, outside the aforementioned contributors is actually largely a homogeneous global phenomenon.

6.3 Separation of Components

Satellite altimetry together with GRACE and GOCE data allows the separation of the different components of sea level. With GRACE the volume (steric) and mass changes are separated at large spatial scales of about 300 km, with GOCE the two components of mean sea level, namely the geoid and the mean dynamic topography, can be separated.

6.4 Reconstruction of Past Sea Level

Estimates of sea level variations for the past century are based on tide gauges data or on reconstruction methods that combine tide gauge records with regional sea level variability derived from either satellite altimetry (Church et al. 2004) or general ocean circulation model outputs (Berge-Nguyen et al. 2008).

The aim of the reconstruction is to combine short 2-D signal (with good spatial sampling) with long 1-D signal (with good temporal sampling) in order to reconstruct long-term, 2-D sea level. The general approach uses EOF decomposition of the 2-D time-series to extract the dominant modes of spatial variability of the signal. Then, the spatial modes are fitted to 1-D records to provide reconstructed multidecadal 2-D sea level fields. Such an approach makes two important assumptions: (1) the temporal length of the 2-D fields is high enough to capture the dominant modes of interannual/decadal variability and (2) errors are Gaussian. Assumption (1) is crucial for past sea level reconstruction.

6.5 Prediction of Sea Level Evolution

The Intergovernmental Panel on Climate Change (IPCC) projections (Bindoff et al. 2007) indicate that sea level should be higher than today by about 40 cm on average by 2100 (within a ±20 cm range).

Considerable uncertainty is associated with these projections, which essentially account for future thermal expansion of the oceans and glaciers melting. The major source of uncertainty is the future behavior of the ice sheets which is almost unknown. Recent observations of the mass balance of the Greenland and West Antarctica ice sheets indicate that a large proportion of ice mass loss observed during the past decade is due to rapid coastal glacier flow into the ocean, with incredible acceleration in the recent years. Such processes due to complex ice dynamics, are not yet fully understood and thus not taken into account in sea level projection models.

6.6 Mathematical Representation of Time-Varying Sea Level

The sea level variability is due to the interaction of traveling waves of different spatial scales and temporal frequencies, the different components are superposed to a noise component and exhibit stationarity or nonstationarity in their statistics.

In the analysis of sea level change, the main components of the sea level heights variability are to be identified and a model that is best suitable to represent the sea level variability constructed.

For its nature, sea level is not completely defined over the sphere. Global and local representations are of interest. The selection depends on the goals.

Fitting a mathematical model to data is usually accomplished by expanding the model as an infinite series of appropriate basis functions. These functions are the solution to the relevant differential equations, subject to relevant boundary conditions. It is generally required that the basis functions form an infinite and complete set and that, for fixed boundary, they are mutually orthogonal over the interval of expansion.

We distinguish between parametric and nonparametric methods, being in parametric method the model chosen a priori and coefficients determined accordingly, while in nonparametric method a data-adaptive basis set is used. Between the nonparametric methods are the multivariate statistical analysis methods, as the singular spectrum analysis (SAS) based on the lag-covariance matrix of the data (Allen and Smith 1994) and the principal component analysis (PCA) in the time domain (TD-PCA) and in the frequency domain (FD-PCA), which perform an optimal decomposition of the data variance respectively in the time and in the frequency domain. In time-domain approach, the base functions are space-dependent functions, generally called empirical orthogonal functions (EOF), and the coefficients, called principal components (PC), are time-dependent functions (Preisendorfer, 1988). The PCA method has no phase information and only detects standing oscillations, not traveling waves.

Commonly used in geodesy is the spherical harmonic set of base functions, where the spherical harmonics can be considered as eigenfunctions of invariant pseudodifferential operators (such as the Beltrami operator) and are expressed in terms of Legendre polynomials in colatitude, while the coefficients are trigonometric functions in latitude. In case of an area of investigation limited to a portion of the Earth's surface, other Legendre polynomials and trigonometric functions are the most appropriate base functions. The method of modeling data over a spherical cap is referred to as spherical cap harmonic analysis (Haines 1985). Whereas the associated Legendre functions for the all sphere have integer degree, those for the spherical cap have noninteger degree.

Propagating phenomena are investigated by the complex PCA (CPCA) in the time domain, where complex time-series are formed from the original time-series and from their Hilbert transform and complex eigenvectors are determined from the cross-covariance or cross-correlation matrices derived from the complex time-series (Horel 1984). This method is therefore used in the study of the El-Nino region.

The extension of PCA to two fields is the CCA statistical method, which identifies pairs of patterns in two fields which fulfill an optimality criterion, we find the linear combination of variables in each set that have maximum correlation (Bretherton et al. 1992). For each field we have therefore a representation in terms of spatial and temporal functions, but, while in the PCA the patterns are orthogonal and therefore appear in both the synthesis and analysis formula, in CCA the spatial functions are not orthogonal and, therefore, the spatial vector of the analysis equation does not necessarily coincide with the spatial patterns of the synthesis equation (Fenoglio-Marc 2001). The independent component analysis method (Hyvarinen et al. 2001) aims at the separation of the components of a signal.

With increasing impact to transient phenomena in geodetic applications alternative representatives deserve more interest. As a first step, the equivalence of multipole with harmonic representations is seen to find various applications in modeling the Earth's interior phenomena. Meanwhile, a large number of highly flexible trial functions serve local and global modeling processes which are basically superior to standard procedures (Groten 2003). With increasing numbers of data sources, such as tide-gauge, altimetric, gravimetric, etc. observations, the optimum estimation based on biased redundant data sets of quite different kind and origin is at stake.

Another common tool for analyzing localized variations of power within a time-serie, is the wavelet method. By breaking down a time-serie into time-frequency space, it allows to determine both the dominant modes of variability and how those modes vary in time.

7 Conclusions

The mean level is variable in time and in space. As sea level rise is a very important component of climate change, it is important to develop techniques for the estimation of sea level trends "cleaned" from decadal variability.

At least 50 year records are needed to separate secular, decadal, and interannual variations and over this long time interval only few tide gauges along the world coastlines are available for the analysis. Moreover, the spatial distribution of tide gauges as well as the existence of interannual and low-frequency signals affects the recovery of secular trends in short records.

Satellite altimetry provides high-precision, high-resolution measurements of sea surface height with global coverage and a revisit time of a few days starting from 1992, i.e., over the last 17 years.

During this time interval, the agreement between altimeter and tide gauge data is good. Global mean sea level rises by an about 3.1 ± 0.4 mm/year between January 1993 and December 2008. Coastal sea level rises at the same rate in the long-term, but with higher interannual variability. Tide gauges conveniently represent the global sea level over the long term. Regionally there is a good agreement between coastal and open-ocean sea level variability from altimeter data, and trends are positive in the main world basins.

Using altimeter data together with space gravity data we can analyze the reason for sea level change and the corresponding components.

At the present time, we are still apart from exact global solutions. Recent data from GOCE and other satellite projects may, however, soon lead to more homogeneously distributed data sets of different kind and of substantially higher accuracy. Consequently, the rigorous treatment in terms of global boundary value problem, as described at the beginning of the chapter, is still far apart. Only with more numerous and redundant datasets unique solutions can be attempted. The approach described in the present chapter will be further applied to the extended data.

References

Allen M, Smith L (1994) Investigating the origins and significance of low frequency modes of climate variability. Geophys Res Lett 21: 883–886

Beckley BD, Lemoine FG, Luthcke SB, Ray RD, Zelensky NP (2007) A reassessment of global and regional mean sea level trends from TOPEX and Jason-1 altimetry based on revised reference frame and orbits. Geophys Res Lett 34:L14608, doi:10.1029/2007GL030002

Berge-Nguyen M, Cazenave A, Lombard A, Llovel W, Viarre J, Cretaux JF (2008) Reconstruction of past decades sea level using thermosteric sea level, tide gauge, satellite altimetry and ocean reanalysis data. Global Planet Change 62: 1–13

Bindoff N, Willebrand J, Artale V, Cazenave A, Gregory J, Gulev S, Hanawa K, Le Quéré C, Levitus S, Nojiri Y, Shum CK, Talley L, Unnikrishnan A (2007) Observations: oceanic climate and sea level. In: Solomon S, Qin D, Manning M, Chen Z, Marquis M, Averyt KB, Tignor M, Miller HL (eds) *Climate change 2007: The physical science basis. Contribution of Working Group I to the fourth assessment report of the Intergovernmental Panel on Climate Change.* Cambridge University Press, Cambridge, UK, pp 385–428

Bretherton C, Smith C, Wallace J (1992) An intercomparison of methods for finding coupled patterns in climate data change. J Clim 5:541–560

Bursa M, Groten E, Kenyon S, Kuba J, Radey K, Vatrt V, Vojtiskova M (2001) Earth dimension specified by geoid potential. Studia Geophys Geodetica 46:1–8

Bursa M, Kenyon S, Kouba J, Sima Z, Vatrt V, Vitek V, Vojtiskova M (2007) The geopotential value W° for specifying the relativistic atomic time scale and a global vertical reference system. J Geod 81:103–110

Cazenave A, Nerem RS (2004) Present-day sea level change. Observations and causes, Rev Geoph 42: RG3001, doi:101029/2003RG000139

Cazenave A, Dominh K, Guinehut S, Berthier E, Llovel W, Ramillien G, Ablain M, Larnicol G (2009) Sea level budget over 2003–2008: A reevaluation from GRACE space gravimetry, satellite altimetry and Argo. Global Planet Change 65(1–2): 83–88

Chelton DB, Ries JC, Haines BJ, Fu LL, Callahan P (2001) Satellite altimetry. In: Fu LL, Cazanave A (eds) Satellite altimetry and earth sciences. Academic, New York, pp 57–64

Church JA, White NJ (2006) A 20th century acceleration in global sea-level rise. Geophys Res Lett 33: L01602, doi: 10.1029/2005GL024826

Church JA, White NJ, Coleman R, Lambeck K, Mitrovica JX (2004) Estimates of the regional distribution of sea level rise over the 1950–2000 period. J Climate 17(13): 2609–2625

Church J, White NJ, Arrup T, Wilson WS, Woodworth PL, Dominigues CM, Hunter JR, Lambeck K (2008) Understanding global sea levels: past, present and future. Sustain Sci 3:9–22, doi:/10.1007/s11625-008-0042-4

Domingues CM, Church J, White NJ, Gleckler PJ, Wijffels SE, Barker PM, Dunn JR (2008) Improved estimates of upper-ocean warming and multi-decadal sea level rise. Nat Lett 453 doi:10.1038/nature07080

Douglas BC (2001) Sea level change in the era of the recording tide gauge. In: Sea level rise, history and consequences. International Geophysics Series, vol 75. Academic, London

Fenoglio-Marc L (2001) Analysis and representation of regional sea level variability fromaltimetry and atmospheric data. Geophys J Int 145:1–18

Fenoglio-Marc L, Tel E (2010) Coastal and global sea level change, J Geodyn 49(3–4): 151–160, doi:10.1016/j.jog.2009.12.003

Fenoglio-Marc L, Groten E, Dietz C (2004) Vertical land motion in the mediterranean Sea from altimetry and tide gauge stations. Mar Geod 27:683–701

Groten E (2003) Ist die Modellbildung in der Geodäsie hinreichend zukunftstauglich? Zeitschr f Vermessungsw 3:192–195

Haines G (1985) Spherical cap harmonic analysis. J Geoph Res 90:2583–2591

Holgate SJ, Woodworth PL (2004) Evidence for enhanced coastal sea level rise during the 1990s. Geophys Res Lett 31: L07305, doi:10.1029/2004GL019626

Holgate SJ (2007) On the decadal rates of sea level change during the twentieth century. Geophys Res Lett 34: L01602, doi:10.1029/2006GL028492

Horel JD (1984) Complex principal component analysis: theory and examples. J Climate Appl Meteor 23:1661–1673

Hyvarinen A, Karhunen J, Oja E (2001) Independent component analysis. Wiley, New York

Jevrejeva S, Grinsted A, Moore JC, Holgate S (2006) Non linear trends and multiyear cycles in sea level records. J Geophys Res 111:C09012, doi:10.1029/2005JC003299

Leuliette E, Nerem RS, Mitchum G (2004) Calibration of Topex/Poseidon and Jason altimeter data to construct a continuous record of mean sea level change. Mar Geod 27: 79–94, doi: 10.1080/01490410490465193

Lombard A, Cazenave A, Le Traon P-Y, Ishii M (2005) Contribution of thermal expansion to present-day sea level change revisited. Global Planet Change 47:11–16

Meier F, Dyurgerov MB, Rick UK, O'Neel S, Pfeffer W, Anderson RS, Anderson SP, Glazovsky AF (2007) Glaciers dominate Eustatic sea level rise in the 21th century, Science 317(5841): 1064–1067, doi: 10.1126/science. 1143906

Mitchum G (2000) An improved calibration of satellite altimetric heights using tide gauge sea level with adjustment for land motion. Mar Geod 23:145–166

Overland JE, Preisendorfer RW (1981) A significance test for principal components applied to a cyclone climatology, Monthly, Weather Review, 110(1):1–4

Paulson A, Zhong S, Wahr J (2007) Inference of mantle viscosity from GRACE and relative sea level data. Geophys J Int 171:497–508, doi:10.111/j.1365-246X.2007.03556.x

Peltier WR (2004) Global glacial isostasy and the surface of the ice-age earth: the ICE-5G (VM2) model and GRACE. Ann Rev Earth Planet Sci 32:111–149

Peltier R (2009) Closure of the budget of global sea level rise over the GRACE era: the importance and magnitudes of the required corrections for global glacial isostatic adjustment. Quaternary Science Reviews 28(17–18): 1658–1674, doi:10.1016/j.quascirev.2009.04.004

Prandi P, Cazenave A, Becker M (2009) Is coastal mean sea level rising faster than the global mean? A comparison between tide gauges and satellite altimetry over 1993–2007. Geophys Res Lett 36, L05602, doi:10.1029/2008GL036564

Preisendorfer RW (1988) Principal component analysis in meteorology and oceanography. In: Developments in atmospheric science, Vol 17. Elsevier, Amsterdam

Rignot E, Box JE, Burgess E, Hanna E (2008) Mass balance of the Greenland ice sheet from 1958 to 2007. Geophys Res Lett 35: L20502, doi:10.1029/2008GL035417

Spada G, Stocchi P (2007) Selen: a Fortran 90 program for solving the sea-level equation. Comput Geosci 33:538–562

Trenberth KE (1984) Signal versus noise in the Southern Oscillation. Month Weather Rev 112:326–332

Trenberth KE, Hurrel JW (1994) Decadal atmosphere-ocean variations in the Pacific. Climate Dyn 9:303–319

White NJ, Church JA, Gregory JM (2005) Coastal and global averaged sea level rise for 1950 to 2000. Geophys Res Lett 32: L01601, doi:101029/2004GL021391

Wilks D (1995) Statistical methods in the atmospheric sciences. Academic, San Diego

Willis JK, Chambers DP, Nerem S (2008) Assessing the global averaged sea level budget on seasonal and interannual timescales. J Geophys Res 113:C06015, doi:10.1029/2007JC004517

Woolf DK, Shaw A, Tsimplis MN (2005) The influence of the North Atlantic Oscillation on sea-level variability in the North Atlantic region. Global Atmosp Ocean Syst 9:145–167, doi: 10.1080/10236730310001633803

13 Unstructured Meshes in Large-Scale Ocean Modeling

Sergey Danilov · Jens Schröter
Alfred Wegener Institute for Polar and Marine Research, Bremerhaven, Germany

W. Freeden, M.Z. Nashed, T. Sonar (Eds.), *Handbook of Geomathematics*, DOI 10.1007/978-3-642-01546-5_13,

Abstract The current status of large-scale ocean modeling on unstructured meshes is discussed in the context of climate applications. Our review is based on FEOM, which is at present the only general circulation model on a triangular mesh with a proven record of global applications. Different setups are considered including some promising alternative finite-element and finite-volume configurations. The focus is on consistency and performance issues which are much easier to achieve with finite-volume methods. On the other hand, they sometimes suffer from numerical modes and require more research before they can be generally recommended for modeling of the general circulation.

1 Introduction

The world ocean is an important component of the Earth's climate system. It essentially contributes to the poleward heat transport and influences dynamics on time scales longer than those of typical weather systems. Numerical models of general ocean circulation present a necessary building block of climate system models. The rapid development of computer technology, advancement in atmospheric modeling, and public demand make ocean modeling a fast developing field of research. The accepted technology almost exclusively uses regular meshes and finite-difference method as the numerical approach. The "large scale" ocean circulation consists of a complex structure of vigorous jet streams (e.g., the Gulf Stream and its extension in the North Atlantic) connected to large sluggish gyres of basin scale. Boundary layers and forcing determine the structure of the flow field, which has a large inertia and evolves only slowly in time. Water masses far from boundaries retain their properties like heat content and chemical composition for hundreds of years. The complexity of coastlines and bottom topography as well as the conservation principles make ocean general circulation modeling a challenging task. The need to resolve narrow passages or simply local refinement of the mesh is commonly solved by brute force, i.e., the global ocean is resolved finer and finer at the cost of increasing the required computational power. However, Dupont et al. (2003) have shown that a simple increase in coastal resolution does not lead to improved dynamics, which in one of the major motivations for using boundary following meshes.

These obstacles make technologies formulated on unstructured meshes very appealing. Indeed, the ability of unstructured meshes to adjust the resolution to specific needs and to capture specific dynamics is their most salient feature.

Driven by this idea there is growing interest toward the development of efficient unstructured-mesh ocean circulation models. Tidal and coastal modeling presents the direction where this effort is already enjoying an obvious success. The size of elements on an unstructured mesh can follow the phase speed of surface gravity waves, $c = (gH)^{1/2}$, where g is the acceleration due to gravity and H the fluid depth. In this way meshes that are uniform in terms of CFL (Courant–Friedrichs–Lewy) criterion can be designed. They allow a natural focus on shallow coastal zones without need for nesting. Numerous applications vividly show benefits introduced by unstructured meshes in this domain (see, i.e., Lynch et al., 1996, 1997; Carrère and Lyard 2003; Chen et al. 2003; Yang and Khangaonkar 2009; Wang et al. 2009a; Zhang and Baptista 2008; Jones and Davies 2008; Hervouet 2000; Walters 2005). Although tidal dynamics are largely barotropic, some of the studies cited above are performed with full 3D models.

The progress in modeling the large-scale ocean circulation on unstructured meshes is slower for several reasons, which include computational load, complexity of dynamics, and the impact of local refinements on global dynamics, details of which are as yet far from being understood. The challenge here is in providing seamlessly working regionally focused setups embedded in an otherwise global coarse mesh. Timmermann et al. (2009) present for the first time results of global ocean simulations performed with an unstructured-mesh finite-element global sea-ice–ocean model, and examples of applications with regional focus are starting to appear (Wang et al. 2009b, 2010).

This review sees its goal in briefly presenting the unstructured mesh approaches, focused at simulating the large-scale ocean circulation, which are either used or were tested by us. A numerous class of barotropic shallow water models, which has rich literature, deserves a special review. Their focus is wave and tidal dynamics whereas the large-scale ocean circulation is dominated by slow dynamics of temperature and salinity fields governed by advective-diffusive processes. Two comprehensive reviews by Pain et al. (2005) and Piggott et al. (2008) highlight potentialities of unstructured mesh methods as applied to ocean modeling in a broader context. They are universally recommended. The status of models formulated on regular meshes is presented in Griffies et al. (2000), while the books by Haidvogel and Beckmann (1999) and Griffies (2004) clarify numerous issues, both fluid dynamical and numerical, which are relevant to the topic.

At present FEOM (Danilov et al. 2004; Wang et al. 2008) is the only general circulation model on unstructured meshes with reasonable performance on integration times of decades (see Timmermann et al. 2009). In the following we will explain its numerics and describe a suite of alternative approaches. All approaches have their own attractions. We give reasons why some of them appear rather appealing and should be subject of future research. By the way of illustration ➊ *Fig. 1* presents the sea surface elevation computed with the North Atlantic setup of FEOM focused on the Carribean basin, Gulf Stream and its extention. The resolution is varying from about 20 km there to about one degree in the rest. The Gulf Stream meanders and vortices are seen in the well resolved part of the mesh. ➊ *Figure 2* zooms into the Loop Current sector of the Gulf Stream to illustrate the formation of an anticyclonic eddy there. The dynamics are much more damped over the coarse part of the mesh, and more than half of mesh nodes are in the fine part occupying only a fraction of total area.

One of the major difficulties in this subject is, in our experience, the communication between mathematicians and oceanographers. Differences in background, language, and approach sometimes lead to misunderstandings. Oceanographers are forced to make the governing Navier–Stokes fluid dynamic equations tractable analytically and numerically. A whole suite of approximations is common (see, e.g., Haidvogel and Beckmann 1999; Griffies 2004; Griffies and Adcroft 2008). The resulting equations then appear reasonably simple and well behaved. It should be understood, however, that the properties of the approximated system of equations are not necessarily also those of the general ocean circulation.

The next two sections contain a brief description of primitive equations (PE) commonly used in ocean modeling and an introduction to finite-element (FE) and finite-volume (FV) methods used on unstructured meshes; they should facilitate further reading. They are followed by a section on consistency requirements of FE approach and two sections on comparative analysis of two FE and two FV setups. We hope that in this way advantages and difficulties of various approaches can be illustrated most easily.

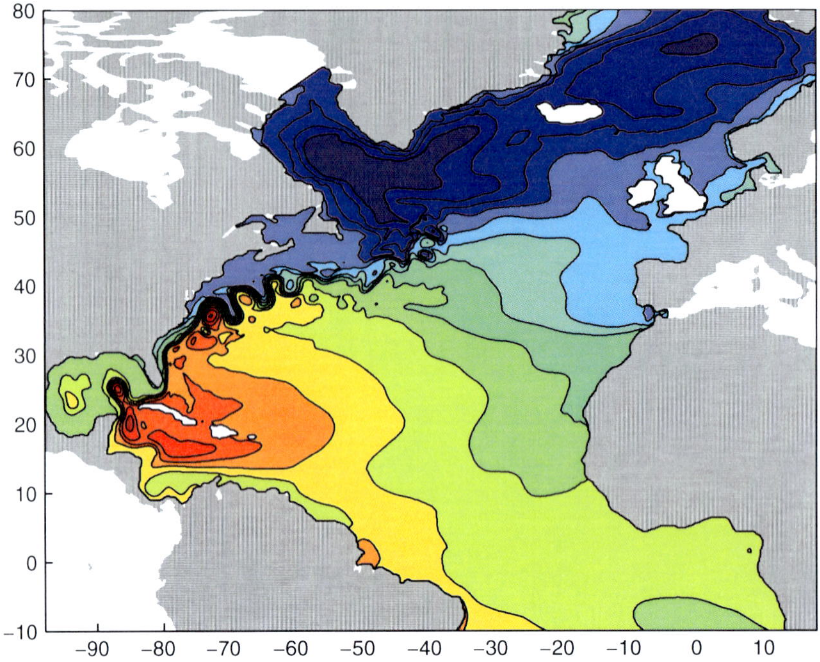

Fig. 1
A snapshot of the sea surface elevation in the North Atlantic (NA) setup of FEOM with focus on the Carribean basin, Gulf Stream, and its extention. Meanders of the Gulf Stream are seen over the well-resolved part of the mesh. The resolution varies from about 20 km in the region of interest and in the vicinity of coastlines to about 120 km in the open ocean. The coarse part of the mesh takes less than half of mesh nodes and is computationally inexpensive. The contour line interval is approximately 11 cm. The setup is relaxed to monthly mean climatology at open boundaries and the surface

2 Dynamic Equations and Typical Approximations

Ocean dynamics on the rotating Earth are described by the standard set of fluid dynamic equations under several approximations. They include the Boussinesq approximation, according to which the full density is replaced by the reference density everywhere except for the gravity term in the momentum equations, and the equation of state. Next, the thin layer approximation acknowledges that the ocean is thin compared to the Earth's radius, so that an elementary volume in spherical coordinates $d\Omega = r^2 \sin\theta dr d\phi d\theta \approx r_e^2 \sin\theta dz d\phi d\theta$. The Earth is assumed to be of spherical shape. The dynamical equations in the Boussinesq and thin layer approximations take the form

$$\partial_t \mathbf{v} + \mathbf{v} \cdot \nabla_3 \mathbf{v} + 2\Omega \times \mathbf{v} + \frac{1}{\rho_0} \nabla_3 p = \mathbf{g} \frac{\rho}{\rho_0} + \nabla \cdot A_h \nabla \mathbf{v} + \partial_z A_v \partial_z \mathbf{v}, \tag{1}$$

$$\nabla_3 \mathbf{v} = 0, \tag{2}$$

where $\mathbf{v} \equiv (\mathbf{u}, w) \equiv (u, v, w)$ represent velocity in the spherical coordinate system, ρ_0 and ρ are the mean density and the deviation from it, respectively, p is the pressure, Ω the earth rotation

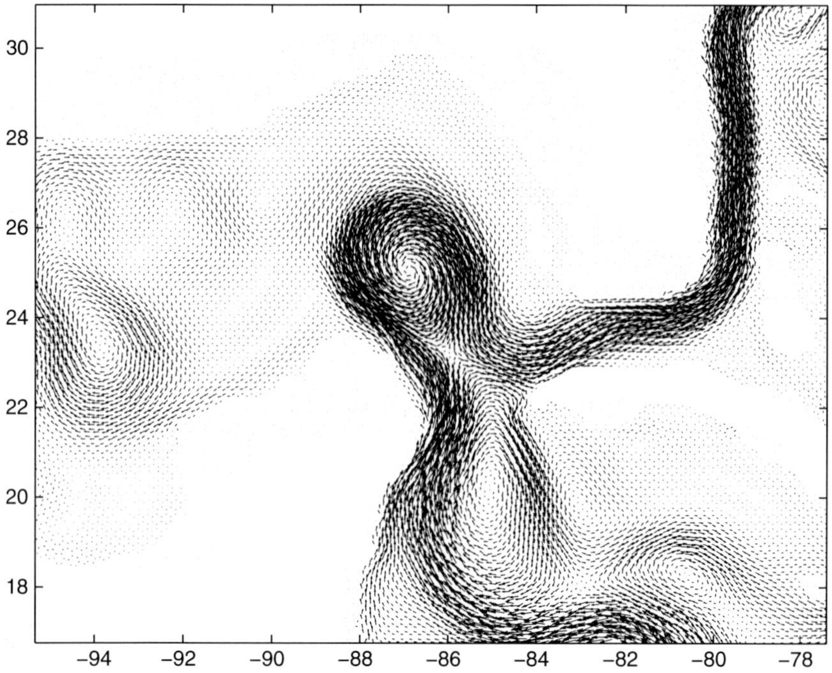

Fig. 2

The horizontal velocity field at 75 m in the Loop Current sector showing the formation of an anticyclonic eddy at the Yucatan and the narrow Florida Current

vector, the lateral and vertical viscosities are denoted by A_h and A_v, $\mathbf{g}$ stands for gravitational acceleration, and ∇_3 and ∇ denote 3D and 2D gradient or divergence operators, respectively. The viscosity is thought of as parameterizing the influence of unresolved scales. It exceeds the molecular viscosity by many orders of magnitude.

In most cases of oceanographic interest the vertical component of momentum equation can further be simplified by noting that vertical acceleration is typically small compared to the acceleration due to gravity. This leads to the hydrostatic approximation

$$\partial_z p = -g\rho, \tag{3}$$

which is assumed below. For the energetic consistency, only the part of the Coriolis force due to the projection of the Earth's angular rotation on the local vertical can be left in the horizontal momentum equations. The omitted terms contain vertical velocity and are therefore small. The resulting momentum equation takes the form

$$\partial_t \mathbf{u} + \mathbf{v} \cdot \nabla_3 \mathbf{u} + f\mathbf{k} \times \mathbf{u} + \frac{1}{\rho_0} \nabla p = \nabla \cdot A_h \nabla \mathbf{u} + \partial_z A_v \partial_z \mathbf{u}, \tag{4}$$

where $\mathbf{k}$ is the local unit vertical vector and $f = 2\Omega \sin\theta$ the Coriolis parameter.

The hydrostatic approximation limits the applicability of dynamical equations. In particular, the propagation of short surface gravity waves and internal gravity waves cannot be properly described. Some of the current ocean circulation models are suggesting a full nonhydrostatic option (MITgcm (Marshall et al. 1997), SUNTANS (Fringer et al. 2006), ICOM (Ford et al.

2004), FEOM), but for the typical applications of large-scale ocean circulation models this is hardly needed. In order to keep both hydrostatic and nonhydrostatic options within the same model the nonhydrostatic part is organized as a perturbation to the hydrostatic one (MITgcm, SUNTANS, FEOM).

In the hydrostatic approximation the total pressure can be integrated to give

$$p = p_a + g\rho_0\eta + p_h, \quad p_h = \int_z^0 g\rho \, \mathrm{d}z. \tag{5}$$

Here p_a is the atmospheric pressure at the free surface, p_h is the pressure due to changes in the weight of the fluid column between $z = 0$ and current depth, and $g\rho_0\eta$ is the part of pressure due to the free surface elevation. In computing it one routinely uses the reference density ρ_0 instead of true density. The incurring error is small. In the hydrostatic approximation only this part of pressure serves as a gauge field and is determined from the vertically integrated continuity equation. Integration of (2) with respect to z gives (see also ➋ Chap. 1)

$$\partial_t\eta + \nabla \cdot \int_{z=-H}^{z=\eta} \mathbf{u} \, \mathrm{d}z = P - E. \tag{6}$$

Surface mass sources $P - E$ (precipitation minus evaporation) are added to the right-hand side of this equation. We emphasize that (6) is not an independent equation.

If the upper integration limit in (6) is set to 0, η is neglected compared to the ocean thickness, giving the linear free surface approximation. This is in many cases a sufficiently accurate option. Neglecting additionally the time derivative is called the rigid lid approximation. The full option is referred to as the nonlinear free surface. The linear free surface and rigid lid approximations imply that the mesh displacement is neglected, which simplifies further computations.

The ocean is bounded by boundaries, and one distinguishes the ocean surface, Γ_1, the bottom, Γ_2, the vertical (lateral) rigid walls Γ_3, and the open boundary Γ_4.

Wind and frictional stresses are applied at the surface and bottom respectively: $A_v\partial_z\mathbf{u} = \boldsymbol{\tau}$ on Γ_1 and $A_v\partial_z\mathbf{u} + A_h\nabla H \cdot \nabla\mathbf{u} = C_\mathrm{d}\mathbf{u}|\mathbf{u}|$ on Γ_2, where $\boldsymbol{\tau}$ is the wind stress, and C_d is the bottom drag coefficient. Lateral walls are impermeable, and can allow either free slip or no slip. Implementation of open boundary conditions is delicate and depends on applications (see, e.g., Marchesiello et al. 2001).

The vertical velocity is diagnosed via the continuity equation:

$$\partial_z w = -\nabla \cdot \mathbf{u}, \tag{7}$$

for which the kinematic boundary conditions at the surface and bottom are:

$$w = \partial_t\eta + \mathbf{u} \cdot \nabla\eta - (P - E),$$

and

$$w = -\nabla H \cdot \mathbf{u} \text{ on } \Gamma_2.$$

They are not independent but linked through (6).

The thermodynamical part consist of equations for potential temperature T and salinity S, and density anomaly ρ is computed via the full equation of state (Jackett and McDougall 1995)

$$\partial_t C + \mathbf{v} \cdot \nabla_3 C - \nabla_3(\mathrm{K}\nabla_3 C) = 0, \tag{8}$$

$$\rho = \rho(T, S, p), \tag{9}$$

where C can be T or S or any tracer, and K is the second-order diffusivity tensor. It can in the general case contain symmetric and antisymmetric part. The former corresponds to ordinary diffusivity (reduces the variance of a tracer field). The latter parameterizes unresolved eddy-induced transport, and is commonly coming through the Gent–McWilliams parameterization (Gent and McWilliams 1990). A detailed exposition can be found in Griffies (2004).

Temperature and salinity can be modified through the surface fluxes. The bottom and lateral walls are insulated. At the open boundary T and S are commonly restored to climatological values.

The main concern with T and S advection–diffusion is to prevent spurious diapycnal mixing associated with either the explicit diffusivity or with implicit diffusivity hidden in the implementation of the advective part of the operator. Tackling the first part of the problem is relatively easy and is achieved through the choice of K constraining mixing to isopycnals planes (planes of constant potential density). The second part of the problem presents the major difficulty for models that do not use isopycnal coordinates. Since residual errors of discrete operators are as a rule larger on unstructured meshes, the performance of advection schemes on unstructured meshes calls for dedicated efforts.

A detailed explanation of approximations and underlying principles is contained, among others, in Griffies (2004) and Griffies and Adcroft (2008).

3 Finite-Element and Finite-Volume Methods

These are the two main methods to treat partial differential equations on unstructured meshes. Typical elements used in ocean modeling are triangular prisms, tetrahedra, or hexagons. Commonly, only the surface mesh (triangular or formed of quadrilaterals) is unstructured whereas nodes are vertically aligned in the body of fluid. Such a choice is dictated by the fact that the ocean is strongly stratified in the vertical direction. Vertical density gradients alias horizontal gradients of pressure on fully unstructured meshes unless a prohibitively fine vertical resolution is used. The well-known problem of pressure gradient errors on meshes with bottom following vertical coordinate has similar roots. This limits in practice the freedom of unstructured meshes to horizontal directions only.

The three-dimensional discretization is obtained by first creating vertical prisms based on the surface mesh and splitting them into smaller prisms by a set of level surfaces. The latter may follow geopotential surfaces (z-coordinate) or bottom topography, there are no limitations other than the admissible pressure gradient errors in the case the levels deviate from the geopotential ones. The bottom following coordinate (sigma-coordinate) is a popular choice in coastal or regional modeling. The common practice of regular-mesh models is rewriting equations with respect to a new set of coordinates (ϕ, θ, σ) where $\sigma = \sigma(\phi, \theta, z, t)$ represents mapping of total water column into interval $[0, 1]$. FV models using sigma-coordinate (like FVCOM, Chen et al. 2003) and some FE models (like TELEMAC, Hervouet 2007) perform this transform. Alternatively, with FE method, one can deal with deformed elements by applying quadrature rules (like FEOM, Wang et al. 2008, or SLIM, White et al. 2008a). In this way the same numerical code can work with any system of levels. The major difference between these two approaches is that horizontal derivatives are computed at constant σ in the former case and at constant z in the latter.

The tetrahedral discretization is used exclusively with the FE method. Tetrahedra are obtained through additionally splitting triangular prisms (a regular prism gives three tetrahedra, but the number reduces to two or one if the prism is cut by bottom topography).

The motivation behind this (almost threefold) increase in the number of elements is that one works with similar elements independent of the choice of level surfaces, which simplifies the implementation.

The sections below serve the introductory purpose. The textbooks by Zienkiewicz and Taylor (2000) or Donea and Huerta (2003) are recommended with respect to the FE method, whereas answers to many questions related to the FV method can be found in Blazek (2001).

3.1 FE Method

The finite element method begins with choosing functional spaces to represent fields on elements of computational mesh. A widely used choice on triangular meshes is a continuous linear representation (P_1, P stands for polynomial, and one for its order), and we start with it. In this case the basis function N_j at node j equals one at the node, goes linearly to zero at neighboring nodes, and stays zero outside the stencil formed by neighboring elements. To simplify the notation we will use Cartesian coordinates (x, y, z) in what follows remembering that full spherical geometry is used in numerical implementations. In the case of tetrahedral representation $N_j = N_j(x, y, z)$. For regular triangular prisms $N_j = N_j(x, y)N_j(z)$, and trilinear representation is used for hexagons. When prisms or hexagons are deformed, this representation is valid in the so-called parent space in which shape of elements is regular. A transform is always performed from the parent to physical space. Any field (for definiteness we will start from tracer C) is expanded in basis functions

$$C(x, y, z, t) = C_j(t)N_j(x, y, z),$$

where summation is implied over repeating indices (the expansion corresponds to linear interpolation). The original Eq. (8) is projected on an appropriate set of test functions M_i, and the term with the second derivatives is integrated by parts:

$$\int M_i \partial_t C \,\mathrm{d}\Omega + \int M_i \mathbf{v} \nabla_3 C \,\mathrm{d}\Omega + \int \nabla_3 M_i \mathrm{K} \nabla_3 C \,\mathrm{d}\Omega = \int M_i q_C \,\mathrm{d}\Gamma_1,$$

where q_C is the flux of tracer C from the ocean through the surface. Substituting the expansion for C one gets a matrix problem

$$\mathrm{M}_{ij}\partial_t C_j + \mathrm{A}_{ij} C_j + D_{ij} C_j = Q_i,$$

where $\mathrm{M}_{ij} = \int M_i N_j \,\mathrm{d}\Omega$ is the mass matrix, $\mathrm{A}_{ij} = \int M_i(\mathbf{v}\nabla)N_j \,\mathrm{d}\Omega$ the advection matrix, and $\mathrm{D}_{ij} = \int \nabla M_i \mathrm{K} \nabla N_i \,\mathrm{d}\Omega$ the diffusion matrix. In many cases the test and basis functions coincide which results in the Galerkin discretization. The space of test functions can be modified to ensure that the equations are stabilized in advection-dominated regimes. The choice $M_i = (1 + \varepsilon(\mathbf{v}\nabla))N_i$, where ε is typically taken as constant on elements, gives the Petrov–Galerkin method. The way how ε is selected influences the properties of transport schemes (see, e.g., Donea and Huerta 2003, for an elementary discussion).

For continuous functions (like P_1) M_{ij} is not diagonal. Although this favorably influences the dispersive properties of transport schemes, it demands solution of large systems of matrix equations which slows down the codes. Additionally, it imposes stronger constraints on the time step size (see, e.g., Donea and Huerta 2003). The diagonal (lumped, or nodal quadrature) approximation can be made to reduce the problem to that typical of finite difference or FV approaches. Computations are routinely organized by cycling through elements.

Returning to the primitive equations the major issue becomes the choice of spaces for the horizontal velocity and scalars (elevation, vertical velocity, temperature, salinity, and pressure should use consistent spaces in the hydrostatic approximation, as will be explained further). This choice influences the overall stability (presence or absence of numerical modes), accuracy, and representation of particular balances (like geostrophy) or processes (like wave propagation). There is ongoing research aimed at analyzing the properties of FE pairs with respect to ocean modeling tasks, but in many cases it remains limited to shallow water equations (exploring their dispersive properties with respect to surface gravity and Rossby waves, see, e.g., Le Roux et al. 2005, 2007). This is helpful. On the other hand, it leaves many 3D issues unexplored, which may partly explain why the existing FE 3D primitive equation ocean models are rather conservative in their selection of spaces.

QUODDY (Lynch et al. 1996), ADCIRC (Westerink et al. 1992), and FEOM (Danilov et al. 2004; Wang et al. 2008; Timmermann et al. 2009) use $P_1 - P_1$ discretization (the space for the horizontal velocity is listed first). TELEMAC (Hervouet 2007) applies a quasi bubble element to approximate the velocity and stays with P_1 for scalars. ICON (Ford et al. 2004) started from $Q_1 - Q_0$ on hexagons (Q_1 denotes trilinear, and Q_0, elementwise constant representations on hexagons) but uses currently $P_1 - P_1$ on tetrahedra (Piggott et al. 2008). The SLIM project begun with the non-conforming P_1^{NC} discretization for the horizontal velocity and P_1 discretization for scalars in horizontal directions, augmented with discontinuous linear representation in the vertical direction (White et al. 2008a), and was limited to uniform density cases. The new direction works with P_1 discontinuous representation (Blaise 2009) and full primitive equations. FEOM followed SLIM in using nonconforming functions, so that a parallel NC branch exists now (Danilov et al. 2008). SELFE (Zhang and Baptista 2008) uses P_1 for elevation, but applies FV methods in the rest.

The $P_1 - P_1$ pair is known to lead to rank deficiency of the gradient operator, so that the codes using it require stabilization. The $Q_1 - Q_0$ pair may lead to similar problems, although it is seldom found in practice. The nonconforming P_1^{NC} functions on triangles (Hua and Thomasset 1984) are associated with edges where they are equal to one, and go linearly to -1 at the opposing node. The advantage brought by them is the diagonal character of mass matrices and stability. The quasi bubble element of TELEMAC includes an additional node at the element center so that the element is split into three smaller ones, with linear representation on them. This helps to overcome the stability issue.

A particular branch of FE models relies on RT0 (Raviart and Thomas (1977)) element which leads to a discretization very close to that given by triangular C-grid (explained below), apart from the mass-matrix part.

This list cannot be complete, but one may notice that the low-order elements are indeed preferred. RT0 and P_1^{NC} elements mentioned here are discontinuous, and formal derivation of weak equations in their case requires adding continuity penalties. These aspects will not be addressed here.

3.2 Finite Volumes

The FV method derives discretized equations by introducing control volumes. They may coincide with the elements of the numerical mesh, but it is not necessarily required. However, the union of all control volumes must provide the tessellation of the domain. The most commonly used placement of variables on triangular meshes (when viewed from the surface) is at centroids (control volumes are based on triangles) or nodes (so-called median-dual control volumes). The common placement of variables in vertical is at mid-levels, so that they characterize the respective control volumes, and only vertical velocity is at full levels. The equations of motion are integrated over control volumes and a time derivative of a quantity is expressed, via the Gauss theorem, in terms of fluxes through the faces of the control volume. Due to this strategy, local and global balances are ensured on the discrete level. To illustrate the FV method it is applied again to the tracer transport Eq. (8) with cell-centered placement of variables. Integrating it over prismatic element (cell) c one obtains

$$\partial_t \int C\,\mathrm{d}\Omega_c + \sum_{j=1}^{5} \mathbf{F}_{jc}\mathbf{n}_{jc} = 0. \tag{10}$$

Here c enumerates the elements, and j the faces for each element; $\mathbf{n}_{jc}$ is the outer normal on the face j of prism c. The discrete tracer values are thought to represent mean over volume quantities: $C_c = \int C\,\mathrm{d}\Omega_c / \int \mathrm{d}\Omega_c$. The essence of the FV approach is in estimating fluxes. The tracer field is unknown on the face, and its reconstruction has to be performed to accurately estimate variables and fluxes at the faces. Linear reconstructions are most popular as they only require information from the nearest neighbors. The reconstructions on two elements across a common face do not necessarily coincide at this face, and so do the direct estimates of the fluxes. To properly tackle these discontinuities one introduces the concept of numerical fluxes that are commonly upwind-biased and limited so as to endow the transport equation with desired properties like monotonicity. Computations are organized in a cycle over faces, and in most cases they are less CPU demanding than with FE.

Among FV based models the branch using triangular C-grid is most numerous and we mention UnTRIM (Casulli and Walters 2000), SUNTANS (Fringer et al. 2006), and the generalization of the method to spherical geometry by Stuhne and Peltier (2006). In this case scalars are placed at circumcenters (viewed from above) and employ prisms as control volumes. The normal component of horizontal velocity is stored at midpoints of vertical faces. The approach used to discretize the momentum equation is a mixture of finite difference and finite volumes.

The method employed by FVCOM (Chen et al. 2003) uses centroids to store velocities and nodes to store scalars (in terms of surface geometry).

3.3 Discontinuous FE

Although ocean circulation models employing discontinuous FE are only appearing (Blaise 2009; the ADCIRC development, see Dawson et al. 2006, should be mentioned too, but thus far it is related to shallow water equations) and will not be discussed here extensively, we briefly outline their main features. One gets the weak formulation by integrating over elements interiors and adding constraints on the fluxes or supplying rules to compute them, as in the FV method. In this case the result is

$$\sum_c \Big(\int (M_i \partial_t C - \mathbf{F}\nabla M_i)\,\mathrm{d}\Omega_c + \int \mathbf{F}\mathbf{n} M_i\,\mathrm{d}\Gamma_c \Big) = 0, \tag{11}$$

where c numbers the elements (cells), volume integration is limited to elements, and Γ_c denotes the boundary of element c. The elements are interconnected through the fluxes, and it is where the FV realm begins.

The new aspect brought by discontinuous Galerkin (DG) FEs is that one can use high-order polynomial representations inside elements. Compared to the FV method this spares one the need of reconstructions required to model advection with high accuracy. High-order representation inside elements allows implementing high-order upwind methods, which is a very valuable feature for parallelization. In contrast to the continuous Galerkin method the mass matrices on discontinuous elements are local and connect only elemental degrees of freedom. They can directly be inverted. The technical difficulty, however, is that these features are achieved through a substantially increased number of degrees of freedom. For instance DG P_1 or P_2 representations on prismatic elements already involve approximately 12 respectively 36 times more degrees of freedom than the number of mesh nodes. Some implementations of DG FE are very close to spectral elements (see, e.g., Zienkiewicz and Taylor 2000, for a general introduction). Successful application of high-order spectral elements has been shown by Iskandarani et al. (2003).

3.4 Brief Summary

It is difficult to give preference to a particular methodology without looking into details of its performance in practical tasks. On the general level, the discontinuous Galerkin method, although very appealing mathematically, is computationally expensive. That is why more efficient low-order FE or FV methods still dominate. The other rationale for using low-order methods is the geometrical complexity of ocean basins. It suggests to invest the available resources in resolving the geometrical structure of the ocean. It is commonly recognized that FV codes are computationally less demanding than FE codes.

Instead of discussing advantages and disadvantages of FE and FV implementations which follow from their mathematical structure, we briefly present, in the next sections, the structure of several approaches used by us. They will highlight practical aspects relevant to large-scale ocean modeling. We begin with explaining space consistency requirements for the FE case. Similar requirements hold for FV case, but they are naturally taken into account through staggering of variables.

4 FE Consistency Requirements

The finite-element method operates with variables belonging to the selected functional spaces. The spaces, however, cannot be chosen independently. We have already mentioned one aspect of this – the rank deficiency of gradient operator and need for stabilization if equal interpolants are used for velocity and pressure ($P_1 - P_1$ case).

4.1 Consistency Between Elevation and Vertical Velocity

The hydrostatic approximation remains a preferred choice in large-scale ocean modeling. It imposes a strong constraint on consistency of functional spaces used to represent the model fields. Assume the functional space for elevation be selected. This immediately implies that the same functional space has to be selected for the vertical velocity w. Indeed, projecting the vertical velocity (divergence) equation on the test functions,

$$\int N_i \partial_z w \, d\Omega = -\int N_i \nabla \mathbf{u} \, d\Omega,$$

and performing integrations by parts, one gets

$$-\int w \partial_z N_i \, d\Omega + \int N_i \partial_t \eta \, d\Gamma_1 = \int \mathbf{u} \nabla N_i \, d\Omega.$$

Here the second integral on the LHS is over the surface ($P - E$ is omitted for brevity), while the integral over the bottom drops out because the bottom is impermeable for the full velocity. Now, summing over vertically aligned nodes (which amounts to replacing 3D test functions with their horizontal projections $N_i = N_{i_s}(x, y)$, where i_s is the index of respective surface node) one obviously arrives at the equation for the elevation

$$\int N_{i_s} \partial_t \eta \, d\Gamma_1 = \int \mathbf{u} \nabla N_{i_s} \, d\Omega,$$

which can be considered as a solvability condition. Thus, the horizontal part of N_i should belong to the functional space selected for the elevation. If, for example, the elevation is interpolated with P_1 functions, then P_1 representation has to be employed for w on tetrahedral elements, and $P_1(x, y) f(z)$ (where $f(z)$ is some convenient vertical representation), on prismatic elements.

An analogous rule that elevation and w be consistently collocated goes without saying in finite-volume or finite-difference codes. Failure to observe this rule spoils the consistency between dynamics (which depends on the elevation) and volume conservation which is unacceptable for long-term ocean modeling.

There are persisting attempts to break this rule (exemplified, for example, by SELFE). Using such models should be limited to short-term coastal applications, and it is so indeed.

4.2 Consistency of Tracer Spaces and Tracer Conservation

It is expected that tracers are conserved by model dynamics. In hydrostatic models maintaining this property hinges on the consistency between w and tracers. Let us now explain how this works.

Consider only the tracer advection term assuming that the mesh is fixed (does not change in time). The term,

$$A_i = \int N_i (\mathbf{u} \nabla + w \partial_z) C \, d\Omega,$$

when summed over test functions transforms to

$$A = \sum A_i = \int (\mathbf{u} \nabla + w \partial_z) C \, d\Omega$$

if Dirichlet boundary conditions are not applied (because sum of all N_i is one in this case; no tracer conservation is expected if the Dirichlet boundary conditions are prescribed). Now realize that equation one solves to determine w, apart from surface integrals due to $\partial_t \eta$, the equation $A = 0$, because $C = C_j N_j$. If the surface term due to $\partial_t \eta$ were absent (as is the case in rigid lid approximation), one would get $A = 0$ provided that N_j used to represent the tracers and test functions in the equation for w are coinciding, because the integral above would be zero for each N_j. Taking a linear combination of such integrals to recover $C = C_j N_j$ does not change the answer. Thus the functional space selected for w defines the representation for tracers.

The presence of surface integrals destroys the true global tracer conservation, which is a known consequence of using linear free surface (fixed mesh), and is shared by all models working in this approximation. There are always uncontrolled fluxes $\int N_i C \partial_t \eta \, d\Gamma_1$ which serve to communicate the tracer C between the volume of fluid below surface $z = 0$ (used in the model) and the layer between $z = 0$ and $z = \eta$. To eliminate these fluxes either rigid lid option should be used or the model must be equipped with full nonlinear free surface (implying that at least the surface of upper cells of the computational mesh is moving). In practice, however, this issue is not critical unless the ocean circulation model is run in a coupled mode with an atmospheric model for long periods of time. See White et al. (2008a) for an FE implementation of nonlinear free surface.

There is one more requirement frequently discussed in connection to the advection schemes. The term A_i should be zero for a uniform tracer field, which is trivially observed.

Now let us look at the conservative advection formulation. In this case, after integration by parts, one has

$$A_i = -\int (\mathbf{u}\nabla N_i + w\partial_z N_i) C \, d\Omega + \int \partial_t \eta C N_i \, d\Gamma_1.$$

The global conservation (again up to the surface flux term) is obvious as the volume term drops out after summing over i. The consistency requirement follows from looking at what happens if $C = \text{const} = 1$. In this case

$$A_i = -\int (\mathbf{u}\nabla N_i + w\partial_z N_i) \, d\Omega + \int \partial_t \eta N_i \, d\Gamma_1.$$

It goes to zero if spaces used to project equation for w and to represent tracers coincide, because in this case A_i is exactly an i equation solved when determining w. White et al. (2008b) discuss this aspect in more detail presenting examples of errors occurring when consistency requirement is violated.

Thus two forms of the tracer advection, the nonconservative and conservative ones, are both possessing necessary properties, but for different reasons. In both cases, however, these properties are rooted into the fact that the same functional spaces have to be used to represent tracers and to solve the continuity equation for w. There is no freedom in selecting spaces for tracers independently of the dynamical part in hydrostatic models.

4.3 Energetic and Pressure Consistency

The work of horizontal pressure forces in the hydrostatic approximation is the way the available potential energy is transferred into the kinetic energy. Numerical schemes should respect this property on the discrete level, which imposes a set of additional consistency requirements on the spaces of functions used and on the way the pressure is computed. Indeed, taking a product of pressure gradients with horizontal velocity and integrating it over the model domain one has

$$\int \mathbf{u} \cdot \nabla p \, d\Omega = -\int p \nabla \cdot \mathbf{u} \, d\Omega = \int p \partial_z w \, d\Omega = -\int \partial_z p w \, d\Omega = \int g\rho w \, d\Omega. \qquad (12)$$

Here we assume for simplicity that lateral walls are rigid. The first equality holds always at the discrete level because the momentum equation is projected on the same functional space as that used for velocities. The second and third require that the pressure is in the space of test functions used to compute w. The forth equality then sets the rule for computing the pressure from the hydrostatic relation. This implies that the pressure has to be obtained as

$$-\int \partial_z p N_i \, d\Omega = \int g\rho N_i \, d\Omega, \tag{13}$$

where N_i are the functions used to expand w. Any deviations from this scheme will lead to inconsistency in the conversion between available potential and kinetic energies.

This way of pressure computation, although demanded by energy conservation, seldom works properly in practice unless a discontinuous representation is selected. In the case of continuous functions N_i the existing horizontal connections leave a mode in p which makes its horizontal derivatives too noisy to be used in oceanographically relevant situations. This difficulty is shared to some extent by all finite-element models based on continuous elevation and is in essence rooted in the structure of the ocean stratification which is much stronger in the vertical direction. One possible way of overcoming it on prismatic meshes lies in applying *horizontal* lumping in the operator parts of equations on w and p, which reduces them to sets of horizontally uncoupled equations for every column of nodes (but note that continuous vertical functions are immediately leading to problems since they lead to matrices with zeros at diagonals).

The horizontal lumping is the preferred option of solving for w in FEOM on prismatic meshes (Wang et al. 2008), and goes through the introduction of potential ϕ with $w = \partial_z \phi$, which formally translates the problem to the second order. White et al. (2008a) do the horizontal lumping too, and use the representation discontinuous in z to alleviate the other difficulty. On tetrahedral meshes, when the potential is introduced, the operator part connects only vertically aligned nodes. The problem, however, remains, this time due to the difference in stencils used to estimate the RHS.

Horizontal lumping on prismatic meshes still does not remove all practical difficulties. The most prominent of them is the pressure gradient error appearing when the vertical levels deviate from geopotential surfaces. Solving this problem through spline interpolations requires to sacrifice the energetic (space) consistency and introduces imbalances in the energy conversion. Wang et al. (2008) present details of the FEOM algorithm, which is largely based on finite differences in the vertical direction; this can be compared with the algorithm of Ford et al. (2004) who split pressure into two contributions belonging to different spaces.

Energetic consistency can only be gained by employing horizontally discontinuous elements (and using Discontinuous Galerkin method) for elevation, tracers, and pressure, which still remains a challenge to modeler's community mostly because of strongly increasing numerical costs. Finite volume and difference methods turn out to be more practical in this respect.

It should be noted that imbalances due to pressure representation are in reality combined with those introduced by time stepping. These latter errors are common to many practical algorithms of solving primitive equations (see discussion in Stuhne and Peltier (2006)). The errors introduced by pressure representation are not necessarily damaging if they remain on the level of energy dissipation due to explicit and implicit (numerical) viscosity. Nevertheless their presence destroys the mathematical beauty of the FE method, as in reality one does not follow the variational formulation rigorously.

5 Nonconforming and Continuous Linear Representations in FEOM

5.1 Preliminary Remarks

The continuous linear representation for the horizontal velocity and scalars (P_1-P_1 discretization) is an obvious choice because it requires minimum memory storage and results in a reasonable number of operations for assembling the RHSs of model equations. Its notorious difficulty is spurious pressure (elevation) modes. These are eliminated by different stabilization techniques such as the Generalized Wave Equation method (used by QUODDY, ADCIRC), Galerkin Least Squares method (Codina and Soto 1997; Danilov et al. 2004) or pressure (elevation) split method (Zienkiewicz and Taylor 2000; Codina and Zienkiewicz 2002; Wang et al. 2008). All methods share the drawback that the horizontal velocity field satisfies a modified, as opposed to exact, vertically integrated continuity equation. The last of them, however, allows a consistent treatment of continuity and tracers which explains why it is selected by Wang et al. (2008).

The velocity space must be larger to get around the rank deficiency, and obvious choices are mini-element (an additional linear or cubic function associated with a node at centroid), quasi bubble (as in TELEMAC), or isoP_2 element (a linear representation obtained by splitting the original triangle into four smaller ones by connecting its mid-edges). Compared to them, the so-called linear non-conforming (NC) representation of velocity (Hua and Thomasset 1984) suggests an advantage of a diagonal mass matrix. It serves as a basis to a model by White et al. (2008) and is also supported as an option by FEOM.

The current standard technology of solving primitive equations for the ocean assumes splitting into the external (barotropic) and baroclinic modes. This is effectively performed through a particular form of the pressure method involving elevation. There are several alternative names such as implicit free surface (introduced by MOM developers), semi-implicit method (Casulli and Walters 2000), pressure correction method, or split method (in fact MITgcm, SUNTANS, FEOM, and ICOM follow it).

SLIM (White et al. 2008a), QUODDY (Lynch et al. 1996), and ADCIRC (Westerink et al. 1992) follow a different technology which introduces the barotropic velocity field (which was used in early versions of FEOM too but is now abandoned). In that case a subproblem for the elevation and vertically integrated velocity is solved first and then the elevation is used to obtain full horizontal velocities. A correction is further required at each time step to maintain consistency with barotropic dynamics. The problem emerges because the depth integrated horizontal velocity cannot be exactly represented in the space of functions used in the horizontal plane in the presence of topography. Indeed, the barotropic velocity $\bar{\mathbf{u}}$ or transport $\mathbf{U} = \int \mathbf{u}\,dz$ in the FE case can only be defined by solving

$$\int N_i \mathbf{u}\, d\Omega = \int N_i \bar{\mathbf{u}}\, d\Omega,$$

where $N_i = N_i(x, y)$ is the horizontal test function, for $\bar{\mathbf{u}} = \bar{\mathbf{u}}_j N_j$ or

$$\int N_i \mathbf{u}\, d\Omega = \int N_i \mathbf{U}\, d\Gamma_1$$

for $\mathbf{U} = \mathbf{U}_j N_j$. It becomes clear then that the divergence term in the vertically integrated continuity equation, $\int N_i \nabla(\int \mathbf{u}\, dz)\, d\Gamma_1 = -\int (\nabla N_i)\mathbf{u}\, d\Omega$ cannot be exactly expressed in terms of barotropic velocity or transport: the horizontal velocity is weighted here with gradients of test functions

$$\int (\nabla N_i)\mathbf{u}\,d\Omega \neq \int (\nabla N_i)\overline{\mathbf{u}}\,d\Omega,$$

and

$$\int (\nabla N_i)\mathbf{u}\,d\Omega \neq \int (\nabla N_i)\mathbf{U}\,d\Gamma_1$$

given the above definitions. Correspondingly, after the barotropic velocity (transport) and elevation are found and elevation is used in computations of the full horizontal velocity, the latter should be corrected. This fact by itself is not new and a similar correction procedure was employed, for instance in MOM-type models. The difference, however, lies in the fact that the reprojection inherent in the FE treatment (and absent in FV and finite difference implementations) can be large. The approach based on the pressure method avoids introducing unnecessary barotropic velocities.

Another technology, the explicit free surface, uses small time steps to solve for external mode. It is getting popular in models formulated on structured grids (see Shchepetkin and McWilliams (2005) for ROMS implementation). In the context of unstructured mesh models this technology is employed by FVCOM. It will not be discussed here.

Since nonconforming elements allow a cleaner implementation of the pressure method, we will start with it in the following section. For brevity we will talk about NC (nonconforming) and CL (continuous linear, i.e., P_1) setups.

5.2 Solving the Dynamical Part with NC Elements

Both NC and CL implementations use continuous linear functions to represent elevation and tracers. The difference is only in the discretization of horizontal velocity. In the NC case, the horizontal velocities are expanded in basis functions that are products of NC linear functions $P_1^{NC}(x, y)$ with continuous linear functions $P_1(z)$. The essential advantage of nonconforming linear functions is orthogonality on triangles. This and the absence of stabilization witness in favor of the NC compared with CL. The disadvantage is the increased number of degrees of freedom which commonly implies computational load.

The orthogonality of NC function is strictly maintained on z-level meshes (vertical and horizontal integration are independent). With generalized vertical coordinates, elements are no longer rectangular prisms and Jacobians of transform from parent to physical domains are functions of horizontal coordinates. This destroys the orthogonality. In order to preserve efficiency, horizontal lumping or edge quadrature rules (see White et al. 2008a) have to be applied, which partly impairs the mathematical elegance of the approach. If elements are gently deformed, the deviations from orthogonality are small and horizontal lumping remains a good compromise. Here we will only discuss the case of z-level meshes.

The momentum equation is discretized using the second-order Adams–Bashforth method for the Coriolis term and the momentum advection. The forward Euler discretization is used for the horizontal viscosity term. Applying to it the Adams–Bashforth method can reduce stability but is convenient and seldom leads to problems. The Coriolis term can be treated semi-implicitly, but this can be beneficial only on coarse meshes to allow for large time steps and is not discussed here. The contribution from vertical viscosity can be optionally treated implicitly (when viscosity is large and CFL limiting may occur). The contribution from the sea surface height η is implicit in typical large-scale applications to suppress inertia-gravity waves. In this case the method remains only first-order accurate in time. One can easily make it second-order if required at almost no cost (by treating η semi-implicitly). The time discretized momentum and vertically integrated continuity equations are

$$\delta(\mathbf{u}^{n+1} - \mathbf{u}^n) + g\nabla\eta^{n+1} - \partial_z A_v \partial_z \mathbf{u}^{n+1} = \mathbf{R}^{n+1/2}, \tag{14}$$

$$\delta(\eta^{n+1} - \eta^n) + \nabla \cdot \int_{-H(x,y)}^{0} \mathbf{u}^{n+1} \, \mathrm{d}z = 0. \tag{15}$$

Here $\delta = 1/\Delta t$, where Δt is the time step, and n enumerates time steps. The RHS of the vertically integrated continuity equation is set to zero for simplicity.

The RHS of (14) contains all terms of the momentum equation other than time derivative, surface pressure, and vertical viscosity. The Coriolis, pressure and viscous terms are computed as

$$\mathbf{R}^{n+1/2} = -\nabla p_h^{n+1/2}/\rho_0 + \nabla A_h \nabla \mathbf{u}^n$$
$$-(3/2+\varepsilon)(\mathbf{f}\times\mathbf{u} + (\mathbf{v}\nabla)\mathbf{u})^n + (1/2+\varepsilon)(\mathbf{f}\times\mathbf{u} + (\mathbf{v}\nabla)\mathbf{u})^{n-1}$$

Here ε is a small constant chosen to stabilize the second-order Adams–Bashforth scheme, and $\mathbf{f} = f\mathbf{k}$.

The ideology of solving the pair (14) and (15) is standard (pressure correction method) and is similar to that used in other models working with implicit free surface (e.g., MITgcm, SUNTANS).

First, during the prediction step one solves for $\mathbf{u}^*$

$$\delta(\mathbf{u}^* - \mathbf{u}^n) + \partial_z A_v \partial_z \mathbf{u}^* + g\nabla\eta^n = \mathbf{R}^{n+1/2}. \tag{16}$$

The predicted velocity is then corrected by solving

$$\delta(\mathbf{u}^{n+1} - \mathbf{u}^*) + g\nabla(\eta^{n+1} - \eta^n) = 0. \tag{17}$$

Formally, combining (16) and (17) one does not recover the original Eq. (14). There is small difference due to the omission of viscous contribution from (17). This difference, however, does not destroy the time accuracy of the method (cf. discussion in Ford et al. 2004). The step reflects the essence of the pressure correction method: only the part of velocity update which projects on elevation gradient update matters as it allows one to satisfy the vertically integrated continuity balance (see further).

Discretizing Eqs. (16) and (17) one gets the following matrix equations

$$(\delta \mathrm{M} + \mathrm{D})\mathbf{u}^* = \delta \mathrm{M}\mathbf{u}^n - g\mathbf{G}\eta^n + \mathbf{R}^{n+1/2}, \tag{18}$$

and

$$(\mathrm{M}\mathbf{u}^{n+1} - \mathrm{M}\mathbf{u}^*) + g\Delta t\mathbf{G}(\eta^{n+1} - \eta^n) = 0. \tag{19}$$

Here the notation used for continuous fields is preserved for their discrete counterparts because the meaning is clear from the context. The matrices introduced above are the mass matrix

$$\mathrm{M}_{ij} = \int N_i N_j \, \mathrm{d}\Omega,$$

the matrix of vertical viscosity

$$\mathrm{D}_{ij} = \int A_v \partial_z N_i \partial_z N_j \, \mathrm{d}\Omega,$$

and of the gradient operator

$$\mathbf{G}_{ij} = \int \mathbf{N}_i \nabla M_j \, d\Omega.$$

Here N_i and M_i are used to denote the basis functions used for the horizontal velocity and elevation, respectively. Due to orthogonality of N_i in the horizontal plane matrices M and D contain only links between vertically aligned nodes. This implies that the problem of matrix inversion is split into E_{2D} (the number of 2D edges) subproblems each of which can be inverted effectively by the sweep algorithm. Since the number of edges is three times that of the nodes, this inversion is relatively expensive, but not so as applying iterative solvers to invert stiffness matrices in the CL case in similar circumstances.

If the selected time step Δt is not CFL limited by the vertical viscosity, its contribution can be estimated at time step n (included into the **R** term). If, additionally, the mass matrix is lumped in the vertical direction, it becomes diagonal, and a very effective numerical algorithm follows.

Expressing velocity from (19) one gets

$$\mathbf{u}^{n+1} = \mathbf{u}^* - g\Delta t \mathrm{M}^{-1}\mathbf{G}(\eta^{n+1} - \eta^n). \tag{20}$$

Now we *first* discretize the vertically integrated continuity Eq. (15) and then substitute (20) to obtain

$$\delta \mathrm{M}_\eta \Delta\eta + g\Delta t \mathbf{G}^T \mathrm{M}^{-1}\mathbf{G}\Delta\eta = \mathbf{G}^T \mathbf{u}^*. \tag{21}$$

Here $\Delta\eta = \eta^{n+1} - \eta^n$ and M_η is the mass matrix of the elevation problem. The discretized vertically integrated continuity equation will be satisfied by $\mathbf{u}^{n+1}$ on completing the time step (solving (21) and updating the horizontal velocity via (20)). Clearly, the velocity field and elevation are consistent and no barotropic velocity is needed for that.

Assembling matrix $\mathbf{G}^T\mathrm{M}^{-1}\mathbf{G}$ has to be done only once during the initialization phase. If M is vertically lumped the assembly is substantially simplified.

In summary, the solution algorithm goes through the following steps:

- Compute $\mathbf{u}^*$ from (18) by inverting $\delta\mathrm{M} + \mathrm{D}$. This requires solving E_{2D} subsystems of equations linking velocities at vertically aligned edges. This step becomes elementary if M is vertically lumped and vertical viscosity is explicit.
- Compute elevation from (21) by iteratively solving the matrix problem defined by $\delta\mathrm{M}_\eta + g\Delta t\mathbf{G}^T\mathrm{M}^{-1}\mathbf{G}$. It has dimension N_{2D} (the number of surface nodes).
- Update the velocity according to (20). This step is elementary if the mass matrix is vertically lumped, and is associated with the sweep algorithm otherwise.

The momentum advection in the NC setup can be a source of difficulty in flows with rich dynamics. A solution that works well in all circumstances filters **u** in $(\mathbf{v}\nabla)\mathbf{u}$ by projecting it on continuous linear functions. Danilov et al. (2008) give details of implementation. An excessively high horizontal viscosity is required otherwise.

Keeping the horizontal velocities continuous in vertical direction is rather a source of nuisance than an advantage. The main drawback is that on z-meshes the velocity degrees of freedom may be located in corners, which is only safe when no-slip boundary condition is used. Assuming P_0 vertical representation removes this difficulty and simultaneously essentially reduces the number of operations required to compute the RHS, and it proves to be a much better choice indeed.

5.3 Solving the Dynamical Part with the CL Approach

In order to remove the spurious pressure modes, an analog of (18,19) is written as

$$(\delta \mathrm{M} + \mathrm{D})\mathbf{u}^* = \delta \mathrm{M}\mathbf{u}^n - g\gamma \mathbf{G}\eta^n + \mathbf{R}^{n+1/2}, \tag{22}$$

$$(\mathrm{M}\mathbf{u}^{n+1} - \mathrm{M}\mathbf{u}^*) + g\Delta t \mathbf{G}(\eta^{n+1} - \gamma\eta^n) = 0. \tag{23}$$

Here the difference to the NC case is in adding a multiplier γ to the η^n term. The strength of stabilization turns out to be proportional to $(1-\gamma)$. In simple terms, γ offsets $\mathbf{u}^*$ from approaching the solution too closely in quasi-stationary regimes. In practical applications $\gamma = 0.95$–0.97 works well. $\mathbf{R}^{n+1/2}$ can be estimated in any appropriate way, see Wang et al. (2008) for FEOM implementation.

The major difficulty in solving (22) is the presence of non-diagonal matrix $\delta\mathrm{M} + \mathrm{D}$ (horizontal basis functions are not orthogonal on elements). Numerical efficiency can be reached in two ways: (i) Use horizontal lumping on $\delta\mathrm{M} + \mathrm{D}$ on prismatic meshes or full lumping on M on tetrahedral meshes (D links only vertically aligned nodes in this case), followed with applying a three-diagonal solver; (ii) Adjust vertical viscosity so that CFL limiting is not occurring and put it to the RHS. Solving for the consistent mass matrix is then done by performing two or three iterations for $\mathrm{M} = \mathrm{M}_L + (\mathrm{M} - \mathrm{M}_L)$, and estimating $(\mathrm{M} - \mathrm{M}_L)\mathbf{u}^*$ from the previous iteration. This is commonly much faster than calling a solver. If neither (i) nor (ii) can be selected, iterative solvers have to be applied and the algorithm becomes slow (loosing easily a factor 2 to 3 in performance).

The other essential distinction from the NC case is that in order to solve for the elevation, one does not use (23). Instead, one first expresses $\mathbf{u}^{n+1}$ from (17) modified by including γ and substitutes it into (15), and *then* discretizes the result. The difference to the NC algorithm is the replacement of operator $-\mathbf{G}^T\mathrm{M}^{-1}\mathbf{G}$ by the Laplacian operator which does not support the pressure modes of operator $\mathbf{G}$. The price for this (necessary) modification is that $\mathbf{u}^{n+1}$ as found from (23) does not satisfy the vertically integrated continuity equation exactly. It is its unprojected counterpart from (17) that does. The latter is used to solve for vertical velocity and to advect tracers while the former is used in the momentum equation. The difference between them is small (it is only due to reprojection of the part due to elevation gradient on linear functions), but important for the consistency of the code. The need to keep two types of horizontal velocity is the major conceptual disadvantage of the CL code. One of these velocities ($\mathbf{u}^{n+1}$) satisfies the boundary conditions but does not exactly satisfy the discrete vertically integrated continuity equation. The other one, $\mathbf{u}^* - g\Delta t\nabla(\eta^{n+1} - \gamma\eta^n)$, on the contrary, does, but satisfies only weak impermeability boundary condition.

It should be reminded that all stabilization methods used in models employing $P_1 - P_1$ discretization suffer from similar issues because the essence of stabilization lies in regularizing the vertically integrated continuity equation. They all introduce the Laplacian operator (instead of or in addition to the true operator $-\mathbf{G}^T\mathrm{M}^{-1}\mathbf{G}$). The approach outlined here has the advantage of explicitly providing the expression for the horizontal velocity which ensures consistency with the vertical velocity and temperature and salinity equations.

The solution of the dynamical part follows the same three basic steps as in the NC case, but the prediction and correction steps require the inversion of the mass matrix or (mass + vertical viscosity) matrix. The inversion of the mass matrices can be done effectively and does not slow down the algorithm as much as the inversion of full $\delta\mathrm{M} + \mathrm{D}$ matrix.

Note that even if stable continuous elements were used for the horizontal velocity, implementing the pressure correction method cleanly would still be difficult: the inverse of sparse mass matrix is a full matrix, and there is no way of assembling $-\mathbf{G}^T\mathrm{M}^{-1}\mathbf{G}$. In this case one has to apply lumping to be practical, or follow the strategy of CL FEOM ($\gamma = 1$ for stable elements). Both strategies degrade the mathematical appeal of the FE approach. Using discontinuous representation for the horizontal velocity alleviates this problem, as mass matrix can easily be inverted elementwise. The nonconforming functions implement an intermediate approach: the support of a basis function is not limited to element, but the functions are orthogonal on elements.

Of strategies based on discontinuous velocity spaces one may mention $P_0 - P_1$ and its higher-order sibling $P_1^{DG} - P_2$. Both are well suited to represent the geostrophic balance, as the pressure gradient can exactly be expressed in the velocity space. The first one can approximately be interpreted in the FV sense, which gives the arrangement of variables used by FVCOM (see ➲ Sect. 6.2). The second one is a promising subject of current research on the level of shallow water equations. However, solving the advective-diffusive equations and determining pressure and vertical velocity in the case of primitive equations may turn to be too expensive for practical 3D applications.

5.4 Vertical Velocity, Pressure, and Tracers

Here we highlight several practical aspects related to the vertical velocity and tracers. The equation for the vertical velocity discretized in the FE way is

$$\int \partial_z N_i w \,\mathrm{d}\Omega - \int N_i \partial_t \eta \,\mathrm{d}\Gamma_1 = - \int \nabla N_i \mathbf{u} \,\mathrm{d}\Omega,$$

where the integration by parts has been performed. The solvability condition for this equation is the vertically integrated continuity equation, so consistency between elevation and vertical velocity is required. If one continues with $w = N_j w_j$, the stiffness matrix becomes $\mathrm{S}_{ij} = \int N_j \partial_z N_i \,\mathrm{d}\Omega$, which presents a problem to iterative solvers if N_i is continuous in vertical direction (zero diagonal entries). Thus, either N_i should be discontinuous in vertical direction (S_{ij} will be modified through the continuity penalties added to the LHS) as proposed in White et al. (2008a), or w has to be represented as $w = \partial_z \phi$ with $\phi = \phi_j N_j$, which transforms the system matrix to $\mathrm{S}_{ij} = \int \partial_z N_i \partial_z N_j \,\mathrm{d}\Omega$. In the latter case the consistency is maintained as it depends on test functions only.

The horizontal connections introduced by horizontally continuous functions still present a difficulty on prismatic meshes, and the practical approach is horizontal lumping (White et al. 2008a; Wang et al. 2008). In case $w = \partial_z \phi$ is solved with linear elements on tetrahedral meshes, S_{ij} connects only vertically aligned nodes and no lumping is needed. When lumping is used, tracer equations have to be modified accordingly to ensure consistency.

We note again that discontinuous representation for scalars would remove the issues mentioned above, but this commonly proves to be much more expensive. Computing pressure in hydrostatic approximation leads to same questions as for w. When a vertically discontinuous representation is used, and horizontal lumping is applied, the pressure can be employed to compute horizontal gradients. For $w = \partial_z \phi$, the consistent pressure can be obtained by projecting the hydrostatic balance on $\partial_z N_i$. Our experience with it on tetrahedral meshes is negative, as there is noise due to the difference in stencils of vertically neighboring nodes. On generalized meshes any attempt to minimize pressure gradient errors would require to sacrifice the consistency. For this reason FEOM (Wang et al. 2008) uses finite differences and cubic spline interpolation. In the opinion of the authors the pressure part is the most unsatisfactory aspect of FE codes using hydrostatic approximation and horizontally continuous representation.

Transport equations for tracers can be solved with any appropriate algorithm providing necessary accuracy and performance. FEOM in most cases applies the FCT scheme by Löhner et al. (1987), which uses the Taylor–Galerkin (TG) scheme as a high-order one (Wang et al. 2008). Consistent mass matrices are retained. They are approximately taken into account through the procedure described above. Applying Petrov–Galerkin type schemes together with the implicit method is found to be too expensive numerically. There is one place, however, where implicitness is often required – the vertical diffusion is frequently made high locally to describe stirring by the wind or in order to remove static instability (overturning). Taking it into account on a rigorous basis requires solving full 3D matrix equations, which is an expensive choice. Again, approximate (compromise) solutions are possible.

The implicit vertical diffusion can be solved for separately after the advection and horizontal diffusion. If the diffusivity tensor is rotated, the respective components are treated together with the horizontal diffusion. The implicit vertical advection substep is formulated as

$$\delta \mathrm{M}(C^{n+1} - C^*) + \mathrm{D}C^{n+1} = 0,$$

where D is the vertical diffusion operator, and C^* is the value updated for all other factors. On prismatic meshes, one lumps M fully and D horizontally. On tetrahedral meshes only M has to be lumped for linear elements. This reduces the full problem to a set of subproblems for vertically aligned nodes, which can be solved effectively. Note that consistent mass matrices are used with advection, which reduces dispersion.

The main conclusions that can be drawn from this section is that requirements to remain practical enforce one to deviate from a rigorous FE treatment in too many places if continuous elements are used for scalars. This would be a strong argument in favor of discontinuous representation if not the incurring computational load (although it might become practical on massive parallel computer architectures). It also points to FV methods as a possible alternative, which treat seamlessly the issues mentioned here. Correspondingly, perspectives suggested by these method should be carefully analyzed.

6 C-Grid and FV Cell-Vertex Type of Discretization

This section presents FV-based methods. We focus on the C-grid and cell-vertex approaches, with more focus on the second one (employed by FVCOM and presenting an FV analog of $P_0 - P_1$ FE discretization). The C-grid approach shows a mode in the horizontal velocity divergence in geostrophically balanced regimes. It performs well in coastal and tidal applications, but not necessarily on large scales of the slowly varying global circulation.

6.1 C-Grid

A suitable code implementation can be obtained by combining the approaches of SUNTANS (time stepping) with that of Stuhne and Peltier (2006) (implementation of spherical geometry).

The surface mesh should satisfy the property of orthogonality implying that circumcenters are contained inside their respective triangles. A small number of defects is commonly accepted, with centroids replacing the circumcenters.

When viewed from the surface, scalar fields are stored at circumcenters, and normal components of horizontal velocity, at midedges. In the vertical direction only the vertical velocity, w, is at full levels, the rest is at midlevels. We will use notation of Stuhne and Peltier (2006) as most concise. Let index c enumerate triangles (cells), and e the edges. A separate index, k, will be used for the vertical direction. For each edge and triangle, one introduces the function $\delta_{e,c}$, which is zero if the triangle c does not contain edge e, 1 if it does and the normal at e is the inner one with respect to c, and -1 if it is the outer one. The spherical equivalent of Perot reconstruction is introduced to map the edge scalar velocity to full velocities at circumcenters

$$A_c \mathbf{u}_c = \sum_e \delta_{e,c} u_e l_e (\mathbf{x}_c - \mathbf{x}_e),$$

and to return back to the edges from the elemental velocities

$$d_e u_e = \sum_c \delta_{e,c} \mathbf{u}_c (\mathbf{x}_c - \mathbf{x}_e).$$

Here the following notation is used: l_e is the length of the edge e, d_e is the distance between the centroids of triangles sharing e, $\mathbf{x}_c$ and $\mathbf{x}_e$ are vectors drawn to the circumcenter and midedge, respectively, from the center of the spherical Earth, and A_c is the area of triangle. A Cartesian reference frame is associated with the Earth center and vector quantities are expressed with respect to it. The distances are measured along great circles, and the areas are those on the sphere. One can symbolically denote these operations as

$$\mathbf{u}_c = (\Gamma u_e)_c$$

and

$$u_e = (\Gamma^{-1} \mathbf{u}_c)_e,$$

where it is assumed that the operator acts on the whole field of edge velocities in the first case and on the whole field of velocity vectors on elements in the second case. The operators are not true inverse of each other, the notation only conveys an idea of what they are used for. Note that Γ^{-1} weights elemental values with quantities belonging to the element. This is a premise condition for preserving symmetry of operators.

The pressure gradient at edge e is estimated as $(\delta_{e,c_1} p_{c_1} + \delta_{e,c_2} p_{c_2})/d_e$, where c_1 and c_2 are elements containing edge e and it is assumed that this rule applies at each velocity level k. This is the finite-difference part of the approach. The rest of momentum equation requires manipulations with vectors. One first applies Γ to reconstruct $\mathbf{u}_c$. Then,

$$\mathbf{F} = -\mathbf{f} \times \mathbf{u}_c - \nabla(\mathbf{u}_c \mathbf{u}_c - A_h \nabla \mathbf{u}_c) - \partial_z (w \mathbf{u}_c - A_v \partial_z \mathbf{u}_c)$$

is estimated. The flux term is computed in the standard finite-volume way. The fluxes are estimated at faces of control volume (a prism defined by c and k). The operation of divergence returns them to centroids. Upwinding of momentum advection can easily be implemented at this stage, but in order to do it properly at least linear reconstruction of full velocity vector to the vertical face is required. The same is true of horizontal faces, but the reconstruction is much easier to implement here. Once $\mathbf{F}$ is computed, one returns to the space of scalar velocities by applying Γ^{-1}. The treatment of w (continuity) and tracer equation is performed in the standard FV way and is very similar to finite difference codes. For example, to solve for w, one starts from the bottom where it is zero on z meshes, and collects volume fluxes through vertical faces, going upward from level to level. Relevant details can be found in Fringer et al. (2006), but in practice more sophisticated flux estimates could be required for tracer advection. Pressure (p_h) computation is finite difference.

To solve for velocities and elevation one may use the procedure outlined above for nonconforming elements. There are no mass matrices involved in the current case. Note that all the places where we were encountering difficulties with finite elements are very straightforward here, and indeed, in terms of computational performance a factor of five can easily be obtained with respect to $P_1^{NC} - P_1$ or $P_1 - P_1$ codes considered above, while preserving consistency.

The weak point of C-grid discretization is its performance in regimes close to the geostrophic balance as discussed in Danilov (2010). We give a simple explanation here. Consider a zonally reentrant flat-bottom channel driven by a time-constant zonal wind. Let density be uniform, so that pressure gradients are due to the elevation only. The stationary equations can be written as

$$\mathrm{C}u_e + g\mathrm{G}\eta = (\tau/h)\theta(z+h) + \mathrm{V}u_e, \quad -\mathrm{G}^{\mathrm{T}}u_n + \partial_z w = 0.$$

Here C, G, V, and $-\mathrm{G}^{\mathrm{T}}$ stand for discrete operators of Coriolis, gradient, viscosity, and divergence, respectively; h is the thickness of the layer experiencing the external wind forcing τ per unit mass, and θ is the theta function. Integrating the equation vertically one obtains $(\mathrm{C} - \mathrm{V})U_e + H\mathrm{G}\eta = \tau$ and $\mathrm{G}^{\mathrm{T}}U_e = 0$, where U_e denotes the vertically integrated u_e and g is absorbed in η. Expressing $U_e = (\mathrm{C}-\mathrm{V})^{-1}(\tau - H\mathrm{G}\eta)$ and requiring its divergence to be zero leads to the equation on η

$$\mathrm{G}^T(\mathrm{C}-\mathrm{V})^{-1}\mathrm{G}\eta = \mathrm{G}^T(\mathrm{C}-\mathrm{V})^{-1}\tau/H.$$

Below the surface layer $u_e = -(\mathrm{C}-\mathrm{V})^{-1}\mathrm{G}\eta$. It turns out that the inversion of $\mathrm{C}-\mathrm{V}$ leads to noise in u_e that has a strong projection on the divergence, which easily exceeds the true divergence by orders of magnitude. Since C is not invertible, we are confronted with a combined issue of Coriolis term and its regularization (viscosity and friction operators). In the time stepping mode the operator $\mathrm{C}-\mathrm{V}$ is estimated on the RHS. The development of divergence noise with time can be diagnosed by computing the divergence of the RHS of momentum equation. It is mainly seen through the systematic noise in the divergence of contributions from the geostrophic balance, which evolves in time until it saturates.

The presence of noise makes the C-grid setup less interesting from the viewpoint of large-scale ocean modeling. We thus continue with the next candidate which does not share this weakness.

6.2 $P_0 - P_1$-Like Discretization

In this case the triangular surface mesh is not necessarily orthogonal. Viewed from the surface, full vectors of horizontal velocity are stored at centroids, and scalar fields are at nodes. Similar to the previous setup (and also to the standard arrangement in finite difference codes), the vertical velocity is allocated at full levels, and all other fields, at midlevels. We will sketch a z-level setup based on implicit treatment of elevation. It follows the same scheme as the NC treatment, once again without complications introduced by the mass matrix. The control volume associated with the velocity vector is the original prismatic element (cell) of the mesh, and one estimates fluxes of momentum through the faces. The contribution of pressure can be computed in the flux form (recalling that the momentum flux tensor can be expressed as $\Pi_{ij} = p\delta_{ij} + \rho_0 v_i v_j - \sigma_{ij}$, with σ_{ij}

the tensor of viscous stresses), giving $\sum_e S_e \mathbf{n}_f(p_1 + p_2)/2$, with sum over vertical faces (e) of prismatic element (c,k). Alternatively, we can assume the linear interpolation in the horizontal plane inside the element and compute the gradient directly. The latter way resembles the FE treatment and provides the identical result.

To accurately estimate the momentum advection part and the contribution from viscous stresses one needs the velocity reconstruction (velocity and its derivatives have to be estimated at faces). It is routinely done with linear least-squares reconstruction based on the information on velocity on three neighbor elements. As a result one obtains the velocity vector on cell c

$$\mathbf{u}_r = \mathbf{u}_c + \mathrm{S}\mathbf{r},$$

where $\mathbf{r} = \mathbf{x} - \mathbf{x}_c$ and $\mathrm{S}_{ij} = \partial_i u_j$ is the matrix of estimated velocity derivatives. Since the reconstruction is performed with respect to the centroid, it is unbiased, i.e., the mean value of velocity on the element is still $\mathbf{u}_c$. The operators providing the reconstruction (the velocity derivatives) are two-dimensional, they are computed, and stored in advance. The reconstruction is sufficient to obtain the second-order estimate for the advective flux on faces. The reconstructions on two elements sharing the face are averaged to compute the normal velocity on this face, and the upwind value is used in the rest. The averaged derivatives can also be used to get the first-order estimate for viscous fluxes as in FVCOM (Chen et al. 2003), which is usually sufficient. The details and proofs can be found in Kobayashi et al. (1999).

Third-order upwinding is used in the vertical direction. Here, the implementation follows finite differences.

In practice, in eddy-resolving simulations of large-scale circulation the horizontal momentum advection part sometimes turns out to be too damping. Its "central" counterpart is not recommended. The scheme obtained by projecting the horizontal velocities first to the P_1 space and employing them to compute the horizontal advection part shows much less damping and can work without upwinding. A frequent problem triggered by the momentum advection is a particular mode manifested through fluctuations of directions of velocity vectors observed at high element Reynolds numbers. Laplacian and biharmonic viscosities are not very efficient in removing it without damping the entire flow too strongly. The Leith modified viscosity of MITgcm can cure the problem (see, i.e., Fox-Kemper and Menemenlis 2008) in most cases.

The control volumes for scalar quantities in the surface plane are obtained by connecting centroids with edge centers. Each such fragment defines a vertical face of 3D control volumes for scalars. The velocity value on each face is known, and central estimate of scalar quantity is straightforward too. What remains is a convenient technology of upwinding in horizontal directions. It is commonly implemented through identifying, for each face, its upwind node. Then an upwind-biased reconstruction is made. It consist in computing mean gradient of scalar quantity over the cluster of elements containing the upwind node and using it to reconstruct the field value at the face. The resulting advection schemes perform stable, but show too strong damping compared with the standard FCT scheme of FEOM in eddy-rich regimes. The other aspect is that reconstructions on median-dual volumes are generally biased (which reduces accuracy), and removing the bias introduces mass matrices.

On the practical side the method presented here performs as fast as the C-grid one, the programming downside is the relative inconveniency of implementing spherical geometry. Yet the way to longitude–latitude free setup is straightforward, as explained in Comblen et al. (2009).

Selecting appropriate advection schemes for both momentum and scalars remains the major issue in eddy-resolving regimes, as it turns out to be much more difficult to find the compromise between damping and stability than with $P_1 - P_1$ or $P_1^{NC} - P_1$ setups presented above.

Our conclusion is that this approach could be tuned for large-scale simulations in the ocean. Similarly to the C-grid setup, it is free of complications with the treatment of w, pressure, and implicit vertical diffusion characteristic of continuous FE representation of scalar quantities. In distinction to it, there are no divergence noise modes. However, divergence-based viscosity is still needed in regimes with high elemental Reynolds numbers.

7 Conclusions

Large-scale ocean circulation modeling is currently dominated by regular grid models. There are several reasons for that, not the least one being the computational efficiency and availability of well-tested algorithms, in particular, for advection.

The methods formulated on unstructured meshes suggest a seamless solution for situations when local focus is required, and in this way make nesting unnecessary. This alone should warrant a variety of applications, and these are beginning to appear. However, the success here is as yet rather moderate, contrary to the coastal and tidal modeling, where it is much more obvious. The explanation is that it is not always easy to find a compromise between consistency and numerical efficiency, whereas long integration times make conservation and maintaining water mass properties a more delicate issue than in coastal applications.

While the question about optimal and robust discretization is still a subject of debate, there already exist solutions showing reasonable performance as applied to large-scale ocean modeling. Our goal was comparing four of them, two belonging to the FE (FEOM) and two to the FV realms. We hope that our practical experience will allow one to easily see major difficulties/advantages associated with particular technologies.

The approaches outlined here do not exploit the full potential of FE and FV technologies, in particular adaptive meshes and high-order elements. In our opinion, at the current stage, there is need in demonstrating the usability of these technologies on practical examples, simultaneously learning from them. Motivated by this idea our studies were focused on low-order elements – $P_1 - P_1$ and $P_1^{NC} - P_1$ FE discretizations (implemented in FEOM), and C-grid and FVCOM- (or $P_0 - P_1$) like FV discretizations. The C-grid case partly uses finite differences, but its core is based on the FV treatment. Its RT0 analog was not tried by us, but it is only different in the existence of mass matrix, which is commonly lumped.

When compared to regular-mesh models, our FE implementations run a factor of approximately 10 (or even more) slower (Danilov et al. 2008). The FCT advection of temperature and salinity takes about 40% of the time step, which indeed becomes expensive when other tracers are added. Both FV setups run about 5 times faster than FE, but this may be slightly corrected down when flux limiters will be added to their transport schemes. The better efficiency is explained by a more convenient data structure and the reduced amount of floating-point operations per element.

The source of major discontent with FE discretizations based on continuous representation of scalar quantities is the presence of horizontal connections and mass matrices. While mass matrices favorably influence the accuracy of algorithms, their presence slows down the performance. A much more important issue, in our opinion, is the necessity of compromises in computing the vertical velocity and pressure and implementing a computationally efficient vertical diffusion. Note that neither of these issues is seen on the level of shallow water equations, which dominate theoretical studies of appropriate discretizations.

Keeping representation for scalars continuous and increasing the order does not seem to be a promising direction, as it inherits all the difficulties seen with P_1 scalars. One indeed needs a discontinuous representation to split vertical and horizontal directions and reach better consistency in treating hydrostatic equations. The latter, to the best of our knowledge, turns out to be prohibitively expensive, yet future developments may modify this statement.

The FV method possesses advantages of the discontinuous FE method and can be thought of as its low-order implementation. It can be much easily tailored to the need of large-scale ocean modeling, as all regular mesh models use it either explicitly (MITgcm) or implicitly (the rest). The major future work should go into design of accurate conservative advection schemes, to be on equal footing with regular mesh models. We have found, however, that C-grid discretization is frequently impaired by the numerical mode in the horizontal velocity divergence, its true origin being rooted in a too small divergence stencil (just triangle), and the fact that the divergence production associated with vorticity dynamics in the presence of Coriolis is not sharing this stencil. It cannot be recommended until solutions suppressing the mode are found. The other discretization method, adapted from FVCOM, performs reasonably well, but is much more demanding with respect to the selection of stabilizing viscosities than our FE implementations.

Despite the current success of FEOM comes at a high computational price, it already serves as a good alternative to nesting methods in regional applications where a high (>10) refinement ratio is required. One has to work hard on its performance on massively parallel computers before the model will be competitive for general purposes. This work should be carried out in parallel with exploring potentialities of other spatial discretization methods, which can provide advantages either on the side of efficiency or numerical accuracy. The FV approach seems to be a promising direction.

References

Blaise S (2009) Development of a finite element marine model. Thesis. École polytechnique de Louvain. Université Catolique de Louvain

Blazek J (2001) Computational fluid dynamics: Principles and applications. Elsevier

Carrère L, Lyard F (2003) Modeling the barotropic response of the global ocean to atmospheric wind and pressure forcing – comparisons with observations. Geophys Res Lett 30:1275

Casulli V, Walters RA (2000) An unstructured grid, three-dimensional model based on the shallow water equations. Int J Numer Meth Fluids 32: 331–348

Chen C, Liu H, Beardsley RC (2003) An unstructured grid, finite-volume, three-dimensional, primitive equations ocean model: Applications to coastal ocean and estuaries. J Atmos Ocean Tech 20:159–186

Codina R, Soto O (1997) Finite element solution of the Stokes problem with dominating Coriolis force. Comput Methods Appl Mech Eng 142: 215–234

Codina R, Zienkiewicz OC (2002) CBS versus GLS stabilization of the incompressible Navier-Stokes equations and the role of the time step as the stabilization parameter. Commun Numer Methods Eng 18:99–112

Comblen R, Legrand S, Deleersnijder E, Legat V (2009) A finite-element method for solving shallow water equations on the sphere. Ocean Modelling 28:12–23

Danilov S, Kivman G, Schröter J (2004) A finite element ocean model: principles and evaluation. Ocean Modell 6:125-150

Danilov S, Wang Q, Losch M, Sidorenko D, Schröter J (2008) Modeling ocean circulation on unstructured meshes: comparison of two horizontal discretizations. Ocean Dynamics 58: 365–374

Danilov S (2010) On utility of triangular C-grid type discretization for numerical modeling of large-scale ocean flows. Submitted to Ocean Dynamics

Dawson CN, Westerink JJ, Feyen JC, Pothina D (2006) Continuous, discontinuous and coupled discontinuous-continuous Galerkin finite element methods for the shallow water equations. Int J Numer Methods Fluids 52:63-88

Donea J, Huerta A (2003) Finite element methods for flow problems. Wiley, New York

Dupont F, Straub DN, Lin CA (2003) Influence of a step-like coastline on the basin scale vorticity budget in mid-latitude gyre models. Tellus 55A: 255–272

Ford R, Pain CC, Piggott MD, Goddard AJH, de Oliveira CRE, Uplemby AP (2004) A non-hydrostatic finite-element model for three-dimensional stratified flows. Part I: model formulation. Mon Wea Rev 132:2832–2844

Fox-Kemper B, Menemenlis D (2008) Can large eddy simulation techniques improve mesoscale rich ocean models? In: Hecht MW, Hasumi H (ed) Ocean modeling in an eddying regime, Geophysical Monograph 177, AGU 2008, pp 319–337

Fringer OB, Gerritsen M, Street RL (2006) An unstructured-grid, finite-volume, nonhydrostatic, parallel coastal ocean simulator. Ocean Modell 14:139–173

Haidvogel DB, Beckmann A (1999): Numerical ocean circulation modeling. Imperial College Press, London

Hua BL, Thomasset F (1984) A noise-free finite-element scheme for the two-layer shallow-water equations. Tellus 36A:157–165

Iskandarani M, Haidvogel DV, Levin JC (2003) A three-dimensional spectral element model for the solution of the hydrostatic primitive equations. J Comp Phys 186:397–425

Gent PR, McWilliams JC (1990) Isopycnal mixing in ocean circulation models. J Phys Oceanogr 20:150–155

Griffies SM, Böning C, Bryan FO, Chassignet EP, Gerdes R, Hasumi H, Hirst A, Treguier A-M, Webb D (2000) Developments in ocean climate modelling. Ocean Modell 2:123–192

Griffies SM (2004) Fundamentals of ocean climate models. Princeton University Press, Princeton

Griffies SM, Adcroft AJ (2008) Formulating the Equations of Ocean Models In Hecht MW, and Hasumi H, (eds) Ocean modeling in an eddying regime, Geophysical Monograph 177, AGU 2008: 281–317

Hervouet J-M (2000) TELEMAC modelling system: an overview. Hydrol Processes 14:2209–2210

Hervouet J-M (2007) Hydrodynamics of Free Surface Flows: Modelling with the Finite Element Method. Wiley, New York

Jackett DR, McDougall TJ (1995) Minimal adjustment of hydrographic profiles to achieve static stability. J Atmos Ocean Techn 12: 381–389

Jones JE, Davies AM (2008) Storm surge computations for the west coast of Britain using a finite element model (TELEMAC). Ocean Dyn 58:337–363

Kobayashi MH, Pereira JMC, Pereira JCF (1999)A conservative finite-volume second-order accurate projection method on hybrid unstructured grids. J Comput Phys 150:40–75

Le Roux DY, Sène A, Rostand V, Hanert E. (2005) On some spurious mode issues in shallow-water models using a linear algebra approach. Ocean Modell 10:83–94

Le Roux DY, Rostand V, Pouliot B (2007) Analysis of numerically induced oscillations in 2D finite-element shallow water models. Part I: Inertia-gravity waves. SIAM J Sci Comput 29:331–360

Löhner R, Morgan K, Peraire J, Vahdati M (1987) Finite-element flux-corrected transport (FEM-FCT) for the Euler and Navier-Stokes equations, Int J Numer Methods Fluids 7:1093–1109

Lynch DR, Ip JTC, Naimie, CE, Werner FE (1996) Comprehensive coastal circulation model with application to the Gulf of Maine. Cont Shelf Res 16:875–906

Lynch DR, Holbroke MJ, Naimie CE (1997) The Maine coastal current: spring climatological circulation. Cont Shelf Res 17:605–634

Marchesiello P, McWilliams JC, Shchepetkin A (2001) Open boundary conditions for long-term integration of regional oceanic models. Ocean Modell 3:1–20

Marshall J, Adcroft A, Hill C, Perelman L, Heisey C (1997) A finite-volume, incompressible Navier-Stokes model for studies of the ocean on parallel computers. J Geophys Res 102:5753–5766

Pain CC, Piggott MD, Goddard AJH, Fang F, Gorman GJ, Marshall DP, Eaton MD, Power PW, de Oliveira CRE (2005) Three-dimensional unstructured mesh ocean modelling, Ocean Modell 10:5–33

Piggott MD, Pain CC, Gorman GJ, Marshall DP, Killworth PD (2008) Unstructured adaptive meshes for ocean modeling. In: Hecht MW, Hasumi H (eds) Ocean modeling in an eddying regime, Geophysical Monograph 177, AGU, pp 383–408

Raviart P, Thomas J (1977) A mixed finite element method for 2nd order elliptic problems. In: Mathematical Aspects of the Finite Element Methods, Lecture Notes in Math. 606, I. Galligani and E. Magenes, Eds., 292–315, Springer-Verlag, Berlin

Shchepetkin AF, McWilliams JC (2005) The regional oceanic modeling system (ROMS):

a split-explicit, free-surface, topography-following-coordinate oceanic model. Ocean Modell 9:347–404

Stuhne GR, Peltier WR (2006) A robust unstructured grid discretization for 3-dimensional hydrostatic flows in spherical geometry: A new numerical structure for ocean general circulation modeling. J Comput Phys 213:704–729

Timmermann R, Danilov S, Schröter J, Böning C, Sidorenko D, Rollenhagen K (2009) Ocean circulation and sea ice distribution in a finite-element global ice–ocean model. Ocean Modell 27: 114–129

Walters RA (2005) Coastal ocean models: two useful finite-element methods. Cont Shelf Res 25: 775–793

Wang Q, Danilov S, Schröter J (2008) Finite Element Ocean circulation Model based on triangular prismatic elements, with application in studying the effect of topography representation. J Geophys Res 113:C05015. doi:10.1029/2007JC004482

Wang B, Fringer OB, Giddings SN, Fong DA (2009a) High-resolution simulations of a macrotidal estuary using SUNTANS. Ocean Modell 28: 167–192

Wang Q, Danilov S, Schröter J (2009b) Bottom water formation in the southern Weddell Sea and the influence of submarine ridges: Idealized numerical simulations. Ocean Modell 28:50–59

Wang Q, Danilov S, Hellmer H, Schröter J (2010) Overflows over the western Ross Sea continental slope, Submitted to J Geophys Res

Westerink JJ, Luettich RA, Blain CA, Scheffner NW (1992) ADCIRC: an advanced three-dimensional circulation model for shelves, coasts and estuaries; Report 2: Users Manual for ADCIRC-2DDI. Contractors Report to the US Army Corps of Engineers. Washington D.C.

White L, Deleersnijder E, Legat V (2008a) A three-dimensional unstructured mesh shallow-water model, with application to the flows around an island and in a wind driven, elongated basin. Ocean Modell 22:26-47

White L, Legat V, Deleersnijder E (2008b) Tracer conservation for three-dimensional, finite element, free-surface, ocean modeling on moving prismatic meshes. Mon Wea Rev 136:420–442. doi:10.1175/2007MWR2137.1

Yang Z, Khangaonkar T (2009) Modeling tidal circulation and stratification in Skagit River estuary using an unstructured grid ocean model. Ocean Modell 28:34–49

Zhang Y, Baptista AM (2008) SELFE: A semi-implicit Eulerian-Lagrangian finite-element model for cross-scale ocean circulation. Ocean Modell 21:71–96

Zienkiewicz OC, Taylor RL (2000) The finite element method. V. 1. Butterworth–Heinemann, Oxford

Zienkiewicz OC, Taylor RL (2000) The finite element method vol. 3: Fluid dynamics. Butterworth–Heinemann, Oxford

14 Numerical Methods in Support of Advanced Tsunami Early Warning

Jörn Behrens
Numerical Methods in Geosciences, KlimaCampus, University of Hamburg, Hamburg, Germany

W. Freeden, M.Z. Nashed, T. Sonar (Eds.), *Handbook of Geomathematics*, DOI 10.1007/978-3-642-01546-5_14,

Abstract After the 2004 Great Sumatra-Andaman Tsunami that devastated vast areas bordering the Indian Ocean and claimed over 230,000 lives many research activities began to improve tsunami early warning capacities. Among these efforts was a large scientific and development project, the German-Indonesian Tsunami Early Warning System (GITEWS) endeavor. Advanced numerical methods for simulating the tsunami propagation and inundation, as well as for evaluating the measurement data and providing a forecast for precise warning bulletins have been developed in that context. We will take the developments of the GITEWS tsunami modeling as a guideline for introducing concepts and existing approaches in tsunami modeling and early warning.

For tsunami propagation and inundation modeling, numerical methods for solving hyperbolic or parabolic partial differential equations play the predominant role. The behavior of tsunami waves is usually modeled by simplifications of the Navier–Stokes equations.

In tsunami early warning approaches, an inverse problem needs to be solved, which can be formulated as follows: *given a number of measurements of the tsunami event, what was the source; and when knowing the source, how do future states look like?* This problem has to be solved within a few minutes in order to be of any use, and the number of available measurements is very small within the first few minutes after the rupture causing a tsunami.

1 Introduction

Tsunami phenomena entered public consciousness after December 26, 2004, when the devastating Great Sumatra–Andaman Tsunami claimed more than 230,000 casualties around the rim of the Indian Ocean. Even though tsunamis happen quite frequently, mostly they are locally limited in their extent and were therefore not noticed by neither the scientific community nor the public until 2004 (Synolakis and Bernard 2006). That event, however, triggered a large number of research endeavors, among them tsunami modeling and early warning system developments. Geomathematical methods play a crucial role in many of the subsystems of current tsunami early warning systems. Geomathematics is involved in geopositioning, included in systems to monitor the Earth crust dislocation, tsunami wave elevation on GPS-equipped buoys, and in new developments to utilize GPS reflectometry to monitor the wave propagation. Mathematical methods are heavily involved in data processing, let it be spaceborne data compression or filtering. In this chapter, we will focus the description to the simulation of tsunami phenomena in order to forecast, hindcast, or assess the impact of tsunamis on coastal regions. In particular, we will demonstrate current approaches to simulate the tsunami wave propagation and inundation on land. Furthermore, we will show how these simulation results can be utilized to forecast tsunami impact in a tsunami early warning system (TEWS).

Tsunami simulation plays a crucial role in assessing in short time the hazard situation (Titov et al. 2005). Usually, a TEWS consists of one or several sensor systems, among them a seismic system to monitor earthquake activity (note that 90% of the tsunamis in the Indian Ocean are caused by subduction zone earthquakes), a tide gauge system, monitoring the wave heights and including deep ocean gauges (buoys), and possibly other systems, like real-time GPS crust deformation monitoring systems, high-frequency (HF) radar systems, etc. (see ➲ *Fig. 1* for a general scheme). All these systems contribute only a small number of individual sensor readings. Therefore, an overall situation perspective relies on simulations in order to "interpolate" between these singular measurements. The problem behind this task is an *assimilation problem.*

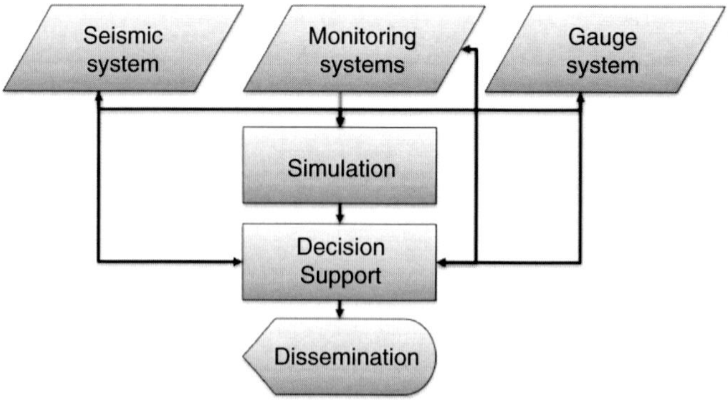

Fig. 1
Generalized architecture of current TEWS. Measurement data (*top*) are either fed into the simulation system or directly evaluated by the decision support unit, before corresponding messages are cast to a dissemination system

Additionally, a TEWS aims to *forecast* the impact of a tsunami, in order to trigger mitigation measures. Therefore, the usual simulation-related steps in tsunami early warning are

1. Assimilate the wave propagation situation, given a (very) small number of measurements.
2. Given the current situation, forecast the impact at the coast or on shore.

In most TEWS the first step tries to assimilate to the source (initial condition) of the wave phenomenon (Wei et al. 2008; Greenslade et al. 2007). We will show later that this is not the only, and possibly not the best way to deal with the problem. In the second step, most of the TEWS give an indication on wave heights at the coast (or even in shallow water off the coast). But information on possible inundation or even risk for settlements and industrial complexes is aimed for.

One important constraint, especially for those areas close to potential rupture areas, is time. In Indonesia, for example, time between earthquake and arrival of a wave at the coast is often less than 30 min. Therefore, a warning should be emitted after only 5–10 min to give at least some time for evacuation or other mitigation measures. A second important constraint is the scarcity of measurements. Generally, the seismic information (location, magnitude, and depth of the earthquake hypocenter) and at most $\mathcal{O}(1)$ other sensor readings are available within the first few minutes after the rupture.

2 Tsunami Propagation and Inundation Modeling

In order to generate tsunami scenarios for hazard assessment of possible future tsunami events, to allow for the assimilation of measurements, or to compute forecasts in case of a submarine earthquake, the tsunami phenomenon needs to be simulated with high degree of accuracy. In the deep ocean, the phenomenon can be modeled very accurately by shallow water theory, since the simplifying assumptions are fulfilled. The (nonlinear) shallow water equations with Coriolis and bottom friction terms

$$\partial_t \mathbf{v} + (\mathbf{v}\cdot\nabla)\mathbf{v} + \mathbf{f}\times\mathbf{v} + g\nabla\zeta + \frac{gn^2\mathbf{v}|\mathbf{v}|}{H^{4/3}} - \nabla\cdot(A_h\nabla\mathbf{v}) = 0, \tag{1}$$

$$\partial_t \zeta + \nabla\cdot(\mathbf{v}H) = 0, \tag{2}$$

hold under the assumptions that $L \gg h$, $h \gg \zeta$, and $W \gg h$, where L is the typical length scale of the domain, W is the typical wave length, h is the mean water depth, and ζ is the wave height over mean sea level. In (1) and (2), $\mathbf{v} = (u(\mathbf{x},t), v(\mathbf{x},t))$ is the horizontal depth-averaged velocity vector, $\mathbf{x} = (\lambda, \phi)$ the coordinate, t the time, f the Coriolis parameter, g the gravitational constant, $H(\mathbf{x},t) = h(\mathbf{x}) + \zeta(\mathbf{x},t)$ the total water depth, composed of the mean water depth h and the surface elevation ζ, n the Manning roughness coefficient, and A_h corresponds to an eddy viscosity.

For tsunami propagation near the shore (where $h \approx \zeta$), these equations are often used in practical applications, but physically they are not valid any more. In these situations, either fully three-dimensional equations could be used, for example, of Boussinesq type or full Navier–Stokes equations. More frequently, nonhydrostatic correction terms are applied, as in Stelling and Zijlema (2003).

2.1 Operational Tsunami Simulation Software

In the framework of the UNESCO International Oceanographic Commission's (IOC) Intergovernmental Coordination Group (ICG) for a Tsunami Warning System for the North-East Atlantic, the Mediterranean and Adjacent Seas (NEAMTWS) effort, the author and coauthors have conducted a survey of currently used operational tsunami simulation systems. No less than 25 different implementations, mostly of the linear or nonlinear shallow water equations are considered operational tsunami simulation models to date. Most of these models use a finite difference discretization scheme, but some use finite elements or finite volumes as their spatial discretization. The most prominent tsunami simulation models are the method of splitting tsunami (MOST) model developed by Titov and Gonzalez (1997) and the TUNAMI model and derivations from it (Goto et al. 1997).

In the author's group an unstructured finite element mesh model, called TsunAWI has been developed (Harig et al. 2008). This model is a typical operational code, in that it is based on Eqs. 1 and 2. It comprises an inundation scheme according to Lynett et al. (2002). While most current tsunami simulation codes use finite difference approximations, the operational code mentioned here uses a finite element discretization with a special pair of elements. Finite elements have been used for a long time in tsunami simulation (see, e.g., Gopalakrishnan and Tung 1980–1981). However, finite element methods have long been unsuccessful in entering the operational model arena. In the next section, we will demonstrate the superior performance of the integrated approach.

2.2 Unstructured Grid Tsunami Simulation

In order to generate an unstructured mesh, we utilize the free mesh generation software (*triangle*) by Shewchuk (1996). The mesh is refined, where either the water is shallow (or where there

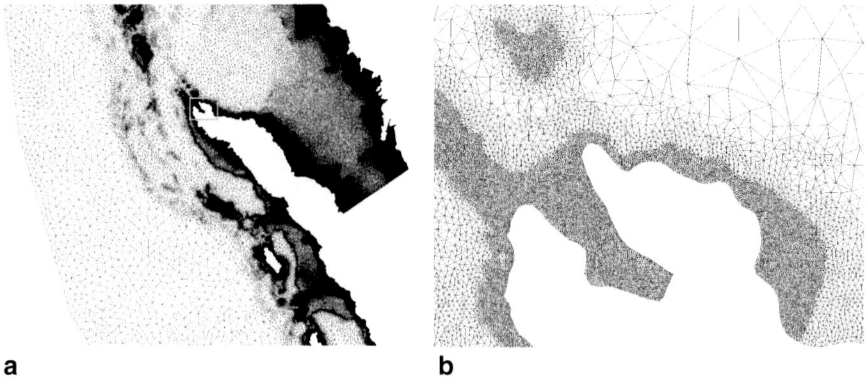

Fig. 2
(a) Locally refined triangular grid for the Indian Ocean close to the northern tip of Sumatra. Note that the grid extends on land and the grid boundary is aligned approximately with the 50 m topography height. (b) Zoom into the Banda Aceh region with an arbitrarily defined refinement area

is land) and where the bathymetry changes rapidly. More precisely, the following criterion for refinement is used:

$$\Delta x \leq \min\left\{c_t\sqrt{gh}, c_\nabla \frac{h}{\nabla h}\right\}, \tag{3}$$

where Δx denotes the mesh size (edge length), c_t and c_∇ are suitably chosen constants, g is the gravitational constant, and h is the mean water depth. By this criterion an almost uniform Courant number can be achieved for the whole simulation domain, even though the mesh size varies dramatically. Given a fixed time-step size Δt the Courant number is defined as

$$\nu = \mathbf{v}\frac{\Delta t}{\Delta x}.$$

Since $\mathbf{v} \approx \sqrt{gh}$, we can chose *one* time-step Δt suitably such that the stability criterion $\nu < 1$ can be fulfilled. A detail of a mesh, for the northern tip of Sumatra is depicted in *Fig. 2*. It is noteworthy that by making the constant $c_t = c_t(\mathbf{x})$ space-dependent, it is possible to arbitrarily refine certain areas of interest (see *Fig. 2*).

The locally refined triangular mesh allows for accurate representation of complex geometries. It has been shown by the author's group that the seamless representation of different scales of waves in deep ocean and in coastal waters gives accurate results in both, the wave propagation and the inundation behavior (Harig et al. 2008).

2.3 Adaptive Mesh Refinement for Tsunami Simulation

In addition to the operational multiscale tsunami simulation code TsunAWI, a research code with adaptive mesh refinement has been developed. Adaptive mesh refinement appears appropriate for online tsunami computations, since the mesh can be defined dynamically in correspondence with the propagating wave, gaining efficiency. A structured mesh approach to adaptive tsunami modeling is described in George (2008). The adaptive mesh generation in our approach is based on amatos, a software library for handling triangular dynamically adapted meshes (Behrens et al. 2005). amatos comprises a number of advanced techniques in order to gain efficiency and support local dynamic mesh adaptation with great ease of use. One of the outstanding features is the built-in space-filling curve ordering of mesh items (nodes, edges, faces), which allows for cache efficiency and simple load balancing in parallel applications (Behrens 2005). This approach relies on a certain mesh refinement strategy and also on an initially consistent mesh (Behrens and Bader 2009).

The fully automatic method starts with a domain boundary (usually a bitmap file with different pixel values for the interior and the exterior of the computational domain) as an input (➲ *Fig. 3*). It then creates an initial mesh of four cells covering the entire bitmap. This initial mesh is locally refined along the boundary. The idea is to finally clear the grid of undesired elements and take the thus created mesh as an initial mesh for further adaptively refined computations. It turns out that the mesh needs to be smoothed and postprocessed in order to obtain triangles, which adhere to certain quality criteria. The boundary nodes shall be aligned with the true boundary (coastline) of the domain, the coastline shall be locally smooth and coastal cells shall not be too distorted. The details are rather technical, but at the end a satisfying mesh can be obtained by this process, without any manual interaction. A different approach to automatic

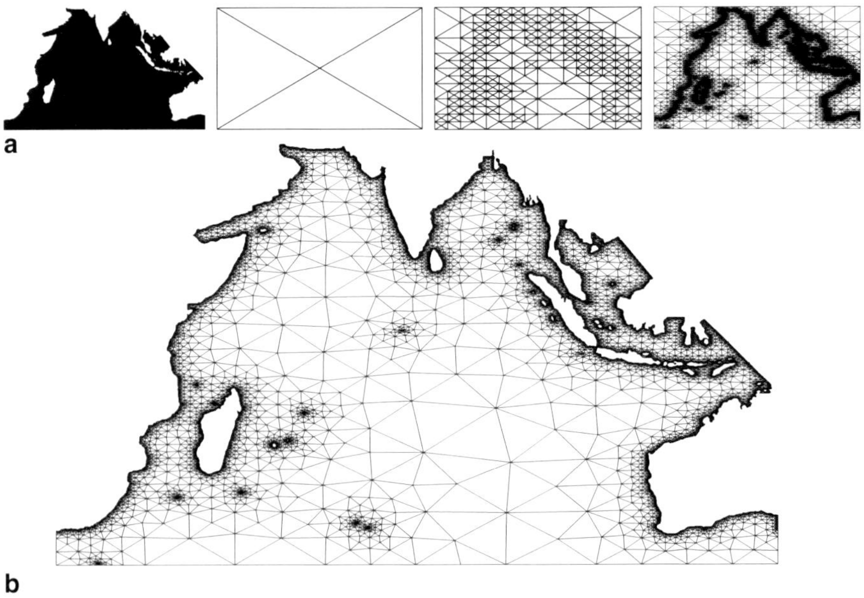

Fig. 3

(a) Four steps in an automatic mesh generation process, which starts with a geometry bitmap and a starting mesh covering the domain with just four cells. Refinement along the bitmap's boundaries yields a locally refined mesh. (b) after edge alignment at the boundary and smoothing the mesh, cells outside the domain are cleared, leaving a numerically well behaved mesh of the complex domain

Table 1
Refinement criteria for adaptive tsunami simulations

Type	Formula	Remark
Indicator	$\eta_\tau = \|\zeta_\tau\|$	Simple value
Indicator	$\eta_\tau = \|\nabla\zeta\|$	Simple gradient
Error estimator	$\eta_\tau = \|\zeta_\tau - \overline{\zeta_{\mathcal{N}(\tau)}}\|$	Averaging discretization error estimator

mesh generation and in particular for automatic shoreline approximation for complex domains like ocean basins is described in Gorman et al. (2008).

By the described procedure, the mesh remains organized in a space-filling curve compatible order. This preserves the optimization properties of amatos even with complex domains. The numerical scheme, used in the adaptive code is essentially that of TsunAWI. In particular, a $P^1 - P^1_{NC}$ conforming/nonconforming finite element pair derived from the work of Hanert et al. (2005) is used. Since the adaptation criterion is not as straight forward as in the fixed mesh approach and does not necessarily preserve a bounded Courant number, a step-size control is applied in the adaptive code.

Special attention is needed for the adaptation criterion. In general, we employ a very simple refinement strategy, based on an l_2-norm approach. A cell is refined, whenever the local error η_τ, τ being the cell under consideration, meets the following inequality:

$$\eta_\tau \geq \theta m^{-1/2} \|\eta_m\|_2,$$

where η_m is the vector of all local errors, θ is a tuning factor, and m is the number of cells. It remains to be defined, how to compute the local error or refinement criterion η_τ.

It turns out that for tsunami wave propagation simulations, the simplest refinement criterion performs best. Several criteria have been tested. They are summarized in *Table 1*. As the simplest indicator, the wave height can be used. The rationale: wherever the wave height ζ is larger than a given threshold, the mesh is refined. It turns out that this criterion is robust and efficient in this special application. Another indicator for refinement is the gradient. If the gradient of the wave is steep, a refinement is needed. However, this criterion is not robust, since the wave becomes very shallow in deep ocean. Finally, a mathematical error estimator, an averaging method—originally introduced by Zienkiewicz and Zhu (1987) for elliptic problems—can be applied. For this type of applications it turns out to be more involved and not as robust as the simple wave height indicator. In the averaging indicator, the computed value within one cell is compared to the average of the surrounding cell's wave heights.

In a preliminary evaluation of the scheme, the minor Andaman tsunami following a magnitude 7.6 earthquake in the Andaman Sea on August 10, 2009 has been simulated. Sea level measurements exist from a nearby deep-ocean assessment and reporting on tsunamis (DART) buoy. *Figure 4* shows the wave propagation on the adaptively refined mesh. Additionally, the wave gauge readings are compared to simulation results by TsunAWI (the operational code) and to the adaptive simulation. Note that the initial conditions are not known exactly for this event. Therefore, the simulations show relatively good correspondence with the measurements (*Fig. 5*).

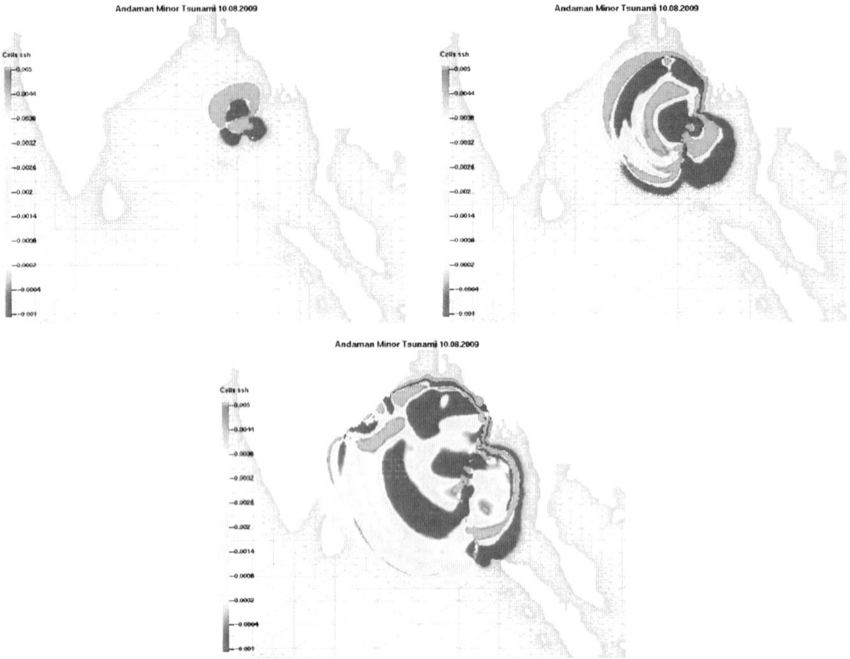

◘ Fig. 4
Wave propagation simulation of the minor Andaman tsunami of August 10, 2009. Snapshots are shown after 28 min, 84 min, and 140 min

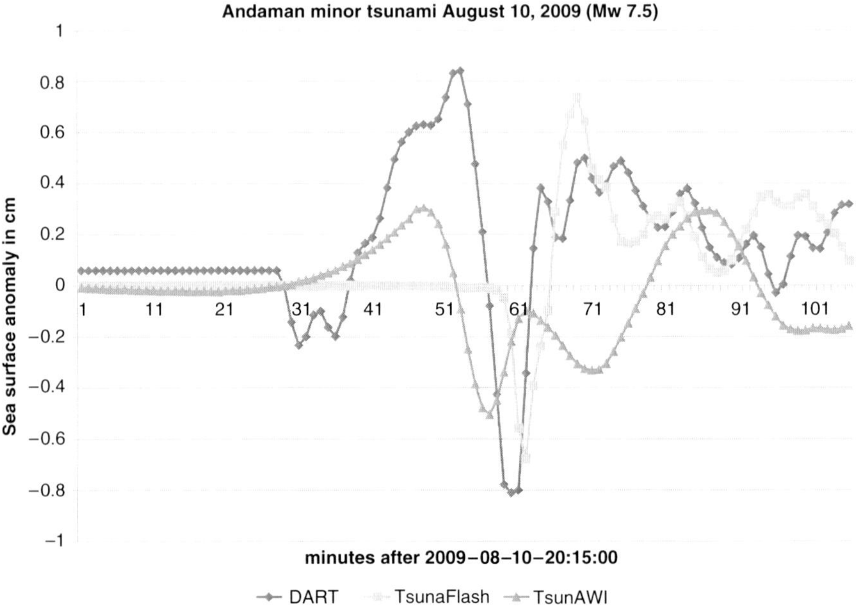

◘ Fig. 5
Comparison of wave gauge measurements, simulation values from TsunAWI and the adaptive code

2.4 Advanced Topics

In this section, a few more advanced methods for simulating tsunamis are listed. We will not give a detailed description, but will only report on these efforts, since they mark elements of future developments.

An important issue in tsunami simulation is to accurately represent initial conditions for the wave propagation. While some convergence has been achieved for subduction zone (normal slip) earthquakes, using uniform plate descriptions (Mansinha and Smylie 1971) or more complex descriptions of the plate rupture (Wang et al. 2003), landslide induced initial conditions are still under active research. About 10–15% of all tsunamis are generated by landslides. It is anticipated that landslide induced tsunamis need a more sophisticated model description (Ward 2001; Berndt et al. 2009).

Inundation modeling still needs scientific attention. In order to simulate the effect of tsunami run-up on land, more sophisticated model formulations are needed. In this case, small-scale turbulent as well as structure–fluid-interaction processes play an important role. Current engineering approaches do not cover these processes and can therefore be misleading when using the modeling results quantitatively. First approaches for improving current standard tsunami simulation models can be found in Yuk et al. (2006).

3 Forecasting Approaches

In order to mitigate the effects of a tsunami event, tsunami early warning systems are in place in several tsunami-prone oceans, and are being installed, where not available. These systems comprise forecasting capabilities with diverse approaches. Most of the current systems rely on a direct relation of earthquakes and tsunamis.

3.1 Existing Tsunami Early Warning Systems

Operational tsunami early warning systems prior to the 2004 Sumatra-Andaman tsunami existed for the Pacific ocean. Two of these systems, namely the one operated by the Japan Meteorological Agency (JMA) and the National Oceanographic and Atmospheric Administration (NOAA) were supported by tsunami simulation (Synolakis and Bernard 2006; Furumoto et al. 1999). Several other systems have since been installed. To be mentioned are the systems operated by Bureau of Meteorology (BoM) in Australia (Greenslade et al. 2007) and Indian National Center for Ocean Information Services (INCOIS). Since the latter systems implement some derivation of the first two, we will describe these as archetypes for different approaches to tsunami early warning.

Since most of the tsunamis threatening US coastal areas are initiated far away from these areas, the NOAA system relies on time. A first and preliminary warning is generated right from earthquake parameters, and based on a decision matrix, which takes earthquake parameters (magnitude, depth, location under the sea) into account. Initial source parameters for tsunami simulations are derived from the preliminary earthquake parameters and pre-computed linear wave propagation scenarios are linearly combined to give a first estimate of wave heights along the coast. A number of DART buoys are positioned far enough from potential tsunami sources so that a clear wave signal can be detected. From the DART wave height and arrival time measurements, a correction to the linear combination of unit sources is inverted to give more accurate forecasts of expected wave arrival times and heights. Additionally, for selected areas of interest, local nonlinear inundation simulations are performed online. This complete system has been applied successfully for the August 2007 Peruvian tsunami (Wei et al. 2008).

While the NOAA system is designed with far-field tsunamis in mind, allowing for some time to pass, before giving accurate and localized tsunami hazard forecasts, the JMA system has been successfully in place to give warnings with only short warning times (near-field tsunamis). Therefore, it gives only estimated information on local wave behavior, but relies on a quick assessment of tsunami wave occurrence with the aim to either confirm or to cancel the warning within short time. However, the simulation is much less accurate, since coastal wave heights are derived from heuristic formulae and wave heights at near-shore control points (Furumoto et al. 1999).

Most of the TEWS installed currently, are derivations in one or the other form of the aforementioned two archetypical systems. In the next section, we will describe a third and different approach to tsunami forecasting, developed by the German–Indonesian Tsunami Early Warning System (GITEWS) consortium and implemented in the Indonesian tsunami early warning system (InaTEWS) operated by the Agency for Meteorology, Climate and Geophysics (BMKG) in Jakarta, Indonesia.

3.2 Multisensor Selection Approach

The start of the development of a near-field tsunami warning system is a thorough sensitivity analysis and uncertainty quantification. From the sensitivity it becomes clear that a precise knowledge of the rupture area would be necessary. One example is given in ❯ *Fig. 6*. Two different rupture areas may correspond to exactly the same seismic parameters. Thus, in a warning situation, they would not be distinguishable. Yet, the effect for a specific coastal area (in this case the West-Sumatran city of Padang) would be critically different: in one case large parts of the city would be inundated, in the other case, only small waves would penetrate the beaches. A false warning—either false positive or, even worse, false negative—would be very likely.

In order to quantify this behavior, a very simple linear uncertainty propagation theory is developed. Let us assume a mapping (forecast functional), which takes any relevant measurement values to a forecast and is usually realized by a simulation model. One example is the linear combination of unit sources to obtain a rupture area, given seismic parameters, as implemented in the warning systems of JMA and NOAA (see previous section). Other mappings are given in ❯ *Table 2*.

In each of these mappings, an uncertainty in the input parameters exists. Additionally, depending on the sensitivity, the mapping amplifies the uncertainty. So, instead of a single input value (data set) for the forecast functional, we have to assume a whole set of equally probable (indistinguishable) input values. More formally, we assume an input set $\mathcal{J}$, an input value ι, and a neighborhood $\mathcal{N}(\iota)$ with a maximum diameter $\Delta\iota$, such that $j \in \mathcal{N}(\iota)$ if $\|j - \iota\| < \Delta\iota$. The mapping acts on $\mathcal{J}$ and maps into the set of forecasts $\mathcal{F}$. It takes the input neighborhood to an output (forecast) set. Thus, let

$$M : \mathcal{J} \longrightarrow \mathcal{F}; \quad \iota \mapsto \phi,$$

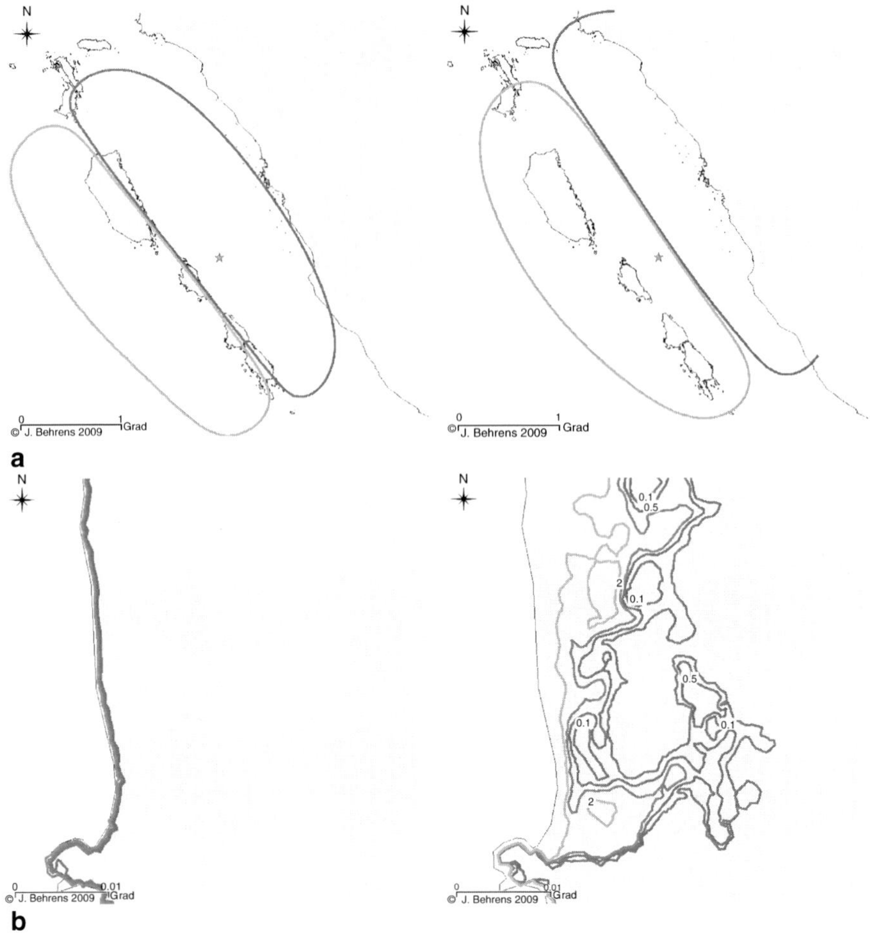

Fig. 6

(a) Two different hypothetical uplift areas corresponding to the exact same seismic parameters (epicenter is shown in map, magnitude is assumed to be Mw 8.5). Left uplift area lies just 50 km further to the west. (b) Flow depth corresponding to the two different uplift areas for Padang showing that the inundation depends sensitively on the exact uplift area

Table 2

Common forecast functionals in tsunami forecasting

Funct.	Description	Input	Output
R	Seismic to rupture	Seismic parameters	Rupture
W	Rupture to wave height	Rupture parameters	Wave
G	GPS dislocation to rupture	Dislocation vectors	Rupture
H	Deep ocean wave height to coastal wave height	Wave height	Wave
S_{GITEWS}	Seismic to scenario	Seismic params.	Scen. ID
G_{GITEWS}	GPS dislocation to scenario	Dislocation vectors	Scen. ID
H_{GITEWS}	Gauge wave height to scenario	Wave height	Scen. ID

then we assume that $M(\mathcal{N}_{\mathcal{J}}(\iota)) = \mathcal{N}_{\mathcal{F}}(\phi)$, with the diameter of $\mathcal{N}_{\mathcal{F}}(\phi)$ being a multiple of the diameter of $\mathcal{N}_{\mathcal{J}}(\iota)$. In other words:

$$\text{if } \|j - \iota\| < \Delta\iota, \text{ then } \|f - \phi\| < \kappa \cdot \Delta\iota, \forall f \in \mathcal{N}_{\mathcal{F}}(\phi). \tag{4}$$

So, κ is the uncertainty amplification factor of the forecast functional. It depends mainly on the sensitivity in the forecast functional. From the example above, we have seen that the sensitivity for the mapping that takes seismic parameters to rupture area is large, i.e., κ for this functional would be large. If κ is large, then false warnings are very likely. It is thus our aim to reduce the uncertainty amplification.

Traditional warning systems use combinations of these forecast functionals. Usually, the seismic parameters are used to derive a rupture area, which then is used for either a propagation simulation or an inundation simulation. This means, usually we have two composed functionals, for example, $W \circ R$. This means we have to compose the uncertainty amplification. Let us assume that $R : \mathcal{S} \longrightarrow \mathcal{R}$ maps from the space $\mathcal{S}$ of seismic parameters to $\mathcal{R}$ of rupture parameters and thus propagates uncertainty neighborhood $\mathcal{N}_{\mathcal{S}}(s) \mapsto \mathcal{N}_{\mathcal{R}}(r)$ with the property $\|\mathcal{N}_{\mathcal{S}}(s)\| < \Delta s$ then $\|\mathcal{N}_{\mathcal{R}}(r)\| < \kappa_R \Delta s$. From here $W : \mathcal{R} \longrightarrow \mathcal{W}$ maps from the space $\mathcal{R}$ of rupture parameters to $\mathcal{W}$ of wave heights with a neighborhood mapping $\mathcal{N}_{\mathcal{R}}(r) \mapsto \mathcal{N}_{\mathcal{W}}(w)$ and an uncertainty propagation $\|\mathcal{N}_{\mathcal{R}}(r)\| < \Delta r$ then $\|\mathcal{N}_{\mathcal{W}}(w)\| < \kappa_W \Delta r$. So, from (4) and the above we can derive

$$\|\mathcal{N}_{\mathcal{S}}(s)\| < \Delta s \text{ then } \|\mathcal{N}_{\mathcal{W}}(w)\| < \kappa_W \Delta r \le \kappa_W \cdot \kappa_R \Delta s;$$

in other words, the uncertainty amplification factors are multiplied, largely increasing uncertainty.

The new approach, implemented in the GITEWS system, works differently. By defining several forecast functionals that map from independent measurements to scenarios, from which wave heights and other data can be extracted. These diverse functionals are not composed but independently combined. Taking the nomenclature from *Table 2*, we can define the GITEWS functionals, all mapping into the same space of scenario data:

$$\begin{aligned} S_{\text{GITEWS}} : \mathcal{S} &\longrightarrow \mathcal{I}, \quad \mathcal{N}_{\mathcal{S}}(s) \mapsto \mathcal{N}_S(i),\\ G_{\text{GITEWS}} : \mathcal{G} &\longrightarrow \mathcal{I}, \quad \mathcal{N}_{\mathcal{G}}(g) \mapsto \mathcal{N}_G(i),\\ H_{\text{GITEWS}} : \mathcal{W} &\longrightarrow \mathcal{I}, \quad \mathcal{N}_{\mathcal{W}}(w) \mapsto \mathcal{N}_H(i). \end{aligned}$$

Even though these functionals all map into the same space, each does so with an individual neighborhood and uncertainty amplification factor. Now, only those scenarios contained in the intersection of these neighborhoods satisfy all independent inputs simultaneously. This is depicted in *Fig. 7*. It is important to note that with this methodology, the uncertainty is reduced by independent measurements of the same physical phenomenon, which are combined in all the available representing scenarios. So, instead of amplifying uncertainty by chaining different forecast functionals, each functional reduces uncertainty by ruling out certain scenarios, which do not match the measured data.

3.3 Implementing the Matching Procedure

So far, we have described how to reduce uncertainty in the forecast process. The forecast functionals are not implemented the same way, they are used for the analysis. The selection of those

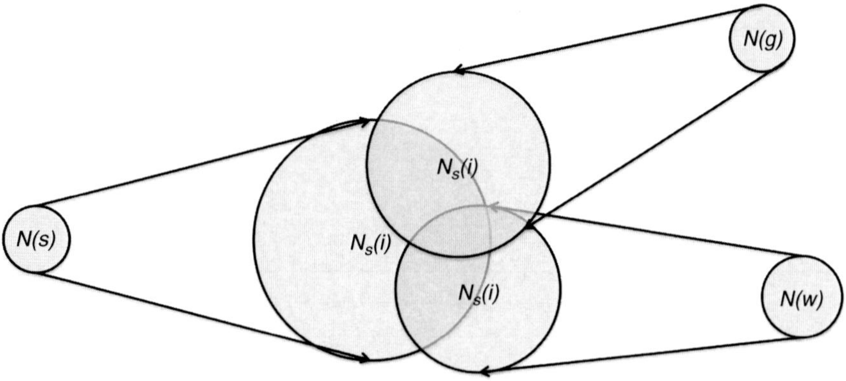

Fig. 7

Combination of GITEWS forecast functionals with reduced uncertainty in the intersection of the mapped uncertainty neighborhoods

scenarios that lie in the intersection of all the forecast functional's image spaces is performed in a matching algorithm, consisting of three essential steps.

First, each of the different types of measurements needs to be compared individually with corresponding values in the precomputed scenarios. This is achieved by individual distance norms. To give an example, the seismic parameter *earthquake location* is given as a pair of coordinates (λ, ϕ). Now, for each scenario $i = 0, \ldots, N$, the individual distance (Euklidean norm)

$$d^i_{\text{loc}} = \|(\lambda, \phi) - (\lambda, \phi)^i\|_2$$

is computed. Similar distance norms are computed for the seismic parameters *magnitude* and *depth*. For each available gauge position the gauge parameters *arrival time* and *wave time series*, and for the GPS sensors the GPS parameter *dislocation vector* are computed analogously.

In a second step, these individual norms are scaled such that the values are normalized, i.e., lie in the unit interval. In this step, small differences are amplified in order to increase resolution and improve the separation between similarly matching scenarios.

The third step combines the individual norms into an aggregated mismatch value. This is achieved by a nested weighted sum approach. In a first substep an aggregate norm is formed for each sensor group (and each scenario). All available arrival times, for example, form a sensor group "arr"; similarly we denote by "ts" the time series, "dis" the dislocation vectors, by "dep" the earthquake depth, and by "mag" the magnitude. Then

$$\begin{aligned} d^i_{\text{dis}} &= \sum_{s \in \{\text{GPS sensors}\}} w_{\text{GPS},s} d^i_{\text{GPS},s}, \\ d^i_{\text{arr}} &= \sum_{s \in \{\text{gauge sensors}\}} w_{\text{gauge},s} d^i_{\text{arr},s}, \\ d^i_{\text{ts}} &= \sum_{s \in \{\text{gauge sensors}\}} w_{\text{gauge},s} d^i_{\text{ts},s}, \end{aligned}$$

where $w_{\text{group},s} \in [0,1]$ is a weight indicating the reliability or availability of the sensor s. In the second substep, for each scenario i the mismatch is aggregated by

$$d^i_{\text{mismatch}} = \sum_{g \in \{\text{loc, dep, mag, arr, ts, dis}\}} W_g \, d^i_g. \tag{5}$$

Note that this aggregated value is a generalized distance norm. It measures the mismatch between a scenario i and a given set of sensor values. If the weights are normalized (i.e., $\sum_g W_g = 1$), then the mismatch is normalized. Thus, a mismatch of 0 means perfect match, and a mismatch of 1 gives the maximum distance between measurement and scenario. Note further that by setting some weights to 0 (and renormalizing), missing values can easily be taken into account.

So far, we have not formulated the notion of uncertainty neighborhoods. In fact, instead of just computing the plain individual distance norms, an uncertainty range is defined for each of the individual norms, in which the difference between the measurement and corresponding scenario values is indistinguishable. Then, a number of scenarios have perfect matches in terms of individual distances or all lie in the uncertainty neighborhood of the image space. With this notion, all those scenarios that have an aggregated mismatch of 0 are in the intersection of the uncertainty neighborhoods as depicted in ➋ *Fig. 7*.

In many realistic cases, and as soon as a sufficiently large number of sensors have detected signals of a tsunami event, the mismatch will not be 0, meaning that the intersection of all forecast functionals is empty. But the mismatch then still gives an objective measure of the best-fitting scenario. Additionally, all those scenarios with a mismatch under a certain threshold could be used for additional statistical analysis and evaluation of uncertainty.

3.4 Experimental Confirmation

In order to confirm the theoretical behavior of the procedure, we define a test case for evaluation. Since we do not have enough relevant real-life data yet, a synthetic test case has been defined. (The author acknowledges the contribution of A. Babeyko in defining and computing the data for this test case.) It has a complex structure, an epicenter location at the rim of the nonuniform rupture area. The assumed magnitude is Mw 8.0 and epicenter location is at (99.93°E, 1.96°S). A sensor network of GPS stations and two deep ocean wave gauges serve with their measurements. And two reference gauges in Padang and Bengkulu are taken as the benchmark. The sensor network is depicted in ➋ *Fig. 8*.

Using only seismic information for the forecast (i.e., the epicenter location and the magnitude), a large uncertainty on the true rupture area remains. Therefore, a large number of possible scenarios match the given data. This is visualized by two different perspectives. Assuming that each scenario can be characterized by a virtual epicenter, which is located in the center of the rupture area, and assuming that the measured given epicenter needs to be located within the rupture area (but not necessarily in the center), all those virtual epicenters corresponding to scenarios with rupture areas containing the measured epicenter, can be plotted. This is depicted in ➋ *Fig. 9* and it shows the uncertainty regarding the rupture area. On the other hand, the uncertainty at the reference gauges can be visualized. A large number of different time-series indicate a large uncertainty and a difficult situation for someone obliged to issue a warning

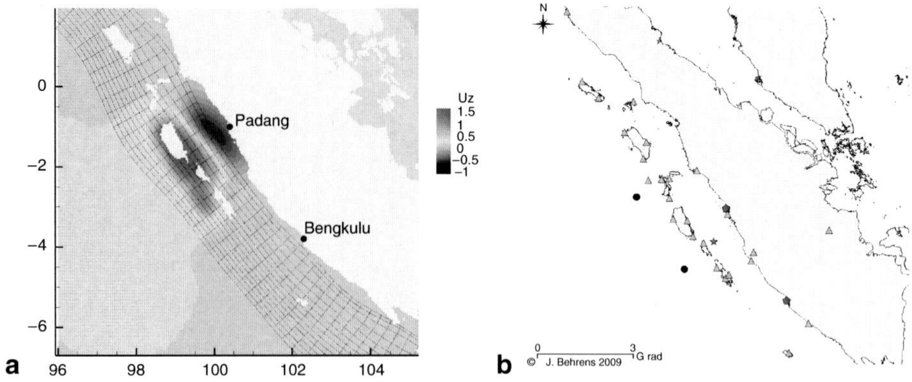

Fig. 8

(a) Initial uplift for the test case; (b) sensor network for the test case. Circles indicate buoys, triangles are GPS stations, the star marks the epicenter, and pentagons are located at the benchmark gauge positions

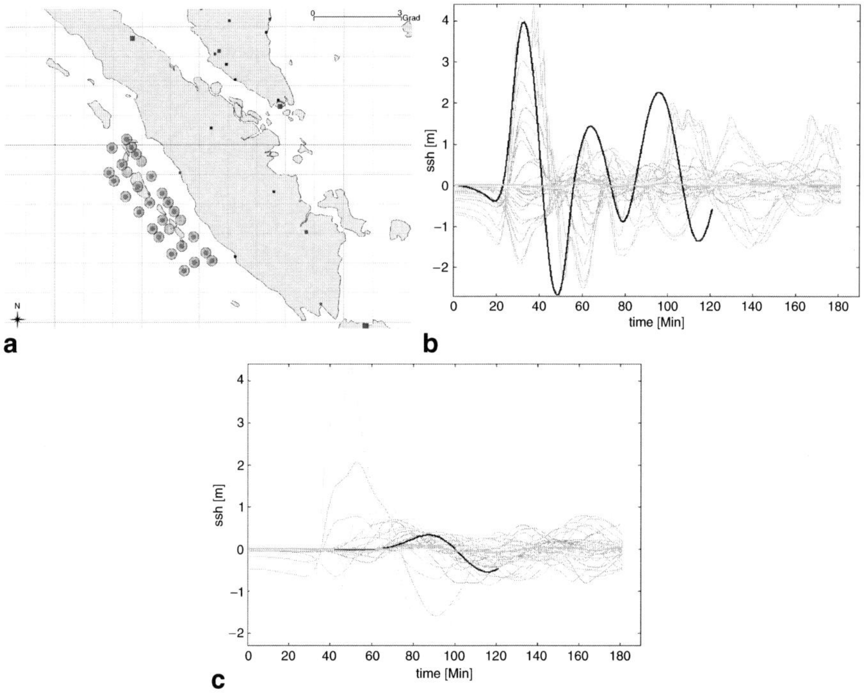

Fig. 9

(a) Possibly matching scenarios for given seismic parameters; (b) wave time series for Padang, for all matching scenarios; (c) wave time series for Bengkulu, for all matching scenarios. The thick black line represents the benchmark "truth"

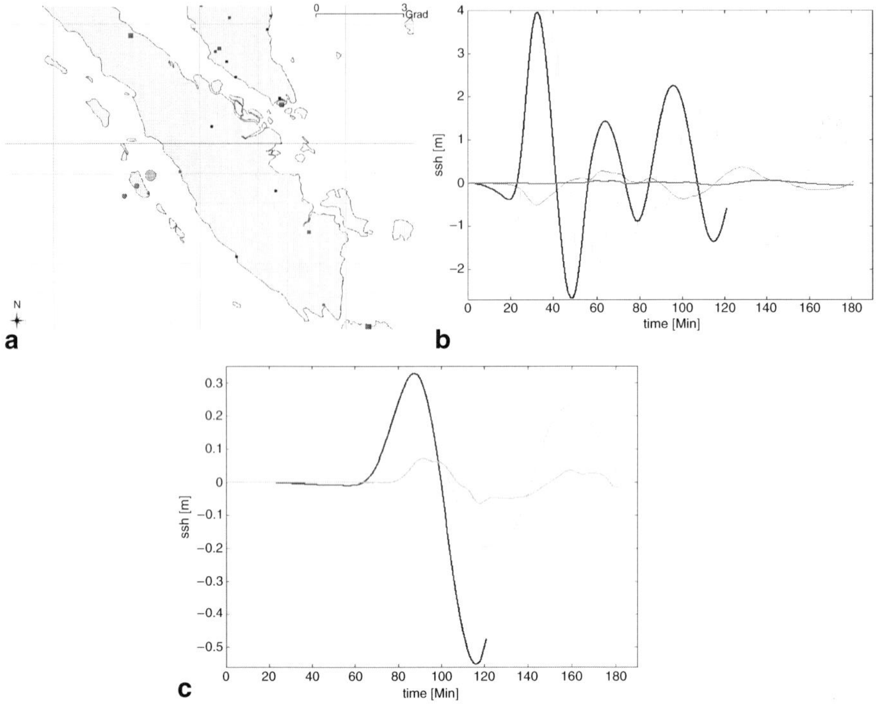

Fig. 10
(a) Possibly matching scenarios for multisensor selection approach; (b) wave time series for Padang; (c) wave time series for Bengkulu

message. Again referring to *Fig. 9*, we see this situation, when only seismic information is used for the forecast. While for the Padang reference gauge, a warning based on the worst case would have been reasonable in spite of the large uncertainty, in Bengkulu this decision would have been a heavy overestimation of the true event, leading to a false positive warning.

If the multisensor selection is used instead, the number of matching scenarios decreases significantly. Only three scenarios remain, that match the given data. There is still some uncertainty remaining, as can be seen from the plots in *Fig. 10*. However, if a warning had been issued based on the worst case, in this situation the estimated arrival times and maximum wave heights would have been quite accurate and definitely correct.

4 Future

Future developments in geomathematical methods for tsunami modeling will concern both application fields, described in this presentation: the inundation modeling and the forecasting. In inundation modeling more sophisticated simulation methods are necessary, including fully three-dimensional models with structure–fluid interaction to account for collapsing buildings, and sedimentation process simulation. A particle method–based approach can be found in Ward (2009).

The multisensor selection–based forecasting method is suitable for several extensions. Once a detection pattern can be identified, in other words suitable input can be measured, and a suitable model can be formulated that maps the detection to a scenario, this can be included into the methodology. For example, an inclusion of land slide mechanics with detection via tilt-meters is viable (Berndt et al. 2009). The inclusion of acoustic sensors might be an option, when applying sophisticated models, including acoustic wave propagation. HF radar based–wave pattern recognition might also be applied to tsunami early warning. However, since the tsunami waves in deep ocean are basically gravity waves, it is still to be demonstrated, how the Doppler-based velocity pattern detection could be used for detecting these types of waves.

5 Conclusion

It has been demonstrated how advanced numerical methods in solving partial differential equations, in combination with an analog assimilation technique, and simple uncertainty quantification and propagation methods can be used to develop accurate and efficient tsunami early warning capabilities. The analog method relies on precomputed tsunami scenarios, which represent the detectable indicators for a tsunami event, namely earthquake parameters, wave heights, and Earth crust dislocation. A rigorous yet simple uncertainty analysis shows how to reduce uncertainty in the near-field tsunami forecasting context.

An adaptive mesh state-of-the-art simulation technology is demonstrated and validated. Operational considerations for tsunami modeling and a number of current operational models are introduced. Current modeling standards adhere to certain validation and verification guidelines as accepted in the community and summarized, for example, in Synolakis et al. (2008). It is shown that unstructured and locally refined meshes show superior accuracy and efficiency within the context of unified tsunami propagation and inundation modeling.

Acknowledgments

The author would like to thank his former group at Alfred Wegener Institute for Polar and Marine Research, Bremerhaven, for substantial contributions and evaluations in developing the simulation software TsunAWI and the simulation system within GITEWS. Widodo Pranowo's contribution to developing and testing the adaptive mesh tsunami code is gratefully acknowledged. This work was partly funded by the Federal Ministry for Education and Research (BMBF) in the GITEWS framework under contract no. 03TSU01.

References

Behrens J (2005) Multilevel optimization by space-filling curves in adaptive atmospheric modeling. In: Hülsemann F, Kowarschik M, Rüde U (eds) Frontiers in simulation—18th symposium on simulation techniques. SCS Publishing House, Erlangen, pp 186–196

Behrens J, Bader M (2009) Efficiency considerations in triangular adaptive mesh refinement. Phil Trans R Soc A 367: 4577–4589

Behrens J, Rakowsky N, Hiller W, Handorf D, Läuter M, Päpke J, Dethloff K (2005) amatos: parallel adaptive mesh generator for atmospheric and oceanic simulation. Ocean Modell 10 (1–2): 171–183

Berndt C, Brune S, Euan S, Zschau J, Sobolev SV (2009) Tsunami modeling of a submarine landslide in the fram strait. Geochem Geophys Geosyst 10: Q04009

Furumoto AS, Tatehata H, Morioko C (1999) Japanese tsunami warning system. Sci Tsunami Hazards 17 (2): 85–105

George DL (2008) Augmented Riemann solvers for the shallow water equations over variable topography with steady states and inundation. J Comput Phys 227 (6): 3089–3113

Gopalakrishnan TC, Tung CC (1980–1981) Run-up of non-breaking waves – a finite-element approach. Coastal Eng 4: 3–22

Gorman GJ, Piggot MD, Wells MR, Pain CC, Allison PA (2008) A systematic approach to unstructured mesh generation for ocean modelling using GMT and Terreno Comput Geosci 34: 1721–1731

Goto C, Ogawa Y, Shuto N, Imamura F (1997) IUGG/IOC TIME project—numerical method of tsunami simulation with the leap-frog scheme. UNESCO, http://ioc3.unesco.org/itic/links.php?go=139

Greenslade DJM, Simanjuntak MA, Burbidge D, Chittleborough J (2007) A first-generation real-time tsunami forecasting system for the Australian region. Research Report 126, Bureau of Meteorology, BMRC, Melbourne, Australia. www.bom.gov.au/pubs/researchreports/researchreports.htm

Hanert E, Le Roux DY, Legat V, Deleersnijder E (2005) An efficient Eulerian finite element method for the shallow water equations. Ocean Modell 10: 115–136

Harig S, Chaeroni C, Pranowo WS, Behrens J (2008) Tsunami simulations on several scales: comparison of approaches with unstructured meshes and nested grids. Ocean Dyn 58: 429–440

Lynett PJ, Wu TR, Liu PLF (2002) Modeling wave runup with depth-integrated equations. Coast Eng 46: 89–107

Mansinha L, Smylie DE (1971) The displacement fields of inclined faults. Bull Seismol Soc Am 61(5): 1433–1440

Shewchuk JR (1996) Triangle: engineering a 2D quality mesh generator and Delaunay triangulator. In: Lin MC, Manocha D (eds) Applied computational geometry: towards geometric engineering. Lecture notes in computer science, vol 1148. Springer, Berlin, pp 203–222

Stelling GS, Zijlema M (2003) An accurate and efficient finite-difference algorithm for non-hydrostatic free-surface flow with application to wave propagation. Int J Numer Meth Fluids 43: 1–23

Synolakis CE, Bernard EN (2006) Tsunami science before and beyond Boxing Day 2004. Phil Trans R Soc A 364: 2231–2265

Synolakis CE, Bernard EN, Titov VV, Kanoglu U, Gonzalez FI (2008) Validation and verification of tsunami numerical models. Pure Appl Geophys 165(11–12): 2197–2228

Titov VV, Gonzalez FI (1997) Implementation and testing of the method of splitting tsunami (most) model. NOAA Technical Memorandum ERL PMEL-112 1927, NOAA, Seattle, WA

Titov VV, Gonzalez FI, Bernard EN, Eble MC, Mofjeld HO, Newman JC, Venturato AJ (2005) Real-time tsunami forecasting: challenges and solutions. Nat Hazards 35: 41–58

Wang R, Martin FL, Roth F (2003) Computation of deformation induced by earthquakes in a multi-layered elastic crust—FORTRAN programs EDGRN/EDCMP. Comput Geosci 29(2): 195–207

Ward SN (2001) Landslide tsunami. J Geophys Res 106(B6): 11,201–11,215

Ward SN (2009) A tsunami ball approach to storm surge and inundation: application to hurricane Katrina, 2005. Int J Geophys (Article ID 324707), 13

Wei Y, Bernard EN, Tang L, Weiss R, Titov VV, Moore C, Spillane M, Hopkins M, Kanoglu U (2008) Real-time experimental forecast of the Peruvian tsunami of August 2007 for U.S. coastlines. Geophss Res Lett 35 (L04609): 1–7

Yuk D, Yim SC, Liu PL-F (2006) Numerical modeling of submarine mass-movement generated waves using RANS model. Comput Geosci 32: 927–935

Zienkiewicz OC, Zhu JZ (1987) A simple error estimator and adaptive procedure for practical engineering analysis. Int J Numer Meth Eng 24: 337–357

15 Efficient Modeling of Flow and Transport in Porous Media Using Multiphysics and Multiscale Approaches

Rainer Helmig · Jennifer Niessner · Bernd Flemisch · Markus Wolff · Jochen Fritz
Department of Hydromechanics and Modeling of Hydrosystems, Institute of Hydraulic Engineering, Universität Stuttgart, Stuttgart, Germany

W. Freeden, M.Z. Nashed, T. Sonar (Eds.), *Handbook of Geomathematics*, DOI 10.1007/978-3-642-01546-5_15,

Abstract Flow and transport processes in porous media including multiple fluid phases are the governing processes in a large variety of geological and technical systems. In general, these systems include processes of different complexity occurring in different parts of the domain of interest. The different processes mostly also take place on different spatial and temporal scales. It is extremely challenging to model such systems in an adequate way accounting for the spatially varying and scale-dependent character of these processes. In this chapter, we give a brief overview of existing upscaling, multiscale, and multiphysics methods, and we present mathematical models and model formulations for multiphase flow in porous media including compositional and non-isothermal flow. Finally, we show simulation results for two-phase flow using a multiphysics method and a multiscale multiphysics algorithm.

1 Introduction

Within hydrological, technical or biological systems, various processes generally occur within different sub-domains. These processes must be considered on different space and time scales, and they require different model concepts and data. Highly complex processes may take place in one part of the system necessitating a fine spatial and temporal resolution, while in other parts of the system, physically simpler processes take place allowing an examination on coarser scales. Considering the porous medium, its heterogeneous structure shows a high dependence on the spatial scale (Niessner and Helmig 2007).

➲ *Figure 1* shows a general physical system which typically consists of certain subregions. The darker regions represent subdomains of the total system dominated by other physical processes than the light part. These processes take place on finer spatial and temporal scales. While traditional models describe this system on one scale—this must be the fine scale if an accurate description is desired—*multiscale* algorithms take the dependence of the processes on both spatial and temporal scales into account, where the connection between both scales is made by up- and downscaling approaches. Much research has been done to upscale either pressure or saturation equation in two-phase flow or include the different scales directly in the numerical scheme by using multiscale finite volumes or elements, see, e.g., E and Engquist (2003b),

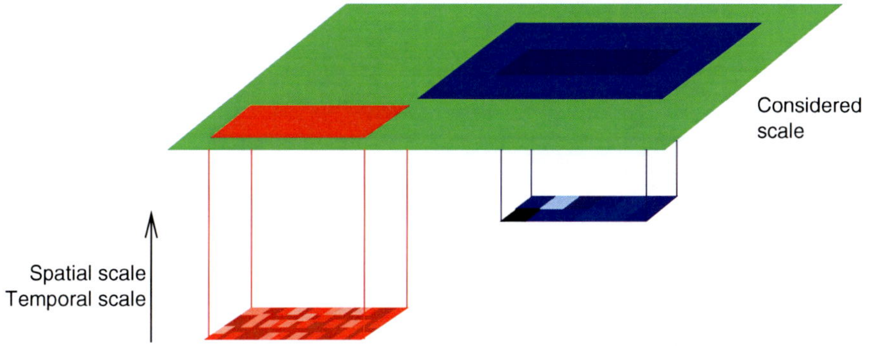

◘ Fig. 1
A general physical system where different processes occur in different parts of a domain and on different scales

E et al. (2003), Chen and Hou (2002), Hou and Wu (1997), Jenny et al. (2003), Durlofsky (1991), Renard and de Marsily (1997), Efendiev et al. (2000), Efendiev and Durlofsky (2002), Chen et al. (2003).

In constrast to the abilities of traditional models, *multiphysics* approaches allow to apply different model concepts in different subdomains. Like that, in the black part of the domain, other processes can be accounted for than in the lighter subdomains. In this respect, research has advanced in the context of domain decomposition techniques (see, e.g., Wheeler et al. 1999; Yotov 2002) and in the context of mortar finite element techniques that allow multiphysics as well as multinumerics coupling (see, e.g., Peszynska et al. 2002).

The advantage of multiscale multiphysics algorithms is that they require much less data as if using the fine-scale model in the whole domain. Additionally, they allow to save computing time or make the computation of very complex and large systems possible that could otherwise not be numerically simulated.

In the following, as an example application, the storage of carbon dioxide (CO_2) in a deep geological formation will be studied and multiscale as well as multiphysics aspects in space and time will be identified. Please note that multiscale and multiphysics aspects are relevant in a large number of additional applications, not only in geological systems, but also in biological (e.g., treatment of brain tumors) and technical (e.g., processes in polymer electrolyte membrane fuel cells) systems. Thus, multiscale multiphysics techniques developed for geological applications can be transferred to a broad range of other problems.

Concerning CO_2 storage, different storage options are commonly considered that are shown in ➋ *Fig. 2*, which is taken from IPCC (2005). According to that figure, possible storage possibilities are given by depleted oil and gas reservoirs, use of CO_2 in the petroleum industry in order to enhance oil and gas recovery or—in a similar spirit—in order to improve the methane production by injection of CO_2. Besides, deep saline formations represent possible storage places,

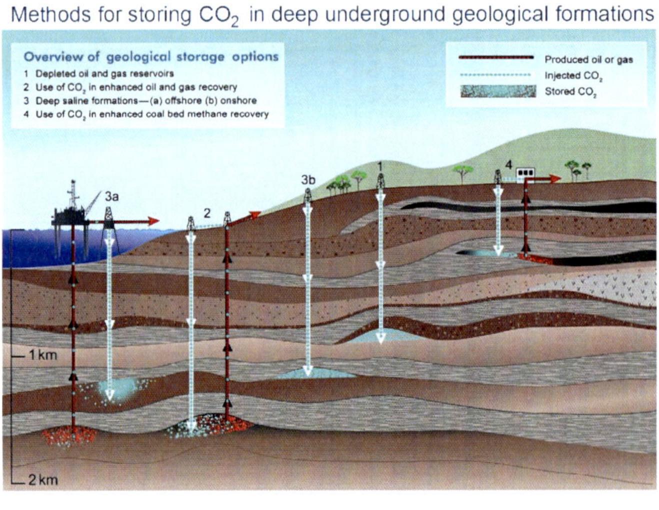

◘ Fig. 2
Carbon dioxide storage scenarios from IPCC special report on carbon capture and storage (From IPCC 2005)

either onshore or offshore. When injecting carbon dioxide, processes take place on highly different spatial and temporal scales. Concerning spatial scales, the processes in the vicinity of the CO_2 plume are very complex including phase change, chemical reactions, etc. But usually, the interest lies on the effect of the CO_2 injection on larger domains, especially if it is to be investigated whether CO_2 is able to migrate to the surface or not. In the vicinity of the CO_2 plume, processes of much higher complexity and much higher fine-scale dependence occur than in the remaining part of the domain. This aspect prescribes both the spatial multiscale and the spatial multiphysics character of this application: around the CO_2 plume, processes have to be resolved on a fine spatial scale in order to be appropriately accounted for. In the rest of the domain of interest, processes may be resolved on a coarser spatial scale. Additionally, the processes occurring in the plume zone and in the non-plume zone are different: While highly complex three-phase three-component processes including reaction need to be considered near the plume, further away from the plume, the consideration of a single-phase system may be sufficient.

With respect to spatial scales we consider ➋ *Fig. 3*, which is again taken from IPCC (2005). In the early time period, i.e., few years after the CO_2 injection ceased, the movement of the CO_2 is determined by advection-dominated multiphase flow (viscous, buoyant, and capillary effects are relevant). In a later time period, when the CO_2 has reached residual saturation everywhere, dissolution and diffusion processes are most decisive for the migration of the carbon dioxide. Eventually, in the very long time range of thousands of years, it is to be expected that the CO_2 will be bound by chemical reactions.

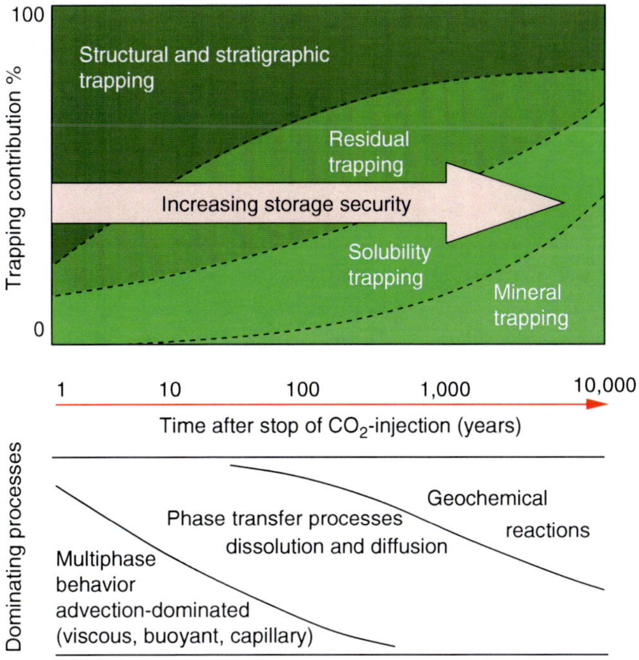

Fig. 3
Time scales of carbon dioxide sequestration (From IPCC 2005)

This chapter is structured as follows: In ➋ Sect. 2, we define the relevant scales considered in this work and give an overview of multiscale and of multiphysics techniques. Next, in ➋ Sect. 3, the mathematical model for flow and transport in porous media is described including non-isothermal flow and different mathematical formulations. In ➋ Sect. 4, the numerical solution procedures for both decoupled and coupled model formulations are explained. In ➋ Sect. 5, we present two different applications of multiphysics and of multiscale multiphysics algorithms. Finally, we conclude in ➋ Sect. 6.

2 State of the Art

We want to give a brief introduction into existing multiscale methods and into methods for scale transfer. First, general definitions of different important scales are given (➋ Sect. 2.1) to point out which are the scales considered in the following sections. Afterward, we give a very general overview of basic approaches for upscaling and different kinds of multiscale methods (➋ Sect. 2.2), and a short introduction to multiphysics methods (➋ Sect. 2.3).

2.1 Definition of Scales

In order to design an appropriate modeling strategy for particular problems, it is important to consider the spatial and temporal scales involved, and how the physical processes and parameters of the system relate to these scales.

A careful definition of relevant length scales can clarify any investigation of scale considerations, although such definitions are a matter of choice and modeling approach (Hristopulos and Christakos 1997). In general, we define the following length scales of concern: the molecular length scale, which is of the order of the size of a molecule; the microscale, or the minimum continuum length scale on which individual molecular interactions can be neglected in favor of an ensemble average of molecular collisions; the local scale, which is the minimum continuum length scale at which the microscale description of fluid movement through individual pores can be neglected in favor of averaging the fluid movement over a representative elementary volume (REV) (this scale is also called REV-scale); the mesoscale, which is a scale on which local scale properties vary distinctly and markedly; and the megascale or field-scale. Measurements or observations can yield representative information across this entire range of scales, depending on the aspect of the system observed and the nature of the instrument used to make the observation. For this reason, we do not specifically define a measurement scale.

➋ *Figure 4* graphically depicts the range of spatial scales of concern in a typical porous medium system. It illustrates two important aspects of these natural systems: several orders of magnitude in potentially relevant length scales exist, and heterogeneity occurs across the entire range of relevant scales. A similar range of temporal scales exists as well, from the picoseconds over which a chemical reaction can occur on a molecular length scale to the centuries or millenniums of concern in the long-term storage of greenhouse gases or atomic waste.

When looking at the REV-scale, we average over both fluid-phase properties and solid-phase properties. In ➋ *Fig. 5*, we schematically show the averaging behavior on the example of the porosity. While averaging over a REV, we assume that the averaged property P does not oscillate significantly. In ➋ *Fig. 5* this is the case in the range of $r_{\min}$ to $r_{\max}$, so an arbitrarily

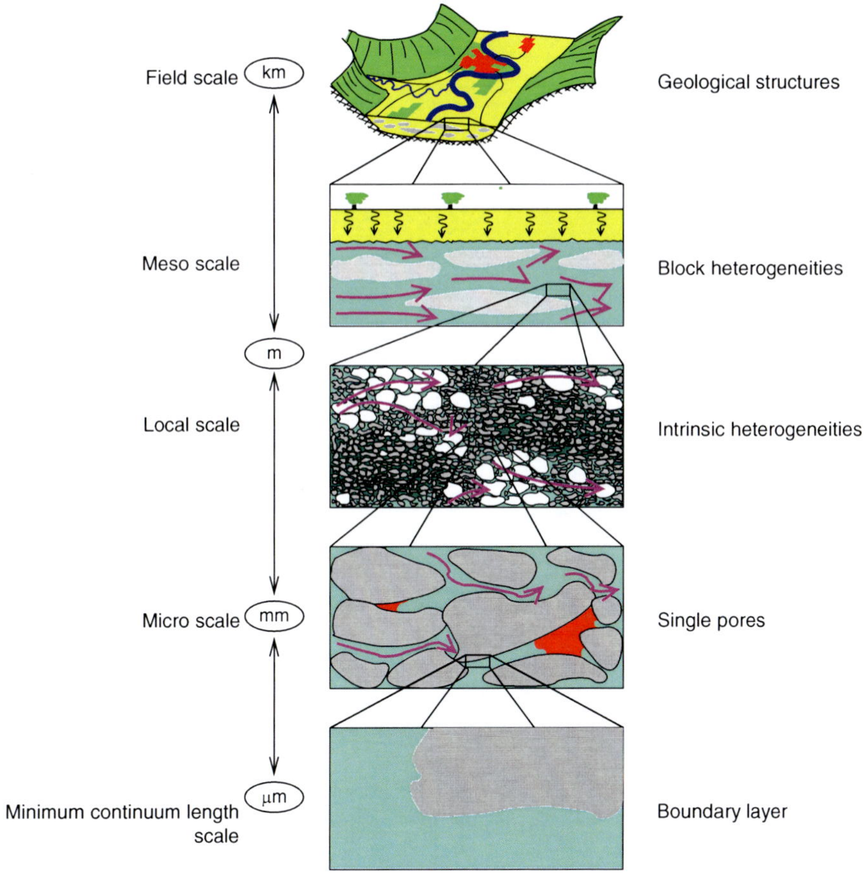

Fig. 4
Different scales for flow in porous media

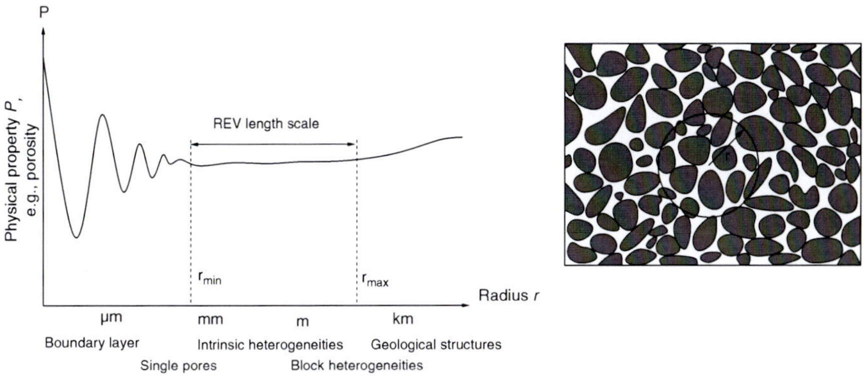

Fig. 5
Different scales for flow in porous media (schematically for *Fig. 4*)

shaped volume V with an inscribed sphere with radius r_{min} and a circumscribed sphere with radius r_{max} can be chosen as REV. Accordingly, we do not assume any heterogeneities on the REV-scale. For our model, we assume that the effects of the sub-REV-scale heterogeneities are taken into account by effective parameters.

The scales of interest in this work are the REV-scale (which we also call fine scale) and the meso-scale (for us, the coarse scale).

2.2 Upscaling and Multiscale Methods

In multiscale modeling, more than one scale is involved in the modeling as the name implies. In general, each pair of scales is coupled in a bidirectional way, where the coarser scale contains the finer scale. This means that upscaling and downscaling methods have to be provided. Upscaling is a transition of the finer to the coarser scale and downscaling vice-versa. Both kinds of operators are generally needed. Only special applications with weak coupling between the scales allow for a mono-directional coupling and thus, only upscaling or only downscaling operators.

Classical upscaling strategies comprise the method of asymptotic expansions (homogenization) and volume averaging. Usually, the fine-scale informations which get lost due to averaging are accounted for by effective parameters in the upscaled equations. For downscaling, the typical methodology is to specify boundary conditions at the boundaries of a coarse-grid block and solve a fine-grid problem in the respective domain. The boundary conditions are obtained either directly from the coarse-scale problem or coarse-scale results are rescaled to fine-scale properties using fine-scale material parameters. In the latter case, fine-grid boundary conditions can be specified along the boundaries of the downscaling domain.

In the following, we provide a brief overview of common upscaling techniques and of multiscale methods.

2.2.1 Upscaling Methods

Effective Coefficients

This method follows the coarse-scale equations to be a priori known, for example, by assuming the same kind of equation than at the fine scale. In that case the problem is to find upscaled or averaged effective model parameters to describe the physical large-scale behavior properly. This method is very commonly applied to upscale the single-phase flow equation in porous media. Effective parameters like the transmissibility are obtained by the solution of local fine-scale problems which can be isolated from the global problem (local upscaling) or coupled to the global problem (local-global upscaling) (e.g., see Durlofsky 1991; Chen et al. 2003).

Pseudo Functions

This method follows the idea of effective coefficients described before. The question is how to determine effective parameters like coarse-scale relative permeabilities, mobilities or fractional flow functions, which depend on the primary variables and therefore vary in time. One possibility, which is for example discussed in Barker and Thibeau (1997) and Darman et al. (2002), is the use of pseudo functions. Most likely, these functions are calculated from the solution of a fine-scale multiphase flow problem (e.g., see, Chen and Durlofsky 2006). Detailed investigations on appropriate boundary conditions for the solution of the local fine-scale problems can be found in Wallstrom et al. (2002a,b).

Volume Averaging/Homogenization Methods

These methods allow to derive new coarse-scale equations from known fine-scale equations. Volume averaging methods applied on porous media flow can, for example, be found in Whitaker (1998) or Gray et al. (1993), homogenization methods are applied in Panfilov (2000). A problem of this averaging approaches is the determination of new constitutive relations appearing in the upscaled coarse-scale equations, which describe the influence of fine-scale fluctuations on the averaged coarse-scale solution. For two-phase flow approaches to calculate terms including fine-scale fluctuations into the coarse-scale equations are developed, for example, by Efendiev et al. (2000) and Efendiev and Durlofsky (2002).

2.2.2 Multiscale Methods

Homogeneous Multiscale Methods

Homogeneous multiscale methods inherently give approximate solutions on the micro-scale. They consist of the traditional numerical approaches to deal with multiscale problems, like, e.g., multi grid methods (Bramble 1993; Briggs et al. 2000; Stüben 2001; Trottenberg et al. 2001), multi resolution wavelet methods (Cattani and Laserra 2003; He and Han 2008; Jang et al. 2004; Urban 2009), multi pole techniques (Giraud et al. 2006; Of 2007; Tornberg and Greengard 2008; Yao et al. 2008), or adaptive mesh refinement (Ainsworth and Oden 2000; Babuska and Strouboulis 2001; Müller 2003). Due to the usually enormous number of degrees of freedom on this scale, this direct numerical solution of real-world multiple scale problems is impossible to realize even with modern supercomputers.

Heterogeneous Multiscale Methods

The heterogeneous multiscale method (HMM), E et al. (2007), proposes general principles for developing accurate numerical schemes for multiple problems, while keeping costs down. It was first introduced in E and Engquist (2003a), and clearly described in E and Engquist (2003b). The general goal of the HMM, as in other multiscale type methods, is to capture the macroscopic behavior of multiscale solutions without resolving all the fine details of the problem. The HMM does this by selectively incorporating the microscale data when needed, and exploiting the characteristics of each particular problem.

Variational Multiscale Method

In Hughes (1995) and Hughes et al. (1998), the authors present the variational multiscale method that serves as a general framework for constructing multiscale methods. An important part of the method is to split the function space into a coarse part, which captures low frequencies, and a fine part, which captures the high frequencies. An approximation of the fine-scale solution is computed and it is used to modify the coarse-scale equations. In recent years, there have been several works on convection–diffusion problems using the variational multiscale framework, see, e.g., Codina (2001), Hauke and García-Olivares (2001), Juanes (2005).

Multiscale Finite-Volume Method

The underlying idea is to construct transmissibilities that capture the local properties of the differential operator. This leads to a multipoint discretization scheme for the finite-volume solution algorithm. The transmissibilities can be computed locally and therefore this step is perfectly suited for massively parallel computers. Furthermore, a conservative fine-scale velocity field can be constructed from the coarse-scale pressure solution. Over the recent years, the method became able to deal with increasingly complex equations (Hajibeygi et al. 2008; Jenny et al. 2006; Lee et al. 2008; Lunati and Jenny 2006, 2007, 2008).

Multiscale Finite-Element Method

Another multiscale method, the multiscale finite element method, was presented in 1997 (Hou and Wu 1997). The theoretical foundation is based on homogenization theory. The main idea is to solve local fine-scale problems numerically in order to use these local solutions to modify the coarse-scale basis functions. There has been a lot of work on this method over the last decade, see, e.g., Aarnes et al. (2006, 2008); Arbogast et al. (2007); Efendiev and Hou (2007); Kim et al. (2007); Kippe et al. (2008).

Multiscale Methods and Domain Decomposition

By comparing the formulations, the authors of Nordbotten and Bjørstad (2008) observe that the multiscale finite-volume method is a special case of a nonoverlapping domain decomposition preconditioner. They go on to suggest how the more general framework of domain decomposition methods can be applied in the multiscale context to obtain improved multiscale estimates.

2.3 Multiphysics Methods

In general, the term multiphysics is used whenever processes which are described by different sets of equations interact and thus are coupled within one global model. The coupling mechanisms can in general be divided into volume (or vertical) coupling and surface (or horizontal) coupling. In this sense, the multiscale approaches introduced before could be interpreted as vertical coupling approaches. Moreover, a large variety of multicontinua models exist. Here, the model domain is physically the same for the different sets of equations, and the exchange is usually performed by means of source and sink terms. Within the context of porous media, most well-known multicontinua models include the double porosity models (Arbogast 1989; Ryzhik 2007), and the MINC method (Multiple INteracting Continua) (Pruess 1985; Smith and Seth 1999).

In contrast to that, horizontal coupling approaches divide the model domain into subdomains sharing common interfaces. The coupling is achieved by enforcing appropriate interface conditions. In physical terms, these interface conditions should state thermodynamic equilibrium (mechanical, thermal, and chemical equilibrium), while in mathematical terms, they often correspond to the continuity of the employed primal and dual variables, like, e.g., pressure and normal velocity. Examples for surface coupling are discrete fracture approaches, Dietrich et al. (2005), or the coupling of porous media flow and free flow domains (Beavers and Joseph 1967; Discacciati et al. 2002; Girault and Rivière 2009; Jäger and Mikelić 2009; Layton et al. 2003). While these two examples couple two different flow regimes, we will in our study concentrate on the coupling of different processes inside one porous media domain. In Albon et al. (1999), the authors present an interface concept to couple two-phase flow processes in different types of porous media. The coupling of different models for one, two or three-phase flow incorporating an iterative non linear solver to ensure the interface coupling conditions was presented in Peszynska et al. (2000).

3 Mathematical Models for Flow and Transport Processes in Porous Media

We introduce the balance equations of flow and transport processes in porous media by means of an REV concept, i.e., on our fine scale. These equations may be upscaled in a subsequent step using one of the techniques of ➋ Sect. 2.2.1 or used in a multiscale technique of ➋ Sect. 2.2.2. After establishing the necessary physical background, the equations for isothermal multiphase flow processes are derived, both for the case of immiscible fluids as well as for miscible fluids. Furthermore, we give an introduction to different decoupled formulations of the balance equations paving the way for specialized solution schemes discussed in the following section. Finally, an extension to non-isothermal processes is provided.

3.1 Preliminaries

After stating the basic definitions of phases and components, the essential fluid and matrix parameters are introduced. Parameters and constitutive relations describing fluid–matrix interactions are discussed, and some common laws for fluid phase equilibria are reviewed.

3.1.1 Basic Definitions

Phases

If two or more fluids fill a volume (e.g., the pore volume), are immiscible and separated by a sharp interface, each fluid is called a phase of the multiphase system. Formally, the solid matrix can also be considered as a phase. If the solubility effects are not negligible, the fluid system has to be considered as a compositional multiphase system.

A pair of two different fluid phases can be divided into a wetting and a non-wetting phase. Here, the important property is the contact angle θ between fluid–fluid interface and solid surface (➋ *Fig.* 6). If the contact angle is acute, the phase has a higher affinity to the solid and is therefore called wetting, whereas the other phase is called non-wetting.

Components

A phase usually consists of several components which can either be pure chemical substances, or consist of several substances which form a unit with constant physical properties, such as air. Thus, it depends on the model problem which substances or mixtures of substances are considered as a component. The choice of the components is essential, as balance equations for compositional flow systems are in general formulated with respect to components.

3.1.2 Fluid Parameters

Compositions and Concentrations

The composition of a phase α is described by fractions of the different components contained in the phase. *Mass fractions* X_α^κ give the ratio of the mass m^κ of one component κ to the total mass of phase α,

$$X_\alpha^\kappa = \frac{m^\kappa}{\sum_\kappa m^\kappa}. \tag{1}$$

From this definition, it is obvious that the mass fractions sum up to unity for each phase,

$$\sum_\kappa X_\alpha^\kappa = 1. \tag{2}$$

A concept which is widely used in chemistry and thermodynamics, are *mole fractions* which phasewise relate the number of molecules of one component to the total number of components. Mole fractions are commonly denoted by lower case letters and can be calculated from mass fractions via the molar mass M^κ by

$$x_\alpha^\kappa = \frac{X_\alpha^\kappa / M^\kappa}{\sum_\kappa X_\alpha^\kappa / M^\kappa} \tag{3}$$

Both, mole fractions and mass fractions are dimensionless quantities. *Concentration* is the mass of a component per volume of the phase, and thus obtained by multiplying the mass fraction of the component by the density of the phase, $C_\alpha^\kappa = \varrho_\alpha X_\alpha^\kappa$, which yields the SI unit $\left[\mathrm{kg/m^3}\right]$.

Density

The density ϱ relates the mass m of an amount of a substance to the volume V which is occupied by it

$$\varrho = \frac{m}{V}. \tag{4}$$

The corresponding unit is $\left[\mathrm{kg/m^3}\right]$. For a fluid phase α, it generally depends on the phase pressure p_α and temperature T, as well as on the composition x_α^κ of the phase,

$$\varrho_\alpha = \varrho_\alpha\left(p_\alpha, T, x_\alpha^\kappa\right). \tag{5}$$

Since the compressibility of the solid matrix as well as its temperature-dependence can be neglected for many applications, one can often assign a constant density to solids.

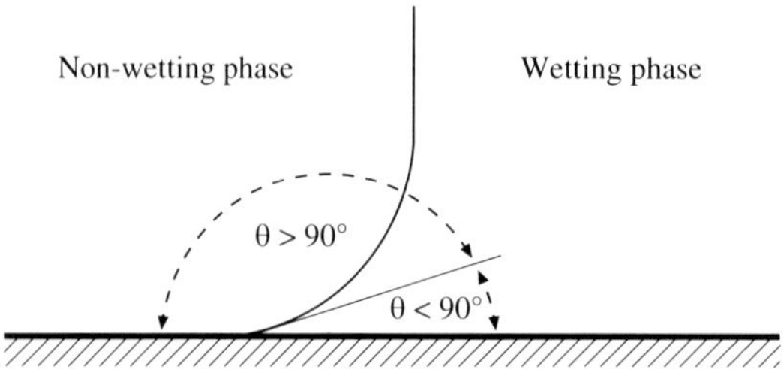

Fig. 6
Contact angle between a wetting and a non-wetting fluid

For liquid phases, the dependence of density on the pressure is usually very low and the contribution by dissolved components is not significant. Thus, the density can be assumed to be only dependent on temperature, $\varrho_\alpha = \varrho_\alpha(T)$. For isothermal systems, the temperature is constant in time and thus, the density of the liquid phase is also constant in time.

The density of gases is highly dependent on temperature as well as on pressure.

Viscosity

Viscosity is a measure for the resistance of a fluid to deformation under shear stress. For Newtonian fluids, the fluid shear stress τ is proportional to the temporal deformation of an angle γ, namely, $\tau = \mu\, \partial\gamma/\partial t$. The proportionality factor μ is called dynamic viscosity with the SI unit $\left[(N \cdot s)/m^2\right] = \left[kg/(m \cdot s)\right]$. In general, the viscosity of liquid phases is primarily determined by their composition and by temperature. With increasing temperature, the viscosity of liquids decreases. Contrarily, the viscosity of gases increases with increasing temperature (see, e.g., Atkins 1994).

3.1.3 Matrix Parameters

Porosity

A porous medium consists of a solid matrix and the pores. The dimensionless ratio of the pore space within the REV to the total volume of the REV is defined as porosity ϕ,

$$\phi = \frac{\text{volume of pore space within the REV}}{\text{total volume of the REV}}. \tag{6}$$

If the solid matrix is assumed to be rigid, the porosity is constant and independent of temperature, pressure or other variables.

Intrinsic Permeability

The intrinsic permeability characterizes the inverse of the resistance of the porous matrix to flow through that matrix. Depending on the matrix type, the permeability may have different values for different flow directions which in general yields a tensor $\mathbf{K}$ with the unit $\left[\mathrm{m}^2\right]$.

3.1.4 Parameters Describing Fluid–Matrix Interaction

Saturation

The pore space is divided and filled by the different phases. In the macroscopic approach (REV-scale), this is expressed by the saturation of each phase α. This dimensionless number is defined as the ratio of the volume of phase α within the REV to the volume of the pore space within the REV:

$$S_\alpha = \frac{\text{volume of phase } \alpha \text{ within the REV}}{\text{volume of the pore space within the REV}}. \tag{7}$$

Assuming that the pore space of the REV is completely filled by the fluid phases α, the sum of the phase saturations must be equal to 1,

$$\sum_\alpha S_\alpha = 1. \tag{8}$$

If no phase transition occurs, the saturations change due to displacement of one phase by another phase. However, a phase can in general not be fully displaced by another, but a certain

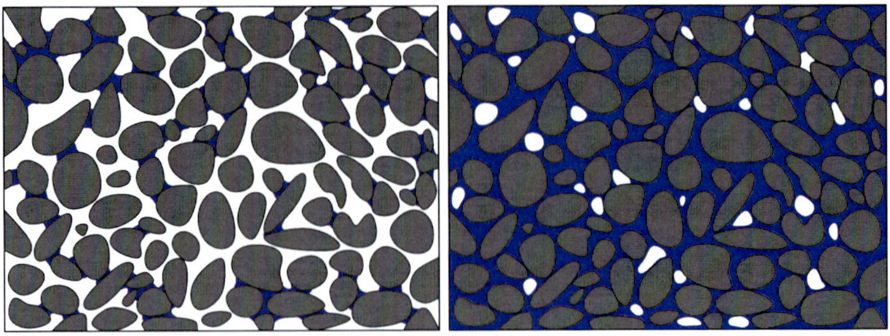

Fig. 7
Residual saturations of the wetting (*left*) and non-wetting (*right*) phase, respectively

saturation will be held back, which is called *residual saturation* . For a wetting phase, a residual saturation occurs if parts of the displaced wetting phase are held back in the finer pore channels during the drainage process (see *Fig. 7*, left-hand side). On the other side, a residual saturation for the non-wetting phase may occur if bubbles of the displaced non-wetting phase are trapped by surrounding wetting phase during the imbibition process (see *Fig. 7*, right-hand side). Therefore, a residual saturation may depend on the pore geometry, the heterogeneity and the displacement process, but also on the number of drainage and imbibition cycles. If the saturation of a phase S_α is smaller than its residual saturation, the relative permeability (Sect. 3.1.4) of phase α is equal to zero which means that no flux of that phase can take place. This implies that a flux can only occur, if the saturation of a phase α lies between the residual saturation and unity ($S_{r\alpha} \leq S_\alpha \leq 1$). With the residual saturation, an effective saturation for a two phase system can be defined in the following way:

$$S_e = \frac{S_w - S_{rw}}{1 - S_{rw}} \qquad S_{rw} \leq S_w \leq 1. \tag{9}$$

Alternatively, in many models the following definition is used:

$$S_e = \frac{S_w - S_{rw}}{1 - S_{rw} - S_{rn}} \qquad S_{rw} \leq S_w \leq 1 - S_{rn}. \tag{10}$$

Which definition has to be used, depends on the way the capillary pressure and the relative permeability curves are obtained, as explained below. Further considerations on the use of effective saturations are made in Helmig (1997).

Capillarity

Due to interfacial tension, forces occur at the interface of two phases. This effect is caused by interactions of the fluids on the molecular scale. Therefore, the interface between a wetting and a non-wetting phase is curved and the equilibrium at the interface leads to a pressure difference between the phases called capillary pressure p_c:

$$p_c = p_n - p_w, \tag{11}$$

where p_n is the non-wetting phase and p_w the wetting phase pressure. In a macroscopic consideration, an increase of the non-wetting phase saturation leads to a decrease of the wetting phase saturation, and, according to microscopic considerations, to the retreat of the wetting fluid to smaller pores. On the REV-scale, it is common to regard the macroscopic capillary pressure as a function of the saturation,

$$p_c = p_c(S_w), \tag{12}$$

the so-called capillary pressure–saturation relation. The simplest way to define a capillary pressure–saturation function is a linear approach:

$$p_c(S_e(S_w)) = p_{c,max}(1 - S_e(S_w)). \tag{13}$$

The most common p_c-S_w-relations for a two-phase system are those of Brooks and Corey and van Genuchten .

In the Brooks–Corey model,

$$p_c(S_e(S_w)) = p_d S_e(S_w)^{-\frac{1}{\lambda}} \qquad p_c \leq p_d, \tag{14}$$

the capillary pressure is a function of the effective saturation S_e. The entry pressure p_d represents the minimum pressure needed for a non-wetting fluid to enter a porous medium initially saturated by a wetting fluid. The parameter λ is called pore-size distribution index and usually lies between 0.2 and 3.0. A very small λ-parameter describes a single size material, while a very large parameter indicates a highly nonuniform material.

The parameters of the Brooks–Corey relation are determined by fitting to experimental data. The effective saturation definition which is used in this parameter fitting is also the one to choose for later application of the respective capillary pressure or relative permeability function.

Relative Permeability

Flow in porous media is strongly influenced by the interaction between the fluid phase and the solid phase. If more than one fluid phase fill the pore space, the presence of one phase also disturbs the flow behavior of another phase. Therefore, the relative permeability $k_{r\alpha}$ which can be considered as a scaling factor is included into the permeability concept. Considering a two fluid-phase system, the space available for one of the fluids depends on the amount of the second fluid within the system. The wetting phase, for example, has to flow around those parts of the porous medium occupied by non-wetting fluid, or has to displace the non-wetting fluid to find new flow paths. In a macroscopic view, this means that the cross-sectional area available for the flow of a phase is depending on its saturation. If the disturbance of the flow of one phase is only due to the restriction of available pore volume caused by the presence of the other fluid, a linear correlation for the relative permeability can be applied,

$$k_{rw}(S_e(S_w)) = S_e(S_w), \tag{15}$$

$$k_{rn}(S_e(S_w)) = 1 - S_e(S_w). \tag{16}$$

This formulation also implies that the relative permeability becomes zero if the residual saturation, representing the amount of immobile fluid, is reached.

In reality, one phase usually not only influences the flow of another phase just by the restriction in available volume, but also by additional interactions between the fluids. If capillary effects occur, the wetting phase, for example, fills the smaller pores if the saturation is small. This means that in case of an increasing saturation of the wetting phase, the relative permeability k_{rw} has to increase slowly if the saturations are still small and it has to increase fast if the saturations become higher, since then the wetting phase begins to fill the larger pores. For the non-wetting phase the opposite situation is the case. Increasing the saturation, the larger pores are filled at first causing a faster rise of k_{rn}. At higher saturations the smaller pores become filled which slows down the increase of the relative permeability. Therefore, correlations for the relative permeabilities can be defined using the known capillary pressure–saturation relationships (details, see Helmig 1997). Besides capillary pressure effects also other effects might occur.

As an example, the Brooks–Corey model is defined as

$$k_{rw}(S_e(S_w)) = S_e(S_w)^{\frac{2+3\lambda}{\lambda}}, \tag{17}$$

$$k_{rn}(S_e(S_w)) = (1 - S_e(S_w))^2 \left[1 - S_e(S_w)^{\frac{2+\lambda}{\lambda}}\right], \tag{18}$$

where λ is the empirical constant from the Brooks–Corey $p_c(S)$-relationship (14). These relative permeabilities do not sum up to unity as for the linear relationship. This is caused by the effects described before, and means that one phase is slowed down stronger by the other phase as it would be only due to the restricted volume available for the flow.

3.1.5 Extended Darcy's Law

In a macroscopic treatment (we are on our fine scale) of porous media, Darcy's law, which was originally obtained experimentally for single phase flow, can be used to calculate averaged velocities using the permeability. For multiphase systems, extended Darcy's law incorporating relative permeabilities is formulated for each phase (details, see Helmig 1997; Scheidegger 1974):

$$\mathbf{v}_\alpha = \frac{k_{r\alpha}}{\mu_\alpha}\mathbf{K}(-\nabla p_\alpha + \varrho_\alpha \mathbf{g}), \tag{19}$$

where $k_{r\alpha}$ is the relative permeability dependent on saturation, and $\mathbf{K}$ the intrinsic permeability dependent on the porous medium, μ the dynamic fluid viscosity, p_α the phase pressure and ϱ_α the phase density, while $\mathbf{g}$ is the gravity vector. The mobility of a phase is defined as $\lambda_\alpha = k_{r\alpha}/\mu_\alpha$. Note that in Eq. 19, the pressure in phase α is used which is important, since the pressure of different phases can differ due to capillarity. The product of the relative and the intrinsic permeability $k_{r\alpha}\mathbf{K}$ is often called total permeability $\mathbf{K}_t$ or effective permeability $\mathbf{K}_e$.

3.1.6 Laws for Fluid Phase Equilibria

We give a short summary of common physical relationships, which govern the equilibrium state between fluid phases and thus, the mass transfer processes, i.e., the exchange of components between phases. While a variety of other relationships can be found in literature, only Dalton's law, Raoult's law, as well as Henry's law are treated here.

Dalton's Law

Dalton's law states that the total pressure of a gas mixture equals the sum of the pressures of the gases that make up the mixture, namely,

$$p_g = \sum_\kappa p_g^\kappa, \tag{20}$$

where p_g^κ is the pressure of a single component κ, the partial pressure, which is by definition the product of the mole fraction of the respective component in the gas phase and the total pressure of the gas phase, i.e.,

$$p_g^\kappa = x_g^\kappa p_g. \tag{21}$$

Raoult's Law

Raoult's law describes the lowering of the vapor pressure of a pure substance in a solution. It relates the vapor pressure of components to the composition of the solution under the simplifying assumption of an ideal solution. The relationship can be derived from the equality of fugacities, see Prausnitz et al. (1967). According to Raoult's law, the vapor pressure of a solution of component κ is equal to the vapor pressure of the pure substance times the mole fraction of component κ in phase α.

$$p_g^\kappa = x_\alpha^\kappa p_{vap}^\kappa \tag{22}$$

Here, p_{vap}^κ denotes the vapor pressure of pure component κ which is generally a function of temperature.

Henry's Law

Henry's law is valid for ideally diluted solutions and ideal gases. It is especially used for the calculation of the solution of gaseous components in liquids. Considering a system with gaseous component κ, a linear relationship between the mole fraction x_α^κ of component κ in the liquid phase and the partial pressure p_g^κ of κ in the gas phase is obtained,

$$x_\alpha^\kappa = H_\alpha^\kappa p_g^\kappa. \tag{23}$$

The parameter H_α^κ denotes the Henry coefficient of component κ in phase α, which is dependent on temperature, $H_\alpha^\kappa = H_\alpha^\kappa(T)$.

➲ *Figure 8* shows the range of applicability of both Henry's law and Raoult's law for a binary system, where component 1 is a component forming a liquid phase, e.g., water, and component 2 is a component forming a gaseous phase, e.g., air. One can see that for low mole fractions of component 2 in the system (small amounts of dissolved air in the liquid phase), Henry's law can be applied whereas for mole fractions of component 1 close to 1 (small amounts of vapor in the gas phase), Raoult's law is the appropriate description. In general, the solvent follows Raoult's law as it is present in excess, whereas the dissolved substance follows Henry's law as it is highly diluted.

3.1.7 The Reynolds Transport Theorem

A common way to derive balance equations in fluid dynamics is to use the Reynolds transport theorem (e.g., White 2003; Helmig 1997), named after the British scientist Osborne Reynolds. Let E be an arbitrary property of the fluid (e.g., mass, energy, momentum) that can be obtained by the integration of a scalar field e over a moving control volume G,

$$E = \int_G e \, \mathrm{d}G\,. \tag{24}$$

The Reynolds transport theorem states that the temporal derivative of the property in a control volume moving with the fluid can be related to local changes of the scalar field by

$$\frac{\mathrm{d}E}{\mathrm{d}t} = \frac{\mathrm{d}}{\mathrm{d}t}\int_G e \, \mathrm{d}G = \int_G \left[\frac{\partial e}{\partial t} + \nabla \cdot (e\mathbf{v})\right] \mathrm{d}G\,, \tag{25}$$

For a general balance equation, we require a conservation of the property E. Thus the property can only change due to sinks and sources, diffusion or dissipation:

$$\frac{\mathrm{d}E}{\mathrm{d}t} = \int_G \frac{\partial e}{\partial t} + \nabla \cdot (e\mathbf{v})\,\mathrm{d}G = \int_G q^e - \nabla \cdot \mathbf{w}\mathrm{d}G\,, \tag{26}$$

where $\mathbf{w}$ is the diffusive flux of e and q^e is the source per unit volume.

3.2 Multiphase Flow

We derive the mass balance equations for the immiscible and the miscible case. Both derivations are based on the general balance equation (26) and the insertion of extended Darcy's law (19) for the involved velocities.

3.2.1 The Immiscible Case

According to the specifications provided in ➋ Sect. 3.1, the mass of a phase α inside a control volume G can be expressed by

$$m_\alpha = \int_G \phi S_\alpha \varrho_\alpha \mathrm{d}G. \tag{27}$$

Under the assumption that the phases are immiscible, the total mass $m_\alpha = \int_G m_\alpha \mathrm{d}G$ is conserved, i.e., $\mathrm{d}m_\alpha/\mathrm{d}t = 0$ in the absence of external sources. Using the general balance equation (26), this mass conservation can be rewritten as

$$\int_G \frac{\partial(\phi \varrho_\alpha S_\alpha)}{\partial t} + \nabla \cdot (\phi S_\alpha \varrho_\alpha \mathbf{v}_{a\alpha})\,\mathrm{d}G = \int_G \varrho_\alpha q_\alpha \mathrm{d}G\,. \tag{28}$$

We emphasize that the diffusive flux is assumed to be zero, since for the motion of phases on an REV scale, no diffusion or dispersion processes are considered. This, however, changes in the most upscaling approaches, where the loss of fine-scale informations is usually counterbalanced by dispersive fluxes on the coarse scale. The control volume considered for the transport

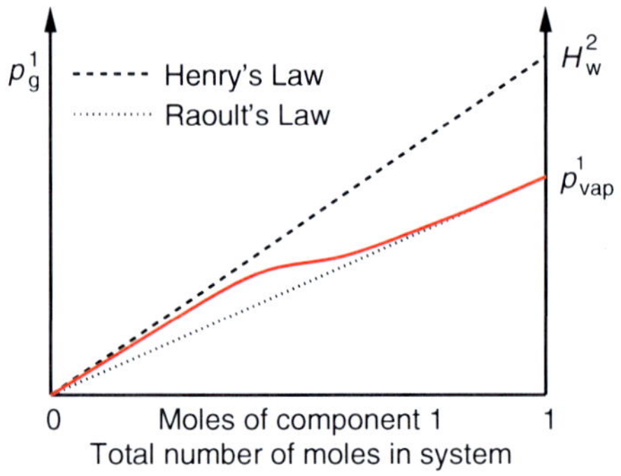

Fig. 8

Applicability of Henry's law and Raoult's law for a binary gas–liquid system (After Lüdecke and Lüdecke 2000)

theorem moves with $\mathbf{v}_{a\alpha}$, which is related to the Darcy velocity $\mathbf{v}_\alpha$ by

$$\mathbf{v}_\alpha = \phi S_\alpha \mathbf{v}_{a\alpha}. \tag{29}$$

Inserting (29) into (28) yields the integral balance equation for a single phase α in a multiphase system

$$\int_G \frac{\partial(\phi \varrho_\alpha S_\alpha)}{\partial t} + \nabla \cdot (\varrho_\alpha \mathbf{v}_\alpha)\, \mathrm{d}G = \int_G \varrho_\alpha q_\alpha \mathrm{d}G\,, \tag{30}$$

Rewriting this equation in differential form and inserting extended Darcy's law (19) yields a system of n_α partial differential equations (with n_α the number of phases),

$$\frac{\partial(\phi \varrho_\alpha S_\alpha)}{\partial t} = -\nabla \cdot (\varrho_\alpha \lambda_\alpha \mathbf{K}(-\nabla p_\alpha + \varrho_\alpha \mathbf{g})) + \varrho_\alpha q_\alpha. \tag{31}$$

Under isothermal conditions, the system (31) of n_α partial differential equations is already closed. In particular, the parameters ϕ, $\mathbf{K}$, $\mathbf{g}$ are intrinsic, and q_α are given source terms. The densities ϱ_α are functions of pressure and the known temperature only, and the mobilities λ_α only depend on the phase saturations. The remaining n_α constitutive relations for the $2n_\alpha$ unknowns S_α, p_α are the closure relation (8) and the $n_\alpha-1$ capillary pressure–saturation relationships (12).

3.2.2 The Miscible Case

We now allow that each phase is made up of different components which can also be dissolved in the other phases. Inserting the total concentration per component,

$$C^\kappa = \phi \sum_\alpha \varrho_\alpha S_\alpha X_\alpha^\kappa, \tag{32}$$

into the general balance equation (26), and applying the same considerations on the velocities as in ➋ Sect. 3.2.1 yields

$$\int_G \frac{\partial C^\kappa}{\partial t} + \sum_\alpha \nabla \cdot (\varrho_\alpha X_\alpha^\kappa \mathbf{v}_\alpha)\, \mathrm{d}G = \int_G q^\kappa \mathrm{d}G\,, \tag{33}$$

Rewriting this in differential form and inserting the extended Darcy law (19), we obtain a set of n_κ partial differential equations (with n_κ the number of components),

$$\frac{\partial C^\kappa}{\partial t} = -\sum_\alpha \nabla \cdot (\varrho_\alpha X_\alpha^\kappa \lambda_\alpha \mathbf{K}(-\nabla p_\alpha + \varrho_\alpha \mathbf{g})) + q^\kappa. \tag{34}$$

We remark that the immiscible case (31) can be easily derived from (34) as a special case. In particular, immiscibility can be equally expressed as X_α^κ being known and constant with respect to space and time. By eventually regrouping and renaming the components with respect to the fixed phase compositions, we can furthermore assume that each component is associated with a distinct phase and $X_\alpha^\kappa = 1$ holds for this particular phase α, whereas it equals zero for all other phases. This directly leads to (31).

In general, we are left with n_κ partial differential equations (34) for the $2n_\alpha + n_\kappa \cdot n_\alpha$ unknowns $p_\alpha, S_\alpha, X_\alpha^\kappa$. Considering (8) and (12) as in the immiscible case, and additionally the closure relations (2), yields $2n_\alpha$ constraints. The remaining $n_\kappa(n_\alpha - 1)$ constraints have to be carefully chosen from the laws for fluid phase equilibria, see ➋ Sect. 3.1.6.

3.3 Decoupled Formulations

It is often advantageous to reformulate the mass balance equations (31) or (34) into one elliptic or parabolic equation for the pressure and one or more hyperbolic–parabolic transport equations for the saturations or concentrations, respectively. In particular, this reformulation allows to employ multiscale or discretization approaches which are especially developed and suited for the corresponding type of equation, and to combine them in various ways. Furthermore, a sequential or iterative solution procedure reduces the amount of unknowns in each solution step. In the following, we introduce these decoupled formulations for the immiscible and for the miscible case.

3.3.1 The Immiscible Case

The reformulation of the multiphase mass balance equations (31) into one pressure equation and one or more saturation equations was primarily derived in Chavent (1976), where the author just called it *a new formulation* for two-phase flow in porous media. Due to the introduction of the idea of fractional flows this formulation is usually called fractional flow formulation.

Pressure Equation

A pressure equation can be derived by addition of the phase mass balance equations. After some reformulation, a general pressure equation can be written as follows:

$$\sum_\alpha S_\alpha \frac{\partial \phi}{\partial t} + \nabla \cdot \mathbf{v}_t + \sum_\alpha \frac{1}{\varrho_\alpha}\left[\phi S_\alpha \frac{\partial \varrho_\alpha}{\partial t} + \mathbf{v}_\alpha \cdot \nabla \varrho_\alpha\right] - \sum_\alpha q_\alpha = 0, \tag{35}$$

with the following definition of a total velocity:

$$\mathbf{v}_t = \sum_\alpha \mathbf{v}_\alpha. \tag{36}$$

Inserting extended Darcy's law (19) into (36) yields

$$\mathbf{v}_t = -\lambda_t \mathbf{K}\left[\sum_\alpha f_\alpha \nabla p_\alpha - \sum_\alpha f_\alpha \varrho_\alpha \mathbf{g}\right]. \tag{37}$$

where $f_\alpha = \lambda_\alpha / \lambda_t$ is the fractional flow function of phase α and $\lambda_t = \sum_\alpha \lambda_\alpha$ is the total mobility. In the following, different possibilities to reformulate this general pressure equation for two-phase flow are shown. There also exist fractional flow approaches for three phases, which are not further considered here. For details, we exemplarily refer to Suk and Yeh (2008).

Global Pressure Formulation for Two-Phase Flow

Defining a global pressure p such that $\nabla p = \sum_\alpha f_\alpha \nabla p_\alpha$ (see below), Eq. 37 can be rewritten as a function of p:

$$\mathbf{v}_t = -\lambda_t \mathbf{K}\left[\nabla p - \sum_\alpha f_\alpha \varrho_\alpha \mathbf{g}\right] \tag{38}$$

Inserting (38) into (35) yields the pressure equation related to the global pressure.

For a domain Ω with boundary $\Gamma = \Gamma_D \cup \Gamma_N$, where Γ_D denotes a Dirichlet and Γ_N a Neumann boundary, the boundary conditions are:

$$\begin{aligned} p &= p_D \quad \text{on} \quad \Gamma_D \qquad \text{and} \\ \mathbf{v}_t \cdot \mathbf{n} &= q_N \quad \text{on} \quad \Gamma_N. \end{aligned} \tag{39}$$

This means that a global pressure has to be found on a Dirichlet boundary which can lead to problems, as the global pressure is no physical variable and thus cannot be measured directly. Following Chavent and Jaffré (1986), the global pressure is defined by

$$p = \frac{1}{2}(p_w + p_n) - \int_{S_c}^{S_w} \left(f_w(S_w) - \frac{1}{2} \right) \frac{dp_c}{dS_w}(S_w)\, dS_w, \tag{40}$$

where S_c is the saturation satisfying $p_c(S_c) = 0$. This definition makes sure that the global pressure is a smooth function and thus is easier to handle from a numerical point of view. However, as shown, e.g., in Binning and Celia (1999), an iterative solution technique is required for more complex (realistic) conditions, where a phase pressure might be known at a boundary. It becomes also clear, that $p = p_w = p_n$, if the capillary pressure between the phases is neglected.

Phase Pressure Formulation for Two-Phase Flow

A pressure equation can also be further derived using a phase pressure which is a physically meaningful parameter in a multiphase system. Investigations of a phase pressure fractional flow formulation can for example be found in Chen et al. (2006) and a formulation including phase potentials has been used in Hoteit and Firoozabadi (2008).

Exploiting Eq. 11, Eq. 37 can be rewritten in terms of one phase pressure. This yields a total velocity in terms of the wetting phase pressure as:

$$\mathbf{v}_t = -\lambda_t \mathbf{K} \left[\nabla p_w + f_n \nabla p_c - \sum_\alpha f_\alpha \varrho_\alpha \mathbf{g} \right], \tag{41}$$

and in terms of a non-wetting phase pressure as:

$$\mathbf{v}_t = -\lambda_t \mathbf{K} \left[\nabla p_n - f_w \nabla p_c - \sum_\alpha f_\alpha \varrho_\alpha \mathbf{g} \right], \tag{42}$$

Substituting $\mathbf{v}_t$ in the general pressure equation (35) by Eq. 41 or 42 yields the pressure equations as function of a phase pressure.

In analogy to the global pressure formulation, the following boundary conditions can be defined:

$$\begin{aligned} p_w &= p_D \quad \text{on} \quad \Gamma_D \qquad \text{or} \\ p_n &= p_D \quad \text{on} \quad \Gamma_D \qquad \text{and} \\ \mathbf{v}_t \cdot \mathbf{n} &= q_N \quad \text{on} \quad \Gamma_N. \end{aligned} \tag{43}$$

It is important to point out that we now have a physically meaningful variable, the phase pressure, instead of the global pressure. So boundary conditions at Dirichlet boundaries can be defined directly, if a phase pressure at a boundary is known.

Saturation Equation

We derive the transport equation for the saturation depending on whether a global or a phase pressure formulation is used. In the first case, a possibly degenerated parabolic–hyperbolic equation is derived, which is quite weakly coupled to the pressure equation. In the second case, a purely hyperbolic equation is obtained with a stronger coupling to the corresponding pressure equation.

Global Pressure Formulation for Two-Phase Flow

In the case of a global pressure formulation a transport equation for saturation related to the total velocity $\mathbf{v}_t$ has to be derived from the general multiphase mass balance equations (35). With the definition of the capillary pressure (11), the extended Darcy's law (19) can be formulated for a wetting and a non-wetting phase as:

$$\mathbf{v}_\mathrm{w} = -\lambda_\mathrm{w}\mathbf{K}(\nabla p_\mathrm{w} - \varrho_\mathrm{w}\mathbf{g}) \tag{44}$$

and

$$\mathbf{v}_\mathrm{n} = -\lambda_\mathrm{n}\mathbf{K}(\nabla p_\mathrm{w} + \nabla p_\mathrm{c} - \varrho_\mathrm{w}\mathbf{g}). \tag{45}$$

Solving (45) for $\mathbf{K}\nabla p_\mathrm{w}$ and inserting it into Eq. 44 yields:

$$\mathbf{v}_\mathrm{w} = \frac{\lambda_\mathrm{w}}{\lambda_\mathrm{n}}\mathbf{v}_\mathrm{n} + \lambda_\mathrm{w}\mathbf{K}[\nabla p_\mathrm{c} + (\varrho_\mathrm{w} - \varrho_\mathrm{n})\mathbf{g}]. \tag{46}$$

With $\mathbf{v}_\mathrm{n} = \mathbf{v}_\mathrm{t} - \mathbf{v}_\mathrm{w}$, (46) can be reformulated as the fractional flow equation for $\mathbf{v}_\mathrm{w}$:

$$\mathbf{v}_\mathrm{w} = \frac{\lambda_\mathrm{w}}{\lambda_\mathrm{w} + \lambda_\mathrm{n}}\mathbf{v}_\mathrm{t} + \frac{\lambda_\mathrm{w}\lambda_\mathrm{n}}{\lambda_\mathrm{w} + \lambda_\mathrm{n}}\mathbf{K}[\nabla p_\mathrm{c} + (\varrho_\mathrm{w} - \varrho_\mathrm{n})\mathbf{g}], \tag{47}$$

which can be further inserted into the wetting phase mass balance equation leading to a transport equation for the wetting phase saturation related to $\mathbf{v}_\mathrm{t}$:

$$\frac{\partial(\phi\varrho_\mathrm{w}S_\mathrm{w})}{\partial t} + \nabla\cdot[\varrho_\mathrm{w}(f_\mathrm{w}\mathbf{v}_\mathrm{t} + f_\mathrm{w}\lambda_\mathrm{n}\mathbf{K}(\nabla p_\mathrm{c} + (\varrho_\mathrm{w} - \varrho_\mathrm{n})\mathbf{g}))] - \varrho_\mathrm{w}q_\mathrm{w} = 0. \tag{48}$$

Some terms of (48) can be reformulated in dependence on the saturation (details, see Helmig 1997). For incompressible fluids and a porosity which does not change in time, the saturation equation of a two-phase system can then be formulated showing the typical character of a transport equation as

$$\begin{aligned}\phi\frac{\partial S_\mathrm{w}}{\partial t} &+ \left[\mathbf{v}_\mathrm{t}\frac{\mathrm{d}f_\mathrm{w}}{\mathrm{d}S_\mathrm{w}} + \frac{\mathrm{d}\,(f_\mathrm{w}\lambda_\mathrm{n})}{\mathrm{d}S_\mathrm{w}}\mathbf{K}(\varrho_\mathrm{w} - \varrho_\mathrm{n})\mathbf{g}\right]\cdot\nabla S_\mathrm{w}\\ &+ \nabla\cdot\left[\bar{\lambda}\mathbf{K}\frac{\mathrm{d}p_\mathrm{c}}{\mathrm{d}S_\mathrm{w}}\nabla S_\mathrm{w}\right] - q_\mathrm{w} + f_\mathrm{w}q_\mathrm{t} = 0,\end{aligned} \tag{49}$$

where $q_\mathrm{t} = q_\mathrm{w} + q_\mathrm{n}$.

Similarly, an equation for the non-wetting phase saturation can be derived which can be written in its final form as:

$$\frac{\partial(\phi\varrho_\mathrm{n}S_\mathrm{n})}{\partial t} + \nabla\cdot[\varrho_\mathrm{n}(f_\mathrm{n}\mathbf{v}_\mathrm{t} - f_\mathrm{n}\lambda_\mathrm{w}\mathbf{K}(\nabla p_\mathrm{c} - (\varrho_\mathrm{n} - \varrho_\mathrm{w})\mathbf{g}))] - \varrho_\mathrm{n}q_\mathrm{n} = 0. \tag{50}$$

Phase Pressure Formulation for Two-Phase Flow

Obviously, for the phase pressure formulation the saturation can be calculated directly from the mass balance equations (35), leading to

$$\frac{\partial\,(\phi\varrho_\mathrm{w}S_\mathrm{w})}{\partial t} + \nabla\cdot(\varrho_\mathrm{w}\mathbf{v}_\mathrm{w}) = q_\mathrm{w} \tag{51}$$

for the wetting phase of a two-phase system and to

$$\frac{\partial (\phi \varrho_n S_n)}{\partial t} + \nabla \cdot (\varrho_n \mathbf{v}_n) = q_n, \tag{52}$$

where the phase velocities can be calculated using extended Darcy's law.

A nice feature of the global pressure formulation is that the two equations (pressure equation and saturation equation) are only weakly coupled through the presence of the total mobility and the fractional flow functions in the pressure equation. These are dependent on the relative permeabilities of the phases and thus dependent on the saturation. This also holds for the phase pressure formulation. However, in this formulation the coupling is strengthened again due to the additional capillary pressure term in the pressure equation.

3.3.2 The Miscible Case

Similarly to the fractional flow formulations for immiscible multiphase flow, decoupled formulations for compositional flow have been developed. However, dissolution and phase changes of components affect the volume of mixtures, compromising the assumption of a divergence-free total velocity field. In Trangenstein and Bell (1989), a pressure equation for compositional flow in porous media based on volume conservation is presented. The basic and physically logical constraint of this equation is that the pore space always has to be filled by some fluid, i.e., the volume of the fluid mixture has to equal the pore volume. Using a similar definition as for component mass fractions in ➋ Sect. 3.1.2, each phase can be assigned a mass fraction that relates the mass of the phase to the mass of all phases present in the control volume and can be related to saturations by $\nu_\alpha = S_\alpha \varrho_\alpha / \sum_\alpha S_\alpha \varrho_\alpha$. Knowing the total mass $m_t = \sum_\alpha S_\alpha \varrho_\alpha = \sum_\kappa C^\kappa$ (with the unit $[\mathrm{kg/m^3}]$ and C^κ as defined in ➋ Sect. 3.2.2) of a mixture inside a control volume and the phase mass fraction ν_α, we can calculate the volume occupied by the phase by $v_\alpha = m_t \nu_\alpha / \varrho_\alpha$. Related to a unit volume, the pore volume is the porosity ϕ and the total mass of a mixture is the sum over the total concentrations and thus the constraint stated above can be expressed as:

$$\left(\sum_\kappa C^\kappa\right) \cdot \left(\sum_\alpha \frac{\nu_\alpha}{\varrho_\alpha}\right) = \phi\,. \tag{53}$$

Since the fluid volume v_t, which is defined by the left-hand side of Eq. 53 always has to fill the pore volume it is not allowed to change, i.e.,

$$\frac{\partial v_t}{\partial t} = 0\,. \tag{54}$$

However, the fluid volume changes due to changes in the composition of the mixture and due to changes in pressure, which can be expressed by derivatives of the volume with respect to pressure and total concentration. These changes have to be compensated by inflow or outflow of mass to or from the control volume which, according to Trangenstein and Bell (1989) or Chen et al. (2000), can be expressed by the pressure equation for compositional flow:

$$-\frac{\partial v_t}{\partial p}\frac{\partial p}{\partial t} + \sum_\kappa \frac{\partial v_t}{\partial C^\kappa} \nabla \cdot \sum_\alpha X_\alpha^\kappa \varrho_\alpha \mathbf{v}_\alpha = \sum_\kappa \frac{\partial v_t}{\partial C^\kappa} q^\kappa\,. \tag{55}$$

In Chen et al. (2000), the authors show a more detailed derivation and also introduce and analyze the fractional flow formulation in terms of phase pressure, weighted phase pressure, and global pressure. However, in this publication, we remain at calculating the phase velocities with extended Darcy's law (Eq. 19) without introducing fractional flows, since—in the opinion of the authors—this is the most convenient.

3.4 Non-Isothermal Flow

The consideration of non-isothermal flow processes involves an additional conservation property: energy. This is expressed as internal energy inside a unit volume which consists of the internal energies of the matrix and the fluids:

$$U = \int_G \phi \sum_\alpha (\varrho_\alpha S_\alpha u_\alpha) + (1-\phi)\,\varrho_s c_s T \,\mathrm{d}G\,, \tag{56}$$

where the internal energy is assumed to be a linear function of temperature T above a reference point. Then c_s denotes the heat capacity of the rock and u_α is the specific internal energy of phase α. The internal energy in a system is increased by heat fluxes into the system and by mechanical work done on the system

$$\frac{\mathrm{d}U}{\mathrm{d}t} = \frac{\mathrm{d}Q}{\mathrm{d}t} + \frac{\mathrm{d}W}{\mathrm{d}t}\,. \tag{57}$$

Heat flows over the control volume boundaries by conduction, which is a linear function of the temperature gradient and occurs in direction of falling temperatures

$$\frac{\mathrm{d}Q}{\mathrm{d}t} = \oint_\Gamma -\mathbf{n}\cdot(-\lambda_s \nabla T)\,\mathrm{d}\Gamma = \int_G \nabla\cdot(\lambda_s \nabla T)\,\mathrm{d}G\,. \tag{58}$$

The mechanical work done by the system (and therefore decreasing its energy) is volume changing work. It is done when fluids flows over the control volume boundaries against a pressure p

$$\frac{\mathrm{d}W}{\mathrm{d}t} = \oint_\Gamma -p\,(\mathbf{n}\cdot\mathbf{v})\,\mathrm{d}\Gamma = \int_G -\nabla\cdot(p\mathbf{v})\,\mathrm{d}G\,. \tag{59}$$

The left-hand side of Eq. 57 can be expressed by the Reynolds transport theorem, where the velocity of the solid phase equals zero

$$\begin{aligned}\frac{\mathrm{d}U}{\mathrm{d}t} = &\int_G \frac{\partial}{\partial t}\left(\phi \sum_\alpha (\varrho_\alpha S_\alpha u_\alpha)\right) + \frac{\partial}{\partial t}((1-\phi)\,\varrho_s c_s T)\,\mathrm{d}G \\ &+ \oint_\Gamma \mathbf{n}\cdot\sum_\alpha (\varrho_\alpha u_\alpha \mathbf{v}_\alpha)\,\mathrm{d}\Gamma\,.\end{aligned} \tag{60}$$

Using the definition of specific enthalpy $h = u + p/\varrho$, the second term on the right-hand side of Eq. 60 and the right-hand side of Eq. 59 can be combined and Eq. 57 can be rewritten to

$$\begin{aligned}&\int_G \frac{\partial}{\partial t}\left(\phi \sum_\alpha (\varrho_\alpha S_\alpha u_\alpha)\right) + \frac{\partial}{\partial t}((1-\phi)\,\varrho_s c_s T)\,\mathrm{d}G \\ &= \oint_\Gamma \mathbf{n}\cdot(\lambda_s \nabla T)\,\mathrm{d}\Gamma - \oint_\Gamma \mathbf{n}\cdot\sum_\alpha (\varrho_\alpha h_\alpha \mathbf{v}_\alpha)\,\mathrm{d}\Gamma\end{aligned} \tag{61}$$

or in differential form

$$\frac{\partial}{\partial t}\left(\phi \sum_{\alpha} (\varrho_{\alpha} S_{\alpha} u_{\alpha})\right) + \frac{\partial}{\partial t}((1-\phi)\,\varrho_{s} c_{s} T)$$
$$= \nabla \cdot (\lambda_{s} \nabla T) - \nabla \cdot \sum_{\alpha} (\varrho_{\alpha} h_{\alpha} \mathbf{v}_{\alpha}) \tag{62}$$

4 Numerical Solution Approaches

After establishing the continuous models, it remains to choose numerical discretization and solution schemes. We will only very briefly address this question for the fully coupled balance equations (31) or (34) in ➋ Sect. 4.1, our main emphasis here is on the decoupled systems, which will be treated in ➋ Sect. 4.2.

4.1 Solution of the Fully Coupled Equations

One possibility to calculate multiphase flow is to directly solve the system of equations given by the balances (31) or (34). These mass balance equations are usually non linear and strongly coupled. Thus, we also call this the fully coupled multiphase flow formulation. After space discretization, the system of equations one has to solve can be written as:

$$\frac{\partial}{\partial t}\mathbf{M}(\underline{u}) + \mathbf{A}(\underline{u}) = \mathbf{R}(\underline{u}), \tag{63}$$

where **M** consists of the accumulation terms, **A** includes the internal flux terms, and **R** is the right-hand side vector which comprises Neumann boundary flux terms as well as source or sink terms. All mass balance equations have to be solved simultaneously due to the strong coupling. Therefore, a linearization technique has to be applied. The most common solution method is the Newton–Raphson-algorithm (Aziz and Wong 1989; Dennis and Schnabel 1996).

Advantages of the fully coupled formulation and the implicit method respectively are that it includes the whole range of physical effects (capillarity, gravity,...) without having additional effort; that it is quite stable; and that it is usually not very sensitive to the choice of the time step size. The disadvantage is that a global system of equations, which is twice as large as for a single phase pressure equation if two-phase flow is calculated (and even larger in the non-isothermal case or including more phases), has to be solved *several* times during each time step, dependent on the number of iterations the linearization algorithm needs to converge.

4.2 Solution of the Decoupled Equations

As before, we split our considerations into the immiscible and the miscible case.

4.2.1 The Immiscible Case

As their name implies, decoupled formulations decouple the system of equations of a multiphase flow formulation to some extent. In the immiscible case, the result is an equation for pressure and additional transport equations for one saturation (see ➋ Sect. 3.3.1) in the case of two-phase flow or several saturations if more phases are considered. The new equations are still weakly coupled due to the saturation-dependent parameters like relative permeabilities or capillary pressure in the pressure equation and the pressure-dependent parameters like density and viscosity in the saturation transport equation. Nevertheless, in many cases it is possible to solve this system of equations sequentially. Numerically, this is usually done by using an IMPES scheme (IMplicit Pressure–Explicit Saturation), which was first introduced in Sheldon and Cardwell (1959) and Stone and Garder (1961). Therefore, the pressure equation is solved first implicitly. From the resulting pressure field the velocity field can be calculated and inserted into the saturation equation which is then solved explicitly.

One major advantage of the decoupled formulation is that it allows for different discretizations of the different equations. For the pressure equation, it is of utmost importance that its solution admits the calculation of a locally conservative velocity field. There are various discretization methods meeting this requirement, like finite volumes with two-point or multi point flux approximation (Aavatsmark 2002; Aavatsmark et al. 2008; Cao et al. 2008; Eigestad and Klausen 2005; Klausen and Winther 2006), mixed finite elements (Allen et al. 1992; Brezzi and Fortin 1991; Huang 2000; Mazzia and Putti 2006; Srinivas et al. 1992), or mimetic finite differences (Berndt et al. 2005; Brezzi et al. 2005a,b; Hyman et al. 2002; Shashkov 1996). Moreover, it is also possible to use discretizations with non conservative standard velocity fields, and employ a post-processing step to reconstruct a locally conservative scheme. This has been investigated for discontinuous Galerkin methods in Bastian and Rivière (2003), while for standard Lagrangian finite elements, it is possible to calculate equilibrated fluxes known from a posteriori error estimation (Ainsworth and Oden 2000).

For the solution of the transport equation(s) also exists a large variety of discretization possibilities, ranging from standard upwind finite volume approaches (Eymard et al. 2000; LeVeque 2002), over higher order discontinuous Galerkin methods with slope limiter (Cockburn et al. 1989; Cockburn and Shu 1989; Ghostine et al. 2009; Hoteit et al. 2004), the modified method of characteristics (Chen et al. 2002; Dawson et al. 1989; Douglas et al. 1999; Ewing et al. 1984), and the Eulerian–Lagrangian localized adjoint method (Ewing and Wang 1994, 1996; Herrera et al. 1993; Russell 1990; Wang et al. 2002), up to streamline methods (Juanes and Lie 2008; Matringe et al. 2006; Oladyshkin et al. 2008).

The IMPES scheme can be very efficient, since a system of equations with only n unknowns, where n is the number of degrees of freedom for the discretization of the pressure equation, has to be solved only once in the pressure step, as opposed to several solutions of a system of equations with m unknowns in the fully coupled scheme, where m is usually at least twice as large as n. However, there are strong restrictions with respect to the choice of the time step size. In a multi dimensional advection dominated porous media flow a time step criterion can be defined as follows:

$$\Delta t_{\mathrm{in}} = \frac{V}{\left|\int\limits_{\Gamma} \sum\limits_{\alpha} (\mathbf{v}_{\alpha_{\mathrm{in}}} \cdot \mathbf{n})\, \mathrm{d}\Gamma\right|} \tag{64}$$

$$\Delta t_{\mathrm{out}} = \frac{V}{\left|\int\limits_{\Gamma} \sum\limits_{\alpha} (\mathbf{v}_{\alpha_{\mathrm{out}}} \cdot \mathbf{n})\, \mathrm{d}\Gamma\right|} \tag{65}$$

$$\Delta t \leq a \min(\Delta t_{\mathrm{in}}, \Delta t_{\mathrm{out}}) \tag{66}$$

where V is the volume within a cell available for the flow, $\mathbf{v}_{\alpha_{\text{in}}}$ is a phase velocity causing flux into a cell, and $\mathbf{v}_{\alpha_{\text{out}}}$ is a phase velocity causing flux out of a cell. To capture the non linear character of two-phase flow it might be necessary to strengthen the time-step restriction. This can be done by choosing a suitable a, where $0 < a \leq 1$. The value of a can either be found heuristically or by a calculation taking into account some more detailed analysis of the non linear behavior (e.g., see Aziz and Settari 1979). In the one-dimensional linear case where $a = 1$, Eqs. 64–66 lead to

$$\Delta t \leq \min\left(\frac{\Delta x}{|\sum_{\alpha} \mathbf{v}_{\alpha_{\text{in}}}|}, \frac{\Delta x}{|\sum_{\alpha} \mathbf{v}_{\alpha_{\text{out}}}|}\right), \tag{67}$$

which is the well-known Courant–Friedrichs–Lewy CFL-condition. Detailed analysis of the stability of an IMPES scheme can be found in Aziz and Settari (1979). There is also shown that the time step size is further restricted if the flow is dominated by diffusive effects. An explicit scheme may then lose its efficiency compared to an implicit scheme, as a large number of time steps can be required. A comparison between an IMPES scheme and an fully implicit scheme can for example be found in Sleep and Sykes (1993).

4.2.2 The Compositional Case

The decoupled compositional multiphase flow equations derived in ➋ Sect. 3.3.2 can be solved sequentially according to the IMPES scheme, where it is commonly referred to as IMPEC scheme, since concentrations are considered in this case. First, the pressure equation (55) is solved implicitly to obtain a pressure field and fluid phase velocities which are used to explicitly solve the transport equation (34). After each of these pressure-transport sequences, the distribution of total concentrations is known. For the next sequence, phase saturation and component mass fractions are needed. These are gained by performing a phase equilibrium- or so-called flash calculation (see below), which forms the last step of the IMPEC scheme.

Flash Calculations

After the evaluation of the transport equation, the total concentrations at each cell or node are known. From these, an overall mass fraction (or feed mass fraction) $z^{\kappa} = C^{\kappa} / \sum_{\kappa} C^{\kappa}$ of each component inside the mixture is calculated. Now the question is, how the phase mass fractions ν_{α} can be calculated from this, i.e., how the different components are distributed among the different phases. Therefore, we introduce at first the equilibrium ratios

$$K_{\alpha}^{\kappa} = \frac{X_{\alpha}^{\kappa}}{X_{\text{r}}^{\kappa}}, \tag{68}$$

which relate the mass fractions of each component in each phase α to its mass fraction in a reference phase r, where K_{r}^{κ} obviously always equals unity. The equilibrium ratios can be obtained by using the laws for fluid phase equilibria in ➋ Sect. 3.1.6, as described in Niessner and Helmig (2007) or by incorporating a thermodynamic equation of state (Aziz and Wong 1989; Nghiem and Li 1984; Michelsen and Mollerup 2007). In the former case, the equilibrium ratios depend only on pressure and are therefore constant for constant pressure and temperature. The feed mass fraction z^{κ} can be related to the phase mass fractions X_{α}^{κ} by

$$z^{\kappa} = \sum_{\alpha} \nu_{\alpha} X_{\alpha}^{\kappa}. \tag{69}$$

Combining and rearranging of (68) and (69) yields

$$X_{\mathrm{r}}^{\kappa} = \sum_{\kappa} \frac{z^{\kappa}}{\sum\limits_{\alpha \neq \mathrm{r}} K_{\alpha}^{\kappa} v_{\alpha} + v_{\mathrm{r}}}, \tag{70}$$

and some more steps, which are elaborately described in Aziz and Wong (1989), yield a set of $n_{\alpha} - 1$ equations, known as the Rachford–Rice equation

$$\sum_{\kappa} \frac{z^{\kappa} \left(K_{\alpha}^{\kappa} - 1\right)}{1 + \sum_{\alpha \neq \mathrm{r}} \left(K_{\alpha}^{\kappa} - 1\right) v_{\alpha}} = 0 \qquad \forall \alpha \neq \mathrm{r}, \tag{71}$$

which generally has to be solved iteratively for the phase mass fraction v_{α}. Only in the case of the same number of phases and components, the Rachford–Rice equation can be solved analytically. Once the phase mass fractions are known, the mass fractions of the components inside the reference phase can be calculated by (70) and then lead to the mass fractions inside the other phases via (68). Other flash calculation approaches which use so-called reduced equation algorithms and which basically use modified forms of the presented equations are presented in Wang and Barker (1995).

5 Application of Multiphysics and Multiscale Methods

Unlike numerical multiscale methods, our ideology is to use a physically based multiscale multiphysics method that uses physical indicators to identify the domains where different processes take place. The idea in mind is always to create a tool which is able to handle complex real-life application. Therefore, the coupling of scales and domains is done in a physically motivated way, e.g., by ensuring continuity of fluxes and pressures across domain boundaries.

5.1 A Multiphysics Example

In this section, we introduce a method to couple compositional two-phase flow with single-phase compositional flow as proposed in Fritz et al. (submitted). The advantage of this coupling is, that for single-phase compositional flow, a simpler pressure equation and only one transport equation have to be solved. Moreover, for single-phase flow, the evaluation of flash calculations can be avoided. This is interesting if these evaluations require lots of computational power such as in many reservoir engineering problems, where flash calculations may occupy up to 70 % of the total CPU time of a model (see Stenby and Wang 1993).

5.1.1 Single-Phase Transport

We want to take a closer look at the equations for the miscible case and assume that only one phase is present. Inserting the definition of the total concentration (32) into the component mass balance equation (34) for one phase α and applying the chain rule yields

$$\phi \varrho_{\alpha} X_{\alpha}^{\kappa} \frac{\partial S_{\alpha}}{\partial t} + S_{\alpha} \varrho_{\alpha} X_{\alpha}^{\kappa} \frac{\partial \phi}{\partial t} + \phi S_{\alpha} \varrho_{\alpha} \frac{\partial X_{\alpha}^{\kappa}}{\partial t} = -\nabla \cdot \left(\mathbf{v}_{\alpha} \varrho_{\alpha} X_{\alpha}^{\kappa}\right) + q^{\kappa} . \tag{72}$$

The first term on the left-hand side equals zero since only one phase is present and thus the saturation always equals unity. Further we assume that phase and matrix compressibilities are of low importance and can be neglected. Then the second and third term on the left-hand side cancel out as well. A volumetric source q_α of the present phase with the composition X_Q^κ is introduced and using the definition $q^\kappa = q_\alpha \varrho_\alpha X_Q^\kappa$, with X_Q^κ in the source-flow, we replace the mass source of component κ and write

$$\phi \varrho_\alpha \frac{\partial X_\alpha^\kappa}{\partial t} = \frac{\partial C^\kappa}{\partial t} = -\varrho_\alpha \nabla \cdot (\mathbf{v}_\alpha X_\alpha^\kappa) + q_\alpha \varrho_\alpha X_Q^\kappa . \tag{73}$$

This is the mass balance equation for the compositional transport in a single phase. The same assumptions can be applied to the compositional pressure equation. For incompressible flow, the derivatives of volume with respect to mass equal the reciprocal of the phase density, i.e., $\partial v_t / \partial C^\kappa = 1/\varrho_\alpha$. Inserting this identity into Eq. 55, setting the derivative of volume with respect to pressure to zero (incompressible flow) and applying the chain rule to the divergence term, we get

$$\sum_\kappa \frac{1}{\varrho_\alpha} \left(X_\alpha^\kappa \varrho_\alpha \nabla \cdot \mathbf{v}_\alpha + \varrho_\alpha \mathbf{v}_\alpha \cdot \nabla X_\alpha^\kappa + X_\alpha^\kappa \mathbf{v}_\alpha \cdot \nabla \varrho_\alpha \right) = \sum_\kappa \frac{1}{\varrho_\alpha} q^\kappa . \tag{74}$$

By definition, the sum of the mass fractions X_α^κ inside one phase equals unity, thus the second term in parenthesis cancels out. Furthermore, the gradient of the density equals zero due to the incompressibility. Again introducing the volumetric source term as above, yields the pressure equation for incompressible single phase compositional flow, which can as well be derived from Eq. 35:

$$\nabla \cdot \mathbf{v}_\alpha = q_\alpha \tag{75}$$

5.1.2 Model Coupling

Consider a spatial integration of Eq. 55 as is done for most numerical discretization. Then the unit of the terms is easily discovered to be volume over time. This basically reveals the physical aspect of the equation, namely the conservation of total fluid volume as also described in ➋ Sect. 3.3.2. The same consideration holds for Eq. 75, which makes sense since it is derived from the compositional pressure equation. The associated transport equations also have a common physical relevance: the conservation of mass.

The clear relation of the multiphase compositional and single phase transport model and the possibility to use the same primary variables (as indicated in Eq. 73) open the way to couple both models. This makes it possible to fit the model to the actual problem and use a sophisticated and accurate model in a subdomain of special interest, whereas a simpler model can be used in the rest of the domain. Consider for example a large hydrosystem which is fully saturated with water except at a spill of nonaqueous phase liquid (NAPL) as displayed in ➋ *Fig. 9*. The components of the NAPL are solvable in water and contaminate it. To model the dissolution of the components in the water, only a small area around the NAPL spill has to be discretized by a compositional two-phase model. The spreading of the contaminants in the larger part of the domain can in contrary be simulated using a single-phase transport model. The advantage in coupling the two models in this domain is that in large parts, the costly equilibrium calculations and evaluation of volume derivatives can be avoided.

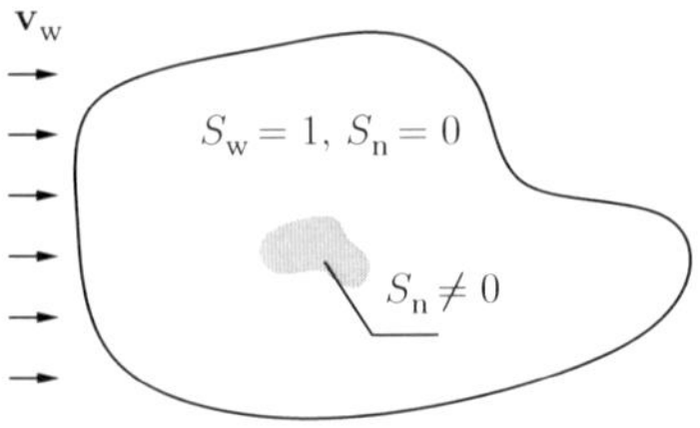

Fig. 9
Multiphysics problem example (From Fritz et al.)

5.1.3 Practical Implementation

The practical implementation of the multiphysics scheme proposed in the preceding sections is done by exploiting the similarity of the equations. Since both pressure equations have the same dimensions and the same unknowns, both can be written into one system of equations. The entries in the stiffness matrix and right-hand side vector are evaluated using either Eq. 55 if the control volume is situated inside the subdomain or Eq. 75 in the other parts of the domain. Equation 55 can also be set up properly at the internal boundary of the subdomain, since all coefficients can be determined. The coefficients concerning the phase which is not present outside the subdomain just have to be set to zero for the outer elements which makes all terms concerning this phase vanish at the boundary. Also the transport equation is well defined at the boundary. Since only one phase is present outside the subdomain, the mobility coefficient in Eq. 34 will equal zero for all other phases and the multiphase compositional mass balance will boil down to Eq. 73 at the interface.

Implicitly, the coupling conditions are already contained in the presented scheme: first, mass fluxes have to be continuous across the subdomain boundary and second, phase velocities have to be continuous across the subdomain boundary. Since only one phase is present outside the subdomain, it is obvious that the second condition requires that only this one phase may flow across the subdomain boundary. Another effect that has to be considered is demixing. Solubilities usually depend on pressure. If a phase is fully saturated with a certain component and then moves further downstream, where the pressure is lower, the solubility decreases and demixing occurs. If the solubility is exceeded outside the subdomain this effect is not represented. These two considerations show the crucial importance of an adequate choice of the subdomain. On the one hand, we want the subdomain to be as small as possible to obtain an economic model, on the other hand, it must be chosen large enough to prevent errors. Especially in large and heterogeneous simulation problems, it is unlikely to determine a proper subdomain in advance, so an automatic adaption is sought. As the most logical scheme, we propose to choose all cells with more than one phase and—since demixing occurs predominantly here—all directly adjacent cells to be part of the subdomain. At the end of each timestep, the choice is checked and superfluous cells are removed and necessary cells are added. This quite easy decomposition can only be expected to be successful in the case of an explicit solution of the transport equations. In particular, the fulfillment of the CFL-condition guarantees that no modeling error will occur, since information is transported at most one cell further in one timestep.

As example for the described subdomain adaptivity see ➲ *Fig. 10*. Displayed is nearly residual air that moves slightly upward before it is dissolved in water. The subdomain is marked by

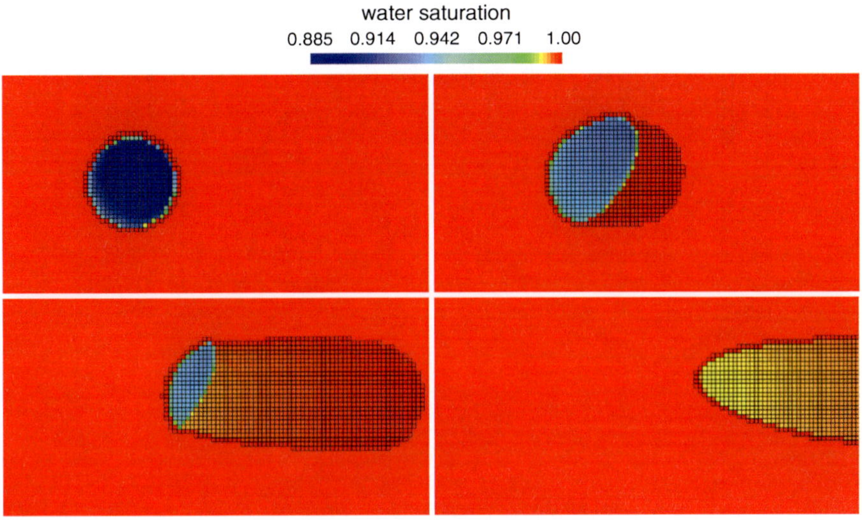

Fig. 10
Dissolution of residual air in water. In black squares: adaptive subdomain at initial conditions, after 100, 200, and 300 timesteps, respectively (From Fritz et al.)

black squares. In the upper row, the initial subdomain and its expansion due to demixing after 100 timesteps can be seen. In the lower row, one can actually see that cells also get removed from the subdomain and that it moves and finally vanishes with the air phase.

5.1.4 Application Example

To demonstrate the performance of the multiphysics approach and to compare it to the full compositional two-phase model on a real-life problem, we chose a benchmark problem from carbon-dioxide sequestration as presented in Eigestad et al. (in press) and Class et al. (in press). Carbon dioxide is injected at a depth of 2,960–3,010 m into a saline aquifer over a time period of 25 years. The given spatial discretization consists of 54,756 control volumes. *Figure 11* shows the results of the simulation runs after 25 years. The result of the multiphysics model on the upper right is obviously in good agreement with the full compositional model. The lower middle displays the subdomain which covers some 4.5% of the model domain in the last timestep. Using the multiphysics model, the number of calls of the flash calculation could be reduced by a factor of 46 as compared to the full compositional model.

5.2 A Multiscale Multiphysics Example

We present the multiscale multiphysics algorithm developed by Niessner and Helmig (2006, 2007, 2009) which couples two different physical models:

- A three-phase–three-component model (fine scale, small subdomain)
- A two-phase model (coarse scale, whole domain).

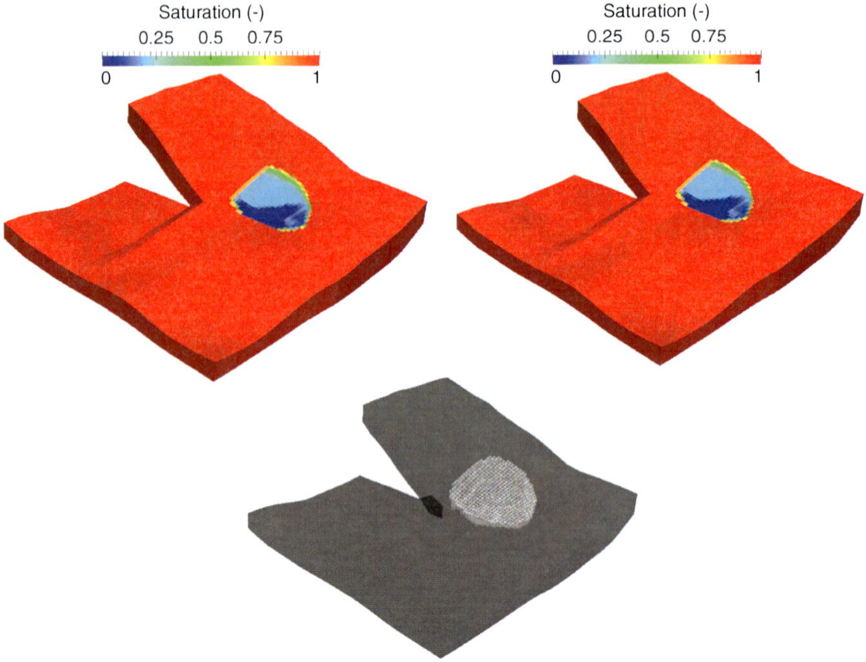

Fig. 11

Results for the Johansen formation benchmark after 25 years. *Upper left*: full compositional two-phase model, *upper right*: multiphysics approach, *lower middle*: subdomain of multiphysics approach

In the following, we present the fine-scale equation system for describing these processes. *Three-phase three-component model*: The three considered phases are the wetting phase w (might be water), an intermediate-wetting phase n (might be an oil), and a non-wetting phase g (which could be a gas). Neglecting gravitational and capillary forces (which are not negligible in many physical systems, of course, but we do so in order to set up the algorithm) and assuming constant densities and viscosities, three-phase three component flow and transport can be described by a pressure equation and two concentration equations in the following way.

The pressure equation is given as

$$\nabla \cdot (\lambda_t \mathbf{K} \nabla p) = 0. \tag{76}$$

Using extended Darcy's law, the Darcy velocity $\mathbf{v}_t$ can be reconstructed as $\mathbf{v}_t = -\lambda_t \mathbf{K} \nabla p$.

The concentration equations yield

$$\frac{\partial C^\kappa}{\partial t} + \mathbf{v}_t \cdot \nabla \sum_\alpha \left(f_\alpha \varrho_\alpha X_\alpha^\kappa \right) = 0 \qquad \kappa \in \{1, 2\}. \tag{77}$$

Note that source and sink terms have been skipped for simplicity. Given the total concentrations and the pressure, saturations can be calculated using flash calculation (see ➋ Sect. 4.2.2).
Two-phase model: We consider a porous medium occupied by a wetting-phase w and a non-wetting gas phase g. The flow of the two phases is described by the pressure equation as given

in Eq. 76 with the only difference that λ_t is fully defined by knowing one of the saturations, $\lambda_t = \lambda_t(S_w)$.

The second equation needed is the saturation equation,

$$\frac{\partial S_w}{\partial t} + \mathbf{v}_t \cdot \nabla f_w = 0, \tag{78}$$

where again, source/sink terms are skipped for sake of simplicity.

5.2.1 Multiscale Multiphysics Algorithm

Figure 12 shows on which scales and in which parts of the domain which equation is solved: pressure and saturation equation are solved in the whole domain and on the coarse scale, while the concentration equations are only solved in a subdomain of interest and on the fine scale. The corresponding multiscale multiphysics algorithm is presented in the context of a discontinuous Galerkin finite element discretization of the partial differential equations using the method given by Bastian (2003) as this scheme is used for calculating the numerical results. However, after slight modifications of the algorithm, other discretization schemes are possible. The elliptic pressure equation is solved using an time-implicit Runge–Kutta scheme. The linear equation system arising from the implicit Runge–Kutta time stepping is preconditioned with a smoother and solved using a linear multigrid solver. Hyperbolic/parabolic saturation and concentration equations are solved time-explicitly using an S-stable three-stage Runge–Kutta method of order three with a modified minmod flux limiter as given in Cockburn and Shu (1998). The overall algorithm consists of the following steps.

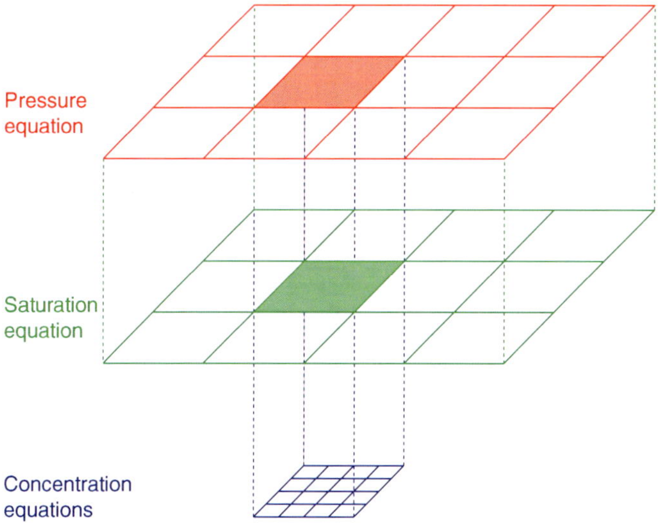

Fig. 12
Scales and solution domain for pressure, saturation, and concentration equations in the multiscale multiphysics algorithm

1. Calculate coarse-scale intrinsic permeabilities.
2. Solve the coarse-scale pressure equation in the total domain.
3. Calculate the coarse-scale velocity field and make it continuous across coarse-element edges.
4. Calculate fine-scale velocities in the subdomain using downscaling and make them continuous across fine-element edges.
5. Solve two concentration equations in the subdomain using Dirichlet boundary conditions and n micro time steps. The latter is necessary in order to fulfill the Courant–Friedrich–Lewy stability criterion for the explicitly solved concentration equations. Here, n is the ratio of coarse-grid discretization length over fine-grid discretization length. Note that the boundary conditions may vary with time as total concentrations depend on saturation.
6. Calculate fine-scale saturations in the subdomain using flash calculations and average them over coarse-scale elements.
7. Solve the coarse-scale saturation equation using these averaged saturations in the subdomain and a source/sink term that makes up for possible mass losses/gains due to changing saturations in the subdomain.
8. Use new saturation field and go to step 2.

Step (7) is necessary as averaged saturations obtained through flash calculations from the solution of local fine-scale concentration equations generally differ from saturations obtained from solving the upscaled saturation equation. This is because in the first case, two-phase two-component processes are accounted for while in the second case, only a two-phase problem is solved. As two-phase two-component processes are the governing processes within the subdomain, these saturations are assumed to be correct and plugged into the solution of the upscaled saturation equation. Now, in order to maintain local mass conservation, a source/sink term within each coarse-scale element is needed that makes up for this difference in mass.

5.2.2 Numerical Results

We consider a domain with a geostatistically generated intrinsic permeability field given on the fine scale. Pressure, saturation, and concentration equations are solved (i.e., mass transfer is taking place in a subdomain) as shown in ➋ *Fig. 12.* ➋ *Figure 13* shows the boundary and initial conditions as well as the spatial extension of the subdomain where mass transfer is accounted for by solving concentration equations and using flash calculations to distribute

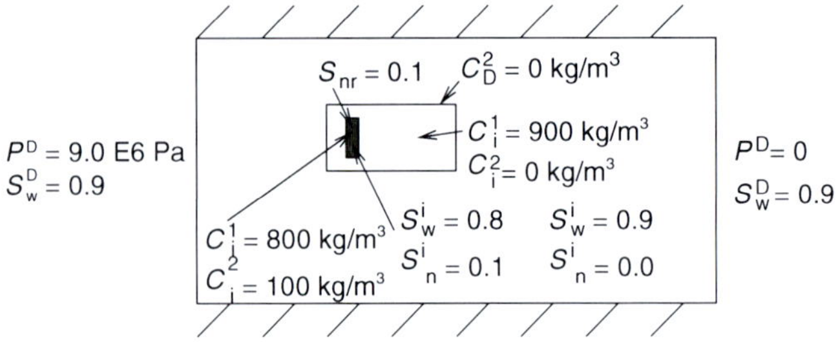

Fig. 13
Setup of example 2. Boundary and initial conditions as well as size and location of the subdomain and the residual LNAPL contamination are shown

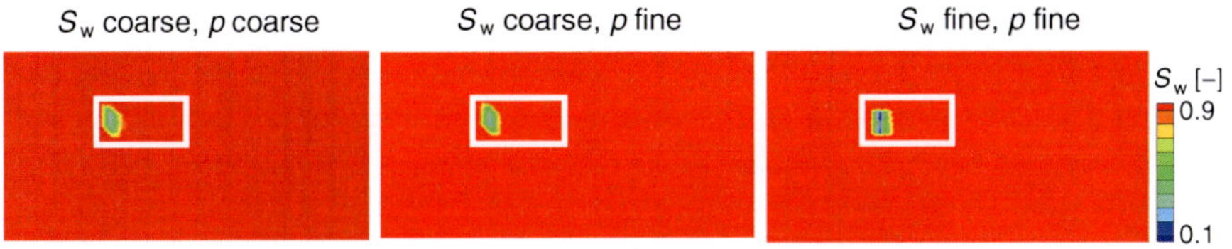

Fig. 14

Water saturation contours for three different test cases. *Left-hand side*: fine-scale reference solution, *middle*: saturation equation is upscaled and pressure equation is solved on the fine scale, *right-hand side*: both pressure and saturation equation are upscaled

components among phases. In a part of the subdomain, there is a residual LNAPL contamination of an initial saturation of $S_n = 0.1$. Again, there is a pressure gradient from left to right and top as well as bottom boundary are closed. Wetting-phase saturation is equal to 0.9 everywhere except for a small part of the subdomain where mass transfer is accounted for (here, wetting-phase saturation is 0.1). Finally, Dirichlet boundary conditions for total concentrations are given such that water and gas phase on the boundary of the subdomain are consisting of water component only, respectively air component only. The motivation for that choice of boundary conditions at the subdomain boundary is that this subdomain has to be chosen sufficiently large in order to capture the whole region where mass transfer is taking place. The residual LNAPL initially consists of LNAPL component only. While three-phase water–LNAPL–gas relative permeability–saturation relationships after Van Genuchten (1980) in combination with Parker et al. (1987) are used in the subdomain, in the remaining part of the domain, two-phase Van Genuchten functions are chosen.

Figure 14 compares water saturation contours at a certain point of time (after 59 h) using the extended multiscale multiphysics algorithm (both pressure and saturation equation are upscaled), the original multiscale multiphysics algorithm (only the saturation equation is upscaled), and the fine-scale reference solution. The location of the subdomain where three-phase–three-component processes are modeled corresponds to the region within the white rectangle. It can be seen that water saturation changes downstream of the residual LNAPL plume and also within the region where the LNAPL is present. As the LNAPL cannot move the change of water saturation must be due to mass transfer processes. Neither the model where the saturation equation is upscaled nor the model where both saturation and pressure equation are upscaled can capture all the details of the fine-scale reference solution. But still, the principal features of the saturation distribution can be captured. Also, the agreement between the results using saturation upscaling only and the results using the model with upscaled pressure and saturation equation is excellent. This indicates again that pressure upscaling and velocity downscaling preserves a high accuracy of the fine-scale velocity field.

6 Conclusion

In this chapter, we have given an overview of multiscale multiphysics methods for flow and transport processes in porous media. Therefore, we defined relevant scales and gave an overview of multiscale and of multiphysics methods. We introduced the mathematical model for compositional non-isothermal multiphase flow and transport in porous media and discussed possible mathematical formulations. Based on that, decoupled and coupled numerical solution strategies are discussed. In a next part, applications examples of multiphysics and of multiscale multiphysics models were given. This work is meant to give an overview of existing multiscale multiphysics approaches for multiphase flow problems in porous media. Examples are given in order to illustrate the effectiveness and applicability of these algorithms.

Future work needs to be done to include more complex processes in the multiscale multiphysics algorithms in order to allow for the modeling of highly complex real-life systems. Also, the development of upscaling techniques and the upscaling of the complex equations is a crucial issue. Coupling techniques need to be improved to allow for a physically based coupling of different multiphysics domains and for the coupling across scales. Numerical methods need to be improved in order to allow for moving meshes if the multiphysics domains move during a simulation and multiscale multiphysics algorithms need to allow for the application of different numerical schemes for the solution of different physical processes (multi numerics).

In conclusion, we state that the development and application of multiscale multiphysics techniques allows to model highly complex physical problems in large domains that could otherwise not be solved numerically.

Acknowledgments

Authors are members of the International Research Training Group NUPUS, financed by the German Research Foundation (DFG , GRK 1398), and of the Cluster of Excellence in Simulation Technology at the University of Stuttgart financed by the German Research Foundation (DFG, EXC 310/1). We thank the DFG for the support.

References

Aarnes JE, Krogstad S, Lie K-A (2006) A hierarchical multiscale method for two-phase flow based upon mixed finite elements and nonuniform coarse grids. Multiscale Model Simul. 5(2): 337–363

Aarnes JE, Krogstad S, Lie K-A (2008) Multiscale mixed/mimetic methods on corner-point grids. Comput Geosci 12(3):297–315

Aavatsmark I (2002) An introduction to multipoint flux approximations for quadrilateral grids. Comput Geosci 6(3-4):405–432. Locally conservative numerical methods for flow in porous media.

Aavatsmark I, Eigestad GT, Mallison BT, Nordbotten JM (2008) A compact multipoint flux approximation method with improved robustness. Numer Methods Partial Differential Equations 24(5):1329–1360

Ainsworth M, Oden JT (2000) A posteriori error estimation in finite element analysis. Pure and Applied Mathematics (New York). Wiley-Interscience [John Wiley & Sons], New York

Albon C, Jaffre J, Roberts J, Wang X, Serres C (1999) Domain decompositioning for some transition problems in flow in porous media. In Chen Z, Ewing R, Shi Z-C (eds), Numerical treatment of multiphase flow in porous media, Lecture Notes in Physics. Springer, Berlin

Allen MB, Ewing RE, Lu P (1992) Well-conditioned iterative schemes for mixed finite-element models of porous-media flows. SIAM J Sci Statist Comput 13(3):794–814

Arbogast T (1989) Analysis of the simulation of single phase flow through a naturally fractured reservoir. SIAM J Numer Anal 26(1):12–29

Arbogast T, Pencheva G, Wheeler MF, Yotov I (2007) A multiscale mortar mixed finite element method. Multiscale Model Simul 6(1):319–346

Atkins P (1994) Physical chemistry. Oxford University Press, 5th edn

Aziz K, Settari A (1979) Petroleum reservoir simulation. Elsevier Applied Science, New York

Aziz K, Wong T (1989) Considerations in the development of multipurpose reservoir simulation models. In: Proceedings first and second international forum on reservoir simulation, pages 77–208. Steiner P

Babuska I, Strouboulis T (2001) The finite element method and its reliability. In: Numerical mathematics and scientific computation. The Clarendon Press Oxford University Press, New York

Barker J, Thibeau S (1997) A critical review of the use of pseudorelative permeabilities for upscaling. SPE Reserv Eng 12(2):138–143

Bastian P (2003) Higher order discontinuous Galerkin methods for flow and transport in Porous media. In Bänsch E (ed), Challenges in scientific computing—*CISC 2002*, number 35 in LNCSE

Bastian P, Rivière B (2003) Superconvergence and $H(\text{div})$ projection for discontinuous Galerkin methods. Int J Numer Methods Fluids 42(10):1043–1057

Beavers GS, Joseph DD (1967) Boundary conditions at a naturally permeable wall. *J Fluid Mech* 30:197–207

Berndt M, Lipnikov K, Shashkov M, Wheeler MF, Yotov I (2005) Superconvergence of the velocity in mimetic finite difference methods on quadrilaterals. *SIAM J Numer Anal* 43(4):1728–1749

Binning P, Celia MA (1999) Practical implementation of the fractional flow approach to multiphase flow simulation. *Adv Water Resour* 22(5): 461–478

Bramble JH (1993) *Multigrid methods*, vol 294 of Pitman research notes in mathematics series. Longman Scientific & Technical, Harlow

Brezzi F, Fortin M (1991) Mixed and hybrid finite element methods, vol 15 of Springer series in computational mathematics. Springer, New York

Brezzi F, Lipnikov K, Shashkov M (2005a) Convergence of the mimetic finite difference method for diffusion problems on polyhedral meshes. SIAM J Numer Anal 43(5):1872–1896

Brezzi F, Lipnikov K, Simoncini V (2005b) A family of mimetic finite difference methods on polygonal and polyhedral meshes. Math Models Methods Appl Sci 15(10):1533–1551

Briggs WL, Henson VE, McCormick SF (2000) A multigrid tutorial, 2nd edn. Society for Industrial and Applied Mathematics (SIAM), Philadelphia

Cao Y, Helmig R, Wohlmuth B (2008) The influence of the boundary discretization on the multipoint flux approximation L-method. In Finite volumes for complex applications V, pp 257–263. ISTE, London

Cattani C, Laserra E (2003) Wavelets in the transport theory of heterogeneous reacting solutes. Int J Fluid Mech Res 30(2):147–152

Chavent G (1976) A new formulation of diphasic incompressible flows in porous media, pp 258–270. Number 503 in Lecture notes in mechanics. Springer, Berlin

Chavent G, Jaffré J (1986) Mathematical models and finite elements for reservoir simulation. North-Holland, Amsterdam

Chen Y, Durlofsky LJ (2006) Adaptive local-global upscaling for general flow scenarios in heterogeneous formations. Transp Porous Media 62(2):157–185

Chen Y, Durlofsky LJ, Gerritsen M, Wen XH (2003) A coupled local-global upscaling approach for simulating flow in highly heterogeneous formations. Adv Water Resour 26(10):1041–1060

Chen Z, Ewing RE, Jiang Q, Spagnuolo AM (2002) Degenerate two-phase incompressible flow. V. Characteristic finite element methods. J Numer Math 10(2):87–107

Chen Z, Guan Q, Ewing R (2000) Analysis of a compositional model for fluid flow in porous media. SIAM J Appl Math 60(3):747–777

Chen Z, Hou TY (2002) A mixed multiscale finite element method for elliptic problems with oscillating coefficients. Math Comput 72(242): 541–576

Chen Z, Huan G, Ma Y (2006) Computational methods for multiphase flows in porous media. SIAM, Computational Science & Engineering, Philadelphia

Class H, Ebigbo A, Helmig R, Dahle H, Nordbotten J, Celia M, Audigane P, Darcis M, Ennis-King J, Fan Y, Flemisch B, Gasda S, Jin M, Krug S, Labregere D, Naderi Beni A, Pawar R, Sbai A, Thomas S, Trenty L, Wei L (in press, DOI:10.1007/s10596-009-9146-x). A benchmark study on problems related to co_2 storage in geologic formations: summary and discussion of the results. Comput Geosci

Cockburn B, Lin SY, Shu C-W (1989) TVB Runge-Kutta local projection discontinuous Galerkin finite element method for conservation laws. III. One-dimensional systems. J Comput Phys 84(1):90–113

Cockburn B, Shu C-W (1989) TVB Runge-Kutta local projection discontinuous Galerkin finite element method for conservation laws. II. General framework. Math Comp 52(186): 411–435

Cockburn B, Shu C-W (1998) The Runge–Kutta discontinuous Galerkin method for conservation laws V: Multidimensional Systems. J Comput Phys 141:199–224

Codina R (2001) A stabilized finite element method for generalized stationary incompressible flows. Comput Methods Appl Mech Eng 190(20–21):2681–2706

Darman NH, Pickup GE, Sorbie KS (2002) A comparison of two-phase dynamic upscaling methods based on fluid potentials. Comput Geosci 6(1):5–27

Dawson CN, Russell TF, Wheeler MF (1989) Some improved error estimates for the modified method of characteristics. *SIAM J Numer Anal* 26(6):1487–1512

Dennis JE, Jr Schnabel RB (1996) Numerical methods for unconstrained optimization and nonlinear equations, vol 16 of *Classics in applied mathematics*. Society for Industrial and Applied Mathematics (SIAM), Philadelphia

Dietrich P, Hemlig R, Sauter M, Hötzl H, Köngeter J, Teutsch G, editors (2005) *Flow and transport in fractured porous media*. Springer, Berlin

Discacciati M, Miglio E, A Q (2002) Mathematical and numerical models for coupling surface and groundwater flows. Appl Num Math, 43:57–74

Douglas J Jr, Huang C-S, Pereira F (1999) The modified method of characteristics with adjusted advection. Numer Math 83(3):353–369

Durlofsky LJ (1991) Numerical calculation of equivalent grid block permeability tensors for heterogeneous porous media. Water Resour Res 27(5):699–708

E W, Engquist B (2003a) The heterogeneous multiscale methods. Commun Math Sci 1(1):87–132

E W, Engquist B (2003b) Multiscale modeling and computation. Notices Am Math Soc 50(9): 1062–1070

E W, Engquist B, Huang Z (2003) Heterogeneous multiscale method: a general methodology for multiscale modeling. Phys Rev 67

E W, Engquist B, Li X, Ren W, Vanden-Eijnden E (2007) Heterogeneous multiscale methods: a review. Commun Comput Phys 2(3):367–450

Efendiev Y, Durlofsky LJ (2002) Numerical modeling of subgrid heterogeneity in two phase flow simulations. Water Resour Res 38(8)

Efendiev Y, Durlofsky LJ, Lee SH (2000) Modeling of subgrid effects in coarse-scale simulations of transport in heterogeneous porous media. Water Resour Res 36(8):2031–2041

Efendiev Y, Hou T (2007) Multiscale finite element methods for porous media flows and their applications. Appl Numer Math 57(5–7): 577–596

Eigestad GT, Klausen RA (2005) On the convergence of the multi-point flux approximation O-method: numerical experiments for discontinuous permeability. Numer Methods Partial Differ Equations 21(6):1079–1098

Eigestad G, Dahle H, Hellevang B, Johansen W, Riis F, Øian E (in press, DOI:10.1007/s10596-009-9153-y). Geological modeling and simulation of co_2 injection in the Johansen formation. Comput Geosci

Ewing RE, Russell TF, Wheeler MF (1984) Convergence analysis of an approximation of miscible displacement in porous media by mixed finite elements and a modified method of characteristics. Comput Methods Appl Mech Engrg 47 (1–2):73–92

Ewing RE, Wang H (1994) Eulerian-Lagrangian localized adjoint methods for variable-coefficient advective-diffusive-reactive equations in groundwater contaminant transport. In: Advances in optimization and numerical analysis (Oaxaca, 1992), vol. 275 of Math Appl pp. 185–205. Kluwer Acad. Publ., Dordrecht

Ewing RE, Wang H (1996) An optimal-order estimate for Eulerian-Lagrangian localized adjoint methods for variable-coefficient advection-reaction problems. SIAM J Numer Anal 33(1): 318–348

Eymard R, Gallouët T, Herbin R (2000) Finite volume methods. In Handbook of numerical analysis, Vol. VII, Handb. Numer Anal, VII, pp 713–1020. North-Holland, Amsterdam

Fritz J, Flemisch B, Helmig R (accepted, preprint on www.nupus.uni-stuttgart.de number 2009/8). Multiphysics modeling of advection-dominated two-phase compositional flow in porous media. Int J Numer Anal Model

Ghostine R, Kesserwani G, Mosé R, Vazquez J, Ghenaim A (2009) An improvement of classical slope limiters for high-order discontinuous Galerkin method. Int J Numer Methods Fluids 59(4):423–442

Giraud L, Langou J, Sylvand G (2006) On the parallel solution of large industrial wave propagation problems. J Comput Acoust 14(1): 83–111

Girault V, Rivière B (2009) Dg approximation of coupled Navier-Stokes and Darcy equations by beaver-Joseph-Saffman interface condition. SIAM J Numer Anal 47:2052–2089

Gray GW, Leijnse A, Kolar RL, Blain CA (1993) Mathematical tools for changing scale in the analysis of physical systems 1st edn. CRC, Boca Rato

Hajibeygi H, Bonfigli G, Hesse MA, Jenny P (2008) Iterative multiscale finite-volume method. J Comput Phys 227(19):8604–8621

Hauke G, García-Olivares A (2001) Variational subgrid scale formulations for the advection-diffusion-reaction equation. Comput Methods Appl Mech Eng 190(51–52):6847–6865

He Y, Han B (2008) A wavelet finite-difference method for numerical simulation of wave propagation in fluid-saturated porous media. Appl Math Mech (English Ed.), 29(11):1495–1504

Helmig R (1997) Multiphase flow and transport processes in the subsurface. Spring

Herrera I, Ewing RE, Celia MA, Russell TF (1993) Eulerian-Lagrangian localized adjoint method: the theoretical framework. Numer. Methods Partial Differential Equations 9(4):431–457

Hoteit H, Ackerer P, Mosé R, Erhel J, Philippe B (2004) New two-dimensional slope limiters for discontinuous Galerkin methods on arbitrary meshes. Int J Numer Methods Engrg 61(14):2566–2593

Hoteit H, Firoozabadi A (2008) Numerical modeling of two-phase flow in heterogeneous permeable media with different capillarity pressures. Adv Water Resour 31(1):56–73

Hou TY, Wu X-H (1997) A multiscale finite element method for elliptic problems in composite materials and porous media. J Comput Phys 134(1):169–189

Hristopulos D, Christakos G (1997) An analysis of hydraulic conductivity upscaling. Nonlinear Anal 30(8):4979–4984

Huang C-S (2000) Convergence analysis of a mass-conserving approximation of immiscible displacement in porous media by mixed finite elements and a modified method of characteristics with adjusted advection. Comput Geosci 4(2):165–184

Hughes TJR (1995) Multiscale phenomena: Green's functions, the Dirichlet-to-Neumann formulation, subgrid scale models, bubbles and the origins of stabilized methods. Comput Methods Appl Mech Eng 127(1–4):387–401

Hughes TJR, Feijóo GR, Mazzei L, Quincy J-B (1998) The variational multiscale method—a paradigm for computational mechanics. Comput Methods Appl Mech Eng 166(1–2):3–24

Hyman J, Morel J, Shashkov M, Steinberg S (2002) Mimetic finite difference methods for diffusion equations. Comput Geosci 6(3–4):333–352. Locally conservative numerical methods for flow in porous media

IPCC (2005) Carbon dioxide capture and storage. Special report of the intergovernmental panel on climate change. Cambridge University Press, Cambridge

Jäger W, Mikelić A (2009) Modeling effective interface laws for transport phenomena between an unconfined fluid and a porous medium using homogenization. Transp Porous Media 78: 489–508

Jang G-W, Kim JE, Kim YY (2004) Multiscale Galerkin method using interpolation wavelets for two-dimensional elliptic problems in general domains. Int J Numer Methods Engrg 59(2): 225–253

Jenny P, Lee SH, Tchelepi H (2003) Multi-scale finite-volume method for elliptic problems in subsurface flow simulations. J Comput Phys 187: 47–67

Jenny P, Lee SH, Tchelepi HA (2006) Adaptive fully implicit multi-scale finite-volume method for multi-phase flow and transport in heterogeneous porous media. J Comput Phys 217(2):627–641

Juanes R (2005) A variational multiscale finite element method for multiphase flow in porous media. Finite Elem Anal Des 41(7–8):763–777

Juanes R, Lie K-A (2008) Numerical modeling of multiphase first-contact miscible flows. II. Front-tracking/streamline simulation. Transp Porous Media 72(1):97–120

Kim M-Y, Park E-J, Thomas SG, Wheeler MF (2007) A multiscale mortar mixed finite element method for slightly compressible flows in porous media. J Korean Math Soc 44(5): 1103–1119

Kippe V, Aarnes JE, Lie K-A (2008) A comparison of multiscale methods for elliptic problems in porous media flow. Comput Geosci 12(3): 377–398

Klausen RA, Winther R (2006) Robust convergence of multi point flux approximation on rough grids. Numer Math 104(3):317–337

Layton WJ, Schieweck F, Yotov I (2003) Coupling fluid flow with porous media flow. SIAM J Numer Anal 40:2195–2218

Lee SH, Wolfsteiner C, Tchelepi HA (2008) Multiscale finite-volume formulation of multiphase flow in porous media: black oil formulation of compressible, three-phase flow with gravity. Comput Geosci 12(3):351–366

LeVeque RJ (2002) *Finite volume methods for hyperbolic problems*. Cambridge Texts in Applied Mathematics. Cambridge University Press, Cambridge

Lüdecke C, Lüdecke D (2000) Thermodynamik. Springer, Berlin

Lunati I, Jenny P (2006) Multiscale finite-volume method for compressible multiphase flow in porous media. J Comput Phys 216(2):616–636

Lunati I, Jenny P (2007) Treating highly anisotropic subsurface flow with the multiscale finite-volume method. Multiscale Model Simul 6(1): 308–318

Lunati I, Jenny P (2008) Multiscale finite-volume method for density-driven flow in porous media. Comput Geosci 12(3):337–350

Matringe SF, Juanes R, Tchelepi HA (2006) Robust streamline tracing for the simulation of porous media flow on general triangular and quadrilateral grids. J Comput Phys 219(2):992–1012

Mazzia A, Putti M (2006) Three-dimensional mixed finite element-finite volume approach for the solution of density-dependent flow in porous media. J Comput Appl Math 185(2): 347–359

Michelsen M, Mollerup J (2007) Thermodynamic models: fundamentals & computational aspects. Tie-Line Publications, Denmark

Müller S (2003) Adaptive multiscale schemes for conservation laws, vol 27 of Lecture notes in computational science and engineering. Springer, Berlin

Nghiem L, Li Y-K (1984) Computation of multiphase equilibrium phenomena with an equation of state. Fluid Phase Equilib 17(1):77–95

Niessner J, Helmig R (2006) Multi-scale modeling of two-phase-two-component processes in heterogeneous porous media. Numer Linear Algebra Appl 13(9):699–715

Niessner J, Helmig R (2007) Multi-scale modeling of three-phase-three-component processes in heterogeneous porous media. Adv Water Resour 30(11):2309–2325

Niessner J, Helmig R (2009) Multi-physics modeling of flow and transport in porous media using a downscaling approach. Adv Water Resour 32(6):845–850

Nordbotten JM, Bjørstad PE (2008) On the relationship between the multiscale finite-volume method and domain decomposition preconditioners. Comput Geosci 12(3):367–376

Of G (2007) Fast multipole methods and applications. In Boundary element analysis, vol 29 of Lect. notes Appl Comput Mech, pp 135–160. Springer, Berlin

Oladyshkin S, Royer J-J, Panfilov M (2008) Effective solution through the streamline technique and HT-splitting for the 3D dynamic analysis of the compositional flows in oil reservoirs. Transp Porous Media 74(3):311–329

Panfilov M (2000) Macroscale models of flow through highly heterogeneous porous media. Kluwer Academic Publishers, Doodrecht

Parker JC, Lenhard RJ, Kuppusami T (1987) A parametric model for constitutive properties governing multiphase flow in porous media. Water Resour Res 23(4):618–624

Peszynska M, Lu Q, Wheeler M (2000) Multiphysics coupling of codes. In: Computational methods in water resources, pp 175–182. A. A. Balkema

Peszynska M, Wheeler MF, Yotov I (2002) Mortar upscaling for multiphase flow in porous media. Comput Geosci 6:73–100

Prausnitz JM, Lichtenthaler RN, Azevedo EG (1967) Molecular thermodynamics of fluid-phase equilibria. Prentice-Hall

Pruess K (1985) A practical method for modeling fluid and heat flow in fractured porous media. SPE J 25(1):14–26

Renard P, de Marsily G (1997) Calculating effective permeability: a review. Adv Water Resour 20:253–278

Russell TF (1990) Eulerian-Lagrangian localized adjoint methods for advection-dominated problems. In: Numerical analysis 1989 (Dundee, 1989), vol 228 of Pitman Res. notes Math Ser, pp 206–228. Longman Sci. Tech., Harlow

Ryzhik V (2007) Spreading of a NAPL lens in a double-porosity medium. Comput Geosci 11(1):1–8

Scheidegger A (1974) The physics of flow through porous media, 3rd edn University of Toronto Press, Toronto

Shashkov M (1996) Conservative finite-difference methods on general grids. Symbolic and Numeric Computation Series. CRC Press, Boca Raton

Sheldon JW, Cardwell WT (1959) One-dimensional, incompressible, noncapillary, two-phase fluid in a porous medium. Petrol Trans AIME 216: 290–296

Sleep BE, Sykes JF (1993) Compositional simulation of groundwater contamination by organic-compounds. 1. Model development and verification. Water Resour Res 29(6):1697–1708

Smith EH Seth MS (1999) Efficient solution for matrix-fracture flow with multiple interacting continua. Int J Numer Anal Methods Geomech 23(5):427–438

Srinivas C, Ramaswamy B, Wheeler MF (1992) Mixed finite element methods for flow through unsaturated porous media. In: Computational methods in water resources, IX, vol. 1 (Denver, CO, 1992), pp 239–246. Comput. Mech., Southampton

Stenby E, Wang P (1993) Noniterative phase equilibrium calculation in compositional reservoir simulation. SPE 26641

Stone HL, Garder AO Jr (1961) Analysis of gas-cap or dissolved-gas reservoirs. Petrol Trans. AIME, 222:92–104

Stüben K (2001) A review of algebraic multigrid. J Comput Appl Math 128(1–2):281–309. Numerical analysis 2000, vol VII, Partial differential equations

Suk H, Yeh G-T (2008) Multiphase flow modeling with general boundary conditions and automatic phase-configuration changes using a fractional-flow approach. Comput Geosci 12(4):541–571

Tornberg A-K, Greengard L (2008) A fast multipole method for the three-dimensional Stokes equations. J Comput Phys 227(3):1613–1619

Trangenstein J Bell J (1989) Mathematical structure of compositional reservoir simulation. SIAM J Sci Stat Comput 10(5):817–845

Trottenberg U, Oosterlee CW, Schüller A (2001) Multigrid. Academic, San Diego

Urban K (2009) Wavelet methods for elliptic partial differential equations. Numerical Mathematics and Scientific Computation. Oxford University Press, Oxford

Van Genuchten MT (1980) A closed-form equation for predicting the hydraulic conductivity of unsaturated soils. Am J Soil Sci 44:892–898

Wallstrom TC, Christie MA, Durlofsky LJ, Sharp DH (2002a) Effective flux boundary conditions for upscaling porous media equations. Transp Porous Media 46(2):139–153

Wallstrom TC, Hou S, Christie MA, Durlofsky LJ, Sharp DH, Zou Q (2002b) Application of effective flux boundary conditions to two-phase upscaling in porous media. Transp Porous Media 46(2):155–178

Wang H, Liang D, Ewing RE, Lyons SL, Qin G (2002) An ELLAM approximation for highly compressible multicomponent flows in porous media. Comput Geosci, 6(3–4):227–251. Locally conservative numerical methods for flow in porous media

Wang P, Barker J (1995) Comparison of flash calculations in compositional reservoir simulation. SPE 30787

Wheeler M, Arbogast T, Bryant S, Eaton J, Lu Q, Peszynska M, Yotov I (1999) A parallel multiblock/multidomain approach to reservoir simulation. In: Fifteenth SPE symposium on reservoir simulation, Houston, Texas, pp 51–62. Society of Petroleum Engineers. SPE 51884

Whitaker S (1998) The method of volume averaging. Kluwer Academic Publishers, Dordretcht

White F (2003) Fluid mechanics. McGraw-Hill, New York

Yao Z-H, Wang H-T, Wang P-B, Lei T (2008) Investigations on fast multipole BEM in solid mechanics. J Univ Sci Technol China 38(1):1–17

Yotov I (2002) Advanced techniques and algorithms for reservoir simulation IV. Multiblock solvers and preconditioners. In: Chadam J, Cunningham A, Ewing RE, Ortoleva P, Wheeler MF (eds), IMA volumes in mathematics and its applications, Vol 131: Resource recovery, confinement, and remediation of environmental hazards. Springer, Berlin

16 Numerical Dynamo Simulations: From Basic Concepts to Realistic Models

Johannes Wicht[1] · *Stephan Stellmach*[2] · *Helmut Harder*[2]
[1]Max-Planck Intitut für Sonnensystemforschung, Kaltenburg-Lindau, Germany
[2]Institut für Geophysik, Westfälische Wilhelms-Universität, Münster, Germany

W. Freeden, M.Z. Nashed, T. Sonar (Eds.), *Handbook of Geomathematics*, DOI 10.1007/978-3-642-01546-5_16,

Abstract The last years have witnessed an impressive growth in the number and quality of numerical dynamo simulations. The numerical models successfully describe many aspects of the geomagnetic field and also set out to explain the various fields of other planets. The success is somewhat surprising since numerical limitation force dynamo modelers to run their models at unrealistic parameters. In particular the Ekman number, a measure for the relative importance of viscous to Coriolis forces, is many orders of magnitude too large: Earth's Ekman number is $E = 10^{-15}$ while even today's most advanced numerical simulations have to content themselves with $E = 10^{-6}$. After giving a brief introduction into the basics of modern dynamo simulations the fundamental force balances are discussed and the question how well the modern models reproduce the geomagnetic field is addressed. First-level properties like the dipole dominance, realistic Elsasser and **magnetic Reynolds numbers**, and an Earth-like reversal behavior are already captured by larger Ekman number simulations around $E = 10^{-3}$. However, low Ekman numbers are required for modeling torsional oscillations which are thought to be an important part of the decadal geomagnetic field variations. Moreover, only low Ekman number models seem to retain the huge dipole dominance of the geomagnetic field once the Rayleigh number has been increased to values where field reversals happen. These cases also seem to resemble the low-latitude field found at Earth's core-mantle boundary more closely than larger Ekman numbers cases.

Keywords: Dynamo; Planetary magnetic fields; Planetary interiors; Numerical model

1 Introduction

Joseph Lamor was the first to suggest that the magnetic fields of the Earth and the sun are maintained by a dynamo mechanism. The underlying physical principle can be illustrated with the homopolar or **disc dynamo** shown in ➋ *Fig. 1*. A metal disc is rotated with speed ω in an initial magnetic field $\mathbf{B}_0$. The motion induces an electromotive force that drives an electric current flowing from the rotation axis to the outer rim of the disc. Sliding contacts close an electric circuit through a coil which produces a magnetic field $\mathbf{B}$. This may subsequently replace $\mathbf{B}_0$ provided the rotation rate exceeds a critical value ω_c that allows to overcome the inherent Ohmic losses. The dynamo is then called **self-exited**. Lorentz forces associated with the induced current break the rotation and limit the magnetic field strength to a value that depends on the torque maintaining the rotation (Roberts 2007).

The disc dynamo like any electrical generator works because of the suitable arrangement of the electrical conductors. In Earth, however, the dynamo process operates in the liquid outer part of Earth's iron core, a homogeneously conducting spherical shell. For these **homogeneous dynamos** to operate, the motion of the liquid dynamo medium and the magnetic field itself must obey a certain complexity. Unraveling the required complexity kept dynamo theoreticians busy for some decades and is still a matter of research. **Cowling's theorem**, for example, states that a perfectly axisymmetric magnetic field cannot be maintained by a dynamo process (Cowling 1957; Ivers and James 1984). The same is true for a purely **toroidal** field (Kaiser et al. 1994), a field whose fieldlines never leave the dynamo region.

Whether or not certain flow configurations would provide dynamo action is typically explored in a kinematic dynamo setup where the flow field is prescribed and steady in time. The magnetic field generation is described by the **induction equation** which can be derived

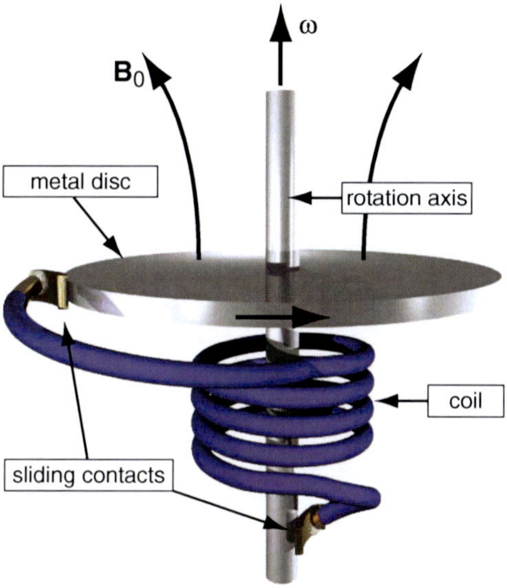

Fig. 1
The homopolar or disc dynamo illustrates the dynamo process

from the pre-Maxwell equations (Roberts 2007):

$$\frac{\partial \mathbf{B}}{\partial t} = \mathrm{Rm}\, \nabla \times (\mathbf{U} \times \mathbf{B}) + \nabla^2 \mathbf{B}. \tag{1}$$

A nondimensional form has been chosen here that uses a length scale ℓ—for example the shell thickness—and the magnetic diffusion time $\tau_\lambda = \ell^2/\lambda$ as the time scale. Here, $\lambda = 1/(\mu\sigma)$ is the magnetic diffusivity with μ the magnetic permeability (of vacuum) and σ the electrical conductivity. Rm is the **magnetic Reynolds number**

$$\mathrm{Rm} = \frac{\mathrm{u}\ell}{\lambda} \tag{2}$$

that depends on the typical flow amplitude u. It measures the ratio of magnetic field production to diffusion via Ohmic decay and has to exceed a certain critical value Rm_c for dynamo action to arise. The induction equation (1) establishes an eigenvalue problem with eigensolutions $\mathbf{B} = \mathbf{B}_0 \exp \epsilon t$. The dynamo is said to work when at least one solution has a positive growth rate, i.e., an eigenvalue ϵ with a positive real part.

Several simple parameterized flows have been found to support dynamo action. Two dimensional flows that depend on only two cartesian coordinates are particularly simple examples. The flow $\mathbf{U} = \sin y\, \hat{\mathbf{x}} + \sin x\, \hat{\mathbf{x}} + (\cos x - \cos y)\, \hat{\mathbf{z}}$ suggested by Roberts (1972) inspired the design of the successful dynamo experiment in Karlsruhe, Germany (Stieglitz and Müller 2001). Here, $\hat{\mathbf{x}}$, $\hat{\mathbf{y}}$, and $\hat{\mathbf{z}}$ are the three cartesian unit vectors. A consistently correlated helicity $H = \mathbf{U}\cdot(\nabla \times \mathbf{U})$ has been shown to be important to provide large-scale dynamo action and the above flow has been constructed so that H promotes the large-scale coherence by having the same sign everywhere.

The convective motions in the dynamo regions of suns and planets are driven by temperature differences. In terrestrial planets with a **freezing inner core**, compositional differences

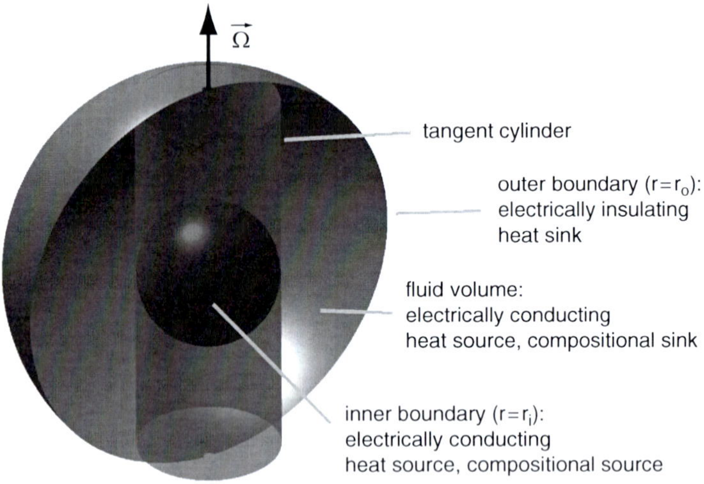

Fig. 2
Principle dynamo setup

provide an additional driving force: The lighter constituents—sulfur, oxygen, silicon—that are mixed into the liquid iron/nickel phase separate from the solid phase and rise from the growing inner core front. Because of the highly supercritical Rayleigh numbers in planetary and solar dynamo regions, the motions are likely turbulent and rather small scale. The feedback of the magnetic field on the flow via the Lorentz force limits the magnetic field growth and thereby determines the magnetic field strength.

Typical state-of-the-art dynamo codes are **self-consistent** and simultaneously solve for the evolution of magnetic field, flow, pressure, temperature, and possibly composition in a rotating spherical shell. The principal setup is shown in *Fig. 2*. Early **self-consistent** models demonstrated the general validity of the concept (Zhang and Busse 1988). The publications by Kageyama and Sato (1995) and Glatzmaier and Roberts (1995), however, are regarded as the first milestones of modern dynamo modeling. Since then, several comparable modern dynamo models have been developed; Christensen and Wicht (2007), Wicht et al. (2009), and Wicht and Tilgner (2010) provide recent overviews.

These models are capable of reproducing the strength and large-scale geometry of the geomagnetic field and also help to understand several smaller-scale features. Many aspects of the geomagnetic field variability are also convincingly captured. Several dynamo models show dipole excursions and reversals very similar to those documented for Earth (Glatzmaier and Roberts 1995; Glatzmaier and Coe 2007; Wicht et al. 2009). Advanced models even show **torsional oscillations** thought to be an important component of the decadal geomagnetic field variation (Zatman and Bloxham 1997; Wicht et al. 2009). In a more recent development, numerical dynamo simulations have also ventured to explain the distinctively different magnetic fields of other planets in the solar system (Stanley and Bloxham 2004; Stanley et al. 2005; Takahashi and Matsushima 2006; Christensen 2006; Wicht et al. 2007; Stanley and Glatzmaier 2010).

The success of modern dynamo simulations seems surprising since the numerical limitations force dynamo modelers to run their codes at parameters that are far from realistic values. For example, the fluid viscosity is generally many orders of magnitude too large in order to damp the small-scale turbulent structures that cannot be resolved numerically. This bears the question whether the numerical dynamos really operate in the same regime as planetary dynamos or reproduce the magnetic field features for the wrong reasons (Glatzmaier 2002). While the numerical simulations certainly do a bad job in modeling the very small-scale flow dynamics they may nevertheless capture the larger-scale dynamo process correctly. This view is supported by recent analysis which show that the numerical results can be extrapolated to the field strength of several planets (Christensen and Aubert 2006) and even fast rotating stars (Christensen et al. 2009) with rather simple **scaling laws**.

The quest for reaching more realistic parameters nevertheless remains and is approached with new numerical methods that will better allow to make use of today's supercomputers as discussed in ➋ Sect. 2.3. Other efforts for improving the realism concern, for example, the treatment of thermal and/or compositional convection including appropriate boundary conditions. These points will be briefly touched in ➋ Sect. 3.

This article concentrates on the question to which degree recent numerical dynamo simulations succeed in modeling Earth's magnetic field and addresses the question which parameter combination helps to reach this goal. Our good knowledge of the geomagnetic field makes it an ideal test case for numerical models. Furthermore the question which changes in the solutions can be expected should the increasing computing power allow to run the models at more realistic parameters in the future is addressed.

The article is organized as follows. The mathematical formulation of the dynamo problem is outlined in ➋ Sect. 2.1 and ➋ 2.2. In ➋ Sect. 2.3, an overview of the diverse numerical methods employed in dynamo simulations is provided. ➋ Section 3 reviews the results from recent spherical dynamo simulations and introduces some new measures to address how well these describe Earth's magnetic field. A conclusion in ➋ Sect. 4 closes the article.

2 Mathematical Formulation

2.1 Basic Equations

Self-consistent dynamo models simultaneously solve for convection and magnetic field generation in a viscous electrically conducting fluid. The convection is driven by temperature differences, and compositional differences also contribute when a **freezing inner core** is modeled. Convection and magnetic field are treated as small disturbances around a **reference state** which is typically assumed to be hydrostatic, adiabatic, and compositionally homogeneous.

Most dynamo simulations of terrestrial planets neglect density and temperature variations of the reference state in the so-called **Boussinesq approximation**. In Earth's outer core both quantities increase by about 25% and similar values can be expected for the other terrestrial planets. These values indicate that the **Boussinesq approximation** may at least serve as a good first approach, all the more so as it involves considerable simplifications of the problem: Viscous heating and Ohmic heating drop out and only the density variations due to temperature and composition fluctuations in the buoyancy term are retained (Braginsky and Roberts 1995).

The following equations comprise the mathematical formulation of the dynamo problem in the **Boussineq approximation**: the **Navier–Stokes equation**

$$\frac{d\mathbf{U}}{dt} = -\nabla P - 2\hat{\mathbf{z}} \times \mathbf{U} + \mathrm{Ra}^{\star} \frac{r}{r_o} C\hat{\mathbf{r}} + (\nabla \times \mathbf{B}) \times \mathbf{B} + \mathrm{E}\nabla^2\mathbf{U}, \tag{3}$$

the **induction equation**

$$\frac{\partial \mathbf{B}}{\partial t} = \nabla \times (\mathbf{U} \times \mathbf{B}) + \frac{\mathrm{E}}{\mathrm{Pm}} \nabla^2 \mathbf{B}, \tag{4}$$

the **codensity evolution equation**

$$\frac{dC}{dt} = \frac{\mathrm{E}}{\mathrm{Pr}} \nabla^2 C + q, \tag{5}$$

the **flow continuity equation**

$$\nabla \cdot \mathbf{U} = 0, \tag{6}$$

and the **magnetic continuity equation**

$$\nabla \cdot \mathbf{B} = 0. \tag{7}$$

Here, d/dt stands for the substantial time derivative $\partial/\partial t + \mathbf{U} \cdot \nabla$, $\mathbf{U}$ is the convective flow, $\mathbf{B}$ the magnetic field, P is a modified pressure that also contains centrifugal effects, and C is the **codensity** explained below.

The equations are given in a nondimensional form that uses the shell thickness $\ell = r_o - r_i$ as a length scale, the rotation period Ω^{-1} as a time scale, the codensity difference Δc across the shell as the codensity scale, and $(\bar{\rho}\mu)^{1/2}\Omega\ell$ as the magnetic scale. Here, $\bar{\rho}$ is the reference state density.

The model is controlled by five dimensionless parameters: the **Ekman number**

$$\mathrm{E} = \frac{\nu}{\Omega \ell^2}, \tag{8}$$

the modified **Rayleigh number**

$$\mathrm{Ra}^\star = \frac{\bar{g}_o \Delta c}{\Omega^2 \ell}, \tag{9}$$

the **Prandtl number**

$$\mathrm{Pr} = \frac{\nu}{\kappa}, \tag{10}$$

the **magnetic Prandtl number**

$$\mathrm{Pm} = \frac{\nu}{\lambda}, \tag{11}$$

and the **aspect ratio**

$$\eta = r_i/r_o. \tag{12}$$

The modified Rayleigh number $\mathrm{Ra}^\star$ is connected to the more classical Rayleigh number

$$\mathrm{Ra} = \frac{\bar{g}_o \Delta c \ell^3}{\kappa \nu} \tag{13}$$

via $\mathrm{Ra}^\star = \mathrm{Ra} E^2 \mathrm{Pr}^{-1}$.

The five dimensionless parameters combine nine physical properties, some of which have not been introduced so far: the kinematic viscosity ν, the thermal and/or compositional diffusivity κ, and the reference state gravity $\bar{g}_0$ at the outer boundary $r = r_o$.

Above equations employ an additional simplification that has been adopted by several dynamo models for terrestrial planets: The density variations due to the **super-adiabatic temperature**—only this component contributes to convection—and due to deviations χ from the homogeneous **reference state** composition $\bar{\chi}$ are combined in the codensity c:

$$c = \alpha T + \gamma \chi, \tag{14}$$

where α and γ are the thermal and compositional expansivity, respectively. For a simplified binary model with heavy constituents (iron, nickel) of total mass m_H and light constituents (sulfur, oxygen, carbon) of total mass m_L the reference state composition is $\bar{\chi} = m_L/(m_L + m_H)$. The compositional expansivity is then given by $\gamma = -\bar{\rho}(\rho_H - \rho_L)/\rho_H\rho_L$, where ρ_H and ρ_L are the densities of heavy and light elements in the liquid core, respectively. In Eq. 14 c, T, and χ refer to dimensional values.

Describing the evolution of temperature and composition by the combined Eq. 5 for codensity $C = c/\Delta c$ assumes that both quantities have similar diffusivities. This seems like a daunting simplification since the chemical diffusivity may be three orders of magnitude smaller than its thermal counterpart (Braginsky and Roberts 1995). The approach is often justified with the argument that the small-scale turbulent mixing, which cannot be resolved by the numerical codes, may result in larger effective **turbulent diffusivities** that are of comparable magnitude (Braginsky and Roberts 1995). This has the additional consequence that the "turbulent" Prandtl number and the magnetic Prandtl number would become of order one (Braginsky and Roberts 1995). Some studies suggest that the differences in diffusivities may have interesting implications (Busse 2002) and the potential effects on planetary dynamo simulations remain to be explored.

Pseudo-spectral codes usually represent flow field and magnetic field by a **poloidal** and a **toroidal** contribution, respectively:

$$\mathbf{U} = \nabla \times \nabla \times (\hat{\mathbf{r}}v) + \nabla \times (\hat{\mathbf{r}}w) , \quad \mathbf{B} = \nabla \times \nabla \times (\hat{\mathbf{r}}g) + \nabla \times (\hat{\mathbf{r}}h). \tag{15}$$

v and g are the poloidal scalar fields, w and h are the toroidal counterparts. Toroidal fields have no radial components and are therefore restricted to spherical shells. The toroidal magnetic field of a planet never leaves the dynamo region and cannot be measured on the planets surface. When using the ansatz 15 the continuity equations for flow (Eq. 6) and magnetic field (Eq. 7) are fulfilled automatically and the number of unknown fields reduces from eight—the three components of **U** and **B**, pressure P, and codensity C—to six. The equations solved for the six unknown fields are typically the radial components of the **Navier–Stokes** and the **induction equation** (3) and (4), respectively, the radial components of the curl of these two equations, the horizontal divergence of the Navier–Stokes equation, and the **codensity evolution equation** (5) (Christensen and Wicht 2007).

The procedure increases the number of spatial derivatives and may thus go along with a loss of precision. This is less an issue for pseudo-spectral codes than for local codes, since the former calculate spatial derivatives in the spectral domain resulting in a higher precision (see ➲ Sect. 2.3).

2.2 Boundary Conditions

The differential equations for the dynamo problem can only be solved when supplemented with appropriate boundary conditions. For the flow **U**, either **rigid** or **free slip boundary conditions** are used. In both cases, the radial flow component is forced to vanish. The horizontal flow components must match the rotation of the boundary for rigid boundary conditions, whereas in the free slip case the horizontal components of the viscous stresses are forced to zero (Christensen and Wicht 2007). While rigid flow conditions seem most appropriate for the dynamo regions of terrestrial planets some authors nevertheless assume free slip boundaries to avoid the thicker Ekman boundary layers that develop at the unrealistically large Ekman numbers used in the numerical simulations (Kuang and Bloxham 1997; Busse and Simitev 2005b).

Most dynamo models assume a purely thermal driving and either employ **fixed temperature** or **fixed heat flux boundary conditions**, mostly for convenience. The latter translates to a fixed radial temperature gradient. For terrestrial planets, the much slower evolving mantle controls how much heat is allowed to leave the core, so that a heat flux condition is appropriate. Lateral variations on the thermal structure of the lower mantle will translate into an inhomogeneous core-mantle boundary heat flux (Aubert et al. 2008a) (see ➋ Sect. 3.3.2). Since the light core elements cannot penetrate the mantle at any significant rate a vanishing flux is the boundary condition of choice for the compositional component (Kutzner and Christensen 2000). The conditions at the boundary to a growing inner core are somewhat more involved. The heat flux originating from the latent heat of the phase transition is proportional to the mass undergoing the phase transition and thus to the compositional flux of light elements. The flux itself depends on the local cooling rate dT/dt which is determined by the convective dynamics and changes with time (Braginsky and Roberts 1995). The resulting conditions have so far only been implemented by Glatzmaier and Roberts (1996).

Since the conductivity of the rocky mantles of terrestrial planets is orders of magnitudes lower than that of the core, the magnetic field can be assumed to match a potential field at the interface $r = r_o$. This matching condition can be formulated into a **magnetic boundary condition** for the individual spherical harmonic filed contributions (Christensen and Wicht 2007). The same applies at the boundary to an electrically insulating inner core assumed in some dynamo simulations for simplicity. A simplified induction equation (4) must be solved for the magnetic field in a conducting inner core which has to match the outer core field at r_i. Refer to Roberts (2007) and Christensen and Wicht (2007) for a more detailed discussion of the magnetic boundary conditions.

2.3 Numerical Methods

Up to now the vast majority of dynamo simulations rely on **pseudo-spectral codes**. Although a spectral approach has been already applied to dynamo problems by Bullard and Gellman (1954), the method gained more widespread popularity in the dynamo community by the pioneering work of Glatzmaier (1984). Since then the method has been optimized and several variants of the original approach are in use today. Here, only the basic ideas of the approach are sketched and for more details refer to the original work (Glatzmaier 1984) or to a recent review (Christensen and Wicht 2007).

The core of the pseudo-spectral codes is the dual representation of the variables in grid space as well as in spectral space. The individual computational steps are performed in whichever representations they are most efficient. Partial derivatives are usually calculated in spectral space whereas the nonlinear terms are calculated in grid space. In the lateral directions (colatitude Θ and longitude Φ) **spherical harmonics** Y_{lm} are the obvious choice as base functions of the spectral expansion. Since spherical harmonics are eigen-functions of the lateral Laplace operator, the calculation of the diffusion terms in the governing equations is particularly simple. Mostly **Chebychev polynomials** C_n are chosen as radial representations, where n is the degree of the polynomial. Chebychev polynomials allow for Fast Fourier transforms and offer the additional advantage of an increased grid resolution near the inner and outer boundaries, where thin boundary layers may develop.

Time stepping is performed in spherical harmonic-radial space (l, m, r) using a mixed implicit/explicit scheme that handles the nonlinear terms as well as the Coriolis force in, for example, an explicit Adams-Bashforth manner. The resulting scheme decouples all spherical harmonic modes (l, m) which allows a fast calculation. A Crank-Nicolson implicit scheme if often used for completing the time stepping for the remaining terms. It is also possible to include the Coriolis force in the implicit discretisation, but this would destroy the decoupling in the latitudinal index l. On the other hand the nonlinear terms are always calculated in local grid space. This requires a transform from a spectral (l, m, r) to a grid (Θ, Φ, r) representation and back. Fast Fourier transforms are applicable in radius and longitude, however, in colatitude Θ the best way to calculate the Legendre transform is by Gauss integration. These Gauss–Legendre transforms are usually the most time-consuming parts of the spectral transform method, and evolve to a severe bottleneck in high-resolution cases.

As stated above, the continuity equations $\nabla \cdot \mathbf{U} = 0$ and $\nabla \cdot \mathbf{B} = 0$ are fulfilled identically by the introduction of poloidal and toroidal scalar fields (15) which is a very convenient feature of the method. Higher-order lateral derivatives of the fields (6th order) are required; however, this creates not a serious problem since the lateral Laplace operator is easily and accurately determined in spectral space. More problematic is the implementation of **parallel computing**. Since derivatives are calculated in spectral space and the spectral base functions – spherical harmonics and Chebychev polynomials – are globally defined, global inter-processor communication is needed. In fact, derivatives are determined in spectral space by recurrence relations (Glatzmaier 1984) which can be regarded as spectral next-neighbor calculation. However, in order to calculate a single-function value at a given point in grid space the complete spectrum must be available for the particular process. Therefore, the calculation of the nonlinear terms during one timestep requires the exchange of the complete spectral dataset between the parallel processes, i.e. global inter-processor communication.

This is a serious obstacle for the efficient use of parallel computing with spectral methods. If only a few parallel processes are required, the solution is a one-dimensional partition of the grid space, for example, in radial or longitudinal direction. This is particularly easy if a shared memory machine is used. For massively parallel computing using several thousands of processes the solution is much more cumbersome. However, as demonstrated by Clune et al. (1999), efficient parallel computing is also possible for spectral methods although the implementation is rather complex. Some approaches replace the Chebychev polynomials in radius by a finite difference approximation (Dormy et al. 1998; Kuang & Bloxham 1999). This adds some more flexibility in the radial discretisation, at the expense of accuracy, but does not help in the parallel computing.

For these reasons, the interest in applying **local methods** to dynamo problems has considerably increased in recent years. Kageyama et al. (1993) and Kageyama and Sato (1997) are early examples for a finite difference compressible approach. However, local approaches, like finite differences, finite volumes, or finite elements, are easier to apply to a compressible flow since in this case the pressure is determined more directly by an equation of state. Since an incompressible flow is a good approximation for the liquid iron core of the terrestrial planets, it seems not appropriate to complicate the physics of the problem further by compressibility. In sharp contrast to spectral methods the use of poloidal and toroidal potentials (15) is not popular in three-dimensional flow simulations by local methods since sixth-order derivatives are then introduced to the problem. (For two-dimensional flows such an approach is standard since only a single potential suffices and only fourth-order derivatives are involved. However 2d flows are irrelevant to dynamo problems which are by necessity three-dimensional.)

Therefore, it is common practice in local methods to retain the primitive variables velocity and pressure. In the last years several local primitive variable approaches have been developed for dynamo simulations. A common feature is that the continuity equation is satisfied iteratively by solving a Poisson-type equation for the pressure or pressure correction although the detailed strategies differ. Usually this correction iteration to fulfill continuity is by far the most compute-intensive part of local approaches. In other aspects, for example, grid structure, the applied methods are extremely diverse. For example, Kageyama and Yoshida (2005) use a YinYang grid approach with two overlapping longitude–latitude grids. For this finite difference method the authors report simulations with nearly up to 10^9 grid points and up to 4,096 parallel processes on the EarthSimulator system. Matsui and Buffett (2005) apply a finite element approach with hexahedral elements based on the parallel GeoFEM thermal-hydraulic subsystem developed for the Japanese EarthSimulator project. Harder and Hansen (2005) use a collocated finite volume method on a cube–sphere grid, i.e., a grid which is projected from the six surfaces of an inscribed cube to the unit sphere. Hejda and Reshetnyak (2003, 2004) also utilize a finite volume method but with a staggered arrangement of variables and a single longitude–latitude grid. Chan et al. (2006) both describe a finite element approach based on a icosahedral triangulation of a spherical surface and a finite difference approach based on a staggered longitude–latitude grid. The method of Fournier et al. (2005) uses a mixture of spectral and local methods: in azimuthal direction a Fourier expansion is used whereas in the meridional plane a spectral element approximation is applied using macro-elements with polynomials of high order.

Wicht et al. (2009) have compiled published **benchmark** data of the various local methods and compared them with the dynamo benchmark of Christensen et al. (2001) which was established solely by spectral methods. Since the benchmark cases were defined at a very moderate Ekman number of $E = 10^{-3}$, the solutions are very smooth and can be approximated with high accuracy by a few eigenfunctions. In this regime fully spectral methods are much more efficient than local methods. Therefore, it is not surprising that of the mentioned local methods only the spectral element approach (Fournier et al. 2005) could compete with the accuracy of fully spectral methods. Presumably the situation would be less clear at lower Ekman numbers, where solutions are less diffusion dominated and the solution spectra are more flat.

As already stated, the motivation for the application of local approaches to the dynamo problem is the attempt to better utilize massively parallel computation. Since simulations at low Ekman number, $E < 10^{-5}$, are desirable for a better representation of the force balance in the Earth's core, and such simulations need as much computational power as possible, the approach seems to be natural. However, although local methods are better suited for parallel computing than spectral methods, local approaches have some other disadvantages which may degrade the performance. For example, in local methods it is much more expensive to fulfill the continuity equations. Another obstacle is the implementation of the **magnetic boundary condition**: usually an insulating exterior is assumed in dynamo simulations. However, this cannot be implemented as a local boundary condition. The easiest way is to match a series expansion of spherical harmonics at the boundary which is easily done with a spectral method. This is also possible with a local approach but requires a transform of the solution at the boundary to spectral space. The amount of extra work depends mainly on the chosen grid structure. An alternative would be to extend the computational domain and solve also for the external field. A more elegant way is to calculate the external field by a boundary element technique (Isakov et al. 2004). The simplest, but somewhat unphysical, approach would be to replace the insulating condition by a local approximation, i.e., apply so-called quasi-vacuum condition which corresponds to vanishing tangential field components at the boundary (Kageyama and Sato 1997; Harder and Hansen 2005). Up to now, it is not clear which is the most suitable alternative.

Due to these limitations local methods are not widely used for production runs; spectral methods are still the workhorse of dynamo research. For example, all models presented in this article were obtained by spectral approaches. The situation may change in the near future for demanding simulations at low Ekman number, or for more specialized cases (lateral variations of material parameters, topographic distortion of the boundaries, etc.) not amenable to spectral methods.

2.4 Force Balances

Before diving into the results from dynamo simulations let us start off with a few considerations about the force balance in the **Navier–Stokes equation**. The fast rotation rate of planets guarantees that the Coriolis force enters the leading order force balance and play a major role in the dynamo dynamics. The Coriolis force therefore typically serves as a reference for judging the relative importance of the other forces. The **Ekman number** scales the ratio of viscous effects to the Coriolis forces. Two additional nondimensional numbers quantify the relative importance of inertial effects and magnetic forces, respectively: the **Rossby number**

$$\mathrm{Ro} = \frac{\mathrm{u}}{\ell\Omega} \tag{16}$$

and the **Elsasser number**

$$\Lambda = \frac{\mathrm{b}^2}{\bar{\rho}\Omega\mu\lambda}. \tag{17}$$

u and b refer to typical dimensional flow and magnetic field amplitudes. The **Rossby number** Ro is identical to the nondimensional flow amplitude in the scaling chosen here. The magnetic Reynolds number $\mathrm{Rm} = \mathrm{u}\ell/\lambda$ has already been introduced in ➲ Sect. 1 as an important measure for the relative importance of magnetic induction to diffusion. Another useful measure is the **Alfvén mach number**

$$M_A = \frac{\mathrm{u}(\bar{\rho}\mu)^{1/2}}{\mathrm{b}}, \tag{18}$$

the ratio of the flow velocity to the typical velocity of Alfvén waves. Note that Ro, Rm, Λ, and M_A are not input parameters for the dynamo simulations but output "parameters" that quantify the solution, namely, the typical (rms) flow amplitude and the typical (rms) magnetic field amplitude. Both the Rossby number and the magnetic Reynolds number measure the flow amplitude: $\mathrm{Ro} = \mathrm{Rm}\,\mathrm{E}/\mathrm{Pm}$. The Alfvén Mach number can be written in terms of the magnetic Reynolds number and the Elsasser number: $M_A = \mathrm{Rm}\mathrm{E}^{1/2}/(\Lambda\mathrm{Pm})^{1/2}$.

Christensen and Aubert (2006) note that the **Rossby number** may not provide a good estimate for the impact of the nonlinear inertial or advection term in the **Navier–Stokes equation** (3). The nonlinearity can give rise to cascades that interchange energy between flows of different length scales. The turbulent transport of energy to increasingly smaller scales in the Kolmogorov cascade is only one example. Other cascades are provided by so-called Reynolds stresses which result from a statistically persistent correlation between different flow components. The fierce zonal winds observed on Jupiter and Saturn are a prominent example for their potential power (Christensen 2002; Heimpel et al. 2005) and demonstrate how small-scale contributions can effectively feed large-scale flows. Christensen and Aubert (2006) suggest to at least incorporate the length scale introduced by the ∇ operator into a refined estimate for the relative importance of the nonlinear inertial forces which they call the **local Rossby number**

$$\mathrm{Ro}_\ell = \frac{\mathrm{u}}{\Omega \ell_u}. \tag{19}$$

The length scale ℓ_u represents a weighted average base on the spherical harmonic decomposition of the flow:

$$\ell_u = \ell\pi / < l > \tag{20}$$

with

$$< l > = \frac{\sum_l l E_k(l)}{\sum_l E_k(l)}, \tag{21}$$

where $E_k(l)$ is the kinetic energy carried by all modes with spherical harmonic degree l.

➲ *Table 1* lists the above introduced parameters for Earth and highlights three characteristics for the force balance in planetary dynamo regions: viscous effects are likely negligible at $\mathrm{E} = 10^{-15}$, inertial forces should play a minor role at $\mathrm{Ro} = 2\times10^{-6}$, and Lorentz forces are of the same order as the Coriolis force since $\Lambda = 1$.

The associated dynamical consequences can be discussed in terms of a bifurcation scenario that starts with the onset of convective motions at a **critical Rayleigh number** Ra_c. Dynamo action sets in at higher Rayleigh numbers where the larger flow amplitude pushes the **magnetic Reynolds number** beyond its critical value. The **small Ekman number** of planetary dynamos suggest that viscous effects are negligible. Close to Ra_c, the Coriolis force is then predominantly balanced by pressure gradients in the so-called **geostrophic** regime. The **Taylor–Proudman theorem** states that the respective flow assumes a two-dimensional configuration, minimizing variations in the direction $\hat{\mathbf{z}}$ along the rotation axis: $\partial \mathbf{U}/\partial z = 0$ for $\mathrm{E} \to 0$.

Buoyancy, however, must be reinstated to allow for convective motions that necessarily involve a radial and thus nongeostrophic component. To facilitate this, viscous effects balance those Coriolis-force contributions which cannot be balanced by pressure gradients (Zhang and Schubert 2000). At lower Ekman numbers the azimuthal flow length scale decreases so that viscous effects can still provide this necessary balance: The azimuthal wave number m_c at onset of convection scales like $\mathrm{E}^{-1/3}$. The impeding effect of the two-dimensionality causes the "classical" **critical Rayleigh number** to rise with decreasing E: $\mathrm{Ra}_c \sim E^{-4/3}$. Inertial effects can contribute to balancing the Coriolis force at larger **Rayleigh numbers** and thus reduce the two-dimensionality of the flow.

➲ *Figure 3* shows the principal convective motions and the effect of decreasing the Ekman number and increasing the Rayleigh number. The **convective columns** are illustrated with isosurfaces of the z-component of vorticity in the upper panels. Red isosurfaces depict cyclones that rotate in the same direction as the overall prograde rotation Ω, blue isosurfaces show anticyclones that rotate in the opposite sense. These columns are restricted to a region outside the so-called tangent cylinder that touches the inner core boundary (see ➲ *Fig. 2*). Inside the tangent cylinder, buoyancy is primarily directed along the axis of rotation so that the **Taylor–Proudman theorem** is even more restricting here. Consequently, the convective motions inside the tangent cylinder start at higher **Rayleigh numbers** a few times Ra_c, and the motions are plume-like rather than column-like. The so-called meridonal circulations, which obviously violate the Taylor–Proudman theorem but are unavoidable in convection, are illustrated by isosurfaces of the z-component of the flow in the lower panels of ➲ *Fig. 2*. They are directed equator-ward in the (red) columns that rotate in the direction of Ω and polar-ward in columns

Table 1

List of parameters and properties for some of the dynamo simulations presented here and for Earth. For model E6 not all the values have been stored during the calculations. Models BM II and E6 have a fourfold azimuthal symmetry so that the dipole tilt is always zero. While this reflects the true solution of the BM II case the symmetry has been imposed in model E6 to save computing time. All models employ rigid flow and fixed temperature boundary conditions. The exception is model E4R106 which mimics purely compositional convection by using a zero flux condition at the outer boundary and a fixed composition condition at the inner boundary. The Prandtl number is unity in all cases and the aspect ratio is fixed to an Earth-like value of $\eta = 0.35$. Some of the cases have been studied elsewhere (Christensen et al. 2001; Wicht et al. 2009; Wicht and Christensen 2009; Wicht and Tilgner 2010). The values provided in these previous publications may differ because of modified definitions. The model names first indicates the negative power of the Ekman number, then the supercriticality Ra/Ra_c. Earth's core Rayleigh number is hard to determine but thought to be large (Gubbins 2001); Earth's local Rossby number has been estimated by Christensen and Aubert (2006), flow amplitudes that enter Ro and Rm are based on secular variation models and match with the scaling law by Christensen and Tilgner (2004). The values *d*, *e*, and *a* are measures to quantify the dipolarity, equatorial symmetry, an axial symmetry of the field at the outer boundary (see Eq. 24–26 in Sect. 3.3.1). Θ refers to the time-averaged dipole tilt

	BM II	E3R9	E4R106	E5R18	E5R36	E5R43	E6	Earth
E	10^{-3}	10^{-3}	3×10^{-4}	3×10^{-5}	3×10^{-5}	3×10^{-5}	3×10^{-6}	10^{-15}
Ra	0.11×10^{-1}	0.55×10^{-1}	1.8	0.09	0.108	0.135	9×10^{-3}	Large
Ra/Ra_c	2	9	106	18	36	43	22	Large
Pm	5	10	3	1	1	1	0.5	3×10^{-7}
Ro	9×10^{-3}	4×10^{-2}	4×10^{-2}	10^{-2}	10^{-2}	2×10^{-2}	2×10^{-3}	2×10^{-6}
Ro_ℓ	0.02	0.12	0.13	0.09	0.10	0.18	0.02	0.09
Rm	46	434	408	356	405	607	261	500
Λ	8	19	5	7	8	4	3	1
M_A	0.2	0.99	1.82	0.74	0.78	1.66	0.40	0.03
d	0.83	0.31	0.41	0.87	0.86	0.20	0.81	0.79
Θ	0	17.6	15.1	2.9	2.2	37.7	0	9.7
e	0	0.58	0.68	0.57	0.57	0.74	–	0.74
a	0.79	0.89	0.94	0.81	0.77	0.94	–	0.94

that rotate in the opposite direction. Inside the tangent cylinder, the **meridional circulation** is mostly outward close to the pole and inward close to the tangent cylinder.

Beyond the onset of dynamo action, the Lorentz force can contribute to balancing the Coriolis force and further releases the two-dimensionality. On theoretic grounds, the Lorentz force is thought to enter the leading order force balance in order to saturate magnetic field growth, and this seems to be confirmed by the geomagnetic value of Λ. The **magnetostrophic balance** assumed to rule planetary dynamo dynamics therefore involves Coriolis forces, pressure gradients, buoyancy, and Lorentz forces, and is characterized by an **Elsasser number** of order one. The influence of Lorentz forces on the flow dynamics will be discussed further in Sect. 3.

Decreasing the **Ekman number** has two principal effects: first an increase of the rotational constraint promoting a more ordered **geostrophic** structure and second a decrease of the length

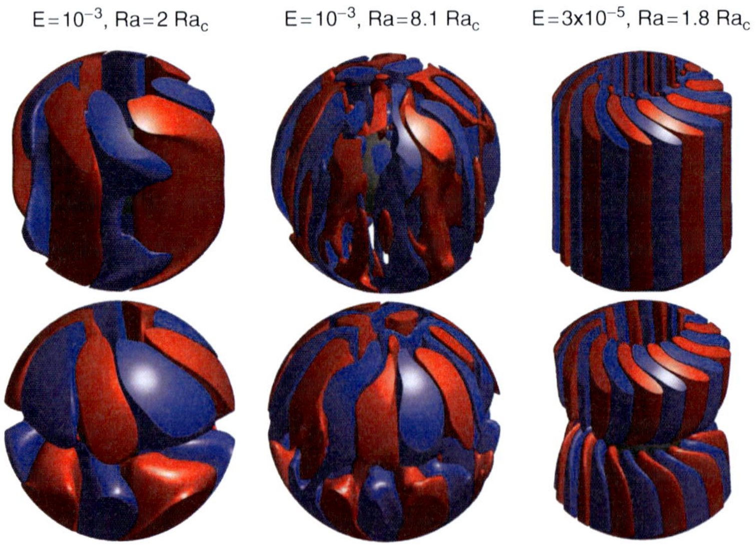

Fig. 3
Flow in nonmagnetic convection simulations at different parameter combinations. The top row shows isosurfaces of the z-components of the vorticity $\nabla \times \mathbf{U}$, the bottom row shows isosurfaces of the z-component of the flow. The Prandtl number is unity in all cases

scales. The dynamics of the smaller flow structures also requires smaller numerical time steps and both effects significantly increase the numerical costs. For example, the time step required for the simulation at $E = 3 \times 10^{-5}$ shown in *Fig. 3* is three orders of magnitude smaller than the time step required at $E = 10^{-3}$. Reaching smaller Ekman numbers is therefore numerically challenging and the available computing power limits the attainable values. An Ekman number of $E = 10^{-6}$ seems to be the current limit for numerical dynamo simulations (Kageyama et al. 2008). Increasing the **Rayleigh number** in order to retain dynamo action and to yield a more realistic structure and time dependence further decreases both length scales and time scales, as discussed in Sect. 3.

The parameters for a few representative dynamo simulations—simple, advanced, and high-end—are listed along with the geophysical values in *Table 1*. Simple dynamos are characterized by moderate Ekman numbers, $E = 10^{-3}$ or even larger, which yield a large-scale solution that can be computed with modest numerical efforts. These dynamos therefore lend themselves to study long-time behavior, to explore the dependencies on the parameters other than E, and to unravel the 3d dynamo mechanism. The simple drifting behavior found at low **Rayleigh numbers** further more recommended the case BM II listed in *Table 1* for a numerical **benchmark** (Christensen et al. 2001). Model E6 represents the high-end of the spectrum at $E = 3\times10^{-6}$. Its solution is very small scale and complex and exhibits a chaotical time dependence. The dynamos at $E = 3 \times 10^{-5}$ listed in *Table 1* represent advanced models that can be simulated on today's mid-range parallel computing systems on a more or less regular basis.

A comparison of the numbers for the most advanced model E6 in *Table 1* with Earth values shows that (1) the Ekman number is still about nine orders of magnitude too large, (2) the Rossby number is three orders of magnitude too large, (3) the magnetic Prandtl number is six orders of magnitude too large, and (4) the Alfvén Mach number is about one order of magnitude too large. On the other hand, (5) the magnetic Reynolds number reaches realistic values, and (6) the Elsasser number is also about right.

In terms of the dynamics this means that (1) viscous effects are much too large, which is necessary to suppress unresolvable small-scale flow features, (2) inertial effects are over-represented, (3) magnetic diffusion is much too low compared to viscous diffusion, (4) the numerical dynamos are too inefficient and produce weaker magnetic fields for a given flow amplitude than Earth. On the positive side, (5) the ratio of magnetic field production to diffusion is realistic, and (6) the relative (rms) impact of the Lorentz force on the flow seems correctly modeled. The role of the local Rossby number will be discussed further below.

3 Toward Realistic Dynamo Simulations

3.1 Dynamo Regimes

One strategy for choosing the parameters for a numerical dynamo model is to fix the Prandtl number at, for example, Pr = 1 to model thermal convection, and to choose E as small as the available numerical computing power permits. The Rayleigh number and the magnetic Prandtl number Pm are then varied until the **critical magnetic Reynolds number** is exceeded and dynamo action starts. Note, however, that some authors have chosen a different approach in selecting their parameters (Glatzmaier 2002). ➋ *Figure 4* shows the dependence of Rm on Ra and Pm for at $E = 10^{-3}$ and Pr = 1. Rigid flow and fixed temperature boundary condition have been assumed. Larger Ra values yield larger flow amplitudes u while larger Pm values are synonymous with lower magnetic diffusivities λ. The increase of either input parameter thus leads to larger magnetic Reynolds numbers $Rm = u\ell/\lambda$ and finally promotes the onset of dynamo action. The minimal critical magnetic Reynolds number is about 50 here, a value typical for spherical dynamo models (Christensen and Aubert 2006).

➋ *Figure 4* shows the four main dynamical regimes that have been identified when exploring the parameter space (Kutzner and Christensen 2002; Christensen and Aubert 2006; Amit and Olson 2008; Takahashi et al. 2008a; Wicht et al. 2009). **C** denotes the purely convective regime and **D** is the regime where **dipole-dominated** magnetic fields are obtained. When u becomes large at values of Ra/Ra_c between 8 and 9, the dynamo changes its character and seems to become less efficient. The dipole component looses its special role which leads to a **multipolar** field configuration—hence regime **M** for **multipolar dynamo regime**—and an overall weaker field as indicated by the smaller Elsasser number (see ➋ *Fig. 4b*). In addition, the critical magnetic Reynolds number increases.

The dipole polarity, which remains stable in regime **D**, frequently changes in regime **M**. Earth-like **reversals**, where the polarity stays constant over long periods and reversals are relatively rare and short events, happen at the transition between the two regimes and define regime **E**. It is somewhat difficult to come up with a clear and unambiguous characterization of "Earth like" reversals. The definition introduced by Wicht et al. (2009) demanding that the magnetic pole should not spend more than 10% of the time in transitional positions farther away than 45° from either pole is followed. Moreover, the dipole field should amount to at least 20% of the total field strength at the outer boundary (see ➋ Sect. 3) on time average.

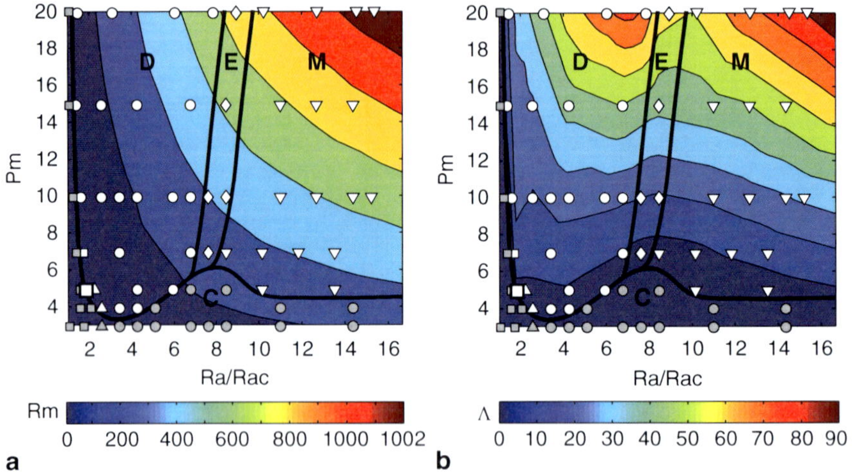

Fig. 4
Regime diagrams that illustrate the transition from stable dipole-dominated dynamos (regime D) to constantly reversing multipolar dynamos (regime M). Models showing Earth-like rare reversals can be found at the transition in regime E. Grey symbols mark nonmagnetic convective solutions (regime C). Different symbols code the time dependence: squares = drifting, upward-pointed triangles = oscillatory, circles = chaotic, diamonds = Earth-like rarely reversing, downward-pointed triangles = constantly reversing. The Ekman number is $E = 10^{-3}$ and the Prandtl number is $Pr = 1$ in all cases

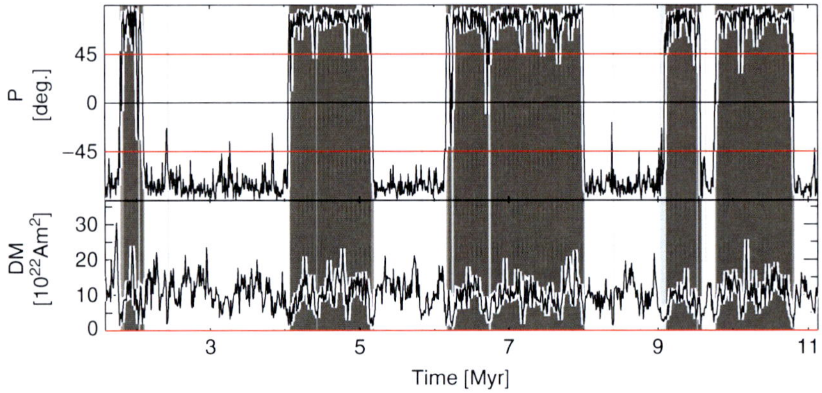

Fig. 5
Reversal sequence in model E3R9 that typifies the behavior of dynamos in regime E (see *Fig. 4*). The *top panel* shows the dipole tilt, the *bottom panel* shows the magnetic dipole moment, rescaled by assuming Earth-like rotation rate, core density, and electrical core conductivity. Excursions where the magnetic pole ventures further away than 45° in latitude from the geographic pole but then returns are marked in light grey. See Wicht et al. (2009) for further explanation

➋ *Figure 5* shows a time sequence for the Earth-like reversing model E3R9 (model T4 in Wicht et al. [2009]) that exhibits ten reversals and several excursions where the magnetic pole also ventures farther away than 45° from the closest geographic pole but then returns. The left boundary of regime **E** is difficult to pin down since it is virtually impossible to numerically prove that a dynamo never reverses (Wicht et al. 2009).

The regime changes from **D** to **E** and further to **M** are attributed to the increased importance of the nonlinear **inertial** effects at larger **Rayleigh numbers** which seem to have reached a critical strength compared to the ordering Coriolis force at the transition to regime **M**. No model for $Ro_\ell < 0.1$ was ever reported to undergo polarity changes (Christensen and Aubert 2006). Even an extremely long run covering an equivalent of 6.8 Myr for the model at $E = 10^{-3}$, $Pm = 10$, and $Ra = 7.5Ra_c$ with $Ro_\ell < 0.09$ showed no reversals. Once in regime **E** the ever-present fluctuation in the **inertial** forces may suffice to drive the dynamo into regime **M** for a short while and thus facilitate **reversals** (Kutzner and Christensen 2002). ➋ *Figure 10*, discussed in more detail in ➋ Sect. 3 below, shows the location of the different regimes with respect to the **local Rossby number** Ro_ℓ.

The scenario outlined for $E = 10^{-3}$ above is repeated at lower Ekman numbers with a few changes. Most notably, the boundary toward the inertia-influenced dynamo regime **M** shifts toward more supercritical Rayleigh numbers. The critical relative strength of **inertial** effects is reached at larger flow amplitudes u since the rotational constraint is stronger at lower E values. The **critical magnetic Reynolds number** is therefore reached at lower magnetic diffusivities λ which means that the minimum **magnetic Prandtl number** Pm_{min} where dynamo action is still retained decreases (Christensen and Aubert 2006). While Pm_{min} is about 4 at $E = 10^{-3}$ (compare ➋ *Fig. 4*), it decreases to roughly 0.1 at $E = 10^{-5}$ (Christensen and Aubert 2006). Christensen and Aubert (2006) suggest that $Pm_{min} = 450\,E^{0.75}$. An extrapolation for Earth's Ekman number yields $Pm_{min} \approx 10^{-8}$ which is safely below the geophysical value of $Pm = 3{\times}10^{-7}$. This predicts that realistic magnetic Prandtl number can indeed be assumed in the numerical models at realistic Ekman numbers. In a few instances it has been reported that both **dipole-dominated** and **multipolar** solutions can be found at identical parameters, at least for Rayleigh numbers close to the boundary of regime **D** and regime **M** (Christensen and Aubert 2006; Simitev and Busse 2009). Simitev and Busse (2009) attribute this to the coexistence of two attractors in a subdomain of the parameter space.

The parameter studies outlined above suggests that the dynamo simulation can be extrapolated to Earth's much lower Ekman number (Christensen and Aubert 2006; Olson and Christensen 2006). The magnetic Prandtl number, the magnetic Reynolds number, and the Elsasser number could still assume realistic values. The Rossby number and the Alfvén Mach number would automatically fall into place so that the field strength and flow amplitude would also assume realistic absolute values. This implies that the dynamical differences play no major role, that viscous effects and **inertial** effects are already small enough to capture at least the primary features of the geodynamo, and that the ratio of viscous to Ohmic dissipation is not essential.

First-level Earth-like dynamos with Elsasser and magnetic Reynolds numbers of roughly the correct size and a **dipole-dominated** field configuration indeed seem to be a very robust feature of the numerical simulations and are found for a wide combination of parameters. The parameter space is much more restricted should the simulation also be required to model additional features that typically have been regarded of secondary significance. Earth-like **reversals** are obtained, for example, in a rather restricted regime. This question is further elaborated in ➋ Sect. 3.

Christensen and Aubert (2006) remark that **dipole-dominated** dynamos, even Earth-like reversing dynamos, can be obtained for a wide range of magnetic Reynolds numbers and Elsasser numbers (see also ➋ *Fig. 4*). Larger Rm and Λ values are, for example, obtained for larger magnetic Prandtl numbers which has to do with the implicit decrease of the magnetic diffusivity λ. Christensen and Aubert (2006) also varied the Ekman number and the Prandtl number and found **dipole-dominated** dynamos for Elsasser numbers between 0.06 and 100. This means that either the **magnetostrophic balance** is not a prerequisite for these dynamo or that the **Elsasser number** is not a good measure for the true impact of the Lorentz force in the **Navier–Stokes equation** (Christensen 2010). A somewhat closer examination of the force balances by Wicht and Christensen (2009) suggest that the latter may actually be the case.

The assumption that Λ has to be of order one is the basis for a classical conjecture that the magnetic field strength should be proportional to the square root of the planetary rotation rate since $b^2 = (\Omega\bar{\rho}\mu\lambda)$. The dimensionless field strengths provided by dynamo simulations are therefore often rescaled by assuming Earth-like values for Ω, core density $\bar{\rho}$, and electrical conductivity $\sigma = 1/(\mu\lambda)$ (see, for example, ➋ *Fig. 5*). Since the results by Christensen and Aubert (2006) challenge this assumption they set out to find a different **scaling law**. A large suit of dynamo models supports the law

$$b^2 \sim f_{Ohm}\, \bar{\rho}^{1/3}\, (Fq_o)^{2/3}\ , \tag{22}$$

where f_{Ohm} is the fraction of the available power that is converted to magnetic energy and Fq_o is the total thermodynamically **available power** (Christensen 2010). The form factor F subsumes the radial dependence of the convective vigor and is typically of order one; q_o is the heat flux through the outer boundary. This **scaling law** not only successfully predicts the field strength for several planets in the solar system (Christensen and Aubert 2006) but also for some fast-rotating stars whose dynamo zones may obey a similar dynamics as those found in the planetary counterparts (Christensen et al. 2009). Refer to Christensen (2010) for a detailed comparison of the different scaling laws that have been proposed over time, not only for the magnetic field strength but also for the flow vigor and other dynamo properties.

The power-based **scaling law** and the force balance scaling $\Lambda = 1$ offer two ways to interpret the field strengths in the numerical simulations, and other alternatives may also be conceivable (Christensen 2010). It thus seems premature to decide whether a particular simulation would yield a realistic field strength until the "true" **scaling law** has been determined, probably with the help of future simulations that further approach the realistic parameters.

The power-dependent **scaling law** for the **local Rossby number** developed by Christensen and Aubert (2006) predicts the value of $Ro_\ell \approx 0.09$ for Earth that is listed in ➋ *Table 1*. This nicely agrees with the fact that the numerical simulations show Earth-like **reversal** in this parameters range. Christensen and Aubert (2006) observe that the transition between dipolar and **multipolar** dynamos always happens at local Rossby numbers around 0.1, independent of the other system parameters. Examples for this behavior are shown in ➋ *Fig. 10a*. This suggests that **inertial** effects play a much larger role in the geodynamo than previously anticipated (Wicht and Tilgner 2010). Scaled to Earth values, the associated length scale ℓ_u (Eq. 20) corresponds to roughly 100 m. It is hard to conceive that such a small length should play an important role in the dynamo process, let alone influence its reversal behavior. The meaning of the **local Rossby number** and the related **scaling laws** needs to be explored further to understand their relevance.

The fact that the power-dependent **scaling law** equation (22) is independent of the Ekman number seems good news for dynamo simulations. However, the numerical results still allow for a weak dependence on the magnetic Prandtl number (Christensen and Aubert 2006; Christensen 2010). This may amount for a significant change when extrapolating the simulation which never go below Pm $\approx$ 0.1 to the planetary value of Pm $= 3\times10^{-7}$. The much smaller viscous diffusion at Pm $\ll$ 1 allows for much smaller scales in the flow than in the magnetic field, causing what is called a scale disparity. The smaller flow scales can therefore have no direct impact on the dynamo process in planetary dynamo regions but may still play a role for the magnetic field generation in dynamo simulations. Unfortunately, lower **magnetic Prandtl numbers** can only be assumed at lower Ekman numbers, as outlined above. A thorough explanation of the possible consequences of the scale disparity on the dynamo process is therefore still missing.

The Prandtl number may also play an important role which has only been explored to some extend so far. Simulations that simultaneously model thermal and compositional convection and allow for the significant differences in the diffusivities of both components are still missing. Several studies that model pure thermal convection but vary the Prandtl number indicate that **inertial** effects decreases upon increase of the Prandtl number in convectively driven flows (Tilgner 1996; Schmalzl et al. 2002; Breuer et al. 2002) as well as in dynamos (Simitev and Busse 2005; Sreenivasan and Jones 2006). Larger Prandtl numbers promote more confined thinner convective flows and lower overall flow amplitudes (Schmalzl et al. 2002; Breuer et al. 2002) and the latter is also responsible for the smaller inertial effects (Sreenivasan and Jones 2006). Higher **magnetic Prandtl numbers** are thus required to sustain dynamo action and keep Rm above the necessary critical value (Simitev and Busse 2005). Simitev and Busse (2005) moreover report that the dipole contribution becomes stronger at larger Prandtl numbers which is in line with the understanding that inertia is responsible for the transition from the dipole to the **multipolar** regime in ➋ *Fig. 4*.

3.2 Dynamo Mechanism

The simple structure of large **Ekman number** and small **Rayleigh number** dynamos still allows to analyze the underlying **dynamo mechanism** (Olson et al. 1999; Wicht and Aubert 2005; Aubert et al. 2008b) which becomes increasingly complex at more realistic parameters. ➋ *Figure 6* illustrates the mechanism at work in the benchmark II (Christensen et al. 2001) dynamo (BMII in ➋ *Table 1*). The solution obeys a four-field azimuthal symmetry and the dynamo process can be illustrated by concentrating on the action of one **cyclone/anticyclone** pair. The mechanism is of the α^2 type, a terminology that goes back to the mean field dynamo theory and refers to the fact that poloidal as well as toroidal magnetic fields are produced by the local action of small-scale flow; small scale refers to the individual convective columns here. The $\alpha\omega$ mechanism where the toroidal field is created by shear in the zonal flow offers an alternative that may operate in models where stress-free boundary conditions allow for stronger zonal flows to develop (Kuang and Bloxham 1997; Busse and Simitev 2005a).

Anticyclones stretch north/south-oriented fieldlines radially outward in the equatorial region. This produces a strong inverse radial field on either side of the equatorial plane which is still clearly visible at the outer boundary r_o. Inverse refers to the radial direction opposing the dominant axial dipole. The field is then wrapped around the **anticyclone** and stretched poleward resulting in a field direction which is now inline with the axial dipole. Advective transport from the **anticyclones** toward the **cyclones** and stretching by the meridional circulation down

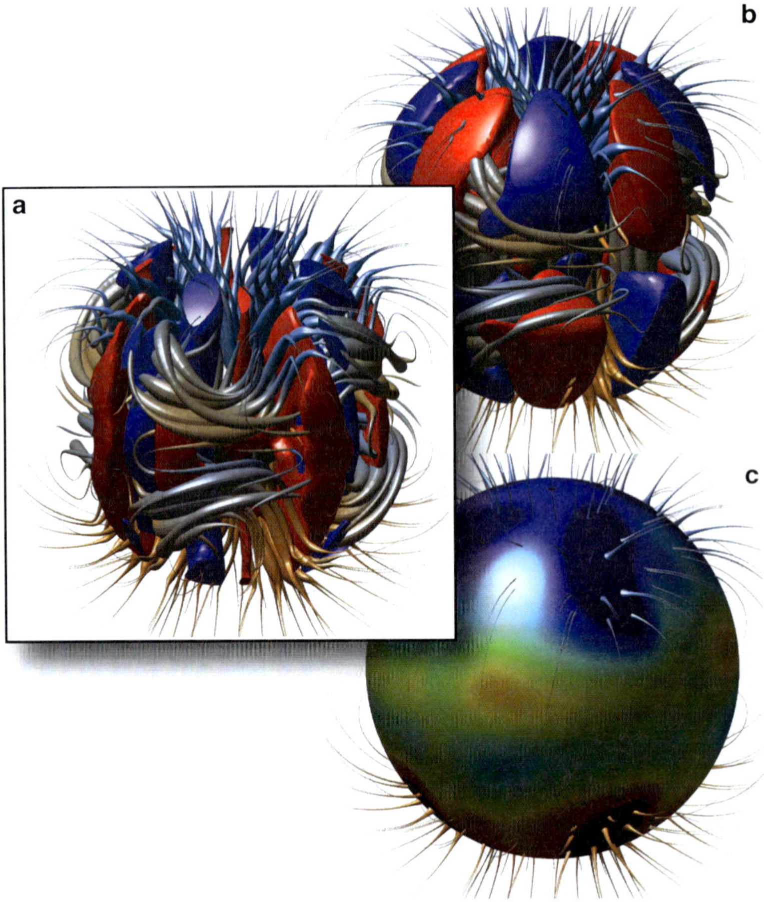

Fig. 6

Illustration of the dynamo mechanism in the benchmark II dynamo. Panel (a) shows isosurfaces of positive (*red*) and negative (*blue*) z-vorticity which illustrate the cyclonic and anticyclonic convective columns. Panel (b) shows isosurfaces of positive (*red*) and negative (*blue*) z-velovity, and panel (c) shows contours of the radial field at the outer boundary. *Red* (*blue*) indicates radially outward (inward) field. The magnetic fieldlines are colored accordingly, their thickness is scaled with the local magnetic field strength

the axis of the **cyclones** closes the magnetic production cycle which maintains the field against Ohmic decay.

The converging flow into the **anticyclones** advectively concentrates that normal polarity field into strong **flux lobes** located at higher latitudes close but outside the **tangent cylinder**. The dipole field is to a good degree established by these flux lobes. Meridional circulation inside the tangent cylinder is responsible for the characteristically weaker magnetic field closer to the poles since it transports fieldlines away from the rotation axis toward the tangent cylinder.

The distinct magnetic features associated with the action of the prograde and retrograde rotating convective columns have been named **magnetic cyclones** and **anticyclones**, respectively, by Aubert et al. (2008b). An identification of the individual elements in the dynamo process becomes increasingly difficult at smaller **Ekman numbers** and larger Rayleigh numbers where the solutions are less symmetric, more small scale, and stronger time dependent. Many typical magnetic structures nevertheless prevail over a wide range of parameters which suggests that the underlying processes may still be similar (Aubert et al. 2008b).

3.3 Comparison with the Geomagnetic Field

A detailed comparison of a numerical solution with the **geomagnetic field** structure and dynamics can serve to judge whether a particular simulation provides a realistic geodynamo model. Global geomagnetic models represent the field in terms of spherical harmonics which allows a downward continuation to the core-mantle boundary. Sufficient resolution in time and space and an even global coverage are key issues here. Refer to Hulot et al. (2010) for a more detailed review of the different geomagnetic field models, only a brief overview is provided here.

A comparison of snapshots always bears the problem that the magnetic field is highly variable. The conclusion may therefore depend on the selected epochs. Time-averaged fields offer additional insight here though certain smaller-scale features vanish in the averaging process. The relatively high resolution and the confidence in data quality make the more modern field models from the past decades an ideal candidate for comparison. However, at least some of their features may not represent the "typical" **geomagnetic field** and they surely do not embrace the full geomagnetic time variability. One should keep this in mind before dismissing a numerical model. So far, no rigorous procedure has been established to judge how well a numerical simulation models the **geomagnetic field**. The assessment presented here may improve the situation but still remains somewhat heuristic.

Modern field models rely on magnetic satellite data that provide a very good spatial coverage. The attainable resolution for the dynamo field is, however, limited by the fact that the crustal magnetic field cannot be separated from the core contribution. The models are therefore restricted to spherical harmonic degrees $l \leq 14$ where the crustal contribution is negligible. Satellite-based models reach back to 1980 and therefore encompass only a very small fraction of the inherent geomagnetic time scales.

Historic field models cover the past 400 years and rely on geomagnetic observatory data and ship measurements among other sources. Naturally, the resolution and precession is inferior to the satellites models and degrades when going back in time. gufml (Jackson et al. 2000), the model used for comparison in the following, provides spherical harmonic degrees up to $l = 14$ in 2.5 year intervals. ➋ *Figure 7* compares the gufml core-mantle boundary (CMB) field model for the epoch 1990 with a snapshot from models E5R36 and E6. The comparison with model E5R32 reveals a striking similarity. An imposed fourfold azimuthal symmetry has been used in model E6 to save computing time. This complicates the comparison but the field seems to show too little structure at low latitudes and lacks the inverse field patches that seem to be typical for the historic **geomagnetic field** in this region. A more detailed discussion of the individual features follows below. Respective comparisons for numerical dynamos at different **Ekman numbers** can be found elsewhere (Christensen and Wicht 2007; Wicht et al. 2009; Wicht and Tilgner 2010).

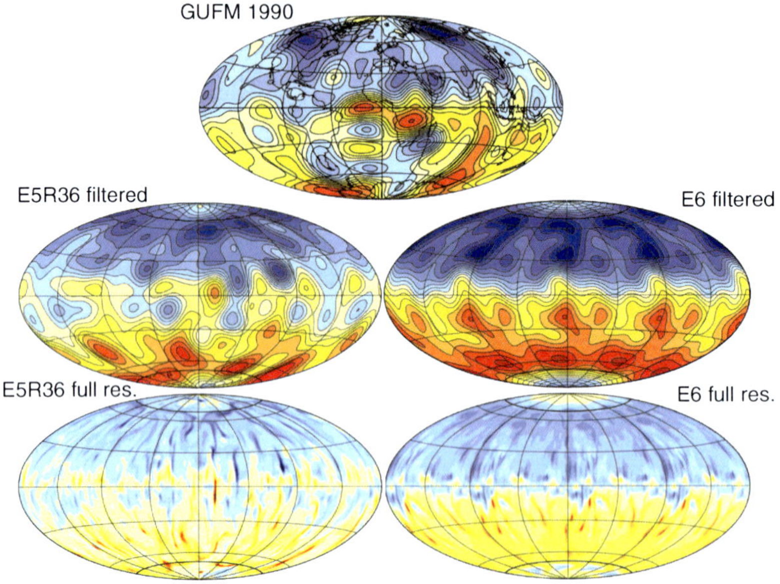

Fig. 7
Comparison of radial magnetic fields at the outer boundary of the dynamo region (core-mantle boundary). The top panel shows the gufm1 field for the year 1990, the other panels show snapshots from models E5R36 and E6, restricted to $l \le 14$ in the middle row and at full numerical resolutions in the bottom row. The color scheme has been inverted for the numerical model to ease the comparison. Generally, the dynamo problem in invariant with respect to a sign change in the magnetic field

The normalized **magnetic energy spectrum** at the core-mantle boundary,

$$\bar{E}_m(l, r = r_0) = \frac{\sum_{m=0}^{m=l} E_m(l, m, r = r_0)}{\sum_{m=0}^{m=1} E_m(l = 1, m, r = r_0)}, \tag{23}$$

provides some information about the importance of the different spherical harmonic degrees in the core field. Here, $E_m(l, m, r)$ is the magnetic energy carried by the mode of spherical harmonic degree l and order m at radius r. *Figure 8* compares the respective gufm1 spectra with spectra from different simulations. The damping employed in the model inversion clearly affects the higher harmonics in gufm1. The spectra for the nondipolar contribution in the numerical simulations remain basically white for low to intermediate degrees with some degrees, notably $l = 5$, sticking out. The relative importance of the dipole contributions grows with decreasing **Ekman number** as long as the solutions belong to regimes **D** or **E**, as further discussed below. The red curve in *Fig. 8* is an example for a multipolar solution where the dipole has lost its prominence all together.

Archeomagnetic models include the magnetic information preserved in recent lava flows and sediments. The widely used CALS7K.2 model by Korte and Constable (Korte et al. 2005; Korte and Constable 2005) is a degree $l = 10$ model that is thought to be reliable up to degree $l = 4$ with a time resolution of about a century. *Figure 9* shows the respective time-averaged CMB field in comparison with time-averaged numerical solutions.

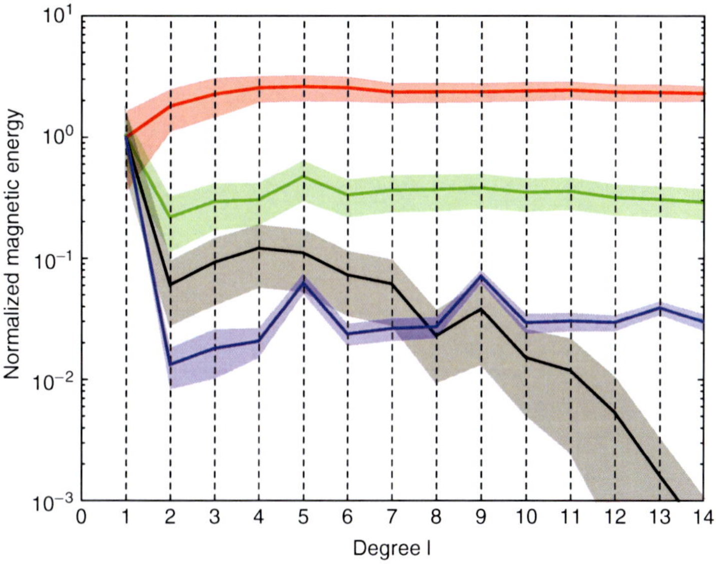

Fig. 8
Time-averaged normalized magnetic energy spectrum at the outer boundary for gufm1 (*black*), the E = 3×10^{-4} model E4R106 (*green*), and the E = 3×10^{-5} models E5R36 (*blue*) and E5R48 (*red*) which belong to regimes D and M, respectively. Colored bars in the width of the standard deviation indicate the time variability

Paleomagnetic data reaching further back in time neither have the spatial nor the temporal resolution and precision to construct faithful models for specific epochs. Statistical interpretations are thus the norm and provide time-averaged fields (TAF) and mean paleo-**secular variation**, for example, in the form of VGP scatter (see below).

In the following the similarity of the geomagnetic field and simulation results are assessed in the context of some key issues that were discussed more heavily in recent years.

3.3.1 Dipole Properties and Symmetries

In the linearized system describing the onset of convection solutions with different **equatorial symmetry** and **azimuthal symmetry** decouple. In terms of spherical harmonics, azimuthal symmetries are related to different orders m, while equatorial symmetric and antisymmetric solutions are described by harmonics where the sum of degree l and order is even or odd, respectively. Equatorial symmetric flows are preferred in rotation-dominated systems and are thus excited at lower **Rayleigh numbers** than antisymmetric contributions. Equatorial symmetric flows support dynamos with either purely equatorial antisymmetric or purely symmetric magnetic fields which in the dynamo context are called the dipolar and quadrupolar families, respectively (Hulot and Bouligand 2005). These names refer to the primary terms in the families, the axial dipole ($l = 1, m = 0$) and the axial quadrupole ($l = 2, m = 0$). Both are coupled by equatorial antisymmetric flows which can transfer energies from one family to the other. The

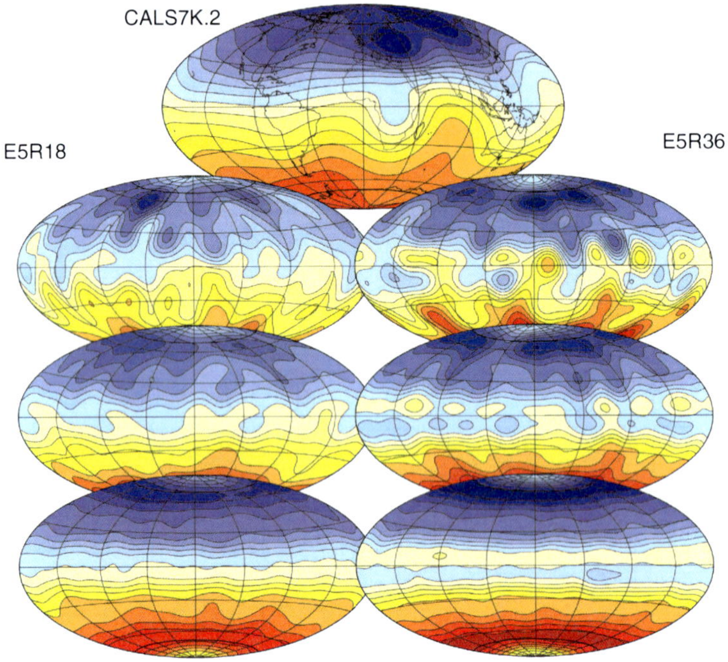

Fig. 9

Time-averaged radial CMB fields for the CALS7K.2 archeomagnetic model by Korte and Constable (2005) and for the numerical models E5R18 (*left*) and E5R36 (*right*). Time averages over intervals corresponding to 500, 5 000, and 50, 000 years, are shown in the top, middle, and bottom panels, respectively

dipolar family is clearly preferred in dynamo solutions at lower **Rayleigh numbers** which correspond to regime **D**. However, a few exceptions have been reported for large **Ekman numbers** (Aubert and Wicht 2004).

Dynamos with perfectly **azimuthal symmetry** and **equatorial symmetry** are found close to the onset of convection where the underlying flow would still retain perfect symmetries in the nonmagnetic case. The respective models have been marked by squares in ➋ *Fig. 4*. These solutions obey a very simple time dependence: a drift of the whole pattern in azimuth. When the **Rayleigh number** or the **magnetic Prandtl number** are increased, the azimuthal symmetry is broken first, then the **equatorial symmetry**. This goes along with a change in time dependence from drifting to chaotic, sometimes via an oscillatory behavior (Wicht 2002; Wicht and Tilgner 2010) (see ➋ *Figs. 4* and ➋ *10*).

The symmetry properties are quantified with time averages of four measures: The dipole contribution is characterized by time averages of its relative importance at r_o, i.e., by the **dipolarity**

$$d = \frac{\left(\sum_{m=0}^{m=1} E_m(l=1, m, r_0)\right)^{1/2}}{\left(\sum_{l=1}^{l=11}\sum_{m=0}^{m=l} E_m(l, m, r_0)\right)^{1/2}}, \tag{24}$$

and by the mean **dipole tilt** Θ, the minimum angle between the magnetic dipole axis and the rotation axis. The symmetry properties of the nondipolar harmonics are quantified by the

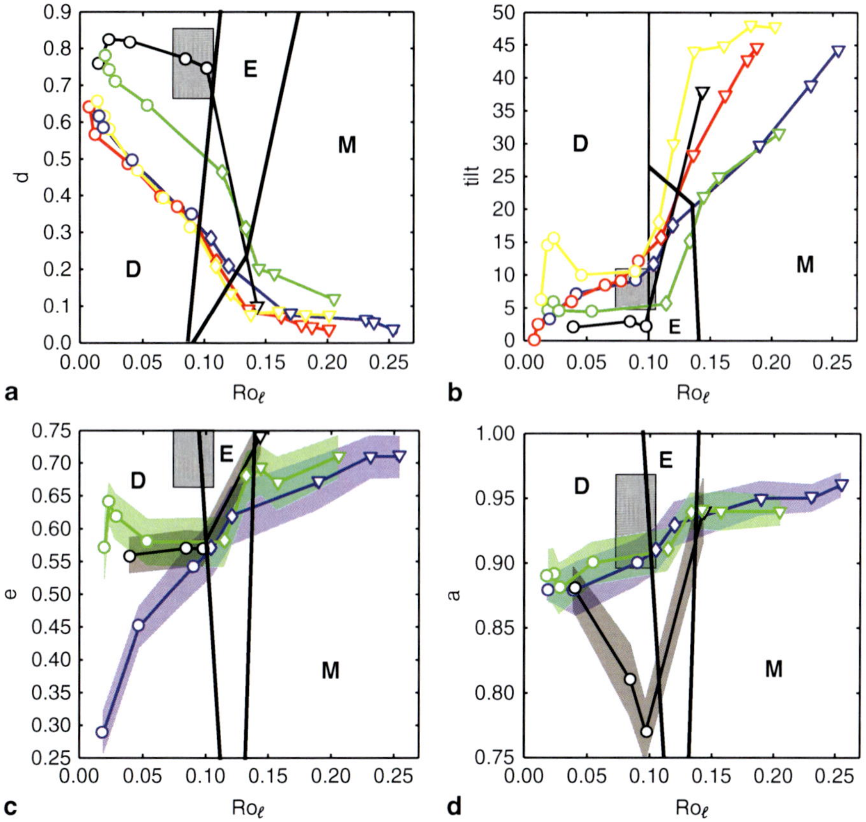

Fig. 10

Dipole properties and symmetry properties of nondipolar contributions for several dynamo models and the gufm1 (*grey boxes*). Different symbols code the time dependence as explained in the caption of *Fig. 4*. Panel (a) shows the time-averaged dipolarity *d* and panel (b) shows the time-averaged dipole tilt. Panels (c) and (d) display the time-averaged equatorial symmetry measure *e* and time-averaged axial symmetry measure *a*, respectively. *Blue* and *red*: models from *Fig. 4* with Pm = 10 and Pm = 20, respectively; *yellow*: identical parameters to the blue model but with chemical boundary conditions; *green*: E = 3×10^{-4}, Pm = 3, chemical boundary conditions; *black*: E = 3×10^{-5}, Pm = 1, fixed temperature conditions. The Prandtl number is unity in all cases. Colored bars in the width of the standard deviation indicate the time variability in panels (c) and (d). The standard deviation amounts to only a few percent in *d* (panel a) and is of the same order as the tilt itself (panel b)

relative strength of the equatorial symmetric contributions, i.e., by the **equatorial symmetry measure**

$$e = \frac{\left(\sum_{l=2}^{l=11}\sum_{m=0}^{m=l,l+m=\text{even}} E_m(l,m,r_0)\right)^{1/2}}{\left(\sum_{l=2}^{l=11}\sum_{m=0}^{m=l} E_m(l,m,r_0)\right)^{1/2}}, \tag{25}$$

and the relative strength of the nonaxially symmetric contributions, i.e., by the **axial symmetry measure**

$$a = \frac{\left(\sum_{l=2}^{l=11} \sum_{m=1}^{m=l} E_m(l, m > 0, r_0)\right)^{1/2}}{\left(\sum_{l=2}^{l=11} \sum_{m=0}^{m=l} E_m(l, m, r_0)\right)^{1/2}} . \quad (26)$$

These four measures are restricted to degrees ($l \leq 11$) to facilitate a comparison with the gufm1 model. The dipole contributions would dominate both measures e and a and are therefore excluded. ➲ *Figure 10* shows the dependence of the four time-averaged measures on the **local Rossby number** Ro_ℓ for several dynamo models. The models span Ekman numbers from $E = 10^{-3}$ to $E = 3{\times}10^{-5}$, magnetic Prandtl numbers between $Pm = 1$ and $Pm = 20$ (see ➲ *Table 1*), and purely thermal as well as purely compositionally driven cases are considered. For each model Ro_ℓ is varied by changing the **Rayleigh number**. The gufm1 model is represented by the gray blocks in ➲ *Fig. 10*. The vertical extension corresponds to minimum and maximum values within the represented 400-year period; the horizontal extension indicates uncertainties in Ro_ℓ (Christensen and Aubert 2006). Note that Earth's **local Rossby number** is based on the scaling derived by Christensen and Aubert (2006) and not on direct measurements (see also the discussion in ➲ Sect. 3.1). ➲ *Table 1* lists the four measures for selected cases.

➲ *Figure 10* demonstrates that the **dipolarity** generally increases with decreasing **Ekman number** while the mean tilt becomes smaller. This can be attributed to the growing influence of rotation at smaller E values; the associated increase in "geostrophic" flow correlation promotes the production of axial dipole field. This is counteracted by the increased influence of inertial forces at larger **Rayleigh numbers**. The strong **dipolarity** of gufm1 and its sizable mean tilt around 10° can only be reached at lower Ekman numbers combined with larger Rayleigh numbers to yield adequate Ro_ℓ values around 0.1. The comparison of the three different models at $E = 10^{-3}$ suggests that neither the magnetic Prandtl number ($Pm = 10$ and $Pm = 20$) nor the driving mode play an important role (see also Wicht et al. (2009)).

The equatorial and axial symmetry measures shown in ➲ *Fig. 10* provide a somewhat inconclusive picture. There is a trend that equatorially symmetric as well as nonaxial contributions become more important with growing **local Rossby number**. The two solutions for intermediate Ro_ℓ values at $E = 3{\times}10^{-5}$, however, do not follow this trend. They are particularly axisymmetric, and seem incompatible with the gufm1 data in this respect. Whether this is typical for low **Ekman number** simulations remains to be established. If true, this would once more suggest that low Ekman number simulations require large **Rayleigh numbers** to yield Earth-like field symmetries. Generally, however, gufm1's e and a values seem within reach of numerical dynamo simulations, in particular, when considering that gufm1 may not represent the true geomagnetic variability here. The degree of dipole dominance and the mean tilt are somewhat better constrained by additional archeomagnetic and paleomagnetic information (Hulot et al. 2010).

3.3.2 Persistent Features and Mantle Influence

Strong normal polarity flux concentrations at higher latitudes are a common feature in all the dynamo simulations. Very similar **flux lobes** can be found in the historic **geomagnetic field**, two in the northern and two in the southern hemisphere, and seem to have changed only little over the last 400 years (Gubbins et al. 2007). Their position and the symmetry with respect to the equator suggest that they are caused by the inflow into convective **cyclones** (see ➲ Sect. 3.2).

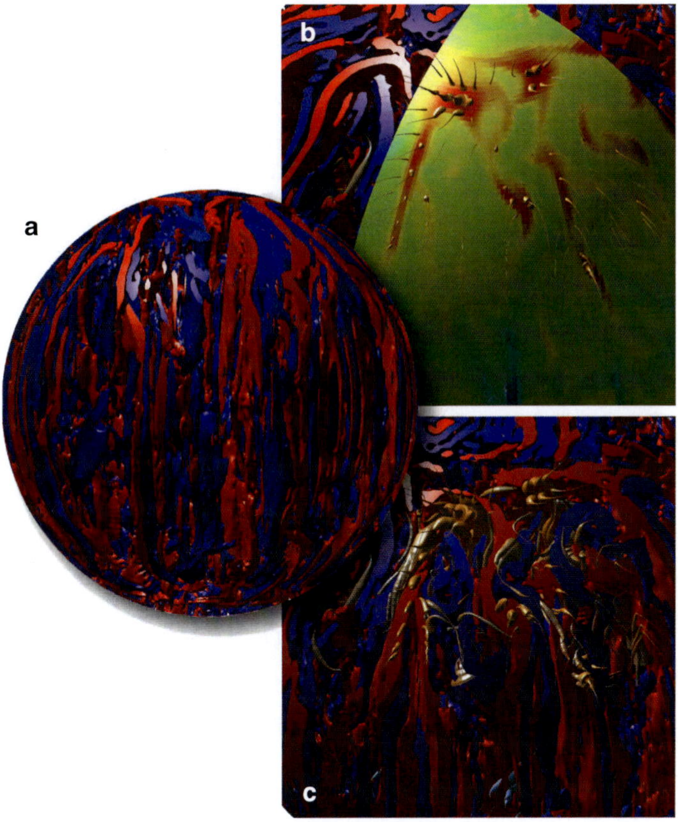

Fig. 11

Panel (a) illustrates the complex sheet-like cyclones (*red*) and anticyclones (*blue*) in model E5R43 with isosurfaces of the z-vorticity. Panels (b) and (c) zoom in on part of the northern hemisphere and illustrate how the normal polarity magnetic field is concentrated by the flow that converges into cyclones. The normal polarity field is radially outward in the northern hemisphere indicated by *red* fieldlines and the *red* radial field contours in panel (b)

Figure 7 demonstrates that the seemingly large-scale **flux lobes** in the filtered field version are the expression of much smaller field concentrations caused by convective features of similar scale. *Figure 11* shows a closeup of the convection around the patches in the dynamo models E5R43. The small-scale correlation of cyclonic features with strong magnetic field patches is evident. As the **Ekman number** decreases, the azimuthal scale of the convective columns shrinks while the scale perpendicular to the rotations axis is much less affected. The columns become increasingly sheet-like which translates into thinner magnetic patches that are stretched into latitudinal direction. In the filtered version, the finer scale is lost and the action of several convective structures appears as one larger magnetic **flux lobe**.

The **flux lobes** are fainter in the time average of the 7,000-year CALS7K.2 model than in the historic field but remain discernable. Even some paleomagnetic TAF models covering 5 Myr report persistent high latitude flux lobes at very similar locations (Gubbins and Kelly 1993; Kelly and Gubbins 1997; Johnson and Constable 1995; Carlut and Courtillot 1998; Johnson et al. 2003).

This has led to speculations whether the lobes represent a more-or-less stationary feature (Gubbins et al. 2007). Since the azimuthal symmetry is not broken in the dynamo models presented above longitudinal features average out over time. ➲ *Figure 9* demonstrates that the historical time span is too short to yield this effect. Even time spans comparable to the period covered by CALS7K.2 may still retain azimuthal structures. The **flux lobes** still appear at comparable location as in snapshots (see ➲ *Fig. 7*) or the historic averages. The solution finally becomes nearly axisymmetric when averaging over periods corresponding to 50,000 years.

This suggests that the persistence over historic or archeomagnetic time scales is nothing special. The persistent azimuthal structures in some 5 Myr TAF models, however, can only be retained when the azimuthal symmetry is broken. The preferred theory is an influence of the lower thermal mantle structure on the dynamo process. It has already been mentioned above that the mantle determines the heat flow out of the dynamo region. Lateral temperature differences in the lower mantle translate into lateral variations in the **CMB heat flux**. The flux is higher where the mantle is colder than average and vice versa. The respective pattern is typically deduced from seismic tomography models that are interpreted in terms of temperature differences yielding so-called **tomographic heat flux models** (Glatzmaier et al. 1999).

Several authors have explored the potential influence of lateral **CMB heat flux** variations on the dynamo process (Glatzmaier et al. 1999; Olson and Christensen 2002; Christensen and Olson 2003; Amit and Olson 2006; Aubert et al. 2007; Gubbins et al. 2007; Willis et al. 2007; Takahashi et al. 2008b; Aubert et al. 2008a) and confirm that they may indeed yield persistent azimuthal features at locations similar to those in the **geomagnetic field**. The time scale over which the lobes become quasi-stationary depends on the model parameters. Willis et al. (2007) find that rather low **Rayleigh numbers** and strong heat flux variations, in the same order as the spherically symmetric total heat flow, are required to lock the patches on a historic time scale. Aubert et al. (2008a), on the other hand, consider a 0.7 Myr average at rather supercritical Rayleigh numbers where the field still varies strongly on historic and even archeomagnetic time scales (Olson and Christensen 2002).

The inhomogeneous **CMB heat flux** may also help to explain some other interesting geophysical features. The nonaxisymmetric structures in paleomagnetic TAF models remain arguable but the axisymmetric properties are much better constrained. Typically, the axial quadrupole amounts to a few percents of the axial dipole while the axial octupole is of similar order but of opposite sign (Johnson 2007). In dynamo simulations, the axial octupole contribution is typically too large. The axial quadrupole contribution, representing an equatorially symmetric mode, is much too low unless the equatorial symmetry is broken by an inhomogeneous **CMB heat flux** condition. The amount of north/south asymmetry required in the boundary condition is somewhat larger than the typical "tomographic" models suggest (Olson and Christensen 2002). The direct translation of seismic velocity differences into temperature differences and subsequently heat flux patterns may oversimplify matters here; for example, compositional variations may also have to be considered.

Amit and Olson (2006) suggest that some features in the flow field at the top of Earth's core that can be deduced from the **secular variation** provided by historic field models may also reflect the influence of the **CMB heat flux** pattern (Amit and Olson, 2006). In a recent analysis Aubert et al. (2008a) evoke the inhomogeneous cooling of the inner core by a quasi-persistent **cyclone** reaching deep into the core to explain seismic inhomogeneities in the outer 100 km of the inner core radius. The preference for magnetic poles to follow two distinct latitude bands during **reversals** (Gubbins and Love 1998) can also be explained with a **tomographic CMB heat flux** model (Coe et al. 2000; Kutzner and Christensen 2004). And finally, the CMB heat flux pattern has been shown to also influence the reversal rate in dynamo simulations and may thus explain the observed changes in the geomagnetic reversal rate (Glatzmaier et al. 1999; Kutzner and Christensen 2004).

3.3.3 Inverse Magnetic Field Production and the Cause for Reversals

The production of strong **inverse magnetic field** on both sides of the equatorial plane is an inherent feature of the fundamental **dynamo mechanism** outlined in ➋ Sect. 3.2. The associated pairwise radial field patches in the outer boundary field are therefore found in many dynamo simulations. Wicht et al. (2009) and Takahashi et al. (2008a), however, report that the patches become less pronounced at lower **Ekman numbers** (see also Sakuraba and Roberts [2009]). Model E6 in ➋ *Fig. 7* and model E5R18 in ➋ *Fig. 9* demonstrate that the patches retreat to a small band around the equator which becomes indiscernible in the filtered field. Normal polarity patches now rule at mid-to lower latitudes but the region around the equator generally shows rather weak field of normal polarity, which is compatible with the strong dipole component in these models. More pronounced inverse patches reappear when the **Rayleigh number** is increased, as is demonstrated by E5R36 in ➋ *Fig. 7* where patches of both polarities may now be found around the equator. The low latitude field remains only weakly inverse when averaged over long time spans (see ➋ *Fig. 9*). The time-averaged archeomagnetic model CALS7K.2 (see ➋ *Fig. 9*) points toward a weakly normal field but the lack of resolution may be an issue here. The larger Rayleigh number in model E5R36 also promotes a stronger breaking of the equatorial symmetry, destroying the pairwise nature of the patches. This contributes to the convincing similarity between the E5R36 solution and the historic **geomagnetic field** where strong and equatorially asymmetric normal polarity patches seem typical at lower latitudes (Jackson 2003; Jackson and Finlay 2007) (see ➋ *Fig. 7*).

The weak magnetic field in the polar regions can give way to inverse field at larger **Rayleigh numbers**, a feature also found in the historic **geomagnetic field**. This inverse field is produced by plume-like convection that rises inside the **tangent cylinder** and the typical associated magnetic structure has been named **magnetic upwellings** (MU) by Aubert et al. (2008b). MUs may also rise at lower to mid-latitudes when the equatorial symmetry is broken to a degree where one leg of the equatorial inverse field production clearly dominates. The symmetry breaking is essential since the inverse fields created on both sides of the equator would cancel otherwise. The MUs may trigger magnetic field **reversals** and excursions when producing enough inverse field to efficiently cancel the prevailing dipole field (Wicht and Olson 2004; Aubert et al. 2008b). This basically resets the polarity and small field fluctuation then decide whether axial dipole field of one or the other direction is amplified after the MUs have ceases (Aubert et al. 2008b; Wicht et al. 2009). MUs are a common feature in larger **Rayleigh number** simulations and vary stochastically in strength, number, and duration. Wicht et al. (2009) suggest that particularly strong or long-lasting MUs are required to trigger reversals. Alternatively, several MUs may constructively team up. Both scenarios remain unlikely at not too large Rayleigh numbers which explains why **reversals** are rare in regime **E** (see ➋ *Figs. 4* and ➋ *10*).

3.3.4 Time Variability

The internal **geomagnetic field** obeys a very rich time variability from short-term variations on the yearly time scale, the geomagnetic jerks, to variations in the reversal frequency on the order of several tens of million years (Hulot et al. 2010). Dynamo simulations are capable of replicating many aspects of the time variability but the relative time scales of the different phenomena may differ from the geomagnetic situation depending on the model parameters.

The fact that not all model parameters are realistic means that there is no unique way to rescale the dimensionless time in the simulations. In order to nevertheless facilitate a direct comparison, the simulation results are often rescaled by assuming an Earth-like magnetic diffusion time of $\tau_\lambda = \ell^2/\lambda \approx 122$ kyr. This approach has been adopted here whenever the time is reported in a simulation in years. Independent of the rescaling problem, the question whether the ratio of different time scales is Earth-like can be addressed. Another important time scale for the dynamo process is the flow time scale $\tau_u = \ell/u$ which is typically called the convective **turnover time** and amounts to roughly 250 years for Earth. τ_u is imprinted on the magnetic field via the induction and is thought to rule the **secular variation**. The **magnetic Reynolds number** provides the ratio of the two time scales in the induction equation (4): $\text{Rm} = \tau_\lambda/\tau_u$. Adopting a realistic **magnetic diffusion time** will therefore also yield a realistic **turnover time** for dynamos with Earth-like magnetic Reynolds numbers.

Christensen and Tilgner (2004) analyzed a suit of dynamo simulations to further elucidate how the typical **secular variation** time scale depends on the degree l of the magnetic component. They find the inverse relationship:

$$\tau_l = 21.7\,\tau_u l^{-1}, \tag{27}$$

which is compatible with the idea that the magnetic field is advected by a large-scale flow and roughly agrees with geomagnetic findings (Hongre et al. 1998; Olsen et al. 2006).

Torsional oscillations (TOs) are a specific form of short-term variations on the decadal time scale (Braginsky 1970). They concern the motion of so-called geostrophic cylinders that are aligned with the planetary rotation axis. TOs are essentially one-dimensional **Alfvén waves** that travel along magnetic field lines connecting the cylinders. The fieldlines act like torsional springs that react to any acceleration of the cylinder with respect to each other. TOs have the correct time scale for explaining the decadal variations in Earth's length of day, which are typically attributed to an exchange of angular momentum between Earth's core and mantle (Jault et al. 1988; Jackson 1997; Jault 2003; Amit and Olson 2006). TOs have reportedly been identified in the geomagnetic **secular variation** signal (Zatman and Bloxham 1997) and are considered a possible origin for the geomagnetic jerks (Bloxham et al. 2002).

The smallness of inertial forces and viscous forces in the magnetostrophic force balance is a prerequisite for **torsional oscillations** to become an important part of the short-term dynamics. Coriolis as well as pressure forces do not contribute to the integrated azimuthal force on geostrophic cylinders for geometrical reasons. This leaves the azimuthal Lorentz forces as the only constituent in the first-order force balance. Taylor (1963) therefore conjectured that the dynamo would assume a configuration where the azimuthal Lorentz forces would cancel along the cylinders until the integrated force can be balanced by viscous or inertial effects. In recognition of Taylor's pioneering work, dynamos where the normalized integrated Lorentz force

$$\mathcal{T}(s) = \frac{\int_{C(s)} \hat{\phi}\cdot((\nabla\times\mathbf{B})\times\mathbf{B})\;dF}{\int_{C(s)} |\hat{\phi}\cdot(\nabla\times\mathbf{B})\times\mathbf{B}|\;dF}\,, \tag{28}$$

is small are said to obey a **Taylor state**. Here, $C(s)$ signifies the geostrophic cylinder of radius s and dF is a respective surface element. **Torsional oscillations** are faster disturbances that ride on the background Taylor state which is established on the **turnover time** scale.

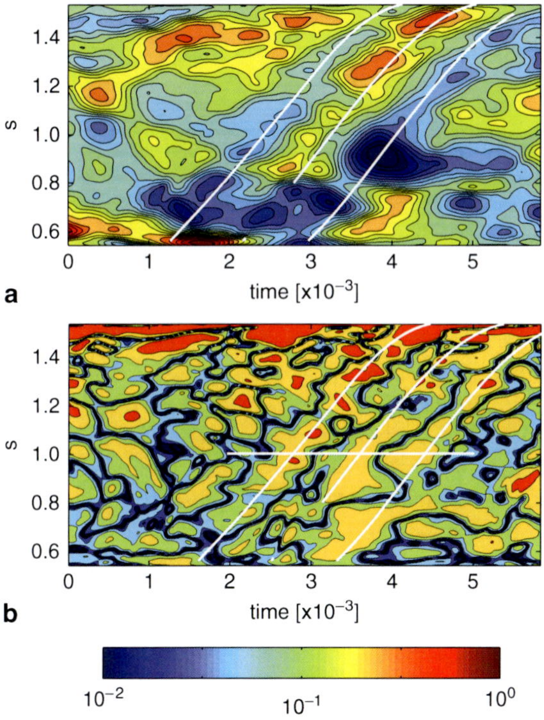

Fig. 12

Panel (a) shows the speed of geostrophic cylinders outside the tangent cylinder for a selected time span in models E6 (Wicht and Christensen 2009). The time is given in units of the magnetic diffusion time here. Torsional oscillations travel from the inner core boundary (*bottom*) toward the outer core boundary (*top*) with the predicted Alfvén velocity (*white lines*). The agreement provides an important clue that the propagating features are indeed torsional oscillations. Panel (b) shows the normalized integrated Lorentz force $\mathcal{T}(s)$ which can reach values down to 10^{-2} where the Taylor state is assumed to a good degree. The Taylor state is broken during the torsional oscillations

Wicht and Christensen (2009) find **torsional oscillations** in dynamo simulations for **Ekman numbers** of $E = 3\times10^{-5}$ or smaller and for relatively low **Rayleigh numbers** where inertial forces remain secondary. ➋ *Figure 12* shows the traveling waves and the $\mathcal{T}(s)$ in their model E6 at $E = 3\times10^{-6}$ where TOs show most clearly (see ➋ *Table 1* for further model parameters and properties). The **Alfvén Mach number** provides the ratio of the flow time scale to the Alfvén time scale characteristic for TOs. **Alfvén waves** travel more than an order of magnitude faster than the typical convective flow speed in Earth's core. In the low Ekman number case where TOs have actually been identified, they were only roughly twice as fast as the flow. Larger ratios can only be expected when the Ekman number is further decreased below $E = 3\times10^{-6}$ (model E6 is 1) where Wicht and Christensen (2009) identified the most Earth-like **torsional oscillations**.

The slowest magnetic time scales are associated with field **reversals**. Geomagnetic reversals typically last some thousand years while simulated reversals seem to take somewhat longer

(Wicht 2005; Wicht et al. 2009). ➋ *Figure 5* shows an example for a reversal sequence in a numerical simulation. The duration of geomagnetic and simulated **reversals** obey a very similar latitudinal dependence (Clement 2004; Wicht 2005; Wicht et al. 2009) with shorter durations at the equator and a gradual increase toward the poles.

During the last several million years the average reversal rate was about four per million years, but there were also long periods in Earth's history when the geodynamo stopped reversing. The most recent of these so-called superchrons happened in the Cretaceous and lasted for about 37 Myr. Whether the variations in reversal rate are caused by changes in Earth's mantle (see ➋ Sect. 3.3.2) (Glatzmaier et al. 1999; Constable 2000) or are an expression of the internal stochastic nature of the dynamo process (Jonkers 2003; Ryan and Sarson 2007) is still debated. Generally, dynamo simulations (Glatzmaier et al. 1999; Wicht et al. 2009) and even rather simple parameterized models seem capable of showing Earth-like variation in reversal frequency. A more thorough analysis of reversal properties and their stochastic nature in geodynamo simulations is still missing. Such an analysis seems impossible at lower Ekman numbers due to the excessive computing times required (Wicht et al. 2009).

Paleomagnetists frequently interpret the local magnetic field provided by a paleomagnetic sample as being caused by a pure dipole contribution which they call a virtual dipole. Consequently, the associated magnetic pole is referred to as a virtual geomagnetic pole (VGP). The **scatter** of the associated virtual geomagnetic pole around its mean position is used to quantify the **secular variation** in paleomagnetic field models. The scatter shows a typical latitudinal dependence with low values around 12° at the equator and rising to about 20° at the poles. Some dynamo simulations show a very similar variation (Kono and Roberts 2002; Wicht 2005; Christensen and Wicht 2007; Wicht et al. 2009). Christensen and Wicht (2007) demonstrate that the amplitude of the scatter depends on the **Rayleigh number** and seems somewhat high at Rayleigh numbers where field reversals happen. Wicht et al. (2009) attribute both the typical latitude dependence of the VGP scatter and of the reversal duration to **magnetic upwellings** rising inside and close to the **tangent cylinder** which increases the general field variability at higher latitudes. Like for reversals, an analysis of the virtual dipole scatter is missing for lower Ekman number cases because of the long time spans required.

3.4 Is There a Distinct Low Ekman Number Regime?

As outlined above, even the most advanced spherical dynamo models are restricted to **Ekman numbers** $E \geq 10^{-6}$ which is still nine orders of magnitude away from Earth's value. Though current dynamos are rather successful and the changes with decreasing Ekman number seemed modest in the models analyzed above there are some indications that more drastic changes may happen at low Ekman numbers. In this section, the respective clues are reviewed and some classical ideas about the dynamics in the low Ekman number regime are also discussed.

3.4.1 The Strong Field Branch

Before direct numerical simulations permitted detailed investigations of fully nonlinear dynamos, **magneto-convection** studies, in which a magnetic field is externally imposed,

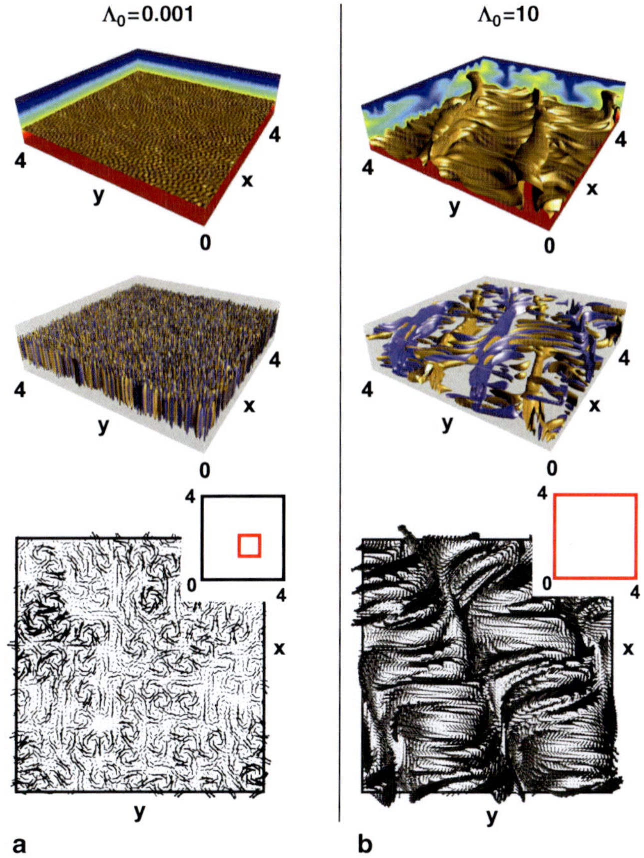

Fig. 13

Snapshots of magneto-convection flows at E = 10^{-5}, Pr = 1, Pm = 0.5 and Ra = 1.2×10^{-2} This should be the same Rayleigh number value as below. for (a) weak and (b) strong imposed horizontal fields in a plane layer model. The top row shows iso-surfaces of temperature at T = 0.7, the middle row shows iso-surfaces of vertical velocity (upflows yellow, downflows blue) while vectors depicting the velocity field at the upper boundary are displayed in the lowermost plots. Note that only the domain indicated by the red square is shown in the bottom panels. Stress-free, perfectly conducting, isothermal boundary conditions have been used. All variables are assumed to be periodic in the horizontal direction, with a fundamental wave length four times larger than the layer height

seemed to suggest that strong magnetic fields would affect convective flows in a drastic way if the **Ekman number** is small. To illustrate this effect, a simple **Cartesian model** is considered here, in which a rapidly rotating layer of electrically conducting fluid is heated from below. A uniform, horizontal magnetic field is externally imposed, and its intensity is measured by the **Elsasser number** $\Lambda_0 = b_0^2/2\Omega\bar{\rho}\mu\lambda$ here, where b_0 denotes the dimensional imposed field strength. Gravity acts downwards, and the rotation axis is assumed to be vertical. The Ekman number defined by using the layer depth as the length scale, is fixed to E = 10^{-5}, and Ra = 1.2×10^{-2}, Pr = 1

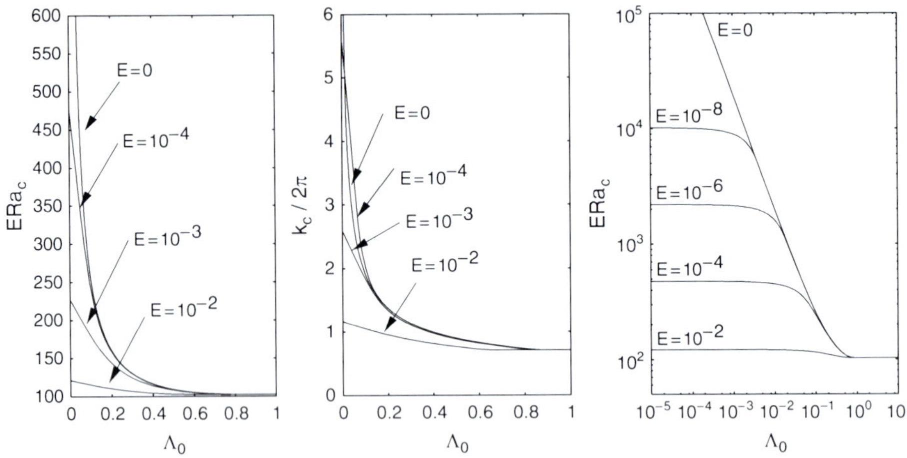

Fig. 14
Critical Rayleigh and wave number for the onset of magneto-convection for $q := \kappa/\eta = 1$ and infinite Prandtl number. The marginal velocity field components are proportional to $\exp(i\ \mathbf{k}\cdot\mathbf{r})$ and $k = |\mathbf{k}|$

and Pm = 0.5 is chosen for the remaining parameters. *Figure 13* illustrates the dynamical consequences of the imposed field for small and large field intensities. If the imposed field is weak, the dynamics is similar to the nonmagnetic case, leading to a flow field consisting of hundreds of individual columns of small $O(\mathrm{E}^{1/3})$ horizontal length scales. As already explained in Sect. 2.4 for the spherical case, the columns have to be this thin in order to provide enough viscous friction to balance the part of the Coriolis forces that cannot be balanced by pressure forces alone (Chandrasekhar 1961). In sharp contrast, for a strong imposed field with $\Lambda_0 = 10$, large convection cells dominate, and both the heat transport and the convective velocities are larger. A detailed analysis reveals that Lorentz forces now chiefly balance the nongeostrophic part of the Coriolis forces, which eliminates the need for small scales in the flow field. Viscous dissipation is reduced to a negligible level, with ohmic losses accounting for more than 98% of the overall dissipation.

Linear stability results for the above configuration (Eltayeb 1972, 1975; Roberts and Jones 2000; Jones and Roberts 2000) provide further information about the asymptotic parameter dependence of the system. Generally, the stability depends on Ra, E, Pr, Pm, and Λ_0. Only one typical example is considered here that serves to illustrate the general behavior. *Figure 14* shows the variation of the critical Rayleigh number Ra_c and wavenumber k_c with Λ_0 for the special case $q = \kappa/\lambda = 1$ and infinite Prandtl number. For $\mathrm{E} \ll 1$, Ra_c drops from $O(\mathrm{E}^{-4/3})$ for weak imposed fields to $\mathrm{Ra}_c = O(\mathrm{E}^{-1})$ for a strong field with $\Lambda_0 = O(1)$. This transition is accompanied by a considerable increase of the spatial flow scales which grow from the purely convective scale, $l_c = O(\mathrm{E}^{1/3})$, to Ekman number-independent scales comparable to the layer depth, $l_c = O(1)$, similar to the behavior observed in the fully nonlinear simulations described above. Note that at low Ekman number, even relatively weak magnetic fields with $\Lambda_0 < O(1)$ have a considerable impact on both l_c and Ra_c, and indeed the fully nonlinear simulations reveal that the value of Λ_0 at which the imposed field starts to noticeably affect the convective flow decreases with decreasing E (not shown here).

The simple **Cartesian model** considered above is certainly over-simplified. Linear studies using spherical shell geometries and more general large-scale magnetic field topologies in fact reveal the existence of more complicated convective modes (Fearn 1979; Sakuraba 2002; Sakuraba and Kono 2000; Zhang and Gubbins 2000b,a), but generally confirm the basic result that strong imposed fields tend to promote large-scale flows and decrease the critical Rayleigh number. For planetary cores, this suggests a huge flow-scale disparity between weakly and strongly magnetic states. Typical estimates based on linear theory suggest that nonmagnetic columnar convection (as illustrated in ➒ *Fig. 3*) would have azimuthal length scales ranging from about 30 m to 1 km, depending on whether molecular or turbulent values of viscosity are used (Zhang and Schubert 2000). The planetary-scale convection cells suggested by **magneto-convection** studies for strong, Earth-like fields would thus be three to five orders of magnitude larger.

Before the advent of direct numerical simulations of the dynamo process, a widespread expectation was that dynamo-generated fields at low Ekman number would act to release the system from the rotational constraint, just as they do in the above **magneto-convection** problem. A large-scale dynamo field, growing from a small initial perturbation, was thought to gradually take over the dynamical role of viscous friction in the momentum equation. It seems reasonable to expect that saturation does not occur before the **Elsasser number** reaches an $O(1)$ value, at which viscous forces become negligible, Lorentz forces completely control the flow scales and a **magnetostrophic** force balance is established. Once the system reaches this so-called **strong field branch**, it might even be possible to reduce the Rayleigh number to values below the $O(E^{-4/3})$ critical Rayleigh number for the onset of nonmagnetic convection without shutting off the dynamo, leading to so-called *subcritical dynamo action*. The reduction of viscous friction to insignificant levels caused by the transition to large-scale flows further suggests that strong field dynamos should be very close to a **Taylor state** if inertial forces remain small.

The **magneto-convection** results described above suggest that the scale separation between weak and strong field cases increases with decreasing E, as does the sensitivity of the convective flow to even weak magnetic fields. Therefore, the strong field scenario might be expected to gain in importance as the **Ekman number** is pushed toward more realistic values in numerical simulations.

3.4.2 Simulations Results at Low Ekman Numbers

Some results from plane layer dynamo simulations seem to support this expectation. Studies by Rotvig and Jones (2002) and Stellmach and Hansen (2004) indicate that dynamo generated fields can significantly affect the dominant flow scales as compared to nonmagnetic convection. Rotvig and Jones (2002), who neglected inertial forces and used a tilted rotation axis, also found clear evidence that their model is close to a **Taylor state** at low Ekman number values. They demonstrated convincingly that viscous forces become irrelevant in the bulk of the convective region and contribute to the force balance only in thin Ekman boundary layers. Convincing evidence for subcritical dynamo action has remained elusive, although studies by St. Pierre (1993) and Stellmach and Hansen (2004) indicate that dynamos with Rayleigh numbers below the nonmagnetic critical value may indeed survive for time spans of the order of a typical magnetic diffusion time.

In most spherical dynamo simulations, however, the magnetic field in general does not seem to chiefly control the dominant flow scales. While this is to be expected for moderate Ekman numbers, the fact that the effect seems to remain small even at $E = O(10^{-5})$ or lower casts doubt on the direct applicability of **magneto-convection** results to the dynamo case. In a recent study at $E = 2 \times 10^{-6}$, Takahashi et al. (2008a) report an increase of ℓ_u by about 20% as compared to the nonmagnetic case, which does not seem to be a huge effect but is perhaps notable as the Elsasser number in their simulations is small compared to unity.

A recent paper by Sakuraba and Roberts (2009) also addresses the issue of whether core flows are dominated by small or large-scale features and raises interesting questions concerning the role of thermal boundary conditions. The vast majority of dynamo simulations has so far been carried out for a fixed, prescribed temperature at the outer core boundary. As mentioned in ➲ Sect. 2.2, this seems somewhat unrealistic as in reality it is mantle convection that largely controls the radial heat flux leaving the core. Specifying the heat flux at the core-mantle boundary in dynamo simulations might thus be more appropriate than assuming a fixed temperature there. Sakuraba and Roberts (2009) compare two simulations at $E = 10^{-6}$, one with a uniform temperature boundary condition applied at the core-mantle boundary, and one with a fixed, uniform heat flux condition instead. Both simulations assume a uniform heat source throughout the core, and include a further localized heat source at the inner core boundary to mimic the release of latent heat due to crystallization.

Surprisingly, the choice of thermal boundary conditions at the core-mantle boundary seems to have a large impact on the system dynamics, as shown in ➲ *Fig. 15*. While fixed temperatures at the outer boundary lead to a flow field that is dominated by small-scale turbulence everywhere, the case with a uniform heat flux boundary condition shows a much larger ability in creating large-scale circulations. In regions where the magnetic field is strong, the velocity field has a large-scale structure, with a dominant azimuthal wavenumber (m) of about six. Small-scale turbulence is still observed in regions where the magnetic field is weak. As Sakuraba and Roberts (2009) argue, it thus seems that the classical idea that a strong magnetic field gives rise to planetary-scale convection is indeed relevant if a fixed heat flux is assumed at the outer boundary. Such a strong effect of the thermal boundary conditions has not been observed in previous simulations (Busse and Simitev 2006), and might occur only at low Ekman number. Future studies of the resulting force balance, along with a more systematic exploration of the parameter space for both dynamos and nonmagnetic convection, are needed to better understand the dynamics causing the observed behavior and to determine if this effect is a robust feature at low E.

Another recent publication reports a substantial change in the system dynamics as more realistic control parameters are approached. Kageyama et al. (2008) observed that the sheet-like convective columns described above are restricted to the inner part of the shell in their simulations at $E = 10^{-6}$ while a strong westward zonal flow develops in the outer part. The solution is shown in ➲ *Fig. 16*. The magnetic field that is generated in the simulation of Kageyama et al. (2008) is not predominantly dipolar, and the dynamo process seems to differ substantially from the simple picture outlined in ➲ Sect. 3.2. Whether the multipolar nature of the field is caused by a larger Rayleigh number or the dipole component simply did not have a chance to develop in their relatively short computational run remains unclear. Also, Kageyama et al. (2008) retain weak compressibility effects which makes the comparison with Boussinesq models difficult.

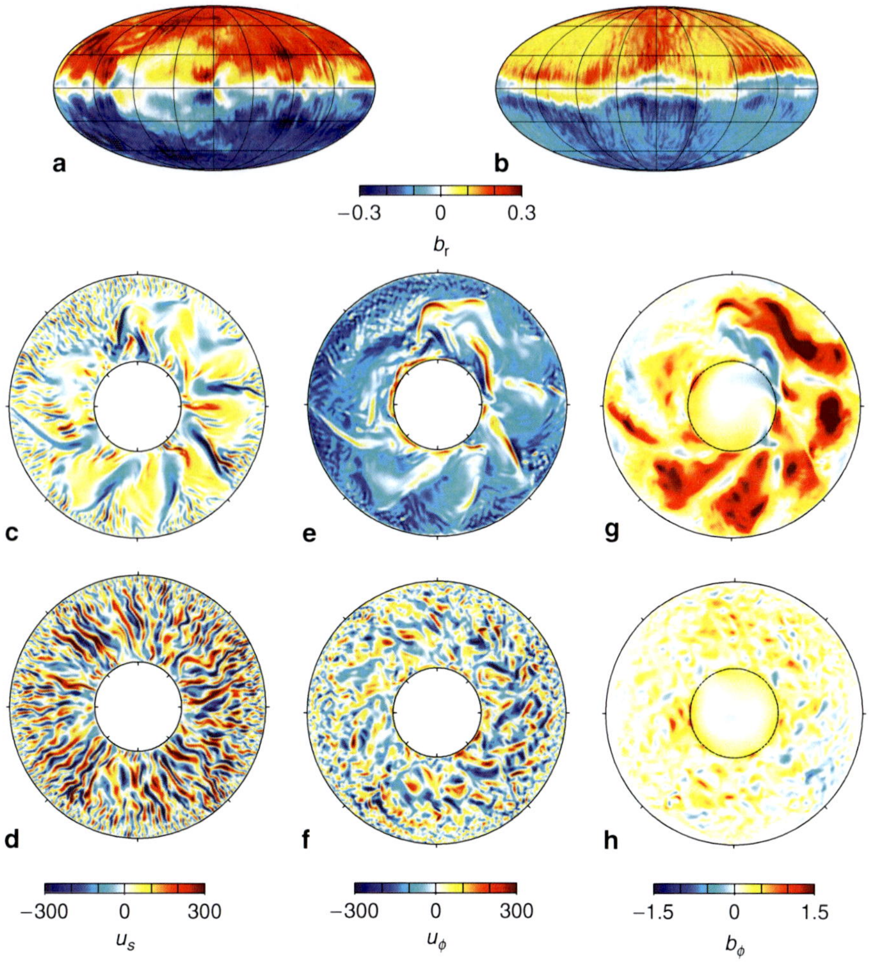

Fig. 15

Snapshots of the velocity and of the magnetic field at E = 10^{-6} in the Sakuraba and Roberts (2009) model. Panels a,c,e and g show results obtained for a simulation with fixed heat flux boundary conditions on the outer shell, while a simulation with prescribed uniform surface temperature gives results as shown in b,d,f and h. The radial magnetic field on the CMB is shown in a and b. The remaining panels are velocity and magnetic field plots in the $z = 0.1r_o$ plane viewed from the north. The radial component of velocity (c,d), the azimuthal component of velocity (e,f) and the azimuthal magnetic field (g,h) are respectively shown. Note that the velocity and magnetic field scales differ from the scaling used in the present paper, see Sakuraba and Roberts (2009) for details

The recent simulations of Sakuraba and Roberts (2009) and Kageyama et al. (2008) challenge the view that current simulations have already reached a regime where only secondary changes can be expected on further decreasing the Ekman number. Due to the enormous difficulties encountered in simulations exploring the limits of what is currently possible, both studies provide information for one specific parameter combination only, and a lot of questions concerning the details of the observed dynamics remain unanswered. Future simulation

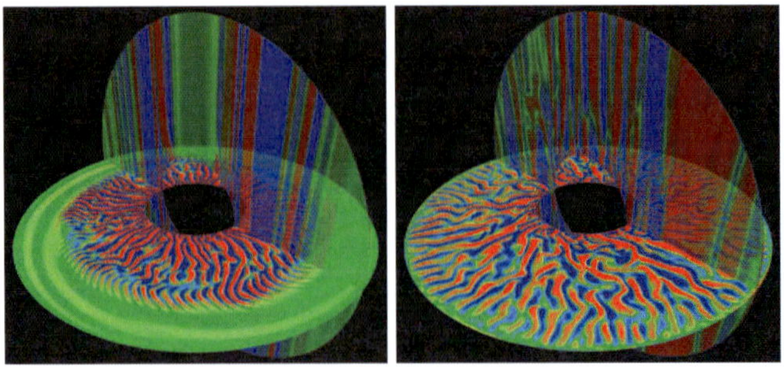

Fig. 16
Flow visualization from the Kageyama et al. (2008) model. An equatorial and a meridional cross-section of the axial component of vorticity are shown. Thin "sheet plumes" with little variations along the rotation axis develop and are restricted to the inner part of the shell when the Ekman number is very low. The Ekman number is 10^{-6} in a and 1.3×10^{-5} in b

efforts targeted at exploring the low Ekman number regime more systematically are clearly needed.

After more that a decade of direct numerical simulations of planetary dynamos, several open questions concerning the low Ekman number regime continue to remain unanswered. It is still not known whether core flows are restricted to small length scales, as most spherical shell simulations currently seem to suggest, or whether planetary-scale convection cells dominate the core flow instead, as suggested by more classical ideas and by the recent results of Sakuraba and Roberts (2009) for a mantle-controlled, uniform heat flux leaving the core. As the results of Kageyama et al. (2008) illustrate, it is also questionable if the vortex columns observed in most simulations and the typical dynamo mechanism caused by them remain dominant features as more realistic Ekman and Rayleigh numbers are reached. It is hoped that the next generation of dynamo models will finally enable us to resolve these open issues.

4 Conclusion

The ingredients of modern numerical dynamo models that so successfully reproduce many features of the geomagnetic field have been outlined. These models seem to correctly capture important aspects of the fundamental dynamics and very robustly produce dipole-dominated fields with Earth-like Elsasser numbers and magnetic Reynolds numbers over a wide range of parameters. Even Earth-like reversals are found when the Rayleigh number is chosen higher enough to guarantee a sufficiently large impact of inertia. Simple scaling laws allow to directly connect simulations with Ekman numbers as large as $E = 10^{-3}$ to the geodynamo at $E = 10^{-15}$, to the dynamos of other planets in the solar system, and even to the dynamo in fast-rotating stars (Christensen and Aubert 2006; Olson and Christensen 2006; Takahashi et al. 2008a; Christensen et al. 2009). All this strongly suggest that the models get the basic dynamo process right.

Lower Ekman number simulations certainly do a better job in modeling the small-scale turbulent flow in the dynamo regions but is the associated excessive increase in numerical costs really warranted for modeling dynamo action? A closer comparison with the geomagnetic field suggests that some features indeed become more Earth-like. Reversing dynamos are inevitable too little "dipolar" at Ekman numbers $E \geq 3\times10^{-4}$. At $E = 3\times10^{-5}$, however, the relative importance of the dipole contribution remains compatible with the historic magnetic field even at Rayleigh numbers where reversals are expected. Torsional oscillations, which may play an important role in the dynamics on decadal times scales, only start to become significant at Ekman numbers $E \leq 3\times10^{-5}$ (Wicht and Christensen 2009). Their time scale still remains too slow at $E = 3\times10^{-6}$ and a further decrease in Ekman number will likely improve matters here. Low Ekman number simulation therefore seem in order to faithfully model the dynamics on decadal time scale where satellite data continue to provide the most reliable geomagnetic field models.

Somewhat contrary to larger Ekman number solutions, the lower Ekman number cases also tend to produce rather simple fields with too small dipole tilts and too little field complexity at low latitudes. The situation, however, changes when the Rayleigh number is increased to the magnitude required for reversals. Here, the low latitude field seems to be even more Earth-like than for simulations at larger Ekman numbers. The need for a combination of low Ekman numbers and large Rayleigh numbers is bad news for the required costs of the respective numerical simulations.

The drastic regime change which marks the transition from the weak field branch to the strong field branch in cartesian magneto-convection problems has never been found in spherical dynamo simulations. Takahashi et al. (2008a) find only a mild 20% increase of the mean flow scale ℓ_u when comparing purely convective and dynamo simulations at $E = 2\times10^{-6}$. Whether the effect may become more significant at smaller Ekman numbers remains an open question.

Two other recent publications suggest that low Ekman number dynamo models may behave even more differently than the limited analysis conducted here revealed. The reason for the nondipolar field and particular convection pattern found by Kageyama et al. (2008) at $E = 10^{-6}$ remains to be clarified. Likewise, the very interesting influence of the thermal boundary condition found by Sakuraba and Roberts (2009) at $E = 10^{-6}$ needs to be explored further.

Dynamos at Ekman numbers larger than say $E = 10^{-5}$ will continue to be the focus of numerical dynamo simulations that aim at understanding the fundamental dynamics and mechanisms. For exploring the long-term behavior relatively large Ekman numbers will remain a necessity. These should be supplemented by increased efforts to explore lower Ekman number models which show interesting distinct features that remain little understood.

The use of boundary conditions that implement the laterally varying heat flux at Earth's core-mantle boundary is another example how dynamo modelers try to improve their numerical codes. The heat flux can influence the reversal behavior (Glatzmaier et al. 1999), helps to model some aspects of the long-term geomagnetic field, and even has the potential to explain a seismic anisotropy of Earth's inner core (Aubert et al. 2008a).

Modern dynamo simulations have already proven useful for exploring the dynamics of planetary interiors. Model refinements and increasing numerical power will further increase their applicability in the future. Several space missions promise to map the interior magnetic fields of Earth (ESA's Swarm mission), Mercury (NASA's MESSENGER and ESA's BepiColombo missions), and Jupiter (NASA's Juno mission) with previously unknown precision. High-end numerical dynamo simulations will be indispensable for translating these measurements into interior properties and dynamical processes.

Acknowledgments

Johannes Wicht thanks Uli Christensen for useful discussions and hints.

References

Amit H, Olson P (2006) Time-averaged and time dependent parts of core flow. Phys Earth Planet Inter 155:120–139

Amit H, Olson P (2008) Geomagnetic dipole tilt changes induced by core flow. Phys Earth Planet Inter 166:226–238

Aubert J, Wicht J (2004) Axial versus equatorial dynamo models with implications for planetary magnetic fields. Earth Planet Sci Lett 221: 409–419

Aubert J, Amit H, Hulot G (2007) Detecting thermal boundary control in surface flows from numerical dynamos. Phys Earth Planet Inter 160:143–156

Aubert J, Amit H, Hulot G, Olson P (2008a) Thermochemical flows couple the Earth's inner core growth to mantle heterogeneity. Nature 454: 758–761

Aubert J, Aurnou J, Wicht J (2008b) The magnetic structure of convection-driven numerical dynamos. Geophys J Int 172:945–956

Bloxham J, Zatman S, Dumberry M (2002) The origin of geomagnetic jerks. Nature 420:65–68

Braginsky S (1970) Torsional magnetohydrodynamic vibrations in the Earth's core and variation in day length. Geomag Aeron 10:1–8

Braginsky S, Roberts P (1995) Equations governing convection in Earth's core and the geodynamo. Geophys Astrophys Fluid Dyn 79: 1–97

Breuer M, Wesseling S, Schmalzl J, Hansen U (2002) Effect of inertia in Rayleigh-B'enard convection. Phys Rev E 69:026320/1–10

Bullard EC, Gellman H (1954) Homogeneous dynamos and terrestrial magnetism. Proc R Soc Lond A A 247:213–278

Busse FH (2002) Is low Rayleigh number convection possible in the Earth's core? Geophys Res Lett 29:070000–1

Busse FH, Simitev R (2005a) Convection in rotating spherical shells and its dynamo states. In: Soward AM, Jones CA, Hughes DW, Weiss NO (eds) Fluid dynamics and dynamos in astrophysics and geophysics. CRC Press, Boca Raton, FL, pp 359–392

Busse FH, Simitev R (2005b) Dynamos driven by convection in rotating spherical shells. Atronom Nachr 326: 231–240

Busse FH, Simitev RD (2006) Parameter dependences of convection-driven dynamos in rotating spherical fluid shells. Geophys Astrophys Fluid Dyn 100:341–361

Carlut J, Courtillot V (1998) How complex is the time-averaged geomagnetic field over the past 5 Myr? Geophys J Int 134:527–544

Chan K, Li L, Liao X (2006) Modelling the core convection using finite element and finite difference methods. Phys Earth Planet Inter 157:124–138

Chandrasekhar S (1961) Hydrodynamic and hydromagnetic stability. Clarendon, Oxford

Christensen U, Aubert J (2006) Scaling properties of convection-driven dynamos in rotating spherical shells and applications to planetary magnetic fields. Geophys J Int 116:97–114

Christensen U, Olson P (2003) Secular variation in numerical geodynamo models with lateral variations of boundary heat flow. Phys Earth Planet Inter 138:39–54

Christensen U, Tilgner A (2004) Power requirement of the geodynamo from Ohmic losses in numerical and laboratory dynamos. Nature 429:169—171

Christensen U, Wicht J (2007) Numerical dynamo simulations. In: Olson P (ed) Treatise on Geophysics, vol 8 (Core dynamics). Elsevier, New York, pp 245–282

Christensen UR (2002) Zonal flow driven by strongly supercritical convection in rotating spherical shells. Numerical Dynamo Simulations "Publisher:" Elsevier, New York J Fluid Mech 470:115–133

Christensen UR (2006) A deep rooted dynamo for Mercury. Nature 444:1056–1058

Christensen UR (2010) Dynamo Scaling Laws and Applications to the Planets. *accepted for publication* at Space. Sci Rev

Christensen UR, Aubert J, Busse FH et al (2001) A numerical dynamo benchmark. Phys Earth Planet Inter 128:25–34

Christensen UR, Holzwarth V, Reiners A (2009) Energy flux determines magnetic field strength of planets and stars. Nature 457:167–169

Clement B (2004) Dependency of the duration of geomagnetic polarity reversals on site latitude. Nature 428:637–640

Clune T, Eliott J, Miesch M, Toomre J, Glatzmaier G (1999) Computational aspects of a code to study rotating turbulent convection in spherical shells. Parallel Comp 25:361–380

Coe R, Hongre L, Glatzmaier A (2000) An examination of simulated geomagnetic reversals from a paleomagnetic perspective. Phil Trans R Soc Lond A358:1141–1170

Constable C (2000) On the rate of occurence of geomagnetic reversals. Phys Earth Planet Inter 118:181–193

Cowling T (1957) The dynamo maintainance of steady magnetic fields. Quart J Mech App Math 10:129–136

Dormy E, Cardin P, Jault D (1998) Mhd flow in a slightly differentially rotating spherical shell, with conducting inner core, in a dipolar magnetic field. Earth Planet Sci Lett 158:15–24

Eltayeb I (1972) Hydromagnetic convection in a rapidly rotating fluid layer. Proc R Soc Lond A 326:229–254

Eltayeb I (1975) Overstable hydromagnetic convection in a rotating fluid layer. J Fluid Mech 71:161–179

Fearn D (1979) Thermal and magnetic instabilities in a rapidly rotating fluid sphere. Geophys Astrophys Fluid Dyn 14:103–126

Fournier A, Bunge H-P, Hollerbach R, Vilotte J-P (2005) A Fourier-spectral element algorithm for thermal convection in rotating axisymmetric containers. J Comp Phys 204:462–489

Glatzmaier G (1984) Numerical simulation of stellar convective dynamos. 1. The model and methods. J Comput Phys 55:461–484

Glatzmaier G (2002) Geodynamo simulations — how realistic are they? Ann Rev Earth Planet Sci 30:237–257

Glatzmaier G, Coe R (2007) Magnetic Polarity Reversals in the Core. In: Olson P (ed) Treatise on Geophysics, Vol 8, (Core dynamics). Elsevier, New York, pp 283–297

Glatzmaier G, Roberts P (1995) A three-dimensional convective dynamo solution with rotating and finitely conducting inner core and mantle. Phys Earth Planet Inter 91:63–75

Glatzmaier G, Roberts P (1996) An anelastic evolutionary geodynamo simulation driven by compositional and thermal convection. Physica D 97:81–94

Glatzmaier G, Coe R, Hongre L, Roberts P (1999) The role of the Earth's mantle in controlling the frequency of geomagnetic reversals. Nature 401:885–890

Gubbins D (2001) The Rayleigh number for convection in the Earth's core. Phys Earth Planet Inter 128:3–12

Gubbins D, Kelly P (1993) Persistent patterns in the geomagnetic field over the past 2.5 ma. Nature 365:829–832

Gubbins D, Love J (1998) Preferred vgp paths during geomagnetic polarity reversals: Symmetry considerations. Geophys Res Lett 25: 1079–1082

Gubbins D, Willis AP, Sreenivasan B (2007) Correlation of Earth's magnetic field with lower mantle thermal and seismic structure. Phys Earth Planet Inter 162:256–260

Harder H, Hansen U (2005) A finite-volume solution method for thermal convection and dynamo problems in spherical shells. Geophys J Int 161:522–532

Heimpel M, Aurnou J, Wicht J (2005) Simulation of equatorial and high-latitude jets on Jupiter in a deep convection model. Nature 438:193–196

Hejda P, Reshetnyak M (2003) Control volume method for the dynamo problem in the sphere with the free rotating inner core. Stud Geophys Geod 47:147–159

Hejda P, Reshetnyak M (2004) Control volume method for the thermal convection problem in a rotating spherical shell: test on the benchmark solution. Stud Geophys Geod, 48, 741–746

Hongre L, Hulot G, Khokholov A (1998) An analysis of the geomangetic field over the past 2000 years. Phys Earth Planet Inter 106:311–335

Hulot G, Bouligand C (2005) Statistical paleomagnetic field modelling and symmetry considerations. Geophys J Int 161, doi:10.1111/j.1365-246X.2005.02612.

Hulot G, Finlay C, Constable C, Olsen N, Mandea M (2010) The Magnetic Field of Planet Earth. *accepted for publication* at Space. Sci Rev

Isakov A, Descombes S, Dormy E (2004) An integro-differential formulation of magnet induction in bounded domains:boundary element-finite volume method. J Comp Phys 197:540–554

Ivers D, James R (1984) Axisymmetric antidynamo theorems in non-uniform compressible fluids. Phil Trans R Soc Lond A 312:179–218

Jackson A (1997) Time dependence of geostrophic core-surface motions. Phys Earth Planet Inter 103:293–311

Jackson A (2003) Intense equatorial flux spots on the surface of the Earth's core. Nature 424:760–763

Jackson A, Finlay C (2007) Geomagnetic secular variation and applications to the core. In: Kono M (ed) Treatise on geophysics, vol 5, (Geomagnetism). Elsevier, New York, pp 147–193

Jackson A, Jonkers A, Walker M (2000) Four centuries of geomagnetic secular variation from historical records. Phil Trans R Soc Lond A 358:957–990

Jault D (2003) Electromagnetic and topographic coupling, and LOD variations. In: Jones CA, Soward AM, Zhang K (eds) Earth's core and lower mantle. Taylor & Francis, London, pp 56–76

Jault D, Gire C, LeMouël J-L (1988) Westward drift, core motion and exchanges of angular momentum between core and mantle. Nature 333:353–356

Johnson C, Constable C (1995) Time averaged geomagnetic field as recorded by lava flows over the past 5 Myr. Geophys J Int 122:489–519

Johnson C, Constable C, Tauxe L (2003) Mapping long-term changed in Earth's magnetic field. Science 300:2044–2045

Johnson CL, McFadden P (2007) Time-averaged field and paleosecular variation. In: Kono M (ed) Treatise on geophysics, vol 5, (Geomagnetism). Elsevier, New York, pp 217–254

Jones C, Roberts P (2000) The onset of magnetoconvection at large Prandtl number in a rotating layer II. Small magnetic diffusion. Geophys Astrophys Fluid Dyn 93:173–226

Jonkers A (2003) Long-range dependence in the cenozoic reversal record. Phys Earth Planet Inter 135:253–266

Kageyama A, Sato T (1995) Computer simulation of a magnetohydrodynamic dynamo. II. Phys Plasmas 2:1421–1431

Kageyama A, Sato T (1997) Generation mechanism of a dipole field by a magnetohydrodynamic dynamo. Phys Rev E 55:4617–4626

Kageyama A, Watanabe K, Sato T (1993) Simulation study of a magnetohydrodynamic dynamo: Convection in a rotating shell. Phys Fluids B 24:2793–2806

Kageyama A, Miyagoshi T, Sato T (2008) Formation of current coils in geodynamo simulations. Nature 454:1106–1109

Kageyama A, Yoshida M (2005) Geodynamo and mantle convection simulations on the earth simulator using the yin-yang grid. J Phys Conf Ser 16:325–338

Kaiser R, Schmitt P, Busse F (1994) On the invisible dynamo. Geophys Astrophys Fluid Dyn 77:93–109

Kelly P, Gubbins D (1997) The geomagnetic field over the past 5 million years. Geophys J Int 128:315–330

Kono M, Roberts P (2002) Recent geodynamo simulations and observations of the geomagnetic field. Rev Geophys 40:1013, doi:10.1029/2000RG000102

Korte M, Constable C (2005) Continuous geomagnetic field models for the past 7 millennia: 2. cals7k. Geochem Geophys Geosys 6, Art No Q02H16

Korte M, Genevey A, Constable C, Frank U, Schnepp E (2005) Continuous geomagnetic field models for the past 7 millennia: 1. a new global data compilation. Geochem Geophys Geosys 6, Art No Q02H15

Kuang W, Bloxham J (1997) An Earth-like numerical dynamo model. Nature 389:371–374

Kuang W, Bloxham J (1999) Numerical modeling of magnetohydrodynamic convection in a rapidly rotating spherical shell: weak and strong field dynamo action. J Comp Phys 153:51–81

Kutzner C, Christensen U (2000) Effects of driving mechanisms in geodynamo models. Geophys Res Lett 27:29–32

Kutzner C, Christensen U (2002) From stable dipolar to reversing numerical dynamos. Phys Earth Planet Inter 131:29–45

Kutzner C, Christensen U (2004) Simulated geomagnetic reversals and preferred virtual geomagnetic pole paths. Geophys J Int 157: 1105–1118

Matsui H, Buffett B (2005) Sub-grid scale model for convection-driven dynamos in a rotating plane layer. submitted to Elsevier

Olsen N, Haagmans R, Sabaka TJ et al (2006) The Swarm End-to-End mission simulator study: A demonstration of separating the various contributions to Earth's magnetic field using synthetic data. Earth Planets Space 58:359–370

Olson P, Christensen U (2002) The time-averaged magnetic field in numerical dynamos with nonuniform boundary heat flow. Geophys J Int 151:809–823

Olson P, Christensen U (2006) Dipole moment scaling for convection-driven planetary dynamos. Earth Planet Sci Lett 250:561–571

Olson P, Christensen U, Glatzmaier G (1999) Numerical modeling of the geodynamo: Mechanism of field generation and equilibration. J Geophys Res 104:10,383–10,404

Roberts P (1972) Kinematic dynamo models. Phil Trans R Soc Lond A 271:663–697

Roberts P (2007) Theory of the geodynamo. In: Olson P (ed) Treatise on geophysics, vol 8, (Core dynamics). Elsevier, New York, pp 245–282

Roberts P, Jones C (2000) The onset of magneto-convection at large Prandtl number in a rotating layer I. Finite magnetic diffusion. Theory of the Geodynamo Geophs Astrophys Fluid Dyn 92:289–325

Rotvig J, Jones C (2002) Rotating convection-driven dynamos at low ekman number. Phys Rev E 66:056308

Ryan DA, Sarson GR (2007) Are geomagnetic field reversals controlled by turbulence within the Earth's core? Geophys Res Lett 34:2307

Sakuraba A (2002) Linear magnetoconvection in rotating fluid spheres permeated by a uniform axial magnetic field. Geophys Astrophys Fluid Dyn 96:291–318

Sakuraba A, Kono M (2000) Effect of a uniform magnetic field on nonlinear magnetocenvection in a rotating fluid spherical shell. Geophys Astrophys Fluid Dyn 92:255–287

Sakuraba A, Roberts P (2009) Generation of a strong magnetic field using uniform heat flux at the surface of the core. Nature Geosci 2:802–805

Schmalzl J, Breuer M, Hansen U (2002) The influence of the Prandtl number on the style of vigorous thermal convection. Geophys Astrophys Fluid Dyn 96:381–403

Simitev R, Busse F (2005) Prandtl-number dependence of convection-driven dynamos in rotating spherical fluid shells. J Fluid Mech 532:365–388

Simitev RD, Busse FH (2009) Bistability and hysteresis of dipolar dynamos generated by turbulent convection in rotating spherical shells. Europhys Lett 85:19001

Sreenivasan B, Jones CA (2006) The role of inertia in the evolution of spherical dynamos. Geophys J Int 164:467–476

St Pierre M (1993) The strong-field branch of the childress-soward dynamo. In Proctor MRE et al (eds) Solar and planetary dynamos, pp 329–337

Stanley S, Bloxham J (2004) Convective-region geometry as the cause of Uranus' and Neptune's unusual magnetic fields. Nature 428: 151–153

Stanley S, Glatzmaier G (2010) Dynamo models for planets other than earth. Space Science Reviews, DOI: 10.1007/s11214-009-9573-y, Only online so far.

Stanley S, Bloxham J, Hutchison W, Zuber M (2005) Thin shell dynamo models consistent with mercury's weak observed magnetic field. Earth Planet Sci Lett DOI: 10.1007/s11214-009-9573-y 234:341–353

Stellmach S, Hansen U (2004) Cartesian convection-driven dynamos at low ekman number. Phys Rev E 70:056312

Stieglitz R, Müller U (2001) Experimental demonstration of the homogeneous two-scale dynamo. Phys Fluids 1:561–564

Takahashi F, Matsushima M (2006) Dipolar and non-dipolar dynamos in a thin shell geometry with implications for the magnetic field of Mercury. Geophys Res Lett 33:L10202

Takahashi F, Matsushima M, Honkura Y (2008a) Scale variability in convection-driven MHD dynamos at low Ekman number. Phys Earth Planet Inter 167:168–178

Takahashi F, Tsunakawa H, Matsushima M, Mochizuki N, Honkura Y (2008b) Effects of thermally heterogeneous structure in the lowermost mantle on the geomagnetic field strength. Earth Planet Sci Lett 272:738–746

Taylor J (1963) The magneto-hydrodynamics of a rotating fluid and the Earth's dynamo problem. Proc R Soc Lond A 274:274–283

Tilgner A (1996) High-Rayleigh-number convection in spherical shells. Phys Rev E 53:4847–4851

Wicht J (2002) Inner-core conductivity in numerical dynamo simulations. Phys Earth Planet Inter 132:281–302

Wicht J (2005) Palaeomagnetic interpretation of dynamo simulations. Geophys J Int 162:371–380

Wicht J, Aubert J (2005) Dynamos in action. GWDG-Bericht 68:49–66

Wicht J, Christensen U (2010) Taylor state and torsional oscillations in numerical dynamo models. Geophys. J. Int. DOI: 10.1111/j.1365-246x.2010.04581.x, Published online only.

Wicht J, Olson P (2004) A detailed study of the polarity reversal mechanism in a numerical dynamo model. Geochem Geophys Geosyst 5, doi:10.1029/2003GC000602

Wicht J, Mandea M, Takahashi F et al (2007) The Origin of Mercury's Internal Magnetic Field. Space Sci Rev 132:261–290

Wicht J, Stellmach S, Harder H (2009) Numerical models of the geodynamo: From fundamental Cartesian models to 3d simulations of field reversals. In: Glassmeier K, Soffel H, Negendank J (eds) Geomagnetic field variations – Space–time structure, processes, and effects on system Earth. Springer, Berlin/Heidelberg/New York, pp 107–158

Wicht J, Tilgner A (2010) Theory and modeling of planetary dynamos. Space Science Review, DOI: 10.1007/s11214-010-9638-y, Published online only.

Willis AP, Sreenivasan B, Gubbins D (2007) Thermal core mantle interaction: Exploring regimes for locked dynamo action. Phys Earth Planet Inter 165:83–92 DOI:10.1007/s11214-010-9638-y

Zatman S, Bloxham J (1997) Torsional oscillations and the magnetic field within the Earth's core. Nature 388:760–761

Zhang K, Gubbins D (2000a) Is the geodynamo process intrinsically unstable? Geophys J Int 140:F1–F4

Zhang K, Gubbins D (2000b) Scale disparities and magnetohydrodynamics in the Earth's core. Phil Trans R Soc Lond A 358: 899–920

Zhang K, Schubert G (2000) Magnetohydrodynamics in rapidly rotating spherical systems. Ann Rev Fluid Mech 32:409–433

Zhang K-K, Busse F (1988) Finite amplitude convection and magnetic field generation in in a rotating spherical shell. Geophys Astrophys Fluid Dyn 44:33–53

17 Mathematical Properties Relevant to Geomagnetic Field Modeling

Terence J. Sabaka[1] · *Gauthier Hulot*[2] · *Nils Olsen*[3]
[1]Planetary Geodynamics Laboratory, Code 698, Greenbelt, MD, USA
[2]Equipe de Géomagnétisme, Institut de Physique du Globe de Paris, Institut de recherche associé au CNRS et *à* l'Université Paris 7, Paris, France
[3]Copenhagen, Denmark

W. Freeden, M.Z. Nashed, T. Sonar (Eds.), *Handbook of Geomathematics*, DOI 10.1007/978-3-642-01546-5_17,

Abstract Geomagnetic field modeling consists in converting large numbers of magnetic observations into a linear combination of elementary mathematical functions that best describes those observations. The set of numerical coefficients defining this linear combination is then what one refers to as a geomagnetic field model. Such models can be used to produce maps. More importantly, they form the basis for the geophysical interpretation of the geomagnetic field, by providing the possibility of separating fields produced by various sources and extrapolating those fields to places where they cannot be directly measured. In this chapter, the mathematical foundation of global (as opposed to regional) geomagnetic field modeling is reviewed, and the spatial modeling of the field in spherical coordinates is focussed. Time can be dealt with as an independent variable and is not explicitly considered. The relevant elementary mathematical functions are introduced, their properties are reviewed, and how they can be used to describe the magnetic field in a source-free (such as the Earth's neutral atmosphere) or source-dense (such as the ionosphere) environment is explained. Completeness and uniqueness properties of those spatial mathematical representations are also discussed, especially in view of providing a formal justification for the fact that geomagnetic field models can indeed be constructed from ground-based and satellite-born observations, provided those reasonably approximate the ideal situation where relevant components of the field can be assumed perfectly known on spherical surfaces or shells at the time for which the model is to be recovered.

1 Introduction

The magnetic field measured at or near the Earth's surface is the superposition of contributions from a variety of sources, as discussed in the accompanying chapter by Olsen et al. in this handbook (PART 2, ➲ Chap. 5). The sophisticated separation of the various fields produced by these sources on the basis of magnetic field observations is a major scientific challenge which requires the introduction of adequate mathematical representations of those fields (see, e.g., Hulot et al. 2007).

Here a synopsis of general properties relevant to such mathematical representations on a planetary scale is provided, which is the scale that is mainly dealt with here (regional representations of the field will only briefly be discussed, and the reader is referred to, e.g., Purucker and Whaler (2007) and to classical texts as Harrison (1987), Blakely (1995), Langel and Hinze (1998)). Such representations are of significant utility since they encode the physics of the magnetic field and allow for a means of deducing these fields from measurements via inverse theory. It should be noted that these representations are spatial in nature, i.e., they provide instantaneous descriptions of the fields. The physical cause of the time variability of such fields is a complicated subject and in its true sense, relies on the electromagnetic dynamics of the environment. Induced magnetism, for instance, is a well-understood subject which is covered in several good texts (e.g., Merrill et al. 1998) as is dynamo theory (cf. ➲ Chap. 16). However, while the physics is known, the incorporation of the dynamics into inverse problems is in its embryonic stages, especially for the Earth's core dynamo , and entails the subject of data assimilation, which is beyond the scope of this discussion. In what follows, time is thus considered as an independent implicit variable.

This chapter is divided into five parts. First, the general equations governing the behavior of any magnetic field are briefly recalled. Then naturally the concepts of *potential* and *non-potential* magnetic fields are introduced, the mathematical representation of which is discussed in the following two sections. Next a short section introduces the useful concept of spatial power spectra. The last section finally deals with uniqueness issues raised by the limited availability of magnetic observations; a topic of paramount importance when defining observational and modeling strategies to recover complete mathematical representations of the field. The chapter concludes with a few words with respect to the practical use of such mathematical representations, guiding the reader to further reading.

2 Helmholtz's Theorem and Maxwell's Equations

A convenient starting point for describing the spatial structure of any vector field is the *Helmholtz theorem* which states that if the divergence and curl of a vector field are known in a particular volume, as well as its normal component over the boundary of that volume, then the vector field is uniquely determined (e.g., Backus et al. 1996). When measurements of the Earth's magnetic field are taken they usually reflect some aspect of the magnetic induction vector **B**. Hence, statements about its divergence and curl will define many of its spatial properties.

Maxwell's equations then provide all the information necessary, sans boundary conditions, to describe the spatial behavior of both **B** and the electric field intensity **E**. They apply to electromagnetic phenomena in media which are at rest with respect to the coordinate system used.

$$\nabla \cdot \mathbf{E} = \frac{\rho_t}{\epsilon_0}, \tag{1}$$

$$\nabla \cdot \mathbf{B} = 0, \tag{2}$$

$$\nabla \times \mathbf{E} = -\frac{\partial \mathbf{B}}{\partial t}, \tag{3}$$

$$\nabla \times \mathbf{B} = \mu_0 \mathbf{J}. \tag{4}$$

In SI units **E** is expressed in V/m and **B** is expressed in *tesla* denoted T ($T = V \cdot s/m^2$). The remaining quantities are the total electric charge density ρ_t in C/m^3, the permittivity of free space $\epsilon_0 = 8.85 \times 10^{-12}\ F/m$, the permeability of free space $\mu_0 = 4\pi \times 10^{-7}\ H/m$, and the total volume current density **J** in A/m^2.

The total volume current density is actually the sum of volume densities from free charge currents $\mathbf{J}_f$, equivalent currents $\mathbf{J}_e$, and displacement currents $\frac{\partial \mathbf{D}}{\partial t}$ such that

$$\mathbf{J} = \mathbf{J}_f + \mathbf{J}_e + \frac{\partial \mathbf{D}}{\partial t}. \tag{5}$$

While $\mathbf{J}_f$ reflects steady free currents in nonmagnetic materials, $\mathbf{J}_e$ represents an equivalent current effect due to magnetized material whose local properties are described by the net dipole moment per unit volume **M** such that

$$\mathbf{J}_e = \nabla \times \mathbf{M}. \tag{6}$$

The displacement current density can be omitted if the time scales of interest are much longer than those required for light to traverse a typical length scale of interest (e.g., Backus et al. 1996). This criterion can be justified by the fact that, for instance, in a linear isotropic medium where $\mathbf{D} = \epsilon \mathbf{E}$ (ϵ is the permittivity of the medium) and $\mathbf{J}_f$ and $\mathbf{J}_e$ are absent, taking the curl of Eq. 4, substituting Eq. 3 and making use of Eq. 2 gives the homogeneous magnetic wave equation:

$$\frac{\partial^2 \mathbf{B}}{\partial t^2} = c_l^2 \nabla^2 \mathbf{B}, \tag{7}$$

where the wave propagates at the speed of light c_l. When time scales of interest are longer than those required for the light to traverse length scales of interest, the left-hand side of Eq. 7 can be neglected, amounting to neglect displacement currents in Eq. 4. Geomagnetism deals with frequencies up to only a few Hz and with length scales smaller than the radius of the Earth, which the light traverses in about 20 ms. Neglecting the displacement current is thus appropriate for most purposes in geomagnetism. Ampere's law is then given in a form which reflects the state of knowledge before Maxwell, thus yielding the *pre-Maxwell equations* where

$$\nabla \times \mathbf{B} = \mu_0 \mathbf{J}, \tag{8}$$

and

$$\mathbf{J} = \mathbf{J}_f + \mathbf{J}_e. \tag{9}$$

What is clear from Maxwell's equations is the absence of magnetic monopoles; magnetic field lines are closed loops. Indeed, because Eq. 2 is valid everywhere and under all circumstances, the magnetic field is solenoidal: its net flux through any closed surface must be zero. By contrast, Ampere's law (Eq. 8) shows that the magnetic field is irrotational only in the absence of free currents and magnetized material. It is this presence or absence of $\mathbf{J}$ that naturally divides magnetic field representations into two classes: *potential* and *non-potential* magnetic fields.

3 Potential Fields

3.1 Magnetic Fields in a Source-Free Shell

Let space be divided into three regions delineated by two shells of lower radius a and higher radius c so that regions I, II, and III are defined as $r \leq a$, $a < r < c$, and $c \leq r$. Imagine that current systems are confined to only regions I and III, and that one wants to describe the field in the source-free region II where $\mathbf{J} = 0$, such as for instance, the neutral atmosphere.

From Ampere's law $\nabla \times \mathbf{B} = 0$ in region II. It is well known that the magnetic field is then conservative and can be expressed as the gradient of a scalar potential V (Lorrain and Corson 1970; Jackson 1998; Backus et al. 1996):

$$\mathbf{B} = -\nabla V. \tag{10}$$

These types of fields are known as *potential* magnetic fields. In addition, from Eq. 2, $\mathbf{B}$ must still be solenoidal which implies

$$\nabla^2 V = 0. \tag{11}$$

i.e., that the potential V be *harmonic*. This is Laplace's equation, which can be solved for V in several different coordinated systems via separation of variables. The spherical system is most natural for the near-Earth environment and its coordinates are (r, θ, ϕ), where r is radial distance from the origin, θ is the polar angle rendered from the north polar axis (colatitude), and

ϕ is the azimuthal angle rendered in the equatorial plane from a prime meridian (longitude). The Laplacian operator may be written in spherical coordinates as

$$\nabla^2 = \frac{2}{r}\frac{\partial}{\partial r} + \frac{\partial^2}{\partial r^2} + \frac{1}{r^2 \sin\theta}\frac{\partial}{\partial\theta}\left(\sin\theta\frac{\partial}{\partial\theta}\right) + \frac{1}{r^2\sin^2\theta}\frac{\partial^2}{\partial\phi^2}. \tag{12}$$

Two linearly independent sets of solutions exist, depending on whether the source currents reside in either region I, V^i, or region III, V^e, i.e., are interior or exterior to the measurement shell, respectively. These solutions correspond to negative and positive powers of r, respectively, and are given by (e.g., Langel 1987; Backus et al. 1996)

$$V^i(r,\theta,\phi) = a\sum_{n=1}^{\infty}\left(\frac{a}{r}\right)^{n+1}\sum_{m=0}^{n}\left(g_n^m\cos m\phi + h_n^m\sin m\phi\right)P_n^m(\cos\theta), \tag{13}$$

$$V^e(r,\theta,\phi) = a\sum_{n=1}^{\infty}\left(\frac{r}{a}\right)^{n}\sum_{m=0}^{n}\left(q_n^m\cos m\phi + s_n^m\sin m\phi\right)P_n^m(\cos\theta), \tag{14}$$

which leads to

$$\mathbf{B}(r,\theta,\phi) = \mathbf{B}_i(r,\theta,\phi) + \mathbf{B}_e(r,\theta,\phi), \tag{15}$$

with

$$\mathbf{B}_i(r,\theta,\phi) = \sum_{n=1}^{\infty}\left(\frac{a}{r}\right)^{n+2}\sum_{m=0}^{n}\left(g_n^m\mathbf{\Pi}_{ni}^{mc}(\theta,\phi) + h_n^m\mathbf{\Pi}_{ni}^{ms}(\theta,\phi)\right), \tag{16}$$

$$\mathbf{B}_e(r,\theta,\phi) = \sum_{n=1}^{\infty}\left(\frac{r}{a}\right)^{n-1}\sum_{m=0}^{n}\left(q_n^m\mathbf{\Pi}_{ne}^{mc}(\theta,\phi) + s_n^m\mathbf{\Pi}_{ne}^{ms}(\theta,\phi)\right), \tag{17}$$

where

$$\mathbf{\Pi}_{ni}^{mc}(\theta,\phi) = (n+1)P_n^m(\cos\theta)\cos m\phi\,\hat{\mathbf{r}} - \frac{dP_n^m(\cos\theta)}{d\theta}\cos m\phi\,\hat{\theta} + \frac{m}{\sin\theta}P_n^m(\cos\theta)\sin m\phi\,\hat{\phi}, \tag{18}$$

$$\mathbf{\Pi}_{ni}^{ms}(\theta,\phi) = (n+1)P_n^m(\cos\theta)\sin m\phi\,\hat{\mathbf{r}} - \frac{dP_n^m(\cos\theta)}{d\theta}\sin m\phi\,\hat{\theta} - \frac{m}{\sin\theta}P_n^m(\cos\theta)\cos m\phi\,\hat{\phi}, \tag{19}$$

$$\mathbf{\Pi}_{ne}^{mc}(\theta,\phi) = -nP_n^m(\cos\theta)\cos m\phi\,\hat{\mathbf{r}} - \frac{dP_n^m(\cos\theta)}{d\theta}\cos m\phi\,\hat{\theta} + \frac{m}{\sin\theta}P_n^m(\cos\theta)\sin m\phi\,\hat{\phi}, \tag{20}$$

$$\mathbf{\Pi}_{ne}^{ms}(\theta,\phi) = -nP_n^m(\cos\theta)\sin m\phi\,\hat{\mathbf{r}} - \frac{dP_n^m(\cos\theta)}{d\theta}\sin m\phi\,\hat{\theta} - \frac{m}{\sin\theta}P_n^m(\cos\theta)\cos m\phi\,\hat{\phi}, \tag{21}$$

and $(\hat{\mathbf{r}},\hat{\theta},\hat{\phi})$ are the unit vectors associated with the spherical coordinates (r,θ,ϕ).

In these equations, a is the reference radius and in this case corresponds to the lower shell, $P_n^m(\cos\theta)$ is the *Schmidt quasi-normalized associated Legendre function* of degree n and order m (both being integers), and g_n^m and h_n^m are internal, and q_n^m and s_n^m external, constants known as the *Gauss coefficients*. Scaling these equations by a yields coefficients whose units are those of **B**, i.e., magnetic induction, which in the near-Earth environment is usually expressed in nanoteslas (nT). Note that although they formally appear in Eqs. 13–17, h_n^0 and s_n^0 Gauss coefficients are in fact not needed (because $\sin m\phi = 0$ when $m = 0$) and therefore not defined.

One should also notice the omission of the $n = 0$ terms in both solutions V^i and V^e (and therefore $\mathbf{B}_i$ and $\mathbf{B}_e$). This is a consequence of the fact that, for V^i, this term leads to an internal field $\mathbf{B}_i$ not satisfying magnetic monopole exclusion at the origin (required by Eq. 2), while for V^e, this term is just a constant (since $P_0^0(\cos\theta) = 1$, see below) which produces no field. It may thus be set to zero.

Finally, the reader should also be warned that the choice of the Schmidt quasi-normalization for the $P_n^m(\cos\theta)$ in Eqs. 13 and 14 is very specific to geomagnetism. It dates back to the early work of Schmidt (1935) and has since been adopted as the conventional norm to be used in geomagnetism, following a resolution of the *International Association of Terrestrial Magnetism and Electricity* (IATME) of the *International Union of Geophysics and Geodesy* (IUGG) (Goldie and Joyce 1940).

3.2 Surface Spherical Harmonics

The angular portions of the terms in Eqs. 13 and 14 are often denoted $Y_n^{m,c}(\theta,\phi) \equiv P_n^m(\cos\theta)\cos m\phi$ and $Y_n^{m,s}(\theta,\phi) \equiv P_n^m(\cos\theta)\sin m\phi$ and are known as *Schmidt quasi-normalized real surface spherical harmonics*. They are related to analogous complex functions $Y_{n,m}(\theta,\phi)$ known as *complex surface spherical harmonics*, which have been extensively studied in the mathematical and physical literature. For any couple of integers (n,m) with $n \geq 0$ and $-n \leq m \leq n$, those $Y_{n,m}(\theta,\phi)$ are defined as complex functions of the form (e.g. Edmonds 1996; Backus et al. 1996):

$$Y_{n,m}(\theta,\phi) = (-1)^m \sqrt{(2n+1)\frac{(n-m)!}{(n+m)!}}\, P_{n,m}(\cos\theta) e^{im\phi}, \tag{22}$$

where the $P_{n,m}(\cos\theta)$ are again the associate Legendre functions of degree n and order m, but now satisfying the much more common *Ferrers normalization* (Ferrers 1877), and defined by (with $\mu = \cos\theta$)

$$P_{n,m}(\mu) = \frac{1}{2^n n!}(1-\mu^2)^{\frac{m}{2}} \frac{d^{n+m}}{d\mu^{n+m}}(\mu^2-1)^n. \tag{23}$$

Note that this definition holds for $-n \leq m \leq n$ and leads to the important property that

$$P_{n,-m}(\mu) = (-1)^m \frac{(n-m)!}{(n+m)!} P_{n,m}(\mu). \tag{24}$$

which then also implies that

$$Y_{n,-m}(\theta,\phi) = (-1)^m \bar{Y}_{n,m}(\theta,\phi). \tag{25}$$

where the overbar denotes complex conjugation. The $Y_{n,m}(\theta,\phi)$ are eigenfunctions of the angular portion of the Laplacian operator ∇_S^2

$$\nabla_S^2 \equiv \frac{1}{\sin\theta}\frac{\partial}{\partial\theta}\left(\sin\theta\frac{\partial}{\partial\theta}\right) + \frac{1}{\sin^2\theta}\frac{\partial^2}{\partial\phi^2}, \tag{26}$$

such that

$$\nabla_S^2 Y_{n,m}(\theta,\phi) = -n(n+1)Y_{n,m}(\theta,\phi). \tag{27}$$

They represent a complete, orthogonal set of complex functions on the unit sphere.

The cumbersome prefactor to be found in Eq. 22 is chosen so that the inner products of these complex surface spherical harmonics over the sphere have the form:

$$\langle Y_{n,m}, Y_{l,k}\rangle \equiv \frac{1}{4\pi}\int_0^{2\pi}\int_0^{\pi} \bar{Y}_{n,m}(\theta,\phi) Y_{l,k}(\theta,\phi)\sin\theta\, d\theta\, d\phi = \delta_{nl}\delta_{mk}, \tag{28}$$

where δ_{ij} is the Kronecker delta. Those complex surface spherical harmonics are then said to be *fully normalized* . The reader should however be aware that not all authors introduce the $1/4\pi$ factor in the definition (Eq. 28) of the inner product, in which case fully normalized complex surface spherical harmonics are not exactly defined as in Eq. 22 (a $1/\sqrt{4\pi}$ factor then needs to be introduced). Here, the convention used in most previous books dealing with geomagnetism (such as, e.g., Merrill and McElhinny (1983); Langel (1987); Merrill et al. (1998), and especially Backus et al. (1996), where many useful mathematical properties satisfied by those functions can be found) is simply chosen. Similarly, the $(-1)^m$ factor in the definition of Eq. 22 is not always introduced.

The Schmidt quasi-normalized $P_n^m(\cos\theta)$ introduced in Eqs. 13 and 14 are then related to the Ferrers normalized $P_{n,m}(\cos\theta)$ through

$$P_n^m(\cos\theta) = \begin{cases} P_{n,m}(\cos\theta) & \text{for} \quad m = 0, \\ \sqrt{\dfrac{2(n-m)!}{(n+m)!}}\, P_{n,m}(\cos\theta) & \text{for} \quad m > 0. \end{cases} \tag{29}$$

and the Schmidt quasi-normalized real surface spherical harmonics $Y_n^{m,c}(\theta,\phi)$ and $Y_n^{m,s}(\theta,\phi)$ are related to the fully normalized complex surface spherical harmonics $Y_{n,m}(\theta,\phi)$ through

$$Y_n^{m,c}(\theta,\phi) = (-1)^m \sqrt{\frac{2}{(2n+1)(1+\delta_{m0})}}\, \Re[Y_{n,m}(\theta,\phi)], \tag{30}$$

$$Y_n^{m,s}(\theta,\phi) = (-1)^m \sqrt{\frac{2}{(2n+1)}}\, \Im[Y_{n,m}(\theta,\phi)], \tag{31}$$

for $n \geq 0$ and $0 \leq m \leq n$.

Note that whereas the $P_{n,m}(\cos\theta)$ are defined for $n \geq 0$ and $-n \leq m \leq n$, which are all needed for the $Y_{n,m}(\theta,\phi)$ to form a complete orthogonal set of complex functions on the unit sphere, the $P_n^m(\cos\theta)$ are only used for $n \geq 0$ and $0 \leq m \leq n$, which is then enough for the $Y_n^{m,c}(\theta,\phi)$ and $Y_n^{m,s}(\theta,\phi)$ to form a complete, orthogonal set of real functions on the unit sphere. They satisfy

$$\left\langle Y_n^{m,c}, Y_l^{k,c} \right\rangle = \left\langle Y_n^{m,s}, Y_l^{k,s} \right\rangle = \frac{1}{2n+1}\delta_{nl}\delta_{mk}, \tag{32}$$

and

$$\left\langle Y_n^{m,c}, Y_l^{k,s} \right\rangle = 0. \tag{33}$$

Finally, because of Eqs. 30 and 31, they too are eigenfunctions of ∇_S^2 and satisfy

$$\nabla_S^2 Y_n^{m,(c,s)}(\theta,\phi) = -n(n+1)\, Y_n^{m,(c,s)}(\theta,\phi). \tag{34}$$

where $Y_n^{m,(c,s)}(\theta,\phi)$ stands for either $Y_n^{m,c}(\theta,\phi)$ or $Y_n^{m,s}(\theta,\phi)$.

The following recursion relationships allow the Schmidt quasi-normalized series to be generated for a given value of θ (Langel 1987):

$$P_n^n(\cos\theta) = \sqrt{\frac{2n-1}{2n}}\, \sin\theta\, P_{n-1}^{n-1}(\cos\theta), \qquad n > 1, \tag{35}$$

$$P_n^m(\cos\theta) = \frac{2n-1}{\sqrt{n^2-m^2}} \cos\theta\, P_{n-1}^m(\cos\theta) - \sqrt{\frac{(n-1)^2-m^2}{n^2-m^2}}\, P_{n-2}^m(\cos\theta), \qquad n > m \geq 0. \tag{36}$$

The first few terms of the series are

$$P_0^0(\cos\theta) = 1, \qquad P_2^2(\cos\theta) = \frac{\sqrt{3}}{2}\sin^2\theta,$$

$$P_1^0(\cos\theta) = \cos\theta, \qquad P_3^0(\cos\theta) = \frac{1}{2}(5\cos^3\theta - 3\cos\theta),$$

$$P_1^1(\cos\theta) = \sin\theta, \qquad P_3^1(\cos\theta) = \frac{\sqrt{3}}{2\sqrt{2}}\sin\theta(5\cos^2\theta - 1),$$

$$P_2^0(\cos\theta) = \frac{1}{2}(3\cos^2\theta - 1), \qquad P_3^2(\cos\theta) = \frac{\sqrt{15}}{2}\sin^2\theta\cos\theta,$$

$$P_2^1(\cos\theta) = \sqrt{3}\cos\theta\sin\theta, \qquad P_3^3(\cos\theta) = \frac{\sqrt{5}}{2\sqrt{2}}\sin^3\theta$$

Plots of the $P_6^m(\cos\theta)$ functions are shown in ➋ *Fig. 1* as a function of θ.

Real surface spherical harmonic functions $Y_n^{m,c}(\theta,\phi)$ and $Y_n^{m,s}(\theta,\phi)$ have $n - m$ zeros in the interval $[0,\pi]$ along meridians and $2m$ zeros along lines of latitude. When $m = 0$, only the $Y_n^{0,c}(\theta,\phi)$ exist, which is then denoted $Y_n^0(\theta,\phi)$. They exhibit annulae of constant sign in longitude and are referred to as *zonal harmonics*. When $n = m$, there are lunes of constant sign in latitude and these are referred to as *sectorial harmonics*. The general cases are termed *tesseral*

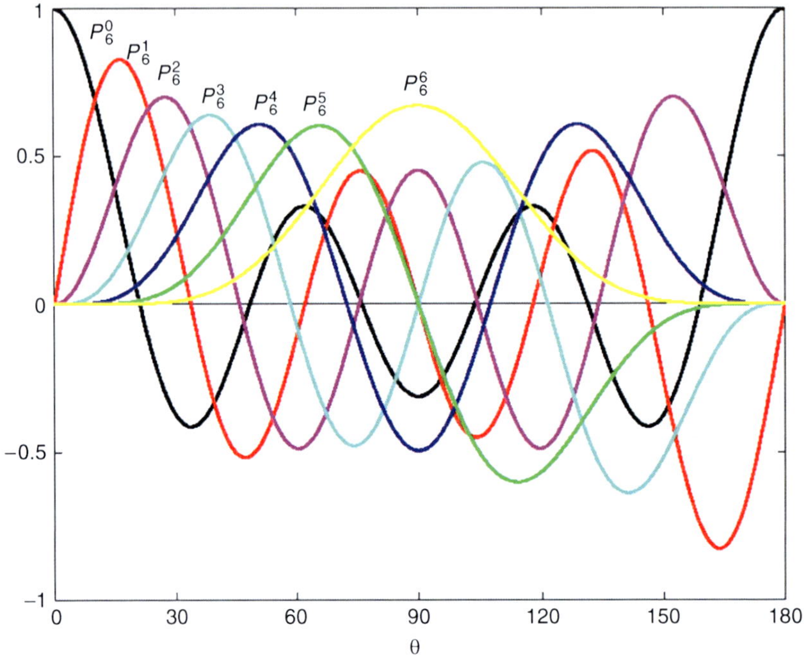

Fig. 1
Schmidt quasi-normalized associated Legendre functions $P_6^m(\cos\theta)$ as a function of θ

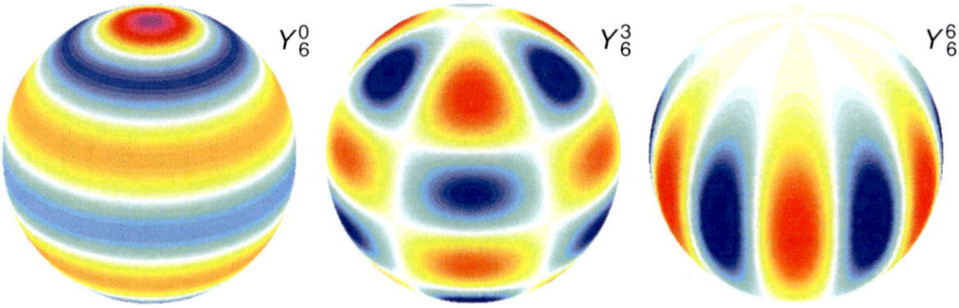

Fig. 2
Color representations (*red* positive, *blue* negative) of real surface harmonics Y_6^0, Y_6^3, and Y_6^6 (Note that $Y_n^{m,c}$ and $Y_n^{m,s}$ are identical to within a $\frac{\pi}{2m}$ longitudinal phase shift)

harmonics. Figure 2 illustrates examples of each from the $n = 6$ family. Finally, note that since $Y_{0,0} = P_{0,0} = P_0^0 = Y_0^0 = 1$, using either Eq. 28 or Eqs. 32–33 for $n = m = 0$ leads to the important additional property that

$$\frac{1}{4\pi}\int_0^{2\pi}\int_0^{\pi} Y_{n,m}(\theta,\phi)\sin\theta\,d\theta\,d\phi = \frac{1}{4\pi}\int_0^{2\pi}\int_0^{\pi} Y_n^m(\theta,\phi)\sin\theta\,d\theta\,d\phi = \delta_{n0}\delta_{m0}, \quad (37)$$

For many more useful properties satisfied by associate Legendre functions and real or complex surface spherical harmonics, the reader is referred to, e.g., Langel (1987), Backus et al. (1996), and Dahlen and Tromp (1998). But beware conventions and normalizations!

3.3 Magnetic Fields from a Spherical Sheet Current

Let space again be divided into three regions, but in a different way. Introduce a single shell of radius b (where $a < b < c$) so that regions I, II, and III are now defined as $r < b$, $r = b$, and $b < r$. This time imagine that the current system is confined to only region II (the spherical shell surface), and that one wants to describe the field produced by those sources in source-free regions I and III, where $\mathbf{J} = 0$. Here, sources previously assumed to lie either below $r = a$ or above $r = c$ are thus ignored and only the description of a field produced by a *spherical sheet current* is considered. This is very applicable to the Earth environment where currents associated with the ionospheric dynamo reside in the E-region and peak near 115 km (cf. Chap. 5).

In region III those sources are seen as internal, producing a field with a potential of the form of Eq. 13:

$$V^i(r,\theta,\phi) = a\sum_{n=1}^{\infty}\left(\frac{a}{r}\right)^{n+1}\sum_{m=0}^{n}\left(a_n^m\cos m\phi + b_n^m\sin m\phi\right)P_n^m(\cos\theta), \quad (38)$$

where a_n^m and b_n^m are now the constants. In region I, by contrast, those sources are seen as external, producing a field with a magnetic potential of the form of Eq. 14. However, because $\mathbf{B}$ is solenoidal its radial component must be continuous across the sheet so that

$$\left.\frac{\partial V^i}{\partial r}\right|_{r=b} = \left.\frac{\partial V^e}{\partial r}\right|_{r=b}. \quad (39)$$

Unlike the independent internal and external magnetic fields found in a shell sandwiched between source bearing regions, the fields in regions I and III here are not independent. They are now coupled through Eq. 39. This also means that the expansion coefficients for V^e are now related to those a_n^m and b_n^m for V^i (Sabaka et al. 2002; Granzow 1983):

$$V^e(r,\theta,\phi) = -a\sum_{n=1}^{\infty}\frac{n+1}{n}\left(\frac{a}{b}\right)^{2n+1}\left(\frac{r}{a}\right)^n\sum_{m=0}^{n}\left(a_n^m\cos m\phi + b_n^m\sin m\phi\right)P_n^m(\cos\theta). \tag{40}$$

Although the radial component of the field is the same just above and below the sheet current, the horizontal components are not. However, they are in the same vertical plane perpendicular to the sheet and if Ampere's circuital law (the integral form of Ampere's law) is applied to the area containing the sheet, then the surface current density is seen to be (e.g., Granzow 1983)

$$\mathbf{J}_s = \frac{1}{\mu_0}\hat{\mathbf{r}}\times(\mathbf{B}_e - \mathbf{B}_i), \tag{41}$$

which in component form is

$$\begin{pmatrix} J_{s\theta} \\ J_{s\phi} \end{pmatrix} = \begin{pmatrix} \frac{1}{\mu_0}\sum_{n=1}^{\infty}\frac{2n+1}{n}\left(\frac{a}{b}\right)^{n+2}\sum_{m=1}^{n}\frac{m}{\sin\theta}\left(-a_n^m\sin m\phi + b_n^m\cos m\phi\right)P_n^m(\cos\theta) \\ -\frac{1}{\mu_0}\sum_{n=1}^{\infty}\frac{2n+1}{n}\left(\frac{a}{b}\right)^{n+2}\sum_{m=0}^{n}\left(a_n^m\cos m\phi + b_n^m\sin m\phi\right)\frac{dP_n^m(\cos\theta)}{d\theta} \end{pmatrix}. \tag{42}$$

Finally, note that $\mathbf{J}_s$ can also be written in the often used form:

$$\mathbf{J}_s = -\hat{\mathbf{r}}\times\nabla\Psi. \tag{43}$$

where Ψ is known as the *sheet current function*. The SI unit for surface current density is A/m, while that of the sheet current function Ψ is A.

4 Non-Potential Fields

In the previous section, mathematical representations were developed for magnetic fields in regions where $\mathbf{J} = 0$. Satellite, surface, and near-surface surveys used in near-Earth magnetic field modeling do not sample the source regions of the core, crust, magnetosphere, or typically the ionospheric E-region. The representations developed so far can therefore be used to describe the fields produced by those sources in regions were observations are made. However, there are additional currents which couple the magnetosphere with the ionosphere in the F-region where satellite measurements are commonly made (cf. ➋ Chap. 5). These additional currents need to be considered. In this section, the condition $\nabla\times\mathbf{B} = 0$ is therefore relaxed. This leads to field forms which are described by two scalar potentials rather than one. While it is true that if $\nabla\cdot\mathbf{B} = 0$, then there exists a vector potential, $\mathbf{A}$, such that $\nabla\times\mathbf{A} = \mathbf{B}$, the magnetic fields of this section, which cannot be written in the form of the gradient of a scalar potential (as in Eq. 10), will be referred to as *nonpotential* fields.

4.1 Helmholtz Representations and Vector Spherical Harmonics

To begin with, consider a very general vector field $\mathbf{F}$. Recall that the Helmholtz theorem then states that if the divergence and curl of $\mathbf{F}$ are known in a particular volume, as well as its normal component over the boundary of that volume, then **F** is uniquely determined. This is also true if **F** decays to zero as $r \to \infty$ (for a rigorous proof and statement, see, e.g., Blakely 1995; Backus et al. 1996). In addition, **F** can then always be written in the form:

$$\mathbf{F} = \nabla S + \nabla \times \mathbf{A}, \tag{44}$$

where S and **A** are however then not uniquely determined. In fact this degree of freedom further makes it possible to choose **A** of the form $\mathbf{A} = T\mathbf{r} + \nabla \times P\mathbf{r}$, so that **F** can also be written as (e.g., Stern 1976)

$$\mathbf{F} = \nabla S + \nabla \times T\mathbf{r} + \nabla \times \nabla \times P\mathbf{r}, \tag{45}$$

This representation has the advantage that if **F** satisfies the vector Helmholtz equation:

$$\nabla^2 \mathbf{F} + k^2 \mathbf{F} = 0, \tag{46}$$

then the three scalar potentials S, T, and P also satisfy the associated scalar Helmholtz equation:

$$\nabla^2 Q + k^2 Q = 0 \tag{47}$$

where Q is either S, T, or P.

The solution to this scalar Helmholtz equation can be achieved through separation of variables in spherical coordinates. Expanding the angular dependence of Q in terms of real surface spherical harmonics, leads to

$$Q(r,\theta,\phi) = \sum_{n=0}^{\infty} \sum_{m=0}^{n} \left(Q_n^{m,c}(r) Y_n^{m,c}(\theta,\phi) + Q_n^{m,s}(r) Y_n^{m,s}(\theta,\phi) \right), \tag{48}$$

and Eq. 47 then implies

$$\left[\frac{d^2}{dr^2} + \frac{2}{r}\frac{d}{dr} + k^2 - \frac{n(n+1)}{r^2} \right] Q_n^{m(c,s)}(r) = 0, \tag{49}$$

which depends only on n. This may be further transformed into Bessel's equation such that the solution of Eq. 49 can be written in the form:

$$Q_n^{m(c,s)}(r) = C_n^{m(c,s)} j_n(kr) + D_n^{m(c,s)} n_n(kr), \tag{50}$$

where $j_n(kr)$ and $n_n(kr)$ are spherical Bessel functions of the first and second kind, respectively (Abramowitz and Stegun 1964), and $C_n^{m(c,s)}$ and $D_n^{m(c,s)}$ are constants. If $k \to 0$, then it can be shown that $j_n(kr)$ and $n_n(kr)$ approach the $r^{-(n+1)}$ internal and r^n external potential forms, respectively (Granzow 1983). This is consistent with the fact that, as seen in the previous section, if the field **F** is potential, Eq. 45 reduces to Eq. 10, and if the corresponding potential is harmonic, Eq. 47 reduces to Eq. 11, in which case it can be expanded in the form of Eqs. 13 and 14.

More generally, if the vector field **F** is to be defined within the spherical shell $a < r < c$, the general solutions (Eq. 50) of Eq. 49 for nonzero k can be used for the purpose of writing expansions of S, T, P, and therefore **F**, in terms of elementary fields. One can then indeed take advantage of the fact that any well-behaved function defined on (a, c) can always be expanded in terms of a sum (over i, or an integral if $c \to \infty$) of spherical Bessel functions of the first kind $j_n(k_i r)$ where $k_i = \frac{x_i}{c}$ and the x_i are the positive roots of $j_n(x) = 0$ (Watson 1966; Granzow

1983). It thus follows that in the spherical shell of interest, any scalar function $Q(r,\theta,\phi)$ (and, in particular, S, T, and P) can be written in the form of the following expansion:

$$Q(r,\theta,\phi) = \sum_{i=1}^{I}\sum_{n=0}^{\infty} j_n(k_i r) \sum_{m=0}^{n} \left(Q_{n,i}^{m,c} Y_n^{m,c}(\theta,\phi) + Q_{n,i}^{m,s} Y_n^{m,s}(\theta,\phi)\right). \tag{51}$$

Using such expansions for S, T, and P in Eq. 45 then provides an expansion of any vector field $\mathbf{F}$ within the spherical shell $a < r < c$ in terms of elementary vector fields, where advantage can be taken of the fact that each $j_n(k_i r) Y_n^{m,c}(\theta,\phi)$ and $j_n(k_i r) Y_n^{m,s}(\theta,\phi)$ will satisfy Eq. 47, with $k = k_i$. More details about how this can be applied to describe any nonpotential magnetic field can be found in Granzow (1983).

Rather than using Eq. 45, one may also use the alternative *Helmholtz representation* (e.g., Backus et al. 1996; Dahlen and Tromp 1998):

$$\mathbf{F} = U\hat{\mathbf{r}} + \nabla_S V - \hat{\mathbf{r}} \times \nabla_S W, \tag{52}$$

where the angular portion

$$\nabla_S = r\nabla - \mathbf{r}\frac{\partial}{\partial r} \tag{53}$$

of the ∇ operator has been introduced (note that $\nabla_S \cdot \nabla_S = \nabla_S^2$ as defined by Eq. 26). Equation 52 amounts to decomposing $\mathbf{F}$ in terms of a purely radial vector field $U\hat{\mathbf{r}}$ and a purely tangent (to the sphere) vector field $\nabla_S V - \hat{\mathbf{r}} \times \nabla_S W$. Still considering $\mathbf{F}$ to be a general (well-behaved) vector field defined within the spherical shell $a < r < c$, this representation can be shown to be unique, provided one requires that for any value of r within the shell, the average values of V and W over the sphere of radius r (denoted $\langle\rangle_{S(r)}$) is such that

$$\langle V \rangle_{S(r)} = \langle W \rangle_{S(r)} = 0. \tag{54}$$

Of course, each scalar function U, V, W can then also be expanded in terms of real surface spherical harmonics of the form:

$$U(r,\theta,\phi) = \sum_{n=0}^{\infty}\sum_{m=0}^{n} \left(U_n^{m,c}(r) Y_n^{m,c}(\theta,\phi) + U_n^{m,s}(r) Y_n^{m,s}(\theta,\phi)\right), \tag{55}$$

and similarly for V and W (for which the $n = 0$ term must however be set to zero because of Eq. 54). This then has the advantage that in implementing Eq. 55 and the equivalent expansions for V and W in Eq. 52, the radial dependence of each $U_n^{m,c}(r)$, $U_n^{m,s}(r)$, etc., is not affected by the ∇_S operator. This leads to

$$\begin{aligned} \mathbf{F}(r,\theta,\phi) = & \sum_{n=0}^{\infty}\sum_{m=0}^{n} \left(U_n^{m,c}(r)\mathbf{P}_n^{m,c}(\theta,\phi) + U_n^{m,s}(r)\mathbf{P}_n^{m,s}(\theta,\phi)\right) \\ & + \sum_{n=1}^{\infty}\sum_{m=0}^{n} \left(V_n^{m,c}(r)\mathbf{B}_n^{m,c}(\theta,\phi) + V_n^{m,s}(r)\mathbf{B}_n^{m,s}(\theta,\phi)\right) \\ & + \sum_{n=1}^{\infty}\sum_{m=0}^{n} \left(W_n^{m,c}(r)\mathbf{C}_n^{m,c}(\theta,\phi) + W_n^{m,s}(r)\mathbf{C}_n^{m,s}(\theta,\phi)\right), \end{aligned} \tag{56}$$

where the various $U_n^{m,(c,s)}(r)$, $V_n^{m,(c,s)}(r)$, and $W_n^{m,(c,s)}(r)$ functions can independently be expanded with the help of any relevant representation, and the following elementary vector functions have been introduced

$$\mathbf{P}_n^{m,(c,s)}(\theta,\phi) = Y_n^{m,(c,s)}(\theta,\phi)\hat{\mathbf{r}}, \tag{57}$$

$$\mathbf{B}_n^{m,(c,s)}(\theta,\phi) = \nabla_S Y_n^{m,(c,s)}(\theta,\phi), \tag{58}$$

$$\mathbf{C}_n^{m,(c,s)}(\theta,\phi) = -\hat{\mathbf{r}} \times \nabla_S Y_n^{m,(c,s)}(\theta,\phi) = \nabla \times Y_n^{m,(c,s)}(\theta,\phi)\mathbf{r}. \tag{59}$$

These are known as *real vector spherical harmonics* (see, e.g., Dahlen and Tromp 1998). Just like the surface spherical harmonics from which they derive, those vectors can also be introduced in complex form and with various norms (see, e.g., Morse and Feshbach 1953; Stern 1976; Granzow 1983; Jackson 1998; Dahlen and Tromp 1998). Here, for simplicity, real quantities and Schmidt quasi-normalizations are considered. Introducing the following inner product for two real vector fields **K** and **L** defined on the unit sphere:

$$\langle \mathbf{K}, \mathbf{L} \rangle \equiv \frac{1}{4\pi} \int_0^{2\pi} \int_0^{\pi} \mathbf{K}(\theta,\phi) \cdot \mathbf{L}(\theta,\phi) \sin\theta \, d\theta \, d\phi, \tag{60}$$

it can be shown that (see, e.g., Dahlen and Tromp 1998)

$$\left\langle \mathbf{K}_n^{m,(c,s)}, \mathbf{L}_l^{k,(c,s)} \right\rangle = 0, \tag{61}$$

as soon as $\mathbf{K}_n^{m,(c,s)}$ and $\mathbf{L}_l^{k,(c,s)}$ are not strictly identical real vector spherical harmonics as defined by Eqs. 57–59. For each $\mathbf{P}_n^{m,(c,s)}$, $\mathbf{B}_n^{m,(c,s)}$, and $\mathbf{C}_n^{m,(c,s)}$, it can further be shown that (see again, e.g., Dahlen and Tromp 1998, but beware the choice of inner product (Eq. 60), and normalization (Eq. 32)):

$$\left\langle \mathbf{P}_n^{m,c}, \mathbf{P}_n^{m,c} \right\rangle = \left\langle \mathbf{P}_n^{m,s}, \mathbf{P}_n^{m,s} \right\rangle = \frac{1}{2n+1}, \tag{62}$$

and

$$\left\langle \mathbf{B}_n^{m,c}, \mathbf{B}_n^{m,c} \right\rangle = \left\langle \mathbf{B}_n^{m,s}, \mathbf{B}_n^{m,s} \right\rangle = \left\langle \mathbf{C}_n^{m,c}, \mathbf{C}_n^{m,c} \right\rangle = \left\langle \mathbf{C}_n^{m,s}, \mathbf{C}_n^{m,s} \right\rangle = \frac{n(n+1)}{2n+1}. \tag{63}$$

Note the discrepancy by a factor of $n(n+1)$ between Eq. 62 and Eq. 63, which can be avoided by introducing an additional factor $(n(n+1))^{-1/2}$ in the right-hand side of the definitions Eqs. 58 and 59 (as is most often done in the literature). Those real vector spherical harmonics are thus mutually orthogonal within each family and between families with respect to this inner product. They provide a convenient general basis for expanding any vector field **F** in the spherical shell $a < r < c$, as is made explicit by Eq. 56.

Of course, many other basis can be built by making appropriate linear combinations of the $\mathbf{P}_n^{m,(c,s)}$, $\mathbf{B}_n^{m,(c,s)}$, and $\mathbf{C}_n^{m,(c,s)}$. Such linear combinations naturally arise when, for instance, implementing expansions of the type Eq. 51 for S, T, and P in Eq. 45. Also, it should be noted that other linear combinations of the kind have already been encountered when potential vector fields were considered (as described in ➋ Sect. 3). This indeed led to Eqs. 16 and 17, where the elementary vector functions $\mathbf{\Pi}_{ni}^{m,(c,s)}(\theta,\phi)$ and $\mathbf{\Pi}_{ne}^{m,(c,s)}(\theta,\phi)$ naturally arose. These can now be rewritten in the form:

$$\mathbf{\Pi}_{ni}^{m,(c,s)} = (n+1)\mathbf{P}_n^{m,(c,s)} - \mathbf{B}_n^{m,(c,s)}, \tag{64}$$

$$\mathbf{\Pi}_{ne}^{m,(c,s)} = -n\mathbf{P}_n^{m,(c,s)} - \mathbf{B}_n^{m,(c,s)}. \tag{65}$$

Together with the $\mathbf{C}_n^{m,(c,s)}$, they again form a general basis of mutually orthogonal vector fields (i.e. satisfying Eq. 61, as one can easily check). They also satisfy

$$\langle \mathbf{\Pi}_{ni}^{m,c}, \mathbf{\Pi}_{ni}^{m,c} \rangle = \langle \mathbf{\Pi}_{ni}^{m,s}, \mathbf{\Pi}_{ni}^{m,s} \rangle = n + 1, \tag{66}$$

$$\langle \mathbf{\Pi}_{ne}^{m,c}, \mathbf{\Pi}_{ne}^{m,c} \rangle = \langle \mathbf{\Pi}_{ne}^{m,s}, \mathbf{\Pi}_{ne}^{m,s} \rangle = n. \tag{67}$$

An interesting discussion of how this alternative basis can be used for the purpose of describing the Earth's magnetic field within a shell where sources with simplified geometry exist, can be found in Winch et al. (2005).

4.2 Mie Representation

Both Helmholtz representations, Eqs. 45 and 52, apply to any well-behaved vector field **F**. Another more restrictive representation can be derived if one requests the field **F** to be solenoidal, as the magnetic field is requested to be because of Eq. 2. Indeed, if a solenoidal field is written in the form of Eq. 45, then S must be harmonic, i.e., it must satisfy Laplace's equation (Eq.11). However, consider the following identity for the $\nabla \times \nabla \times$ operator in spherical coordinates

$$\nabla \times \nabla \times P\mathbf{r} = \nabla \left[\frac{\partial}{\partial r} (rP) \right] - \mathbf{r} \nabla^2 P. \tag{68}$$

This suggests that if an additional scalar potential of the form:

$$P_s = \frac{1}{r} \int S \, dr \tag{69}$$

is added to the original scalar potential P, then ∇S may be absorbed into Eq. 68. This is true since P_s is then also harmonic and

$$S = \frac{\partial}{\partial r} (rP_s). \tag{70}$$

What this means is that any solenoidal field such as **B** may be written as the curl of a vector potential, which is to say, the last two terms of Eq. 45. The condition of $\nabla \cdot \mathbf{B}$ has eliminated the need for one of the three original scalar potentials and leaves **B** in the form:

$$\mathbf{B} = \nabla \times T\mathbf{r} + \nabla \times \nabla \times P\mathbf{r}. \tag{71}$$

This representation can be shown to be unique provided one requests

$$\langle T \rangle_{S(r)} = \langle P \rangle_{S(r)} = 0, \tag{72}$$

for all values of r within the spherical shell of interest $a < r < c$. It is known as the *Mie representation* for a solenoidal vector (Mie 1908; Backus 1986; Backus et al. 1996; Dahlen and Tromp 1998).

The first term in Eq. 71 is known as the *toroidal* part of **B**, denoted as $\mathbf{B}_{tor}$, and T is known as the *toroidal scalar potential*

$$\mathbf{B}_{tor} = \nabla \times T\mathbf{r}, \tag{73}$$

$$= \frac{1}{\sin\theta} \frac{\partial T}{\partial \phi} \hat{\theta} - \frac{\partial T}{\partial \theta} \hat{\phi}. \tag{74}$$

This part has no radial component and its surface divergence is zero

$$\nabla_S \cdot \mathbf{B}_{tor} = 0, \tag{75}$$

where the surface divergence operates on a given vector field $\mathbf{F}$ as (recall the definition Eq. 53 of ∇_S):

$$\nabla_S \cdot \mathbf{F} = 2F_r + \frac{1}{\sin\theta}\left[\frac{\partial}{\partial\theta}(F_\theta \sin\theta) + \frac{\partial F_\phi}{\partial\phi}\right]. \tag{76}$$

The second term in Eq. 71 is known as the *poloidal* part of $\mathbf{B}$, denoted as $\mathbf{B}_{pol}$, and P is known as the *poloidal scalar potential*

$$\mathbf{B}_{pol} = \nabla \times \nabla \times P\mathbf{r}, \tag{77}$$

$$= -\frac{1}{r}\nabla_S^2 P\,\hat{\mathbf{r}} + \frac{1}{r}\frac{\partial}{\partial\theta}\left[\frac{\partial}{\partial r}(rP)\right]\hat{\theta} + \frac{1}{r\sin\theta}\frac{\partial}{\partial\phi}\left[\frac{\partial}{\partial r}(rP)\right]\hat{\phi}. \tag{78}$$

This part has a vanishing surface curl on the sphere, i.e., it satisfies

$$\hat{\mathbf{r}} \cdot (\nabla_S \times \mathbf{B}_{pol}) = \Lambda_S \cdot \mathbf{B}_{pol} = 0, \tag{79}$$

where the operator $\Lambda_S \equiv \hat{\mathbf{r}} \times \nabla_S$ known as the *surface curl* operator has been introduced (Backus et al. 1996; Dahlen and Tromp 1998). This operator satisfies for any vector field $\mathbf{F}$:

$$\Lambda_S \cdot \mathbf{F} = \frac{1}{\sin\theta}\left[\frac{\partial}{\partial\theta}\left(F_\phi \sin\theta\right) - \frac{\partial F_\theta}{\partial\phi}\right]. \tag{80}$$

The terms *toroidal* and *poloidal* were coined by Elsasser (1946) and so the Mie representation is sometimes also referred to as the *toroidal–poloidal decomposition* .

Just like for the Helmholtz representations, one can again expand the toroidal and poloidal scalar fields in terms of real spherical harmonics:

$$T(r,\theta,\phi) = \sum_{n=1}^{\infty}\sum_{m=0}^{n}\left(T_n^{m,c}(r)Y_n^{m,c}(\theta,\phi) + T_n^{m,s}(r)Y_n^{m,s}(\theta,\phi)\right), \tag{81}$$

$$P(r,\theta,\phi) = \sum_{n=1}^{\infty}\sum_{m=0}^{n}\left(P_n^{m,c}(r)Y_n^{m,c}(\theta,\phi) + P_n^{m,s}(r)Y_n^{m,s}(\theta,\phi)\right), \tag{82}$$

where the sum starts at $n = 1$, and not at $n = 0$, because of Eq. 72. This then leads to (recalling Eq. 59):

$$\mathbf{B}_{tor}(r,\theta,\phi) = \sum_{n=1}^{\infty}\sum_{m=0}^{n}\left(T_n^{m,c}(r)\mathbf{C}_n^{m,c}(\theta,\phi) + T_n^{m,s}(r)\mathbf{C}_n^{m,s}(\theta,\phi)\right), \tag{83}$$

and (recalling Eq. 57 and 58, and making use of Eq. 27):

$$\mathbf{B}_{pol}(r,\theta,\phi) = \sum_{n=1}^{\infty}\sum_{m=0}^{n} n(n+1)\left(P_n^{m,c}(r)\mathbf{P}_n^{m,c}(\theta,\phi) + P_n^{m,s}(r)\mathbf{P}_n^{m,s}(\theta,\phi)\right)$$

$$+\sum_{n=1}^{\infty}\sum_{m=0}^{n}\left(\frac{d}{dr}\left(rP_n^{m,c}(r)\right)\mathbf{B}_n^{m,c}(\theta,\phi) + \frac{d}{dr}\left(rP_n^{m,s}(r)\right)\mathbf{B}_n^{m,s}(\theta,\phi)\right), \tag{84}$$

which shows that $\mathbf{B}_{tor}$ and $\mathbf{B}_{pol}$ are expressible in terms of different families of vector spherical harmonics. As a result, it also follows that for any value of r within the shell of interest $a < r < c$,

$\mathbf{B}_{tor}$ and $\mathbf{B}_{pol}$ are orthogonal over the sphere of radius r, i.e., with respect to the inner product defined by Eq. 60:

$$\left\langle \mathbf{B}_{tor}(r), \mathbf{B}_{pol}(r) \right\rangle = 0. \tag{85}$$

4.3 Relationship of B and J Mie Representations

Recall from the pre-Maxwell form of Ampere's law that $\mathbf{J}$ is also the curl of a vector. This means that it also is a solenoidal field

$$\nabla \cdot \mathbf{J} = 0 \tag{86}$$

and so also possesses a toroidal–poloidal decomposition. Let T_B and P_B be the toroidal and poloidal scalars representing the magnetic field $\mathbf{B}$ and let T_J and P_J be the same for the volume current density $\mathbf{J}$. It follows from Ampere's law Eq. 8 and Eqs. 68 and 71 that

$$\mu_0 \mathbf{J} = \nabla \times \nabla \times T_B \mathbf{r} + \nabla \times \nabla \times \nabla \times P_B \mathbf{r}, \tag{87}$$

$$= \nabla \times \left(-\nabla^2 P_B\right) \mathbf{r} + \nabla \times \nabla \times T_B \mathbf{r}. \tag{88}$$

Taking advantage of the uniqueness of the Mie representation when Eq. 72 is satisfied, one can then identify the following relationships between the various scalar functions

$$T_J = -\frac{1}{\mu_0} \nabla^2 P_B, \tag{89}$$

$$P_J = \frac{1}{\mu_0} T_B. \tag{90}$$

This then shows that poloidal magnetic fields are associated with toroidal current densities and toroidal magnetic fields are associated with poloidal current densities.

Conversely, one may solve Eqs. 89 and 90 for T_B and P_B in terms of T_J and P_J. The first equation yields Poisson's equation for P_B at position $\mathbf{r}$

$$\nabla^2 P_B(\mathbf{r}) = -\mu_0 T_J(\mathbf{r}) \tag{91}$$

whose classical solution (see, e.g., Jackson 1998) is given by

$$P_B(\mathbf{r}) = \frac{\mu_0}{4\pi} \int_{\tau'} \frac{T_J(\mathbf{r}')}{|\mathbf{r} - \mathbf{r}'|} d\tau', \tag{92}$$

where τ' is a volume enclosing all currents, $d\tau'$ is the differential volume element, and $\mathbf{r}'$ is the position within the volume. The second equation yields simply

$$T_B(\mathbf{r}) = \mu_0 P_J(\mathbf{r}). \tag{93}$$

Substituting these into Eqs. 73 and 77 shows the dependence of the toroidal and poloidal parts of $\mathbf{B}$ with respect to those of $\mathbf{J}$

$$\mathbf{B}_{tor}(\mathbf{r}) = \nabla \times [\mu_0 P_J(\mathbf{r})]\, \mathbf{r}, \tag{94}$$

$$\mathbf{B}_{pol}(\mathbf{r}) = \nabla \times \nabla \times \left[\frac{\mu_0}{4\pi} \int_{\tau'} \frac{T_J(\mathbf{r}')}{|\mathbf{r} - \mathbf{r}'|} d\tau'\right] \mathbf{r}. \tag{95}$$

This is an interesting result considering that ∇ is operating at the position $\mathbf{r}$ where $\mathbf{B}$ is being calculated. If this position happens to be in a source-free region where $\mathbf{J} = 0$, then $P_J(\mathbf{r}) = 0$

and the toroidal magnetic field $\mathbf{B}_{tor}(\mathbf{r}) = 0$. This implies that *toroidal magnetic fields only exist within a conductor or magnetized material where the associated poloidal* **J** *is present*. At the same location **r**, the poloidal magnetic field $\mathbf{B}_{pol}$, however, does not identically vanish. This is because $\mathbf{B}_{pol}$ is sensitive to the toroidal current scalar $T_J(\mathbf{r}')$ evaluated within the distant current-carrying volume, and not only to its local value. Of course at this point Eq. 95 collapses to the usual potential form:

$$\mathbf{B}_{pol}(\mathbf{r}) = \nabla\left[\frac{\partial}{\partial r}\left(rP_B(\mathbf{r})\right)\right] = \nabla\left[\frac{\partial}{\partial r}\left(r\frac{\mu_0}{4\pi}\int_{\tau'}\frac{T_J(\mathbf{r}')}{|\mathbf{r}-\mathbf{r}'|}\,d\tau'\right)\right]. \tag{96}$$

This is because of Eq. 68 and of the fact that

$$\nabla^2 P_B(\mathbf{r}) = \begin{cases} -\mu_0 T_J(\mathbf{r}) & \text{for } \mathbf{r} \text{ inside } \tau', \\ 0 & \text{for } \mathbf{r} \text{ outside } \tau'. \end{cases} \tag{97}$$

This result then also provides the possibility of relating the present formalism to the one previously described in ➲ Sect. 3 when considering potential fields that arise in a source-free shell $a < r < c$. Then indeed, the internal (below $r = a$) and external (above $r = c$) current sources lead to *internal* P_B^i and *external* P_B^e magnetic poloidal scalar potentials, which can be related to the harmonic scalar potentials V^i and V^e defined by Eqs. 13 and 14 through

$$V^i = -\frac{\partial}{\partial r}\left(rP_B^i\right), \tag{98}$$

$$V^e = -\frac{\partial}{\partial r}\left(rP_B^e\right), \tag{99}$$

and leading to expansions of the form:

$$P_B^i(r,\theta,\phi) = a\sum_{n=1}^{\infty}\left(\frac{a}{r}\right)^{n+1}\sum_{m=0}^{n}\left(G_n^m\cos m\phi + H_n^m\sin m\phi\right)P_n^m(\cos\theta), \tag{100}$$

$$P_B^e(r,\theta,\phi) = a\sum_{n=1}^{\infty}\left(\frac{r}{a}\right)^{n}\sum_{m=0}^{n}\left(Q_n^m\cos m\phi + S_n^m\sin m\phi\right)P_n^m(\cos\theta), \tag{101}$$

where

$$\begin{cases} G_n^m = \frac{1}{n}g_n^m \\ H_n^m = \frac{1}{n}h_n^m \end{cases}, \qquad \begin{cases} Q_n^m = -\frac{1}{n+1}q_n^m \\ S_n^m = -\frac{1}{n+1}s_n^m \end{cases}. \tag{102}$$

This then amounts to state that within a source-free spherical shell $a<r<c$, the r dependence of the $P_n^{m,c}(r)$ and $P_n^{m,s}(r)$ in Eq. 82 is entirely specified, and of the form:

$$P_n^{m,c}(r) = a\left[\frac{g_n^m}{n}\left(\frac{a}{r}\right)^{n+1} - \frac{q_n^m}{n+1}\left(\frac{r}{a}\right)^{n}\right], \tag{103}$$

$$P_n^{m,s}(r) = a\left[\frac{h_n^m}{n}\left(\frac{a}{r}\right)^{n+1} - \frac{s_n^m}{n+1}\left(\frac{r}{a}\right)^{n}\right], \tag{104}$$

while of course $T_n^{m,c}(r) = T_n^{m,s}(r) = 0$.

4.4 Magnetic Fields in a Current-Carrying Shell

The toroidal–poloidal decomposition provides a convenient way of describing the magnetic field in a current-carrying shell $a < r < c$. The magnetic field then has four parts

$$\mathbf{B} = \mathbf{B}^{i}_{pol} + \mathbf{B}^{e}_{pol} + \mathbf{B}^{sh}_{pol} + \mathbf{B}^{sh}_{tor}, \tag{105}$$

where $\mathbf{B}^{i}_{pol}$ is the potential field due to toroidal currents $\mathbf{J}^{i}_{tor}$ in the region $r < a$ (in fact, due to all currents in the region $r < a$, since poloidal currents in this region do not produce any field in the shell $a < r < c$), $\mathbf{B}^{e}_{pol}$ is the potential field due to toroidal currents $\mathbf{J}^{e}_{tor}$ in the region $c < r$ (in fact, due to all currents in the region $c < r$, since poloidal currents in this region do not produce any field in the shell $a < r < c$), and $\mathbf{B}^{sh}_{pol}$ and $\mathbf{B}^{sh}_{tor}$ are the nonpotential poloidal and toroidal fields due to in situ toroidal $\mathbf{J}^{sh}_{tor}$ and poloidal $\mathbf{J}^{sh}_{pol}$ currents in the shell. From what was just seen, it is now known that the behavior of $\mathbf{B}^{i}_{pol}$ and $\mathbf{B}^{e}_{pol}$ is entirely dictated by the fact that their respective poloidal scalars P^{i}_{B} and P^{e}_{B} must take the forms Eq. 100 and 101. Those potential fields are thus entirely defined by the knowledge of the G^{m}_{n}, H^{m}_{n}, Q^{m}_{n}, and S^{m}_{n} coefficients (or equivalently, of the Gauss g^{m}_{n}, h^{m}_{n}, q^{m}_{n}, and s^{m}_{n} coefficients). By contrast, all that is known about $\mathbf{B}^{sh}_{pol}$ and $\mathbf{B}^{sh}_{tor}$ is that they are associated with toroidal and poloidal scalars which can be written in the form Eq. 81 and 82, so that they themselves can be written in the form Eq. 83 and 84. The radial dependence of the corresponding $P^{m,(c,s)}_{n}(r)$ and $T^{m,(c,s)}_{n}(r)$ functions is unknown, but dictated by the distribution of the current sources $\mathbf{J}^{sh}_{tor}$ and $\mathbf{J}^{sh}_{pol}$ within the shell, because of Eq. 92 and 93.

It is interesting at this stage to introduce an *inner shell* $a' < r < c'$ within the spherical shell $a < r < c$ considered so far (and which is referred here as the *outer shell*). Space can then naturally be divided into five regions (➲ *Fig. 3*): region I, for $r < a$, region II for $a < r < a'$, region III for $a' < r < c'$ (the inner shell), region IV for $c' < r < c$, and region V for $r > c$. As was just seen, everywhere within the outer shell $a < r < c$, the field $\mathbf{B}$ can be written in the form of Eq. 105. But in exactly the same way, everywhere within the inner shell $a' < r < c'$, the field can also be

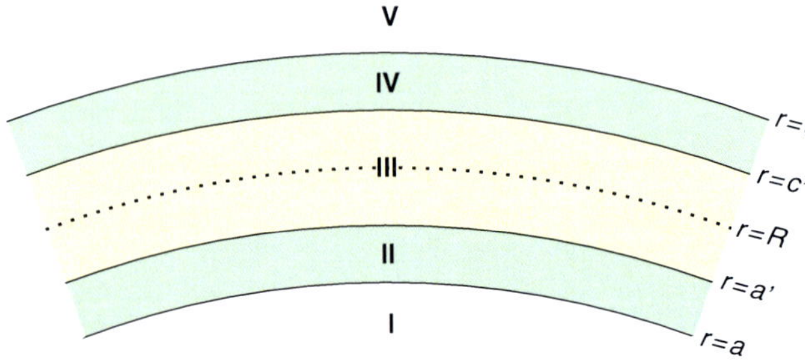

Fig. 3
Schematic showing a meridional cross-section of a general current-carrying shell $a < r < c$, within which an *inner shell* $a' < r < c'$ is defined. This then defines five regions, as identified by their numbers (I, II, III, IV, and V), within which both toroidal and poloidal sources can be found. These will contribute differently to the field, depending on where the field is observed. Of particular interest is the way these contribute when the field is observed in the inner shell and this shell progressively shrinks to the sphere $r = R$ (see text for details)

written in the form:

$$\mathbf{B} = \mathbf{B}^{i}_{pol}{}' + \mathbf{B}^{e}_{pol}{}' + \mathbf{B}^{sh}_{pol}{}' + \mathbf{B}^{sh}_{tor}{}', \tag{106}$$

where $\mathbf{B}^{i}_{pol}{}'$ is the potential field due to toroidal currents $\mathbf{J}^{i}_{tor}{}'$ in the region $r < a'$, $\mathbf{B}^{e}_{pol}{}'$ is the potential field due to toroidal currents $\mathbf{J}^{e}_{tor}{}'$ in the region $c' < r$, and $\mathbf{B}^{sh}_{pol}{}'$ and $\mathbf{B}^{sh}_{tor}{}'$ are the nonpotential poloidal and toroidal fields due to in situ toroidal $\mathbf{J}^{sh}_{tor}{}'$ and poloidal $\mathbf{J}^{sh}_{pol}{}'$ currents in the inner shell. It is important to note that $\mathbf{B}^{i}_{pol}{}' \neq \mathbf{B}^{i}_{pol}$ and $\mathbf{B}^{e}_{pol}{}' \neq \mathbf{B}^{e}_{pol}$. Toroidal currents flowing in regions I (resp. V), which already contributed to $\mathbf{B}^{i}_{pol}$ (resp. $\mathbf{B}^{e}_{pol}$), still contribute to $\mathbf{B}^{i}_{pol}{}'$ (resp. $\mathbf{B}^{e}_{pol}{}'$). But toroidal currents flowing in region II (resp. IV), which contributed to $\mathbf{B}^{sh}_{pol}$ and not to $\mathbf{B}^{i}_{pol}$ (resp. $\mathbf{B}^{e}_{pol}$), now contribute to $\mathbf{B}^{i}_{pol}{}'$ (resp. $\mathbf{B}^{e}_{pol}{}'$). Only the toroidal currents flowing in region III, which already contributed to $\mathbf{B}^{sh}_{pol}$, still contribute to $\mathbf{B}^{sh}_{pol}{}'$. By contrast, $\mathbf{B}^{sh}_{tor}{}' = \mathbf{B}^{sh}_{tor}$, both because the toroidal–poloidal decomposition is unique and because the toroidal field is only sensitive to the local behavior of the poloidal currents (recall Eq. 94). Obviously, the smaller the thickness $h = c' - a'$ of the inner shell, the less sources contributing to $\mathbf{B}^{sh}_{pol}$, still contribute to $\mathbf{B}^{sh}_{pol}{}'$. In fact, it can be shown that if this inner shell eventually shrinks to a sphere of radius $r = R$ (➋ *Fig. 3*), then $\mathbf{B}^{sh}_{pol}{}'$ goes to zero as $h/R \to 0$, while $\mathbf{B}^{sh}_{tor}{}'$ remains equal to $\mathbf{B}^{sh}_{tor}$ (Backus 1986). Further introducing $\mathbf{B}_{iR}$ (resp. $\mathbf{B}_{eR}$) as the limit of $\mathbf{B}^{i}_{pol}{}'$ (resp. $\mathbf{B}^{e}_{pol}{}'$) when $h/R \to 0$ then makes it possible to introduce the following unique *decomposition of a magnetic field* **B** *on a sphere of radius* $r = R$ *surrounded by sources*:

$$\mathbf{B}(R,\theta,\phi) = \mathbf{B}_{iR}(R,\theta,\phi) + \mathbf{B}_{eR}(R,\theta,\phi) + \mathbf{B}^{sh}_{tor}(R,\theta,\phi), \tag{107}$$

where (recall Eq. 83)

$$\mathbf{B}^{sh}_{tor}(R,\theta,\phi) = \sum_{n=1}^{\infty}\sum_{m=0}^{n}\left(T_n^{m,c}(R)\mathbf{C}_n^{m,c}(\theta,\phi) + T_n^{m,s}(R)\mathbf{C}_n^{m,s}(\theta,\phi)\right), \tag{108}$$

is the toroidal field produced on the sphere $r = R$ by the local poloidal currents, and $\mathbf{B}_{iR}(R,\theta,\phi)$ and $\mathbf{B}_{eR}(R,\theta,\phi)$ are the potential fields produced on the sphere $r = R$ by all sources, respectively, below and above $r = R$. These fields are in fact the values taken for $r = R$ of the potential fields $\mathbf{B}_{iR}(r,\theta,\phi)$ and $\mathbf{B}_{eR}(r,\theta,\phi)$ which may be defined more generally for, respectively, $r \geq R$ and $r \leq R$, with the help of (recall Eqs. 98-102 and Eq. 16-17):

$$\mathbf{B}_{iR}(r,\theta,\phi) = \sum_{n=1}^{\infty}\left(\frac{a}{r}\right)^{n+2}\sum_{m=0}^{n}\left(g_n^m(R)\mathbf{\Pi}_{ni}^{mc}(\theta,\phi) + h_n^m(R)\mathbf{\Pi}_{ni}^{ms}(\theta,\phi)\right), \tag{109}$$

$$\mathbf{B}_{eR}(r,\theta,\phi) = \sum_{n=1}^{\infty}\left(\frac{r}{a}\right)^{n-1}\sum_{m=0}^{n}\left(q_n^m(R)\mathbf{\Pi}_{ne}^{mc}(\theta,\phi) + s_n^m(R)\mathbf{\Pi}_{ne}^{ms}(\theta,\phi)\right), \tag{110}$$

where the R dependence of the Gauss coefficients is here to recall that these Gauss coefficients describe the fields produced by all sources, respectively, below $(g_n^m(R), h_n^m(R))$ and above $(q_n^m(R), s_n^m(R))$ $r = R$.

4.5 Thin-Shell Approximation

Finally, if h/R is not zero, but small enough, one may still write within the (then thin-shell) $a' < r < c'$ around $r = R$:

$$\mathbf{B}(r,\theta,\phi) = \mathbf{B}_i(r,\theta,\phi) + \mathbf{B}_e(r,\theta,\phi) + \mathbf{B}_{tor}^{sh}(R,\theta,\phi), \tag{111}$$

where $\mathbf{B}_i(r,\theta,\phi)$, $\mathbf{B}_e(r,\theta,\phi)$, and $\mathbf{B}_{tor}^{sh}(R,\theta,\phi)$ are given by Eqs. 108–110. Equation 111 is then correct to within $\mathbf{B}_{pol}^{sh}$ and $\mathbf{B}_{tor}^{sh}$ corrections of order h/R (Backus 1986). This approximation is known as the *thin-shell approximation*.

5 Spatial Power Spectra

In discussing the various components of the Earth's magnetic field, and in particular the way each component contributes on average to the observed magnetic field, it will prove useful to deal with the concept of *spatial power spectra* . This concept was introduced by Mauersberger (1956) and popularized by Lowes (1966, 1974), both in the case of potential fields. However, it is quite straightforward to introduce these also in the case of nonpotential fields.

Indeed, consider the sphere $S(R)$ of radius $r = R$ and assume the most general case when this sphere is surrounded by sources. Then, as seen previously, the field can be written in the form of Eq. 107, and its average squared magnitude over $S(R)$ can be written in the form:

$$\begin{aligned}\langle \mathbf{B}^2(R,\theta,\phi)\rangle_{S(R)} &= \frac{1}{4\pi}\int_0^{2\pi}\int_0^{\pi} \mathbf{B}(R,\theta,\phi)\cdot\mathbf{B}(R,\theta,\phi)\sin\theta\, d\theta\, d\phi \\ &= W^i(R) + W^e(R) + W^T(R),\end{aligned} \tag{112}$$

where

$$W^i(R) = \sum_{n=1}^{\infty} W_n^i(R), \qquad W^e(R) = \sum_{n=1}^{\infty} W_n^e(R), \qquad W^T(R) = \sum_{n=1}^{\infty} W_n^T(R), \tag{113}$$

with

$$W_n^i(R) = (n+1)\left(\frac{a}{R}\right)^{2n+4} \sum_{m=0}^{n} \left[(g_n^m(R))^2 + (h_n^m(R))^2\right], \tag{114}$$

$$W_n^e(R) = n\left(\frac{R}{a}\right)^{2n-2} \sum_{m=0}^{n} \left[(q_n^m(R))^2 + (s_n^m(R))^2\right], \tag{115}$$

$$W_n^T(R) = \frac{n(n+1)}{2n+1} \sum_{m=0}^{n} \left[(T_n^{m,c}(R))^2 + (T_n^{m,s}(R))^2\right], \tag{116}$$

all of which follows from Eqs. 108–110 and the orthogonality properties of the $\mathbf{C}_n^{m,(c,s)}$ and $\mathbf{\Pi}_{n,(i,e)}^{m,(c,s)}$ spherical harmonic vectors (recall Eqs. 60–62).

Equations 112–113 then show that each type of field—the potential field produced by all sources above $r = R$, the potential field produced by all sources below $r = R$, and the nonpotential (toroidal) field produced by the local (poloidal) sources on $r = R-$, and within each type of field, each degree n (in fact, each elementary field of degree n and order m, as is further shown by Eqs. 114–116) contributes independently to the average squared magnitude $\langle \mathbf{B}^2(R,\theta,\phi)\rangle_{S(R)}$ on the sphere $r = R$. Hence, plotting $W_n^i(R)$ (resp. $W_n^e(R)$, $W_n^T(R)$) as a function of n provides a very convenient mean of identifying which sources, and within each type of source, which degrees n, most contribute on average to the magnetic field $\mathbf{B}(R,\theta,\phi)$ on the sphere $r = R$. Such plots are known as *spatial power spectra*.

In the more restrictive (and better known) case when the field is potential within a source-free shell $a < r < c$, the field takes the form (Eq. 15) $\mathbf{B}(r,\theta,\phi) = \mathbf{B}_i(r,\theta,\phi) + \mathbf{B}_e(r,\theta,\phi)$, where $\mathbf{B}_i(r,\theta,\phi)$ and $\mathbf{B}_e(r,\theta,\phi)$ are respectively defined by Eq. 16 and Eq. 17. In exactly the same way, it can again be shown that for any value of r within the shell, each degree n of the field $\mathbf{B}_i(r,\theta,\phi)$ of internal origin (with sources below $r = a$) and of the field $\mathbf{B}_e(r,\theta,\phi)$ of external origin (with sources above $r = c$) contribute to the average squared magnitude $\langle \mathbf{B}^2(r,\theta,\phi)\rangle_{S(r)}$ on the sphere $S(r)$ by, respectively,

$$W_n^i(r) = (n+1)\left(\frac{a}{r}\right)^{2n+4}\sum_{m=0}^{n}\left[\left(g_n^m\right)^2 + \left(h_n^m\right)^2\right], \tag{117}$$

$$W_n^e(r) = n\left(\frac{r}{a}\right)^{2n-2}\sum_{m=0}^{n}\left[\left(q_n^m\right)^2 + \left(s_n^m\right)^2\right], \tag{118}$$

which again define spatial power spectra. Note that in that case, the only r-dependence is the one due to the geometric factors $(a/r)^{2n+4}$ and $(r/a)^{2n-2}$. The spectrum for the field of internal origin $W_n^i(r)$ is what is often referred to as the *Lowes–Mauersberger* spectrum.

6 Mathematical Uniqueness Issue

Sections 2 to 4 introduced several mathematical ways of representing magnetic fields in spherical shells, when sources of those fields lie either below, within, or above those shells. The goal is now to take advantage of those representations to explain how the best possible description of the Earth's magnetic field can be recovered from available observations. Obviously, the more the observations, the better the description. However, even an infinite number of observations might not be enough to guarantee that one does eventually achieve a proper description of the field. Observations do not only need to be numerous; they also need to provide adequate information. This is an important issue since, in practice, observations cannot be made anywhere. Historical observations have all been made at the Earth's surface. Aeromagnetic surveys provided additional observations after 1950, and satellite missions only started in the 1960s. Assuming that all those observations could have been made in an infinitely dense way (here, the issue of the limited density of observations is not discussed) and instantly (temporal issues are not discussed either), this means that the best information historical observations and aeromagnetic surveys can provide is the knowledge of $\mathbf{B}$ or of some derived quantities (components, direction, or intensity) over the entire surface of the Earth. Satellites bring additional information, which at best is a complete knowledge of $\mathbf{B}$ or of some derived quantities within a thin shell where sources can be found. To what extent is this enough to recover a complete mathematical description of the Earth's magnetic field ?

6.1 Uniqueness of Magnetic Fields in a Source-Free Shell

First consider the situation when the sources of the magnetic field are a priori known to be either internal to an inner surface Σ_i, or external to an outer surface Σ_e, so that the shell in-between

Σ_i and Σ_e is source-free. Within this shell, Eqs. 2 and 8 show that both $\nabla\cdot\mathbf{B}(=0)$ and $\nabla\times\mathbf{B}(=0)$ are known. Applying Helmholtz theorem then shows that the field is completely characterized within the shell, provided the normal component of the field is known everywhere on both Σ_i and Σ_e.

This very general result, also known in potential theory as the *uniqueness theorem for Neumann boundary conditions* (e.g., Kellogg 1954; Blakely 1995), can be made more explicit in the case considered in ➋ Sect. 3 when the shell is spherical, and Σ_i and Σ_e are defined by $r = a$ and $r = c$. In that case, it is known that within the shell, $\mathbf{B}$ is described by $\mathbf{B}(r,\theta,\phi) = \mathbf{B}_i(r,\theta,\phi) + \mathbf{B}_e(r,\theta,\phi)$ (Eq. 15) where $\mathbf{B}_i(r,\theta,\phi)$ and $\mathbf{B}_e(r,\theta,\phi)$ are given by Eqs. 16 and 17, the radial components of which can easily be inferred from Eqs. 18–21:

$$B_{ir}(r,\theta,\phi) = \sum_{n=1}^{\infty}(n+1)\left(\frac{a}{r}\right)^{n+2}\sum_{m=0}^{n}\left(g_n^m\cos m\phi + h_n^m\sin m\phi\right)P_n^m(\cos\theta), \tag{119}$$

$$B_{er}(r,\theta,\phi) = \sum_{n=1}^{\infty}(-n)\left(\frac{r}{a}\right)^{n-1}\sum_{m=0}^{n}\left(q_n^m\cos m\phi + s_n^m\sin m\phi\right)P_n^m(\cos\theta), \tag{120}$$

This can then be used together with Eqs. 32 and 33 to show that

$$\frac{2n+1}{4\pi}\int_{\theta=0}^{\pi}\int_{\phi=0}^{2\pi}B_r(r,\theta,\phi)P_n^m(\cos\theta)\cos m\phi\sin\theta\,d\theta\,d\phi = (n+1)\left(\frac{a}{r}\right)^{n+2}g_n^m - n\left(\frac{r}{a}\right)^{n-1}q_n^m, \tag{121}$$

$$\frac{2n+1}{4\pi}\int_{\theta=0}^{\pi}\int_{\phi=0}^{2\pi}B_r(r,\theta,\phi)P_n^m(\cos\theta)\sin m\phi\sin\theta\,d\theta\,d\phi = (n+1)\left(\frac{a}{r}\right)^{n+2}h_n^m - n\left(\frac{r}{a}\right)^{n-1}s_n^m. \tag{122}$$

Then, assuming that the normal component of $\mathbf{B}$ is known on both Σ_i $(r = a)$ and Σ_e $(r = c)$, amounts to assume that $\mathbf{B}(a,\theta,\phi)$ and $\mathbf{B}(c,\theta,\phi)$ are both completely known; making use of Eqs. 121 and 122 once for $r = a$ and again for $r = c$, leads to a set of linear equations from which all Gauss coefficients $(g_n^m, h_n^m, q_n^m, s_n^m)$ can be inferred; and recalling Eqs. 15, 16, and 17, shows that the field $\mathbf{B}$ is then indeed completely defined within the spherical shell $a < r < c$.

More generally, it is important to note that the field $\mathbf{B}$ can be characterized just as well as soon as $B_r(r,\theta,\phi)$ is completely known for two different values r_1 and r_2 of r, provided $a \le r_1 < r_2 \le c$ (i.e., provided the two spherical surfaces defined by $r = r_1$ and $r = r_2$ lie within the source-free spherical shell).

Similar conclusions can be reached if the potential $V(r,\theta,\phi)$ in place of the radial component $B_r(r,\theta,\phi)$ is assumed to be known on both Σ_i and Σ_e. This very general result, known in potential theory as the *uniqueness theorem for Dirichlet boundary conditions* (e.g., Kellogg 1954; Blakely 1995), can also be made more explicit in the case when the shell is spherical, and Σ_i and Σ_e are defined by $r = a$ and $r = c$. In that case, it is known that within the shell, $V(r,\theta,\phi) = V_i(r,\theta,\phi) + V_e(r,\theta,\phi)$, where $V_i(r,\theta,\phi)$ and $V_e(r,\theta,\phi)$ are given by Eqs. 13 and 14, which can be used together with Eqs. 32 and 33 to show that

$$\frac{2n+1}{4\pi a}\int_{\theta=0}^{\pi}\int_{\phi=0}^{2\pi}V(r,\theta,\phi)P_n^m(\cos\theta)\cos m\phi\sin\theta\,d\theta\,d\phi = \left(\frac{a}{r}\right)^{n+1}g_n^m + \left(\frac{r}{a}\right)^{n}q_n^m, \tag{123}$$

$$\frac{2n+1}{4\pi a}\int_{\theta=0}^{\pi}\int_{\phi=0}^{2\pi}V(r,\theta,\phi)P_n^m(\cos\theta)\sin m\phi\sin\theta\,d\theta\,d\phi = \left(\frac{a}{r}\right)^{n+1}h_n^m + \left(\frac{r}{a}\right)^{n}s_n^m. \tag{124}$$

Then, assuming that $V(r,\theta,\phi)$ is known on both Σ_i $(r = a)$ and Σ_e $(r = c)$, makes it possible to use Eqs. 123 and 124, once for $r = a$, and again for $r = c$, to form another set of linear equations from which all Gauss coefficients $(g_n^m, h_n^m, q_n^m, s_n^m)$ can again be inferred. The same reasoning as in the previous case then follows, leading to the same conclusions (and generalization to the case when $V(r,\theta,\phi)$ is known for $r = r_1$ and r_2, provided $a \leq r_1 < r_2 \leq c$).

In practice however, only components of the magnetic field **B** and not its potential V are directly accessible to observations. The previous result can nevertheless be used to show yet another, directly useful, uniqueness property now applying to the $B_\theta(r,\theta,\phi)$ component of the field, when this component is assumed to be known on the two spherical surfaces defined by $r = r_1$ and r_2, where $a \leq r_1 < r_2 \leq c$. This component is such that

$$B_\theta(r,\theta,\phi) = -\frac{1}{r}\frac{\partial V(r,\theta,\phi)}{\partial \theta}. \tag{125}$$

Integrating along a meridian (with fixed values of r and ϕ) starting from $\theta = 0$ thus leads to

$$V(r,\theta,\phi) = C(r) - r\int_0^\theta B_\theta(r,\theta',\phi)d\theta', \tag{126}$$

which shows that if $B_\theta(r,\theta,\phi)$ is known, so is $V(r,\theta,\phi)$, to within a function $C(r)$. But it is known from Eqs. 13, 14, and 37, that the average value of $V(r,\theta,\phi)$ over the sphere $S(r)$ of radius r is such that $\langle V(r,\theta,\phi)\rangle_{S(r)} = 0$ (recall that this is true, only because magnetic fields do not have monopole sources). It thus follows that

$$C(r) = r\left\langle \int_0^\theta B_\theta(r,\theta',\phi)d\theta'\right\rangle_{S(r)}, \tag{127}$$

and that as soon as $B_\theta(r,\theta,\phi)$ is known for a given value of r, so is $V(r,\theta,\phi)$. $V(r,\theta,\phi)$ can then again be used to compute the set of linear equations (Eqs. 123 and 124) for two different values r_1 and r_2 of r such that $a \leq r_1 < r_2 \leq c$, from which all Gauss coefficients $(g_n^m, h_n^m, q_n^m, s_n^m)$ can finally again be inferred.

It is important to note that by contrast, no similar conclusion applies when $B_\phi(r,\theta,\phi)$, rather than $B_\theta(r,\theta,\phi)$, is considered. This is because, as is clear from, e.g., Eqs. 16–21, $B_\phi(r,\theta,\phi)$ is totally insensitive to zonal fields (described by spherical harmonic terms of order $m = 0$).

Of course, and as must now be obvious to the reader, other useful uniqueness properties can also be derived by combining the knowledge of $B_r(r,\theta,\phi)$ for a given value r_1 of r, and of $B_\theta(r,\theta,\phi)$ for another value r_2 of r. Of particular relevance to the historical situation (when observations are only available at the Earth's surface) is the case when both $B_r(r,\theta,\phi)$ and $B_\theta(r,\theta,\phi)$ are simultaneously known for a given value R of r, where $a \leq R \leq c$. In that case, Eqs. 121 and 122 on one hand, and Eqs. 123 and 124 on another hand, hold for $r = R$. This again leads to a set of linear equations from which all the Gauss coefficients $(g_n^m, h_n^m, q_n^m, s_n^m)$ can be inferred, and the field once again fully determined.

Finally, it is important to note that independently of the method used to define the field in a unique way within the source-free shell $a < r < c$, all Gauss coefficients are then recovered, and it therefore becomes possible to identify $\mathbf{B}_i$ (Eq. 16) and $\mathbf{B}_e$ (Eq. 17).

6.2 Uniqueness Issues Raised by Directional-Only Observations

Now focus a little more on the possibility of completely defining a magnetic field when information is only available on a single surface within the source-free shell. It has been shown that if the field has both internal and external sources, knowing both $B_r(r, \theta, \phi)$ and $B_\theta(r, \theta, \phi)$ over a sphere defined by $r = R$ within the source-free shell, is enough to achieve uniqueness and identify the fields of internal and external origin. Implicit is also the conclusion that just knowing one component of the field (be it B_r, B_θ, or B_ϕ) would not be enough. At least two components are needed, and in fact not just any two components, since, as previously seen, B_ϕ does not provide as much information as B_r or B_θ. Of course, knowing the entire field **B** is even better. The field is then overdetermined (at least its nonzonal component), which is very useful since, in practice, the field is known just at a finite number of sites, and not everywhere at the Earth's surface (cf. ➋ Chap. 5). This in fact is the way Gauss first proved that the Earth's magnetic field is mainly of internal origin (Gauss 1839, for a detailed account of how Gauss did proceed in practice, see, e.g., Langel 1987).

But before Gauss first introduced a way to measure the magnitude of a magnetic field (which he did in 1832, see, e.g., Malin 1987), only inclination and declination observations were made. Such observations have a serious drawback. They cannot tell the difference between the real magnetic field and the same magnetic field multiplied by some arbitrary positive constant λ. But if such *directional-only* observations are available everywhere at the Earth's surface, could it be that they nevertheless provide enough information for the Earth's magnetic field to be completely characterized, to within the global positive factor λ ?

Until recently, most authors felt that this was indeed the case, at least when a priori assuming the field to be of internal origin (e.g., Kono 1976). However directional-only observations are not linearly related to the Gauss coefficients, and the answer turns out to be more subtle, as first recognized by Proctor and Gubbins (1990).

Relying on complex-variables, mathematical tools very different from those used in this Chapter, Proctor and Gubbins (1990) investigated axisymmetric fields (i.e., zonal fields, with only $m = 0$ Gauss coefficients) of internal origin and succeeded in exhibiting a family of different fields all sharing exactly the same direction everywhere on a spherical surface $r = R$ enclosing all sources. Clearly, even a perfect knowledge of the direction of the field on the sphere $r = R$ would not be enough to fully characterize a field belonging to such a family, even to within a global positive factor.

The fields exhibited by Proctor and Gubbins (1990) are however special in several respects: they are axisymmetric, antisymmetric with respect to the equator, and of octupole type (displaying four loci of magnetic poles on the sphere $r = R$, one magnetic pole at each geographic pole, and two mid-latitude axisymmetric lines of magnetic poles, a magnetic pole being defined as a point where the field is perpendicular to the surface). The nonuniqueness property they reveal may therefore very well not transpose to Earth-like situations involving magnetic fields strongly dominated by their dipole component.

This led Hulot et al. (1997) to reconsider the problem in a more general context. These authors again assumed the field to entirely be of internal origin, but only requested it to be *regular enough* (i.e., physical). Using a potential theory type of approach (which is again not given in any detail here), they were able to show that if the direction of the field is known everywhere on a *smooth enough* surface Σ_i enclosing all sources, and if those directions do not reveal anymore than N loci of magnetic poles on Σ_i, then all fields satisfying these boundary conditions belong to an open cone (in such an open cone, any nonzero positive linear combination of solutions is a solution) of dimension $N - 1$. Applying this result to the fields exhibited by Proctor and Gubbins (1990), which display $N = 4$ loci of poles on Σ_i (which is then the sphere $r = R$), shows that those fields would belong to open cones of dimension 3 (in fact, 2, when the equatorial antisymmetry is taken into account, see Hulot et al. 1997), leaving enough space for fields from this family to share the same directions on Σ_i, and yet not be proportional to each other. But applying this result to the historical magnetic field, which only displays $N = 2$ poles on Σ_i (which is then the Earth's surface), leads to a different conclusion. The Earth's field belongs to an open cone of dimension 1, which shows that it can indeed be recovered from directional-only data on Σ_i (the Earth's surface) to within the already discussed global positive factor λ. Note however that this result only holds when all contributions from external sources are ignored.

Similar results can be derived in the case, less-relevant to the Earth, when the field is assumed to have all its sources *outside* the surface on which its direction is assumed to be known (Hulot et al. 1997). But to the authors knowledge, no results applying to the most general situation when both internal and external sources are simultaneously considered, have yet been derived. Only a relatively weak statement can easily be made in the much more trivial case when the direction of the field is assumed to be known in a subshell of the source-free shell separating external and internal sources (and not just on a surface, Bloxham 1985; Proctor and Gubbins 1990; Lowes et al. 1995). In that case indeed, if two fields $\mathbf{B}(\mathbf{r})$ and $\mathbf{B}'(\mathbf{r})$ share the same direction within the subshell, then a scalar function $\lambda(\mathbf{r})$ exists such that $\mathbf{B}'(\mathbf{r}) = \lambda(\mathbf{r})\mathbf{B}(\mathbf{r})$ within it. But within that subshell, $\mathbf{B}(\mathbf{r})$ and $\mathbf{B}'(\mathbf{r})$ must satisfy $\nabla \cdot \mathbf{B} = 0$, $\nabla \times \mathbf{B} = 0$, $\nabla \cdot \mathbf{B}' = 0$, and $\nabla \times \mathbf{B}' = 0$. This implies $\nabla\lambda \cdot \mathbf{B} = 0$ and $\nabla\lambda \times \mathbf{B} = 0$, i.e., that $\lambda(\mathbf{r})$ is a constant λ within the subshell. Since both fields $\mathbf{B}(\mathbf{r})$ and $\mathbf{B}'(\mathbf{r})$ can be written in a unique way in the form Eqs. 15–21 within that source-free subshell, this means that all Gauss coefficients describing $\mathbf{B}'(\mathbf{r})$ are then proportional (by a factor λ) to those describing $\mathbf{B}(\mathbf{r})$. Hence $\mathbf{B}'(\mathbf{r}) = \lambda\mathbf{B}(\mathbf{r})$ not only within the subshell within which observations are available, but also beyond this shell, provided one remains within the source-free shell.

6.3 Uniqueness Issues Raised by Intensity-Only Observations

Measuring the full magnetic field $\mathbf{B}$ requires a lot of care. Nowadays, by far the most demanding step turns out to be the orientation of the measured field with respect to the geocentric reference frame. This is especially true in the context of satellite measurements (cf. ➲ Chap. 5). By contrast, measuring the intensity $F = |\mathbf{B}|$ of the field , is comparatively easier, and when satellites first started making magnetic measurements from space, they only measured the intensity of the field. But to what extent can a magnetic field be completely determined (to within a global sign, of course) when only its intensity is measured ?

First consider the now familiar case when a source-free shell can be defined, and information (i.e., intensity F) is only available on a single surface within that shell. Though no explicit results have yet been published (at least to the authors knowledge) in the general case when the field has both external and internal sources, it can be easily anticipated that this will be a very unfavorable situation: it is already known that the knowledge of only one of the B_r, B_θ, or B_ϕ components everywhere on this surface is not enough and that at least B_r and B_θ need to be simultaneously known.

In fact, even in the case when the field is further assumed to only have internal sources (i.e., enclosed within the surface where intensity is assumed known), no general conclusion can be drawn. Some important specific results are however available. In particular, Backus (1968) showed that if the field is of internal origin and further assumed to be a finite sum of vector spherical harmonics (i.e., if all Gauss coefficients g_n^m and h_n^m are a priori known to be zero for $n > N$, where N is a finite integer, and the sum in Eq. 16 is therefore finite), then the field can indeed be completely determined (to within a global sign) by the knowledge of F everywhere on a spherical surface $r = R$ enclosing all sources. But this result will generally not hold if the field is *not* a finite sum of vector spherical harmonics.

This was first shown by Backus (1970), which exhibited what are now known as the *Backus series* (see also the comment by Walker 1992). These are fields $\mathbf{B}_M(r, \theta, \phi)$ of internal origin and order M (i.e., defined by Gauss coefficients with $g_n^m = h_n^m = 0$ if $m \neq M$), where M can be any positive integer. The Gauss (internal) coefficients of $\mathbf{B}_M(r, \theta, \phi)$ are defined by a recursion relation, which needs not be explicited here. Suffices to say that this relation is chosen so as to ensure that each field $\mathbf{B}_M(r, \theta, \phi)$, (1) has Gauss coefficients that converge fast enough with increasing degree n, for the convergence of the infinite sum (Eq. 16) defining $\mathbf{B}_M(r, \theta, \phi)$ to be ensured for all values of $r \geq R$, and, (2) satisfies $\mathbf{B}_M(r, \theta, \phi) \cdot \mathbf{B}_D(r, \theta, \phi) = 0$ everywhere on the surface $r = R$, where $\mathbf{B}_D(r, \theta, \phi)$ is the axial dipole of internal origin defined by the single Gauss coefficient $g_1^0 = 1$ in Eq. 16. These Backus series can then be used to define an infinite number of pairs of magnetic field of internal origin $\mathbf{B}_{M+} = \alpha \mathbf{B}_D + \beta \mathbf{B}_M$ and $\mathbf{B}_{M-} = \alpha \mathbf{B}_D - \beta \mathbf{B}_M$, where (α, β) can be any pair of real (nonzero) values. These pairs will automatically satisfy (as one can easily check) $|\mathbf{B}_{M-}(r, \theta, \phi)| = |\mathbf{B}_{M+}(r, \theta, \phi)|$ on the sphere $r = R$, while obviously $\mathbf{B}_{M-} \neq \mathbf{B}_{M+}$. A perfect knowledge of the intensity of the field on the sphere $r = R$ would therefore not be enough to characterize a field belonging to any such pairs, even to within a global sign.

An interesting generalization of this result to the case when $\mathbf{B}_D$ does not need to be an axial dipole field has more recently been published by Alberto et al. (2004). If $\mathbf{B}_N$ is any arbitrary field of internal origin defined by a finite number of Gauss coefficients with maximum degree N, a field $\mathbf{B'}_N$ can again always be found such that $\mathbf{B'}_N(r, \theta, \phi) \cdot \mathbf{B}_N(r, \theta, \phi) = 0$ everywhere on the surface $r = R$. Many additional pairs of magnetic fields $\mathbf{B}_{N+} = \alpha \mathbf{B}_N + \beta \mathbf{B'}_N$ and $\mathbf{B}_{N-} = \alpha \mathbf{B}_N - \beta \mathbf{B'}_N$ again sharing the same intensity on the sphere $r = R$, can thus be found.

Those results are interesting. However they do not provide a general answer to the question of the uniqueness of arbitrary fields of internal origin only constrained by intensity data on a surface enclosing all sources (when the field is not a finite sum of vector spherical harmonics). But does this really matter in practice?

It may indeed be argued that the best model of the Earth's magnetic field of internal origin that will ever be recovered, will anyway be in the form of a finite number of Gauss coefficients (those compatible with the resolution matching the spatial distribution of the limited number of observations), and that because of the Backus (1968) uniqueness result previously discussed, this model would necessarily be determined in a unique way (to within a sign) by the knowledge of the intensity of the field at the Earth's surface (assuming that some strategy of measurement has been used so as to minimize any contribution of the field of external origin). In practice, this indeed is the case. However, this practical uniqueness turns out to be very relative and misleading, as the study of Stern et al. (1980) illustrates. In this study, the authors use a data set of full vector magnetic field observations collected by the 1980 Magsat satellite (and carefully selected to minimize any local or external sources, which they then consider as a source of noise) once to produce a field model A best explaining the observed intensity, and once again to produce a model B best explaining the observed vector field. They found that the two models differ very significantly. Both models predict similar intensity at the Earth's surface (to within measurement error, typically a few nT), but they strongly disagree when predicting the full vector field, with model A leading to errors (up to 2,000 nT !) far more than tolerated by measurement errors on the measured vector field, contrary to model B. This disagreement can be traced back to the fact that the difference $\mathbf{B}_B - \mathbf{B}_A$ between the predictions $\mathbf{B}_A$ and $\mathbf{B}_B$ of the two models tend to satisfy $(\mathbf{B}_B - \mathbf{B}_A) \cdot (\mathbf{B}_B + \mathbf{B}_A) = 0$.

One way of interpreting this result is to note that any practical optimizing procedure used to look for a model A best fitting the intensity (and just the intensity), will be much less sensitive to an error $\delta\mathbf{B}$ perpendicular to the observed field $\mathbf{B}$ (which then produces a second-order error $(\delta\mathbf{B})^2/|\mathbf{B}|$ in the intensity) than to a comparable error along the observed field (which then produces a first-order error $\delta\mathbf{B}$ in the intensity). By contrast, the optimizing procedure used to look for a model B best fitting the full vector $\mathbf{B}$, will make sure this error is kept small, whatever its direction. As a result, the difference $\mathbf{B}_B - \mathbf{B}_A$ will be largest in the direction perpendicular to the observed field, the value of which, to lowest order, is close to $(\mathbf{B}_B + \mathbf{B}_A)/2$. The two models are thus bound to lead to predictions such that $(\mathbf{B}_B - \mathbf{B}_A) \cdot (\mathbf{B}_B + \mathbf{B}_A) = 0$. Since model A makes erroneous predictions in the direction perpendicular to the observed field, this effect is often referred to as the *perpendicular error effect* (Lowes 1975; Langel 1987).

Some authors however also refer to this effect as the *Backus effect* (e.g., Stern et al. 1980; Langel 1987), quite correctly, since this error also seems to be closely related to the type of nonuniqueness exhibited by the fields constructed with the help of the Backus series. To see this, first recall that model B is constrained by full vector field observations. Were those perfect and available everywhere at the Earth's surface, model B would be perfectly and uniquely determined. Improving the data set would thus lead model B to eventually match the *true Earth* model. Model A differs from model B in a macroscopic way, and it is very likely that improving the intensity data set would *not* lead model A to converge toward the *true Earth* model. This suggests that at least another model than the *true Earth* model could be found in the limit perfect intensity data is available everywhere at the Earth's surface, and that when only a finite imperfect intensity data-set is available, a model recovered by optimizing the fit just to this intensity data set, such as model A, is a truncated and approximate version of this alternative model. Indeed, the magnitude and geographical distribution of the difference $\mathbf{B}_B - \mathbf{B}_A$ is very comparable to that of the Backus series (e.g., Stern and Bredekamp 1975).

This undesired nonuniqueness property is problematic. To alleviate the resulting Backus/perpendicular-error effect, several practical solutions have therefore been proposed, such as adding a minimum of vector field data to the intensity-only data set (as first investigated by Barraclough and Nevitt 1976), or taking into account even poor determinations of the field direction, which indeed brings considerable improvement (Holme and Bloxham 1995). From a more formal point of view, the next conceptual improvement was however brought by Khokhlov et al. (1997). Again relying on a potential theory type of approach, these authors showed that a simple and unambiguous way of characterizing a *regular enough* field of internal origin only defined by the knowledge of its intensity F on a *smooth enough* surface Σ_i enclosing all its sources, is to locate its (possibly several) magnetic dip equators on Σ_i (defined as curves across which the component of $\mathbf{B}$ normal to Σ_i changes sign). In other words, if a field of internal origin is defined by both its intensity everywhere on Σ_i and the location of its dip equator(s) on Σ_i, it then is completely and uniquely determined (to within a global sign). Note that indeed, any two fields of pairs built with the help of the Backus series (or their generalization by Alberto et al. 2004), do not share the same equators. The practical usefulness of this theoretical result was subsequently investigated by Khokhlov et al. (1999), and demonstrated in real situations by Ultré-Guérard et al. (1998) and, more recently, by Holme et al. (2005). However, the most obvious practical conclusion that should be drawn from these various results is that intensity-only strategies of measuring the Earth's magnetic field on a planetary scale is a dangerous one. Indeed all advanced near-Earth magnetic field missions are now designed to make sure the full field **B** is measured (cf. Hulot et al., 2007). Besides, and as shown later, this turns out to be mandatory also because near-Earth satellites do not orbit in a source-free shell.

Finally, and for completeness, the situation when the intensity of the field is no longer known on a surface, but within a volume within the source-free shell should be considered (in which case, again a field with both internal and external sources may be considered). Then, as shown by Backus (1974), provided the source-free shell can be considered as connected (any two points in the shell can be joined by a smooth curve within the shell, which is the case of the spherical shell $a < r < c$), knowing the intensity of the field in an open region contained entirely within the source-free shell, is enough to determine the field entirely up to its global sign.

6.4 Uniqueness of Magnetic Fields in a Shell Enclosing a Spherical Sheet Current

So far, the possibility of recovering a complete description of a magnetic field has only been addressed when some information is available on some surface or in some volume within a source-free shell, the sources of the field lying above and/or below the shell. This is typically the situation when one attempts to describe the Earth's magnetic field within the neutral atmosphere, using ground-based, shipborne and aeromagnetic observations. It was also noted that once the field is fully determined within such a source-free shell, then both the field of internal origin $\mathbf{B}_i$ and the field of external origin $\mathbf{B}_e$ can be identified.

Now, what if one considers a shell within which currents can be found? This is a situation that must be considered when attempting to also analyze satellite data, since satellites fly above the E-region where the ionospheric dynamo resides, and within the F-region where additional currents can be found (cf. ➋ Chap. 5).

Here, currents in the F-region are first ignored, and a spherical shell $a < r < c$ is assumed which only contains a spherical sheet current (which would typically describe the currents produced by the ionospheric dynamo in the E-region) at radius $r = b$. Then the sources of the field are assumed to lie below $r = a$ (sources referred to as $\mathbf{J}(r < a)$ sources), on $r = b$ ($\mathbf{J}_s(r = b)$ sources), and above $r = c$ ($\mathbf{J}(r > c)$ sources). What kind of information does one then need to make sure the field produced by those sources can be completely described everywhere within that shell?

To address this question, first note that since $\mathbf{J}(r < a)$ sources lie below $r = a$, the potential field they produce can be described by a set of Gauss coefficients (g_n^m, h_n^m) which may be used in Eq. 16 to predict this field for any value $r > a$. In the same way, since $\mathbf{J}(r > c)$ sources lie above $r = c$, the potential field they produce can be described by a set of Gauss coefficients (q_n^m, s_n^m) which may be used in Eq. 17 to predict this field for any value $r < c$. Finally, since the $\mathbf{J}_s(r = b)$ sources correspond to a spherical sheet current, (a_n^m, b_n^m) Gauss coefficients can also be introduced and used to predict the potential field produced by those sources thanks to Eq. 38 for $r > b$ and Eq. 40 for $r < b$.

Now, consider the lower subshell $a < r < b$. This shell is a source-free shell of the type considered so far. All the previous results relevant to the case when both internal and external sources are to be found therefore apply. Take the most relevant case when information about the field is available only on a surface (say, the Earth's surface) within the subshell $a < r < b$. Then one must be able to know at least two components of the field everywhere on that surface, typically $B_r(r, \theta, \phi)$ and $B_\theta(r, \theta, \phi)$ to fully characterize the field within that subshell. Once one knows these, one can recover all Gauss coefficients describing the field within that subshell. In particular, one will recover the (g_n^m, h_n^m) Gauss coefficients of the field produced by the $\mathbf{J}(r < a)$ sources, which is seen as the field of internal origin for that subshell. One can also recover the Gauss coefficients of the field of external origin for that subshell. But this field is the one produced by both $\mathbf{J}_s(r = b)$ sources and $\mathbf{J}(r > c)$ sources. One will therefore recover the sum of their Gauss coefficients, i.e. (recall Eq. 40), $q_n^m - ((n+1)/n)(a/b)^{2n+1} a_n^m$ and $s_n^m - ((n+1)/n)(a/b)^{2n+1} b_n^m$.

Next, consider the upper subshell $b < r < c$. Obviously, the same reasoning as above can be made, and good use can again be made of, for instance, the knowledge of two components of the field on a surface within that subshell (say the surface covered by an orbiting satellite, assuming local sources can be neglected). But it must be acknowledged that the $\mathbf{J}_s(r = b)$ sources are now seen as sources of internal origin. This then leads to the conclusion that the following sums of Gauss coefficients will also be recovered, $g_n^m + a_n^m$ and $h_n^m + b_n^m$ (recall Eq. 40), together with the Gauss coefficients q_n^m and s_n^m.

The reader will then notice that for each degree n and order m, four quantities (g_n^m, $q_n^m - ((n+1)/n)(a/b)^{2n+1} a_n^m$, $g_n^m + a_n^m$, and q_n^m) will have been recovered to constrain the three Gauss coefficients g_n^m, q_n^m, and a_n^m, while four additional quantities (h_n^m, $s_n^m - ((n+1)/n)(a/b)^{2n+1} b_n^m$, $h_n^m + b_n^m$ and s_n^m) will have been recovered to constrain the three Gauss coefficients h_n^m, s_n^m, and b_n^m. In each case, one has one constraint too many. One constraint could have therefore been dropped. Indeed, and as the reader can easily check, in such a situation when only a spherical sheet current is to be found in an otherwise source-free spherical shell $a < r < c$, the knowledge of two components (typically $B_r(r, \theta, \phi)$ and $B_\theta(r, \theta, \phi)$) on a surface within one of the subshell, and of only one component (typically $B_r(r, \theta, \phi)$ or $B_\theta(r, \theta, \phi)$) on a surface in the other subshell, is enough to fully characterize the field everywhere within the shell $a < r < c$. Beware however that all the limitations already identified in ➲ Sects. 6.1 and ➲ 6.3 when attempting to make use of either the B_ϕ component or the intensity also hold in the present case.

It is important to stress that the above results hold only because it was assumed that all sources to be found within the shell $a < r < c$ lie on a spherical infinitely thin sheet. What if the sheet is itself a subshell of some thickness (defined by say $b - e < r < b + e$)? In that case one may still use Eq. 38 to describe the field the sources within that shell would produce for $r > b + e$. However, the corresponding Gauss coefficients (a_n^m, b_n^m) may then no longer be used in Eq. 40 to predict the field the same sources would produce for $r < b - e$, because the continuity equation (Eq. 39) no longer holds. As a result, if the spherical sheet current $r = b$ does have some non negligible thickness, one must back-up from the previous conclusions. In that case, one really does need to have two independent sources of information (such as two components of the field on a surface) within each subshell, to fully characterize the field within each subshell. Also, the reader should note that the field is then still *not* fully defined within the current-carrying shell $b - e < r < b + e$. This finally brings one to the issue of uniquely defining a magnetic field within a current-carrying shell.

6.5 Uniqueness of Magnetic Fields in a Current-Carrying Shell

If the most general case of a current-carrying shell $a < r < c$ is now considered within which any type of currents can be found, the field may no longer be written in the form $\mathbf{B}(r, \theta, \phi)$

$= \mathbf{B}_i(r,\theta,\phi) + \mathbf{B}_e(r,\theta,\phi)$ (Eq. 15). But it may be written in the form (Eq. 105, recall ➱ Sect. 4):

$$\mathbf{B} = \mathbf{B}^i_{pol} + \mathbf{B}^e_{pol} + \mathbf{B}^{sh}_{pol} + \mathbf{B}^{sh}_{tor}, \tag{128}$$

where $\mathbf{B}^i_{pol}$ describes the field produced by all internal sources (referred to as the $\mathbf{J}(r < a)$ sources just above, only the toroidal type of which contribute to $\mathbf{B}^i_{pol}$, as seen in ➱ Sect. 4), $\mathbf{B}^e_{pol}$ describes the field produced by all external sources (the $\mathbf{J}(r > c)$ sources, only the toroidal type of which contribute to $\mathbf{B}^e_{pol}$), and $\mathbf{B}^{sh}_{pol}$ and $\mathbf{B}^{sh}_{tor}$ are the nonpotential poloidal and toroidal fields due to in situ toroidal $\mathbf{J}^{sh}_{tor}$ and poloidal $\mathbf{J}^{sh}_{pol}$ currents in the shell. Because of those currents, the field $\mathbf{B}$ may generally no longer be defined in a unique way everywhere in such a shell. However, if appropriate additional assumptions with respect to the nature of those local currents are introduced, some uniqueness results can again be derived. To see this, it is useful to start from the decomposition Eq. 107 of the magnetic field $\mathbf{B}(R,\theta,\phi)$ on the sphere of radius $r = R$ (recall ➱ Sect. 4):

$$\mathbf{B}(R,\theta,\phi) = \mathbf{B}_{iR}(R,\theta,\phi) + \mathbf{B}_{eR}(R,\theta,\phi) + \mathbf{B}^{sh}_{tor}(R,\theta,\phi) \tag{129}$$

where $\mathbf{B}^{sh}_{tor}(R,\theta,\phi)$, $\mathbf{B}_{iR}(R,\theta,\phi)$, and $\mathbf{B}_{eR}(R,\theta,\phi)$ are given by respectively Eqs. 108, 109, and 110. Because of the orthogonality of the vector spherical harmonics $\mathbf{C}^{m,(c,s)}_n$, $\mathbf{\Pi}^{m,(c,s)}_{ni}$, and $\mathbf{\Pi}^{m,(c,s)}_{ne}$, one may then write (recall Eqs. 63, 66, and 67):

$$\begin{aligned}\left\langle \mathbf{B}(R,\theta,\phi), \mathbf{\Pi}^{m,c}_{ni}(\theta,\phi)\right\rangle &= \frac{1}{4\pi}\int_{\theta=0}^{\pi}\int_{\phi=0}^{2\pi} \mathbf{B}(R,\theta,\phi)\cdot\mathbf{\Pi}^{m,c}_{ni}(\theta,\phi)\sin\theta\, d\theta\, d\phi\\ &= (n+1)\left(\frac{a}{R}\right)^{n+2} g^m_n(R),\end{aligned} \tag{130}$$

$$\left\langle \mathbf{B}(R,\theta,\phi), \mathbf{\Pi}^{m,s}_{ni}(\theta,\phi)\right\rangle = (n+1)\left(\frac{a}{R}\right)^{n+2} h^m_n(R), \tag{131}$$

$$\left\langle \mathbf{B}(R,\theta,\phi), \mathbf{\Pi}^{m,c}_{ne}(\theta,\phi)\right\rangle = n\left(\frac{R}{a}\right)^{n-1} q^m_n(R), \tag{132}$$

$$\left\langle \mathbf{B}(R,\theta,\phi), \mathbf{\Pi}^{m,s}_{ne}(\theta,\phi)\right\rangle = n\left(\frac{R}{a}\right)^{n-1} s^m_n(R), \tag{133}$$

$$\left\langle \mathbf{B}(R,\theta,\phi), \mathbf{C}^{m,c}_{n}(\theta,\phi)\right\rangle = \frac{n(n+1)}{2n+1} T^{m,c}_n(R), \tag{134}$$

$$\left\langle \mathbf{B}(R,\theta,\phi), \mathbf{C}^{m,s}_{n}(\theta,\phi)\right\rangle = \frac{n(n+1)}{2n+1} T^{m,s}_n(R), \tag{135}$$

which shows that the complete knowledge of $\mathbf{B}(R,\theta,\phi)$ on the spherical surface $r = R$ already makes it possible to identify the Gauss coefficients $(g^m_n(R), h^m_n(R))$ of the potential field $\mathbf{B}_{iR}(r,\theta,\phi)$ produced above $r = R$ by all sources below $r = R$, the Gauss coefficients $(q^m_n(R), s^m_n(R))$ of the potential field $\mathbf{B}_{eR}(r,\theta,\phi)$ produced below $r = R$ by all sources above $r = R$, and the coefficients $T^{m,c}_n(R)$ and $T^{m,s}_n(R)$ defining the toroidal field $\mathbf{B}^{sh}_{tor}(R,\theta,\phi)$ produced by the local (poloidal) sources at $r = R$. This uniqueness theorem, due to Backus (1986) is however not powerful enough in general to reconstruct the field $\mathbf{B}(r,\theta,\phi)$ in a unique way everywhere within the shell $a < r < c$. But one may introduce additional assumptions that reasonably apply to the near-Earth environment (see ➱ *Fig. 4* for a schematic sketch, and ➱ Chap. 5 in the handbook for more details).

Assume that the shell $a < r < c$ can be divided into two subshells $a < r < b$ and $b < r < c$ separated by a spherical sheet current at $r = b$. Assume that the lower subshell describes the

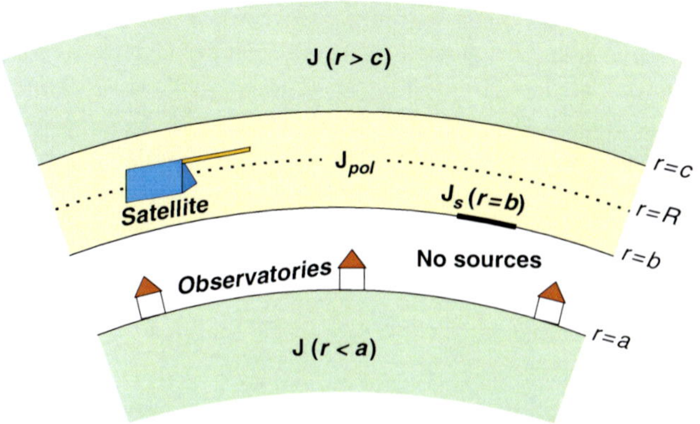

Fig. 4

Uniqueness of a magnetic field recovered from partial information within a current-carrying shell. In this special case relevant to geomagnetism, it is assumed that any source can lie below $r = a$ (*internal* $J(r < a)$ sources), and above $r = c$ (*external* $J(r > c)$ sources), no sources can lie within the lower subshell ($a < r < b$, the neutral atmosphere), a spherical sheet current can lie at $r = b$ (the *E*-region $J_s(r = b)$ sources), and only poloidal sources can lie within the upper subshell ($b < r < c$, the *F*-region ionosphere). The knowledge of B on a sphere $r = R$ in the upper subshell (as provided by, e.g., a satellite) and of enough components of B on the sphere $r = a$ (as provided by, e.g., observatories at the Earth's surface), is then enough to recover the field produced by most sources in many places (see text for details)

neutral atmosphere and is therefore source-free, while the spherical sheet current describes the ionospheric *E*-region (which is again considered as infinitely thin). Finally assume that the upper subshell describes the ionospheric current-carrying *F*-region within which near-Earth satellites orbit. Since those currents are known to be mainly the so-called *field-aligned currents* at polar latitudes (i.e., aligned with the dominant poloidal field, see, e.g., Olsen 1997), it is not unreasonable to further assume that those currents have no toroidal components (they are mainly in radial direction). In other words, assume that no $\mathbf{J}^{sh}_{tor}$ sources, but only $\mathbf{J}^{sh}_{pol}$ sources lie in the $b < r < c$ upper subshell. These assumptions, together with the previous uniqueness theorem can then be combined in a very powerful way.

Indeed, since $\mathbf{B}(R, \theta, \phi)$ is already assumed to be known on a sphere $r = R$ in the upper subshell, the Gauss coefficients $(g_n^m(R), h_n^m(R), q_n^m(R), s_n^m(R))$ can be inferred from Eqs. 130–133. The $(q_n^m(R), s_n^m(R))$ Gauss coefficients then describe the field produced below $r = R$ by all sources above $r = R$. But since between $r = R$ and $r = c$ only $\mathbf{J}^{sh}_{pol}$ sources are to be found which only produce local toroidal fields, the only sources contributing to the field described by the $(q_n^m(R), s_n^m(R))$ coefficients are the $\mathbf{J}(r > c)$ sources. It thus follows that the $(q_n^m(R), s_n^m(R))$ are the (q_n^m, s_n^m) Gauss coefficients describing the potential field produced below $r = c$ by the $\mathbf{J}(r > c)$ sources. A similar use of the Gauss coefficients $(g_n^m(R), h_n^m(R))$ recovered from Eqs. 130 and 131 can then also be made. Those describe the field produced above $r = R$ by all sources below $r = R$. But no sources between $r = b$ and $r = R$ contribute. Only $\mathbf{J}(r < a)$ and $\mathbf{J}_s(r = b)$ sources do. It then also follows that the $(g_n^m(R), h_n^m(R))$ are the sums $(g_n^m + a_n^m, h_n^m + b_n^m)$ of the Gauss coefficients describing the field produced above $r = a$ by the $\mathbf{J}(r < a)$ sources and above $r = b$ by the $\mathbf{J}_s(r = b)$ sources. This then brings one back to a situation similar to the one previously encountered when considering a shell $a < r < c$ just enclosing a spherical sheet current at $r = b$. If, in addition to knowing the field $\mathbf{B}(R, \theta, \phi)$ on the spherical surface $r = R$, enough information is also available at the Earth's surface (say at least B_r), all Gauss coefficients (g_n^m, h_n^m), (q_n^m, s_n^m), and (a_n^m, b_n^m) can again be recovered. In which case, the field produced by the $\mathbf{J}(r < a)$ sources can be predicted everywhere for $r > a$, the one produced by the $\mathbf{J}(r > c)$ sources can be predicted everywhere for $r < c$, and the field produced by $\mathbf{J}_s(r = b)$ sources can be predicted everywhere (except $r = b$). In particular, the total field $\mathbf{B}(r, \theta, \phi)$ is then completely defined within the lower subshell $a < r < b$. Note, however, and this is important, that the field $\mathbf{B}_{tor}^{sh}(r, \theta, \phi)$ (which is zero in the source-free lower subshell), is then still only known for $r = R$ and cannot be predicted elsewhere in the upper subshell (defined by $b < r < c$).

Of course, this is not the only set of assumptions one may introduce. One may first relax the infinitely thin assumption made for the spherical sheet current, which could then extend from $r = b - e$ to $r = b + e$, as was assumed earlier. Provided enough information is available within the lower subshell $a < r < b$ (say the two components $B_r(r, \theta, \phi)$ and $B_\theta(r, \theta, \phi)$ at the Earth's surface) then again, $\mathbf{J}(r < a)$ sources can be predicted everywhere for $r > a$, the one produced by the $\mathbf{J}(r > b)$ sources can be predicted everywhere for $r < b$ and the field produced by the spherical (thick) sheet sources can be predicted everywhere except within $b - e < r < b + e$. Of course, the field $\mathbf{B}_{tor}^{sh}(r, \theta, \phi)$ would still only be known for $r = R$ and would not be predicted elsewhere in the upper subshell.

In fact, and as must now be obvious to the reader, this last point is precisely one of the several issues that make satellite data difficult to take advantage of. In particular, satellites are never on a rigorously circular orbit, and do not exactly sample $\mathbf{B}(R, \theta, \phi)$ (apart from the difficulties related to the proper sampling in space and time of the temporal variations of these currents). Rather they sample drifting elliptic shells within a spherical shell of average radius R but with some thickness h. When this shell is thin enough, one may however rely on the thin-shell approximation introduced in ➲ Sect. 4, in which case one may use Eq. 111. As was then noted, this approximation is correct to within $\mathbf{B}_{pol}^{sh}$ and $\mathbf{B}_{tor}^{sh}$ corrections of order h/R. If the satellite indeed orbits within a region were only *poloidal currents* are to be found, $\mathbf{B}_{pol}^{sh} = 0$ and this correction only affects $\mathbf{B}_{tor}^{sh}$. If h/R is indeed small, this correction is small enough, and all the reasoning above may be repeated.

The practical applicability of such an approach for satellite magnetic measurements has been investigated by Backus (1986) and Olsen (1997). Olsen (1997) points out that for the Magsat 1980 mission, these numbers are $h \approx 100$ km $\ll R \approx 6{,}821$ km, which seems to justify the thin-shell approximation. Then, indeed, magnetic signatures from both field-aligned currents in the polar latitudes and meridional coupling currents associated with the equatorial electrojet (EEJ) (Maeda et al., 1982) are detected in the Magsat data, each examples of $\mathbf{B}_{tor}^{sh}$ produced by local $\mathbf{J}_{pol}^{sh}$ sources.

Of course, additional or alternative assumptions can also be used. Olsen (1997) for instance considers only radial poloidal currents in the sampling shell of Magsat; an assumption which is basically valid except for mid-latitude inter-hemispheric currents . From Eq. 68 one can obtain purely radial currents if the radial dependency of P_J is proportional to $1/r$, thus eliminating the first term. This idea can also be extended to purely meridional currents ($J_\phi = 0$), either in the standard geocentric coordinates (Olsen 1997) or in a *quasi-dipole (QD) coordinate system* (Richmond 1995), as has been done by Sabaka et al. (2004) (the QD system is a warped coordinate system useful in describing phenomena which are organized according to the Earth's main field, see Richmond (1995) and ➲ Sect. 3.1 of the chapter by Olsen et al. in the handbook).

Two classes of admissible scalar functions are then found to contribute to meridional currents: (1) those which are purely radial, and (2) those which are QD zonal, i.e. $m = 0$. Clearly only the second class contribute to the horizontal components of $\mathbf{J}$, and so a nonvanishing first term in Eq. 68 for P_J is required.

Finally, it should also be mentioned that advanced investigations of the CHAMP satellite (Maus 2007) data have recently provided several examples of situations revealing the presence of some $\mathbf{B}^{sh}_{pol}$ fields (and $\mathbf{J}^{sh}_{tor}$ sources) at satellite altitude, contradicting the assumptions described above (e.g., Lühr et al. 2002; Maus and Lühr 2006; Stolle et al. 2006). Although those fields are usually small (on the order of a few nT at most) they can be of comparable magnitude to the weakest signals produced by the smallest scales of the field of internal origin (the $\mathbf{J}(r < a)$ sources), which sets a limit to satellites ability to recover this field, despite the high quality of the measurements. No doubt that this limit is one of the greatest challenge the soon to be launched (2011) ESA's *Swarm* mission will have to face (Friis-Christensen et al. 2006, 2009).

7 Concluding Comments: From Theory to Practice

The present review was intended to provide the reader with the mathematical background relevant to geomagnetic field modeling. Often, mathematical rigor required that a number of simplifying assumptions be introduced with respect to the location of the various magnetic field sources, and to the type and distribution of magnetic observations. In particular, these observations were systematically assumed to continuously sample idealized regions (be it an idealized spherical "Earth" surface, or an idealized spherical "ionospheric" layer or shell). Also, all of the observations were implicitly assumed to be error free and synchronous in time, thereby avoiding the issue of the mathematical representation of the time variation of the various fields. Yet, those fields vary in time, sometimes quite fast in the case of the field produced by external sources, and in practice, observations are limited in number, affected by measurement errors and not always synchronous (satellites take some time to complete their orbits). These departures from the ideal situations considered in this chapter are a significant source of concern for the practical computation of geomagnetic field models based on the various mathematical properties derived here. But they can fortunately be handled. Provided relevant temporal parameterizations are introduced, appropriate data selection used, and adequate so-called inverse methods employed, geomagnetic field models defined in terms of time-varying Gauss coefficients of the type described here can indeed be computed. Details about the way this is achieved is however very much dependent on the type of observations analyzed and on the field contribution one is more specifically interested in. Examples of geomagnetic field modeling based on historical ground-based observations , with special emphasis on the main field produced within the Earth's core can be found in, e.g., Jackson et al. (2000) and Jackson and Finlay (2007). A recent example of a geomagnetic field model based on satellite data and focusing on the field produced by the magnetization within the Earth's crust is provided by Maus et al. (2008). Additional examples based on the joint use of contemporary ground-based and satellite-born observations for the modeling of both the field of internal and external origin, can otherwise be found in, e.g., Sabaka et al. (2004), Thomson and Lesur (2007), Lesur et al. (2008), Olsen et al. (2009), and in the review paper by Hulot et al. (2007), where many more references are provided. For approximation of the geomagnetic field the conventional system of vector spherical harmonics is used. An approach based on locally supported vector wavelets is studied in the next chapter (➋ Chap. 18).

Acknowledgment

This is IPGP contribution 2596.

References

Abramowitz M, Stegun I A (1964) *Handbook of mathematical functions*. Dover, New York, 1964

Alberto P, Oliveira O, Pais MA (2004) On the non-uniqueness of main geomagnetic field determined by surface intensity measurements: the Backus problem. *Geophys J Int* 159: 558–554. doi: 10.1111/j.1365-246X.2004.02413.x

Backus GE (1968) Applications of a non-linear boundary value problem for Laplace's equation to gravity and geomagnetic intensity surveys. *Q J Mech Appl Math* 21: 195–221

Backus GE (1970) Non-uniqueness of the external geomagnetic field determined by surface intensity measurements. *J Geophys Res* 75 (31): 6339–6341

Backus GE (1974) Determination of the external geomagnetic field from intensity measurements. *Geophys Res Lett* 1 (1): 21

Backus G (1986) Poloidal and toroidal fields in geomagnetic field modeling. *Rev Geophys* 24: 75–109

Backus G, Parker R, Constable C (1996) *Foundations of geomagnetism*. Cambridge Univ. Press, New York

Barraclough DR, Nevitt C (1976) The effect of observational errors on geomagnetic field models based solely on total-intensity measurements. *Phys Earth Planet Int* 13: 123–131

Blakely RJ (1999) *Potential theory in gravity and magnetic applications*. Cambridge Press, Cambridge

Bloxham J (1985) *Geomagnetic secular variation*. PhD thesis, Cambridge University

Dahlen F, Tromp J (1998) *Theoretical global seismology*. Princeton University Press, Princeton

Edmonds A (1996) *Angular momentum in quantum mechanics*. Princeton University Press, Princeton

Elsasser W (1946) Induction effects in terrestrial magnetism. Part I. Theory. *Phys Rev* 69 (3-4): 106–116

Ferrers NM (1877) *An elementary treatise on spherical harmonics and subjects connected with them*. Macmillan, London

Friis-Christensen E, Lühr H, Hulot G (2006) *Swarm*: a constellation to study the Earth's magnetic field. *Earth Planets Space* 58: 351–358

Friis-Christensen E, Lühr H, Hulot G, Haagmans R, Purucker M (2009) Geomagnetic research from space. *Eos* 90: 25

Gauss CF (1839) *Allgemeine Theorie des Erdmagnetismus*. Resultate aus den Beobachtungen des Magnetischen Vereins im Jahre 1838. Göttinger Magnetischer Verein, Leipzig

Goldie AHR, Joyce JW (1940) In: *Proceedings of the 1939 Washington Assembly of the Association of Terrestrial Magnetism and Electricity of the International Union of Geodesy and Geophysics*, vol 11 (6). Neill & Co, Edinburgh

Granzow DK (1983) Spherical harmonic representation of the magnetic field in the presence of a current density. *Geophys J R Astron Soc* 74: 489–505

Harrison CGA (1987) The crustal field. In: Jacobs JA (ed) *Geomagnetism*, vol 1. Academic, London, pp 513–610

Holme R, Bloxham J (1995) Alleviation of the backus effect in geomagnetic field modelling. *Geophys Res Lett* 22: 1641–1644

Holme R, James MA, Lühr H (2005) Magnetic field modelling from scalar-only data: resolving the Backus effect with the equatorial electrojet. *Earth Planets Space* 57: 1203–1209

Hulot G, Khokhlov A, Le Mouël JL (1997) Uniqueness of mainly dipolar magnetic fields recovered from directional data. *Geophys J Int* 129: 347–354

Hulot G, Sabaka TJ, Olsen N (2007) The present field. In: Kono M (ed) *Treatise on geophysics*, vol 5. Elsevier, Amsterdam

Jackson J (1998) *Classical electrodynamics*. Wiley, New York

Jackson A, Finlay CC (2007) Geomagnetic secular variation and its application to the core. In: Kono M, (ed) *Treatise on geophysics*, vol 5. Elsevier, Amsterdam

Jackson A, Jonkers ART, Walker MR (2000) Four centuries of geomagnetic secular variation from historical records. *Phil Trans R Soc Lond* A 358: 957–990

Kellogg OD (1954) *Foundations of potential theory*. Dover, New York

Khokhlov A, Hulot G, Le Mouël JL (1997) On the Backus effect - I. *Geophys J Int* 130: 701–703

Khokhlov A, Hulot G, Le Mouël JL (1999) On the Backus effect - II. *Geophys J Int* 137: 816–820

Kono M (1976) Uniqueness problems in the spherical analysis of the geomagnetic field direction data. *J Geomagn Geoelectr* 28: 11–29

Langel RA (1987) The main field. In: Jacobs JA (ed) *Geomagnetism*, vol 1. Academic, London, pp 249–512

Langel RA, Hinze WJ (1998) *The magnetic field of the Earth's lithosphere: the satellite perspective.* Cambridge University Press, Cambridge

Lesur V, Wardinski I, Rother M, Mandea M (2008) GRIMM: the GFZ reference internal magnetic model based on vector satellite and observatory data. *Geophys J Int* 173: 382–294

Lorrain P, Corson D (1970) *Electromagnetic fields and waves.* WH Freeman, San Francisco

Lowes FJ (1966) Mean-square values on sphere of spherical harmonic vector fields. *J Geophys Res* 71: 2179

Lowes FJ (1974) Spatial power spectrum of the main geomagnetic field, and extrapolation to the core. *Geophys J R Astr Soc* 36: 717–730

Lowes FJ (1975) Vector errors in spherical harmonic analysis of scalar data. *Geophys J R Astron Soc* 42: 637–651

Lowes FJ, De Santis A, Duka B (1999) A discussion of the uniqueness of a Laplacian potential when given only partial information on a sphere. *Geophys J Int* 121: 579–584

Lühr H, Maus S, Rother M (2002) First in-situ observation of night-time *F* region currents with the CHAMP satellite. *Geophys Res Lett* 29 ((10):127.1–127.4. doi: 10.1029/2001 GL 013845)

Maeda H, Iyemori T, Araki T, Kamei T (1982) New evidence of a meridional current system in the equatorial ionosphere. *Geophys Res Lett* 9: 337–340

Malin S (1987) Historical introduction to geomagnetism. In: Jacobs JA (ed) *Geomagnetism*, vol 1. Academic, London, pp 1–49

Mauersberger P (1956) Das Mittel der Energiedichte des geomagnetischen Hauptfeldes an der Erdoberfläche und seine säkulare Änderung. *Gerl Beitr Geophys* 65: 207–215

Maus S (2007) CHAMP magnetic mission. In: Gubbins D, Herrero-Bervera E (eds) *Encyclopedia of geomagnetism and paleomagnetism.* Springer, Heidelberg

Maus S, Lühr H (2006) A gravity-driven electric current in the earth's ionosphere identified in champ satellite magnetic measurements. *Geophys Res Lett* 33: L02812. doi:10.1029/2005GL024436

Maus S, Yin F, Lühr H, Manoj C, Rother M, Rauberg J, Michaelis I, Stolle C, Müller R (2008) Resolution of direction of oceanic magnetic lineations by the sixth-generation lithospheric magnetic field model from CHAMP satellite magnetic measurements. *Geochem Geophys Geosyst* 9 (7): Q07021

Merrill R, McElhinny M (1983) *The Earth's magnetic field.* Academic, London

Merrill R, McFadden P, McElhinny M (1998) *The magnetic field of the Earth: paleomagnetism, the core, and the deep mantle.* Academic, London

Mie G (1908) Considerations on the optic of turbid media, especially colloidal metal sols. *Ann Phys(Leipzig)* 25: 377–442

Morse P, Feshbach H (1953) *Methods of theoretical physics.* International series in pure and applied physics. McGraw-Hill, New York

Olsen N (1997) Ionospheric *F* region currents at middle and low latitudes estimated from Magsat data. *J Geophys Res* 102 (A3): 4563–4576

Olsen N, Mandea M, Sabaka TJ, Tøffner-Clausen L (2009) CHAOS-2—a geomagnetic field model derived from one decade of continuous satellite data. *Geophys J Int* 199(3): 1477–1487. doi: doi:10.1111/j.1365-246X.2009.04386.x

Proctor MRE, Gubbins D (1990) Analysis of geomagnetic directional data. *Geophys J Int* 100: 69–77

Purucker M, Whaler K (2007) Crustal magnetism. In: Kono M (ed) *Treatise on geophysics*, vol 5. Elsevier, Amsterdam, pp 195–235

Richmond AD (1995) Ionospheric electrodynamics using magnetic Apex coordinates. *J Geomagn Geoelectr* 47: 191–212

Sabaka TJ, Olsen N, Langel RA (2002) A comprehensive model of the quiet-time near-Earth magnetic field: Phase 3. *Geophys J Int* 151: 32–68

Sabaka TJ, Olsen N, Purucker ME (2004) Extending comprehensive models of the Earth's magnetic field with Ørsted and CHAMP data. *Geophys J Int* 159: 521–547. doi: 10.1111/j.1365-246X.2004.02421.x

Schmidt A (1935) *Tafeln der Normierten Kugelfunktionen.* Engelhard-Reyher Verlag, Gotha

Stern DP (1976) Representation of magnetic fields in space. *Rev Geophys* 14: 199–214

Stern DP, Bredekamp JH (1975) Error enhancement in geomagnetic models derived from scalar data. *J Geophys Res* 80: 1776–1782

Stern DP, Langel RA, Mead GD (1980) Backus effect observed by Magsat. *Geophys Res Lett* 7: 941–944

Stolle C, Lühr H, Rother M, Balasis G (2006) Magnetic signatures of equatorial spread *F*, as observed by the CHAMP satellite. *J Geophys Res* 111: A02304. doi:10.1029/2005JA011184

Thomson AWP, Lesur V (2007) An improved geomagnetic data selection algorithm for global geomagnetic field modelling. *Geophys J Int* 169 (3): 951–963

Ultré-Guérard P, Hamoudi M, Hulot G (1998) Reducing the Backus effect given some knowledge of the dip-equator. *Geophys Res Lett* 22 (16): 3201–3204

Walker AD (1992) Comment on "Non-uniqueness of the external geomagnetic field determined by surface intensity measurements" by Georges E. Backus. *J Geophys Res* 97 (B10): 13991

Watson GN (1966) *A treatise on the theory of Bessel function*. Cambridge University Press, London

Winch D, Ivers D, Turner J, Stening R (2005) Geomagnetism and Schmidt quasi-normalization. *Geophys J Int* 160 (2): 487–504

18 Multiscale Modeling of the Geomagnetic Field and Ionospheric Currents

Christian Gerhards
Geomathematics Group, Department of Mathematics, University of Kaiserslautern, Kaiserslautern, Germany

W. Freeden, M.Z. Nashed, T. Sonar (Eds.), *Handbook of Geomathematics*, DOI 10.1007/978-3-642-01546-5_18,

Abstract This chapter reports on the recent application of multiscale techniques to the modeling of geomagnetic problems. Two approaches are presented: a spherical harmonics-oriented one, using frequency packages, and a spatially oriented one, using regularizations of the single layer kernel and Green's function with respect to the Beltrami operator. As an example both approaches are applied to the separation of the magnetic field with respect to interior and exterior sources and the reconstruction of radial ionospheric currents.

1 Introduction

The last decade has severely improved the understanding of the Earth's magnetic field due to high-precision vector satellite data delivered from the Ørsted and CHAMP satellite missions that were launched in 1999 and 2000, respectively. A new satellite constellation (Swarm) is anticipated to be launched in 2011 to deliver even more accurate measurements. Satellite data generally contain contributions from various sources of the Earth's magnetic field, which are mainly due to dynamo processes in the Earth's interior, electric currents in the iono- and magnetosphere, and static magnetization in the Earth's crust in combination with induction processes (cf. ➋ *Fig. 1*). A major task is the separation of these sources (for more details the reader is referred to ➋ Chaps. 5 and ➋ 19). Beside adequate data pre-selection (e.g., night time data and magnetically quiet data to avoid iono- and magnetospheric contributions), this requires a great effort in modeling the different contributions (e.g., the constantly revised IGRF for the Earth's main field). Classical texts on these topics are, for example, Hulot et al. (2007), Langel (1987), and Langel and Hinze (1998).

In this chapter we mainly regard the magnetic field of the Earth's exterior that is governed by Maxwell's equations. Length and time scales of Earth's magnetic field, however, are such that the displacement currents can be neglected (cf. Backus et al. 1996). Therefore, it is reasonable to concentrate the modeling effort on the pre-Maxwell equations

$$\nabla \wedge b = j, \tag{1}$$

$$\nabla \cdot b = 0, \tag{2}$$

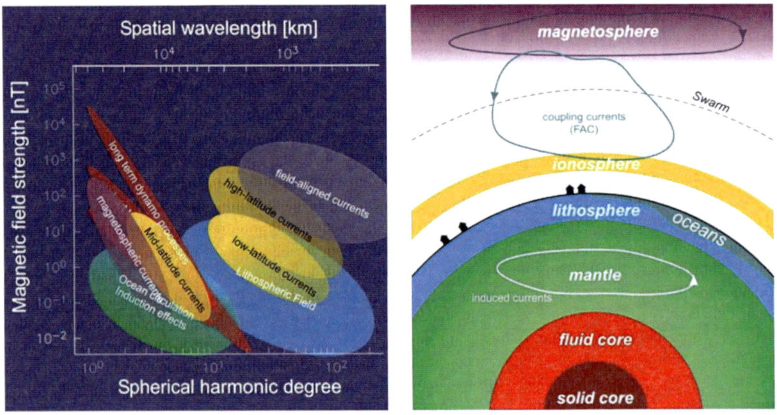

Fig. 1
Schematic description of Earth's magnetic field contributions (cf. Haagmans et al. (2008))

where b denotes the magnetic field and j the electrical current density (the unit system is chosen such that the vacuum permeability μ_0 occurring in the SI system is equal to 1). In regions where no such currents are present, for example the spherical shell $\Omega_{(R_0,R_1)}$ between the Earth's surface Ω_{R_0} and the lower bound of the ionosphere Ω_{R_1} (here Ω_{R_i} denote spheres of radius R_i, $i = 0, 1$, with center at zero), the classical Gauss representation, basically introduced in Gauss (1839), yields b to be a potential field ∇U, with U a scalar harmonic function in $\Omega_{(R_0,R_1)}$. Denoting by $\{Y_{n,k}\}_{n=0,1,\ldots,\,k=1,\ldots,2n+1}$ an orthonormal system of spherical harmonics, the scalar potential can then be represented as

$$U(x) = \sum_{n=0}^{\infty}\sum_{k=1}^{2n+1} a_{n,k}(r)Y_{n,k}(\xi), \quad R_0 < r = |x| < R_1,\ x = r\xi, \tag{3}$$

with adequate scalars $a_{n,k}$ depending on the radius r. More precisely, the radial dependence is given by

$$a_{n,k}(r) = c_{n,k}\, r^n + d_{n,k}\frac{1}{r^{n+1}}, \tag{4}$$

representing the two physically relevant solutions for the potential U: those that are harmonic in the interior $\Omega_{R_1}^{int}$ and those harmonic in the exterior $\Omega_{R_0}^{ext}$. The constants $c_{n,k}$, $d_{n,k}$ can be uniquely determined from vector data given on a single sphere Ω_r contained in the spherical shell $\Omega_{(R_0,R_1)}$ (cf. Backus et al. 1996).

Satellite measurements, however, are generally conducted in regions of ionospheric currents, that is, $j \neq 0$. So Gauss' representation breaks down for modeling from such data sources. As a substitute, Backus (1986) and Gerlich (1972) suggested the Mie representation for geomagnetic applications. Any solenoidal vector field in a shell $\Omega_{(R_1,R_2)}$, such as the magnetic field b, can be decomposed into

$$b = p_b + q_b = \nabla \wedge LP_b + LQ_b, \tag{5}$$

with uniquely defined vector fields p_b, q_b and corresponding scalars P_b, Q_b that are uniquely determined by

$$\frac{1}{4\pi}\int_\Omega P_b(r\eta)d\omega(\eta) = \frac{1}{4\pi}\int_\Omega Q_b(r\eta)d\omega(\eta) = 0, \quad R_1 < r < R_2. \tag{6}$$

(Here ∇ denotes the gradient and $L_x = x \wedge \nabla_x$ the curl gradient.) The term p_b is known as the toroidal part of the magnetic field and has a vanishing surface curl, $L^* \cdot p_b = 0$. The toroidal part q_b, on the other hand, has vanishing surface divergence, $\nabla^* \cdot q_b = 0$, and is purely tangential. (∇^* denotes the surface gradient, L_ξ^* the surface curl gradient $\xi \wedge \nabla_\xi^*$, and Δ^* the Beltrami operator $\nabla^* \cdot \nabla^*$.) Due to (1), the corresponding current density allows a Mie representation as well,

$$j = p_j + q_j = \nabla \wedge LP_j + LQ_j. \tag{7}$$

In combination with the pre-Maxwell equations, this yields a fundamental connection for the Mie scalars of the magnetic field and its source currents,

$$Q_b = P_j, \tag{8}$$

$$\Delta P_b = -Q_j. \tag{9}$$

In other words, toroidal magnetic fields are caused by poloidal currents and poloidal fields by toroidal currents. Furthermore, the Mie scalars represent the analogon to the scalar potential U, although they are not necessarily harmonic, and can be expanded in terms of spherical harmonics.

It is the aim of this chapter, however, to present alternatives to the classical Fourier expansion in terms of spherical harmonics. In ➋ Sect. 2, we present two sets of vectorial counterparts to scalar spherical harmonics avoiding the detour via scalar potentials, as well as Green's function for the Beltrami operator as a more space-oriented approach. ➋ Section 3 introduces general concepts of multiscale techniques to circumvent the problems arising from scalar or vectorial Fourier expansions. These techniques are exemplarily applied to the separation of the magnetic field with respect to interior and exterior sources and the reconstruction of radial currents in ➋ Sect. 4. The remaining two sections give a short outlook on future perspectives and a final conclusion.

2 Scientifically Relevant Function Systems

Fourier expansions on the sphere in terms of spherical harmonics are the classical approach of potential field modeling in geosciences. In geomagnetism, this relates to the potential U in source-free regions and the poloidal and toroidal scalars P_b and Q_b in non-source-free regions. An extensive overview on the mathematics of magnetic field modeling in terms of spherical harmonics is, for example, given in the contribution Sabaka et al. (2009) of this handbook. Any square integrable function F of class $L^2(\Omega)$ can be represented as

$$F(\xi) = \sum_{n=0}^{\infty} \sum_{k=1}^{2n+1} F^{\wedge}(n,k) Y_{n,k}(\xi), \quad \xi \in \Omega, \tag{10}$$

with Fourier coefficients

$$F^{\wedge}(n,k) = \int_{\Omega} F(\eta) Y_{n,k}(\eta)\, d\omega(\eta). \tag{11}$$

A vectorial counterpart that is forming a complete orthonormal system in the space $l^2(\Omega)$ of all square integrable vector valued functions on the sphere can avoid the detour via scalar potentials in the generally vectorial framework of geomagnetics. The main motivation for the definition of a vectorial basis system is the Helmholtz representation, decomposing any continuously differentiable vector field f of class $c^{(1)}(\Omega)$ into a normal and a tangential component,

$$f(\xi) = \xi F_1(\xi) + \nabla_\xi^* F_2(\xi) + L_\xi^* F_3(\xi), \quad \xi \in \Omega, \tag{12}$$

with uniquely defined F_1 of class $C^{(1)}(\Omega)$ and F_2, F_3 of class $C^{(2)}(\Omega)$ satisfying

$$\frac{1}{4\pi} \int_{\Omega} F_2(\eta)\, d\omega(\eta) = \frac{1}{4\pi} \int_{\Omega} F_3(\eta)\, d\omega(\eta) = 0. \tag{13}$$

The connection to geomagnetic modeling is given by a combination of the Helmholtz and the Mie representation of the magnetic field,

$$b(x) = \xi \frac{\Delta_\xi^* P_b(r\xi)}{r} - \nabla_\xi^* \frac{\frac{\partial}{\partial r}(r P_b(r\xi))}{r} + L_\xi^* Q_b(r\xi), \quad r = |x|,\ x = r\xi. \tag{14}$$

A Fourier expansion of b with respect to Helmholtz representation oriented vector spherical harmonics, as introduced in ➋ Sect. 2.1, then directly delivers the poloidal and toroidal part of the magnetic field. The calculation of the actual Mie scalars, however, requires to solve the differential equations

$$\Delta_\xi^* P_b(r\xi) = r\xi \cdot b(r\xi), \quad \Delta_\xi^* Q_b(r\xi) = L_\xi^* \cdot b(r\xi), \quad \xi \in \Omega. \tag{15}$$

Solutions to these equations can as well be expressed in terms of spherical harmonics. Green's function with respect to the Beltrami operator, however, offers simple integral representations. Since it has a closed representation in space domain, it is a reasonable foundation for geomagnetic modeling from a less frequency-oriented point of view. More details on this can be found in ➲ Sect. 2.2.

2.1 Vector Spherical Harmonics

For further applications we define the three mutually orthogonal surface operators $o_\xi^{(1)} F(\xi) = \xi F(\xi)$, $o_\xi^{(2)} F(\xi) = \nabla_\xi^* F(\xi)$ and $o_\xi^{(3)} F(\xi) = L_\xi^* F(\xi)$, with F a sufficiently smooth scalar function. A system of vector spherical harmonics of type i is then defined by

$$y_{n,k}^{(i)} = \left(\mu_n^{(i)}\right)^{-\frac{1}{2}} o^{(i)} Y_{n,k}, \quad i = 1,2,3, \tag{16}$$

for $n = 0_i, 0_i + 1, \ldots, k = 1, \ldots, 2n+1$, with $0_1 = 0$, $0_2 = 0_3 = 1$, and normalization constants $\mu_n^{(1)} = 1$, $\mu_n^{(2)} = \mu_n^{(3)} = n(n+1)$. Due to the Helmholtz theorem, orthonormality and completeness carry over from the scalar case. In other words, any function f of class $l^2(\Omega)$ can be represented as a Fourier series

$$f(\xi) = \sum_{i=1}^{3} \sum_{n=0_i}^{\infty} \sum_{k=1}^{2n+1} (f^{(i)})^\wedge(n,k)\, y_{n,k}^{(i)}(\xi), \quad \xi \in \Omega, \tag{17}$$

with Fourier coefficients

$$(f^{(i)})^\wedge(n,k) = \int_\Omega f(\eta) \cdot y_{n,k}^{(i)}(\eta)\, d\omega(\eta). \tag{18}$$

This way we have defined a vectorial orthonormal basis system for $l^2(\Omega)$ that decomposes a vector field with respect to the Helmholtz representation. Fundamental properties such as the addition theorem and the Funk–Hecke formula can be similarly formulated in a vectorial framework. A detailed overview on this is found in Freeden et al. (1998), Freeden and Schreiner (2009) and Freeden and Schreiner (2010).

There are, however, other ways to define vector basis systems. A system especially useful to separation of sources requires the operators $\tilde{o}^{(1)} = o^{(1)}\left(D + \frac{1}{2}\right) - o^{(2)}$, $\tilde{o}^{(2)} = o^{(1)}\left(D - \frac{1}{2}\right) + o^{(2)}$, $\tilde{o}^{(3)} = o^{(3)}$, with

$$D = \left(-\Delta^* + \frac{1}{4}\right)^{\frac{1}{2}}. \tag{19}$$

The operator D is characterized by the spherical harmonics $Y_{n,k}$ being eigenfunctions to the eigenvalues $n + \frac{1}{2}$. Its inverse is also called the single layer operator. We define an alternative system of vector spherical harmonics of type i by

$$\tilde{y}_{n,k}^{(i)} = \left(\tilde{\mu}_n^{(i)}\right)^{-\frac{1}{2}} \tilde{o}^{(i)} Y_{n,k}, \quad i = 1,2,3, \tag{20}$$

for $n = 0_i, 0_i + 1, \ldots, k = 1, \ldots, 2n+1$, and constants $\tilde{\mu}_n^{(1)} = (n+1)(2n+1)$, $\tilde{\mu}_n^{(2)} = n(2n+1)$, and $\tilde{\mu}_n^{(3)} = n(n+1)$. The completeness property of the scalar spherical harmonics is again conserved but the orthogonality gets lost. The advantage of this system is its connection to outer and inner harmonics. More precisely,

$$\nabla_x H_{n,k}^{int}(R;x) = \frac{1}{R^2}\left(\frac{r}{R}\right)^{n-1}\left(\tilde{\mu}_n^{(2)}\right)^{\frac{1}{2}} \tilde{y}_{n,k}^{(2)}(\xi), \quad r = |x| < R, x = r\xi, \tag{21}$$

$$-\nabla_x H_{n,k}^{ext}(R;x) = \frac{1}{R^2}\left(\frac{R}{r}\right)^{n+2}\left(\tilde{\mu}_n^{(1)}\right)^{\frac{1}{2}} \tilde{y}_{n,k}^{(1)}(\xi), \quad r = |x| > R, x = r\xi. \tag{22}$$

By $H_{n,k}^{int}(R;\cdot)$ we denote the unique solution to the Laplace equation $\Delta U = 0$ in Ω_R^{int} with boundary value $U = \frac{1}{R} Y_{n,k}\left(\frac{\cdot}{R}\right)$ on Ω_R, and $H_{n,k}^{ext}(R;\cdot)$ the solution to the corresponding boundary value problem in the exterior Ω_R^{ext}. For more details on the above described vectorial basis system, the reader is referred to Backus (1986), Edmonds (1957), and Mayer (2003).

2.2 Green's Function for the Beltrami Operator

As Green's function with respect to the Beltrami operator we define the rotationally invariant function $G(\Delta^*;\cdot)$ satisfying

$$\Delta_\xi^* G(\Delta^*;\xi\cdot\eta) = -\frac{1}{4\pi}, \quad -1 \le \xi\cdot\eta < 1, \tag{23}$$

for $\xi, \eta \in \Omega$, with a singularity of type

$$G(\Delta^*;\xi\cdot\eta) = O(\ln(1-\xi\cdot\eta)). \tag{24}$$

Uniqueness is guaranteed by the claim that $G(\Delta^*;\cdot)$ has vanishing integral mean value on the sphere. Observing that the spherical harmonics $Y_{n,k}$ are eigenfunctions to the Beltrami operator with eigenvalues $-n(n+1)$, one can derive the Fourier representation

$$G(\Delta^*;\xi\cdot\eta) = \sum_{n=1}^{\infty}\sum_{k=1}^{2n+1} \frac{1}{-n(n+1)} Y_{n,k}(\xi) Y_{n,k}(\eta), \tag{25}$$

for $\xi, \eta \in \Omega$, $-1 \le \xi\cdot\eta < 1$. It should be remarked that the missing zeroth order term is responsible for $\eta \mapsto \Delta_\xi^* G(\Delta^*;\xi\cdot\eta)$ differing by the constant $\frac{1}{4\pi}$ from the Dirac kernel. On the other hand, this property allows the comparison of a twice continuously function on the unit sphere with its integral mean value and an error term involving Green's function. This is stated in

Theorem 1
Let F be of class $C^{(2)}(\Omega)$ Then

$$F(\xi) = \frac{1}{4\pi}\int_\Omega F(\eta)d\omega(\eta) + \int_\Omega G(\Delta^*;\xi\cdot\eta)\,\Delta_\eta^* F(\eta)\,d\omega(\eta), \quad \xi \in \Omega.$$

As a consequence of the above representation, it is straight forward to see that

$$F(\xi) = \int_\Omega G(\Delta^*;\xi\cdot\eta)H(\eta)\,d\omega(\eta), \quad \xi \in \Omega. \tag{26}$$

is the uniquely defined solution with vanishing integral mean value to the Beltrami equation

$$\Delta_\xi^* F(\xi) = H(\xi), \quad \xi \in \Omega, \tag{27}$$

where H is a continuous function satisfying $\frac{1}{4\pi}\int_\Omega H(\eta)\,d\omega(\eta) = 0$. For a detailed overview on Green's function the reader is again referred to Freeden et al. (1998) and Freeden and Schreiner (2009). What makes Green's function interesting in comparison to a Fourier expansion is its closed representation

$$G(\Delta^*;\xi\cdot\eta) = \frac{1}{4\pi}\ln(1-\xi\cdot\eta) + \frac{1}{4\pi}(1-\ln(2)), \tag{28}$$

for $\xi,\eta \in \Omega$, $-1 \le \xi\cdot\eta < 1$. This way one can represent the modeled quantities in a more spatially oriented framework than it would be possible by any kind of spherical harmonic expansion. It does not deliver a basis system like spherical harmonics do but gives simple integral representations for quantities that are governed by differential equations involving the surface operators ∇^*, L^*, and Δ^*, to which many geomagnetic problems (e.g., Mie scalars, radial currents, separation of sources) can be reduced.

3 Key Issues for Multiscale Techniques

Characteristic for spherical harmonics, scalar as well as vectorial, is their perfect localization in frequency domain. In fact, each spherical harmonic $Y_{n,k}$ corresponds to a fixed frequency $\frac{2\pi}{n}$. The uncertainty principle, though, states that exact frequency and space localization are mutually exclusive. For our modeling effort, this implies that small local changes in the magnetic field affect every Fourier coefficient in a spherical harmonic expansion. That effect is especially severe for data that is only irregularly distributed (as is the case for geomagnetic ground data). As a remedy, spherical cap harmonic analysis has been introduced in Haines (1985) and revised in Thebault et al. (2006). General spherical harmonics are substituted by an orthonormal system of harmonic functions on a spherical cap satisfying certain boundary conditions on the function itself and its first latitudinal derivative. This way local Fourier expansions become possible, but at the price of neglecting any data outside the region under consideration, which can lead to unrealistic errors in the approximation. An alternative is the transition from spherical harmonics to zonal kernel functions of improved spatial localization. More precisely, we define

$$\Phi(\xi,\eta) = \sum_{n=0}^{\infty}\sum_{k=1}^{2n+1}\Phi^\wedge(n)Y_{n,k}(\xi)Y_{n,k}(\eta), \quad \xi,\eta\in\Omega, \tag{29}$$

with coefficients $\Phi^\wedge(n)$ satisfying

$$\sum_{n=0}^{\infty}\frac{2n+1}{4\pi}|\Phi^\wedge(n)|^2 < \infty. \tag{30}$$

One can distinguish four main types of kernels: band-limited and non-band-limited, space-limited and non-space-limited ones (➋ *Fig. 2*). Band-limited kernels are characterized by $\Phi^\wedge(n) = 0$, $n \ge N$, for some sufficiently large N. In other words, they are strongly localized in frequency domain. Perfect frequency localization would be given by the Legendre kernels, defined via the symbols $\Phi^\wedge(n) = \frac{4\pi}{2n+1}$, for $n = m$ with a fixed integer m, and $\Phi^\wedge(n) = 0$ otherwise. The non-band-limited counterparts generally show a much stronger spatial localization. In fact, the uncertainty principle demands any space-limited kernel (i.e., $\Phi(\xi,\cdot)$ has locally compact support) to be non-band-limited. Perfect spatial localization is given by the Dirac kernel with $\Phi^\wedge(n) = 1$ for all $n = 0,1,\ldots$ (which, however, is only to be understood

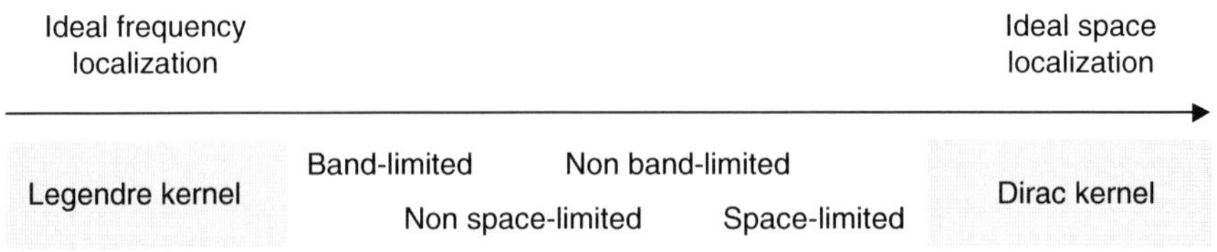

Fig. 2
Frequency/space localization of zonal kernels

in a distributional sense since (30) is not satisfied). Non-space-limited kernels, on the other hand, can be both, band-limited or non-band-limited. For a more precise quantitative categorization of space and frequency localization, the reader is referred to Freeden (1998) and the contribution Freeden and Schreiner (2010) of this handbook.

A scale-dependent transition from frequency localizing to space localizing kernels can be achieved by symbols $\Phi_J^\wedge(n)$ that are tending to one as the scale J tends to infinity. This allows the analysis of given data at different scales of spatial resolution. A band-limited example is given by the cubic polynomial (CP) kernel

$$\Phi_J^\wedge(n) = \begin{cases} \left(1 - n2^{-J}\right)^2 \left(1 + n2^{-(J-1)}\right), & n \le 2^J - 1, \\ 0, & \text{else}, \end{cases} \tag{31}$$

a non-band-limited one by the Abel–Poisson kernel

$$\Phi_J^\wedge(n) = e^{-n\,2^{-J}}, \quad n = 0, 1, \dots . \tag{32}$$

Figure 3 illustrates the evolution of space and frequency localization of these two kernels at different scales J. Such kernels can be used to deliver rough global approximations for small scales and sparse data sets, while higher, better space localizing scales deliver high resolution reconstructions in regions of dense data coverage.

A second problem occurring with spherical harmonics is, however, not fixed by the above approach: for high degrees, spherical harmonics become strongly oscillating which complicates the numerical evaluation of occurring convolutions. Kernels that have a closed representation in terms of elementary functions are more advantageous here. They are especially suitable if they have local support since this reduces the evaluation of the occurring convolutions to integrations over small bounded regions on the sphere (cf. Michel (2006)). Approaches to such functions can be found in Sect. 3.2, while Sect. 3.1 concentrates on the definition of a more frequency-oriented multiscale analysis with scale-dependent kernels of type (29).

3.1 Wavelets as Frequency Packages

We now give a more precise description of how the above outlined method of superposing spherical harmonics to kernels/frequency packages of better spatial localization can be used for a multiscale reconstruction of square integrable functions. Directly regarding the vectorial case (as it is desirable for geomagnetic modeling), we set

$$\phi_J^{(i)}(\xi,\eta) = \sum_{n=0_i}^{\infty} \sum_{k=1}^{2n+1} \left(\phi_J^{(i)}\right)^\wedge(n)\, Y_{n,k}(\xi)\, y_{n,k}^{(i)}(\eta), \quad \xi, \eta \in \Omega, \tag{33}$$

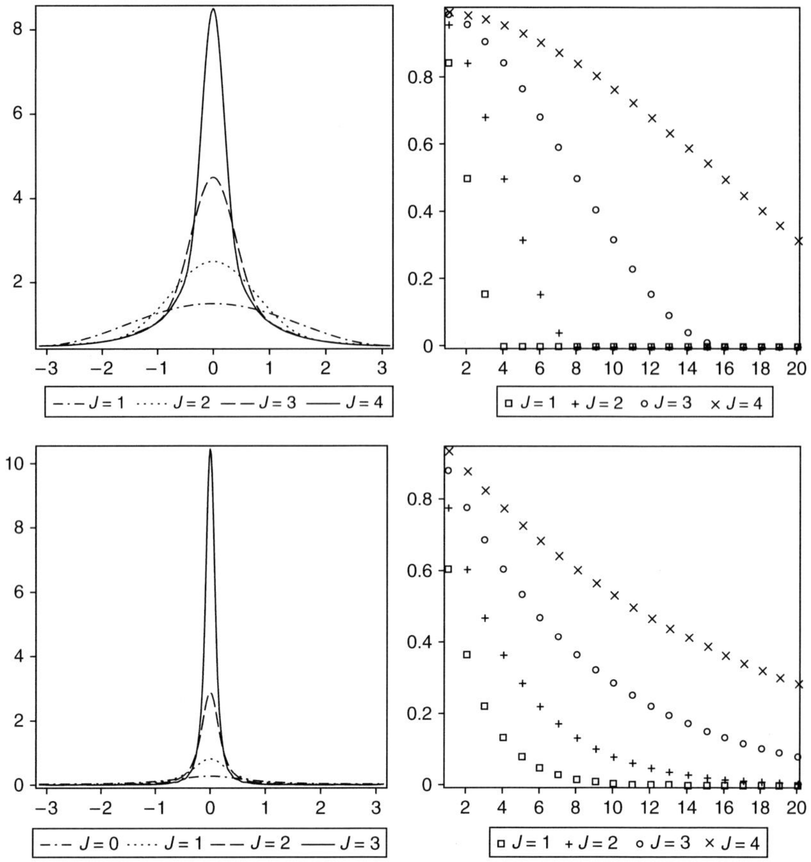

Fig. 3
CP kernel with respect to angular distance (*top left*) and corresponding Fourier coefficients (*top right*). Analogous plots for the Abel–Poisson kernel (*bottom*)

for $i = 1, 2, 3$. This kernel is called a scaling function of type i if the symbols satisfy

$$\text{(i)}\quad \lim_{J\to\infty}\left(\left(\phi_J^{(i)}\right)^{\wedge}(n)\right)^2 = 1, \quad n = 0_i, 0_i + 1, \ldots, \tag{34}$$

$$\text{(ii)}\quad \sum_{n=0}^{\infty} \frac{2n+1}{4\pi}\left(\left(\phi_J^{(i)}\right)^{\wedge}(n)\right)^2 < \infty, \quad J \in \mathbb{Z}, \tag{35}$$

$$\text{(iii)}\quad \left(\left(\phi_{J+1}^{(i)}\right)^{\wedge}(n)\right)^2 \geq \left(\left(\phi_J^{(i)}\right)^{\wedge}(n)\right)^2, \quad J \in \mathbb{Z},\ n = 0_i, 0_i + 1, \ldots. \tag{36}$$

The CP as well as the Abel–Poisson symbols satisfies all three conditions. Condition (i) is responsible for the actual approximation property, which states for a function f of class $l^2(\Omega)$

that $f = f^{(1)} + f^{(2)} + f^{(3)}$ with

$$f^{(i)} = \lim_{J\to\infty} \phi_J^{(i)} \star \phi_J^{(i)} * f = \lim_{J\to\infty} \sum_{n=0_i}^{\infty} \sum_{k=1}^{2n+1} \left(\left(\phi_J^{(i)}\right)^{\wedge}(n) \right)^2 \left(f^{(i)}\right)^{\wedge}(n,k)\, y_{n,k}^{(i)}. \tag{37}$$

For brevity, we defined the scalar valued convolution $\phi_J * f$ of a vectorial scaling kernel against a vectorial function f of class $l^2(\Omega)$ by $\int_\Omega \phi_J(\cdot,\eta)\cdot f(\eta)d\omega(\eta)$, and the vector valued convolution $\phi_J \star F$ of a vectorial scaling kernel against a scalar function F of class $L^2(\Omega)$ by $\int_\Omega \phi_J(\eta,\cdot)F(\eta)d\omega(\eta)$. Condition (iii) implies that the scale spaces

$$\mathcal{V}_J^{(i)} = \left\{ \phi_J^{(i)} \star \phi_J^{(i)} * f \,|\, f \in l^2(\Omega) \right\}, \quad J \in \mathbb{Z},\ i = 1,2,3, \tag{38}$$

are nested in the sense

$$\mathcal{V}_J^{(i)} \subset \mathcal{V}_{J+1}^{(i)} \subset \cdots \subset l^2(\Omega), \quad J \in \mathbb{Z},\ i = 1,2,3. \tag{39}$$

In other words, the approximation error is monotonically decreasing. This property, where the set of scale spaces $\mathcal{V}_J^{(i)}$ is also called a multiresolution analysis, is generally only satisfied for non-space-limited kernels (at least all known geophysically relevant locally supported kernels do not satisfy condition (iii)). Approximation (37) can be further split up to pay tribute to irregularly distributed data. The goal is a unified representation that can supply high resolution approximations in regions of high data density and less resolving approximations in regions of sparse data coverage. Therefore, we define so called wavelet kernels of type i and scale J,

$$\psi_J^{(i)}(\xi,\eta) = \sum_{n=0_i}^{\infty} \sum_{k=1}^{2n+1} \left(\psi_J^{(i)}\right)^{\wedge}(n) Y_{n,k}(\xi) y_{n,k}^{(i)}(\eta), \quad \xi,\eta \in \Omega, \tag{40}$$

via the symbols

$$\left(\psi_J^{(i)}\right)^{\wedge}(n) = \left(\left(\left(\phi_{J+1}^{(i)}\right)^{\wedge}(n) \right)^2 - \left(\left(\phi_J^{(i)}\right)^{\wedge}(n) \right)^2 \right)^{\frac{1}{2}}, \quad J \in \mathbb{Z},\ i = 1,2,3. \tag{41}$$

From this definition it is easily seen that

$$\phi_{J+1}^{(i)} \star \phi_{J+1}^{(i)} * f = \phi_J^{(i)} \star \phi_J^{(i)} * f + \psi_J^{(i)} \star \psi_J^{(i)} * f, \quad J \in \mathbb{Z},\ i = 1,2,3, \tag{42}$$

enabling us to state the following decomposition theorem.

Theorem 2
Let $\left\{\phi_J^{(i)}\right\}_{i=1,2,3,\,J\in\mathbb{Z}}$ *be a set of scaling kernels and* $J_0 \in \mathbb{Z}$. *Then* f *of class* $l^2(\Omega)$ *can be decomposed into* $f = f^{(1)} + f^{(2)} + f^{(3)}$ *with*

$$f^{(i)} = \phi_{J_0}^{(i)} \star \phi_{J_0}^{(i)} * f + \sum_{J=J_0}^{\infty} \psi_J^{(i)} \star \psi_J^{(i)} * f, \quad i = 1,2,3.$$

This is a multiscale decomposition satisfying the desired properties. A global trend approximation of f is given by $\phi_{J_0}^{(i)} \star \phi_{J_0}^{(i)} * f$ and can be calculated from a sparse data set. It can be improved by adding detail information of the type $\psi_J^{(i)} \star \psi_J^{(i)} * f$. Observing the increasing spatial localization of the kernels $\psi_J^{(i)}$, this means that information of more and more local origin is added at each step. In regions of high data density, detail information can be added up to high scales with only causing minor (scale-dependent) deteriorations outside the region under consideration. In other words, a high resolution "zooming-in" into regions of better data coverage becomes possible without neglecting data outside these regions. Furthermore, the connection of this approach to (vector) spherical harmonics preserves at least to a certain degree the physical meaningful interpretation of features of long, medium, and short wavelengths. On the other hand, it is also this connection that not completely eliminates the global properties of spherical harmonics and still requires the numerically awkward evaluations of spherical harmonics at high degrees. This is where a spatially oriented approach as in the upcoming section can help. It should finally only be remarked that the above presented method works for any basis system of $l^2(\Omega)$ instead of $\left\{y_{n,k}^{(i)}\right\}$, in particular the system $\left\{\tilde{y}_{n,k}^{(i)}\right\}$.

3.2 Locally Supported Wavelets

A multiscale approach addressing the problems occurring from the global nature of spherical harmonics has just been presented by forming frequency packages to gain better spatial localizations. This, we have seen, still involves the evaluation of spherical harmonics. From a numerical point of view it would be desirable to have kernels with closed representations in space domain. Kernels with locally compact support are especially suitable since they reduce the computation effort for the occurring convolutions to integrations over small bounded regions. We start from the assumption that we already have a set of continuous vectorial functions $\phi_\rho : \Omega \times \Omega \to \mathbb{R}^3$ or a set of continuous tensorial functions $\boldsymbol{\phi}_\rho : \Omega \times \Omega \to \mathbb{R}^{3\times 3}$, with scaling parameter $\rho > 0$, satisfying

$$\text{(i)}\quad \phi_\rho(\xi,\eta) = \phi_\sigma(\xi,\eta), \quad \boldsymbol{\phi}_\rho(\xi,\eta) = \boldsymbol{\phi}_\sigma(\xi,\eta), \quad \xi,\eta \in \Omega,\ 1-\xi\cdot\eta \geq \rho \geq \sigma, \tag{43}$$

$$\text{(ii)}\quad \lim_{\rho\to 0+} \sup_{\xi\in\Omega} \left| \int_\Omega \phi_\rho(\xi,\eta)\cdot f(\eta)d\omega(\eta) - G(\xi) \right| = 0, \tag{44}$$

$$\lim_{\rho\to 0+} \sup_{\xi\in\Omega} \left| \int_\Omega \boldsymbol{\phi}_\rho(\xi,\eta) f(\eta)d\omega(\eta) - g(\xi) \right| = 0, \tag{45}$$

where G or g denote the modeled scalar or vectorial quantities, respectively, and f the accessible data, which is assumed to be at least continuous on the sphere instead of square integrable as in the previous section. Tensorial kernels are required for the reconstruction of the tangential components (or the quantity as a whole) while vectorial kernels are sufficient for reconstructions where one is only interested in the radial component (e.g., field-aligned currents). Those kernels, however, do generally not have locally compact support. Locality is achieved by taking the difference at different parameters ρ. More precisely, for discrete scales $2 \geq \rho_0 > \rho_1 > \ldots > 0$ with $\lim_{J\to\infty} \rho_J = 0$ we set the scaling function (of scale J) to be $\phi_J = \phi_{\rho_J}$ or $\boldsymbol{\phi}_J = \boldsymbol{\phi}_{\rho_J}$. The corresponding difference wavelet kernel (of scale J) is defined by

$$\psi_J(\xi,\eta) = \phi_{J+1}(\xi,\eta) - \phi_J(\xi,\eta), \quad \xi,\eta \in \Omega, \tag{46}$$

$$\boldsymbol{\psi}_J(\xi,\eta) = \boldsymbol{\phi}_{J+1}(\xi,\eta) - \boldsymbol{\phi}_J(\xi,\eta), \quad \xi,\eta \in \Omega, \tag{47}$$

From property (i) it is obvious that $\psi_J(\xi,\eta) = 0$ and $\boldsymbol{\psi}_J(\xi,\eta) = 0$ for $1-\xi\cdot\eta \geq \rho_J$, thus, having a scale dependent locally compact support on a spherical cap of radius ρ_J. Denoting by $\boldsymbol{\phi}_J \star f$ the vector valued convolution $\int_\Omega \boldsymbol{\phi}_J(\cdot,\eta)f(\eta)d\omega(\eta)$ of a tensorial kernel against a vectorial function f, it is now easily seen that the following multiscale decomposition holds true.

Theorem 3
Let $\{\phi_J\}_{J=0,1...}$ *and* $\{\boldsymbol{\phi}_J\}_{J=0,1,...}$ *be scaling kernels satisfying properties (i), (ii) and* $J_0 \in \mathbb{Z}$*. Then we have for sufficiently smooth* f *and modeled quantities* G, g,

$$G(\xi) = (\phi_{J_0} \star f)(\xi) + \sum_{J=J_0}^{\infty} (\psi_J \star f)(\xi), \quad \xi \in \Omega,$$

$$g(\xi) = (\boldsymbol{\phi}_{J_0} \star f)(\xi) + \sum_{J=J_0}^{\infty} (\boldsymbol{\psi}_J \star f)(\xi), \quad \xi \in \Omega.$$

The above representation allows "zooming-in" into regions of higher data density like the frequency approach does. The local support of the wavelets avoids any deteriorations outside the spherical cap under consideration, global data is only needed for the trend approximations $\phi_{J_0} \star F$ and $\boldsymbol{\phi}_{J_0} \star f$. Furthermore, it implies that a Fourier expansion of the wavelet kernels contains contributions at arbitrarily high spherical harmonic degrees. This way, such a multiscale approach accounts for small-wavelength features at every step of the "zooming-in" procedure, which makes it especially suitable for reconstructing local phenomena (e.g., crustal field anomalies or field-aligned currents, cf. ➋ *Fig. 1*). A closed representation of the kernels in space domain, however, makes it often difficult to find an explicit representation of the Fourier coefficients in a spherical harmonic expansion, so that one looses the easy physical interpretability that has been preserved by the construction of frequency packages. On the other hand, this circumvents the problem of the evaluation of high-degree spherical harmonics. Elementary functions as they occur in the representations of many of such space domain-oriented kernels usually offer a good numerical handling. The main task in a multiscale framework as presented in this section is the actual construction of scaling kernels satisfying conditions (i) and (ii), and finding a closed representation for them. Green's function for the Beltrami operator, we have seen, is a good foundation for geomagnetic modeling. A regularization around its singularity enables the construction of suitable scaling kernels and is treated in the remaining paragraphs of this section. Finally, it should be remarked that different from the frequency package approach, the spatially oriented kernels generally do not constitute a multiresolution analysis in the sense of nested scale spaces. Condition (ii) implies that the wavelet expansions converge toward the modeled quantity, but the approximation does not necessarily have to improve at each step of the multiscale reconstruction.

3.2.1 Regularized Green's Function and Single Layer Kernel

As mentioned, the crucial step for an application of the above presented multiscale technique is the construction of geophysically relevant sets of scaling kernels satisfying conditions (i) and (ii). Regularizations of Green's function for the Beltrami operator and the single layer kernel, together with comparing them to the original quantities, constitute a sufficient foundation for that purpose. A detailed overview and proofs of the stated limit relations for Green's function can be found in Freeden and Schreiner (2009) and Freeden and Gerhards (2010). By the single layer kernel we denote the function

$$S(\xi \cdot \eta) = \frac{1}{\sqrt{2}}(1 - \xi \cdot \eta)^{-\frac{1}{2}}, \quad -1 \leq \xi \cdot \eta < 1, \tag{48}$$

for $\xi, \eta \in \Omega$. The single layer operator as defined in ➲ Sect. 2.1 can then be expressed as an integral operator

$$D_\xi^{-1} F(\xi) = \int_\Omega S(\xi \cdot \eta) F(\eta)\, d\omega(\eta), \quad \xi \in \Omega, \tag{49}$$

for F of class $C^{(0)}(\Omega)$. It is not as naturally connected to the governing geomagnetic equations as Green's function and the surface operators ∇^*, L^*, and Δ^* are, but is required for a decomposition of the magnetic field with respect to the sources involving the operators $\tilde{o}^{(i)}$, $i = 1, 2, 3$.

A regularization of the weakly singular Green function is simply achieved by substituting the original function on a spherical cap around the singularity by a sufficiently smooth function matching the boundary conditions. For the separation of sources a twice continuously differentiable regularization is satisfactory, so that we choose

$$G^\rho(\Delta^*; \xi \cdot \eta) = \begin{cases} \frac{1}{4\pi} \ln(1 - \xi \cdot \eta) + \frac{1}{4\pi}(1 - \ln(2)), & 1 - \xi \cdot \eta \geq \rho, \\ -\frac{1}{8\pi\rho^2}(1 - \xi \cdot \eta)^2 + \frac{1}{2\pi\rho}(1 - \xi \cdot \eta) & \\ +\frac{1}{4\pi}(\ln(\rho) - \ln(2)) - \frac{1}{8\pi}, & 1 - \xi \cdot \eta < \rho. \end{cases} \tag{50}$$

For the reconstruction of radial currents as in ➲ Sect. 4.2 one has to choose the regularization on the spherical cap $\{\eta \in \Omega | 1 - \xi \cdot \eta\}$ such that $G^\rho(\Delta^*; \cdot)$ is three times continuously differentiable. Analogously a continuously differentiable regularization of the single layer kernel is achieved by

$$S^\rho(\xi \cdot \eta) = \begin{cases} \frac{1}{\sqrt{2}}(1 - \xi \cdot \eta)^{-\frac{1}{2}}, & 1 - \xi \cdot \eta \geq \rho, \\ -\frac{1}{2\sqrt{2}}\rho^{-\frac{3}{2}}(1 - \xi \cdot \eta) + \frac{3}{2\sqrt{2}}\rho^{-\frac{1}{2}}, & 1 - \xi \cdot \eta < \rho. \end{cases} \tag{51}$$

Both regularized kernels are plotted in ➲ *Fig. 4*. It is obvious that $G^\rho(\Delta^*; \xi \cdot \eta) = G^\sigma(\Delta^*; \xi \cdot \eta)$ and $S^\rho(\xi \cdot \eta) = S^\sigma(\xi \cdot \eta)$ for $1 - \xi \cdot \eta \geq \rho \geq \sigma$, therefore, already satisfying condition (i). Elementary calculations using the integrability of Green's function and the single layer kernel deliver for functions F of class $C^{(0)}(\Omega)$,

$$\lim_{\rho \to 0+} \sup_{\xi \in \Omega} \left| \int_\Omega G^\rho(\Delta^*; \xi \cdot \eta) F(\eta) d\omega(\eta) - \int_\Omega G(\Delta^*; \xi \cdot \eta) F(\eta) d\omega(\eta) \right| = 0, \tag{52}$$

$$\lim_{\rho \to 0+} \sup_{\xi \in \Omega} \left| \int_\Omega S^\rho(\xi \cdot \eta) F(\eta) d\omega(\eta) - \int_\Omega S(\xi \cdot \eta) F(\eta) d\omega(\eta) \right| = 0. \tag{53}$$

Vectorial kernels result from the application of the surface gradient and the surface curl gradient. Since the surface gradient and the surface curl gradient of $G(\Delta^*; \cdot)$ are till integrable on the sphere, a similar argumentation yields

$$\lim_{\rho \to 0+} \sup_{\xi \in \Omega} \left| \int_\Omega \mathbf{\Lambda}_\xi G^\rho(\Delta^*; \xi \cdot \eta) \cdot f(\eta) d\omega(\eta) - \int_\Omega \mathbf{\Lambda}_\xi G(\Delta^*; \xi \cdot \eta) \cdot f(\eta) d\omega(\eta) \right| = 0,$$

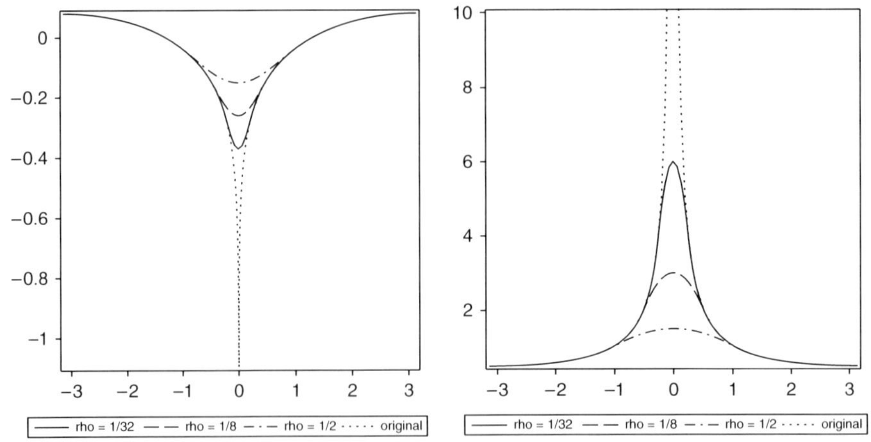

Fig. 4
Regularized Green's function (*left*) and single layer kernel (*right*), plotted with respect to the angular distance

for $\boldsymbol{\Lambda}$ one of the operators ∇^* or L^* and f of class $c^{(0)}(\Omega)$. The derivatives of the single layer kernel, on the other hand, are not integrable anymore, so that one requires more smoothness on the convolved function. Using that L^* and D^{-1} commutate, one can show that

$$\lim_{\rho\to 0+}\sup_{\xi\in\Omega}\left|\int_\Omega \boldsymbol{\Lambda}_\xi S^\rho(\xi\cdot\eta)F(\eta)d\omega(\eta)-\boldsymbol{\Lambda}_\xi\int_\Omega S(\xi\cdot\eta)F(\eta)d\omega(\eta)\right|=0, \tag{54}$$

for F of class $C^{(1)}(\Omega)$. Using Gauss' theorem on the sphere, twice application of the surface differential operators delivers tensorial kernels satisfying

$$\begin{aligned}\lim_{\rho\to 0+}\sup_{\xi\in\Omega}\Bigg|&\int_\Omega\left(\boldsymbol{\Lambda}^{(1)}_\xi\otimes\boldsymbol{\Lambda}^{(2)}_\eta G^\rho(\Delta^*;\xi\cdot\eta)\right)f(\eta)d\omega(\eta)\\&-\boldsymbol{\Lambda}^{(1)}_\xi\int_\Omega\left(\boldsymbol{\Lambda}^{(2)}_\eta G(\Delta^*;\xi\cdot\eta)\right)\cdot f(\eta)d\omega(\eta)\Bigg|=0,\end{aligned} \tag{55}$$

for f of class $c^{(1)}(\Omega)$ and $\boldsymbol{\Lambda}^{(1)}$, $\boldsymbol{\Lambda}^{(2)}$ one of the operators ∇^*, L^*. Assuming that the not-regularized terms denote the modeled quantities G, g, we thus obtain relations as required in (ii). Various further assertions involving Green's function and the single layer kernel can be shown under sufficient smoothness assumptions on the data F and f. Those limit relations provided here should, however, be sufficient to motivate the convergence of the multiscale decompositions exemplarily treated in the next section.

4 Application to Geomagnetic Problems

In this section we want to present results on how the two previously defined multiscale techniques can be applied to the problem of separating the magnetic field with respect to its sources and to the reconstruction of radial ionospheric currents. In both cases, the frequency package as well as the space-oriented approach will be described.

4.1 Separation of Sources

A separation of external and internal contributions to the magnetic field can be based on physically motivated models of the external sources such as field aligned currents, magnetotail currents, magnetospheric ring currents. Besides the pure separation, this gives more detailed information about the structure of the external sources. For a model parametrization of ionospheric sources the reader is referred to, for example, Lühr et al. (2009), for a model of magnetospheric sources to Tsyganenko and Sitnov (2007). An empirical approach of separating the core field from exterior sources through the observation of a change in the power spectrum of Earth's magnetic field around a spherical harmonic degree of 13–15 is due to Lowes (1966) and Mauersberger (1956). Extensive references on the subject of separation of sources can also be found in Olsen et al. (2009). From a mathematical point of view, however, these approaches are somewhat incomplete. First attempts to separate inner and outer sources for gradient fields in a mathematically rigorous manner have been made in Gauss (1839). The coefficients $a_{n,k}(r)$ in representation (3) yield a separation in the sense that the coefficients $c_{n,k}$ represent the part of the magnetic field due to sources outside Ω_{R_1} and $d_{n,k}$ the part due to sources inside Ω_{R_0}. In non-source-free regions, we have seen, this representation breaks down and the more general Mie decomposition has to be used. Using Biot Savart's law of classical electrodynamics (cf. Jackson 1975), one can show that the poloidal part p_b of the magnetic field can be split into a part p_b^{int} due to sources in Ω_R^{int}, satisfying

$$\nabla_x \wedge p_b^{int}(R;x) = \begin{cases} q_j(x), & x \in \Omega_R^{int}, \\ 0, & x \in \Omega_R^{ext}, \end{cases} \tag{56}$$

and a part p_b^{ext} due to external sources, satisfying

$$\nabla_x \wedge p_b^{ext}(R;x) = \begin{cases} 0, & x \in \Omega_R^{int}, \\ q_j(x), & x \in \Omega_R^{ext}, \end{cases} \tag{57}$$

with q_j denoting the toroidal part of the corresponding source currents j. A detailed derivation of this split-up is given in Backus et al. (1996). The remaining toroidal part q_b of the magnetic field can be interpreted as the part due to poloidal sources crossing the sphere Ω_R. Summarized, a separation of the magnetic field with respect to its sources is given by

$$b = p_b^{int} + p_b^{ext} + q_b. \tag{58}$$

A formulation of this decomposition in terms of spherical harmonics can also be found in Olsen et al. (2009). We concentrate on multiscale expansions, and first present the frequency package approach which can be found in a more detailed treatment in Mayer (2003) and Mayer and Maier (2006).

4.1.1 Wavelets as Frequency Packages

Observing from (56) that p_b^{int} is divergence-free and curl-free in Ω_R^{ext}, it can be represented as a gradient field ∇U^{ext} in Ω_R^{ext} with a potential U^{ext} that is harmonic in the exterior space. Analogously, p_b^{ext} is a gradient field ∇U^{int} in Ω_R^{int} with U^{int} harmonic in the interior. The potentials can therefore be expanded in terms of inner and outer harmonics, so that relations

(21) and (22) deliver

$$p_b^{int}(R;x) = \sum_{n=0}^{\infty}\sum_{k=1}^{2n+1} (\tilde{b}^{(1)})^{\wedge}(R;n,k)\,\tilde{y}^{(1)}_{n,k}(\xi), \quad |x| = R,\ x = R\xi, \tag{59}$$

$$p_b^{ext}(R;x) = \sum_{n=1}^{\infty}\sum_{k=1}^{2n+1} (\tilde{b}^{(2)})^{\wedge}(R;n,k)\,\tilde{y}^{(2)}_{n,k}(\xi), \quad |x| = R,\ x = R\xi, \tag{60}$$

$$q_b(R;x) = \sum_{n=1}^{\infty}\sum_{k=1}^{2n+1} (\tilde{b}^{(1)})^{\wedge}(R;n,k)\,\tilde{y}^{(3)}_{n,k}(\xi), \quad |x| = R,\ x = R\xi. \tag{61}$$

In other words, the vector spherical harmonics $\left\{\tilde{y}^{(i)}_{n,k}\right\}$ form a basis system that decomposes the magnetic field with respect to its sources. A multiscale reconstruction can then be achieved as described in ➲ Sect. 3.1 by setting the scaling kernel to

$$\phi_J^{(i)}(\xi,\eta) = \sum_{n=0_i}^{\infty}\sum_{k=1}^{2n+1} \left(\phi_J^{(i)}\right)^{\wedge}(n)\, Y_{n,k}(\xi)\,\tilde{y}^{(i)}_{n,k}(\eta), \quad i = 1,2,3,\ \xi,\eta \in \Omega, \tag{62}$$

for $\xi,\eta \in \Omega$, with the symbols $\left(\phi_J^{(i)}\right)^{\wedge}(n)$, for example, those corresponding to the CP kernel. Choosing a sufficiently large scale J_{max}, we get the approximation

$$\begin{aligned} p_b^{int}(R;x) \approx{}& R^{-2}\left(\tilde{\mu}_n^{(1)}\right)^{\frac{1}{2}}\left(\phi_{J_0}^{(1)} \star \phi_{J_0}^{(1)} \ast b(R\cdot)\right)(\xi) \\ &+ R^{-2}\left(\tilde{\mu}_n^{(1)}\right)^{\frac{1}{2}}\sum_{J=J_0}^{J_{max}}\left(\psi_J^{(1)} \star \psi_J^{(1)} \ast b(R\cdot)\right)(\xi) \end{aligned} \tag{63}$$

$$\begin{aligned} p_b^{ext}(R;x) \approx{}& R^{-2}\left(\tilde{\mu}_n^{(2)}\right)^{\frac{1}{2}}\left(\phi_{J_0}^{(2)} \star \phi_{J_0}^{(2)} \ast b(R\cdot)\right)(\xi) \\ &+ R^{-2}\left(\tilde{\mu}_n^{(2)}\right)^{\frac{1}{2}}\sum_{J=J_0}^{J_{max}}\left(\psi_J^{(2)} \star \psi_J^{(2)} \ast b(R\cdot)\right)(\xi) \end{aligned} \tag{64}$$

for the poloidal parts, and

$$\begin{aligned} q_b(R;x) \approx{}& \left(\tilde{\mu}_n^{(3)}\right)^{\frac{1}{2}}\left(\phi_{J_0}^{(3)} \star \phi_{J_0}^{(3)} \ast b(R\cdot)\right)(\xi) \\ &+ \left(\tilde{\mu}_n^{(3)}\right)^{\frac{1}{2}}\sum_{J=J_0}^{J_{max}}\left(\psi_J^{(3)} \star \psi_J^{(3)} \ast b(R\cdot)\right)(\xi) \end{aligned} \tag{65}$$

for the toroidal part, with $|x| = R$, $x = R\xi$. In Mayer and Maier (2006) the above method has been applied to a pre-processed set of 7 months of CHAMP data (➲ *Fig. 5*). ➲ *Figures* 6 and ➲ 7 show the reconstruction of p_b^{int} up to scale $J_{max} = 5$ and the remaining external and satellite orbit contributions p_b^{ext} and q_b. In the external contributions one can actually recognize the structure of the source field, for example a polar field-aligned current systems which will be treated in more detail in the next section. Local reconstructions over the North pole and Africa can as well be found in the above mentioned publication.

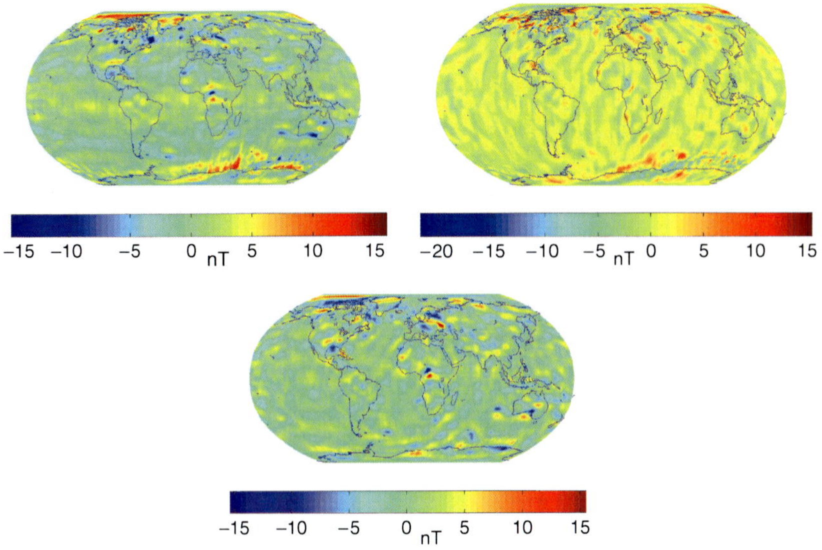

Fig. 5
Original CHAMP data set averaged over a regular grid of 130 × 130 points: X-component (*top left*), Y-component (*top right*), Z-component (*bottom*)

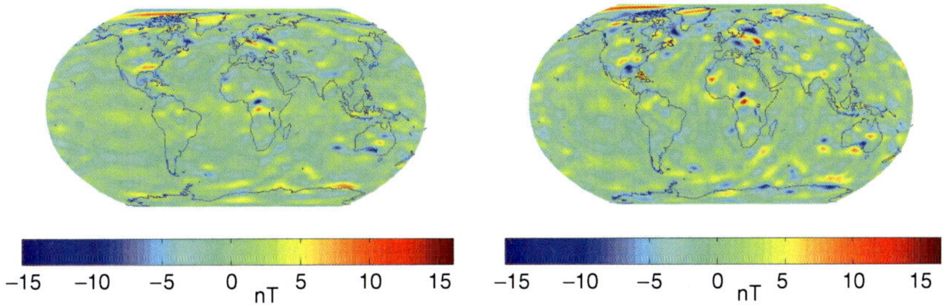

Fig. 6
X-component (*left*) and Z-component (*right*) of the magnetic field due to sources inside the satellite's orbit, reconstructed with CP scaling kernels at scale $J_{max} = 5$

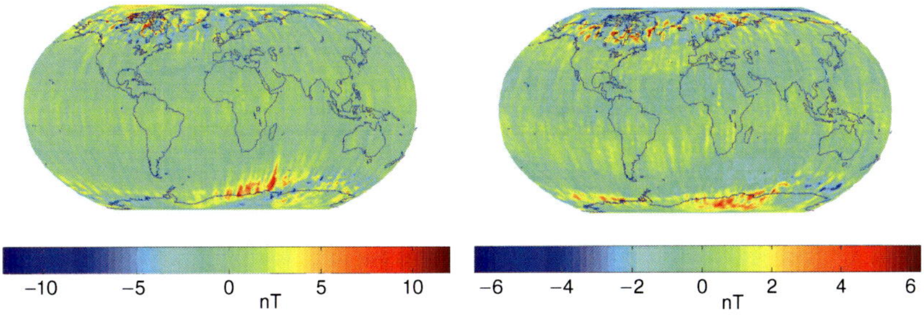

Fig. 7
X-component (*left*) and Z-component (*right*) of the magnetic field due to sources in the exterior or crossing the satellite's orbit, calculated as the difference of the original data and the reconstructed part due to interior sources

4.1.2 Locally Supported Wavelets

The connection of the operators $\tilde{o}^{(i)}$, $i = 1, 2, 3$, to the vector spherical harmonics $\tilde{y}^{(i)}_{n,k}$, $i = 1, 2, 3$, yields that a representation

$$b(x) = \tilde{o}^{(1)}_{\xi}\tilde{B}_1(R\xi) + \tilde{o}^{(2)}_{\xi}\tilde{B}_2(R\xi) + \tilde{o}^{(3)}_{\xi}\tilde{B}_3(R\xi), \quad |x| = R,\ x = R\xi, \tag{66}$$

of a twice continuously differentiable magnetic field b decomposes it with respect to the sources. More precisely, $\tilde{o}^{(1)}\tilde{B}_1$ denotes the part due to sources in Ω_R^{int}, $\tilde{o}^{(2)}\tilde{B}_2$ the part due to exterior sources in Ω_R^{ext} and $\tilde{o}^{(3)}\tilde{B}_3$ the part due to sources on the sphere or crossing the sphere Ω_R. Comparing this representation to the orthogonal Helmholtz representation, one can derive that the scalars $\tilde{B}_1$, $\tilde{B}_2$, $\tilde{B}_3$ are uniquely defined by the condition

$$\frac{1}{4\pi}\int_{\Omega}\tilde{B}_1(R\eta) - \tilde{B}_2(R\eta)\,d\omega(\eta) = \frac{1}{4\pi}\int_{\Omega}\tilde{B}_3(R\eta)\,d\omega(\eta) = 0, \tag{67}$$

and that they can be expressed as

$$\tilde{B}_1 = \frac{1}{2}D^{-1}B_1 + \frac{1}{4}D^{-1}B_2 - \frac{1}{2}B_2, \tag{68}$$

$$\tilde{B}_2 = \frac{1}{2}D^{-1}B_1 + \frac{1}{4}D^{-1}B_2 + \frac{1}{2}B_2, \tag{69}$$

$$\tilde{B}_3 = B_3, \tag{70}$$

where B_1, B_2, B_3 correspond to the Helmholtz decomposition $b = o^{(1)}B_1 + o^{(2)}B_2 + o^{(3)}B_3$. Theorem 1 and its consequences for solutions to the Beltrami equation imply a representation of the Helmholtz scalars in terms of Green's function with respect to the Beltrami operator, so that (66) and (68) lead after some lengthy but elementary calculations to

$$\begin{aligned} p_b^{int}(R;x) = {} & \frac{1}{2}\xi(\xi\cdot b(R\xi)) + \frac{1}{4}\xi D_{\xi}^{-1}(\xi\cdot b(R\xi)) + \frac{1}{2}\xi D_{\xi}^{-1}\nabla_{\xi}^{*}\cdot b(R\xi) \\ & - \frac{1}{2}\nabla_{\xi}^{*}D_{\xi}^{-1}(\xi\cdot b(R\xi)) - \frac{1}{2}\nabla_{\eta}^{*}\int_{\Omega}\nabla_{\xi}^{*}G(\Delta^{*};\xi\cdot\eta)\cdot b(R\eta)d\omega(\eta) \\ & + \frac{1}{4}\nabla_{\xi}^{*}\int_{\Omega}\nabla_{\eta}^{*}D_{\xi}^{-1}G(\Delta^{*};\xi\cdot\eta)\cdot b(R\eta)d\omega(\eta), \end{aligned} \tag{71}$$

for $|x| = R$, $x = R\xi$. Substituting the occurring Green functions and single layer kernels by their regularized counterparts, we eventually arrive at the following definition of scaling kernels,

$$\phi_J^{(int)}(\xi,\eta) = \frac{1}{2}\Delta_{\xi}^{*}G^{\rho_J}(\Delta^{*};\xi\cdot\eta)\eta + \frac{1}{8\pi}S^{\rho_J}(\xi\cdot\eta)\eta - \frac{1}{4\pi}\nabla_{\eta}^{*}S^{\rho_J}(\xi\cdot\eta), \tag{72}$$

$$\boldsymbol{\phi}_J^{(int)}(\xi,\eta) = \frac{1}{2}\nabla_{\xi}^{*}\otimes\nabla_{\eta}^{*}\left(\frac{1}{2}D_{\xi}^{-1}G^{\rho_J}(\Delta^{*};\xi\cdot\eta) - G^{\rho_J}(\Delta^{*};\xi\cdot\eta)\right) - \frac{1}{4\pi}\nabla_{\xi}^{*}S^{\rho_J}(\xi\cdot\eta)\otimes\eta. \tag{73}$$

Similarly one can derive scaling kernels $\phi_J^{(ext)}$, $\boldsymbol{\phi}_J^{(ext)}$, and $\boldsymbol{\phi}_J^{(q)}$ for the poloidal field due to external sources and the toroidal field due to currents crossing the sphere. Their explicit representation is omitted for brevity, a visualization of $\phi_J^{(int)}$ and $\phi_J^{(ext)}$ and the corresponding locally supported wavelets at different scales can be found in ➲ *Fig. 8*. The limit relations from the end of ➲ Sect. 3.2 and some slight modifications now tell us that a sufficiently large J_{max}

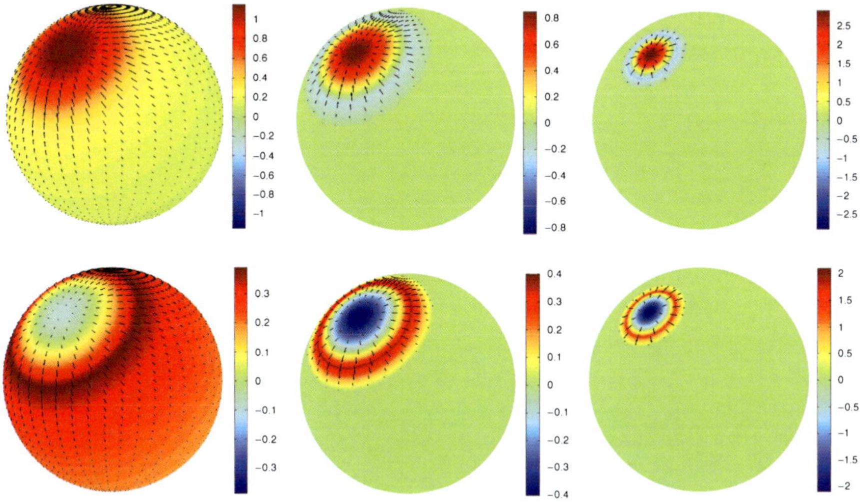

Fig. 8

Scaling kernel $\phi_J^{(int)}$ at scale $J = 2$ with corresponsding wavelets at scales $J = 2$ and $J = 4$ (top, left to right), and scaling kernel $\phi_J^{(int)}$ at scale $J = 3$ with corresponding wavelets at scales $J = 2$ and $J = 4$ (bottom, left to right). Colors indicate the radial component, arrows the tangential one

can be chosen, so that

$$p_b^{(int)}(R;x) \approx \xi\left(\phi_{J_0}^{(int)} \star b(R\cdot)\right)(\xi) + \xi \sum_{J=J_0}^{J_{max}} \left(\psi_J^{(int)} \star b(R\cdot)\right)(\xi) + \left(\phi_{J_0}^{(int)} \star b(R\cdot)\right)(\xi) + \sum_{J=J_0}^{J_{max}} \left(\psi_J^{(int)} \star b(R\cdot)\right)(\xi), \tag{74}$$

$$p_b^{(ext)}(R;x) \approx \xi\left(\phi_{J_0}^{(ext)} \star b(R\cdot)\right)(\xi) + \xi \sum_{J=J_0}^{J_{max}} \left(\psi_J^{(ext)} \star b(R\cdot)\right)(\xi) + \left(\phi_{J_0}^{(ext)} \star b(R\cdot)\right)(\xi) + \sum_{J=J_0}^{J_{max}} \left(\psi_J^{(ext)} \star b(R\cdot)\right)(\xi), \tag{75}$$

$$q_b(R;x) \approx \left(\phi_{J_0}^{(q)} \star b(R\cdot)\right)(\xi) + \sum_{J=J_0}^{J_{max}} \left(\psi_J^{(q)} \star b(R\cdot)\right)(\xi), \tag{76}$$

for $|x| = R$, $x = R\xi$. This approach has yet to be tested on adequate data sets, but summarizing, we have derived a frequency package oriented and a spatially oriented multiscale separation of the Earth's magnetic field with respect to its sources, based on the same physical principles. In combination with the classical Fourier approach, this supplies three different tools for an adequate reconstruction of the long-, medium- as well as short-wavelength contributions to the magnetic field.

4.2 Reconstruction of Radial Currents

As a second application, we treat the reconstruction of radial currents from magnetic fields given on a sphere of fixed radius. Field-aligned currents in the polar regions were first treated in Birkeland (1908) while a system of field-aligned currents at low latitudes due to the equatorial electrojet was proposed in Untied (1967). Both current systems are of regional nature (i.e., short wave-lengths corresponding to spherical harmonic degrees greater than 10) and can be assumed to be approximately radial. A spherical harmonic approach to their calculation applying the Mie representation is given in Olsen (1997) where also more references on this topic can be found. Here we give representations in terms of the two multiscale approaches presented in ➋ Sect. 3. First, we deal with the approach via frequency packages, which can be found in more detail in Maier (2005), then we turn to the construction of locally supported wavelets, as treated in Freeden and Gerhards (2010). General point of departure is the observation that the combination of (8), (9) with the Helmholtz and Mie representation (14) yields

$$j(x) = \xi \frac{\Delta_\xi^* Q_b(r\xi)}{r} - \nabla_\xi^* \frac{\frac{\partial}{\partial r}(rQ_b(r\xi))}{r} - L_\xi^* \Delta_{r\xi} P_b(r\xi), \quad r = |x|,\ x = r\xi, \tag{77}$$

connecting the Helmholtz representation of the current density j with the Mie scalars of the magnetic field b.

4.2.1 Wavelets as Frequency Packages

The toroidal part of the magnetic field is entirely due to the L^*-contributions of b, so that it can be expanded solely in terms of scaling kernels of type 3. In other words, if defining

$$\phi_J^{(i)}(\xi,\eta) = \sum_{n=0_i}^{\infty} \sum_{k=1}^{2n+1} \phi_J^\wedge(n) Y_{n,k}(\xi) y_{n,k}^{(i)}(\eta), \quad i = 1,2,3, \tag{78}$$

for $\xi, \eta \in \Omega$, with $\left(\phi_J^{(i)}\right)^\wedge(n)$, for example, the symbols of the CP kernel, we get

$$q_b(x) = \left(\phi_{J_0}^{(3)} \star \phi_{J_0}^{(3)} \star b(r\cdot)\right)(\xi) + \sum_{J=J_0}^{\infty} \left(\psi_J^{(3)} \star \psi_J^{(3)} \star b(r\cdot)\right)(\xi), \tag{79}$$

for $r = |x|$, $x = r\xi$. For a representation of the toroidal scalar Q_b, satisfying $L^* Q_b = q_b$, we need the scalar kernel

$$\Phi_J(\xi,\eta) = \sum_{n=1}^{\infty} \sum_{k=1}^{2n+1} \left(\phi_J^{(3)}\right)^\wedge(n) Y_{n,k}(\xi) Y_{n,k}(\eta), \quad \xi, \eta \in \Omega. \tag{80}$$

It satisfies $L_\xi^* \Phi_J(\eta,\xi) = \phi_J^{(3)}(\xi,\eta)$, which yields the representation

$$Q_b(x) = \left(\Phi_{J_0} \star \phi_{J_0}^{(3)} \star b(r\cdot)\right)(\xi) + \sum_{J=J_0}^{\infty} \left(\Psi_J \star \psi_J^{(3)} \star b(r\cdot)\right)(\xi), \quad \xi \in \Omega. \tag{81}$$

Equation (77) now states that the radial currents can simply be calculated by $j_{rad}(x) = \frac{1}{r}\xi \Delta_\xi^* Q_b(r\xi)$ for $r = |x|$, $x = r\xi$. The vectorial kernel

$$\phi_J^{(1),\Delta^*}(\eta,\xi) = \xi \Delta_\xi^* \Phi_J(\xi,\eta) = \sum_{n=1}^{\infty} \sum_{k=1}^{2n+1} -n(n+1)\left(\phi_J^{(3)}\right)^\wedge(n) Y_{n,k}(\eta) y_{n,k}^{(1)}(\xi) \tag{82}$$

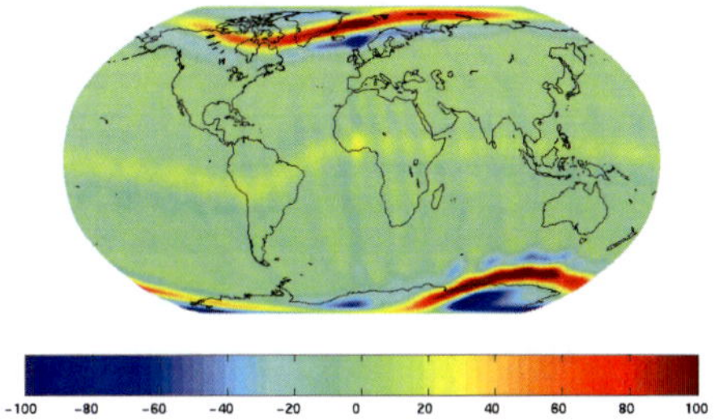

Fig. 9
Radial component of the currents calculated from MAGSAT evening data with CP kernels at scale $J_{max} = 6$ (units in nA/m^2)

finally delivers the desired multiscale representation of the radial currents. For a sufficiently large scale J_{max} we have

$$j_{rad}(x) \approx \frac{1}{r}\big(\phi_{J_0}^{(1),\Delta^*} \star \phi_{J_0}^{(3)} * b(r\cdot)\big)(\xi) + \frac{1}{r}\sum_{J=J_0}^{J_{max}} \big(\psi_J^{(1),\Delta^*} \star \psi_J^{(3)} * b(r\cdot)\big)(\xi), \tag{83}$$

with $r = |x|$, $x = r\xi$. Figure 9 shows a reconstruction of radial currents from MAGSAT data as published in Maier (2005). One can well recognize the before mentioned currents in the polar regions as well as a weaker upward direct current system along the dip equator. It should be remarked that the data sets used for such calculations requires a precedent treatment of the satellite data to obtain the magnetic field due to such current systems. This can be done by the subtraction of main field models and models of the crustal field or by separation techniques as presented in the previous section.

4.2.2 Locally Supported Wavelets

The space-oriented method starts from the same situation as above, that is, we need a representation of the toroidal scalar Q_b. Since $q_b = L^* Q_b$ implies $L^* \cdot b = \Delta^* Q_b$, Green's function for the Beltrami operator delivers

$$Q_b(x) = \int_\Omega L_\xi^* G(\Delta^*; \xi \cdot \eta) \cdot b(r\eta)\, d\omega(\eta), \quad r = |x|, x = r\xi. \tag{84}$$

Relation (77) again implies

$$j_{rad}(x) = \frac{1}{r}\xi\Delta_\xi^* \int_\Omega L_\xi^* G(\Delta^*; \xi \cdot \eta) \cdot b(r\eta)\, d\omega(\eta), \quad r = |x|, x = r\xi. \tag{85}$$

Substituting Green's function by its regularized version, and defining the scaling kernel

$$\phi_J(\xi, \eta) = L_\xi^* \Delta_\xi^* G^{\rho_J}(\Delta^*; \xi \cdot \eta), \quad \xi, \eta \in \Omega, \tag{86}$$

with $\rho_0 > \rho_1 > \ldots > 0$ a series of scales satisfying $\lim_{J\to\infty} \rho_J = 0$, limit relations like those in Sect. 3.2 deliver

$$j_{rad}(x) \approx \frac{1}{r}\xi(\phi_{J_0} * b(r\cdot))(\xi) + \frac{1}{r}\xi \sum_{J=J_0}^{J_{max}} (\psi_J * b(r\cdot))(\xi), \quad r = |x|, x = r\xi, \tag{87}$$

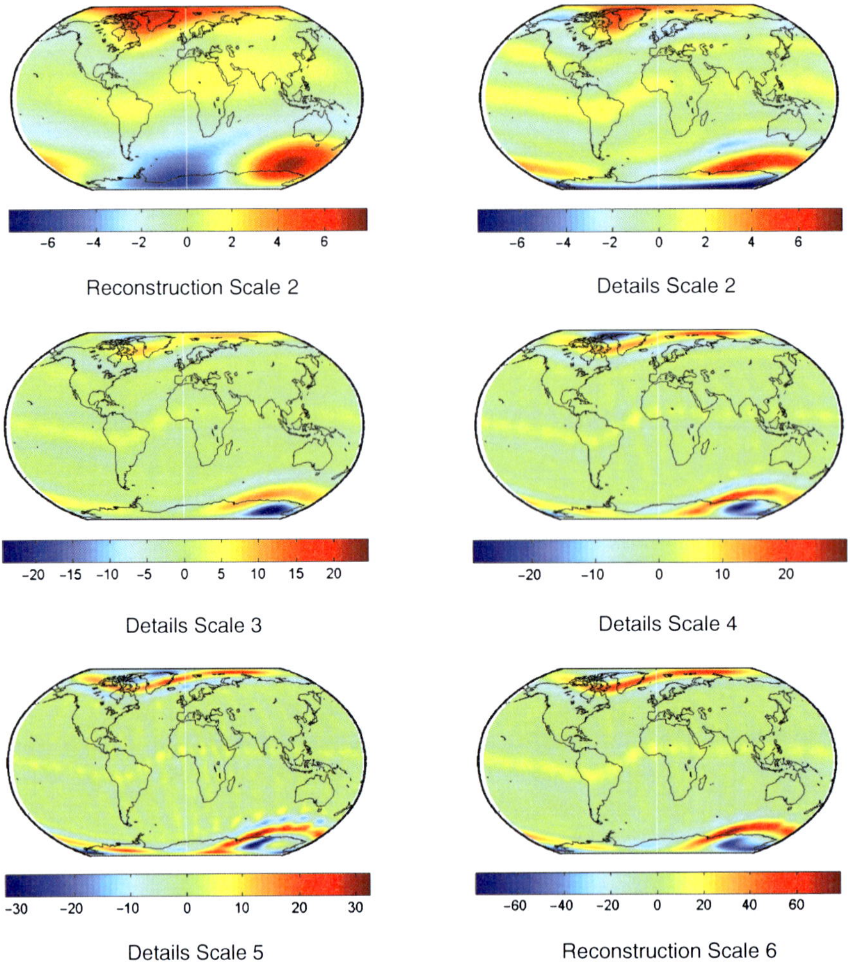

Fig. 10

Radial component of the currents calculated from MAGSAT evening magnetic field data up to scale $J_{max} = 6$ (bottom right). Scale $J_0 = 2$ represents a first trend approximation (top left). The remaining figures (top right to bottom left) show the detail information added by the wavelet convolutions at each scale (units nA/m^2)

for a sufficiently large J_{max}. *Figure 10* shows the reconstruction from MAGSAT evening magnetic field measurements at various spatial resolutions (cf. Moehring (2010)). One can well recognize that the detail information added at each step of the multiscale procedure reveals more and more local features at growing scales. The data set used is slightly different from that in Maier (2005) but produces analogous polar and equatorial current systems. Interesting in this approach is that the properties of $G(\Delta^*;\cdot)$ imply that already the scaling kernels are having locally compact support (i.e., $\phi_J(\xi,\eta) = 0$ on the spherical cap given by $1 - \xi \cdot \eta < \rho_J$). This implies that radial currents can be calculated from purely local data. Furthermore, if one is not interested in the evolution over the different scales, the calculation of $\phi_{J_{max}} * b(r\cdot)$, for a large J_{max}, offers a fast and precise approximation of the radial currents.

5 Future Directions

Although research on the Earth's magnetic field has been conducted for far more than a century, it is especially recent satellite missions like Ørsted and CHAMP that made the need of adequate mathematical methods more and more apparent to handle the large amount of accessible data as well as to improve the accuracy of the modeling. Swarm, the new ESA geomagnetic satellite mission (cf. Friis-Christensen et al. 2006) with an intended launch in 2011, will provide even more accurate data and due to the constellation of the three involved satellites offer new modeling approaches. The two topics addressed here only present a fraction of the complex problems arising from the desire to understand the Earth's magnetic field. Especially the magnetohydrodynamic phenomena affecting core processes are intensively studied (e.g., Amit and Olsen 2004; Jackson 1997; Pais et al. 2004; Zatman and Bloxham 1997 to name only a few) but far from being satisfactory understood. Also an accurate global mapping of the Earth's mantle conductivity is more and more object of recent research as adequate satellite data becomes available (e.g., Kuvshinov et al. 2006; Olsen 1999).

Concerning the multiscale methods presented in this chapter, near future research will be including the numerical implementation of locally supported wavelet methods. The typical integration approach as presented in Driscoll and Healy (1994) is well representing the current satellite data situation with denser data towards the poles. However, other integration methods might be more advantageous for the structure of locally supported wavelets. Beyond this, approaches that are solely limited to bounded regions of the sphere could offer multiscale methods that do not need a first global trend approximation. This might, on the other hand, be at the price of a necessity of boundary information.

A point that Fourier methods as well as the presented space and frequency-oriented multiscale methods with application to geomagnetic modeling have in common is their limitation to spherical geometries. To comprise the altitude variations in satellite measurements one might wish for modeling on surfaces that better describe the satellite's true orbit. Approaches toward this have, for example, been made in Freeden and Mayer (2003). The limit and jump relations from classical potential theory on parallel surfaces were used to define scaling and wavelet functions on arbitrary geometries with application to boundary value problems for the Laplacian equation $\Delta U = 0$, thus, allowing the calculation of the scalar potential in source-free regions.

6 Conclusion

The conventional approximation in geomagnetism is based on spherical harmonics (see ➲ Chap. 17). Spherical wavelets have been a major research focus only for several years (e.g., Dahlke et al. 1995; Freeden and Schreiner 1995; Freeden et al. 1998; Holschneider 1996; Schröder and Swelden 1995). Their application to geomagnetic modeling, however, is rather recent. In Chambodut et al. (2005) and Holschneider et al. (2003), for example, Poisson wavelets were introduced for the modeling of gradient fields. In this chapter we have given an overview of two different multiscale approaches. The more general formulation in terms of frequency packages (cf. ➲ Sect. 3.1) has been introduced to geomagnetic modeling in Bayer et al. (2001) and further applied, for example, in Maier (2005), Maier and Mayer (2003), and Mayer and Maier (2006), also concerning geomagnetically relevant problems not treated in this chapter. A first application to geomagnetic modeling where locally supported wavelets have been constructed for the calculation of radial currents (cf. ➋ Sect. 4.2) can be found in Freeden and Gerhards (2010) and Moehring (2010). Analogous concepts of spatially oriented multiscale methods as described in ➋ Sect. 3.2 have already been successfully applied to local modeling of geostrophic ocean flow and Earth's disturbing potential in Fehlinger et al. (2007) and Fehlinger et al. (2008).

Summarizing, multiscale methods are useful wherever unevenly distributed data sets occur or where the quantity under consideration is of regional nature. In general, we might come to the conclusion that, with the tools at hand, an optimal procedure for the approximation of geomagnetic quantities could look as follows: (i) a trend approximation for small wavelengths via a spherical harmonic Fourier expansion, (ii) band-limited scale dependent kernels for the reconstruction of medium-wavelengths features, and (iii) locally supported kernels for the analysis of seriously space localizing phenomena.

References

Amit H, Olsen P (2004) Helical core flow from geomagnetic secular variation. Phys Earth Planet Inter 147:125

Backus GE (1986) Poloidal and toroidal fields in geomagnetic field modeling. Rev Geophys 24: 75–109

Backus GE, Parker R, Constable C (1996) Foundations of geomagnetism. Cambridge University Press, Cambridge

Bayer M, Freeden W, Maier T (2001) A vector wavelet approach to iono- and magnetospheric geomagnetic satellite data. J Atm Sol-Ter Phys 63:581–597

Birkeland K (1908) The Norwegian aurora polaris expedition 1902–1903, vol 1. H. Aschehoug, Oslo

Chambodut A, Panet I, Mandea M, Diament M, Holschneider M (2005) Wavelet frames: an alternative to spherical harmonic representation of potential fields. J Geophys Int 163: 875–899

Dahlke S, Dahmen W, Schmitt W, Weinreich I (1995) Multiresolution analysis and wavelets on S^2 and S^3. Numer Funct Anal Optim 16:19–41

Driscoll J, Healy D (1994) Computating Fourier transforms and convolutions on the 2-sphere. Adv Appl Math 15:202–250

Edmonds AR (1957) Angular momentum in quantum mechanics. Princeton University Press, Princeton

Fehlinger T, Freeden W, Gramsch S, Mayer C, Michel D, Schreiner M (2007). Local modelling of sea surface topography from (geostrophic) ocean flow. ZAMM 87:775–791

Fehlinger T, Freeden W, Mayer C, Schreiner M (2008) On the local multiscale determination of the earths disturbing potential from discrete deflections of the vertical. Comput Geosci 12: 473–490

Freeden W (1998) The uncertainty principle and its role in physical geodesy. In: Freeden W (ed) Progress in geodetic science. Shaker, Aachen

Freeden W, Gerhards C (2010) Poloidal–toroidal modeling by locally supported vector wavelets. Math Geosc, doi:10.1007/s11004-009-9262-0

Freeden W, Mayer C (2008) Wavelets generated by layer potentials. App Comp Harm Ana 14: 195–237

Freeden W, Schreiner M (1995) Non-orthogonal expansion on the sphere. Math Meth Appl Sci 18:83–120

Freeden W, Schreiner M (2010) Special functions in mathematical geosciences - an attempt of categorization. In: Freeden W, Nashed Z, Sonar T (eds), Handbook of geomathematics. Springer, Heidelberg

Freeden W, Schreiner M (2009) Spherical functions of mathematical (geo-) sciences. Springer, Heidelberg

Freeden W, Gervens T, Schreiner M (1998) Constructive approximation on the sphere – with applications to geosciences. Oxford University Press, New York

Friis-Christensen E, Lühr H, Hulot G (2006) Swarm: A constellation to study the Earth's magnetic field. Earth Planets Space 58:351–358

Gauss CF (1839) Allgemeine Theorie des Erdmagnetismus. Resultate aus den Beobachtungen des Magnetischen Vereins im Jahre 1838. Göttinger Magnetischer Verein, Leipzig

Gerlich G (1972) Magnetfeldbeschreibung mit Verallgemeinerten Poloidalen und Toroidalen Skalaren. Z Naturforsch 8:1167–1172

Haagmanns R, Kern M, Plank G (2008) Swarm - the earth's magnetic field and environment explorers. RSSD Seminar ESTEC.

Haines GV (1985) Spherical cap harmonic analysis. J Geophys Res 90:2583–2591

Holschneider M (1996) Continuous wavelet transforms on the sphere. J Math Phys 37:4156–4165

Holschneider M, Chambodut A, Mandea M (2003) From global to regional analysis of the magnetic field on the sphere using wavelet frames. Phys Earth Planet Inter 135:107–124

Hulot G, Sabaka TJ, Olsen N (2007) The present field. In: Kono M (ed), Treatise on Geophysics, vol 5. Elsevier, Amsterdam

Jackson JD (1975) Classical electrodynamics. Wiley, New York

Jackson A (1997) Time-dependency of tangentially geostrophic core surface motions. Earth Planet Inter 103:293–311

Kuvshinov A, Sabaka TJ, Olsen N (2006) 3-D electromagnetic induction studies using the swarm constellation: mapping conductivity anomalies in the Earth's mantle. Earth Planets Space 58: 417–427

Langel RA (1987) The main field. In: Jacobs JA (ed), Geomagnetism, vol 1. Academic, London

Langel RA, Hinze WJ (1998) The magnetic field of the Earth's lithosphere: the satellite perspective. Cambridge University Press, Cambridge

Lowes FJ (1966) Mean square values on sphere of spherical harmonic fields. J Geophys Res 71:2179

Lühr H, Korte M, Mandea M (2009) The recent magnetic field. In: Glassmeier K-H, Soffel H, Negendank JW (eds), Geomagnetic variations. Springer, Berlin

Maier T (2005) Wavelet-Mie-representation for solenoidal vector fields with applications to ionospheric geomagnetic data. SIAM J Appl Math 65:1888–1912

Maier T, Mayer C (2003) Multiscale downward continuation of the crustal field from CHMAP FGM data. In: First CHAMP mission results for gravity, magnetic and atmospheric studies. Springer, Berlin

Mauersberger P (1956) Das Mittel der Energiedichte des Geomagnetischen Hauptfeldes an der Erdoberfläche und Seine Sekuläre Änderung. Gerl Beitr Geophys 65:207–215

Mayer C (2003) Wavelet modeling of ionospheric currents and induced magnetic fields from satellite data. PhD Thesis, University of Kaiserslautern

Mayer C, Maier T (2006) Separating inner and outer Earth's magnetic field from CHAMP satellite measurements by means of vector scaling functions and wavelets. Geophys. J Int 167: 1188–1203

Michel V (2006) Fast approximation on the 2-sphere by optimally localized approximate identities. Schriften zur Funktionalanalysis und Geomathematik 29, University of Kaiserslautern

Moehring S (2010) Multiscale modeling of ionospheric current systems. Diploma Thesis, University of Kaiserslautern.

Olsen N (1997) Ionospheric F-region currents at middle and low latitudes estimated from MAGSAT data. J Geophys Res 102:4563–4576

Olsen N (1999) Induction studies with satellite data. Surv Geophys 20:309–340

Olsen N, Glassmeier K-H, Jia X (2009) Separation of the magnetic field into external and internal parts. Space Sci Rev, doi:10.1007/s11214-009-9563-0

Pais MA, Oliviera O, Nogueira F (2004) Nonuniqueness of inverted core-mantle boundary flows and deviations from tangential geostrophy. J Geophys Res 109:doi:10.1029/2004JB003012

Sabaka T, Hulot G, Olsen N (2010) Mathematical properties relevant to geomagnetic field modeling. In: Freeden W, Nashed Z, Sonar T (eds), Handbook of geomathematics. Springer, Heidelberg

Schröder P, Swelden W (1995) Spherical wavelets on the sphere. In: Approximation theory, vol VIII. World Scientific, Singapore

Thebault E, Schott JJ, Mandea M (2006) Revised spherical cap harmonics analysis (R-SCHA): validation and properties. J Geophys Res 111:doi:10.1029/2005JB003836

Tsyganenko NA, Sitnov MI (2007) Magnetospheric configurations from a high-resolution data-based magnetic field model. J Geophys Res 112:doi:10.1029/2007JA012260

Untied J (1967) A model of the equatorial electrojet involving meridional currents. J Geophys Res 72:5799–5810

Zatman S, Bloxham J (1997) Torsional oscillations and the magnetic field within the Earth's core. Nature 388:760763

19 The Forward and Adjoint Methods of Global Electromagnetic Induction for CHAMP Magnetic Data

Zdeněk Martinec[1,2]
[1]Dublin Institute for Advanced Studies, Dublin, Ireland
[2]Department of Geophysics, Faculty of Mathematics and Physics, Charles University, Prague, Czech Republic

W. Freeden, M.Z. Nashed, T. Sonar (Eds.), *Handbook of Geomathematics*, DOI 10.1007/978-3-642-01546-5_19,

Abstract Detailed mathematical derivations of the forward and adjoint sensitivity methods are presented for computing the electromagnetic induction response of a 2-D heterogeneous conducting sphere to a transient external electric current excitation. The forward method is appropriate for determining the induced spatiotemporal electromagnetic signature at satellite altitudes associated with the upper and mid-mantle conductivity heterogeneities, while the adjoint method provides an efficient tool for computing the sensitivity of satellite magnetic data to the conductivity structure of the Earth's interior. The forward and adjoint initial boundary-value problems, both solved in the time domain, are identical, except for the specification of the prescribed boundary conditions. The respective boundary-value data at the satellite's altitude are the X magnetic component measured by the CHAMP vector magnetometer along the satellite track for the forward method and the difference between the measured and predicted Z magnetic component for the adjoint method. Both methods are alternatively formulated for the case when the time-dependent, spherical harmonic Gauss coefficients of the magnetic field generated by external equatorial ring currents in the magnetosphere and the magnetic field generated by the induced eddy currents in the Earth, respectively, are specified.

Before applying these methods, the CHAMP vector magnetic data are modeled by a two-step, track-by-track spherical harmonic analysis. As a result, the X and Z components of CHAMP magnetic data are represented in terms of series of Legendre polynomial derivatives. Four examples of the two-step analysis of the signals recorded by the CHAMP vector magnetometer are presented. The track-by-track analysis is applied to the CHAMP data recorded in the year 2001, yielding a 1-year time series of spherical harmonic coefficients.

The output of the forward modeling of electromagnetic induction, that is, the predicted Z component at satellite altitude, can then be compared with the satellite observations. The squares of the differences between the measured and predicted Z component summed up over all CHAMP tracks determine the misfit. The sensitivity of the CHAMP data, that is, the partial derivatives of the misfit with respect to mantle conductivity parameters, are then obtained by the scalar product of the forward and adjoint solutions, multiplied by the gradient of the conductivity and integrated over all CHAMP tracks. Such exactly determined sensitivities are checked against the numerical differentiation of the misfit, and a good agreement is obtained. The attractiveness of the adjoint method lies in the fact that the adjoint sensitivities are calculated for the price of only an additional forward calculation, regardless of the number of conductivity parameters. However, since the adjoint solution proceeds backwards in time, the forward solution must be stored at each time step, leading to memory requirements that are linear with respect to the number of steps undertaken.

Having determined the sensitivities, the conjugate gradient inversion is run to infer 1-D and 2-D conductivity structures of the Earth based on the CHAMP residual time series (after the subtraction of the static field and secular variations as described by the CHAOS model) for the year 2001. It is shown that this time series is capable of resolving both 1-D and 2-D structures in the upper mantle and the upper part of the lower mantle, while it is not sufficiently long to reliably resolve the conductivity structure in the lower part of the lower mantle.

Keywords 2-D electrical conductivity; Adjoint sensitivity method; CHAMP magnetic data; Electromagnetic induction; Finite elements; Inverse problem; Spherical harmonics; Time-domain approach

1 Introduction

Global-scale electromagnetic (EM) induction processes in the Earth, induced by external magnetic storms, are traditionally used to determine the electrical conductivity of the Earth's interior. During large magnetic storms, charged particles in the near-Earth plasma sheet are energized and injected deeper into the magnetosphere, producing the storm-time ring current at roughly 3–4 Earth radii (Kivelson and Russell, 1995; Daglis et al., 1999). At mid-latitudes on the Earth's surface, the magnetic potential due to this magnetospheric current has a nearly axisymmetric structure. It changes with colatitude predominantly as the cosine of colatitude, that is, as the Legendre function $P_{10}(\cos \vartheta)$ (e.g., Eckhardt et al., 1963). The time evolution of a magnetic storm also follows a characteristic pattern. Its initial phase is characterized by a rapid intensification of the ring current over time scales of several hours. The following main phase of a storm, which can last as long as 2 to 3 days in the case of severe storms, is characterized by the occurrence of multiple intense substorms, with the associated auroral and geomagnetic effects. The final recovery phase of a storm is characterized by an exponential relaxation of the ring current to its usual intensity with a characteristic period of several days (Hultqvist, 1973).

The instantaneous strength of the magnetospheric ring currents is conventionally monitored by the Dst index, an average of the horizontal magnetic field at a collection of geomagnetic observatories located at equatorial latitudes. The EM induction signal, defined herein as the residual magnetic induction vector $\boldsymbol{B}$ that remains after core, lithospheric and ionospheric contributions are removed, is of the order of 20–200 nT (Langel et al., 1996). The total magnetospheric signal consists of a primary term owing to the ring current itself, plus an induced term from eddy currents inside the Earth responding to fluctuations in ring current intensity. In geomagnetic dipole coordinates, the magnetic-field signal induced by a zonal external current source in a radially symmetric Earth follows a zonal distribution, adding roughly 20–30% to the primary signal at the dipole equator, and canceling up to 80% of the primary signal near the geomagnetic poles. This is easily demonstrated by an analytic calculation of the response of a conducting sphere in a uniform magnetic field (Everett and Martinec, 2003). Even when the source has a simple spatial structure, induced currents in reality are more complicated since they are influenced by the Earth's heterogeneous conductivity structure. The resulting 3-D induction effects at satellite altitudes caused by the spatially complicated induced current flow can be large and dependent on upper-mantle electrical conductivity.

Electrical conductivity is an important deep-Earth physical property, spatial variations of which provide fundamental information concerning geodynamic processes such as the subduction of slabs, the ascent of mantle plumes, and the convection of anomalously hot mantle material. The electrical conductivity of the upper to mid-mantle is conventionally studied using frequency-domain geomagnetic induction techniques. The traditional approach involves the estimation of surface impedances from land-based observatory recordings of geomagnetic time variations of external origin in the period range of several hours to several days (e.g., Eckhardt et al., 1963; Banks, 1969; Schultz and Larsen, 1987, 1990). The underlying electrical conductivity is extracted from the measured impedances by means of forward modeling and inversion. The task is difficult, however, since the global distribution of magnetic observatories is sparse and irregular, the quality of the magnetic time series is variable, and assumptions about the spatial and temporal variability of external magnetic sources must often be introduced to be able to carry out electromagnetic induction modeling (e.g., Langel and Estes, 1985a).

The recent high-precision magnetic missions, Ørsted (launched 1999 February) and CHAMP (2000 July), may have the ability to help these problems to be overcome. Unlike land-based observatories, satellites acquire data with no regard for oceans and continents, hemispheres, or political boundaries. On the other hand, however, the combined spatial and temporal character of satellite signals makes their analysis more difficult than that of their terrestrial counterparts, which manifest only temporal variations. Nonetheless, significant progress has already been made in separating the signals due to EM induction in the Earth from satellite magnetic data (Didwall, 1984; Oraevsky et al., 1993, Olsen, 1999; Tarits and Grammatica, 2000; Korte et al., 2003; Constable and Constable, 2004; Kuvshinov and Olsen, 2006).

In order to model 3-D EM induction effects in the geomagnetic field at satellite altitudes quantitatively, a transient EM induction in a 3-D heterogeneous sphere needs to be simulated. Several techniques are available to model the geomagnetic response of a 3-D heterogeneous sphere in the Fourier-frequency domain, each based on a different numerical method: the spherical thin-sheet method (Fainberg et al., 1990; Kuvshinov et al., 1999a), the finite-element method (Everett and Schultz, 1996; Weiss and Everett, 1998), the integral-equation method (Kuvshinov et al., 1999b), the finite-difference method (Uyeshima and Schultz, 2000) and the spectral finite-element method (Martinec, 1999). It is, however, inconvenient to study the geomagnetic induction response to a transient excitation, such as a magnetic storm, in the Fourier-frequency domain. Moreover, the complicated spatial and temporal variability of satellite data favors a time-domain approach. Several time-domain methods for computing the geomagnetic induction response to a transient external source have recently been developed (Hamano, 2002; Everett and Martinec, 2003; Martinec et al., 2003; Velímský and Martinec, 2005).

On a planetary scale, the electrical conductivity of the Earth's interior is determined from permanent geomagnetic observatory recordings by applying inverse theory (e.g., Banks, 1969; Schultz and Larsen, 1987, 1990; and others). Although electrical conductivity depends on the temperature, pressure, and chemical composition of the Earth's interior with many degrees of freedom, one is always forced to parameterize the conductivity by only a few parameters so that the inverse modeling can be carried out with a certain degree of uniqueness. A major difficulty in the choice of model parameterization is to introduce those parameters that are most important for interpreting the data. Strictly, this information cannot be known a priori, but can be inferred from the analysis of the sensitivities, that is, partial derivatives of the data with respect to model parameters. Once the sensitivities to all parameters are available, they can subsequently be used for ranking the relative importance of conductivity parameters to a forward-modeled response, for refining an initial conductivity model to improve the fit to the observed data, and for assessing the uncertainty of the inverse-modeled conductivity distribution due to the propagation of errors contaminating the data.

A straightforward approach to compute the sensitivities is the so-called brute-force method, whereby the partial derivatives with respect to model parameters are approximated by the centered difference of two forward model runs. Although the brute-force method is not particularly elegant, it is useful for computing the sensitivities with respect to a small number of model parameters, or for testing the accuracy of faster algorithms (which is the case in this chapter). However, it becomes impractical for a conductivity model with a large number of parameters.

There are two advanced formal techniques for calculating the sensitivities: the forward sensitivity method and the adjoint sensitivity method. In the forward sensitivity method, the forward model is differentiated with respect to model parameters and the resulting forward sensitivity equations are solved for the partial derivatives of the field variables. If there are M model parameters, then the solutions of M forward sensitivity equations are required. Although this is an excellent method when M is small, it, however, becomes computationally expensive for larger values of M. The mathematical foundations of the forward and adjoint sensitivities for linear and nonlinear dynamical systems are presented by Marchuk (1995), Cacuci (2003), and Sandu et al. (2003), while its application to EM induction modeling is described by Jupp and Vozoff (1977), Rodi (1976), Oldenburg (1990), and McGillivray et al. (1994).

The adjoint sensitivity method, applied also hereafter, is a powerful complement to the forward sensitivity method. In this method, the adjoint sensitivity equations are solved by making use of nearly identical forward modeling code, but running it backwards in time. For a physical system that is linearly dependent on its model parameters, the adjoint sensitivity equations are independent of the original (forward) equations (Cacuci, 2003), and hence the adjoint sensitivity equations are solved only once in order to obtain the adjoint solution. The sensitivities to all model parameters are then obtained by a subsequent integration of the product of the forward and adjoint solutions. Thus, there is no need to solve repeatedly the forward sensitivity equations, as in the forward sensitivity method. The adjoint sensitivity method is thus the most efficient for sensitivity analysis of models with large numbers of parameters. The adjoint sensitivity method for a general physical system was, for example, described by Morse and Feshbach (1953), Lanczos (1961), Marchuk (1995), Cacuci (2003), and Tarantola (2005), and its application to EM induction problems was demonstrated by Weidelt (1975), Madden and Mackie (1989), McGillivray and Oldenburg (1990), Oldenburg (1990), Farquharson and Oldenburg (1996), Newman and Alumbaugh (1997, 2000), Dorn et al. (1999), Rodi and Mackie (2001), Avdeev and Avdeeva (2006), and Kelbert et al. (2008).

In this chapter, the forward and adjoint methods for data recorded by the vector flux gate magnetometer on board of the CHAMP is presented. The satellite was launched on July 15, 2000, into a near-polar orbit with an inclination approximately 87.3° and initial altitude of 454 km. Magnetic-storm signals which are generated by equatorial ring currents in the magnetosphere are concentrated on here. The configuration of the CHAMP orbit and the distribution of the magnetospheric ring currents allows one to model EM induction processes in an axisymmetric geometry. The main aim here is to rigorously formulate the forward and adjoint sensitivity methods and use them for interpreting the CHAMP magnetic data for the year 2001.

2 Basic Assumptions on EM Induction Modeling for CHAMP Magnetic Data

In the following, the magnetic signals induced by equatorial ring currents in the magnetosphere and measured by the CHAMP vector flux gate magnetometer are considered. To obtain these signals, the CHAMP magnetic data are processed in a specific manner, and several assumptions for the EM induction modeling are made. The data processing steps and assumptions (see Martinec and McCreadie [2004], and Velímský et al. [2006] for more details) are as follows.

1. The CHAOS model of the Earth's magnetic field (Olsen et al., 2006a) is first subtracted from the CHAMP vector magnetic data. The residual magnetic time series are only considered along night-side satellite tracks at low and mid-latitudes, allowing the reduction of the magnetic effects of ionospheric currents and field-aligned currents in the polar regions.
2. The satellite is assumed to orbit the Earth sufficiently fast (one CHAMP orbit takes approximately 90 min) compared to the time variations of the ring current to allow the separation of the spatial and temporal changes of the magnetic field observed by the single satellite

to be performed in a simple way. Each night-side satellite track is considered to sample a snapshot of the magnetic field at the time when the satellite crosses the magnetic equator.

3. Since the CHAMP satellite orbit is nearly polar, magnetic signals sampled along a track are dominantly influenced by latitudinal changes in the electrical conductivity of the Earth's mantle. The effect of longitudinal variations in electrical conductivity on track data is not considered and it is assumed that the electrical conductivity σ of the Earth is axisymmetric, that is,

$$\sigma = \sigma(r, \vartheta) \qquad \text{in } \mathcal{B}, \tag{1}$$

where $\mathcal{B}$ is a conducting sphere approximating a heterogeneous Earth, r is the radial distance from the center of $\mathcal{B}$ and ϑ is the colatitude.

4. Since ring-current magnetospheric excitation has nearly an axisymmetric geometry, it is assumed that, for a given satellite track, the inducing and induced magnetic fields possess an axisymmetry property, that is,

$$G_{jm}^{(e)}(t) = G_{jm}^{(i)}(t) = 0 \qquad \text{for } m \neq 0, \tag{2}$$

where $G_{jm}^{(e)}(t)$ and $G_{jm}^{(i)}(t)$ are the time-dependent, spherical harmonic Gauss coefficients of the magnetic field generated by external equatorial ring currents in the magnetosphere and the magnetic field generated by the induced eddy currents in the Earth, respectively.

3 Forward Method of Global EM Induction

In this section, the forward method of EM induction is formulated for a 2-D case when the electrical conductivity and external sources of electromagnetic variations are axisymmetrically distributed and when the external current excitation has a transient feature similar to that of a magnetic storm. Most of considerations in this section follow the papers by Martinec (1997, 1999).

3.1 Formulation of EM Induction for a 3-D Inhomogeneous Earth

An initial, boundary-value problem (IBVP) for EM induction is first formulated for a 3-D inhomogeneous Earth. The conducting sphere $\mathcal{B}$ is excited by a specified source originating in the ionosphere and magnetosphere. From a broad spectrum of temporal variations in the geomagnetic field, long-period geomagnetic variations with periods ranging from several hours to tens of days is considered. The EM induction within $\mathcal{B}$ for this range of periods is governed by Maxwell's equations taken in the quasi-static approximation:

$$\operatorname{curl} \mathbf{B} = \mu\sigma \boldsymbol{E}, \tag{3}$$

$$\operatorname{curl} \boldsymbol{E} = -\frac{\partial \boldsymbol{B}}{\partial t}, \tag{4}$$

$$\operatorname{div} \boldsymbol{B} = 0, \tag{5}$$

$$\operatorname{div} \boldsymbol{E} = \frac{\varrho}{\epsilon}, \tag{6}$$

where $\boldsymbol{E}$ is the electric intensity, $\boldsymbol{B}$ is the magnetic induction, ϱ is the volume charge density, ϵ and μ are the permitivity and permeability, respectively, and σ is the electrical conductivity. The vacuum values for μ and ϵ are used for the whole space.

Since a 3-D inhomogeneous Earth is being considered, the volume charge density ϱ does not vanish exponentially as in a homogeneous medium (e.g., Stratton, 1941), but electric charges may accumulate in regions where the electrical conductivity has a nonzero gradient. The assumption of the quasi-static approximation of the electromagnetic field then yields the volume charge density ϱ in the form (Pěč and Martinec, 1986; Weaver, 1994)

$$\varrho = -\frac{\epsilon}{\sigma}\left(\operatorname{grad}\sigma \cdot \boldsymbol{E}\right). \tag{7}$$

From this, Eq. 6 may be written in the form

$$\operatorname{div}\left(\sigma \boldsymbol{E}\right) = 0, \tag{8}$$

showing that the electric current density $\boldsymbol{i} = \sigma \boldsymbol{E}$ is divergence-free.

Substituting Eqs. 3 into 4, the magnetic diffusion equation for $\boldsymbol{B}$ is obtained:

$$\operatorname{curl}\left(\frac{1}{\sigma}\operatorname{curl}\boldsymbol{B}\right) + \mu\frac{\partial \boldsymbol{B}}{\partial t} = 0 \qquad \text{in } \mathcal{B}. \tag{9}$$

Note that the vector field $\boldsymbol{B}$ is automatically forced to be divergence-free by Eq. 9. Therefore, if Eq. 9 is not violated, the divergence-free condition (5) must not be explicitly imposed on $\boldsymbol{B}$.

Furthermore, it is assumed that the region external to $\mathcal{B}$ is a perfect insulator that models the near-space environment to the Earth's surface. As seen from the Biot-Savart law, expressed in Eq. 3, the magnetic induction $\boldsymbol{B}_0$ in this insulator (the subscript 0 indicating a vacuum) is taken to be irrotational,

$$\operatorname{curl}\boldsymbol{B}_0 = 0. \tag{10}$$

Hence, $\boldsymbol{B}_0$ can be expressed as the negative gradient of magnetic scalar potential U,

$$\boldsymbol{B}_0 = -\operatorname{grad} U. \tag{11}$$

The constraint $\operatorname{div}\boldsymbol{B}_0 = 0$ then implies that the potential U satisfies the scalar Laplace equation

$$\nabla^2 U = 0. \tag{12}$$

Applying the toroidal-spheroidal decomposition of a vector to the magnetic induction $\boldsymbol{B}_0$, that is, writing $\boldsymbol{B}_0 = \boldsymbol{B}_{0,T} + \boldsymbol{B}_{0,S}$, and using the fact that the gradient of a scalar generates a spheroidal vector, Eq. 11 shows that the toroidal component of the magnetic induction in a vacuum vanishes,

$$\boldsymbol{B}_{0,T} = 0. \tag{13}$$

To complete the specifications, boundary conditions are prescribed on the external surface $\partial\mathcal{B}$ of sphere $\mathcal{B}$. Decomposing the magnetic induction $\boldsymbol{B}$ into its toroidal and spheroidal parts, $\boldsymbol{B} = \boldsymbol{B}_T + \boldsymbol{B}_S$, it is required, in view of Eq. 13, that the toroidal part of $\boldsymbol{B}$ vanishes on $\partial\mathcal{B}$,

$$\boldsymbol{B}_T = 0 \qquad \text{on } \partial\mathcal{B}, \tag{14}$$

and that the tangential components of $\boldsymbol{B}$ are continuous across $\partial\mathcal{B}$,

$$\boldsymbol{n} \times \left(\boldsymbol{B} - \boldsymbol{B}_0\right) = 0 \qquad \text{on } \partial\mathcal{B}, \tag{15}$$

where $\boldsymbol{n}$ is the outward unit normal to $\partial\mathcal{B}$ and "×" stands for the cross-product of vectors. In view of Eq. 13, the last condition requires that the tangential components of the spheroidal part of $\boldsymbol{B}$ are fixed on $\partial\mathcal{B}$,

$$\boldsymbol{n} \times \boldsymbol{B}_S = \boldsymbol{b}_t \qquad \text{on } \partial\mathcal{B}, \tag{16}$$

where $\boldsymbol{b}_t$ is the tangential component of the magnetic field $\boldsymbol{B}_0$ in the atmosphere taken at the surface $\partial\mathcal{B}$ at time $t \geq 0$,

$$\boldsymbol{b}_t := \boldsymbol{n} \times \boldsymbol{B}_0\big|_{\partial\mathcal{B}}. \tag{17}$$

Without additional specifications, $\boldsymbol{b}_t$ is assumed to be derivable from ground observations of the geomagnetic field variations. Equation 9 is yet subject to the initial condition

$$\boldsymbol{B}\big|_{t=0} = \boldsymbol{B}^0 \qquad \text{in } \mathcal{B}, \tag{18}$$

where $\boldsymbol{B}^0$ is the initial value of the magnetic induction.

The other two boundary conditions require the continuity of the tangential components of the electric intensity (e.g., Stratton, 1941):

$$\boldsymbol{n} \times (\boldsymbol{E} - \boldsymbol{E}_0) = 0 \qquad \text{on } \partial\mathcal{B}, \tag{19}$$

and the continuity of the normal component of the magnetic induction:

$$\boldsymbol{n} \cdot (\boldsymbol{B} - \boldsymbol{B}_0) = 0 \qquad \text{on } \partial\mathcal{B}. \tag{20}$$

A solution of Eq. 9 with the two boundary conditions (14) and (17) is now proved to be unique. Suppose that there exists two solutions, $\boldsymbol{B}_1$ and $\boldsymbol{B}_2$, that satisfy the problem. The difference vector $\boldsymbol{B} := \boldsymbol{B}_1 - \boldsymbol{B}_2$ is zero at the initial time $t = 0$ and satisfies a homogeneous boundary-value problem for $t > 0$:

$$\operatorname{curl}\left(\frac{1}{\sigma}\operatorname{curl}\boldsymbol{B}\right) + \mu\frac{\partial \boldsymbol{B}}{\partial t} = 0 \qquad \text{in } \mathcal{B}, \tag{21}$$

$$\boldsymbol{B}_T = 0 \qquad \text{on } \partial\mathcal{B}, \tag{22}$$

$$\boldsymbol{n} \times \boldsymbol{B}_S = 0 \qquad \text{on } \partial\mathcal{B}. \tag{23}$$

Multiplying Eq. 21 by $\boldsymbol{B}$ and making use of Green's theorem

$$\int_{\mathcal{B}} \operatorname{curl}\left(\frac{1}{\sigma}\operatorname{curl}\boldsymbol{a}\right)\cdot \boldsymbol{b}\, dV = \int_{\mathcal{B}} \frac{1}{\sigma}(\operatorname{curl}\boldsymbol{a}\cdot\operatorname{curl}\boldsymbol{b})\, dV - \int_{\partial\mathcal{B}} \frac{1}{\sigma}\operatorname{curl}\boldsymbol{a}\cdot(\boldsymbol{n}\times\boldsymbol{b})\, dS, \tag{24}$$

where $\boldsymbol{a}$ and $\boldsymbol{b}$ are differentiable vector fields and "·" stands for the scalar product of vectors, results in

$$\int_{\mathcal{B}} \frac{1}{\sigma}(\operatorname{curl}\boldsymbol{B}\cdot\operatorname{curl}\boldsymbol{B})\, dV + \frac{\mu}{2}\frac{\partial}{\partial t}\int_{\mathcal{B}}(\boldsymbol{B}\cdot\boldsymbol{B})\, dV = \int_{\partial\mathcal{B}}\frac{1}{\sigma}\operatorname{curl}\boldsymbol{B}\cdot(\boldsymbol{n}\times\boldsymbol{B})\, dS. \tag{25}$$

Decomposing $\boldsymbol{B}$ into its toroidal and spheroidal parts, $\boldsymbol{B} = \boldsymbol{B}_T + \boldsymbol{B}_S$, the surface integral on the right-hand side then reads as

$$\int_{\partial\mathcal{B}}\frac{1}{\sigma}\operatorname{curl}\boldsymbol{B}\cdot(\boldsymbol{n}\times\boldsymbol{B})\, dS = \int_{\partial\mathcal{B}}\frac{1}{\sigma}\operatorname{curl}\boldsymbol{B}\cdot[(\boldsymbol{n}\times\boldsymbol{B}_T)+(\boldsymbol{n}\times\boldsymbol{B}_S)]\, dS. \tag{26}$$

Both of the integrals on the right-hand side are equal to zero because of conditions (22) and (23). Moreover, integrating Eq. 25 over time and using the initial condition $\boldsymbol{B} = 0$ at $t = 0$, one obtains

$$\frac{\mu}{2}\int_{\mathcal{B}}(\boldsymbol{B}\cdot\boldsymbol{B})\, dV = -\int_0^T\left[\int_{\mathcal{B}}\frac{1}{\sigma}(\operatorname{curl}\boldsymbol{B}\cdot\operatorname{curl}\boldsymbol{B})\, dV\right]dt. \tag{27}$$

Since $\sigma > 0$ in $\mathcal{B}$, the term on the right-hand side is always equal to or less than zero. On the other hand, the energy integral on the left-hand side is positive or equal to zero. Equation 27 can therefore only be satisfied if the difference field $\boldsymbol{B}$ is equal to zero at any time $t > 0$. Hence, Eq. 9, together with the boundary conditions (14) and (17) and the initial condition (18) satisfying divergence-free condition (5), ensures the uniqueness of the solution.

3.2 Special Case: EM Induction in an Axisymmetric Case

Adopting the above assumptions for the case of EM induction for CHAMP magnetic data (see ➋ Sect. 2), the discussion is confined to the problem with axisymmetric electrical conductivity $\sigma = \sigma(r, \vartheta)$ and assumed that the variations of magnetic field recorded by the CHAMP magnetometer are induced by a purely zonal external source. Under these two restrictions, both the inducing and induced parts of magnetic induction $\boldsymbol{B}$ are an axisymmetric vector field that may be represented in terms of zonal spherical vector harmonics $\boldsymbol{Y}_j^\ell(\vartheta)$ (Varshalovich et al., 1989, ➋ Sect. 7.3). Using these harmonics, the toroidal-spheroidal decomposition of magnetic induction $\boldsymbol{B}$ can be expressed in the form

$$\begin{aligned}\boldsymbol{B}(r,\vartheta) &= \boldsymbol{B}_T(r,\vartheta) + \boldsymbol{B}_S(r,\vartheta)\\ &= \sum_{j=1}^{\infty}\sum_{\ell=j-1}^{j+1} B_j^\ell(r)\boldsymbol{Y}_j^\ell(\vartheta)\,,\end{aligned} \tag{28}$$

where $\boldsymbol{Y}_j^j(\vartheta)$ and $\boldsymbol{Y}_j^{j\pm1}(\vartheta)$ are zonal toroidal and spheroidal vector spherical harmonics, respectively, and $B_j^j(r)$ and $B_j^{j\pm1}(r)$ are spherical harmonic expansion coefficients of the toroidal ($\boldsymbol{B}_T$) and spheroidal ($\boldsymbol{B}_S$) parts of $\boldsymbol{B}$, respectively. Moreover, the r and ϑ components of $\boldsymbol{Y}_j^j(\vartheta)$ and the φ component of $\boldsymbol{Y}_j^{j\pm1}(\vartheta)$ are identically equal to zero, which means that the toroidal-spheroidal decomposition of $\boldsymbol{B}$ is also decoupled with respect to the spherical components of $\boldsymbol{B}_T$ and $\boldsymbol{B}_S$. The toroidal-spheroidal decomposition (28) can be introduced such that the toroidal part $\boldsymbol{B}_T$ is divergence-free and the r component of the curl of the spheroidal part $\boldsymbol{B}_S$ vanishes:

$$\operatorname{div}\boldsymbol{B}_T = 0\,, \qquad \boldsymbol{e}_r\cdot\operatorname{curl}\boldsymbol{B}_S = 0. \tag{29}$$

It should be emphasized that the axisymmetric geometry of the problem allows one to abbreviate the notation and drop the angular-order index $m = 0$ for scalar and vector spherical harmonics.

The IBVP formulated in the previous section is now examined for the axisymmetric case. First, the product $\sigma\boldsymbol{B}$ is decomposed into the toroidal and spheroidal parts. Considering two differential identities

$$\begin{aligned}\operatorname{div}(\sigma\boldsymbol{B}) &= \sigma\operatorname{div}\boldsymbol{B} + \operatorname{grad}\sigma\cdot\boldsymbol{B},\\ \operatorname{curl}(\sigma\boldsymbol{B}) &= \sigma\operatorname{curl}\boldsymbol{B} + \operatorname{grad}\sigma\times\boldsymbol{B}\end{aligned} \tag{30}$$

for $\boldsymbol{B}_T$ and $\boldsymbol{B}_S$, respectively, and realizing that (i) for an axisymmetric electrical conductivity σ, the φ component of the vector grad σ is identically equal to zero, and (ii) for purely zonal behavior of $\boldsymbol{B}_T$ and $\boldsymbol{B}_S$, the r and ϑ components of $\boldsymbol{B}_T$ and the φ component of $\boldsymbol{B}_S$ are identically equal to zero, it is found that the products $\operatorname{grad}\sigma\cdot\boldsymbol{B}_T = 0$ and $\boldsymbol{e}_r\cdot(\operatorname{grad}\sigma\times\boldsymbol{B}_S) = 0$. Then

$$\operatorname{div}(\sigma\boldsymbol{B}_T) = 0\,, \qquad \boldsymbol{e}_r\cdot\operatorname{curl}(\sigma\boldsymbol{B}_S) = 0\,. \tag{31}$$

In other words, the product of σ with $\boldsymbol{B}_T$ and $\boldsymbol{B}_S$ results again in toroidal and spheroidal vectors, respectively. Note that such a "decoupled" spheroidal-toroidal decomposition of $\sigma\boldsymbol{B}$ is broken once either of the two basic assumptions of the axisymmetric geometry of the problem is violated.

Furthermore, since the rotation of a spheroidal vector is a toroidal vector and vice versa, the IBVP for an axisymmetric case can be split into two decoupled IBVPs: (1) the problem formulated for the spheroidal magnetic induction $\boldsymbol{B}_S$:

$$\operatorname{curl}\left(\frac{1}{\sigma}\operatorname{curl}\boldsymbol{B}_S\right)+\mu\frac{\partial\boldsymbol{B}_S}{\partial t}=0 \qquad \text{in } \mathcal{B}, \tag{32}$$

$$\operatorname{div}\boldsymbol{B}_S=0 \qquad \text{in } \mathcal{B}, \tag{33}$$

$$\boldsymbol{n}\times\boldsymbol{B}_S=\boldsymbol{b}_t \qquad \text{on } \partial\mathcal{B}. \tag{34}$$

with an inhomogeneous initial condition $\boldsymbol{B}_S=\boldsymbol{B}_S^0$ at $t=0$, and (2) the problem for the toroidal magnetic induction $\boldsymbol{B}_T$:

$$\operatorname{curl}\left(\frac{1}{\sigma}\operatorname{curl}\boldsymbol{B}_T\right)+\mu\frac{\partial\boldsymbol{B}_T}{\partial t}=0 \qquad \text{in } \mathcal{B}, \tag{35}$$

with homogeneous initial and boundary conditions. Attention is hereafter turned to the first IBVP, since the latter one has only a trivial solution $\boldsymbol{B}_T=0$ in $\mathcal{B}$. Note again that such a "decoupled" spheroidal-toroidal decomposition of the boundary-value problem for EM induction cannot be achieved if the conductivity σ depends also on the longitude φ and/or when the external excitation source has not only zonal, but also tesseral and sectoral spherical components.

For convenience, the toroidal vector potential $\boldsymbol{A}$ ($\boldsymbol{A}$ is not labeled by subscript T since its spheroidal counterpart is not used in this text) that generates the spheroidal magnetic induction $\boldsymbol{B}_S$ is introduced:

$$\boldsymbol{B}_S=\operatorname{curl}\boldsymbol{A}, \qquad \operatorname{div}\boldsymbol{A}=0. \tag{36}$$

By this prescription, the divergence-free constraint (33) is automatically satisfied and the IBVP for the spheroidal magnetic induction $\boldsymbol{B}_S$ is transformed to the IBVP for the toroidal vector potential $\boldsymbol{A}=\boldsymbol{A}(r,\vartheta)$. In the classical mathematical formulation, the toroidal vector potential $\boldsymbol{A}\in C^2(\overline{\mathcal{B}})\times C^1(\langle 0,\infty))$ is searched for such that $\boldsymbol{B}_S=\operatorname{curl}\boldsymbol{A}$ and

$$\frac{1}{\mu}\operatorname{curl}\operatorname{curl}\boldsymbol{A}+\sigma\frac{\partial\boldsymbol{A}}{\partial t}=0 \qquad \text{in } \mathcal{B}, \tag{37}$$

$$\operatorname{div}\boldsymbol{A}=0 \qquad \text{in } \mathcal{B}, \tag{38}$$

$$\boldsymbol{n}\times\operatorname{curl}\boldsymbol{A}=\boldsymbol{b}_t \qquad \text{on } \partial\mathcal{B}, \tag{39}$$

$$\boldsymbol{A}|_{t=0}=\boldsymbol{A}^0 \qquad \text{in } \mathcal{B}, \tag{40}$$

where the conductivity $\sigma\geq 0$ is a continuous function in $\mathcal{B}$, $\sigma\in C(\mathcal{B})$, $\mu>0$ is the constant permitivity of a vacuum, $\boldsymbol{b}_t(\vartheta,t)\in C^2(\partial\mathcal{B})\times C^1(\langle 0,\infty))$ and $\boldsymbol{A}^0$ is the generating potential for the initial magnetic induction $\boldsymbol{B}_S^0$ such that $\boldsymbol{B}_S^0=\operatorname{curl}\boldsymbol{A}^0$. At the internal interfaces, where the electrical conductivity changes discontinuously, the continuity of the tangential components of magnetic induction and electric intensity is required. The various functional spaces used in this approach are listed in ➋ *Table 1*.

The magnetic diffusion equation (37) for $\boldsymbol{A}$ follows from the Maxwell equation (3). To satisfy Faraday's law (4), the electric intensity has to be a toroidal vector, $\boldsymbol{E}=\boldsymbol{E}_T$, of the form

$$\boldsymbol{E}_T=-\frac{\partial\boldsymbol{A}}{\partial t}. \tag{41}$$

Note that the electric intensity $\boldsymbol{E}_T$ is not generated by the gradient of a scalar electromagnetic potential, since the gradient of a scalar results in a spheroidal vector that would contradict the requirement that $\boldsymbol{E}_T$ is a toroidal vector. Under the prescription (41), the continuity condition

Table 1
List of functional spaces used

$C(\overline{\mathcal{D}})$	Space of the continuous functions defined in the domain $\overline{\mathcal{D}}$ ($\overline{\mathcal{D}}$ is the closure of $\mathcal{D}$)
$C^1(\langle 0,\infty))$	Space of the functions for which the classical derivatives up to first order are continuous in the interval $\langle 0,\infty)$
$C^2(\overline{\mathcal{D}})$	Space of the functions for which the classical derivatives up to second order are continuous in $\overline{\mathcal{D}}$
$L_2(\mathcal{D})$	Space of square-integrable functions in $\mathcal{D}$

(19) of the tangential components of the electric intensity can be ensured by the continuity of the toroidal vector potential $\boldsymbol{A}$,

$$\boldsymbol{A} = \boldsymbol{A}_0 \qquad \text{on } \partial\mathcal{B}, \tag{42}$$

since $\boldsymbol{A}$ has only nonzero tangential components.

3.3 Gauss Representation of Magnetic Induction in the Atmosphere

As mentioned above, the Earth's atmosphere in the vicinity of the Earth is assumed to be nonconducting, with the magnetic induction $\boldsymbol{B}_0$ generated by the magnetic scalar potential U, which is a harmonic function satisfying Laplace's equation (12). Under the assumption of axisymmetric geometry (see ➋ Sect. 2), its solution is given in terms of zonal solid scalar spherical harmonics $r^j Y_j(\vartheta)$ and $r^{-j-1}Y_j(\vartheta)$:

$$U(r,\vartheta,t) = a\sum_{j=1}^{\infty}\left[\left(\frac{r}{a}\right)^j G_j^{(e)}(t) + \left(\frac{a}{r}\right)^{j+1} G_j^{(i)}(t)\right] Y_j(\vartheta) \qquad \text{for } r \geq a, \tag{43}$$

where a is the radius of a conducting sphere $\mathcal{B}$ which is equal to a mean radius of the Earth, and $G_j^{(e)}(t)$ and $G_j^{(i)}(t)$ are the time-dependent, zonal spherical harmonic Gauss coefficients of the external and internal magnetic fields, respectively.

Using the following formula for the gradient of a scalar function $f(r)Y_j(\vartheta)$ in spherical coordinates (Varshalovich et al., 1989, p. 217),

$$\text{grad}\left[f(r)Y_j(\vartheta)\right] = \sqrt{\frac{j}{2j+1}}\left(\frac{d}{dr}+\frac{j+1}{r}\right)f(r)\boldsymbol{Y}_j^{j-1}(\vartheta) - \sqrt{\frac{j+1}{2j+1}}\left(\frac{d}{dr}-\frac{j}{r}\right)f(r)\boldsymbol{Y}_j^{j+1}(\vartheta), \tag{44}$$

the magnetic induction in a vacuum ($r \geq a$) may be expressed in terms of solid vector spherical harmonics as

$$\boldsymbol{B}_0(r,\vartheta,t) = -\sum_{j=1}^{\infty}\left[\sqrt{j(2j+1)}\left(\frac{r}{a}\right)^{j-1} G_j^{(e)}(t)\boldsymbol{Y}_j^{j-1}(\vartheta) + \sqrt{(j+1)(2j+1)}\left(\frac{a}{r}\right)^{j+2} G_j^{(i)}(t)\boldsymbol{Y}_j^{j+1}(\vartheta)\right]. \tag{45}$$

This formula again demonstrates the fact that the toroidal component of the magnetic field in a vacuum vanishes, $\boldsymbol{B}_{0,T} = 0$.

For the following considerations, it is convenient to express the magnetic induction $\boldsymbol{B}_0$ in terms of the toroidal vector potential $\boldsymbol{A}_0$ such that $\boldsymbol{B}_0 = \operatorname{curl} \boldsymbol{A}_0$. Using the rotation formulae (226) and (227) for vector spherical harmonics, the spherical harmonic representation of the toroidal vector potential in a vacuum reads as (see also ➋ Chaps. 17 and ➋ 18)

$$\boldsymbol{A}_0(r, \vartheta, t) = a \sum_{j=1}^{\infty} \left[\sqrt{\frac{j}{j+1}} \left(\frac{r}{a}\right)^{j} G_j^{(e)}(t) - \sqrt{\frac{j+1}{j}} \left(\frac{a}{r}\right)^{j+1} G_j^{(i)}(t) \right] \boldsymbol{Y}_j^j(\vartheta). \tag{46}$$

The representation (46) of $\boldsymbol{A}_0$ by solid spherical harmonics $r^j \boldsymbol{Y}_j^j(\vartheta)$ and $r^{-j-1}\boldsymbol{Y}_j^j(\vartheta)$ is consistent with the fact that $\boldsymbol{A}_0$ satisfies the vector Laplace equation, $\nabla^2 \boldsymbol{A}_0 = 0$, as seen from Eqs. 37 and (38) for $\sigma = 0$.

As introduced above, $\partial\mathcal{B}$ is a sphere (of radius a) with the external normal $\boldsymbol{n}$ coinciding with the spherical base vector $\boldsymbol{e}_r$, that is, $\boldsymbol{n} = \boldsymbol{e}_r$. Taking into account expression (217) for the polar components of the vector spherical harmonics, the horizontal northward X component of the magnetic induction vector $\boldsymbol{B}_0$ at radius $r \geq a$ is

$$X(r, \vartheta, t) := -\boldsymbol{e}_\vartheta \cdot \boldsymbol{B}_0 = \sum_{j=1}^{\infty} X_j(r, t) \frac{\partial Y_j(\vartheta)}{\partial \vartheta}, \tag{47}$$

where $\boldsymbol{e}_\vartheta$ is the spherical base vector in the colatitudinal direction. The spherical harmonic coefficients X_j are expressed in the form

$$X_j(r, t) = \left(\frac{r}{a}\right)^{j-1} G_j^{(e)}(t) + \left(\frac{a}{r}\right)^{j+2} G_j^{(i)}(t). \tag{48}$$

Similarly, the spherical harmonic representation of the vertical downward Z component of the magnetic induction vector $\boldsymbol{B}_0$ at radius $r \geq a$ is

$$Z(r, \vartheta, t) := -\boldsymbol{e}_r \cdot \boldsymbol{B}_0 = \sum_{j=1}^{\infty} Z_j(r, t) Y_j(\vartheta), \tag{49}$$

where the spherical harmonic coefficients Z_j are

$$Z_j(r, t) = j \left(\frac{r}{a}\right)^{j-1} G_j^{(e)}(t) - (j+1) \left(\frac{a}{r}\right)^{j+2} G_j^{(i)}(t). \tag{50}$$

Equations 48 and 50 show that the coefficients X_j and Z_j are composed of two different linear combinations of the spherical harmonics $G_j^{(e)}$ of the external electromagnetic sources and the spherical harmonics $G_j^{(i)}$ of the induced electromagnetic field inside the Earth. Consequently, there is no need to specify these coefficients separately when X_j and Z_j are used as the boundary-value data for the forward and adjoint modeling of EM induction, respectively.

Making use of Eq. 45 and formula (223) for the cross product of $\boldsymbol{e}_r$ with the spheroidal vector spherical harmonics $\boldsymbol{Y}_j^{j\pm1}(\vartheta)$, the tangential component of magnetic induction $\boldsymbol{B}_0$ in a vacuum ($r \geq a$) has the form

$$\boldsymbol{e}_r \times \boldsymbol{B}_0(r, \vartheta, t) = -\sum_{j=1}^{\infty} \sqrt{j(j+1)} \left[\left(\frac{r}{a}\right)^{j-1} G_j^{(e)}(t) + \left(\frac{a}{r}\right)^{j+2} G_j^{(i)}(t) \right] \boldsymbol{Y}_j^j(\vartheta), \tag{51}$$

which, in view of Eq. 48, can be rewritten in terms of the spherical harmonic coefficients $X_j(r, t)$:

$$\boldsymbol{e}_r \times \boldsymbol{B}_0(r, \vartheta, t) = -\sum_{j=1}^{\infty} \sqrt{j(j+1)} X_j(r, t) \boldsymbol{Y}_j^j(\vartheta). \tag{52}$$

In the other words, the axisymmetric geometry allows the determination of $\boldsymbol{e}_r \times \boldsymbol{B}_0$ from the horizontal northward X component of the magnetic induction vector $\boldsymbol{B}_0$.

In particular, the ground magnetic observation vector $\boldsymbol{b}_t$, defined by Eq. 17, can be expressed as

$$\boldsymbol{b}_t(\vartheta, t) = -\sum_{j=1}^{\infty} \sqrt{j(j+1)} \left[G_j^{(e)}(t) + G_j^{(i)}(t) \right] \boldsymbol{Y}_j^j(\vartheta) \tag{53}$$

$$= -\sum_{j=1}^{\infty} \sqrt{j(j+1)} X_j(a,t) \boldsymbol{Y}_j^j(\vartheta). \tag{54}$$

Likewise, the satellite magnetic observation vector $\boldsymbol{B}_t$, defined by

$$\boldsymbol{B}_t := \boldsymbol{n} \times \boldsymbol{B}_0 |_{\partial\mathcal{A}}, \tag{55}$$

where $\partial\mathcal{A}$ is the mean-orbit sphere of radius $r = b$, can be expressed in terms of the external and internal Gauss coefficients and spherical harmonic coefficients $X_j(b,t)$, respectively, as

$$\boldsymbol{B}_t(\vartheta, t) = -\sum_{j=1}^{\infty} \sqrt{j(j+1)} \left[\left(\frac{b}{a}\right)^{j-1} G_j^{(e)}(t) + \left(\frac{a}{b}\right)^{j+2} G_j^{(i)}(t) \right] \boldsymbol{Y}_j^j(\vartheta) \tag{56}$$

$$= -\sum_{j=1}^{\infty} \sqrt{j(j+1)} X_j(b,t) \boldsymbol{Y}_j^j(\vartheta). \tag{57}$$

4 Forward Method of EM Induction for the *X* Component of CHAMP Magnetic Data

The forward method of EM induction can, at least, be formulated for two kinds of the boundary-value data. Either the X component of the CHAMP magnetic data (considered in this section) or the Gauss coefficients of the external magnetic field (the next section) along the track of CHAMP satellite is specified. Most of considerations in this section follow the papers by Martinec (1997), Martinec et al. (2003), and Martinec and McCreadie (2004).

4.1 Classical Formulation

The IBVP (37–40) assumes that magnetic data $\boldsymbol{b}_t$ are prescribed on the Earth's surface. For satellite measurements, this requires the continuation of magnetic data from satellite-orbit altitudes down to the Earth's surface. Since the downward continuation of satellite magnetic data poses a fundamental problem, a modification of the IBVP (37)–(40) such that the X component of CHAMP magnetic data is used directly as boundary values at satellite altitudes is given in this section.

The solution domain is extended by the atmosphere $\mathcal{A}$ surrounding the conducting sphere $\mathcal{B}$. Since the magnetic signals from night-time, mid-latitude tracks only are considered, it is assumed that there are no electric currents in $\mathcal{A}$. This assumption is not completely correct, but it is still a good approximation (Langel and Estes, 1985b). Moreover, $\mathcal{A}$ is treated as a nonconducting spherical layer with the inner boundary coinciding with the surface $\partial\mathcal{B}$ of the conducting sphere $\mathcal{B}$ with radius $r = a$ and the outer boundary coinciding with the mean-orbit sphere $\partial\mathcal{A}$ of radius $r = b$.

The classical mathematical formulation of the IBVP of global EM induction for satellite magnetic data is as follows. Find the toroidal vector potential $\boldsymbol{A}$ in the conducting sphere $\mathcal{B}$ and the toroidal vector potential $\boldsymbol{A}_0$ in the nonconducting atmosphere $\mathcal{A}$ such that the magnetic induction vectors in $\mathcal{B}$ and $\mathcal{A}$ are expressed in the forms $\boldsymbol{B} = \operatorname{curl}\boldsymbol{A}$ and $\boldsymbol{B}_0 = \operatorname{curl}\boldsymbol{A}_0$, respectively, and, for $t > 0$, it holds that

$$\frac{1}{\mu}\operatorname{curl}\operatorname{curl}\boldsymbol{A} + \sigma\frac{\partial \boldsymbol{A}}{\partial t} = 0 \qquad \text{in } \mathcal{B}, \tag{58}$$

$$\operatorname{div}\boldsymbol{A} = 0 \qquad \text{in } \mathcal{B}, \tag{59}$$

$$\operatorname{curl}\operatorname{curl}\boldsymbol{A}_0 = 0 \qquad \text{in } \mathcal{A}, \tag{60}$$

$$\operatorname{div}\boldsymbol{A}_0 = 0 \qquad \text{in } \mathcal{A}, \tag{61}$$

$$\boldsymbol{A} = \boldsymbol{A}_0 \qquad \text{on } \partial\mathcal{B}, \tag{62}$$

$$\boldsymbol{n}\times\operatorname{curl}\boldsymbol{A} = \boldsymbol{n}\times\operatorname{curl}\boldsymbol{A}_0 \qquad \text{on } \partial\mathcal{B}, \tag{63}$$

$$\boldsymbol{n}\times\operatorname{curl}\boldsymbol{A}_0 = \boldsymbol{B}_t \qquad \text{on } \partial\mathcal{A}, \tag{64}$$

$$\boldsymbol{A}|_{t=0} = \boldsymbol{A}^0 \qquad \text{in } \mathcal{B}\cup\mathcal{A}, \tag{65}$$

where the mathematical assumptions imposed on the functions $\boldsymbol{A}$, $\boldsymbol{A}_0$, σ, μ, $\boldsymbol{B}_t$, and $\boldsymbol{A}^0$ are the same as for the IBVP (37)–(40), see ➤ *Table 1*. The continuity condition (62) on $\boldsymbol{A}$ is imposed on a solution since the intention is to apply a different parameterization of $\boldsymbol{A}$ in the sphere $\mathcal{B}$ and the spherical layer $\mathcal{A}$. The term $\boldsymbol{B}_t$ represents the tangential components of the magnetic induction $\boldsymbol{B}_0$ at the satellite altitudes and $\boldsymbol{n}$ is the unit normal to $\partial\mathcal{A}$. The axisymmetric geometry allows (see ➤ Sect. 7) to determine $\boldsymbol{B}_t$ from the horizontal northward X component of the magnetic induction vector $\boldsymbol{B}$ measured by the CHAMP vector magnetometer.

4.2 Weak Formulation

4.2.1 Ground Magnetic Data

The IBVP (37)–(40) for the ground magnetic data $\boldsymbol{b}_t$ is now reformulated in a weak sense. The solution space is introduced as

$$V := \{\boldsymbol{A}|\boldsymbol{A}\in L_2(\mathcal{B}),\ \operatorname{curl}\boldsymbol{A}\in L_2(\mathcal{B}),\ \operatorname{div}\boldsymbol{A} = 0 \text{ in } \mathcal{B}\}, \tag{66}$$

where the functional space $L_2(\mathcal{B})$ is introduced in ➤ *Table 1*. The weak formulation of the IBVP (37)–(40) consists of finding $\boldsymbol{A}\in V\times C^1(\langle 0,\infty))$ such that at a fixed time it satisfies the following variational equation:

$$a(\boldsymbol{A},\delta\boldsymbol{A}) + b(\boldsymbol{A},\delta\boldsymbol{A}) = f(\delta\boldsymbol{A}) \qquad \forall\delta\boldsymbol{A}\in V, \tag{67}$$

where the bilinear forms $a(\cdot,\cdot)$, $b(\cdot,\cdot)$ and the linear functional $f(\cdot)$ are defined as follows:

$$a(\boldsymbol{A}, \delta \boldsymbol{A}) := \frac{1}{\mu} \int_{\mathcal{B}} (\operatorname{curl} \boldsymbol{A} \cdot \operatorname{curl} \delta \boldsymbol{A}) dV, \tag{68}$$

$$b(\boldsymbol{A}, \delta \boldsymbol{A}) := \int_{\mathcal{B}} \sigma(r, \vartheta) \Big(\frac{\partial \boldsymbol{A}}{\partial t} \cdot \delta \boldsymbol{A} \Big) dV, \tag{69}$$

$$f(\delta \boldsymbol{A}) := -\frac{1}{\mu} \int_{\partial \mathcal{B}} (\boldsymbol{b}_t \cdot \delta \boldsymbol{A}) dS. \tag{70}$$

It can be seen that the assumptions imposed on the potential $\boldsymbol{A}$ are weaker in the weak formulation than in the classical formulation. Moreover, the assumptions concerning the electrical conductivity σ and the boundary data $\boldsymbol{b}_t$ can also be made weaker in the latter formulation. It is sufficient to assume that the electrical conductivity is a square-integrable function in $\mathcal{B}$, $\sigma \in L_2(\mathcal{B})$, and the boundary data at a fixed time is a square-integrable function on $\partial\mathcal{B}$, $\boldsymbol{b}_t \in L_2(\partial\mathcal{B}) \times C^1(\langle 0, \infty))$.

To show that the weak solution generalizes the classical solution to the problem (37)–(40), it is for the moment assumed that the weak solution $\boldsymbol{A}$ is sufficiently smooth and belongs to $\boldsymbol{A} \in C^2(\overline{\mathcal{B}})$. Then, the following Green's theorem is valid:

$$\int_{\mathcal{B}} (\operatorname{curl} \boldsymbol{A} \cdot \operatorname{curl} \delta \boldsymbol{A}) dV = \int_{\mathcal{B}} (\operatorname{curl} \operatorname{curl} \boldsymbol{A} \cdot \delta \boldsymbol{A}) dV - \int_{\partial \mathcal{B}} (\boldsymbol{n} \times \operatorname{curl} \boldsymbol{A}) \cdot \delta \boldsymbol{A} \, dS. \tag{71}$$

In view of this, the variational equation (67) can be rewritten as follows:

$$\frac{1}{\mu} \int_{\mathcal{B}} (\operatorname{curl} \operatorname{curl} \boldsymbol{A} \cdot \delta \boldsymbol{A}) dV - \frac{1}{\mu} \int_{\partial \mathcal{B}} (\boldsymbol{n} \times \operatorname{curl} \boldsymbol{A}) \cdot \delta \boldsymbol{A} \, dS + \int_{\mathcal{B}} \sigma \Big(\frac{\partial \boldsymbol{A}}{\partial t} \cdot \delta \boldsymbol{A} \Big) dV = -\frac{1}{\mu} \int_{\partial \mathcal{B}} (\boldsymbol{b}_t \cdot \delta \boldsymbol{A}) dS. \tag{72}$$

Taking first Eq. 72 only for the test functions $\delta \boldsymbol{A} \in C_0^\infty(\mathcal{B})$, where $C_0^\infty(\mathcal{B})$ is the space of infinitely differentiable functions with compact support in $\mathcal{B}$, and making use of the implication

$$\boldsymbol{f} \in L_2(\mathcal{B}), \quad \int_{\mathcal{B}} (\boldsymbol{f} \cdot \delta \boldsymbol{A}) dV = 0 \quad \forall \delta \boldsymbol{A} \in C_0^\infty(\mathcal{B}) \Rightarrow \boldsymbol{f} = 0 \quad \text{in } \mathcal{B}, \tag{73}$$

Eq. 37 is proved. To obtain the boundary condition (39), the following implication is used

$$\boldsymbol{f} \in L_2(\partial \mathcal{B}), \quad \int_{\partial \mathcal{B}} (\boldsymbol{f} \cdot \delta \boldsymbol{A}) dS = 0 \quad \forall \delta \boldsymbol{A} \in C^\infty(\overline{\mathcal{B}}) \Rightarrow \boldsymbol{f} = 0 \quad \text{on } \partial \mathcal{B}, \tag{74}$$

where $C^\infty(\overline{\mathcal{B}})$ is the space of infinitely differentiable functions in $\overline{\mathcal{B}}$. It can be seen that if a weak solution of the problem exists and is sufficiently smooth, for instance, if $\boldsymbol{A} \in C^2(\overline{\mathcal{B}})$, then this solution satisfies the differential equation (37) and the boundary condition (39), all taken at time t. Thus, the weak solution generalizes the classical solution $C^2(\overline{\mathcal{B}})$ since the weak solution may exist even though the classical solution does not exist. However, if the classical solution exists, it is also the weak solution (Křížek and Neittaanmäki, 1990).

4.2.2 Satellite Magnetic Data

Turning the attention now to the weak formulation of IBVP (58)–(65) for satellite magnetic data $\boldsymbol{B}_t$, the intention is to apply different parameterizations of the potentials $\boldsymbol{A}$ and $\boldsymbol{A}_0$. In addition to the solution space V for the conducting sphere $\mathcal{B}$, the solution space V_0 for the nonconducting atmosphere $\mathcal{A}$ is introduced:

$$V_0 := \left\{ \boldsymbol{A}_0 | \boldsymbol{A}_0 \in C^2(\mathcal{A}), \ \operatorname{div} \boldsymbol{A}_0 = 0 \text{ in } \mathcal{A} \right\}. \tag{75}$$

Note that the continuity condition (62) is not imposed on either of the solution spaces V and V_0. Instead, the Lagrange multiplier vector $\boldsymbol{\lambda}$ and a solution space for it are introduced:

$$V_\lambda := \{\boldsymbol{\lambda} | \boldsymbol{\lambda} \in L_2(\partial\mathcal{B})\}. \tag{76}$$

The weak formulation of the IBVP (58)–(65) consists of finding $(\boldsymbol{A}, \boldsymbol{A}_0, \boldsymbol{\lambda}) \in (V, V_0, V_\lambda) \times C^1(\langle 0, \infty))$ such that at a fixed time they satisfy the following variational equation:

$$\begin{aligned} &a(\boldsymbol{A}, \delta\boldsymbol{A}) + b(\boldsymbol{A}, \delta\boldsymbol{A}) + a_0(\boldsymbol{A}_0, \delta\boldsymbol{A}_0) + c(\delta\boldsymbol{A} - \delta\boldsymbol{A}_0, \boldsymbol{\lambda}) + c(\boldsymbol{A} - \boldsymbol{A}_0, \delta\boldsymbol{\lambda}) = F(\delta\boldsymbol{A}_0) \\ &\forall \delta\boldsymbol{A} \in V,\ \forall \delta\boldsymbol{A}_0 \in V_0,\ \forall \delta\boldsymbol{\lambda} \in V_\lambda, \end{aligned} \tag{77}$$

where the bilinear forms $a(\cdot,\cdot)$ and $b(\cdot,\cdot)$ are defined by Eqs. 68 and 69, and the additional bilinear forms $a_0(\cdot,\cdot)$ and $c(\cdot,\cdot)$ and the new linear functional $F(\cdot)$ are defined as follows:

$$a_0(\boldsymbol{A}_0, \delta\boldsymbol{A}_0) := \frac{1}{\mu}\int_{\mathcal{A}} (\operatorname{curl}\boldsymbol{A}_0 \cdot \operatorname{curl}\delta\boldsymbol{A}_0)\,dV, \tag{78}$$

$$c(\boldsymbol{A} - \boldsymbol{A}_0, \boldsymbol{\lambda}) := \int_{\partial\mathcal{B}} (\boldsymbol{A} - \boldsymbol{A}_0) \cdot \boldsymbol{\lambda}\,dS, \tag{79}$$

$$F(\delta\boldsymbol{A}_0) := -\frac{1}{\mu}\int_{\partial\mathcal{A}} (\boldsymbol{B}_t \cdot \delta\boldsymbol{A}_0)\,dS. \tag{80}$$

To show that the weak solution generalizes the classical solution to the problem (58)–(65), it is again assumed that the weak solution $\boldsymbol{A}$ is sufficiently smooth and belongs to $\boldsymbol{A} \in C^2(\overline{\mathcal{B}})$. By making use of Green's theorem (71), the variational equation (77) can be written as

$$\begin{aligned} &\frac{1}{\mu}\int_{\mathcal{B}} (\operatorname{curl}\operatorname{curl}\boldsymbol{A} \cdot \delta\boldsymbol{A})\,dV - \frac{1}{\mu}\int_{\partial\mathcal{B}} (\boldsymbol{n} \times \operatorname{curl}\boldsymbol{A}) \cdot \delta\boldsymbol{A}\,dS + \int_{\mathcal{B}} \sigma\Big(\frac{\partial\boldsymbol{A}}{\partial t} \cdot \delta\boldsymbol{A}\Big)dV \\ &+ \frac{1}{\mu}\int_{\mathcal{A}} (\operatorname{curl}\operatorname{curl}\boldsymbol{A}_0 \cdot \delta\boldsymbol{A}_0)\,dV - \frac{1}{\mu}\int_{\partial\mathcal{A}} (\boldsymbol{n} \times \operatorname{curl}\boldsymbol{A}_0) \cdot \delta\boldsymbol{A}_0\,dS \\ &+ \frac{1}{\mu}\int_{\partial\mathcal{B}} (\boldsymbol{n} \times \operatorname{curl}\boldsymbol{A}_0) \cdot \delta\boldsymbol{A}_0\,dS + \int_{\partial\mathcal{B}} (\delta\boldsymbol{A} - \delta\boldsymbol{A}_0) \cdot \boldsymbol{\lambda}\,dS \\ &+ \int_{\partial\mathcal{B}} (\boldsymbol{A} - \boldsymbol{A}_0) \cdot \delta\boldsymbol{\lambda}\,dS = -\frac{1}{\mu}\int_{\partial\mathcal{A}} (\boldsymbol{B}_t \cdot \delta\boldsymbol{A}_0)\,dS. \end{aligned} \tag{81}$$

Now, taking Eq. 81 for the test functions $\delta\boldsymbol{A} \in C_0^\infty(\mathcal{B})$ and making use of the implication (73), Eq. 58 is proved. Likewise, taking Eq. 81 for the test functions $\delta\boldsymbol{A}_0 \in C_0^\infty(\mathcal{A})$ and using the implication (73) for the domain $\mathcal{A}$, Eq. 60 is proved. To obtain the continuity conditions (62) and (63), the implication (74) is used for the test functions $\delta\boldsymbol{A} \in C^\infty(\overline{\mathcal{B}})$. The boundary condition (64) can be obtained by an analogous way if the implication (74) is considered for $\partial\mathcal{A}$. It can therefore be concluded that if a weak solution of the problem exists and is sufficiently smooth, for instance, if $\boldsymbol{A}$ belongs to the space of functions whose second-order derivatives are continuous in $\bar{\mathcal{B}}$, then this solution satisfies the differential equations (58) and (60), the interface conditions (62) and (63), and the boundary conditions (64), all taken at time t.

4.3 Frequency-Domain and Time-Domain Solutions

Two approaches of solving the IBVPs of EM induction with respect to the time variable t are now presented.

The variational equation (67) is first solved in the Fourier-frequency domain, assuming that all field variables have a harmonic time dependence of the form $e^{i\omega t}$. Denoting the Fourier image of $\boldsymbol{A}$ by $\hat{\boldsymbol{A}}$, the weak formulation of EM induction for ground magnetic data in the frequency domain is described by the variational equation:

$$a(\hat{\boldsymbol{A}}, \delta\hat{\boldsymbol{A}}) + i\omega\, b_1(\hat{\boldsymbol{A}}, \delta\hat{\boldsymbol{A}}) = f(\delta\hat{\boldsymbol{A}}) \qquad \forall \delta\hat{\boldsymbol{A}} \in V, \tag{82}$$

where the new bilinear form $b_1(\hat{\boldsymbol{A}}, \delta\hat{\boldsymbol{A}})$ is defined by

$$b_1(\hat{\boldsymbol{A}}, \delta\hat{\boldsymbol{A}}) := \int_{\mathcal{B}} \sigma(r, \vartheta)(\hat{\boldsymbol{A}} \cdot \delta\hat{\boldsymbol{A}})\,dV. \tag{83}$$

Having solved Eq. 82 for $\hat{\boldsymbol{A}}$, the solution is transformed back to the time domain by applying the inverse Fourier transform.

Alternatively, the IBVP for ground magnetic data can be solved directly in the time domain, which is the approach applied in the following. There are several choices for representing the time derivative of the toroidal vector potential $\boldsymbol{A}$ in the bilinear form $b(\cdot,\cdot)$. For simplicity, the explicit Euler differencing scheme will be chosen and $\partial\boldsymbol{A}/\partial t$ will be approximated by the differences of $\boldsymbol{A}$ at two subsequent time levels (Press et al., 1992):

$$\frac{\partial \boldsymbol{A}}{\partial t} \approx \frac{\boldsymbol{A}(r, \vartheta, t_{i+1}) - \boldsymbol{A}(r, \vartheta, t_i)}{t_{i+1} - t_i} =: \frac{{}^{i+1}\boldsymbol{A} - {}^{i}\boldsymbol{A}}{\Delta t_i}, \tag{84}$$

where ${}^{i}\boldsymbol{A}$ denotes the values of $\boldsymbol{A}$ at discrete time levels $0 = t_0 < t_1 < \cdots < t_{i+1} < \cdots$. The variational equation (67), which is now solved at each time level t_i, $i = 0, 1, \ldots$, has the form

$$a({}^{i+1}\boldsymbol{A}, \delta\boldsymbol{A}) + \frac{1}{\Delta t_i} b_1({}^{i+1}\boldsymbol{A}, \delta\boldsymbol{A}) = \frac{1}{\Delta t_i} b_1({}^{i}\boldsymbol{A}, \delta\boldsymbol{A}) + f({}^{i+1}\boldsymbol{b}_t, \delta\boldsymbol{A}) \qquad \forall \delta\boldsymbol{A} \in V, \tag{85}$$

where the bilinear form $b_1(\cdot,\cdot)$ is defined by Eq. 83.

The same two approaches can be applied to the IBVP of EM induction for satellite magnetic data. Here, the time-domain approach is only presented. The variational equation (77) is discretized with respect to time and solved at each time level t_i:

$$\begin{aligned} &a({}^{i+1}\boldsymbol{A}, \delta\boldsymbol{A}) + \frac{1}{\Delta t_i} b_1({}^{i+1}\boldsymbol{A}, \delta\boldsymbol{A}) + a_0({}^{i+1}\boldsymbol{A}_0, \delta\boldsymbol{A}_0) + c(\delta\boldsymbol{A} - \delta\boldsymbol{A}_0, {}^{i+1}\boldsymbol{\lambda}) + c({}^{i+1}\boldsymbol{A} - {}^{i+1}\boldsymbol{A}_0, \delta\boldsymbol{\lambda}) \\ &= \frac{1}{\Delta t_i} b_1({}^{i}\boldsymbol{A}, \delta\boldsymbol{A}) + F({}^{i+1}\boldsymbol{B}_t, \delta\boldsymbol{A}_0) \qquad \forall \delta\boldsymbol{A} \in V,\ \forall \delta\boldsymbol{A}_0 \in V_0,\ \forall \delta\boldsymbol{\lambda} \in V_\lambda. \end{aligned} \tag{86}$$

4.4 Vector Spherical Harmonics Parameterization Over Colatitude

For the axisymmetric geometry of external sources and the conductivity model, it has been shown that the induced electromagnetic field is axisymmetric and the associated toroidal vector potential is an axisymmetric vector. It may be represented in terms of zonal toroidal vector spherical harmonics $\boldsymbol{Y}_j^j(\vartheta)$. Their explicit forms are as follows (more details are given the Appendix):

$$\boldsymbol{Y}_j^j(\vartheta) := P_{j1}(\cos\vartheta)\boldsymbol{e}_\varphi, \tag{87}$$

where $P_{j1}(\cos\vartheta)$ is the associated Legendre function of degree j and order $m = 1$, and $\boldsymbol{e}_\varphi$ is the spherical base vector in the longitudinal direction. An important property of the functions $\boldsymbol{Y}_j^j(\vartheta)$ is that they are divergence-free:

$$\operatorname{div}\left[f(r)\boldsymbol{Y}_j^j(\vartheta)\right] = 0, \tag{88}$$

where $f(r)$ is a differentiable function.

The required toroidal vector potential $\boldsymbol{A}$ and test functions $\delta\boldsymbol{A}$ inside the conducting sphere $\mathcal{B}$ can be represented as a series of the functions $\boldsymbol{Y}_j^j(\vartheta)$:

$$\begin{Bmatrix} \boldsymbol{A}(r,\vartheta,t) \\ \delta\boldsymbol{A}(r,\vartheta) \end{Bmatrix} = \sum_{j=1}^{\infty} \begin{Bmatrix} A_j^j(r,t) \\ \delta A_j^j(r) \end{Bmatrix} \boldsymbol{Y}_j^j(\vartheta), \tag{89}$$

where $A_j^j(r,t)$ and $\delta A_j^j(r)$ are spherical harmonic expansion coefficients. The divergence-free property of functions $\boldsymbol{Y}_j^j(\vartheta)$ implies that both the toroidal vector potential $\boldsymbol{A}$ and test functions $\delta\boldsymbol{A}$ are divergence-free. Therefore, the parameterization (89) of potentials $\boldsymbol{A}$ and $\delta\boldsymbol{A}$ automatically satisfies the requirement that the functions from the solution space V be divergence-free. The parameterization (89) is also employed for the Lagrange multipliers $\boldsymbol{\lambda}(\vartheta,t)$ and the associated test functions $\delta\boldsymbol{\lambda}(\vartheta)$ with the expansion coefficients $\lambda_j^j(t)$ and $\delta\lambda_j^j$, respectively.

Introducing the spherical harmonic representation of the zonal toroidal vector $\boldsymbol{A}$, the curl of $\boldsymbol{A}$ is a zonal spheroidal vector:

$$\operatorname{curl}\boldsymbol{A} = \sum_{j=1}^{\infty}\sum_{\ell=j-1}^{j+1,2} R_j^\ell(\boldsymbol{A};r)\boldsymbol{Y}_j^\ell(\vartheta), \tag{90}$$

where $R_j^\ell(\boldsymbol{A};r)$ are given by Eq. 227 in the Appendix. The substitution of Eqs. 89 and 90 into Eqs. 68 and 83 leads to the spherical harmonic representation of the bilinear forms $a(\cdot,\cdot)$ and $b_1(\cdot,\cdot)$:

$$a(\boldsymbol{A},\delta\boldsymbol{A}) = \frac{1}{\mu}\sum_{j=1}^{\infty}\sum_{\ell=j-1}^{j+1,2}\int_0^a R_j^\ell(\boldsymbol{A};r)R_j^\ell(\delta\boldsymbol{A};r)r^2\,dr, \tag{91}$$

$$b_1(\boldsymbol{A},\delta\boldsymbol{A}) = \int_0^a \mathcal{E}(\boldsymbol{A},\delta\boldsymbol{A};r)r^2\,dr, \tag{92}$$

where the orthogonality property (221) of vector spherical harmonics has been employed and $\mathcal{E}$ denotes the angular part of the ohmic energy,

$$\mathcal{E}(\boldsymbol{A},\delta\boldsymbol{A};r) = 2\pi\int_0^\pi \sigma(r,\vartheta)\sum_{j_1=1}^{\infty} A_{j_1}^{j_1}(r,t)P_{j_1,1}(\cos\vartheta)\sum_{j_2=1}^{\infty}\delta A_{j_2}^{j_2}(r)P_{j_2,1}(\cos\vartheta)\sin\vartheta\,d\vartheta. \tag{93}$$

Likewise, substituting Eqs. 53 and 89 into Eq. 70 results in the spherical harmonic representation of the linear functional $f(\cdot)$:

$$f(\delta\boldsymbol{A}) = \frac{a^2}{\mu}\sum_{j=1}^{\infty}\sqrt{j(j+1)}\left[G_j^{(e)}(t)+G_j^{(i)}(t)\right]\delta A_j^j(a). \tag{94}$$

4.5 Finite-Element Approximation over the Radial Coordinate

Inside the conducting sphere $\mathcal{B}$, the range of integration $\langle 0,a\rangle$ is divided over the radial coordinate into P subintervals by the nodes $0 = r_1 < r_2 < \cdots < r_P < r_{P+1} = a$. The piecewise linear basis functions defined at the nodes by the relation $\psi_k(r_i) = \delta_{ki}$ can be used as the basis function of the Sobolev functional space $W_2^1(0,a)$. Note that only two basis functions are nonzero in the interval $r_k \le r \le r_{k+1}$, namely,

$$\psi_k(r) = \frac{r_{k+1}-r}{h_k}, \qquad \psi_{k+1}(r) = \frac{r-r_k}{h_k}, \tag{95}$$

where $h_k = r_{k+1} - r_k$. Since both the unknown solution $A_j^j(r,t)$ and test functions $\delta A_j^j(r)$ are elements of this functional space, they can be parameterized by piecewise linear finite elements $\psi_k(r)$ such that

$$\left\{ \begin{array}{c} A_j^j(r,t) \\ \delta A_j^j(r) \end{array} \right\} = \sum_{k=1}^{P+1} \left\{ \begin{array}{c} A_j^{j,k}(t) \\ \delta A_j^{j,k} \end{array} \right\} \psi_k(r). \tag{96}$$

The finite-element representation of curl $\boldsymbol{A}$ coefficients then reads as

$$R_j^{j-1}(\boldsymbol{A};r) = -\sqrt{\frac{j+1}{2j+1}} \left[\left(-\frac{1}{h_k} + \frac{j+1}{r}\psi_k(r) \right) A_j^{j,k} + \left(\frac{1}{h_k} + \frac{j+1}{r}\psi_{k+1}(r) \right) A_j^{j,k+1} \right],$$

$$R_j^{j+1}(\boldsymbol{A};r) = -\sqrt{\frac{j}{2j+1}} \left[\left(-\frac{1}{h_k} - \frac{j}{r}\psi_k(r) \right) A_j^{j,k} + \left(\frac{1}{h_k} - \frac{j}{r}\psi_{k+1}(r) \right) A_j^{j,k+1} \right], \tag{97}$$

where $r_k \le r \le r_{k+1}$.

Since the electrical conductivity $\sigma(r,\vartheta) \in L_2(\mathcal{B})$, the radial dependence of σ can be approximated by piecewise constant functions,

$$\sigma(r,\vartheta) = \sigma_k(\vartheta), \qquad r_k \le r \le r_{k+1}, \tag{98}$$

where $\sigma_k(\vartheta)$ does not depend on the radial coordinate r and may be further approximated by piecewise constant functions in colatitude ϑ. (However, this approximation will not be denoted explicitly.) The integrals over r in Eqs. 91 and 92 can be divided into P subintervals:

$$a(\boldsymbol{A},\delta\boldsymbol{A}) = \frac{1}{\mu} \sum_{j=1}^{\infty} \sum_{\ell=j-1}^{j+1,2} \sum_{k=1}^{P} \int_{r_k}^{r_{k+1}} R_j^{\ell}(\boldsymbol{A};r) R_j^{\ell}(\delta\boldsymbol{A};r) r^2 dr, \tag{99}$$

$$b_1(\boldsymbol{A},\delta\boldsymbol{A}) = \sum_{k=1}^{P} \int_{r_k}^{r_{k+1}} \mathcal{E}(\boldsymbol{A},\delta\boldsymbol{A};r) r^2 dr, \tag{100}$$

and the integration over r is reduced to the computation of integrals of the type

$$\int_{r_k}^{r_{k+1}} \psi_i(r)\psi_j(r) r^2 dr, \tag{101}$$

where the indices i and j are equal to k and/or $k+1$. These integrals can be evaluated numerically, for example, by means of the two-point Gauss–Legendre numerical quadrature with the weights equal to 1 and the nodes $x_{1,2} = \pm 1/\sqrt{3}$ (Press et al., 1992, ➋ Sect. 4.5). For instance, the quadrature formula for the integral in Eq. 100 can be written in the form

$$b_1(\boldsymbol{A},\delta\boldsymbol{A}) = \sum_{k=1}^{P} \sum_{\alpha=1}^{2} \mathcal{E}\Big(\boldsymbol{A}(r_\alpha), \delta\boldsymbol{A}(r_\alpha); r_\alpha \Big) \frac{r_\alpha^2 h_k}{2}, \tag{102}$$

where $r_\alpha := \frac{1}{2}(h_k x_\alpha + r_k + r_{k+1})$, $\alpha = 1,2$. The integration over colatitude ϑ in the term $\mathcal{E}$, see Eq. 93, can also be carried out numerically by the Gauss–Legendre quadrature formula. Computational details of this approach can be found in Orszag (1970) or Martinec (1989).

4.6 Solid Vector Spherical Harmonics Parameterization of A_0

Our attention is now turned to the parameterization of the toroidal vector potential $\boldsymbol{A}_0$ and test functions $\delta\boldsymbol{A}_0$ in an insulating atmosphere $\mathcal{A}$. Equation 46 shows that $\boldsymbol{A}_0$, and also $\delta\boldsymbol{A}_0$, can be represented as a series of the zonal toroidal vector spherical harmonics $\mathbf{Y}_j^j(\vartheta)$ with the spherical expansion coefficients $A_{0,j}^j$ and $\delta A_{0,j}^j$, respectively, of the form

$$\begin{Bmatrix} A_{0,j}^j(r,t) \\ \delta A_{0,j}^j(r) \end{Bmatrix} = a\left[\sqrt{\frac{j}{j+1}}\left(\frac{r}{a}\right)^j \begin{Bmatrix} G_j^{(e)}(t) \\ \delta G_j^{(e)} \end{Bmatrix} - \sqrt{\frac{j+1}{j}}\left(\frac{a}{r}\right)^{j+1} \begin{Bmatrix} G_j^{(i)}(t) \\ \delta G_j^{(i)} \end{Bmatrix}\right] \quad \text{for } a \le r \le b. \tag{103}$$

The zonal scalar-magnetic Gauss coefficients $G_j^{(e)}(t)$ and $G_j^{(i)}(t)$ are considered known in the case of the IBVP for ground magnetic data as they constitute the ground magnetic data $\boldsymbol{b}_t$ on $\mathcal{B}$, see (53). However, for satellite magnetic data, the coefficients $G_j^{(e)}(t)$ and $G_j^{(i)}(t)$ in the insulating atmosphere $\mathcal{A}$ are, in addition to coefficients $A_j^{j,k}(t)$, unknowns and are sought by solving the IBVP (58–65). The associated test-function coefficients are denoted by $\delta G_j^{(e)}$ and $\delta G_j^{(i)}$.

Applying the operator curl on the parameterization (103) and substituting the result into Eq. 78 results in the parameterization of the bilinear form $a_0(\cdot,\cdot)$:

$$a_0(\mathbf{A}_0, \delta\mathbf{A}_0) = \frac{a^3}{\mu}\sum_{j=1}^{\infty}\left\{ j\left[\left(\frac{b}{a}\right)^{2j+1} - 1\right] G_j^{(e)}(t)\delta G_j^{(e)} - (j+1)\left[\left(\frac{a}{b}\right)^{2j+1} - 1\right] G_j^{(i)}(t)\delta G_j^{(i)}\right\}, \tag{104}$$

where a and b are the radii of the spheres $\partial\mathcal{B}$ and $\partial\mathcal{A}$, respectively.

The continuity condition (62), that is, $\mathbf{A} = \mathbf{A}_0$ on $\partial\mathcal{B}$, is now expressed in terms of spherical harmonics. Substituting for the spherical harmonics representations (89) of $\mathbf{A}$ and (103) of $\mathbf{A}_0$, respectively, into Eq. 62 results in the constraint between the external coefficients $G_j^{(e)}(t)$, the internal coefficients $G_j^{(i)}(t)$ of the toroidal vector potential $\mathbf{A}_0$ in the atmosphere $\mathcal{A}$, and the coefficients $A_j^j(a,t)$ of the toroidal vector potential $\mathbf{A}$ in the conducting sphere $\mathcal{B}$:

$$A_j^j(a,t) = a\left[\sqrt{\frac{j}{j+1}}\, G_j^{(e)}(t) - \sqrt{\frac{j+1}{j}}\, G_j^{(i)}(t)\right]. \tag{105}$$

This continuity condition can be used to express the bilinear form $c(\cdot,\cdot)$, defined by Eq. 79, in terms of spherical harmonics as follows:

$$c(\mathbf{A} - \mathbf{A}_0, \boldsymbol{\lambda}) = \sum_{j=1}^{\infty}\left[\frac{1}{a}\sqrt{\frac{j}{j+1}} A_j^j(a,t) + G_j^{(e)}(t) - \frac{j}{j+1} G_j^{(i)}(t)\right]\lambda_j^j(t), \tag{106}$$

where $\lambda_j^j(t)$ are zonal toroidal vector spherical harmonic expansion coefficients of the Lagrange multiplier $\boldsymbol{\lambda}$.

Finally, making use of Eqs. 57 and 103, the linear functional $F(\cdot)$ defined by Eq. 80 can be expressed in the form

$$F(\delta\mathbf{A}_0) = \frac{b^2}{\mu}\sum_{j=1}^{\infty}\sqrt{j(j+1)}X_j(b,t)\delta A_{0,j}^j(b), \tag{107}$$

where the spherical harmonic coefficients $\delta A_{0,j}^j(b)$ of the test functions $\delta\mathbf{A}_0(b,\vartheta)$ are given by Eq. 103 for $r = b$:

$$\delta A_{0,j}^j(b) = a\left[\sqrt{\frac{j}{j+1}}\left(\frac{b}{a}\right)^j \delta G_j^{(e)} - \sqrt{\frac{j+1}{j}}\left(\frac{a}{b}\right)^{j+1} \delta G_j^{(i)}\right]. \tag{108}$$

In view of this, the functional $F(\cdot)$ thus reads as

$$F(\delta \mathbf{A}_0) = \frac{ab^2}{\mu} \sum_{j=1}^{\infty} X_j(b,t) \left[j \left(\frac{b}{a}\right)^j \delta G_j^{(e)} - (j+1) \left(\frac{a}{b}\right)^{j+1} \delta G_j^{(i)} \right]. \tag{109}$$

5 Forward Method of EM Induction for the External Gauss Coefficients

The case where the time series of the CHAMP-derived coefficients $X_j^{(\mathrm{obs})}(t)$ and $Z_j^{(\mathrm{obs})}(t)$ coefficients (see ➋ Sect. 7) are converted to a time series of spherical harmonic coefficients of external and internal fields at the CHAMP satellite altitude is now considered. To obtain these coefficients, denoted by $G_j^{(e,\mathrm{obs})}(t)$ and $G_j^{(i,\mathrm{obs})}(t)$, the Gaussian expansion of the external magnetic potential is undertaken at the satellite orbit of radius $r = b$, which results in Eqs. 48 and 50 where the radius r equals to b. The straightforward derivation then yields

$$\begin{aligned} G_j^{(e,\mathrm{obs})}(t) &= \frac{1}{2j+1}\left[(j+1)X_j^{(\mathrm{obs})}(t) + Z_j^{(\mathrm{obs})}(t)\right], \\ G_j^{(i,\mathrm{obs})}(t) &= \frac{1}{2j+1}\left[jX_j^{(\mathrm{obs})}(t) - Z_j^{(\mathrm{obs})}(t)\right]. \end{aligned} \tag{110}$$

The satellite observables $G_j^{(e,\mathrm{obs})}(t)$ and $G_j^{(i,\mathrm{obs})}(t)$ are related to the original, ground-based Gauss coefficients $G_j^{(e)}(t)$ and $G_j^{(i)}(t)$ by

$$\begin{aligned} G_j^{(e,\mathrm{obs})}(t) &= (b/a)^{j-1} G_j^{(e)}(t), \\ G_j^{(i,\mathrm{obs})}(t) &= (a/b)^{j+2} G_j^{(i)}(t). \end{aligned} \tag{111}$$

When the Gauss coefficients $G_j^{(e)}(t)$ and $G_j^{(i)}(t)$ are computed from the satellite observables $G_j^{(e,\mathrm{obs})}(t)$ and $G_j^{(i,\mathrm{obs})}(t)$ by inverting Eq. 111, the aim is to solve the downward continuation of satellite magnetic data from the satellite's orbit to the Earth's surface. It is, in principal, a numerically unstable problem, in particular for higher-degree spherical harmonic coefficients, since noise contaminated the satellite observables $G_j^{(e,\mathrm{obs})}(t)$ and $G_j^{(i,\mathrm{obs})}(t)$ is amplified by a factor of $(b/a)^j$ when computing the ground-based Gauss coefficients $G_j^{(e)}(t)$ and $G_j^{(i)}(t)$. Hence, the IBVP of EM induction is assumed to be solved only for low-degree spherical harmonics, typically up to spherical harmonic degree $j_{\max} = 5$. In this case, Martinec and McCreadie (2004) showed that the downward continuation of the satellite-determined coefficients from the CHAMP satellite orbit to the ground is numerically stable. This fact will be adopted and low-degree $G_j^{(e)}(t)$ and $G_j^{(i)}(t)$ are assumed to be calculated from the satellite observables by inverting (111). However, it should be noted that future satellite missions, such as SWARM (Olsen et al., 2006b), may provide reliable information about higher-degree spherical harmonic coefficients. Then, their downward continuation from a satellite's altitude to the ground will become numerically unstable and the forward and adjoint IBVP of EM induction will need to be formulated directly for $G_j^{(e,\mathrm{obs})}(t)$ and $G_j^{(i,\mathrm{obs})}(t)$.

5.1 Classical Formulation

Given the Gauss coefficients $G_j^{(e)}(t)$ and $G_j^{(i)}(t)$ as observations, the forward IBVP of EM induction can be reformulated. From several possible combinations of these coefficients, it is natural to consider that the external Gauss coefficients $G_j^{(e)}(t)$ are used as the boundary-value data for the forward EM induction method, while the internal Gauss coefficients $G_j^{(i)}(t)$ are used for the adjoint EM induction method. The modification of the forward method is now derived.

The first modification concerns the solution domain. While in the previous case for the X and Z components of the CHAMP magnetic data, the solution domain consists of a conducting sphere $\mathcal{B}$ surrounded by an insulating atmosphere $\mathcal{A}$, in the present case where the Gauss coefficients $G_j^{(e)}(t)$ and $G_j^{(i)}(t)$ are used as observations, it is sufficient to consider only the conducting sphere $\mathcal{B}$ as the solution domain. Note, however, that the solution domain will again consist of the unification of $\mathcal{B}$ and $\mathcal{A}$ when the satellite observables $G_j^{(e,\mathrm{obs})}(t)$ and $G_j^{(i,\mathrm{obs})}(t)$ are taken as observations.

Another modification concerns the boundary condition (16). Making use of Eq. 45 and formulae (222) and (223) for the scalar and vector products of $\boldsymbol{e}_r$ with the spheroidal vector spherical harmonics $\boldsymbol{Y}_j^{j\pm1}(\vartheta)$, the continuity of the normal and tangential components of the magnetic induction vector $\boldsymbol{B}$ on the boundary $\partial\mathcal{B}$, see Eqs. 15 and 20, is of the form

$$\left[\boldsymbol{n}\cdot\operatorname{curl}\boldsymbol{A}(a)\right]_j = -jG_j^{(e)} + (j+1)G_j^{(i)}, \tag{112}$$

$$\left[\boldsymbol{n}\times\operatorname{curl}\boldsymbol{A}(a)\right]_j^j = -\sqrt{j(j+1)}\left(G_j^{(e)} + G_j^{(i)}\right). \tag{113}$$

Combining these equations such that the external and internal Gauss coefficients are separated, and making the scalar product of $\boldsymbol{e}_r$ with Eq. 90 for $r = a$, that is,

$$\left[\boldsymbol{n}\cdot\operatorname{curl}\boldsymbol{A}(a)\right]_j = -\frac{\sqrt{j(j+1)}}{a}A_j^j(a), \tag{114}$$

results in

$$-\frac{j}{a}A_j^j(a) + \left[\boldsymbol{n}\times\operatorname{curl}\boldsymbol{A}(a)\right]_j^j = -\sqrt{\frac{j}{j+1}}(2j+1)G_j^{(e)}, \tag{115}$$

$$\frac{j+1}{a}A_j^j(a) + \left[\boldsymbol{n}\times\operatorname{curl}\boldsymbol{A}(a)\right]_j^j = -\sqrt{\frac{j+1}{j}}(2j+1)G_j^{(i)}. \tag{116}$$

The last two equations represent the boundary conditions, which will be used in the forward and adjoint IBVP of EM induction, respectively.

The forward IBVP of EM induction for the Gauss coefficient $G_j^{(e)}(t)$ can now be formulated as follows. Given the conductivity $\sigma(r,\vartheta)$ of sphere $\mathcal{B}$, the toroidal vector potential $\boldsymbol{A}$ is searched for such that, for $t > 0$, it holds that

$$\frac{1}{\mu}\operatorname{curl}\operatorname{curl}\boldsymbol{A} + \sigma\frac{\partial\boldsymbol{A}}{\partial t} = \boldsymbol{0} \quad \text{in } \mathcal{B} \tag{117}$$

with the boundary condition

$$\left[\boldsymbol{n}\times\operatorname{curl}\boldsymbol{A}(a)\right]_j^j - \frac{j}{a}A_j^j(a) = -\sqrt{\frac{j}{j+1}}(2j+1)G_j^{(e)} \quad \text{on } \partial\mathcal{B} \tag{118}$$

and the inhomogeneous initial condition

$$\boldsymbol{A}|_{t=0} = \boldsymbol{A}^0 \quad \text{in } \mathcal{B}. \tag{119}$$

5.2 Weak Formulation

The IBVP (117)–(119) can again be reformulated in a weak sense. By this it is meant that $\boldsymbol{A} \in V \times C^1((0,\infty))$ is searched for such that at a fixed time it satisfies the following variational equation:

$$a_1(\boldsymbol{A}, \delta\boldsymbol{A}) + b(\boldsymbol{A}, \delta\boldsymbol{A}) = f_1(\delta\boldsymbol{A}) \qquad \forall \delta\boldsymbol{A} \in V, \tag{120}$$

where the solution space V is defined by Eq. 66. The new bilinear form $a_1(\cdot,\cdot)$ and the new linear functional $f_1(\cdot)$ are expressed in terms of the original bilinear form $a(\cdot,\cdot)$ and the coefficient $A_j^j(a,t)$ as follows:

$$a_1(\boldsymbol{A}, \delta\boldsymbol{A}) = a(\boldsymbol{A}, \delta\boldsymbol{A}) - \frac{a}{\mu}\sum_{j=1}^{\infty} jA_j^j(a,t)\delta A_j^j(a), \tag{121}$$

$$f_1(\delta\boldsymbol{A}) = \frac{a^2}{\mu}\sum_{j=1}^{\infty}\sqrt{\frac{j}{j+1}}(2j+1)G_j^{(e)}(t)\delta A_j^j(a). \tag{122}$$

It should be emphasized that there is a difference in principle between the original variational equation 67 and the modification (120) in prescribing the boundary data on the surface $\partial\mathcal{B}$. Equation 67 requires the prescription of the tangential components of the total magnetic induction in a vacuum on $\partial\mathcal{B}$. Inspecting the functional $f(\cdot)$ in Eq. 94 shows that this requirement leads to the necessity to define the linear combinations $G_j^{(e)}(t) + G_j^{(i)}(t)$ for $j = 1, 2, \ldots$, as input boundary data for solving Eq. 67. In contrast to this scheme, the functional $f_1(\cdot)$ on the right-hand side of Eq. 120 only contains the spherical harmonic coefficients $G_j^{(e)}(t)$. Hence, to solve Eq. 120, only the spherical harmonic coefficients $G_j^{(e)}(t)$ of the external electromagnetic source need to be prescribed while the spherical harmonic coefficients $G_j^{(i)}(t)$ of the induced magnetic field within the earth are determined after solving Eq. 120 by means of Eqs. 115 and 116:

$$G_j^{(i)} = \frac{j}{j+1}G_j^{(e)} - \frac{1}{a}\sqrt{\frac{j}{j+1}}A_j^j(a). \tag{123}$$

The former scheme is advantageous in the case where there is no possibility of separating the external and internal parts of magnetic induction observations by spherical harmonic analysis. The latter scheme can be applied if such an analysis can be carried out, or in the case when the external magnetic source is defined by a known physical process.

6 Time-Domain, Spectral Finite-Element Solution

Finally, the spectral finite-element solution to the IBVP of EM induction for CHAMP magnetic data is introduced. For the sake of simplicity, the case where the spherical harmonic coefficients

$G_j^{(e)}(t)$ of the external electromagnetic source are considered as input observations is treated first. Introducing the finite-dimensional functional space as

$$V_h := \left\{ \delta \boldsymbol{A} = \sum_{j=1}^{j_{\max}} \sum_{k=1}^{P+1} \delta A_j^{j,k} \psi_k(r) \boldsymbol{Y}_j^j(\vartheta) \right\}, \tag{124}$$

where $j_{\max}$ and P are finite cut-off degrees, the Galerkin method for approximating the solution of variational equation (120) at a fixed time t_{i+1} consists in finding ${}^{i+1}\boldsymbol{A}_h \in V_h$ such that

$$a_1({}^{i+1}\boldsymbol{A}_h, \delta \boldsymbol{A}_h) + \frac{1}{\Delta t_i} b_1({}^{i+1}\boldsymbol{A}_h, \delta \boldsymbol{A}_h) = \frac{1}{\Delta t_i} b_1({}^{i}\boldsymbol{A}_h, \delta \boldsymbol{A}_h) + f_1\left({}^{i+1}G_j^{(e)}, \delta \boldsymbol{A}_h\right) \qquad \forall \delta \boldsymbol{A}_h \in V_h. \tag{125}$$

The discrete solution ${}^{i+1}\boldsymbol{A}_h$ of this system of equations is called the *time-domain, spectral finite-element solution*. For a given angular degree j (and a fixed time t_{i+1}), there are $P+1$ unknown coefficients ${}^{i+1}A_j^{j,k}$ in the system (125) that describe the solution in the conducting sphere $\mathcal{B}$. Once this system is solved, the coefficient ${}^{i+1}G_j^{(i)}$ of the induced magnetic field is computed by means of the continuity condition (123).

The time-domain, spectral finite-element solution can similarly be introduced to the IBVP of EM induction for the CHAMP magnetic data in the case where the spherical harmonic expansion coefficients $X_j(t)$ of the X component of the magnetic induction vector $\boldsymbol{B}_0$ measured at satellite altitudes are considered as input observations. Beside the functional space V_h, the finite-dimensional functional subspaces of the spaces V_0 and V_λ is constructed by the following prescriptions:

$$V_{0,h} := \left\{ \delta \boldsymbol{A}_0 = a \sum_{j=1}^{j_{\max}} \left[\sqrt{\frac{j}{j+1}} \left(\frac{r}{a}\right)^j \delta G_j^{(e)} - \sqrt{\frac{j+1}{j}} \left(\frac{a}{r}\right)^{j+1} \delta G_j^{(i)} \right] \boldsymbol{Y}_j^j(\vartheta) \right\}, \tag{126}$$

$$V_{\lambda,h} := \left\{ \delta \boldsymbol{\lambda} = \sum_{j=1}^{j_{\max}} \delta \lambda_j^j \, \boldsymbol{Y}_j^j(\vartheta) \right\}. \tag{127}$$

The Galerkin method for approximating the solution of the variational equation (77) at a fixed time t_{i+1} consists in finding ${}^{i+1}\boldsymbol{A}_h \in V_h$, ${}^{i+1}\boldsymbol{A}_{0,h} \in V_{0,h}$ and ${}^{i+1}\boldsymbol{\lambda}_h \in V_{\lambda,h}$, satisfying the variational equation

$$\begin{aligned} & a({}^{i+1}\boldsymbol{A}_h, \delta \boldsymbol{A}_h) + \frac{1}{\Delta t_i} b_1({}^{i+1}\boldsymbol{A}_h, \delta \boldsymbol{A}_h) + a_0({}^{i+1}\boldsymbol{A}_{0,h}, \delta \boldsymbol{A}_{0,h}) + c(\delta \boldsymbol{A}_h - \delta \boldsymbol{A}_{0,h}, {}^{i+1}\boldsymbol{\lambda}_h) \\ & \quad + c({}^{i+1}\boldsymbol{A}_h - {}^{i+1}\boldsymbol{A}_{0,h}, \delta \boldsymbol{\lambda}_h) = \frac{1}{\Delta t_i} b_1({}^{i}\boldsymbol{A}_h, \delta \boldsymbol{A}_h) + F({}^{i+1}\boldsymbol{B}_t, \delta \boldsymbol{A}_{0,h}) \\ & \forall \delta \boldsymbol{A}_h \in V_h, \ \forall \delta \boldsymbol{A}_{0,h} \in V_{0,h}, \ \forall \delta \boldsymbol{\lambda}_h \in V_{\lambda,h}. \end{aligned} \tag{128}$$

For a given angular degree j (and a fixed time t_{i+1}), the unknowns in Eq. 128 consist of $P+1$ coefficients ${}^{i+1}A_j^{j,k}$ describing the solution in the conducting sphere $\mathcal{B}$, the coefficients ${}^{i+1}G_j^{(e)}$ and ${}^{i+1}G_j^{(i)}$ describing the solution in a nonconducting spherical layer $\mathcal{A}$, and ${}^{i+1}\lambda_j^j$ ensuring the continuity of potentials ${}^{i+1}\boldsymbol{A}$ and ${}^{i+1}\boldsymbol{A}_0$ on the Earth's surface $\partial\mathcal{B}$. In total, there are $P+4$ unknowns in the system for a given j.

Martinec et al. (2003) tested the time-domain, spectral finite-element method for the spherical harmonic coefficients $G_j^{(e)}(t)$, described by the variational equation (125), by comparing the results with the analytical and semi-analytical solutions to EM induction in two concentrically and eccentrically nested spheres of different, but constant electrical conductivities. They showed that the numerical code implementing the time-domain, spectral finite-element method for $G_j^{(e)}(t)$ performs correctly, and the time-domain, spectral finite-element method is particularly appropriate when the external current excitation is transient. Later on, Martinec and McCreadie (2004) made use of these results and tested the time-domain, spectral finite-element method for satellite magnetic data, described by the variational equation (128), by comparing it with the time-domain, spectral finite-element method for ground magnetic data. They showed that agreement between the numerical results of the two methods for synthetic data is excellent.

7 CHAMP Data Analysis

7.1 Selection and Processing of Vector Data

The data analyzed in this chapter were recorded by the three-component vector magnetometer on board of CHAMP. To demonstrate the performance of the forward method, from all records spanning more than 8 years, the 1-year-long time series from January 1, 2001, (track No. 2610) to January 10, 2002 (track. No. 8402) has been selected. Judging from the Dst index (➋ *Fig. 4*), there were about ten events when the geomagnetic field was significantly disturbed by magnetic storms or substorms. In order to minimize the effect of strong day-side ionospheric currents, night-side data recorded by the satellite between 18:00 and 6:00 local solar time are only used.

In the first step of the data processing, the CHAOS model of the Earth's magnetic field (Olsen et al., 2006a) is used to separate the signals corresponding to EM induction by storm-time magnetospheric currents. Based on the CHAOS model, the main and crustal fields up to degree 50 and the secular variation up to degree 18 are removed from the CHAMP data.

In the next step, the horizontal magnetic components (X, Y) are rotated from geographic coordinates to dipole coordinates, assuming that the north geomagnetic pole is at 78.8° N, 70.7° W. Since an axisymmetric geometry of external currents and mantle electrical conductivity is assumed, the dipolar longitudinal component Y is not considered hereafter and X and Z are used to describe the northward and downward magnetic components in dipolar coordinates, respectively. ➋ *Figure 1* shows an example of the original and rotated data from CHAMP track No. 6755.

7.2 Two-Step, Track-by-Track Spherical Harmonic Analysis

The input data of the two-step, track-by-track spherical harmonic analysis are the samples of the X component of the residual magnetic signal along an individual satellite track, that is, data set (ϑ_i, X_i), $i = 1, \ldots, N$, where ϑ_i is the geomagnetic colatitude of the ith measurement side, and N is the number of data points. The magnetic data from low and mid-latitudes within the interval $(\vartheta_1, \vartheta_2)$ are only considered in accordance with the assumption that global EM induction is driven by the equatorial ring currents in the magnetosphere. Hence, observations from the polar regions, which are contaminated by signals from field-aligned currents and polar electrojets, are excluded from the analyzed time series. The satellite track data (ϑ_i, X_i) are referenced to the time when CHAMP passes the magnetic equator.

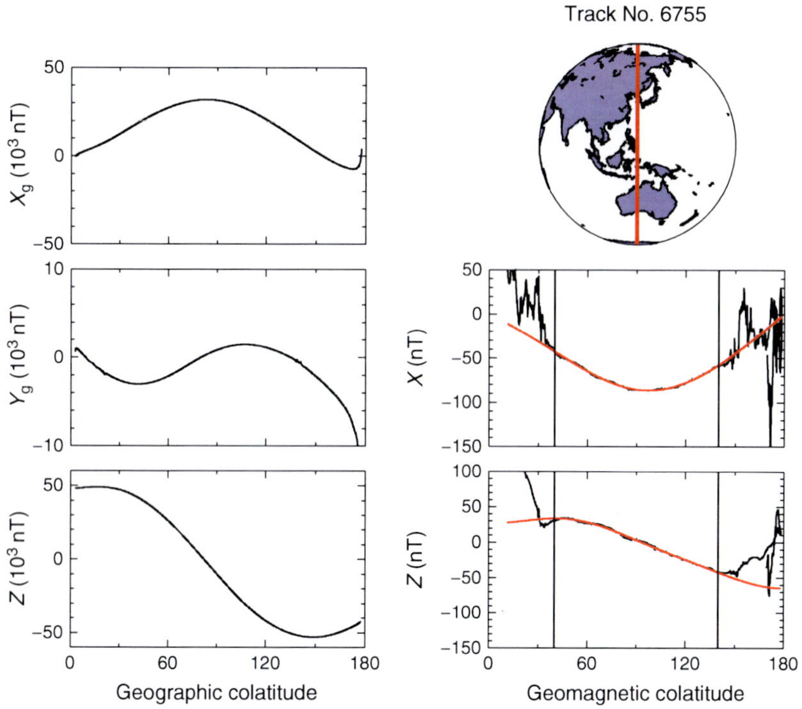

Fig. 1

CHAMP satellite magnetic data along track No. 6755 (*red line* on global map shows the satellite track), which samples the initial phase of a magnetic storm on September 26, 2001, above the East Pacific Ocean. *Left panels*: the original CHAMP data plotted along geographical colatitude. X_g, Y_g, and Z components point, respectively, to the geographic north, the geographic east and downwards. *Right panels*: *Black lines* denote X and Z CHAMP components after the removal of the CHAOS model and the rotation of the residual field to dipole coordinates. The *red lines* show the results of the two-step, track-by-track spherical harmonic analysis, including the extrapolation into the polar regions using data from the mid-colatitude interval (40°, 140°), as marked by dotted lines

In view of parameterization (47), N observational equations for data X_i are considered in the form

$$\sum_{j=1}^{j_{\max}} X_j(t)\frac{\partial Y_j(\vartheta_k)}{\partial \vartheta} + e_i = X_i, \qquad i = 1, \cdots, N, \tag{129}$$

where $X_j(t)$ are the expansion coefficients to be determined by a least-squares method and $j_{\max}$ is the cut-off degree. The measurement errors e_k are assumed to have zero means, uniform variances σ^2, and are uncorrelated:

$$\begin{aligned} Ee_i &= 0, \\ \operatorname{var} e_i &= \sigma^2, \\ \operatorname{cov}(e_i, e_j) &= 0 \qquad \text{for } i \neq j, \end{aligned} \tag{130}$$

where E, var, and cov are the statistical expectancy, the variance, and the covariance operator, respectively. The spherical harmonic analysis of satellite-track magnetic measurements of the X component of the magnetic induction vector is performed in two steps.

7.2.1 Change of the Interval of Orthogonality

In the first step, the data X_i are mapped from the mid-latitude interval $\vartheta \in (\vartheta_1, \vartheta_2)$ onto the half-circle interval $\vartheta' \in (0, \pi)$ by the linear transformation

$$\vartheta'(\vartheta) = \pi \frac{\vartheta - \vartheta_1}{\vartheta_2 - \vartheta_1}, \tag{131}$$

and then adjusted by a series of Legendre polynomials:

$$X(\vartheta') = \sum_{j=0}^{N'} X'_j Y_j(\vartheta'). \tag{132}$$

Likewise, the samples of the Z component of the residual magnetic signal along an individual satellite track, that is, data set (ϑ_i, Z_i) are first mapped from the mid-latitude interval $\vartheta \in (\vartheta_1, \vartheta_2)$ onto the half-circle interval $\vartheta' \in (0, \pi)$ and then expanded into a series of Legendre polynomials:

$$Z(\vartheta') = \sum_{j=0}^{N'} Z'_j Y_j(\vartheta'). \tag{133}$$

The expansion coefficients X'_j and Z'_j are determined by fitting the models (132) and (133) to mid-latitude magnetic data X_i and Z_i, respectively. Since the accuracy of the CHAMP magnetic measurements is high, both long-wavelength and short-wavelength features of the mid-latitude data are adjusted. That is why the cut-off degree N' is chosen to be large. In the following numerical examples, $N' = 25$, while the number of datum points is $N = 1{,}550$. Because of data errors, the observational equations based on the models (132) and (133) are inconsistent and an exact solution to these systems does not exist. The solution to each system of equations is estimated by a least-squares method. Since this method is well documented in the literature (e.g., Bevington, 1969), no details are given.

7.2.2 Extrapolation of Magnetic Data from Mid-Latitudes to Polar Regions

When the analysis of mid-latitude data X_i is complete, the signal that best fits the mid-latitude data is extrapolated to the polar regions. To do it, it is required that the original parameterization (129) of the X component matches that found in the previous step:

$$\sum_{j=1}^{j_{\max}} X_j(t) \left.\frac{\partial Y_j(\vartheta)}{\partial \vartheta}\right|_{\vartheta(\vartheta')} = \sum_{k=0}^{N'} X'_k Y_k(\vartheta'), \tag{134}$$

where $\vartheta = \vartheta(\vartheta')$ denotes the inverse mapping to (131) and the coefficients X'_k are known from the previous step. To determine $X_j(t)$, the orthonormality property of $Y_k(\vartheta')$ is used and the extrapolation condition (134) is rewritten as a system of linear algebraic equations

$$2\pi \sum_{j=1}^{j_{\max}} X_j(t) \int_{\vartheta'=0}^{\pi} \left. \frac{\partial Y_j(\vartheta)}{\partial \vartheta} \right|_{\vartheta(\vartheta')} Y_k(\vartheta') \sin \vartheta' d\vartheta' = X'_k \tag{135}$$

for $k = 0, 1, \cdots, N'$. In a similar way, the extrapolation condition for the Z component can be expressed as

$$2\pi \sum_{j=1}^{j_{\max}} Z_j(t) \int_{\vartheta'=0}^{\pi} Y_j(\vartheta(\vartheta')) Y_k(\vartheta') \sin \vartheta' d\vartheta' = Z'_k. \tag{136}$$

In contrast to the previous step, only long-wavelength features of mid-latitude data are extrapolated to the polar regions; thus, $j_{\max} \ll N'$. In the following numerical examples, only the range $2 \leq j_{\max} \leq 6$ is considered, depending on the character of the mid-latitude data. This choice implies that both systems of linear equations are overdetermined and are solved by a least-squares method. The least-squares estimates of the coefficients $X_j(t)$ and $Z_j(t)$ will be denoted by $X_j^{(\mathrm{obs})}(t)$ and $Z_j^{(\mathrm{obs})}(t)$, respectively. Respective substitutions of $X_j^{(\mathrm{obs})}(t)$ and $Z_j^{(\mathrm{obs})}(t)$ into Eqs. 47 and 49 yield smooth approximations of the X and Z components inside the colatitude interval $(\vartheta_1, \vartheta_2)$ as well as undisturbed extrapolations into the polar regions $(0^\circ, \vartheta_1) \cup (\vartheta_2, 180^\circ)$.

7.2.3 Selection Criteria for Extrapolation

The crucial points of the extrapolation are the choice of the truncation degree $j_{\max}$ of the parameterization (129) and the determination of the colatitude interval $(\vartheta_1, \vartheta_2)$ where the data are not disturbed by the polar currents. Martinec and McCreadie (2004) and Velímský et al. (2006) imposed three criteria to determine these two parameters. First, the power of the magnetic field from the external ring currents is concentrated in the low-degree harmonic coefficients, particularly in the $j = 1$ term, and the leakage of electromagnetic energy into higher-degree terms caused by the Earth's conductivity and electric-current geometry monotonically decreases. This criterion is applied in such a way that the analysis begins with degree $j_{\max} = 1$, increases it by one and plots the degree-power spectrum of the coefficients $X_j^{(\mathrm{obs})}(t)$. While the degree-power spectrum is a monotonically decreasing function of angular degree j, increasing the cut-off degree $j_{\max}$ is continued. Once the degree-power spectrum of $X_j^{(\mathrm{obs})}(t)$ no longer decreases monotonically, the actual cut-off degree is taken from the previous step for which the degree-power spectrum still monotonically decreases. The degree-power spectrum of coefficients $X_j^{(\mathrm{obs})}(t)$ for the final choice of cut-off degree $j_{\max}$ is shown in the third-row panels of ➲ *Fig. 5*.

This criterion can be interpreted as follows. The largest proportion of the magnetospheric ring-current excitation energy is concentrated in the low-degree harmonic coefficients, particularly in the $j = 1$ term. The leakage of the electromagnetic energy from degree $j = 1$ to higher degrees is caused by lateral heterogeneities in the electrical conductivity of the Earth's mantle. The more pronounced the lateral heterogeneities, the larger the transport of energy from degree $j = 1$ to higher degrees. Accepting the criterion of a monotonically decreasing degree-power spectrum means therefore that the Earth's mantle is regarded as only weakly laterally heterogeneous.

Second, the first derivative of the X component with respect to colatitude does not change sign in the polar regions. This criterion excludes unrealistic oscillatory behavior of the X component in these regions caused by a high-degree extrapolation. Third, if the least-squares estimate of the X component of CHAMP data over the colatitude interval $(\vartheta_1 - 5^\circ, \vartheta_2 + 5^\circ)$ differs by more than 10 nT compared to the estimate over the interval $(\vartheta_1, \vartheta_2)$, the field due to the polar currents is assumed to encroach upon the field produced by near-equatorial ring currents, and the narrower colatitude interval $(\vartheta_1, \vartheta_2)$ is considered to contain only the signature generated by the near-equatorial currents. Applying these criteria to the CHAMP-track data iteratively, starting from degree $j = 1$ and the colatitude interval $(10^\circ, 170^\circ)$ and proceeding to higher degrees and shorter colatitude intervals, it is found that the maximum cut-off degree varies from track to track, but does not exceed $j_{\max} = 6$ and the colatitude interval is usually $(40^\circ, 140^\circ)$.

The extrapolation of the Z component from the field at low and mid-latitudes is more problematic than that for the X component. This is because (i) the second selection criterion cannot be applied since the Z component does not approach zero at the magnetic poles as seen from parameterization (49) and (ii) the Z component of CHAMP magnetic data contains a larger portion of high-frequency noise than the X component, which, in principle, violates the assumption of the third selection criterion. ➋ *Figure 5* shows that the leakage of electromagnetic energy from $j = 1$ to higher-degree terms is not monotonically decreasing for the Z component. That is why the least-squares estimates $Z_j^{(\mathrm{obs})}(t)$ are extrapolated to polar regions from the colatitude interval $(\vartheta_1, \vartheta_2)$ and up to the spherical degree $j_{\max}$ determined for the X component.

7.2.4 Examples of Spherical Harmonic Analysis of the CHAMP Magnetic Data

Presented here are four examples of the spherical harmonic analysis of the CHAMP magnetic data recorded in the period from September 25 to October 7, 2001. This period is chosen because it includes a magnetic storm followed by a magnetic substorm, as seen from the behavior of the *Dst* index (see ➋ *Fig. 2*). For demonstration purposes, four CHAMP-track data sets are chosen: the data recorded along track No. 6732 as an example of data analysis before a magnetic storm occurs; track No. 6755 represents the main phase of a magnetic storm; track No. 6780 represents the recovery phase of a storm; and track No. 6830 represents the appearance of a substorm.

In ➋ *Fig. 3*, the X component of the original CHAMP magnetic data reduced by the main magnetic field and the lithospheric magnetic field is shown. The top panels show the residual magnetic signals for the night-time mid-latitudes and the filtered signals after the first step of the spherical harmonic analysis has been performed. The mid-latitude data X_i are adjusted by the model (132) rather well by choosing $N' = 25$. For the sake of completeness, the second-row panels of ➋ *Fig. 3* show the degree-power spectrum of the coefficients X'_j. The degree-power spectrum of the coefficients $X_j^{(\mathrm{obs})}(t)$ for the cut-off degree $j_{\max}$ chosen according to the first selection criterion is shown in the third-row panels of ➋ *Fig. 3*.

The bottom panels of ➋ *Fig. 3* show the residual signals over the whole night-time track derived from the CHAMP observations and the signals extrapolated from low-latitude and mid-latitude data. First, the well-known fact can be seen that the original magnetic data are disturbed at the polar regions by sources other than equatorial ring currents in the magnetosphere.

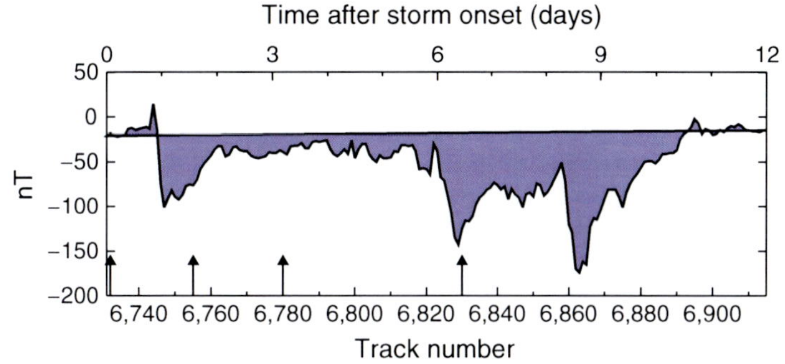

Fig. 2
The *Dst* index for the magnetic storm that occurred between September 25 and October 7, 2001. The arrows mark the satellite tracks chosen to demonstrate the two-step, track-by-track spherical harmonic analysis of satellite magnetic data

Second, since there is no objective criterion for evaluating the quality of the extrapolation of the X component to the polar regions, it is regarded subjectively. For the track data shown here, but also for the other data for the magnetic storm considered, the extrapolation of the X component from mid-latitudes to the polar regions works reasonably well, provided that the cut-off degree $j_{\max}$ and the colatitude interval $(\vartheta_1, \vartheta_2)$ are chosen according to the criteria introduced above.

The procedure applied to the 2001-CHAMP-track data results in time series of spherical harmonic coefficients $X_j^{(\text{obs})}(t)$ and $Z_j^{(\text{obs})}(t)$ for $j = 1, \cdots, 4$. As an example, the resulting coefficients for degree $j = 1$ are plotted in ➋ *Fig. 4* as functions of time after January 1, 2001. As expected, there is a high correlation between the first-degree harmonics $X_1^{(\text{obs})}(t)$ and $Z_1^{(\text{obs})}(t)$ and the *Dst* index for the days that experienced a magnetic storm.

7.3 Power-Spectrum Analysis

Although the method applied in this chapter is based on the time-domain approach, it is valuable to inspect the spectra of the $X_j^{(\text{obs})}(t)$ and $Z_j^{(\text{obs})}(t)$ time series. ➋ *Figure 5* shows the maximum-entropy power-spectrum estimates (Press et al., 1992, Sect. 13.7) of the first four spherical harmonics of the horizontal and vertical components. It can be seen that the magnitudes of the power spectra of the X component monotonically decrease with increasing harmonic degree, which is a consequence of the first selection criterion applied in the two-step, track-by-track analysis. For instance, the power spectrum of the second-degree terms is about two orders of magnitude smaller than that of the first-degree terms. As already mentioned, and also seen in ➋ *Fig. 5*, this is not the case for the Z component, where the magnitude of the maximum-entropy power-spectrum of the Z component is larger than that of the X component for $j > 1$, which demonstrates that the Z component of the CHAMP magnetic data contains a larger portion of high-frequency noise than the X component.

Despite analyzing only night-side tracks, there is a significant peak at the period of 1 day in the power spectra of the higher-degree harmonics ($j \geq 2$), but, surprisingly, missing in the spectra of the first-degree harmonic. To eliminate the induction effect of residual dawn/dusk ionospheric electric currents, the night-side local-solar time interval is shrunk from (18:00, 6:00) to (22:00, 4:00). However, a 1-day period signal remains present in the CHAMP residual

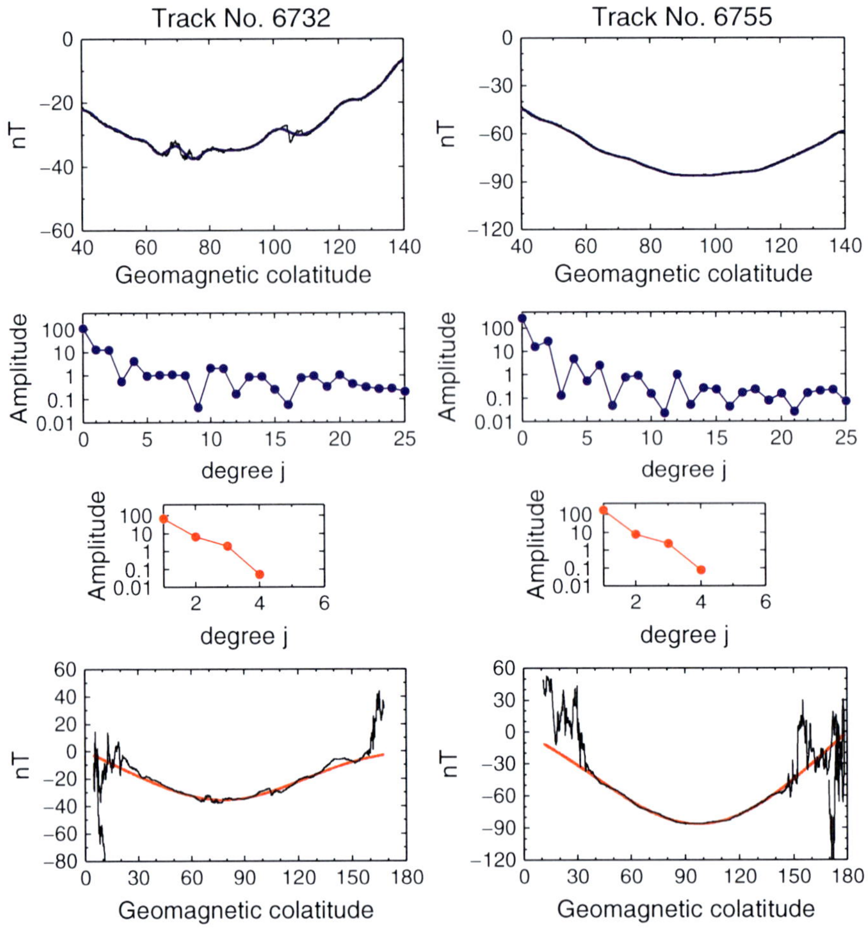

Fig. 3

Examples of the two-step, track-by-track spherical harmonic analysis of magnetic signals along four satellite tracks. The *top panels* show the X component of the residual magnetic signals at the night-time mid-latitudes derived from the CHAMP magnetic observations (*thin lines*) and the predicted signals after the first step of the spherical harmonic analysis has been completed (*thick lines*). The number of samples in the original signals is $N = 1,550$. The *second-* and *third-row panels* show the degree-power spectrum of the coefficients X_j' and $X_j(t)$, respectively. The cut-off degree of the coefficients X_j' is fixed to $N' = 25$, while the cut-off degree j_{max} of the coefficients $X_j(t)$ is found by the criteria discussed in the text. The *bottom panels* show the X component of the residual magnetic signals over the whole night-time tracks (*thin lines*) and the signals extrapolated from mid-latitude data according to the second step of the spherical harmonic analysis (*thick lines*). The longitude when the CHAMP satellite crosses the equator of the geocentric coordinate system is −55.19°, 127.19°, −97.15°, and 174.23° for tracks No. 6732, 6755, 6780 and 6830, respectively

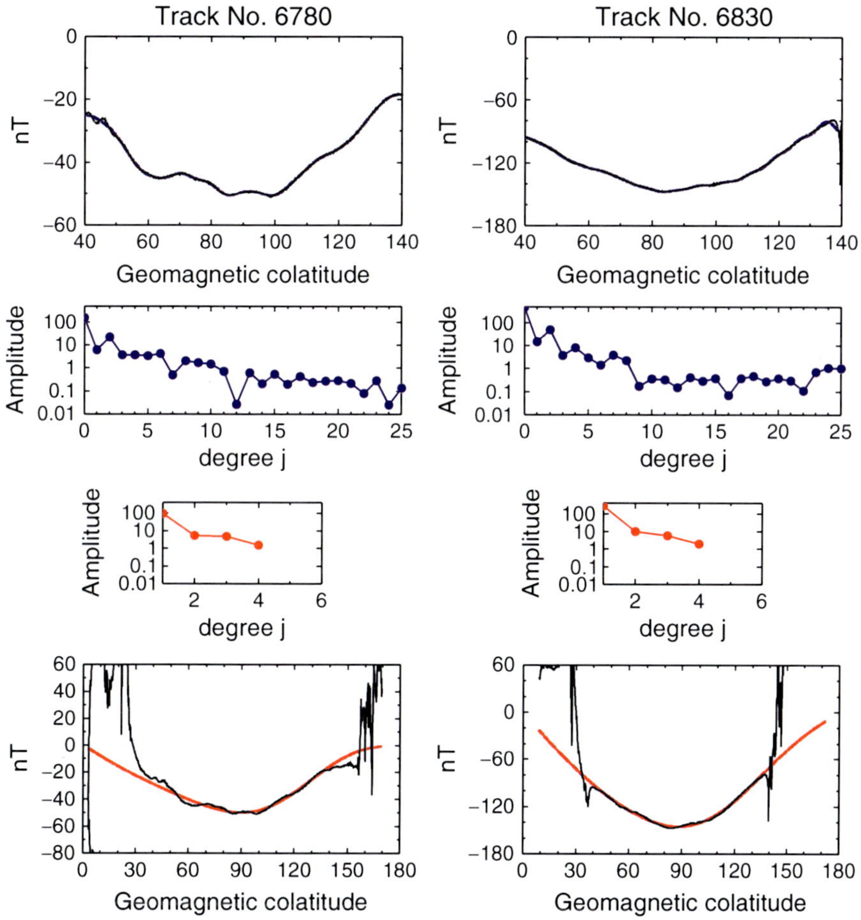

Fig. 3 (continued)

signal (not shown here). To locate a region of potential inducing electric currents, time series of $X_j^{(\mathrm{obs})}(t)$ and $Z_j^{(\mathrm{obs})}(t)$ coefficients are converted to time series of spherical harmonic coefficients $G_j^{(e,\mathrm{obs})}(t)$ and $G_j^{(i,\mathrm{obs})}(t)$ of the external and internal fields counted with respect to the CHAMP satellite altitude by applying Eq. 110. The maximum-entropy power-spectrum estimates of the external and internal coefficients $G_j^{(e,\mathrm{obs})}(t)$ and $G_j^{(i,\mathrm{obs})}(t)$ are shown in *Fig. 6*. It can be seen that these spectra for degrees $j = 2$ to 4 also have a peak at a period of 1 day. This means that at least part of $G_j^{(e,\mathrm{obs})}(t)$ originates in the magnetosphere or even the magnetopause and magnetic tail, while $G_j^{(i,\mathrm{obs})}(t)$ may originate from the residual night-side ionospheric currents and/or the electric currents induced in the Earth by either effect.

Figure 6 also shows that, while the periods of peak values in the external and internal magnetic fields for degree $j = 1$ correspond to each other, for the higher-degree spherical harmonic coefficients, such a correspondence is only valid for some periods, for instance, 6.8, 5.6, or 4.8 days. However, the peak for the period of 8.5 days in the internal component for $j = 2$ is hardly detectable in the external field. This could be explained by a three-dimensionality effect

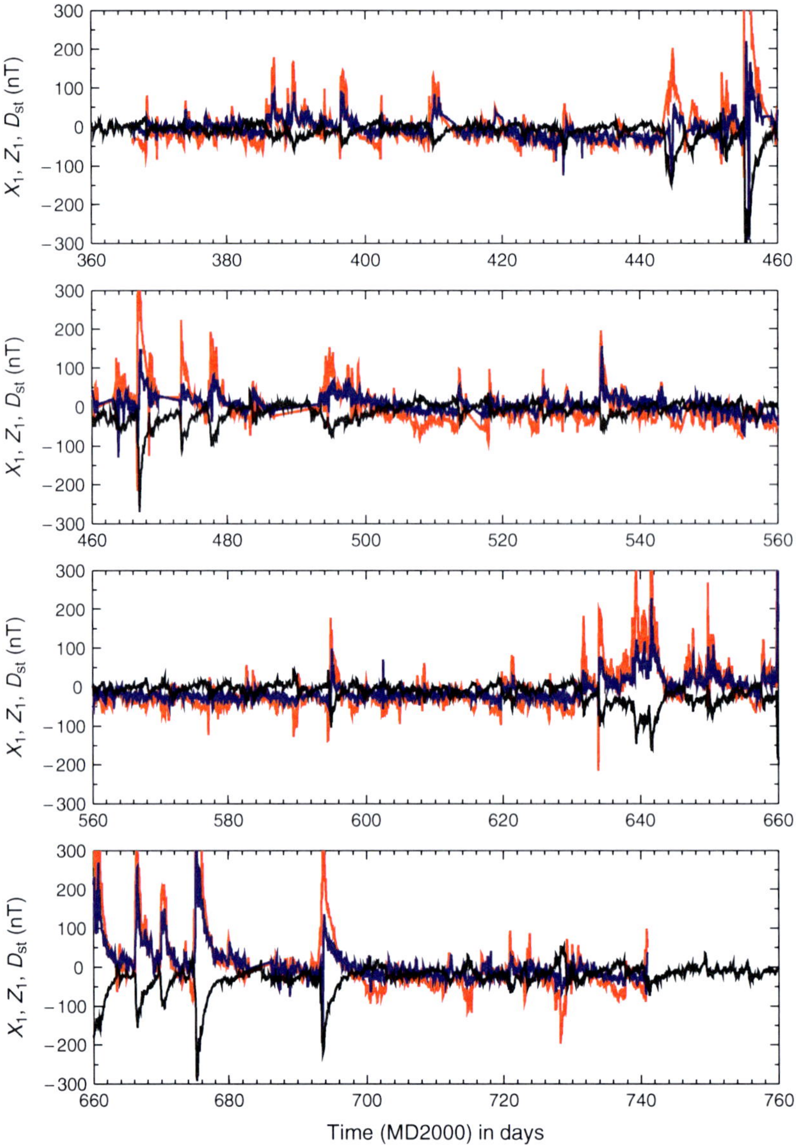

■ Fig. 4

Time series of the spherical harmonic coefficients $X_1^{(\mathrm{obs})}(t)$ (*red*) and $Z_1^{(\mathrm{obs})}(t)$ (*blue*) of horizontal and vertical components obtained by the two-step, track-by-track spherical harmonic analysis of CHAMP data for the year 2001. A mean and linear trend have been removed following Olsen et al. (2005). The coefficients from the missing tracks are filled by cubic spline interpolation applied to the detrended time series. Note that the sign of the X_1 component is opposite to that of the *Dst* index (*black line*). Time on the horizontal axis is measured from midnight of January 1, 2000

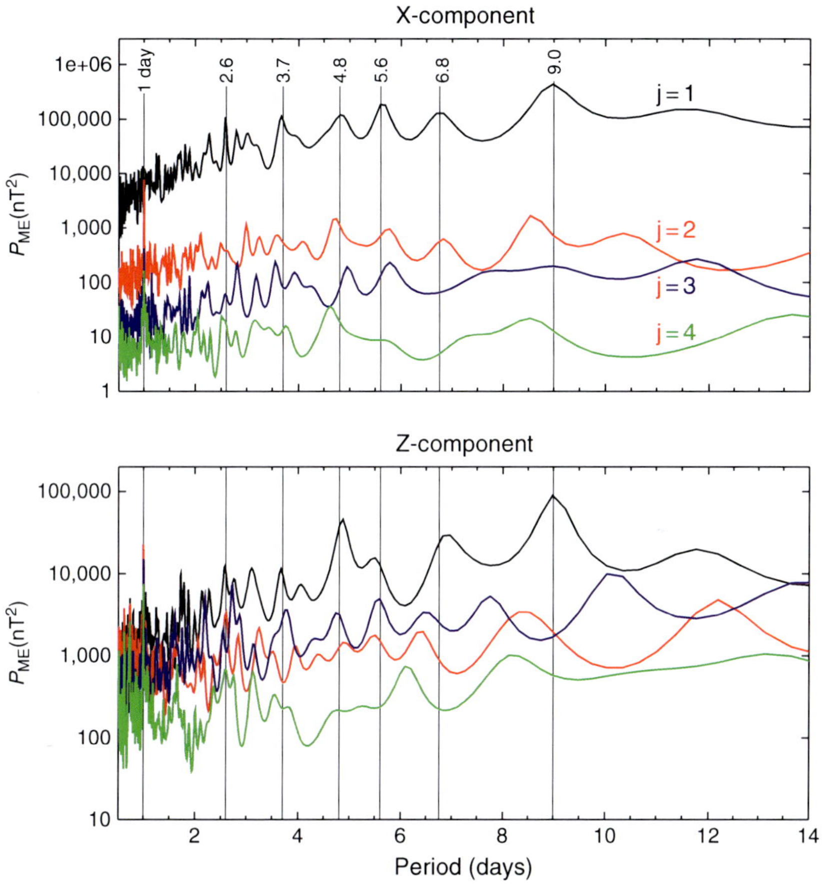

Fig. 5

The maximum-entropy power-spectrum estimates of the spherical harmonic coefficients of $X_j^{(obs)}(t)$ (*top panel*) and $Z_j^{(obs)}(t)$ (*bottom panel*) components. Degrees $j = 1, 2, 3$, and 4 are shown by *black, red, blue*, and *green* lines, respectively, The spectra have peaks at higher harmonics of the 27-day solar rotation period, that is at periods of 9, 6.8, 5.6, 4.8 days, etc

in the electrical conductivity of the Earth's mantle that causes the leakage of electromagnetic energy from degree $j = 1$ to the second and higher-degree terms. This leakage may partly shift the characteristic periods in the resulting signal due to interference between signals with various spatial wavelengths and periods.

8 Adjoint Sensitivity Method of EM Induction for the *Z* Component of CHAMP Magnetic Data

In this section, the adjoint sensitivity method of EM induction for computing the sensitivities of the *Z* component of CHAMP magnetic data with respect to the mantle's conductivity

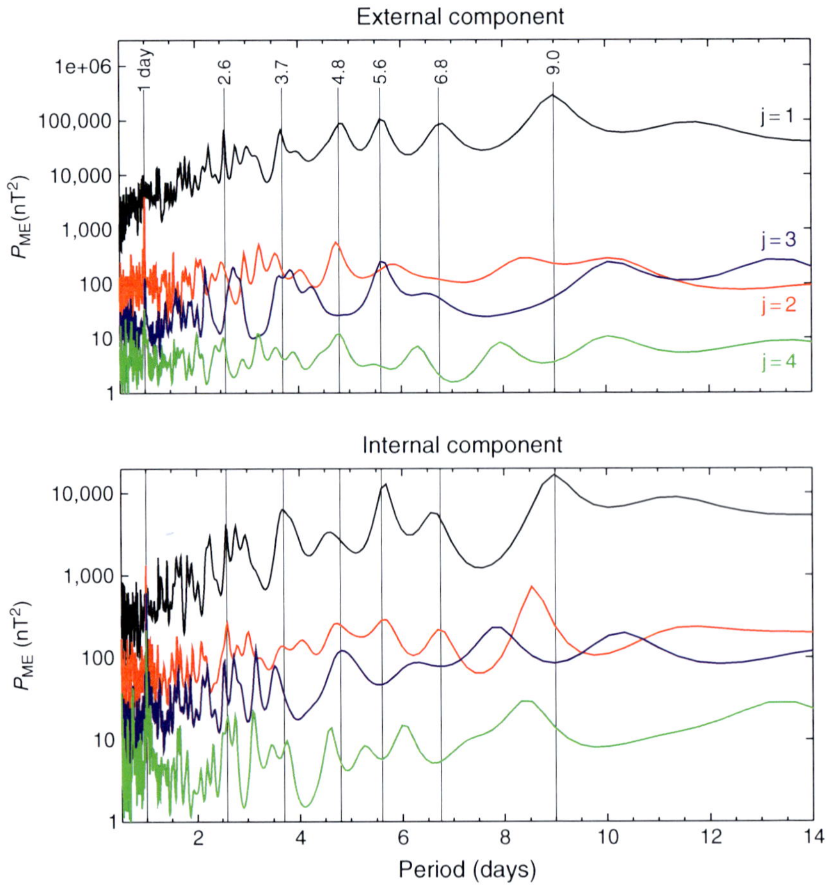

Fig. 6
As Fig. 5, but for the external and internal Gauss coefficients $G_j^{(e,obs)}(t)$ and $G_j^{(i,obs)}(t)$, counted with respect to the CHAMP satellite altitude

structure is formulated. Most of considerations in this section follow the paper by Martinec and Velímský (2009).

8.1 Forward Method

The forward method of EM induction for the X component of CHAMP magnetic data was formulated in Sect. 4. In this case, the solution domain G for EM induction modeling is the unification of the conducting sphere $\mathcal{B}$ and the insulating spherical layer $\mathcal{A}$, that is, $G = \mathcal{B} \cup \mathcal{A}$ with the boundary ∂G coinciding with the mean-orbit sphere, that is, $\partial G = \partial \mathcal{A}$. The forward IBVP (58–65) for the toroidal vector potential $\boldsymbol{A}$ in G can then be written in the abbreviated form as

$$\frac{1}{\mu} \operatorname{curl} \operatorname{curl} \boldsymbol{A} + \sigma \frac{\partial \boldsymbol{A}}{\partial t} = \boldsymbol{0} \quad \text{in } G \tag{137}$$

with the boundary condition

$$\boldsymbol{n} \times \operatorname{curl} \boldsymbol{A} = \boldsymbol{B}_t \quad \text{on } \partial G, \tag{138}$$

and the inhomogeneous initial condition

$$\boldsymbol{A}|_{t=0} = \boldsymbol{A}^0 \quad \text{in } G. \tag{139}$$

Note that the conductivity $\sigma = 0$ in the insulating atmosphere $\mathcal{A}$ implies that the second term in Eq. 137 vanishes in $\mathcal{A}$ and Eq. 137 reduces to Eq. 60.

8.2 Misfit Function and Its Gradient in the Parameter Space

Let the conductivity $\sigma(r, \vartheta)$ of the conducting sphere $\mathcal{B}$ now be represented in terms of an M-dimensional system of r- and ϑ-dependent base functions and denote the expansion coefficients of this representation to be $\sigma_1, \sigma_2, \ldots, \sigma_M$. Defining the conductivity parameter vector $\vec{\sigma} := (\sigma_1, \sigma_2, \ldots, \sigma_M)$, the dependence of the conductivity $\sigma(r, \vartheta)$ on the parameters $\vec{\sigma}$ can be made explicit as

$$\sigma = \sigma(r, \vartheta; \vec{\sigma}). \tag{140}$$

In ➋ Sect. 4, it is shown that the solution of the IBVP for CHAMP magnetic data enables the modeling of the time evolution of the normal component $B_n := \boldsymbol{n} \cdot \boldsymbol{B}$ of the magnetic induction vector on the mean-orbit sphere ∂G along the satellite tracks. These predicted data $B_n(\vec{\sigma})$ can be compared with the observations $B_n^{(\text{obs})} = -Z^{(\text{obs})}$ of the normal component of the magnetic induction vector by the CHAMP onboard magnetometer. The differences between observed and predicted values can then be used as a misfit for the inverse EM induction modeling. The adjoint method of EM induction presented hereafter calculates the sensitivity of the forward-modeled data $B_n(\vec{\sigma})$ on the conductivity parameters $\vec{\sigma}$ by making use of the differences $B_n^{(\text{obs})} - B_n(\vec{\sigma})$ as boundary-value data.

Let the observations $B_n^{(\text{obs})}$ be made for times $t \in (0, T)$ such that, according to assumption (2) in ➋ Sect. 2, $B_n^{(\text{obs})}(\vartheta, t_i)$ at a particular time $t_i \in (0, T)$ corresponds to the CHAMP observations along the ith satellite track. The least-squares misfit is then defined as

$$\chi^2(\vec{\sigma}) := \frac{b}{2\mu} \int_0^T \int_{\partial G} w_b^2 \left[B_n^{(\text{obs})} - B_n(\vec{\sigma})\right]^2 dS\, dt, \tag{141}$$

where the weighting factor $w_b = w_b(\vartheta, t)$ is chosen to be dimensionless such that the misfit has the SI unit $\mathrm{m^3 s T^2}/[\mu]$, $[\mu]=\mathrm{kg\, m\, s^{-2} A^{-2}}$. If the observations $B_n^{(\text{obs})}$ contain random errors which are statistically independent, the statistical variance of the observations may be substituted for the reciprocal value of w_b^2 (e.g., Bevington, 1969, Sects. 6–4). The ϑ dependence of w_b allows the elimination of the track data from the polar regions which are contaminated by signals from field-aligned currents and polar electrojets, while the time dependence of w_b allows the elimination of the track data for time instances when other undesirable magnetic effects at low and mid-latitudes contaminate the signal excited by equatorial ring currents in the magnetosphere.

The sensitivity analysis or inverse modeling requires the computation of the partial derivative of the misfit with respect to the model parameters, that is, the derivatives $\partial\chi^2/\partial\sigma_m$, $m = 1, \ldots, M$, often termed the sensitivities of the misfit with respect to the model parameters σ_m (e.g., Sandu et al., 2003). To abbreviate the notation, the partial derivatives with respect to the conductivity parameters are ordered in the gradient operator in the M-dimensional parameter space,

$$\nabla_{\vec{\sigma}} := \sum_{m=1}^{M} \hat{\sigma}_m \frac{\partial}{\partial \sigma_m}, \tag{142}$$

where the hat in $\hat{\sigma}_m$ indicates a unit vector.

Realizing that the observations $B_n^{(\text{obs})}$ are independent of the conductivity parameters $\vec{\sigma}$, that is, $\nabla_{\vec{\sigma}} B_n^{(\text{obs})} = \vec{0}$, the gradient of $\chi^2(\vec{\sigma})$ is

$$\nabla_{\vec{\sigma}} \chi^2 = -\frac{b}{\mu} \int_0^T \int_{\partial G} \Delta B_n(\vec{\sigma}) \nabla_{\vec{\sigma}} B_n \, dS \, dt, \tag{143}$$

where $\Delta B_n(\vec{\sigma})$ are the weighted residuals of the normal component of the magnetic induction vector:

$$\Delta B_n(\vec{\sigma}) := w_b^2 \left[B_n^{(\text{obs})} - B_n(\vec{\sigma}) \right]. \tag{144}$$

The straightforward approach to find $\nabla_{\vec{\sigma}} \chi^2$ is to approximate $\partial \chi^2 / \partial \sigma_m$ by a numerical differentiation of forward model runs. Due to the size of the parameter space, this procedure is often extremely computationally expensive.

8.3 The Forward Sensitivity Equations

The forward sensitivity analysis computes the sensitivities of the forward solution with respect to the conductivity parameters, that is, the partial derivatives $\partial \boldsymbol{A} / \partial \sigma_m$, $m = 1, \ldots, M$. Using them, the forward sensitivities $\nabla_{\vec{\sigma}} B_n$ are computed and substituted into Eq. 143 for $\nabla_{\vec{\sigma}} \chi^2$.

To form the forward sensitivity equations, also called the linear tangent equations of the model (e.g., McGillivray et al., 1994; Cacuci, 2003; Sandu et al., 2003, 2005), the conductivity model (140) is considered in the forward model Eqs. 137–139. Differentiating them with respect to the conductivity parameters $\vec{\sigma}$ yields

$$\frac{1}{\mu} \operatorname{curl} \operatorname{curl} \nabla_{\vec{\sigma}} \boldsymbol{A} + \sigma \frac{\partial \nabla_{\vec{\sigma}} \boldsymbol{A}}{\partial t} + \nabla_{\vec{\sigma}} \sigma \frac{\partial \boldsymbol{A}}{\partial t} = \boldsymbol{0} \quad \text{in } G \tag{145}$$

with homogeneous boundary condition

$$\boldsymbol{n} \times \operatorname{curl} \nabla_{\vec{\sigma}} \boldsymbol{A} = \boldsymbol{0} \quad \text{on } \partial G \tag{146}$$

and homogeneous initial condition

$$\nabla_{\vec{\sigma}} \boldsymbol{A}|_{t=0} = \boldsymbol{0} \quad \text{in } G, \tag{147}$$

where $\nabla_{\vec{\sigma}} \boldsymbol{B}_t = \nabla_{\vec{\sigma}} \boldsymbol{A}^0 = \boldsymbol{0}$ have been substituted because the boundary data $\boldsymbol{B}_t$ and the initial condition $\boldsymbol{A}^0$ are independent of the conductivity parameters $\vec{\sigma}$. In the forward sensitivity analysis, for each parameter σ_m and associated forward solution $\boldsymbol{A}$, a new source term $\nabla_{\vec{\sigma}} \sigma \, \partial \boldsymbol{A} / \partial t$ is created and the forward sensitivity equations (145–147) are solved to compute the partial derivative $\partial \boldsymbol{A} / \partial \sigma_m$. The forward sensitivity analysis is known to be very effective when the sensitivities of a larger number of output variables are computed with respect to a small number of model parameters (Sandu et al., 2003; Petzold et al., 2006).

In ➋ Sect. 9, the adjoint sensitivity method of EM induction for the case when the Gauss coefficients are taken as observations will be dealt with. In this case, the boundary condition (146) has a more general form:

$$\boldsymbol{n} \times \operatorname{curl} \nabla_{\vec{\sigma}} \boldsymbol{A} - \boldsymbol{L}(\nabla_{\vec{\sigma}} \boldsymbol{A}) = \boldsymbol{0} \quad \text{on } \partial G, \tag{148}$$

where $\boldsymbol{L}$ is a linear vector operator acting on a vector function defined on the boundary ∂G. For the case studied now, however, $\boldsymbol{L} = \boldsymbol{0}$.

8.4 The Adjoint Sensitivity Equations

The adjoint method provides an efficient alternative to the forward sensitivity analysis for evaluating $\nabla_{\vec{\sigma}} \chi^2$ without explicit knowledge of $\nabla_{\vec{\sigma}} \boldsymbol{A}$, that is, without solving the forward sensitivity equations. Hence, the adjoint method is more efficient for problems involving a large number of model parameters. Because the forward sensitivity equations are linear in $\nabla_{\vec{\sigma}} \boldsymbol{A}$, an adjoint equation exists (Cacuci, 2003).

The adjoint sensitivity analysis proceeds by forming the inner product of Eqs. 145 and 148 with an yet unspecified adjoint function $\hat{\boldsymbol{A}}(r, \vartheta, t)$, then integrated over G and ∂G, respectively, and subtracted from each other:

$$\begin{aligned} &\frac{1}{\mu} \int_G \operatorname{curl}\operatorname{curl} \nabla_{\vec{\sigma}} \boldsymbol{A} \cdot \hat{\boldsymbol{A}}\, dV - \frac{1}{\mu} \int_{\partial G} (\boldsymbol{n} \times \operatorname{curl} \nabla_{\vec{\sigma}} \boldsymbol{A}) \cdot \hat{\boldsymbol{A}}\, dS + \frac{1}{\mu} \int_{\partial G} \boldsymbol{L}(\nabla_{\vec{\sigma}} \boldsymbol{A}) \cdot \hat{\boldsymbol{A}}\, dS \\ &+ \int_G \sigma \frac{\partial \nabla_{\vec{\sigma}} \boldsymbol{A}}{\partial t} \cdot \hat{\boldsymbol{A}}\, dV + \int_G \nabla_{\vec{\sigma}} \sigma \frac{\partial \boldsymbol{A}}{\partial t} \cdot \hat{\boldsymbol{A}}\, dV = 0, \end{aligned} \tag{149}$$

where the dot stands for the scalar product of vectors.

In the next step, the integrals in Eq. 149 are transformed such that $\nabla_{\vec{\sigma}} \boldsymbol{A}$ and $\hat{\boldsymbol{A}}$ interchange. To achieve this, the Green's theorem is considered for two sufficiently smooth functions $\boldsymbol{f}$ and $\boldsymbol{g}$ in the form

$$\int_G \operatorname{curl} \boldsymbol{f} \cdot \operatorname{curl} \boldsymbol{g}\, dV = \int_G \operatorname{curl}\operatorname{curl} \boldsymbol{f} \cdot \boldsymbol{g}\, dV - \int_{\partial G} (\boldsymbol{n} \times \operatorname{curl} \boldsymbol{f}) \cdot \boldsymbol{g}\, dS. \tag{150}$$

Interchanging the functions $\boldsymbol{f}$ and $\boldsymbol{g}$ and subtracting the new equation from the original one results in the integral identity

$$\int_G \operatorname{curl}\operatorname{curl} \boldsymbol{f} \cdot \boldsymbol{g}\, dV - \int_{\partial G} (\boldsymbol{n} \times \operatorname{curl} \boldsymbol{f}) \cdot \boldsymbol{g}\, dS = \int_G \operatorname{curl}\operatorname{curl} \boldsymbol{g} \cdot \boldsymbol{f}\, dV - \int_{\partial G} (\boldsymbol{n} \times \operatorname{curl} \boldsymbol{g}) \cdot \boldsymbol{f}\, dS. \tag{151}$$

By this, the positions of $\nabla_{\vec{\sigma}} \boldsymbol{A}$ and $\hat{\boldsymbol{A}}$ can be exchanged in the first two integrals in Eq. 149:

$$\begin{aligned} &\frac{1}{\mu} \int_G \operatorname{curl}\operatorname{curl} \hat{\boldsymbol{A}} \cdot \nabla_{\vec{\sigma}} \boldsymbol{A}\, dV - \frac{1}{\mu} \int_{\partial G} (\boldsymbol{n} \times \operatorname{curl} \hat{\boldsymbol{A}}) \cdot \nabla_{\vec{\sigma}} \boldsymbol{A}\, dS + \frac{1}{\mu} \int_{\partial G} \boldsymbol{L}(\nabla_{\vec{\sigma}} \boldsymbol{A}) \cdot \hat{\boldsymbol{A}}\, dS \\ &+ \int_G \sigma \frac{\partial \nabla_{\vec{\sigma}} \boldsymbol{A}}{\partial t} \cdot \hat{\boldsymbol{A}}\, dV + \int_G \nabla_{\vec{\sigma}} \sigma \frac{\partial \boldsymbol{A}}{\partial t} \cdot \hat{\boldsymbol{A}}\, dV = 0, \end{aligned} \tag{152}$$

To perform the same transformation in the fourth integral, Eq. 152 is integrated over the time interval $t \in (0, T)$, that is,

$$\frac{1}{\mu}\int_0^T \int_G \operatorname{curl}\operatorname{curl}\hat{\boldsymbol{A}} \cdot \nabla_{\vec{\sigma}}\boldsymbol{A}\, dV dt - \frac{1}{\mu}\int_0^T \int_{\partial G} (\boldsymbol{n} \times \operatorname{curl}\hat{\boldsymbol{A}}) \cdot \nabla_{\vec{\sigma}}\boldsymbol{A}\, dS dt$$
$$+ \frac{1}{\mu}\int_0^T \int_{\partial G} \boldsymbol{L}(\nabla_{\vec{\sigma}}\boldsymbol{A}) \cdot \hat{\boldsymbol{A}}\, dS dt + \int_0^T \int_G \sigma \frac{\partial \nabla_{\vec{\sigma}}\boldsymbol{A}}{\partial t} \cdot \hat{\boldsymbol{A}}\, dV dt$$
$$+ \int_0^T \int_G \nabla_{\vec{\sigma}}\sigma \frac{\partial \boldsymbol{A}}{\partial t} \cdot \hat{\boldsymbol{A}}\, dV dt = 0. \tag{153}$$

Then the order of integration is exchanged over the spatial variables and time in the fourth integral and perform the time integration by partes:

$$\int_0^T \frac{\partial \nabla_{\vec{\sigma}}\boldsymbol{A}}{\partial t} \cdot \hat{\boldsymbol{A}}\, dt = \nabla_{\vec{\sigma}}\boldsymbol{A} \cdot \hat{\boldsymbol{A}}\Big|_{t=T} - \nabla_{\vec{\sigma}}\boldsymbol{A} \cdot \hat{\boldsymbol{A}}\Big|_{t=0} - \int_0^T \nabla_{\vec{\sigma}}\boldsymbol{A} \cdot \frac{\partial \hat{\boldsymbol{A}}}{\partial t}\, dt. \tag{154}$$

The second term on the right-hand side is equal to zero because of the homogeneous initial condition (147). Finally, Eq. 153 takes the form

$$\frac{1}{\mu}\int_0^T \int_G \operatorname{curl}\operatorname{curl}\hat{\boldsymbol{A}} \cdot \nabla_{\vec{\sigma}}\boldsymbol{A}\, dV dt - \frac{1}{\mu}\int_0^T \int_{\partial G} (\boldsymbol{n} \times \operatorname{curl}\hat{\boldsymbol{A}}) \cdot \nabla_{\vec{\sigma}}\boldsymbol{A}\, dS dt$$
$$+ \frac{1}{\mu}\int_0^T \int_{\partial G} \boldsymbol{L}(\nabla_{\vec{\sigma}}\boldsymbol{A}) \cdot \hat{\boldsymbol{A}}\, dS dt + \int_G \sigma \nabla_{\vec{\sigma}}\boldsymbol{A} \cdot \hat{\boldsymbol{A}}\Big|_{t=T} dV$$
$$- \int_0^T \int_G \sigma \nabla_{\vec{\sigma}}\boldsymbol{A} \cdot \frac{\partial \hat{\boldsymbol{A}}}{\partial t}\, dV dt + \int_0^T \int_G \nabla_{\vec{\sigma}}\sigma \frac{\partial \boldsymbol{A}}{\partial t} \cdot \hat{\boldsymbol{A}}\, dV dt = 0. \tag{155}$$

Remembering that $\nabla_{\vec{\sigma}} B_n$ is the derivative that is to be eliminate from $\nabla_{\vec{\sigma}}\chi^2$, the homogeneous equation (155) is added to Eq. 143 (note the physical units of Eq. 155 are the same as $\nabla_{\vec{\sigma}}\chi^2$, namely, $\mathrm{m^3 s T^2}/[\mu\sigma]$ provided that the physical units of $\hat{\boldsymbol{A}}$ are the same as of $\boldsymbol{A}$, namely, Tm):

$$\nabla_{\vec{\sigma}}\chi^2 = \frac{1}{\mu}\int_0^T \int_G \operatorname{curl}\operatorname{curl}\hat{\boldsymbol{A}} \cdot \nabla_{\vec{\sigma}}\boldsymbol{A}\, dV dt - \frac{1}{\mu}\int_0^T \int_{\partial G} (\boldsymbol{n} \times \operatorname{curl}\hat{\boldsymbol{A}}) \cdot \nabla_{\vec{\sigma}}\boldsymbol{A}\, dS\, dt$$
$$+ \frac{1}{\mu}\int_0^T \int_{\partial G} \boldsymbol{L}(\nabla_{\vec{\sigma}}\boldsymbol{A}) \cdot \hat{\boldsymbol{A}}\, dS dt + \int_G \sigma \nabla_{\vec{\sigma}}\boldsymbol{A} \cdot \hat{\boldsymbol{A}}\Big|_{t=T} dV - \int_0^T \int_G \sigma \frac{\partial \hat{\boldsymbol{A}}}{\partial t} \cdot \nabla_{\vec{\sigma}}\boldsymbol{A}\, dV dt$$
$$+ \int_0^T \int_G \nabla_{\vec{\sigma}}\sigma \frac{\partial \boldsymbol{A}}{\partial t} \cdot \hat{\boldsymbol{A}}\, dV dt - \frac{b}{\mu}\int_0^T \int_{\partial G} \Delta B_n \nabla_{\vec{\sigma}} B_n\, dS\, dt. \tag{156}$$

The adjoint function $\hat{\boldsymbol{A}}$ has been considered arbitrary so far. The aim is now to impose constraints on it such that the originally arbitrary $\hat{\boldsymbol{A}}$ transforms to the well-defined *adjoint toroidal vector potential*. The volume integrals over G proportional to $\nabla_{\vec{\sigma}}\boldsymbol{A}$ are first eliminated by requiring that

$$\frac{1}{\mu}\operatorname{curl}\operatorname{curl}\hat{\boldsymbol{A}} - \sigma \frac{\partial \hat{\boldsymbol{A}}}{\partial t} = \boldsymbol{0} \quad \text{in } G, \tag{157}$$

with the terminal condition on $\hat{\boldsymbol{A}}$:

$$\hat{\boldsymbol{A}}\Big|_{t=T} = \boldsymbol{0} \qquad \text{in } G. \tag{158}$$

The boundary condition for $\hat{A}$ on ∂G is derived from the requirement that the surface integrals over ∂G in Eq. 156 cancel each other, that is,

$$\int_{\partial G} (\boldsymbol{n} \times \operatorname{curl} \hat{\boldsymbol{A}}) \cdot \nabla_{\vec{\sigma}} \boldsymbol{A} \, dS - \int_{\partial G} \boldsymbol{L}(\nabla_{\vec{\sigma}} \boldsymbol{A}) \cdot \hat{\boldsymbol{A}} \, dS + b \int_{\partial G} \Delta B_n \nabla_{\vec{\sigma}} B_n \, dS = 0 \tag{159}$$

at any time $t \in (0, T)$. This condition will be elaborated on in the next section. Under these constraints, the gradient of $\chi^2(\vec{\sigma})$ takes the form

$$\nabla_{\vec{\sigma}} \chi^2 = \int_0^T \int_G \nabla_{\vec{\sigma}} \sigma \frac{\partial \boldsymbol{A}}{\partial t} \cdot \hat{\boldsymbol{A}} \, dV dt. \tag{160}$$

8.5 Boundary Condition for the Adjoint Potential

To relate $\nabla_{\vec{\sigma}} \boldsymbol{A}$ and $\nabla_{\vec{\sigma}} B_n$ in the constraint described by Eq. 159 and, subsequently, to eliminate $\nabla_{\vec{\sigma}} \boldsymbol{A}$ from it, $\boldsymbol{A}$ needs to be parameterized. In the colatitudinal direction, $\boldsymbol{A}$ will be represented as a series of the zonal toroidal vector spherical harmonics $\boldsymbol{Y}_j^j(\vartheta)$ in the form given by Eq. 89, which is also employed for the adjoint potential $\hat{\boldsymbol{A}}$:

$$\begin{Bmatrix} \boldsymbol{A}(r, \vartheta, t) \\ \hat{\boldsymbol{A}}(r, \vartheta, t) \end{Bmatrix} = \sum_{j=1}^{\infty} \begin{Bmatrix} A_j^j(r, t) \\ \hat{A}_j^j(r, t) \end{Bmatrix} \boldsymbol{Y}_j^j(\vartheta). \tag{161}$$

In the radial direction, inside a conducting sphere $\mathcal{B}$ of radius a, the spherical harmonic expansion coefficients $A_j^j(r, t)$ are parameterized by $P + 1$ piecewise-linear finite elements $\psi_k(r)$ on the interval $0 \leq r \leq a$ as shown by Eq. 96:

$$A_j^j(r, t) = \sum_{k=1}^{P+1} A_j^{j,k}(t) \psi_k(r). \tag{162}$$

In an insulating atmosphere $\mathcal{A}$, the spherical harmonic expansion coefficients $A_j^j(r, t)$ are parameterized in the form given by Eq. 103:

$$A_j^j(r, t) = a \left[\sqrt{\frac{j}{j+1}} \left(\frac{r}{a} \right)^j G_j^{(e)}(t) - \sqrt{\frac{j+1}{j}} \left(\frac{a}{r} \right)^{j+1} G_j^{(i)}(t) \right]. \tag{163}$$

The same parameterizations as shown by Eqs. 162 and 163 are taken for the coefficients $\hat{A}_j^j(r, t)$.

The first aim is to express the gradient $\nabla_{\vec{\sigma}} \chi^2$ in terms of spherical harmonics. Since the upper-boundary ∂G of the solution domain G is the mean-orbit sphere of radius b, the external normal $\boldsymbol{n}$ to ∂G coincides with the unit vector $\boldsymbol{e}_r$, that is $\boldsymbol{n} = \boldsymbol{e}_r$. Applying the gradient operator $\nabla_{\vec{\sigma}}$ on the equation $B_n = \boldsymbol{e}_r \cdot \operatorname{curl} \boldsymbol{A}$ and using Eq. 228 yields

$$\nabla_{\vec{\sigma}} B_n(r, \vartheta, t) = -\frac{1}{r} \sum_{j=1}^{\infty} \sqrt{j(j+1)} \, \nabla_{\vec{\sigma}} A_j^j(r, t) Y_j(\vartheta). \tag{164}$$

Moreover, applying a two-step, track-by-track spherical harmonic analysis on the residual satellite-track data ΔB_n defined by Eq. 144, these observables can, at a particular time $t \in (0, T)$, be represented as a series of the zonal scalar spherical harmonics

$$\Delta B_n(\vartheta, t; \vec{\sigma}) = \sum_{j=1}^{\infty} \Delta B_{n,j}(t; \vec{\sigma}) Y_j(\vartheta) \tag{165}$$

with spherical harmonic coefficients of the form

$$\Delta B_{n,j}(t;\vec{\sigma}) = \frac{1}{b^2}\int_{\partial G} w_b^2\left[B_n^{(\mathrm{obs})}(\vartheta,t) - B_n(b,\vartheta,t;\vec{\sigma})\right] Y_j(\vartheta)\, dS. \tag{166}$$

Substituting Eqs. 164 and 165 into Eq. 143 and employing the orthonormality property (216) of the zonal scalar spherical harmonics $Y_j(\vartheta)$, the gradient of the misfit χ^2 becomes

$$\nabla_{\vec{\sigma}}\chi^2 = \frac{b^2}{\mu}\int_0^T \sum_{j=1}^{\infty}\sqrt{j(j+1)}\,\Delta B_{n,j}(t;\vec{\sigma})\nabla_{\vec{\sigma}}A_j^j(b,t)\,dt. \tag{167}$$

The constraint (159) with $\boldsymbol{L} = \boldsymbol{0}$, that is, for the case of the boundary condition (146) is now expressed in terms of spherical harmonics. By the parameterization (161) and the assumption $\boldsymbol{n} = \boldsymbol{e}_r$, the differential relation (230) applied to $\hat{\boldsymbol{A}}$ yields

$$\boldsymbol{n}\times\operatorname{curl}\hat{\boldsymbol{A}} = \sum_{j=1}^{\infty}\left[\boldsymbol{n}\times\operatorname{curl}\hat{\boldsymbol{A}}(r,t)\right]_j^j \boldsymbol{Y}_j^j(\vartheta), \tag{168}$$

where

$$\left[\boldsymbol{n}\times\operatorname{curl}\hat{\boldsymbol{A}}(r,t)\right]_j^j = -\left(\frac{d}{dr}+\frac{1}{r}\right)\hat{A}_j^j(r). \tag{169}$$

The first constituent in the first integral of the constraint (159) is expressed by Eq. 168, while the second constituent can be obtained by applying the gradient operator $\nabla_{\vec{\sigma}}$ to Eq. 161. The two constituents in the second integral of the constraint (159) are expressed by Eqs. 164 and (165), respectively. Performing all indicated substitutions, one obtains

$$\begin{aligned}&\int_{\varphi=0}^{2\pi}\int_{\vartheta=0}^{\pi}\sum_{j_1=1}^{\infty}\left[\boldsymbol{n}\times\operatorname{curl}\hat{\boldsymbol{A}}(b,t)\right]_{j_1}^{j_1}\boldsymbol{Y}_{j_1}^{j_1}(\vartheta)\cdot\sum_{j_2=1}^{\infty}\nabla_{\vec{\sigma}}A_{j_2}^{j_2}(b,t)\,\boldsymbol{Y}_{j_2}^{j_2}(\vartheta)\,b^2\sin\vartheta\, d\vartheta\, d\varphi\\ &= b\int_{\varphi=0}^{2\pi}\int_{\vartheta=0}^{\pi}\sum_{j_1=1}^{\infty}\Delta B_{n,j_1}(t;\vec{\sigma})Y_{j_1}(\vartheta)\frac{1}{b}\sum_{j_2=1}^{\infty}\sqrt{j_2(j_2+1)}\,\nabla_{\vec{\sigma}}A_{j_2}^{j_2}(b,t)Y_{j_2}(\vartheta)\,b^2\sin\vartheta\, d\vartheta\, d\varphi.\end{aligned} \tag{170}$$

Interchanging the order of integration over the full solid angle and summations over j's, and making use of the orthonormality properties (216) and (221) of the zonal scalar and vector spherical harmonics, respectively, Eq. 170 reduces to

$$\sum_{j=1}^{\infty}\left[\boldsymbol{n}\times\operatorname{curl}\hat{\boldsymbol{A}}(b,t)\right]_j^j\nabla_{\vec{\sigma}}A_j^j(b,t) = \sum_{j=1}^{\infty}\sqrt{j(j+1)}\,\Delta B_{n,j}(t;\vec{\sigma})\nabla_{\vec{\sigma}}A_j^j(b,t), \tag{171}$$

which is to be valid at any time $t \in (0,T)$. To satisfy this constraint independently of $\nabla_{\vec{\sigma}}A_j^j(b,t)$, one last condition is imposed upon the adjoint potential $\hat{\boldsymbol{A}}$, namely,

$$\left[\boldsymbol{n}\times\operatorname{curl}\hat{\boldsymbol{A}}(b,t)\right]_j^j = \sqrt{j(j+1)}\,\Delta B_{n,j}(t;\vec{\sigma}) \quad \text{on } \partial G \tag{172}$$

at any time $t \in (0,T)$.

8.6 Adjoint Method

The formulation of the adjoint method of EM induction for the Z component of CHAMP satellite magnetic data can be summarized as follows.

Given the electrical conductivity model $\sigma(r,\vartheta)$ in the sphere $\mathcal{B}$, the forward solution $A(r,\vartheta,t)$ in $\mathcal{B}$ and the atmosphere $\mathcal{A}$ for $t \in (0,T)$ and the observations $B_n^{(\text{obs})}(t)$ on the mean-orbit sphere ∂G of radius $r = b$, with uncertainties quantified by weighting factor w_b, find the adjoint potential $\hat{A}(r,\vartheta,t)$ in $G = \mathcal{B} \cup \mathcal{A}$ by solving the adjoint problem:

$$\frac{1}{\mu}\operatorname{curl}\operatorname{curl}\hat{\boldsymbol{A}} - \sigma\frac{\partial \hat{\boldsymbol{A}}}{\partial t} = \boldsymbol{0} \quad \text{in } G \tag{173}$$

with the boundary condition

$$\left[\boldsymbol{n} \times \operatorname{curl}\hat{\boldsymbol{A}}(b,t)\right]_j^j = \sqrt{j(j+1)}\,\Delta B_{n,j}(t) \quad \text{on } \partial G \tag{174}$$

and the terminal condition

$$\hat{\boldsymbol{A}}\big|_{t=T} = \boldsymbol{0} \qquad \text{in } G. \tag{175}$$

The gradient of the misfit $\chi^2(\vec{\sigma})$ is then expressed as

$$\nabla_{\vec{\sigma}}\chi^2 = \int_0^T \int_G \nabla_{\vec{\sigma}}\sigma\frac{\partial \boldsymbol{A}(t)}{\partial t} \cdot \hat{\boldsymbol{A}}(t)\, dV dt. \tag{176}$$

The set of Eqs. 173–175 is referred to as the *adjoint problem* of the forward problem specified by Eqs. 137–139. Combining the forward solution $\boldsymbol{A}$ and the adjoint solution $\hat{\boldsymbol{A}}$ according to Eq. 176 thus gives the exact derivative of the misfit χ^2.

8.7 Reverse Time

The numerical solution of Eq. (173), solved backwards in time from $t = T$ to $t = 0$, is inherently unstable. Unlike the case of the forward model equation and the forward sensitivity equation, the adjoint equation effectively includes negative diffusion, which enhances numerical perturbations instead of smoothing them, leading to an unstable solution. To avoid such numerical instability, the sign of the diffusive term in Eq. 173 is changed by reversing the time variable. Let the reverse time $\tau = T - t$, $\tau \in (0,T)$, and the reverse-time adjoint potential $\check{\boldsymbol{A}}(\tau)$ be introduced such that

$$\hat{\boldsymbol{A}}(t) = \hat{\boldsymbol{A}}(T-\tau) =: \check{\boldsymbol{A}}(\tau). \tag{177}$$

Hence

$$\frac{\partial \hat{\boldsymbol{A}}}{\partial t} = -\frac{\partial \check{\boldsymbol{A}}}{\partial \tau}, \tag{178}$$

and Eq. 173 transforms to the diffusion equation for the reverse-time adjoint potential $\check{\boldsymbol{A}}(\tau)$:

$$\frac{1}{\mu}\operatorname{curl}\operatorname{curl}\check{\boldsymbol{A}} + \sigma\frac{\partial \check{\boldsymbol{A}}}{\partial \tau} = \boldsymbol{0} \quad \text{in } G \tag{179}$$

with the boundary condition

$$\left[\boldsymbol{n} \times \operatorname{curl}\check{\boldsymbol{A}}(b,\tau)\right]_j^j = \sqrt{j(j+1)}\,\Delta B_{n,j}(T-\tau) \quad \text{on } \partial G. \tag{180}$$

The terminal condition (175) for $\hat{\boldsymbol{A}}$ becomes the initial condition for the potential $\check{\boldsymbol{A}}$:

$$\check{\boldsymbol{A}}\big|_{\tau=0} = \boldsymbol{0} \qquad \text{in } G. \tag{181}$$

With these changes, the adjoint equations become similar to those of the forward method, and thus nearly identical numerical methods can be applied.

In addition, the gradient (176) transforms to

$$\nabla_{\vec{\sigma}} \chi^2 = \int_0^T \int_G \nabla_{\vec{\sigma}} \sigma \frac{\partial \boldsymbol{A}(t)}{\partial t} \cdot \check{\boldsymbol{A}}(T-t)\, dV dt. \tag{182}$$

The importance of Eq. 182 is that, once the forward problem (137)–(139) is solved and the misfit χ^2 is evaluated from Eq. 141, the gradient $\nabla_{\vec{\sigma}} \chi^2$ may be evaluated for little more than the cost of a single solution of the adjoint system (179)–(181) and a single scalar product in Eq. 182, regardless of the dimension of the conductivity parameter vector $\vec{\sigma}$. This is compared to other methods of evaluating $\nabla_{\vec{\sigma}} \chi^2$ that typically require the solution of the forward problem (137)–(139) per component of $\vec{\sigma}$.

The specific steps involved in the adjoint computations are now explained. First, the forward solutions $\boldsymbol{A}(t_i)$ are calculated at discrete times $0 = t_0 < t_1 < \cdots < t_n = T$ by solving the forward problem (137)–(139), and each solution $\boldsymbol{A}(t_i)$ *must be stored.* Then, the reverse-time adjoint solutions $\check{\boldsymbol{A}}(t_i)$, $i = 0, \cdots, n$, are calculated, proceeding again forwards in time according to Eqs. 179–181. As each adjoint solution is computed, the misfit and its derivative are updated according to Eqs. 141 and 182, respectively. When $\check{\boldsymbol{A}}(T)$ has finally been calculated, both χ^2 and $\nabla_{\vec{\sigma}} \chi^2$ are known. The forward solutions $\boldsymbol{A}(t_i)$ are stored because Eqs. 180 and 182 depend on them for the adjoint calculation. As a result, the numerical algorithm has memory requirements that are linear with respect to the number of time steps. This is the main drawback of the adjoint method.

8.8 Weak Formulation

The adjoint IBVP (179)–(181) can again be reformulated in a weak sense. Creating an auxiliary boundary-value vector

$$\begin{aligned} \boldsymbol{B}_t^{(\mathrm{adj})}(\vartheta, \tau) &:= \sum_{j=1}^{\infty} \sqrt{j(j+1)}\, \Delta B_{n,j}(T-\tau) \boldsymbol{Y}_j^j(\vartheta) \\ &= -\sum_{j=1}^{\infty} \sqrt{j(j+1)}\, \Delta Z_j(T-\tau) \boldsymbol{Y}_j^j(\vartheta), \end{aligned} \tag{183}$$

where the negative vertical downward Z component of the magnetic induction vector $\boldsymbol{B}_0$ has been substituted for the normal upward B_n component of $\boldsymbol{B}_0$, the boundary condition (180) can be written as

$$\boldsymbol{n} \times \operatorname{curl} \check{\boldsymbol{A}} = \boldsymbol{B}_t^{(\mathrm{adj})} \quad \text{on } \partial G. \tag{184}$$

It can be seen that the adjoint problem (179), (181), and (184) for the reverse-time adjoint potential $\check{\boldsymbol{A}}(\tau)$ has the same form as the forward problem (137)–(139) for the forward potential $\boldsymbol{A}$. Hence, the weak formulation of the adjoint problem is given by the variational equation (77), where the forward boundary-data vector $\boldsymbol{B}_t$ is to be replaced by the adjoint boundary-data vector $\boldsymbol{B}_t^{(\mathrm{adj})}$. In addition, the form-similarity between the expression (57) for $\boldsymbol{B}_t$ and the expression (183) for $\boldsymbol{B}_t^{(\mathrm{adj})}$ enables one to express the spherical harmonic representation (109) of the linear functional $F(\cdot)$ in a unified form:

$$F(\delta \boldsymbol{A}_0) = \frac{ab^2}{\mu} \sum_{j=1}^{\infty} D_j(t) \left[j \left(\frac{b}{a}\right)^j \delta G_j^{(e)} - (j+1) \left(\frac{a}{b}\right)^{j+1} \delta G_j^{(i)} \right], \tag{185}$$

where

$$D_j(t) = \begin{cases} X_j(t) & \text{for the forward method,} \\ \Delta Z_j(T-t) & \text{for the adjoint method,} \end{cases} \tag{186}$$

and $\Delta Z_j(t)$ is the residual between the Z component of the CHAMP observations and the forward-modeled data, that is, $\Delta Z_j(t) = Z_j^{(\text{obs})}(t) - Z_j(t;\vec{\sigma})$.

9 Adjoint Sensitivity Method of EM Induction for the Internal Gauss Coefficients of CHAMP Magnetic Data

In this section, the adjoint sensitivity method of EM induction for computing the sensitivities of the internal Gauss coefficients of CHAMP magnetic data with respect to mantle-conductivity structure is formulated.

9.1 Forward Method

The forward method of EM induction for the external Gauss coefficients of CHAMP magnetic was formulated in ➋ Sect. 5. As discussed, the solution domain for EM induction modeling is the conducting sphere $\mathcal{B}$ with the boundary $\partial\mathcal{B}$ coinciding with the mean Earth surface with radius $r = a$. Since both the external and internal Gauss coefficients are associated with the spherical harmonic expansion of the magnetic scalar potential U in a near-space atmosphere to the Earth's surface, the boundary condition for $G_j^{(e)}(t)$ can only be formulated in terms of spherical harmonic expansion coefficients of the sought-after toroidal vector potential $\boldsymbol{A}$. First, the forward IBVP of EM induction for the Gauss coefficient $G_j^{(e)}(t)$ is briefly reviewed: The toroidal vector potential $\boldsymbol{A}$ inside the conductive sphere $\mathcal{B}$ with a given conductivity $\sigma(r,\vartheta)$ is sought such that, for $t > 0$, it holds that

$$\frac{1}{\mu}\operatorname{curl}\operatorname{curl}\boldsymbol{A} + \sigma\frac{\partial \boldsymbol{A}}{\partial t} = \boldsymbol{0} \quad \text{in } \mathcal{B} \tag{187}$$

with the boundary condition

$$\left[\boldsymbol{n}\times\operatorname{curl}\boldsymbol{A}(a)\right]_j^j - \frac{j}{a}A_j^j(a) = -\sqrt{\frac{j}{j+1}}(2j+1)G_j^{(e)} \quad \text{on } \partial\mathcal{B} \tag{188}$$

and the inhomogeneous initial condition

$$\boldsymbol{A}|_{t=0} = \boldsymbol{A}^0 \quad \text{in } \mathcal{B}. \tag{189}$$

9.2 Misfit Function and Its Gradient in the Parameter Space

In comparison with the adjoint sensitivity method for the Z component of CHAMP magnetic data, yet another modification concerns the definition of a misfit function. Let the observations $G_j^{(i,\text{obs})}(t)$, $j = 1,2,\cdots,j_{\max}$, be made over the time interval $(0,T)$. The least-squares misfit is then defined as

$$\chi^2(\vec{\sigma}) := \frac{a^3}{2\mu}\int_0^T \sum_{j=1}^{j_{\max}} \left[G_j^{(i,\text{obs})}(t) - G_j^{(i)}(t;\vec{\sigma})\right]^2 dt, \tag{190}$$

where the forward-modeled data $G_j^{(i)}(t;\vec{\sigma})$ are computed according to Eq. 123 after solving the forward IBVP of EM induction (187)–(189). To bring the misfit function (190) to a form that is analogous to Eq. 141, two auxiliary quantities are introduced

$$\begin{Bmatrix} G^{(i,\mathrm{obs})}(\vartheta,t) \\ G^{(i)}(\vartheta,t;\vec{\sigma}) \end{Bmatrix} = \sum_{j=1}^{j_{\max}} \begin{Bmatrix} G_j^{(i,\mathrm{obs})}(t) \\ G_j^{(i)}(t;\vec{\sigma})(t) \end{Bmatrix} Y_j(\vartheta), \tag{191}$$

by means of which, and considering the orthonormality property of spherical harmonics $Y_j(\vartheta)$, the misfit (190) can be written as

$$\chi^2(\vec{\sigma}) = \frac{a}{2\mu} \int_0^T \int_{\partial\mathcal{B}} \left[G^{(i,\mathrm{obs})} - G^{(i)}(\vec{\sigma})\right]^2 dS\, dt. \tag{192}$$

In contrast to Eq. 141, the weighting factor w_b^2 does not appear in the integral (192) since possible inconsistencies in the CHAMP magnetic data are already considered in data processing for $G_j^{(i,\mathrm{obs})}(t)$ (see ➋ Sect. 7).

Realizing that the observations $G^{(i,\mathrm{obs})}$ are independent of the conductivity parameters $\vec{\sigma}$, that is, $\nabla_{\vec{\sigma}} G^{(i,\mathrm{obs})} = \vec{0}$, the gradient of $\chi^2(\vec{\sigma})$ is

$$\nabla_{\vec{\sigma}} \chi^2 = -\frac{a}{\mu} \int_0^T \int_{\partial\mathcal{B}} \Delta G^{(i)}(\vec{\sigma}) \nabla_{\vec{\sigma}} G^{(i)}\, dS\, dt, \tag{193}$$

where $\Delta G^{(i)}(\vec{\sigma})$ are the residuals of the internal Gauss coefficients:

$$\Delta G^{(i)}(\vec{\sigma}) := G^{(i,\mathrm{obs})} - G^{(i)}(\vec{\sigma}). \tag{194}$$

9.3 Adjoint Method

Differentiating eqs (187) and (189) for the forward solution with respect to conductivity parameters $\vec{\sigma}$ results in the sensitivity equations of the same form as Eqs. 145 and 147, but now valid inside the sphere $\mathcal{B}$. The appropriate boundary condition for the sensitivities $\nabla_{\vec{\sigma}} \boldsymbol{A}$ is obtained by differentiating Eq. (188) with respect to the parameters $\vec{\sigma}$:

$$\left[\boldsymbol{n} \times \mathrm{curl}\, \nabla_{\vec{\sigma}} \boldsymbol{A}(a)\right]_j^j - \frac{j}{a} \nabla_{\vec{\sigma}} A_j^j(a) = 0 \quad \text{on } \partial\mathcal{B}. \tag{195}$$

Multiplying the last equation by the zonal toroidal vector spherical harmonics and summing up the result over j, the sensitivity equation (195) can be written in the form of Eq. 148, where the linear vector boundary operator $\boldsymbol{L}$ has the form

$$\boldsymbol{L}(\nabla_{\vec{\sigma}} \boldsymbol{A}) = \frac{1}{a} \sum_{j=1}^{\infty} j \nabla_{\vec{\sigma}} A_j^j(a) \boldsymbol{Y}_j^j(\vartheta). \tag{196}$$

In view of the form-similarity between the expressions (143) and (193), the boundary condition for $\hat{\boldsymbol{A}}$ on $\partial\mathcal{B}$ can be deduced from the condition (159):

$$\int_{\partial\mathcal{B}} (\boldsymbol{n} \times \mathrm{curl}\, \hat{\boldsymbol{A}}) \cdot \nabla_{\vec{\sigma}} \boldsymbol{A}\, dS - \int_{\partial\mathcal{B}} \boldsymbol{L}(\nabla_{\vec{\sigma}} \boldsymbol{A}) \cdot \hat{\boldsymbol{A}}\, dS + a \int_{\partial\mathcal{B}} \Delta G^{(i)} \nabla_{\vec{\sigma}} G^{(i)}\, dS = 0, \tag{197}$$

which must be valid at any time $t \in (0, T)$.

To express the partial derivatives of the forward-modeled data $G_j^{(i)}$ with respect to the conductivity parameters $\vec{\sigma}$, that is, the gradient $\nabla_{\vec{\sigma}} G_j^{(i)}$ in terms of the sensitivities $\nabla_{\vec{\sigma}} \boldsymbol{A}$, Eq. 123 is differentiated with respect to the conductivity parameters $\vec{\sigma}$:

$$\nabla_{\vec{\sigma}} G_j^{(i)} = -\frac{1}{a}\sqrt{\frac{j}{j+1}}\,\nabla_{\vec{\sigma}} A_j^j(a), \tag{198}$$

where $\nabla_{\vec{\sigma}} G_j^{(e)} = 0$ has been considered because the forward-model boundary data $G_j^{(e)}$ are independent of the conductivity parameters $\vec{\sigma}$.

The constraint (197) can finally be expressed in terms of spherical harmonics. The first constituent in the first integral of the constraint (197) is expressed by Eq. 168, while the second constituent can be obtained by applying the gradient operator $\nabla_{\vec{\sigma}}$ to Eq. 161. The two constituents in the second and third integrals of the constraint (197) are expressed by Eqs. 196 and 161, and by Eqs. (194) and (198), respectively. Performing all indicated substitutions results in

$$\begin{aligned}&\int_{\varphi=0}^{2\pi}\int_{\vartheta=0}^{\pi}\sum_{j_1=1}^{\infty}\left[\boldsymbol{n}\times\operatorname{curl}\hat{\boldsymbol{A}}(a,t)\right]_{j_1}^{j_1}\boldsymbol{Y}_{j_1}^{j_1}(\vartheta)\cdot\sum_{j_2=1}^{\infty}\nabla_{\vec{\sigma}}A_{j_2}^{j_2}(a,t)\,\boldsymbol{Y}_{j_2}^{j_2}(\vartheta)\,a^2\sin\vartheta\,d\vartheta\,d\varphi\\&-\frac{1}{a}\int_{\varphi=0}^{2\pi}\int_{\vartheta=0}^{\pi}\sum_{j_1=1}^{\infty}j_1\nabla_{\vec{\sigma}}A_{j_1}^{j_1}(a,t)\boldsymbol{Y}_{j_1}^{j_1}(\vartheta)\cdot\sum_{j_2=1}^{\infty}\hat{A}_{j_2}^{j_2}(a,t)\,\boldsymbol{Y}_{j_2}^{j_2}(\vartheta)\,a^2\sin\vartheta\,d\vartheta\,d\varphi\\&=a\int_{\varphi=0}^{2\pi}\int_{\vartheta=0}^{\pi}\sum_{j_1=1}^{\infty}\Delta G_{j_1}^{(i)}(t;\vec{\sigma})Y_{j_1}(\vartheta)\frac{1}{a}\sum_{j_2=1}^{\infty}\sqrt{\frac{j_2}{j_2+1}}\,\nabla_{\vec{\sigma}}A_{j_2}^{j_2}(a,t)Y_{j_2}(\vartheta)\,a^2\sin\vartheta\,d\vartheta\,d\varphi.\end{aligned} \tag{199}$$

Interchanging the order of integration over the full solid angle and summations over j's, and making use of the orthonormality properties (216) and (221) of the zonal scalar and vector spherical harmonics, respectively, Eq. 199 reduces to

$$\sum_{j=1}^{\infty}\left\{\left[\boldsymbol{n}\times\operatorname{curl}\hat{\boldsymbol{A}}(a,t)\right]_j^j-\frac{j}{a}\hat{A}_j^j(a,t)\right\}\nabla_{\vec{\sigma}}A_j^j(a,t)=\sum_{j=1}^{\infty}\sqrt{\frac{j}{j+1}}\,\Delta G_j^{(i)}(t;\vec{\sigma})\nabla_{\vec{\sigma}}A_j^j(a,t). \tag{200}$$

To satisfy this constraint independent of $\nabla_{\vec{\sigma}}A_j^j(a,t)$, one last condition is imposed upon the adjoint potential $\hat{\boldsymbol{A}}$, namely,

$$\left[\boldsymbol{n}\times\operatorname{curl}\hat{\boldsymbol{A}}(a,t)\right]_j^j-\frac{j}{a}\hat{A}_j^j(a,t)=\sqrt{\frac{j}{j+1}}\,\Delta G_j^{(i)}(t;\vec{\sigma})\quad\text{on }\partial\mathcal{B} \tag{201}$$

for $j = 1, 2, \cdots, j_{\max}$, and at any time $t \in (0, T)$.

The formulation of the adjoint method of EM induction for the internal Gauss coefficients is now summarized. Given the electrical conductivity $\sigma(r,\vartheta)$ in the conducting sphere $\mathcal{B}$, the forward solution $\boldsymbol{A}(r,\vartheta,t)$ in $\mathcal{B}$ and the observations $G_j^{(i,\text{obs})}(t)$, $j = 1, 2, \cdots, j_{\max}$, on the mean sphere $\partial\mathcal{B}$ of radius $r = a$ for the time interval $(0, T)$, find the adjoint potential $\hat{\boldsymbol{A}}(r,\vartheta,t)$ in $\mathcal{B}$, such that, for $t > 0$, it satisfies the magnetic diffusion equation

$$\frac{1}{\mu}\operatorname{curl}\operatorname{curl}\hat{\boldsymbol{A}}-\sigma\frac{\partial\hat{\boldsymbol{A}}}{\partial t}=\boldsymbol{0}\quad\text{in }\mathcal{B} \tag{202}$$

with the boundary condition

$$\left[\boldsymbol{n} \times \operatorname{curl} \check{\boldsymbol{A}}(a,t)\right]_j^j - \frac{j}{a}\check{A}_j^j(a,t) = \sqrt{\frac{j}{j+1}}\,\Delta G_j^{(i)}(t) \quad \text{on } \partial\mathcal{B} \tag{203}$$

for $j = 1, 2, \cdots, j_{\max}$, and the terminal condition

$$\check{\boldsymbol{A}}\Big|_{t=T} = \boldsymbol{0} \qquad \text{in } \mathcal{B}. \tag{204}$$

9.4 Weak Formulation

To find a stable solution of diffusion equation (202), the reverse time $\tau = T - t$ and the reverse-time adjoint potential $\check{A}(\tau)$ are introduced in the same manner as in ➋ Sect. 8.7. By this transformation, the negative sign at the diffusive term in Eq. 202 is inverted to a positive sign. The adjoint IBVP for the $\check{\boldsymbol{A}}$ can be reformulated in a weak sense and described by the variational equation

$$a_1(\check{\boldsymbol{A}}, \delta\boldsymbol{A}) + b(\check{\boldsymbol{A}}, \delta\boldsymbol{A}) = f_2(\delta\boldsymbol{A}) \qquad \forall \delta\boldsymbol{A} \in V, \tag{205}$$

where the solution space V, the bilinear forms $a_1(\cdot,\cdot)$ and $b(\cdot,\cdot)$ are given by Eqs. 66, 121, and 69, respectively, and the new linear functional $f_2(\cdot)$ is defined by

$$f_2(\delta\boldsymbol{A}) = -\frac{a^2}{\mu}\sum_{j=1}^{\infty}\sqrt{\frac{j}{j+1}}\,\Delta G_j^{(i)}(T-\tau)\,\delta A_j^j(a). \tag{206}$$

Here, $\Delta G_j^{(i)}(t)$ are the residuals between the coefficients $G_j^{(i,\mathrm{sur})}(t)$ determined from the CHAMP observations of the X and Z components of the magnetic induction vector and continued downwards from the satellite's altitude to the Earth's surface according to Eq. 111,

$$G_j^{(i,\mathrm{sur})}(t) = (b/a)^{j+2} G_j^{(i,\mathrm{obs})}(t), \tag{207}$$

and the forward-modeled coefficients $G_j^{(i)}(t;\vec{\sigma})$:

$$\Delta G_j^{(i)}(t) = G_j^{(i,\mathrm{sur})}(t) - G_j^{(i)}(t;\vec{\sigma}). \tag{208}$$

Having determined the forward solution $\boldsymbol{A}$ and the reverse-time adjoint solution $\check{\boldsymbol{A}}$, the gradient of the misfit $\chi^2(\vec{\sigma})$ with respect to the conductivity parameters $\vec{\sigma}$ can be computed by

$$\nabla_{\vec{\sigma}}\chi^2 = \int_0^T \int_G \nabla_{\vec{\sigma}}\sigma \frac{\partial \boldsymbol{A}(t)}{\partial t} \cdot \check{\boldsymbol{A}}(T-t)\, dV dt. \tag{209}$$

9.5 Summary

The forward and adjoint IBVPs of EM induction for the CHAMP satellite data can be formulated in a unified way. Let $\boldsymbol{F}$ denote either the toroidal vector potential $\boldsymbol{A}$ or the reverse-time adjoint toroidal vector potential $\check{\boldsymbol{A}}$ for the forward and the adjoint problems, respectively. $\boldsymbol{F}$ is sought inside the conductive sphere $\mathcal{S}$ with a given conductivity $\sigma(r,\vartheta)$ such that, for $t > 0$, it satisfies the magnetic diffusion equation

$$\frac{1}{\mu}\operatorname{curl}\operatorname{curl}\boldsymbol{F} + \sigma\frac{\partial \boldsymbol{F}}{\partial t} = \boldsymbol{0} \quad \text{in } \mathcal{S} \tag{210}$$

Table 2
Boundary conditions for the forward and adjoint methods

Method	Satellite magnetic components ($r = b$)	Ground-based Gauss coefficients ($r = a$)
Forward	$\left[n \times \operatorname{curl} F(t)\right]_j^j = -\sqrt{j(j+1)}X_j(t)$	$\left[n \times \operatorname{curl} F(t)\right]_j^j - \frac{j}{a}F_j^j(t) = -\sqrt{\frac{j}{j+1}}(2j+1)G_j^{(e)}(t)$
Adjoint	$\left[n \times \operatorname{curl} F(\tau)\right]_j^j = -\sqrt{j(j+1)}\,\Delta Z_j(T-\tau)$	$\left[n \times \operatorname{curl} F(\tau)\right]_j^j - \frac{j}{a}F_j^j(\tau) = \sqrt{\frac{j}{j+1}}\Delta G_j^{(i)}(T-\tau)$

with the inhomogeneous initial condition

$$\boldsymbol{F}|_{t=0} = \boldsymbol{F}^0 \quad \text{in } \mathcal{S} \tag{211}$$

and an appropriate boundary condition chosen from the set of boundary conditions summarized for convenience in *Table 2.*

10 Sensitivity Analysis for CHAMP Magnetic Data

The forward and adjoint solutions are now computed for the 2001-CHAMP data (see Sect. 7) with spherical-harmonic cut-off degree $j_{\max} = 4$ and time step $\Delta t = 1\,\text{h}$. The sensitivity analysis of the data will be performed with respect to two different conductivity models: a three-layer, 1-D conductivity model and a two-layer, 2-D conductivity model. For each case, the approximation error of the adjoint sensitivity method is first investigated and then the conjugate gradient method is run to search for an optimal conductivity model by adjusting the Z component of CHAMP data in a least-squares sense.

10.1 Brute-Force Sensitivities

Sensitivities generated with the adjoint sensitivity method (ASM), called hereafter as the adjoint sensitivities, will be compared to those generated by direct numerical differentiation of the misfit, the so-called brute-force method (BFM) (e.g., Bevington, 1969), in which the partial derivative of misfit with respect to σ_m at the point $\vec{\sigma}^0$ is approximated by the second-order-accuracy-centered difference of two forward model runs:

$$\left[\frac{\partial \chi^2}{\partial \sigma_m}\right]_{\vec{\sigma}^0} \approx \frac{\chi^2\left(\sigma_1^0, \dots, \sigma_m^0 + \varepsilon, \dots, \sigma_M^0\right) - \chi^2\left(\sigma_1^0, \dots, \sigma_m^0 - \varepsilon, \dots, \sigma_M^0\right)}{2\varepsilon}, \tag{212}$$

where ε refers to a perturbation applied to the nominal value of σ_m^0.

10.2 Model Parameterization

To parameterize the electrical conductivity, the radial interval $< 0, a >$ is divided into L subintervals by the nodes $0 = R_1 < R_2 < \cdots < R_L < R_{L+1} = a$ such that the radial dependence of the electrical conductivity $\sigma(r, \vartheta)$ is approximated by piecewise constant functions:

$$\sigma(r, \vartheta) = \sigma_\ell(\vartheta), \qquad R_\ell \le r \le R_{\ell+1}, \tag{213}$$

where $\sigma_\ell(\vartheta)$ for a given layer $\ell = 1, \cdots, L$ does not depend on the radial coordinate r. Moreover, let $\sigma_\ell(\vartheta)$ be parameterized by the zonal scalar spherical harmonics $Y_j(\vartheta)$. As a result, the logarithm of the electrical conductivity is considered in the form

$$\log \sigma(r, \vartheta; \vec{\sigma}) = \sqrt{4\pi} \sum_{\ell=1}^{L} \sum_{j=0}^{J} \sigma_{\ell j} \xi_\ell(r) Y_j(\vartheta), \tag{214}$$

where $\xi_\ell(r)$ is equal to 1 in the interval $R_\ell \le r \le R_{\ell+1}$ and 0 elsewhere. The number of conductivity parameters $\sigma_{\ell j}$, that is, the size of conductivity parameter vector $\vec{\sigma}$, is $M = L(J+1)$.

10.3 Three-Layer, 1-D Conductivity Model

Consider a 1-D conducting sphere $\mathcal{B}$ consisting of the lithosphere, the upper mantle (UM), the upper (ULM), and lower (LLM) parts of the lower mantle, and the core. The interfaces between the conductivity layers are kept fixed at depths of 220, 670, 1,500 and 2,890 km, respectively. The conductivities of the lithosphere and the core are 0.001 and 10,000 S/m, respectively, and fixed at these values for all computation runs, hence the number of conductivity parameters $\sigma_{\ell 0}$ is $L = 3$. The nominal values of the conductivity parameters are $\sigma_{10}^0 = 1$ (hence $\sigma_{\mathrm{LLM}} = 10$ S/m), $\sigma_{20}^0 = 0$ ($\sigma_{\mathrm{ULM}} = 1$ S/m), and $\sigma_{30}^0 = -1$ ($\sigma_{\mathrm{UM}} = 0.1$ S/m).

10.3.1 Sensitivity Comparison

The results of the sensitivity tests computed for the three-layer, 1-D conductivity model are summarized in ➋ *Fig. 7*, where the top panels show the misfit χ^2 as a function of one conductivity parameter $\sigma_{\ell 0}$, with the other two equal to the nominal values. The bottom panels compare the derivatives of the misfit obtained by the ASM with the BFM. From these results, two conclusions can be drawn. First, the differences between the derivatives of the misfit obtained by the ASM and BFM (the dashed lines in the bottom panels) are about one order (for σ_{30}) and at least two orders (for σ_{10} and σ_{20}) of magnitude smaller than the derivatives themselves, which justifies the validity of the ASM. The differences between the adjoint and brute-force sensitivities are caused by the approximation error in the time numerical differentiation (84). This error can be reduced by low-pass filtering of CHAMP time series (Martinec and Velímský, 2009).

Second, both the top and bottom panels show that the misfit χ^2 is most sensitive to the conductivity changes in the upper mantle and decreases with increasing depth of the conductivity layer, being least sensitive to conductivity changes in the lower part of the lower mantle.

10.3.2 Conjugate Gradient Inversion

The sensitivity results in ➋ *Fig. 7* are encouraging with regards to the solution of the inverse problem for a 1-D mantle conductivity structure. The conjugate gradient (CG) minimization with bracketing and line searching is employed using Brent's method with derivatives

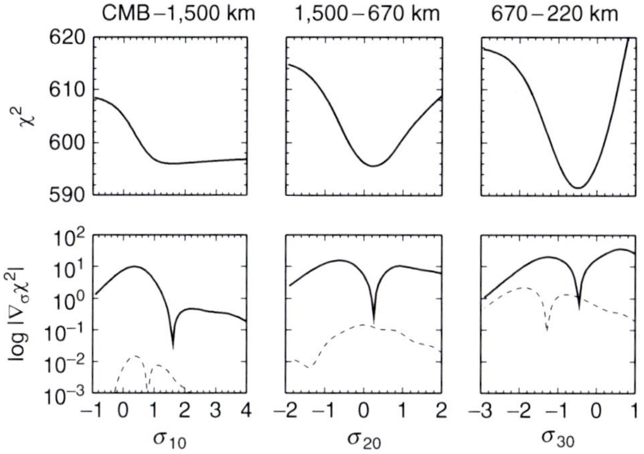

Fig. 7
The misfit χ^2 (*top panels*) and the magnitude of its sensitivities $\nabla_{\vec{\sigma}}\chi^2$ (*bottom panels*) as functions of the conductivity parameters $\sigma_{\ell 0}$ for the three-layer, 1-D conductivity model consisting of the lower and upper parts of the lower mantle ($\ell = 1, 2$) and the upper mantle ($\ell = 3$). Two panels in a column show a cross-section through the respective hypersurface χ^2 and $|\nabla_{\vec{\sigma}}\chi^2|$ in the 3-D parameter space along one parameter, while the other two model parameters are kept fixed and equal to nominal values $\vec{\sigma}^0 = (2, 0, -1)$. The adjoint sensitivities computed for $\Delta = 1$ h (the *solid lines* in the *bottom panels*) are compared with the brute-force sensitivities ($\varepsilon = 0.01$) and their differences are shown (*the dashed lines*)

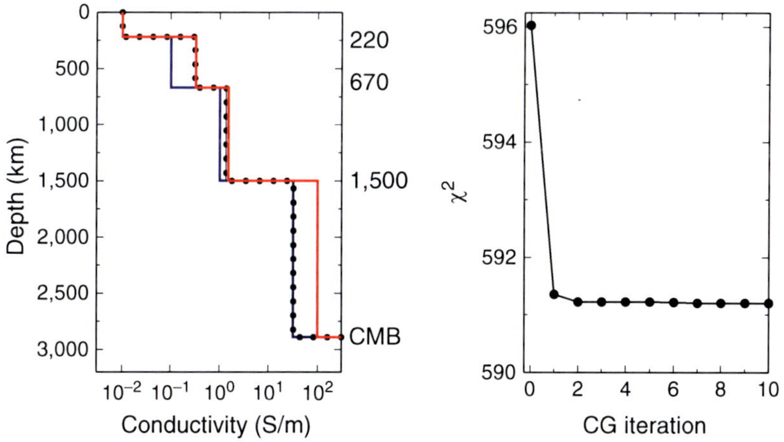

Fig. 8
Three-layer, 1-D conductivity model (*left panel*) best fitting the 2001-CHAMP data (*red line*), the starting model for the CG minimization (*blue line*) and the model after the first iteration (*dotted line*). The right panel shows the misfit χ^2 as a function of CG iterations

(Press et al., 1992, Sect. 10.3) obtained by the ASM. The inverse problem is solved for the three parameters $\sigma_{\ell 0}$, with starting values equal to (1.5, 0, −1).

Figure 8 shows the results of the inversion, where the left panel displays the conductivity structure in the three-layer mantle and the right panel the misfit χ^2 as a function of the

CG iterations. The blue line shows the starting model of the CG minimization, the dotted line the model after the first iteration and the red line the model after ten iterations. As expected from the sensitivity tests, the minimization first modifies the conductivities of the UM and ULM, to which the misfit χ^2 is the most sensitive. When the UM and ULM conductivities are improved, the CG minimization also changes the LLM conductivity. The optimal values of the conductivity parameters after ten iterations are $(\sigma_{10}, \sigma_{20}, \sigma_{30}) = (1.990, 0.186, -0.501)$. This corresponds to the conductivities $\sigma_{\mathrm{ULM}} = 1.53$ S/m and $\sigma_{\mathrm{UM}} = 0.32$ S/m for ULM and UM, which are considered to be well resolved, while the conductivity $\sigma_{\mathrm{LLM}} = 97.8$ S/m should be treated with some reservation, because of its poor resolution. A CHAMP time series longer than 1 year would be necessary to increase the sensitivity of CHAMP data to the LLM conductivity.

10.4 Two-Layer, 2-D Conductivity Model

10.4.1 Sensitivity Comparison

The adjoint sensitivities are now computed for the 2-D conductivity model, again consisting of the lithosphere, the upper mantle, and the upper and lower parts of the lower mantle, with the interfaces at depths of 220, 670, 1,500, and 2,890 km, respectively. The conductivities of the UM and ULM are now considered to be ϑ-dependent, such that the cut-off degree J in the conductivity parameterization (214) is equal to $J = 1$. The conductivity of the lithosphere is again fixed to 0.001 S/m. Because of the rather poor resolution of the LLM conductivity, this conductivity is chosen to be equal to the optimal value obtained by the CG minimization, that is, 97.8 S/m, and is kept fixed throughout the sensitivity tests and subsequent inversion. Complementary to the sensitivity tests for the zonal coefficients $\sigma_{\ell 0}$ shown in ➋ *Fig. 7*, the sensitivity tests for non-zonal coefficients $\sigma_{\ell 1}$ of the ULM ($\ell = 1$) and UM ($\ell = 2$) are now carried out in a manner similar to that applied in ➋ Sect. 10.3.1, with the same nominal values for the zonal coefficients $\sigma_{\ell 0}$ and $\sigma_{11}^0 = \sigma_{21}^0 = 0$. The forward and adjoint solutions are again computed for the 2001-CHAMP data (see ➋ Sect. 7) with spherical-harmonic cut-off degree $j_{\max} = 4$ and time step $\Delta t = 1$ h. The earth model is again divided into 40 finite-element layers with layer thicknesses increasing with depth.

➋ *Figure 9* summarizes the results of the sensitivity tests. The top panels show the misfit χ^2 as a function of the parameters σ_{11} and σ_{21}, where only one conductivity parameter is varied and the other to zero. The bottom panels compare the derivatives of the misfit obtained by the ASM and the BFM. It can be seen that the adjoint sensitivities show very good agreement with the brute-force results, with differences not exceeding 0.01% of the magnitude of the sensitivities themselves. Moreover, the sensitivities to latitudinal dependency of conductivity are significant, again more pronounced in the upper mantle than in the lower mantle. This tells that the CHAMP data are capable of revealing lateral variations of conductivity in the upper and lower mantle.

10.4.2 Conjugate Gradient Inversion

The sensitivity results in ➋ *Fig. 9* are encouraging to attempt to solve the inverse problem for lateral variations of conductivity in the mantle. For this purpose, the CG minimization with

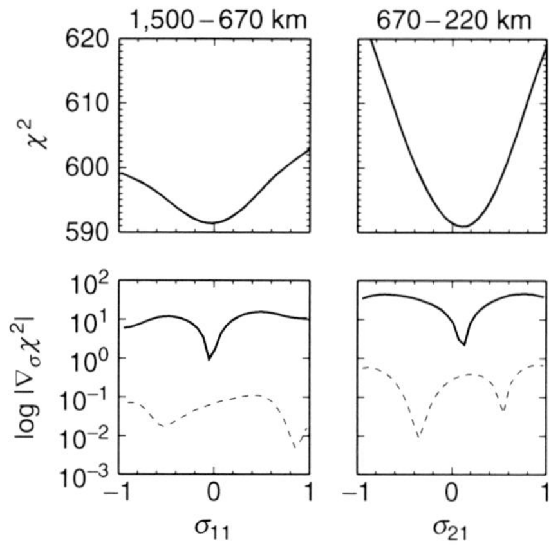

Fig. 9

As for Fig. 7, but with respect to the conductivity parameters $\sigma_{\ell 1}$ of the latitudinally dependent conductivities of the upper part of the lower mantle (ℓ = 1) and the upper mantle (ℓ = 2). The nominal values of the conductivity parameters $(\sigma^0_{\ell 0}, \sigma^0_{\ell 1})$ = (0, 0, −1, 0). The results apply to conductivities of the lithosphere, the lower part of the lower mantle and the core equal to 0.001, 97.8, and 10^4 S/m, respectively

derivatives obtained by the ASM is again employed. The inverse problem is solved for four parameters, $\sigma_{\ell 0}$ and $\sigma_{\ell 1}$, $\ell = 1, 2$. The starting values of $\sigma_{\ell 0}$ are the nominal values of the three-layer, 1-D conductivity model (see Sect. 10.3), while the values of $\sigma_{\ell 1}$ are set equal to zero at the start of minimization.

The results of the inversion are summarized in *Fig. 10*, where the left and center panels show the conductivity structure in the ULM and UM, while the right panel shows the misfit χ^2 as a function of CG iterations. The blue lines show the starting model of minimization, the dotted lines the model of minimization after the first iteration and the red lines the final model of minimization after eight iterations. These models are compared with the optimal three-layer, 1-D conductivity model (black lines) found in Sect. 10.3.2. Again, as indicated by the sensitivity tests, the minimization, at the first stage, adjusts the conductivity in the upper mantle, to which the misfit χ^2 is the most sensitive, and then varies the ULM conductivity, to which the misfit is less sensitive. The optimal values of the conductivity parameters after eight iterations are $(\sigma_{10}, \sigma_{11}, \sigma_{20}, \sigma_{21}) = (0.192, -0.008, -0.476, 0.106)$. It is concluded that the mantle conductivity variations in the latitudinal direction reach about 20% of the mean value in the upper mantle, and about 4% in the upper part of the lower mantle. Comparing the optimal values of the zonal coefficients σ_{10} and σ_{20} with those found in Sect. 10.3.2 for a 1-D conductivity model, it is concluded that the averaged optimal 2-D conductivity structure closely approaches the optimal 1-D structure. This is also indicated in *Fig. 10*, where the final 2-D conductivity profile (red lines) intersects the optimal 1-D conductivity profile (black lines) at the magnetic equator.

11 Conclusions

This chapter has been motivated by efforts to give a detailed presentation of the advanced mathematical methods available for interpreting the time series of CHAMP magnetic data such that

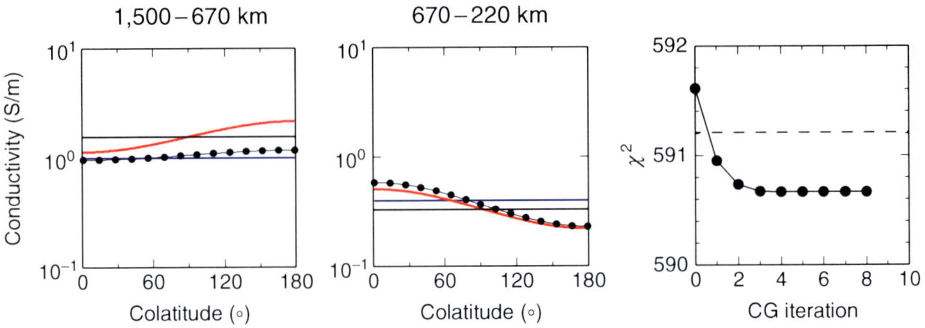

Fig. 10

Two-layer, latitudinally dependent conductivity model of the upper part of the lower mantle and the upper mantle (*left* and *middle panels*). The model best fitting the 2001-CHAMP data (*red lines*), the starting model for the CG minimization (*blue lines*) and the model after the first iteration (*dotted line*) are compared to the best 1-D conductivity model from Fig. 8 (*black lines*). The *right panel* shows the misfit χ^2 as a function of the number of CG iterations, the *dashed line* shows the misfit χ^2 for the best 1-D conductivity model

the complete time series, not only their parts, can be considered in forward and inverse modeling and still be computationally feasible. It turned out that these criteria are satisfied by highly efficient methods of forward and adjoint sensitivity analysis that are numerically based on the time-domain, spectral finite-element method. This has been demonstrated for the year 2001 CHAMP time series with a time step of 1 h. To apply the forward and adjoint sensitivity methods to longer time series is straightforward, leading to memory and computational time requirements that are linear with respect to the number of time steps undertaken. The analysis of the complete, more than 8-year long, CHAMP time series is ongoing with the particular objective of determining the lower-mantle conductivity.

The achievement of the present approach is its ability to use satellite data directly, without continuing them from the satellite altitude to the ground level or without decomposing them into the exciting and induced parts by spherical harmonic analysis. This fact is demonstrated for a 2-D configuration, for which the electrical conductivity and the external sources of the electromagnetic variations are axisymmetrically distributed and for which the external current excitation is transient, as for a magnetic storm. The 2-D case corresponds to the situation where vector magnetic data along each track of a satellite, such as the CHAMP satellite, is used. The present approach can be extended to the transient electromagnetic induction in a 3-D heterogeneous sphere if the signals from multiple satellites, simultaneously supplemented by ground-based magnetic observations, are available in the future.

The presented sensitivity analysis has shown that the 2001-CHAMP data are clearly sensitive to latitudinal variations in mantle conductivity. This result suggests the need to modify the forward and adjoint methods for an axisymmetric distribution of mantle conductivity to the case where the CHAMP data will only be considered over particular areas above the Earth's surface, for instance, the Pacific Ocean, allowing the study of how latitudinal variations in conductivity differs from region to region. This procedure would enable one to find not only conductivity variations in the latitudinal direction, but also longitudinally. This idea warrants further investigation, because it belongs to the category of problems related to data assimilation and methods of constrained minimization can be applied. Similar methods can also be applied to the assimilation of the recordings at permanent geomagnetic observatories into the conductivity models derived from satellite observations.

Acknowledgments

The author thanks Kevin Fleming for his comments on the manuscript. The author acknowledges support from the Grant Agency of the Czech Republic through Grant No. 205/09/0546.

Appendix: Zonal Scalar and Vector Spherical Harmonics

In this section, we define the zonal scalar and vector spherical harmonics, introduce their orthonormality properties and give some other relations. All considerations follow the book by Varshalovich et al. (1989), which is referenced in the following.

The zonal scalar spherical harmonics $Y_j(\vartheta)$ can be defined in terms of the Legendre polynomials $P_j(\cos\vartheta)$ of degree j (ibid., p. 134, Eq. 6):

$$Y_j(\vartheta) := \sqrt{\frac{2j+1}{4\pi}} P_j(\cos\vartheta), \tag{215}$$

where $j = 0, 1, \ldots$. The orthogonality property of the Legendre polynomials over the interval $0 \le \vartheta \le \pi$ (ibid., p. 149, Eq. 10) results in the orthonormality property of the zonal scalar spherical harmonics $Y_j(\vartheta)$ over the full solid angle $(0 \le \vartheta \le \pi, 0 \le \varphi < 2\pi)$:

$$\int_{\varphi=0}^{2\pi} \int_{\vartheta=0}^{\pi} Y_{j_1}(\vartheta) Y_{j_2}(\vartheta) \sin\vartheta d\vartheta d\varphi = \delta_{j_1 j_2}, \tag{216}$$

where δ_{ij} stands for the Kronecker delta symbol. Note that the integration over longitude φ can be performed analytically, resulting in the multiplication by a factor of 2π. However, the form of the double integration will be kept since it is consistent with surface integrals considered in the main text.

The zonal vector spherical harmonics $\boldsymbol{Y}_j^\ell(\vartheta)$, $j = 0, 1, \ldots, \ell = j \pm 1, j$, can be defined (see also ➊ Chaps. 18 and ➊ 19) via their polar components (ibid., p. 211, Eq. (10); pp. 213–214, Eqs. 25–27):

$$\begin{aligned}
\sqrt{j(2j+1)}\, \boldsymbol{Y}_j^{j-1}(\vartheta) &= j\, Y_j(\vartheta)\, \boldsymbol{e}_r + \frac{\partial Y_j(\vartheta)}{\partial\vartheta} \boldsymbol{e}_\vartheta, \\
\sqrt{(j+1)(2j+1)}\, \boldsymbol{Y}_j^{j+1}(\vartheta) &= -(j+1)\, Y_j(\vartheta)\, \boldsymbol{e}_r + \frac{\partial Y_j(\vartheta)}{\partial\vartheta} \boldsymbol{e}_\vartheta, \\
\sqrt{j(j+1)}\, \boldsymbol{Y}_j^{j}(\vartheta) &= -i\, \frac{\partial Y_j(\vartheta)}{\partial\vartheta} \boldsymbol{e}_\varphi,
\end{aligned} \tag{217}$$

where $i = \sqrt{-1}$, and $\boldsymbol{e}_r$, $\boldsymbol{e}_\vartheta$, and $\boldsymbol{e}_\varphi$ are spherical base vectors. The vector functions $\boldsymbol{Y}_j^{j\pm1}(\vartheta)$ are called the zonal spheroidal vector spherical harmonics and $\boldsymbol{Y}_j^j(\vartheta)$ are the zonal toroidal vector spherical harmonics. A further useful form of the zonal toroidal vector spherical harmonics can be obtained considering $\partial Y_j(\vartheta)/\partial\vartheta = \sqrt{j(j+1)} P_{j1}(\cos\vartheta)$ (ibid., p. 146, Eq. 5), where $P_{j1}(\cos\vartheta)$ is fully normalized associated Legendre functions of order $m = 1$:

$$\boldsymbol{Y}_j^j(\vartheta) = -i\, P_{j1}(\cos\vartheta) \boldsymbol{e}_\varphi. \tag{218}$$

The orthonormality property of the spherical base vectors and the zonal scalar spherical harmonics combine to give the orthonormality property of the zonal vector spherical harmonics (ibid., p. 227, Eq. 121):

$$\int_{\varphi=0}^{2\pi}\int_{\vartheta=0}^{2\pi} \boldsymbol{Y}_{j_1}^{\ell_1}(\vartheta)\cdot\left[\boldsymbol{Y}_{j_2}^{\ell_2}(\vartheta)\right]^{*}\sin\vartheta\, d\vartheta\, d\varphi = \delta_{j_1 j_2}\delta_{\ell_1\ell_2}, \tag{219}$$

where the dot stands for the scalar product of vectors and the asterisk denotes complex conjugation.

Since both the zonal scalar spherical harmonics and the spherical base vectors are real functions, Eq. 217 shows that the spheroidal vector harmonics $\boldsymbol{Y}_j^{j\pm1}(\vartheta)$ are real functions, whereas the toroidal vector harmonics $\boldsymbol{Y}_j^{j}(\vartheta)$ are pure imaginary functions. To avoid complex arithmetics, $\boldsymbol{Y}_j^{j}(\vartheta)$ is redefined in such a way that they become real functions (of colatitude ϑ):

$$\boldsymbol{Y}_j^{j}(\vartheta) := P_{j1}(\cos\vartheta)\boldsymbol{e}_\varphi. \tag{220}$$

At this stage, a remark about this step is required. To avoid additional notation, the same notation is used for the real and complex versions of $\boldsymbol{Y}_j^{j}(\vartheta)$, since the real version of $\boldsymbol{Y}_j^{j}(\vartheta)$ is exclusively used throughout this chapter. It is in contrast to Martinec (1997), Martinec et al. (2003), and Martinec and McCreadie (2004), where the complex functions $\boldsymbol{Y}_j^{j}(\vartheta)$, defined by Eq. 218, have been used. However, the re-definition (220) only makes sense for studying a phenomenon with an axisymmetric geometry. For a more complex phenomenon, the original definition (218) is to be used.

The orthonormality property (219) for the real zonal vector spherical harmonics now reads as

$$\int_{\varphi=0}^{2\pi}\int_{\vartheta=0}^{2\pi} \boldsymbol{Y}_{j_1}^{\ell_1}(\vartheta)\cdot\boldsymbol{Y}_{j_2}^{\ell_2}(\vartheta)\sin\vartheta\, d\vartheta\, d\varphi = \delta_{j_1 j_2}\delta_{\ell_1\ell_2}. \tag{221}$$

The formulae for the scalar and vector products of the radial unit vector $\boldsymbol{e}_r$ and the zonal vector spherical harmonics $\boldsymbol{Y}_j^{\ell}(\vartheta)$ follow from Eq. 217:

$$\begin{aligned}
\boldsymbol{e}_r\cdot\boldsymbol{Y}_j^{j-1}(\vartheta) &= \sqrt{\frac{j}{2j+1}}\,Y_j(\vartheta),\\
\boldsymbol{e}_r\cdot\boldsymbol{Y}_j^{j+1}(\vartheta) &= -\sqrt{\frac{j+1}{2j+1}}\,Y_j(\vartheta),\\
\boldsymbol{e}_r\cdot\boldsymbol{Y}_j^{j}(\vartheta) &= 0,
\end{aligned} \tag{222}$$

and

$$\begin{aligned}
\boldsymbol{e}_r\times\boldsymbol{Y}_j^{j-1}(\vartheta) &= \sqrt{\frac{j+1}{2j+1}}\,\boldsymbol{Y}_j^{j}(\vartheta),\\
\boldsymbol{e}_r\times\boldsymbol{Y}_j^{j+1}(\vartheta) &= \sqrt{\frac{j}{2j+1}}\,\boldsymbol{Y}_j^{j}(\vartheta),\\
\boldsymbol{e}_r\times\boldsymbol{Y}_j^{j}(\vartheta) &= -\sqrt{\frac{j+1}{2j+1}}\,\boldsymbol{Y}_j^{j-1}(\vartheta)-\sqrt{\frac{j}{2j+1}}\,\boldsymbol{Y}_j^{j+1}(\vartheta).
\end{aligned} \tag{223}$$

Any vector $\boldsymbol{A}(\vartheta)$ which depends on colatitude ϑ and which is square-integrable over the interval $0 \le \vartheta \le \pi$ may be expanded in a series of the zonal vector spherical harmonics, that is,

$$\boldsymbol{A}(\vartheta) = \sum_{j=0}^{\infty} \sum_{\ell=|j-1|}^{j+1} A_j^\ell \boldsymbol{Y}_j^\ell(\vartheta) \tag{224}$$

with the expansion coefficients given by

$$A_j^\ell = \int_{\varphi=0}^{2\pi} \int_{\vartheta=0}^{2\pi} \boldsymbol{A}(\vartheta) \cdot \boldsymbol{Y}_j^\ell(\vartheta) \sin\vartheta d\vartheta d\varphi. \tag{225}$$

The curl of vector $\boldsymbol{A}(r,\vartheta)$ is then

$$\operatorname{curl} \boldsymbol{A} = \sum_{j=1}^{\infty} \sum_{\ell=j-1}^{j+1} R_j^\ell(r) \boldsymbol{Y}_j^\ell(\vartheta), \tag{226}$$

where (ibid, p. 217, Eq. 55)

$$\begin{aligned} R_j^{j-1}(r) &= -\sqrt{\frac{j+1}{2j+1}}\left(\frac{d}{dr} + \frac{j+1}{r}\right) A_j^j(r), \\ R_j^{j+1}(r) &= -\sqrt{\frac{j}{2j+1}}\left(\frac{d}{dr} - \frac{j}{r}\right) A_j^j(r), \\ R_j^j(r) &= \sqrt{\frac{j+1}{2j+1}}\left(\frac{d}{dr} - \frac{j-1}{r}\right) A_j^{j-1}(r) + \sqrt{\frac{j}{2j+1}}\left(\frac{d}{dr} + \frac{j+2}{r}\right) A_j^{j+1}(r). \end{aligned} \tag{227}$$

The radial and tangential components of curl $\boldsymbol{A}$ may be evaluated as

$$\boldsymbol{e}_r \cdot \operatorname{curl} \boldsymbol{A} = -\frac{1}{r} \sum_{j=1}^{\infty} \sqrt{j(j+1)}\, A_j^j(r) Y_j(\vartheta), \tag{228}$$

and

$$\boldsymbol{e}_r \times \operatorname{curl} \boldsymbol{A} = -\sum_{j=1}^{\infty} \left(\frac{d}{dr} + \frac{1}{r}\right) A_j^j(r) \boldsymbol{Y}_j^j(\vartheta) + \sum_{j=1}^{\infty} R_j^j(r) \left(\boldsymbol{e}_r \times \boldsymbol{Y}_j^j(\vartheta)\right). \tag{229}$$

In particular, for a toroidal vector $\boldsymbol{A}(\vartheta)$, the coefficients $A_j^{j\pm1}(r) = 0$ and Eq. 229 reduces to

$$\boldsymbol{e}_r \times \operatorname{curl} \boldsymbol{A} = -\sum_{j=1}^{\infty} \left(\frac{d}{dr} + \frac{1}{r}\right) A_j^j(r) \boldsymbol{Y}_j^j(\vartheta). \tag{230}$$

References

Avdeev DB, Avdeeva AD (2006) A rigorous three-dimensional magnetotelluric inversion. PIER 62:41–48

Banks R (1969) Geomagnetic variations and the electrical conductivity of the upper mantle. Geophys J R Astr Soc 17:457–487

Banks RJ, Ainsworth JN (1992) Global induction and the spatial structure of mid-latitude geomagnetic variations. Geophys J Int 110:251–266

Bevington PR (1969) Data reduction and error analysis for the physical sciences. McGraw-Hill, New York

Cacuci DG (2003) Sensitivity and uncertainty analysis. Volume I. Theory. Chapman & Hall/CRC, Boca Raton, FL

Constable S, Constable C (2004) Observing geomagnetic induction in magnetic satellite measurements and associated implications for mantle conductivity. Geochem Geophys Geosyst 5:Q01006. doi:10.1029/2003GC000634

Daglis IA, Thorne RM, Baumjohann W, Orsini S (1999) The terrestrial ring current: origin, formation and decay. Rev Geophys 37:407–438

Didwall EM (1984) The electrical conductivity of the upper mantle as estimated from satellite magnetic field data. J Geophys Res 89:537–542

Dorn O, Bertete-Aquirre H, Berryman JG, Papanicolaou GC (1999) A nonlinear inversion method for 3-D electromagnetic imaging using adjoint fields. Inverse Prob 15:1523–1558

Eckhardt D, Larner K, Madden T (1963) Long periodic magnetic fluctuations and mantle conductivity estimates. J Geophys Res 68:6279–6286

Everett ME, Martinec Z (2003) Spatiotemporal response of a conducting sphere under simulated geomagnetic storm conditions. Phys Earth Planet Inter 138:163–181

Everett ME, Schultz A (1996) Geomagnetic induction in a heterogeneous sphere: Az-imuthally symmetric test computations and the response of an undulating 660-km discontinuity. J Geophys Res 101:2765–2783

Fainberg EB, Kuvshinov AV, Singer BSh (1990) Electromagnetic induction in a spherical Earth with non-uniform oceans and continents in electric contact with the underlying medium – I. Theory, method and example. Geophys J Int 102:273–281

Farquharson CG, Oldenburg DW (1996) Approximate sensitivities for the electromagnetic inverse problem. Geophys J Int 126:235–252

Hamano Y (2002) A new time-domain approach for the electromagnetic induction problem in a three-dimensional heterogeneous earth. Geophys J Int 150:753–769

Hultqvist B (1973) Perturbations of the geomagnetic field. In: Egeland A, Holter O, Omholt A (eds) Cosmical geophysics. Universitetsforlaget, Oslo, pp 193–201

Jupp DLB, Vozoff K (1977) Two-dimensional magnetotelluric inversion. Geophys J R Astr Soc 50:333–352

Kelbert A, Egbert GD, Schultz A (2008) Nonlinear conjugate gradient inversion for global EM induction: resolution studies. Geophys J Int 173:365–381

Kivelson MG, Russell CT (1995) Introduction to space physics, Cambridge University Press, Cambridge.

Korte M, Constable S, Constable C (2003) Separation of external magnetic signal for induction studies. In: Reigber Ch, Lühr H, Schwintzer P (eds) First CHAMP mission results for gravity, magnetic and atmospheric studies. Springer, Berlin, pp 315–320

Křížek M, Neittaanmäki P (1990) Finite element approximation of variational problems and applications. Longmann Scientific and Technical/Wiley, New York

Kuvshinov A, Olsen N (2006) A global model of mantle conductivity derived from 5 years of CHAMP, Ørsted, and SAC-C magnetic data. Geophys Res Lett 33:L18301. doi:10.1029/2006GL027083

Kuvshinov AV, Avdeev DB, Pankratov OV (1999a) Global induction by Sq and Dst sources in the presence of oceans: bimodal solutions for non-uniform spherical surface shells above radially symmetric earth models in comparison to observations. Geophys J Int 137:630–650

Kuvshinov AV, Avdeev DB, Pankratov OV, Golyshev SA (1999b) Modelling electromagnetic fields in 3-D spherical earth using fast integral equation approach. Expanded abstract of the 2nd International Symposium on 3-D Electromagnetics, pp 84–88

Lanczos C (1961) Linear differential operators. Van Nostrand, Princeton, NJ

Langel RA, Estes RH (1985a) Large-scale, near-field magnetic fields from external sources and the corresponding induced internal field. J Geophys Res 90:2487–2494

Langel RA, Estes RH (1985b) The near-Earth magnetic field at 1980 determined from Magsat data. J Geophys Res 90:2495–2510

Langel RA, Sabaka TJ, Baldwin RT, Conrad JA (1996) The near-Earth magnetic field from magnetospheric and quiet-day ionospheric sources and how it is modeled. Phys Earth Planet Inter 98:235–268

Madden TM, Mackie RL (1989) Three-dimensional magnetotelluric modelling and inversion. Proc Inst Electron Electric Eng 77:318–333

Marchuk GI (1995) Adjoint equations and analysis of complex systems. Kluwer, Dordrecht

Martinec Z (1989) Program to calculate the spectral harmonic expansion coefficients of the two scalar fields product. Comp Phys Commun 54:177–182

Martinec Z (1997) Spectral–finite element approach to two-dimensional electromagnetic induction in a spherical earth. Geophys J Int 130:583–594

Martinec Z (1999) Spectral–finite element approach to three-dimensional electromagnetic induction in a spherical earth. Geophys J Int 136:229–250

Martinec Z, McCreadie H (2004) Electromagnetic induction modelling based on satellite magnetic vector data. Geophys J Int 157:1045–1060

Martinec Z, Velímský J (2009) The adjoint sensitivity method of global electromagnetic induction for CHAMP magnetic data. Geophys J Int 179:1372–1396. doi: 10.1111/j.1365-246X.2009.04356.x

Martinec Z, Everett ME, Velímský J (2003) Time-domain, spectral-finite element approach to transient two-dimensional geomagnetic induction in a spherical heterogeneous earth. Geophys J Int 155:33–43

McGillivray PR, Oldenburg DW (1990) Methods for calculating Fréchet derivatives and sensitivities for the non-linear inverse problems: a comparative study. Geophys Prosp 38:499–524

McGillivray PR, Oldenburg DW, Ellis RG, Habashy TM (1994) Calculation of sensitivities for the frequency-domain electromagnetic problem. Geophys J Int 116:1–4

Morse PW, Feshbach H (1953) Methods of theoretical physics. McGraw-Hill, New York

Newman GA, Alumbaugh DL (1997) Three-dimensional massively parallel electromagnetic inversion – I. Theory Geophys J Int 128:345–354

Newman GA, Alumbaugh DL (2000) Three-dimensional magnetotelluric inversion using non-linear conjugate on induction effects of geomagnetic daily variations from equatorial gradients. Geophys J Int 140:410–424

Oldenburg DW (1990) Inversion of electromagnetic data: an overview of new techniques. Surv Geophys 11:231–270

Olsen N (1999) Induction studies with satellite data. Surv Geophys 20:309–340

Olsen N, Sabaka TJ, Lowes F (2005) New parameterization of external and induced fields in geomagnetic field modeling, and a candidate model for IGRF 2005. Earth Planets Space 57:1141–1149

Olsen N, Lühr H, Sabaka TJ, Mandea M, Rother M, Toffner-Clausen L, Choi S (2006a) CHAOS – a model of the Earth's magnetic field derived from CHAMP, Ørsted & SAC-C magnetic satellite data. Geophys J Int 166:67–75

Olsen N, Haagmans R, Sabaka T, Kuvshinov A, Maus S, Purucker M, Rother M, Lesur V, Mandea M (2006b) The swarm end-to-end mission simulator study: separation of the various contributions to earths magnetic field using synthetic data. Earth Planets Space 58:359–370

Oraevsky VN, Rotanova NM, Semenov VYu, Bondar TN, Abramova DYu (1993) Magnetovariational sounding of the Earth using observatory and MAGSAT satellite data. Phys Earth Planet Inter 78:119–130

Orszag SA (1970) Transform method for the calculation of vector-coupled sums: application to the spectral form of the vorticity equation. J Atmos Sci 27:890

Pěč K, Martinec Z (1986) Spectral theory of electromagnetic induction in a radially and laterally inhomogeneous Earth. Studia Geoph et Geod 30:345–355

Petzold L, Li ST, Cao Y, Serban R (2006) Sensitivity analysis of differential-algebraic equations and partial differential equations. Comp Chem Eng 30:1553–1559

Press WH, Teukolsky SA, Vetterling WT, Flannery BP (1992) Numerical recipes in Fortran. The Art of Scientific Computing. Cambridge University Press, Cambridge

Rodi WL (1976) A technique for improving the accuracy of finite element solutions of MT data. Geophys J R Astr Soc 44:483–506

Rodi WL, Mackie RL (2001) Nonlinear conjugate gradients algorithm for 2-D magnetotel-luric inversion. Geophysics 66:174–187

Sandu A, Daescu DN, Carmichael GR (2003) Direct and adjoint sensitivity analysis of chemical kinetic systems with KPP: I–theory and software tools. Atmos Environ 37:5083–5096

Sandu A, Daescu DN, Carmichael GR, Chai T (2005) Adjoint sensitivity analysis of regional air quality models. J Comput Phys 204:222–252

Schultz A, Larsen JC (1987) On the electrical conductivity of the mid-mantle, I, Calculation of equivalent scalar magnetotelluric response functions. Geophys J R Astr Soc 88:733–761

Schultz A, Larsen JC (1990) On the electrical conductivity of the mid-mantle, II, Delineation of heterogeneity by application of extremal inverse solutions. Geophys J Int 101:565–580

Stratton JA (1941) Electromagnetic theory. Wiley, New Jersey (reissued in 2007)

Tarantola A (2005) Inverse problem theory and methods for model parameter estimation. SIAM, Philadelphia, PA

Tarits P, Grammatica N (2000) Electromagnetic induction effects by the solar quiet magnetic field at satellite altitude. Geophys Res Lett 27:4009–4012

Uyeshima M, Schultz A (2000) Geoelectromagnetic induction in a heterogeneous sphere: a new three-dimensional forward solver using a conservative staggered-grid finite difference method. Geophys J Int 140:636–650

Varshalovich DA, Moskalev AN, Khersonskii VK (1989) Quantum theory of angular momentum. World Scientific, Singapore

Velímský J, Martinec Z (2005) Time-domain, spherical harmonic-finite element approach to transient three-dimensional geomagnetic induction in a spherical heterogeneous Earth. Geophys J Int 161:81–101

Velímský J, Martinec Z, Everett ME (2006) Electrical conductivity in the Earth's mantle inferred from CHAMP satellite measurements – I. Data processing and 1-D inversion. Geophys J Int 166:529–542

Weaver JT (1994) Mathematical methods for geo-electromagnetic induction, research studies press. Wiley, New York

Weidelt P (1975) Inversion of two-dimensional conductivity structure. Phys Earth Planet Inter 10:282–291

Weiss CJ, Everett ME (1998) Geomagnetic induction in a heterogeneous sphere: fully three-dimensional test computations and the response of a realistic distribution of oceans and continents. Geophys J Int 135: 650–662

20 Asymptotic Models for Atmospheric Flows

Rupert Klein
FB Mathematik & Informatik, Institut für Mathematik, Freie Universität Berlin, Berlin, Germany

W. Freeden, M.Z. Nashed, T. Sonar (Eds.), *Handbook of Geomathematics*, DOI 10.1007/978-3-642-01546-5_20,

Abstract Atmospheric flows feature length and time scales from 10^{-5} to 10^{5} m and from microseconds to weeks and more. For scales above several kilometers and minutes, there is a natural scale separation induced by the atmosphere's thermal stratification together with the influences of gravity and Earth's rotation, and the fact that atmospheric flow Mach numbers are typically small. A central aim of theoretical meteorology is to understand the associated scale-specific flow phenomena, such as internal gravity waves, baroclinic instabilities, Rossby waves, cloud formation and moist convection, (anti-)cyclonic weather patterns, hurricanes, and a variety of interacting waves in the tropics. Such understanding is greatly supported by analyses of reduced sets of model equations which capture just those fluid mechanical processes that are essential for the phenomenon in question while discarding higher-order effects. Such reduced models are typically proposed on the basis of combinations of physical arguments and mathematical derivations, and are not easily understood by the meteorologically untrained. This chapter demonstrates how many well-known reduced sets of model equations for specific, scale-dependent atmospheric flow phenomena may be derived in a unified and transparent fashion from the full compressible atmospheric flow equations using standard techniques of formal asymptotics. It also discusses an example for the limitations of this approach.

➲ Sections 3–5 of this contribution are a recompilation of the author's more comprehensive article "Scale-dependent models for atmospheric flows", Annual Reviews of Fluid Mechanics, 42 (2010).

1 Introduction

Atmospheric flows feature length and time scales from 10^{-5} to 10^{5} m and from microseconds to weeks and more, and many different physical processes persistently interact across these scales. A central aim of theoretical meteorology is to understand the associated scale-specific flow phenomena, such as internal gravity waves, baroclinic instabilities, Rossby waves, cloud formation and moist convection, (anti-)cyclonic weather patterns, hurricanes, and a variety of interacting waves in the tropics. To cope with the associated complexity, theoretical meteorologists have developed a large number of reduced mathematical models which capture the essence of these scale-dependent processes while consciously discarding effects that would merely amount to small, nonessential corrections. See, e.g., the texts by Pedlosky (1987), Gill (1982), Zeytounian (1990), White (2002), and Majda (2002) for detailed derivations and for discussions of the usefulness and validity of such reduced model equations.

➲ Section 2 of this contribution presents two classical examples of such reduced model equations and discusses possible lines of thought for their derivations with emphasis on the relation between *meteorological scale analysis* and *formal asymptotics*.

➲ Section 3, introduces several universal physical characteristics of Earth and its atmosphere that are associated with the atmosphere's thermal stratification, with the influences of gravity and Earth's rotation, and with the fact that atmospheric flow Mach numbers are typically small. These characteristics give rise to three nondimensional parameters, two of which are small and motivate a unified approach to meteorological modeling via formal asymptotics. In pursuing this route, a particular distinguished limit among these parameters is introduced, which leaves us with only one small reference parameter, and this is used in the sequel as the basis for asymptotic expansions.

In ➋ Sect. 4 we find that many classical reduced scale-dependent models of theoretical meteorology can in fact be rederived through formal asymptotics based on this remaining small parameter in a unified fashion directly from the full three-dimensional compressible flow equations. The distinguished limit adopted in ➋ Sect. 3 is thus justified in hindsight as one that supports a substantial part of the established meteorological theory.

The reduced models rederived in ➋ Sect. 4 are all well-known, so that this section should be considered a reorganization of a family of classical models under a common systematic mathematical framework. The models discussed are all single-scale models in the sense that they incorporate a single horizontal and a single vertical spacial scale, and a single timescale only. In contrast, ➋ Sect. 5 briefly sketches more recent developments for multiscale problems.

➋ Section 6 draws some conclusions.

2 Examples of Reduced Models, Scale Analysis, and Asymptotics

Arguments from formal asymptotic analysis are the principal tool used in this chapter to generate an overview of the multitude of reduced mathematical models for atmospheric flows. Yet, formal asymptotic expansions do have their limitations. In this section, we consider two examples of established meteorological models: the quasi-geostrophic (QG) model, which does have a straightforward derivation through formal (single-scale) asymptotics, and the class of anelastic and pseudo-incompressible *sound-proof models*, which does not. See (White 2002; Salmon 1983; Névir 2004; Oliver 2006) for discussions of alternative approaches to the construction of reduced models.

2.1 QG-Theory

A prominent example of a reduced model of theoretical meteorology is the *QG model*: (see ➋ Sect. 4.3.2 for its derivation)

$$(\partial_t + \boldsymbol{u} \cdot \nabla_\xi)\, q = 0, \tag{1}$$

where

$$\begin{aligned} q &= \nabla_\xi^2 \widetilde{\pi} + \beta\,\eta + \frac{f}{\overline{\rho}}\frac{\partial}{\partial z}\left(\frac{\overline{\rho}}{d\overline{\theta}/dz}\frac{\partial \widetilde{\pi}}{\partial z}\right), \\ \boldsymbol{u} &= \boldsymbol{k} \times \nabla_\xi \widetilde{\pi}. \end{aligned} \tag{2}$$

Here $\widetilde{\pi}, \boldsymbol{u}$ are a pressure variable and the horizontal flow velocity, $\overline{\rho}(z), \overline{\theta}(z)$ are the vertical background stratifications of density and potential temperature (the "potential temperature" is closely related to entropy: it is the temperature which a parcel of air would attain if compressed or expanded adiabatically to sea-level pressure), $\boldsymbol{\xi} = (\xi, \eta)$ are horizontal spacial coordinates, ∇_ξ denotes the horizontal components of the gradient, $\boldsymbol{k}$ is the local vertical unit vector at $(\xi, \eta = 0)$, $f = 2\boldsymbol{k} \cdot \boldsymbol{\Omega}$ is the Coriolis parameter, with $\boldsymbol{\Omega}$ the Earth rotation vector, and β is its derivative w.r.t. η.

The central dynamical variable in the QG theory is the *potential vorticity*, q, which according to $(2)_1$ is a superposition of the vertical component of the relative vorticity, $\nabla_\xi^2 \widetilde{\pi} \equiv \boldsymbol{k} \cdot \nabla_\xi \times \boldsymbol{u}$, a contribution from meridional (north–south) variations of the Coriolis parameter, and of a third term that represents the effects of vortex stretching due to the compression and extension of vertical columns of air.

The QG theory has been derived originally to explain the formation and evolution of the large-scale vortical atmospheric flow patterns that constitute the familiar high and low pressure systems seen regularly on mid latitude weather maps. It also served as the basis for Richardson's first attempt at a numerical weather forecasting, (see Lynch 2006). Its classical derivation (see, e.g., Pedlosky 1987) can be reproduced one-to-one within the formal asymptotic framework to be summarized in subsequent sections.

2.2 Sound-Proof Models

This is in contrast, e.g., to the situation with the classical derivations of reduced *sound-proof* models for atmospheric flows on much smaller scales, such as the extended anelastic models summarized by Bannon (1996), or the pseudo-incompressible model by Durran (1989). In the simplest case of an inviscid, adiabatic flow in a nonrotating frame of reference, the latter model is given by

$$\begin{aligned} \rho_\tau + \nabla\cdot(\rho\boldsymbol{v}) &= 0 \\ (\rho\boldsymbol{v})_\tau + \nabla\cdot(\rho\boldsymbol{v}\circ\boldsymbol{v}) + \overline{P}\,\nabla\widetilde{\pi} &= \rho g\,\boldsymbol{k}. \\ \nabla\cdot(\overline{P}\boldsymbol{v}) &= 0. \end{aligned} \tag{3}$$

Here $\rho, \boldsymbol{v}, \widetilde{\pi}$ are the density, flow velocity, and pressure field for this model, g is the acceleration of gravity, and

$$\overline{P}(z) \equiv \overline{\rho}(z)\overline{\Theta}(z) \tag{4}$$

is the density-weighted background stratification of potential temperature.

As shown in Davies et al. (2003) and Klein (2009), this model results formally from the full compressible flow equations for an ideal gas with constant specific heat capacities in a straightforward fashion. The point of departure are the compressible Euler equations with gravity in the form

$$\begin{aligned} \rho_t + \nabla\cdot(\rho\boldsymbol{v}) &= 0 \\ (\rho\boldsymbol{v})_t + \nabla\cdot(\rho\boldsymbol{v}\circ\boldsymbol{v}) + P\,\nabla\pi &= \rho g\,\boldsymbol{k}. \\ p_t + \boldsymbol{v}\cdot\nabla + \gamma p\nabla\cdot\boldsymbol{v} &= 0, \end{aligned} \tag{5}$$

where p is the thermodynamic pressure, and

$$P = p^{\frac{1}{\gamma}} \quad \text{and} \quad \pi = p^{\Gamma}/\Gamma \quad \text{with} \quad \Gamma = \frac{\gamma - 1}{\gamma}, \tag{6}$$

and γ the isentropic exponent. Now we rewrite $(5)_3$ as

$$P_t + \nabla\cdot(P\boldsymbol{v}) = 0, \tag{7}$$

and assume (i) that all pressure variations are systematically small so that $P_t \approx 0$, and (ii) that $P \equiv \overline{P}(z)$ corresponds to the hydrostatic pressure distribution of an atmosphere a rest. With these two assumptions, the pseudo-incompressible model in (4) emerges from (5). Close inspection of Durran's original derivation of (4) reveals that his only essential assumption consists, in fact, of small deviations of the thermodynamic pressure from a hydrostatically balanced background distribution.

Interestingly, there is no derivation of this reduced model via classical single or multiple scales asymptotic analysis unless one adopts the additional assumption of weak stratification of the potential temperature, so that

$$\overline{\Theta}(z) \equiv \frac{\overline{P}(z)}{\overline{\rho}(z)} = 1 + Ma^2 \overline{\theta}(z), \tag{8}$$

where *Ma* is the flow Mach number. The reason is that the compressible flow equations from (5) describe three processes with asymptotically different characteristic time scales as indicated in ➋ *Table 1* (Klein 2009).

In the table, $u_{\text{ref}} \sim 10 \text{ m s}^{-1}$ is a typical atmospheric flow velocity, $p_{\text{ref}} \sim 10^5$ Pa is a reference sea level pressure, $\rho_{\text{ref}} \sim 1 \text{ kg m}^{-3}$ is an associated reference density, and $h_{\text{sc}} = \frac{p_{\text{ref}}}{\rho_{\text{ref}} g} \sim 10$ km is the pressure scale height, i.e., a characteristic vertical distance over which the atmospheric pressure changes appreciably as a response to the diminishing mass of air that it has to balance against the force of gravity at higher altitudes. Notice that the quantity $\sqrt{g h_{\text{sc}}} = \sqrt{p_{\text{ref}}/\rho_{\text{ref}}}$, the speed of *external gravity waves* of the atmosphere, equals the speed of sound at reference conditions, except for a factor of $\sqrt{\gamma}$. Recall that $\overline{\Theta}(z)$ is the background stratification of the potential temperature defined in (4).

The table reveals that, generally, compressible flows under the influence of gravity feature three distinct characteristic time scales: those of advection, and of internal wave and sound propagation. The pseudo-incompressible model from (3) does not support acoustic waves due to the divergence constraint imposed on the velocity field in $(3)_3$. Yet it does feature the processes of advection and internal wave propagation. According to the preceeding discussion of characteristic time scales, the pseudo-incompressible model will thus be a two-timescale model unless the characteristic frequency of internal waves is slowed down to the time scale of advection by assuming $\frac{h_{\text{sc}}}{\overline{\Theta}} \frac{d\overline{\Theta}}{dz} = O(Ma^2)$ as $Ma \to 0$. In fact, Ogura and Phillips (1962) adopted this assumption in the derivation of their original *anelastic* model (see also ➋ Sect. 4.3.3). Yet, their model was criticized because of this assumption, because stratifications of order $Ma^2 \sim 10^{-3}$ imply potential temperature variations across the troposphere, i.e., the lower 10 – 15 km of the atmosphere, of less than 1 K, which is far too small in comparison with the observed 30 – 50 K.

Table 1

Dimensional and nondimensional inverse time scales for the three physical processes described by the inviscid, adiabatic compressible flow equations with gravity in (6)

Process	Inverse timescale	Dimensionless inverse timescale
Advection	$\dfrac{u_{\text{ref}}}{h_{\text{sc}}}$	1
Internal gravity waves	$N = \sqrt{\dfrac{g}{\overline{\Theta}} \dfrac{d\overline{\Theta}}{dz}}$	$\dfrac{1}{\gamma Ma} \sqrt{\dfrac{h_{\text{sc}}}{\overline{\Theta}} \dfrac{d\overline{\Theta}}{dz}}$
Sound propagation	$\dfrac{h_{\text{sc}}}{\sqrt{\gamma p_{\text{ref}}/\rho_{\text{ref}}}}$	$\dfrac{1}{Ma}$

Generalized *sound-proof* models proposed later (see Durran 1989; Bannon 1996, and references therein), are thus genuine two-scale models.

These observations allow us to exemplify explicitly one remark by White (2002) who takes the somewhat extreme stance that "Such expansion methods *[single-scale asymptotic expansions; explanation added by the author]* may be suspected of lending a cosmetic veneer to what is rather crude and restricted scale analysis." In fact, formal asymptotic methods when faced with a two-timescale problem will allow us to filter out the fast scale and produce a reduced balanced model that captures the remaining slower process(es) only. Faced with a three-scale problem, such as the compressible atmospheric flows discussed in this section, asymptotic multiple scales analysis will provide us with separate equations for the processes taking place on each scale, and with additional coupling constraints that guarantee the validity of the coupled model over the longest of the involved time scales. Standard techniques of asymptotic analysis will, however, *not* allow us to eliminate from a three-scale problem only the fastest scale while retaining a reduced two-scale pde model that captures the remaining two slower time scales only. In this sense, the meteorological scale analyses presented by Durran (1989) and Bannon (1996) are not equivalent to formal asymptotic approaches. Their analyses are, however, also not entirely satisfactory in the author's view, because they do not at all address the asymptotic two-scale properties of the remaining pseudo-incompressible and anelastic model equations.

Mathematically speaking, an interesting challenge remains: There are many textbooks on methods of formal asymptotic analysis, (see, e.g., Kevorkian and Cole, 1996), and there is also a multitude of examples where mathematicians have rigorously proven the validity of approximate equation systems derived with these techniques, (see, e.g., Majda 2002; Schochet 2005, and references therein). In contrast, a mathematically sound classification of the methods of meteorological scale analysis as referred to, e.g., by White (2002), and an explicit characterization of what distinguishes them from formal asymptotics seems outstanding. Similarly, rigorous proofs of validity of the resulting reduced models such as the extended anelastic and pseudo-incompressible models are still missing.

3 Dimensional Considerations

3.1 Characteristic Scales and Dimensionless Parameters

Here we borrow from Keller and Ting (1951) and Klein (2008) to argue for an inherent scale separation in large-scale atmospheric flows that will motivate the asymptotic arguments to be presented subsequently. ➋ *Table 2* lists eight universal characteristics of atmospheric motions involving the physical dimensions of length, time, mass, and temperature.

Earth's radius, its rotation rate, and the acceleration of gravity, are obviously universal. The sea level pressure is set by the mass (weight) of the atmosphere, which is essentially constant in time. The water freezing temperature is a good reference value for the large-scale, long-time averaged conditions on Earth. See also Rahmstorf et al. (2004) for the related paleo climatic record. The equator-to-pole potential temperature difference is maintained by the inhomogeneous irradiation from the sun and its magnitude appears to be stable over very long periods of time. An average vertical potential temperature difference across the troposphere, i.e., across the lower 10–15 km of the atmosphere, is of the same order of magnitude as the equator-to-pole temperature difference (see, e.g., Frierson 2008, and references therein). Finally the dry air gas

Table 2
Universal characteristics of atmospheric motions

Earth's radius	$a = 6 \cdot 10^6$ m
Earth's rotation rate	$\Omega \sim 10^{-4}\,\mathrm{s}^{-1}$
Acceleration of gravity	$g = 9.81\,\mathrm{ms}^{-2}$
Sea level pressure	$p_{\mathrm{ref}} = 10^5\,\mathrm{kgm}^{-1}\mathrm{s}^{-2}$
H_2O freezing temperature	$T_{\mathrm{ref}} \sim 273$ K
Equator–pole pot. temp. diff. Tropospheric vertical pot. temp. diff.	$\Delta\Theta \sim 40$ K
Dry gas constant	$R = 287\,\mathrm{m}^2\,\mathrm{s}^{-2}\,\mathrm{K}^{-1}$
Dry isentropic exponent	$\gamma = 1.4$

Table 3
Auxiliary quantities of interest derived from those in Table 2

Sea-level air density	$\rho_{\mathrm{ref}} = p_{\mathrm{ref}}/(R\,T_{\mathrm{ref}}) \sim 1.25\,\mathrm{kg\,m}^{-3}$
Pressure scale height	$h_{\mathrm{sc}} = p_{\mathrm{ref}}/(g\,\rho_{\mathrm{ref}}) \sim 8$ km
Sound speed	$c_{\mathrm{ref}} = \sqrt{\gamma p_{\mathrm{ref}}/\rho_{\mathrm{ref}}} \sim 330\,\mathrm{m\,s}^{-1}$
Internal wave speed	$c_{\mathrm{int}} = \sqrt{g h_{\mathrm{sc}} \dfrac{\Delta\Theta}{T_{\mathrm{ref}}}} \sim 110\,\mathrm{m\,s}^{-1}$
Thermal wind velocity	$u_{\mathrm{ref}} = \dfrac{2}{\pi} \dfrac{g h_{\mathrm{sc}}}{\Omega a} \dfrac{\Delta\Theta}{T_{\mathrm{ref}}} \sim 12\,\mathrm{m\,s}^{-1}$

constant is a good approximation to local values, because the mass fractions of water vapor and greenhouse gases are very small.

These seven dimensional characteristics allow for three independent dimensionless combinations in addition to the isentropic exponent γ. A possible choice is

$$
\begin{aligned}
\Pi_1 &= \frac{h_{\mathrm{sc}}}{a} \sim 1.6 \cdot 10^{-3}, \\
\Pi_2 &= \frac{\Delta\Theta}{T_{\mathrm{ref}}} \sim 1.5 \cdot 10^{-1}, \\
\Pi_3 &= \frac{c_{\mathrm{ref}}}{\Omega a} \sim 4.7 \cdot 10^{-1},
\end{aligned}
\tag{9}
$$

with h_{sc} the density scale height and c_{ref} of the order of a characteristic speed of sound or of barotropic (external) gravity waves as given in the first three items of *Table 3*. The last two items in *Table 3* are further characteristic signal speeds derived from the quantities of *Table 2*: c_{int} corresponds to the horizontal phase speed of linearized internal gravity waves in the long-wavelength limit (see Gill 1982, Chap. 6). In estimating a typical horizontal flow velocity, we have used the so-called thermal wind relation which provides an estimate for the vertical variation of the horizontal velocity in an atmosphere that is simultaneously in hydrostatic and geostrophic balance. Here "geostrophy" refers to a balance of the horizontal pressure gradient and Coriolis force. See the discussion of Eq. 23 for more details. Importantly, internal gravity waves are dispersive, their phase speed depending strongly on the wave number vector, so that c_{int} is merely an upper estimate and, in practice, internal wave signals may move at speeds as low as u_{ref}.

Table 4
Hierarchy of physically distinguished scales in the atmosphere

Planetary scale	$L_P = \frac{\pi}{2} a \sim 10{,}000$ km
Obukhov radius	$L_{Ob} = \frac{c_{ref}}{\Omega} \sim 3{,}300$ km
Synoptic scale	$L_{Ro} = \frac{c_{int}}{\Omega} \sim 1{,}100$ km
Meso-β scale	$L_{meso} = \frac{u_{ref}}{\Omega} \sim 150$ km
Meso-γ scale	$h_{sc} = \frac{\gamma p_{ref}}{g \rho_{ref}} \sim 11$ km

Consider now the distances which sound waves, internal waves, and particles advected by the thermal wind would travel during the characteristic time of Earth's rotation, $1/\Omega \sim 10^4$ s. Together with the scale height, h_{sc}, and the equator-to-pole distance, L_P, we find the hierarchy of characteristic lengths displayed in *Table 4*.

The technical terms in the left column are often used for these scales in the meteorological literature. The synoptic reference scale, L_{Ro}, is also called the "Rossby radius," and the Obukhov scale is frequently termed "external Rossby radius." These length scales are naturally induced by fluid dynamical processes in the atmosphere and they give rise to multiple scales regimes when the characteristic signal speeds differ significantly. If we fix, in turn, a length scale to be considered, then the different characteristic signal speeds give rise to multiple times instead of multiple lengths. In general situations, one will be faced with combined multiple length–multiple time regimes.

3.2 Distinguished Limits

Even for the very simple problem of the linear oscillator with small mass and small damping, an asymptotic expansion that allows for independent variation of the two parameters is bound to fail, because limits taken in the space of the mass and damping parameters turn out to be path dependent (Klein 2010). If the mass vanishes faster than the damping, non-oscillatory, purely damped solutions prevail; if the damping vanishes faster we obtain rapidly oscillating limit solutions. If that is so even for a simple linear problem there is little hope for independent multiple parameter expansions in more complex fluid dynamical problems. Therefore, being faced with multiple small parameters as in (9), we proceed by introducing distinguished limits, or coupled limit processes: the parameters are functionally related and asymptotic analyses proceed in terms of a single expansion parameter only.

The characteristic signal speeds from *Table 3* are compatible with the scalings

$$\frac{c_{int}}{c_{ref}} \sim 1/3 \sim \sqrt{\varepsilon}, \quad \frac{u_{ref}}{c_{int}} \sim 1/9 \sim \varepsilon, \quad \text{and} \quad \frac{u_{ref}}{c_{ref}} \sim \varepsilon^{\frac{3}{2}}, \tag{10}$$

and this corresponds to letting

$$\Pi_1 = c_1 \varepsilon^3, \quad \Pi_2 = c_2 \varepsilon, \quad \Pi_3 = c_3 \sqrt{\varepsilon} \tag{11}$$

with $c_i = O(1)$ as $\varepsilon \to 0$ for the parameters in (9). The length scales in ➋ *Table 4* then obey

$$L_{\text{meso}} = \frac{h_{\text{sc}}}{\varepsilon}, \quad L_{\text{Ro}} = \frac{h_{\text{sc}}}{\varepsilon^2}, \quad L_{\text{Ob}} = \frac{h_{\text{sc}}}{\varepsilon^{\frac{5}{2}}}, \quad L_{\text{p}} = \frac{h_{\text{sc}}}{\varepsilon^3}, \tag{12}$$

and we find that all these familiar meteorologically relevant scales can be accessed naturally through asymptotic scalings in the small parameter ε.

3.3 Remarks of Caution

We will restrict our discussion here to length scales larger or equal to the density scale height. Of course, on the much smaller length and time scales comparable to those of typical engineering applications, turbulence will induce a continuous range of scales. Analyses that exclusively rely on the assumption of scale separations will be of limited value in studying such flows. The interested reader may want to consult (Oberlack 2006) for theoretical foundations and references.

There are serious voices in the literature (see, e.g., White 2002, sec. 9.3) who would not necessarily agree that formal asymptotics as promoted here are the best way forward in developing approximate model equations for atmospheric flow applications. Salmon (1998) and Norbury and Roulstone (2002) are excellent starting points for the interested reader.

Other serious voices in the literature, e.g., Lovejoy et al. (2008), would dispute our basic proposition that there be a natural scale separation for atmospheric flows of sufficiently large scale. Their arguments are based on the fact that it is very difficult to detect such scale separations in the observational record. To our defense we refer to our previous discussion, and to Lundgren (1982), or Hunt and Vassilicos (1991). These authors demonstrate that some localized, i.e., ideally scale-separated, and entirely realizable flow structures exhibit full spectra without gaps and with exponents akin to those of turbulence spectra. In other words, there may very well be clean scale separations even if they are not visible in spectral and similar global decompositions.

4 Classical Single-Scale Models

In this section, we summarize the hierarchy of models that obtains from the full compressible flow equations when we assume solutions to depend on a single horizontal, a single vertical, and a single time scale only. See ➋ Sect. 5 for a brief discussion of multiscale regimes.

➋ Section 4.1 will set the stage for the subsequent discussions by introducing the nondimensional governing equations and a general scale transformation that will be used excessively later on.

There is common agreement among meteorologists that, on length scales small compared to the planetary scale, acoustic modes contribute negligibly little to the atmospheric dynamics. As a consequence, there has been long-standing interest in analogs to the classical incompressible flow equations that would be valid for meteorological applications. Such models would capture the effects of advection, vertical stratification, and of the Coriolis force, while discarding acoustic modes. According to our previous discussion, given a horizontal characteristic length, acoustic waves, internal waves, and advection are all associated with their individual characteristic time scales. Therefore, after having decided to eliminate the sound waves we still have the option of adjusting a reduced model to the internal wave dynamics (➋ Sect. 4.2) or to capture the effects of advection (➋ Sect. 4.3).

Near-equatorial motions are addressed in ➋ Sects. 4.4 and ➋ 4.5 summarizes the *hydrostatic primitive equations* which are at the core of most current global weather forecast and climate models.

To simplify notations, the tangent plane approximation is used throughout this section. That is, we neglect the influence of Earth's sphericity, except for including the horizontal variation of the Coriolis parameter. See standard textbooks, such as Gill (1982) and Pedlosky (1987), for discussions of the regime of validity of this approximation.

4.1 Scalings of the Governing Equations

4.1.1 Governing Equations

The compressible flow equations for an ideal gas with constant specific heat capacities and including gravity, rotation, and generalized source terms will serve as the basis for the subsequent discussions. We nondimensionalize using $\ell_{\text{ref}} = h_{\text{sc}}$ and $t_{\text{ref}} = h_{\text{sc}}/u_{\text{ref}}$ for space and time, and $p_{\text{ref}}, T_{\text{ref}}, R, u_{\text{ref}}$ from ➋ *Tables 2* and ➋ *3* for the thermodynamic variables and velocity. In addition, we employ the distinguished limits from (10), (11). The governing equations for the dimensionless horizontal and vertical flow velocities, $\mathbf{v}_{\parallel}, w$, pressure, p, density, ρ, and potential temperature, $\Theta = p^{1/\gamma}/\rho$, as functions of the horizontal and vertical coordinates, $\boldsymbol{x}, z$, and time, t, then read

$$
\begin{aligned}
\left(\frac{\partial}{\partial t} + \mathbf{v}_{\parallel}\cdot\nabla_{\parallel} + w\frac{\partial}{\partial z}\right)\mathbf{v}_{\parallel} + \varepsilon\left(2\boldsymbol{\Omega}\times\mathbf{v}\right)_{\parallel} + \frac{1}{\varepsilon^3\rho}\nabla_{\parallel}p &= \mathbf{Q}_{\mathbf{v}_{\parallel}},\\
\left(\frac{\partial}{\partial t} + \mathbf{v}_{\parallel}\cdot\nabla_{\parallel} + w\frac{\partial}{\partial z}\right)w + \varepsilon\left(2\boldsymbol{\Omega}\times\mathbf{v}\right)_{\perp} + \frac{1}{\varepsilon^3\rho}\frac{\partial p}{\partial z} &= Q_w - \frac{1}{\varepsilon^3},\\
\left(\frac{\partial}{\partial t} + \mathbf{v}_{\parallel}\cdot\nabla_{\parallel} + w\frac{\partial}{\partial z}\right)\rho + \rho\,\nabla\cdot\mathbf{v} &= 0,\\
\left(\frac{\partial}{\partial t} + \mathbf{v}_{\parallel}\cdot\nabla_{\parallel} + w\frac{\partial}{\partial z}\right)\Theta &= Q_{\Theta}.
\end{aligned}
\tag{13}
$$

Here $\boldsymbol{b}_{\parallel}, \boldsymbol{b}_{\perp}$ denote the horizontal and vertical components of a vector $\boldsymbol{b}$, respectively. The terms $Q_{[\cdot]}$ in (13) stand for general source terms, molecular transport, and further unresolved-scale closure terms. In the sequel we assume that these terms will not affect the leading-order asymptotic balances to be discussed. If in one or the other practical situation it turns out that this assumption is void, these terms can be handled within the same framework as the rest of the terms. See, e.g., Klein and Majda (2006), for examples involving moist processes in the atmosphere.

4.1.2 Some Revealing Transformations

The Exner Pressure

For the analysis of atmospheric flows, the Exner pressure $\pi = p^{\Gamma}/\Gamma$ with $\Gamma = (\gamma - 1)/\gamma$ turns out to be a very convenient variable. The pressure is always in hydrostatic balance to leading order in the atmosphere. For a given horizontally averaged mean stratification of the potential temperature, $\overline{\Theta}(z)$, the corresponding hydrostatic pressure will satisfy $d\overline{p}/dz = -\overline{\rho} = -\overline{p}^{1/\gamma}/\overline{\Theta}$, (see the $O(1/\varepsilon^3)$–terms in $(13)_2$) with exact solution

$$\overline{p}(z) = (\Gamma\overline{\pi}(z))^{\frac{1}{\Gamma}} \quad \text{where} \quad \overline{\pi}(z) = \Gamma^{-1} - \int_0^z \frac{1}{\overline{\Theta}(z')} dz'. \tag{14}$$

Thus the Exner pressure captures the essential vertical variation of the hydrostatic background pressure due to the atmosphere's stratification without exhibiting the very strong drop of the pressure $\overline{p}$ due to the compressibility of air. For example, for a neutrally stratified atmosphere with $\overline{\Theta} \equiv$ const, $\overline{\pi}$ is linear in z, whereas $\overline{p}$ is a strongly nonlinear function. Most atmospheric flow phenomena may be understood as perturbations away from such a hydrostatically balanced background state.

A General Scale Transformation

Suppose we are interested in solutions to (13) with a characteristic time $T = t_{\text{ref}}/\varepsilon^{\alpha_t}$ and a horizontal characteristic length $L = h_{\text{sc}}/\varepsilon^{\alpha_x}$ with $\alpha_t, \alpha_x > 0$. Appropriate rescaled coordinates are

$$\tau = \varepsilon^{\alpha_t} t \quad \text{and} \quad \boldsymbol{\xi} = \varepsilon^{\alpha_x} \boldsymbol{x}. \tag{15}$$

Furthermore, we take into account that the relative (vertical) variation of the potential temperature in the atmosphere is of order $O(\Delta\Theta/T_{\text{ref}}) = O(\varepsilon)$ and that the vertical and horizontal velocities in most of the regimes to be considered scale as $|w|/|\mathbf{v}_\parallel| \sim h_{\text{sc}}/L = O(\varepsilon^{\alpha_x})$. This suggests new dependent variables,

$$\begin{aligned}
\Theta(\tau, \boldsymbol{\xi}, z) &= 1 + \varepsilon\overline{\theta}(z) + \varepsilon^{\alpha_\pi}\widetilde{\theta}(\tau, \boldsymbol{\xi}, z),\\
\pi(\tau, \boldsymbol{\xi}, z) &= \overline{\pi}(z) + \varepsilon^{\alpha_\pi}\Gamma\widetilde{\pi}(\tau, \boldsymbol{\xi}, z),\\
w(\tau, \boldsymbol{\xi}, z) &= \varepsilon^{\alpha_x}\,\widetilde{w}(\tau, \boldsymbol{\xi}, z),
\end{aligned} \tag{16}$$

with $\overline{\pi}(z)$ from (14) for $\overline{\Theta}(z) = 1 + \varepsilon\overline{\theta}(z)$. Then the governing equations become

$$\begin{aligned}
&\left(\frac{\varepsilon^{\alpha_t}}{\varepsilon^{\alpha_x}}\frac{\partial}{\partial\tau} + \mathbf{v}_\parallel\cdot\nabla_\xi + \widetilde{w}\frac{\partial}{\partial z}\right)\mathbf{v}_\parallel + \frac{(2\boldsymbol{\Omega}\times\mathbf{v})_\parallel}{\varepsilon^{\alpha_x-1}} + \frac{\varepsilon^{\alpha_\pi}}{\varepsilon^3}\,\Theta\,\nabla_\xi\widetilde{\pi} = \mathbf{Q}^\varepsilon_{\mathbf{v}_\parallel}\\
&\left(\frac{\varepsilon^{\alpha_t}}{\varepsilon^{\alpha_x}}\frac{\partial}{\partial\tau} + \mathbf{v}_\parallel\cdot\nabla_\xi + \widetilde{w}\frac{\partial}{\partial z}\right)\widetilde{w} + \frac{(2\boldsymbol{\Omega}\times\mathbf{v})_\perp}{\varepsilon^{2\alpha_x-1}} + \frac{\varepsilon^{\alpha_\pi}}{\varepsilon^{3+2\alpha_x}}\left(\Theta\frac{\partial\widetilde{\pi}}{\partial z} - \frac{\widetilde{\theta}}{\overline{\Theta}}\right) = Q^\varepsilon_w\\
&\left(\frac{\varepsilon^{\alpha_t}}{\varepsilon^{\alpha_x}}\frac{\partial}{\partial\tau} + \mathbf{v}_\parallel\cdot\nabla_\xi + \widetilde{w}\frac{\partial}{\partial z}\right)\widetilde{\pi} + \frac{\gamma\pi}{\varepsilon^{\alpha_\pi}}\left(\nabla_\xi\cdot\mathbf{v}_\parallel + \frac{\partial\widetilde{w}}{\partial z} + \frac{\widetilde{w}}{\gamma\Gamma\pi}\frac{d\overline{\pi}}{dz}\right) = \frac{\gamma\pi Q^\varepsilon_\Theta}{\Theta}\\
&\left(\frac{\varepsilon^{\alpha_t}}{\varepsilon^{\alpha_x}}\frac{\partial}{\partial\tau} + \mathbf{v}_\parallel\cdot\nabla_\xi + \widetilde{w}\frac{\partial}{\partial z}\right)\widetilde{\theta} + \frac{\varepsilon}{\varepsilon^{\alpha_\pi}}\,\widetilde{w}\,\frac{d\overline{\theta}}{dz} = Q^\varepsilon_\Theta
\end{aligned} \tag{17}$$

Below we will assume that the source terms, $Q^\varepsilon_{[\,\cdot\,]}$, are of sufficiently high order in ε to not affect the leading-order asymptotics.

Importantly, in writing down the equations in (17) we have merely adopted a transformation of variables, but no approximations. Together with the definitions in (15), (16), they are equivalent to the original version of the compressible flow equations in (13). However, below we will employ judicious choices for the scaling exponents, $\alpha_{[\,\cdot\,]}$, and assume solutions that adhere to the implied scalings in that $\mathbf{v}_\parallel, \widetilde{w}, \widetilde{\pi}, \widetilde{\theta} = O(1)$ and that the partial derivative operators $\partial_\tau, \nabla_\xi, \partial_z$ yield $O(1)$ results when applied to these variables as $\varepsilon \to 0$. This will allow us to

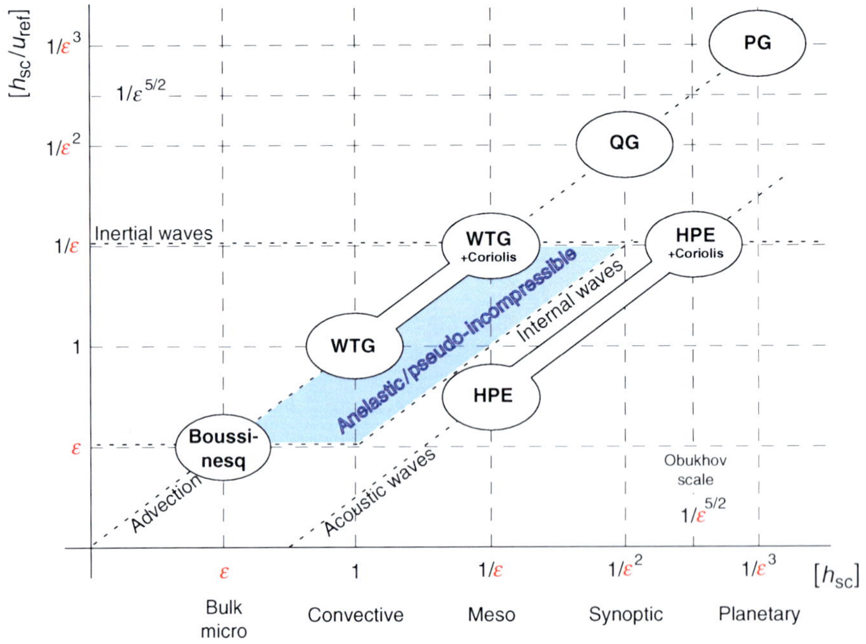

Fig. 1
Scaling regimes and model equations for atmospheric flows. *WTG* weak temperature gradient approximations, *QG* quasi-geostrophic, *PG* planetary geostrophic, *HPE* hydrostatic primitive equations. The WTG and HPE models cover a wide range of spacial scales assuming the associated advective and acoustic times scales, respectively. The anelastic and pseudo-incompressible models for realistic flow regimes cover multiple spatiotemporal scales (Sect. 2). See similar graphs for near-equatorial flows in Majda and Klein (2003) and Majda (2007b)

efficiently carve out the essence of various limit regimes for atmospheric flows without having to go through the details of the asymptotic expansions—an exercise that *can* be completed, of course, see, e.g., Majda and Klein (2003) and Klein (2004), but whose detailed exposition would be too tedious and space-consuming for the present purposes. *Figure 1* summarizes the mid latitude flow regimes to be discussed in this way below.

Considering (17), the classical Strouhal, Mach, Froude, and Rossby numbers are now related to ε and the spatiotemporal scaling exponents via

$$St^{-1} = \frac{L}{UT} = \frac{\varepsilon^{\alpha_t}}{\varepsilon^{\alpha_x}}, \qquad Fr^2 = \gamma \mathrm{Ma}^2 = \frac{\gamma U^2}{C^2} = \varepsilon^3, \qquad Ro = \varepsilon^{\alpha_x - 1}, \tag{18}$$

where T, L, U, C are the characteristic time and length scale, horizontal flow velocity, and sound speed in the rescaled variables from Eqs. 16 and 17, respectively.

4.2 Midlatitude Internal Gravity Wave Models

Here we consider (17) with $\alpha_t = \alpha_x - 1$, so that temporal variations are fast compared with trends due to advection, and we restrict ourselves to length scales up to the synoptic scale, $L_{\mathrm{Ro}} = h_{sc}/\varepsilon^2$, so that $\alpha_x \in \{0, 1, 2\}$. The resulting leading-order models will include the time derivative at $O(1/\varepsilon)$ but not the advection terms. The Coriolis terms will not dominate the time derivative unless $\alpha_x > 2$. As a consequence, for any horizontal scale $L \le h_{sc}/\varepsilon^2$, the level of pressure perturbations can be assessed from the horizontal momentum equation, $(17)_1$, by balancing the pressure gradient and time derivative term and we find $\alpha_\pi = 2$. With this information at hand, we observe that the velocity divergence term will dominate the pressure equation $(17)_3$ so that a divergence constraint arises at leading order. Collecting these observations, we obtain a generic leading-order model for horizontal scales $h_{sc} \le L \le h_{sc}/\varepsilon^2 = L_{Ro}$,

$$\begin{aligned} \frac{\partial \mathbf{v}_\parallel}{\partial \tau} + a\, f\, \boldsymbol{k} \times \mathbf{v}_\parallel + \nabla_\xi \widetilde{\pi} &= \mathbf{Q}^\varepsilon_{\mathbf{v}_\parallel}, \\ b\, \frac{\partial \widetilde{w}}{\partial \tau} + \frac{\partial \widetilde{\pi}}{\partial z} &= \widetilde{\theta}, \\ \nabla \cdot (\overline{P} \mathbf{v}) &= 0, \\ \frac{\partial \widetilde{\theta}}{\partial \tau} + \widetilde{w}\, \frac{d\overline{\theta}}{dz} &= Q^\varepsilon_\Theta. \end{aligned} \tag{19}$$

Here we have introduced the vertical component of the Earth rotation vector, $f = \boldsymbol{k} \cdot 2\boldsymbol{\Omega}$, with $\boldsymbol{k}$ the vertical unit vector, and we have abbreviated

$$\nabla \cdot (\overline{P} \mathbf{v}) = \overline{P} \left(\nabla_\xi \cdot \mathbf{v}_\parallel + \frac{\partial \widetilde{w}}{\partial z} + \frac{\widetilde{w}}{\gamma \Gamma \overline{\pi}} \frac{d\overline{\pi}}{dz} \right),$$

where $\overline{P}(z) = \overline{\pi}(z)^{1/\gamma\Gamma} = \overline{p}^{1/\gamma}$. The parameters a, b in (19) depend on the characteristic horizontal scale and are defined as

$$a = \begin{cases} 1 & (\alpha_x = 2) \\ 0 & (\alpha_x \in \{0, 1\}) \end{cases}, \qquad b = \begin{cases} 1 & (\alpha_x = 0) \\ 0 & (\alpha_x \in \{1, 2\}) \end{cases}. \tag{20}$$

For any pair (a, b), the model equations in (19) are "sound-proof" due to the velocity divergence constraint $(19)_3$. They support internal gravity waves through the interplay of vertical advection of the background stratification in $(19)_4$ and the buoyancy force on the right hand side of the vertical momentum equation, $(19)_2$. When $b = 0$, i.e., for horizontal scales $L \gg h_{sc}$, we have $\partial\widetilde{\pi}/\partial z = \widetilde{\theta}$ and these waves are called hydrostatic; for $b = 1$ they are non-hydrostatic with the inertia of the vertical motion participating in the dynamics.

When $a = 0$, i.e., for length scales small compared with the synoptic scale, $L_{Ro} \sim h_{sc}/\varepsilon^2$, the internal waves propagate freely. Hydrostatic states at rest with horizontally homogeneous $\widetilde{\theta}$ are then the only equilibrium states of the system. For length scales comparable to the synoptic scale, however, $a = 1$, and the Coriolis effect affects the dynamics. As a consequence, there are new nontrivial steady states involving the geostrophic balance of the Coriolis and horizontal pressure gradient terms, i.e., $f\, \boldsymbol{k} \times \mathbf{v}_\parallel + \nabla_\xi \widetilde{\pi} = 0$. Instead of assuming hydrostatic states at rest, such a system will adjust to the geostrophic state compatible with its initial data over long times, and it will generally release only a fraction of the full potential energy that it held initially (see Gill, 1982, Chap. 7).

4.3 Balanced Models for Advection Time Scales

Here we let $\alpha_t = \alpha_x$ in (17), i.e., we consider processes that will always include the effects of advection in their temporal evolution. Now we will have to distinguish between the meso scales,

$L = h_{sc}$ and $L = L_{\text{meso}} = h_{sc}/\varepsilon$ on the one hand, and the synoptic scale, $L = L_{\text{Ro}} = h_{sc}/\varepsilon^2$, respectively. For the latter, the Coriolis term becomes asymptotically dominant.

4.3.1 Weak Temperature Gradient Models for the Meso Scales

On the meso scales, i.e., for $\alpha_x \in \{0, 1\}$, the horizontal momentum balance in $(17)_1$ yields $\alpha_\pi = 3$ and we find the "weak temperature gradient" or "quasi-nondivergent" approximation at leading order (Held and Hoskins 1985; Sobel et al. 2001; Zeytounian 1990),

$$\begin{aligned} \left(\frac{\partial}{\partial\tau} + \mathbf{v}_{\parallel}\cdot\nabla_{\xi} + \widetilde{w}\frac{\partial}{\partial z}\right)\mathbf{v}_{\parallel} + a\,f\,\boldsymbol{k}\times\mathbf{v}_{\parallel} + \nabla_{\xi}\widetilde{\pi} &= \mathbf{Q}^{\varepsilon}, \\ \nabla\cdot(\overline{P}\mathbf{v}) &= 0, \\ \widetilde{w}\,\frac{d\overline{\theta}}{dz} &= Q_{\Theta}^{\varepsilon}. \end{aligned} \tag{21}$$

where $a = 0$ for $L = h_{sc}$ and $a = 1$ for $L = h_{sc}/\varepsilon = L_{\text{meso}}$.

The third equation in (21) describes how diabatic heating in a sufficiently strongly stratified medium induces vertical motions so that air parcels adjust quasi-statically to their individual vertical levels of neutral buoyancy. Replacing the potential temperature transport equation from $(17)_4$ with $(21)_3$ suppresses internal gravity waves, just as the divergence constraints suppress the acoustic modes. With $(21)_3$, the vertical velocity is fixed, given the diabatic source term. The second equation then becomes a divergence constraint for the horizontal velocity, and the perturbation pressure $\widetilde{\pi}$ is responsible for guaranteeing compliance with this constraint. The vertical momentum equation, $(17)_2$, in this regime becomes the determining equation for potential temperature perturbations, $\widetilde{\theta}$. These are passive unless they influence the source terms in (21).

Weak temperature gradient models are frequently employed for studies in tropical meteorology, see Klein (2006); [and references therein].

4.3.2 Synoptic Scales and the Quasi-Geostrophic Approximation

On the larger synoptic length and advective time scales, $T = h_{sc}/\varepsilon^2 u_{\text{ref}} \sim 12$ h and $L = h_{sc}/\varepsilon^2 \sim 1{,}100$ km, for which $\alpha_t = \alpha_x = 2$, we find that the Coriolis and pressure gradient terms form the *geostrophic balance* at leading order, and this changes the dynamics profoundly in comparison with the weak temperature gradient dynamics from the last section. Here we encounter the likely most prominent classical model of theoretical meteorology, the quasi-geostrophic model (see Gill 1982; Pedlosky 1987; Zeytounian 1990) and references therein. See also Muraki et al. (1999) for a higher-order accurate QG theory.

Matching the Coriolis and pressure gradient terms, which are $O(\varepsilon^{-1})$ and $O(\varepsilon^{\alpha_\pi - 3})$, provides $\alpha_\pi = 2$. The horizontal and vertical momentum equations reduce to the geostrophic and hydrostatic balances, respectively, viz.,

$$f_0\,\boldsymbol{k}\times\mathbf{v}_{\parallel} + \nabla_{\xi}\widetilde{\pi} = 0, \qquad \frac{\partial\widetilde{\pi}}{\partial z} = \widetilde{\theta}, \tag{22}$$

where $f_0 = \boldsymbol{k} \cdot 2\boldsymbol{\Omega}|_{\xi=0}$. By taking the curl of $(22)_1$ and inserting $(22)_2$ we obtain

$$f_0 \begin{pmatrix} -\partial v/\partial z \\ \partial u/\partial z \\ \nabla_\xi \cdot \mathbf{v}_\parallel \end{pmatrix} = \begin{pmatrix} -\partial^2 \widetilde{\pi}/\partial z \partial \eta \\ \partial^2 \widetilde{\pi}/\partial z \partial \xi \\ 0 \end{pmatrix} = \begin{pmatrix} -\partial \widetilde{\theta}/\partial \eta \\ \partial \widetilde{\theta}/\partial \xi \\ 0 \end{pmatrix}, \tag{23}$$

with ξ, η the east and northward components of $\boldsymbol{\xi}$, respectively. The first two components in (23) represent the "thermal wind relation" which was introduced earlier in the introduction and was used in ➋ *Table 3* to define the universal reference flow velocity, u_{ref}. The third component in (23) states that the horizontal flow is divergence free at leading order, so that $\widetilde{\pi}$ may be considered a stream function for the horizontal flow according to $(22)_1$.

Equation 22 does not allow us to determine the temporal evolution of the flow. An evolution equation is obtained by going to the next order in ε and eliminating the arising higher-order perturbation functions using the vertical momentum, pressure, and potential temperature equations (see Pedlosky 1987, sec. 6.5). The result is, as one may have expected, a vorticity transport equation and one finds

$$\left(\frac{\partial}{\partial \tau} + \mathbf{v}_\parallel \cdot \nabla_\xi\right) q = Q_q^\varepsilon, \tag{24}$$

where

$$q = \zeta + \beta\,\eta + \frac{f}{\overline{\rho}} \frac{\partial}{\partial z}\left(\frac{\overline{\rho}\,\widetilde{\theta}}{d\overline{\theta}/dz}\right) \tag{25}$$

is the leading-order potential vorticity with ζ, β, η defined through

$$\zeta = \boldsymbol{k} \cdot \nabla_\xi \times \mathbf{v}_\parallel, \qquad \eta = \boldsymbol{\xi} \cdot \vec{n}^{\text{north}}, \qquad \boldsymbol{k} \cdot 2\boldsymbol{\Omega} = f + \beta\eta. \tag{26}$$

Thus ζ is the vertical component of vorticity, η the northward component of $\boldsymbol{\xi}$, and β the derivative of the Coriolis parameter with respect to η. The last term in (25) captures the effect on vorticity of the vertical stretching of a column of air induced by nonzero first-order horizontal divergences and the constraint of mass continuity. The potential vorticity source term, Q_q^ε, in (24) is a combination of the heat and momentum source terms and their gradients.

Using the geostrophic and hydrostatic balances in (22), the definition of potential vorticity in (25) becomes

$$q = \nabla_\xi^2 \widetilde{\pi} + \beta\,\eta + \frac{f}{\overline{\rho}} \frac{\partial}{\partial z}\left(\frac{\overline{\rho}}{d\overline{\theta}/dz} \frac{\partial \widetilde{\pi}}{\partial z}\right). \tag{27}$$

This, together with $(22)_1$, reveals that the transport equation from (24) is a closed equation for the perturbation pressure field, $\widetilde{\pi}$, given appropriate definitions of the source terms and boundary conditions.

The interpretation of the potential vorticity transport equation in (24) is as follows: Total vorticity is advected along (horizontal) pathlines of the leading-order flow. In the process, vertical vorticity, ζ, adjusts to compensate for local changes of the planetary rotation, $\beta\eta$, when there are northward or southward excursions, and feels the effects of vortex stretching as captured by the third term on the right in (27). Important manifestations of these processes are planetary-scale meandering motions, the Rossby waves, and baroclinic instabilities which induce the sequence of cyclones and anti cyclones that constitute much of the weather statistics in the middle latitudes, see Pedlosky (1987, Chap. 7) for a comprehensive discussion. Importantly, in the QG theory both acoustic and internal gravity waves are asymptotically filtered out, as the characteristic time scale considered is that of advection only.

4.3.3 Ogura and Phillips' Anelastic Model for Weak Stratification

Thus far we faced the alternative of adopting either the internal wave or the advection time scale in constructing reduced single-scale models. A compact description of the combined effects of both would not transpire. Ogura and Phillips (1962) realized that this separation may be overcome for weak stratification and diabatic heating. To summarize their development, we let $\left(\overline{\theta}, Q_{\Theta}^{\varepsilon}\right) = \varepsilon^2 \left(\overline{\theta}^*, Q_{\Theta}^*\right) = O(\varepsilon^2)$ in $(17)_4$ but otherwise follow the developments for the advective time scale in ➋ Sect. 4.3.1. This leads to the classical anelastic model,

$$
\begin{aligned}
\left(\frac{\partial}{\partial \tau} + \mathbf{v}_{\parallel} \cdot \nabla_{\xi} + \widetilde{w} \frac{\partial}{\partial z}\right) \mathbf{v}_{\parallel} + \nabla_{\xi} \widetilde{\pi} &= \mathbf{Q}_{\mathbf{v}_{\parallel}}^{\varepsilon}, \\
\left(\frac{\partial}{\partial \tau} + \mathbf{v}_{\parallel} \cdot \nabla_{\xi} + \widetilde{w} \frac{\partial}{\partial z}\right) \widetilde{w} + \frac{\partial \widetilde{\pi}}{\partial z} &= \widetilde{\theta} + Q_{w}^{\varepsilon}, \\
\nabla \cdot (\overline{P} \mathbf{v}) &= 0, \\
\left(\frac{\partial}{\partial \tau} + \mathbf{v}_{\parallel} \cdot \nabla_{\xi} + \widetilde{w} \frac{\partial}{\partial z}\right) \widetilde{\theta} + \widetilde{w} \frac{d \overline{\theta}^*}{dz} &= Q_{\Theta}^*.
\end{aligned}
\tag{28}
$$

The velocity divergence constraint in $(28)_3$ is responsible for suppressing sound waves. For the present weak variations of potential temperature it is equivalent to the more common formulation, $\nabla \cdot (\overline{\rho} \mathbf{v}) = 0$, see the definition of Θ below (13). The model involves a Boussinesq-type approximation for buoyancy in $(28)_2$, and the buoyancy evolution equation $(28)_4$ accounts explicitly for the background stratification. As a consequence, the system does support both internal gravity waves and the nonlinear effects of advection.

The total variation of potential temperature allowed for in the Ogura and Phillips' anelastic model is of order $O(\varepsilon^3) = O(Ma^2) \sim 10^{-3}$ (see (16), (18), and let $\overline{\theta} = O(\varepsilon^2)$). This amounts to temperature variations of merely a few Kelvin in dimensional terms. This is in contrast with the observed variations of 30...50 K across the troposphere. To address this issue, various generalizations of the classical anelastic model allowing for stronger stratifications have been proposed (see, e.g., Bannon 1996; Durran 1989, and references therein). See also the previous discussion (➋ Sect. 2).

4.4 Scalings for Near-Equatorial Motions

In the tropics, the same processes as encountered so far contribute to the atmospheric fluid dynamics, i.e., we have advection, inertial effects due to the Coriolis force, and internal gravity and sound waves. However, the fact that the Coriolis force for the horizontal motion changes sign (and is zero) at the equator, lets the tropical dynamics exhibit a range of very different phenomena as compared to middle and high latitude motions. There is, in particular, a spectrum of near-equatorial trapped waves which combine aspects of internal gravity and Rossby waves and which move predominantly in the zonal direction along the equator.

For the asymptotic scaling in ε of the Coriolis term in the horizontal momentum balance near the equator we have $(2\mathbf{\Omega} \times \mathbf{v})_{\parallel} = f \boldsymbol{k} \times \mathbf{v}_{\parallel} + w 2\mathbf{\Omega}_{\parallel} \times \boldsymbol{k}$, where $f = 2\mathbf{\Omega} \cdot \boldsymbol{k} = 2\Omega \sin(\phi) = 2\Omega\phi + O(\phi^3)$ for small latitudes, $\phi \ll 1$. For scales smaller than the planetary scale, we will place our coordinate system onto the equator and let the space coordinate y point to the north. Then, as y is nondimensionalized by $h_{sc} = \varepsilon^3 a$, we have $\phi = \varepsilon^3 y$ and in $(17)_1$ find

$$
\varepsilon(2\mathbf{\Omega} \times \mathbf{v})_{\parallel} = \varepsilon^4 y \, 2\Omega \, \boldsymbol{k} \times \mathbf{v}_{\parallel} + O(\varepsilon w). \tag{29}
$$

For spatiotemporal scales that are too small for the Coriolis effect to play a leading-order role, we will find essentially the same small-scale equations discussed in ➋ Sect. 4.2 for internal waves and in ➋ Sect. 4.3 for the balanced weak temperature gradient flows, except that the Coriolis parameter will be a linear function of the meridional coordinate.

A prominent difference between tropical and midlatitude dynamics arises at the equatorial synoptic scales, i.e., at horizontal spacial scales for which the Coriolis term again dominantes the advection terms in the momentum balance even under the equatorial scaling from (29). We will find the governing equations for a variety of dispersive combined internal–inertial waves that are confined to the near-equatorial region and, on longer time scales, a geostrophically balanced model relevant to the tropics.

In analogy with our discussion of the scales in ➋ *Table 4* we assess the equatorial synoptic scale, L_{es}, by equating the time an internal gravity wave would need to pass this characteristic distance with the inverse of the Coriolis parameter. The latter is, however, now proportional to the scale we are interested in, and we obtain in dimensional terms, $c_{\mathrm{int}}/L_{\mathrm{es}} = L_{\mathrm{es}}\Omega/a$, i.e., $L_{\mathrm{es}} \sim \sqrt{c_{\mathrm{int}}\, a/\Omega}$. Using (9–11), we recover the scaling developed by Majda and Klein (2003)

$$\frac{L_{\mathrm{es}}}{h_{\mathrm{sc}}} \sim \sqrt{\frac{c_{\mathrm{ref}}}{\Omega\, a}\,\frac{c_{\mathrm{int}}}{c_{\mathrm{ref}}}\,\frac{a}{h_{\mathrm{sc}}}} \sim \varepsilon^{-\frac{5}{2}}. \tag{30}$$

Comparing this with (12) we observe that the equatorial synoptic scale matches the asymptotic scaling of the midlatitude Obukhov or external Rossby radius, and this corresponds to an intriguing connection between baroclinic near-equatorial waves and barotropic midlatitude modes discussed by Wang and Xie (1996) and Majda and Biello (2003).

4.4.1 Equatorial Sub-Synoptic Flow Models

For scales smaller than the equatorial synoptic scale with $L \sim h_{\mathrm{sc}}/\varepsilon^2$ or smaller, the situation is very similar to that encountered for sub-synoptic scales in the midlatitudes. On the internal wave time scale, the leading-order model is again that of internal wave motions without influence of the Coriolis force as in ➋ Sect. 4.2. On advective time scales, weak temperature gradient models are recovered in analogy with those found in ➋ Sect. 4.3.1, yet with a non constant Coriolis parameter $f = 2\Omega\,\eta$. For related derivations see Browning et al. (2000), Sobel et al. (2001) and Majda and Klein (2003). When the zonal characteristic length is allowed to be large compared to the meridional one, then a modified set of equations is obtained that still retains the advection of zonal momentum, but that is geostrophically balanced in the meridional direction, Majda and Klein (2003). We omit listing all these equations here as they are easily obtained from those in ➋ Sect. 4.3.1 taking into account the equatorial representation of the Coriolis term in (29).

4.4.2 Equatorial Synoptic and Planetary Models

In the tropics, there is a strong anisotropy between the zonal and meridional directions due to the variation of the Coriolis parameter. Here we distinguish the zonal and meridional velocity components, u and v, respectively, so that $\mathbf{v}_{\parallel} = u\,\boldsymbol{i} + v\,\boldsymbol{j}$. In the meridional direction, we allow for waves extending over the equatorial synoptic distance, L_{es}, whereas in the zonal direction, we have $L_x = L_{\mathrm{es}}$ for synoptic scale waves, but $L_x \sim L_{\mathrm{es}}/\varepsilon$ for equatorial long waves (Majda and

Klein 2003). Just as we scaled the vertical velocity with the vertical-to-horizontal aspect ratio before, we rescale the meridional velocity here with the meridional-to-zonal aspect ratio, i.e., we let $\eta = \varepsilon^{5/2} y$ throughout this section, but $v = \varepsilon^{\alpha_v} \widetilde{v}$, with $\alpha_v = 0$ for synoptic scale waves and $\alpha_v = 1$ for equatorial long waves. Also, we focus on the internal wave or advection time scale, so that $\alpha_t = \alpha_x - 1$ or $\alpha_t = \alpha_x$, respectively, with $\alpha_x \in \{5/2, 7/2\}$. In the usual way the pressure scaling is found by balancing the dominant terms in the (zonal) momentum equation and we have $\alpha_\pi = 2$. These scalings yield the generalized linear equatorial (wave) equations

$$
\begin{aligned}
b\,\frac{\partial u}{\partial \tau} - 2\Omega\,\eta\,\widetilde{v} + \frac{\partial \widetilde{\pi}}{\partial \xi} &= \mathbf{Q}^{\varepsilon}_{u},\\
b\,a\,\frac{\partial \widetilde{v}}{\partial \tau} + 2\Omega\,\eta\,u + \frac{\partial \widetilde{\pi}}{\partial \eta} &= \mathbf{Q}^{\varepsilon}_{\widetilde{v}},\\
\frac{\partial \widetilde{\pi}}{\partial z} &= \widetilde{\theta},\\
\nabla\cdot(\overline{P}\mathbf{v}) &= 0,\\
b\,\frac{\partial \widetilde{\theta}}{\partial \tau} + \widetilde{w}\,\frac{d\overline{\theta}}{dz} &= Q^{\varepsilon}_{\Theta}.
\end{aligned}
\tag{31}
$$

For synoptic scale waves with isotropic scaling in the horizontal directions $a = 1$, whereas for equatorial long waves there is geostrophic balance in the meridional direction and $a = 0$. To describe processes advancing on the slower advection time scale, $b = 0$, and we obtain the Matsuno–Webster–Gill model for geostrophically balanced near-equatorial motions, Matsuno (1966).

The equations in (31) support, for $b \neq 0$, a rich family of essentially different traveling waves (Gill 1982; Pedlosky 1987; Majda 2002; Wheeler and Kiladis 1999). There are fast equatorially trapped internal gravity waves that correspond to the midlatitude internal waves from ➲ Sect. 4.2, slowly westward moving equatorially trapped Rossby waves, the Yanai waves which are special solutions behaving like internal gravity waves at short wavelengths but like Rossby waves in the long wavelength limit, and finally there are Kelvin waves, which travel exclusively eastward and are associated with purely zonal motions, i.e., with $\widetilde{v} \equiv 0$.

In the quasi-steady balanced case, $b = 0$, the equations in (31) reduce to a linear model for the geostrophically balanced flow for which exact solutions are available, given the source terms $Q^{\varepsilon}_{[\cdot]}$ (see, e.g., Matsuno 1966). These exact solutions are determined, however, only up to a purely zonal, i.e., x-independent mean flow. Such a mean flow must either be prescribed or obtained from an analysis of the next larger zonal scale (Majda and Klein, 2003). The mathematical character of near-equatorial geostrophically balanced flows is very different from that of their midlatitude analogs from ➲ Sect. 4.3.2.

4.5 The Hydrostatic Primitive Equations

In the construction of computational global weather forecasting or climate models, it is important that the basic fluid flow model used is uniformly valid on the entire globe and for the most extreme flow conditions found. As we have seen, the large-scale model equations discussed in the previous sections are adapted to either the middle latitudes or the near-equatorial region. At the same time, one observes relatively large mean flow velocities of the order of $|\mathbf{v}| \sim 50$ m/s, corresponding to Mach numbers $Ma \sim 0.15$, in strong jet streams near the tropopause (see, e.g., Schneider 2006), and it turns out that global scale barotropic wave perturbations may be equally considered as external gravity or as planetary sound waves.

These observations lead us to consider fully compressible flows with Mach numbers of order unity on horizontal scales large compared with the scale height, h_{sc}. Going back to our general rescaled compressible flow equations, we nondimensionalize the flow velocity by the sound speed, c_{ref} instead of by u_{ref} recalling that $u_{ref}/c_{ref} = Ma \sim \varepsilon^{\frac{3}{2}}$ according to (10). Then we let

$$\widehat{\boldsymbol{v}}_{\parallel} = \varepsilon^{\frac{3}{2}} \mathbf{v}_{\parallel}, \qquad \alpha_x \in \left\{1, \ldots, \frac{5}{2}\right\}, \qquad \alpha_t = \alpha_x - \frac{3}{2} \tag{32}$$

and obtain, after reverting to the variables $p, \rho, \widehat{\boldsymbol{v}}_{\parallel}, \widehat{w}$, and dropping the “hats” for convenience of notation,

$$\begin{aligned}
\left(\frac{\partial}{\partial t} + \mathbf{v}_{\parallel} \cdot \nabla_{\xi} + w \frac{\partial}{\partial z}\right) \mathbf{v}_{\parallel} + a\, f \boldsymbol{k} \times \mathbf{v} + \frac{1}{\rho}\, \nabla_{\xi} p &= \mathbf{Q}_{\mathbf{v}_{\parallel}},\\
\frac{1}{\rho} \frac{\partial p}{\partial z} &= -1,\\
\left(\frac{\partial}{\partial t} + \mathbf{v}_{\parallel} \cdot \nabla_{\xi} + w \frac{\partial}{\partial z}\right) \rho + \rho\, \nabla \cdot \mathbf{v} &= 0,\\
\left(\frac{\partial}{\partial t} + \mathbf{v}_{\parallel} \cdot \nabla_{\xi} + w \frac{\partial}{\partial z}\right) \Theta &= Q_{\Theta},
\end{aligned} \tag{33}$$

where $f = 2\boldsymbol{\Omega} \cdot \boldsymbol{k}$ is the vertical component of the Earth rotation vector and $a = 0$ for $\alpha_x < 5/2$, whereas $a = 1$ for $\alpha_x = 5/2$.

The only approximations relative to the full compressible flow equations from (13) result from the hydrostatic approximation for the pressure in the second equation, from the neglect of higher-order contributions to the Coriolis terms both in the vertical and horizontal momentum balances, i.e., from the “traditional approximation,” (see, e.g., White 2002), and possibly from our assumption regarding the nondominance of the source terms $Q_{[\cdot]}$.

The main consequence of the hydrostatic approximation is that the vertical velocity becomes a diagnostic variable that can be determined from the horizontal velocity, pressure, and potential temperature fields at any given time by integrating a second-order ordinary differential equation in z. This can be demonstrated, e.g., by differentiating $(33)_2$ with respect to time and collecting all expressions involving w. A more elegant way to reveal the diagnostic nature of vertical velocity is to move to pressure coordinates, i.e., to introduce the time dependent coordinate transformation, $(t, \boldsymbol{\xi}, z) \to (t, \boldsymbol{\xi}, p)$, where the transformation rule is given by the hydrostatic relation in $(33)_2$. After this transformation, one obtains a divergence constraint

$$\nabla_{\xi}^{p} \cdot \widehat{\boldsymbol{v}}_{\parallel} + \frac{\partial \omega}{\partial p} = 0 \tag{34}$$

where $\omega = Dp/Dt$ is the analog of vertical velocity in the pressure coordinates (see, e.g., Salmon 1998; White 2002).

The traditional approximation for the Coriolis terms results automatically from the present asymptotic considerations. At the same time, it is extremely important for energetic reasons: unless the traditional approximation is introduced together with the hydrostatic one, the resulting equations will not feature an energy conservation principle, and this would render them useless for long-term integrations (White 2002).

In terms of physical processes represented, the hydrostatic primitive equations cover advection, hydrostatic internal gravity waves, baroclinic and barotropic inertial (Rossby) waves, and barotropic large-scale external gravity waves. The latter are the remnant of compressibility in this system and may thus equivalently be considered as its sound waves. Due to the fact that the Coriolis term does not become dominant in (33), there is no degeneration of these equations near the equator, and the equatorial region is represented with uniform accuracy.

The hydrostatic primitive equations form the fluid dynamical basis of most global weather forecasting and climate models today (see, e.g., Lorenz 1967; Zeytounian 1990; Salmon 1998; Houghton 2002; Shaw and Shepherd 2009). In regional weather forecasting, they are being replaced today with non-hydrostatic models (Clark et al. 2009; Koppert et al. 2009).

5 Developments for Multiple Scales Regimes

Reduced single-scale models for atmospheric motions as discussed in the previous section constitute the scientific basis of theoretical meteorology. Most of them have been derived long ago and have stood the test of time. In contrast, much of today's research aims at modeling interactions across multiple length and/or timescales. Here we briefly sketch just two recent examples, both of which address the ubiquitous question of how computational simulation techniques can be improved systematically by employing multi-scale modeling approaches. The reader may consult (Klein 2010) for further examples and discussions.

5.1 Shaw and Shepherd's Parameterization Framework

One important recent application of multiple scales theories is the development of sound bases for the *parameterization* of the net effects of unresolved scales in computational atmospheric flow models. An excellent example is the recent work by Shaw and Shepherd (2009). Burkhardt and Becker (2006) discuss the potentially detrimental effect of subgrid-scale parameterizations of effective diffusion, friction, and dissipation in global climate models that are not energetically and thermodynamically consistent. Shaw and Shepherd (2009) employ multiple scales asymptotics to reveal how mesoscale internal wave processes (see ➲ Sect. 4.2) would cumulatively affect planetary scale motions as described by the hydrostatic primitive equations (➲ Sect. 4.5). Using a different set of ideas associated with wave activity conservation laws developed in Shepherd (1990), they move on to derive simplified expressions for the discovered interaction terms that would allow them not to solve the meso-scale wave model in detail, as the formal multiple scales asymptotics procedure would require, but would provide effective closures that are energetically and thermodynamically consistent.

Shaw and Shepherd restrict their framework, in this first layout, to the characteristic internal gravity wave time for the meso scale, and to the planetary acoustic time scale. Within their framework, there is thus great potential for incorporation of wide-ranging results on wave–meanfield interactions as developed in recent years (see, e.g., Bühler 2010, and references therein). It is as yet not clear, however, how the closure schemes addressing wave activities will carry over to situations where nonlinear advection and (moist) convection play an important role on the small scales.

5.2 Superparameterization

It is, of course, by no means guaranteed that the fast or small-scale equations in a multiple scales asymptotic analysis are analytically tractable. For example, in a two-scale model that involves the length scales $\ell \sim h_{sc}$ and $L \sim h_{sc}/\varepsilon$ and covers the advective time scale h_{sc}/u_{ref} for the scale ℓ, the small-scale model would be either the weak temperature gradient flow equations from ➋ Sect. 4.3.1 or some version of the anelastic model from ➋ Sect. 4.3.3 (Klein and Majda 2006). In both cases one will generally have to resort to numerical computation in order to determine the small-scale dynamics. In such a situation, the technique of multiple scales analysis can nevertheless yield valuable guidelines regarding the reduced or averaged information that is to be extracted from the small-scale model and communicated to the large-scale equations. E and Engquist have systematized these procedures and proposed a framework for the construction of computational "heterogeneous multiscale models" (see Engquist et al. 2007).

In the context of meteorological modeling, there is a related very active recent development aiming at overcoming one of the major deficiencies of today's climate models in an efficient way. Cloud distributions in the atmosphere feature a wide spectrum of length and time scales (Tessier et al. 1993). Therefore, it is quite generally not possible to even remotely resolve cloud processes in general circulation models for the climate, which feature horizontal grid sizes of 100 km and more. Grabowski (2001, 2004) proposed the superparameterization approach, which involves operating a two-dimensional cloud resolving model, based in his case on an extended version of the anelastic model from ➋ Sect. 4.3.3, within each of the climate model's grid boxes, and extracting the net effects of the small-scale processes on the large-scale flow through appropriate averaging procedures. In today's terms, the approach may be considered a heterogeneous multiscale model for cloud processes in climate applications. Majda (2007a) formulated a general framework for the construction of such hierarchies of embedded models based on the general multiple scales approach from Majda and Klein (2003) and Klein (2004). In particular, he accounts explicitly for the fact that between the convective length scale of the order of h_{sc} and the planetary (climate) scale there are two more groups of characteristic length and time scales to be accounted for as explained here in ➋ Sect. 4.1. In this context it is worth noting that a direct translation of multiple scales techniques into a working computational geo-fluid dynamics code was shown to be feasible already by Nadiga et al. (1997).

6 Conclusions

This chapter has summarized a unified view on the construction of reduced models in theoretical meteorology that is based on formal asymptotic arguments. A wide range of established models of theoretical meteorology can be reproduced in this way, and this approach also paves the way for systematic studies of multiscale interactions through more sophisticated techniques of asymptotic analysis. Yet, through one explicit example—the class of sound-proof models—we have also outlined the limitations of the approach in that there are well-established reduced models which do not lend themselves to formal asymptotic derivation in a straightforward fashion.

This review has also consciously addressed only the fluid mechanically induced scale separations and their consequences for the construction of reduced flow models. Ubiquitous source terms in the equations for real-life applications associated with diabatic effects, turbulence closures and the like may impose additional characteristics both in terms of spatio temporal scales and in terms of their response to variations of the flow variables. As is well known, e.g., from combustion theory, this can strongly affect the resulting simplified models (see, e.g., Peters 2000; Klein and Majda 2006).

There is a rich literature on the mathematically rigorous justifications of reduced asymptotic models. Schochet (2005) reviews work on the classical problem of low Mach number flows and, in particular, includes recent developments to systematically addressing multiple length and time-scale problems. There are also many examples of rigorous justifications of asymptotic models for geophysical flows, and here the reader may consult (Embid and Majda 1998; Babin et al. 2002; Majda 2002; Zeitlin 2007; Bresch and Gérard-Varet 2007; Dutrifoy et al. 2009; Levermore et al. 1996; Bresch and Desjardins 2003; Cao and Titi 2003, 2007; Masmoudi 2007; Feireisl et al. 2008, and references therein).

Acknowledgments

The author thanks Ulrich Achatz, Dargan Frierson, Juan Pedro Mellado, Norbert Peters, Heiko Schmidt, and Bjorn Stevens for very helpful discussions and suggestions concerning the content and structure of this manuscript, and Ulrike Eickers for her careful proofreading.

References

Babin A, Mahalov A, Nicolaenko B (2002) Fast singular limits of stably stratified 3d Euler and Navier-Stokes equations and ageostrophic wave fronts. In: Norbury J, Roulstone I (eds) Large-scale atmosphere-ocean dynamics 1: analytical methods and numerical models. Cambridge University Press, Cambridge

Bannon PR (1996) On the anelastic approximation for a compressible atmosphere. J Atmos Sci 53: 3618–3628

Bresch D, Desjardins B (2003) Existence of global weak solutions for a 2d viscous shallow water equations and convergence to the quasi-geostrophic model. Comm Math Phys 238:211–223

Bresch D, Gérard-Varet D (2007) On some homogenization problems from shallow water theory. Appl Math Lett 20:505–510

Browning G, Kreiss HO, Schubert WH (2000) The role of gravity waves in slowly varying in time tropospheric motions near the equator. J Atmos Sci 57:4008–4019

Bühler O (2010) Wave-mean interactions in fluids and superfluids. Ann Rev Fluid Mech 42:205–228

Burkhardt U, Becker E (2006) A consistent diffusion-dissipation parameterization in the ECHAM climate model. Mon Wea Rev 134:1194–1204

Cao C, Titi E (2003) Global well-posedness and finite dimensional global attractor for a 3-d planetary geostrophic viscous model. Comm Pure Appl Math 56:198–233

Cao C, Titi E (2007) Global well-posedness of the three-dimensional viscous primitive equations of large scale ocean and atmosphere dynamics. Ann Math 166:245–267

Clark P et al (2009) The weather research & forecasting model. http://www.wrf-model.org/

Davies T, Staniforth A, Wood N, Thuburn J (2003) Validity of anelastic and other equation sets as inferred from normal-mode analysis. Q J R Meteorol Soc 129:2761–2775

Durran DR (1989) Improving the anelastic approximation. J Atmos Sci 46:1453–1461

Dutrifoy A, Schochet S, Majda AJ (2009) A simple justification of the singular limit for equatorial shallow-water dynamics. Comm Pure Appl Math LXI:322–333

Embid P, Majda AJ (1998) Averaging over fast gravity waves for geophysical flows with unbalanced initial data. Theor Comp Fluid Dyn 11: 155–169

Engquist B, E W, Vanden-Eijnden E (2007) Heterogeneous multiscale methods: A review. Comm Comput Phys 2:367–450

Feireisl E, Málek J, Novtoný A, Stravskraba I (2008) Anelastic approximation as a singular limit of the compressible Navier-Stokes system. Comm Part Diff Equat 33:157–176

Frierson DMW (2008) Midlatitude static stability in simple and comprehensive general circulation models. J Atmos Sci 65:1049–1062

Gill AE (1982) Atmosphere-ocean dynamics, vol 30 of Intl Geophysics Series. Academic, San Diego

Grabowski WW (2001) Coupling cloud processes with the large-scale dynamics using the cloud-resolving convection parameterization (CRCP). J Atmos Sci 58:978–997

Grabowski WW (2004) An improved framework for superparameterization. J Atmos Sci 61: 1940–1952

Held IM, Hoskins BJ (1985) Large-scale eddies and the general circulation of the troposphere. Adv Geophys 28:3–31

Houghton J (ed) (2002) The physics of atmospheres. Cambridge University Press, Cambridge

Hunt JCR, Vassilicos JC (1991) Kolmogoroffs contributions to the physical and geometrical understanding of small-scale turbulence and recent developments. Proc R Soc London A 434: 183–210

Keller J, Ting L (1951) Approximate equations for large scale atmospheric motions. Internal Report, Inst. for Mathematics & Mechanics (renamed to Courant Institute of Mathematical Sciences in 1962), NYU, (http://www.arxiv.org/abs/physics/0606114v2)

Kevorkian J, Cole J (1996) Multiple scale and singular perturbation methods. Springer, New York

Klein R (2004) An applied mathematical view of theoretical meteorology. In: Applied mathematics entering the 21st century; invited talks from the ICIAM 2003 congress, vol 116 of SIAM proceedings in applied mathematics

Klein R (ed) (2006) Theoretical developments in tropical meteorology. Special issue, Theor. Comp. Fluid Dyn., vol 20. Springer, Berlin

Klein R (2008) An unified approach to meteorological modelling based on multiple-scales asymptotics. Adv Geosci 15:23–33

Klein R (2009) Asymptotics, structure, and integration of sound-proof atmospheric flow equations. Theor Comp Fluid Dyn 23:161–195

Klein R (2010) Scale-dependent asymptotic models for atmospheric flows. Ann Rev Fluid Mech 42: 249–274

Klein R, Majda AJ (2006) Systematic multiscale models for deep convection on mesoscales. Theor Comp Fluid Dyn 20:525–551

Koppert HJ et al (2009) Consortium for small-scale modelling. http://www.cosmo-model.org/

Levermore CD, Oliver M, Titi ES (1996) Global well-posedness for models of shallow water in a basin with a varying bottom. Indiana Univ Math J 45:479–510

Lorenz EN (1967) The nature and theory of the general circulation of the atmosphere. World Meteorological Organization, Geneva

Lovejoy S, Tuck AF, Hovde SJ, Schertzer D (2008) Do stable atmospheric layers exist? Geophys Res Lett 35:L01802

Lundgren TS (1982) Strained spiral vortex model for turbulent fine structure. Phys Fluids 25: 2193–2203

Lynch P (2006) The emergence of numerical weather prediction: Richardson's dream. Cambridge University Press, Cambridge

Majda AJ (2002) Introduction to P.D.E.'s and waves for the atmosphere and ocean. Courant Lecture Notes vol 9. American Mathematical Society & Courant Institute of Mathematical Sciences

Majda AJ (2007a) Multiscale models with moisture and systematic strategies for superparameterization. J Atmos Sci 64:2726–2734

Majda AJ (2007b) New multiscale models and self-similarity in tropical convection. J Atmos Sci 64:1393–1404

Majda AJ, Biello JA (2003) The nonlinear interaction of barotropic and equatorial baroclinic Rossby waves. J Atmos Sci 60:1809–1821

Majda AJ, Klein R (2003) Systematic multi-scale models for the tropics. J Atmos Sci 60:393–408

Masmoudi N (2007) Rigorous derivation of the anelastic approximation. J Math Pures et Appliquées 3:230–240

Matsuno T (1966) Quasi-geostrophic motions in the equatorial area. J Met Soc Jap 44:25–43

Muraki DJ, Snyder C, Rotunno R (1999) The next-order corrections to quasigeostrophic theory. J Atmos Sci 56:1547–1560

Nadiga BT, Hecht MW, Margolin LG, Smolarkiewicz PK (1997) On simulating flows with multiple time scales using a method of averages. Theor Comp Fluid Dyn 9:281–292

Névir P (2004) Ertel's vorticity theorems, the particle relabelling symmetry and the energy-vorticity theory of fluid mechanics. Meteorologische Zeitschrift 13:485–498

Norbury J, Roulstone I (eds) (2002) Large scale atmosphere-ocean dynamics I: analytical methods and numerical models. Cambridge University Press, Cambridge

Oberlack M (2006) Symmetries, invariance and self-similarity in turbulence. Springer, Berlin

Ogura Y, Phillips NA (1962) Scale analysis of deep moist convection and some related numerical calculations. J Atmos Sci 19:173–179

Oliver M (2006) Variational asymptotics for rotating shallow water near geostrophy: a transformational approach. J Fluid Mech 551:197–234

Pedlosky J (1987) Geophysical fluid dynamics. Springer, 2nd edn, Berlin

Peters N (2000) Turbulent combustion. Cambridge University Press, Cambridge

Rahmstorf S et al (2004) Cosmic rays, carbon dioxide and climate. EOS 85:38–41

Salmon R (1983) Practical use of Hamilton's principle. J Fluid Mech 132:431–444

Salmon R (1998) Lectures on geophysical fluid dynamics. Oxford Univ. Press, Oxford

Schneider T (2006) The general circulation of the atmosphere. Ann Rev Earth Planet Sci 34: 655–688

Schochet S (2005) The mathematical theory of low Mach number flows. M2AN 39:441–458

Shaw TA, Shepherd TG (2009) A theoretical framework for energy and momentum consistency in subgrid-scale parameterization for climate models. J Atmos Sci 66:3095–3114

Shepherd T (1990) Symmetries, conservation laws, and Hamiltonian structure in geophysical fluid dynamics. Adv Geophys 32:287–338

Sobel A, Nilsson J, Polvani L (2001) The weak temperature gradient approximation and balanced tropical moisture waves. J Atmos Sci 58: 3650–3665

Tessier Y, Lovejoy S, Schertzer D (1993) Universal multi-fractals: theory and observations for rain and clouds. J Appl Meteorol 32:223–250

Wang B, Xie X (1996) Low-frequency equatorial waves in vertically sheared zonal flow. Part I: stable waves. J Atmos Sci 53:449–467

Wheeler M, Kiladis GN (1999) Convectively coupled equatorial waves analysis of clouds and temperature in the wavenumber-frequency domain. J Atmos Sci 56:374–399

White AA (2002) A view of the equations of meteorological dynamics and various approximations. In: Norbury J, Roulstone I (eds) Large-scale atmosphere-ocean dynamics 1: analytical methods and numerical models. Cambridge University Press, Cambridge

Zeitlin V (ed) (2007) Nonlinear dynamics of rotating shallow water: methods and advances. Elsevier Science Publishers B.V.

Zeytounian RK (1990) Asymptotic modeling of atmospheric flows. Springer: Heidelberg

21 Modern Techniques for Numerical Weather Prediction: A Picture Drawn from *Kyrill*

Nils Dorband · Martin Fengler · Andreas Gumann · Stefan Laps
Gais, Schweiz

W. Freeden, M.Z. Nashed, T. Sonar (Eds.), *Handbook of Geomathematics*, DOI 10.1007/978-3-642-01546-5_21,

Abstract This chapter gives a short overview on modern numerical weather prediction (NWP): The chapter sketches the mathematical formulation of the underlying physical problem and its numerical treatment and gives an outlook on statistical weather forecasting (MOS). Special emphasis is given to the *Kyrill* event in order to demonstrate the application of the different methods.

1 Introduction

On 18th and 19th of January 2007, one of the most severe storms during the last decades came across Europe: *Kyrill* as it was called by German meteorologists. With its hurricane-force winds *Kyrill* was leaving a trail of destruction in its wake as it traveled across Northern and Central Europe. Public life broke down completely: schools, universities, and many companies have been closed beforehand; parts of the energy supply and public transport came to a virtual standstill; hundreds of flights have been canceled and finally more than several dozens found their death due to injuries in accidents. The total economical damage has been estimated to EUR 2.4 billion.

Starting from this dramatic event this chapter tries to sketch the strengths, weaknesses, and challenges of modern weather forecasting. First, we will introduce the basic setup in tackling the underlying physical problem. This leads us to the governing dynamical equations and the approach used by European Centre for Medium-Range Weather Forecasts (ECMWF, www.ecmfw.int) for solving them. To complete the picture we also show how ensemble prediction techniques help to identify potential risks of a forthcoming event. And finally we show how model output statistics (MOS) techniques—a statistical post-processing method—help to estimate the specific impact of such weather phenomena at specific locations. Clearly, in each of these steps mathematics plays the fundamental role. However, in order to clarify the relation between mathematics and meteorology we demonstrate several of the aspects discussed below on the example of the winter storm *Kyrill*.

1.1 What Is a Weather Forecast?

Weather forecasting is nothing else than telling someone how the weather is going to develop. However, this definition does not explain the mechanics that are necessary for doing this. At the beginning of the twenty-first century it is a lot more than reading the future from stars or throwing chicken bones as it might have been hundreds of years ago—although some of the readers might suggest this.

Today the demands for precise weather forecasts are manifold. Classically, the main driver for modern meteorology came from marine and aeronautical purposes. But next to these aspects it is clear that for agriculture, insurance business, or energy producers the weather is an increasingly important economic factor. Actually, there are studies estimating that 80% of the value-added chain is directly or at least indirectly depending on weather. Last but not least, the human itself has an intrinsic interest in weather when clothing for the day. Currently, the market for weather forecasts is still heavily developing. At the moment one estimates the turnover in the weather business to be $10 billion.

If one talks about meteorology and weather forecasts one is usually talking about forecasts lasting for the next couple of days. One distinguishes between real-time forecasts of the next few hours—commonly called *nowcasting*—and the classical weather forecast for several days (usually up to 10). When going beyond 10 days it becomes more convenient to speak of trend analyses for the next several weeks, a terminology that indicates that even the scientists know about the strong challenges in doing forecasts for such a time interval. Forecasts that cover months or years are actually no common topics in meteorology but rather in climatology. However, it should be outlined that meteorology pushes the virtual border to climatology step by step from days to weeks—and currently, one is heavily working on monthly forecasts by introducing atmosphere and ocean coupled systems to meteorology which was once the hobby-horse of climatologists.

Finally, there is a rather small community that is specialized to analyze the dynamical forecasts derived from above. By applying statistical techniques starting from *linear regression* and ending up at complex *nonlinear models* between historical model data and stations' measurements, one is able to refine the forecasts for specific stations. This immediately distills the real benefit of the forecast.

Now, let us have a closer look at the different ingredients of a weather forecast. At first glance, we are confronted with an initial value problem. Thus, once endowed with the initial state of the atmosphere and the complete set of all physical processes describing the world outside, we are able to compute in a deterministic way all future states of the atmosphere, such as temperature, precipitation, wind, etc. Unfortunately, in practice we know little about this initial state which introduces a significant uncertainty right from the beginning. Due to the nonlinear nature of the dynamical problem, this uncertainty can lead to very large errors in the prediction. In the most extreme cases, it can even drive the numerical model into a completely wrong atmospheric state, which can lead to missing or not properly predicting important events like *Kyrill*. Statistical methods can be applied for dealing with such uncertainties. We sketch out so-called ensemble techniques immediately after having discussed the deterministic case.

2 Data Assimilation Methods: The Journey from 1d-Var to 4d-Var

Numerical weather models are central tools for modern meteorology. With rapidly increasing computer power during the last decade, decreasing cost of hardware and improvements in weather and climate codes and numerical methods, it has become possible to model the global and mesoscale dynamics of the atmosphere with accurate physics and well-resolved dynamics. However, the predictive power of all models is still limited due to some very fundamental problems. Arguably the most severe one is our ignorance of the initial condition for a simulation. While, in many areas of the world, we do have lots of data from ground based weather stations, for the higher layers of the atmosphere we must rely either on very sparse direct measurements, like radio soundings, or on remote sensing observations, that are obtained for example from radar stations or satellites. The amount and quality of such remote sensing data is increasing rapidly, but they are often in a form that is not particularly useful for numerical weather prediction (NWP). It is highly nontrivial to properly inject information into the NWP models, when the observed quantities (as for example radar reflectivity) are only indirectly related to the model parameters (typically temperature, pressure, humidity, and wind velocities). At the same time, assimilation of such data is a key element for creating realistic initial states of the atmosphere. Some of the basic techniques for assimilating data are described in this section.

2.1 Observational Nudging

A simple but effective approach to data assimilation is to modify the background analysis by terms proportional to the difference between the model state and the observational data. An example is the widely used Cressman analysis scheme. If $\mathbf{x}_b$ is the background analysis and $\mathbf{y}_i$ a vector of i observations, the model state $\mathbf{x}$ provided by a simple Cressman analysis would be

$$\mathbf{x} = \mathbf{x}_b + \frac{\sum_{i=1}^{n} w(i,j)(\mathbf{y}_i - \mathbf{x}_{b,i})}{\sum_{i=1}^{n} w(i,j)}, \tag{1}$$

where the weights $w(i,j)$ are a function of the distance $d_{i,j}$ between the points i and j and take on the value 1 for $i = j$ (Daley 1991). There are different possible definitions for these weights. In methods that are commonly referred to as observational nudging, the condition $w_{i=j} = 1$ is dropped, so that a weighted average between the background state and the observations is performed. Observational nudging can be used as a four-dimensional analysis method, that is, observational data from different points in time are considered. Instead of modifying the background state directly at an initial time, source terms are added to the evolution equations, so that the model is forced dynamically toward the observed fields. Effectively, this is equivalent to changing the governing evolution equations. Therefore the source term has to be chosen small enough to prevent the model from drifting into unrealistic physical configurations.

Observational nudging is still used in lots of operational NWP systems. It is a straightforward method, with a lot of flexibility. A common criticism is that the modifications are done without respecting the consistency of the atmospheric state which might lead to unrealistic configurations.

2.2 Variational Analysis

In variational data assimilation, a cost function is defined from the error covariances and an atmospheric state that minimizes that cost function is constructed. If model errors can be neglected and assuming that background and observation errors are normal, unbiased distributions, the cost function is

$$J = \frac{1}{2}(\mathbf{x}(0) - \mathbf{x}_b(0))^{\mathrm{T}}\mathbf{B}^{-1}(\mathbf{x}(0) - \mathbf{x}_b(0)) + \frac{1}{2}(\mathbf{y} - \mathbf{H}(\mathbf{x}))^{\mathrm{T}}\mathbf{R}^{-1}(\mathbf{y} - \mathbf{H}(\mathbf{x})), \tag{2}$$

where $\mathbf{x}_b$ is the background field, $\mathbf{y}$ the vector of observations, and $\mathbf{H}(\mathbf{x})$ the observation operator that translates the model fields to the observed quantities. $\mathbf{B}$ and $\mathbf{R}$ are the background error covariance and observation error covariance, respectively. The operator $\mathbf{H}(\mathbf{x})$ provides a mechanism for assimilating any observational quantities that can be derived from the model parameters, without the necessity of solving the inverse problem. An example is radar reflectivity: it is much more difficult to match the atmospheric conditions to a radar image than to compute a reflectivity out of model parameters. The latter is what $\mathbf{H}(\mathbf{x})$ does and is all that is needed for defining the cost function.

Having defined a suitable cost function, variational data analysis is reduced to a high-dimensional linear or nonlinear (depending on the properties of $\mathbf{H}(\mathbf{x})$) minimization problem (see Lorenc 1986; Menke 1984; Tarantola 1987 for more detailed discussions and Baker et al. 2004 or Courtier et al. 1998 as examples for implementations of such algorithms).

In order to find solutions to the minimization problem, a number of simplifying assumptions can be made. One simplification has already been introduced, when we neglected the model errors in Eq. 2.

Another common simplification is to evaluate all observation operators at a fixed time only, neglecting the time dependence of the observations. This leads to the so-called 3D-VAR scheme (see Parrish and Derber 1992, for more details). The resulting cost function is then minimized using variational methods. A method closely related to 3D-VAR is Optimal Interpolation, which uses the same approximation, but solves the minimization problem not via variational methods but by direct inversion (see Bouttier and Coutier 2001, and references given therein).

In 4D-VAR methods, the time parameter of the observations is taken into account. Therefore the assimilation of data is not only improving the initial state of the model, but also the dynamics during some period of time. The minimization of the cost function is then considerably more difficult. Commonly applied methods for finding a solution are iterative methods of solving in a linearized regime, and using a sequence of linearized solutions for approximating the solution to the full nonlinear problem (Bouttier and Coutier 2001).

While the variational methods outlined above are aiming toward optimizing the state vector of the full three-dimensional atmosphere, they are general enough to be applied to simpler problems. An example that is frequently encountered is 1D-VAR, where assimilation is done only for a vertical column at a fixed coordinate and time. This method plays an important role in the analysis of satellite data.

There are other assimilation methods that can be combined with the previously discussed methods, or even replace them. One of the more widely used ones is the Kalman filter, which can be used to assimilate 4D data (space and time), and is capable of taking into account the time dependence of the model errors.

3 Basic Equations

This section is dedicated to the set of basic, physical, and rather complex equations that are commonly used for numerical weather prediction. Before entering details of the partial differential equations it is necessary to talk about coordinate systems. While the horizontal one (λ, θ) is quite common in mathematics we keep a special eye on the vertical one.

3.1 Vertical Coordinate System

3.1.1 σ-Coordinates

Any quantity that exhibits a one-to-one relation to height z may be used as vertical coordinate. If the hydrostatic approximation is made, pressure is such a quantity, since $\rho > 0$ lets $\partial p/\partial z = -\rho g$ be negative everywhere. In these so-called pressure coordinates, the independent variables are λ, θ, p, t instead of λ, θ, z, t and the height z becomes a dependent variable (Norbury and Roulstone 2002).

The physical height is rarely used as the vertical coordinate in atmospheric simulation models. Some models use pressure as vertical coordinate, because it simplifies the equations, at least if the atmosphere is in hydrostatic balance, which is generally true for synoptic and mesoscale motion. In such models, the 500 hPa isobaric surface (which undulates in space and time) for instance is a fixed reference level. For complex terrain it is better to use *sigma*-coordinates instead of pressure, because a sigma (or *terrain-following*) coordinate system allows for a high resolution just above ground level, whatever altitude the ground level may be. σ-coordinates are defined by

$$\sigma = \frac{p_{\mathrm{sfc}} - p}{p_{\mathrm{sfc}}},$$

where p_{sfc} is the ground-level pressure, and p the variable pressure. The σ-coordinates range from 1 at the ground to 0 at the top of the atmosphere.

The sigma coordinate found the basis for an essential modification that is introduced by the eta coordinates.

3.1.2 η-Coordinates

The fundamental base in the eta system is not at the ground surface, but at mean sea level (Simmons and Burridge 1981). The eta coordinate system has surfaces that remain relatively horizontal at all times. At the same time, it retains the mathematical advantages of the pressure-based system that does not intersect the ground. It does this by allowing the bottom atmospheric layer to be represented within each grid box as a flat "step." The eta coordinate system defines the vertical position of a point in the atmosphere as a ratio of the pressure difference between that point and the top of the domain to that of the pressure difference between a fundamental base below the point and the top of the domain. The ETA coordinate system varies from 1 at the base to 0 at the top of the domain. Because it is pressure based and normalized, it is easy to mathematically cast governing equations of the atmosphere into a relatively simple form.

There are several advantages of eta coordinates compared with the sigma ones, which should be mentioned:

(i) Eta models do not need to perform the vertical interpolations that are necessary to calculate the pressure gradient force (PGF) in sigma models (Mesinger and Janji, 1985). This reduces the error in PGF calculations and improves the forecast of wind and temperature and moisture changes in areas of steeply sloping terrain.

(ii) Although the numerical formulation near the surface is more complex, the low-level convergence in areas of steep terrain are far more representative of real atmospheric conditions than in the simpler formulations in sigma models (Black 1994). Especially, precipitation forecasts improve in these areas significantly, which more than compensates for the slightly increased computer run time.

(iii) Compared with sigma models, eta models can often improve forecasts of cold air outbreaks, damming events, and lee-side cyclogenesis. For example, in cold-air damming events, the inversion in the real atmosphere above the cold air mass on the east side of a mountain are preserved almost exactly in an eta model.

Unfortunately eta coordinates also introduce some drawbacks and come along with certain limitations, for example:

(i) The step nature of the eta coordinate makes it difficult to retain detailed vertical structure in the boundary layer over the entire model domain, particularly over elevated terrain.
(ii) Gradually sloping terrain is not reflected within the Eta models. Since all terrain is represented in discrete steps, gradual slopes that extend over large distances can be concentrated within as few as one step. This unrealistic compression of the slope into a small area can be compensated, in part, by increasing the vertical and/or horizontal resolution.
(iii) By its step nature Eta models have difficulty predicting extreme downslope wind events.

For models using eta coordinates the user is referred to the ETA Model (Black 1994) and, naturally, to the ECMWF model as introduced below.

3.2 The Eulerian Formulation of the Continuous Equations

In the following we gather the set of equations used at ECMWF for describing the atmospherical flow. In detail we follow exactly the extensive documentation provided by IFS (2006a).

To be more specific, we introduce a spherical coordinate system given by (λ, θ, η), where λ denotes the longitude, θ the latitude, and η the so-called hybrid vertical coordinate as introduced above. Then vertical coordinate η could be considered as a monotonic function of the pressure p and the surface pressure p_{sfc}, that is, $\eta(p, p_{sfc})$ such that

$$\eta(0, p_{sfc}) = 0 \qquad \text{and} \qquad \eta(p_{sfc}, p_{sfc}) = 1. \tag{3}$$

Then the equations of momentum can be written as

$$\begin{aligned}
&\frac{\partial U}{\partial t} + \frac{1}{a\cos^2\theta}\left(U\frac{\partial U}{\partial \lambda} + V\cos\theta\frac{\partial U}{\partial \theta}\right) + \dot{\eta}\frac{\partial U}{\partial \eta} \\
&\qquad - fV + \frac{1}{a}\left(\frac{\partial \phi}{\partial \lambda} + R_{\mathrm{dry}}T_v\frac{\partial}{\partial \lambda}\ln p\right) = P_U + K_U \\
&\frac{\partial V}{\partial t} + \frac{1}{a\cos^2\theta}\left(U\frac{\partial V}{\partial \lambda} + V\cos\theta\frac{\partial V}{\partial \theta} + \sin\theta(U^2 + V^2)\right) + \dot{\eta}\frac{\partial V}{\partial \eta} \\
&\qquad + fU + \frac{\cos\theta}{a}\left(\frac{\partial \phi}{\partial \theta} + R_{\mathrm{dry}}T_v\frac{\partial}{\partial \theta}\ln p\right) = P_V + K_V,
\end{aligned} \tag{4}$$

where a is the Earth's radius, $\dot{\eta}$ is the vertical velocity $\dot{\eta} = \frac{\mathrm{d}\eta}{\mathrm{d}t}$, ϕ is the geopotential, R_{dry} is the gas constant of dry air and T_v is the virtual temperature defined by

$$T_v = T\left(1 + \left(\frac{R_{\mathrm{vap}}}{R_{\mathrm{dry}}} - 1\right)q\right),$$

where T is the temperature, q is the specific humidity, and R_{vap} is the gas constant of water vapor. The terms P_U and P_V represent contributions of additional physical background processes that are discussed later on. K_U and K_V denote horizontal diffusion.

Equation (4) is coupled with the thermodynamic equation given by

$$\frac{\partial T}{\partial t} + \frac{1}{a\cos^2\theta}\left(U\frac{\partial T}{\partial \lambda} + V\cos\theta\frac{\partial T}{\partial \theta}\right) + \dot{\eta}\frac{\partial T}{\partial \eta} - \frac{\kappa T_v \omega}{(1 + (\delta - 1)q)p} = P_T + K_T, \tag{5}$$

with $\kappa = \frac{R_{\text{dry}}}{c_{p_{\text{dry}}}}$ (with $c_{p_{\text{dry}}}$ the specific heat of dry air at constant pressure), ω is the pressure coordinate vertical velocity $\omega = \frac{\mathrm{d}p}{\mathrm{d}t}$, and $\delta = \frac{c_{p_{\text{vap}}}}{c_{p_{\text{dry}}}}$ with $c_{p_{\text{vap}}}$ the specific heat of water vapor at constant pressure. Again P_T abbreviates additional physical background processes, whereas K_T denotes horizontal diffusion terms.

The moisture equation reads as

$$\frac{\partial q}{\partial t} + \frac{1}{a\cos^2\theta}\left(U\frac{\partial q}{\partial \lambda} + V\cos\theta\frac{\partial q}{\partial \theta}\right) + \dot{\eta}\frac{\partial q}{\partial \eta} = P_q + K_q, \tag{6}$$

where P_q and K_q are again background process and diffusion terms.

The set of Eqs. (4)–(6) get closed by the continuity equation

$$\frac{\partial}{\partial t}\left(\frac{\partial p}{\partial \eta}\right) + \nabla\cdot\left(\mathbf{v}_H\frac{\partial p}{\partial \eta}\right) + \frac{\partial}{\partial \eta}\left(\dot{\eta}\frac{\partial p}{\partial \eta}\right) = 0, \tag{7}$$

where $\mathbf{v}_H$ is the vector (u, v) of the horizontal wind speed.

Now, under the assumption of an hydrostatic flow the geopotential ϕ in (4) can be written as

$$\frac{\partial \phi}{\partial \eta} = -\frac{R_{\text{dry}}T_v}{p}\frac{\partial p}{\partial \eta}.$$

Then the vertical velocity ω in (5) is given by

$$\omega = -\int_0^{\eta}\nabla\cdot\left(\mathbf{v}_h\frac{\partial p}{\partial \eta}\right)\mathrm{d}\eta + \mathbf{v}_H\cdot\nabla p.$$

By integrating Eq. (7) with boundary conditions $\dot{\eta} = 0$ taken at the levels from Eq. (3), we end up with an expression for the change in surface pressure

$$\frac{\partial p_{\text{sfc}}}{\partial t} = -\int_0^{1}\nabla\cdot\left(\mathbf{v}_H\frac{\partial p}{\partial \eta}\right)\mathrm{d}\eta$$

$$\dot{\eta}\frac{\partial p}{\partial \eta} = -\frac{\partial p}{\partial t} - \int_0^{\eta}\nabla\cdot\left(\mathbf{v}_H\frac{\partial p}{\partial \eta}\right)\mathrm{d}\eta.$$

3.3 Physical Background Processes

When we discuss about physical background processes we talk, for example, about radiation, turbulent diffusion, and interactions with the surface; subgrid-scale orographic drag, convection, clouds and large-scale precipitation, surface parametrization, methane oxidation, ozone chemistry parametrization, climatological data, etc. All of the above-mentioned processes have in common that they are parameterized and triggered subsequently after the computation of the prognostic equations. To make this idea more evident we have in the following a closer look to the generation of clouds and precipitation.

3.3.1 Clouds and Precipitation

The described equations allow for modeling subsequent physical processes that influence via the introduced forcing terms P_x the prognostic equations. For convenience we keep an eye on two important processes in order to demonstrate the general purpose: cloud modeling and large-scale (stratiform) precipitation. We follow the representation given in IFS (2006b).

3.3.2 Clouds

For simplicity we focus only on stratiform (non-convective) clouds.

Having once implicitly introduced a vertical and horizontal grid in space (the Gauss–Legendre transform implies a regular grid) one can define the cloud and ice water content of a specific grid volume (cell) as

$$l = \frac{1}{V} \int_V \frac{\rho_w}{\rho} \, \mathrm{d}V,$$

where ρ_w is the density of cloud water, ρ is the density of moist air, and V is the volume of the grid box. The fraction of the grid box that is covered by clouds is given by a.

Then the time change of cloud water and ice can be obtained by

$$\frac{\partial l}{\partial t} = A(l) + S_{\mathrm{conv}} + S_{\mathrm{strat}} - E_{\mathrm{cld}} - G_{\mathrm{prec}}$$

together with

$$\frac{\partial a}{\partial t} = A(a) + \delta a_{\mathrm{conv}} + \delta a_{\mathrm{strat}} - \delta a_{\mathrm{evap}},$$

where $A(l)$ and $A(a)$ denote the transport of cloud water/ice and cloud area through the boundaries of the grid volume. $S_{\mathrm{conv}}, \delta a_{\mathrm{conv}}$ are the formation of cloud water/ice and cloud area by convective processes, resp. $S_{\mathrm{strat}}, \delta a_{\mathrm{strat}}$ by stratiform condensation processes. E_{cld} is the evaporation rate of cloud water/ice. G_{prec} is the rate of precipitation falling out of the cloud. And finally we denote by δa_{evap} the rate of decrease of the cloud area by evaporation.

For the formation of clouds one distinguishes two cases, namely the processes in case of already existing clouds and the formation of new clouds. Details can be found in IFS (2006b), but trivially spoken, new clouds are assumed to form when the relative humidity is larger than a certain threshold that depends on the pressure level, that is, tropospheric clouds are generated if the relative humidity exceeds 100%. Finally, it is important to mention that the formation of new clouds comes along with evaporative processes which introduces reversibility into the system.

3.3.3 Precipitation

Similarly, we sketch the procedure for estimating the amount of precipitation in some grid-box. The precipitation

$$P = \frac{1}{A} \int PH(l) \, \mathrm{d}A,$$

where the step function $H(l)$ depends on the portion of the cell containing clouds at condensate specific humidity l and A denotes the volume of the grid-box. The precipitation fraction can then be expressed as

$$a = \frac{1}{A} \int H(l)H(P) \, \mathrm{d}A.$$

The autoconversion from liquid cloud-water to rain and also from ice to snow is parametrized in Sundqvist (1978) and can be written as

$$G = ac_0\left(1 - e^{-\frac{l_{\text{cld}}}{l_{\text{crit}}}^2}\right).$$

The reader should be aware that there is also an additional, completely different process contributing to the large-scale precipitation budget, namely in case of clear-sky conditions. Again we refer to IFS (2006b), which also describes ice sedimentation, evaporation of precipitation, and melting of snow.

Rain and snow is removed from the atmospheric column immediately but can evaporate, melt, and interact with the cloud-water in all layers through which it passes.

3.4 The Discretization

There are many ways of tackling the above-mentioned coupled set of partial differential equations. However, for meteorological reasons, one uses a scheme introduced by Simmons and Burridge (1981) based on frictionless adiabatic flow. It is designed such that it conserves angular momentum, which helps avoiding timing problems in traveling fronts. Therefore, one introduces a fixed number of vertical layers at fixed pressure levels—the so-called vertical (finite element) discretization build of cubic B-splines. The prognostic horizontal variables T, u, v, ϕ, q, and p are represented in terms of scalar spherical harmonics (see Freeden et al. 1998, for an extensive introduction). At the moment ECMWF uses a representation of 92 layers in the vertical and horizontally spherical harmonics of degree one to 1279. All (nonlinear) differential operators acting on the spherical harmonics are applied after the transformation from Fourier into space domain on the grid. Then, physical, parameterized background processes are applied in space and one projects back to Fourier domain, where finally the diffusion terms are applied.

For the discretization one leaves the Eulerian representation and uses a Semi-Lagrangian formulation. This is for two reasons. First, Eulerian schemes often require small time steps to avoid numerical instability (CFL condition): that is, the prognostic variable must not be advected more than one grid length per time step. The maximum time step is therefore defined by the strongest winds. To overcome this problem one uses a Lagrangian numerical scheme where the prognostic variable is assumed to be conserved for an individual particle in the advection process along its trajectory. The drawback is that with a pure Lagrangian framework it would be impossible to maintain uniform resolution over the forecast region. A set of marked particles, would ultimately result in dense congestion at some geographical locations, complete absence in other. To overcome this difficulty a semi-Lagrangian scheme has been developed.

In this numerical scheme at every time-step the grid-points of the numerical mesh are representing the arrival points of backward trajectories at the future time. The point reached during this back-tracking defines where an air parcel was at the beginning of the time-step. During the transport the particle is subject to various physical and dynamical forcing. Essentially, all prognostic variables are then found through interpolation (using values at the previous time-step for the interpolation grid) to this departure point.

In contrast to the Eulerian framework the semi-Lagrangian scheme allows the use of large time-steps without limiting the stability. One limitation for stability is that trajectories should not cross each other. Another, that particles should not overtake another. Therefore, the choice of the time-step in the semi-Lagrangian scheme is only limited by numerical accuracy. However, despite its stability properties severe truncation errors may cause misleading results.

Interestingly, one should note when talking about accuracy that the convergence order of the underlying Galerkin method could be massively improved by switching to a nonlinear formulation as proposed in Fengler (2005).

Finally, we would like to outline that the horizontal discretization used for the Gauss–Legendre transformation (e.g., see Fengler 2005) is—due to performance issues—slightly modified. Originally the Gauss–Legendre grid converges massively to the poles, which is due to the zeros of the Legendre polynomials that accumulate at the boundary. This introduces naturally a work overload in polar regions, where little is known about the atmospheric conditions and numerical noise due to the pole convergence/singularity of the underlying vector spherical harmonics. To overcome these problems one integrates over reduced lattices that drop points on each row such that one keeps powers of $2^n 3^m 5^k$, which allow for fast Fourier transforms. Experimentally one was able to show that this modification introduced only minor artifacts that are negligible in comparison to the effort to be spent for avoiding them.

4 Ensemble Forecasts

Now, having sketched the basic dynamics behind numerical weather prediction we may draw from this a bit more frowning picture. Indeed, numerical weather prediction generally suffers from two types of uncertainty: First, the initial state of the atmosphere is known only to an approximate extent and, second, the numerical weather prediction models themselves exhibit an intrinsic uncertainty. In modern numerical weather prediction systems, assessing this double uncertainty employing *ensemble forecasting* has, on the one hand, become a major challenge and, on the other, provides a set of tools for probability-based decision-making. Ultimately, ensemble forecasting allows to quantitatively estimate the potential environmental and entrepreneurial risks of a forthcoming severe weather event. Over the past 20 years, ensemble forecasting has been implemented and further developed at the main weather prediction centers. For an overview of historic and recent developments, see Lewis (2005) and Leutbecher and Palmer (2008).

Historically and practically, there are different ways to tackle the uncertainty issue inherent in numerical weather forecasts. The most basic idea is the one that probably any professional forecaster employs in his daily work: he compares the forecasts of different numerical models, which is sometimes referred to as the "poor man's ensemble." The more sophisticated version is an ensemble simulation, for which a certain numerical model is evaluated many times using different sets of initial conditions as well as different parameter sets for the parametrizations of the atmospheric physics.

Due to the increased computational requirements compared to a single deterministic run, ensemble simulations are usually carried out using a lower horizontal resolution and a smaller number of vertical levels than the main deterministic run of the respective model. In addition to the perturbed ensemble members, one usually launches a control run, with the resolution of the ensemble members and still the best initialization available, the one which is in use for the main deterministic run. The different ensemble members ideally represent the different possible ways in which the current state of the atmosphere might possibly evolve. The variance or spread of the different members of the ensemble as well as the deviation from the control run provide useful information on the reliability of the forecast and on its future development.

The two sources of uncertainty present in numerical weather prediction cannot really be distinguished in the final output of a numerical forecast. A numerical weather prediction model is a highly nonlinear dynamical system living in a phase space of about 10^6 to 10^8 dimensions. Nevertheless, the underlying evolution equations are well defined and deterministic and, accordingly, the system exhibits deterministic chaotic behavior. This implies that small variations in the initial state of the system may rapidly grow and lead to diverging final states. At the same time, the errors from the initial state blend with the errors caused by the model itself, which stem from the choice of the parametrization coefficients, from truncation errors and from discretization errors. Thus, the errors in the final state are flow-dependent and change from one run of the model to another. Technically speaking, the purpose of ensemble forecasting is to appropriately sample the phase space of the numerical model in order to estimate the probability density function of the final outcome.

There are several methods which are commonly used to create the perturbations of the initial state. The perturbations of the initial state have to be set up in such a way that they are propagated during the model run and thus lead to significant deviations of the final state of the ensemble members. The perturbations which grow strongly during the dynamical evolution identify the directions of initial uncertainty which lead to the largest forecast errors.

The first group of methods to create the perturbations of the initial state is based on ensemble-specific data assimilation techniques. The ensemble Kalman filter, which adds pseudo-random numbers to the assimilated observations, used by the Canadian Meteorological Service belongs to this category (Houtekamer et al. 2005).

The second group of methods is based on the so-called bred vector technique. This technique is based on the idea to repeatedly propagate and rescale a random initial perturbation in order to breed the perturbations which are the most important ones in the dynamical evolution. A bred vector technique is being employed by the US National Center for Enironmental Prediction (NCEP) (Toth and Kalnay 1997).

The third group of methods is based on the identification of the leading singular vectors of the operator which is responsible for the propagation of the perturbations. The leading singular vectors have to be identified for each initial state and different ensemble members can then be initialized with different linear combinations of the leading singular vectors. The singular vector technique is employed by ECMWF (Molteni et al. 1996).

The physical effects living on spatial scales which are not resolved by numerical weather prediction models are usually represented by parametrizations as sketched above. The most common approach to introducing model uncertainty is to perturb the parameters of the model's parametrizations. Other less commonly used approaches are multi-model ensembles and stochastic-dynamic parametrizations. For a review of the current methods, which are used in order to represent model error, see Palmer et al. (2005).

It is a highly nontrivial task to adjust an ensemble simulation such that it is neither over- nor under-dispersive. The spread of the members of an ideal ensemble should be such that its probability density function perfectly matches the probability distribution of the possible atmospheric configurations. This can only be reached by properly adjusting both the perturbations of the initial state as well as the perturbations of the model. Interestingly, numerical weather prediction models rather tend to be under-dispersive and rapidly converge toward the climate normal if the perturbations of the model are insufficient. Calibration techniques based on statistics of past ensemble forecasts can be used in a post-processing step in order to improve the forecast skill and in order to adjust the statistical distribution of an ensemble simulation.

Interpreting the outcome of an ensemble simulation is much more sophisticated than interpreting a single deterministic run. The first step is usually to compare the main deterministic run of the respective model with the control run of the ensemble. Both these runs are initialized using the best guess for the initial state of the atmosphere which is available. Large deviations indicate that the model resolution has a strong impact on the outcome for the given atmospheric configuration. In the second step, one usually investigates the spread of the members of the ensemble, their median, and the deviation from the control run. A small spread of the ensemble members indicates a comparatively predictable state of the atmosphere, whereas a large spread indicates an unstable and less predictable state. Finally, the ensemble members allow for probabilistic weather predictions. If, for example, only a fraction of the ensemble members predicts a specific event for a certain region, the actual probability of the occurrence of the event can be derived from this fraction. For a more detailed overview and further references on measuring the forecast skill of ensemble simulations, see Candille and Talagrand (2005)

There are many different ways to depict the outcome of an ensemble simulation. The most prominent one is probably the ensemble plume for a certain location in the simulation domain. For an ensemble plume, the forecasts of all members for a certain parameter are plotted against the lead time. Additional information can be incorporated by including the control run, the deterministic run, the ensemble median, and optionally the climate characteristics. Ensemble plumes allow for a quick overview of the spread of the ensemble members and the easy comparison to the ensemble median, the control run, the deterministic run, and the climate normal (see, e.g., ❯ *Fig. 17*).

5 Statistical Weather Forecast (MOS)

Once endowed with the algorithms and techniques sketched above one can do what is known as (dynamical) numerical weather prediction.

Dropping for a moment any thoughts on physics and mathematics, that is, convergence, stability, formulation of the equations, technical difficulties, and accuracy, a weather model provides us with nothing else but some numerical output that is usually given either as gridded data on different layers and some (native) mesh or by some spherical harmonic coefficients. Either one could be mapped onto some regular grid for drawing charts and maps. However, any kind of regularity in the data immediately gives a convenient access of time series from model data if one archives model runs from the past. Clearly, this opens the way to answer questions concerning accuracy and model performance at some specific location but also allows to refine the model in a so-called post-processing step. This leads us to MOS (Model Output Statistics), which relates the historical model information to measurements that have been taken at a certain coordinate by linear or nonlinear regression. Hence, a dense station network helps to improve the outcome of a numerical weather prediction tremendously. For example, regional and especially local effects that are either physically not modeled or happen at some scale that is not resolved by the underlying model could be made visible in this statistically improved weather forecast. Such *local effects* could be for example luv- and lee-effects, some cold air basin, exposition to special wind systems in valleys, Foehn, inversion, and so on. ❯ *Figure 1* shows the comparison between accuracy of the ECMWF direct model output (DMO) and the ECMWF-MOS at a station that is located on Hiddensee in the Baltic Sea. The MOS system improves significantly the

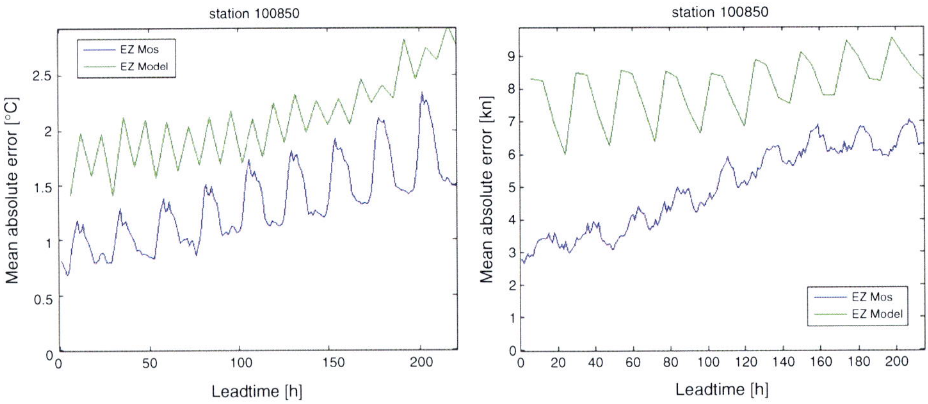

Fig. 1
Comparison of MOS and DMO error in temperature and wind speed

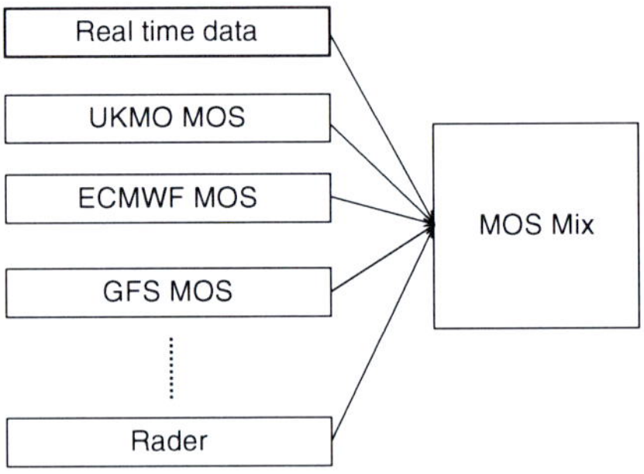

Fig. 2
Combination of different MOS systems

model performance by detecting station specific characteristics like sea breezes, sea-land wind circulation, sea warming in autumn, and so forth *Fig. 1*.

Finally, the accuracy of statistical post-processing systems can be improved by combining the MOS systems of different models: For example, such as a ECMWF-MOS together with a GFS-MOS derived from the NOAA/NCEP GFS model, UKMO-/UKNAE-MOS derived from the Met Office's global and meso-scale (NAE) models, and additional different MOS systems. The combination done by an expert system (see *Fig. 2*) adjusts the weighting by the current forecast skill and reduces the error variances tremendously and, thus, further improves the performance as shown in *Fig. 3*.

6 Applying the Techniques to *Kyrill*

Keeping in mind the methods introduced so far we will have a closer look of how they apply to the *Kyrill* event. We use ECMWF and Met Office NAE model data for an analysis of the

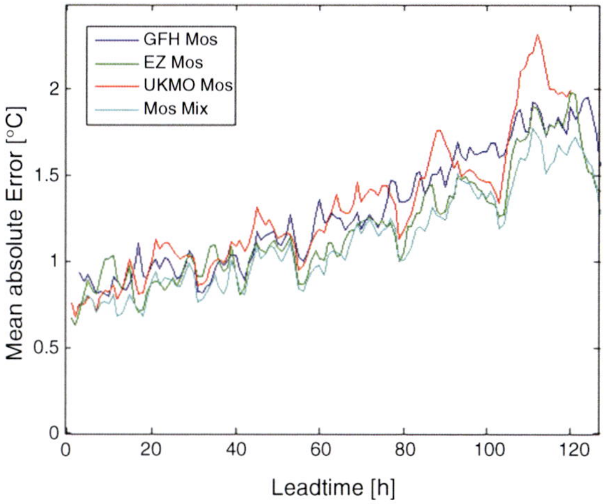

Fig. 3
Increase of accuracy when combining the different forecasts

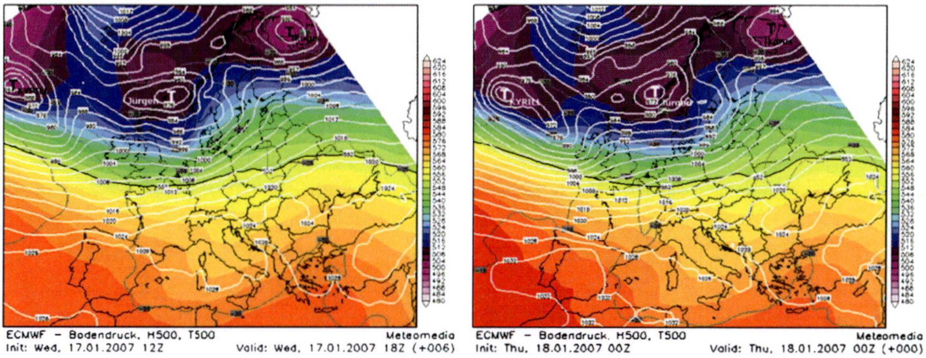

Fig. 4
Geopotential height [Dm] and temperature at 500 hPa at 17 January 2007 18z (left) and 18 January 2007 0z (right)

weather conditions around January 17th and 18th, shown in *Fig. 4*. A detailed synoptic analysis of this severe storm has been done in Fink et al. (2009). The reason for us to choose this event for an analysis based on model output is due to two facts. First, the mesoscale flow patterns of this winter storm are well captured — even 8–10 days before landfall. This outlines the good formulation of the flow and its parametrization done by ECMWF. On the other hand, we observe certain effects that are not resolved by a spherical harmonic representation and which shows the limitations of this formulation, namely sharp and fast traveling cold fronts coming with convectively embedded rainfall. Therefore, at first we have a closer look at the ECMWF analysis to describe the geopotential height field and afterward we use the Met Office NAE model that uses a finite difference representation to resolve very local effects. This model is nested into a global one but operates with lead times of only up to the next 36 h. The reader should note that all times are given in Greenwich Mean Time, also known as UTC, Zulu-time, or z-time.

6.1 Analysis of the Air Pressure and Temperature Fields

The images in ➋ *Fig. 4* show a strong westerly flow that has dominated the first weeks of January. In the evening of January 17th the depression Jürgen, as it was dubbed by German meteorologists, influenced Central Europe with some windy and rainy weather. The first signals of *Kyrill* can be seen on the western edge of the domain over the Atlantic ocean. This map shows also that *Kyrill* is accompanied by very cold air on its backside in great heights, which indicates the high potential of a quickly intensifying low-pressure cell.

In the night of the 18th we observe from ➋ *Fig. 4*, a strong separation of cold air masses in northern and milder ones in southern parts of Central Europe. This line of separation developed over the British Islands to the frontal zone of *Kyrill*. Meanwhile *Kyrill* has shown a pressure minimum of 962 hPa at mean sea level and traveled quickly eastward due to the strong flow at 500 hPa (about 278 to 315 km/h). In the early morning of the 18th *Kyrill* developed strong gradients to a height laying above Spain and northern parts of Africa. This distinctive gradient lead — in the warm sector of *Kyrill* — to the first damages in Ireland, South England, and northern parts of France with gusts of more than 65 kn. The pressure gradients further intensified and *Kyrill* was classified as a strong winter storm (see ➋ *Fig. 5*). While in southern and south-westerly parts of Germany the pressure gradients started rising, the eastern parts were influenced by relatively calm parts of the height ridge between *Kyrill* and Jürgen. The embedded warm front of *Kyrill* brought strong rainfall in western parts of Germany, especially in the low mountain range. The warm front drove mild air into South England, Northern France, and West Germany. Till noon of the 18th *Kyrill* has developed into a large storm depression yielding gusts at wind speeds of more than 120 km/h in wide areas of South England, North France, the Netherlands, Belgium, Luxembourg, Germany, and Switzerland (see ➋ *Fig. 5*). Even parts of Austria were affected.

In the late noon till the early evening of the 18th the heavy and impressively organized cold front of *Kyrill* traveled from North West of Germany to the South East bringing heavy rain, hail, strong gusts, and thunderstorms (see ➋ *Fig. 6*). Due to the strong gradient in pressure and the cold front, gusts at velocities between 120 km/h-160 km/h have been measured even in the plains and lowlands. Finally, it should be mentioned that after the cold front has passed, a convergence and backward oriented occlusion connected to the depression brought heavy rainfall in the North West of Germany.

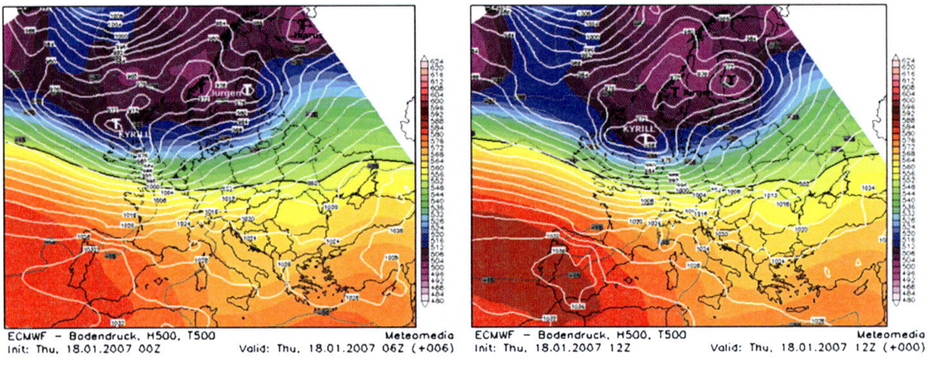

◘ Fig. 5

Geopotential height (dm) and temperature at 500 hPa on 18 January 2007 6z (left) and 12z (right)

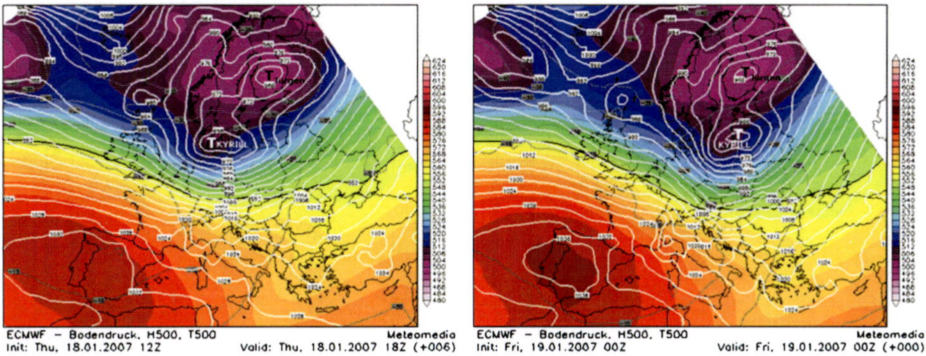

Fig. 6
Geopotential height (dm) and temperature at 500 hPa on 18 January 2007 18z (left) and 19 January 2007 0z (right)

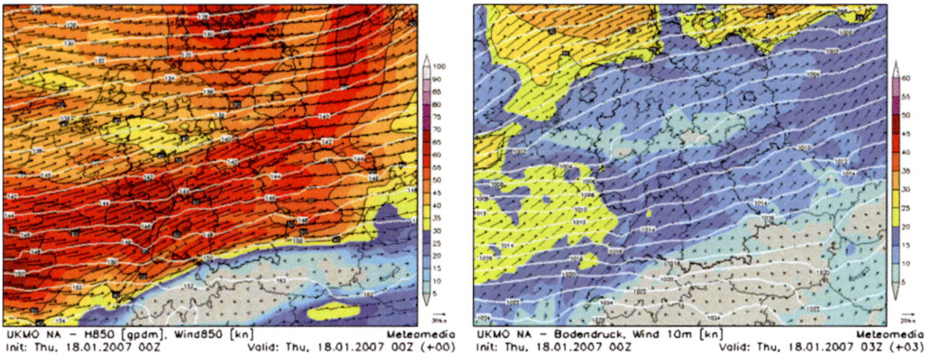

Fig. 7
Wind speed 10m above ground at 18 January 2007 0z (left) and 3z (right)

In the back of *Kyrill*, which was passing through quickly, the weather calmed down and heavy winds only occurred in the South and the low mountain range.

6.2 Analysis of *Kyrills* Surface Winds

The following figures have been computed from the UK Met Office NAE model that uses a finite difference formulation at a horizontal resolution of ca. 12 km. They show the model surface winds at a height of 10 m above West France, Germany, Switzerland, Austria, and Eastern parts of Poland. The reader should note that the colors indicate the strength of the wind speed at the indicated time and not the absolute value of gusts in a certain time interval. The overlay of white isolines show the pressure field corrected to mean sea level.

Starting at midnight of the 18th in *Figs. 7* and *8*, we observe a strong intensification of the surface winds over the North Sea and the described landfall. The strong winds inshore are due to the Channel acting like a nozzle.

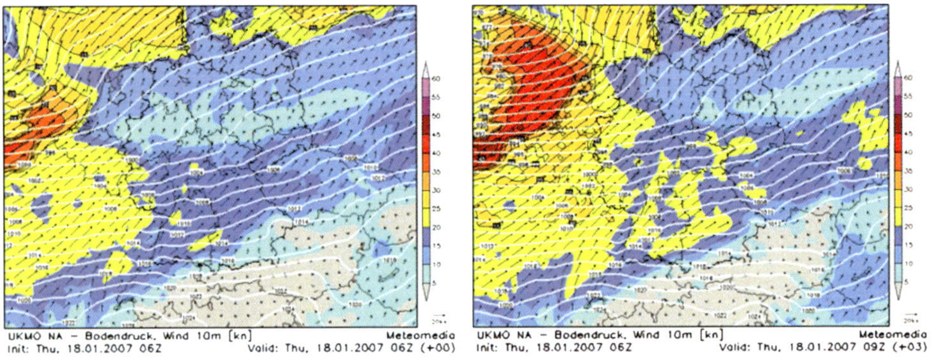

Fig. 8

Wind speed 10 m above ground on 18 January 2007 6z (left) and 9z (right)

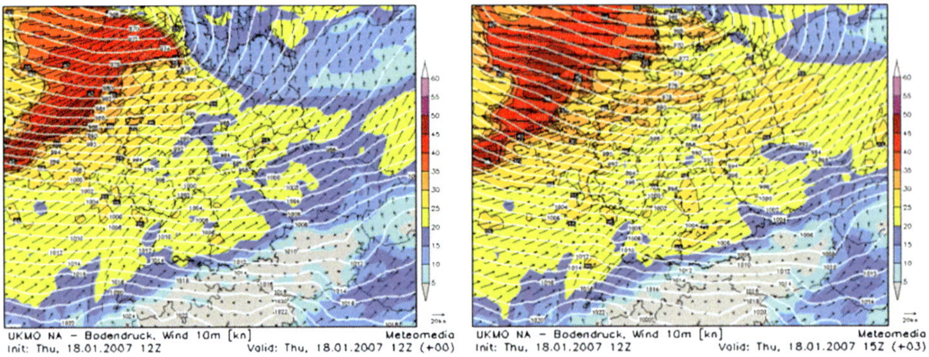

Fig. 9

Wind speed 10 m above ground on 18 January 2007, 12z (left) and 15z (right)

Noteworthy when comparing *Figs. 9–11* is the change in wind direction when the cold front entered northern parts of Germany.

6.3 Analysis of *Kyrill's* 850hPa Winds

To understand the heaviness of *Kyrill's* gusts we have closer look at the model on the 850 hPa pressure level. In these layers, due to the convective nature of the cold front, heavy rainfall causes transport of the strong horizontal momentum into the vertical. Caused by this kind of mixing, fast traveling air at heights of 1200 m and 1500 m is pushed down to earth. These so-called *down bursts* are commonly responsible for the heavy damages of such a storm.

From the images shown in *Fig. 12* we observe a westerly flow at strong but not too heavy wind speeds. The reader should note that the wind speeds are given in knots. Moreover, it should be outlined that the "calm" regions in the Alps are numerical artifacts that have their origin in the fact that the 850 hPa layer is in these mountainous regions in the ground. In the following images shown from *Figs. 13–16* we now observe the cold front with these strong winds passing over Germany.

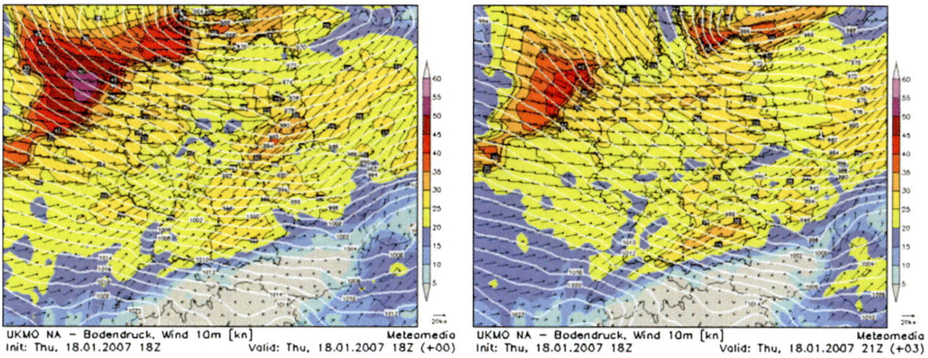

Fig. 10

Wind speed 10 m above ground on 18 January 2007, 18z (left) and 21z (right)

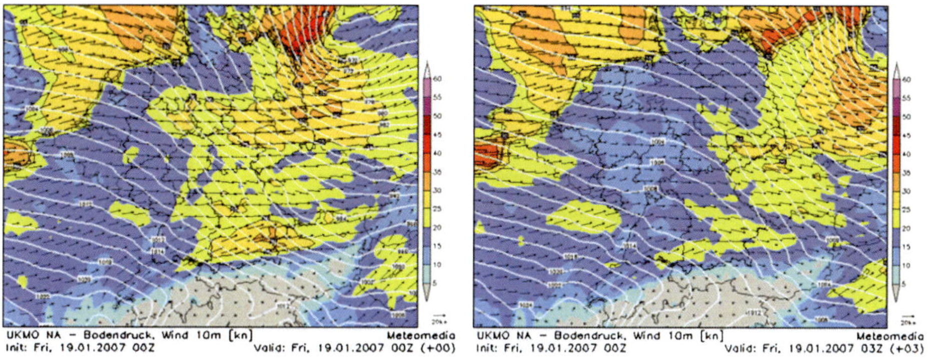

Fig. 11

Wind speed 10 m above ground on 19 January 2007, 0z (left) and 3z (right)

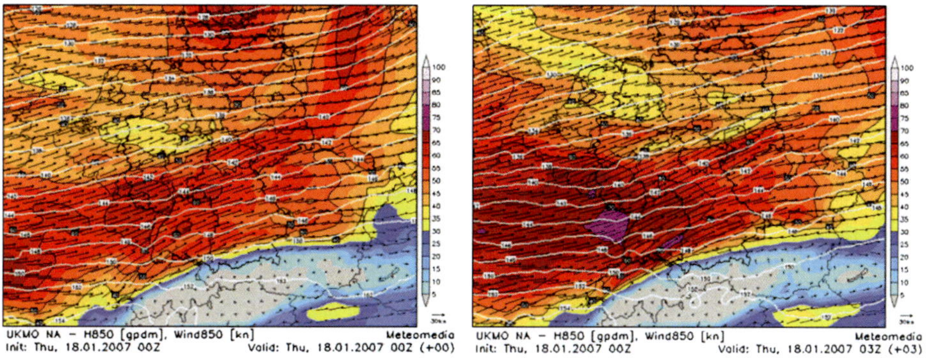

Fig. 12

850 hPa-Wind on 18 January 2007 0z (left) and 3z (right)

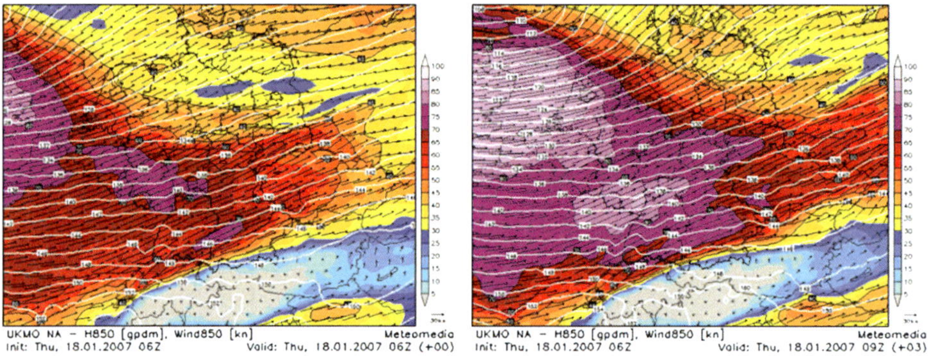

Fig. 13
850 hPa-Wind on 18 January 2007 6z (left) and 9z (right)

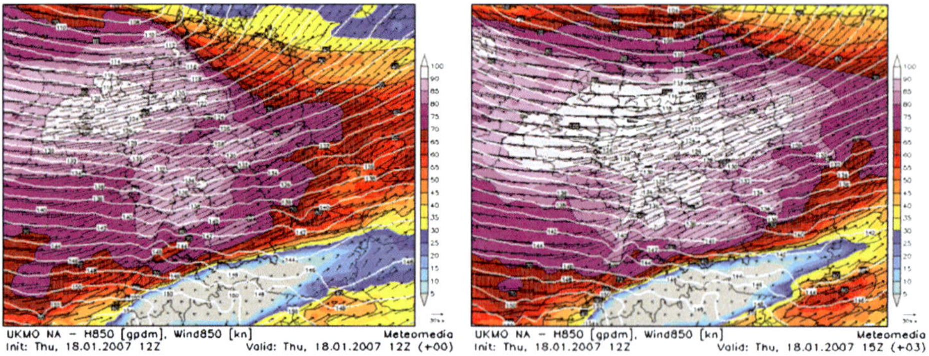

Fig. 14
850 hPa-Wind on 18 January 2007, 12z (left) and 15z (right)

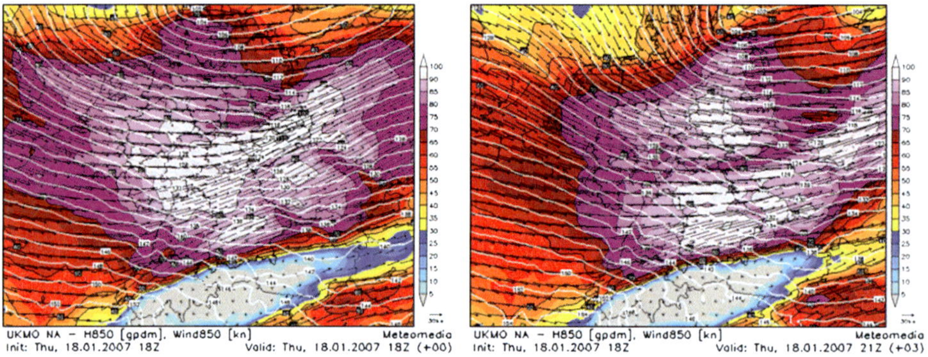

Fig. 15
850 hPa-Wind on 18 January 2007, 18z (left) and 21z (right)

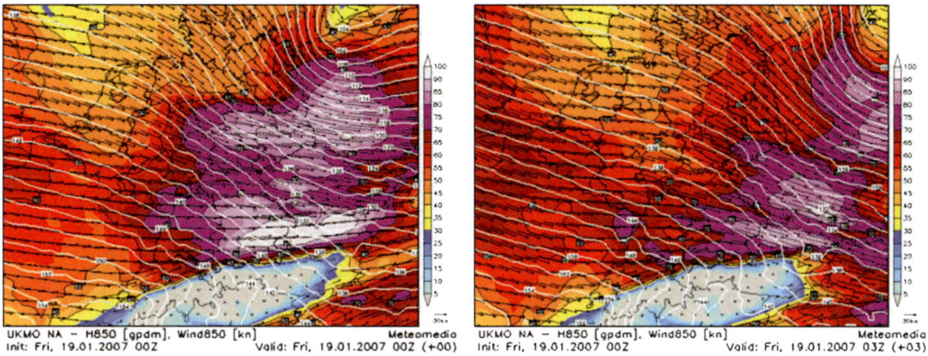

Fig. 16
850 hPa-Wind on 19 January 2007, 0z (left) and 3z (right)

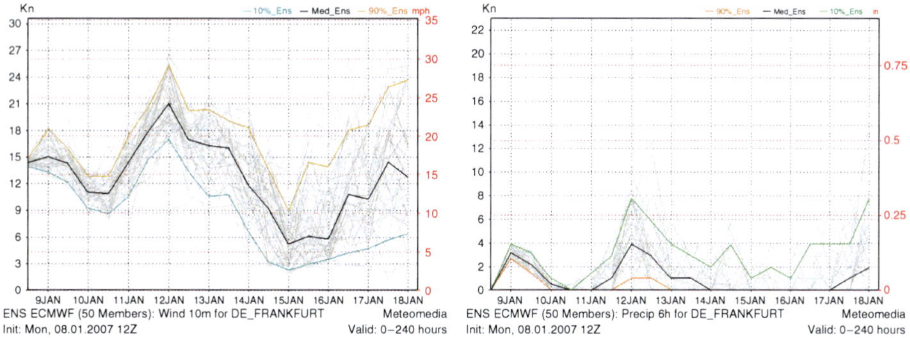

Fig. 17
Ensemble plumes for Frankfurt from 8 January 2007, 12z. 10 m Wind (left) and precipitation (right)

6.4 Ensemble Forecasts

In the scope of the 50-member ECMWF ensemble forecasting system, first signals for a severe winter storm have shown up as early as 10 days in advance. In ➋ *Fig. 17*, we show a so-called ensemble plume plot for the 10-m wind for the city of Frankfurt, Germany. Plotted over the lead-time, the plume plot contains the results for the 50 ensemble members, the ensemble median as well as the 10% and the 90% quantile. In the ensemble run initialized on Monday 8.12., 12z, at least a fraction of the ensemble members exceeded wind speeds of 24 kn for January 18. At the same time, a fraction of the ensemble members exceeded 14 mm for the 6-h precipitation for the same forecast time (see ➋ *Fig. 17*).

It should be noted that, despite the underestimated strength of the event, the timing was already highly precise at that time. However, on January 8, the different members of the ensemble did not exhibit a very consistent outcome for the future *Kyrill* event.

In the following, we will track the ensemble forecasts for the *Kyrill* winter storm for the city of Frankfurt, Germany, while approaching the time of the pass of the main front.

In ➋ *Figs. 18* and ➋ *19*, we show according ensemble plume plots for the wind and the 6-h precipitation for the ensemble run initialized on January 10, 12z.

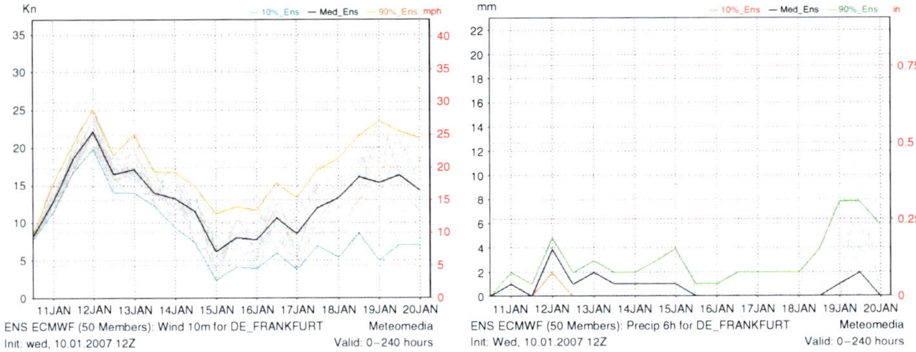

Fig. 18

Ensemble plumes for Frankfurt from 10 January 2007, 12z. 10-m Wind (left) and precipitation (right)

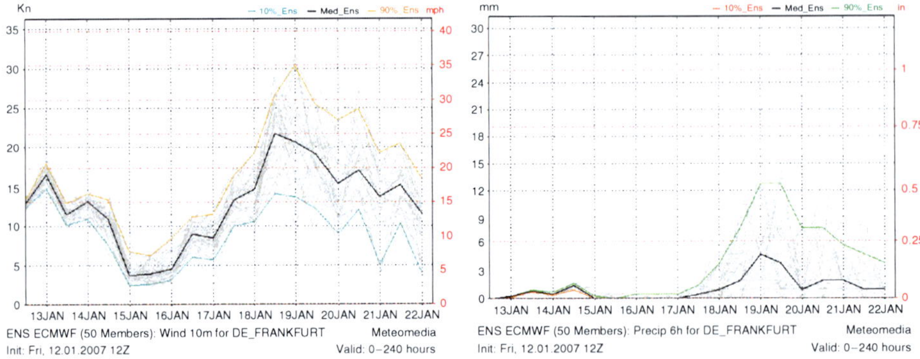

Fig. 19

Ensemble plumes for Frankfurt from 12 January 2007, 12z. Wind (left) and precipitation (right)

The signals for the 10-m wind considerably increased compared to the previous run. A fraction of the ensemble members now exceeds 30 kn and the ensemble median clearly exceeds 15 kn. At the same time, a distinct peak in the expected precipitation develops, with an increased number of the ensemble members exceeding 14 mm for the 6-h precipitation.

The timing for the main event is almost unchanged compared to the previous run. The strongest winds are still expected to occur around 19 January 0z.

In the plots for the ensemble run initialized January 12, 12z shown in ➲ *Figs. 19* and ➲ *20*, the signals for the severe winter storm *Kyrill* become more pronounced. At that time, about 7 days in advance, almost all ensemble members exceed 10m winds of 15kn and the 90% quantile exceeds 30kn. At the same time, the distinct peak in the 6-h precipitation becomes more pronounced. The timing of the event is hardly altered, with the strongest winds still expected to break their way around January 19, 0z.

About 5 days before the main event, the 10% quantile for the 10-m winds almost reaches 18 kn and a very consistent picture develops in the scope of the ECMWF ensemble forecasts (see ➲ *Fig. 20*). Based on the very consistent picture, 10-m winds of up to 30 kn can be expected for the time between January 18, 12z and January 19, 0z. The 6-h precipitation can be expected to reach about 5 mm based on the ensemble median.

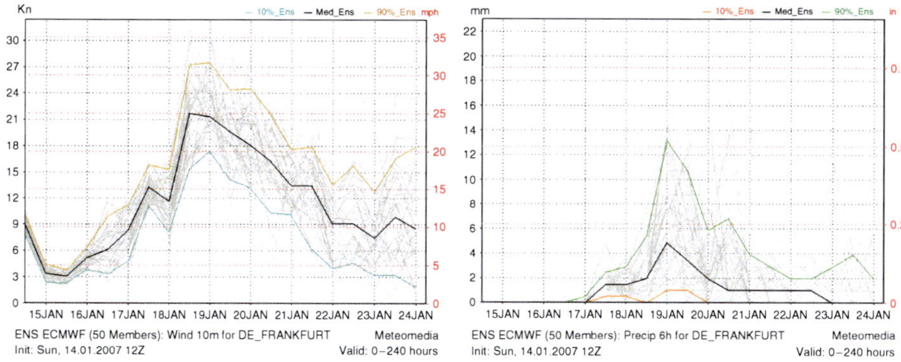

Fig. 20

Ensemble plumes for Frankfurt from 14 January 2007, 12z. Wind (left) and Precipitation (right)

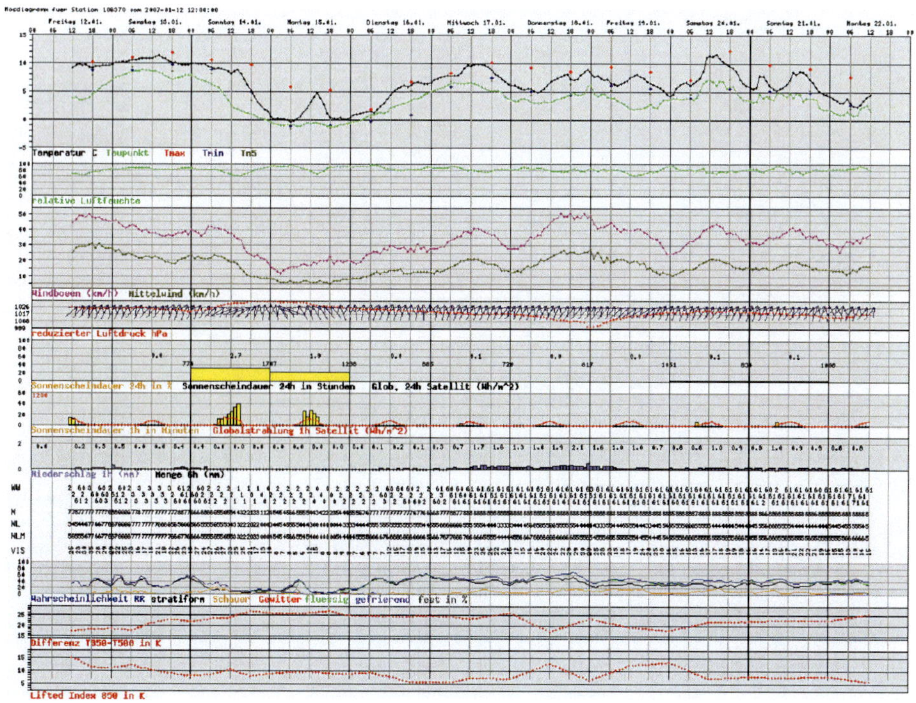

Fig. 21

MOS chart for Frankfurt from 12 January 2007, 12z

6.5 MOS Forecasts

In order to stress the importance of local forecasts which take into account local effects, we will show in the following three MOS forecast charts for the city of Frankfurt, Germany (*Figs. 21–23*).

The MOS forecasts shown in the following are based on the deterministic run of the ECMWF model. Already from the MOS run based on the model output from January 12, 12z,

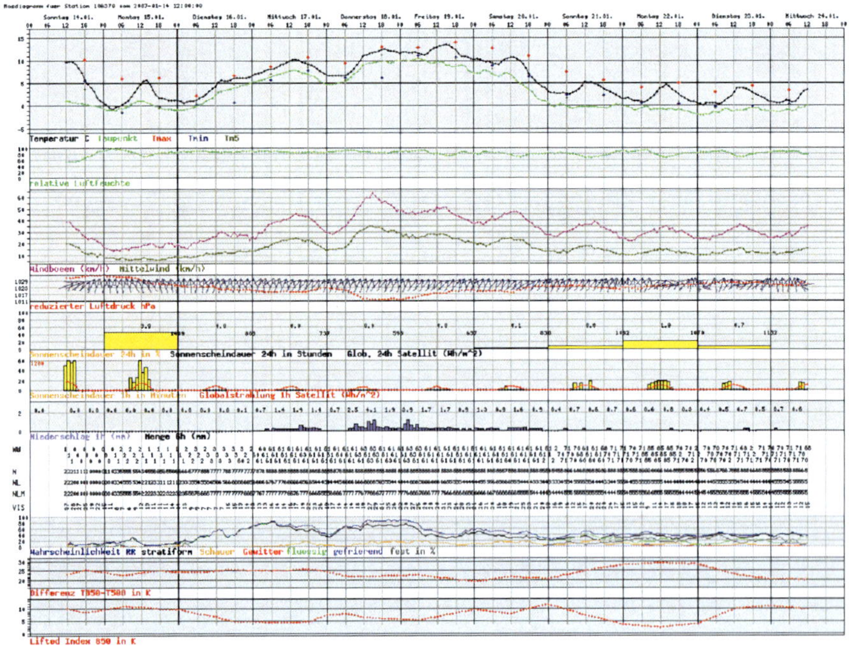

Fig. 22
MOS chart for Frankfurt from 14 January 2007, 12z

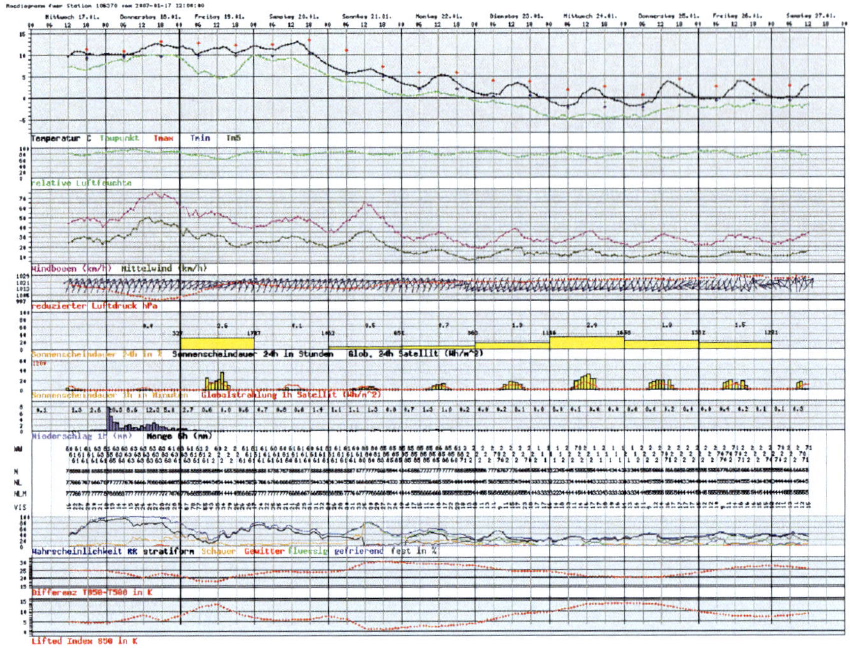

Fig. 23
MOS chart for Frankfurt from 17 January 2007, 12z

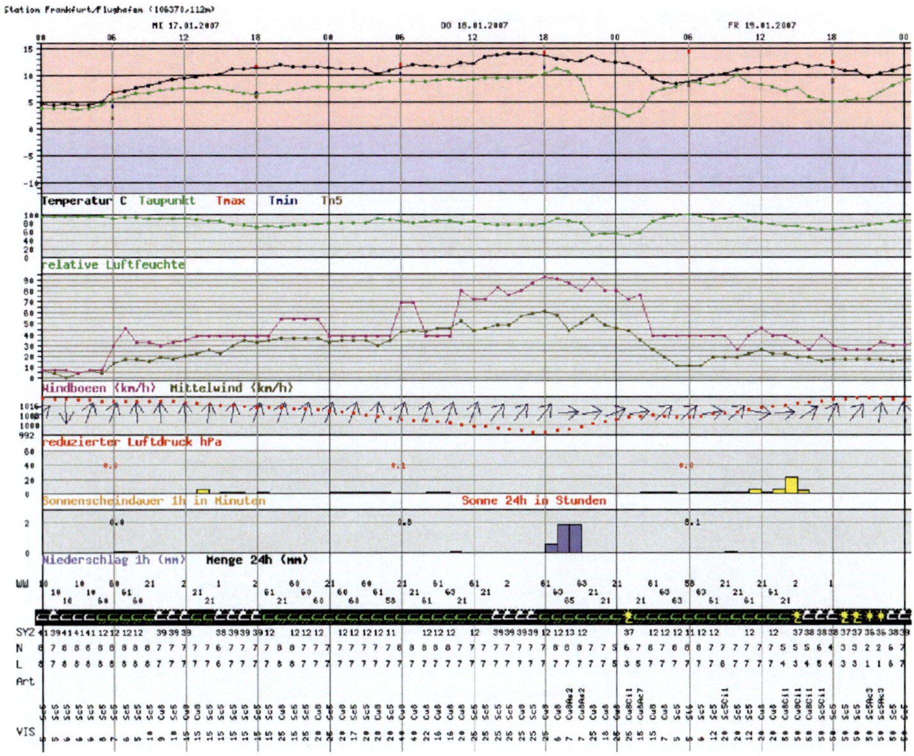

Fig. 24

Observation chart for Frankfurt for the *Kyrill* event

a detailed and very precise picture of the *Kyrill* winter storm can be drawn. All the main features of the event are contained as well as the precise timing, which had already been found in the ensemble forecasts. The strongest gusts with more than 50 km/h are expected for the late evening of January 19. At the same time, the reduced pressure is expected to drop to about 999 hPa at this specific location.

In the MOS run based on the January 14, 12z model output, the pass of the cold front with the following peak of the gusts and the wind becomes more pronounced than in the previous run. The peak in the precipitation was to be expected after the pass of the cold front.

In the MOS run based on the January 17, 12z ECMWF model output, the speed of the maximum gusts increased even more and the expected precipitation rises. From these MOS charts, the importance of local forecasts can easily be understood. Comparing the MOS charts with the ensemble forecasts shown in the previous section, it is obvious that local effects play a crucial role.

In Fig. 24, we show an observation chart in order to verify the MOS forecast from January 17, 12z. The precipitation has been overestimated by the forecasts, but the sea level pressure, the wind, the gusts, and the temperature profile have been captured to a very high precision.

6.6 Weather Radar

For the sake of completeness we should also have a closer look at the radar images for the same time period. The radar images in ➋ *Figs. 25–29* show shaded areas where one observes strong rainfall or hail. The stronger the precipitation event the brighter the color. These images provide a deep insight to the strong, narrow cold front traveling from North Germany to the South. In ➋ *Figs. 27* and ➋ *28* we see an extreme sharpness and an impressively strong organization in the fast traveling cold front. This vastly damaging front had a cross diameter of 4–6 km that the

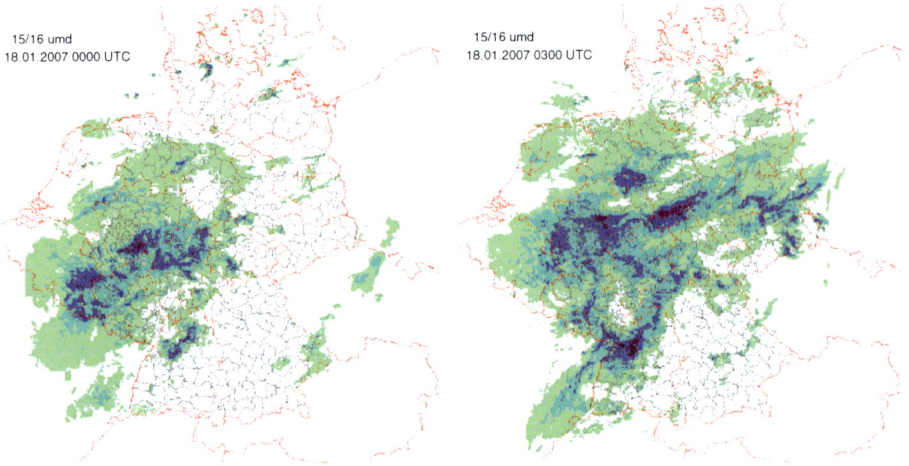

Fig. 25
Radar images from 18.1.2007 0z (left) and 3z (right)

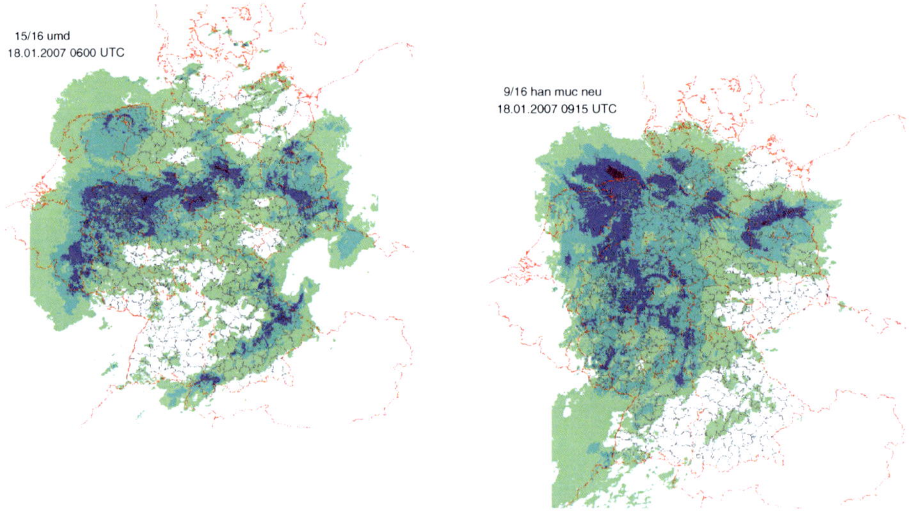

Fig. 26
Radar images from 18.1.2007 6z (left) and 9z (right)

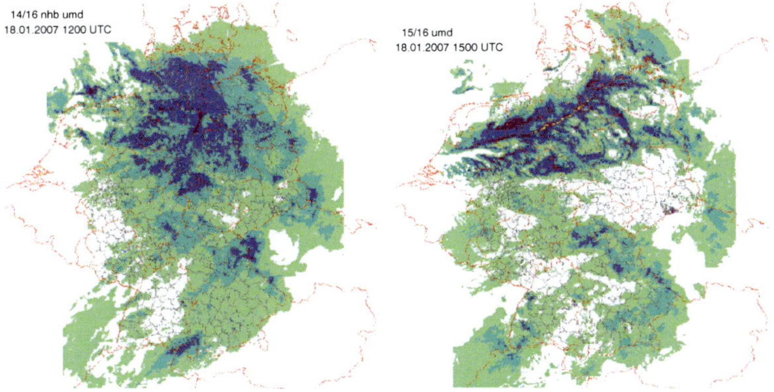

Fig. 27
Radar images from 18.1.2007 12z (left) and 15z (right)

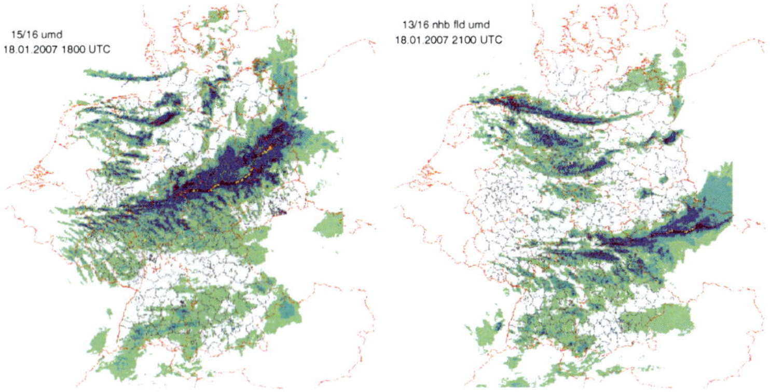

Fig. 28
Radar images from 18.1.2007 18z (left) and 21z (right)

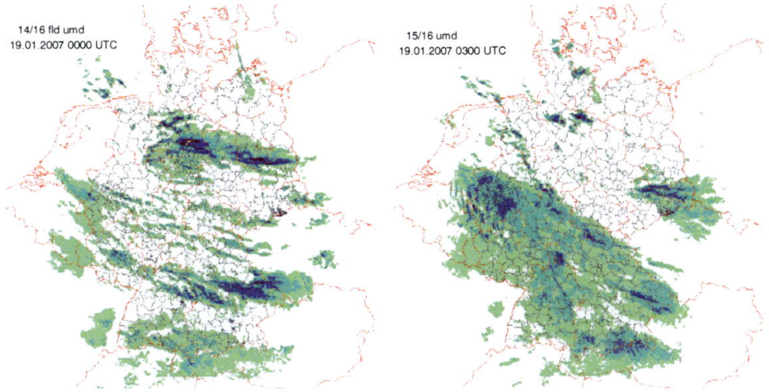

Fig. 29
Radar images from 19.1.2007 0z (left) and 19.1. 3z (right)

models we talked above, unfortunately, did not show due to their resolution. To resolve these pattern properly is a challenging task for the current research.

7 Conclusion

Since the early days of computer simulations scientist have been interested in using them for modeling the atmosphere and predicting the weather. Nowadays these efforts have evolved into essential tools for meteorologists. There still is and will be rapid development in the future, due to the access to fast enough computers and the availability of highly sophisticated mathematical methods and algorithms for solving the underlying nonlinear problems. Some of the most important methods that are used in current operational forecasting codes have been reviewed in this chapter.

After the fundamental set of governing partial differential equations was defined, a formulation and discretization suitable for solving these equations numerically on three-dimensional grids was introduced. On the example of cloud formation and precipitation, we showed how microphysical processes are coupled to such a model. Some commonly used methods for assimilating information from observational data to improve the accuracy of predictions have been outlined.

It is crucial to understand, that even though — as a well defined initial value problem — the models are deterministic, the complexity and nonlinearity of the underlying mathematics as well as our ignorance of exact initial conditions make it difficult to predict the quality of a single forecast. To tackle that problem, statistical methods are developed, so-called Ensemble Forecasts. Finally statistical postprocessing techniques, known as Model Output Statistics (MOS) can be used for further improving the forecast quality at specific locations.

During the last two decades, through methods as the ones described here, numerical weather models opened a window for observing and investigating the atmosphere at an unprecedented level of detail and contribute significantly to our ability to understand and predict the weather and its dynamics. As an example for the type and quality of information we can extract from atmospheric simulations, in combination with statistical analysis and observational data (weather station reports and radar maps) we analyzed the winter storm *Kyrill* and the processes that were leading to this event. For such an analysis, model data provide us with detailed temperature, pressure, and wind maps, that would not be available at a comparable frequency through observational data alone. The dynamics at different pressure levels that was eventually leading to this devastating storm were described and analyzed in detail based on data obtained from the ECMWF and the UK Met Office NAE models. We then looked at the event through ensemble plumes and MOS diagrams, which give us first hints at the storm more than a week in advance and rather precise quantitative predictions about wind speeds and precipitation a couple of days before the event at specific locations.

The comparison to Radar images reveals limitations of the model predictions, due to fine, localized structures that are not accurately resolved on the model grids. This demonstrates the need for better local models with very high spatial resolutions and more reliable coupling to all available observational data. Despite the overall accurate picture we already obtain, such high resolution models can be very valuable when preparing for a severe weather situation, for example for emergency teams that have to decide where to start evacuations or move manpower and machinery.

Acknowledgments

The authors would like to acknowledge Meteomedia, especially Markus Pfister and Mark Vornhusen for many fruitful discussions and help with the ensemble and MOS charts. Moreover, the authors' gratitude goes to ECMWF for providing an extensive documentation and scientific material to their current systems.

References

Baker DM, Huang W, Guo YR, Bourgeois A, Xiao XN. (2004) A three-dimensional variational data assimilation system for MM5: Implementation and initial results. Mon Wea Rev 132: 897.

Black TL. (1994) The new NMC mesoscale eta model: Description and forecast examples. Wea Forecasting 9: 265.

Bouttier F, Coutier P. (2001) Meteorological training course lecture series, ECMWF.

Candille G, Talagrand O. (2005) Evaluation of probabilistic prediction systems for a scalar variable. Quart J Roy Meteor Soc 131: 2131.

Courtier P et al. (1998) The ECMWF implementation of three-dimensional variational assimilation (3DVAR). I: Formulation. Quart J Roy Meteor Soc 124: 1783.

Daley R. (1991) Atmospheric data analysis. Cambridge University Press.

ECMWF Webpage: www.ecmwf.int.

Fengler M. (2005) Vector spherical harmonic and vector wavelet based non-linear galerkin schemes for solving the incompressible Navier-Stokes equation on the sphere. Shaker Verlag, Maastricht, Germany.

Fink AH, Brücher T, Ermert V, Krüger A, Pinto JG. (2009), The European storm kyrill in January 2007: synoptic evolution and considerations with respect to climate change. Nat Hazards Earth Syst Sci 9:405-423.

Freeden W, Gervens T, Schreiner M. (1998) Constructive approximation on the sphere (with applications to geomathematics). Oxford Science Publications, Clarendon, Oxford.

Houtekamer PL, Mitchell HL, Pellerin G, Buehner M, Charron M, Spacek L, Hansen B. (2005) Atmospheric data assimilation with an ensemble kalman filter: Results with real observations. Mon Weather Rev 133: 604.

IFS Documentation (2006a) Cy31r1 operational implementation 12 September 2006; Part III: Dynamics and Numerical Procedures.

IFS Documentation (2006b) Cy31r1 operational implementation 12 September 2006; Part IV: Physical Processes.

Lewis JM. (2005) Roots of ensemble forecasting. Mon Weather Rev 133: 1865.

Leutbecher M, Palmer TN. (2008) Ensemble prediction of tropical cyclones using targeted diabatic singular vectors. J Comput Phys 227: 3515.

Lorenc AC. (1986) Analysis methods for numerical weather prediction. Mon Weather Rev 112: 1177.

Menke W. (1984) Geophysical data analysis: discrete inverse theory. Academic Press, New-York

Mesinger F, Janji Z. (1985) Problems and numerical methods of incorporation of mountains in atmospheric models. Lect Appl Math, 22: 81–120.

Molteni F, Buizza R, Palmer TN. (1996) The ECMWF ensemble prediction system: Methodology and validation. Quart J Roy Meteor Soc 122: 73.

Norbury J, Roulstone I. (2002) Large-scale atmosphere ocean dynamics, vol I. Cambridge, University Press.

Palmer TN, Shutts GJ, Hagedorn R, Doblas-Reyes FJ, Jung T, Leutbecher M. (2005) Representing model uncertainty in weather and climate prediction. Annu Rev Earth Planet Sci 33: 163.

Parrish D, Derber J. (1992) The National Meteorological Center's spectral statistical interpolation analysis system. Mon Weather Rev 120: 1747.

Simmons AJ, Burridge (1981) An energy and angular momentum conserving vertical finite difference scheme and hybrid vertical coordinates. Mon Weather Rev 109:758–766.

Sundqvist H. (1978) A parametrization scheme for non-convective condensation including prediction of cloud water content. Q J R Meteorol Soc 104:677–690.

Tarantola A. (1987) Inverse problem theory. Methods for data fitting and model parameter estimation. Elsevier, Amsterdam.

Toth Z, Kalnay E. (1997) Ensemble forecasting at NCEP: The breeding method. Mon Weather Rev 125:3297.

22 Modeling Deep Geothermal Reservoirs: Recent Advances and Future Problems

Maxim Ilyasov · Isabel Ostermann · Alessandro Punzi
Fraunhofer ITWM, Kaiserslautern, Germany

W. Freeden, M.Z. Nashed, T. Sonar (Eds.), *Handbook of Geomathematics*, DOI 10.1007/978-3-642-01546-5_22,

Abstract Due to the increasing demand of renewable energy production facilities, modeling geothermal reservoirs is a central issue in today's engineering practice. After over 40 years of study, many models have been proposed and applied to hundreds of sites worldwide. Nevertheless, with increasing computational capabilities new efficient methods are becoming available. The aim of this chapter is to present recent progress on seismic processing as well as fluid and thermal flow simulations for porous and fractured subsurface systems. The commonly used methods in industrial energy exploration and production such as forward modeling, seismic migration, and inversion methods together with continuum and discrete flow models for reservoir monitoring and management are reviewed. Furthermore, for two specific features numerical examples are presented. Finally, future fields of studies are described.

1 Introduction

Due to the climate change and the increase of energy production, the need for renewable energy is a rising factor in today's economy, and it features geothermal energy as one of its most promising representatives because of its independence of external influences such as seasonal climatic behavior. The complexity of the entire energy production chain makes the modeling of geothermal reservoirs a difficult challenge for many categories of scientists. In fact, for the geothermal energy production system to be efficient it is of essential importance not only to have a deep understanding of the geologic, thermal, and mechanical configuration of the potential site, but also to be able to predict possible consequences of this invasive procedure. For this reason geologists, engineers, physicists, and mathematicians must share their expertise and work together in order to provide an efficient solution for this ambitious task.

The idea behind industrial geothermal energy production is to use the intrinsic heat stored in the Earth's crust which can be extracted via convection by deep circulation of water. Cold water is pumped into 3–5 km deep reservoirs, warmed up by the underlying heat, and then extracted at a temperature between 100°C and 300°C (e.g., DiPippo 2008). The gained energy can be employed directly for the heat market or it can be used indirectly for electricity generation.

Countries with active volcanism such as Iceland have a longstanding tradition of industrial use of geothermal energy gained by reservoirs with high enthalpy (>180°C), which are easily accessible due to their shallow depths. Nevertheless, the absence of these sources does not exclude the substantial geothermal potential in other regions or countries (for details on the geothermal potential in Germany see, e.g., Jung 2007; Schulz 2009) in the shape of deep reservoirs with low enthalpy (<180°C) occurring in sediment basins or in the vicinity of reservoirs with high enthalpy. For a detailed categorization of geothermal reservoirs due to their temperature, see, e.g., Sanyal (2005). Basically, there are three different types of systems for deep geothermal power production: thermowells, hydrothermal reservoirs, and petrothermal reservoirs. In a thermowell the heat transfer medium circulates in a closed cycle within a u-pipe or a coaxial heat exchanger, therefore only one borehole has to be drilled. Although there is the advantage of no contact with the groundwater, the relatively low productivity gives rise to focus on the aforementioned alternatives representing open systems. The concept behind hydrothermal systems is to let thermal water found in deep reservoirs circulate between two or three drilled deep wells through a previously existing aquifer. Typically, these reservoirs consist of a porous medium layer heated from below by a hot stratum of impermeable material.

By contrast, in petrothermal systems the water flows through fractured rock the porosity of which can be enhanced by hydraulic stimulation. In the latter case water is artificially pumped into the reservoir. In this chapter we focus on hydro- and petrothermal systems.

The key issues that mathematicians have to face are the detection of potential reservoirs along with the surrounding subsurface structure including temperature, capacity, and hydraulic characteristics of the reservoir and the development of a comprehensive model describing the dynamics of the production process, in particular as far as flow, temperature, and composition of the fluid are concerned. Recent events such as earthquakes (e.g., Phillips et al. 2002) show that another fundamental mathematical problem is the understanding of inner stress conformation and dynamics. A key issue is the determination of the fault patterns and its accompanying fractures in the area of interest in the deeper underground. Regarding the geophysical detection of rock morphology, seismic migration methods (see, e.g., Biondi 2006; Yilmaz 2001 and the references therein) based on the acoustic and elastic-isotropic wave equation are currently used. Additionally, methods concerning seismic inversion (velocity analysis) like traveltime reflection tomography (e.g., Symes 1995; Bleistein et al. 2001; Yilmaz 2001) are applied to define approximate velocity and density models of the explorative area. The main lack of these methods is that they provide no information about porosity and permeability without available geological logs.

Based on the knowledge gained by seismic modeling (location, orientation, and aperture of cracks) a mathematical description of the stress field prior to production can be provided. Also, due to the danger of fluid flow in a highly stressed system it is of crucial importance to understand the evolution of the stress during the production process. Another core issue is modeling the actual underground flow of water, which must take into account several aspects such as thermal flow, chemicals flow, evolution of the pressure gradient, and eventual consequences on the highly stressed rock conformation. This means solving coupled equations for elasticity as well as for two-, three- or multiphase fluid, heat, and mass flow in porous or fractured media (see, e.g., Pruess 1990 for the flow problems).

The complexity of the topic and the wide variety of strategies used to model these highly problematic situations, make it necessary to give an outline of the tools actually employed. For this reason, the aim of this chapter is to provide a comprehensive summary of mathematical knowledge and techniques today in use for modeling geothermal reservoirs. In ➲ Sect. 2 we present the main physical problems that arise in geothermal reservoirs, as far as seismic imaging and inversion as well as flow modeling are concerned, while ➲ Sect. 3 reviews the actual strategies in use to understand and solve these problems. In ➲ Sect. 4 some examples of numerical models for seismic and heat transport problems are presented, whereas in ➲ Sect. 5 ideas and trends for further studies are suggested. Finally, we draw some conclusions in ➲ Sect. 6.

2 Scientific Relevance

Independent of the actual type of a deep geothermal system, the process that a geothermal site and the associated power plant traverses (from exploration via simulation and prediction to actual operation) is strongly influenced by mathematical modeling of the core problems prior to and during the operation of the geothermal power plant. The key issues that we address are the seismic exploration of the site and the subsequent modeling of transport and stress processes inside the detected reservoir.

2.1 Reservoir Detection

Each seismic problem starts with data acquisition. For this purpose some kind of source (explosion, vibrator, airgun, etc.) is propagated through the area of interest. These waves are reflected at the places of impedance contrasts (rapid changes of the medium density), propagated back, and then recorded on the surface or/and in available boreholes by receivers of seismic energy. The recorded seismograms are carefully processed to detect fractures along with their location, orientation, and aperture, that are needed for interpretation and definition of the target reservoir.

Nowadays no specific methods for geothermal energy exploration are available. In order to determine the structure, depth, and thickness of the target reservoir, the standard seismic methods from the hydrocarbon industry are applied to the two- or three-dimensional seismic sections. This approach often has rather poor quality in the sense of folding and resolution. Since the interest of geothermal projects is not only focused on structure heights and traps as in oil field practice, but also on fault zones and karst structures opened under recent stress conditions, the interpretation for geothermal needs is significantly complicated.

All methods can be distinguished between time- or depth-migration strategies and between application to post- or pre-stack data sets. The time-migration strategy is used to resolve conflicts in dipping events with different velocities. The depth-migration strategy handles strong lateral velocity variations associated with complex overburden structures.

The numerical techniques used to solve the migration problem can generally be separated into three broad categories: (i) summation or integral methods based on the integral solution, e.g., Kirchhoff Migration (see Yilmaz 2001 and the references therein); (ii) methods based on the finite-difference schemes, e.g., reverse-time migration (see Baysal et al. 1983; Yilmaz 2001; Popov et al. 2008); (iii) transform methods based on frequency–wavenumber implementations, e.g., frequency-space and frequency-wavenumber migration (see Yilmaz 2001; Claerbout 2009). All these migration methods rely on some approximation of the scalar acoustic or elastic-isotropic wave equation.

According to the geothermal requirements, highly accurate approximations and efficient numerical techniques must be chosen in order to handle steep dipping events and complex velocity models with strong lateral and vertical variations, as well as to construct the subsurface image in a locally defined region with high resolution on available computational resources.

These migration algorithms require an accurate velocity model. Adapting the interval velocity using an inversion by comparing the measured travel times with simulated travel times is called reflection tomography. There are many versions of reflection tomography, but they all use ray-tracing techniques and are formulated mathematically as an optimization problem. The most popular and efficient methods are ray-based travel time tomography (e.g., Bishop et al. 1985), waveform- and full-wave inversion tomography (FWI) (see Tarantola 1984), and Gaussian beam tomography (e.g., Popov et al. 2008; Semtchenok et al. 2009).

The "true" velocity estimation is usually obtained by an iterative process called migration velocity analysis (MVA) (see Biondi 2006), which uses the kinematic information gained by the migration and consists of the following steps:

initial step perform reflection tomography of the coarse velocity structure using a priori knowledge about the subsurface

iterative step migrate seismic data sets and apply the imaging condition; update the velocity function by tomography inversion

The global convergence of this optimization problem is a first-order concern. Moreover, the problem is non convex and consequently has several local minima (see Bleistein et al. 2001; Biondi 2006 and the references therein).

2.2 Stress and Flow Problems

A central problem in modeling the behavior of a deep geothermal reservoir, is to predict the possible consequences of injecting cold fluid into a porous or fractured medium. For this reason, it is crucial to determine the specific reservoir geometry, the mechanical properties of the rock, and the inner potential mechanical energy prior to production. As a matter of fact, these parameters influence the borehole stability, the orientation of hydraulically induced fractures, the fluid flow anisotropies, and the slip and dilation tendency of existing faults and fractures.

In enhanced geothermal systems (EGS), that are one category of petrothermal systems, the possible efficiency of the geothermal reservoirs is too low for industrial purposes. Hence, the usual method to improve the productivity is to use fluid pressure to increase the aperture of the fractures and their conductivity and, thus, the permeability of the reservoir. With the help of hydraulic stimulation (e.g., Ghassemi 2003; Zubkov et al. 2007) the fractures propagate along the maximum principle stress direction. The fracture path is governed by the actual stress field in the reservoir acting at the fracture tips (e.g., Moeck et al. 2009). Often the main difficulty of the mechanical models that predict the fracture growth during the stimulation process is to couple the applied pressure of the fluid with the fracture aperture and the rock stress, since the parameters that govern this coupling (such as the rock stiffness and the shear stiffness) are very difficult to detect.

Fluid diffusion into the rock formation, the so-called leak-off, causes the pore pressure to rise and therefore the surrounding rock to expand (e.g., Biot 1941; Rice and Cleary 1976). Thus, the relation between pore pressure and stress (see Engelder and Fischer 1994; Addis 1997; Hillis 2000, 2001, 2003; Zhou and Ghassemi 2009 and the references therein) does not only lead to changes in permeability and porosity due to crack growth, but also to changes in the velocity regime of seismic waves, microseismicity, reactivation of slips and faults, disturbance of borehole stability, and changes in the flow paths of the fluid through the reservoir (e.g., Altmann et al. 2008).

Along with the poroelastic effects on the rock matrix, the other major sources of possible rock displacement, even after terminating the injection of fluid, are thermoelastic processes due to the temperature difference between the injected fluid and the medium (see Hicks et al. 1996; Ghassemi 2003; Ghassemi and Tarasovs 2004; Ghassemi and Zhang 2004; Brouwer et al. 2005; Yin 2008). As a matter of fact, the hot rock is cooled down by the injection of cold fluid and, as a consequence, the rock shrinks contrasting the poroelastic effects of dilation (e.g., Nakao and Ishido 1998). Normally, on a short-time scale the mechanical effects are dominant and the thermoelastic effects can be neglected, but in a long-term simulation they have to be taken into account. The details concerning the modeling of the stress field in geothermal applications are beyond the scope of this chapter. For a more specific treatment of this problem the reader is referred to Ghassemi (2003), Rutqvist and Stephansson (2003) as well as to Evans et al. (1999), Jing and Hudson (2002), and the references therein for comprehensive reviews.

The problem of fluid flow in a geothermal reservoir is very complex for several reasons. First of all, the data regarding geometry and composition of the domain are usually lacking or incomplete. Moreover, the direct feedback from the physical situation needed to compare theoretical results with practical measurements is difficult to obtain. Furthermore, the large number of parameters and coupled processes that have to be taken into account make the development of a comprehensive model a very hard task.

The three main phenomena that have to be modeled are fluid flow, thermal flow, and chemical flow. Chemicals can dissolute and precipitate during fluid circulation eventually reducing the permeability of the medium and influencing the flow (see Cheng and Yeh 1998; Durst and Vuataz 2000; Jing et al. 2002; Kühn and Stöfen 2005; Kühn 2009 and the references therein), but the contribution of this process to thermal flow is limited, therefore most of the coupled models just neglect it. Depending on the physical conditions of the reservoir the problems arising during modeling fluid and heat flow differ.

As far as hydrothermal systems are concerned the mainstream assumption is to represent the reservoir as a porous medium. The theory of porous media was first introduced by Darcy in the nineteenth century and has been extensively studied since. The main idea is that the instantaneous discharge rate through a porous medium is proportional to the viscosity of the fluid and the pressure drop over a given distance. This general idea can be coupled with several heat transport formulations such as the advection diffusion equation.

There are two basic categories of models for petrothermal systems consisting of a fractured medium: continuum methods and discrete methods. The continuum model approach can be subdivided into three methods, namely the effective continuum method (ECM), the dual-continuum or the generalized multiple-continuum model, and the stochastic-continuum model. The ECM (e.g., Wu 2000) is based on the assumption of local thermodynamic equilibrium between fracture and matrix at all times. Apart from this assumption, the most important condition this model includes is the existence and the determination of a representative elementary volume (REV) for the fractured medium for which equivalent hydraulic coefficients, such as the permeability tensor, have to be determined, and for which the porous medium theory can be subsequently used. Since the applicability of this method is restricted due to the fact that there is generally no guarantee that a REV exists for a given site (cf. Long et al. 1982), or that the assumption of local thermodynamic equilibrium between fracture and matrix is violated by rapid flow and transport processes (see Pruess et al. 1986), alternatives such as the dual-continuum method or the more general multiple interacting continua (MINC) method (see Barenblatt et al. 1960; Warren and Root 1963; Kazemi 1969; Pruess and Narasimhan 1985; Wu and Qin 2009 and the references therein) are needed. The dual-porosity method as one possible form of dual-continuum methods was first introduced by Barenblatt et al. (1960), applying the idea of two separate but overlapping continua, one modeling the fracture system and the other modeling the porous matrix blocks. The method takes into account the fluid flow not only within the media, as the general Darcy model does, but also between the fractures and the matrix blocks. Typically, the flow between the fracture system and the matrix is modeled as pseudosteady, assuming pressure equilibration between the fracture system and the matrix blocks within the duration of each time step. This assumption fails in presence of large matrix blocks, low matrix permeability or silification of fracture surfaces. For other models concerning fracture–matrix interaction the reader is referred to Berkowitz (2002). The stochastic-continuum approach represents the fractured medium as a single continuum but, as opposed to the ECM, it uses geostatistical parameters. The concept was first introduced for fractured media by Neuman and Depner (1988) and Tsang et al. (1996). Altogether, the continuum approach is a simple method, since it ignores the complex geometry of fractured systems and employs effective parameters to describe their behavior.

In contrast to this, the discrete approach is based on the explicit determination of the fractures, thus the need for a detailed knowledge of the geometry of the fractures arises. Hence, there are two different tasks which discrete models can fulfill, namely modeling the flow when the continuum methods cannot be applied, and specifying the effective parameters needed in the continuum approach, if they are applicable. Similar to the continuum models the discrete models can be divided into three different methods: single-fracture models, fracture-network models, fracture–matrix models. In single-fracture models only one fracture is considered. There are different realizations for this concept, both deterministic and stochastic. Usually, the fracture surface, shape, and orientation are simplified and the flow is described, e.g., by the parallel plate concept (e.g., Tang et al. 1981) or inside a planar fracture in an infinite reservoir (e.g., Ghassemi et al. 2003). An early statistical approach dealing with preferential flow paths based on the variation of fracture aperture can be found in Tsang et al. (1996). The concept of discrete fracture network (DFN) models is based on the idea of restricting the fluid flow only to the fractures and assuming the surrounding rock matrix to be impermeable (e.g., Dershowitz et al. 2004). To model the flow, simplifying methods such as the parallel plate concept (see Lee et al. 1999) or the channeling-flow concept (see Tsang and Tsang 1987) are often applied. The last approach in the discrete category is the fracture–matrix model, also called explicit discrete-fracture and matrix method (see Snow 1965; Sudicky and McLaren 1992; Stothoff and Or 2000; Reichenberger et al. 2006 and the references therein), which uses a discrete fracture network but regards the rock matrix as a continuum as opposed to the DFN model. In contrast to the other discrete models, this approach allows to represent fluid potential gradients and fluxes between the fracture system and the rock matrix on a physical level. Nonetheless, this advantage is often reduced by the fact that the models are restricted to a vertical or horizontal orientation of the fractures (e.g., Travis 1984) or that the influence of the real geometry of the fractures such as their tortuosity is neglected. Since not only a detailed knowledge of the matrix geometric properties is rarely available at a given site, but also the application of this method is computationally intensive, the less demanding dual-continuum or the more general MINC method are widely used.

3 Current Strategies

In this section we present the actual strategies today in use to solve the key issues in seismic and flow modeling of geothermal reservoirs, which we presented in ➲ Sect. 2.

3.1 Mathematical Models of Seismic Migration and Inversion

Since until now known standard seismic methods are commonly used in geothermal energy projects, which was briefly mentioned in ➲ Sect. 2.1, and are well described in the literature, in this section we give only a short description of some models, that yield the most accurate migration results. Furthermore, we describe a model that can be used to detect fractures and also to update the velocity model.

As mentioned before, all seismic methods are based on the acoustic or elastic-isotropic wave equation. For the sake of simplicity we consider the scalar acoustic wave equation

$$\left(\Delta - \frac{1}{C^2}\frac{\partial^2}{\partial t^2}\right) U = 0, \tag{1}$$

where Δ is the Laplace operator, $C \in \mathcal{C}^1(V)$ is a velocity function in the given medium $V \subset \mathbb{R}^2$ or $V \subset \mathbb{R}^3$, and $U \in \mathcal{C}^1(V \times [0, t_{\text{end}}])$ is a pressure function. Substituting the function C in Eq. (1) by the compressional velocity function and the shear velocity function, the elastic-isotropic wave equation is obtained. In order to reduce unwanted internal reflections and artifacts, Eq. (1) can be modified, such that the nonreflecting wave equation is used (see, e.g., Baysal et al. 1984), damping terms in the boundary layers are added (see, e.g., Fletcher et al. 2006), or possible filtering techniques like least-squares attenuation are applied (see, e.g., Guitton and Kaelin 2006). The construction of an appointed seismic model depends on the approximation methods used to construct a numerical solution of Eq. (1).

Moreover, data sets are usually sorted by the geometrical positions of sources and corresponding receivers. One of the most used sorting techniques is the common-shot-gather in which all seismograms from the same shot are sequentially saved as shown in ➲ *Fig. 1*. Another sorting method is the common-midpoint-gather which sorts all seismic traces above the same midpoint (➲ *Fig. 2*). Other sorting techniques are described in Yilmaz (2001) and Biondi (2006).

The downward-continuation migration method starts at the surface and computes the wave field from the current depth level to the next. As an example we consider the hybrid depth migration method (see Zhang and Zhang 2004), which splits Eq. (1) into two one-way wave equations for the down-going and up-going waves, namely

$$\frac{\partial U}{\partial x_3} = \pm \frac{i\omega}{C}\sqrt{1 + \frac{C^2}{\omega^2}\left(\frac{\partial^2}{\partial x_1^2} + \frac{\partial^2}{\partial x_2^2}\right)}U, \tag{2}$$

where ω is the angular frequency and $U \in \mathcal{C}^1(V \times [0; +\infty))$ is the pressure function. The source wave field is simulated as a down-going wave field, and the wave field of the receivers

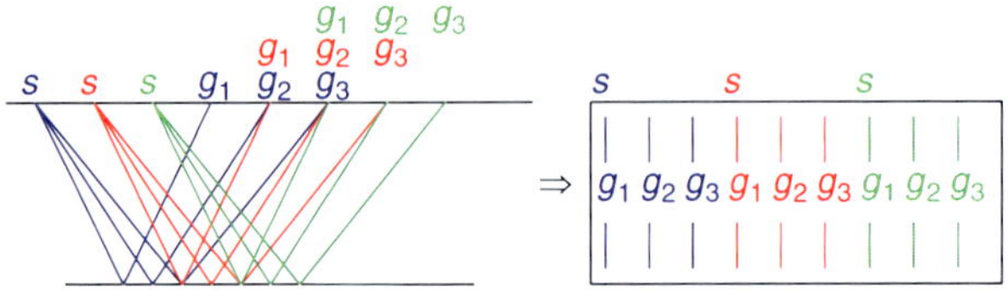

Fig. 1
Common-shot-gather

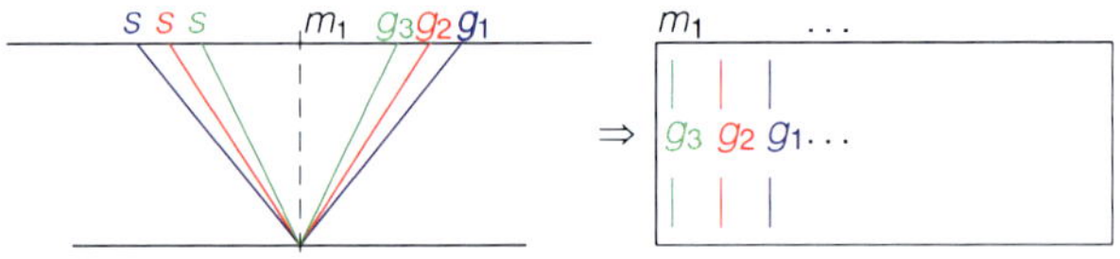

Fig. 2
Common-midpoint-gather

is extrapolated as an up-going wave field. The migration image is obtained by convolution (imaging condition, see, e.g., Biondi 2006) of both wave fields over all frequencies and over all source positions. Equation (2) can be numerically solved in frequency-space as well as frequency-wavenumber domain.

One of the most used methods is the reverse-time migration introduced by Baysal et al. (1983), since it is capable of resolving all kinds of waves. Furthermore, common-shot-gather data sorting is commonly used. The source wave field is simulated by Eq. (1) in direct time, whereas the wave field of the corresponding receivers is extrapolated in reverse time by Eq. (1). The migration result is gained by convolution of both wave fields at time zero. For forward modeling and migration the methods that are currently used are Runge–Kutta schemes (see, e.g., Bonomi et al. 1998), alternating direction implicit schemes (see, e.g., Mitchell and Griffiths 1980; Mufti 1985), and additive schemes (see, e.g., Samarskii 2001; Samarskii and Vabishchevich 2001), however, they are very sensitive to the accuracy of the velocity model. To define an adequate velocity model, the methods described in ➲ Sect. 2.1 are applied. The tomographic methods are based on the high-frequency approximation of the wave equation

$$\left(\frac{\partial T}{\partial x_1}\right)^2 + \left(\frac{\partial T}{\partial x_2}\right)^2 + \left(\frac{\partial T}{\partial x_3}\right)^2 = \frac{1}{C^2}, \tag{3}$$

where $T \in C^1(V \times [0; +\infty))$ defines the travel time surface. Equation (3) is called the eikonal equation (see Tarantola 1984; Bishop et al. 1985). To define or to update the velocity model, the waveform inversion is also employed (see, e.g., Bunks et al. 1995; Pratt et al. 1998; Pratt 1999). The main principle of tomography and waveform inversion can be formulated as a classical optimization problem, e.g., by the Karush–Kuhn–Tucker theory. For example, travel time tomography optimizes the slowness vector by minimizing the difference between the simulated travel time based on the eikonal equation and the recorded travel time. The waveform inversion adapts the velocity model by minimizing the misfit of simulated and observed amplitudes.

In contrast to the methods described above, the integral method can be used for forward modeling and migration as well as for FWI. For this purpose, we apply the Fourier transform to Eq. (1) and obtain the Helmholtz equation

$$\Delta U(x) + K^2(x)U(x) = 0, \text{ in } V, \tag{4}$$

that fulfills the Sommerfeld radiation condition

$$|\nabla U(x) - \mathrm{i}K(x)U(x)| = O(|x|^{-2}),\ |x| \to \infty, \tag{5}$$

where $K \in C^1(V)$ is the wavenumber function, and $U \in C^1(V)$ is the pressure function. In order to use the classical Helmholtz theory, we rewrite Eq. (4) with variable parameter K as the following system

$$\Delta U_I + \kappa^2 U_I = 0 \tag{6}$$

$$\Delta U_S + \kappa^2 U_S = -\kappa^2 \Lambda U \tag{7}$$

$$U = U_I + U_S, \tag{8}$$

where $\kappa \in \mathbb{C}$ is a fixed parameter describing the “background” wavenumber profile, U_I denotes the “incident” wave field in the background medium, and U_S is the scattered wave field. For κ we define the corresponding perturbation function

$$\Lambda(x) = \frac{K^2(x)}{\kappa^2} - 1. \tag{9}$$

Assuming $U_I|_{\partial V} = F \in C^1(\partial V)$, $U_S|_{\partial V} = 0$, and $G(x, \cdot)|_{\partial V} = G(\cdot, y)|_{\partial V} = 0$ for $x, y \in V \cup \partial V$, the solution of system (6)–(8) can be represented as

$$U_I(x) = \int_{\partial V} \frac{\partial}{\partial \nu(y)} G(x, y) F(y) d\omega(y), \tag{10}$$

$$U_S(x) = \kappa^2 \int_V G(x, y) \Lambda(y) \left(U_I(y) + U_S(y)\right) dy, \tag{11}$$

where $\nu(y)$ denotes the (inner unit) surface normal field. The function G defined by

$$G(x, y) = \frac{e^{i\kappa|x-y|}}{4\pi|x-y|} \tag{12}$$

denotes the fundamental solution of Eq. (6).

Finally, forward modeling and migration problems can be mathematically explained as follows. Using seismograms F, which are observed on the surface, and approximate values of the velocity field C ($K = \omega/C$, where ω is the angular frequency), the pressure wave field U in the defined volume V needs to be found. Because of the nonlinearity of the second integral in Eq. (11), some kind of linearization has to be applied, e.g., Born (see Bleistein et al. 2001), Rytov, or De Wolf approximation (see Wu et al. 2006).

The inverse problem can be formulated as an update of the velocity function C (or the perturbation function Λ) by the given pressure wave field U in the volume V and the observed seismograms F.

3.2 Flow Models

In this section the two main approaches and their inherent methods for modeling coupled flow problems in geothermal reservoirs are described referring to either hydrothermal or petrothermal reservoirs.

3.2.1 Hydrothermal Systems

As mentioned in ➋ Sect. 2.2 Darcy's law is the basic equation for laminar flow of a fluid in a porous medium representing a hydrothermal reservoir, namely

$$\mathbf{v} = -\frac{\mathbf{k}}{\mu}(\nabla P + \rho_f \mathbf{g}), \tag{13}$$

where $\mathbf{v}$ is the velocity vector, $\mathbf{k}$ is the hydraulic permeability tensor, μ is the viscosity of the fluid, P is the pressure, ρ_f is the density of the fluid, and $\mathbf{g}$ is the body acceleration vector. Although this equation was first derived by empirical measurements, a mathematical justification can be provided, e.g., by the homogenization method described in Ene and Poliševski (1987) or the volume averaging method described by Sahimi (1995).

Based on the law of mass conservation the continuity equation reads as

$$0 = \frac{\partial(\rho_f \Phi)}{\partial t} + \nabla \cdot (\rho_f \mathbf{v}) + \rho_f W, \tag{14}$$

with porosity Φ and source or sink term W.

The equation for heat transport incorporating advection and diffusion can also be derived (see Kolditz 1997) in a similar fashion from the law of energy conservation and can be written as

$$\frac{\partial(\rho c T)}{\partial t} = \nabla \cdot (\lambda \nabla T - (\rho c)_f T \mathbf{v}) + Q, \tag{15}$$

where ρc is the product of the density and the thermal capacity of the saturated porous medium, T is the temperature, λ is the thermal conductivity of the porous medium, $(\rho c)_f$ is the product of the density and the thermal capacity of the fluid, and Q is the heat generation rate.

In order to solve the fluid and thermal flow problem for a porous medium representing a geothermal reservoir, the coupled Eqs. (13)–(15) have to be solved. Commonly, numerical solution methods are based on finite element, finite volume, or finite difference methods (e.g., Zyvoloski 1983; Zhao et al. 1999; O'Sullivan et al. 2001; Chen et al. 2006).

From a mathematical point of view the problem has been challenged by Ene and Poliševski (1987) who proved the existence and uniqueness of a solution for the incompressible fluid case, i.e., $\nabla \cdot \mathbf{v} = 0$, in both bounded and unbounded domains.

3.2.2 Petrothermal Systems

We already described the two categories for modeling fractured geothermal reservoirs (petrothermal reservoirs) in ➋ Sect. 2.2. In this section we want to give more detailed information on the corresponding continuum and discrete models.

Continuum Models

Effective Continuum Method (ECM). The equations that have to be solved for coupled fluid and thermal flow in the ECM are the same as in the porous medium approach (see ➋ Sect. 3.2.1) since the fractured medium is treated as a single homogeneous continuum with effective coefficients. Determining these effective coefficients, especially the permeability and the dispersion tensor, is the most difficult task since they depend nonlinearly on the pressure. Since there is a large variety of different types of models using ECM, we first present two basic models for determining effective coefficients in fractured rocks introduced by Pruess et al. (1986) and Wu (2000) based on a deterministic approach.

Pruess and his coworkers developed a semi-empirical model to represent an unsaturated fracture/matrix system as a single equivalent continuum, therein using the simple arithmetic sum of the conductivity of the rock matrix and the fracture system as the approximation of the unsaturated conductivity of the fractured rock

$$\mathbf{k}(P) = \mathbf{k}_{\text{matrix}}(P) + \mathbf{k}_{\text{fracture}}(P). \tag{16}$$

A similar approximation of the isotropic permeability of a fracture/matrix system was used by (Peters and Klavetter 1988). The approximation (16) is based on the above mentioned assumption of the local thermodynamic equilibrium between the fracture system and the rock matrix blocks (cf. ➋ Sect. 2.2) breaking down in the case of fast transients.

One of the examples of including solute transport in models for fractured media is given by Wu (2000), who presents a generalized ECM formulation to model multiphase, non-isothermal flow and solute transport. The governing equations for compositional transport and energy conservation contain the key effective correlations such as capillary pressures, relative permeability, dispersion tensor, and thermal conductivity. Introducing a fracture/matrix combined capillary pressure curve as a function of an effective liquid saturation, the corresponding fracture and matrix saturations can be determined by inversion of the capillary pressure functions of fracture and matrix, respectively. Consequently, the effective relative permeabilities of the corresponding fluid phase can be determined as the weighted sum (via the absolute permeabilities of the fracture system and the matrix) of the relative effective permeabilities of the fracture system and the matrix evaluated at the respective saturations. Similarly, the thermal conductivity can be specified. For the determination of the dispersion tensor the Darcy velocities in the fracture system and the matrix are needed additionally to the corresponding saturations. Note that approximation (16) is also incorporated in this model for the effective continuum permeability.

An attempt to categorize the different mathematical concepts beyond the problem to find good estimates for the matrix-dependent effective parameters can be found in Sahimi (1995).

Dual-Continuum and MINC Methods. In the special case of dual-porosity two conservation laws are needed, one for the fracture and one for the rock matrix blocks, respectively, (see Barenblatt et al. 1960)

$$0 = \frac{\partial(\rho_f \Phi_1)}{\partial t} + \nabla \cdot (\rho_f \mathbf{v}_1) - w \tag{17}$$

$$0 = \frac{\partial(\rho_f \Phi_2)}{\partial t} + \nabla \cdot (\rho_f \mathbf{v}_2) + w, \tag{18}$$

where w is the mass of the liquid which flows from the rock matrix into the fractures per unit time and unit rock volume. The subscripts 1 and 2 represent the fracture phase and the rock matrix phase, respectively. Note that Eqs. (17) and (18) are in the same form as that for a single porous medium, cf. (14).

Combining Eq. (17) with Darcy's law and Eq. (18) with the assumption of the dependence of the porosity in the rock matrix only on the corresponding pressure in the rock matrix we obtain, assuming a slightly compressible fluid in a homogeneous medium and neglecting the small high-order terms,

$$\frac{\partial p_1}{\partial t} - \eta \frac{\partial \Delta p_1}{\partial t} = \kappa \Delta p_1, \tag{19}$$

where p_1 is the pressure in the fracture phase, η and κ are constants depending on the system. The first numerical implementation of a finite difference multi-phase, dual-porosity scheme including gravity and imbibition was presented by Kazemi et al. (1976). It can be shown (see Arbogast 1989) that models of this kind are well-posed, given appropriate boundary and initial conditions. For the derivation of the double-porosity model due to homogenization, see Arbogast et al. (1990).

An extension of the dual-porosity model has been proposed by Pruess and Narasimhan (1985), the so-called multiple interacting continua (MINC) method. The idea is to divide the domain into regions in which the thermal dynamic equilibrium can be assumed and to treat every block of the rock matrix not as a porous medium as in the dual-porosity method, but as a set of different porous media with different constituting properties all following Darcy's law. In this way, the transient interaction between matrix and fractures can be shown to be more realistic. A coherent representation for the thermal flow has also been incorporated into the model using the integral finite difference method. Examples on how this method can be adapted to different situations can be found in Kimura et al. (1992) and Wu and Pruess (2005).

Further information on multi-physics and multi-scale approaches can be found in this book in Helmig et al. (2010) and the references therein.

Stochastic-Continuum Method. As mentioned in ➋ Sect. 2.2 the stochastic-continuum method is related to the ECM but is based on modeling the site specific hydraulic parameters, such as the hydraulic conductivity, as random variables via stochastic methods, e.g., the Monte Carlo simulation. However, measuring the hydraulic conductivity in this manner is problematic since it is a scale-dependent parameter that usually has a higher variance at smaller scales, possesses a varying support volume, and is derived from well-tests whose scales have to match the required support scale of the model (see, e.g., Odén and Niemi 2006).

The pioneering work by Neuman and Depner (1988) verifies the dependence of the effective principal hydraulic conductivity solely on mean, variance, and integral scales of local log hydraulic conductivities for ellipsoidal covariance functions. The proposed method requires that the medium is locally isotropic and that the variance is small.

The stochastic-continuum model of Tsang et al. (1996) is based on a nonparametric approach using the sequential indicator simulation method. Hydraulic conductivity data from point injection tests serve the purpose of deriving the needed input for this simulation, namely the thresholds dividing the possible range of values of hydraulic conductivity in the stochastic-continuum into classes, and the corresponding fractions within each of these classes. In order to reflect the fractures and the rock matrix they introduced a long-range correlation for the high hydraulic conductivity part of the distribution in the preferred planes of fractures. Thus, both the fractures and the rock matrix contribute to the hydraulic conductivity even though they are not treated as two different continua as in the dual-continuum model described above. Due to the inherent restrictions of the model and the reduction of the uncertainty they propose to employ spatially integrated quantities to model flow and transport processes in a strongly heterogeneous reservoir reflecting the continuum quality (spatial invariability) of their model.

Based on the concept of a stochastic representative elementary volume (REV) using multiple realizations of stochastic discrete fracture network (DFN) models (see below) simulated via the Monte Carlo method Min et al. (2004) determine the equivalent permeability tensor with the help of the two-dimensional UDEC code by Itasca Consulting Group Inc. (2000). The central relation that is used in this code is a generalized Darcy law for anisotropic and homogeneous porous media (see Bear 1972)

$$\mathbf{v}_i = F \sum_{j=1}^{2} \frac{\mathbf{k}_{ij}}{\mu} \frac{\partial P}{\partial x_j}, \tag{20}$$

where F is the cross-section area of the DFN model. One of the most important steps in the analysis is to prove the existence of the resulting permeability as a tensor, which was done by comparing the derived results with the ellipse equation of the directional permeability. The second important step is to determine whether a REV can be established for a specific site or not. Min and his coworkers presented two criteria, "coefficient of variation" and "prediction error", to show the existence of the REV and to specify its size for Sellafield, UK.

Discrete Models

Single-Fracture Models. The analysis of an accurate behavior of a single fracture is crucial in the understanding of situations in which most of the flow occurs in a few dominant paths. The classical idea for modeling a rock fracture is to consider it as a pair of smooth parallel planes (see Lomize 1951). These kinds of models are interesting from a mathematical point of view, since they often offer an analytical or semi-analytical solution for the flow (see, e.g., Wu et al. 2005), and they are widely used in reservoir modeling, since they can be useful for a quantitative analysis, but they are far from realistic.

The reality is, the surface of the fracture is usually rough to the point that the flow can fail to satisfy the cubic law (see Bear et al. 1993)

$$v = \frac{g}{12\mu} a_{\mathrm{p}}^{3} \frac{\Delta h}{l}, \tag{21}$$

where v is the flow rate, g is the gravity acceleration, a_{p} is the aperture, Δh is the head loss, and l is the fracture length. Several authors have tried to characterize this deviation from the cubic law with fractal (e.g., Brown 1987; Fomin et al. 2003) and statistical (e.g., Tsang and Tsang 1989) modeling of the fracture roughness. In general it is acknowledged that the fractures themselves are two-dimensional networks of variable aperture. Most of the time, the trick is not to identify every single roughness in the surface, even though a number of technical methods are used to compute a precise profilometric analysis, but rather the scale of the roughness which has a dominant influence on the fluid flow (see Berkowitz 2002).

Recently, another method realizing the irregularity of the fracture surface was introduced to model (Navier–Stokes) flow through a single fracture, namely the lattice Boltzmann method (see Kim et al. 2003; Eker and Akin 2006 and the references therein). In contrast to the traditional "top-down" methods employing partial differential equations, the idea of this "bottom-up" method is to use simple rules to represent fluid flow based on the Boltzmann equation.

Additionally, to problems caused by roughness the situation is complicated by the influence of deformation processes due to flow and pressure gradients which should be considered at this scale (see, e.g., Auradou 2009).

For heat extraction from hot dry rock (HDR) systems that are a member of petrothermal systems, e.g., Heuer et al. (1991) and Ghassemi et al. (2003) developed mathematical models for a single fracture. The essential feature of the approach by Heuer et al. (1991) is that the presented (one-dimensional) model can be solved by analytical methods. Furthermore, a generalization to an infinite number of parallel cracks is also given. A three-dimensional model of heat transport in a planar fracture in an infinite reservoir is given by Ghassemi et al. (2003), who derive an integral equation formulation with a Green's function effectively sidestepping the need to discretize the geothermal reservoir.

Discrete Fracture Network (DFN). Among the methods for modeling a geothermal reservoir, the DFN approach is one of the most accurate, but also one of the most difficult to implement. This model restricts fluid flow to the fractures and regards the surrounding rock as impermeable (e.g., Dershowitz et al. 2004). Thus, as described for the single-fracture models, most of the time fluid flow through fractures is compared to flow between parallel plates with smooth walls or to flow through pipes. In contrast to the dual-porosity model, the fluid flow in the DFN model is governed by the cubic law. Based on this law (momentum equation) and the continuity relationship, unknown heads at intersections of the fracture network can be determined in the DFN model. A comparison of the dual-porosity method and the DFN method can be found in Lee et al. (1999). They identified the fracture volume fraction and the aperture to be the most significant parameters in the dual-porosity and the DFN model, respectively. Furthermore, they derived a one-dimensional analytical solution of the dual-porosity model for a confined fractured aquifer problem with the help of Fourier and Laplace transforms based on an earlier solution by Ödner (1998). Different approaches to characterize a reservoir fracture network and the flow in such a network were introduced, e.g., based on stochastic models, fractal models, fuzzy logic, and neural networks (incorporating field data), a combination of these models, or on percolation theory (see Watanabe and Takahashi 1995; Mo et al. 1998; Jing et al. 2000; Ouenes 2000; Maryška et al. 2004; Tran and Rahman 2006 and the references therein). Note that percolation theory can also be used to determine the connectivity of a DFN and its effective permeability (see Berkowitz 1995; Mo et al. 1998; Masihi et al. 2007). Analytical models for the determination of the permeability of an anisotropic DFN are presented by Chen et al. (1999).

The idea beyond the stochastic models is that not the entire fracture network can be located via seismic and geological means since an important amount of flow can occur in fractures too small to be detected. Thus, these fractures are generated stochastically. Information on the morphology of the system has to be gathered first (via Monte Carlo sampling or geological analysis) in order to assign the correct probability distributions, and then a realistic fracture network can be generated, providing a fairly good approximation of the real underground situation. The statistical distribution of fracture orientation is often described as a Fisher distribution (see Fischer et al. 1993), whereas fracture aperture and size can be sampled from log-normal or Gaussian distributions. For more details see Assteerawatt (2008).

Fracture–Matrix Models. The fracture–matrix models are an extension of the above described DFN models in terms of considering the rock matrix as a porous medium. Thus, the influence of the interaction between fractures and surrounding rock matrix on the physical processes during fluid and thermal flow through the reservoir can be captured and analyzed on a more realistic level. For this reason, these models can also be used to determine the parameters needed in the continuum methods (see, e.g., Lang 1995; Lang und Helmig 1995).

Due to its computational complexity, the problem of combining an accurate description of the fracture system with a permeable matrix has only recently been studied more extensively (e.g., by R. Helmig and his workgroup, University of Stuttgart, Germany). In fact, with the progress of technology, it is now possible to model a very complex set of fractures, and nonetheless take into account the fact that every fracture can exchange fluid with the surrounding rock matrix. An attempt for this kind of study has been done with success by Reichenberger et al. (2006). The main idea is that the capillary pressure and the flow must be continuous across the fracture boundary, therefore a proper interface condition should be given. This method is still at an early stage, and at the moment can be rarely used due to its strong dependency on a precise and complete knowledge of the real on-site fracture–matrix configuration.

For further information on (numerical) methods concerned with flow and transport through fractured media the reader is referred to the recent assessments in Adler and Thovert (1999), Sanyal et al. (2000), Dietrich et al. (2005), and Neuman (2005). A schematic overview of the presented methods used to model geothermal reservoirs is given in ➋ *Fig. 3* introducing the following abbreviations for the respective methods: DC (double-continuum), SC (stochastic-continuum), SF (single-fracture), and FM (fracture–matrix).

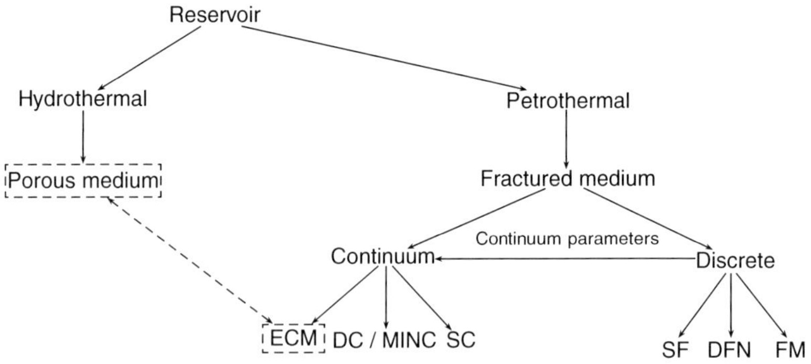

Fig. 3
Scheme of geothermal reservoir models characterizing today's situation

4 Numerical Examples

In this section we present specific numerical examples for seismic migration and heat transport.

4.1 Modeling of Seismic Wave Propagation

4.1.1 Reverse-Time Migration Using Compact Differences Schemes

In this section we consider the numerical construction of the reverse-time migration (see Baysal et al. 1983) in more detail. There are many numerical approaches (see also Chaps. 25 and 43 of this work) to get a stable numerical scheme: explicit finite differences methods (e.g., Bording and Liner 1998), implicit finite differences methods (e.g., Mufti 1985), Galerkin methods (e.g., Du and Bancroft 2004), migration methods using Gaussian beams (e.g., Popov et al. 2006), and spectral finite differences schemes (e.g., Bonomi et al. 1998). We introduce here an alternative approach based on the compact additive finite differences schemes.

As starting point we consider the nonreflecting wave equation (see Baysal et al. 1984) and reformulate it in operator form

$$\frac{\partial^2 U}{\partial t^2} + \mathcal{A}U = 0, \quad \mathcal{A} = -\nabla\left(C^2 \nabla U\right), \quad \mathcal{A} = \sum_{\alpha=1}^{p} \mathcal{A}^{(\alpha)}, \tag{22}$$

where $\mathcal{A}^{(\alpha)}$ can be used for domain decomposition into p domains (usually applied for three-dimensional models) as well as for direction splitting into p directions (usually applied for two-dimensional models). For the sake of simplicity, we consider the two-dimensional version of Eq. (22) and define

$$\mathcal{A}^{(\alpha)} U = -\frac{\partial}{\partial x_\alpha}\left(C^2 \frac{\partial}{\partial x_\alpha} U\right) \tag{23}$$

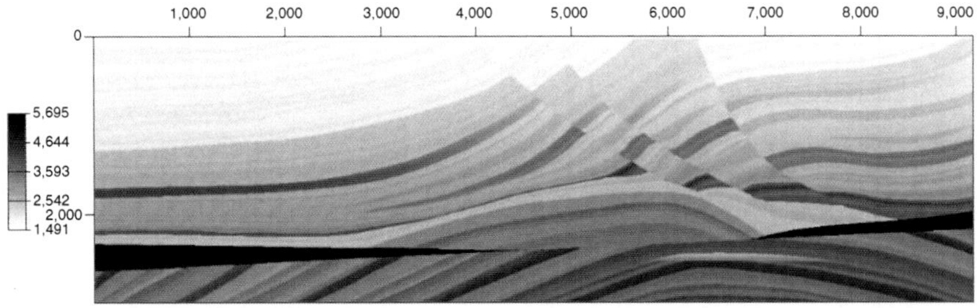

Fig. 4
Strong velocity model for the "Marmousi" data set

for $\alpha = 1, 2$. According to the theorem introduced in Samarskii and Vabishchevich (2001), for all $\sigma \geq \frac{p}{4}$, $(\mu = \sigma\tau^2)$, and for all τ (discretization step for the time direction), the additive scheme

$$\frac{U_{n+1} - 2U_n + U_{n-1}}{\tau^2} + \sum_{\alpha=1}^{p} \left(E + \mu\mathcal{A}^{(\alpha)}\right)^{-1} \mathcal{A}^{(\alpha)} U_n = 0, \tag{24}$$

where E is the identity matrix, is unconditionally stable for problem (22). Equation (24) can be also described as an explicit–implicit finite difference scheme and is solved in two steps. The implicit step consists of solving the following p equations

$$\left(E + \mu\mathcal{A}^{(\alpha)}\right) \tilde{U}^{(\alpha)} = \mathcal{A}^{(\alpha)} U_n, \quad \alpha = 1, \ldots, p. \tag{25}$$

Subsequently, the explicit step uses the solution $\tilde{U}^{(\alpha)}$ to find U_{n+1} by solving

$$\frac{U_{n+1} - 2U_n + U_{n-1}}{\tau^2} + \sum_{\alpha=1}^{p} \tilde{U}^{(\alpha)} = 0. \tag{26}$$

We apply this numerical scheme to the "Marmousi" data set (see Versteeg 1994), that is a synthetic data set used for verification of migration algorithms. We simulate the source wave field in direct time and denote it by U^s (the snapshots of the wave propagation after 0.6 s, 1.1 s, 1.6 s, 2.1 s, and 2.6 s are illustrated in *Figs. 5–9*. The receiver wave field is extrapolated in reverse time and denoted by U^g.

The resulting image is then obtained by convolution of both wave fields at time zero. To achieve a higher level of accuracy one can reduce the noise and the artifacts by imposing the imaging condition

$$I(x) = \frac{\sum_t U_t^s(x) U_t^g(x)}{\sum_t \left(U_t^s(x)\right)^2} + \frac{\sum_t U_t^s(x) U_t^g(x)}{\sum_t \left(U_t^g(x)\right)^2}. \tag{27}$$

The result of the reverse-time migration applied to the "Marmousi" data set using the strong velocity model (*Fig. 4*) is illustrated in *Fig. 10*.

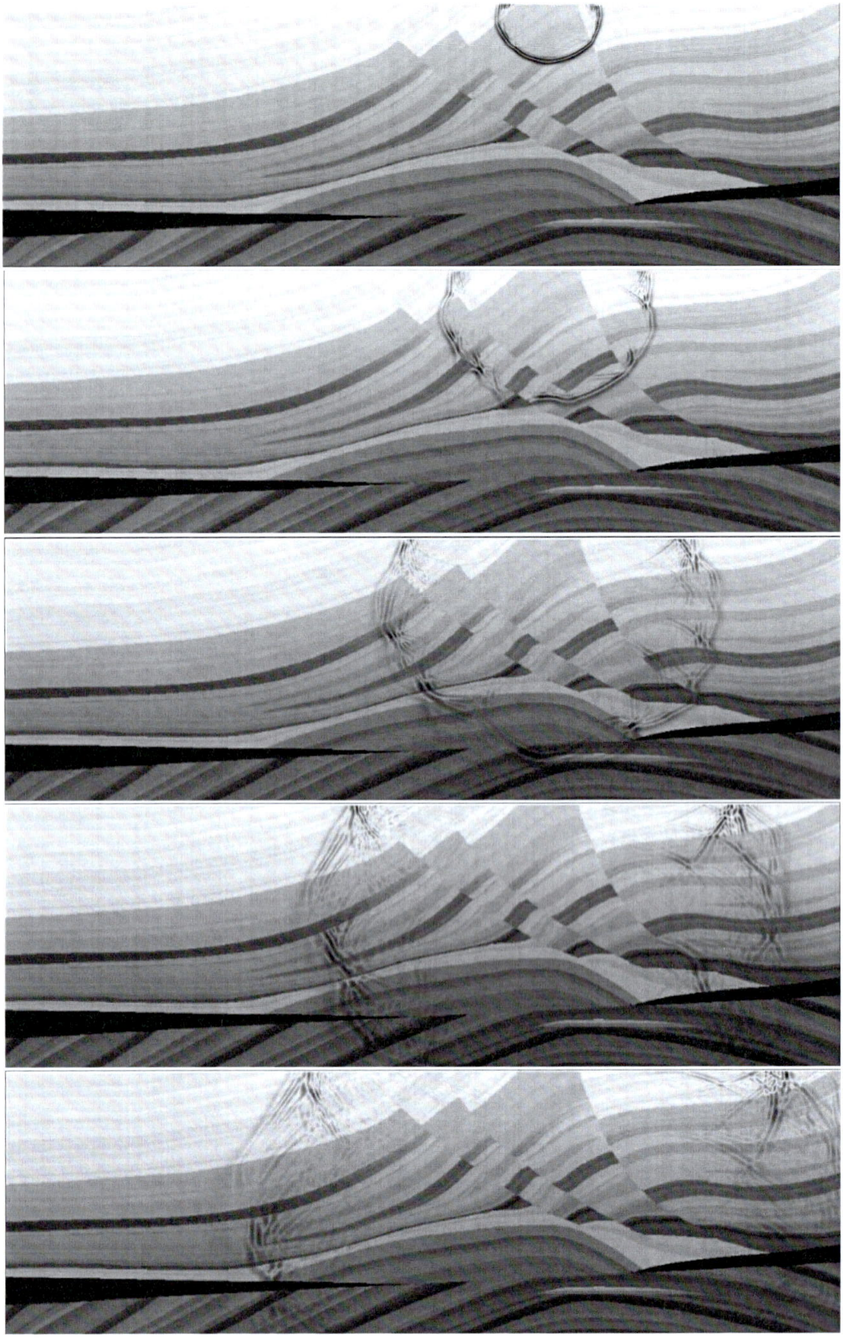

Figs. 5–9

Result of wave propagation in the "Marmousi" strong velocity model (*Fig. 4*) after 0.6 s, 1.1 s, 1.6 s, 2.1 s, and 2.6 s, respectively

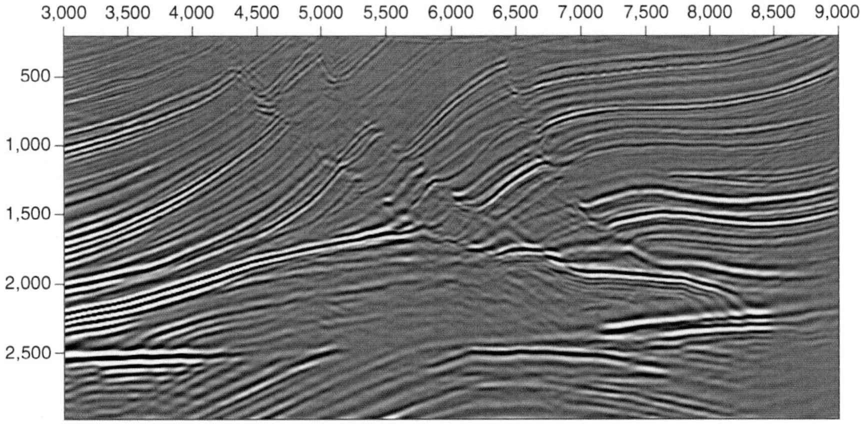

Fig. 10
"Marmousi" migration image

4.1.2 Wavelet Approach for Seismic Wave Propagation

The basic development of the Helmholtz theory dates back to Sommerfeld (1947) and Müller (1969). By use of regularization techniques applied to metaharmonic functions, introduced and expanded in Freeden et al. (2003) and Freeden and Mayer (2007), we construct a new approach to solve the Helmholtz equation (4) according to the integral equations (10) and (11). For $\tau, \sigma \in \mathbb{R}$, $\tau \neq 0$, the operator $P(\tau, \sigma; \kappa) : \mathcal{L}^2(\partial V) \to C(\partial V)$ defined by

$$P(\tau, \sigma; \kappa)F(x) = \int_{\partial V} F(y) \frac{e^{i\kappa|x+\tau\mu(x)-(y+\sigma\nu(y))|}}{|x + \tau\mu(x) - (y + \sigma\nu(y))|} d\omega(y), \tag{28}$$

is called the Helmholtz potential operator for the continuous function F on ∂V. We note, that the kernel function in the integral equation (28) is the fundamental solution $G(x + \tau\mu(x), y + \sigma\nu(y))$ of Eq. (10). From this definition we obtain the single layer potential on ∂V for values on $\Sigma(\tau) = \{x \in \mathbb{R}^3 \,|x = y + \tau\mu(y),\ y \in \partial V\}$, $P(\tau, 0; \kappa) : \mathcal{L}^2(\partial V) \to C(\partial V)$, which is defined by

$$P(\tau, 0; \kappa)F(x) = \int_{\partial V} F(y) \frac{e^{i\kappa|x+\tau\mu(x)-y|}}{|x + \tau\mu(x) - y|} d\omega(y). \tag{29}$$

Analogously, we can represent the double layer potential on ∂V for the values on $\Sigma(\tau)$, $P_{|2}(\tau, 0; \kappa) : \mathcal{L}^2(\partial V) \to C(\partial V)$, given by

$$\begin{aligned} P_{|2}(\tau, 0; \kappa)F(x) &= \frac{\partial}{\partial\sigma} P(\tau, \sigma; \kappa)F(x)\Big|_{\sigma=0} \\ &= \int_{\partial V} F(y) \left(\frac{\partial}{\partial\nu(y)} \frac{e^{i\kappa|x+\tau\mu(x)-(y+\sigma\nu(y))|}}{|x + \tau\mu(x) - (y + \sigma\nu(y))|} \right)\Bigg|_{\sigma=0} d\omega(y) \\ &= \int_{\partial V} F(y) \frac{\nu(y)(x + \tau\mu(x) - y)e^{i\kappa|x+\tau\mu(x)-y|}}{|x + \tau\mu(x) - y|^2} \\ &\quad \times \left(\frac{1}{|x + \tau\mu(x) - y|} - i\kappa \right) d\omega(y), \end{aligned} \tag{30}$$

and the operator $P_{|1}(\tau,0;\kappa) : \mathcal{L}^2(\partial V) \to \mathcal{C}(\partial V)$ representing the normal derivative of the single layer potential on ∂V for the values on $\Sigma(\tau)$ defined by

$$\begin{aligned} P_{|1}(\tau,0;\kappa)F(x) &= \frac{\partial}{\partial\tau}P(\tau,\sigma;\kappa)F(x)\Big|_{\sigma=0} \\ &= -\int_{\partial V} F(y)\frac{\mu(x)(x+\tau\mu(x)-y)e^{\mathrm{i}\kappa|x+\tau\mu(x)-y|}}{|x+\tau\mu(x)-y|^2} \\ &\quad\times\left(\frac{1}{|x+\tau\mu(x)-y|}-\mathrm{i}\kappa\right)d\omega(y). \end{aligned} \tag{31}$$

If $\sigma = \tau = 0$, the kernels of the formally defined potentials $P(0,0;\kappa)$ and $P_{|i}(0,0;\kappa)$; $i = 1,2$; have weak singularities. In order to define the classical jump and limit relations, we have to introduce the adjoint operators. By $P^*(\tau,\sigma;\kappa) : \mathcal{L}^2(\partial V) \to \mathcal{C}(\partial V)$ we denote the adjoint operator to $P(\tau,\sigma;\kappa)$, which satisfies

$$(F_1, P(\tau,\sigma;\kappa)F_2)_{\mathcal{L}^2(\partial V)} = \left(P^*(\tau,\sigma;\kappa)F_1, F_2\right)_{\mathcal{L}^2(\partial V)} \tag{32}$$

for all functions $F_1, F_2 \in \mathcal{L}^2(\partial V)$ with respect to the scalar product in $\mathcal{L}^2(\partial V)$. Using Fubini's theorem we can define the adjoint operators explicitly as

$$P^*(\tau,0;\kappa)F(x) = \int_{\partial V} F(y)\frac{e^{-\mathrm{i}\overline{\kappa}|x-y-\tau\mu(y)|}}{|x-y-\tau\mu(y)|}d\omega(y), \tag{33}$$

$$\begin{aligned} P^*_{|1}(\tau,0;\kappa)F(x) &= \int_{\partial V} F(y)\frac{\mu(y)(x-y-\tau\mu(y))e^{-\mathrm{i}\overline{\kappa}|x-y-\tau\mu(y)|}}{|x-y-\tau\mu(y)|^2} \\ &\quad\times\left(\mathrm{i}\overline{\kappa}+\frac{1}{|x-y-\tau\mu(y)|}\right)d\omega(y), \end{aligned} \tag{34}$$

$$\begin{aligned} P^*_{|2}(\tau,0;\kappa)F(x) &= -\int_{\partial V} F(y)\frac{\nu(x)(x-y-\tau\mu(y))e^{\mathrm{i}\overline{\kappa}|x-y-\tau\mu(y)|}}{|x-y-\tau\mu(y)|^2} \\ &\quad\times\left(\mathrm{i}\overline{\kappa}+\frac{1}{|x-y-\tau\mu(y)|}\right)d\omega(y), \end{aligned} \tag{35}$$

where $\overline{\kappa}$ denotes the complex conjugate of κ.

According to the theory described in Ilyasov (2010), we define the classical jump relations and limit formulae not only for smooth regions but also for regions $V \subset \mathbb{R}^3$ including edges and vertices. Let I denote the identity operator. Furthermore, the operator $\tilde{I} : \mathcal{L}^2(\partial V) \to \mathcal{C}(\partial V)$ is defined by $\tilde{I}F(x) = (\mu(x)\nu(x))\,F(x)$, $F \in \mathcal{L}^2(\partial V)$. Then, for all sufficiently small $\tau > 0$ and functions $F \in \mathcal{L}^2(\partial V)$ the jump and limit relations may be formulated for $i = 1,2,3$ by

$$\lim_{\substack{\tau\to 0\\ \tau>0}} \|J_i(\tau;\kappa)F\|_{\mathcal{L}^2(\partial V)} = 0, \quad \lim_{\substack{\tau\to 0\\ \tau>0}} \left\|J_i(\tau;\kappa)^*F\right\|_{\mathcal{L}^2(\partial V)} = 0, \tag{36}$$

$$\lim_{\substack{\tau\to 0\\ \tau>0}} \left\|L_i^{\pm}(\tau;\kappa)F\right\|_{\mathcal{L}^2(\partial V)} = 0, \quad \lim_{\substack{\tau\to 0\\ \tau>0}} \left\|L_i^{\pm}(\tau;\kappa)^*F\right\|_{\mathcal{L}^2(\partial V)} = 0, \tag{37}$$

where the operators J_i and $L_i^{\pm}$ are formally defined by

$$J_1(\tau;\kappa) = P(\tau,0;\kappa) - P(-\tau,0;\kappa), \tag{38}$$
$$J_2(\tau;\kappa) = P_{|1}(\tau,0;\kappa) - P_{|1}(-\tau,0;\kappa) + 4\pi\tilde{I}, \tag{39}$$
$$J_3(\tau;\kappa) = P_{|2}(\tau,0;\kappa) - P_{|2}(-\tau,0;\kappa) - 4\pi I, \tag{40}$$
$$L_1^{\pm}(\tau;\kappa) = P(\pm\tau,0;\kappa) - P(0,0;\kappa), \tag{41}$$
$$L_2^{\pm}(\tau;\kappa) = P_{|1}(\pm\tau,0;\kappa) - P_{|1}(0,0;\kappa) \pm 2\pi\tilde{I}, \tag{42}$$
$$L_3^{\pm}(\tau;\kappa) = P_{|2}(\pm\tau,0;\kappa) - P_{|2}(0,0;\kappa) \mp 2\pi I. \tag{43}$$

According to the explicit representation of the limit and jump relations, we can rewrite the formally defined potentials $P_{|2}(0,0;\kappa)$ resp. $P(0,0;\kappa)$, that describe the solution (10) resp. (11), for $F \in \mathcal{L}^2(\partial V)$ and $U \in \mathcal{L}^2(V)$ by

$$\lim_{\substack{\tau\to 0\\ \tau>0}} \int_{\partial V} \Phi_\tau^I(x,y)F(y)\,d\omega(y) = P_{|2}(0,0;\kappa)F, \tag{44}$$

$$\lim_{\substack{\tau\to 0\\ \tau>0}} \int_{V} \Phi_\tau^{II}(x,y)U(y)\,dy = P(0,0;\kappa)U, \tag{45}$$

where the kernel functions are defined by

$$\begin{aligned}\Phi_\tau^I(x,y) &= \frac{e^{i\kappa|x+\tau\mu(x)-y|}}{2|x+\tau\mu(x)-y|^2}\left(i\kappa + \frac{1}{|x+\tau\mu(x)-y|}\right)\\ &\quad - \frac{e^{i\kappa|x-\tau\mu(x)-y|}}{2|x-\tau\mu(x)-y|^2}\left(i\kappa + \frac{1}{|x-\tau\mu(x)-y|}\right),\end{aligned} \tag{46}$$

$$\Phi_{\pm\tau}^{II}(x,y) = \frac{e^{i\kappa|x\pm\tau\mu(x)-y|}}{|x\pm\tau\mu(x)-y|}. \tag{47}$$

The family $\{\Phi_\tau^i\}_{\tau>0}$ of kernels $\Phi_\tau^i : \partial V \times \partial V \to \mathbb{R}$ is called a ∂V-scaling function of type i, for $\tau > 0$ and $i = I, II$ (see Freeden et al. 2003). For $\tau = 1$, $\Phi_1^i . \partial V \times \partial V \to \mathbb{R}$ is called the mother kernel of the ∂V-scaling function of type i. Correspondingly, for $\tau > 0$, an invertible weight function α (e.g. $\alpha(\tau) = \tau^{-1}$), and $i = I, II$, the family $\{\Psi_\tau^i\}_{\tau>0}$ of kernels $\Psi_\tau^i : \partial V \times \partial V \to \mathbb{R}$ given by

$$\Psi_\tau^i(x,y) = -\alpha^{-1}(\tau)\frac{d}{d\tau}\Phi_\tau^i(x,y),\ x,y \in \partial V, \tag{48}$$

is called a ∂V-wavelet function of type i. The differential equation (48) is called the (scale continuous) ∂V-scaling equation of type i. An illustration of the scaling function $\Phi_\tau^I(x,y)$ and the corresponding wavelet function $\Psi_\tau^I(x,y)$ for different scales is shown in ➋ *Fig. 11.*

Observing the discretized ∂V-scaling equation of type i we obtain for the function $F \in \mathcal{L}^2(\partial V)$ and integers $J \in \mathbb{Z}$ and $N \in \mathbb{N}$ the relation

$$\begin{aligned}\int_{\partial V} \Phi_{J+N}^i(x,y)F(y)\,d\omega(y) &= \int_{\partial V} \Phi_J^i(x,y)F(y)\,d\omega(y)\\ &\quad + \sum_{j=J}^{J+N-1} \int_{\partial V} \Psi_j^i(x,y)F(y)\,d\omega(y),\end{aligned} \tag{49}$$

that describes the principle of a recursive reconstruction scheme (a tree algorithm) (see ➋ *Fig. 12*).

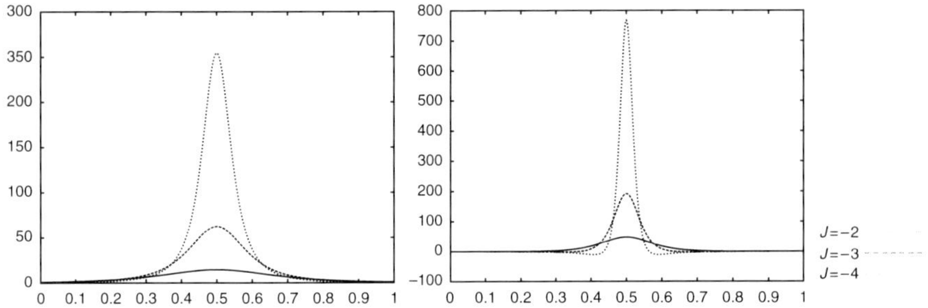

Fig. 11
(1D) Scaling function $\Phi^I_\tau(x,y)$ (*left*) and (1D) wavelet function $\Psi^I_\tau(x,y)$ (*right*) for different scales $\tau = 2^J$

Then, defining the (scale) discretized ∂V-scaling function of type i, viz. $\left\{\Phi^i_{\tau_j}\right\}_{j\in\mathbb{Z},}$ $\tau_j = 2^{-j}$, by $\Phi^i_j = \Phi^i_{\tau_j}$, and the corresponding (scale) discretized ∂V-wavelet function $\Psi^i_j = \Phi^i_{j+1} - \Phi^i_j$, we can rewrite the solution in discrete form for given $J_0 \in \mathbb{Z}$, namely

$$U_I(x) = \int_{\partial V} \Phi^I_{J_0}(x,y)F(y)d\omega(y) + \sum_{j=J_0}^{+\infty} \int_{\partial V} \Psi^I_j(x,y)F(y)d\omega(y), \tag{50}$$

$$U_S(x) = \kappa^2 \int_V \Phi^{II}_{J_0}(x,y)\Lambda(y)U(y)dy + \kappa^2 \sum_{j=J_0}^{+\infty} \int_V \Psi^{II}_j(x,y)\Lambda(y)U(y)dy. \tag{51}$$

Now, for sufficiently large $J \in \mathbb{N}$, we obtain the discrete reconstruction formulas that regularize the singular integral Eqs. (10) and (11) by integral equations of regular type

$$U^J_I(x) = \sum_{k=1}^{N_{J_0}} \alpha_k^{N_{J_0}} \Phi^I_{J_0}\left(x, y_k^{N_{J_0}}\right) + \sum_{j=J_0}^{J-1}\sum_{k=1}^{N_j} \alpha_k^{N_j} \Psi^I_j\left(x, y_k^{N_j}\right), \tag{52}$$

$$U^J_S(x) = \sum_{k=1}^{M_{J_0}} \beta_k^{M_{J_0}} \Phi^{II}_{J_0}\left(x, y_k^{M_{J_0}}\right) + \sum_{j=J_0}^{J-1}\sum_{k=1}^{M_j} \beta_k^{M_j} \Psi^{II}_j\left(x, y_k^{M_j}\right), \tag{53}$$

where the vectors $\alpha^{N_j} \in \mathbb{C}^{N_j}$ and $\beta^{M_j} \in \mathbb{C}^{M_j}$ are defined for given weights $w_k^{N_j}, \tilde{w}_k^{M_j} \in \mathbb{C}$ by $\alpha_k^{N_j} = w_k^{N_j} F\left(y_k^{N_j}\right)$ and $\beta_k^{M_j} = \tilde{w}_k^{M_j} \Lambda\left(y_k^{M_j}\right) U^J_I\left(y_k^{M_j}\right)$.

The migration process is formulated as a forward problem: given function values F on the boundary and the perturbation function Λ (cf. (9)), find the pressure wave field $U_I + U_s$ in the region V. The inverse problem consists of finding the function Λ based on the given boundary function F and the pressure wave field $U_I + U_S$ in the region V. The use of the wavelet approach allows a multi-scale reconstruction of the solution of both the forward as well as the inverse problem via a tree algorithm (see Freeden et al. 2003), which describes a fast wavelet transform for scientific computation. Moreover, the main advantage of this technique is the space localizing property of the wavelets, that makes the refinement of a certain evaluation point possible by only using points from the "near field" of the wavelet kernel. The contribution of the remaining points to the approximation is already incorporated after the application of the initial step of the tree algorithm.

The proposed wavelet approach for migration, applied to a simple post-stack data set including a scattering point and a v-shaped reflector, is shown in ➋ *Fig. 13.*

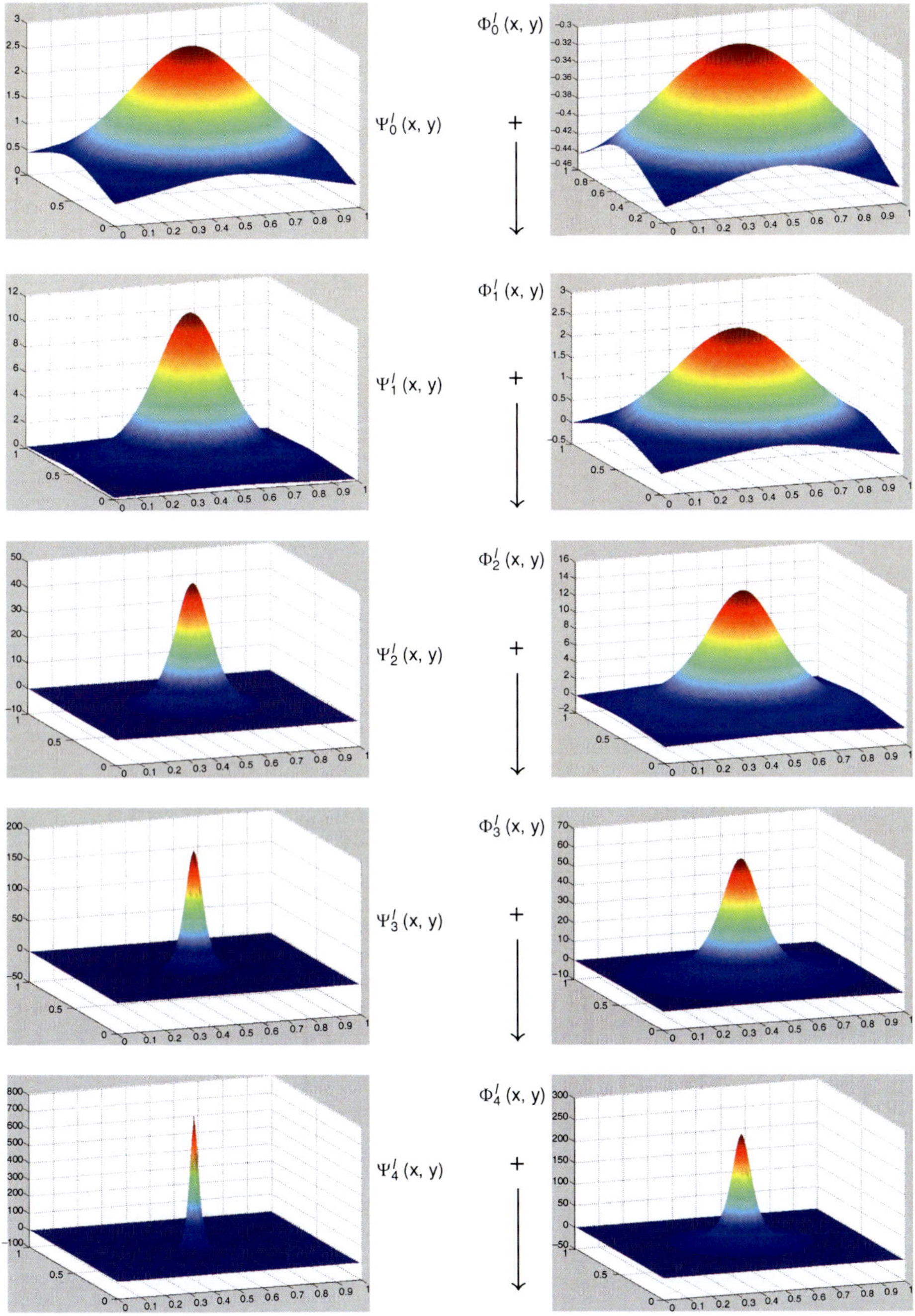

Fig. 12
Reconstruction scheme (tree algorithm) showing strong space localization of (2D) wavelets

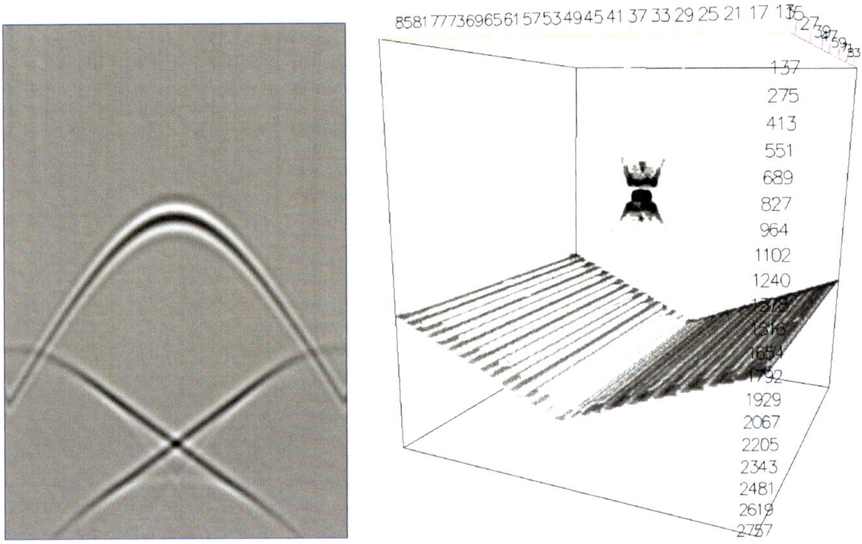

Fig. 13
2D section of a 3D input data set (*left*) and migration result using the wavelet reconstruction for the scale $J = 7$ (*right*)

4.2 Heat Transport

Regarding the heat extraction from a hydrothermal reservoir, we study the initial-boundary-value-problem (IBVP) (54)–(56) consisting of the transient advection diffusion equation representing the heat transport in a porous medium composed of a fluid and a solid phase, the initial temperature T_0 and the Neumann boundary condition F

$$\rho c \frac{\partial T}{\partial t} = \nabla \cdot (\lambda \nabla T) - \Phi(\rho c)_{\mathrm{f}} \mathbf{v} \cdot \nabla T + Q \quad \text{in } \mathcal{B} \times (0, t_{\mathrm{end}}], \tag{54}$$

$$T(\cdot, 0) = T_0 \quad \text{in } \mathcal{B}, \tag{55}$$

$$\frac{\partial T}{\partial \nu} = F \quad \text{on } \partial\mathcal{B} \times (0, t_{\mathrm{end}}], \tag{56}$$

where the bounded, open region $\mathcal{B} \subset \mathbb{R}^3$ denotes the reservoir, the source and sink term Q describes the influence of injection of cold water, ν is the outer unit normal to the boundary of $\mathcal{B}$, denoted by $\partial\mathcal{B}$, and $0 < t_{\mathrm{end}} < \infty$ represents the duration of the temporal simulation.

Multiplying Eq. (54) with test functions $U \in \mathcal{H}^1(\mathcal{B})$, applying the first Green theorem and incorporating the Neumann boundary condition (56) we obtain the weak formulation of the IBVP. The necessary classes of parameters, that yield continuity and modified coercivity of the quadratic form a given by

$$\begin{aligned} a(t; T, U) := {} & \frac{1}{\rho c} \int_{\mathcal{B}} \lambda(x,t) \nabla_x T(x,t) \cdot \nabla_x U(x) dx \\ & + \frac{\Phi(\rho c)_{\mathrm{f}}}{\rho c} \int_{\mathcal{B}} \mathbf{v}(x,t) \cdot (\nabla_x T(x,t)) U(x) dx \end{aligned} \tag{57}$$

are identified in Ostermann (2009). Defining the source-boundary-term R by

$$(R,U) := \frac{1}{\rho c}\int_{\mathcal{B}} Q(x,t)U(x)dx + \frac{1}{\rho c}\int_{\partial\mathcal{B}} \lambda(x,t)F(x,t)U(x)d\omega(x), \tag{58}$$

using the notation $(\cdot,\cdot)$ for the pairing between $\mathcal{H}^1(\mathcal{B})$ and its dual $(\mathcal{H}^1(\mathcal{B}))^\star$, and introducing the bounded, linear mapping $A(t) : \mathcal{H}^1(\mathcal{B}) \to (\mathcal{H}^1(\mathcal{B}))^\star$ satisfying

$$a(t;T,U) = -(A(t)T,U)_{\mathcal{L}^2(\mathcal{B})}, \quad T,U \in \mathcal{H}^1(\mathcal{B}), \tag{59}$$

the weak formulation of the IBVP can be represented as the initial-value-problem

$$\frac{dT}{dt} = A(t)T + R, \quad T(0) = T_0. \tag{60}$$

Note that this initial-value-problem holds in the sense of $(\mathcal{H}^1(\mathcal{B}))^\star$-valued distributions.

Existence and uniqueness of the solution of Eq. (60) follows for $R \in \mathcal{L}^2((0,t_{\text{end}});(\mathcal{H}^1(\mathcal{B}))^\star)$ and $T_0 \in \mathcal{L}^2(\mathcal{B})$ by application of the theoretical framework given in Renardy and Rogers (1993). The continuity of this unique solution is a direct consequence of Sobolev's lemma.

For the numerical implementation it is vital to approximate the problem with a finite-dimensional one, whose solution converges in the limit to the solution of the original problem. This is realized by introducing a linear Galerkin scheme and scalar kernels $K_J(\cdot,y_i)$, $i = 1,\dots,y_{N_J}$, with respect to a system Y of pairwise disjoint points $y_1,\dots,y_{N_J} \in \mathcal{B}$. Defining the subspace

$$\mathcal{H}_J := \text{span}\{K_J(\cdot,y_1),\dots,K_J(\cdot,y_{N_J})\} \tag{61}$$

and requiring for the limit $J \to \infty$ that

$$\bigcup_J^{\infty} \mathcal{H}_J \text{ is dense in } \mathcal{H}^1(\mathcal{B}), \tag{62}$$

we consider the approximation of the weak solution of IBVP (54)–(56) given by

$$T_J : \mathbb{R}_0^+ \to \mathcal{H}_J, \quad t \mapsto T_J(t) = \sum_{i=1}^{N_J} T_i^J(t)K_J(\cdot,y_i) \tag{63}$$

and satisfying the weak formulation of the original problem for test functions in $\mathcal{H}_J$. Strong and weak-$\star$ convergence of the proposed scheme was proven in different $\mathcal{L}^p$-spaces in Ostermann (2009) in a similar fashion as in Lions and Magenes (1972) and Fengler (2005).

In *Fig. 14* we show the temporal evolution of the temperature difference $T - T_0$ on the equatorial plane inside the unit ball, our computational domain, for an example that is based on the parameters given in *Table 1*.

Using the weighted balance between the injected temperature $T_{\text{inj}} = 343.15$ K (70°C) (here temporally and spatially constant) and the reconstructed temperature at the injection point $T(x_{\text{inj}},t)$ the source and sink term Q is modeled as a point source/sink located at the injection point explicitly reading as

$$Q(x,t) = \Phi(\rho c)_f \tilde{Q}(T_{\text{inj}} - T(x_{\text{inj}},t))\delta(x - x_{\text{inj}}), \tag{64}$$

where $\tilde{Q} = 0.005\ \frac{\text{m}^3}{\text{s}}$, $x_{\text{inj}} = (0.467758, -0.170250, 0)^T$, and δ describe the discharge rate, the location of the injection point, and the delta distribution, respectively.

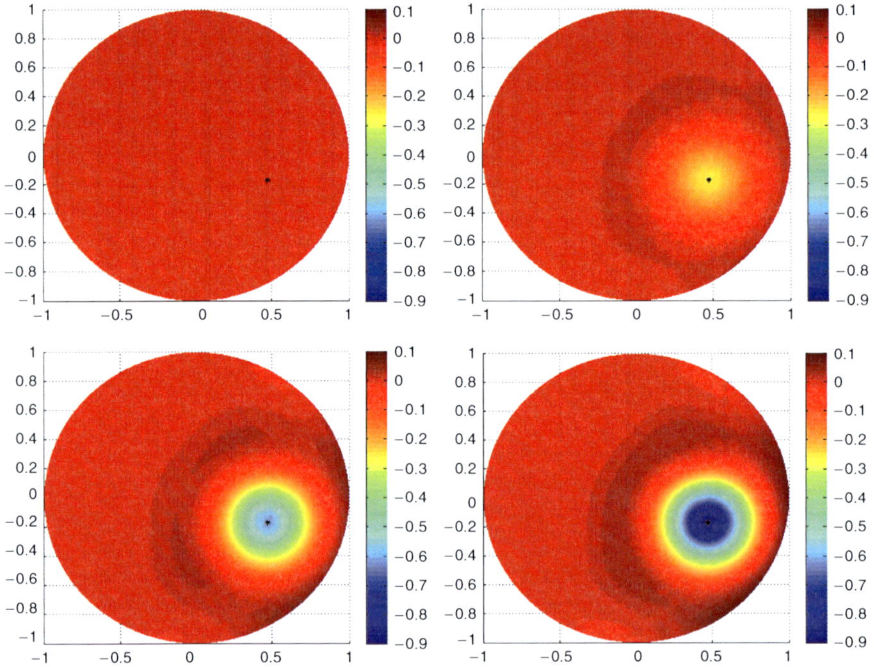

Fig. 14

Temporal evolution of the temperature difference $T - T_0$ on the equatorial plane after 0 (*top left*), 150, 000 (*top right*), 350, 000 (*down left*), and 500, 000 (*down right*) timesteps ($*$ injection point)

Table 1

Rock parameters and fluid velocity of a porous medium composed of sandstone and water

$\rho c = 2296.486803 \frac{\mathrm{kJ}}{\mathrm{m^3 \cdot K}}$,	$\lambda = 1.17246574 \cdot 10^{-3} \frac{\mathrm{kJ}}{\mathrm{m \cdot K \cdot s}}$,	$\Phi = 0.254$
$(\rho c)_f = 4219.8945 \frac{\mathrm{kJ}}{\mathrm{m^3 \cdot K}}$,	$v = (0, -10^{-5}, 0)^T \frac{\mathrm{m}}{\mathrm{s}}$	

Furthermore, we choose a temporally and spatially constant initial temperature $T_0 = 393.15$ K (120°C) and Neumann boundary condition $F = 0.1 \frac{\mathrm{K}}{\mathrm{s}}$. The linear kernels needed in the Galerkin scheme are chosen based on the strictly biharmonic kernel

$$K(x, y) = \frac{1}{4\pi} \frac{|x - y|}{2} \tag{65}$$

and an equidistributed grid inside the computational domain. As expected it can be seen in *Fig. 14* that a cold front expands from the injection point.

5 Future Directions

After more than 40 years of study, the science of modeling geothermal reservoirs has been applied to many field problems by engineers. Thereby, it has pioneered many fields of studies in geology, hydrology (flow transport, solute transport), and so on. Nevertheless, there are still problems that have to be solved and, consequently, several directions future studies can develop, namely:

- The numerical realization of all existing seismic methods yields more or less adequate results, even though the resolution and accuracy are not high enough for the interpretation of images of geothermal reservoirs. Because of the success of wavelets in data compression, solution of differential equations and recursive computation of local events, that are very important for the interpretation of geological subsurface images, we assume that the wavelet approach introduced in Sec. 4.1 enables us to construct numerical schemes for seismic migration and inversion that yield results with locally high resolution. To obtain more information relevant for geothermal projects, seismic data should be observed with different parameters (peak frequency, wave length) as for the hydrocarbon industry. Moreover, the wavelet approach can be used to locally improve the results gained by standard seismic migration and inversion methods.
- Due to the increasing detail with which seismic methods detect the position and the composition of the fractures, and to the rising computational efficiency in solving complex numerical systems of PDEs, it is natural to expect that valid formulations of explicit discrete fracture-matrix problems will arise.
- It seems that only a few authors have treated the problem of coupling the flow equations for DFN with an appropriate description of the heat transport, probably because the problem is of minor interest when modeling oil reservoirs and nuclear waste disposal. Furthermore, when the heat is considered, the difficulty of the problem increases considerably, since in this case there is another term, energy, to be accounted for. The mainstream approach (see, e.g., Chen et al. 2006) is to state the well-known conservation of energy equation for porous media

$$\frac{\partial}{\partial t}\left(\phi \sum_{\alpha} \rho_\alpha S_\alpha U_\alpha + (1-\phi)\rho_s C_s T\right) + \nabla \cdot \sum_{\alpha} \rho_\alpha \mathbf{v}_\alpha H_\alpha - \nabla \cdot (\lambda_T \nabla T) = Q, \qquad (66)$$

  where S_α, μ_α, $\mathbf{v}_\alpha$, U_α, H_α are the saturation, viscosity, (Darcy) velocity, internal energy, and enthalpy of the phase α, respectively, whereas λ_T, ρ_s, C_s and Q are the total thermal conductivity, the density of the matrix, the heat capacity of the matrix, and the heat source term. An accurate mathematical description of this process (especially across the fracture boundaries) has yet to be provided.

6 Conclusion

In this chapter the main methods for mathematical modeling of reservoir detection and flow simulation related to geothermal projects have been presented. The whole spectrum of seismic inversion and migration methods is reviewed, focusing the attention on an efficient classification of the currently used numerical techniques. It is the opinion of the authors that one of the most promising strategies of seismic data processing is reverse-time migration, the principle of which is exemplarily described by a compact differences scheme implying an efficient and parallel implementation, since it allows direction splitting or domain decomposition. The validity of the proposed scheme is shown by the result of its application to the synthetic "Marmousi" data set. Additionally, we introduced a wavelet approach for seismic migration and inversion in regions with non-smooth boundaries. Subsequently, a simple three-dimensional data set was migrated via the space localized wavelets that allow an efficient computation by means of a tree algorithm.

As far as underground water flow in geothermal reservoirs is concerned, the continuum and discrete concepts have been described. A three-dimensional numerical method for heat transport in a single continuum has been studied in detail and numerical results have been presented. Although in engineering practice the usual approach is to represent fractured rock using multi-continuum models, in recent time a shift toward discrete fracture–matrix modeling has been observed in consideration of its more realistic approximation of field characteristics. The aim of future studies will be to include a description for heat transport in the aforementioned models.

Acknowledgments

This work and the work of the project group "Geomathematik," ITWM, Kaiserslautern, is supported by the "Ministerium für Umwelt, Forsten und Verbraucherschutz, Rheinland-Pfalz" within the scope of the project "MathGeotherm" (Kapitel 14 02, Titel 686 72) and by the research center "Center for Mathematical and Computational Modelling ((CM)^2)," TU Kaiserslautern, within the scope of the project "EGMS" (PI: Prof. Dr. W. Freeden). The authors thank the "Competence Center High Performance Computing and Visualization," ITWM, Kaiserslautern, for providing seismic data sets and visualization tools.

References

Addis MA (1997) The stress-depletion response of reservoirs. SPE 38720

Adler PM, Thovert JF (1999) Fractures and fracture networks. Theory and applications of transport in porous media, vol 15, Springer

Altmann JB, Dorner A, Schoenball M, Müller BIR, Müller TM (2008) Modellierung von porendruckinduzierten Änderungen des Spannungsfelds in Reservoiren. Kongressband, Geothermiekongress 2008, Karlsruhe, Germany, November 11–13, 2008

Arbogast T (1989) Analysis of the simulation of single phase flow through a naturally fractured reservoir. SIAM J Numer Anal 26:12–29

Arbogast T, Douglas J, Hornung U (1990) Derivation of the double porosity model of single phase flow via homogenization theory. SIAM J Math Anal 21(4):823–836

Assteerawatt A (2008) Flow and transport modelling of fractured aquifers based on a geostatistical approach. PhD dissertation, Institute of Hydraulic Engineering, University of Stuttgart

Auradou H (2009) Influence of wall roughness on the geometrical, mechanical and transport properties of single fractures. J Phys D Appl Phys 42:214015

Barenblatt GI, Zheltov IP, Kochina IN (1960) Basic concepts in the theory of seepage of homogeneous liquids in fissured rocks. PMM, Sov Appl Math Mech 24:852–864

Baysal E, Kosloff DD, Sherwood JWC (1983) Reverse time migration. Geophysics 48(1):1514–1524

Baysal E, Kosloff DD, Sherwood JWC (1984) A two-way nonreflecting wave equation. Geophysics 49(2):132–141

Bear J (1972) Dynamics of fluids in porous media. Elsevier, New York

Bear J, Tsang CF, de Marsily G (1993) Flow and contaminant transport in fractured rock. Academic, San Diego

Berkowitz B (1995) Analysis of fracture network connectivity using percolation theory. Math Geol 27(4):467–483

Berkowitz B (2002) Characterizing flow and transport in fractured geological media: a review. Adv Water Res 25:852–864

Biondi BL (2006) Three-dimensional seismic imaging. Society of Exploration Geophysicists, Tulsa

Biot MA (1941) General theory of three-dimensional consolidation. J Appl Phys 12:155–164

Bishop TN, Bube KP, Cutler RT, Langam RT, Love PL, Resnick JR, Shuey RT, Spinder DA, Wyld HW (1985) Tomographic determination of velocity and depth in laterally varying media. Geophysics 50:903–923

Bleistein N, Cohen JK, Stockwell JW Jr (2001) Mathematics of multidimensional seismic imaging, migration, and inversion. Springer, New York

Bonomi E, Brieger L, Nardone C, Pieroni E (1998) 3D spectral reverse time migration with no-wraparound absorbing conditions. 68th Annual International Meeting, SEG, Expanded Abstracts

Bording RP, Liner CL (1998) Theory of 2.5D reverse time migration. 64th Annual International Meeting, SEG, Expanded Abstracts, pp 692–694

Brouwer GK, Lokhorst A, Orlic B (2005) Geothermal heat and abandoned gas reservoirs in the Netherlands. Proceedings, World Geothermal Congress 2005, Antalya, Turkey, April 24–29, 2005

Brown SR (1987) Fluid flow through rock joints: The effect of surface roughness. J Geophys Res 92(B2):1337–1347

Bunks C, Saleck FM, Zaleski S, Chavent G (1995) Multiscale seismic waveform inversion. Geophysics 60(5):1457–1473

Chen M, Bai M, Roegiers JC (1999) Permeability tensors of anisotropic fracture networks. Math Geol 31(4):355–373

Chen Z, Huan G, Ma Y (2006) Computational methods for multiphase flows in porous media. Computational Science & Engineering, SIAM, Philadelphia

Cheng HP, Yeh GT (1998) Development and demonstrative application of a 3-D numerical model of subsurface flow, heat transfer, and reactive chemical transport: 3DHYDROGEOCHEM. J Contam Hydrol 34:47–83

Claerbout J (2009) Basic earth imaging. Standford University, Standford

Dershowitz WS, La Pointe PR, Doe TW (2004) Advances in discrete fracture network modeling. In: Proceedings, US EPA/NGWA fractured rock conference, Portland, pp 882–894

Dietrich P, Helmig R, Sauter M, Hötzl H, Köngeter J, Teutsch G (2005) Flow and transport in fractured porous media. Springer, Berlin

DiPippo R (2008) Geothermal power plants: principles, applications, case studies and environmental impact, 2nd edn. Butterworth-Heinemann, Amsterdam

Du X, Bancroft JC (2004) 2-D wave equation modeling and migration by a new finite difference scheme based on the Galerkin method. 74th Annual International Meeting, SEG, Expanded Abstracts, pp 1107–1110

Durst P, Vuataz FD (2000) Fluid-rock interactions in hot dry rock reservoirs. A review of the HDR sites and detailed investigations of the Soultz-Sous-Forets system. Proceedings, World Geothermal Congress 2000, Kyushu-Tohoku, Japan, May 28–June 10, 2000

Eker E, Akin S (2006) Lattice Boltzmann simulation of fluid flow in synthetic fractures. Transp Porous Media 65:363–384

Ene HI, Poliševski D (1987) Thermal flow in porous media. D Reidel Publishing Company, Dordrecht, Holland

Engelder T, Fischer MP (1994) Influence of poroelastic behavior on the magnitude of minimum horizontal stress, Sh, in overpressured parts of sedimentary basins. Geology 22: 949–952

Evans KF, Cornet FH, Hashida T, Hayashi K, Ito T, Matsuki K, Wallroth T (1999) Stress and rock mechanics issues of relevance to HDR/HWR engineered geothermal systems: review of developments during the past 15 years. Geothermics 28:455–474

Fengler MJ (2005) Vector spherical harmonic and vector wavelet based non-linear Galerkin schemes for solving the incompressible Navier–Stokes equation on the sphere. PhD dissertation, Geomathematics Group, TU Kaiserslautern, Shaker

Fischer N, Lewis T, Embleton B (1993) Statistical analysis of spherical data. Cambridge University Press, Cambridge

Fletcher RP, Fowler PJ, Kitchenside P, Albertin U (2006) Suppressing unwanted internal reflections in prestack reverse-time migration. Geophysics 71(6):E79–E82

Fomin S, Hashida T, Shimizu A, Matsuki K, Sakaguchi K (2003) Fractal concept in numerical simulation of hydraulic fracturing of the hot dry rock geothermal reservoir. Hydrol Process 17:2975–2989

Freeden W, Mayer C (2007) Modeling tangential vector fields on regular surfaces by means of Mie potentials. Wavelets Multiresol Inform Process 5(3):417–449

Freeden W, Mayer C, Schreiner M (2003) Tree algorithms in wavelet approximations by Helmholtz potential operators. Numer Funct Anal Optim 24(7,8):747–782

Ghassemi A (2003) A thermoelastic hydraulic fracture design tool for geothermal reservoir development. Final report, Department of Geology & Geological Engineering, University of North Dakota

Ghassemi A, Tarasovs S (2004) Three-dimensional modeling of injection induced thermal stresses with an example from Coso. Proceedings, 29th workshop on geothermal reservoir engineering, Stanford University, Stanford, CA, January 26–28, 2004, SGP-TR-173

Ghassemi A, Zhang Q (2004) Poro-thermoelastic mechanisms in wellbore stability and reservoir stimulation. Proceedings, 29th workshop on geothermal reservoir engineering, Stanford University, Stanford, CA, January 26–28, 2004, SGP-TR-173

Ghassemi A, Tarasovs S, Cheng AHD (2003) An integral equation solution for three-dimensional heat extraction from planar fracture in hot dry rock. Int J Numer Anal Meth Geomech 27:989–1004

Guitton A, Kaelin B (2006) Least-square attenuation of reverse time migration artefacts. Annual Meeting, SEG, New Orleans, pp 1514–1524

Helmig R, Niessner J, Flemisch B, Wolff M, Fritz J (2010) Efficient modelling of flow and transport in porous media using multi-physics and multi-scale approaches. In: Freeden W, Nashed Z, Sonar T (eds) Handbook of GeoMathematics, Springer, Berlin

Heuer N, Küpper T, Windelberg D (1991) Mathematical model of a hot dry rock system. Geophys J Int 105:659–664

Hicks TW, Pine RJ, Willis-Richards J, Xu S, Jupe AJ, Rodrigues NEV (1996) A hydro-thermo-mechanical numerical model for HDR geothermal reservoir evaluation. Int J Rock Mech Min Sci Geomech Abstr 33(5):499–511

Hillis RR (2000) Pore pressure/stress coupling and its implications for seismicity. Explor Geophys 31:448–454

Hillis RR (2001) Coupled changes in pore pressure and stress in oil fields and sedimentary basins. Pet Geosci 7:419–425

Hillis RR (2003) Pore pressure/stress coupling and its implications for rock failure. In: Vanrensbergen P, Hillis RR, Maltman AJ, Morley CK (eds) Subsurface sediment mobilization. Geological Society of London, London, pp 359–368

Ilyasov M (2010) On the modeling of metaharmonic functions in regions with non-smooth boundaries and its application to seismics. PhD dissertation, Geomathematics Group, TU Kaiserslautern, in preparation

Itasca Consulting Group Inc. (2000) UDEC users's guide. Minnesota

Jing L, Hudson JA (2002) Numerical methods in rock mechanics. Int J Rock Mech Min Sci 39:409–427

Jing Z, Willis-Richards J, Watanabe K, Hashida T (2000) A three-dimensional stochastic rock mechanics model of engineered geothermal systems in fractured crystalline rock. J Geophys Res 105(B10):23.663–23.679

Jing Z, Watanabe K, Willis-Richards J, Hashida T (2002) A 3-D water/rock chemical interaction model for prediction of HDR/HWR geothermal reservoir performance. Geothermics 31:1–28

Jung R (2007) Stand und Aussichten der Tiefengeothermie in Deutschland. Erdöl, Erdgas, Kohle 123(2):1–7

Kazemi H (1969) Pressure transient analysis of naturally fractured reservoirs with uniform fracture distribution. Soc Pet Eng J Trans AIME 246:451–461

Kazemi H, Merril LS, Porterfield KL, Zeman PR (1976) Numerical simulation of water-oil flow in naturally fractured reservoirs. Proceedings, SPE-AIME 4th symposium on numerical simulation of reservoir performance, Los Angeles, CA, February 19–20, 1976, SPE 5719

Kim I, Lindquist WB, Durham WB (2003) Fracture flow simulation using a finite-difference lattice Boltzmann method. Phys Rev E 67:046708

Kimura S, Masuda Y, Hayashi K (1992) Efficient numerical method based on double porosity model to analyze heat and fluid flows in fractured rock formations. JSME Int J, Ser 2 35(3):395–399

Kolditz O (1997) Strömung, Stoff- und Wärmetransport im Kluftgestein. Borntraeger, Berlin, Stuttgart

Kühn M (2009) Modelling feedback of chemical reactions on flow fields in hydrothermal systems. Surv Geophys 30(3):233–251

Kühn M, Stöfen H (2005) A reactive flow model of the geothermal reservoir Waiwera, New Zealand. Hydrol J 13(4):606–626

Lang U (1995) Simulation regionaler Strömungs- und Transportvorgänge in Karstaquiferen mit Hilfe des Doppelkontinuum-Ansatzes: Methodenentwicklung und Parameteridentifikation. PhD dissertation, University Stuttgart, Germany, Mitteilungen Heft 85

Lang U, Helmig R (1995) Numerical modeling in fractured media—identification of measured field data. In: Krasny J, Mls J (eds) Groundwater quality: remediation and protection. IAHS and University Karlova, Prague, pp 203–212

Lee J, Choi SU, Cho W (1999) A comparative study of dual-porosity model and discrete fracture network model. KSCE J Civ Eng 3(2):171–180

Lions JL, Magenes E (1972) Non-homogeneous boundary value problems and applications I. Die Grundlehren der mathematischen Wissenschaften in Einzeldarstellungen, Band 181, Springer

Lomize GM (1951) Seepage in fissured rocks. State Press, Moscow

Long J, Remer J, Wilson C, Witherspoon P (1982) Porous media equivalents for networks of discontinuous fractures. Water Resour Res 18(3):645–658

Maryška J, Severýn O, Vohralík M (2004) Numerical simulation of fracture flow with a mixed-hybrid FEM stochastic discrete fracture network model. Comput Geosci 8:217–234

Masihi M, King P, Nurafza P (2007) Fast estimation of connectivity in fractured reservoirs using percolation theory. SPE J 12(2):167–178

Min KB, Jing L, Stephansson O (2004) Determining the equivalent permeability tensor for fractured rock masses using a stochastic REV approach: method and application to the field data from Sellafield, UK. Hydrogeol J 12: 497–510

Mitchell AR, Griffiths DF (1980) The finite difference method in partial differential equations. Wiley, Chichester

Mo H, Bai M, Lin D, Roegiers JC (1998) Study of flow and transport in fracture network using percolation theory. Appl Math Modell 22: 277–291

Moeck I, Kwiatek G, Zimmermann G (2009) The in-situ stress field as a key issue for geothermal field development—a case study from the NE German Basin. Proceedings, 71st EAGE conference & exhibition, Amsterdam, T025

Mufti IR (1985) Seismic modeling in the implicit mode. Geophys Prospect 33:619–656

Müller C (1969) Foundations of the mathematical theory of electromagnetic waves. Springer, Berlin

Nakao S, Ishido T (1998) Pressure-transient behavior during cold water injection into geothermal wells. Geothermics 27(4):401–413

Neuman S (2005) Trends, prospects and challenges in quantifying flow and transport through fractured rocks. Hydrogeol J 13:124–147

Neuman S, Depner J (1988) Use of variable-scale pressure test data to estimate the log hydraulic conductivity covariance and dispersivity of fractured granites near Oracle, Arizona. J Hydrol 102:475–501

Odén M, Niemi A (2006) From well-test data to input to stochastic continuum models: effect of the variable support scale of the hydraulic data. Hydrogeol J 14:1409–1422

Ödner H (1998) One-dimensional transient flow in a finite fractured aquifer system. Hydrol Sci J 43(2):243–265

Ostermann I (2009) Wärmetransportmodellierung in tiefen geothermischen Systemen. Bericht 45, Schriften zur Funktionalanalysis und Geomathematik, Department of Mathematics, TU Kaiserslautern

O'Sullivan MJ, Pruess K, Lippmann MJ (2001) State of the art of geothermal reservoir simulation. Geothermics 30:395–429

Ouenes A (2000) Practical application of fuzzy logic and neural networks to fractured reservoir characterization. Comput Geosci 26:953–962

Peters RR, Klavetter EA (1988) A continuum model for water movement in an unsaturated fractured rock mass. Water Resour Res 24:416–430

Phillips WS, Rutledge JT, House LS, Fehler MC (2002) Induced microearthquake patterns in hydrocarbon and geothermal reservoirs: six case studies. Pure Appl Geophys 159:345–369

Popov MM, Semtchenok NM, Popov PM, Verdel AR (2006) Gaussian beam migration of multivalued zero-offset data. Proceedings, International Conference, Days on Diffraction, St Petersburg, Russia, May 30–June 2, 2006, pp 225–234

Popov MM, Semtchenok NM, Verdel AR, Popov PM (2008) Reverse time migration with Gaussian beams and velocity analysis applications. Extended Abstracts, 70th EAGE Conference & Exhibitions, Rome, F048

Pratt RG (1999) Seismic waveform inversion in the frequency domain, Part I: theory and verification in a physical scale model. Geophysics 64(3):888–901

Pratt RG, Shin C, Hicks GJ (1998) Gauss-Newton and full Newton methods in frequency-space seismic waveform inversion. Geophysics 133(2): 341–362

Pruess K (1990) Modeling of geothermal reservoirs: fundamental processes, computer simulation and field applications. Geothermics 19(1):3–15

Pruess K, Narasimhan TN (1985) A practical method for modeling fluid and heat flow in fractured porous media. Soc Pet Eng J 25:14–26

Pruess K, Wang JSY, Tsang YW (1986) Effective continuum approximation for modeling fluid and heat flow in fractured porous tuff. Sandia National Laboratories Report SAND86-7000, Albuquerque

Reichenberger V, Jakobs H, Bastian P, Helmig R (2006) A mixed-dimensional finite volume method for two-phase flow in fractured porous media. Adv Water Resour 29(7):1020–1036

Renardy M, Rogers RC (1993) An introduction to partial differential equations. Texts in applied mathematics, vol 13, Springer, Berlin

Rice JR, Cleary MP (1976) Some basic stress diffusion solutions for fluid-saturated elastic porous media with compressible constituents. Rev Geophys Space Phys 14(2):227–241

Rutqvist J, Stephansson O (2003) The role of hydromechanical coupling in fractured rock engineering. Hydrogeol J 11:7–40

Sahimi M (1995) Flow and transport in porous media and fractured rock: from classical methods to modern approaches. VCH, Weinheim

Samarskii AA (2001) The theory of difference schemes. Marcel Dekker Inc., New York

Samarskii AA, Vabishchevich PN (2001) Additive schemes for problems of mathematical physics. Nauka, Moscow

Sanyal S (2005) Classification of geothermal systems—a possible scheme. Proceedings, 30th Workshop on Geothermal Reservoir Engineering, Stanford University, Stanford, California, January 31–February 2, SGP-TR-176 85–92

Sanyal SK, Butler SJ, Swenson D, Hardeman B (2000) Review of the state-of-the-art of numerical simulation of enhanced geothermal systems. Proceedings, World Geothermal Congress 2000, Kyushu–Tohoku, Japan, May 28–June 10, 2000

Schulz R (2009) Aufbau eines geothermischen Informationssystems für Deutschland, Endbericht. Leibniz-Institut für Angewandte Geophysik Hannover

Semtchenok NM, Popov MM, Verdel AR (2009) Gaussian beam tomography. Extended Abstracts, 71st EAGE Conference & Exhibition, Amsterdam, U032

Snow DT (1965) A parallel plate model of fractured permeable media. PhD dissertation, University of California, Berkeley

Sommerfeld A (1947) Partielle Differentialgleichungen der Physik. Akademische Verlagsgesellschaft Geest & Portig KG

Stothoff S, Or D (2000) A discrete-fracture boundary integral approach to simulating coupled energy and moisture transport in a fractured porous medium. In: Faybishenko B, Witherspoon PA, Benson SM (eds) Dynamics of fluids in fractured rocks, concepts and recent advances. AGU Geophysical Monograph 122, American Geophysical Union, Washington, pp 269–279

Sudicky EA, McLaren RG (1992) The Laplace transform Galerkin technique for large-scale simulation of mass transport in discretely fractured porous formations. Water Resour Res 28(2):499–514

Symes WW (1995) Mathematics of reflection tomography: the Rice inversion project. Department of Computational and Applied Mathematics, Rice University, Houston

Tang D, Frind E, Sudicky E (1981) Contaminant transport in fractured porous media: analytical solution for a single fracture. Water Resour Res 17(3):555–564

Tarantola A (1984) Inversion of seismic reflection data in the acoustic approximation. Geophysics 49:1259–1266

Tran NH, Rahman SS (2006) Modelling discrete fracture networks using neuro-fractal-stochastic simulation. J Eng Applied Sci 1(2):154–160

Travis BJ (1984) TRACR3D: a model of flow and transport in porous/fractured media. Los Alamos National Laboratory LA-9667-MS, Los Alamos

Tsang Y, Tsang C (1987) Chanel flow model through fractured media. Water Resour Res 23(3):467–479

Tsang Y, Tsang C (1989) Flow chaneling in a single fracture as a two-dimensional strongly heterogeneous permeable medium. Water Resour Res 25(9):2076–2080

Tsang Y, Tsang C, Hale F, Dverstop B (1996) Tracer transport in a stochastic continuum model of fractured media. Water Resour Res 32(10):3077–3092

Versteeg R (1994) The Marmousi experience: velocity model determination on a synthetic complex data set. The Leading Edge, Rice University, Houston

Warren JE, Root PJ (1963) The behavior of naturally fractured reservoirs. Soc Pet Eng J Trans AIME 228:245–255

Watanabe K, Takahashi T (1995) Fractal geometry characterization of geothermal reservoir fracture networks. J Geophys Res 100(B1): 521–528

Wu YS (2000) On the effective continuum method for modeling multiphase flow, multicomponent transport and heat transfer in fractured rock. In: Faybishenko B, Witherspoon PA, Benson SM (eds) Dynamics of fluids in fractured rocks, concepts and recent advances. AGU Geophysical Monograph 122, American Geophysical Union, Washington, pp 299–312

Wu YS, Pruess K (2005) A physically based numerical approach for modeling fracture-matrix interaction in fractured reservoirs. Proceedings, World Geothermal Congress 2005, Antalya, Turkey, April 24–29, 2005

Wu YS, Qin G (2009) A generalized numerical approach for modeling multiphase flow and transport in fractured porous media. Comm Comput Phys 6(1):85–108

Wu X, Pope GA, Shook, GM, Srinivasan S (2005) A semi-analytical model to calculate energy production in single fracture geothermal reservoirs. Geothermal Resour Council Trans 29:665–669

Wu RS, Xie XB, Wu XY (2006) One-way and one-return approximations (de Wolf approximation) for fast elastic wave modeling in complex media. Adv Geophys 48(5):265–322

Yilmaz Ö (2001) Seismic data analysis: processing, inversion, and interpretation of seismic data. Society of Exploration Geophysicists, Tulsa

Yin S (2008) Geomechanics-reservoir modeling by displacement discontinuity-finite element method. PhD dissertation, University of Waterloo, Ontario, Canada

Zhang W, Zhang G (2004) 3D hybrid depth migration and four-way splitting scheme. J Comp Math 24:463–474

Zhao C, Hobbs BE, Baxter K, Mühlhaus HB, Ord A (1999) A numerical study of pore-fluid, thermal and mass flow in fluid-saturated porous rock basins. Eng Comput 16(2):202–214

Zhou XX, Ghassemi A (2009) Three-dimensional poroelastic simulation of hydraulic and natural fractures using the displacement discontinuity method. Proceedings, 34th Workshop on Geothermal Reservoir Engineering, Stanford University, Stanford, California, February 9–11, 2009, SGP-TR-187

Zubkov VV, Koshelev VF, Lin'kov AM (2007) Numerical modeling of hydraulic fracture initiation and development. J Min Sci 43(1):40–56

Zyvoloski G (1983) Finite element methods for geothermal reservoir simulation. Int J Numer Anal Methods Geomech 7:75–86

23 Phosphorus Cycles in Lakes and Rivers: Modeling, Analysis, and Simulation

Andreas Meister[1] · *Joachim Benz*[2]
[1]Work-Group of Analysis and Applied Mathematics, Department of Mathematics, University of Kassel, Kassel, Germany
[2]Faculty of Organic Agricultural Sciences, Work-Group Data-Processing and Computer Facilities, University of Kassel, Witzenhausen, Germany

W. Freeden, M.Z. Nashed, T. Sonar (Eds.), *Handbook of Geomathematics*, DOI 10.1007/978-3-642-01546-5_23,

Abstract From spring to summer period, a large number of lakes are laced with thick layers of algae implicitly representing a serious problem with respect to the fish stock as well as other important organisms and at the end for the complete biological diversity of species. Consequently, the investigation of the cause-and-effect chain represents an important task concerning the protection of the natural environment. Often such situations are enforced by an oversupply of nutrient. As phosphorus is the limiting nutrient element for most of all algae growth processes an advanced knowledge of the phosphorus cycle is essential. In this context the chapter gives a survey on our recent progress in modeling and numerical simulation of plankton spring bloom situations caused by eutrophication via phosphorus accumulation. Due to the underlying processes we employ the shallow water equations as the fluid dynamic part coupled with additional equations describing biogeochemical processes of interest within both the water layer and the sediment. Depending on the model under consideration one is faced with significant requirements like positivity as well as conservativity in the context of stiff source terms. The numerical method used to simulate the dynamic part and the evolution of the phosphorus and different biomass concentrations is based on a second-order finite volume scheme extended by a specific formulation of the modified Patankar approach to satisfy the natural requirements to be unconditionally positivity preserving as well as conservative due to stiff transition terms. Beside a mathematical analysis, several test cases are shown which confirm both the theoretical results and the applicability of the complete numerical scheme. In particular, the flow field and phosphorus dynamics for the West Lake in Hangzhou, China are computed using the previously stated mass and positivity preserving finite volume scheme.

1 Introduction

The main objective of this chapter is to present a stable and accurate numerical method for a wide range of applications in the field of complex ecosystem models. Thereby, we focus on the process of eutrophication, which represents a serious problem through giving rise to excessive algae blooms in all eutrophic fresh water ecosystems. Thus, it is crucial in ecological science to understand and predict the coherences of the underlying dynamics. Even taking into account the systematical difficulties of modeling and simulation (Poethke 1994), elaborate modeling provides a very useful tool for deeper understanding and long range prediction (Sagehashi et al. 2000).

In the 1970s computer-based modeling and simulation in ecological science with a special focus on lake eutrophication examination started to achieve good results. Deterministic models were proposed in Park et al. (1974), Jørgensen (1975), and Straškraba and Gnauk (1985).

Up to now due to both deeper knowledge of the underlying principles and growing computational power the examined models have been developed more sophisticated. The evolved models nowadays pose severe demands on the used numerical schemes. Models including matter cycles, where the material, for example, atoms, can only change their configuration and thus are neither created nor destroyed, require the algorithm to maintain the mass and number of atoms in the system, respectively. Another computational nontrivial but obvious demand is to ensure positivity for all examined material constituents. Consequently, in recent years, one can observe increasing interest in the design of positivity preserving schemes.

A survey on positive advection transport methods can be found in Smolarkiewicz (2006). With respect to shallow water flows one is often faced with the transition between dry and wet areas such that the construction of positive schemes is widespread for this kind of application, see Berzins (2001), Audusse and Bristeau (2005), Chertock and Kurganov (2008), Ricchiuto and Bollermann (2009) and the references therein. Concerning stiff reaction terms a modification of common Runge–Kutta schemes which ensures the positivity was originally suggested by Patankar in the context of turbulent mixing (Patankar 1980). Unfortunately the schemes obtained were not able to retain the characteristic of mass conservation, even though mass conservativity is a feature of the original Runge–Kutta schemes. The scheme was improved in Burchard et al. (2003) retaining the positivity but also reacquiring the conservation of mass. Based on this idea, a more complex form of conservation was achieved in a series of publications (Bruggeman et al. 2007; Broekhuizen et al. 2008). First applications of the modified Patankar-approach in combination with classical discretization schemes for the solution of advection-diffusion-reaction equations are presented in Burchard et al. (2005, 2006) with a specific focus on marine ecosystem dynamics in the North and the Baltic Sea. But being more applicable to flexible definitions of conservation they lost crucial numerical stability. A detailed overview of those schemes can be found in Zardo (2005). Due to the superior numerical stability the modified Patankar ansatz is used. Also the ensured conservation of mass is appropriate for the model under consideration.

One major and essential aspect for growth is nutrient supply. To predict the growth of an algae reliably one has to consider the limiting factor (Schwoerbel and Brendelberger 2005). The basis for the implemented ecological model is presented by Hongping and Jianyi in Hongping and Jianyi (2002). As the model under consideration describes algae growth in the West Lake, China with a ratio of more then 1:14 for phosphorus to nitrogen the limiting nutrient is phosphorus (Lampert and Sommer 1999). Therefore it is sufficient to restrict the considered nourishment to phosphorus for modeling purpose.

The model features the description of four different groups of algae species and zooplankton with the particularity of mapping specific phosphorus contents for each of these organisms. Furthermore, phosphorus is presented in solute and organic form in the water body as well as in the sediment.

Beside the biological and chemical processes the flow field of the lake itself is of major importance. The current leads to an unbalanced distribution of matter inside a lake and thus creates significant differences in phosphorus concentrations.

The two-dimensional shallow water equations form an adequate formulation of shallow water flow. By means of sophisticated high resolution schemes (Toro 2001; Vater 2004), one is able to simulate complicated phenomena like, for example, a dam break problem or others, see Vázquez-Cendón (2007), Stoker (1957). The two-dimensional shallow water equations thus serve as basis formulation for the flow phenomena of interest.

The phosphorus cycle model together with the two-dimensional shallow water equations forms the complete set of convection-diffusion-reaction equations and is a set of hyperbolic parabolic partial differential equations.

2 Mathematical Modeling

Modeling ecological process in lakes, urban rivers, or channels possess two main demands. First, a reliable mathematical model has to be formulated which includes fundamental properties and effects of eutrophication including local biochemical phenomena as well as transport processes due to convection and diffusion. Thus, beside the consideration of the dynamics of biomass and phosphorus concentrations, one has to take account of the flow situation in the water body. Consequently, the equations governing the process of interest are given by a combination of the Saint-Venant system for the flow part and an advection-diffusion-reaction system for the biochemical part.

Neglecting the influence of rain and evaporation the underlying fluid dynamic part represented by the shallow water equations can be written in the case of a constant bottom topography in form of the conservation law

$$\partial_t \mathbf{u}_s + \sum_{m=1}^{2} \partial_{x_m} \mathbf{f}^c_{m,s}(\mathbf{u}_s) = \mathbf{0}.$$

Thereby, $\mathbf{u}_s = (H, \Phi v_1, \Phi v_2)^T$ and $\mathbf{f}^c_{m,s}(\mathbf{u}_s) = \left(H v_m, \Phi v_m v_1 + \frac{1}{2}\delta_1^m \Phi^2, \Phi v_m v_2 + \frac{1}{2}\delta_2^m \Phi^2\right)^T$ are referred to as the vector of conserved quantities and the flux function in which $H, \mathbf{v} = (v_1, v_2)^T$ denote the water height and the velocity, respectively. Furthermore, $\Phi = gH$ represents the geopotential with the gravity force g and the Kronecker delta is denoted by δ_i^m.

The ecological part consists of processes which describe the behavior of biomass and organic phosphorus of four different groups of algae species (BA, PA) and zooplankton (BZ, PZ). To describe the complete phosphorus cycle including solute phosphorus in the water body (PS), organic phosphorus in the detritus (PZ) and organic and inorganic phosphorus in the sediment (PE_O, PE_I) are considered, see ➋ *Fig. 2*. This model is based on the *West Lake Model*, which is published in Hongping and Jianyi (2002). The dynamics of biomass demands positivity and the cycle of phosphorus positivity and conservativity as well. The interactions between these ecological system elements are nonlinear, for example, growth is formulated as *Michaelis–Menten* kinetics. In addition, the particulate components (biomass, organic phosphorus, and detritus) in the water body are influenced by advective transport. The solute phosphorus in the water body is affected by advective and diffusive transport. For the elements inside of the sediment no horizontal transport is considered. The vertical exchange between phosphorus in the sediment and in the water body is driven by diffusion.

Consequently, the governing equations for the phosphorus cycle and the biomass dynamics (➋ *Fig. 1*) represents a system of advection-diffusion-reaction equations of the form

$$\partial_t \mathbf{u}_p + \sum_{m=1}^{2} \partial_{x_m} \mathbf{f}^c_{m,p}(\mathbf{u}_s, \mathbf{u}_p) = \sum_{m=1}^{2} \partial_{x_m} \mathbf{f}^v_{m,p}(\mathbf{u}_p) + \mathbf{q}_p(\mathbf{u}_s, \mathbf{u}_p) \tag{1}$$

where the components read

$$\mathbf{u}_p = (BA, BZ, PA, PZ, PD, PS, PE_I, PE_O)^T$$

and the convective and viscous fluxes are

$$\mathbf{f}^c_{m,p}(\mathbf{u}_s, \mathbf{u}_p) = v_m (BA, BZ, PA, PZ, PD, PS, 0, 0)^T$$

and

$$\mathbf{f}^v_{m,p}(\mathbf{u}_p) = (0, \ldots, 0, \lambda_{PS}\partial_{x_m} PS, \lambda_{PE_I}\partial_{x_m} PE_I, 0)^T,$$

respectively. Note, that BA and PA denotes a vector of four constituents by it selves. With respect to the spatial domain $\mathcal{D} \subset \mathbf{R}^2$, which is assumed to be polygonal bounded, i.e., $\partial\mathcal{D} = \cup_{k=1}^n \partial\mathcal{D}_k$,

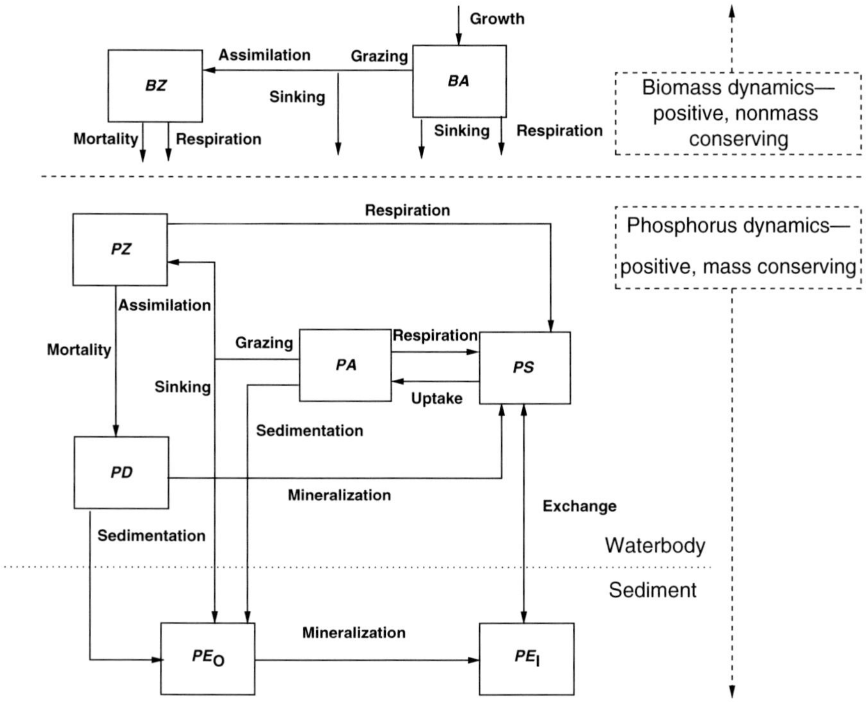

Fig. 1
Phosphorus and biomass dynamic

the complete system consisting of the fluid and biochemical part can be written as

$$\partial_t \mathbf{U} + \sum_{m=1}^{2} \partial_{x_m} \mathbf{F}_m^c(\mathbf{U}) = \sum_{m=1}^{2} \partial_{x_m} \mathbf{F}_m^v(\mathbf{U}) + \mathbf{Q}(\mathbf{U}) \text{ in } \mathcal{D} \times \mathbf{R}^+, \tag{2}$$

where $\mathbf{U} = (\mathbf{u}_s, \mathbf{u}_p)^T$, $\mathbf{F}_m^c = (\mathbf{f}_{m,s}^c, \mathbf{f}_{m,p}^c)^T$, $\mathbf{F}_m^v = (0, \mathbf{f}_{m,p}^v)^T$, and $\mathbf{Q} = (0, \mathbf{q}_p)^T$.

A detailed description of the biochemical part will be presented in Sec. 3.3.

3 Numerical Method

The discretization of the mathematical model is based on a conventional finite volume scheme. To satisfy the specific requirements of positivity with respect to the involved biogeochemical system it is necessary to employ an appropriate method for the corresponding local transition terms. Thus, for the sake of simplicity, we first present the finite volume approach and investigate its applicability in the context of the fluid dynamic part of the governing equation. Thereafter, we will discuss positivity preserving, conservative scheme for stiff and non-stiff ordinary differential equations. Similar to the finite-volume approach, we will discuss the properties of this numerical scheme not only theoretically but also by means of different test cases. Finally, we combine both parts to simulate the flow field and the phosphorus cycle with respect to the West lake as a comprehensive practical application.

3.1 Finite Volume Method

Finite volume schemes are formulated on arbitrary control volumes and well-known successful time stepping schemes and spatial discretization techniques can easily be employed in the general framework. This approximation technique perfectly combines the needs concerned with robustness and accuracy as well as the treatment of complicated geometries. We start with the description of a state-of-the-art finite volume scheme. The development of the method is presented in various articles (Meister 1998, 2003; Meister and Sonar 1998; Meister and Oevermann 1998; Meister and Vömel 2001) which also proof the validity of the algorithm in the area of inviscid and viscous flow fields.

Smooth solutions of the system of the shallow water equations exist in general only for short time and well-known phenomena like shock waves and contact discontinuities develop naturally. Hence we introduce the concept of weak solutions which represents the basis for each finite volume scheme. Every single finite volume scheme can be derived using the concept of weak solutions. A bounded set $\sigma \subset \mathbf{R}^2$ is said to be a control volume if Gauss' integral theorem is applicable to functions defined on σ. The mapping $\widetilde{\mathbf{u}}$ is called a weak solution of the system (2) if

$$\frac{d}{dt}\int_{\sigma} \mathbf{U}\,d\mathbf{x} = -\int_{\partial\sigma}\sum_{m=1}^{2}\mathbf{F}_m^c(\mathbf{U})n_m\,ds + \int_{\partial\sigma}\sum_{m=1}^{2}\mathbf{F}_m^v(\mathbf{U})n_m\,ds + \int_{\sigma}\mathbf{Q}(\mathbf{U})\,d\mathbf{x} \tag{3}$$

is valid on every control volume $\sigma \subset \mathcal{D}$ with outer unit normal vector $\mathbf{n} = (n_1, n_2)^T$. Note that this formula can be derived from the system (2) by integrating over a control volume and applying Gauss' integral theorem. The solution class considered can be described by the function space $BV\left(\mathbf{R}_0^+; L^1 \cap L^\infty(\mathbf{R}^2;\mathbf{R}^{17})\right)$, i.e., the mapping $t \mapsto \mathbf{U}(\cdot,t)$ is of bounded variation and the image is an integrable function of the space variables which is bounded almost everywhere.

The numerical approximation of Eq. 3 requires an appropriate discretization of the space part $\overline{\mathcal{D}}$ as well as the time part $\mathbf{R}^+$.

We start from an arbitrary conforming triangulation $\mathcal{T}^h$ of the domain $\overline{\mathcal{D}}$ which is called the primary mesh consisting of finitely many (say $\#\mathcal{T}^h$) triangles $\mathcal{T}_i$, $i = 1,\ldots,\#\mathcal{T}^h$. Thereby, the parameter h corresponds to a typical one-dimensional geometrical measure of the triangulation, as, for example, the maximum diameter of the smallest in-closing circle of the triangle $\mathcal{T}_i$, $i = 1,\ldots,\#\mathcal{T}^h$. For a comprehensive definition of a primary grid, we refer to Sonar (1997a). Furthermore, $\mathcal{N}^h$ denotes the index set of all nodes of the triangulation $\mathcal{T}^h$ and is subdivided by $\mathcal{N}^h = \mathcal{N}^{h,\mathcal{D}} \cup \mathcal{N}^{h,\partial\mathcal{D}}$, where $\mathcal{N}^{h,\mathcal{D}}$ is associated with the inner points and $\mathcal{N}^{h,\partial\mathcal{D}}$ includes the indices of the boundary points. We set $N := \#\mathcal{N}^h$ and denote the three edges of the triangle $\mathcal{T}$ by $e_{\mathcal{T},k}$, $k = 1,2,3$. Furthermore, we define

$$E(i) = \left\{e_{\mathcal{T},k} \mid k \in \{1,2,3\}, \mathcal{T} \in \mathcal{T}^h, \text{ node } \mathbf{x}_i \in e_{\mathcal{T},k}\right\},$$

$$V(i) = \left\{\mathcal{T} \mid \text{node } \mathbf{x}_i \text{ is vertex of } \mathcal{T} \in \mathcal{T}^h\right\},$$

and

$$C(\mathcal{T}) = \{i \mid i \in \{1,\ldots,N\}, \text{ node } \mathbf{x}_i \text{ is vertex of } \mathcal{T}\}.$$

For the calculation of the triangulation we employ an algorithm developed by Friedrich (Friedrich 1993) which provides mostly isotropic grids. As we see in the following, the occurrence of second-order derivatives within the partial differential equation requires the evaluation of first order derivative on the boundary of each control volume used in the finite volume scheme. Due to this fact it is advantageous to employ a box-type method where the computation of a first-order derivative at the boundary of each box is straightforward. We define a discrete control volume σ_i as the open subset of $\mathbf{R}^2$ including the node $\mathbf{x}_i = (x_{i1}, x_{i2})^T$ and bounded by the straight lines, which are defined by the connection of the midpoint of the edge $e_{\mathcal{T},k} \in E(i)$ with the point

$$\mathbf{x}_s = (x_{s1}, x_{s2})^T = \sum_{i \in C(\mathcal{T})} \alpha_i \mathbf{x}_i$$

of the corresponding triangle $\mathcal{T}$, see Fig. 2. In the case that $\mathbf{x}_i$ is at the boundary of the computational domain, the line defined by the connection of the midpoint of the boundary edge and the node itself is also a part of $\partial\sigma_i$. For the calculation of the weights we employ

$$\alpha_i = \frac{1}{2 \sum\limits_{m \in C(\mathcal{T})} |l_m|} \sum_{\substack{m \in C(\mathcal{T}) \\ m \neq i}} |l_m| \text{ with } |l_i| = \|\mathbf{x}_j - \mathbf{x}_k\|_2 \text{ for } i, j, k \in C(\mathcal{T}). \tag{4}$$

This definition exhibits the advantage that the deformation of the control volume with respect to distorted triangles is much smaller compared to those achieved by the use of $\mathbf{x}_s$ as the barycenter of the triangle $\mathcal{T}$.

The union $\mathcal{B}^h$ of all boxes σ_i, $i \in \mathcal{N}^h$ is called the secondary mesh. Let $N(i)$ denote the index set of all nodes neighboring node $\mathbf{x}_i$, i.e., those nodes $\mathbf{x}_j$, $j \neq i$, for which $\int_{\partial\sigma_i \cap \partial\sigma_j} 1\, ds \neq 0$ is valid. In general, for $j \in N(i)$ the boundary between the control volume σ_i and σ_j consists of two line segments which are denoted by l_{ij}^k, $k = 1, 2$. Furthermore, $\mathbf{n}_{ij}^k$, $k = 1, 2$, represent the accompanying unit normal vectors. Note that in the case of a boundary box σ_i there exist two adjacent cells σ_j, $j \in \mathcal{N}^{h,\partial\mathcal{D}}$, such that $\partial\sigma_i \cap \partial\sigma_j$ consists of one line segment l_{ij}^k only. In order to achieve a unique representation we interpret the lacking line segment as having the length zero.

Introducing the cell average

$$\mathbf{U}_i(t) := \frac{1}{|\sigma_i|} \int_{\sigma_i} \mathbf{U}(\mathbf{x}, t) d\mathbf{x}$$

the integral form with regard to an inner box σ_i can be written as

$$\frac{d}{dt}\mathbf{U}_i(t) = -\left(\mathcal{L}_i^c \mathbf{U}\right)(t) + \left(\mathcal{L}_i^v \mathbf{U}\right)(t) + (\mathcal{Q}_i \mathbf{U})(t)$$

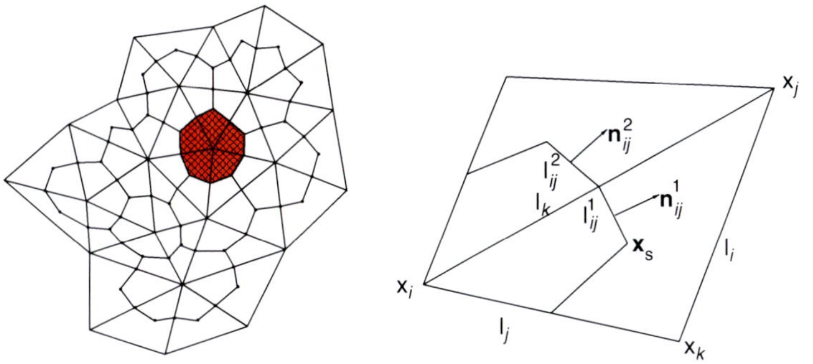

Fig. 2
General form of a control volume (*left*) and its boundary (*right*)

$p_{i,j}(\mathbf{c})$ and $d_{i,j}(\mathbf{c})$ must satisfy $p_{i,j}(\mathbf{c}) = d_{j,i}(\mathbf{c})$. In addition to these transition terms, we consider for the ith constituent local production by $p_{i,i}(\mathbf{c}) \geq 0$ and similar local destruction by $d_{i,i}(\mathbf{c}) \geq 0$. Thus, the system we start to investigate here can be written as

$$\frac{d}{dt}c_i = \sum_{\substack{j=1\\ j\neq i}}^{I} p_{i,j}(\mathbf{c}) - \sum_{\substack{j=1\\ j\neq i}}^{I} d_{i,j}(\mathbf{c}) + p_{i,i}(\mathbf{c}) - d_{i,i}(\mathbf{c}),\ i = 1,\ldots,I, \tag{6}$$

where $\mathbf{c} = \mathbf{c}(t) = (c_1(t),\ldots,c_I(t))^T$ denotes the vector of the I constituents. This system has to be solved under the initial conditions

$$\mathbf{c}^0 = \mathbf{c}(t=0) > \mathbf{0}. \tag{7}$$

Definition 3.1

The system (6) is called fully conservative, if the production and destruction terms satisfy

$$p_{i,j}(\mathbf{c}) = d_{j,i}(\mathbf{c}) \ \text{ for } \ j \neq i, \ \ j,i \in \{1,\ldots,I\}$$

and

$$p_{i,i}(\mathbf{c}) = d_{i,i}(\mathbf{c}) = 0 \ \text{ for } \ i \in \{1,\ldots,I\}.$$

Furthermore, we always consider ecosystems where the constituents are by nature positive. From a mathematical point of view one can easily prove by means of a simple contradiction argument, that for non negative initial conditions $c_i(0) \geq 0$, $i = 1,\ldots,I$, the condition

$$d_{i,j}(\mathbf{c}) \longrightarrow 0 \ \text{ for } \ c_i \longrightarrow 0$$

for all $j \in \{1,\ldots,I\}$, guarantees that the quantities $c_i(t)$, $i = 1,\ldots,I$ remain non negative for all $t \in \mathbf{R}^+$. Consequently, the properties mentioned above have to be maintained by the discretization scheme which means that no gain or loss of mass should occur for numerical reasons and that the concentration of all constituents have to remain positive independent of the time step size used. Based on the following standard formulation of a discretization scheme as

$$\mathbf{c}^{n+1} = \mathbf{c}^n + \Delta t \Phi(\mathbf{c}^n, \mathbf{c}^{n+1}; \Delta t)$$

we introduce some notations and definitions which are used in the subsequent parts of the chapter.

Definition 3.2

For a given discretization scheme Φ *we call*

$$\mathbf{e} = \mathbf{c}^{n+1} - \mathbf{c}\left(t^{n+1}\right) \ \text{ with } \ \mathbf{c}^{n+1} = \mathbf{c}\left(t^n\right) + \Delta t \Phi\left(\mathbf{c}\left(t^n\right), \mathbf{c}\left(t^{n+1}\right); \Delta t\right)$$

the local discretization error vector, where $\mathbf{c}(t)$ *represents the exact solution of the initial value problem (6), (7) and* $\Delta t = t^{n+1} - t^n$. *Moreover, we write*

$$\mathbf{M} = \mathcal{O}\left(\Delta t^p\right) \ \text{ as } \ \Delta t \to 0, \ \ p \in \mathbf{N}_0,$$

in terms of $m_{i,j} = \mathcal{O}\left(\Delta t^p\right)$ *as* $\Delta t \to 0$, $p \in \mathbf{N}_0$ *for all elements* $m_{i,j}$, $i = 1,\ldots,r$, $j = 1,\ldots,k$ *of the matrix* $\mathbf{M} \in \mathbf{R}^{r\times k}$.

It is worth mentioning that the production and destruction terms $p_{i,j}$, $d_{i,j}$, $i,j = 1,\ldots,I$, are considered to be sufficiently smooth and we require the solution $\mathbf{c}$ of the initial value problem (6) and (7) to be sufficiently differentiable. In the following, we always consider the case of a vanishing time step Δt and thus the supplement $\Delta t \to 0$ will be neglected for simplification and we use the expression accuracy in the sense of consistency.

Definition 3.3
A discretization scheme Φ is called

- *Consistent of order p with respect to the ordinary differential equation (6), if*

$$\mathbf{e} = \mathcal{O}\left(\Delta t^{p+1}\right),$$

- *Unconditionally positive, if $\mathbf{c}^{n+1} > 0$ for any given $\mathbf{c}^n := \mathbf{c}(t^n) > 0$ and any arbitrary large time step $\Delta t \geq 0$ independent of the specific definition of the production and destruction terms within the ordinary differential equation (6)*
- *Conservative, if*

$$\sum_{i=1}^{I}\left(c_i^{n+1} - c_i^n\right) = 0 \tag{8}$$

for any fully conservative ordinary differential equation (6) and $\mathbf{c}^n := \mathbf{c}(t^n)$.

At first, we consider the well-known forward Euler scheme

$$c_i^{n+1} = c_i^n + \Delta t\left(\sum_{\substack{j=1\\j\neq i}}^{I} p_{i,j}(\mathbf{c}^n) - \sum_{\substack{j=1\\j\neq i}}^{I} d_{i,j}(\mathbf{c}^n) + p_{i,i}(\mathbf{c}^n) - d_{i,i}(\mathbf{c}^n)\right) \tag{9}$$

for $i = 1,\ldots,I$. Quite obviously, we obtain the following result concerning the positivity and conservativity of the numerical method.

Theorem 3.4
The forward Euler scheme (9) is conservative but not unconditional positivity preserving.

Proof: By means of the abbreviations $P_i = \sum_{\substack{j=1\\j\neq i}}^{I} p_{i,j}$ and $D_i = \sum_{\substack{j=1\\j\neq i}}^{I} d_{i,j}$ and utilizing the properties of the production and destruction rates we deduce

$$\sum_{i=1}^{I}\left(c_i^{n+1} - c_i^n\right) = \underbrace{\Delta t\sum_{i=1}^{I}\left(P_i\left(\mathbf{c}^n\right) - D_i\left(\mathbf{c}^n\right)\right)}_{=0} + \Delta t\sum_{i=1}^{I}\left(\underbrace{p_{i,i}(\mathbf{c}^n)}_{=0} - \underbrace{d_{i,i}(\mathbf{c}^n)}_{=0}\right) = 0,$$

which proves that the method is conservative. Due to the fact that the property to be unconditionally positive is independent of the production and destruction terms, we consider a fully conservative system with non vanishing right-hand side. Thus, there exists at least one index $i \in \{1,\ldots,I\}$ such that $P_i\left(\mathbf{c}^n\right) - D_i\left(\mathbf{c}^n\right) < 0$ for a given $\mathbf{c}^n > 0$. Thus, using

$$\Delta t > \frac{c_i^n}{D_i\left(\mathbf{c}^n\right) - P_i\left(\mathbf{c}^n\right)} \quad \text{yields} \quad c_i^{n+1} = c_i^n + \Delta t\left(P_i\left(\mathbf{c}^n\right) - D_i\left(\mathbf{c}^n\right)\right) < 0,$$

which proves the statement. □

To overcome this disadvantage, Patankar (Patankar 1980) suggested to weight the destruction terms $d_{i,j}(\mathbf{c})$ and $d_{i,i}(\mathbf{c})$ by the factor $\frac{c_i^{n+1}}{c_i^n}$.

Theorem 3.5
The Patankar scheme

$$c_i^{n+1} = c_i^n + \Delta t \left(\sum_{\substack{j=1 \\ j \neq i}}^{I} p_{i,j}(\mathbf{c}^n) - \sum_{\substack{j=1 \\ j \neq i}}^{I} d_{i,j}(\mathbf{c}^n) \frac{c_i^{n+1}}{c_i^n} + p_{i,i}(\mathbf{c}^n) - d_{i,i}(\mathbf{c}^n) \frac{c_i^{n+1}}{c_i^n} \right)$$

for $i = 1, \ldots, I$ *is unconditional positivity preserving but not conservative.*

Proof: We simply rewrite the Patankar scheme in the form

$$\underbrace{\left(1 + \frac{\Delta t}{c_i^n} \left(\sum_{\substack{j=1 \\ j \neq i}}^{I} d_{i,j}(\mathbf{c}^n) + d_{i,i}(\mathbf{c}^n) \right) \right)}_{\geq 1} c_i^{n+1} = \underbrace{c_i^n + \Delta t \left(\sum_{\substack{j=1 \\ j \neq i}}^{I} p_{i,j}(\mathbf{c}^n) + p_{i,i}(\mathbf{c}^n) \right)}_{\geq c_i^n > 0}$$

for $i = 1, \ldots, I$. Thus, we can immediately conclude the positivity of the method. However, one can easily see that even in the case of the simple system

$$c_1'(t) = c_2(t) - 2c_1(t)$$
$$c_2'(t) = 2c_1(t) - c_2(t)$$

and initial conditions $\mathbf{c}^0 = \mathbf{c}(t = 0) = (1,1)^T$ one gets

$$c_1^1 = \frac{1 + \Delta t}{1 + 2\Delta t}, \quad c_2^1 = \frac{1 + 2\Delta t}{1 + \Delta t}$$

such that

$$\sum_{i=1}^{2} \left(c_i^1 - c_i^0 \right) = \frac{\Delta t^2}{1 + 3\Delta t + 2\Delta t^2} \neq 0 \text{ for all } \Delta t > 0.$$

□

Theorem 3.5 shows that the so-called Patankar-trick represents a cure with respect to the positivity constraint but this method suffers from the fact that the conservativity is violated since production and destruction terms are handled in a different manner. Inspired by the Patankar-trick, Burchard et al. Burchard et al. (2003) introduced a modified Patankar approach where source as well as sink terms are treated in the same way. However, this procedure can only directly be applied to conservative systems. Thus, an extension of this modified Patankar scheme to take account of additional non conservative reaction terms as appearing within the biomass dynamics has been presented in Benz et al. (2009). With respect to the Euler scheme as the underlying basic solver, this extended modified Patankar approach can be written in the form

$$c_i^{n+1} = c_i^n + \Delta t \left(\sum_{\substack{j=1 \\ j \neq i}}^{I} p_{i,j}\left(\mathbf{c}^n\right) \frac{c_j^{n+1}}{c_j^n} - \sum_{\substack{j=1 \\ j \neq i}}^{I} d_{i,j}\left(\mathbf{c}^n\right) \frac{c_i^{n+1}}{c_i^n} + \left(p_{i,i}(\mathbf{c}^n) - d_{i,i}(\mathbf{c}^n)\right)\omega_i^n \right) \quad (10)$$

for $i = 1, \ldots, I$, where

$$\omega_i^n = \begin{cases} \frac{c_i^{n+1}}{c_i^n}, & \text{if } p_{i,i}(\mathbf{c}^n) - d_{i,i}(\mathbf{c}^n) < 0, \\ 1, & \text{otherwise.} \end{cases}$$

Theorem 3.6

The extended modified Patankar scheme (10) is conservative in the sense of Definition 3.3.

Proof: It is easily seen by straightforward calculations that

$$\sum_{i=1}^{I}\left(c_i^{n+1}-c_i^n\right)=\Delta t\sum_{i=1}^{I}\left(\sum_{\substack{j=1\\j\neq i}}^{I} p_{i,j}\left(\mathbf{c}^n\right)\frac{c_j^{n+1}}{c_j^n}-\sum_{\substack{j=1\\j\neq i}}^{I} d_{i,j}\left(\mathbf{c}^n\right)\frac{c_i^{n+1}}{c_i^n}\right)$$

$$+\Delta t\sum_{i=1}^{I}\underbrace{\left(p_{i,i}(\mathbf{c}^n)-d_{i,i}(\mathbf{c}^n)\right)}_{=0}\omega_i^n$$

$$=\Delta t\sum_{\substack{i,j=1\\j\neq i}}^{I}\underbrace{\left(p_{i,j}\left(\mathbf{c}^n\right)-d_{j,i}\left(\mathbf{c}^n\right)\right)}_{=0}\frac{c_j^{n+1}}{c_j^n}=0,$$

which proves the statement. □

Theorem 3.7

The extended modified Patankar scheme (10) applied to the system of differential equations (6) is unconditionally positivity preserving.

Proof: The Patankar-type approach (10) can be written in the form

$$\mathbf{A}\mathbf{c}^{n+1}=\mathbf{b}^n, \tag{11}$$

where $\mathbf{A}=(a_{i,j})\in\mathbf{R}^{I\times I}$ with

$$a_{i,i}=1+\frac{\Delta t}{c_i^n}\left(\max\{0,d_{i,i}(\mathbf{c}^n)-p_{i,i}(\mathbf{c}^n)\}+\sum_{\substack{k=1\\k\neq i}}^{I} d_{i,k}(\mathbf{c}^n)\right)>0,\quad i=1,\ldots,I,$$

$$a_{i,j}=\ \Delta t\frac{p_{i,j}(\mathbf{c}^n)}{c_j^n}\le 0,\quad i,j-1,\ldots,I,\quad i\neq j.$$

and

$$b_i^n=c_i^n+\Delta t\max\{0,p_{i,i}(\mathbf{c}^n)-d_{i,i}(\mathbf{c}^n)\}\ge c_i^n>0,\quad i=1,\ldots,I.$$

Hence, for $i=1,\ldots,I$ we have

$$|a_{i,i}|>\Delta t\sum_{\substack{k=1\\k\neq i}}^{I}\frac{d_{i,k}(\mathbf{c}^n)}{c_i^n}=-\Delta t\sum_{\substack{k=1\\k\neq i}}^{I}\frac{p_{k,i}(\mathbf{c}^n)}{c_i^n}=\sum_{\substack{k=1\\k\neq i}}^{I}(-a_{k,i})=\sum_{\substack{k=1\\k\neq i}}^{I}|a_{k,i}|$$

which directly shows that the point Jacobi matrix $\mathbf{B}=\mathbf{I}-\mathbf{D}^{-1}\mathbf{A}^T$ defined by means of the diagonal matrix $\mathbf{D}^{-1}=diag\left\{a_{1,1}^{-1},\ldots,a_{I,I}^{-1}\right\}$ satisfies $\rho(\mathbf{B})\le\|\mathbf{B}\|_\infty<1$. Thus, the matrix $\mathbf{B}$ is convergent. Regarding the fact that the matrix $\mathbf{A}$ contains only nonpositive off-diagonal elements and positive diagonal entries we can conclude with a standard statement from the numerical linear algebra that $\mathbf{A}^T$ and therefore $\mathbf{A}$ are M-matrices. Thus, $\mathbf{A}^{-1}$ exists and is nonnegative, i.e., $\mathbf{A}\ge 0$. This fact implies that

$$\mathbf{c}^{n+1}=\mathbf{A}^{-1}\mathbf{b}^n\ge\mathbf{A}^{-1}\mathbf{c}^n>0$$

since at least one entry per row of the matrix $\mathbf{A}^{-1}$ is positive. □

The Patankar scheme as well as the modified and extended modified version can be interpreted as a perturbed Euler scheme due to the incorporation of the weights. Thus, it is quite not obvious that these schemes are still first order accurate. Fortunately, the order of accuracy of the underlying Euler scheme transmits to each variant discussed above. Similar to the error analysis presented in Burchard et al. (2003) for the modified Patankar scheme we will now prove this important property for the extended formulation.

Theorem 3.8

The extended modified Patankar scheme (10) is first-order accurate in the sense of the local discretization error.

Proof: Since the time step inside the extended modified Patankar scheme (10) is performed by the solution of a linear system of equations, it is advantageous to investigate the entries of the inverse matrix $\mathbf{A}^{-1} = (\tilde{a}_{i,j}) \in \mathbf{R}^{I\times I}$. Introducing $\mathbf{e} = (1,\ldots,1)^T \in \mathbf{R}^I$ one can easily verify $\mathbf{e}^T\mathbf{A} \geq \mathbf{e}^T$. Within the proof of Theorem 3.7 we have seen that $\mathbf{A}$ is regular and $\mathbf{A}^{-1} \geq 0$. Thus, we obtain $\mathbf{e}^T \geq \mathbf{e}^T\mathbf{A}^{-1}$ that yields

$$0 \leq \tilde{a}_{i,j} \leq 1, \quad i,j = 1,\ldots,I \tag{12}$$

independent of the time step size $\Delta t > 0$ used. Starting from the formulation of the time step in the form of the linear system (11) one can conclude

$$\frac{c_i^{n+1}}{c_i^n} = \sum_{j=1}^{I} \underbrace{\tilde{a}_{i,j}}_{\in[0,1]} \frac{b_j^n}{c_i^n} = \mathcal{O}(1), \quad i = 1,\ldots,I,$$

from the estimation (12). Introducing this property into the extended modified Patankar scheme (10) and determining $\mathbf{c}^n := \mathbf{c}(t^n)$ for simplification leads to

$$\begin{aligned} c_i^{n+1} - c_i^n &= \Delta t \underbrace{\left(\sum_{\substack{j=1\\ j\neq i}}^{I} p_{i,j}(\mathbf{c}^n) \frac{c_j^{n+1}}{c_j^n} - \sum_{\substack{j=1\\ j\neq i}}^{I} d_{i,j}(\mathbf{c}^n) \frac{c_i^{n+1}}{c_i^n} + \left(p_{i,i}(\mathbf{c}^n) - d_{i,i}(\mathbf{c}^n)\right)\omega_i^n \right)}_{=\mathcal{O}(1)} \\ &= \mathcal{O}(\Delta t). \end{aligned}$$

Thus we obtain

$$\frac{c_i^{n+1}}{c_i^n} - 1 = \frac{c_i^{n+1} - c_i^n}{c_i^n} = \mathcal{O}(\Delta t) \tag{13}$$

for $i = 1,\ldots,I$. A similar results is valid for the weight ω_i^n due to

$$\omega_i^n - 1 = \begin{cases} \dfrac{c_i^{n+1} - c_i^n}{c_i^n} = \mathcal{O}(\Delta t), & \text{if } p_{i,i}(\mathbf{c}^n) - d_{i,i}(\mathbf{c}^n) < 0, \\ 0 = \mathcal{O}(\Delta t), & \text{otherwise.} \end{cases} \tag{14}$$

A combination of (13) and (14) with a simple Taylor series expansion yields

$$
\begin{aligned}
& c_i\left(t^{n+1}\right) \\
&= c_i\left(t^n\right) + \Delta t \frac{dc_i}{dt}\left(t^n\right) + \mathcal{O}\left(\Delta t^2\right) \\
&= c_i\left(t^n\right) + \Delta t \left(\sum_{\substack{j=1\\ j\neq i}}^{I} p_{i,j}\left(\mathbf{c}\left(t^n\right)\right) - \sum_{\substack{j=1\\ j\neq i}}^{I} d_{i,j}\left(\mathbf{c}\left(t^n\right)\right) + p_{i,i}\left(\mathbf{c}\left(t^n\right)\right) - d_{i,i}\left(\mathbf{c}\left(t^n\right)\right) \right) \\
&\qquad + \mathcal{O}\left(\Delta t^2\right) \\
&= c_i^n + \Delta t \left(\sum_{\substack{j=1\\ j\neq i}}^{I} p_{i,j}\left(\mathbf{c}^n\right) \frac{c_j^{n+1}}{c_j^n} - \sum_{\substack{j=1\\ j\neq i}}^{I} d_{i,j}\left(\mathbf{c}^n\right) \frac{c_i^{n+1}}{c_i^n} + \left(p_{i,i}\left(\mathbf{c}^n\right) - d_{i,i}\left(\mathbf{c}^n\right)\right) \omega_i^n \right) \\
&\qquad - \underbrace{\Delta t \left(\sum_{\substack{j=1\\ j\neq i}}^{I} p_{i,j}\left(\mathbf{c}^n\right) \frac{c_j^{n+1} - c_j^n}{c_j^n} - \sum_{\substack{j=1\\ j\neq i}}^{I} d_{i,j}\left(\mathbf{c}^n\right) \frac{c_i^{n+1} - c_i^n}{c_i^n} \right)}_{=\mathcal{O}(\Delta t)} \\
&\qquad - \underbrace{\Delta t \left(\left(p_{i,i}\left(\mathbf{c}^n\right) - d_{i,i}\left(\mathbf{c}^n\right)\right)\left(\omega_i^n - 1\right)\right)}_{=\mathcal{O}(\Delta t)} + \mathcal{O}\left(\Delta t^2\right) \\
&= c_i^{n+1} + \mathcal{O}\left(\Delta t^2\right),
\end{aligned}
$$

which completes the proof. □

In order to increase the accuracy one can easily integrate the idea describe in the context of the extended modified Patankar scheme within a standard second-order Runge–Kutta method. Similar to the proof presented in Burchard et al. (2003), it can be shown that such an extension is second-order accurate, unconditionally positivity preserving and conservative in the sense of Definition 3.3.

3.2.1 Numerical Results for Positive Ordinary Differential Equations

The first test cases presents a simple linear system of ordinary differential equations. The results are taken from the original paper by Burchard, Deleersnijder, and Meister (Burchard et al. 2003). The system can be written as

$$d_t c_1 = c_2 - 5c_1, \quad \text{and} \quad d_t c_2 = 5c_1 - c_2 \tag{15}$$

with initial values set to be $c_1(0) = 0.9$ and $c_2(0) = 0.1$. The analytic solution is $c_1(t) = (1 + 4.4\exp(-6t))/6$ and $c_2(t) = 1 - c_1(t)$.

Using the step size $\Delta t = 0.25$ one obtains negative concentrations for the standard forward Euler scheme, whereas the modified Patankar approach still preserves the positivity of the solution. Furthermore, both schemes are conservative, which can be seen from the horizontal line representing the sum of both concentrations (➲ *Fig. 4*).

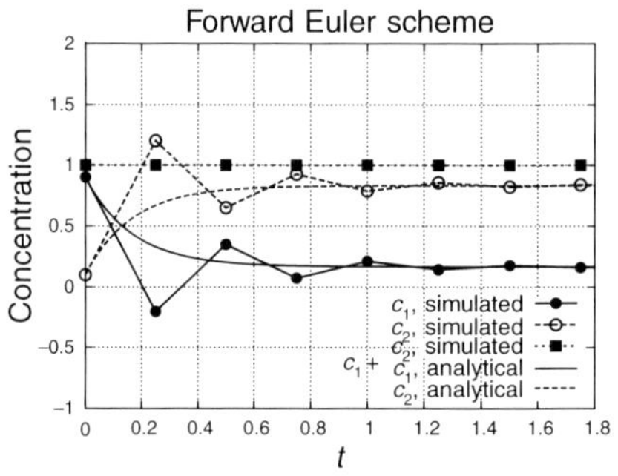

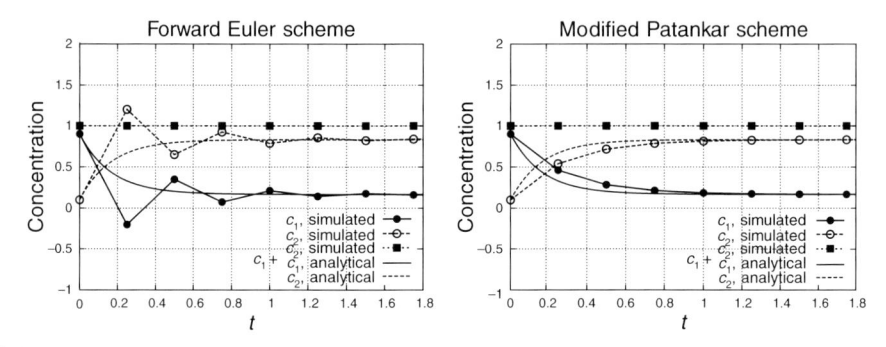

Fig. 4

Numerical approximation (Δt = 0.25) and analytical solution of the simple linear system with the forward Euler scheme (*left*) and the (extended) modified Patankar scheme (*right*)

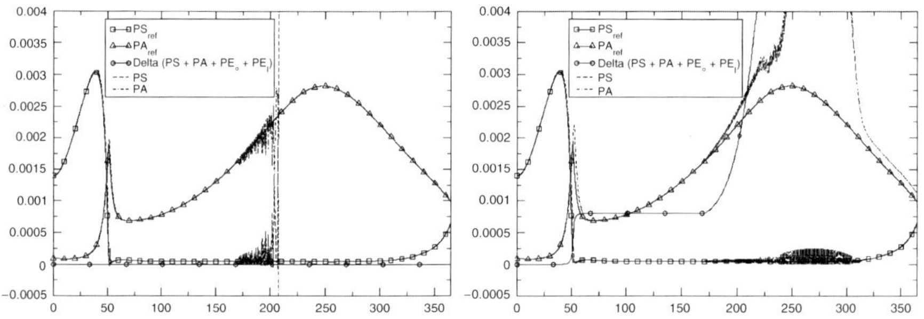

Fig. 5

Results of the forward Euler scheme (*left*) and the Patankar scheme (*right*) concerning the second test case

As a second test cases we consider the simplified conservative biochemical system

$$\frac{d}{dt}\begin{pmatrix} PA \\ PE_I \\ PE_O \\ PS \end{pmatrix} = \begin{pmatrix} uptba - setpa - resp \\ minpe - exchp \\ setpa - minpe \\ exchp - uptba + resp \end{pmatrix},$$

where $uptba$, $setpa$, $resp$, $exchp$, and $minpe$ denote the phosphorus uptake, phosphorus of setting phytoplankton, phosphorus due to respiration, exchange between water and sediment and the mineralization of organic phosphorus, respectively. The results obtained by the standard forward Euler scheme, the Patankar scheme and the modified Patankar scheme for a constant time step size $\Delta t = 1/3\ d$ are compared with a high resolution numerical result using the time step size $\Delta t = 1/100\ d$. *Figures 5* and 6 confirm the analytic statements concerning the three schemes used. The modified Patankar scheme is positivity preserving and conservative while the Patankar method suffers from non-conservativity and the standard Euler approach yields meaningless negative values for both the solute phosphorus concentration in the water body and the phosphorus within the biomass. Note that the conservativity can be

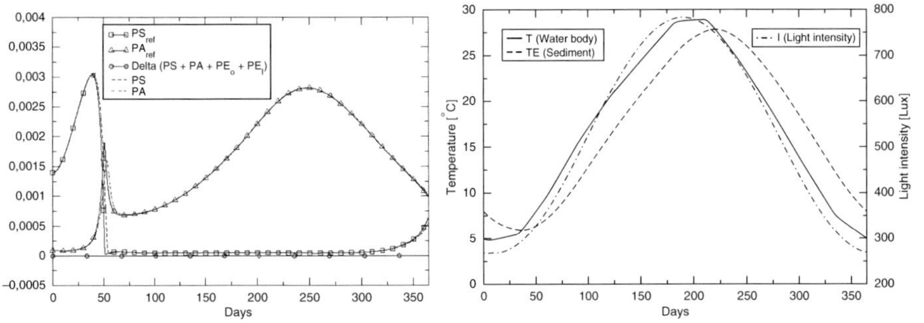

Fig. 6
Results of the modified Patankar scheme (*left*) concerning the second test case. Distribution of sediment temperature (*TE*), water temperature (*T*) and light intensity (*I*) over the course of the year (*right*) with respect to the West Lake

Table 1
Constituents of the phosphorus cycle

Description	Abbreviation	Unit
Solute phosphorus/PO_4^-	*PS*	g/m^3
Phosphorus in detritus	*PD*	g/m^3
Inorganic and solute phosphorus in sediment	PE_I	g/m^3
Organic phosphorus in sediment	PE_O	g/m^3
Biomass of zooplankton	*BZ*	g/m^3
and its content of phosphorus	*PZ*	g/m^3
Biomasses of four different groups of algae species	BA_{A-D}	g/m^3
with their respective content of phosphorus	PA_{A-D}	g/m^3

observed by $\text{Delta}(PS + PA + PE_I + PE_O) := (PS + PA + PE_I + PE_O)^n - (PS + PA + PE_I + PE_O)^0$.

3.3 Practical Applications

The ecological model presented here is an enhancement of the phosphorus cycle model proposed in Hongping and Jianyi (2002). All considered processes are depicted in *Fig. 2*. For the sake of simplicity only one biomass algae (*BA*) and respectively only one phosphorus content (*PA*) are included in *Fig. 2*. The remaining three biomasses and corresponding phosphorus can easily be included since their evolution is quite alike. The examined constituents of the biological system are summarized with their abbreviations and their units respectively in *Table 1*.

The impact of the current on the different kinds of biomass and phosphorus constituents within the governing equations (2) is performed as follows. Regarding the biochemical system the flux $\mathbf{f}^c_{m,p}$ describes the passive advective transport with the existing current for all constituents of the phosphorus cycle, which are in the water body. Furthermore, the additional term $\mathbf{f}^v_{m,p}$ includes diffusive effects for the solute components *PS* and PE_I with appropriate diffusion coefficients Diff_{PS} and Diff_{PE_I}. The expansion of the right-hand side $\mathbf{q}_P(\mathbf{U})$ is given by

$$\mathbf{q}_P(\mathbf{U}) = \begin{pmatrix} exchp - \sum_i uptba_i + minpd + zresp + \sum_i resp_i \\ -minpd - setpd + zmorp \\ -exchp + minpe \\ -minpe + setpd + \sum_i (setpa_i + gsinkp_i) \\ uptba_A - resp_A - setpa_A - gsinkp_A - assimp_A \\ uptba_B - resp_B - setpa_B - gsinkp_B - assimp_B \\ uptba_C - resp_C - setpa_C - gsinkp_C - assimp_C \\ uptba_D - resp_D - setpa_D - gsinkp_D - assimp_D \\ -zresp - zmorp + \sum_i assimp_i \\ \sum_i assim_i - zres - zmor \\ growth_A - res_A - sink_A - graz_A \\ growth_B - res_B - sink_B - graz_B \\ growth_C - res_C - sink_C - graz_C \\ growth_D - res_D - sink_D - graz_D \end{pmatrix}.$$

A detailed description of the terms is presented in Appendix 4

For the proper computability of the above given equations the temperature of the water T and of the sediment TE must be supplied. Based on the air temperature stated in Weather (2008a,b,c), these temperatures are solutions of a one-dimensional heat equation for the proper distribution shown in ➋ *Fig. 6*. The figure also shows the assumed light intensity over the course of 1 year. For the rain q also annual developments are used from the same source of information.

Most of the constants are adopted from Hongping and Jianyi (2002). All constants used are assembled in ➋ *Table 2*, Appendix A.

Even though the model described above is based on Hongping and Jianyi (2002), it has many important new features. An extremely important fact is that the underlying flow field of the water coupled with the biological and chemical equations are simultaneously solved. This gives a much more detailed image of the modeled occurrences in the lake then any prediction made under the assumption of even distributed materia can hope for.

Another major difference between the model proposed in Hongping and Jianyi (2002) and our model is the splitting of the phosphorus in sediment PE to organic phosphorus in the sediment PE_O and inorganic phosphorus in the sediment PE_I. A temperature-driven mineralization process from sedimented matter to solute phosphorus is introduced. As only the second form of phosphorus can contribute to the phosphorus supply in the water body, this offers a better reproduction of the phosphorus accumulation in sediment during the winter period. The fixation of phosphorus in the sediment is also an important process from the ecological point of view. Nevertheless, due to the missing pH and O_2-concentration the fixation is not considered in this model.

Not complete uptaken biomass of algae by grazing is assimilated. The model by Hongping and Jianyi assumes a complete assimilation. Our model expects a gain of 80% of the biomass for the zooplankton biomass and a loss of 20%.

A critical mathematical and also biological certainty is that all processes (with one exception: $exchp$) are positive. This is the mathematical formulation for the fact that processes are not reversible. For example, zooplankton grazing cannot grant a mass win for the grazed algae or respiration cannot end up with the algae gathering phosphorus from the water body. This should be modeled accordingly.

Table 2

Description and numerical values for all constants

Constant	Value	Unit	Description
F_{min}	0.05	$\frac{g}{m^3}$	Min feeding concentration of phytoplankton
F_s	0.25	$\frac{g}{m^3}$	Menten feeding constant phytoplankton
GR_{max}	0.09	$\frac{1}{d}$	Max graze rate of zooplankton
R_i	0.18, 1, 1, 1	—	Preference factor for grazing i
DPs_{S_i}	0.027, 0.016,		
	0.016, 0.018	$\frac{g}{m^3}$	Menten constant for i phosphorus uptake
TOP_i	30, 23, 20, 21	°C	Optimal temperature for growth i
K_1	1.5	$\frac{1}{d}$	Extinction coefficient of water
K_2	1	$\frac{1}{d}$	Extinction coefficient of phytoplankton
LOP_i	310, 340, 350, 310	Lux	Optimal light radiation for growth i
$growth_{max_i}$	3.3, 2.35, 2.43, 2.37	$\frac{1}{d}$	Max growth rate i
RO_i	0.007, 0.003,		
	0.003, 0.003	$\frac{1}{d}$	Respiration rate i
K_{set_i}	0.025, 0.016,		
	0.016, 0.017	$\frac{m}{d}$	Velocity of sedimentation i
Pin_{min_i}	0.005, 0.005,		
	0.005, 0.005	$\frac{g}{m^3}$	Min content of phosphorus in i
Pin_{max_i}	0.015, 0.020,		
	0.015, 0.015	$\frac{g}{m^3}$	Max content of phosphorus in i
PUP_{max_i}	0.07, 0.1,		
	0.07, 0.07	$\frac{1}{d}$	Max phosphorus uptake rate of i
mor	0.0004	$\frac{1}{d}$	Death rate of zooplankton
ZRO	0.03	$\frac{1}{d}$	Respiration rate of zooplankton
ZRM	0.04	—	Respiration multiplier of zooplankton
K_{ex}	0.03	$\frac{1}{d}$	Phosphorus exchange coefficient
K_{m1}	0.01	$\frac{1}{d}$	Mineralization rate of PD
S_{m1}	0.8	—	Temp. coeff. of mineralization rate PD
K_{m2}	0.178	$\frac{1}{d}$	Mineralization rate of PE
S_{m2}	1.08	—	Temp. coeff. of mineralization rate PE
V_s	0.125	$\frac{m}{d}$	Sedimentation rate
rd	0.38	—	Percentage of organic phosphorus in detritus
$util_i$	0.8, 0.8, 0.8, 0.8	—	Usage of feed material
$TCOEF$	0.38	$\frac{1}{°C}$	Q10 coefficient

Nevertheless, the original model allows sign changes for example in $graz_i$, $uptba_i$ or $asim$. It also allows greater uptakes then the maximum uptake rate. This flaws are corrected by using equations oriented at equations proposed in Hongping and Jianyi (2002) with the proper constraints. An example for this is the determination of PC_i.

Since no measured data are available for a comparison with the numerical simulation, the results are valid only under the assumptions included in the model.

To demonstrate the applicability of the scheme for the discussed model we show various numerical results. The first test case presents a course of 1 year after the computation of several years for the whole phosphorus cycle model. In ❯ *Fig. 7*, the changes of the phosphorus of one algae and there biomass together with the solute phosphorus in the water body and the

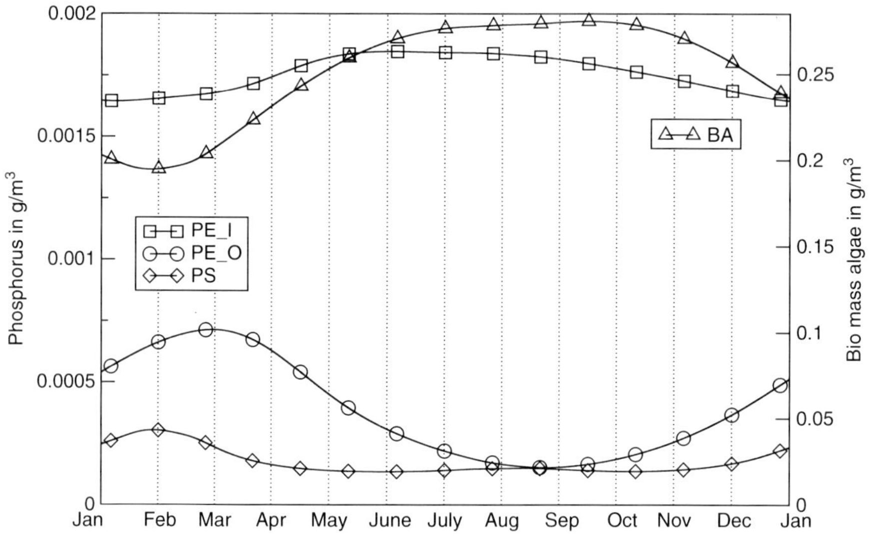

Fig. 7
Course of 1 year for phosphorus in water and sediment with one biomass algae

phosphorus in the sediment are presented. The scheme gives reasonable results for the change of the phosphorus content over the seasonal changes as can be seen in *Fig. 7*. The phosphorus concentration PE_O augments in spring. With the rising of the temperature the bacteria in the sediment start to convert the organic-bound phosphorus, which was accumulated over the winter, into inorganic solute phosphorus.

Through the diffusive exchange between the solute phosphorus in water and sediment the solute phosphorus in the water augments and is directly consumed by the starting algae growth BA.

At end of the year, the phosphorus concentration PE_O and PS increases due to the decease of the algae BA. PE_O is augmented by the dead biological mass, PS through lesser consumption and exchange between PE_I and PS.

As second test case shown in *Figs. 8* and *9* solute phosphorus inflow has been examined. It was assumed that all boundaries of the West Lake except the small channels in the south end (lower left part of the figure) and in the north eastern corner (upper right part of the figure) are fixed walls, i.e., no particles can pass this boundary. The southern boundary is considered the inflow, the north eastern boundary the outflow.

For a clear visualization of the proportions of the current, high inflow velocities have been assumed. One can observe the curvature of the current in the lake after the straight inflow following the structure of the lake. Then the current splits into a western and an eastern part. The west flowing current circles back into the lake behind the three islands and through the small openings into the nearly cut off western part of the lake. The eastern directed part of the flow pours out of the West Lake.

The West Lake was assumed to contain only traces of phosphorus PS at the beginning of the calculation. This allows the undisturbed observation of the distribution of in-flowed phosphorus. One can see clearly how the phosphorus is distributed. High concentrations of phosphorus are found directly along the strong currents, the inflow area and also the triangular dead area

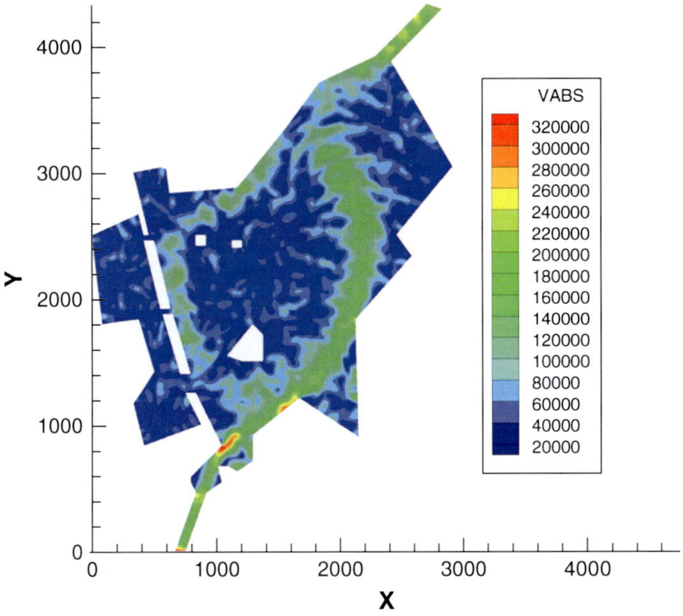

Fig. 8
Absolute value of the velocity $v_{abs} = \|v\|$ in the West Lake

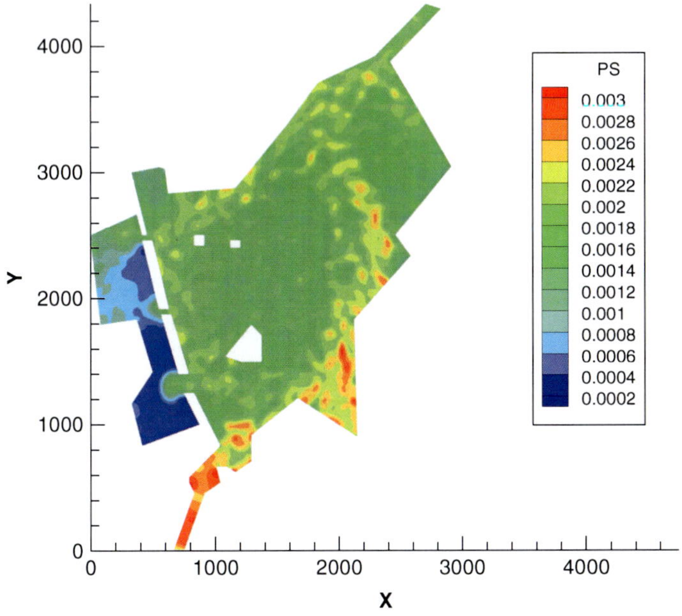

Fig. 9
Solute phosphorus floating into the West Lake

along the south eastern boundary of the lake. Also one can note that through the absence of a strong current into the nearly cut off western part of the lake the phosphorus is entering very slowly.

4 Conclusion

A complex phosphorus cycle containing four different groups of algae species and their respective phosphorus content was presented. Furthermore, a higher order finite volume scheme was introduced.

Through various numerical results the scheme has proven to be able to solve simultaneously a real life fluid dynamics problem and a sophisticated system of positive and conservative phosphor cycle describing biological and chemical equations.

The task is accomplished via a splitting ansatz combining higher order flux solving methods with problem adopted ordinary differential equation techniques, the extended modified Patankar ansatz. Constructed in this way, the scheme preserves the important characteristics of conservativity and positivity necessary to obtain meaningful numerical results.

Thus state-of-the-art modeling and numerical techniques have been successfully combined in the context of an eutrophication lake modeling.

Acknowledgment

First of all, we would like to express our thanks to W. Freeden for initiating this *Handbook of Geo-mathematics* and giving us the possibility to participate. Several mathematical parts of this chapter are developed in a joint work with H. Burchard and E. Deelersnijder. A. Meister is grateful for this productive cooperation. Furthermore, A. Meister would like to thank Th. Sonar for many years of fruitful collaborations in the field of computational fluid dynamics.

Appendix A: Ecological Model

In order to give a deeper insight in the structure of the source term $\mathbf{q}_p(\mathbf{u}_s, \mathbf{u}_p)$, we now concentrate on the specific expression for the biomass of ith algae group:

1. Temperature impact on growth of algae group i

$$TT_i = \frac{T}{TOP_i} e^{\frac{TOP_i - T}{TOP_i}}.$$

2. Phosphor availability factor for algae group i

$$PP_i = \frac{PS}{DP_{S_i} + PS}.$$

3. Light extinction $K = K_1 + K_2 \sum_j BA_j$.
4. Depth averaged available light intensity $L = I\left(\frac{1-e^{-KH}}{KH}\right)$.

5. Light intensity impact on growth of algae group i

$$LL_i = \frac{L}{LOP_i} e^{\frac{LOP_i - L}{LOP_i}}.$$

6. Growth of algae i

$$growth_i = growth_{max_i}\ TT_i\ PP_i\ LL_i\ BA_i.$$

7. Respiration loss of biomass of algae group i

$$res_i = RO_i\ e^{TCOEF\ T}\ BA_i.$$

8. Sinking, i.e., dying, of biomass of algae group i

$$sink_i = BA_i\ \frac{K_{set_i}}{H}.$$

9. Total food preference weighted biomass algae $F = \sum_j R_j\ BA_j$.
10. Available-food factor for zooplankton

$$FF = \begin{cases} 0.0 & F_{min} > F \\ \frac{F - F_{min}}{F_s + F - F_{min}} & F_{min} \leq F \end{cases}.$$

11. Grazing of biomass algae i through zooplankton

$$graz_i = GR_{max}\ FF \frac{R_i\ BA_i}{F} BZ.$$

The biomass zooplankton dynamics are given by

1. Biomass gain for zooplankton through grazing algae i

$$assim_i = util_i\ graz_i.$$

2. Respiration of zooplankton

$$zres = ZRO\ e^{TCOEF\ T}\ BZ + ZRM \sum_j graz_j.$$

3. Death rate of zooplankton $zmor = mor\ BZ$.

The phosphorus dynamics are given as:

1. Diffusive exchange between the solute phosphorus in sediment and water

$$exchp = K_{ex}\ (PE_I - PS).$$

2. Mineralization of PD: $minpd = K_{m1} S_{m1}^{T-20}\ PD$.
3. Mineralization from PE_O to PE_I: $minpe = K_{m2} S_{m2}^{TE-20}\ PE_O$.
4. Sedimentation of PD: $setpd = \frac{PD}{H}\ (1 - rd)\ V_s$.

The exchange dynamics between phosphorus in living cells and the phosphorus without associated biomass are:

1. Phosphorus fraction from biomass of algae group i

$$PCON_i = \frac{PA_i}{BA_i}.$$

2. Influence of the phosphorus already contained in the biomass of algae i for its growth

$$PC_i = \begin{cases} 1 & PCON_i < Pin_{min_i} \\ 0 & PCON_i > Pin_{max_i} \\ \frac{Pin_{max_i} - PCON_i}{Pin_{max_i} - Pin_{min_i}} & otherwise \end{cases}$$

3. Phosphorus uptake from algae group i

$$uptba_i = PUP_{max_i}\, PC_i\, BA_i\, PP_i.$$

4. Respiration loss of phosphorus of algae i

$$resp_i = res_i\, PCON_i.$$

5. Sinking, i.e., dying, of phosphorus of algae group i

$$setpa_i = sink_i\, PCON_i.$$

6. Phosphorus loss of algae group i through zooplankton grazing

$$grazp_i = graz_i\, PCON_i.$$

7. Biomass loss of algae group i through zooplankton grazing not assimilated by zooplankton

$$gsink_i = (1 - util_i)\, graz_i.$$

8. PD gain through zooplankton grazing algae group i

$$gsinkp_i = gsink_i\, PCON_i.$$

9. Phosphorus gain for zooplankton through grazing algae group i

$$assimp_i = assim_i\, PCON_i.$$

10. Phosphorus fraction from biomass of zooplankton

$$PCON_z = \frac{PZ}{BZ}.$$

11. Respiration amount from zooplankton

$$zresp = zres\, PCON_z.$$

12. PZ loss through mortality

$$zmorp = zmor\, PCON_z.$$

13. PZ loss through respiration

$$zresp = zres\, PCON_z.$$

References

Ansorge R, Sonar Th (2009) Mathematical models of fluid dynamics. Wiley-VCH, New York

Audusse E, Bristeau M-O (2005) A well-balanced positivity preserving second-order scheme for shallow water flows on unstructured meshes. J Comput Phys 206:311–333

Barth TJ, Jesperson DC (1989) The design and application of upwind schemes on unstructured meshes. AIAA paper 89-0366

Benz J, Meister A, Zardo PA (2009) A conservative, positivity preserving scheme for advection-diffusion-reaction equations in biochemical applications. In: Hyperbolic Problems - Proceedings of Symposia in Applied Mathematics, Eds: E. Tadmor, J.-G. Liu, A. E. Tzavaras; Amer. Math. Society, 2009

Berzins M (2001) Modified mass matrices and positivity preservation for hyperbolic and parabolic PDEs. Commun Num Meth Eng 9:659–666

Broekhuizen N, Rickard GJ, Bruggeman J, Meister A (2008) An improved and generalized second order, unconditionally positive, mass conserving integration scheme for biochemical systems. Appl Num Math 58:319–340

Bruggeman J, Burchard H, Kooi BW, Sommeijer B (2007) A second-order, unconditionally positive, mass-conserving integration scheme for biochemical systems. Appl Num Math 57:36–58

Burchard H, Deleersnijder E, Meister A (2003) A high-order conservative Patankar-type discretisation for stiff systems of production–destruction equations. Appl Num Math 47:1–30

Burchard H, Deleersnijder E, Meister A (2005) Application of modified Patankar schemes to stiff biogeochemical models of the water column. Ocean Dyn 55(3-4):326-337

Burchard H, Bolding K, Kühn W, Meister A, Neumann T, Umlauf L (2006) Description of a flexible and extendable physical-biogeochemical model system for the water column. J Mar Syst 61:180–211

Chertock A, Kurganov A (2008) A second-order positivity preserving central upwind scheme for Chemotaxis and Haptotaxis Models. Num Math 111:169–205

Friedrich O (1993) A new method for generating inner points of triangulations in two dimensions. Comput Meth Appl Mech Eng 104:77–86

Hirsch C (1988a) Numerical computation of internal and external flows, Vol 1. Wiley, New York

Hirsch C (1988b) Numerical computation of internal and external flows, Vol 2. Wiley, New York

Hongping P, Jianyi M (2002) Study on the algal dynamic model for West Lake, Hangzhou. Ecol Model 148:67–77

Jørgensen SE (1975) A eutrophication model for a lake. Ecol Model 2:147–165

Lampert W, Sommer U (1999) Limnoökologie. Georg Thieme, Stuttgart

LeVeque RJ (1990) Numerical methods for conservation laws. Birkhäuser, Boston

Meister A (1998) Comparison of different Krylov subspace methods embedded in an implicit finite volume scheme for the computation of viscous and inviscid flow fields on unstructured grids. J Comput Phys 140:311–345

Meister A (2003) Viscous flow fields at all speeds: analysis and numerical simulation. J Appl Math Phys 54:1010–1049

Meister A, Oevermann M (1998) An implicit finite volume approach of the $k - \epsilon$ turbulence model on unstructured grids. ZAMM 78(11): 743–757

Meister A, Sonar Th (1998) Finite-volume schemes for compressible fluid flow. Surv Math Ind 8: 1–36

Meister A, Vömel C (2001) Efficient preconditioning of linear systems arising from the discretization of hyperbolic conservation laws. Adv Comput Math 14(1):49–73

Park RA et al (1974) A generalized model for simulating lake ecosystems. Simulation 21:33–50

Patankar SV (1980) Numerical heat transfer and fluid flows. McGraw-Hill, New York

Poethke H-J (1994) Analysieren, Verstehen und Prognostizieren. PhD thesis, Johannes Gutenberg-Universität Mainz, Mainz

Ricchiuto M, Bollermann A (2009) Stabilized residual distribution for shallow water simulations. J Comput Phys 228(4):1071–1115

Sagehashi M, Sakoda A, Suzuki M (2000) A predictive model of long-term stability after biomanipulation of shallow lakes. Water Res 34(16):4014–4028

Schwoerbel J, Brendelberger H (2005) Einführung in die Limnologie. Elsevier Spektrum Akademischer Munich

Smolarkiewicz PK (2006) Multidimensional positive definite advection transport algorithm: an overview. Int J Num Meth Fluids 50:1123–1144

Sonar Th (1997a) On the construction of essentially non-oscillatory finite volume approximations to hyperbolic conservation laws on general triangulations: polynomial recovery, accuracy, and stencil selection. Comp Meth Appl Mech Eng 140:157–181

Sonar Th (1997b) Mehrdimensionale ENO-Verfahren. Teubner, Stuttgart

Stoker JJ (1957) Water waves. Interscience Publisher, New York

Straškraba M, Gnauk A (1985) Freshwater ecosystems. Elsevier, Amsterdam

Toro EF (1999) Riemann solvers and numerical methods for fluid dynamics. Springer, Berlin

Toro EF (2001) Shock-capturing methods for free-surface shallow flows. Wiley, New York

Vater S (2004) A new projection method for the zero Froud number shallow water equations. Master's thesis, Freie Universität Berlin, Berlin

Vázquez-Cendón M-E (2007) Depth averaged modelling of turbulent shallow water flow with wet-dry fronts. Arch Comput Meth Eng 14(3): 303–341

Weather Hangzhou (2008a) Internet, May 23, http://www.chinatoday.com.cn/english/chinatours/hangzhou.htm

Weather Hangzhou (2008b) Internet, May 23, http://www.ilec.or.jp/database/asi/asi-53.html

Weather Hangzhou (2008c) Internet, May 23, http://www.chinatoday.com.cn/english/chinatours/hangzhou.htm

Zardo PA (2005) Konservative und positive Verfahren für autonome gewöhnliche Differentialgleichungssysteme. Master's thesis, University Kassel